The Tidal Disruption of Stars by Massive Black Holes

Peter G. Jonker • Iair Arcavi • E. Sterl Phinney •
Elena M. Rossi • Nicholas C. Stone •
Sjoert van Velzen
Editors

The Tidal Disruption of Stars by Massive Black Holes

Previously published in *Space Science Reviews* in the Topical Collection "The Tidal Disruption of Stars by Massive Black Holes"

Springer

Editors
Peter G. Jonker
SRON, Utrecht
Radboud University Nijmegen
Nijmegen, Gelderland, The Netherlands

Iair Arcavi
Tel Aviv University
Tel Aviv, Israel

E. Sterl Phinney
California Institute of Technology
Pasadena, CA, USA

Elena M. Rossi
Leiden Observatory
Leiden, The Netherlands

Nicholas C. Stone
Racah Institute of Physics
The Hebrew University of Jerusalem
Jerusalem, Israel

Sjoert van Velzen
Leiden University
Amsterdam, The Netherlands

ISSN 1385-7525 Space Sciences Series of ISSI
ISBN 978-94-024-2145-3

Cover Image: Cover image: A view of the accretion disc around the supermassive black hole, with jet-like structures flowing away from the disc. The extreme mass of the black hole bends spacetime, allowing the far side of the accretion disc to be seen as an image above and below the black hole. Credit: © Deutsches Elektronen-Synchrotron (DESY); illustration by the Science Communication Lab, Germany.

This Springer imprint is published by the registered company Springer Nature B.V.
The registered company address is: Van Godewijckstraat 30, 3311 GX Dordrecht, The Netherlands

The day returns, but nevermore
Returns the traveller to the shore,
And the tide rises, the tide falls.

Henry Wadsworth Longfellow,
"The Tide Rises, The Tide Falls"

Preface to the Topical Collection: The Tidal Disruption of Stars by Massive Black Holes

For several decades, astronomers have speculated that a hapless star could wander too close to a super massive black hole (SMBH) and be torn apart by tidal forces. It is only with the recent advent of numerous wide-field transient surveys that such events have been detected in the form of giant-amplitude, luminous flares of electromagnetic radiation from the centers of otherwise quiescent galaxies. These discoveries span the entire electromagnetic spectrum, from γ-rays through X-rays, ultra-violet, optical, infrared, and radio. A small number of events launch relativistic jets. These tidal disruption events (TDEs) have caused widespread excitement as they can be used to study the properties of quiescent, otherwise undetectable, SMBHs; the populations and dynamics of stars in galactic nuclei; the physics of black hole accretion including the potential to detect relativistic effects near the SMBH; and the physics of (radio) jet formation and evolution in a pristine environment. For scientific questions concerning quiescent SMBHs, TDEs are unique probes beyond the local universe. TDEs can also occur around active galactic nuclei (AGN), although uniquely identifying such an event on top of a bright AGN is difficult.

Whereas AGNs (such as quasars) host SMBHs that are supplied by steady streams of fuel for thousands to even millions of years, TDEs offer a unique opportunity to study individual SMBHs with accretion rates that change by orders of magnitude on human timescales of days to months or sometimes years. In addition, because the rate of TDEs can be substantially enhanced in binary black hole systems, TDEs could point us to galaxies that are likely to host so far elusive compact binary SMBHs. Gaining knowledge on their presence and properties is of paramount importance for the ESA-led mission LISA, that will detect gravitational waves from the binary's final stages of coalescence. After the merger, a recoil kick imparted on the newly formed SMBH can cause a second wave of enhanced rate of TDEs, which can potentially be located off-centre with respect to the nucleus of the host galaxy.

Currently, the diverse emission properties of flares associated with TDEs is not fully understood. This challenge is being addressed by a sharp increase in observational work and theoretical modelling. Over the next few years, the largest growth areas will likely be in the greatly expanded surveys of the transient sky, and in new numerical modeling techniques. Together these will reveal how SMBHs shine by ripping apart orbiting stars and swallowing the stellar debris.

In light of this foreseen growth, many new researchers are expected to enter the field. Therefore, the time was deemed ripe to compose a comprehensive overview of the state of the art in this rapidly-evolving field. This book was planned and launched at a workshop held at the International Space Science Institute (ISSI) in Bern on 8–12 October, 2018.

The book starts off with a description of the observational properties of TDEs. Saxton et al. discuss the multi-wavelength properties and rates of those TDEs and TDE candidates selected via X-ray observations, and van Velzen et al. for those discovered in the optical and ultra-violet part of the spectrum. Alexander et al. present the radio properties of TDEs and show that only a small fraction of TDEs launches radio-luminous, mildly relativistic jets. A second van Velzen et al. contribution delves into cross-wavelength correlations, as well as echoes and reverberation from TDEs. Maguire et al. review the properties of TDEs from white dwarfs including the prospects for their detection across different wavelengths. Zabludoff et al. discuss how to distinguish TDEs from possible impostors such as supernovae, AGN flares, and gamma-ray bursts. These authors also discuss several candidate TDEs whose nature is still contended. Finally, French et al. summarize the host galaxy properties of known TDEs, and what can be learned from them, including the prospects for using host galaxy classification to assist in identifying TDEs in upcoming large transient surveys.

The book then turns to the theory behind TDEs and the emission mechanisms at play. Stone et al. present the theory that determines the rate of tidal disruptions, and overview theoretical TDE rate estimates. Rossi et al. describe the analytic theory of the tidal disruption process itself, followed by a presentation of existing numerical hydrodynamics simulations of stellar tidal disruption. Although the theory of the disruption process is reasonably well-understood, subsequent aspects of a TDE are substantially more uncertain. Bonnerot & Stone describe a particularly unsettled aspect of TDE physics: the dissipational processes by which highly eccentric stellar debris streams lose orbital energy before settling into some type of accretion flow. Next, Dai et al. describe the physics of more fully circularized TDE discs, where the accretion rates may be highly super- or sub-Eddington. Many stages of TDE evolution present problems in 3D hydrodynamics that can only be modeled precisely with numerical simulations. Lodato et al. review the numerical techniques and results that have been used so far in the literature. Roth et al. then review existing theories of radiative emission processes in TDEs, focusing on the current debates regarding the origin of optical/ultra-violet light in TDEs. Coughlin et al. present the unique features – both dynamical and hydrodynamical – of TDEs occurring in *binary* SMBH systems. Finally, Krolik et al. overview future prospects for simulations of TDEs, emphasizing both existing physical challenges and possible computational advances that may overcome them.

While there is a loose ordering of topics, the observational chapters are presented in the rough order that different wavelengths or other observational aspects of TDEs entered the literature, and theoretical chapters are mostly presented in a temporal sequence following the disruption process and its aftermath. Nevertheless, each chapter in this book has been designed as a stand-alone review, with appropriate references to the other chapters, where more details on a specific topic may be found. As such, there is no single "correct" order in which to proceed, and the reader should feel free to flip directly to the chapter of greatest interest to them.

Due to an error, the name of the editor Nicholas C. Stone was inadvertently left out of the list of editors for the collection of papers published in Space Science Reviews that form the chapters of this book. This error has been corrected in the book.

Acknowledgements: The editors would like to thank all authors and coordinators for their hard work in creating this book, the staff at ISSI for their friendly and generous support, and the future readers book, who will no doubt answer many of the TDE puzzles that so far remain unresolved.

Peter G. Jonker
Department of Astrophysics/IMAPP, Radboud University, P.O. Box 9010, 6500 GL, Nijmegen, The Netherlands
SRON, Netherlands Institute for Space Research, Sorbonnelaan 2, 3584 CA, Utrecht, The Netherlands

Iair Arcavi
The School of Physics and Astronomy, Tel Aviv University, Tel Aviv, 69978, Israel
CIFAR Azrieli Global Scholars program, CIFAR, Toronto, Canada

E. Sterl Phinney
Theoretical Astrophysics, 350-17 California Institute of Technology, Pasadena CA 91125, USA

Elena M. Rossi
Leiden Observatory, Leiden University, PO Box 9513, 2300, RA Leiden, The Netherlands

Nicholas C. Stone
Racah Institute of Physics, The Hebrew University, Jerusalem, 91904, Israel

Sjoert van Velzen
Leiden Observatory, Leiden University, PO Box 9513, 2300, RA Leiden, The Netherlands

Contents

Space Sci Rev (2020) 216:85
https://doi.org/10.1007/s11214-020-00708-4

X-Ray Properties of TDEs

R. Saxton[1] · S. Komossa[2] · K. Auchettl[3,4] · P.G. Jonker[5,6]

Received: 29 October 2019 / Accepted: 16 June 2020 / Published online: 1 July 2020

Abstract Observational astronomy of tidal disruption events (TDEs) began with the detection of X-ray flares from quiescent galaxies during the ROSAT all-sky survey of 1990–1991. The flares complied with theoretical expectations, having high peak luminosities (L_x up to $\geq 4 \times 10^{44}$ erg/s), a thermal spectrum with $kT \sim$ few $\times 10^5$ K, and a decline on timescales of months to years, consistent with a diminishing return of stellar debris to a black hole of mass 10^{6-8} $M_\odot$. These measurements gave solid proof that the nuclei of quiescent galaxies are habitually populated by a super-massive black hole. Beginning in 2000, XMM-*Newton*, Chandra and *Swift* have discovered further TDEs which have been monitored closely at multiple wavelengths. A general picture has emerged of, initially near-Eddington accretion, powering outflows of highly-ionised material, giving way to a calmer sub-Eddington phase,

The Tidal Disruption of Stars by Massive Black Holes
Edited by Peter G. Jonker, Sterl Phinney, Elena Maria Rossi, Sjoert van Velzen, Iair Arcavi and Maurizio Falanga

✉ R. Saxton
rsaxton@esa.int

S. Komossa
astrokomossa@gmx.de

K. Auchettl
katie.auchettl@nbi.ku.dk

P.G. Jonker
p.jonker@astro.ru.nl

1 TPZ-VEGA for ESA, XMM-Newton SOC, ESAC, Apartado 78, 28691 Villanueva de la Cañada, Madrid, Spain

2 Max Planck Institut für Radioastronomie, Auf dem Huegel 69, 53121 Bonn, Germany

3 DARK, Niels Bohr Institute, University of Copenhagen, Lyngbyvej 2, 2100, Copenhagen, Denmark

4 Present address: School of Physics, The University of Melbourne, Parkville, VIC 3010, Australia

5 SRON, Netherlands Institute for Space Research, Sorbonnelaan 2, 3584 CA, Utrecht, The Netherlands

6 Department of Astrophysics/IMAPP, Radboud University, P.O. Box 9010, 6500 GL, Nijmegen, The Netherlands

where the flux decays monotonically, and finally a low accretion rate phase with a harder X-ray spectrum indicative of the formation of a disk corona. There are exceptions to this rule though which at the moment are not well understood. A few bright X-ray TDEs have been discovered in optical surveys but in general X-ray TDEs show little excess emission in the optical band, at least at times coincident with the X-ray flare. X-ray TDEs are powerful new probes of accretion physics down to the last stable orbit, revealing the conditions necessary for launching jets and winds. Finally we see that evidence is mounting for nuclear and non-nuclear intermediate mass black holes based on TDE flares which are relatively hot and/or fast.

Keywords X-ray · TDE · Black holes · Accretion disks

1 Introduction

A key unsolved question in extra-galactic astrophysics in the 70s and 80s was, whether black holes exist at the centers of most or all galaxies. While 10% of galaxies were Active Galactic Nuclei (AGN) and believed to be powered by accretion onto supermassive black holes (SMBH), the remaining 90% were quiescent, inactive galaxies with no signs of permanent activity. Did they still harbour black holes at their centers? If so, how could we actually detect such "dormant" black holes? Only in a handful of the closest (quiescent) galaxies, could SMBHs be discerned by their impact on stellar rotation curves (Sargent et al. 1978). In order to tell, whether SMBHs existed not only in AGN, but in the majority of quiescent galaxies, Rees (1988) suggested that, if they disrupted a star and subsequently accreted the matter, then they would reveal their presence by a *temporary* X-ray flare, lasting a few months.

The integrated X-ray output of the stellar population of a galaxy is approximately 10^{39}–10^{41} erg s^{-1} (Fabbiano 1986) whereas the X-ray flare produced by a TDE emits at a significant fraction of the Eddington luminosity, 10^{42}–10^{45} erg s^{-1} (e.g. Rees 1988), and is hence relatively easy to detect thanks to the high contrast. While it is straight-forward to distinguish an AGN from a quiescent, in-active galaxy by means of optical spectroscopy (see next section), other X-ray source populations like flare stars, novae, or ULXs embedded in external galaxies are too distant to be resolved from their host galaxies by optical spectroscopy. However, they can still be readily distinguished by the X-ray peak luminosity of a TDE, which is orders of magnitude higher than the luminosity of other X-ray transient phenomena such as Novae, Supernovae, flares stars, accreting Galactic binaries and ULX.

It was shown that X-ray emission from TDEs should follow the rate of return of stellar debris to the black hole, which to a first approximation decays with a power-law index of $-5/3$ (Rees 1988; Phinney 1989; Evans and Kochanek 1989; Lodato and Rossi 2011) such that the luminosity is given by:

$$L_X = A(\frac{t - t_0}{1\ \mathrm{yr}})^{-5/3} \tag{1}$$

The first detections of TDE X-ray flares were made with the ROSAT satellite (see Sect. 3), which found transient sources that displayed the predicted TDE characteristics; a short rise to peak, a steady decline, high peak luminosities, a soft X-ray spectrum, and, importantly, occurred in quiescent, in-active galaxies. With the advent of Chandra, XMM-*Newton*, the Neil Gehrels *Swift* satellite (hereafter *Swift*; Gehrels et al. 2004) and multi-wavelength follow-up, differences in the spectra and light curve of each event both compared to AGN and within the class of TDEs detected became more apparent. In particular *Swift*, with its ability to perform high cadence, long-term monitoring of the X-ray and optical/UV light curve and spectrum, has had a major impact in the field of TDEs.

The X-rays generated during a TDE are believed to be produced from the innermost stable orbits of the black hole and experience the strong gravitational field. This means they can be used to probe the Kerr and Schwarzschild metrics via precession, quasi-periodic oscillations (QPOs) and reverberation (see the Echo Chapter and Rees 1990).

At the time of writing,[1] the number of X-ray emitting TDEs and X-ray emitting TDE candidates is small enough that each one can be discussed individually, which we do in the following sections. A summary of the properties of each object is given in Table 1.

2 TDE Identification: How to Distinguish Between a Quiescent Galaxy and an AGN by Optical Spectroscopy

Astronomers distinguish between quiescent galaxies and Active Galactic Nuclei (AGN). They are routinely identified and distinguished by means of optical spectroscopy. AGN are galaxies which harbor a permanent accretion disk that emits a strong multi-wavelength continuum including X-rays. This photoionises ambient gas, and produces strong characteristic, long-lived emission lines. Quiescent galaxies are those which *do not harbor a permanent accretion disk* and therefore *do not emit any luminous X-rays from their cores*, neither variable nor constant, and they *lack the characteristic optical emission lines* (Osterbrock 1989).

The two of them can therefore be distinguished by means of *optical spectroscopy*: AGN have characteristic, high-ionization *narrow emission lines*, which arise in low-density gas at large nuclear distances (the so-called 'Narrow-line Region'; NLR). The NLR is photoionized by the photons from the *permanent* accretion disk, and it *traces the activity over a long time interval of 1000s of years*. In contrast, quiescent galaxies *lack these characteristic NLR emission lines*. The majority of quiescent galaxies have no emission lines at all while star-formation galaxies have low-ionization lines, very different from NLRs (Baldwin et al. 1981; Osterbrock 1989; Kewley et al. 2001; Kauffmann et al. 2003). Therefore, a crucial and very reliable distinction between an AGN and a quiescent galaxy is to take an optical spectrum, and determine the presence or absence of the characteristic NLR emission lines. Identifying quiescent galaxies by optical spectroscopy has been routinely carried out for decades (see Osterbrock 1989, for a good textbook). Optical spectroscopy to check for a quiescent, inactive host galaxy has therefore been an important step in the identification of any new TDE (e.g. Komossa and Greiner 1999), following the procedures suggested by theoreticians (e.g. Rees 1988, 1990) irrespective of the waveband in which it was discovered.[2]

Finally, we would like to note, that (1) low-mass AGN and (2) AGN in LINERs (low-ionization nuclear emission regions) are more difficult to identify spectroscopically (e.g. Greene 2012). High S/N optical spectroscopy is required in case (1), while (2) LINER-like emission lines, which populate a separate region in diagnostic diagrams, are known to be powered by several different physical mechanisms; shocks, photoionization by old stellar populations, and/or AGN (e.g. Schulz and Fritsch 1994).

[1] In this chapter we include publications written up until mid-2019.

[2] Note, that broad and narrow optical emission lines can also be *temporarily* excited by the TDE itself (e.g. Komossa et al. 2008; Wang et al. 2012). However, these lines differ from classical NLRs and can be distinguished if there is more than one post-flare optical spectrum, since they are not permanent but will change and fade away quickly.

3 ROSAT Soft X-Ray TDEs

Given the low disruption rate of TDEs (about 1 event every 10^{4-5} years per galaxy; Sect. 12), large-area sky surveys are best-suited to detect these events. The X-ray mission ROSAT was an ideal experiment to systematically search for, and detect, TDEs. Launched in June 1990, it carried out the first imaging X-ray survey of the entire sky in the soft X-ray band, at energies between (0.1–2.4) keV (e.g. Trümper 1994). The ROSAT all-sky survey (RASS) lasted for about a year and was then followed by an eight-year phase of pointed observations of selected targets, including deep follow-ups of exceptional sources and transients discovered during the RASS. ROSAT was equipped with two prime instruments, a high-resolution imager (HRI) providing a spatial resolution of 5 arcsec, and a positionally-sensitive proportional counter (PSPC) which allowed broad-band X-ray spectroscopy and achieved a spatial resolution of 25 arcsec on-axis.

Several luminous, high-amplitude X-ray flares from quiescent galaxies, matching remarkably well the predictions of tidal disruption theory (e.g. Rees 1988, 1990), have first been identified by ROSAT. Four main events were discovered, from the galaxies NGC 5905 (Bade et al. 1996; Komossa and Bade 1999), RXJ1242-1119 (Komossa and Greiner 1999), RXJ1624+7554 (Grupe et al. 1999) and RXJ1420+5334 (Greiner et al. 2000). Among these, NGC 5905 and RXJ1242-1119 are the best-covered events in terms of their multi-wavelength follow-up observations and their long-term X-ray lightcurves, spanning time intervals of more than a decade, with amplitudes of decline larger than a factor of 1000 (Komossa et al. 2004; Halpern et al. 2004; Komossa 2005). We therefore summarize some main properties of these two events in the subsections below.

In addition, the ROSAT archive was used to identify new TDEs after the end of the mission; either sources which were bright during the ROSAT epoch, and had faded substantially when re-observed with present X-ray missions (Cappelluti et al. 2009; Maksym et al. 2014a; Khabibullin and Sazonov 2014), or events which were found to be bright in later observations, but much fainter or undetected with ROSAT. The ROSAT database has therefore played an important role in the identification of most soft X-ray TDEs known today.

3.1 NGC 5905

NGC 5905 was first noticed due to its luminous, soft ($kT = 0.06$ keV) X-ray outburst with (lower limit on the) peak luminosity in the soft X-ray band of $L_{\mathrm{x,peak}} = 7 \times 10^{42}$ erg/s during the RASS[3] (Komossa and Bade 1999). It remained bright for at least 5 days (the time interval while its position was repeatedly scanned during the RASS) increasing in luminosity to the observed peak (Fig. 1). Given its daily coverage for almost a week, it has one of the best-covered early lightcurves of the non-radio-emitting X-ray TDEs.

X-rays then declined on the timescale of months to years, as observed in multiple re-observations with ROSAT (Fig. 2). Within the 5 arcsec positional error (ROSAT HRI), the X-rays came from the center of this nearby barred spiral galaxy ($z = 0.011$; Fig. 1). While the X-ray spectrum was initially very soft, it had hardened significantly ($\Gamma_{\mathrm{x}} = 2.4$) 3 years later, when re-observed with ROSAT. The decline of its X-ray lightcurve is consistent with the predicted $t^{-5/3}$ law (Fig. 2), as first reported based on its ROSAT observations (Komossa and Bade 1999) and then confirmed with Chandra (Halpern et al. 2004). All observations of this event are in very good agreement with the predictions (Rees 1990) from tidal disruption theory (Bade et al. 1996; Komossa and Bade 1999).

[3]This luminosity is even higher, if a powerlaw model is fit, and if there is absorption from the event's host galaxy.

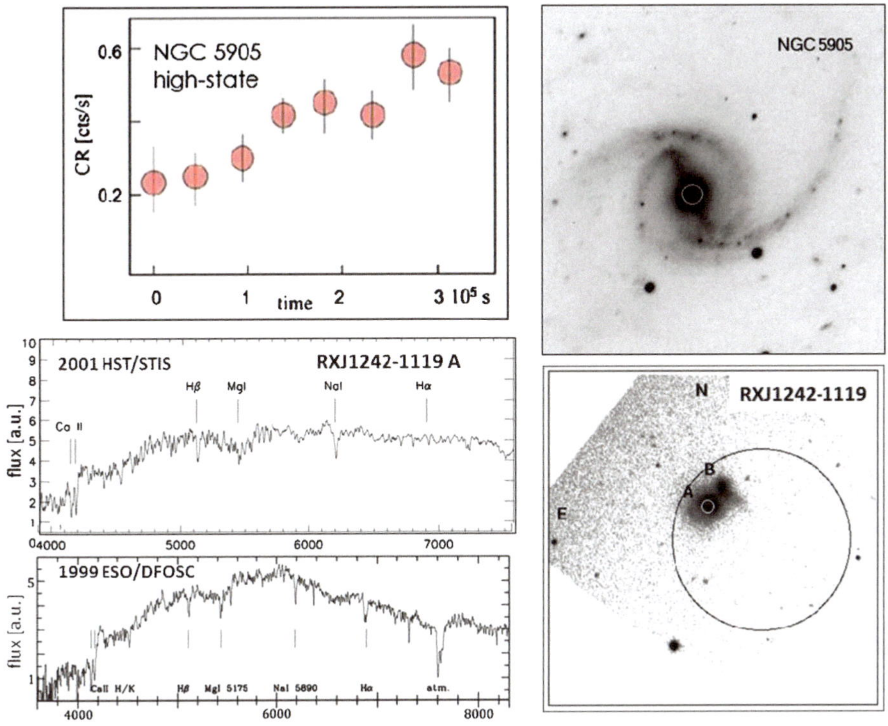

Fig. 1 Upper left: Early X-ray lightcurve of NGC 5905 showing its rise to the observed maximum (Komossa and Bade 1999; Bade et al. 1996) when the galaxy position was repeatedly covered during the RASS scans. Upper right: Optical image of NGC 5905 (Komossa 2002); a nearby giant barred spiral galaxy. The X-ray flare was observed from the centre of the galaxy. The white circle corresponds to the ROSAT HRI error circle of 5″. Lower right: Optical image of RXJ1242-1119. The inital ROSAT X-ray error circle (black) contained two galaxies, both inactive. The X-ray error circle was narrowed down with Chandra (white), which detected fading X-ray emission from the core of RXJ1242-1119A (Komossa et al. 2004). Lower left: Optical spectrum of RXJ1242-1119A [upper panel: HST STIS spectrum from August 2001 (Gezari et al. 2003); lower panel: groundbased ESO spectrum from January 1999 (Komossa and Greiner 1999)]. These spectra completely lack emission lines and reveal a quiescent, inactive galaxy

A very important first step in the identification of every single TDE from a *quiescent* galaxy, is to take an optical spectrum of the host galaxy, and search for the absence of AGN activity down to deep limits. Ground-based optical spectroscopy before and after the flare has led to a starburst (HII-type) classification of the host galaxy of NGC 5905 (Komossa and Bade 1999). An HST observation after the flare showed evidence for faint enhanced [OIII] emission from the core of the galaxy (Gezari et al. 2003) – either long-lived or excited by the flare itself. Using scaling relations between host galaxy and central SMBH mass for elliptical galaxies (Ferrarese and Merritt 2000), an upper limit on the SMBH mass of NGC 5905 of a few times $10^8\ M_\odot$ was estimated; near the upper limit for the tidal disruption of a solar-type star and as such could imply a spinning SMBH. However, the host of NGC 5905 is a spiral galaxy and these tend to lie below the scaling relations of ellipticals. Using instead the relation of Salucci et al. (2000) for spirals gives a BH mass estimate of $\sim 10^7\ M_\odot$ (Komossa 2002).

Integrating over its (observed) lightcurve (Eq. (1)), only a small fraction of a solar mass needed to be accreted to power this flare (Komossa 2002; Li et al. 2002). We look at possible

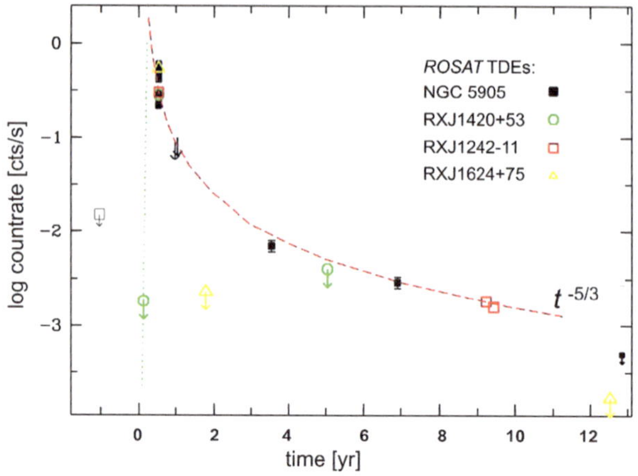

Fig. 2 Joint X-ray lightcurve of four TDEs identified with ROSAT, all shifted to the same peak time. The decline is consistent with a $t^{-5/3}$ law (dashed lined). This point was first made based on the ROSAT data of NGC 5905 (Komossa and Bade 1999), and later for the overall luminosity evolution of the sources displayed above (e.g., Fig. 1 of Komossa 2004). RXJ1242-1119 shows a further drop in X-ray emission at late times (not shown here), deviating from the early phase decline law, implying a total amplitude of decline of a factor > 1000 (Komossa 2005)

explanations for this in Sect. 4.4. Note that all such estimates provide a lower limit on the accreted stellar mass, since the events may not have been caught exactly at peak luminosity, and since a significant fraction of the luminosity may be emitted outside the soft X-ray band.

The morphology of the host galaxy of NGC 5905 is that of a barred spiral galaxy of type SB; one of the largest spirals known and particularly well resolved in optical imaging since nearby (Fig. 1). Its multiple triaxial structures with a secondary bar might aid the efficient re-filling of its stellar loss-cone orbits, thereby boosting the stellar tidal disruption rate (see Sect. 3.3.3 of Komossa 2002).

3.2 RXJ1242-1119

A very luminous X-ray outburst was discovered from the inactive galaxy RX J1242-1119 (at $z = 0.05$) with ROSAT during a serendipitous pointed observation. Its X-ray spectrum was extremely soft, one of the steepest ever identified among galaxies, and among the steepest in the ROSAT data base, with a photon index of $\Gamma_x = 5.1 \pm 0.9$ (Komossa and Greiner 1999). At a soft X-ray luminosity of 4×10^{44} erg/s (a conservative lower limit, assuming no absorption in the host galaxy itself, and without applying a bolometric correction), this event was exceptionally luminous.

Optical spectroscopy of the host galaxy, both ground-based and with the HST, reveals a quiescent, in-active galaxy, with no emission lines detected at all down to deep limits (Komossa and Greiner 1999; Gezari et al. 2003).

During the RASS itself, RXJ1242-1119 was not detected, implying an initial amplitude of variability of a factor of > 20, and a rise-time of < 1.5 years. Deep Chandra follow-ups (cf. next section) then revealed the fading of the X-ray emission from the TDE by a factor of up to > 1000 (Komossa et al. 2004; Halpern et al. 2004; Komossa 2005) more than a decade after its high-state. Given its extreme properties and deep follow-ups, this event continues to be one of the best cases of a TDE identified so far (Komossa et al. 2004).

Integrating only over the *observed* (0.1–2.4) X-ray lightcurve then gave a strict lower limit on the total emitted energy of 1.6×10^{51} erg/s (Eq. (1)) and on the accreted mass of $M > 0.01\eta_{0.1}^{-1}\ M_\odot$ for the TDE in RXJ1242-1119 (Sect. 3 of Komossa et al. 2004).

Based on the host galaxy blue magnitude measured with the optical monitor OM onboard XMM-Newton, $m_\mathrm{b} = 17.56 \pm 0.05$, the mass of the black hole at the center of the galaxy was estimated, based on the correlation between the absolute blue magnitude of the bulge of an elliptical galaxy and its SMBH mass (Ferrarese and Merritt 2000). This yielded an SMBH mass on the order of $M_\mathrm{BH} \approx 10^8\ M_\odot$ (Komossa et al. 2004).

3.3 More Recently Identified TDEs in the ROSAT Archive

During the ROSAT mission, an X-ray catalogue of > 100,000 X-ray sources was created (Voges et al. 1999). This data base can still be used, in combination with the X-ray data bases more recently created with the missions XMM-Newton and Chandra, to search for new TDEs bright during the ROSAT observation(s). Cappelluti et al. (2009) identified the X-ray outburst TDXF1347-3254 ($z = 0.037$) that way, while Maksym et al. (2014a) and Khabibullin and Sazonov (2014) reported the detection of an outburst from RBS 1032 ($z = 0.026$).

The event TDXF1347-3254 was the first one identified in a galaxy cluster, in Abell 3571 (Cappelluti et al. 2009). X-ray emission from one of the cluster galaxies, LEDA 095953, was bright in ROSAT but decayed by a factor 650 between 1992 and 2004. The black hole mass, M_BH, was estimated to be $10^7\ M_\odot$ and the integrated luminosity was $> 2 \times 10^{50}$ erg/s, implying an accreted mass between 0.001 and 0.03 solar masses. The ROSAT PSPC spectrum had an effective $kT = 120$ eV. Multi-band optical/NIR photometry, taken at the time, indicated an inactive host galaxy (Cappelluti et al. 2009) which has subsequently been confirmed by optical spectroscopy (Wevers et al. 2019a).

RBS 1032 was a supersoft ($\Gamma_\mathrm{x} \sim 5$) and luminous ($\sim 10^{43}$ erg/s) ROSAT PSPC source, which later had faded by a factor $\sim$ 100–300 when re-observed with XMM-Newton. It is hosted by an inactive dwarf galaxy (Ghosh et al. 2006; Maksym et al. 2014a). From the shape of the light curve, Khabibullin and Sazonov (2014) deduced that the event had been first seen well after peak and that the peak bolometric luminosity was more likely to have been a few $\times 10^{44}\ \mathrm{erg\,s^{-1}}$. This in turn implies that $M_\mathrm{BH} > 10^6\ M_\odot$. The event remained quite soft ($\Gamma_\mathrm{x} = 3.4 \pm 0.3$) 20 years after discovery.

3.4 Summary of the Properties of the ROSAT TDEs

In summary, the ROSAT TDEs are characterized by:

- Soft X-ray (0.1–2.4 keV) peak luminosities up to several $10^{44}\ \mathrm{erg\,s^{-1}}$.
- Very soft X-ray spectra near peak, with black-body temperatures in the range $kT_\mathrm{bb} =$ 0.04–0.12 keV (or, alternatively, with powerlaw indices in the range $\Gamma_\mathrm{x} = 4$–5), followed by a spectral hardening within years.
- A decline law of NGC 5905 and RXJ1242-1119 consistent with $t^{-5/3}$, and a drop below that decline law after $t > 10$ years.
- Total amplitudes of decline up to factors > 1000–6000, measured in deep dedicated Chandra follow-up observations (Sect. 4).
- Host galaxies, which are optically inactive, radio-inactive, and also X-ray inactive at low-state.
- The host galaxy morphology of the nearest event, in NGC 5905, is a giant barred spiral galaxy.

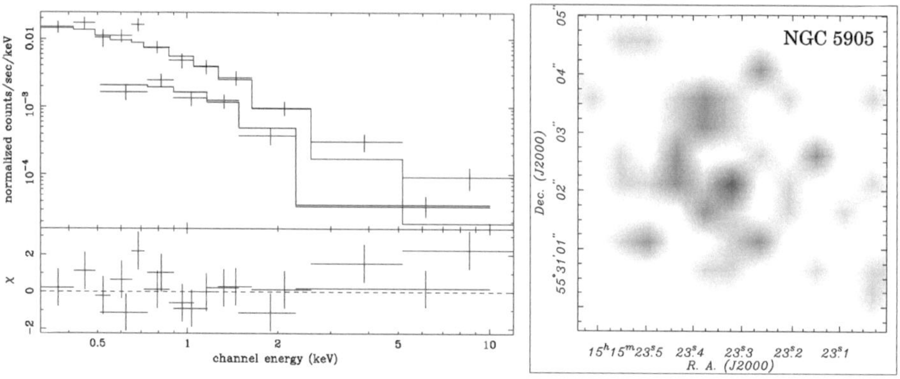

Fig. 3 Left: X-ray spectrum of RXJ1242-1119, the first TDE followed up with XMM-Newton. The observations revealed a strong hardening of the X-ray spectrum compared to the high-state observation (Komossa et al. 2004). Right: Deep Chandra X-ray image of NGC 5905 (Halpern et al. 2004) 12 years after the flare. Most (or all) of the low-flux state emission is widely extended and from the host galaxy

- All of them occurred in the nearby universe ($z = 0.011$–0.147).
- Black hole mass estimates are in the range $10^{6-8}\ M_{\odot}$.

Given their extreme properties, particularly the absence of an optical AGN down to deep limits, huge peak luminosities and total amplitudes of variability (the largest seen in galaxies so far), they continue to be among the most reliably identified TDEs to date.

4 Using Chandra and XMM-Newton to Follow Up and Detect TDEs

At the end of the 20th century the launch of the XMM-Newton (Jansen et al. 2001) and Chandra (Weisskopf et al. 2000) satellites provided the first high spatial and spectral resolution X-ray observation that covered the soft and medium energy bands (0.2–12 keV). XMM-*Newton* is flying three CCD cameras, with energy resolution of 50–100 eV at 1 keV and a reflection grating spectrometer (RGS) with a spectral resolution from 100 to 500 (FWHM) in the energy range 0.33–2.5 keV, as well as an Optical Monitor (OM) hosting visible and UV filters. Chandra hosts the ACIS camera, the High-Resolution camera (HRC) giving sub-arcsecond spatial resolution and low-energy (resolution > 1000) and high-energy (resolution = 200–800) transmission gratings.

These observatories were used for the first deep follow-up campaigns of the TDEs initially discovered with ROSAT. Chandra and XMM-Newton observations of RXJ 1242-1119 detected faint and fading X-ray emission a decade after the initial TDE high-state consistent with the optical position of the core of the host galaxy within the ~ 1 arcsecond spatial error of Chandra, supporting the TDE interpretation (Komossa et al. 2004). With XMM-Newton, a first high-quality post-flare spectrum of the faint late-state X-ray emission from a TDE (RXJ 1242-1119) was obtained (Fig. 3), which showed a significant spectral hardening of the event, from $\Gamma_x \sim 5.1 \pm 0.9$ during high-state to $\Gamma_x = 2.5 \pm 0.2$ with XMM-*Newton* (Komossa et al. 2004), possibly related to the formation of an accretion-disk corona, initially absent. We look further at the long-term evolution of the X-ray spectra of TDEs in Sect. 4.5.3.

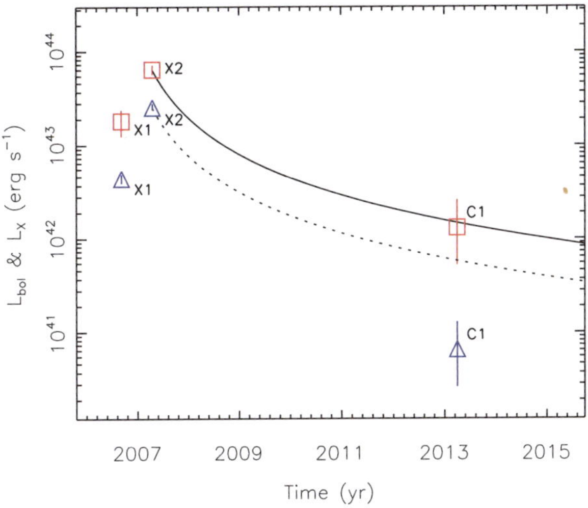

Fig. 4 The soft X-ray (blue) and bolometric (red) light curve of 2XMMi J184725.1-631724, with X1, X2 and C1 representing the first and second XMM-*Newton* observations and the first Chandra observation respectively (adapted from Lin et al. 2018b). The solid line plots a fit to the bolometric luminosity of the function $(t - t_0)^{-5/3}$ with t_0 set at 1 month before the date of X1, while the dashed line fits the same function to the soft X-ray luminosity

Chandra and XMM-*Newton* have been used for catalogue searches of TDEs, and also for the quick identification of new TDEs at peak by comparison with the RASS X-ray catalogues.

XMM-Newton generally makes observations of a few 10s of ks detecting around 50 sources within its quarter square degree field-of-view. These 'serendipitous' sources have been collated into a catalogue, at the time of writing in its third incarnation (3XMM; Rosen et al. 2016) and containing upwards of 500,000 independent sources from 2% of the sky. Chandra has produced a catalogue (CSC 2.0; Evans et al. 2010) with 315,000 sources from 1% of the sky.

With a mean flux of 10^{-14} (10^{-15}) $\mathrm{erg\,s^{-1}\,cm^{-2}}$ for XMM-Newton (Chandra) in the 0.2–2 keV energy band the catalogued sources are generally too faint to be compared with historical observations in a search for variability. However, many fields have been observed on multiple occasions and it is possible to look at the flux history of a significant fraction of the catalogue for up to two decades. These comparisons have yielded a number of candidate TDEs.

4.1 TDEs Discovered Using XMM-Newton and Chandra

4.1.1 2XMMi J184725.1-631724

One such object was 2XMMi J184725.1-631724 (Lin et al. 2011), detected in an XMM-*Newton* pointed observation with a peak observed soft X-ray flux of 2×10^{-12} $\mathrm{erg\,s^{-1}\,cm^{-2}}$ in 2007 but not detected by a further XMM-Newton observation in 2013 ($F_X < 2.5 \times 10^{-14}$ $\mathrm{erg\,s^{-1}\,cm^{-2}}$), with Chandra confirming a drop in flux by a factor 1000 (Lin et al. 2018b) in 2013 (Fig. 4). The source whose position is consistent with the nucleus of the optically-inactive galaxy IC 4765-f01-1504 ($z = 0.0353$), was initially flagged due to its very soft spectrum (equivalent black-body temperature of $kT < 100$ eV).

A fainter, but still flux-enhanced, observation of 2XMMi J184725.1-631724 in 2006 allows the disruption time to be constrained to within about one month enabling the index of the flux decay law to be well measured in this TDE. Surprisingly it turns out to be considerably steeper than $-5/3$. However, we must bear in mind that this is an in-band flux

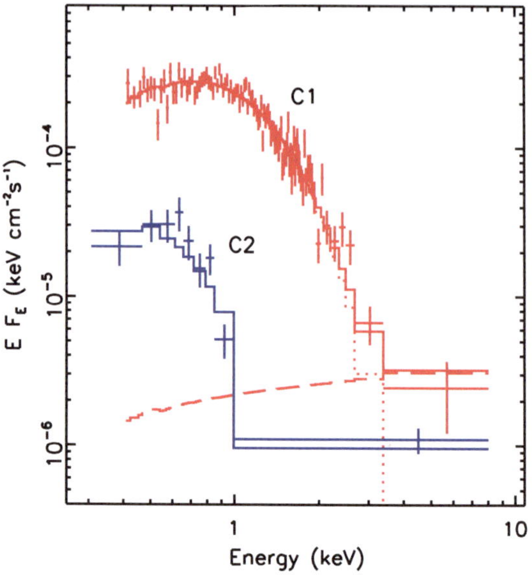

Fig. 5 Chandra spectra of 3XMM J215022.4-55108 from 2006-05-05 (red) and 2016-09-14 (blue) fit with a thermal disk model, of $kT = 280(140)$ eV, plus a weak power-law (adapted from Lin et al. 2018a)

whereas we should be comparing the bolometric radiation output, under the assumption that this traces the fall back rate of the stellar debris. The peak bolometric luminosity of the TDE was 6×10^{43} erg s^{-1}, and the BH mass $M_{\mathrm{BH}} = 1 \times 10^{6-7}\ M_{\odot}$, based on the host galaxy V and K magnitudes, with the peak soft X-ray spectrum dominated by a thermal model of $kT \sim 80$ eV. As the spectrum appears to be dominated by black-body radiation, the emission temperature should decrease as $T \propto L^{1/4}$ (see Sect: 4.5) and hence in the Chandra observation of 2013 will have cooled to $kT \sim 35$ eV. At this temperature, much of the flux is shifted out of the X-ray band and the bolometric correction is correspondingly larger than that needed for $kT = 80$ eV. Accounting for this effect brings the index of the *bolometric* flux decay into good agreement with $-5/3$ (Lin et al. 2018b).

4.1.2 3XMM J152130.7+074916 and 3XMM J215022.4-55108: Possible TDEs from Intermediate Mass Black Holes

The X-ray spectra of TDEs are usually soft in the early phase. If this temperature is related to the black hole mass (see Eq. (4)), then in principle this allows us to identify elusive intermediate mass black holes (IMBH) within the TDE population from their X-ray spectrum, something that is not possible in AGN surveys where the spectrum is dominated by a power-law with a mass-independent slope. The TDE 3XMM J152130.7+074916 is a good example. Discovered in an XMM-*Newton* observation in 2000 with a flux 100x higher than that seen in Chandra 4 months earlier, the spectrum could be fit with a disk model with an inner temperature of $kT \sim 170$ eV and M_{BH} between 0.19 and $1.4 \times 10^{6}\ M_{\odot}$ depending on BH spin (Lin et al. 2015).

An even lower mass was inferred in 3XMM J215022.4-55108 which lies about 12.5 kpc from the nucleus of a redshift 0.055 galaxy (Lin et al. 2018a). Here the effective temperature peaked at $kT = 280$ eV, reducing to 140 eV as the luminosity decayed from $L_X = 10^{43}$ to 10^{42} erg s^{-1} (Fig. 5). If these temperatures really come from the inner edge of the accretion disk then the BH in this case has a mass of a few $\times 10^{4}\ M_{\odot}$.

This method of estimating black hole mass can only work if the correct spectral model is applied and should be used with caution. For example, a corona (see Sect. 4.5), if present, would push X-rays to higher energy, artificially increasing the measured temperature and causing the black hole mass to be underestimated. The effect of black hole spin also needs to be taken into account, causing an order of magnitude variation even in simple models (e.g. Lin et al. 2015).

4.1.3 TDEs in Clusters of Galaxies

One way of maximising the chances of finding TDEs using Chandra and XMM-*Newton* pointed observations is to repeatedly observe a rich cluster of galaxies and hence simultaneously monitor many galaxies (Wang and Merritt 2004). Maksym et al. (2010) observed the Abell class 4 cluster, A1689, which at $z \approx 0.18$, has a radius of 10 arcminutes nicely fitting into the Chandra field of view. They observed the cluster, which has an estimated 2100 galaxy members, 6 times with Chandra and once with XMM-*Newton* over an 8 year period. One galaxy, SDSS J131122.15-012345.6, varied by a factor 30 over the observations, displaying a soft spectrum ($kT \sim 100$ eV) and peak $L_X > 5 \times 10^{42}$ erg s^{-1}. From relations with the galaxy magnitude they estimated the black hole mass to be $M_{\rm BH} = 1\text{–}7 \times 10^6\ M_\odot$. The light curve could be reasonably well fit with the canonical $t^{-5/3}$ law, from whence the total X-ray luminosity can be found by integrating over Eq. (1) for the duration of the flare. They estimate the bolometric correction factor to be 1.4 and found the total luminosity emitted over the event to be $L_{\rm bol} \sim 10^{50}$ erg/s. After correcting for the expected emission before the first detection, the equivalent total mass accreted was found from Eq. (2) to be ~ 0.01 solar masses, a surprisingly low value which we address in Sect. 4.4.

Another cluster which received a lot of attention from Chandra was A1795, observed a total of 17 times between 1999 and 2010 (Maksym et al. 2013). One dwarf galaxy, WINGS J1348, with a mass of just $3 \times 10^8\ M_\odot$ (Maksym et al. 2014b), showed a bright X-ray source in the first observation which subsequently decayed by a factor > 50. It appears likely that this flare was first captured by the EUVE satellite, in the 0.016–0.163 keV band, a year before the launch of Chandra, making it the only known TDE detected in the EUV band to date. The spectrum in the first Chandra observation was soft ($\Gamma_{\rm x} = 4.1$ or $kT = 84$ eV) with evidence for further softening over the following year. The sparse light curve again can be reasonably fit to a canonical $t^{-5/3}$ decay index.

4.2 TDEs Discovered in the XMM-Newton Slew Survey

XMM-Newton slews between targets at a rate of 90 degrees per hour and keeps its most sensitive camera, EPIC-pn (Strüder et al. 2001), open to record the sky. While a given source passes through the camera in just 15 seconds, the large mirror effective area and short frame time of the detector gives a detection limit of $F_{0.2\text{–}2} = 6 \times 10^{-13}$ erg s^{-1} cm^{-2} and a positional accuracy of ~ 8 arcseconds (Saxton et al. 2008). Again at the time of writing, 85% of the sky has been covered by slews and > 20000 sources are contained in the XMMSL2 catalogue. The slew sensitivity is well matched with that of the RASS and meaningful comparisons can be made between these surveys to find the bright population of X-ray TDEs. This was first done by Esquej et al. (2007) who found five, previously anonymous, galaxies that showed a factor > 20 increase in flux between the RASS and slew surveys. Follow-up optical spectra showed that two of these were likely to be due to AGN variability, two (NGC 3599 and SDSS J132341.97+482701.3) were good candidates for TDEs, while XMMSL1 J024916.6-041244 was apparently a persistent Seyfert 1.9 galaxy but showed TDE traits, namely a very soft spectrum and a factor 100 flux decay (Strotjohann et al. 2016; Auchettl et al. 2017; Wevers et al. 2019b).

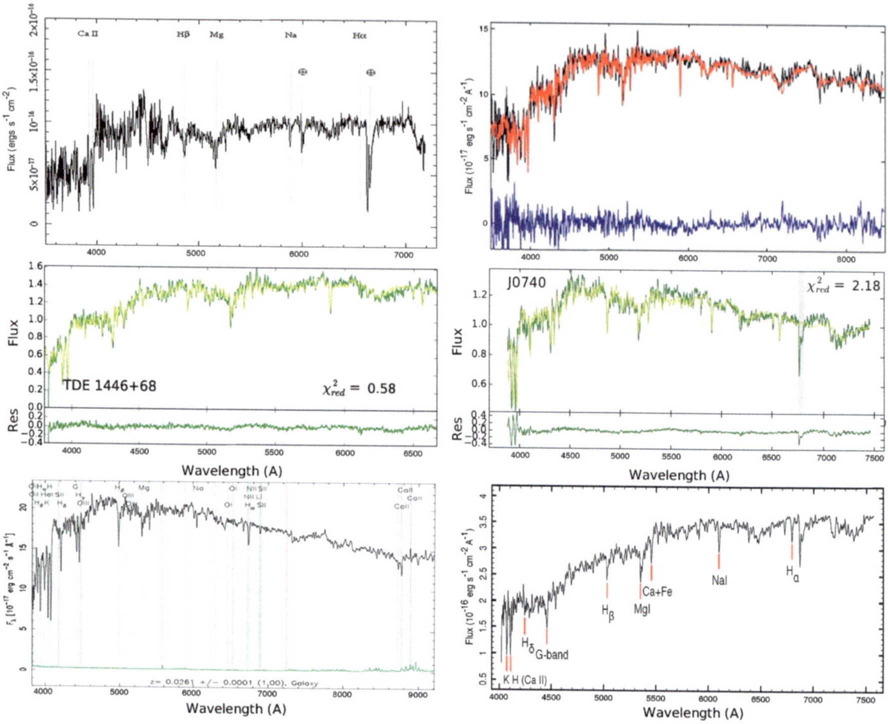

Fig. 6 Optical spectra of TDEs detected by XMM-*Newton*. Top left: SDSS J120136.02+300305.5 (Saxton et al. 2012a), top right: SDSS J132341.97+482701.3 (Esquej et al. 2007); middle left: XMMSL2 J144605.0+685735 (Saxton et al. 2019), middle right: 2MASX 07400785-8539307 (Saxton et al. 2017), bottom left: RBS 1032 (Ghosh et al. 2006), bottom right: 2XMMi J184725.1-631724 (Lin et al. 2011)

4.2.1 SDSS J120136.02+300305.5

The large sky coverage and relatively fast processing of XMM-Newton slews opened up the possibility of investigating events while they were close to the peak of their luminosity. This was first applied to SDSS J120136.02+300305.5 (hereafter SDSS J1201+30) spotted in a slew in 2010 (Saxton et al. 2012a,b). Figure 7 shows the X-ray light curve of SDSS J1201+30 as seen in XMM-Newton observations and short exposures taken with *Swift*. Intriguingly the initially high-luminosity emission from this event (10^{45} erg s^{-1}) disappeared 3 weeks after discovery, reducing by at least a factor 50 between 2010-06-30 and 2010-07-07. The possibility that this was caused by a temporary local absorption event is unlikely given the long, >21 day, duration of the drop in luminosity. Another possibility is that the return of material to the BH was interrupted by the passage of a secondary black hole (Liu et al. 2014). This scenario is further explored in Sect. 11. The emission from this TDE was soft, but wider than a single black-body, being empirically well fit by a Bremsstrahlung model (Saxton et al. 2012a). During the year of observations, the event got softer showing no evidence for the development of a hard tail.

4.2.2 2MASX 0740-85

2MASX 07400785-8539307 (hereafter 2MASX 0740-85) was detected in a slew in 2014 a factor 20 brighter than an upper limit from ROSAT (Saxton et al. 2017). Monitoring by

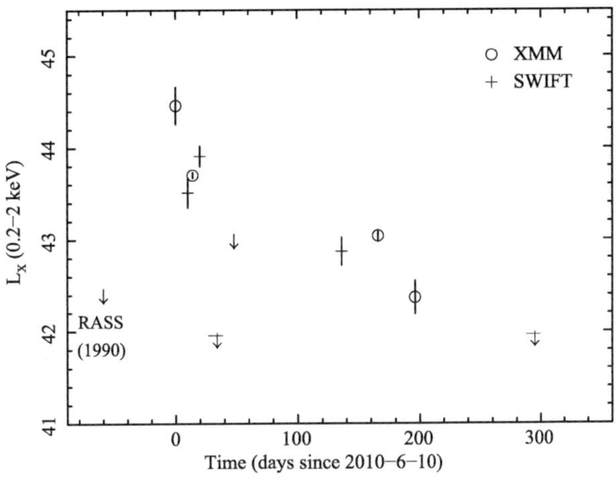

Fig. 7 Soft X-ray light curve of SDSS J1201+30. Upper limits are from the RASS or Swift-XRT (adapted from Saxton et al. 2012a)

Swift and XMM-*Newton* over the following 550 days revealed a drop in X-ray flux by a factor of 70 (Fig. 8). The host galaxy was shown to be optically inactive (Fig. 6). The first X-ray observations were highly variable, with a flux doubling time of 400 seconds (Fig. 9). Using a variability-mass scaling relationship (Ponti et al. 2012) the black hole mass was estimated to be $M_{\rm BH} = 3.5^{+6.5}_{-2.4} \times 10^6\ M_\odot$. Light travel-time arguments then place the size of the emitting region within $73R_g$ of the black hole.

The X-ray spectrum of the event in 2MASS 0740-85 is unusual because it consists of a power-law of $\Gamma_{\rm X} \sim 2$ which dominates in the "hard" 2–10 keV X-ray band and contributes $\sim 25\%$ of the total energy. While other TDEs often show a weak hard-tail to the soft emission, which can be approximated by a (usually) steep power-law (e.g. Lin et al. 2018b; Holoien et al. 2016b; Kara et al. 2018) this feature is akin to the dominant power-law seen almost ubiquitously in AGN (Nandra and Pounds 1994). In other sources, SWIFT J164449.3+573451, Swift J2058.4+0516 and IGR J12580+0134 (see the Gamma-ray Chapter), the hard X-rays were likely generated in a jet as evidenced by the strong accompanying radio emission. The radio emission from 2MASX 0740-85 was rather weak, 1 mJy at 1.53 GHz, and may have been generated from shocks in an outflow rather than a collimated jet (Alexander et al. 2017).

2MASX 0740-85 had a UV flare contemporaneous with the X-ray flare (Fig. 8) which decayed with a flatter slope (Saxton et al. 2017). The relationship between the X-ray and UV emission of X-ray selected TDEs is explored further in Sect. 9.1.

The X-ray and UV emission from 2MASX 0740-85 could be reasonably well connected using a multi-coloured disk model (see Sect. 4.5) suggesting that the wide-band emission may be coming from a single coherent structure (Saxton et al. 2017).

4.2.3 XMMSL2 J1446+68

XMMSL2 J144605.0+685735 (hereafter XMMSL2 J1446+68) is another TDE detected in an XMM-*Newton* slew, this time in 2016. In this event the X-ray flux remained stable for the first ~ 100 days after discovery, before experiencing a drop of a factor 100 over the following 500 days (Fig. 8 Saxton et al. 2019). The host galaxy is optically inactive (Fig. 6) and with the peak bolometric luminosity, $L_{\rm bol} \sim 10^{43}$ erg s^{-1}, interpretations other than a TDE are unlikely (Saxton et al. 2019).

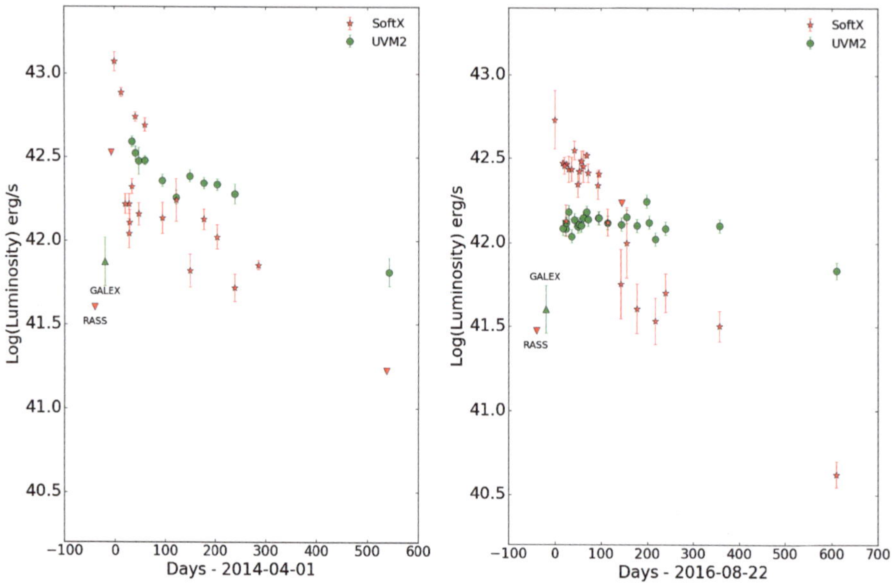

Fig. 8 The soft (0.2–2 keV) X-ray (red) and UVM2 (2340 Å; green) light curves for 2MASX 0740-85 (left) and XMMSL2 J1446+68 (right). Luminosity has been corrected for galactic extinction and includes the contribution from the host galaxy. GALEX-nuv filter measurements from 2007 have been rescaled to the UVM2 filter in both plots, which are adapted from Saxton et al. (2017) and Saxton et al. (2019)

Fig. 9 A one-day EPIC-pn light curve of 2MASX 0740-85 taken by XMM-*Newton* soon after discovery, showing large short-term variability (adapted from Saxton et al. 2017)

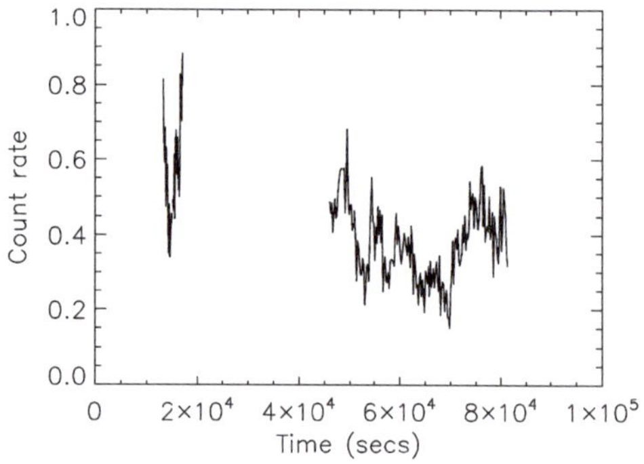

It exhibited X-ray spectra consistent with a single power-law, stretching from 0.3–10 keV with slope, $\Gamma_{\rm x} \sim 2.6$. Even in high signal-to-noise spectra there is no significant evidence of a low-energy thermal component in addition to this power-law. The black-hole mass estimated from optical absorption-line widths is quite high, $M_{\rm BH} = 7^{+17}_{-5} \times 10^7\ M_\odot$ (Wevers et al. 2019b), and thermal emission may be too cool to enter the X-ray band (see Sect. 4.5). The spectral shape remained constant over time in this event, even while the X-ray flux fell by a factor 100.

The event was accompanied by UV emission which remained roughly constant for 400 days before decaying by a magnitude over the next few hundred days. This disconnect between the UV and X-ray flux is unusual and is discussed further in Sect. 9.1.

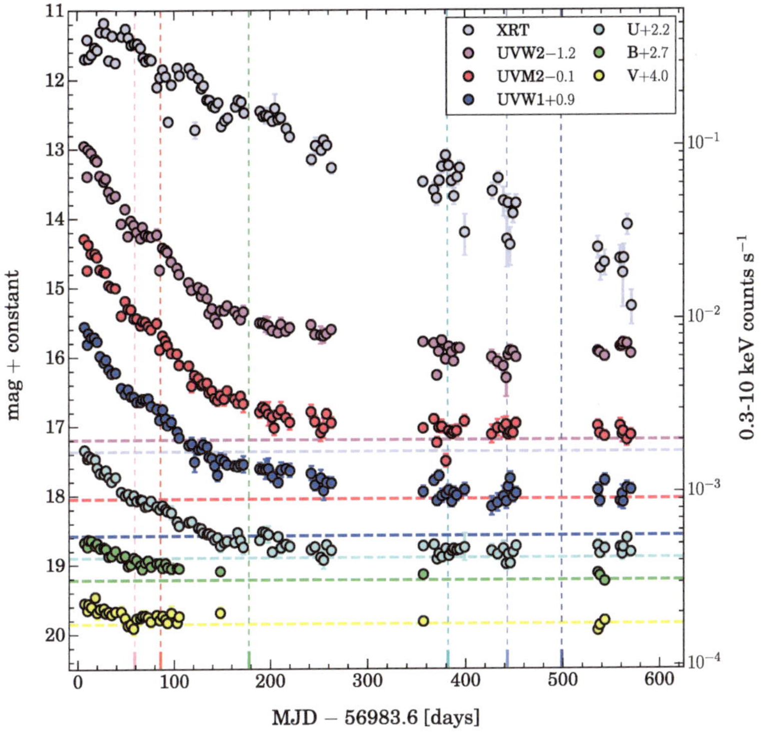

Fig. 10 The X-ray, UV and optical light-curves of ASASSN-14li (taken from Brown et al. 2017)

4.3 X-Ray Bright Events Discovered in Optical Surveys

For reasons which are currently not well understood, optically and UV-bright TDEs tend to show little or no X-ray emission (Gezari et al. 2009; Arcavi et al. 2014; Jonker et al. 2019, and see the Optical Chapter). Nevertheless there are some notable exceptions, two of which are examined below.

4.3.1 ASASSN-14li

ASASSN-14li was first discovered by the All-Sky Automated Survey for SuperNovae (ASASSN: Shappee et al. 2014) on 22nd of November 2014. This source was coincident with the center of galaxy, PGC043234 which is located at a distance of ~ 90 Mpc and its optical spectra exhibit characteristics consistent with that of a TDE (see the Optical Chapter). An extensive multi-wavelength monitoring campaign using *Swift* revealed in addition to its UV and optical emission, strong X-ray emission arising from this event (see Fig. 10: Holoien et al. 2016b; Brown et al. 2017). Holoien et al. (2016b) found that the X-ray luminosity decayed at a much slower rate than that seen in the optical/UV wavelengths, with X-rays becoming the dominant source of emission approximately 40 days after peak. Using Swift observations spanning 600 days, Brown et al. (2017) found that ASASSN-14li remained bright in both UV and X-ray wavelengths even at late times, while the total energy radiated in X-ray and UV/optical was comparable.

Due to its X-ray brightness, ASASSN-14li was a source of many target of opportunity observations by Chandra and XMM-Newton. Both X-ray grating and CCD spectra showed that the X-ray emission could be described by a simple black-body with a temperature $kT \sim$ 50–60 eV (Miller et al. 2015; Holoien et al. 2016b; Brown et al. 2017; Bright et al. 2018), which cooled during the decay (Brown et al. 2017).

Using deep grating spectra taken close to the peak of the flare, Miller et al. (2015) found that the black-body emission of ASASSN-14li was modified by absorption from photoionised species of N, O, S, Ar, and Ca. The absorption lines were blue-shifted by $v_{\rm shift} = -360 \pm 50$ km/s initially, slowing at later times (Miller et al. 2015). This led the authors to suggest that the X-ray absorption either arises in strong outflows of highly ionized, low velocity X-ray gas, from a super-Eddington wind or from stellar debris.

More recently, Kara et al. (2018) found the early time X-ray spectra exhibit a broad, P-Cygni-like absorption feature around ~ 0.7 keV, which fades with time. Using photoionisation modelling, they find that this feature is consistent with absorption by OVIII in a very fast (0.2c) and highly ionised outflow. Compared to the low velocity outflow detected by Miller et al. (2015), Kara et al. (2018) suggest that this high velocity component is produced much closer to the black hole. Since the ionisation parameters of these two components are similar, it is possible that the lower-velocity component arises from the fast outflow decelerating as it collides with the debris stream or another medium.

4.3.2 ASASSN-15oi

ASASSN-15oi was another X-ray bright TDE discovered by ASASSN on the 14th of August 2015. Located at a distance of 214 Mpc, this source exhibited similar spectral features to other optically detected TDEs (see the Optical Chapter) such as declining strong helium features, black-body emission and a declining light curve (Holoien et al. 2016a, 2018b). However, compared to ASASSN-14li, the X-ray emission from ASASSN-15oi was much weaker. At inital discovery, the detected X-ray emission from the source was lower than an upper limit derived using ROSAT. However, follow-up Swift observations revealed behaviour unseen at the time in any other TDE.[4] Typically, the X-ray emission of a TDE decays following a simple powerlaw (see Sect. 5), however, in this case the X-ray emission of ASASSN-15oi brightened by nearly an order of magnitude before fading again Gezari et al. (2017), Holoien et al. (2018b). As the timescale of this increase was approximately one year, Gezari et al. (2017) suggested that the behaviour was the result of delayed accretion due to inefficient circularization of the stellar debris stream, while Holoien et al. (2018b) suggested that this behaviour resulted from material surrounding the accretion disk becoming optically thin to X-ray radiation a few months after discovery. Using XMM-Newton, Gezari et al. (2017) found that the X-ray emission from the source was best described by a black-body plus a weak powerlaw component. While at both early and late times they find the temperature of black-body component does not change significantly (~ 45 eV), the powerlaw component may become softer with time ($\Gamma_{\rm initial} \sim 2.5$ to $\Gamma_{\rm later} \sim 3.3$). Using the Swift observations, Holoien et al. (2018b) find that the black-body component is an order of magnitude stronger than the powerlaw component, and is responsible for the flux variations.

[4] Similar behaviour has now been detected in AT2018fyk (Wevers et al. 2019a) and AT2019azh (Liu et al. 2019).

4.4 Accreted Mass

The mass accreted during an event can be estimated by integrating the total emitted X-ray luminosity

$$\Delta M = \frac{k_{\mathrm{Bol}}}{\eta c^2} \int_t^{\infty} L_X(t)dt \tag{2}$$

where η is the efficiency of conversion of gravitational energy into radiation, generally taken to be 10%, and k_{Bol} is the factor to correct X-ray to bolometric luminosity (see Netzer 2019, for a recent description). From this calculation, the accreted mass often appears to be ~ 0.01 solar masses (Brown et al. 2016; Auchettl et al. 2017; Saxton et al. 2017, and references in Sects. 3 and 4.1.3); a low value which potentially leads to a missing energy problem (Piran et al. 2015b).

We expect from stellar population analysis that the average mass of a disrupted main sequence star will be between 0.1 and 1 solar mass (e.g. Guillochon et al. 2014; van Velzen et al. 2019b; Mockler et al. 2018). If the fraction of stellar debris returning to the black hole is $\sim$ 20–50%, as dynamical studies predict (Ayal et al. 2000; Bonnerot and Lu 2019) then, at face value, $\lesssim 10\%$ of the available accretion energy is being converted into radiation in these events. There are several possible explanations for this:

- These flares are actually caused by the stripping of the atmosphere of an evolved star (MacLeod et al. 2012) and hence much less material is available for accretion.
- The conversion of mass to light factor (η) is around 0.01, i.e. just 10% of that seen in AGN.
- We always miss the peak of the emission and underestimate the total luminosity.
- The initial super-Eddington accretion produces strong winds which push a large fraction of the material away from the black hole (e.g. Metzger and Stone 2016; Dai et al. 2018).
- A large fraction of the radiation is actually emitted in the unobservable EUV and the bolometric correction is seriously underestimated (e.g. Komossa et al. 2004).
- The returning matter forms an accretion structure which evolves viscously and hence drains into the black hole at a much lower rate than $t^{-5/3}$ (Cannizzo et al. 1990; van Velzen et al. 2019a).
- We are underestimating the column density of material surrounding the event (Auchettl et al. 2017).
- A combination of the above or a further unknown factor.

4.5 X-Ray Spectra

The spectral features imprinted by the environment of an AGN are constant for decades, even if the details may vary on timescales as short as minutes. These are well described in e.g. Netzer (2013) and in essence consist of a pseudo-power-law, dominant from 2–100 keV, with a slope of roughly 2 (Turner and Pounds 1989; Nandra and Pounds 1994), a soft emission component, which may be due to Compton-upscattering by cool electrons ($kT_e \sim 0.2$ keV; Done et al. 2012) of photons generated in the disk and a pure thermal soft X-ray component, usually hidden by the stronger Comptonised component. This emission is modified by reflection off distant neutral material and off the ionised disk and various absorption features with various ionisation states and column densities. In radio-loud objects a jet produces very strong power-law emission which tends to mask the other features. This complex mix requires very high signal-to-noise spectra to begin to deconvolve the individual components.

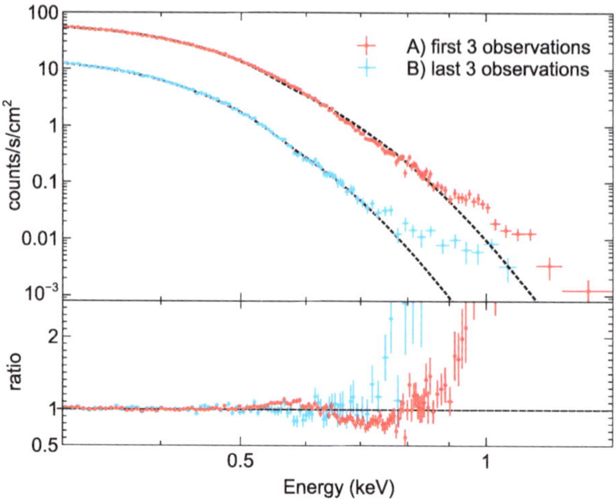

Fig. 11 X-ray spectrum of ASSASN-14li from early (red) and late (blue) XMM-*Newton* observations. Figure adapted from Kara et al. (2018). The best fit with a single black-body is shown as a dotted black line. A hard excess is evident in both fits

The prediction is that in the simplified case of a TDE, where we add material at a well-determined date to a SMBH, we ought to be able to determine the temporal onset of each of the components mentioned above and hence achieve a better understanding of persistent accreting SMBH and the timescales involved in the distribution of matter about the black hole.

A wide range of empirical models have been used to fit the broad-band XMM-*Newton* and Chandra spectra of TDEs. A single-temperature black-body, with luminosity given by

$$L = \sigma A T^4 \tag{3}$$

where σ is the Boltzmann constant, A is the surface area of the emitting region and T is the temperature in Kelvin, is a reasonable description in a number of cases (e.g., early emission of ASASSN-14li; Sect. 4.3.1). As the stellar debris returns to the BH it should form a hot, optically-thick structure, with an effective black-body temperature dependent on the BH mass and the mass accretion rate, expressed in terms of the Eddington accretion rate, $\dot{M}_{\rm Edd} = L_{\rm Edd}/\eta c^2$ (Ulmer 1999)

$$kT \sim 40 M_6^{-1/4} \dot{M}_{\rm Edd}^{1/4}\ (\rm eV) \tag{4}$$

where M_6 is the mass of the black hole in units of $10^6\ M_\odot$ and $\dot{M}_{\rm Edd}$ is the Eddington-limited accretion rate.

However, these fits often leave a high-energy tail (e.g. 2XMMi J184725.1-631724: Lin et al. 2011, SDSS J1201+30: Saxton et al. 2012b, ASASSN-14li: Kara et al. 2018, 3XMM J150052.0+015452: Lin et al. 2017a, ASASSN-15oi: Holoien et al. 2018b) which needs further explanation. A high-statistic example is shown in Fig. 11. Several explanations have been offered for this in the literature which we run through below.

A: A multi-temperature structure: This assumes that the material is in a thin-disk configuration, where the emissivity index of the material is strongly weighted towards the centre so that most of the emission occurs at the higher temperatures. A common implementation is the *diskbb* model (Makishima et al. 1986). The range of temperatures produces a broader spectrum than a single-T black-body and can fit the observed spectra in some cases (Lin et al. 2011, 2015, 2018a). This model leads to a high inner disk temperature and, from Eq. (4), a correspondingly lower black-hole mass.

B: An empirical power-law: A single power-law usually proves to be a poor-fit when used to model the whole of a high-statistic TDE spectrum. When it is used, then the slope is usually very steep, ($\Gamma_x > 4$; see Sects. 3, 4 and 10), in excess of the index value of 1.5 to 2.3 (Turner and Pounds 1989; Nandra and Pounds 1994) ubiquitously found in persistent AGN and believed to be produced by Comptonisation of disk photons by a high temperature ($kT_e > 100$ keV), optically thin plasma. It is more successful in fitting the *hard-excess* in good quality spectra of TDEs, but even here the slope tends to be steep (e.g. $\Gamma \sim 3.5$ in 2XMM 1847 (Lin et al. 2011) and also in ASASSN-15oi Gezari et al. 2017; Holoien et al. 2018b or those summarised in Auchettl et al. 2017).

C: Bremsstrahlung: This is emission from an optically-thin plasma which produces a wider spectrum than a black-body. It was a good fit to SDSS J1201+30 with a temperature of 390 eV, reducing to 280 eV in a later observation (Saxton et al. 2012a). The main draw-back of this model is that, being optically-thin, at this temperature it should produce substantial narrow line X-ray emission from N, O, Fe (e.g. Mewe et al. 1985) which was not seen in SDSS J1201+30 or in other TDEs.

D. Inverse Compton: A nearby population of electrons, with significant kinetic energy compared to the photon energies, will give energy to a fraction of the photons in a process known as Compton-upscattering, thus creating a hard tail to the thermal spectrum. This mechanism has been proposed for the soft-X-ray excess regularly seen in the spectrum of AGN (e.g. Done et al. 2012). In the soft X-ray regime, the final energy of an upscattered photon in collision with a thermal electron population of temperature kT_e, is related to its initial energy by

$$E_f \sim \left(1 + \frac{4kT_e}{mc^2}\right) E_i \tag{5}$$

The hard excess produced by the Inverse Compton effect has a shape which is well approximated by a power-law with slope dependent on the temperature and optical depth of the electrons (see Nishimura et al. 1986). This model provides a more physical fit to SDSS 1201+30 (Saxton et al. 2012b) and has been used to fit the excess in XMM-*Newton* and Chandra spectra of TDEs such as 3XMM J150052.0+015452 (Lin et al. 2017a). Two popular spectral models are (compbb; Nishimura et al. 1986) and (comptt; Titarchuk 1994). In these models the typical hard-excess slope of $\Gamma = 3$–4 is provided by a Maxwellian electron population of temperature, $kTe \sim 5$–10 keV and optical depth $\tau \sim 1$.

4.5.1 Hard Power-Law Emission

Another emission feature which has been seen is a hard ($\Gamma \sim 2$) power-law, equivalent to the dominant emission mechanism in AGN. In some cases this is associated with a relativistic jet (see the Gamma-ray Chapter) accompanied by strong emission at radio wavelengths. In IGR J12580+0134[5] and 2MASX 0740-85 the power-law is dominant, even at low energies, but radio emission is modest (~ 1 mJy or $L_R = 10^{37-39}$ erg s^{-1}). This component seems to be the analogue of the X-ray power-law seen in radio-quiet AGN,[6] which is generally explained by Compton-upscattering of disk photons by a population of very high energy electrons ($kTe > 100$ keV) located at a few R_g from the black hole, a proximity which leads

[5] Note that the classification of IGR J12580+0134 as a TDE has been questioned based on its WISE colours, pre-flare data and hardness ratio evolution (see A.17 of Auchettl et al. 2017).

[6] Although Irwin et al. (2015) make a case for the 2–10 keV emission coming from the inverse Compton component of the jet in IGR J12580+0134.

to variations on time scales of minutes to hours. In fact, it is this fast variability, also seen in 2MASX 0740-85 (Fig. 9), which locates the X-ray emission close to the BH.

4.5.2 Absorption Features

In the spectra of some events, a good fit can only be obtained by adding one or more *ionised* absorption features. This is the case in the CCD spectra of 3XMM J1521 and SDSS J1201+30 which show evidence for absorption features from apparently outflowing material (Saxton et al. 2012a; Lin et al. 2015). A more detailed analysis of this material can be found from the grating instrument on-board XMM-*Newton*, the RGS, in the bright TDE ASSASN-14li (see Sect. 4.3.1 and Miller et al. 2015; Kara et al. 2018). Outflows have been regularly found in BHs accreting close to the Eddington limit, in solar mass systems as well as SMBH (Arav et al. 1994; Crenshaw et al. 1999; Pounds et al. 2003).

Any *neutral* absorption (N_H), in excess of that of our galaxy and the TDE host galaxy, found in TDE spectra would be crucial for constraining the geometry of the debris and accreting material. For example, the reprocessing model (Dai et al. 2018; Metzger and Stone 2016) seeks to explain the difference between optical and X-ray TDEs in terms of a viewing angle (see the Emission Mechanisms Chapter). The very soft spectra of X-ray selected TDEs are highly sensitive to cold absorption and their very presence argues against large columns of neutral gas in the line of sight (e.g. Komossa and Bade 1999). As an example, a TDE at $z = 0.05$, with a Galactic column, $N_H = 1 \times 10^{20}$ cm^{-2} and black-body emission of $kT = 60$ eV has its 0.2–2 keV flux reduced by a factor 6 by an absorption of $N_H = 1 \times 10^{21}$ cm^{-2} intrinsic to the host galaxy of the event. In many detailed fits, neutral absorption in excess of that of our own galaxy is not required (e.g. Saxton et al. 2017, 2012a; Lin et al. 2015, 2018a). There are exceptions: SWIFT J164449.3+573451 ($N_H = 2 \times 10^{22}$ cm^{-2}; Burrows et al. 2011);[7] IGR J12580+0134 ($N_H = 7 \times 10^{22}$ cm^{-2}; Nikołajuk and Walter 2013), 3XMM J150052.0+015452 ($N_H = 4.2 \times 10^{21}$ cm^{-2}; Lin et al. 2017a) and ASSASN-14li ($N_H = 1.4 \times 10^{20}$ cm^{-2}; Miller et al. 2015), but in none of these cases has the excess neutral absorption been seen to change between observations. This either means that the absorption comes from the host galaxy, or that the material was produced by the TDE but in a form which did not decrease in density and was not significantly ionised by the nuclear radiation, during the event.

The derived absorbing column is of course dependent on the emission model used in the spectral fit. Auchettl et al. (2017) fitted a large sample of TDE spectra assuming a single absorbed power-law model, finding that a large fraction of X-ray TDEs have column densities (N_H) that are at least two times greater than the Galactic column density measured along the line of sight to these events. This result holds when a power-law is the correct emission model for the TDE and is useful for investigating variations with time. It does not return the correct absolute value though, for the different spectral models which are commonly seen in TDEs, e.g. the multi-component spectrum of ASASSN-14li (Fig. 11 and Miller et al. 2015; Kara et al. 2018).

4.5.3 Long-Term Spectral Evolution

It is interesting to see what happens to the accreting material after many years when the accretion rate has dropped well below the Eddington limit. The late-time spectrum of the

[7] Note the highly reddened, absorbed host galaxy of this event.

ROSAT TDE RXJ 1242-1119 was measured with XMM-Newton and implied a strong hardening of the X-ray spectrum from $\Gamma_X = 5.1$ to $\Gamma_X = 2.5$ (Komossa et al. 2004). A dedicated study of the late-time spectra of ROSAT TDEs by Chandra, taken > 10 years after the disruption, showed relatively hard spectra ($\Gamma < 2.5$) in all cases (Halpern et al. 2004; Vaughan et al. 2004). The Chandra study is complicated by the very low statistics (< 25 photons in each spectrum) and by the low-luminosity of the residual emission ($L_X = 5 \times 10^{39}$ to 2×10^{41} erg s^{-1}) which is comparable to that seen from the binary population of the galaxy, at least in the case of NGC 5905, where most or all of the low-state emission is extended and clearly not associated with the nucleus (Halpern et al. 2004). In NGC 5905 a spectral hardening was already measured with ROSAT itself, 3 years after the maximum ($\Gamma_X = 4.0$ to $\Gamma_X = 2.4$; Komossa and Bade 1999). There is some theoretical expectations that accreting debris will collapse into a thin disk leaving a long-lived low-level emission for many years. Such emission has accretion rates of $\dot{M} \sim 10^{-4}$ and an index $\Gamma_X \sim 2$ (Cannizzo et al. 1990). In the case of RXJ1242-1119, the latest deep Chandra follow-ups have shown a deviation from the early-phase decline law, in form of a deep dip below the early-phase decline law, and may indicate a change in accretion mode (Komossa 2005), as predicted by e.g. Rees (1990).

In NGC 3599 the spectrum was still relatively soft 6 years after the peak emission (Saxton et al. 2015) with an equivalent fit of a power-law of $\Gamma = 2.7 \pm 0.3$ plus a $kT = 50$ eV black-body. In RBS 1032 the spectrum after 20 years had hardened from $\Gamma = 5$ to $\Gamma = 3.4$ (Maksym et al. 2014a). Interestingly, Jonker et al. (2019) found late-time X-ray emission in three optically-selected TDEs. The spectra of these could be modeled with power-laws of $\Gamma = 2.5$–3.9, compatible with those of the X-ray selected events.

The flux of the soft black-body component certainly drops over the years but due to the paucity of high quality early and late-time spectra, it is not yet clear whether the harder component remains roughly constant or increases over time. More observations, with higher statistics, are needed to decipher what the long-term spectral hardening means in physical terms.

5 Light Curves

The evolution of a TDE's light curve depends heavily on a number of factors such as the stellar structure (e.g., Guillochon and Ramirez-Ruiz 2013), and whether the emission arises from fall-back (e.g., Rees 1988; Evans and Kochanek 1989; Phinney 1989), disk emission (e.g., Rees 1988; Cannizzo and Gehrels 2009; Lodato and Rossi 2011) or super-Eddington accretion (e.g., Cannizzo et al. 2011). X-ray light curves of TDEs are rarely well enough sampled to be able to identify the time of disruption, or even the time of peak flux. Commonly, when fitting the light curve, the two parameters, t_0 and the index α are found to be degenerate and α is fixed to the canonical value of $-5/3$ to obtain an estimate of the date of disruption. However, studies of the optical emission from individual sources (e.g. Wevers et al. 2019a), and detailed global studies in the Far-UV (van Velzen et al. 2019a) have shown that this behaviour is not necessarily universal, especially at late-times. Auchettl et al. (2017), took a different approach. Under the assumption that t_0 is the time of the first, peak flux, detection, they showed that the X-ray emission from TDEs then seems to decay with a power-law index that is shallower than $t^{-5/3}$, implying that the viscous timescale[8] is long for

[8]Here the viscous timescale is defined as the time it takes for material to accrete onto a black hole, and depends on the height and radius of the disk and the orbital period (Guillochon and Ramirez-Ruiz 2015).

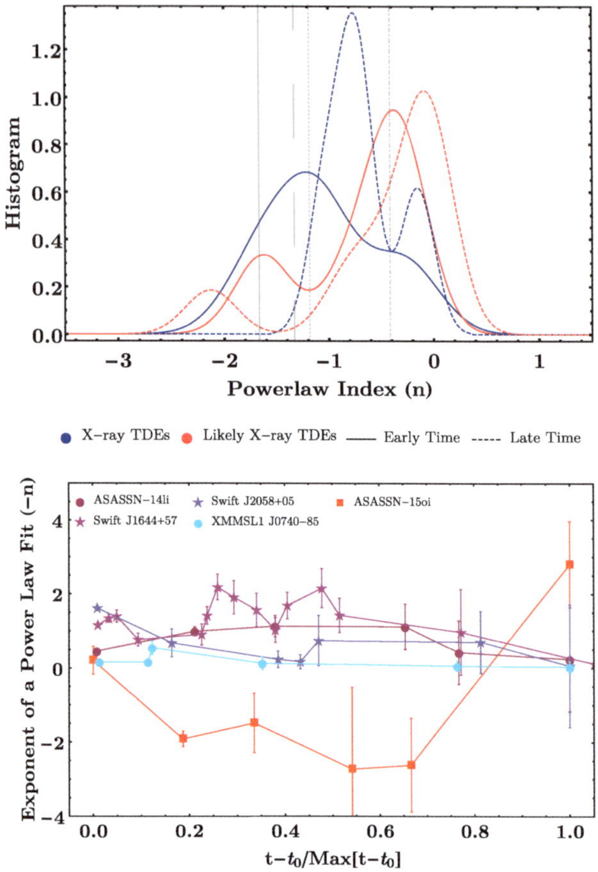

Fig. 12 *Upper:* Histogram of powerlaw indexes as seen at early (solid) and late (dashed) times during the decay of the X-ray emission from a sample of X-ray TDE candidates. Overlaid as the vertical solid, large dashed, dotted and dash-dotted grey lines are the powerlaw indexes for fallback ($-5/3$), advective, super-Eddington slim disk accretion ($-4/3$), viscous disk accretion ($-19/16$) and disk emission ($-5/12$). Figure taken from Auchettl et al. (2017). *Lower:* Best fit powerlaw index and its uncertainty for the X-ray TDE sample of Auchettl et al. (2017)

these events (Guillochon and Ramirez-Ruiz 2013). At early times, TDEs were found to have power-law indexes that are consistent with both fall-back ($t^{-5/3}$) and disk emission ($t^{-5/12}$) (see Lodato and Rossi 2011), while at late times most, if not all, events tend to have decay rates consistent with disk emission (see Fig. 12 upper panel). The transition between these different emission processes is not necessarily smooth, with small timescale variations in the power-law index seen as each event evolves (Fig. 12 lower panel), while the time it takes for each event to undergo this transition varies between sources. Within the class of X-ray emitting TDEs, those whose emission is dominated by a strong jet (e.g., jetted TDEs such as Swift J1644+57: Bloom et al. 2011; Levan et al. 2011; Burrows et al. 2011; Levan et al. 2016) showed multiple transitions between different emission processes, while those whose emission is not dominated by a jet (thermal TDEs such as ASASSN-14li: Miller et al. 2015; Holoien et al. 2016b; Brown et al. 2017; Bright et al. 2018) tend to show less variation in their power-law index which tends to fluctuate around the index associated with one type of emission process.

Even though the X-ray light curves of TDEs tend to decay following a powerlaw index that is consistent with fall back, or disk emission, a handful of TDEs show strong deviations from this behaviour. Even though the light curve of the jetted TDE Swift J1644+57 globally exhibits an approximately $t^{-5/3}$ decay (which is punctuated by flaring, variability and dips) (Bloom et al. 2011; Levan et al. 2011; Burrows et al. 2011; Levan et al. 2016; Mangano et al.

Fig. 13 X-ray light curve of 3XMM J150052.0+015452 from Chandra (blue circles and upper limit), XMM-Newton (red squares) and Swift-XRT (green diamond). Data taken from Lin et al. (2017a)

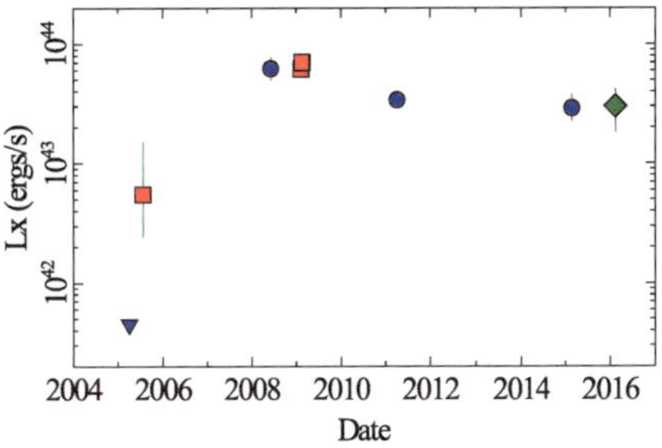

2016), a striking feature of its light curve is the dramatic decrease in X-ray flux ~500 days after its initial detection. Within ~ 4 days, the X-ray flux dropped by a factor of > 50, which corresponds to a decay index steeper than t^{-70} (Levan et al. 2016). This behaviour suggests that the accretion underwent a state change; either it suddenly stopped, possibly consistent with the star being completely accreted onto the black hole (Quataert and Kasen 2012), or the accretion became sub-Eddington and radiatively efficient which dramatically represses power-law jets (e.g., van Velzen et al. 2011b; Russell et al. 2011; Zauderer et al. 2013).

6 Indirectly Identified X-Ray Events from Reprocessing into High-Ionization Emission Lines, and X-Ray Follow-Up

Some luminous X-ray TDE candidates were not detected directly in the X-ray regime, but the flaring X-ray emission was indirectly inferred based on the presence of luminous, then fading, high-ionization iron coronal emission lines in optical spectra. These lines need a strong incident X-ray continuum in order to be created (Komossa et al. 2008, 2009; Wang et al. 2011, 2012). Only one of these events, SDSS J095209.56+214313.3, had X-ray follow-up spectroscopy which revealed a relatively flat X-ray spectrum, and a strong decline in X-ray luminosity with $L_{\mathrm{x,low}} = 4 \times 10^{40}$ erg/s between 2–10 keV (Komossa et al. 2009). Selected non-X-ray properties of these events will be further discussed in the Echo Chapter.

7 Long-Lived Events

We have seen in the previous sections that some TDEs (e.g. NGC 5905, 2MASX 0740-85, SDSS J120136.02+300305.5 and IGR J12580+0134) decay quickly from their peak X-ray luminosity. Other events, however, maintain their peak emission for longer; NGC 3599 had a plateau of at least 18 months before decaying (Saxton et al. 2015) and XMMSL2 J1446+68 at least 100 days. The champion though is 3XMM J150052.0+015452 (Lin et al. 2017a), a TDE from a dwarf galaxy, which rose within 4 months in 2005 and has been decaying very slowly for more than ten years (Fig. 13). This event has maintained a soft spectrum, being well modelled with a low-temperature, $kT \sim 40$ eV, black-body, heavily Comptonised by optically-thick, low-energy electrons (`COMPTT` model), for the whole of its plateau phase. This model, in stellar-mass black holes, is believed to signify a super-Eddington accretion state (e.g. Middleton et al. 2013). The last Chandra observation, has a relative deficit of hard flux which can be interpreted as either; that the accretion mode has changed to a super-soft (sub-Eddington) state or that an ionised absorber was in the line-of-sight which caused the sharp spectral drop towards higher energies. The black hole mass is estimated to be

$\sim 10^6\ M_\odot$ based on the mass of the host dwarf galaxy, consistent with the observed $L_{\rm bol} \sim 10^{44}$ erg s^{-1}. Lin et al. (2017a) interpreted the slow evolution of the bolometric light curve as indicating a distant circularisation of the stellar debris, leading to higher viscosity and a consequently slow fallback of the material to the BH. The mass of the disrupted object is then around 2 $M_\odot$ with $\sim 0.9\ M_\odot$ being converted into radiation. Two unusual persistent AGN have shown similar long high-luminosity, soft X-ray emission; GSN 069 (Miniutti et al. 2013) and 2XMM J123103.2+110648 (Terashima et al. 2012). Both of these have been interpreted by some authors as a TDE occurring in an AGN (Lin et al. 2017b; Shu et al. 2018).

There are several theoretical reasons for why a TDE may be long-lived. For example, a long super-Eddington accretion phase, perhaps involving a disrupted object with a large mass (e.g. Lin et al. 2017a); a partially stripped evolved star atmosphere (MacLeod et al. 2012) or late, distant circularisation (Guillochon and Ramirez-Ruiz 2015; Shiokawa et al. 2015).

8 Very Fast Events

Some TDEs are expected to lead to fast X-ray flares. For instance, a TDE involving an intermediate-mass black hole (IMBH), defined here as a black hole with a mass less than $10^5\ M_\odot$ (e.g. Evans et al. 2015). Especially if the disrupted star was a compact star, such as a white dwarf (see the White Dwarf Chapter), then the associated orbital time scales are short and the relativistic periastron precession is large, potentially leading to a short circularization time (see the Disruption Chapter) and, therefore, a short rise time for the accretion flare (e.g. Clausen and Eracleous 2011; MacLeod et al. 2016; Shen 2018). Alternatively, it is predicted that in some cases a shock occurs the star upon disruption which breaks out of the star and gives rise to a brief X-ray flare (e.g. Yalinewich et al. 2018). Finally, blazar-like variability in those TDEs where a relativistic jet launched from the near-vicinity of the black hole is pointing close to our line of sight can give rise to fast X-ray variability even for black holes more massive than IMBHs (e.g. Bloom et al. 2011; Levan et al. 2011).

Below we discuss the fast X-ray events that have been suggested to be caused by tidal disruption events, we do note, however, that relatively few things are known about these events, making their interpretation as being caused by a tidal disruption event much less secure than some of the other events of longer duration in this manuscript.

First of all, we need to be aware that several aspects can make an event appear "fast". For instance, if given the limited sensitivity of any instrument only the peak of a (longer duration) outburst is detected an event can appear as "short" or "fast". This effect probably affected earlier detections of fast events (e.g. Pye and McHardy 1983; Grindlay 1999) more than recent detections with more sensitive instruments, although of course events at large(r) distances will still cause this effect even in modern detectors (as all detectors have a sensitivity limit). Furthermore, several minute-to-hour scale X-ray flares are known that have nothing to do with tidal disruption events, such as M-star flares (Heise et al. 1975) and more generally, flares due to stellar activity often induced by binary interactions such as those of RS CVn stars (Pye and McHardy 1983). And finally, accretion flares from Galactic low- and high-mass X-ray binaries (Liu et al. 2000, 2001) and thermonuclear explosions on the surface of an accreting neutron star (so called Type I X-ray bursts) can appear as fast X-ray flares (e.g. Galloway et al. 2008).

In order to weed out flares from such events multi-wavelength data is crucial. For that a source localization accurate to the order of arcseconds is important: this astrometric accuracy comes naturally with *Chandra*, XMM-*Newton* and *Swift*-XRT-discovered transients.

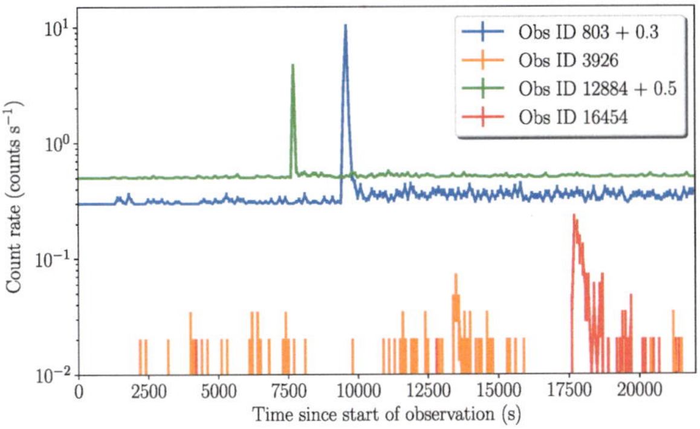

Fig. 14 The light curves of the four fast X-ray transients discovered in *Chandra* data. The data has been binned in time intervals of 100 seconds. The time zero on the X-axis corresponds to the start of the observation in each case. The observation ID of the *Chandra* data is indicated in the top right side of the figure. Clearly, the detected count rate varies significantly between the four events. Furthermore, the duration of the flares differs as well. The flare detected in observation ID 12884 is the shortest

XRT 000519: The first of this new type of fast transient X-ray sources (XRT 000519) was found in an archival *Chandra* observation (Observation ID 803; Jonker et al. 2013; see Fig. 14). The source position lies 12.16 arcminutes from the centre of the Virgo Cluster galaxy M 86, but it does not fall in the M 86 $\mu_B = 25$ magnitude per arcsec^2 isophote area. Optical Isaac Newton Telescope images show a tidal stream stripped off the galaxy SDSS J122541.29+130251.2, suggesting that M 86 is undergoing a minor merger. The small projected distance on the sky between the position of the transient and that of the tidal stream suggests that the transient is associated with M 86. Uniquely to this source (when compared to the others, see below) is that the main flare is double peaked.

Deep observations with the William Herschel Telescope (WHT) $r' \approx 25$ and $K \approx 20$-band observations rule out an M-star flare and also globular cluster hosts brighter than $M_V \sim -6$ at the location of M 86 for this event.

XRT 110203: A second transient which has properties that strongly resemble those of XRT 000519 is XRT 110203 (*Chandra* observation ID 12884; Glennie et al. 2015). Also this event lies close to a cluster of galaxies (ACO 3581; $z = 0.023$) but no host galaxy is detected so far.

CDFS-XT1: Bauer et al. (2017) report the discovery of a third fast X-ray transient in the *Chandra* Deep Field South survey (CDFS-XT1; *Chandra* observation ID 16454). The light curve indicates a rise time of $\sim$ 70–160 s. In this case they found a possible host dwarf galaxy in the 3D-HST field (with AB magnitude $R \sim 27.5$), but there is no cluster of galaxies nearby. Unfortunately, there is no spectroscopic redshift of this source. The photometric redshift of ~ 2 assigned to this host is much larger than the redshifts associated with the first two events. However, the redshift estimation is still very uncertain as the host was only detected in two HST filters (ACS/F606W $\sim$V-band and WFC3/F125W $\sim$J-band). In addition, the host-event association also needs to be confirmed as the chance alignment probability for such deep images is non-negligible and finally, the astrometric alignment between the host and the transient event is not perfect.

Given that these three events share the same timescale, have no clear or only a faint host, using Occam's razor, we suggest that the three events are drawn from the same parent population. Glennie et al. (2015) estimate a rate of $\sim 10^5$ events per year over the whole sky with a peak X-ray flux greater than 10^{-10} erg cm^{-2} s^{-1}.

Irwin et al. (2016) found one source to flare to a peak L_X of 9×10^{40} erg s^{-1} and 5 repeat flares to $L_X \approx 10^{40}$ erg s^{-1} and the probable detection of persistent/long term X-ray emission from optical sources consistent with being a globular cluster or ultra-compact dwarf host. These flares have a similar time scale and peak luminosities $L_X > 10^{39}$ erg s^{-1}. Repeat flares have not been observed for the three X-ray flares XRT 000519, XRT 110203 and CDFS-XT1 discussed above, although the repetition time scale is > 4 days for one and ~ 1.8 days for the second repeating source in Irwin et al. (2016) and given the sparseness of the X-ray observations of the fields of the three X-ray flares above one cannot rule out that all events repeat with timescales of days to weeks. Whereas this might argue against cataclysmic events, such as TDEs, models predicting multiple flares due to for instance partial disruptions of a white dwarf on an eccentric bound orbit or binary black holes might well still be consistent with the observations (Zalamea et al. 2010). However, we do note that the predicted periods would be of order of hundreds to perhaps thousands of seconds, not days for the bound white dwarf orbiting an IMBH.

For the flares where *Chandra* detected sufficient number of counts for a spectral decomposition, it was found that the spectra of the flares are well-fit with a power law shape (Jonker et al. 2013; Glennie et al. 2015), although in the latter case the value of the power law index depends strongly on the assumed interstellar extinction present. In the case of the flare called "source 1" in Irwin et al. (2016), the spectrum was also well-fit by a power law with index 1.6 ± 0.3 (90% confidence), fixing the Galactic column density to the value of 1.8×10^{20} cm^{-2}. The power law index varied slightly between the two peaks in the case of the XRT 000519 flare, with the second peak having a slightly softer power law index (1.95 ± 0.05; 68% confidence) than the first peak (1.6 ± 0.1; 68% confidence). A power law index of 1.4 ± 0.2 (68% confidence) provided a good fit for the flare reported in Bauer et al. (2017), although those authors also warn that a softer index (closer to 2) can not be ruled out as the power law index and (Galactic) extinction are to a large degree degenerate given the relatively low number of counts detected.

Other fast X-ray flares, often with time scales of thousands of seconds, hence slightly longer than the ~ 100 s time scale for the main flares above, have been reported as ultra-long gamma-ray bursts (Levan et al. 2014), although the energy bands used to measure the duration of the flares is different, making their comparison more difficult.

A particularly interesting case is CDFS-XT2 (Xue et al. 2019). In that paper it was interpreted as the X-ray signal from a binary neutron star merger. However, (Peng et al. 2019) interpret it as a white dwarf TDE. We note that the light curve (Xue et al. 2019) is more like the ultra-long GRBs (Levan et al. 2014), than the faster-still X-ray transients we report on here. However, clearly there is overlap in the properties of these events.

The energies and timescales associated with these events imply that compact objects such as massive black holes must be involved. Given the expected rate of 10^5 over the whole sky per year, a conservative assumed log N– log S (similar to other observed X-ray sources where per 2 decades in luminosity ten times more sources are found [e.g. Mineo et al. 2012]), and the instantaneous sensitivity of the eROSITA satellite should find ~ 1 of these fast X-ray events per day.

9 Multi-Waveband Properties of X-Ray Selected TDEs

9.1 UV and Optical

Among the first X-ray TDEs identified, NGC 5905 had quasi-simultaneous optical photometry, thanks to photographic plate archives which covered a timespan of several decades

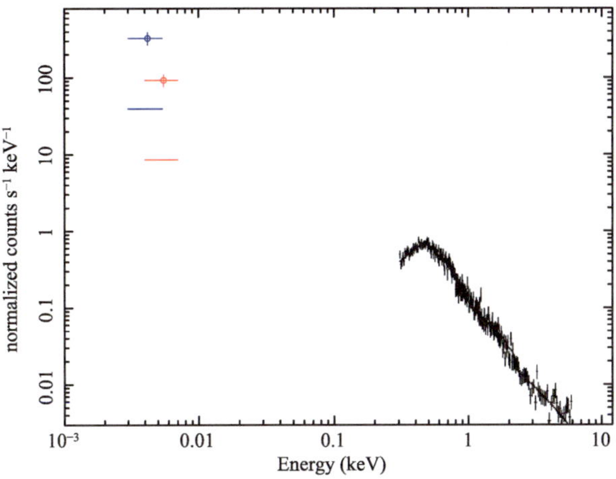

Fig. 15 An extrapolation of the best fit X-ray model, a power-law plus a $kT = 86$ eV black-body, to the XMM-OM, UVW1 and UVM2 filter data of 2MASX 0740-85. The single temperature thermal model significantly underpredicts the observed UV flux

(Fig. 2 of Komossa and Bade 1999). No long-term optical variability of this HII-type galaxy was discovered, and no optical flaring quasi-simultaneous with the X-ray flare was detected.

The Swift and XMM-*Newton* satellites both host an UV/optical telescope in their payload which allows these bands to be monitored simultaneously with the X-ray emission. Initially, it was thought that any emission in these bands would be from the Rayleigh-Jeans tail of the hot plasma which produces the soft X-ray emission and would therefore rise and decay simultaneously with the X-rays (e.g. Ulmer 1999). In 2MASX 0740-85 the UV and X-ray flux did decay from peak quasi-simultaneously over 550 days (Fig. 8; Saxton et al. 2017). Nevertheless, the galaxy-subtracted UV emission sits well in excess of a simple extrapolation of the thermal component from the X-ray spectrum of this source to UV wavebands (Fig. 15). The UV and X-ray data can be well modelled by a structure which includes emission from a range of temperatures, such as a thin accretion disk (Saxton et al. 2017). This finding agrees with the low temperature emission of $\sim 20{,}000$ K ubiquitously found in TDEs discovered in the optical or UV bands (Gezari et al. 2009; van Velzen et al. 2011a; Holoien et al. 2016a, and see the Optical Chapter). The similar temporal behaviour of the X-ray and UV emission is mirrored in ASASSN-14li (Sect. 4.3.1 and Holoien et al. 2016b; Brown et al. 2017) and indicates prompt accretion (or efficient circulation: Guillochon and Ramirez-Ruiz 2015). In this event, the high-density, multi-wavelength monitoring indicated a possible delay of 32 days between the X-ray emission and that of the UV (Pasham et al. 2017, and see the Echo Chapter). If the UV is produced by shocks in colliding streams of debris (Piran et al. 2015a; Shiokawa et al. 2015) then it would naturally precede the X-ray emission which is generated when that same material falls down to the black hole. The timescale for the infall is indeed expected to be a few weeks (Piran et al. 2015a; Shiokawa et al. 2015; Bonnerot et al. 2017, and the Disruption Chapter and Accretion Disc Chapter).

In ASSASN-15oi the behaviour was quite different (Sect. 4.3.2). Here the UV light fell by 5 magnitudes over 200 days while the soft X-ray luminosity increased by a factor 10. The X-rays subsequently declined back to their initial level after a further 400 days. Gezari et al. (2017) and Holoien et al. (2018b) presented late-time observations of ASASSN-15oi. Their studies revealed that the thermal X-ray emission from the source brightened by an order of magnitude during its first year of evolution, rather than following a powerlaw decline as seen from its optical/UV light curves. After ~ 600 days, the X-ray emission had faded back to the levels originally detected at peak. Gezari et al. (2017) suggested that the ~ 1 year it takes in the rise to peak for the X-ray emission is the result of delayed accretion on a $10^6\ M_\odot$ black hole. This delayed accretion is a result of inefficient circularisation of the debris disk due to a delay in the time it takes for the material to accrete onto the black hole (Lodato 2012; Piran et al. 2015a; Guillochon and Ramirez-Ruiz 2015). However, as the circularisation timescale is proportional to $M_{\rm BH}^{-7/6}$ (Bonnerot et al. 2017), Gezari et al. (2017) also suggested that if the black hole of ASASSN-15oi was closer to $10^7\ M_\odot$ as originally estimated by Holoien et al. (2016a) using the host galaxy mass and the $M_{\rm BH} - M_{\rm bulge}$ relation of McConnell and Ma (2013), then the circularisation timescale would be much shorter, and may not be able to explain the observed behaviour. Another possibility is that ASASSN-15oi is heavily absorbed by material surrounding the accretion disk which is optically thick to X-ray radiation at early times (e.g., Metzger and Stone 2016). Even though, observationally Holoien et al. (2016a), Gezari et al. (2017) and Holoien et al. (2018b) were unable to constrain the column density in the direction of the source to confirm this suggestion, the possibility that some X-ray TDEs are surrounded by dense material may be supported by the study of Auchettl et al. (2017).

A still different picture is present in the light curve of XMMSL2 J1446+68 (Fig. 8). Here the UV flux from the galaxy increased by ~ 1 magnitude, prior to the first X-ray detection and then stayed flat for 400 days before fading by $\sim$ a magnitude over the next 200 days. Meanwhile, the X-ray emission was constant for 100 days after discovery before fading by a factor ~ 100 over the next 500 days (Saxton et al. 2019). The delayed decay of the UV emission in this event suggests that a reservoir of relatively cool material was maintained for about a year. One possible explanation is that the material formed an accretion disk which drained viscously until running out of material (van Velzen et al. 2019a).

In SDSS J1201+30 the UV emission from the galaxy was apparently unaffected by the disruption event, while the X-rays faded by a factor 100. While we should note that a constant UV/optical flux can simply mean that the flare is obscured in these bands or the contrast with the bright host galaxy is poor, the diversity of relative behaviour between the optical, UV and X-ray bands represents a challenge to current models of TDE evolution. A full understanding of the diversity of optical/UV emission behaviour from X-ray selected TDEs awaits a better understanding of the mechanisms which are responsible for the optical/UV emission. This theme is addressed in the Optical Chapter.

9.2 Radio

Dedicated radio follow-up was performed on the TDE in NGC 5905, in order to exclude the (very unlikely) scenario of a blazar hiding in this nearby starburst galaxy, and to search for the first time for jet emission associated with a TDE itself (Komossa 2002). Based on an observation with the VLA A array carried out in 1996, no nuclear radio emission was detected, with a 5σ upper limit for the presence of a central point source of < 150 μJy (Komossa and Dahlem 2001; Komossa 2002). At a distance of 75.4 Mpc of NGC 5905, this corresponds to a limit of $L_{8.46\ \rm GHz} < 9 \times 10^{36}\ \rm erg\,s^{-1}$. Extended low-level radio emission at lower frequencies is present in NGC 5905, and consistent with its HII-type nature (see Sect. 2.3.2 of Komossa 2002). A search for late-time radio emission from the nucleus of NGC 5905 was carried out by Bower et al. (2013), who did not detect any to an upper limit of 200 μJy. Radio emission from RXJ 1242-1119 was searched for in the FIRST VLA sky survey at 20 cm (Komossa 2002) but none was detected. In order to search for very late-time radio emission, Bower et al. (2013) also carried out follow-ups of the other ROSAT TDEs. No radio emission was found from RXJ 1242-1119, while a second source in the error circle of RXJ 1420.4+5334 does emit faintly (114 ± 24 μJy) in the radio regime.

Following the Swift detection of a jetted TDE, radio follow-up was more routinely carried out for newly identified TDEs. Radio upper limits of 100–200 μJy were reported for

SDSS J1201+30 (Saxton et al. 2012a) and 10 μJy for XMMSL2 J1446+68 (Saxton et al. 2019). Faint radio emission (1.2 mJy at 1.5 GHz; $L_{1.5\,\mathrm{GHz}} = 10^{37}$ erg s^{-1}) was detected 1 year after discovery from 2MASX 0740-85, which faded over the following months (Alexander et al. 2017), while slightly stronger emission was monitored in ASASSN-14li (Alexander et al. 2016).

In summary, the thermal X-ray TDEs are not strong radio emitters. For a full analysis of the radio properties of these and other TDEs see the Radio Chapter.

10 Interpreting X-Ray TDEs with a Reprocessing Model

Various authors have explored the effect on the emitted radiation of an optically-thick, static or expanding envelope of material created during the event (Ulmer 1999; Strubbe and Quataert 2009; Metzger and Stone 2016; Dai et al. 2018; Roth and Kasen 2018). In this section we compare this model with X-ray TDE properties. The X-ray emission from a (non-jetted) TDE is generally quite soft in nature, and is well described by a black-body with a temperature between 10–100 eV (see the previous sections) or a powerlaw with photon index of $\Gamma > 4$ (Auchettl et al. 2017, and earlier work). These temperatures are generally consistent with the picture suggested by Rees (1988) of a black-body with a temperature between 10^{5-6} Kelvin arising from an accretion disk (e.g., Ulmer 1999; Bonning et al. 2007). However, the temperatures derived from optical/UV studies of these events and optical/UV only events are an order of magnitude less than those measured from X-rays (see the Optical Chapter) and may feasibly represent emission from the reprocessing of nuclear radiation (e.g. Loeb and Ulmer 1997).

Those that have a strong jet such as Swift J1644+57 and Swift J2058+05 (Cenko et al. 2012) generally exhibit much harder X-ray emission, which can be best described by a simple powerlaw with a photon index of $\Gamma \sim 1$–2. There are also cases of non-jetted TDE, such as ASASSN-15oi, XMMSL1 J0740-85 and PS18kh (Holoien et al. 2018a; van Velzen et al. 2019b) which exhibit both the soft black-body component and a weaker powerlaw component with a temperature and photon index similar to that listed above (see Sect. 4.5). This diversity in the observational characteristics of these events has been suggested to be a natural result of the viewing-angle with respect to the orientation of the accretion disk (e.g., Dai et al. 2018), or that X-ray TDEs and optical/UV only TDEs could result from a separate class of events that have compact debris disk due to large apsidal precession of the self-intersecting streams of the disrupted star (e.g., Dai et al. 2015). Jonker et al. (2019) discuss three optically-selected TDEs which were not detected in X-rays during the optical flare but 8–10 years later had luminosities of $L_X \sim 10^{41-42}$. In the reprocessing model, the X-ray emission from these events will have been massively suppressed at early times and only become visible when the density of the absorbing material decreased.

Observationally, it has been shown that the softness of non-jetted events is intrinsic to the source, with the spectral energy distribution of these events peaking in the UV/soft X-ray band (Auchettl et al. 2017, and references in Sects. 3, 4), consistent with theoretical expectations for black holes with masses $< 10^7\ M_\odot$ (e.g., Ulmer 1999; Dai et al. 2018). Using the spectral energy distributions of these events, Auchettl et al. (2017) saw that X-ray selected TDEs have high X-ray to optical ratios (see Fig. 16). The fact that these events produce significant amounts of both X-ray and optical/UV emission opens the possibility that a considerable fraction of their emission is being reprocessed into optical/UV wavelengths,[9]

[9] An alternative explanation was given by van Velzen et al. (2019a) who suggested that the properties of these sources are not a result of reprocessing but are due to a viscously spreading, unobscured accretion disk. This

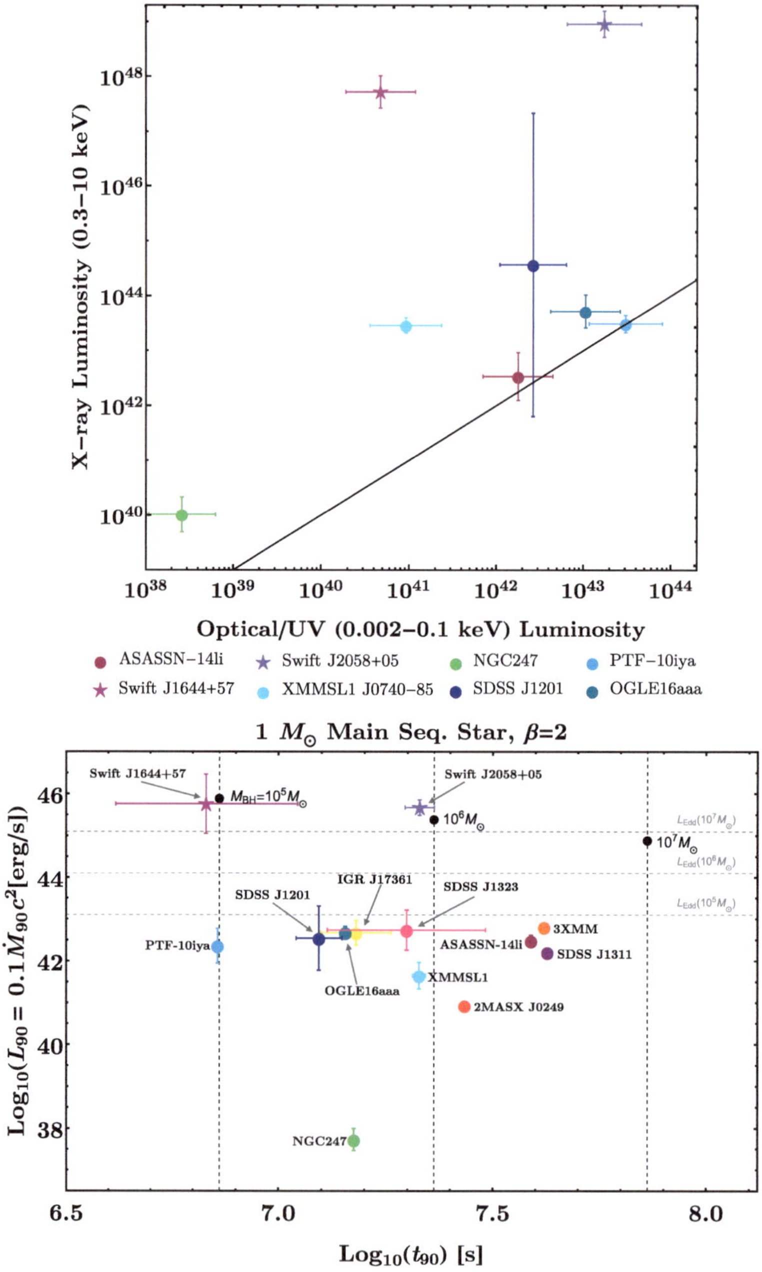

Fig. 16 Upper: Peak X-ray v peak optical/UV luminosity for a selection of X-ray bright TDEs. Lower: The mean isotropic luminosity plotted against the time taken for the TDE to emit between 5 and 90% of its total output. Both plots are taken from Auchettl et al. (2017)

work was extended into the X-ray regime in Jonker et al. (2019), who infer the existence of a long-lived accretion disk to explain the relatively high late-time X-ray luminosity of three optically-selected TDEs.

with the variation seen in Fig. 16 perhaps implying that these events experience significantly different reprocessing rates.

A separation between jetted and non-jetted events is seen when one attempts to derive the isotropic luminosity of each source (Auchettl et al. 2017, Fig. 16). Even though TDEs exhibit a wide range of isotropic luminosities, jetted events tend to have $L_{\rm iso} \sim 10^{44}$ erg s^{-1}, while non-jetted X-ray events have $L_{\rm iso} \lesssim 10^{42}$ erg s^{-1}.[10] TDEs which have been detected in optical/UV wavelengths only, known as "veiled X-ray TDEs" in the nomenclature of Auchettl et al. (2017), could be surrounded by very dense material (e.g., PS1-10jh: Guillochon et al. 2014), which would lead to the X-rays being reprocessed completely into optical/UV wavelengths (e.g., Dai et al. 2018). These optical/UV sources have isotropic luminosities that fall between the jetted and non-jetted X-ray TDEs, in what Auchettl et al. (2017) refer to as a "reprocessing valley" (see e.g., Gezari et al. 2012; Chornock et al. 2014; Guillochon et al. 2014; Gezari et al. 2015; Piran et al. 2015b). As such, these events could emit X-rays but have all their emission reprocessed into lower wavelengths.

Assuming that a main sequence star is being fully disrupted, Auchettl et al. (2017) find that the light curves of X-ray TDEs are consistent with a disruption from a black hole with mass between 10^{5-7} $M_\odot$. These results are consistent with that derived from e.g., modelling the light curves of the individual events (e.g., Mockler et al. 2018), from bulge-black hole mass relations of e.g., McConnell and Ma (2013), Reines and Volonteri (2015), and optical spectroscopy (Wevers et al. 2017).

11 TDEs in Binary SMBHs and Recoiling SMBHs

11.1 TDEs in Binary SMBHs

The appearance of TDEs which occur in binary SMBHs can be different from TDEs of single SMBHs, including lightcurves which look characteristically different (Liu et al. 2009; Coughlin et al. 2017), and including rates which can be boosted by up to several orders of magnitude in some stages of binary evolution (e.g. Chen et al. 2009).

The lightcurve of the TDE from SDSS J120136.02+300305.5 (Sect. 4.2.1; Fig. 7) does not show a smooth decline, but exhibits episodes of dipping. One month after the peak, the X-ray emission suddenly dropped by a factor of > 50 within a week and the source was no longer detected by Swift. X-rays re-appeared after 115 d, and then dropped a second time. While such a behaviour could arise due to beaming in jetted sources, no radio emission was detected from this TDE (Saxton et al. 2012a). Instead, the characteristic intermittence and recovery of the lightcurve of SDSS J1201+30 is reminiscent of predictions by Liu et al. (2009), who computed TDE lightcurves in binary SMBHs. In that case, the second SMBH acts as a perturber and the accretion stream on the primary is temporarily interrupted. Simulations by Liu et al. (2014) have shown, that the lightcurve of SDSS J1201+30 is consistent with a binary SMBH model with a primary mass of 10^6 $M_\odot$, a mass ratio $q \sim 0.1$ and a semi-major axis of 0.6 milli-pc.

This was the first supermassive binary BH (SMBBH) candidate identified in a non-active host galaxy, and the one with the most compact orbit among the known SMBBH candidates (review by Komossa and Zensus 2016). It has overcome the so-called "final parsec problem"

[10] Here $L_{\rm iso}$ is defined as the mean isotropic luminosity, after correcting for beaming, emitted by the event in the interval where the light curve contains between 5% and 95% of the total emitted luminosity (see Auchettl et al. 2017 and references therein for an explanation of the derivation of isotropic luminosity).

(e.g. Colpi 2014). Upon coalescence, it will be a strong source of gravitational wave emission in the sensitivity regime of the upcoming generation space-based gravitational wave detectors. If significant numbers of SMBBHs exist at the cores of non-active galaxies, we expect to see more such events in well-sampled lightcurves of TDEs with *Swift* or the future Einstein Probe (EP; Yuan et al. 2015, 2016) mission. A good lightcurve coverage is essential for constraining the system parameters.

11.2 TDEs as Signposts of Recoiling SMBHs

Luminous X-ray flares from TDEs which occur *off-nuclear* are possible signposts of recoiling SMBHs (Komossa and Merritt 2008). At X-ray peak luminosities in the quasar regime ($L_X \sim 10^{42-46}$ erg/s), there is no other mechanism, which could produce a long-lived off-nuclear X-ray flare.

Potentially, a flaring recoiling SMBH could hide among the population of ultraluminous X-ray sources (ULX) which have been identified in nearby galaxies, however, the known ULXs have much lower X-ray luminosities than AGN, and they are by far (i.e., by orders of magnitude) too abundant to be explained by TDEs from recoiling SMBHs (see, e.g., the discussion by Strateva and Komossa 2009; Jonker et al. 2012).

12 X-Ray TDE Rates

To calculate the frequency of tidal disruptions it is necessary to perform a uniform analysis of a well-controlled sample of events. Donley et al. (2002), performed a systematic survey of galaxies observed in both the *ROSAT* All-Sky Survey and subsequent *ROSAT* pointed observations, re-detecting 3 TDEs which had been previously reported (see Sect. 3). From this work, which covered 9% of the sky, they calculated a TDE rate of 9×10^{-6} gal^{-1} yr^{-1}. Esquej et al. (2008) found 2 TDEs in the XMM-*Newton* slew survey by comparing fluxes with *ROSAT* observations taken 10–15 years earlier and derived a rate of 2.3×10^{-4} gal^{-1} yr^{-1}. A further search for events was made by comparing RASS data with deep XMM-*Newton* pointed observations (Khabibullin and Sazonov 2014) giving a baseline of 10–20 years. They discovered three events in the 2% of the sky covered by XMM-*Newton*, obtaining a rate of 3×10^{-5} gal^{-1} yr^{-1}. Finally, from multiple observations of the large cluster of galaxies, A1689, Maksym et al. (2010) derived a rate of 6×10^{-5} TDEs gal^{-1} yr^{-1}.

A survey measures three main quantities: the number of detected TDE, the number of square degrees covered and the flux limit reached. A set of assumptions are then adopted to find the volume of sky which has been sampled, and hence the number of galaxies observed, and the fraction of a one-year light curve which has been observed. To convert these into a rate, assumptions have to be made about the peak luminosity (L_P) of the event, the galaxy density and the shape of the light curve (for a full description of the derivation see the Rates Chapter). In Table 2 we list the measured values and adopted assumptions for each of the soft X-ray TDE rate calculations. We see that the assumptions vary greatly between each calculation. L_P has been taken to be between 2.8×10^{43} (Donley et al. 2002) and $\sim 1 \times 10^{44}$ (Khabibullin and Sazonov 2014; Esquej et al. 2008) erg s^{-1}. Similarly the light curve shape has been taken as effectively flat for one year (Donley et al. 2002), flat for 0.19 years and then dropping to zero (Khabibullin and Sazonov 2014) or $t^{-5/3}$ (Esquej et al. 2008; Maksym et al. 2010). Integrating over the latter gives a surveyed volume equivalent to observing the peak luminosity for 0.013 years. Not surprisingly these different assumptions produce large differences in the final calculated rates, which range from 9×10^{-6} to 2.3×10^{-4} gal^{-1} yr^{-1}.

Table 1 Summary of soft X-ray TDEs referred to in this chapter

Name	Date[a]	z	$L_{X,peak}$[b]	Spectrum[c]	Ref[d]
NGC 5905	7/1990	0.011	7×10^{42}	$kT = 60$ eV	(Bade et al. 1996)
RXJ1624+7554	8/1990	0.064	2×10^{44}	$\Gamma = 3$	(Grupe et al. 1999)
RBS 1032	11/1990	0.026	1×10^{43}	$kT = 120$ eV	Ghosh et al. (2006)[e]
RXJ1420+5334	12/1990	0.147	2.5×10^{44}	$kT = 38$ eV	(Greiner et al. 2000)
RXJ 1242-1119	7/1992	0.05	4×10^{44}	$kT = 60$ eV	(Komossa and Greiner 1999)
TDXF1347-3254	12/1992	0.037	7×10^{42}	$kT = 120$ eV	(Cappelluti et al. 2009)
WINGS J1348	12/1999	0.063	1×10^{42}	$kT = 84$ eV	(Maksym et al. 2013)
3XMM J152130.7+074916	8/2000	0.179	5×10^{43}	$kT \sim 170$ eV	(Lin et al. 2015)
NGC 3599	11/2003	0.002	5×10^{41}	$kT = 95$ eV	(Esquej et al. 2007)
SDSS J132341.97+482701.3	12/2003	0.0875	5×10^{43}	–	(Esquej et al. 2007)
SDSS J131122.15-012345.6	2/2004	0.18	5×10^{42}	$kT \sim 100$ eV	Maksym et al. (2010)
XMMSL1 J024916.6-041244	7/2004	0.0186	2×10^{42}	$kT = 110$ eV	(Esquej et al. 2007)
3XMM J150052.0+015452	7/2005	0.145	6×10^{43}	$kT \sim 40$ eV	(Lin et al. 2017a)
SDSS J095209.56+214313.3	12/2005	0.079	–	–	(Komossa et al. 2008)
3XMM J215022.4-55108	5/2006	0.055	7×10^{42}	$kT \sim 280$ eV	(Lin et al. 2018a)
2XMMi J184725.1-631724	9/2006	0.0353	3×10^{43}	$kT \sim 80$ eV	(Lin et al. 2011)
SDSS J120136.02+300305.5	6/2010	0.146	3×10^{44}	$kT = 70$ eV	(Saxton et al. 2012a)
2MASX 0740-85	4/2014	0.0173	2×10^{43}	$kT \sim 80$ eV, $\Gamma_X \sim 2$	(Saxton et al. 2017)
ASASSN-14li	11/2014	0.0206	3×10^{43}	$kT = 50$ eV	(Holoien et al. 2016b)
ASASSN-15oi	8/2015	0.0479	6×10^{42}	~ 45 eV	Gezari et al. (2017)
XMMSL2 J1446+68	8/2016	0.029	6×10^{42}	$\Gamma_X \sim 2.6$	(Saxton et al. 2019)

[a]Date of first detection in X-rays

[b]The highest unabsorbed X-ray luminosity measured. Units of erg s^{-1}

[c]Dominant spectral component(s) at peak. "kT" refers to a black-body temperature, Γ refers to a power-law slope

[d]The first reference to the X-ray measurements

[e]Identified as a TDE by Maksym et al. (2014a) and Khabibullin and Sazonov (2014)

Table 2 Summary of TDE survey rates

Survey[a]	Factor[b]	Lum^{c}_{X} (erg/s)	Decay[d] model	Baseline (Years)	ρ_{gal}[e] (Mpc^{-3})	F_{lim}[f] ($erg\,s^{-1}\,cm^{-2}$)	D_{lim}[g] (Mpc)	Nsrc[h]	Area[i] (% sky)	Rate[j] $gal^{-1}\,yr^{-1}$	Rate[k] $gal^{-1}\,yr^{-1}$
RASSvRXP	20	2.8×10^{43}	1 yr	0.5–8	0.014	2.0×10^{-12}	342	3	9	9×10^{-6}	3.4×10^{-5}
RASSvXMMP	10	10^{44}[l]	0.19 yrs	10–20	0.02	3.5×10^{-13}	877	3	2	3×10^{-5}	2.1×10^{-4}
XMMSvRASS	20	10^{44}	$t^{-5/3}$	10–18	0.023	5.0×10^{-12}	406	2	15	2.3×10^{-4}	1.1×10^{-4}
Abell 1689	4–5	$\sim 10^{44}$	$t^{-5/3}$	1–7	–	$\sim 1 \times 10^{-13}$	–	1	–	6×10^{-5}	6×10^{-5}

[a]RASS v ROSAT pointed (0.2–2.4 keV) (Donley et al. 2002); RASS bright source catalogue v XMM-Newton pointed (0.2–2 keV) data (Khabibullin and Sazonov 2014); XMM, EPIC-pn slew (0.2–2 keV) from catalogue XMMS1d3 v RASS (Esquej et al. 2008); repeat observations of 2100 galaxies in Abell 1689 over 7 years (Maksym et al. 2010)

[b]Minimum variability factor needed for a detection

[c]Assumed peak X-ray luminosity

[d]Assumed light curve duration or decay function; flat for 1 year (Donley et al. 2002), flat for 0.19 years then dropping to zero (Khabibullin and Sazonov 2014) or $t^{-5/3}$

[e]Density of galaxies in the local universe

[f]Detection flux limit (observed) in the 0.2–2 keV band. Khabibullin and Sazonov (2014) use the ROSAT Bright Source Catalogue which contains sources with a minimum count rate of 0.05 c/s in the ROSAT PSPC camera. This translates to a 0.2–2 keV flux of 3.5×10^{-13} $erg\,s^{-1}\,cm^{-2}$ using their spectral model of a $kT = 50$ eV black body absorbed by a Galactic column of $N_H = 5 \times 10^{20}$ cm^{-2}

[g]Derived limiting distance for TDE detection

[h]Number of TDEs found in the survey. Note that (Donley et al. 2002) consider sources which have faded between the RASS and ROSAT pointed observations and hence exclude RXJ 1242-1119

[i]Percentage of the sky covered in the survey. In the case of Abell 1689, 2100 galaxies were repeatedly observed

[j]The original reported TDE rate based on the given assumptions

[k]The TDE rate calculated under common assumptions; $L_{X,peak} = 2.8 \times 10^{43}$, an average visibility time of 0.19 years and a galaxy density of $\rho = 0.02$ Mpc^{-3} (see text)

[l]Khabibullin and Sazonov (2014) use a bolometric luminosity of 7×10^{44} $erg\,s^{-1}$ without explicitly giving the correction factor to the X-ray band. They refer to Khabibullin et al. (2014) from where an X-ray peak luminosity of 1×10^{44} $erg\,s^{-1}$ can be derived based on the Eddington luminosity and K-correction for a 10^6 $M_\odot$ black hole

To show the importance of the details of the calculation, we take as a common set of assumptions, the peak luminosity from Donley et al. (2002), $L_P = 2.8 \times 10^{43}$ erg s^{-1}, the visibility time from Khabibullin and Sazonov (2014) (0.19 years) and a galaxy density of $\rho = 0.02$ Mpc^{-3} and apply it to the surveys.[11] Results are shown in Table 2 where the rates now vary between 3.4 and 21 $\times 10^{-5}$ gal^{-1} yr^{-1}, within a factor 7 of each other.[12] This immediately shows that the *measurements* in these surveys are actually in better agreement than they first appear.

For an accurate calculation of the absolute value of the TDE rate, a larger and less-biased sample of TDEs are needed to constrain their properties. In particular the TDE peak luminosity function is not currently well constrained (although see Sun et al. 2015; Auchettl et al. 2018, for first attempts) and the large variety of TDE light curves (e.g. Fig. 8) introduce considerable uncertainty in the rates which will only be resolved when an unbiased estimate of the median light curve is available.

In summary, the *reported* soft-X-ray-selected TDE rates in the literature lie between $\sim 1 \times 10^{-5}$ and 2×10^{-4} TDEs gal^{-1} yr^{-1}. If we adopt a common set of assumptions, then the surveys agree to better than a factor 7. Note that the measured rate is a lower limit due to the fraction of TDEs which are not observed in X-rays due to absorption by gas in the host galaxy, tidal debris or outflowing material from the accretion process, e.g. Sembay and West (1993) estimate that 40% of soft X-ray TDEs will be located in edge-on galaxies and their soft X-rays absorbed away within the host galaxy.

13 Conclusions and Future Prospects

The bulk of the TDEs discussed in Sect. 2–5 share the following characteristics:

- X-ray peak luminosities between 10^{42} and a few times 10^{44} erg s^{-1}.
- Very soft X-ray spectra near peak, with black-body temperatures in the range $kT_{bb} =$ 0.04–0.12 keV (or, alternatively, with powerlaw indices in the range $\Gamma_x = 3$–5).
- A spectral hardening within years.
- A decline on a timescale of years down to a quiescent level.
- An absence of X-ray emission lines
- Fast variability in several events (minutes to hours).
- Host galaxies which are quiescent and in-active both before and after the disruption.
- A decline in flux by factors up to 1000–6000.

While there are certainly exceptions to these properties, the overall picture is of X-ray radiation emitted by optically-thick material within a few R_g of the black hole, which fades with the diminishing return of tidal debris.

The limited duration of events has allowed us to follow the passage from super-Eddington to sub-Eddington to low-level (RIAF/ADAF) accretion on year-to-decade timescales and witness the onset and cessation of disk winds which are responsible for driving material into the host galaxy.

[11] Note that the cluster survey of (Maksym et al. 2010) uses a self-consistent set of assumptions which are not affected by this change.

[12] All the surveys use very strong TDE candidates when calculating the rates except for Khabibullin and Sazonov (2014) which identified one very likely TDE (RBS 1032) and two possible TDEs in their sample. If only RBS 1032 had been adopted here then the survey TDE rates with common assumptions would agree to a factor ≈ 3.

One of the exciting prospects of X-ray TDE observations is the possibility of identifying IMBH from their distinctive disruptive properties. This was broached in Sect. 4.1.2 and also in the section on the enigmatic *very fast events*, whose light curves may indicate the disruption of a compact star by a BH of mass $\leq 10^5\ M_\odot$.

Deep follow-up observations of TDEs near their peak are allowing us to probe the extremes of accretion physics in relatively clean environments. They allow us to follow the evolution of disk winds and coronae, search for relativistic (precession) effects in the Kerr metric, estimate BH spin, carry out absorption/emission-line spectroscopy of ionized matter in outflow (either stellar debris or accretion disk winds), and study the jet-disk coupling and jet evolution in jetted events.

Although more than 20 years old, the science of X-ray emitting TDEs can be considered to be still in its infancy, with each new event presenting traits which modify our understanding of the disruption process. With the launch of the eRosita telescope (Predehl et al. 2010) on board the Spectrum-Roentgen-Gamma (SRG) mission, it is expected that hundreds of new TDEs will be found (Khabibullin et al. 2014; Jonker et al. 2019). If well monitored in dedicated follow-ups by other missions, these will fill in the parameter space and provide us with a more complete picture of the phenomenon. Within a few years, the Einstein Probe will provide excellent light curves of the rise, peak and initial decay phases of hundreds of X-ray TDEs and allow detailed modeling of the fall back process of a large number of TDEs.

Acknowledgements Dacheng Lin and Erin Kara are warmly thanked for providing updated figures of their work. RS would like to thank Peter Maksym for early help with the rates calculation. Sjoert Van Velzen and an anonymous second referee are thanked for comments and suggestions which improved the manuscript.

References

K.D. Alexander, E. Berger, J. Guillochon, B.A. Zauderer, P.K.G. Williams, Discovery of an outflow from radio observations of the tidal disruption event ASASSN-14li. Astrophys. J. Lett. **819**, 25 (2016). https://doi.org/10.3847/2041-8205/819/2/L25

K.D. Alexander, M.H. Wieringa, E. Berger, R.D. Saxton, S. Komossa, Radio observations of the tidal disruption event XMMSL1 J0740-85. Astrophys. J. **837**, 153 (2017). https://doi.org/10.3847/1538-4357/aa6192

N. Arav, Z.-Y. Li, M.C. Begelman, Radiative acceleration in outflows from broad absorption-line quasi-stellar objects. II. Wind models. Astrophys. J. **432**, 62 (1994). https://doi.org/10.1086/174549

I. Arcavi, A. Gal-Yam, M. Sullivan, Y.-C. Pan, S.B. Cenko, A. Horesh, E.O. Ofek, A. De Cia, L. Yan, C.-W. Yang, D.A. Howell, D. Tal, S.R. Kulkarni, S.P. Tendulkar, S. Tang, D. Xu, A. Sternberg, J.G. Cohen, J.S. Bloom, P.E. Nugent, M.M. Kasliwal, D.A. Perley, R.M. Quimby, A.A. Miller, C.A. Theissen, R.R. Laher, A continuum of H- to He-rich tidal disruption candidates with a preference for E+A galaxies. Astrophys. J. **793**, 38 (2014). https://doi.org/10.1088/0004-637X/793/1/38

K. Auchettl, J. Guillochon, E. Ramirez-Ruiz, New physical insights about tidal disruption events from a comprehensive observational inventory at X-ray wavelengths. Astrophys. J. **838**, 149 (2017). https://doi.org/10.3847/1538-4357/aa633b

K. Auchettl, E. Ramirez-Ruiz, J. Guillochon, A comparison of the X-ray emission from tidal disruption events with those of active galactic nuclei. Astrophys. J. **852**, 37 (2018). https://doi.org/10.3847/1538-4357/aa9b7c

S. Ayal, M. Livio, T. Piran, Tidal disruption of a solar-type star by a supermassive black hole. Astrophys. J. **545**, 772–780 (2000). https://doi.org/10.1086/317835

N. Bade, S. Komossa, M. Dahlem, Detection of an extremely soft X-ray outburst in the HII-like nucleus of NGC 5905. Astron. Astrophys. **309**, L35 (1996). http://adsabs.harvard.edu/abs/1996A%26A...309L..35B

J.A. Baldwin, M.M. Phillips, R. Terlevich, Classification parameters for the emission-line spectra of extragalactic objects. Publ. Astron. Soc. Pac. **93**, 5–19 (1981). https://doi.org/10.1086/130766

F.E. Bauer, E. Treister, K. Schawinski, S. Schulze, B. Luo, D.M. Alexander, W.N. Brandt, A. Comastri, F. Forster, R. Gilli, D.A. Kann, K. Maeda, K. Nomoto, M. Paolillo, P. Ranalli, D.P. Schneider, O. Shemmer, M. Tanaka, A. Tolstov, N. Tominaga, P. Tozzi, C. Vignali, J. Wang, Y. Xue, G. Yang, A new, faint population of X-ray transients. Mon. Not. R. Astron. Soc. **467**, 4841–4857 (2017). https://doi.org/10.1093/mnras/stx417

J.S. Bloom, D. Giannios, B.D. Metzger, S.B. Cenko, D.A. Perley, N.R. Butler, N.R. Tanvir, A.J. Levan, P.T. O'Brien, L.E. Strubbe, F. De Colle, E. Ramirez-Ruiz, W.H. Lee, S. Nayakshin, E. Quataert, A.R. King, A. Cucchiara, J. Guillochon, G.C. Bower, A.S. Fruchter, A.N. Morgan, A.J. van der Horst, A possible relativistic jetted outburst from a massive black hole fed by a tidally disrupted star. Science **333**, 203 (2011). https://doi.org/10.1126/science.1207150

C. Bonnerot, W. Lu, Simulating realistic disc formation in tidal disruption events. ArXiv e-prints, 1906-05865 (2019)

C. Bonnerot, E.M. Rossi, G. Lodato, Long-term stream evolution in tidal disruption events. Mon. Not. R. Astron. Soc. **464**, 2816–2830 (2017). https://doi.org/10.1093/mnras/stw2547

E.W. Bonning, L. Cheng, G.A. Shields, S. Salviander, K. Gebhardt, Accretion disk temperatures and continuum colors in QSOs. Astrophys. J. **659**, 211–217 (2007). https://doi.org/10.1086/510712

G.C. Bower, B.D. Metzger, S.B. Cenko, J.M. Silverman, J.S. Bloom, Late-time radio emission from X-ray-selected tidal disruption events. Astrophys. J. **763**(2), 84 (2013). https://doi.org/10.1088/0004-637X/763/2/84

J.S. Bright, R.P. Fender, S.E. Motta, K. Mooley, Y.C. Perrott, S. van Velzen, S. Carey, J. Hickish, N. Razavi-Ghods, D. Titterington, P. Scott, K. Grainge, A. Scaife, T. Cantwell, C. Rumsey, Long-term radio and X-ray evolution of the tidal disruption event ASASSN-14li. Mon. Not. R. Astron. Soc. **475**(3), 4011–4019 (2018). https://doi.org/10.1093/mnras/sty077

P.J. Brown, Y. Yang, J. Cooke, M. Olaes, R.M. Quimby, D. Baade, N. Gehrels, P. Hoeflich, J. Maund, J. Mould, F. Patat, L. Wang, J.C. Wheeler, ASASSN-15lh: a superluminous ultraviolet rebrightening observed by Swift and Hubble. ArXiv e-prints (2016)

J.S. Brown, T.W.-S. Holoien, K. Auchettl, K.Z. Stanek, C.S. Kochanek, B.J. Shappee, J.L. Prieto, D. Grupe, The long term evolution of ASASSN-14li. Mon. Not. R. Astron. Soc. **466**, 4904–4916 (2017). https://doi.org/10.1093/mnras/stx033

D.N. Burrows, J.A. Kennea, G. Ghisellini, V. Mangano, B. Zhang, K.L. Page, M. Eracleous, P. Romano, T. Sakamoto, A.D. Falcone, J.P. Osborne, S. Campana, A.P. Beardmore, A.A. Breeveld, M.M. Chester, R. Corbet, S. Covino, J.R. Cummings, P. D'Avanzo, V. D'Elia, P. Esposito, P.A. Evans, D. Fugazza, J.M. Gelbord, K. Hiroi, S.T. Holland, K.Y. Huang, M. Im, G. Israel, Y. Jeon, Y.-B. Jeon, H.D. Jun, N. Kawai, J.H. Kim, H.A. Krimm, F.E. Marshall, P. Mészáros, H. Negoro, N. Omodei, W.-K. Park, J.S. Perkins, M. Sugizaki, H.-I. Sung, G. Tagliaferri, E. Troja, Y. Ueda, Y. Urata, R. Usui, L.A. Antonelli, S.D. Barthelmy, G. Cusumano, P. Giommi, A. Melandri, M. Perri, J.L. Racusin, B. Sbarufatti, M.H. Siegel, N. Gehrels, Relativistic jet activity from the tidal disruption of a star by a massive black hole. Nature **476**, 421–424 (2011). https://doi.org/10.1038/nature10374

J.K. Cannizzo, N. Gehrels, A new paradigm for gamma-ray bursts: long-term accretion rate modulation by an external accretion disk. Astrophys. J. **700**, 1047–1058 (2009). https://doi.org/10.1088/0004-637X/700/2/1047

J.K. Cannizzo, H.M. Lee, J. Goodman, The disk accretion of a tidally disrupted star onto a massive black hole. Astrophys. J. **351**, 38 (1990). http://adsabs.harvard.edu/abs/1990ApJ...351...38C

J.K. Cannizzo, E. Troja, G. Lodato, GRB 110328A/Swift J164449.3+573451: the tidal obliteration of a deeply plunging star? Astrophys. J. **742**, 32 (2011). https://doi.org/10.1088/0004-637X/742/1/32

N. Cappelluti, M. Ajello, P. Rebusco, S. Komossa, A. Bongiorno, C. Clemens, M. Salvato, P. Esquej, T. Aldcroft, J. Greiner, H. Quintana, A candidate tidal disruption event in the Galaxy cluster Abell 3571. Astron. Astrophys. **495**(2), 9–12 (2009). http://www.aanda.org/10.1051/0004-6361/200811479

S.B. Cenko, J.S. Bloom, S.R. Kulkarni, L.E. Strubbe, A.A. Miller, N.R. Butler, R.M. Quimby, A. Gal-Yam, E.O. Ofek, E. Quataert, L. Bildsten, D. Poznanski, D.A. Perley, A.N. Morgan, A.V. Filippenko, D.A. Frail, I. Arcavi, S. Ben-Ami, A. Cucchiara, C.D. Fassnacht, Y. Green, I.M. Hook, D.A. Howell, D.J. Lagattuta, N.M. Law, M.M. Kasliwal, P.E. Nugent, J.M. Silverman, M. Sullivan, S.P. Tendulkar, O. Yaron, Ptf10iya: a short-lived, luminous flare from the nuclear region of a star-forming galaxy. Mon. Not. R. Astron. Soc. **420**(3), 2684–2699 (2012). http://adsabs.harvard.edu/abs/2012MNRAS.420.2684C

X. Chen, P. Madau, A. Sesana, F.K. Liu, Enhanced tidal disruption rates from massive black hole binaries. Astrophys. J. Lett. **697**, 149 (2009). http://adsabs.harvard.edu/abs/2009ApJ...697L.149C

R. Chornock, E. Berger, S. Gezari, B.A. Zauderer, A. Rest, L. Chomiuk, A. Kamble, A.M. Soderberg, I. Czekala, J. Dittmann, M. Drout, R.J. Foley, W. Fong, M.E. Huber, R.P. Kirshner, A. Lawrence, R. Lunnan, G.H. Marion, G. Narayan, A.G. Riess, K.C. Roth, N.E. Sanders, D. Scolnic, S.J. Smartt,

K. Smith, C.W. Stubbs, J.L. Tonry, W.S. Burgett, K.C. Chambers, H. Flewelling, K.W. Hodapp, N. Kaiser, E.A. Magnier, D.C. Martin, J.D. Neill, P.A. Price, R. Wainscoat, The ultraviolet-bright, slowly declining transient PS1-11af as a partial tidal disruption event. Astrophys. J. **780**, 44 (2014). https://doi.org/10.1088/0004-637X/780/1/44

D. Clausen, M. Eracleous, Probing intermediate-mass black holes with optical emission lines from tidally disrupted white dwarfs. Astrophys. J. **726**(1), 34 (2011). http://adsabs.harvard.edu/abs/2011ApJ...726...34C

M. Colpi, Massive binary black holes in galactic nuclei and their path to coalescence. Space Sci. Rev. **183**(1-4), 189–221 (2014). https://doi.org/10.1007/s11214-014-0067-1

E.R. Coughlin, P.J. Armitage, C. Nixon, M.C. Begelman, Tidal disruption events from supermassive black hole binaries. Mon. Not. R. Astron. Soc. **465**, 3840–3864 (2017). https://doi.org/10.1093/mnras/stw2913

D.M. Crenshaw, S.B. Kraemer, A. Boggess, S.P. Maran, R.F. Mushotzky, C.-C. Wu, Intrinsic absorption lines in Seyfert 1 galaxies. I. Ultraviolet spectra from the Hubble Space Telescope. Astrophys. J. **516**(2), 750–768 (1999). https://doi.org/10.1086/307144

L. Dai, J.C. McKinney, M.C. Miller, Soft X-ray temperature tidal disruption events from stars on deep plunging orbits. Astrophys. J. Lett. **812**, 39 (2015). https://doi.org/10.1088/2041-8205/812/2/L39

L. Dai, J.C. McKinney, N. Roth, E. Ramirez-Ruiz, M.C. Miller, A unified model for tidal disruption events. Astrophys. J. Lett. **859**, 20 (2018). https://doi.org/10.3847/2041-8213/aab429

C. Done, S.W. Davis, C. Jin, O. Blaes, M. Ward, Intrinsic disc emission and the soft X-ray excess in active galactic nuclei. Mon. Not. R. Astron. Soc. **420**, 1848–1860 (2012). https://doi.org/10.1111/j.1365-2966.2011.19779.x

J.L. Donley, W.N. Brandt, M. Eracleous, T. Boller, Large-amplitude X-ray outbursts from galactic nuclei: a systematic survey using ROSAT archival data. Astron. J. **124**, 1308 (2002). http://adsabs.harvard.edu/abs/2002AJ....124.1308D

P. Esquej, R.D. Saxton, M.J. Freyberg, A.M. Read, B. Altieri, M. Sanchez-Portal, G. Hasinger, Candidate tidal disruption events from the XMM-Newton slew survey. Astron. Astrophys. **462**, 49 (2007). http://adsabs.harvard.edu/abs/2007A%26A...462L..49E

P. Esquej, R.D. Saxton, S. Komossa, A.M. Read, M.J. Freyberg, G. Hasinger, D.A. García-Hernández, H. Lu, J. Rodriguez Zaurín, M. Sánchez-Portal, H. Zhou, Evolution of tidal disruption candidates discovered by XMM-Newton. Astron. Astrophys. **489**, 543 (2008). http://adsabs.harvard.edu/abs/2008A%26A...489..543E

C.R. Evans, C.S. Kochanek, The tidal disruption of a star by a massive black hole. Astrophys. J. Lett. **346**, 13–16 (1989). https://doi.org/10.1086/185567

I.N. Evans, F.A. Primini, K.J. Glotfelty, C.S. Anderson, N.R. Bonaventura, J.C. Chen, J.E. Davis, S.M. Doe, J.D. Evans, G. Fabbiano, E.C. Galle, D.G. Gibbs II, J.D. Grier, R.M. Hain, D.M. Hall, P.N. Harbo, X.H. He, J.C. Houck, M. Karovska, V.L. Kashyap, J. Lauer, M.L. McCollough, J.C. McDowell, J.B. Miller, A.W. Mitschang, D.L. Morgan, A.E. Mossman, J.S. Nichols, M.A. Nowak, D.A. Plummer, B.L. Refsdal, A.H. Rots, A. Siemiginowska, B.A. Sundheim, M.S. Tibbetts, D.W. Van Stone, S.L. Winkelman, P. Zografou, The Chandra source catalog. Astrophys. J. Suppl. Ser. **189**, 37–82 (2010). https://doi.org/10.1088/0067-0049/189/1/37

C. Evans, P. Laguna, M. Eracleous, Ultra-close encounters of stars with massive black holes: tidal disruption events with prompt hyperaccretion. Astrophys. J. Lett. **805**, 19 (2015). https://doi.org/10.1088/2041-8205/805/2/L19

G. Fabbiano, The X-ray properties of normal galaxies. Publ. Astron. Soc. Pac. **98**, 525–543 (1986). https://doi.org/10.1086/131790

L. Ferrarese, D. Merritt, A fundamental relation between supermassive black holes and their host galaxies. Astrophys. J. Lett. **539**(1), 9–12 (2000). https://doi.org/10.1086/312838

D.K. Galloway, M.P. Muno, J.M. Hartman, D. Psaltis, D. Chakrabarty, Thermonuclear (type I) X-ray bursts observed by the Rossi X-Ray Timing Explorer. Astrophys. J. Suppl. Ser. **179**, 360–422 (2008). https://doi.org/10.1086/592044

N. Gehrels, G. Chincarini, P. Giommi, K.O. Mason, J.A. Nousek, A.A. Wells, N.E. White, S.D. Barthelmy, D.N. Burrows, L.R. Cominsky, K.C. Hurley, F.E. Marshall, P. Mészáros, P.W.A. Roming, L. Angelini, L.M. Barbier, T. Belloni, S. Campana, P.A. Caraveo, M.M. Chester, O. Citterio, T.L. Cline, M.S. Cropper, J.R. Cummings, A.J. Dean, E.D. Feigelson, E.E. Fenimore, D.A. Frail, A.S. Fruchter, G.P. Garmire, K. Gendreau, G. Ghisellini, J. Greiner, J.E. Hill, S.D. Hunsberger, H.A. Krimm, S.R. Kulkarni, P. Kumar, F. Lebrun, N.M. Lloyd-Ronning, C.B. Markwardt, B.J. Mattson, R.F. Mushotzky, J.P. Norris, J. Osborne, B. Paczynski, D.M. Palmer, H.-S. Park, A.M. Parsons, J. Paul, M.J. Rees, C.S. Reynolds, J.E. Rhoads, T.P. Sasseen, B.E. Schaefer, A.T. Short, A.P. Smale, I.A. Smith, L. Stella, G. Tagliaferri, T. Takahashi, M. Tashiro, L.K. Townsley, J. Tueller, M.J.L. Turner, M. Vietri, W. Voges, M.J. Ward, R.

Willingale, F.M. Zerbi, W.W. Zhang, The Swift gamma-ray burst mission. Astrophys. J. **611**(2), 1005–1020 (2004). https://doi.org/10.1086/422091

S. Gezari, J.P. Halpern, S. Komossa, D. Grupe, K.M. Leighly, Follow-up Hubble Space Telescope/Space Telescope Imaging Spectroscopy of three candidate tidal disruption events. Astrophys. J. **592**(1), 42–51 (2003). https://doi.org/10.1086/375553

S. Gezari, T. Heckman, S.B. Cenko, M. Eracleous, K. Forster, T.S. Gonçalves, D.C. Martin, P. Morrissey, S.G. Neff, M. Seibert, D. Schiminovich, T.K. Wyder, Luminous thermal flares from quiescent supermassive black holes. Astrophys. J. **698**, 1367–1379 (2009). https://doi.org/10.1088/0004-637X/698/2/1367. http://adsabs.harvard.edu/abs/2009ApJ...698.1367G

S. Gezari, R. Chornock, A. Rest, M.E. Huber, K. Forster, E. Berger, P.J. Challis, J.D. Neill, D.C. Martin, T. Heckman, A. Lawrence, C. Norman, G. Narayan, R.J. Foley, G.H. Marion, D. Scolnic, L. Chomiuk, A. Soderberg, K. Smith, R.P. Kirshner, A.G. Riess, S.J. Smartt, C.W. Stubbs, J.L. Tonry, W.M. Wood-Vasey, W.S. Burgett, K.C. Chambers, T. Grav, J.N. Heasley, N. Kaiser, R.-P. Kudritzki, E.A. Magnier, J.S. Morgan, P.A. Price, An ultraviolet-optical flare from the tidal disruption of a helium-rich stellar core. Nature **485**, 217–220 (2012). https://doi.org/10.1038/nature10990

S. Gezari, R. Chornock, A. Lawrence, A. Rest, D.O. Jones, E. Berger, P.M. Challis, G. Narayan, PS1-10jh continues to follow the fallback accretion rate of a tidally disrupted star. Astrophys. J. Lett. **815**, 5 (2015). https://doi.org/10.1088/2041-8205/815/1/L5

S. Gezari, S.B. Cenko, I. Arcavi, X-ray brightening and UV fading of tidal disruption event ASASSN-15oi. Astrophys. J. Lett. **851**, 47 (2017). https://doi.org/10.3847/2041-8213/aaa0c2

K.K. Ghosh, V. Suleymanov, I. Bikmaev, S. Shimansky, N. Sakhibullin, RBS 1032: a dwarf-nucleated spheroidal galaxy with an intermediate-mass black hole hosted in a globular cluster. Mon. Not. R. Astron. Soc. **371**(4), 1587–1593 (2006). https://doi.org/10.1111/j.1365-2966.2006.10723.x

A. Glennie, P.G. Jonker, R.P. Fender, T. Nagayama, M.L. Pretorius, Two fast X-ray transients in archival Chandra data. Mon. Not. R. Astron. Soc. **450**, 3765–3770 (2015). https://doi.org/10.1093/mnras/stv801

J.E. Greene, Low-mass black holes as the remnants of primordial black hole formation. Nat. Commun. **3**, 1304 (2012). https://doi.org/10.1038/ncomms2314

J. Greiner, R. Schwarz, S. Zharikov, M. Orio, Rx j1420.4+5334—another tidal disruption event? Astron. Astrophys. **362**, 25 (2000). http://adsabs.harvard.edu/abs/2000A%26A...362L..25G

J.E. Grindlay, Fast X-ray transients and gamma-ray bursts: constraints on beaming. Astrophys. J. **510**(2), 710–714 (1999). https://doi.org/10.1086/306617

D. Grupe, H.-C. Thomas, K.M. Leighly, RX J1624.9+7554: a new X-ray transient AGN. Astron. Astrophys. **350**, 31–34 (1999)

J. Guillochon, E. Ramirez-Ruiz, Hydrodynamical simulations to determine the feeding rate of black holes by the tidal disruption of stars: the importance of the impact parameter and stellar structure. Astrophys. J. **767**, 25 (2013). https://doi.org/10.1088/0004-637X/767/1/25

J. Guillochon, E. Ramirez-Ruiz, A dark year for tidal disruption events. Astrophys. J. **809**, 166 (2015). https://doi.org/10.1088/0004-637X/809/2/166

J. Guillochon, H. Manukian, E. Ramirez-Ruiz, PS1-10jh: the disruption of a main-sequence star of near-solar composition. Astrophys. J. **783**, 23 (2014). https://doi.org/10.1088/0004-637X/783/1/23

J.P. Halpern, S. Gezari, S. Komossa, Follow-up Chandra observations of three candidate tidal disruption events. Astrophys. J. **604**, 572–578 (2004). https://doi.org/10.1086/381937

J. Heise, A.C. Brinkman, J. Schrijver, R. Mewe, E.H.B.M. Gronenschild, A.J.F. den Boggende, J. Grindlay, Evidence for X-ray emission from flare stars observed by ANS. Astrophys. J. Lett. **202**, 73–76 (1975). https://doi.org/10.1086/181984

T.W.-S. Holoien, C.S. Kochanek, J.L. Prieto, D. Grupe, P. Chen, D. Godoy-Rivera, K.Z. Stanek, B.J. Shappee, S. Dong, J.S. Brown, U. Basu, J.F. Beacom, D. Bersier, J. Brimacombe, E.K. Carlson, E. Falco, E. Johnston, B.F. Madore, G. Pojmanski, M. Seibert, ASASSN-15oi: a rapidly evolving, luminous tidal disruption event at 216 Mpc. Mon. Not. R. Astron. Soc. **463**, 3813–3828 (2016a). https://doi.org/10.1093/mnras/stw2272

T.W.-S. Holoien, C.S. Kochanek, J.L. Prieto, K.Z. Stanek, S. Dong, B.J. Shappee, D. Grupe, J.S. Brown, U. Basu, J.F. Beacom, D. Bersier, J. Brimacombe, A.B. Danilet, E. Falco, Z. Guo, J. Jose, G.J. Herczeg, F. Long, G. Pojmanski, G.V. Simonian, D.M. Szczygieł, T.A. Thompson, J.R. Thorstensen, R.M. Wagner, P.R. Woźniak, Six months of multiwavelength follow-up of the tidal disruption candidate ASASSN-14li and implied TDE rates from ASAS-SN. Mon. Not. R. Astron. Soc. **455**, 2918–2935 (2016b). https://doi.org/10.1093/mnras/stv2486

T.W.-S. Holoien, M.E. Huber, B.J. Shappee, M. Eracleous, K. Auchettl, J.S. Brown, M.A. Tucker, K.C. Chambers, C.S. Kochanek, K.Z. Stanek, A. Rest, D. Bersier, R.S. Post, G. Aldering, K.A. Ponder, J.D. Simon, E. Kankare, D. Dong., G. Hallinan, J. Bulger, T.B. Lowe, E.A. Magnier, A.S.B. Schultz, C.Z. Waters, M. Willman, D. Wright, D.R. Young, S. Dong, J.L. Prieto, T.A. Thompson, L. Denneau, H.

Flewelling, A.N. Heinze, S.J. Smartt, K.W. Smith, B. Stalder, J.L. Tonry, H. Weiland, PS18kh: a new tidal disruption event with a non-axisymmetric accretion disk. ArXiv e-prints (2018a)

T.W.-S. Holoien, J.S. Brown, K. Auchettl, C.S. Kochanek, J.L. Prieto, B.J. Shappee, J. Van Saders, The unusual late-time evolution of the tidal disruption event ASASSN-15oi. Mon. Not. R. Astron. Soc. **480**, 5689–5703 (2018b). https://doi.org/10.1093/mnras/sty2273

J.A. Irwin, R.N. Henriksen, M. Krause, Q.D. Wang, T. Wiegert, E.J. Murphy, G. Heald, E. Perlman, CHANG-ES V: nuclear outflow in a Virgo cluster spiral after a tidal disruption event. Astrophys. J. **809**(2), 172 (2015). https://doi.org/10.1088/0004-637X/809/2/172

J.A. Irwin, W.P. Maksym, G.R. Sivakoff, A.J. Romanowsky, D. Lin, T. Speegle, I. Prado, D. Mildebrath, J. Strader, J. Liu, J.M. Miller, Ultraluminous X-ray bursts in two ultracompact companions to nearby elliptical galaxies. Nature **538**, 356–358 (2016). https://doi.org/10.1038/nature19822

F. Jansen, D. Lumb, B. Altieri, J. Clavel, M. Ehle, C. Erd, C. Gabriel, M. Guainazzi, P. Gondoin, R. Much, R. Munoz, M. Santos, N. Schartel, D. Texier, G. Vacanti, XMM-Newton observatory. I. The spacecraft and operations. Astron. Astrophys. **365**, 1–6 (2001). https://doi.org/10.1051/0004-6361:20000036

P.G. Jonker, M. Heida, M.A.P. Torres, J.C.A. Miller-Jones, A.C. Fabian, E.M. Ratti, G. Miniutti, D.J. Walton, T.P. Roberts, The nature of the bright ULX X-2 in NGC 3921: a Chandra position and HST candidate counterpart. Astrophys. J. **758**(1), 28 (2012). https://doi.org/10.1088/0004-637X/758/1/28

P.G. Jonker, A. Glennie, M. Heida, T. Maccarone, S. Hodgkin, G. Nelemans, J.C.A. Miller-Jones, M.A.P. Torres, R. Fender, Discovery of a new kind of explosive X-ray transient near M86. Astrophys. J. **779**, 14 (2013). https://doi.org/10.1088/0004-637X/779/1/14

P.G. Jonker, N.C. Stone, A. Generozov, S. van Velzen, B. Metzger, Implications from late-time X-ray detections of optically selected tidal disruption events: state changes, unification, and detection rates. ArXiv e-prints, 1906-12236 (2019)

E. Kara, L. Dai, C.S. Reynolds, T. Kallman, Ultrafast outflow in tidal disruption event ASASSN-14li. Mon. Not. R. Astron. Soc. **474**, 3593–3598 (2018). https://doi.org/10.1093/mnras/stx3004

G. Kauffmann, T.M. Heckman, C. Tremonti, J. Brinchmann, S. Charlot, S.D.M. White, S.E. Ridgway, J. Brinkmann, M. Fukugita, P.B. Hall, Ž. Ivezić, G.T. Richards, D.P. Schneider, The host galaxies of active galactic nuclei. Mon. Not. R. Astron. Soc. **346**(4), 1055–1077 (2003). https://doi.org/10.1111/j.1365-2966.2003.07154.x

L.J. Kewley, C.A. Heisler, M.A. Dopita, S. Lumsden, Optical classification of southern warm infrared galaxies. Astrophys. J. Suppl. Ser. **132**(1), 37–71 (2001). https://doi.org/10.1086/318944

I. Khabibullin, S. Sazonov, Stellar tidal disruption candidates found by cross-correlating the ROSAT bright source catalogue and XMM-Newton observations. Mon. Not. R. Astron. Soc. **444**(2), 1041–1053 (2014). https://doi.org/10.1093/mnras/stu1491

I. Khabibullin, S. Sazonov, R. Sunyaev, SRG/eROSITA prospects for the detection of stellar tidal disruption flares. Mon. Not. R. Astron. Soc. **437**(1), 327–337 (2014). https://doi.org/10.1093/mnras/stt1889

S. Komossa, Ludwig Biermann award lecture: X-ray evidence for supermassive black holes at the centers of nearby, non-active galaxies. Rev. Mod. Astron. **15**, 27 (2002)

S. Komossa, The extremes of (X-ray) variability among galaxies: flares from stars tidally disrupted by supermassive black holes, in *The Interplay Among Black Holes, Stars and ISM in Galactic Nuclei*, ed. by T. Storchi-Bergmann, L.C. Ho, H.R. Schmitt. IAU Symposium, vol. 222 (2004), pp. 45–48. https://doi.org/10.1017/S1743921304001425

S. Komossa, Growing black holes: observational evidence for stellar tidal disruption events, in *Growing Black Holes: Accretion in a Cosmological Context*, ed. by A. Merloni, S. Nayakshin, R.A. Sunyaev (2005), pp. 159–163. https://doi.org/10.1007/11403913_27

S. Komossa, N. Bade, The giant X-ray outbursts in NGC 5905 and IC 3599: follow-up observations and outburst scenarios. Astron. Astrophys. **343**, 775–787 (1999)

S. Komossa, M. Dahlem, X-ray outbursts from nearby 'normal' and active galaxies. A review, new radio observations, and an X-ray search for further tidal disruption flares. ArXiv Astrophysics e-prints (2001)

S. Komossa, J. Greiner, Discovery of a giant and luminous X-ray outburst from the optically inactive galaxy pair RX J1242.6-1119. Astron. Astrophys. **349**, 45–48 (1999)

S. Komossa, D. Merritt, Tidal disruption flares from recoiling supermassive black holes. Astrophys. J. **683**(1), 21–24 (2008). http://stacks.iop.org/1538-4357/683/i=1/a=L21

S. Komossa, J.A. Zensus, Compact object mergers: observations of supermassive binary black holes and stellar tidal disruption events, in *Star Clusters and Black Holes in Galaxies Across Cosmic Time*, ed. by Y. Meiron, S. Li, F.-K. Liu, R. Spurzem. IAU Symposium, vol. 312 (2016), pp. 13–25. https://doi.org/10.1017/S1743921315007395

S. Komossa, J. Halpern, N. Schartel, G. Hasinger, M. Santos-Lleo, P. Predehl, A huge drop in the X-ray luminosity of the nonactive galaxy RX J1242.6-1119A, and the first postflare spectrum: testing the tidal disruption scenario. Astrophys. J. Lett. **603**, 17 (2004). http://adsabs.harvard.edu/abs/2004ApJ...603L..17K

S. Komossa, H. Zhou, T. Wang, M. Ajello, J. Ge, J. Greiner, H. Lu, M. Salvato, R. Saxton, H. Shan, D. Xu, W. Yuan, Discovery of superstrong, fading, iron line emission and double-peaked Balmer lines of the galaxy SDSS J095209.56+214313.3: the light echo of a huge flare. Astrophys. J. **678**(1), 13–16 (2008). http://adsabs.harvard.edu/abs/2008ApJ...678L..13K

S. Komossa, H. Zhou, A. Rau, M. Dopita, A. Gal-Yam, J. Greiner, J. Zuther, M. Salvato, D. Xu, H. Lu, NTT, Spitzer, and Chandra spectroscopy of SDSSJ095209.56+214313.3: the most luminous coronal-line supernova ever observed, or a stellar tidal disruption event? Astrophys. J. **701**(1), 105–121 (2009). https://doi.org/10.1088/0004-637X/701/1/105

A.J. Levan, N.R. Tanvir, S.B. Cenko, D.A. Perley, K. Wiersema, J.S. Bloom, A.S. Fruchter, A.d.U. Postigo, P.T. O'Brien, N. Butler, A.J. van der Horst, G. Leloudas, A.N. Morgan, K. Misra, G.C. Bower, J. Farihi, R.L. Tunnicliffe, M. Modjaz, J.M. Silverman, J. Hjorth, C. Thöne, A. Cucchiara, J.M.C. Cerón, A.J. Castro-Tirado, J.A. Arnold, M. Bremer, J.P. Brodie, T. Carroll, M.C. Cooper, P.A. Curran, R.M. Cutri, J. Ehle, D. Forbes, J. Fynbo, J. Gorosabel, J. Graham, D.I. Hoffman, S. Guziy, P. Jakobsson, A. Kamble, T. Kerr, M.M. Kasliwal, C. Kouveliotou, D. Kocevski, N.M. Law, P.E. Nugent, E.O. Ofek, D. Poznanski, R.M. Quimby, E. Rol, A.J. Romanowsky, R. Sánchez-Ramírez, S. Schulze, N. Singh, L. van Spaandonk, R.L.C. Starling, R.G. Strom, J.C. Tello, O. Vaduvescu, P.J. Wheatley, R.A.M.J. Wijers, J.M. Winters, D. Xu, An extremely luminous panchromatic outburst from the nucleus of a distant galaxy. Science **333**, 199 (2011). https://doi.org/10.1126/science.1207143

A.J. Levan, N.R. Tanvir, R.L.C. Starling, K. Wiersema, K.L. Page, D.A. Perley, S. Schulze, G.A. Wynn, R. Chornock, J. Hjorth, S.B. Cenko, A.S. Fruchter, P.T. O'Brien, G.C. Brown, R.L. Tunnicliffe, D. Malesani, P. Jakobsson, D. Watson, E. Berger, D. Bersier, B.E. Cobb, S. Covino, A. Cucchiara, A. de Ugarte Postigo, D.B. Fox, A. Gal-Yam, P. Goldoni, J. Gorosabel, L. Kaper, T. Krühler, R. Karjalainen, J.P. Osborne, E. Pian, R. Sánchez-Ramírez, B. Schmidt, I. Skillen, G. Tagliaferri, C. Thöne, O. Vaduvescu, R.A.M.J. Wijers, B.A. Zauderer, A new population of ultra-long duration gamma-ray bursts. Astrophys. J. **781**, 13 (2014). https://doi.org/10.1088/0004-637X/781/1/13

A.J. Levan, N.R. Tanvir, G.C. Brown, B.D. Metzger, K.L. Page, S.B. Cenko, P.T. O'Brien, J.D. Lyman, K. Wiersema, E.R. Stanway, A.S. Fruchter, D.A. Perley, J.S. Bloom, Late time multi-wavelength observations of Swift J1644+5734: a luminous optical/IR bump and quiescent X-ray emission. Astrophys. J. **819**, 51 (2016). https://doi.org/10.3847/0004-637X/819/1/51

L.-X. Li, R. Narayan, K. Menou, The giant X-ray flare of NGC 5905: tidal disruption of a star, a brown dwarf, or a planet? Astrophys. J. **576**, 753 (2002). http://adsabs.harvard.edu/abs/2002ApJ...576..753L

D. Lin, E.R. Carrasco, D. Grupe, N.A. Webb, D. Barret, S.A. Farrell, Discovery of an ultrasoft X-ray transient source in the 2XMM catalog: a tidal disruption event candidate. Astrophys. J. **738**, 52 (2011). https://doi.org/10.1088/0004-637X/738/1/52

D. Lin, P.W. Maksym, J.A. Irwin, S. Komossa, N.A. Webb, O. Godet, D. Barret, D. Grupe, S.D.J. Gwyn, An ultrasoft X-ray flare from 3XMM J152130.7+074916: a tidal disruption event candidate. Astrophys. J. **811**, 43 (2015). https://doi.org/10.1088/0004-637X/811/1/43

D. Lin, J. Guillochon, S. Komossa, E. Ramirez-Ruiz, J.A. Irwin, W.P. Maksym, D. Grupe, O. Godet, N.A. Webb, D. Barret, B.A. Zauderer, P.-A. Duc, E.R. Carrasco, S.D.J. Gwyn, A likely decade-long sustained tidal disruption event. Nat. Astron. **1**, 0033 (2017a). https://doi.org/10.1038/s41550-016-0033

D. Lin, O. Godet, L.C. Ho, D. Barret, N.A. Webb, J.A. Irwin, Large decay of X-ray flux in 2XMM J123103.2+110648: evidence for a tidal disruption event. Mon. Not. R. Astron. Soc. **468**, 783–789 (2017b). https://doi.org/10.1093/mnras/stx489

D. Lin, J. Strader, E.R. Carrasco, D. Page, A.J. Romanowsky, J. Homan, J.A. Irwin, R.A. Remillard, O. Godet, N.A. Webb, H. Baumgardt, R. Wijnands, D. Barret, P.-A. Duc, J.P. Brodie, S.D.J. Gwyn, A luminous X-ray outburst from an intermediate-mass black hole in an off-centre star cluster. Nat. Astron. **2**, 656–661 (2018a). https://doi.org/10.1038/s41550-018-0493-1

D. Lin, J. Strader, E.R. Carrasco, O. Godet, D. Grupe, N.A. Webb, D. Barret, J.A. Irwin, Multiwavelength follow-up observations of the tidal disruption event candidate 2XMMi J184725.1-631724. Mon. Not. R. Astron. Soc. **474**, 3000–3008 (2018b). https://doi.org/10.1093/mnras/stx2940

Q.Z. Liu, J. van Paradijs, E.P.J. van den Heuvel, A catalogue of high-mass X-ray binaries. Astron. Astrophys. Suppl. Ser. **147**, 25–49 (2000). https://doi.org/10.1051/aas:2000288

Q.Z. Liu, J. van Paradijs, E.P.J. van den Heuvel, A catalogue of low-mass X-ray binaries. Astron. Astrophys. **368**, 1021–1054 (2001). https://doi.org/10.1051/0004-6361:20010075

F.K. Liu, S. Li, X. Chen, Interruption of tidal-disruption flares by supermassive black hole binaries. Astrophys. J. **706**(1), 133–137 (2009). https://doi.org/10.1088/0004-637X/706/1/L133

F.K. Liu, S. Li, S. Komossa, A milliparsec supermassive black hole binary candidate in the galaxy SDSS J120136.02+300305.5. Astrophys. J. **786**, 103 (2014). https://doi.org/10.1088/0004-637X/786/2/103

X.-L. Liu, L.-M. Dou, R.-F. Shen, J.-H. Chen, The UV/optical peak and X-ray brightening in TDE candidate AT2019azh: a case of stream-stream collision and delayed accretion. ArXiv e-prints, 1912-06081 (2019)

G. Lodato, Challenges in the modeling of tidal disruption events lightcurves, in *European Physical Journal Web of Conferences*. European Physical Journal Web of Conferences, vol. 39 (2012), p. 01001. https://doi.org/10.1051/epjconf/20123901001
G. Lodato, E.M. Rossi, Multiband light curves of tidal disruption events. Mon. Not. R. Astron. Soc. **410**, 359–367 (2011). https://doi.org/10.1111/j.1365-2966.2010.17448.x
A. Loeb, A. Ulmer, Optical appearance of the debris of a star disrupted by a massive black hole. Astrophys. J. **489**(2), 573–578 (1997). https://doi.org/10.1086/304814
M. MacLeod, J. Guillochon, E. Ramirez-Ruiz, The tidal disruption of giant stars and their contribution to the flaring supermassive black hole population. Astrophys. J. **757**(2), 134 (2012). https://doi.org/10.1088/0004-637X/757/2/134. http://adsabs.harvard.edu/abs/2012ApJ...757..134M
M. MacLeod, J. Guillochon, E. Ramirez-Ruiz, D. Kasen, S. Rosswog, Optical thermonuclear transients from tidal compression of white dwarfs as tracers of the low end of the massive black hole mass function. Astrophys. J. **819**, 3 (2016). https://doi.org/10.3847/0004-637X/819/1/3
K. Makishima, Y. Maejima, K. Mitsuda, H.V. Bradt, R.A. Remillard, I.R. Tuohy, R. Hoshi, M. Nakagawa, Simultaneous X-ray and optical observations of GX 339-4 in an X-ray high state. Astrophys. J. **308**, 635–643 (1986). https://doi.org/10.1086/164534
W.P. Maksym, M.P. Ulmer, M. Eracleous, A tidal disruption flare in A1689 from an archival X-ray survey of galaxy clusters. Astrophys. J. **722**(2), 1035–1050 (2010). http://adsabs.harvard.edu/abs/2010ApJ...722.1035M
W.P. Maksym, M.P. Ulmer, M.C. Eracleous, L. Guennou, L.C. Ho, A tidal flare candidate in Abell 1795. Mon. Not. R. Astron. Soc. **435**(3), 1904–1927 (2013). http://adsabs.harvard.edu/abs/2013MNRAS.435.1904M
W.P. Maksym, D. Lin, J.A. Irwin, RBS 1032: a tidal disruption event in another dwarf galaxy? Astrophys. J. Lett. **792**, 29 (2014a). https://doi.org/10.1088/2041-8205/792/2/L29
W.P. Maksym, M.P. Ulmer, K.C. Roth, J.A. Irwin, R. Dupke, L.C. Ho, W.C. Keel, C. Adami, Deep spectroscopy of the $M_V \sim -14.8$ host galaxy of a tidal disruption flare in A1795. Mon. Not. R. Astron. Soc. **444**, 866–873 (2014b). https://doi.org/10.1093/mnras/stu1485
V. Mangano, D.N. Burrows, B. Sbarufatti, J.K. Cannizzo, The definitive X-ray light curve of Swift J164449.3+573451. Astrophys. J. **817**, 103 (2016). https://doi.org/10.3847/0004-637X/817/2/103
N.J. McConnell, C.-P. Ma, Revisiting the scaling relations of black hole masses and host galaxy properties. Astrophys. J. **764**, 184 (2013). https://doi.org/10.1088/0004-637X/764/2/184
B.D. Metzger, N.C. Stone, A bright year for tidal disruptions. Mon. Not. R. Astron. Soc. **461**, 948–966 (2016). https://doi.org/10.1093/mnras/stw1394
R. Mewe, E.H.B.M. Gronenschild, G.H.J. van den Oord, Calculated X-radiation from optically thin plasmas. V. Astron. Astrophys. Suppl. Ser. **62**, 197–254 (1985)
M.J. Middleton, J.C.A. Miller-Jones, S. Markoff, R. Fender, M. Henze, N. Hurley-Walker, A.M.M. Scaife, T.P. Roberts, D. Walton, J. Carpenter, Bright radio emission from an ultraluminous stellar-mass microquasar in M 31. Nature **493**(7431), 187–190 (2013). https://doi.org/10.1038/nature11697
J.M. Miller, J.S. Kaastra, M.C. Miller, M.T. Reynolds, G. Brown, S.B. Cenko, J.J. Drake, S. Gezari, J. Guillochon, K. Gultekin, J. Irwin, A. Levan, D. Maitra, W.P. Maksym, R. Mushotzky, P. O'Brien, F. Paerels, J. de Plaa, E. Ramirez-Ruiz, T. Strohmayer, N. Tanvir, Flows of X-ray gas reveal the disruption of a star by a massive black hole. Nature **526**, 542–545 (2015). https://doi.org/10.1038/nature15708
S. Mineo, M. Gilfanov, R. Sunyaev, X-ray emission from star-forming galaxies – I. High-mass X-ray binaries. Mon. Not. R. Astron. Soc. **419**, 2095–2115 (2012). https://doi.org/10.1111/j.1365-2966.2011.19862.x
G. Miniutti, R.D. Saxton, P.M. Rodríguez-Pascual, A.M. Read, P. Esquej, M. Colless, P. Dobbie, M. Spolaor, A high Eddington-ratio, true Seyfert 2 galaxy candidate: implications for broad-line region models. Mon. Not. R. Astron. Soc. **433**, 1764–1777 (2013). https://doi.org/10.1093/mnras/stt850
B. Mockler, J. Guillochon, E. Ramirez-Ruiz, Weighing black holes using tidal disruption events. ArXiv e-prints (2018)
K. Nandra, K.A. Pounds, GINGA observations of the X-ray spectra of Seyfert galaxies. Mon. Not. R. Astron. Soc. **268**, 405 (1994). https://doi.org/10.1093/mnras/268.2.405
H. Netzer, *The Physics and Evolution of Active Galactic Nuclei* (2013)
H. Netzer, Bolometric correction factors for active galactic nuclei. Mon. Not. R. Astron. Soc. **488**(4), 5185–5191 (2019). https://doi.org/10.1093/mnras/stz2016
M. Nikołajuk, R. Walter, Tidal disruption of a super-Jupiter by a massive black hole. Astron. Astrophys. **552**, 75 (2013). http://adsabs.harvard.edu/abs/2013A%26A...552A..75N
J. Nishimura, K. Mitsuda, M. Itoh, Comptonization of soft X-ray photons in an optically thin hot plasma. Publ. Astron. Soc. Jpn. **38**, 819–830 (1986)
D.E. Osterbrock, *Astrophysics of Gaseous Nebulae and Active Galactic Nuclei* (1989)

D.R. Pasham, S.B. Cenko, A. Sadowski, J. Guillochon, N.C. Stone, S. van Velzen, J.K. Cannizzo, Optical/UV-to-X-ray echoes from the tidal disruption flare ASASSN-14li. Astrophys. J. Lett. **837**, 30 (2017). https://doi.org/10.3847/2041-8213/aa6003

Z.-K. Peng, Y.-S. Yang, R.-F. Shen, L.-J. Wang, J.-H. Zou, B.-B. Zhang, CDF-S XT1 and XT2: white dwarf tidal disruption events by intermediate-mass black holes? Astrophys. J. Lett. **884**(2), 34 (2019). https://doi.org/10.3847/2041-8213/ab481b

E.S. Phinney, Cosmic merger mania. Nature **340**, 595–596 (1989). https://doi.org/10.1038/340595a0

T. Piran, A. Sądowski, A. Tchekhovskoy, Jet and disc luminosities in tidal disruption events. Mon. Not. R. Astron. Soc. **453**, 157–165 (2015a). https://doi.org/10.1093/mnras/stv1547

T. Piran, G. Svirski, J. Krolik, R.M. Cheng, H. Shiokawa, Disk formation versus disk accretion—what powers tidal disruption events? Astrophys. J. **806**, 164 (2015b). https://doi.org/10.1088/0004-637X/806/2/164

G. Ponti, I. Papadakis, S. Bianchi, M. Guainazzi, G. Matt, P. Uttley, N.F. Bonilla, CAIXA: a catalogue of AGN in the XMM-Newton archive. III. Excess variance analysis. Astron. Astrophys. **542**, 83 (2012). https://doi.org/10.1051/0004-6361/201118326

K.A. Pounds, J.N. Reeves, A.R. King, K.L. Page, P.T. O'Brien, M.J.L. Turner, A high-velocity ionized outflow and XUV photosphere in the narrow emission line quasar PG1211+143. Mon. Not. R. Astron. Soc. **345**(3), 705–713 (2003). https://doi.org/10.1046/j.1365-8711.2003.07006.x

P. Predehl, H. Böhringer, H. Brunner, M. Brusa, V. Burwitz, N. Cappelluti, E. Churazov, K. Dennerl, M. Freyberg, P. Friedrich, eROSITA on SRG, in *X-Ray Astronomy 2009; Present Status, Multi-Wavelength Approach and Future Perspectives*, ed. by A. Comastri, L. Angelini, M. Cappi. American Institute of Physics Conference Series, vol. 1248 (2010), pp. 543–548. https://doi.org/10.1063/1.3475336

J.P. Pye, I.M. McHardy, The Ariel V sky survey of fast-transient X-ray sources. Mon. Not. R. Astron. Soc. **205**, 875–888 (1983). https://doi.org/10.1093/mnras/205.3.875

E. Quataert, D. Kasen, Swift 1644+57: the longest gamma-ray burst? Mon. Not. R. Astron. Soc. **419**, 1–5 (2012). https://doi.org/10.1111/j.1745-3933.2011.01151.x

M.J. Rees, Tidal disruption of stars by black holes of 10 to the 6th-10 to the 8th solar masses in nearby galaxies. Nature **333**, 523–528 (1988). https://doi.org/10.1038/333523a0

M.J. Rees, Black holes in galactic centers. Sci. Am. **263**, 56–60 (1990). https://doi.org/10.1038/scientificamerican1190-56

A.E. Reines, M. Volonteri, Relations between central black hole mass and total galaxy stellar mass in the local universe. Astrophys. J. **813**, 82 (2015). https://doi.org/10.1088/0004-637X/813/2/82

S.R. Rosen, N.A. Webb, M.G. Watson, J. Ballet, D. Barret, V. Braito, F.J. Carrera, M.T. Ceballos, M. Coriat, R. Della Ceca, G. Denkinson, P. Esquej, S.A. Farrell, M. Freyberg, F. Grisé, P. Guillout, L. Heil, F. Koliopanos, D. Law-Green, G. Lamer, D. Lin, R. Martino, L. Michel, C. Motch, A. Nebot Gomez-Moran, C.G. Page, K. Page, M. Page, M.W. Pakull, J. Pye, A. Read, P. Rodriguez, M. Sakano, R. Saxton, A. Schwope, A.E. Scott, R. Sturm, I. Traulsen, V. Yershov, I. Zolotukhin, The XMM-Newton serendipitous survey. VII. The third XMM-Newton serendipitous source catalogue. Astron. Astrophys. **590**, 1 (2016). https://doi.org/10.1051/0004-6361/201526416

N. Roth, D. Kasen, What sets the line profiles in tidal disruption events? Astrophys. J. **855**(1), 54 (2018). https://doi.org/10.3847/1538-4357/aaaec6

D.M. Russell, J.C.A. Miller-Jones, T.J. Maccarone, Y.J. Yang, R.P. Fender, F. Lewis, Testing the jet quenching paradigm with an ultradeep observation of a steadily soft state black hole. Astrophys. J. Lett. **739**, 19 (2011). https://doi.org/10.1088/2041-8205/739/1/L19

P. Salucci, C. Ratnam, P. Monaco, L. Danese, The masses of black holes in the nuclei of spirals. Mon. Not. R. Astron. Soc. **317**(2), 488–496 (2000). https://doi.org/10.1046/j.1365-8711.2000.03622.x

W.L.W. Sargent, P.J. Young, A. Boksenberg, K. Shortridge, C.R. Lynds, F.D.A. Hartwick, Dynamical evidence for a central mass concentration in the galaxy M87. Astrophys. J. **221**, 731–744 (1978). https://doi.org/10.1086/156077

R.D. Saxton, A.M. Read, P. Esquej, M.J. Freyberg, B. Altieri, D. Bermejo, The first XMM-Newton slew survey catalogue: XMMSL1. Astron. Astrophys. **480**, 611–622 (2008). https://doi.org/10.1051/0004-6361:20079193

R.D. Saxton, A.M. Read, P. Esquej, S. Komossa, S. Dougherty, P. Rodriguez-Pascual, D. Barrado, A tidal disruption-like X-ray flare from the quiescent galaxy SDSS J120136.02+300305.5. Astron. Astrophys. **541**, 106 (2012a). https://doi.org/10.1051/0004-6361/201118367

R.D. Saxton, A.M. Read, S. Komossa, P. Esquej, A well-monitored, X-ray selected, tidal disruption event, in *European Physical Journal Web of Conferences*. European Physical Journal Web of Conferences, vol. 39 (2012b), p. 02002. https://doi.org/10.1051/epjconf/20123902002

R.D. Saxton, S.E. Motta, S. Komossa, A.M. Read, Was the soft X-ray flare in NGC 3599 due to an AGN disc instability or a delayed tidal disruption event? Mon. Not. R. Astron. Soc. **454**, 2798–2803 (2015). https://doi.org/10.1093/mnras/stv2160

R.D. Saxton, A.M. Read, S. Komossa, P. Lira, K.D. Alexander, M.H. Wieringa, XMMSL1 J074008.2-853927: a tidal disruption event with thermal and non-thermal components. Astron. Astrophys. **598**, 29 (2017). https://doi.org/10.1051/0004-6361/201629015

R.D. Saxton, A.M. Read, S. Komossa, P. Lira, K.D. Alexander, I. Steele, F. Ocaña, E. Berger, P. Blanchard, XMMSL2 J144605.0+685735: a slow tidal disruption event. ArXiv e-prints, 1908-01065 (2019)

H. Schulz, C. Fritsch, Possible continuum-diluting processes to generate Liner-like spectra. Astron. Astrophys. **291**, 713–719 (1994)

S. Sembay, R.G. West, Black hole remnants: soft X-ray flares from tidally disrupted stars. Mon. Not. R. Astron. Soc. **262**, 141–150 (1993). https://doi.org/10.1093/mnras/262.1.141

B.J. Shappee, J.L. Prieto, D. Grupe, C.S. Kochanek, K.Z. Stanek, G. De Rosa, S. Mathur, Y. Zu, B.M. Peterson, R.W. Pogge, S. Komossa, M. Im, J. Jencson, T.W.-S. Holoien, U. Basu, J.F. Beacom, D.M. Szczygieł, J. Brimacombe, S. Adams, A. Campillay, C. Choi, C. Contreras, M. Dietrich, M. Dubberley, M. Elphick, S. Foale, M. Giustini, C. Gonzalez, E. Hawkins, D.A. Howell, E.Y. Hsiao, M. Koss, K.M. Leighly, N. Morrell, D. Mudd, D. Mullins, J.M. Nugent, J. Parrent, M.M. Phillips, G. Pojmanski, W. Rosing, R. Ross, D. Sand, D.M. Terndrup, S. Valenti, Z. Walker, Y. Yoon, The man behind the curtain: X-rays drive the UV through NIR variability in the 2013 active galactic nucleus outburst in NGC 2617. Astrophys. J. **788**, 48 (2014). https://doi.org/10.1088/0004-637X/788/1/48

R.-F. Shen, Fast, ultra-luminous X-ray bursts from tidal stripping of white dwarfs by intermediate-mass black holes. ArXiv e-prints (2018)

H. Shiokawa, J.H. Krolik, R.M. Cheng, T. Piran, S.C. Noble, General relativistic hydrodynamic simulation of accretion flow from a stellar tidal disruption. Astrophys. J. **804**, 85 (2015). https://doi.org/10.1088/0004-637X/804/2/85

X.W. Shu, S.S. Wang, L.M. Dou, N. Jiang, J.X. Wang, T.G. Wang, A long decay of X-ray flux and spectral evolution in the supersoft active galactic nucleus GSN 069. Astrophys. J. Lett. **857**, 16 (2018). https://doi.org/10.3847/2041-8213/aaba17

I.V. Strateva, S. Komossa, The X-ray point-source population of NGC 1365: the puzzle of two highly-variable ultraluminous X-ray sources. Astrophys. J. **692**, 443 (2009). https://doi.org/10.1088/0004-637X/692/1/443

N.L. Strotjohann, R.D. Saxton, R.L.C. Starling, P. Esquej, A.M. Read, P.A. Evans, G. Miniutti, Highly variable AGN from the XMM-Newton slew survey. Astron. Astrophys. **592**, 74 (2016). https://doi.org/10.1051/0004-6361/201628241

L.E. Strubbe, E. Quataert, Optical flares from the tidal disruption of stars by massive black holes. Mon. Not. R. Astron. Soc. **400**, 2070 (2009). http://adsabs.harvard.edu/abs/2009MNRAS.400.2070S

L. Strüder, U. Briel, K. Dennerl, R. Hartmann, E. Kendziorra, N. Meidinger, E. Pfeffermann, C. Reppin, B. Aschenbach, W. Bornemann, H. Bräuninger, W. Burkert, M. Elender, M. Freyberg, F. Haberl, G. Hartner, F. Heuschmann, H. Hippmann, E. Kastelic, S. Kemmer, G. Kettenring, W. Kink, N. Krause, S. Müller, A. Oppitz, W. Pietsch, M. Popp, P. Predehl, A. Read, K.H. Stephan, D. Stötter, J. Trümper, P. Holl, J. Kemmer, H. Soltau, R. Stötter, U. Weber, U. Weichert, C. von Zanthier, D. Carathanassis, G. Lutz, R.H. Richter, P. Solc, H. Böttcher, M. Kuster, R. Staubert, A. Abbey, A. Holland, M. Turner, M. Balasini, G.F. Bignami, N. La Palombara, G. Villa, W. Buttler, F. Gianini, R. Lainé, D. Lumb, P. Dhez, The European photon imaging camera on XMM-Newton: the pn-CCD camera. Astron. Astrophys. **365**, 18–26 (2001). https://doi.org/10.1051/0004-6361:20000066

H. Sun, B. Zhang, Z. Li, Extragalactic high-energy transients: event rate densities and luminosity functions. Astrophys. J. **812**(1), 33 (2015). https://doi.org/10.1088/0004-637X/812/1/33

Y. Terashima, N. Kamizasa, H. Awaki, A. Kubota, Y. Ueda, A candidate active galactic nucleus with a pure soft thermal X-ray spectrum. Astrophys. J. **752**, 154 (2012). https://doi.org/10.1088/0004-637X/752/2/154

L. Titarchuk, Generalized comptonization models and application to the recent high-energy observations. Astrophys. J. **434**, 570 (1994). https://doi.org/10.1086/174760

J. Trümper, Highlights of ROSAT, in *Frontiers of Space and Ground-Based Astronomy*, ed. by W. Wamsteker, M.S. Longair, Y. Kondo. Astrophysics and Space Science Library, vol. 187 (1994), p. 47. https://doi.org/10.1007/978-94-011-0794-5_5

T.J. Turner, K.A. Pounds, The EXOSAT spectral survey of AGN. Mon. Not. R. Astron. Soc. **240**, 833–880 (1989). https://doi.org/10.1093/mnras/240.4.833

A. Ulmer, Flares from the tidal disruption of stars by massive black holes. Astrophys. J. **514**, 180–187 (1999). https://doi.org/10.1086/306909

S. van Velzen, G.R. Farrar, S. Gezari, N. Morrell, D. Zaritsky, L. Östman, M. Smith, J. Gelfand, A.J. Drake, Optical discovery of probable stellar tidal disruption flares. Astrophys. J. **741**(2), 73 (2011a). https://doi.org/10.1088/0004-637X/741/2/73. http://adsabs.harvard.edu/abs/2011ApJ...741...73V

S. van Velzen, E. Körding, H. Falcke, Radio jets from stellar tidal disruptions. Mon. Not. R. Astron. Soc. **417**(1), 51–55 (2011b). https://doi.org/10.1111/j.1745-3933.2011.01118.x

S. van Velzen, N.C. Stone, B.D. Metzger, S. Gezari, T.M. Brown, A.S. Fruchter, Late-time UV observations of tidal disruption flares reveal unobscured, compact accretion disks. Astrophys. J. **878**(2), 82 (2019a). https://doi.org/10.3847/1538-4357/ab1844

S. van Velzen, S. Gezari, S.B. Cenko, E. Kara, J.C.A. Miller-Jones, T. Hung, J. Bright, N. Roth, N. Blagorodnova, D. Huppenkothen, L. Yan, E. Ofek, J. Sollerman, S. Frederick, C. Ward, M.J. Graham, R. Fender, M.M. Kasliwal, C. Canella, R. Stein, M. Giomi, V. Brinnel, J. van Santen, J. Nordin, E.C. Bellm, R. Dekany, C. Fremling, V.Z. Golkhou, T. Kupfer, S.R. Kulkarni, R.R. Laher, A. Mahabal, F.J. Masci, A.A. Miller, J.D. Neill, R. Riddle, M. Rigault, B. Rusholme, M.T. Soumagnac, Y. Tachibana, The first tidal disruption flare in ZTF: from photometric selection to multi-wavelength characterization. Astrophys. J. **872**(2), 198 (2019b). https://doi.org/10.3847/1538-4357/aafe0c

S. Vaughan, R. Edelson, R.S. Warwick, Chandra observations of five X-ray transient galactic nuclei. Mon. Not. R. Astron. Soc. **349**, 1–5 (2004). https://doi.org/10.1111/j.1365-2966.2004.07615.x

W. Voges, B. Aschenbach, T. Boller, H. Bräuninger, U. Briel, W. Burkert, K. Dennerl, J. Englhauser, R. Gruber, F. Haberl, The ROSAT all-sky survey bright source catalogue. Astron. Astrophys. **349**, 389–405 (1999)

J. Wang, D. Merritt, Revised rates of stellar disruption in galactic nuclei. Astrophys. J. **600**, 149 (2004). http://adsabs.harvard.edu/abs/2004ApJ...600..149W

T.-G. Wang, H.-Y. Zhou, L.-F. Wang, H.-L. Lu, D. Xu, Transient superstrong coronal lines and broad bumps in the galaxy SDSS J074820.67+471214.3. Astrophys. J. **740**(2), 85 (2011). https://doi.org/10.1088/0004-637X/740/2/85

J.-M. Wang, C. Cheng, Y.-R. Li, Clumpy accretion onto black holes. I. Clumpy-advection-dominated accretion flow structure and radiation. Astrophys. J. **748**(2), 147 (2012). http://adsabs.harvard.edu/abs/2012ApJ...748..147W

M.C. Weisskopf, H.D. Tananbaum, L.P. Van Speybroeck, S.L. O'Dell, Chandra X-ray Observatory (CXO): overview, in *X-Ray Optics, Instruments, and Missions III*, ed. by J.E. Truemper, B. Aschenbach. Society of Photo-Optical Instrumentation Engineers (SPIE) Conference Series, vol. 4012 (2000), pp. 2–16. https://doi.org/10.1117/12.391545

T. Wevers, S. van Velzen, P.G. Jonker, N.C. Stone, T. Hung, F. Onori, S. Gezari, N. Blagorodnova, Black hole masses of tidal disruption event host galaxies. Mon. Not. R. Astron. Soc. **471**, 1694–1708 (2017). https://doi.org/10.1093/mnras/stx1703

T. Wevers, D.R. Pasham, S. van Velzen, G. Leloudas, S. Schulze, J.C.A. Miller-Jones, P.G. Jonker, M. Gromadzki, E. Kankare, S.T. Hodgkin, Ł. Wyrzykowski, Z. Kostrzewa-Rutkowska, S. Moran, M. Berton, K. Maguire, F. Onori, S. Mattila, M. Nicholl, Evidence for rapid disc formation and reprocessing in the X-ray bright tidal disruption event candidate AT 2018fyk. Mon. Not. R. Astron. Soc. **488**(4), 4816–4830 (2019a). https://doi.org/10.1093/mnras/stz1976

T. Wevers, N.C. Stone, S. van Velzen, P.G. Jonker, T. Hung, K. Auchettl, S. Gezari, F. Onori, D.M. Sánchez, Z. Kostrzewa-Rutkowska, Black hole masses of tidal disruption event host galaxies II. Mon. Not. R. Astron. Soc. **487**, 4136 (2019b). https://doi.org/10.1093/mnras/stz1602

Y.Q. Xue, X.C. Zheng, Y. Li, W.N. Brandt, B. Zhang, B. Luo, B.-B. Zhang, F.E. Bauer, H. Sun, B.D. Lehmer, X.-F. Wu, G. Yang, X. Kong, J.Y. Li, M.Y. Sun, J.-X. Wang, F. Vito, A magnetar-powered X-ray transient as the aftermath of a binary neutron-star merger. Nature **568**(7751), 198–201 (2019). https://doi.org/10.1038/s41586-019-1079-5

A. Yalinewich, J. Guillochon, R. Sari, A. Loeb, Shock breakouts from tidal disruption events. ArXiv e-prints (2018)

W. Yuan, C. Zhang, H. Feng, S.N. Zhang, Z.X. Ling, D. Zhao, J. Deng, Y. Qiu, J.P. Osborne, P. O'Brien, Einstein Probe – a small mission to monitor and explore the dynamic X-ray Universe. ArXiv e-prints, 1506-07735 (2015)

W. Yuan, S. Komossa, C. Zhang, H. Feng, Z.-X. Ling, D.H. Zhao, S.-N. Zhang, J.P. Osborne, P. O'Brien, R. Willingale, Detecting tidal disruption events of massive black holes in normal galaxies with the Einstein Probe, in *Star Clusters and Black Holes in Galaxies Across Cosmic Time*, ed. by Y. Meiron, S. Li, F.-K. Liu, R. Spurzem IAU Symposium, vol. 312 (2016), pp. 68–70. https://doi.org/10.1017/S1743921315007516

I. Zalamea, K. Menou, A.M. Beloborodov, White dwarfs stripped by massive black holes: sources of coincident gravitational and electromagnetic radiation. Mon. Not. R. Astron. Soc. **409**, 25–29 (2010). https://doi.org/10.1111/j.1745-3933.2010.00930.x

B.A. Zauderer, E. Berger, R. Margutti, G.G. Pooley, R. Sari, A.M. Soderberg, A. Brunthaler, M.F. Bietenholz, Radio monitoring of the tidal disruption event Swift J164449.3+573451. II. The relativistic jet shuts off and a transition to forward shock X-ray/radio emission. Astrophys. J. **767**(2), 152 (2013). https://doi.org/10.1088/0004-637X/767/2/152

Space Sci Rev (2021) 217:18
https://doi.org/10.1007/s11214-020-00759-7

CORRECTION

Correction to: X-Ray Properties of TDEs

R. Saxton[1] · S. Komossa[2] · K. Auchettl[3,4] · P.G. Jonker[5,6]

Accepted: 21 October 2020 / Published online: 2 February 2021

Abstract Observational astronomy of tidal disruption events (TDEs) began with the detection of X-ray flares from quiescent galaxies during the ROSAT all-sky survey of 1990–1991. The flares complied with theoretical expectations, having high peak luminosities (L_x up to $\geq 4 \times 10^{44}$ erg/s), a thermal spectrum with $kT \sim$ few $\times$ 10^5 K, and a decline on timescales of months to years, consistent with a diminishing return of stellar debris to a black hole of mass 10^{6-8} $M_\odot$. These measurements gave solid proof that the nuclei of quiescent galaxies are habitually populated by a super-massive black hole. Beginning in 2000, XMM-*Newton*,

The Tidal Disruption of Stars by Massive Black Holes
Edited by Peter G. Jonker, Sterl Phinney, Elena Maria Rossi, Sjoert van Velzen, Iair Arcavi and Maurizio Falanga

The article by Saxton et al., Space. Sci. Rev. (2020) 216:85, contains a number of errors in the references and a missing copyright statement, which have been corrected in this version and should be regarded as the final version by the reader.

The original article can be found online at https://doi.org/10.1007/s11214-020-00708-4

✉ R. Saxton
rsaxton@esa.int

S. Komossa
astrokomossa@gmx.de

K. Auchettl
katie.auchettl@nbi.ku.dk

P.G. Jonker
p.jonker@astro.ru.nl

1 TPZ-Vega for ESA, XMM-Newton SOC, ESAC, Apartado 78, 28691 Villanueva de la Cañada, Madrid, Spain

2 Max Planck Institut für Radioastronomie, Auf dem Huegel 69, 53121 Bonn, Germany

3 DARK, Niels Bohr Institute, University of Copenhagen, Lyngbyvej 2, 2100, Copenhagen, Denmark

4 Present address: School of Physics, The University of Melbourne, Parkville, VIC 3010, Australia

5 SRON, Netherlands Institute for Space Research, Sorbonnelaan 2, 3584 CA, Utrecht, The Netherlands

Chandra and *Swift* have discovered further TDEs which have been monitored closely at multiple wavelengths. A general picture has emerged of, initially near-Eddington accretion, powering outflows of highly-ionised material, giving way to a calmer sub-Eddington phase, where the flux decays monotonically, and finally a low accretion rate phase with a harder X-ray spectrum indicative of the formation of a disk corona. There are exceptions to this rule though which at the moment are not well understood. A few bright X-ray TDEs have been discovered in optical surveys but in general X-ray TDEs show little excess emission in the optical band, at least at times coincident with the X-ray flare. X-ray TDEs are powerful new probes of accretion physics down to the last stable orbit, revealing the conditions necessary for launching jets and winds. Finally we see that evidence is mounting for nuclear and non-nuclear intermediate mass black holes based on TDE flares which are relatively hot and/or fast.

Keywords X-ray · TDE · Black holes · Accretion disks

Correction to: Space Sci. Rev. (2020) 216:85 https://doi.org/10.1007/s11214-020-00708-4

The article by Saxton et al., Space Sci Rev (2020) 216:85, contains a number of errors in the references and a missing copyright statement, which we correct below.

- Reference Brown et al. 2016 in the text (Sect. 4.4) should have referred to Brown et al. 2017. Brown et al. 2016 should be removed from the bibliography.
- Reference "Cannizzo and Gehrels (2009)" in the text (Sect. 5) should have referred to "Cannizzo et al. (1990)". Cannizzo and Gehrels (2009) should be removed from the bibliography.
- Reference Alexander et al. 2016 in Sect. 4.2.2 should read Alexander et al. 2017.
- Reference Truemper et al. (1994) should be replaced by Truemper et al. (1982) in the text (Sect. 3), and in the references with J. Truemper, The ROSAT mission. Adv. Space Res. 2(4), 241–249 (1982). https://doi.org/10.1016/0273-1177(82)90070-9.
- Reference S.B. Cenko et al. 2012 is incorrect. The correct one is: S.B. Cenko, H.A. Krimm, A. Horesh, A. Rau, D.A. Frail, J.A. Kennea, A.J. Levan, S.T. Holland, N.R. Butler, R.M. Quimby, J.S. Bloom, A.V. Filippenko, A. Gal-Yam, J. Greiner, S.R. Kulkarni, E.O. Ofek, F. Olivares E., P. Schady, J.M. Silverman, N.R. Tanvir, D. Xu, Swift J2058.4+0516: discovery of a possible second relativistic tidal disruption flare? Astrophys. J. 753, 77 (2012). https://doi.org/10.1088/0004-637X/753/1/77.
- Reference E.S. Phinney 1989 is incorrect. The correct one is: E.S. Phinney, Manifestations of a massive black hole in the galactic center, in *The Center of the Galaxy*, ed. by M. Morris IAU Symposium, vol. 136, 1989, p. 543. https://ui.adsabs.harvard.edu/abs/1989IAUS..136..543P.
- Reference M.J. Rees 1990 is incorrect. The correct one is: M.J. Rees, Dead quasars in nearby galaxies? Science 247, 817 (1990). https://doi.org/10.1126/science.247.4944.817.
- Reference J.M. Wang et al. 2012 is incorrect. The correct one is: T.G. Wang, H.Y. Zhou, S. Komossa, H. Wang, W. Yuan, C. Yang, Extreme coronal line emitters: tidal disruption of stars by massive black holes in galactic nuclei? Astrophys. J. 749, 115 (2012). https://doi.org/10.1088/0004-637X/749/2/115.

[6] Department of Astrophysics/IMAPP, Radboud University, P.O. Box 9010, 6500 GL, Nijmegen, The Netherlands

The following references in the bibliography were incomplete and should have read:

N. Cappelluti, M. Ajello, P. Rebusco, S. Komossa, A. Bongiorno, C. Clemens, M. Salvato, P. Esquej, T. Aldcroft, J. Greiner, H. Quintana, A candidate tidal disruption event in the Galaxy cluster Abell 3571. Astron. Astrophys. 495(2), L9 (2009). https://doi.org/10.1051/0004-6361/200811479

J. Greiner, R. Schwarz, S. Zharikov, M. Orio, RX J1420.4+5334 – another tidal disruption event? Astron. Astrophys. 362, L25 (2000). http://ui.adsabs.harvard.edu/abs/2000A%26A...362L..25G

D. Grupe, H.-C. Thomas, K.M. Leighly, RX J1624.9+7554: A new X-ray transient AGN. Astron. Astrophys. 350, L31 (1999). https://ui.adsabs.harvard.edu/abs/1999A%26A...350L..31G

T.W.-S. Holoien, M.E. Huber, B.J. Shappee, M. Eracleous, K. Auchettl, J.S. Brown, M.A. Tucker, K.C. Chambers, C.S. Kochanek, K.Z. Stanek, A. Rest, D. Bersier, R.S. Post, G. Aldering, K.A. Ponder, J.D. Simon, E. Kankare, D. Dong., G. Hallinan, J. Bulger, T.B. Lowe, E.A. Magnier, A.S.B. Schultz, C.Z. Waters, M. Willman, D. Wright, D.R. Young, S. Dong, J.L. Prieto, T.A. Thompson, L. Denneau, H. Flewelling, A.N. Heinze, S.J. Smartt, K.W. Smith, B. Stalder, J.L. Tonry, H. Weiland, PS18kh: A New Tidal Disruption Event with a Non-Axisymmetric Accretion Disk. ArXiv e-prints, arXiv:1808.02890 (2018a)

S. Komossa, Ludwig Biermann award lecture: X-ray evidence for supermassive black holes at the centers of nearby, non-active galaxies. Rev. Mod. Astron. 15, 27 (2002). https://ui.adsabs.harvard.edu/abs/2002RvMA...15...27K

S. Komossa, N. Bade, The giant X-ray outbursts in NGC5905 and IC 3599: follow-up observations and outburst scenarios. Astron. Astrophys. 343, 775–787 (1999). https://ui.adsabs.harvard.edu/abs/1999A%26A...343..775K

S. Komossa, M. Dahlem, X-ray outbursts from nearby 'normal' and active galaxies. A review, new radio observations, and an X-ray search for further tidal disruption flares. MAXI workshop on AGN. ArXiv e-prints, astro-ph/0106422 (2001)

S. Komossa, J. Greiner, Discovery of a giant and luminous X-ray outburst from the optically inactive galaxy pair RX J1242.6-1119. Astron. Astrophys. 349, L45, (1999). https://ui.adsabs.harvard.edu/abs/1999A%26A...349L..45K

S. Komossa, D. Merritt, Tidal disruption flares from recoiling supermassive black holes. Astrophys. J. Lett. 683(1), 21–24 (2008). https://doi.org/10.1086/591420

S. Komossa, H. Zhou, T. Wang, M. Ajello, J. Ge, J. Greiner, H. Lu, M. Salvato, R. Saxton, H. Shan, D. Xu, W. Yuan, Discovery of superstrong, fading, iron line emission and double-peaked Balmer lines of the galaxy SDSS J095209.56+214313.3: the light echo of a huge flare. Astrophys. J. Lett. 678(1), 13–16 (2008). https://doi.org/10.1086/588281

R. Mewe, E.H.B.M. Gronenschild, G.H.J. van den Oord, Calculated X-radiation from optically thin plasmas. V. Astron. Astrophys. Suppl. Ser.62, 197–254 (1985). http://ui.adsabs.harvard.edu/abs/1985A%26AS...62..197M

B. Mockler, J. Guillochon, E. Ramirez-Ruiz, Weighing black holes using tidal disruption events. ArXiv eprints, 1801.08221 (2018)

H. Netzer, The Physics and Evolution of Active Galactic Nuclei (Cambridge University Press, Cambridge, 2013). https://doi.org/10.1017/CBO9781139109291

D.E. Osterbrock Astrophysics of Gaseous Nebulae and Active Galactic Nuclei (University Science Books, Mill Valley, 1989).

H. Schulz, C. Fritsch, Possible continuum-diluting processes to generate Liner-like spectra. Astron. Astrophys. 291, 713–719 (1994). https://ui.adsabs.harvard.edu/abs/1994A%26A...291..713S

R.-F. Shen, Fast, ultra-luminous X-ray bursts from tidal stripping of White Dwarfs by Intermediate-mass Black Holes. ArXiv e-prints, 1809.09359 (2018)
W. Voges, B. Aschenbach, T. Boller, H. Bräuninger, U. Briel, W. Burkert, K. Dennerl, J. Englhauser, R. Gruber, F. Haberl, The ROSAT all-sky survey bright source catalogue. Astron. Astrophys. 349, 389–405 (1999). https://ui.adsabs.harvard.edu/abs/1999A%26A...349..389V
A. Yalinewich, J. Guillochon, R. Sari, A. Loeb, Shock Breakouts from Tidal Disruption Events. ArXiv e-prints, 1808.10447 (2018)

– missing copyright/reference: Fig. 1: The proper copyright citation for a subimage of Fig. 1 was deleted before submission. The sentence should correctly read: "Lower right: Optical image of RXJ1242-1119, adapted from Komossa and Greiner (1999)."

1 Introduction

A key unsolved question in extra-galactic astrophysics in the 70s and 80s was, whether black holes exist at the centers of most or all galaxies. While 10% of galaxies were Active Galactic Nuclei (AGN) and believed to be powered by accretion onto supermassive black holes (SMBH), the remaining 90% were quiescent, inactive galaxies with no signs of permanent activity. Did they still harbour black holes at their centers? If so, how could we actually detect such "dormant" black holes? Only in a handful of the closest (quiescent) galaxies, could SMBHs be discerned by their impact on stellar rotation curves (Sargent et al. 1978). In order to tell, whether SMBHs existed not only in AGN, but in the majority of quiescent galaxies, Rees (1988) suggested that, if they disrupted a star and subsequently accreted the matter, then they would reveal their presence by a *temporary* X-ray flare, lasting a few months.

The integrated X-ray output of the stellar population of a galaxy is approximately 10^{39}–10^{41} erg s^{-1} (Fabbiano 1986) whereas the X-ray flare produced by a TDE emits at a significant fraction of the Eddington luminosity, 10^{42}–10^{45} erg s^{-1} (e.g. Rees 1988), and is hence relatively easy to detect thanks to the high contrast. While it is straight-forward to distinguish an AGN from a quiescent, in-active galaxy by means of optical spectroscopy (see next section), other X-ray source populations like flare stars, novae, or ULXs embedded in external galaxies are too distant to be resolved from their host galaxies by optical spectroscopy. However, they can still be readily distinguished by the X-ray peak luminosity of a TDE, which is orders of magnitude higher than the luminosity of other X-ray transient phenomena such as Novae, Supernovae, flares stars, accreting Galactic binaries and ULX.

It was shown that X-ray emission from TDEs should follow the rate of return of stellar debris to the black hole, which to a first approximation decays with a power-law index of −5/3 (Rees 1988; Phinney 1989; Evans and Kochanek 1989; Lodato and Rossi 2011) such that the luminosity is given by:

$$L_X = A\left(\frac{t - t_0}{1\,\mathrm{yr}}\right)^{-5/3} \tag{1}$$

The first detections of TDE X-ray flares were made with the ROSAT satellite (see Sect. 3), which found transient sources that displayed the predicted TDE characteristics; a short rise to peak, a steady decline, high peak luminosities, a soft X-ray spectrum, and, importantly, occurred in quiescent, in-active galaxies. With the advent of Chandra, XMM-*Newton*, the Neil Gehrels *Swift* satellite (hereafter *Swift*; Gehrels et al. 2004) and multi-wavelength follow-up, differences in the spectra and light curve of each event both compared to AGN and within the class of TDEs detected became more apparent. In particular *Swift*, with its ability to perform high cadence, long-term monitoring of the X-ray and optical/UV light curve and spectrum, has had a major impact in the field of TDEs.

The X-rays generated during a TDE are believed to be produced from the innermost stable orbits of the black hole and experience the strong gravitational field. This means they can be used to probe the Kerr and Schwarzschild metrics via precession, quasi-periodic oscillations (QPOs) and reverberation (see the Echo Chapter and Rees 1990).

At the time of writing,[1] the number of X-ray emitting TDEs and X-ray emitting TDE candidates is small enough that each one can be discussed individually, which we do in the following sections. A summary of the properties of each object is given in Table 1.

2 TDE Identification: How to Distinguish Between a Quiescent Galaxy and an AGN by Optical Spectroscopy

Astronomers distinguish between quiescent galaxies and Active Galactic Nuclei (AGN). They are routinely identified and distinguished by means of optical spectroscopy. AGN are galaxies which harbor a permanent accretion disk that emits a strong multi-wavelength continuum including X-rays. This photoionises ambient gas, and produces strong characteristic, long-lived emission lines. Quiescent galaxies are those which *do not harbor a permanent accretion disk* and therefore *do not emit any luminous X-rays from their cores*, neither variable nor constant, and they *lack the characteristic optical emission lines* (Osterbrock 1989).

The two of them can therefore be distinguished by means of *optical spectroscopy*: AGN have characteristic, high-ionization *narrow emission lines*, which arise in low-density gas at large nuclear distances (the so-called 'Narrow-line Region'; NLR). The NLR is photoionized by the photons from the *permanent* accretion disk, and it *traces the activity over a long time interval of 1000s of years*. In contrast, quiescent galaxies *lack these characteristic NLR emission lines*. The majority of quiescent galaxies have no emission lines at all while star-formation galaxies have low-ionization lines, very different from NLRs (Baldwin et al. 1981; Osterbrock 1989; Kewley et al. 2001; Kauffmann et al. 2003). Therefore, a crucial and very reliable distinction between an AGN and a quiescent galaxy is to take an optical spectrum, and determine the presence or absence of the characteristic NLR emission lines. Identifying quiescent galaxies by optical spectroscopy has been routinely carried out for decades (see Osterbrock 1989, for a good textbook). Optical spectroscopy to check for a quiescent, inactive host galaxy has therefore been an important step in the identification of any new TDE (e.g. Komossa and Greiner 1999), following the procedures suggested by theoreticians (e.g. Rees 1988; Rees 1990) irrespective of the waveband in which it was discovered.[2]

Finally, we would like to note, that (1) low-mass AGN and (2) AGN in LINERs (low-ionization nuclear emission regions) are more difficult to identify spectroscopically (e.g. Greene 2012). High S/N optical spectroscopy is required in case (1), while (2) LINER-like emission lines, which populate a separate region in diagnostic diagrams, are known to be powered by several different physical mechanisms; shocks, photoionization by old stellar populations, and/or AGN (e.g. Schulz and Fritsch 1994).

[1] In this chapter we include publications written up until mid-2019.

[2] Note, that broad and narrow optical emission lines can also be *temporarily* excited by the TDE itself (e.g. Komossa et al. 2008; Wang et al. 2012). However, these lines differ from classical NLRs and can be distinguished if there is more than one post-flare optical spectrum, since they are not permanent but will change and fade away quickly.

Table 1 Summary of soft X-ray TDEs referred to in this chapter

Name	Date[a]	z	$L_{X,peak}$[b]	Spectrum[c]	Ref.[d]
NGC 5905	7/1990	0.011	7×10^{42}	$kT = 60$ eV	(Bade et al. 1996)
RXJ1624+7554	8/1990	0.064	2×10^{44}	$\Gamma = 3$	(Grupe et al. 1999)
RBS 1032	11/1990	0.026	1×10^{43}	$kT = 120$ eV	Ghosh et al. (2006)[e]
RXJ1420+5334	12/1990	0.147	2.5×10^{44}	$kT = 38$ eV	(Greiner et al. 2000)
RXJ 1242–1119	7/1992	0.05	4×10^{44}	$kT = 60$ eV	(Komossa and Greiner 1999)
TDXF1347-3254	12/1992	0.037	7×10^{42}	$kT = 120$ eV	(Cappelluti et al. 2009)
WINGS J1348	12/1999	0.063	1×10^{42}	$kT = 84$ eV	(Maksym et al. 2013)
3XMM J152130.7+074916	8/2000	0.179	5×10^{43}	$kT \sim 170$ eV	(Lin et al. 2015)
NGC 3599	11/2003	0.002	5×10^{41}	$kT = 95$ eV	(Esquej et al. 2007)
SDSS J132341.97+482701.3	12/2003	0.0875	5×10^{43}	–	(Esquej et al. 2007)
SDSS J131122.15-012345.6	2/2004	0.18	5×10^{42}	kT$\sim$ 100 eV	Maksym et al. (2010)
XMMSL1 J024916.6-041244	7/2004	0.0186	2×10^{42}	$kT = 110$ eV	(Esquej et al. 2007)
3XMM J150052.0+015452	7/2005	0.145	6×10^{43}	kT$\sim$ 40 eV	(Lin et al. 2017a)
SDSS J095209.56+214313.3	12/2005	0.079	–	–	(Komossa et al. 2008)
3XMM J215022.4-55108	5/2006	0.055	7×10^{42}	$kT \sim 280$ eV	(Lin et al. 2018a)
2XMMi J184725.1-631724	9/2006	0.0353	3×10^{43}	kT$\sim$ 80 eV	(Lin et al. 2011)
SDSS J120136.02+300305.5	6/2010	0.146	3×10^{44}	$kT = 70$ eV	(Saxton et al. 2012a)
2MASX 0740-85	4/2014	0.0173	2×10^{43}	kT$\sim$ 80 eV, $\Gamma_X \sim 2$	(Saxton et al. 2017)
ASASSN-14li	11/2014	0.0206	3×10^{43}	$kT = 50$ eV	(Holoien et al. 2016b)
ASASSN-15oi	8/2015	0.0479	6×10^{42}	$\sim$45 eV	Gezari et al. (2017)
XMMSL2 J1446+68	8/2016	0.029	6×10^{42}	$\Gamma_X \sim 2.6$	(Saxton et al. 2019)

[a]Date of first detection in X-rays

[b]The highest unabsorbed X-ray luminosity measured. Units of erg s^{-1}

[c]Dominant spectral component(s) at peak. "kT" refers to a black-body temperature, Γ refers to a power-law slope

[d]The first reference to the X-ray measurements

[e]Identified as a TDE by Maksym et al. (2014a) and Khabibullin and Sazonov (2014)

3 ROSAT Soft X-Ray TDEs

Given the low disruption rate of TDEs (about 1 event every 10^{4-5} years per galaxy; Sect. 12), large-area sky surveys are best-suited to detect these events. The X-ray mission ROSAT was an ideal experiment to systematically search for, and detect, TDEs. Launched in June 1990, it carried out the first imaging X-ray survey of the entire sky in the soft X-ray band, at energies between (0.1–2.4) keV (e.g. Truemper 1982). The ROSAT all-sky survey (RASS) lasted for about a year and was then followed by an eight-year phase of pointed observations of selected targets, including deep follow-ups of exceptional sources and transients discovered during the RASS. ROSAT was equipped with two prime instruments, a high-resolution imager (HRI) providing a spatial resolution of 5 arcsec, and a positionally-sensitive proportional counter (PSPC) which allowed broad-band X-ray spectroscopy and achieved a spatial resolution of 25 arcsec on-axis.

Several luminous, high-amplitude X-ray flares from quiescent galaxies, matching remarkably well the predictions of tidal disruption theory (e.g. Rees 1988; Rees 1990), have first been identified by ROSAT. Four main events were discovered, from the galaxies NGC5905 (Bade et al. 1996; Komossa and Bade 1999), RXJ1242-1119 (Komossa and Greiner 1999), RXJ1624+7554 (Grupe et al. 1999) and RXJ1420+5334 (Greiner et al. 2000). Among these, NGC5905 and RXJ1242-1119 are the best-covered events in terms of their multi-wavelength follow-up observations and their long-term X-ray lightcurves, spanning time intervals of more than a decade, with amplitudes of decline larger than a factor of 1000 (Komossa et al. 2004; Halpern et al. 2004; Komossa 2005). We therefore summarize some main properties of these two events in the subsections below.

In addition, the ROSAT archive was used to identify new TDEs after the end of the mission; either sources which were bright during the ROSAT epoch, and had faded substantially when re-observed with present X-ray missions (Cappelluti et al. 2009; Maksym et al. 2014a; Khabibullin and Sazonov 2014), or events which were found to be bright in later observations, but much fainter or undetected with ROSAT. The ROSAT database has therefore played an important role in the identification of most soft X-ray TDEs known today.

3.1 NGC 5905

NGC 5905 was first noticed due to its luminous, soft ($kT = 0.06$ keV) X-ray outburst with (lower limit on the) peak luminosity in the soft X-ray band of $L_{\rm x,peak} = 7 \times 10^{42}$ erg/s during the RASS[3] (Komossa and Bade 1999). It remained bright for at least 5 days (the time interval while its position was repeatedly scanned during the RASS) increasing in luminosity to the observed peak (Fig. 1). Given its daily coverage for almost a week, it has one of the best-covered early lightcurves of the non-radio-emitting X-ray TDEs.

X-rays then declined on the timescale of months to years, as observed in multiple re-observations with ROSAT (Fig. 2). Within the 5 arcsec positional error (ROSAT HRI), the X-rays came from the center of this nearby barred spiral galaxy ($z = 0.011$; Fig. 1). While the X-ray spectrum was initially very soft, it had hardened significantly ($\Gamma_{\rm x} = 2.4$) 3 years later, when re-observed with ROSAT. The decline of its X-ray lightcurve is consistent with the predicted $t^{-5/3}$ law (Fig. 2), as first reported based on its ROSAT observations (Komossa and Bade 1999) and then confirmed with Chandra (Halpern et al. 2004). All observations of this event are in very good agreement with the predictions (Rees 1990) from tidal disruption theory (Bade et al. 1996; Komossa and Bade 1999).

[3]This luminosity is even higher, if a powerlaw model is fit, and if there is absorption from the event's host galaxy.

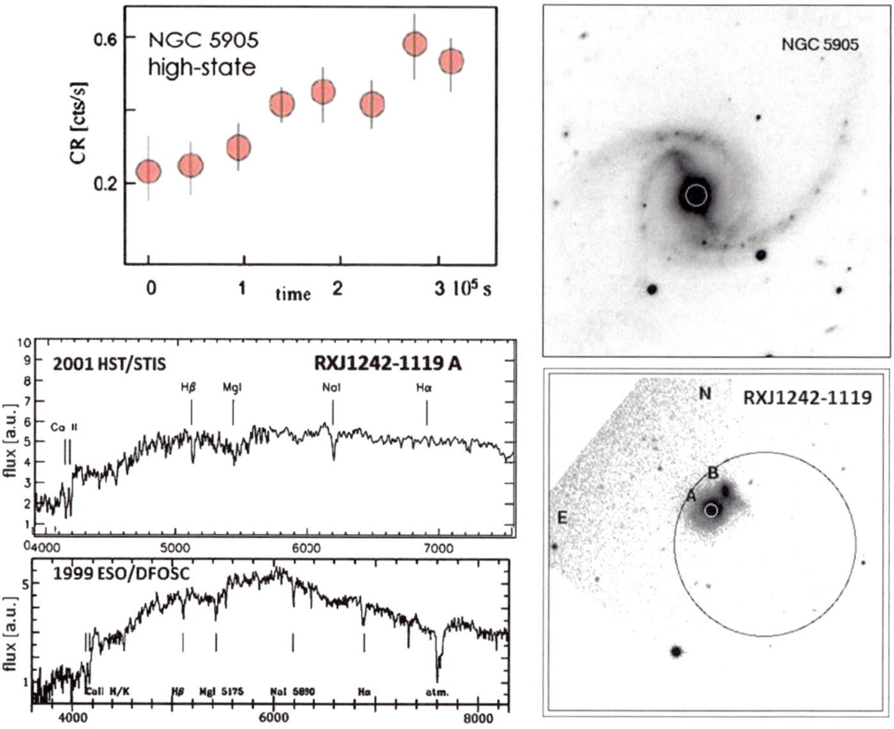

Fig. 1 Upper left: Early X-ray lightcurve of NGC 5905 showing its rise to the observed maximum (Komossa and Bade 1999; Bade et al. 1996) when the galaxy position was repeatedly covered during the RASS scans. Upper right: Optical image of NGC 5905 (Komossa 2002); a nearby giant barred spiral galaxy. The X-ray flare was observed from the centre of the galaxy. The white circle corresponds to the ROSAT HRI error circle of 5″. Lower right: Optical image of RXJ1242-1119, adapted from Komossa and Greiner (1999). The initial ROSAT X-ray error circle (black) contained two galaxies, both inactive. The X-ray error circle was narrowed down with Chandra (white), which detected fading X-ray emission from the core of RXJ1242-1119A (Komossa et al. 2004). Lower left: Optical spectrum of RXJ1242-1119A [upper panel: HST STIS spectrum from August 2001 (Gezari et al. 2003); lower panel: groundbased ESO spectrum from January 1999 (Komossa and Greiner 1999)]. These spectra completely lack emission lines and reveal a quiescent, inactive galaxy

A very important first step in the identification of every single TDE from a *quiescent* galaxy, is to take an optical spectrum of the host galaxy, and search for the absence of AGN activity down to deep limits. Ground-based optical spectroscopy before and after the flare has led to a starburst (HII-type) classification of the host galaxy of NGC 5905 (Komossa and Bade 1999). An HST observation after the flare showed evidence for faint enhanced [OIII] emission from the core of the galaxy (Gezari et al. 2003) – either long-lived or excited by the flare itself. Using scaling relations between host galaxy and central SMBH mass for elliptical galaxies (Ferrarese and Merritt 2000), an upper limit on the SMBH mass of NGC 5905 of a few times 10^8 $M_\odot$ was estimated; near the upper limit for the tidal disruption of a solar-type star and as such could imply a spinning SMBH. However, the host of NGC 5905 is a spiral galaxy and these tend to lie below the scaling relations of ellipticals. Using instead the relation of Salucci et al. (2000) for spirals gives a BH mass estimate of $\sim 10^7$ $M_\odot$ (Komossa 2002).

Integrating over its (observed) lightcurve (Eq. (1)), only a small fraction of a solar mass needed to be accreted to power this flare (Komossa 2002; Li et al. 2002). We look at possible

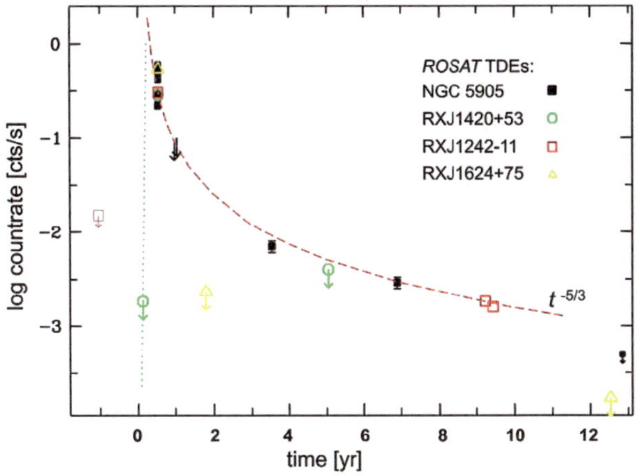

Fig. 2 Joint X-ray lightcurve of four TDEs identified with ROSAT, all shifted to the same peak time. The decline is consistent with a $t^{-5/3}$ law (dashed lined). This point was first made based on the ROSAT data of NGC 5905 (Komossa and Bade 1999), and later for the overall luminosity evolution of the sources displayed above (e.g., Fig. 1 of Komossa 2004). RXJ1242-1119 shows a further drop in X-ray emission at late times (not shown here), deviating from the early phase decline law, implying a total amplitude of decline of a factor >1000 (Komossa 2005)

explanations for this in Sect. 4.4. Note that all such estimates provide a lower limit on the accreted stellar mass, since the events may not have been caught exactly at peak luminosity, and since a significant fraction of the luminosity may be emitted outside the soft X-ray band.

The morphology of the host galaxy of NGC 5905 is that of a barred spiral galaxy of type SB; one of the largest spirals known and particularly well resolved in optical imaging since nearby (Fig. 1). Its multiple triaxial structures with a secondary bar might aid the efficient re-filling of its stellar loss-cone orbits, thereby boosting the stellar tidal disruption rate (see Sect. 3.3.3 of Komossa 2002).

3.2 RXJ1242–1119

A very luminous X-ray outburst was discovered from the inactive galaxy RX J1242-1119 (at $z = 0.05$) with ROSAT during a 'serendipitous' pointed observation. Its X-ray spectrum was extremely soft, one of the steepest ever identified among galaxies, and among the steepest in the ROSAT data base, with a photon index of $\Gamma_x = 5.1 \pm 0.9$ (Komossa and Greiner 1999). At a soft X-ray luminosity of 4×10^{44} erg/s (a conservative lower limit, assuming no absorption in the host galaxy itself, and without applying a bolometric correction), this event was exceptionally luminous.

Optical spectroscopy of the host galaxy, both ground-based and with the HST, reveals a quiescent, in-active galaxy, with no emission lines detected at all down to deep limits (Komossa and Greiner 1999; Gezari et al. 2003).

During the RASS itself, RXJ1242-1119 was not detected, implying an initial amplitude of variability of a factor of >20, and a rise-time of <1.5 years. Deep Chandra follow-ups (cf. next section) then revealed the fading of the X-ray emission from the TDE by a factor of up to >1000(Komossa et al. 2004; Halpern et al. 2004; Komossa 2005) more than a decade after its high-state. Given its extreme properties and deep follow-ups, this event continues to be one of the best cases of a TDE identified so far (Komossa et al. 2004).

Integrating only over the *observed* (0.1–2.4) X-ray lightcurve then gave a strict lower limit on the total emitted energy of 1.6×10^{51} erg/s (Eq. (1)) and on the accreted mass of $M > 0.01\eta_{0.1}^{-1}$ $M_\odot$ for the TDE in RXJ1242-1119 (Sect. 3 of Komossa et al. 2004).

Based on the host galaxy blue magnitude measured with the optical monitor OM onboard XMM-Newton, $m_b = 17.56 \pm 0.05$, the mass of the black hole at the center of the galaxy was estimated, based on the correlation between the absolute blue magnitude of the bulge of an elliptical galaxy and its SMBH mass (Ferrarese and Merritt 2000). This yielded an SMBH mass on the order of $M_{BH} \approx 10^8$ $M_\odot$ (Komossa et al. 2004).

3.3 More Recently Identified TDEs in the ROSAT Archive

During the ROSAT mission, an X-ray catalogue of >100,000 X-ray sources was created (Voges et al. 1999). This data base can still be used, in combination with the X-ray data bases more recently created with the missions XMM-Newton and Chandra, to search for new TDEs bright during the ROSAT observation(s). Cappelluti et al. (2009) identified the X-ray outburst TDXF1347-3254 ($z = 0.037$) that way, while Maksym et al. (2014a) and Khabibullin and Sazonov (2014) reported the detection of an outburst from RBS 1032 ($z = 0.026$).

The event TDXF1347-3254 was the first one identified in a galaxy cluster, in Abell 3571 (Cappelluti et al. 2009). X-ray emission from one of the cluster galaxies, LEDA 095953, was bright in ROSAT but decayed by a factor 650 between 1992 and 2004. The black hole mass, M_{BH}, was estimated to be 10^7 $M_\odot$ and the integrated luminosity was $>2 \times 10^{50}$ erg/s, implying an accreted mass between 0.001 and 0.03 solar masses. The ROSAT PSPC spectrum had an effective $kT = 120$ eV. Multi-band optical/NIR photometry, taken at the time, indicated an inactive host galaxy (Cappelluti et al. 2009) which has subsequently been confirmed by optical spectroscopy (Wevers et al. 2019a).

RBS 1032 was a supersoft ($\Gamma_x \sim 5$) and luminous ($\sim 10^{43}$ erg/s) ROSAT PSPC source, which later had faded by a factor $\sim$ 100–300 when re-observed with XMM-Newton. It is hosted by an inactive dwarf galaxy (Ghosh et al. 2006; Maksym et al. 2014a). From the shape of the light curve, Khabibullin and Sazonov (2014) deduced that the event had been first seen well after peak and that the peak bolometric luminosity was more likely to have been a few $\times 10^{44}$ erg s^{-1}. This in turn implies that $M_{BH} > 10^6$ $M_\odot$. The event remained quite soft ($\Gamma_x = 3.4 \pm 0.3$) 20 years after discovery.

3.4 Summary of the Properties of the ROSAT TDEs

In summary, the ROSAT TDEs are characterized by:

- Soft X-ray (0.1–2.4 keV) peak luminosities up to several 10^{44} erg s^{-1} .
- Very soft X-ray spectra near peak, with black-body temperatures in the range $kT_{bb} =$ 0.04–0.12 keV (or, alternatively, with powerlaw indices in the range $\Gamma_x = 4$–5), followed by a spectral hardening within years.
- A decline law of NGC 5905 and RXJ1242-1119 consistent with $t^{-5/3}$, and a drop below that decline law after $t > 10$ years.
- Total amplitudes of decline up to factors >1000–6000, measured in deep dedicated Chandra follow-up observations (Sect. 4).
- Host galaxies, which are optically inactive, radio-inactive, and also X-ray inactive at low-state.
- The host galaxy morphology of the nearest event, in NGC 5905, is a giant barred spiral galaxy.

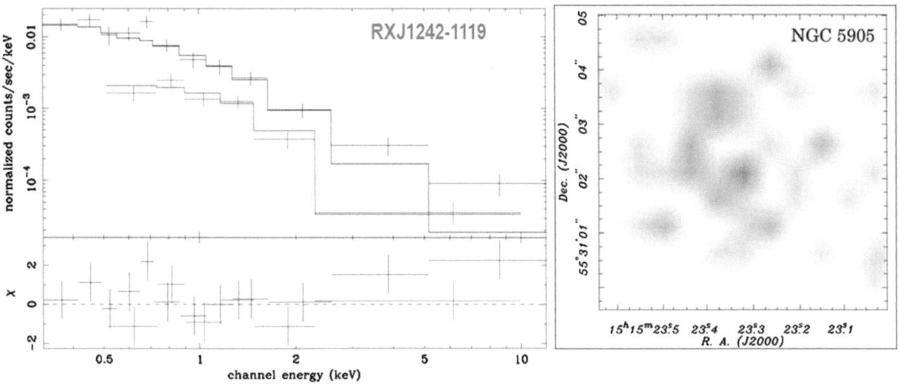

Fig. 3 Left: X-ray spectrum of RXJ1242-1119, the first TDE followed up with XMM-Newton. The observations revealed a strong hardening of the X-ray spectrum compared to the high-state observation (Komossa et al. 2004). Right: Deep Chandra X-ray image of NGC 5905 (Halpern et al. 2004) 12 years after the flare. Most (or all) of the low-flux state emission is widely extended and from the host galaxy

- All of them occurred in the nearby universe ($z = 0.011$–0.147).
- Black hole mass estimates are in the range 10^{6-8} $M_\odot$.

Given their extreme properties, particularly the absence of an optical AGN down to deep limits, huge peak luminosities and total amplitudes of variability (the largest seen in galaxies so far), they continue to be among the most reliably identified TDEs to date.

4 Using Chandra and XMM-Newton to Follow up and Detect TDEs

At the end of the 20th century the launch of the XMM-Newton (Jansen et al. 2001) and Chandra (Weisskopf et al. 2000) satellites provided the first high spatial and spectral resolution X-ray observation that covered the soft and medium energy bands (0.2–12 keV). XMM-*Newton* is flying three CCD cameras, with energy resolution of 50–100 eV at 1 keV and a reflection grating spectrometer (RGS) with a spectral resolution from 100 to 500 (FWHM) in the energy range 0.33–2.5 keV, as well as an Optical Monitor (OM) hosting visible and UV filters. Chandra hosts the ACIS camera, the High-Resolution camera (HRC) giving sub-arcsecond spatial resolution and low-energy (resolution > 1000) and high-energy (resolution $= 200$–800) transmission gratings.

These observatories were used for the first deep follow-up campaigns of the TDEs initially discovered with ROSAT. Chandra and XMM-Newton observations of RXJ 1242-1119 detected faint and fading X-ray emission a decade after the initial TDE high-state consistent with the optical position of the core of the host galaxy within the $\sim$1 arcsecond spatial error of Chandra, supporting the TDE interpretation (Komossa et al. 2004). With XMM-Newton, a first high-quality post-flare spectrum of the faint late-state X-ray emission from a TDE (RXJ 1242–1119) was obtained (Fig. 3), which showed a significant spectral hardening of the event, from $\Gamma_x \sim 5.1 \pm 0.9$ during high-state to $\Gamma_x = 2.5 \pm 0.2$ with XMM-*Newton* (Komossa et al. 2004), possibly related to the formation of an accretion-disk corona, initially absent. We look further at the long-term evolution of the X-ray spectra of TDEs in Sect. 4.5.3.

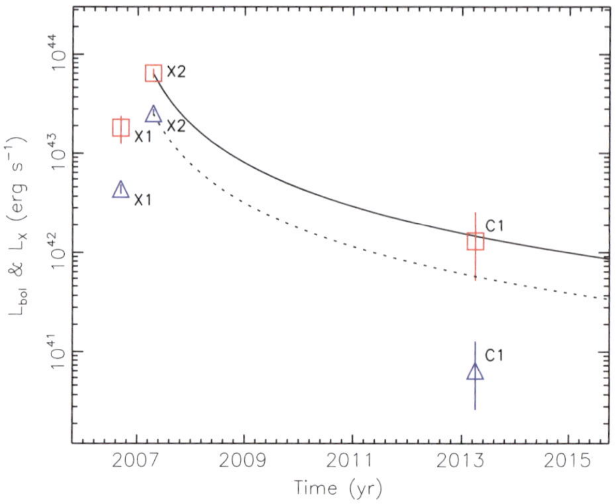

Fig. 4 The soft X-ray (blue) and bolometric (red) light curve of 2XMMi J184725.1-631724, with X1, X2 and C1 representing the first and second XMM-*Newton* observations and the first Chandra observation respectively (adapted from Lin et al. 2018b). The solid line plots a fit to the bolometric luminosity of the function $(t - t_0)^{-5/3}$ with t_0 set at 1 month before the date of X1, while the dashed line fits the same function to the soft X-ray luminosity

Chandra and XMM-*Newton* have been used for catalogue searches of TDEs, and also for the quick identification of new TDEs at peak by comparison with the RASS X-ray catalogues.

XMM-Newton generally makes observations of a few 10s of ks detecting around 50 sources within its quarter square degree field-of-view. These 'serendipitous' sources have been collated into a catalogue, at the time of writing in its third incarnation (3XMM; Rosen et al. 2016) and containing upwards of 500,000 independent sources from 2% of the sky. Chandra has produced a catalogue (CSC 2.0; Evans et al. 2010) with 315,000 sources from 1% of the sky.

With a mean flux of 10^{-14} (10^{-15}) $\text{erg}\,\text{s}^{-1}\,\text{cm}^{-2}$ for XMM-Newton (Chandra) in the 0.2–2 keV energy band the catalogued sources are generally too faint to be compared with historical observations in a search for variability. However, many fields have been observed on multiple occasions and it is possible to look at the flux history of a significant fraction of the catalogue for up to two decades. These comparisons have yielded a number of candidate TDEs.

4.1 TDEs Discovered Using XMM-Newton and Chandra

4.1.1 2XMMi J184725.1-631724

One such object was 2XMMi J184725.1-631724 (Lin et al. 2011), detected in an XMM-*Newton* pointed observation with a peak observed soft X-ray flux of 2×10^{-12} $\text{erg}\,\text{s}^{-1}\,\text{cm}^{-2}$ in 2007 but not detected by a further XMM-Newton observation in 2013 ($F_X < 2.5 \times 10^{-14}$ $\text{erg}\,\text{s}^{-1}\,\text{cm}^{-2}$), with Chandra confirming a drop in flux by a factor 1000 (Lin et al. 2018b) in 2013 (Fig. 4). The source whose position is consistent with the nucleus of the optically-inactive galaxy IC 4765-f01-1504 ($z = 0.0353$), was initially flagged due to its very soft spectrum (equivalent black-body temperature of $kT < 100$ eV).

A fainter, but still flux-enhanced, observation of 2XMMi J184725.1-631724 in 2006 allows the disruption time to be constrained to within about one month enabling the index of the flux decay law to be well measured in this TDE. Surprisingly it turns out to be considerably steeper than $-5/3$. However, we must bear in mind that this is an in-band flux whereas we should be comparing the bolometric radiation output, under the assumption that

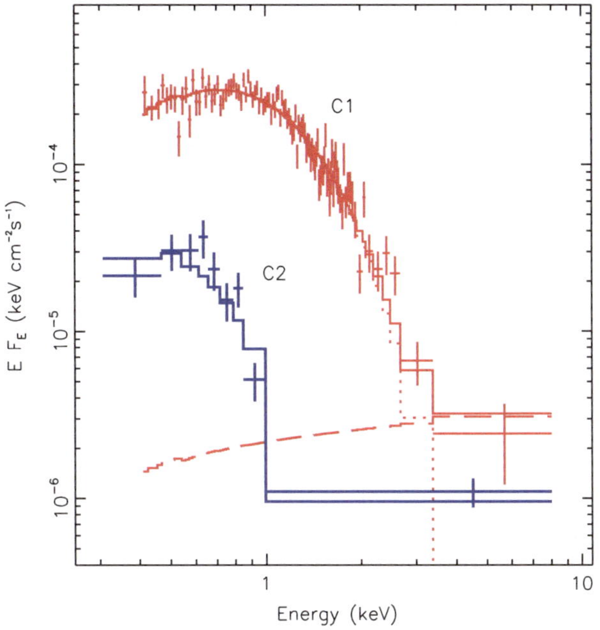

Fig. 5 Chandra spectra of 3XMM J215022.4-55108 from 2006-05-05 (red) and 2016-09-14 (blue) fit with a thermal disk model, of $kT = 280(140)$ eV, plus a weak power-law (adapted from Lin et al. 2018a)

this traces the fall back rate of the stellar debris. The peak bolometric luminosity of the TDE was 6×10^{43} erg s^{-1} , and the BH mass $M_{\rm BH} = 1 \times 10^{6-7}$ $M_\odot$, based on the host galaxy V and K magnitudes, with the peak soft X-ray spectrum dominated by a thermal model of $kT \sim 80$ eV. As the spectrum appears to be dominated by black-body radiation, the emission temperature should decrease as $T \propto L^{1/4}$ (see Sect: 4.5) and hence in the Chandra observation of 2013 will have cooled to $kT \sim 35$ eV. At this temperature, much of the flux is shifted out of the X-ray band and the bolometric correction is correspondingly larger than that needed for $kT = 80$ eV. Accounting for this effect brings the index of the *bolometric* flux decay into good agreement with $-5/3$ (Lin et al. 2018b).

4.1.2 3XMM J152130.7+074916 and 3XMM J215022.4-55108: Possible TDEs from Intermediate Mass Black Holes

The X-ray spectra of TDEs are usually soft in the early phase. If this temperature is related to the black hole mass (see Eq. (4)), then in principle this allows us to identify elusive intermediate mass black holes (IMBH) within the TDE population from their X-ray spectrum, something that is not possible in AGN surveys where the spectrum is dominated by a power-law with a mass-independent slope. The TDE 3XMM J152130.7+074916 is a good example. Discovered in an XMM-*Newton* observation in 2000 with a flux 100x higher than that seen in Chandra 4 months earlier, the spectrum could be fit with a disk model with an inner temperature of $kT \sim 170$ eV and $M_{\rm BH}$ between 0.19 and 1.4 $\times 10^6$ $M_\odot$ depending on BH spin (Lin et al. 2015).

An even lower mass was inferred in 3XMM J215022.4-55108 which lies about 12.5 kpc from the nucleus of a redshift 0.055 galaxy (Lin et al. 2018a). Here the effective temperature peaked at $kT = 280$ eV, reducing to 140 eV as the luminosity decayed from $L_X = 10^{43}$ to 10^{42}erg s^{-1} (Fig. 5). If these temperatures really come from the inner edge of the accretion disk then the BH in this case has a mass of a few $\times 10^4$ $M_\odot$.

This method of estimating black hole mass can only work if the correct spectral model is applied and should be used with caution. For example, a corona (see Sect. 4.5), if present, would push X-rays to higher energy, artificially increasing the measured temperature and causing the black hole mass to be underestimated. The effect of black hole spin also needs to be taken into account, causing an order of magnitude variation even in simple models (e.g. Lin et al. 2015).

4.1.3 TDEs in Clusters of Galaxies

One way of maximising the chances of finding TDEs using Chandra and XMM-*Newton* pointed observations is to repeatedly observe a rich cluster of galaxies and hence simultaneously monitor many galaxies (Wang and Merritt 2004). Maksym et al. (2010) observed the Abell class 4 cluster, A1689, which at $z \approx 0.18$, has a radius of 10 arcminutes nicely fitting into the Chandra field of view. They observed the cluster, which has an estimated 2100 galaxy members, 6 times with Chandra and once with XMM-*Newton* over an 8 year period. One galaxy, SDSS J131122.15-012345.6, varied by a factor 30 over the observations, displaying a soft spectrum ($kT \sim 100$ eV) and peak $L_X > 5 \times 10^{42}$ erg s^{-1}. From relations with the galaxy magnitude they estimated the black hole mass to be $M_{\mathrm{BH}} = 1 - 7 \times 10^6$ M$_\odot$. The light curve could be reasonably well fit with the canonical $t^{-5/3}$ law, from whence the total X-ray luminosity can be found by integrating over Eq. (1) for the duration of the flare. They estimate the bolometric correction factor to be 1.4 and found the total luminosity emitted over the event to be $L_{\mathrm{bol}} \sim 10^{50}$ ergs. After correcting for the expected emission before the first detection, the equivalent total mass accreted was found from Eq. (2) to be ~ 0.01 solar masses, a surprisingly low value which we address in Sect. 4.4.

Another cluster which received a lot of attention from Chandra was A1795, observed a total of 17 times between 1999 and 2010 (Maksym et al. 2013). One dwarf galaxy, WINGS J1348, with a mass of just 3×10^8 M$_\odot$ (Maksym et al. 2014b), showed a bright X-ray source in the first observation which subsequently decayed by a factor > 50. It appears likely that this flare was first captured by the EUVE satellite, in the 0.016–0.163 keV band, a year before the launch of Chandra, making it the only known TDE detected in the EUV band to date. The spectrum in the first Chandra observation was soft ($\Gamma_{\mathrm{x}} = 4.1$ or $kT = 84$ eV) with evidence for further softening over the following year. The sparse light curve again can be reasonably fit to a canonical $t^{-5/3}$ decay index.

4.2 TDEs Discovered in the XMM-Newton Slew Survey

XMM-Newton slews between targets at a rate of 90 degrees per hour and keeps its most sensitive camera, EPIC-pn (Strüder et al. 2001), open to record the sky. While a given source passes through the camera in just 15 seconds, the large mirror effective area and short frame time of the detector gives a detection limit of $F_{0.2\text{–}2} = 6 \times 10^{-13}$ erg s^{-1} cm^{-2} and a positional accuracy of ~ 8 arcseconds (Saxton et al. 2008). Again at the time of writing, 85% of the sky has been covered by slews and $> 20\,000$ sources are contained in the XMMSL2 catalogue. The slew sensitivity is well matched with that of the RASS and meaningful comparisons can be made between these surveys to find the bright population of X-ray TDEs. This was first done by Esquej et al. (2007) who found five, previously anonymous, galaxies that showed a factor >20 increase in flux between the RASS and slew surveys. Follow-up optical spectra showed that two of these were likely to be due to AGN variability, two (NGC 3599 and SDSS J132341.97+482701.3) were good candidates for TDEs, while XMMSL1 J024916.6-041244 was apparently a persistent Seyfert 1.9 galaxy but showed TDE traits, namely a very soft spectrum and a factor 100 flux decay (Strotjohann et al. 2016; Auchettl et al. 2017; Wevers et al. 2019b).

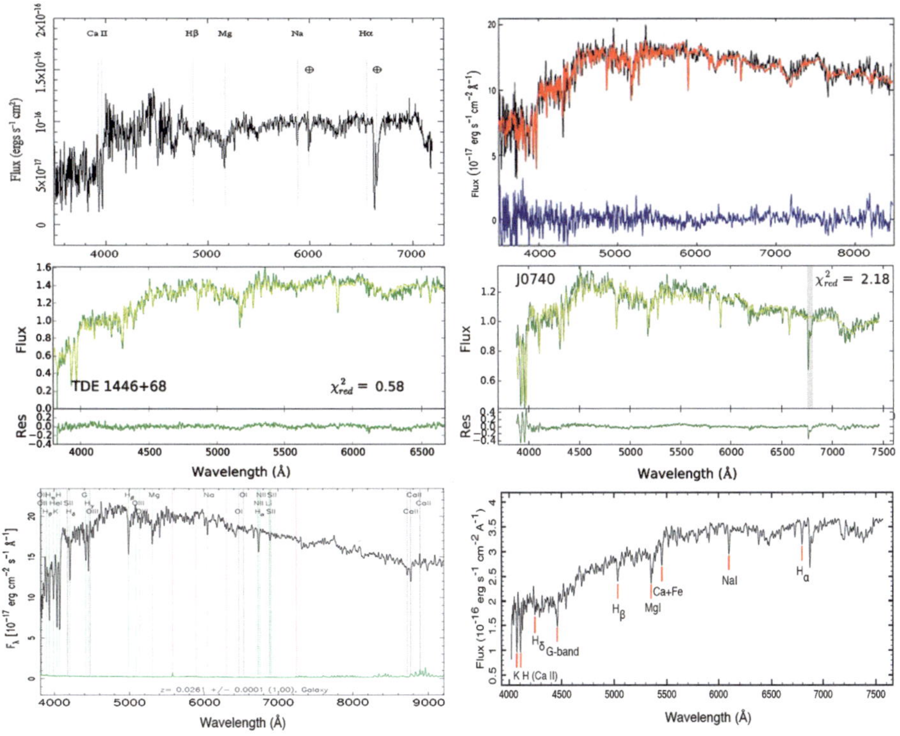

Fig. 6 Optical spectra of TDEs detected by XMM-*Newton*. Top left: SDSS J120136.02+300305.5 (Saxton et al. 2012a), top right: SDSS J132341.97+482701.3 (Esquej et al. 2007); middle left: XMMSL2 J144605.0+685735 (Saxton et al. 2019), middle right: 2MASX 07400785-8539307 (Saxton et al. 2017), bottom left: RBS 1032 (Ghosh et al. 2006), bottom right: 2XMMi J184725.1-631724 (Lin et al. 2011)

4.2.1 SDSS J120136.02+300305.5

The large sky coverage and relatively fast processing of XMM-Newton slews opened up the possibility of investigating events while they were close to the peak of their luminosity. This was first applied to SDSS J120136.02+300305.5 (hereafter SDSS J1201+30) spotted in a slew in 2010 (Saxton et al. 2012a,b). Figure 7 shows the X-ray light curve of SDSS J1201+30 as seen in XMM-Newton observations and short exposures taken with *Swift*. Intriguingly the initially high-luminosity emission from this event (10^{45} erg s^{-1}) disappeared 3 weeks after discovery, reducing by at least a factor 50 between 2010-06-30 and 2010-07-07. The possibility that this was caused by a temporary local absorption event is unlikely given the long, >21 day, duration of the drop in luminosity. Another possibility is that the return of material to the BH was interrupted by the passage of a secondary black hole (Liu et al. 2014). This scenario is further explored in Sect. 11. The emission from this TDE was soft, but wider than a single black-body, being empirically well fit by a Bremsstrahlung model (Saxton et al. 2012a). During the year of observations, the event got softer showing no evidence for the development of a hard tail.

4.2.2 2MASX 0740-85

2MASX 07400785-8539307 (hereafter 2MASX 0740-85) was detected in a slew in 2014 a factor 20 brighter than an upper limit from ROSAT (Saxton et al. 2017). Monitoring by

Fig. 7 Soft X-ray light curve of SDSS J1201+30. Upper limits are from the RASS or Swift-XRT (adapted from Saxton et al. 2012a)

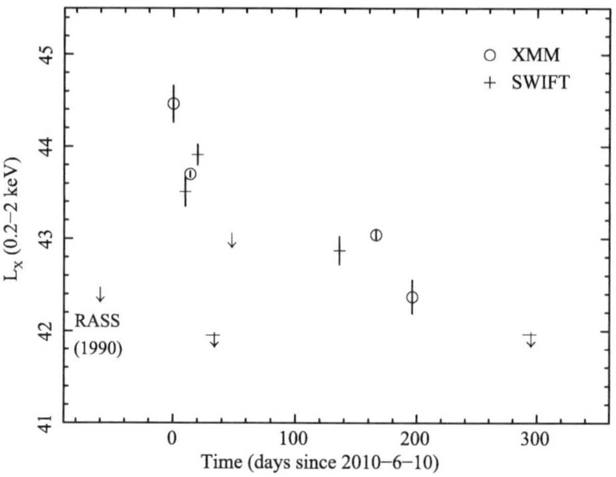

Swift and XMM-*Newton* over the following 550 days revealed a drop in X-ray flux by a factor of 70 (Fig. 8). The host galaxy was shown to be optically inactive (Fig. 6). The first X-ray observations were highly variable, with a flux doubling time of 400 seconds (Fig. 9). Using a variability-mass scaling relationship (Ponti et al. 2012) the black hole mass was estimated to be $M_{\rm BH} = 3.5^{+6.5}_{-2.4} \times 10^6\ {\rm M}_\odot$. Light travel-time arguments then place the size of the emitting region within 73 R_g of the black hole.

The X-ray spectrum of the event in 2MASS 0740-85 is unusual because it consists of a power-law of $\Gamma_{\rm x} \sim 2$ which dominates in the "hard" 2–10 keV X-ray band and contributes $\sim 25\%$ of the total energy. While other TDEs often show a weak hard-tail to the soft emission, which can be approximated by a (usually) steep power-law (e.g. Lin et al. 2018b; Holoien et al. 2016a; Kara et al. 2018) this feature is akin to the dominant power-law seen almost ubiquitously in AGN (Nandra and Pounds 1994). In other sources, SWIFT J164449.3+573451, Swift J2058.4+0516 and IGR J12580+0134 (see the Gamma-ray Chapter), the hard X-rays were likely generated in a jet as evidenced by the strong accompanying radio emission. The radio emission from 2MASX 0740-85 was rather weak, 1 mJy at 1.53 GHz, and may have been generated from shocks in an outflow rather than a collimated jet (Alexander et al. 2017).

2MASX 0740-85 had a UV flare contemporaneous with the X-ray flare (Fig. 8) which decayed with a flatter slope (Saxton et al. 2017). The relationship between the X-ray and UV emission of X-ray selected TDEs is explored further in Sect. 9.1.

The X-ray and UV emission from 2MASX 0740-85 could be reasonably well connected using a multi-coloured disk model (see Sect. 4.5) suggesting that the wide-band emission may be coming from a single coherent structure (Saxton et al. 2017).

4.2.3 XMMSL2 J1446+68

XMMSL2 J144605.0+685735 (hereafter XMMSL2 J1446+68) is another TDE detected in an XMM-*Newton* slew, this time in 2016. In this event the X-ray flux remained stable for the first ~ 100 days after discovery, before experiencing a drop of a factor 100 over the following 500 days (Fig. 8; Saxton et al. 2019). The host galaxy is optically inactive (Fig. 6) and with the peak bolometric luminosity, $L_{\rm bol} \sim 10^{43}\ {\rm erg\,s^{-1}}$, interpretations other than a TDE are unlikely (Saxton et al. 2019).

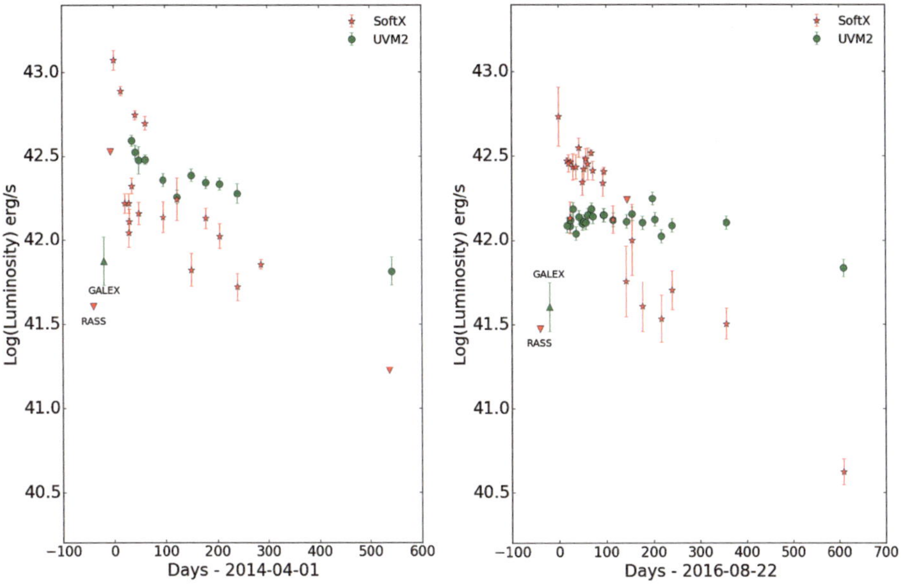

Fig. 8 The soft (0.2–2 keV) X-ray (red) and UVM2 (2340 Å; green) light curves for 2MASX 0740-85 (left) and XMMSL2 J1446+68 (right). Luminosity has been corrected for galactic extinction and includes the contribution from the host galaxy. GALEX-nuv filter measurements from 2007 have been rescaled to the UVM2 filter in both plots, which are adapted from Saxton et al. (2017) and Saxton et al. (2019)

It exhibited X-ray spectra consistent with a single power-law, stretching from 0.3–10 keV with slope, $\Gamma_x \sim 2.6$. Even in high signal-to-noise spectra there is no significant evidence of a low-energy thermal component in addition to this power-law. The black-hole mass estimated from optical absorption-line widths is quite high, $M_{\rm BH} = 7^{+17}_{-5} \times 10^7\ \rm M_\odot$ (Wevers et al. 2019b), and thermal emission may be too cool to enter the X-ray band (see Sect. 4.5). The spectral shape remained constant over time in this event, even while the X-ray flux fell by a factor 100.

The event was accompanied by UV emission which remained roughly constant for 400 days before decaying by a magnitude over the next few hundred days. This disconnect between the UV and X-ray flux is unusual and is discussed further in Sect. 9.1.

4.3 X-Ray Bright Events Discovered in Optical Surveys

For reasons which are currently not well understood, optically and UV-bright TDEs tend to show little or no X-ray emission (Gezari et al. 2009; Arcavi et al. 2014; Jonker et al. 2019, and see the Optical Chapter). Nevertheless there are some notable exceptions, two of which are examined below.

4.3.1 ASASSN-14li

ASASSN-14li was first discovered by the All-Sky Automated Survey for SuperNovae (ASASSN: Shappee et al. 2014) on 22nd of November 2014. This source was coincident with the center of galaxy, PGC043234 which is located at a distance of $\sim$ 90 Mpc and its optical spectra exhibit characteristics consistent with that of a TDE (see the Optical Chapter). An extensive multi-wavelength monitoring campaign using *Swift* revealed in addition

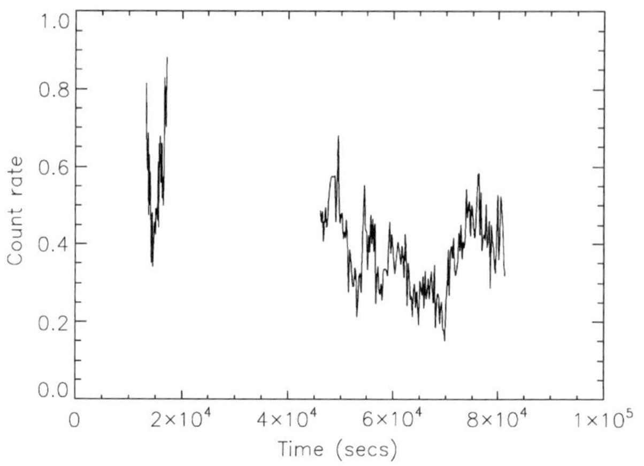

Fig. 9 A one-day EPIC-pn light curve of 2MASX 0740-85 taken by XMM-*Newton* soon after discovery, showing large short-term variability (adapted from Saxton et al. 2017)

to its UV and optical emission, strong X-ray emission arising from this event (see Fig. 10: Holoien et al. 2016b, Brown et al. 2017). Holoien et al. (2016b) found that the X-ray luminosity decayed at a much slower rate than that seen in the optical/UV wavelengths, with X-rays becoming the dominant source of emission approximately 40 days after peak. Using Swift observations spanning 600 days, Brown et al. (2017) found that ASASSN-14li remained bright in both UV and X-ray wavelengths even at late times, while the total energy radiated in X-ray and UV/optical was comparable.

Due to its X-ray brightness, ASASSN-14li was a source of many target of opportunity observations by Chandra and XMM-Newton. Both X-ray grating and CCD spectra showed that the X-ray emission could be described by a simple black-body with a temperature $kT \sim$ 50–60 eV (Miller et al. 2015; Holoien et al. 2016b; Brown et al. 2017; Bright et al. 2018), which cooled during the decay (Brown et al. 2017).

Using deep grating spectra taken close to the peak of the flare, Miller et al. (2015) found that the black-body emission of ASASSN-14li was modified by absorption from photoionised species of N, O, S, Ar, and Ca. The absorption lines were blue-shifted by $v_{\rm shift} = -360 \pm 50$ km/s initially, slowing at later times (Miller et al. 2015). This led the authors to suggest that the X-ray absorption either arises in strong outflows of highly ionized, low velocity X-ray gas, from a super-Eddington wind or from stellar debris.

More recently, Kara et al. (2018) found the early time X-ray spectra exhibit a broad, P-Cygni-like absorption feature around ~0.7 keV, which fades with time. Using photoionisation modelling, they find that this feature is consistent with absorption by OVIII in a very fast (0.2c) and highly ionised outflow. Compared to the low velocity outflow detected by Miller et al. (2015), Kara et al. (2018) suggest that this high velocity component is produced much closer to the black hole. Since the ionisation parameters of these two components are similar, it is possible that the lower-velocity component arises from the fast outflow decelerating as it collides with the debris stream or another medium.

4.3.2 ASASSN-15oi

ASASSN-15oi was another X-ray bright TDE discovered by ASASSN on the 14th of August 2015. Located at a distance of 214 Mpc, this source exhibited similar spectral features to other optically detected TDEs (see the Optical Chapter) such as declining strong helium features, black-body emission and a declining light curve (Holoien et al. 2016a, 2018b). However, compared to ASASSN-14li, the X-ray emission from ASASSN-15oi was much

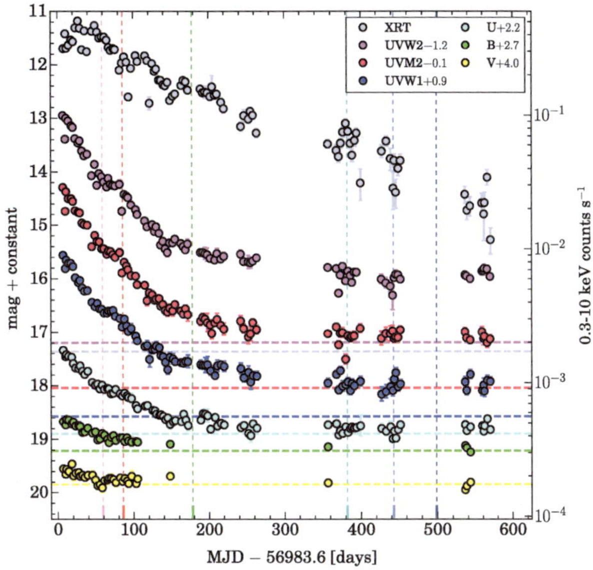

Fig. 10 The X-ray, UV and optical light-curves of ASASSN-14li (taken from Brown et al. (2017))

weaker. At initial discovery, the detected X-ray emission from the source was lower than an upper limit derived using ROSAT. However, follow-up Swift observations revealed behaviour unseen at the time in any other TDE.[4] Typically, the X-ray emission of a TDE decays following a simple powerlaw (see Sect. 5), however, in this case the X-ray emission of ASASSN-15oi brightened by nearly an order of magnitude before fading again (Gezari et al. 2017; Holoien et al. 2018b). As the timescale of this increase was approximately one year, Gezari et al. (2017) suggested that the behaviour was the result of delayed accretion due to inefficient circularization of the stellar debris stream, while Holoien et al. (2018b) suggested that this behaviour resulted from material surrounding the accretion disk becoming optically thin to X-ray radiation a few months after discovery. Using XMM-Newton, Gezari et al. (2017) found that the X-ray emission from the source was best described by a black-body plus a weak powerlaw component. While at both early and late times they find the temperature of black-body component does not change significantly ($\sim$45 eV), the powerlaw component may become softer with time ($\Gamma_{\rm initial} \sim 2.5$ to $\Gamma_{\rm later} \sim 3.3$). Using the Swift observations, Holoien et al. (2018b) find that the black-body component is an order of magnitude stronger than the powerlaw component, and is responsible for the flux variations.

4.4 Accreted Mass

The mass accreted during an event can be estimated by integrating the total emitted X-ray luminosity

$$\Delta M = \frac{k_{\rm Bol}}{\eta c^2} \int_t^\infty L_X(t)dt \qquad (2)$$

where η is the efficiency of conversion of gravitational energy into radiation, generally taken to be 10%, and $k_{\rm Bol}$ is the factor to correct X-ray to bolometric luminosity (see Netzer 2019,

[4]Similar behaviour has now been detected in AT2018fyk (Wevers et al. 2019a) and AT2019azh (Liu et al. 2019).

for a recent description). From this calculation, the accreted mass often appears to be ~ 0.01 solar masses (Brown et al. 2017; Auchettl et al. 2017; Saxton et al. 2017, and references in Sects. 3 and 4.1.3); a low value which potentially leads to a missing energy problem (Piran et al. 2015b).

We expect from stellar population analysis that the average mass of a disrupted main sequence star will be between 0.1 and 1 solar mass (e.g. Guillochon et al. 2014; van Velzen et al. 2019b; Mockler et al. 2018). If the fraction of stellar debris returning to the black hole is $\sim$ 20–50%, as dynamical studies predict (Ayal et al. 2000; Bonnerot and Lu 2019) then, at face value, $\lesssim 10\%$ of the available accretion energy is being converted into radiation in these events. There are several possible explanations for this:

- These flares are actually caused by the stripping of the atmosphere of an evolved star (MacLeod et al. 2012) and hence much less material is available for accretion.
- The conversion of mass to light factor (η) is around 0.01, i.e. just 10% of that seen in AGN.
- We always miss the peak of the emission and underestimate the total luminosity.
- The initial super-Eddington accretion produces strong winds which push a large fraction of the material away from the black hole (e.g. Metzger and Stone 2016; Dai et al. 2018).
- A large fraction of the radiation is actually emitted in the unobservable EUV and the bolometric correction is seriously underestimated (e.g. Komossa et al. 2004).
- The returning matter forms an accretion structure which evolves viscously and hence drains into the black hole at a much lower rate than $t^{-5/3}$ (Cannizzo et al. 1990; van Velzen et al. 2019a).
- We are underestimating the column density of material surrounding the event (Auchettl et al. 2017).
- A combination of the above or a further unknown factor.

4.5 X-Ray Spectra

The spectral features imprinted by the environment of an AGN are constant for decades, even if the details may vary on timescales as short as minutes. These are well described in e.g. Netzer (2013) and in essence consist of a pseudo-power-law, dominant from 2–100 keV, with a slope of roughly 2 (Turner and Pounds 1989; Nandra and Pounds 1994), a soft emission component, which may be due to Compton-upscattering by cool electrons ($kT_e \sim 0.2$ keV; Done et al. 2012) of photons generated in the disk and a pure thermal soft X-ray component, usually hidden by the stronger Comptonised component. This emission is modified by reflection off distant neutral material and off the ionised disk and various absorption features with various ionisation states and column densities. In radio-loud objects a jet produces very strong power-law emission which tends to mask the other features. This complex mix requires very high signal-to-noise spectra to begin to deconvolve the individual components.

The prediction is that in the simplified case of a TDE, where we add material at a well-determined date to a SMBH, we ought to be able to determine the temporal onset of each of the components mentioned above and hence achieve a better understanding of persistent accreting SMBH and the timescales involved in the distribution of matter about the black hole.

A wide range of empirical models have been used to fit the broad-band XMM-*Newton* and Chandra spectra of TDEs. A single-temperature black-body, with luminosity given by

$$L = \sigma A T^4 \tag{3}$$

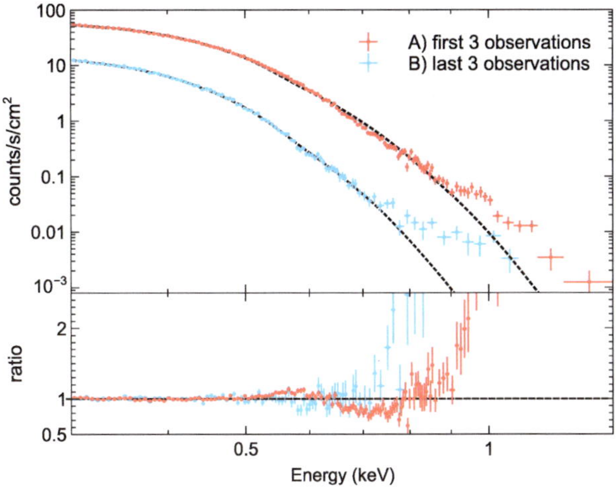

Fig. 11 X-ray spectrum of ASSASN-14li from early (red) and late (blue) XMM-*Newton* observations. Figure adapted from Kara et al. (2018). The best fit with a single black-body is shown as a dotted black line. A hard excess is evident in both fits

where σ is the Boltzmann constant, A is the surface area of the emitting region and T is the temperature in Kelvin, is a reasonable description in a number of cases (e.g., early emission of ASASSN-14li; Sect. 4.3.1). As the stellar debris returns to the BH it should form a hot, optically-thick structure, with an effective black-body temperature dependent on the BH mass and the mass accretion rate, expressed in terms of the Eddington accretion rate, $\dot{M}_{\rm Edd} = L_{\rm Edd}/\eta c^2$ (Ulmer 1999)

$$kT \sim 40 M_6^{-1/4} \dot{M}_{\rm Edd}^{1/4} \text{ (eV)} \tag{4}$$

where M_6 is the mass of the black hole in units of 10^6 $M_\odot$ and $\dot{M}_{\rm Edd}$ is the Eddington-limited accretion rate.

However, these fits often leave a high-energy tail (e.g. 2XMMi J184725.1-631724: Lin et al. (2011), SDSS J1201+30: Saxton et al. (2012b), ASASSN-14li: Kara et al. (2018), 3XMM J150052.0+015452: Lin et al. (2017a), ASASSN-15oi: Holoien et al. (2018b)) which needs further explanation. A high-statistic example is shown in Fig. 11. Several explanations have been offered for this in the literature which we run through below.

A: A multi-temperature structure: This assumes that the material is in a thin-disk configuration, where the emissivity index of the material is strongly weighted towards the centre so that most of the emission occurs at the higher temperatures. A common implementation is the *diskbb* model (Makishima et al. 1986). The range of temperatures produces a broader spectrum than a single-T black-body and can fit the observed spectra in some cases (Lin et al. 2011, 2015, 2018a). This model leads to a high inner disk temperature and, from Eq. (4), a correspondingly lower black-hole mass.

B: An empirical power-law: A single power-law usually proves to be a poor-fit when used to model the whole of a high-statistic TDE spectrum. When it is used, then the slope is usually very steep, ($\Gamma_{\rm x} > 4$; see Sect. 3, 4 and 10), in excess of the index value of 1.5 to 2.3 (Turner and Pounds 1989; Nandra and Pounds 1994) ubiquitously found in persistent AGN and believed to be produced by Comptonisation of disk photons by a high temperature ($kT_e > 100$ keV), optically thin plasma. It is more successful in fitting the *hard-excess* in good quality spectra of TDEs, but even here the slope tends to be steep (e.g. $\Gamma \sim 3.5$ in 2XMM 1847 (Lin et al. 2011) and also in ASASSN-15oi Gezari et al. 2017, Holoien et al. 2018b or those summarised in Auchettl et al. 2017).

C: Bremsstrahlung: This is emission from an optically-thin plasma which produces a wider spectrum than a black-body. It was a good fit to SDSS J1201+30 with a temperature of 390 eV, reducing to 280 eV in a later observation (Saxton et al. 2012a). The main draw-back of this model is that, being optically-thin, at this temperature it should produce substantial narrow line X-ray emission from N, O, Fe (e.g. Mewe et al. 1985) which was not seen in SDSS J1201+30 or in other TDEs.

D. Inverse Compton: A nearby population of electrons, with significant kinetic energy compared to the photon energies, will give energy to a fraction of the photons in a process known as Compton-upscattering, thus creating a hard tail to the thermal spectrum. This mechanism has been proposed for the soft-X-ray excess regularly seen in the spectrum of AGN (e.g. Done et al. 2012). In the soft X-ray regime, the final energy of an upscattered photon in collision with a thermal electron population of temperature kT_e, is related to its initial energy by

$$E_f \sim (1 + \frac{4kT_e}{mc^2})E_i \tag{5}$$

The hard excess produced by the Inverse Compton effect has a shape which is well approximated by a power-law with slope dependent on the temperature and optical depth of the electrons (see Nishimura et al. 1986). This model provides a more physical fit to SDSS 1201+30 (Saxton et al. 2012b) and has been used to fit the excess in XMM-*Newton* and Chandra spectra of TDEs such as 3XMM J150052.0+015452 (Lin et al. 2017a). Two popular spectral models are (compbb; Nishimura et al. 1986) and (comptt; Titarchuk 1994). In these models the typical hard-excess slope of $\Gamma = 3 - 4$ is provided by a Maxwellian electron population of temperature, $kTe \sim 5 - 10$ keV and optical depth $\tau \sim 1$.

4.5.1 Hard Power-Law Emission

Another emission feature which has been seen is a hard ($\Gamma \sim 2$) power-law, equivalent to the dominant emission mechanism in AGN. In some cases this is associated with a relativistic jet (see the Gamma-ray Chapter) accompanied by strong emission at radio wavelengths. In IGR J12580+0134[5] and 2MASX 0740-85 the power-law is dominant, even at low energies, but radio emission is modest (~ 1 mJy or $L_R = 10^{37-39}$ erg s^{-1}). This component seems to be the analogue of the X-ray power-law seen in radio-quiet AGN,[6] which is generally explained by Compton-upscattering of disk photons by a population of very high energy electrons ($kTe > 100$ keV) located at a few R_g from the black hole, a proximity which leads to variations on time scales of minutes to hours. In fact, it is this fast variability, also seen in 2MASX 0740-85 (Fig. 9), which locates the X-ray emission close to the BH.

4.5.2 Absorption Features

In the spectra of some events, a good fit can only be obtained by adding one or more *ionised* absorption features. This is the case in the CCD spectra of 3XMM J1521 and SDSS J1201+30 which show evidence for absorption features from apparently outflowing material (Saxton et al. 2012a; Lin et al. 2015). A more detailed analysis of this material can

[5] Note that the classification of IGR J12580+0134 as a TDE has been questioned based on its WISE colours, pre-flare data and hardness ratio evolution (see A.17 of Auchettl et al. 2017)

[6] although Irwin et al. (2015) make a case for the 2–10 keV emission coming from the inverse Compton component of the jet in IGR J12580+0134.

be found from the grating instrument on-board XMM-*Newton*, the RGS, in the bright TDE ASSASN-14li (see Sect. 4.3.1 and Miller et al. 2015; Kara et al. 2018). Outflows have been regularly found in BHs accreting close to the Eddington limit, in solar mass systems as well as SMBH (Arav et al. 1994; Crenshaw et al. 1999; Pounds et al. 2003).

Any *neutral* absorption (N_H), in excess of that of our galaxy and the TDE host galaxy, found in TDE spectra would be crucial for constraining the geometry of the debris and accreting material. For example, the reprocessing model (Dai et al. 2018; Metzger and Stone 2016) seeks to explain the difference between optical and X-ray TDEs in terms of a viewing angle (see the Emission Mechanisms Chapter). The very soft spectra of X-ray selected TDEs are highly sensitive to cold absorption and their very presence argues against large columns of neutral gas in the line of sight (e.g. Komossa and Bade 1999). As an example, a TDE at $z = 0.05$, with a Galactic column, $N_H = 1 \times 10^{20}$ cm^{-2} and black-body emission of $kT = 60$ eV has its 0.2–2 keV flux reduced by a factor 6 by an absorption of $N_H = 1 \times 10^{21}$ cm^{-2} intrinsic to the host galaxy of the event. In many detailed fits, neutral absorption in excess of that of our own galaxy is not required (e.g. Saxton et al. 2017, 2012a; Lin et al. 2015, 2018a). There are exceptions: SWIFT J164449.3+573451 ($N_H = 2 \times 10^{22}$cm^{-2}; Burrows et al. 2011);[7] IGR J12580+0134 ($N_H = 7 \times 10^{22}$cm^{-2}; Nikołajuk and Walter 2013), 3XMM J150052.0+015452 ($N_H = 4.2 \times 10^{21}$cm^{-2}; Lin et al. 2017a) and ASSASN-14li ($N_H = 1.4 \times 10^{20}$cm^{-2}; Miller et al. 2015), but in none of these cases has the excess neutral absorption been seen to change between observations. This either means that the absorption comes from the host galaxy, or that the material was produced by the TDE but in a form which did not decrease in density and was not significantly ionised by the nuclear radiation, during the event.

The derived absorbing column is of course dependent on the emission model used in the spectral fit. Auchettl et al. (2017) fitted a large sample of TDE spectra assuming a single absorbed power-law model, finding that a large fraction of X-ray TDEs have column densities (N_H) that are at least two times greater than the Galactic column density measured along the line of sight to these events. This result holds when a power-law is the correct emission model for the TDE and is useful for investigating variations with time. It does not return the correct absolute value though, for the different spectral models which are commonly seen in TDEs, e.g. the multi-component spectrum of ASASSN-14li (Fig. 11 and Miller et al. 2015; Kara et al. 2018).

4.5.3 Long-Term Spectral Evolution

It is interesting to see what happens to the accreting material after many years when the accretion rate has dropped well below the Eddington limit. The late-time spectrum of the ROSAT TDE RXJ 1242-1119 was measured with XMM-Newton and implied a strong hardening of the X-ray spectrum from $\Gamma_X = 5.1$ to $\Gamma_X = 2.5$ (Komossa et al. 2004). A dedicated study of the late-time spectra of ROSAT TDEs by Chandra, taken $>$ 10 years after the disruption, showed relatively hard spectra ($\Gamma < 2.5$) in all cases (Halpern et al. 2004; Vaughan et al. 2004). The Chandra study is complicated by the very low statistics ($<$ 25 photons in each spectrum) and by the low-luminosity of the residual emission ($L_X = 5 \times 10^{39}$ to 2×10^{41} erg s^{-1}) which is comparable to that seen from the binary population of the galaxy, at least in the case of NGC 5905, where most or all of the low-state emission is extended and clearly not associated with the nucleus (Halpern et al. 2004). In NGC 5905 a spectral hardening was already measured with ROSAT itself, 3 years after the maximum ($\Gamma_X = 4.0$ to $\Gamma_X =$

[7] Note the highly reddened, absorbed host galaxy of this event.

2.4; Komossa and Bade 1999). There is some theoretical expectations that accreting debris will collapse into a thin disk leaving a long-lived low-level emission for many years. Such emission has accretion rates of $\dot{M} \sim 10^{-4}$ and an index $\Gamma_X \sim 2$ (Cannizzo et al. 1990). In the case of RXJ1242-1119, the latest deep Chandra follow-ups have shown a deviation from the early-phase decline law, in form of a deep dip below the early-phase decline law, and may indicate a change in accretion mode (Komossa 2005), as predicted by e.g. Rees (1990).

In NGC 3599 the spectrum was still relatively soft 6 years after the peak emission (Saxton et al. 2015) with an equivalent fit of a power-law of $\Gamma = 2.7 \pm 0.3$ plus a $kT = 50$ eV black-body. In RBS 1032 the spectrum after 20 years had hardened from $\Gamma = 5$ to $\Gamma = 3.4$ (Maksym et al. 2014a). Interestingly, Jonker et al. (2019) found late-time X-ray emission in three optically-selected TDEs. The spectra of these could be modeled with power-laws of $\Gamma = 2.5$–3.9, compatible with those of the X-ray selected events.

The flux of the soft black-body component certainly drops over the years but due to the paucity of high quality early and late-time spectra, it is not yet clear whether the harder component remains roughly constant or increases over time. More observations, with higher statistics, are needed to decipher what the long-term spectral hardening means in physical terms.

5 Light Curves

The evolution of a TDE's light curve depends heavily on a number of factors such as the stellar structure (e.g., Guillochon and Ramirez-Ruiz 2013), and whether the emission arises from fall-back (e.g., Rees 1988; Evans and Kochanek 1989; Phinney 1989), disk emission (e.g., Rees 1988; Cannizzo et al. 1990; Lodato and Rossi 2011) or super-Eddington accretion (e.g., Cannizzo et al. 2011). X-ray light curves of TDEs are rarely well enough sampled to be able to identify the time of disruption, or even the time of peak flux. Commonly, when fitting the light curve, the two parameters, t_0 and the index α are found to be degenerate and α is fixed to the canonical value of $-5/3$ to obtain an estimate of the date of disruption. However, studies of the optical emission from individual sources (e.g. Wevers et al. 2019a), and detailed global studies in the Far-UV (van Velzen et al. 2019a) have shown that this behaviour is not necessarily universal, especially at late-times. Auchettl et al. (2017), took a different approach. Under the assumption that t_0 is the time of the first, peak flux, detection, they showed that the X-ray emission from TDEs then seems to decay with a power-law index that is shallower than $t^{-5/3}$, implying that the viscous timescale[8] is long for these events (Guillochon and Ramirez-Ruiz 2013). At early times, TDEs were found to have power-law indexes that are consistent with both fall-back ($t^{-5/3}$) and disk emission ($t^{-5/12}$) (see Lodato and Rossi 2011), while at late times most, if not all, events tend to have decay rates consistent with disk emission (see Fig. 12 upper panel). The transition between these different emission processes is not necessarily smooth, with small timescale variations in the power-law index seen as each event evolves (Fig. 12 lower panel), while the time it takes for each event to undergo this transition varies between sources. Within the class of X-ray emitting TDEs, those whose emission is dominated by a strong jet (e.g., jetted TDEs such as Swift J1644+57: Bloom et al. 2011, Levan et al. 2011, Burrows et al. 2011, Levan et al. 2016) showed multiple transitions between different emission processes, while those whose emission is not dominated by a jet (thermal TDEs such as ASASSN-14li: Miller et al. 2015, Holoien et al. 2016b,

[8]Here the viscous timescale is defined as the time it takes for material to accrete onto a black hole, and depends on the height and radius of the disk and the orbital period (Guillochon and Ramirez-Ruiz 2015).

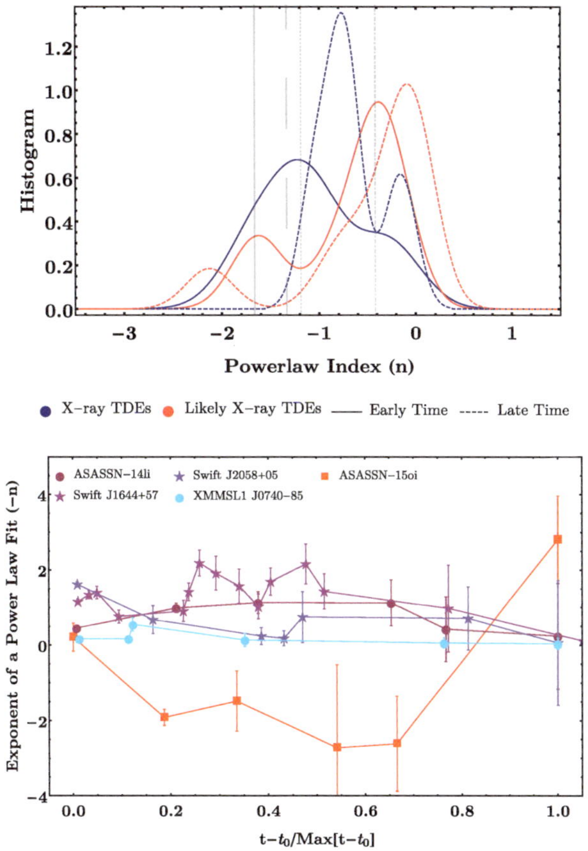

Fig. 12 Upper: Histogram of powerlaw indexes as seen at early (solid) and late (dashed) times during the decay of the X-ray emission from a sample of X-ray TDE candidates. Overlaid as the vertical solid, large dashed, dotted and dash-dotted grey lines are the powerlaw indexes for fallback (−5/3), advective, super-Eddington slim disk accretion (−4/3), viscous disk accretion (−19/16) and disk emission (−5/12). Figure taken from Auchettl et al. (2017). Lower: Best fit powerlaw index and its uncertainty for the X-ray TDE sample of Auchettl et al. (2017)

Brown et al. 2017, Bright et al. 2018) tend to show less variation in their power-law index which tends to fluctuate around the index associated with one type of emission process.

Even though the X-ray light curves of TDEs tend to decay following a powerlaw index that is consistent with fall back, or disk emission, a handful of TDEs show strong deviations from this behaviour. Even though the light curve of the jetted TDE Swift J1644+57 globally exhibits an approximately $t^{-5/3}$ decay (which is punctuated by flaring, variability and dips) (Bloom et al. 2011; Levan et al. 2011; Burrows et al. 2011; Levan et al. 2016; Mangano et al. 2016), a striking feature of its light curve is the dramatic decrease in X-ray flux ∼500 days after its initial detection. Within ∼ 4 days, the X-ray flux dropped by a factor of >50, which corresponds to a decay index steeper than t^{-70} (Levan et al. 2016). This behaviour suggests that the accretion underwent a state change; either it suddenly stopped, possibly consistent with the star being completely accreted onto the black hole (Quataert and Kasen 2012), or the accretion became sub-Eddington and radiatively efficient which dramatically represses power-law jets (e.g., van Velzen et al. 2011a; Russell et al. 2011; Zauderer et al. 2013).

6 Indirectly Identified X-Ray Events from Reprocessing into High-Ionization Emission Lines, and X-Ray Follow-up

Some luminous X-ray TDE candidates were not detected directly in the X-ray regime, but the flaring X-ray emission was indirectly inferred based on the presence of luminous, then fading, high-ionization iron coronal emission lines in optical spectra. These lines need a strong incident X-ray continuum in order to be created (Komossa et al. 2008, 2009; Wang et al. 2011, 2012). Only one of these events, SDSS J095209.56+214313.3, had X-ray follow-up spectroscopy which revealed a relatively flat X-ray spectrum, and a strong decline in X-ray luminosity with $L_{\mathrm{x,low}} = 4 \times 10^{40}$ erg/s between 2–10 keV (Komossa et al. 2009). Selected non-X-ray properties of these events will be further discussed in the Echo Chapter.

7 Long-Lived Events

We have seen in the previous sections that some TDEs (e.g. NGC 5905, 2MASX 0740-85, SDSS J120136.02+300305.5 and IGR J12580+0134) decay quickly from their peak X-ray luminosity. Other events, however, maintain their peak emission for longer; NGC 3599 had a plateau of at least 18 months before decaying (Saxton et al. 2015) and XMMSL2 J1446+68 at least 100 days. The champion though is 3XMM J150052.0+015452 (Lin et al. 2017a), a TDE from a dwarf galaxy, which rose within 4 months in 2005 and has been decaying very slowly for more than ten years (Fig. 13). This event has maintained a soft spectrum, being well modelled with a low-temperature, $kT \sim 40$ eV, black-body, heavily Comptonised by optically-thick, low-energy electrons (`COMPTT` model), for the whole of its plateau phase. This model, in stellar-mass black holes, is believed to signify a super-Eddington accretion state (e.g. Middleton et al. 2013). The last Chandra observation, has a relative deficit of hard flux which can be interpreted as either; that the accretion mode has changed to a super-soft (sub-Eddington) state or that an ionised absorber was in the line-of-sight which caused the sharp spectral drop towards higher energies. The black hole mass is estimated to be $\sim 10^6$ $M_\odot$ based on the mass of the host dwarf galaxy, consistent with the observed $L_{\mathrm{bol}} \sim 10^{44}$ erg s^{-1}. Lin et al. (2017a) interpreted the slow evolution of the bolometric light curve as indicating a distant circularisation of the stellar debris, leading to higher viscosity and a consequently slow fallback of the material to the BH. The mass of the disrupted object is then around 2 $M_\odot$ with $\sim$0.9 $M_\odot$ being converted into radiation. Two unusual persistent AGN have shown similar long high-luminosity, soft X-ray emission; GSN 069 (Miniutti et al. 2013) and 2XMM J123103.2+110648 (Terashima et al. 2012). Both of these have been interpreted by some authors as a TDE occurring in an AGN (Lin et al. 2017b; Shu et al. 2018).

There are several theoretical reasons for why a TDE may be long-lived. For example, a long super-Eddington accretion phase, perhaps involving a disrupted object with a large mass (e.g. Lin et al. 2017a); a partially stripped evolved star atmosphere (MacLeod et al. 2012) or late, distant circularisation (Guillochon and Ramirez-Ruiz 2015; Shiokawa et al. 2015).

8 Very Fast Events

Some TDEs are expected to lead to fast X-ray flares. For instance, a TDE involving an intermediate-mass black hole (IMBH), defined here as a black hole with a mass less than 10^5 $M_\odot$ (e.g. Evans et al. 2015). Especially if the disrupted star was a compact star, such as a white dwarf (see the White Dwarf Chapter), then the associated orbital time scales are short and the relativistic periastron precession is large, potentially leading to a short circularization time (see the Disruption Chapter) and, therefore, a short rise time for the accretion flare (e.g. Clausen and Eracleous 2011, MacLeod et al. 2016, Shen 2018). Alternatively, it is

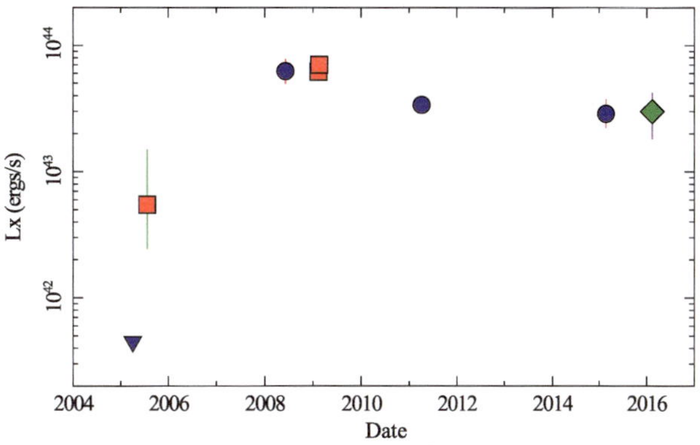

Fig. 13 X-ray light curve of 3XMM J150052.0+015452 from Chandra (blue circles and upper limit), XMM-Newton (red squares) and Swift-XRT (green diamond). Data taken from Lin et al. (2017a)

predicted that in some cases a shock occurs the star upon disruption which breaks out of the star and gives rise to a brief X-ray flare (e.g. Yalinewich et al. 2018). Finally, blazar–like variability in those TDEs where a relativistic jet launched from the near–vicinity of the black hole is pointing close to our line of sight can give rise to fast X-ray variability even for black holes more massive than IMBHs (e.g. Bloom et al. 2011, Levan et al. 2011).

Below we discuss the fast X-ray events that have been suggested to be caused by tidal disruption events, we do note, however, that relatively few things are known about these events, making their interpretation as being caused by a tidal disruption event much less secure than some of the other events of longer duration in this manuscript.

First of all, we need to be aware that several aspects can make an event appear "fast". For instance, if given the limited sensitivity of any instrument only the peak of a (longer duration) outburst is detected an event can appear as "short" or "fast". This effect probably affected earlier detections of fast events (e.g. Pye and McHardy 1983; Grindlay 1999) more than recent detections with more sensitive instruments, although of course events at large(r) distances will still cause this effect even in modern detectors (as all detectors have a sensitivity limit). Furthermore, several minute–to–hour scale X-ray flares are known that have nothing to do with tidal disruption events, such as M–star flares (Heise et al. 1975) and more generally, flares due to stellar activity often induced by binary interactions such as those of RS CVn stars (Pye and McHardy 1983). And finally, accretion flares from Galactic low–and high–mass X-ray binaries (Liu et al. 2000; Liu et al. 2001) and thermonuclear explosions on the surface of an accreting neutron star (so called Type I X-ray bursts) can appear as fast X-ray flares (e.g. Galloway et al. 2008).

In order to weed out flares from such events multi–wavelength data is crucial. For that a source localization accurate to the order of arcseconds is important: this astrometric accuracy comes naturally with *Chandra*, XMM–*Newton* and *Swift*–XRT–discovered transients.

XRT 000519: The first of this new type of fast transient X-ray sources (XRT 000519) was found in an archival *Chandra* observation (Observation ID 803; Jonker et al. 2013; see Fig. 14). The source position lies 12.16 arcminutes from the centre of the Virgo Cluster galaxy M 86, but it does not fall in the M 86 $\mu_B = 25$ magnitude per arcsec^2 isophote area. Optical Isaac Newton Telescope images show a tidal stream stripped off the galaxy SDSS J122541.29+130251.2, suggesting that M 86 is undergoing a minor merger. The small projected distance on the sky between the position of the transient and that of the tidal stream suggests that the transient is associated with M 86. Uniquely to this source (when compared to the others, see below) is that the main flare is double peaked.

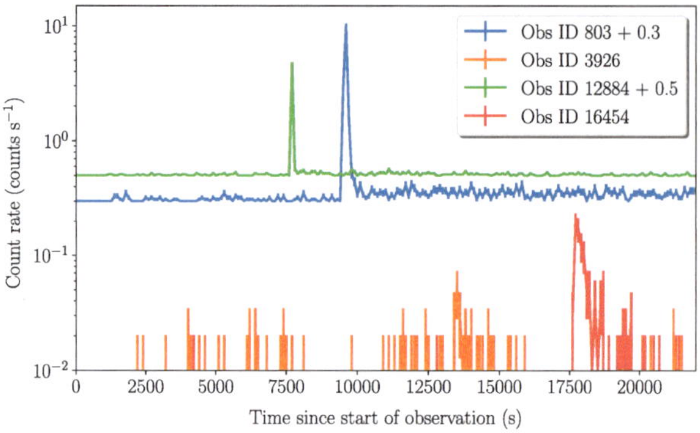

Fig. 14 The light curves of the four fast X-ray transients discovered in *Chandra* data. The data has been binned in time intervals of 100 seconds. The time zero on the X-axis corresponds to the start of the observation in each case. The observation ID of the *Chandra* data is indicated in the top right side of the figure. Clearly, the detected count rate varies significantly between the four events. Furthermore, the duration of the flares differs as well. The flare detected in observation ID 12884 is the shortest

Deep observations with the William Herschel Telescope (WHT) $r' \approx 25$ and $K \approx 20$–band observations rule out an M-star flare and also globular cluster hosts brighter than $M_V \sim -6$ at the location of M 86 for this event.

XRT 110203: A second transient which has properties that strongly resemble those of XRT 000519 is XRT 110203 (*Chandra* observation ID 12884; Glennie et al. 2015). Also this event lies close to a cluster of galaxies (ACO 3581; $z = 0.023$) but no host galaxy is detected so far.

CDFS-XT1: Bauer et al. (2017) report the discovery of a third fast X-ray transient in the *Chandra* Deep Field South survey (CDFS-XT1; *Chandra* observation ID 16454). The light curve indicates a rise time of ∼70–160 s. In this case they found a possible host dwarf galaxy in the 3D-HST field (with AB magnitude R∼27.5), but there is no cluster of galaxies nearby. Unfortunately, there is no spectroscopic redshift of this source. The photometric redshift of ∼2 assigned to this host is much larger than the redshifts associated with the first two events. However, the redshift estimation is still very uncertain as the host was only detected in two HST filters (ACS/F606W ∼V-band and WFC3/F125W ∼J-band). In addition, the host–event association also needs to be confirmed as the chance alignment probability for such deep images is non-negligible and finally, the astrometric alignment between the host and the transient event is not perfect.

Given that these three events share the same timescale, have no clear or only a faint host, using Occam's razor, we suggest that the three events are drawn from the same parent population. Glennie et al. (2015) estimate a rate of $\sim 10^5$ events per year over the whole sky with a peak X-ray flux greater than 10^{-10} erg cm^{-2} s^{-1}.

Irwin et al. (2016) found one source to flare to a peak L_X of 9×10^{40} erg s^{-1} and 5 repeat flares to $L_X \approx 10^{40}$ erg s^{-1} and the probable detection of persistent/long term X-ray emission from optical sources consistent with being a globular cluster or ultra–compact dwarf host. These flares have a similar time scale and peak luminosities $L_X > 10^{39}$ erg s^{-1}. Repeat flares have not been observed for the three X-ray flares XRT 000519, XRT 110203 and CDFS-XT1 discussed above, although the repetition time scale is >4 days for one and ∼1.8 days for the second repeating source in Irwin et al. (2016) and given the sparseness of the X-ray observations of the fields of the three X-ray flares above one cannot rule out that all events repeat with timescales of days to weeks. Whereas this might argue against cataclysmic events, such as TDEs, models predicting multiple flares due to for instance partial disruptions of a white dwarf on an eccentric bound orbit or binary black holes might well still be consistent with the observations (Zalamea et al. 2010). However, we do note that the predicted periods would be of order of hundreds to perhaps thousands of seconds, not days for the bound white dwarf orbiting an IMBH.

For the flares where *Chandra* detected sufficient number of counts for a spectral decomposition, it was found that the spectra of the flares are well–fit with a power law shape (Jonker et al. 2013; Glennie et al. 2015), although in the latter case the value of the power law index depends strongly on the assumed interstellar extinction present. In the case of the flare called "source 1" in Irwin et al. (2016), the spectrum was also well–fit by a power law with index 1.6 ± 0.3 (90% confidence), fixing the Galactic column density to the value of 1.8×10^{20} cm^{-2}. The power law index varied slightly between the two peaks in the case of the XRT 000519 flare, with the second peak having a slightly softer power law index (1.95 ± 0.05; 68% confidence) than the first peak (1.6 ± 0.1; 68% confidence). A power law index of 1.4 ± 0.2 (68% confidence) provided a good fit for the flare reported in Bauer et al. (2017), although those authors also warn that a softer index (closer to 2) can not be ruled out as the power law index and (Galactic) extinction are to a large degree degenerate given the relatively low number of counts detected.

Other fast X-ray flares, often with time scales of thousands of seconds, hence slightly longer than the $\sim$100 s time scale for the main flares above, have been reported as ultra-long gamma-ray bursts (Levan et al. 2014), although the energy bands used to measure the duration of the flares is different, making their comparison more difficult.

A particularly interesting case is CDFS-XT2 (Xue et al. 2019). In that paper it was interpreted as the X-ray signal from a binary neutron star merger. However, (Peng et al. 2019) interpret it as a white dwarf TDE. We note that the light curve (Xue et al. 2019) is more like the ultra-long GRBs (Levan et al. 2014), than the faster-still X-ray transients we report on here. However, clearly there is overlap in the properties of these events.

The energies and timescales associated with these events imply that compact objects such as massive black holes must be involved. Given the expected rate of 10^5 over the whole sky per year, a conservative assumed logN–logS (similar to other observed X-ray sources where per 2 decades in luminosity ten times more sources are found [e.g. Mineo et al. 2012]), and the instantaneous sensitivity of the eROSITA satellite should find $\sim$1 of these fast X-ray events per day.

9 Multi-Waveband Properties of X-Ray Selected TDEs

9.1 UV and Optical

Among the first X-ray TDEs identified, NGC 5905 had quasi-simultaneous optical photometry, thanks to photographic plate archives which covered a timespan of several decades (Fig. 2 of Komossa and Bade 1999). No long-term optical variability of this HII-type galaxy was discovered, and no optical flaring quasi-simultaneous with the X-ray flare was detected.

The Swift and XMM-*Newton* satellites both host an UV/optical telescope in their payload which allows these bands to be monitored simultaneously with the X-ray emission. Initially, it was thought that any emission in these bands would be from the Rayleigh-Jeans tail of the hot plasma which produces the soft X-ray emission and would therefore rise and decay simultaneously with the X-rays (e.g. Ulmer 1999). In 2MASX 0740-85 the UV and X-ray flux

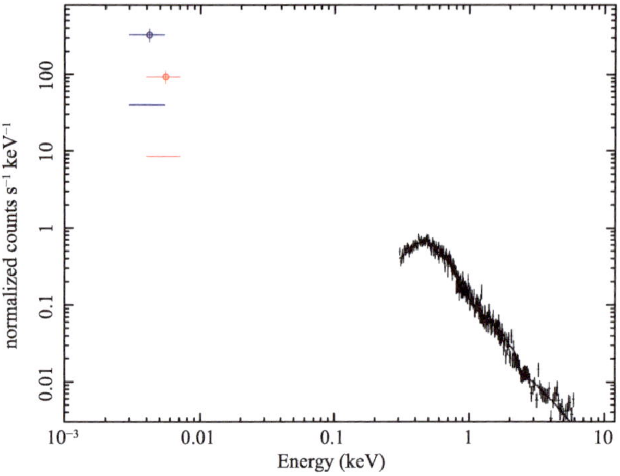

Fig. 15 An extrapolation of the best fit X-ray model, a power-law plus a $kT = 86$ eV black-body, to the XMM-OM, UVW1 and UVM2 filter data of 2MASX 0740-85. The single temperature thermal model significantly underpredicts the observed UV flux

did decay from peak quasi-simultaneously over 550 days (Fig. 8; Saxton et al. 2017). Nevertheless, the galaxy-subtracted UV emission sits well in excess of a simple extrapolation of the thermal component from the X-ray spectrum of this source to UV wavebands (Fig. 15). The UV and X-ray data can be well modelled by a structure which includes emission from a range of temperatures, such as a thin accretion disk (Saxton et al. 2017). This finding agrees with the low temperature emission of $\sim 20{,}000$ K ubiquitously found in TDEs discovered in the optical or UV bands (Gezari et al. 2009; van Velzen et al. 2011b; Holoien et al. 2016a, and see the Optical Chapter). The similar temporal behaviour of the X-ray and UV emission is mirrored in ASASSN-14li (Sect. 4.3.1 and Holoien et al. 2016b; Brown et al. 2017) and indicates prompt accretion (or efficient circulation: Guillochon and Ramirez-Ruiz 2015). In this event, the high-density, multi-wavelength monitoring indicated a possible delay of 32 days between the X-ray emission and that of the UV (Pasham et al. 2017, and see the Echo Chapter). If the UV is produced by shocks in colliding streams of debris (Piran et al. 2015a; Shiokawa et al. 2015) then it would naturally precede the X-ray emission which is generated when that same material falls down to the black hole. The timescale for the infall is indeed expected to be a few weeks (Piran et al. 2015a; Shiokawa et al. 2015; Bonnerot et al. 2017, and the Disruption Chapter and Accretion Disc Chapter).

In ASSASN-15oi the behaviour was quite different (Sect. 4.3.2). Here the UV light fell by 5 magnitudes over 200 days while the soft X-ray luminosity increased by a factor 10. The X-rays subsequently declined back to their initial level after a further 400 days. Gezari et al. (2017) and Holoien et al. (2018b) presented late-time observations of ASASSN-15oi. Their studies revealed that the thermal X-ray emission from the source brightened by an order of magnitude during its first year of evolution, rather than following a powerlaw decline as seen from its optical/UV light curves. After $\sim$600 days, the X-ray emission had faded back to the levels originally detected at peak. Gezari et al. (2017) suggested that the $\sim$1 year it takes in the rise to peak for the X-ray emission is the result of delayed accretion on a 10^6 $M_\odot$ black hole. This delayed accretion is a result of inefficient circularisation of the debris disk due to a delay in the time it takes for the material to accrete onto the black hole (Lodato 2012; Piran et al. 2015a; Guillochon and Ramirez-Ruiz 2015). However, as the circularisation timescale is proportional to $M_{\rm BH}^{-7/6}$ (Bonnerot et al. 2017), Gezari et al. (2017) also suggested that if the black hole of ASASSN-15oi was closer to 10^7 $M_\odot$ as originally estimated by Holoien et al. (2016a) using the host galaxy mass and the $M_{\rm BH} - M_{\rm bulge}$ relation of McConnell and Ma (2013), then the circularisation timescale would be much shorter, and may not be able to explain the observed behaviour. Another possibility is that ASASSN-15oi is heavily absorbed by material surrounding the accretion disk which is optically thick to X-ray radiation at early times (e.g., Metzger and Stone 2016). Even though, observationally Holoien et al. (2016a), Gezari et al. (2017) and Holoien et al. (2018b) were unable to constrain the column density in the direction of the source to confirm this suggestion, the possibility that some X-ray TDEs are surrounded by dense material may be supported by the study of Auchettl et al. (2017).

A still different picture is present in the light curve of XMMSL2 J1446+68 (Fig. 8). Here the UV flux from the galaxy increased by $\sim$1 magnitude, prior to the first X-ray detection and then stayed flat for 400 days before fading by $\sim$ a magnitude over the next 200 days. Meanwhile, the X-ray emission was constant for 100 days after discovery before fading by a factor $\sim$100 over the next 500 days (Saxton et al. 2019). The delayed decay of the UV emission in this event suggests that a reservoir of relatively cool material was maintained for about a year. One possible explanation is that the material formed an accretion disk which drained viscously until running out of material (van Velzen et al. 2019a).

In SDSS J1201+30 the UV emission from the galaxy was apparently unaffected by the disruption event, while the X-rays faded by a factor 100. While we should note that a constant UV/optical flux can simply mean that the flare is obscured in these bands or the contrast with the bright host galaxy is poor, the diversity of relative behaviour between the optical, UV and X-ray bands represents a challenge to current models of TDE evolution. A full understanding of the diversity of optical/UV emission behaviour from X-ray selected TDEs awaits a better understanding of the mechanisms which are responsible for the optical/UV emission. This theme is addressed in the Optical Chapter.

9.2 Radio

Dedicated radio follow-up was performed on the TDE in NGC 5905, in order to exclude the (very unlikely) scenario of a blazar hiding in this nearby starburst galaxy, and to search for the first time for jet emission associated with a TDE itself (Komossa 2002). Based on an observation with the VLA A array carried out in 1996, no nuclear radio emission was detected, with a 5σ upper limit for the presence of a central point source of < 150 μJy (Komossa and Dahlem 2001; Komossa 2002). At a distance of 75.4 Mpc of NGC 5905, this corresponds to a limit of $L_{8.46\,\mathrm{GHz}} < 9 \times 10^{36}$ erg s^{-1}. Extended low-level radio emission at lower frequencies is present in NGC 5905, and consistent with its HII-type nature (see Sect. 2.3.2 of Komossa 2002). A search for late-time radio emission from the nucleus of NGC 5905 was carried out by Bower et al. (2013), who did not detect any to an upper limit of 200 μJy. Radio emission from RXJ 1242-1119 was searched for in the FIRST VLA sky survey at 20 cm (Komossa 2002) but none was detected. In order to search for very late-time radio emission, Bower et al. (2013) also carried out follow-ups of the other ROSAT TDEs. No radio emission was found from RXJ 1242-1119, while a second source in the error circle of RXJ 1420.4+5334 does emit faintly (114 ± 24 μJy) in the radio regime.

Following the Swift detection of a jetted TDE, radio follow-up was more routinely carried out for newly identified TDEs. Radio upper limits of 100–200 μJy were reported for SDSS J1201+30 (Saxton et al. 2012a) and 10 μJy for XMMSL2 J1446+68 (Saxton et al. 2019). Faint radio emission (1.2 mJy at 1.5 GHz; $L_{1.5\mathrm{GHz}} = 10^{37}$ erg s^{-1}) was detected 1 year after discovery from 2MASX 0740-85, which faded over the following months (Alexander et al. 2017), while slightly stronger emission was monitored in ASASSN-14li (Alexander et al. 2016).

In summary, the thermal X-ray TDEs are not strong radio emitters. For a full analysis of the radio properties of these and other TDEs see the Radio Chapter.

10 Interpreting X-Ray TDEs with a Reprocessing Model

Various authors have explored the effect on the emitted radiation of an optically-thick, static or expanding envelope of material created during the event (Ulmer 1999; Strubbe and Quataert 2009; Metzger and Stone 2016; Dai et al. 2018; Roth and Kasen 2018). In this section we compare this model with X-ray TDE properties. The X-ray emission from a (non-jetted) TDE is generally quite soft in nature, and is well described by a black-body with a temperature between 10–100 eV (see the previous sections) or a powerlaw with photon index of $\Gamma > 4$ (Auchettl et al. 2017, and earlier work). These temperatures are generally consistent with the picture suggested by Rees (1988) of a black-body with a temperature between 10^{5-6} Kelvin arising from an accretion disk (e.g., Ulmer 1999; Bonning et al. 2007). However, the temperatures derived from optical/UV studies of these events and optical/UV only events are an order of magnitude less than those measured from X-rays (see the Optical Chapter) and may feasibly represent emission from the reprocessing of nuclear radiation (e.g. Loeb and Ulmer 1997).

Those that have a strong jet such as Swift J1644+57 and Swift J2058+05 (Cenko et al. 2012) generally exhibit much harder X-ray emission, which can be best described by a simple powerlaw with a photon index of $\Gamma \sim 1$–2. There are also cases of non-jetted TDE, such as ASASSN-15oi, XMMSL1 J0740-85 and PS18kh (Holoien et al. 2018a; van Velzen et al. 2019b) which exhibit both the soft black-body component and a weaker powerlaw component with a temperature and photon index similar to that listed above (see Sect. 4.5). This diversity in the observational characteristics of these events has been suggested to be a natural result of the viewing-angle with respect to the orientation of the accretion disk (e.g., Dai et al. 2018), or that X-ray TDEs and optical/UV only TDEs could result from a separate class of events that have compact debris disk due to large apsidal precession of the self-intersecting streams of the disrupted star (e.g., Dai et al. 2015). Jonker et al. (2019) discuss three optically-selected TDEs which were not detected in X-rays during the optical flare but 8–10 years later had luminosities of $L_X \sim 10^{41-42}$. In the reprocessing model, the X-ray emission from these events will have been massively suppressed at early times and only become visible when the density of the absorbing material decreased.

Observationally, it has been shown that the softness of non-jetted events is intrinsic to the source, with the spectral energy distribution of these events peaking in the UV/soft X-ray band (Auchettl et al. 2017, and references in Sect. 3, 4), consistent with theoretical expectations for black holes with masses $<10^7$ $M_\odot$ (e.g., Ulmer 1999; Dai et al. 2018). Using the spectral energy distributions of these events, Auchettl et al. (2017) saw that X-ray selected TDEs have high X-ray to optical ratios (see Fig. 16). The fact that these events produce significant amounts of both X-ray and optical/UV emission opens the possibility that a considerable fraction of their emission is being reprocessed into optical/UV wavelengths,[9] with the variation seen in Fig. 16 perhaps implying that these events experience significantly different reprocessing rates.

A separation between jetted and non-jetted events is seen when one attempts to derive the isotropic luminosity of each source (Auchettl et al. 2017, Fig. 16). Even though TDEs exhibit a wide range of isotropic luminosities, jetted events tend to have $L_{iso} \sim 10^{44}$ erg s^{-1},

[9] An alternative explanation was given by van Velzen et al. (2019a) who suggested that the properties of these sources are not a result of reprocessing but are due to a viscously spreading, unobscured accretion disk. This work was extended into the X-ray regime in Jonker et al. (2019), who infer the existence of a long-lived accretion disk to explain the relatively high late-time X-ray luminosity of three optically-selected TDEs.

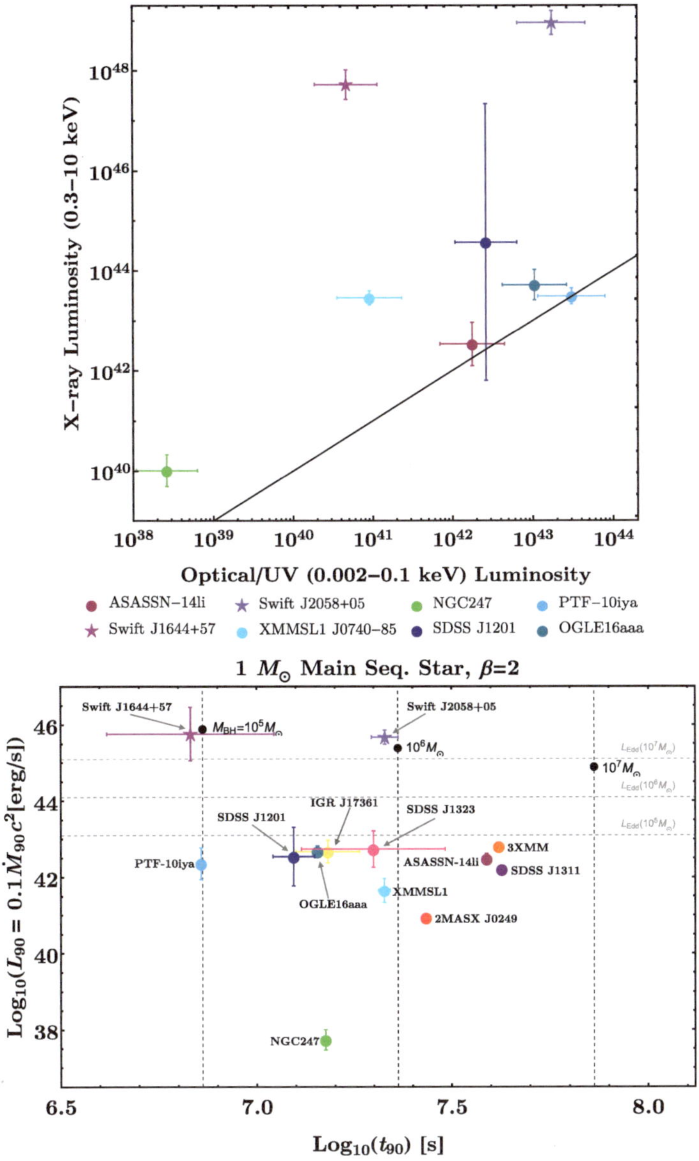

Fig. 16 Upper: Peak X-ray v peak optical/UV luminosity for a selection of X-ray bright TDEs. Lower: The mean isotropic luminosity plotted against the time taken for the TDE to emit between 5 and 90% of its total output. Both plots are taken from Auchettl et al. (2017)

while non-jetted X-ray events have $L_{\mathrm{iso}} \lesssim 10^{42}$ erg s^{-1}.[10] TDEs which have been detected in optical/UV wavelengths only, known as "veiled X-ray TDEs" in the nomenclature of

[10]Here L_{iso} is defined as the mean isotropic luminosity, after correcting for beaming, emitted by the event in the interval where the light curve contains between 5% and 95% of the total emitted luminosity (see Auchettl et al. (2017) and references therein for an explanation of the derivation of isotropic luminosity.).

Auchettl et al. (2017), could be surrounded by very dense material (e.g., PS1-10jh: Guillochon et al. 2014), which would lead to the X-rays being reprocessed completely into optical/UV wavelengths (e.g., Dai et al. 2018). These optical/UV sources have isotropic luminosities that fall between the jetted and non-jetted X-ray TDEs, in what Auchettl et al. (2017) refer to as a "reprocessing valley" (see e.g., Gezari et al. 2012, Chornock et al. 2014, Guillochon et al. 2014, Gezari et al. 2015, Piran et al. 2015b). As such, these events could emit X-rays but have all their emission reprocessed into lower wavelengths.

Assuming that a main sequence star is being fully disrupted, Auchettl et al. (2017) find that the light curves of X-ray TDEs are consistent with a disruption from a black hole with mass between 10^{5-7} $M_\odot$. These results are consistent with that derived from e.g., modelling the light curves of the individual events (e.g., Mockler et al. 2018), from bulge-black hole mass relations of e.g., McConnell and Ma (2013), Reines and Volonteri (2015), and optical spectroscopy (Wevers et al. 2017).

11 TDEs in Binary SMBHs and Recoiling SMBHs

11.1 TDEs in Binary SMBHs

The appearance of TDEs which occur in binary SMBHs can be different from TDEs of single SMBHs, including lightcurves which look characteristically different (Liu et al. 2009; Coughlin et al. 2017), and including rates which can be boosted by up to several orders of magnitude in some stages of binary evolution (e.g. Chen et al. 2009).

The lightcurve of the TDE from SDSS J120136.02+300305.5 (Sect. 4.2.1; Fig. 7) does not show a smooth decline, but exhibits episodes of dipping. One month after the peak, the X-ray emission suddenly dropped by a factor of >50 within a week and the source was no longer detected by Swift. X-rays re-appeared after 115 d, and then dropped a second time. While such a behaviour could arise due to beaming in jetted sources, no radio emission was detected from this TDE (Saxton et al. 2012a). Instead, the characteristic intermittence and recovery of the lightcurve of SDSS J1201+30 is reminiscent of predictions by Liu et al. (2009), who computed TDE lightcurves in binary SMBHs. In that case, the second SMBH acts as a perturber and the accretion stream on the primary is temporarily interrupted. Simulations by Liu et al. (2014) have shown, that the lightcurve of SDSS J1201+30 is consistent with a binary SMBH model with a primary mass of 10^6 $M_\odot$, a mass ratio $q \sim 0.1$ and a semi-major axis of 0.6 milli-pc.

This was the first supermassive binary BH (SMBBH) candidate identified in a non-active host galaxy, and the one with the most compact orbit among the known SMBBH candidates (review by Komossa and Zensus 2016). It has overcome the so-called "final parsec problem" (e.g. Colpi 2014). Upon coalescence, it will be a strong source of gravitational wave emission in the sensitivity regime of the upcoming generation space-based gravitational wave detectors. If significant numbers of SMBBHs exist at the cores of non-active galaxies, we expect to see more such events in well-sampled lightcurves of TDEs with *Swift* or the future Einstein Probe (EP; Yuan et al. 2015, 2016) mission. A good lightcurve coverage is essential for constraining the system parameters.

11.2 TDEs as Signposts of Recoiling SMBHs

Luminous X-ray flares from TDEs which occur *off-nuclear* are possible signposts of recoiling SMBHs (Komossa and Merritt 2008). At X-ray peak luminosities in the quasar regime

($L_X \sim 10^{42-46}$ erg/s), there is no other mechanism, which could produce a long-lived off-nuclear X-ray flare.

Potentially, a flaring recoiling SMBH could hide among the population of ultraluminous X-ray sources (ULX) which have been identified in nearby galaxies, however, the known ULXs have much lower X-ray luminosities than AGN, and they are by far (i.e., by orders of magnitude) too abundant to be explained by TDEs from recoiling SMBHs (see, e.g., the discussion by Strateva and Komossa 2009; Jonker et al. 2012).

12 X-Ray TDE Rates

To calculate the frequency of tidal disruptions it is necessary to perform a uniform analysis of a well-controlled sample of events. Donley et al. (2002), performed a systematic survey of galaxies observed in both the *ROSAT* All-Sky Survey and subsequent *ROSAT* pointed observations, re-detecting 3 TDEs which had been previously reported (see Sect. 3). From this work, which covered 9% of the sky, they calculated a TDE rate of 9×10^{-6} gal^{-1} yr^{-1}. Esquej et al. (2008) found 2 TDEs in the XMM-*Newton* slew survey by comparing fluxes with *ROSAT* observations taken 10–15 years earlier and derived a rate of 2.3×10^{-4} gal^{-1} yr^{-1}. A further search for events was made by comparing RASS data with deep XMM-*Newton* pointed observations (Khabibullin and Sazonov 2014) giving a baseline of 10–20 years. They discovered three events in the 2% of the sky covered by XMM-*Newton*, obtaining a rate of 3×10^{-5} gal^{-1} yr^{-1}. Finally, from multiple observations of the large cluster of galaxies, A1689, Maksym et al. (2010) derived a rate of 6×10^{-5} TDEs gal^{-1} yr^{-1}.

A survey measures three main quantities: the number of detected TDE, the number of square degrees covered and the flux limit reached. A set of assumptions are then adopted to find the volume of sky which has been sampled, and hence the number of galaxies observed, and the fraction of a one-year light curve which has been observed. To convert these into a rate, assumptions have to be made about the peak luminosity (L_P) of the event, the galaxy density and the shape of the light curve (for a full description of the derivation see the Rates Chapter). In Table 2 we list the measured values and adopted assumptions for each of the soft X-ray TDE rate calculations. We see that the assumptions vary greatly between each calculation. L_P has been taken to be between 2.8×10^{43} (Donley et al. 2002) and $\sim 1 \times 10^{44}$ (Khabibullin and Sazonov 2014; Esquej et al. 2008) erg s^{-1}. Similarly the light curve shape has been taken as effectively flat for one year (Donley et al. 2002), flat for 0.19 years and then dropping to zero (Khabibullin and Sazonov 2014) or $t^{-5/3}$ (Esquej et al. 2008; Maksym et al. 2010). Integrating over the latter gives a surveyed volume equivalent to observing the peak luminosity for 0.013 years. Not surprisingly these different assumptions produce large differences in the final calculated rates, which range from 9×10^{-6} to 2.3×10^{-4} gal^{-1} yr^{-1} .

To show the importance of the details of the calculation, we take as a common set of assumptions, the peak luminosity from Donley et al. (2002), $L_P = 2.8 \times 10^{43}$ erg s^{-1}, the visibility time from Khabibullin and Sazonov (2014) (0.19 years) and a galaxy density of $\rho = 0.02$ Mpc^{-3} and apply it to the surveys.[11] Results are shown in Table 2 where the rates now vary between 3.4 and 21×10^{-5} gal^{-1} yr^{-1} , within a factor 7 of each other.[12] This

[11] Note that the cluster survey of (Maksym et al. 2010) uses a self-consistent set of assumptions which are not affected by this change.

[12] All the surveys use very strong TDE candidates when calculating the rates except for Khabibullin and Sazonov (2014) which identified one very likely TDE (RBS 1032) and two possible TDEs in their sample. If only RBS 1032 had been adopted here then the survey TDE rates with common assumptions would agree to a factor ≈ 3.

immediately shows that the *measurements* in these surveys are actually in better agreement than they first appear.

For an accurate calculation of the absolute value of the TDE rate, a larger and less-biased sample of TDEs are needed to constrain their properties. In particular the TDE peak luminosity function is not currently well constrained (although see Sun et al. 2015; Auchettl et al. 2018, for first attempts) and the large variety of TDE light curves (e.g. Fig. 8) introduce considerable uncertainty in the rates which will only be resolved when an unbiased estimate of the median light curve is available.

In summary, the *reported* soft-X-ray-selected TDE rates in the literature lie between $\sim 1 \times 10^{-5}$ and 2×10^{-4} TDEs gal^{-1} yr^{-1}. If we adopt a common set of assumptions, then the surveys agree to better than a factor 7. Note that the measured rate is a lower limit due to the fraction of TDEs which are not observed in X-rays due to absorption by gas in the host galaxy, tidal debris or outflowing material from the accretion process, e.g. Sembay and West (1993) estimate that 40% of soft X-ray TDEs will be located in edge-on galaxies and their soft X-rays absorbed away within the host galaxy.

13 Conclusions and Future Prospects

The bulk of the TDEs discussed in Sect. 2–5 share the following characteristics:

- X-ray peak luminosities between 10^{42} and a few times 10^{44} erg s^{-1} .
- Very soft X-ray spectra near peak, with black-body temperatures in the range $kT_{bb} =$ 0.04–0.12 keV (or, alternatively, with powerlaw indices in the range $\Gamma_x = 3$–5).
- A spectral hardening within years.
- A decline on a timescale of years down to a quiescent level.
- An absence of X-ray emission lines.
- Fast variability in several events (minutes to hours).
- Host galaxies which are quiescent and in-active both before and after the disruption.
- A decline in flux by factors up to 1000–6000.

While there are certainly exceptions to these properties, the overall picture is of X-ray radiation emitted by optically-thick material within a few R_g of the black hole, which fades with the diminishing return of tidal debris.

The limited duration of events has allowed us to follow the passage from super-Eddington to sub-Eddington to low-level (RIAF/ADAF) accretion on year-to-decade timescales and witness the onset and cessation of disk winds which are responsible for driving material into the host galaxy.

One of the exciting prospects of X-ray TDE observations is the possibility of identifying IMBH from their distinctive disruptive properties. This was broached in Sect. 4.1.2 and also in the section on the enigmatic *very fast events*, whose light curves may indicate the disruption of a compact star by a BH of mass $\leq 10^5$ $M_\odot$.

Deep follow-up observations of TDEs near their peak are allowing us to probe the extremes of accretion physics in relatively clean environments. They allow us to follow the evolution of disk winds and coronae, search for relativistic (precession) effects in the Kerr metric, estimate BH spin, carry out absorption/emission-line spectroscopy of ionized matter in outflow (either stellar debris or accretion disk winds), and study the jet-disk coupling and jet evolution in jetted events.

Although more than 20 years old, the science of X-ray emitting TDEs can be considered to be still in its infancy, with each new event presenting traits which modify our understanding of the disruption process. With the launch of the eRosita telescope (Predehl et al. 2010)

Table 2 Summary of TDE survey rates

Survey[a]	Factor[b]	Lum_X[c] (erg/s)	Decay[d] model	Baseline (Years)	ρ_{gal}[e] (Mpc^{-3})	F_{lim}[f] ($erg\,s^{-1}\,cm^{-2}$)	D_{lim}[g] (Mpc)	Nsrc[h]	Area[i] (% sky)	Rate[j] $gal^{-1}\,yr^{-1}$	Rate[k] $gal^{-1}\,yr^{-1}$
RASSvRXP	20	2.8×10^{43}	1 yr	0.5–8	0.014	2.0×10^{-12}	342	3	9	9×10^{-6}	3.4×10^{-5}
RASSvXMMP	10	10^{44}[l]	0.19 yrs	10–20	0.02	3.5×10^{-13}	877	3	2	3×10^{-5}	2.1×10^{-4}
XMMSvRASS	20	10^{44}	$t^{-5/3}$	10–18	0.023	5.0×10^{-12}	406	2	15	2.3×10^{-4}	1.1×10^{-4}
Abell 1689	4–5	$\sim 10^{44}$	$t^{-5/3}$	1–7	–	$\sim 1 \times 10^{-13}$	–	1	–	6×10^{-5}	6×10^{-5}

[a] RASS v *ROSAT* pointed (0.2–2.4 keV) (Donley et al. 2002); RASS bright source catalogue v XMM-*Newton* pointed (0.2–2 keV) data (Khabibullin and Sazonov 2014); XMM, EPIC-pn slew (0.2–2 keV) from catalogue XMMS1d3 v RASS (Esquej et al. 2008); repeat observations of 2100 galaxies in Abell 1689 over 7 years (Maksym et al. 2010)

[b] Minimum variability factor needed for a detection

[c] Assumed peak X-ray luminosity

[d] Assumed light curve duration or decay function; flat for 1 year (Donley et al. 2002), flat for 0.19 years then dropping to zero (Khabibullin and Sazonov 2014) or $t^{-5/3}$

[e] Density of galaxies in the local universe

[f] Detection flux limit (observed) in the 0.2–2 keV band. Khabibullin and Sazonov (2014) use the ROSAT Bright Source Catalogue which contains sources with a minimum count rate of 0.05 c/s in the ROSAT PSPC camera. This translates to a 0.2–2 keV flux of $3.5 \times 10^{-13}\ erg\,s^{-1}\,cm^{-2}$ using their spectral model of a $kT = 50$ eV black body absorbed by a Galactic column of $N_H = 5 \times 10^{20}\ cm^{-2}$

[g] Derived limiting distance for TDE detection

[h] Number of TDEs found in the survey. Note that (Donley et al. 2002) consider sources which have faded between the RASS and *ROSAT* pointed observations and hence exclude RXJ 1242-1119

[i] Percentage of the sky covered in the survey. In the case of Abell 1689, 2100 galaxies were repeatedly observed

[j] The original reported TDE rate based on the given assumptions

[k] The TDE rate calculated under common assumptions; $L_{X,peak} = 2.8 \times 10^{43}$, an average visibility time of 0.19 years and a galaxy density of $\rho = 0.02\ Mpc^{-3}$ (see text)

[l] Khabibullin and Sazonov (2014) use a bolometric luminosity of $7 \times 10^{44}\ erg\,s^{-1}$ without explicitly giving the correction factor to the X-ray band. They refer to Khabibullin et al. (2014) from where an X-ray peak luminosity of $1 \times 10^{44}\ erg\,s^{-1}$ can be derived based on the Eddington luminosity and K-correction for a $10^6\ M_\odot$ black hole

on board the Spectrum-Roentgen-Gamma (SRG) mission, it is expected that hundreds of new TDEs will be found (Khabibullin et al. 2014; Jonker et al. 2019). If well monitored in dedicated follow-ups by other missions, these will fill in the parameter space and provide us with a more complete picture of the phenomenon. Within a few years, the Einstein Probe will provide excellent light curves of the rise, peak and initial decay phases of hundreds of X-ray TDEs and allow detailed modeling of the fall back process of a large number of TDEs.

Acknowledgements Dacheng Lin and Erin Kara are warmly thanked for providing updated figures of their work. RS would like to thank Peter Maksym for early help with the rates calculation. Sjoert Van Velzen and an anonymous second referee are thanked for comments and suggestions which improved the manuscript.

References

K.D. Alexander, E. Berger, J. Guillochon, B.A. Zauderer, P.K.G. Williams, Discovery of an outflow from radio observations of the tidal disruption event ASASSN-14li. Astrophys. J. Lett. **819**, 25 (2016). https://doi.org/10.3847/2041-8205/819/2/L25

K.D. Alexander, M.H. Wieringa, E. Berger, R.D. Saxton, S. Komossa, Radio observations of the tidal disruption event XMMSL1 J0740-85. Astrophys. J. **837**, 153 (2017). https://doi.org/10.3847/1538-4357/aa6192

N. Arav, Z.-Y. Li, M.C. Begelman, Radiative acceleration in outflows from broad absorption-line quasi-stellar objects. II. Wind models. Astrophys. J. **432**, 62 (1994). https://doi.org/10.1086/174549

I. Arcavi, A. Gal-Yam, M. Sullivan, Y.-C. Pan, S.B. Cenko, A. Horesh, E.O. Ofek, A. De Cia, L. Yan, C.-W. Yang, D.A. Howell, D. Tal, S.R. Kulkarni, S.P. Tendulkar, S. Tang, D. Xu, A. Sternberg, J.G. Cohen, J.S. Bloom, P.E. Nugent, M.M. Kasliwal, D.A. Perley, R.M. Quimby, A.A. Miller, C.A. Theissen, R.R. Laher, A continuum of H- to He-rich tidal disruption candidates with a preference for E+A galaxies. Astrophys. J. **793**, 38 (2014). https://doi.org/10.1088/0004-637X/793/1/38

K. Auchettl, J. Guillochon, E. Ramirez-Ruiz, New physical insights about tidal disruption events from a comprehensive observational inventory at X-ray wavelengths. Astrophys. J. **838**, 149 (2017). https://doi.org/10.3847/1538-4357/aa633b

K. Auchettl, E. Ramirez-Ruiz, J. Guillochon, A comparison of the X-ray emission from tidal disruption events with those of active galactic nuclei. Astrophys. J. **852**, 37 (2018). https://doi.org/10.3847/1538-4357/aa9b7c

S. Ayal, M. Livio, T. Piran, Tidal disruption of a solar-type star by a supermassive black hole. Astrophys. J. **545**, 772–780 (2000). https://doi.org/10.1086/317835

N. Bade, S. Komossa, M. Dahlem, Detection of an extremely soft x-ray outburst in the HII-like nucleus of NGC 5905. Astron. Astrophys. **309**, L35–L38 (1996). https://ui.adsabs.harvard.edu/abs/1996A%26A...309L..35B

J.A. Baldwin, M.M. Phillips, R. Terlevich, Classification parameters for the emission-line spectra of extragalactic objects. Publ. Astron. Soc. Pac. **93**, 5–19 (1981). https://doi.org/10.1086/130766

F.E. Bauer, E. Treister, K. Schawinski, S. Schulze, B. Luo, D.M. Alexander, W.N. Brandt, A. Comastri, F. Forster, R. Gilli, D.A. Kann, K. Maeda, K. Nomoto, M. Paolillo, P. Ranalli, D.P. Schneider, O. Shemmer, M. Tanaka, A. Tolstov, N. Tominaga, P. Tozzi, C. Vignali, J. Wang, Y. Xue, G. Yang, A new, faint population of X-ray transients. Mon. Not. R. Astron. Soc. **467**, 4841–4857 (2017). https://doi.org/10.1093/mnras/stx417

J.S. Bloom, D. Giannios, B.D. Metzger, S.B. Cenko, D.A. Perley, N.R. Butler, N.R. Tanvir, A.J. Levan, P.T. O'Brien, L.E. Strubbe, F. De Colle, E. Ramirez-Ruiz, W.H. Lee, S. Nayakshin, E. Quataert, A.R. King, A. Cucchiara, J. Guillochon, G.C. Bower, A.S. Fruchter, A.N. Morgan, A.J. van der Horst, A possible relativistic jetted outburst from a massive black hole fed by a tidally disrupted star. Science **333**, 203 (2011). https://doi.org/10.1126/science.1207150

C. Bonnerot, W. Lu, Simulating realistic disc formation in tidal disruption events. ArXiv e-prints, arXiv:1906.05865 (2019)

C. Bonnerot, E.M. Rossi, G. Lodato, Long-term stream evolution in tidal disruption events. Mon. Not. R. Astron. Soc. **464**, 2816–2830 (2017). https://doi.org/10.1093/mnras/stw2547

E.W. Bonning, L. Cheng, G.A. Shields, S. Salviander, K. Gebhardt, Accretion disk temperatures and continuum colors in QSOs. Astrophys. J. **659**, 211–217 (2007). https://doi.org/10.1086/510712

G.C. Bower, B.D. Metzger, S.B. Cenko, J.M. Silverman, J.S. Bloom, Late-time radio emission from X-ray-selected tidal disruption events. Astrophys. J. **763**(2), 84 (2013). https://doi.org/10.1088/0004-637X/763/2/84

J.S. Bright, R.P. Fender, S.E. Motta, K. Mooley, Y.C. Perrott, S. van Velzen, S. Carey, J. Hickish, N. Razavi-Ghods, D. Titterington, P. Scott, K. Grainge, A. Scaife, T. Cantwell, C. Rumsey, Long-term radio and X-ray evolution of the tidal disruption event ASASSN-14li. Mon. Not. R. Astron. Soc. **475**, 4011–4019 (2018). https://doi.org/10.1093/mnras/sty077

J.S. Brown, T.W.-S. Holoien, K. Auchettl, K.Z. Stanek, C.S. Kochanek, B.J. Shappee, J.L. Prieto, D. Grupe, The long term evolution of ASASSN-14li. Mon. Not. R. Astron. Soc. **466**, 4904–4916 (2017). https://doi.org/10.1093/mnras/stx033

D. Burrows, J. Kennea, G. Ghisellini, V. Mangano, B. Zhang, K. Page, M. Eracleous, P. Romano, T. Sakamoto, A. Falcone, J. Osborne, S. Campana, A. Beardmore, A. Breeveld, M. Chester, R. Corbet, S. Covino, J. Cummings, P. D'avanzo, V. D'elia, P. Esposito, P. Evans, D. Fugazza, J. Gelbord, K. Hiroi, S. Holland, K. Huang, M. Im, G. Israel, Y. Jeon, H. Jun, N. Kawai, J. Kim, H. Krimm, F. Marshall, P. Mészáros, H. Negoro, N. Omodei, W. Park, J. Perkins, M. Sugizaki, H. Sung, G. Tagliaferri, E. Troja, Y. Ueda, Y. Urata, R. Usui, L. Antonelli, S. Barthelmy, G. Cusumano, P. Giommi, A. Melandri, M. Perri, J. Racusin, B. Sbarufatti, M. Siegel, N. Gehrels, Relativistic jet activity from the tidal disruption of a star by a massive black hole. Nature **476**, 421–421424424 (2011). http://adsabs.harvard.edu/abs/2011Natur.476..421B

J.K. Cannizzo, H.M. Lee, J. Goodman, The disk accretion of a tidally disrupted star onto a massive black hole. Astrophys. J. **351**, 38 (1990). https://doi.org/10.1086/168442

J.K. Cannizzo, E. Troja, G. Lodato, GRB 110328A/Swift J164449.3+573451: the tidal obliteration of a deeply plunging star? Astrophys. J. **742**, 32 (2011). https://doi.org/10.1088/0004-637X/742/1/32

N. Cappelluti, M. Ajello, P. Rebusco, S. Komossa, A. Bongiorno, C. Clemens, M. Salvato, P. Esquej, T. Aldcroft, J. Greiner, H. Quintana, A candidate tidal disruption event in the Galaxy cluster Abell 3571. Astron. Astrophys. **495**, L9–L12 (2009). https://doi.org/10.1051/0004-6361/200811479

S.B. Cenko, H.A. Krimm, A. Horesh, A. Rau, D.A. Frail, J.A. Kennea, A.J. Levan, S.T. Holland, N.R. Butler, R.M. Quimby, J.S. Bloom, A.V. Filippenko, A. Gal-Yam, J. Greiner, S.R. Kulkarni, E.O. Ofek, F. Olivares E., P. Schady, J.M. Silverman, N.R. Tanvir, D. Xu, Swift J2058.4+0516: discovery of a possible second relativistic tidal disruption flare? Astrophys. J. **753**, 77 (2012). https://doi.org/10.1088/0004-637X/753/1/77

X. Chen, P. Madau, A. Sesana, F.K. Liu, Enhanced tidal disruption rates from massive black hole binaries. Astrophys. J. Lett. **697**(2), 149–152 (2009). https://doi.org/10.1088/0004-637X/697/2/L149

R. Chornock, E. Berger, S. Gezari, B.A. Zauderer, A. Rest, L. Chomiuk, A. Kamble, A.M. Soderberg, I. Czekala, J. Dittmann, M. Drout, R.J. Foley, W. Fong, M.E. Huber, R.P. Kirshner, A. Lawrence, R. Lunnan, G.H. Marion, G. Narayan, A.G. Riess, K.C. Roth, N.E. Sanders, D. Scolnic, S.J. Smartt, K. Smith, C.W. Stubbs, J.L. Tonry, W.S. Burgett, K.C. Chambers, H. Flewelling, K.W. Hodapp, N. Kaiser, E.A. Magnier, D.C. Martin, J.D. Neill, P.A. Price, R. Wainscoat, The ultraviolet-bright, slowly declining transient PS1-11af as a partial tidal disruption event. Astrophys. J. **780**, 44 (2014). https://doi.org/10.1088/0004-637X/780/1/44

D. Clausen, M. Eracleous, Probing intermediate-mass black holes with optical emission lines from tidally disrupted white dwarfs. Astrophys. J. **726**(1), 34 (2011). https://doi.org/10.1088/0004-637X/726/1/34

M. Colpi, Massive binary black holes in galactic nuclei and their path to coalescence. Space Sci. Rev. **183**(1-4), 189–221 (2014). https://doi.org/10.1007/s11214-014-0067-1

E.R. Coughlin, P.J. Armitage, C. Nixon, M.C. Begelman, Tidal disruption events from supermassive black hole binaries. Mon. Not. R. Astron. Soc. **465**, 3840–3864 (2017). https://doi.org/10.1093/mnras/stw2913

D.M. Crenshaw, S.B. Kraemer, A. Boggess, S.P. Maran, R.F. Mushotzky, C.-C. Wu, Intrinsic absorption lines in Seyfert 1 galaxies. I. Ultraviolet spectra from the Hubble Space Telescope. Astrophys. J. **516**(2), 750–768 (1999). https://doi.org/10.1086/307144

L. Dai, J.C. McKinney, M.C. Miller, Soft X-ray temperature tidal disruption events from stars on deep plunging orbits. Astrophys. J. Lett. **812**, 39 (2015). https://doi.org/10.1088/2041-8205/812/2/L39

L. Dai, J.C. McKinney, N. Roth, E. Ramirez-Ruiz, M.C. Miller, A unified model for tidal disruption events. Astrophys. J. Lett. **859**, 20 (2018). https://doi.org/10.3847/2041-8213/aab429

C. Done, S.W. Davis, C. Jin, O. Blaes, M. Ward, Intrinsic disc emission and the soft X-ray excess in active galactic nuclei. Mon. Not. R. Astron. Soc. **420**, 1848–1860 (2012). https://doi.org/10.1111/j.1365-2966.2011.19779.x

J.L. Donley, W.N. Brandt, M. Eracleous, T. Boller, Large-amplitude X-ray outbursts from galactic nuclei: a systematic survey using ROSAT archival data. Astron. J. **124**(3), 1308–1321 (2002). https://doi.org/10.1086/342280

P. Esquej, R.D. Saxton, M.J. Freyberg, A.M. Read, B. Altieri, M. Sanchez-Portal, G. Hasinger, Candidate tidal disruption events from the XMM-Newton slew survey. Astron. Astrophys. **462**, 49 (2007). http://adsabs.harvard.edu/abs/2007A%26A...462L..49E

P. Esquej, R.D. Saxton, S. Komossa, A.M. Read, M.J. Freyberg, G. Hasinger, D.A. García-Hernández, H. Lu, J. Rodriguez Zaurín, M. Sánchez-Portal, H. Zhou, Evolution of tidal disruption candidates discovered by XMM-Newton. Astron. Astrophys. **489**, 543 (2008). http://adsabs.harvard.edu/abs/2008A%26A...489..543E

C.R. Evans, C.S. Kochanek, The tidal disruption of a star by a massive black hole. Astrophys. J. Lett. **346**, 13–16 (1989). https://doi.org/10.1086/185567

I.N. Evans, F.A. Primini, K.J. Glotfelty, C.S. Anderson, N.R. Bonaventura, J.C. Chen, J.E. Davis, S.M. Doe, J.D. Evans, G. Fabbiano, E.C. Galle, D.G. Gibbs II, J.D. Grier, R.M. Hain, D.M. Hall, P.N. Harbo, X.H. He, J.C. Houck, M. Karovska, V.L. Kashyap, J. Lauer, M.L. McCollough, J.C. McDowell, J.B. Miller, A.W. Mitschang, D.L. Morgan, A.E. Mossman, J.S. Nichols, M.A. Nowak, D.A. Plummer, B.L. Refsdal, A.H. Rots, A. Siemiginowska, B.A. Sundheim, M.S. Tibbetts, D.W. Van Stone, S.L. Winkelman, P. Zografou, The Chandra Source Catalog. Astrophys. J. Suppl. Ser. **189**, 37–82 (2010). https://doi.org/10.1088/0067-0049/189/1/37

C. Evans, P. Laguna, M. Eracleous, Ultra-close encounters of stars with massive black holes: tidal disruption events with prompt hyperaccretion. Astrophys. J. Lett. **805**, 19 (2015). https://doi.org/10.1088/2041-8205/805/2/L19

G. Fabbiano, The X-ray properties of normal galaxies. Publ. Astron. Soc. Pac. **98**, 525–543 (1986). https://doi.org/10.1086/131790

L. Ferrarese, D. Merritt, A fundamental relation between supermassive black holes and their host galaxies. Astrophys. J. Lett. **539**(1), 9–12 (2000). https://doi.org/10.1086/312838

D.K. Galloway, M.P. Muno, J.M. Hartman, D. Psaltis, D. Chakrabarty, Thermonuclear (type I) X-ray bursts observed by the Rossi X-ray Timing Explorer. Astrophys. J. Suppl. Ser. **179**, 360–422 (2008). https://doi.org/10.1086/592044

N. Gehrels, G. Chincarini, P. Giommi, K.O. Mason, J.A. Nousek, A.A. Wells, N.E. White, S.D. Barthelmy, D.N. Burrows, L.R. Cominsky, K.C. Hurley, F.E. Marshall, P. Mészáros, P.W.A. Roming, L. Angelini, L.M. Barbier, T. Belloni, S. Campana, P.A. Caraveo, M.M. Chester, O. Citterio, T.L. Cline, M.S. Cropper, J.R. Cummings, A.J. Dean, E.D. Feigelson, E.E. Fenimore, D.A. Frail, A.S. Fruchter, G.P. Garmire, K. Gendreau, G. Ghisellini, J. Greiner, J.E. Hill, S.D. Hunsberger, H.A. Krimm, S.R. Kulkarni, P. Kumar, F. Lebrun, N.M. Lloyd-Ronning, C.B. Markwardt, B.J. Mattson, R.F. Mushotzky, J.P. Norris, J. Osborne, B. Paczynski, D.M. Palmer, H.-S. Park, A.M. Parsons, J. Paul, M.J. Rees, C.S. Reynolds, J.E. Rhoads, T.P. Sasseen, B.E. Schaefer, A.T. Short, A.P. Smale, I.A. Smith, L. Stella, G. Tagliaferri, T. Takahashi, M. Tashiro, L.K. Townsley, J. Tueller, M.J.L. Turner, M. Vietri, W. Voges, M.J. Ward, R. Willingale, F.M. Zerbi, W.W. Zhang, The Swift gamma-ray burst mission. Astrophys. J. **611**(2), 1005–1020 (2004). https://doi.org/10.1086/422091

S. Gezari, J.P. Halpern, S. Komossa, D. Grupe, K.M. Leighly, Follow-up Hubble Space Telescope/Space Telescope imaging spectroscopy of three candidate tidal disruption events. Astrophys. J. **592**(1), 42–51 (2003). https://doi.org/10.1086/375553

S. Gezari, T. Heckman, S.B. Cenko, M. Eracleous, K. Forster, T.S. Gonçalves, D.C. Martin, P. Morrissey, S.G. Neff, M. Seibert, D. Schiminovich, T.K. Wyder, Luminous thermal flares from quiescent supermassive black holes. Astrophys. J. **698**(2), 1367–1379 (2009). https://doi.org/10.1088/0004-637X/698/2/1367

S. Gezari, R. Chornock, A. Rest, M.E. Huber, K. Forster, E. Berger, P.J. Challis, J.D. Neill, D.C. Martin, T. Heckman, A. Lawrence, C. Norman, G. Narayan, R.J. Foley, G.H. Marion, D. Scolnic, L. Chomiuk, A. Soderberg, K. Smith, R.P. Kirshner, A.G. Riess, S.J. Smartt, C.W. Stubbs, J.L. Tonry, W.M. Wood-Vasey, W.S. Burgett, K.C. Chambers, T. Grav, J.N. Heasley, N. Kaiser, R.-P. Kudritzki, E.A. Magnier, J.S. Morgan, P.A. Price, An ultraviolet-optical flare from the tidal disruption of a helium-rich stellar core. Nature **485**, 217–220 (2012). https://doi.org/10.1038/nature10990

S. Gezari, R. Chornock, A. Lawrence, A. Rest, D.O. Jones, E. Berger, P.M. Challis, G. Narayan, PS1-10jh continues to follow the fallback accretion rate of a tidally disrupted star. Astrophys. J. Lett. **815**, 5 (2015). https://doi.org/10.1088/2041-8205/815/1/L5

S. Gezari, S.B. Cenko, I. Arcavi, X-ray brightening and UV fading of tidal disruption event ASASSN-15oi. Astrophys. J. Lett. **851**, 47 (2017). https://doi.org/10.3847/2041-8213/aaa0c2

K.K. Ghosh, V. Suleymanov, I. Bikmaev, S. Shimansky, N. Sakhibullin, RBS 1032: a dwarf-nucleated spheroidal galaxy with an intermediate-mass black hole hosted in a globular cluster. Mon. Not. R. Astron. Soc. **371**(4), 1587–1593 (2006). https://doi.org/10.1111/j.1365-2966.2006.10723.x

A. Glennie, P.G. Jonker, R.P. Fender, T. Nagayama, M.L. Pretorius, Two fast X-ray transients in archival Chandra data. Mon. Not. R. Astron. Soc. **450**, 3765–3770 (2015). https://doi.org/10.1093/mnras/stv801

J.E. Greene, Low-mass black holes as the remnants of primordial black hole formation. Nat. Commun. **3**, 1304 (2012). https://doi.org/10.1038/ncomms2314

J. Greiner, R. Schwarz, S. Zharikov, M. Orio, RX J1420.4+5334 – another tidal disruption event? Astron. Astrophys. **362**, L25–L28 (2000). https://ui.adsabs.harvard.edu/abs/2000A%26A...362L..25G

J.E. Grindlay, Fast x-ray transients and gamma-ray bursts: constraints on beaming. Astrophys. J. **510**(2), 710–714 (1999). https://doi.org/10.1086/306617.

D. Grupe, H.-C. Thomas, K.M. Leighly, RX J1624.9+7554: a new X-ray transient AGN. Astron. Astrophys. **350**, L31–L34 (1999). https://ui.adsabs.harvard.edu/abs/1999A%26A...350L..31G

J. Guillochon, E. Ramirez-Ruiz, Hydrodynamical simulations to determine the feeding rate of black holes by the tidal disruption of stars: the importance of the impact parameter and stellar structure. Astrophys. J. **767**, 25 (2013). https://doi.org/10.1088/0004-637X/767/1/25

J. Guillochon, E. Ramirez-Ruiz, A dark year for tidal disruption events. Astrophys. J. **809**, 166 (2015). https://doi.org/10.1088/0004-637X/809/2/166

J. Guillochon, H. Manukian, E. Ramirez-Ruiz, PS1-10jh: the disruption of a main-sequence star of near-solar composition. Astrophys. J. **783**, 23 (2014). https://doi.org/10.1088/0004-637X/783/1/23

J.P. Halpern, S. Gezari, S. Komossa, Follow-up Chandra observations of three candidate tidal disruption events. Astrophys. J. **604**, 572–578 (2004). https://doi.org/10.1086/381937

J. Heise, A.C. Brinkman, J. Schrijver, R. Mewe, E.H.B.M. Gronenschild, A.J.F. den Boggende, J. Grindlay, Evidence for X-ray emission from flare stars observed by ANS. Astrophys. J. Lett. **202**, 73–76 (1975). https://doi.org/10.1086/181984

T.W.-S. Holoien, C.S. Kochanek, J.L. Prieto, D. Grupe, P. Chen, D. Godoy-Rivera, K.Z. Stanek, B.J. Shappee, S. Dong, J.S. Brown, U. Basu, J.F. Beacom, D. Bersier, J. Brimacombe, E.K. Carlson, E. Falco, E. Johnston, B.F. Madore, G. Pojmanski, M. Seibert, ASASSN-15oi: a rapidly evolving, luminous tidal disruption event at 216 Mpc. Mon. Not. R. Astron. Soc. **463**, 3813–3828 (2016a). https://doi.org/10.1093/mnras/stw2272

T.W.-S. Holoien, C.S. Kochanek, J.L. Prieto, K.Z. Stanek, S. Dong, B.J. Shappee, D. Grupe, J.S. Brown, U. Basu, J.F. Beacom, D. Bersier, J. Brimacombe, A.B. Danilet, E. Falco, Z. Guo, J. Jose, G.J. Herczeg, F. Long, G. Pojmanski, G.V. Simonian, D.M. Szczygieł, T.A. Thompson, J.R. Thorstensen, R.M. Wagner, P.R. Woźniak, Six months of multiwavelength follow-up of the tidal disruption candidate ASASSN-14li and implied TDE rates from ASAS-SN. Mon. Not. R. Astron. Soc. **455**, 2918–2935 (2016b). https://doi.org/10.1093/mnras/stv2486

T.W.-S. Holoien, M.E. Huber, B.J. Shappee, M. Eracleous, K. Auchettl, J.S. Brown, M.A. Tucker, K.C. Chambers, C.S. Kochanek, K.Z. Stanek, A. Rest, D. Bersier, R.S. Post, G. Aldering, K.A. Ponder, J.D. Simon, E. Kankare, D. Dong, G. Hallinan, J. Bulger, T.B. Lowe, E.A. Magnier, A.S.B. Schultz, C.Z. Waters, M. Willman, D. Wright, D.R. Young, S. Dong, J.L. Prieto, T.A. Thompson, L. Denneau, H. Flewelling, A.N. Heinze, S.J. Smartt, K.W. Smith, B. Stalder, J.L. Tonry, H. Weiland, PS18kh: a New Tidal Disruption Event with a Non-Axisymmetric Accretion Disk. ArXiv e-prints, arXiv:1808.02890 (2018a)

T.W.-S. Holoien, J.S. Brown, K. Auchettl, C.S. Kochanek, J.L. Prieto, B.J. Shappee, J. Van Saders, The unusual late-time evolution of the tidal disruption event ASASSN-15oi. Mon. Not. R. Astron. Soc. **480**, 5689–5703 (2018b). https://doi.org/10.1093/mnras/sty2273

J.A. Irwin, R.N. Henriksen, M. Krause, Q.D. Wang, T. Wiegert, E.J. Murphy, G. Heald, E. Perlman, CHANG-ES V: nuclear outflow in a Virgo Cluster spiral after a tidal disruption event. Astrophys. J. **809**(2), 172 (2015). https://doi.org/10.1088/0004-637X/809/2/172

J.A. Irwin, W.P. Maksym, G.R. Sivakoff, A.J. Romanowsky, D. Lin, T. Speegle, I. Prado, D. Mildebrath, J. Strader, J. Liu, J.M. Miller, Ultraluminous X-ray bursts in two ultracompact companions to nearby elliptical galaxies. Nature **538**, 356–358 (2016). https://doi.org/10.1038/nature19822

F. Jansen, D. Lumb, B. Altieri, J. Clavel, M. Ehle, C. Erd, C. Gabriel, M. Guainazzi, P. Gondoin, R. Much, R. Munoz, M. Santos, N. Schartel, D. Texier, G. Vacanti, XMM-Newton observatory. I. The spacecraft and operations. Astron. Astrophys. **365**, 1–6 (2001). https://doi.org/10.1051/0004-6361:20000036

P.G. Jonker, M. Heida, M.A.P. Torres, J.C.A. Miller-Jones, A.C. Fabian, E.M. Ratti, G. Miniutti, D.J. Walton, T.P. Roberts, The nature of the bright ULX X-2 in NGC 3921: a Chandra position and HST candidate counterpart. Astrophys. J. **758**(1), 28 (2012). https://doi.org/10.1088/0004-637X/758/1/28

P.G. Jonker, A. Glennie, M. Heida, T. Maccarone, S. Hodgkin, G. Nelemans, J.C.A. Miller-Jones, M.A.P. Torres, R. Fender, Discovery of a new kind of explosive X-ray transient near M86. Astrophys. J. **779**, 14 (2013). https://doi.org/10.1088/0004-637X/779/1/14

P.G. Jonker, N.C. Stone, A. Generozov, S. van Velzen, B. Metzger, Implications from Late-Time X-ray Detections of Optically Selected Tidal Disruption Events: State Changes, Unification, and Detection Rates. ArXiv e-prints, arXiv:1906.12236 (2019)

E. Kara, L. Dai, C.S. Reynolds, T. Kallman, Ultrafast outflow in tidal disruption event ASASSN-14li. Mon. Not. R. Astron. Soc. **474**, 3593–3598 (2018). https://doi.org/10.1093/mnras/stx3004

G. Kauffmann, T.M. Heckman, C. Tremonti, J. Brinchmann, S. Charlot, S.D.M. White, S.E. Ridgway, J. Brinkmann, M. Fukugita, P.B. Hall, Ž. Ivezić, G.T. Richards, D.P. Schneider, The host galaxies of active galactic nuclei. Mon. Not. R. Astron. Soc. **346**(4), 1055–1077 (2003). https://doi.org/10.1111/j.1365-2966.2003.07154.x

L.J. Kewley, C.A. Heisler, M.A. Dopita, S. Lumsden, Optical classification of southern warm infrared galaxies. Astrophys. J. Suppl. Ser. **132**(1), 37–71 (2001). https://doi.org/10.1086/318944

I. Khabibullin, S. Sazonov, Stellar tidal disruption candidates found by cross-correlating the ROSAT Bright Source Catalogue and XMM-Newton observations. Mon. Not. R. Astron. Soc. **444**(2), 1041–1053 (2014). https://doi.org/10.1093/mnras/stu1491

I. Khabibullin, S. Sazonov, R. Sunyaev, SRG/eROSITA prospects for the detection of stellar tidal disruption flares. Mon. Not. R. Astron. Soc. **437**(1), 327–337 (2014). https://doi.org/10.1093/mnras/stt1889

S. Komossa, Ludwig Biermann award lecture: X-ray evidence for supermassive black holes at the centers of nearby, non-active galaxies. Rev. Mod. Astron. **15**, 27 (2002). https://ui.adsabs.harvard.edu/abs/2002RvMA...15...27K

S. Komossa, The extremes of (X-ray) variability among galaxies: flares from stars tidally disrupted by supermassive black holes, in *The Interplay Among Black Holes, Stars and ISM in Galactic Nuclei*, ed. by T. Storchi-Bergmann, L.C. Ho, H.R. Schmitt IAU Symposium, vol. 222 (2004), pp. 45–48. https://doi.org/10.1017/S1743921304001425

S. Komossa, Growing black holes: observational evidence for stellar tidal disruption events, in *Growing Black Holes: Accretion in a Cosmological Context*, ed. by A. Merloni, S. Nayakshin, R.A. Sunyaev (2005), pp. 159–163. https://doi.org/10.1007/11403913_27

S. Komossa, N. Bade, The giant X-ray outbursts in NGC 5905 and IC 3599: follow-up observations and outburst scenarios. Astron. Astrophys. **343**, 775–787 (1999). https://ui.adsabs.harvard.edu/abs/1999A%26A...343..775K

S. Komossa, M. Dahlem, X-ray outbursts from nearby 'normal' and active galaxies. A review, new radio observations, and an X-ray search for further tidal disruption flares. MAXI workshop on AGN. ArXiv e-prints, arXiv:astro-ph/0106422 (2001)

S. Komossa, J. Greiner, Discovery of a giant and luminous X-ray outburst from the optically inactive galaxy pair RX J1242.6-1119. Astron. Astrophys. **349**, L45–L48 (1999). https://ui.adsabs.harvard.edu/abs/1999A%26A...349L..45K

S. Komossa, D. Merritt, Tidal disruption flares from recoiling supermassive black holes. Astrophys. J. Lett. **683**(1), 21–24 (2008). https://doi.org/10.1086/591420

S. Komossa, J.A. Zensus, Compact object mergers: observations of supermassive binary black holes and stellar tidal disruption events, in *Star Clusters and Black Holes in Galaxies Across Cosmic Time*, ed. by Y. Meiron, S. Li, F.-K. Liu, R. Spurzem IAU Symposium, vol. 312 (2016), pp. 13–25. https://doi.org/10.1017/S1743921315007395

S. Komossa, J. Halpern, N. Schartel, G. Hasinger, M. Santos-Lleo, P. Predehl, A huge drop in the X-ray luminosity of the nonactive galaxy RX J1242.6-1119A, and the first postflare spectrum: testing the tidal disruption scenario. Astrophys. J. Lett. **603**(1), 17–20 (2004). https://doi.org/10.1086/382046

S. Komossa, H. Zhou, T. Wang, M. Ajello, J. Ge, J. Greiner, H. Lu, M. Salvato, R. Saxton, H. Shan, D. Xu, W. Yuan, Discovery of superstrong, fading, iron line emission and double-peaked Balmer lines of the galaxy SDSS J095209.56+214313.3: the light echo of a huge flare. Astrophys. J. Lett. **678**(1), 13–16 (2008). https://doi.org/10.1086/588281

S. Komossa, H. Zhou, A. Rau, M. Dopita, A. Gal-Yam, J. Greiner, J. Zuther, M. Salvato, D. Xu, H. Lu, NTT, Spitzer, and Chandra spectroscopy of SDSSJ095209.56+214313.3: the most luminous coronal-line supernova ever observed, or a stellar tidal disruption event? Astrophys. J. **701**(1), 105–121 (2009). https://doi.org/10.1088/0004-637X/701/1/105

A.J. Levan, N.R. Tanvir, S.B. Cenko, D.A. Perley, K. Wiersema, J.S. Bloom, A.S. Fruchter, A.d.U. Postigo, P.T. O'Brien, N. Butler, A.J. van der Horst, G. Leloudas, A.N. Morgan, K. Misra, G.C. Bower, J. Farihi, R.L. Tunnicliffe, M. Modjaz, J.M. Silverman, J. Hjorth, C. Thöne, A. Cucchiara, J.M.C. Cerón, A.J. Castro-Tirado, J.A. Arnold, M. Bremer, J.P. Brodie, T. Carroll, M.C. Cooper, P.A. Curran, R.M. Cutri, J. Ehle, D. Forbes, J. Fynbo, J. Gorosabel, J. Graham, D.I. Hoffman, S. Guziy, P. Jakobsson, A. Kamble, T. Kerr, M.M. Kasliwal, C. Kouveliotou, D. Kocevski, N.M. Law, P.E. Nugent, E.O. Ofek, D. Poznanski, R.M. Quimby, E. Rol, A.J. Romanowsky, R. Sánchez-Ramírez, S. Schulze, N. Singh, L. van Spaandonk, R.L.C. Starling, R.G. Strom, J.C. Tello, O. Vaduvescu, P.J. Wheatley, R.A.M.J. Wijers, J.M. Winters, D. Xu, An extremely luminous panchromatic outburst from the nucleus of a distant galaxy. Science **333**, 199 (2011). https://doi.org/10.1126/science.1207143

A.J. Levan, N.R. Tanvir, R.L.C. Starling, K. Wiersema, K.L. Page, D.A. Perley, S. Schulze, G.A. Wynn, R. Chornock, J. Hjorth, S.B. Cenko, A.S. Fruchter, P.T. O'Brien, G.C. Brown, R.L. Tunnicliffe, D. Malesani, P. Jakobsson, D. Watson, E. Berger, D. Bersier, B.E. Cobb, S. Covino, A. Cucchiara, A. de Ugarte Postigo, D.B. Fox, A. Gal-Yam, P. Goldoni, J. Gorosabel, L. Kaper, T. Krühler, R. Karjalainen, J.P. Osborne, E. Pian, R. Sánchez-Ramírez, B. Schmidt, I. Skillen, G. Tagliaferri, C. Thöne, O. Vaduvescu, R.A.M.J. Wijers, B.A. Zauderer, A new population of ultra-long duration gamma-ray bursts. Astrophys. J. **781**(1), 13 (2014). https://doi.org/10.1088/0004-637X/781/1/13

A.J. Levan, N.R. Tanvir, G.C. Brown, B.D. Metzger, K.L. Page, S.B. Cenko, P.T. O'Brien, J.D. Lyman, K. Wiersema, E.R. Stanway, A.S. Fruchter, D.A. Perley, J.S. Bloom, Late time multi-wavelength observations of Swift J1644+5734: a luminous optical/IR bump and quiescent X-ray emission. Astrophys. J. **819**, 51 (2016). https://doi.org/10.3847/0004-637X/819/1/51

L.-X. Li, R. Narayan, K. Menou, The giant X-ray flare of NGC 5905: tidal disruption of a star, a brown dwarf, or a planet? Astrophys. J. **576**(2), 753–761 (2002). https://doi.org/10.1086/341890

D. Lin, E.R. Carrasco, D. Grupe, N.A. Webb, D. Barret, S.A. Farrell, Discovery of an ultrasoft X-ray transient source in the 2XMM catalog: a tidal disruption event candidate. Astrophys. J. **738**, 52 (2011). https://doi.org/10.1088/0004-637X/738/1/52

D. Lin, P.W. Maksym, J.A. Irwin, S. Komossa, N.A. Webb, O. Godet, D. Barret, D. Grupe, S.D.J. Gwyn, An ultrasoft X-ray flare from 3XMM J152130.7+074916: a tidal disruption event candidate. Astrophys. J. **811**, 43 (2015). https://doi.org/10.1088/0004-637X/811/1/43

D. Lin, J. Guillochon, S. Komossa, E. Ramirez-Ruiz, J.A. Irwin, W.P. Maksym, D. Grupe, O. Godet, N.A. Webb, D. Barret, B.A. Zauderer, P.-A. Duc, E.R. Carrasco, S.D.J. Gwyn, A likely decade-long sustained tidal disruption event. Nat. Astron. **1**, 0033 (2017a). https://doi.org/10.1038/s41550-016-0033

D. Lin, O. Godet, L.C. Ho, D. Barret, N.A. Webb, J.A. Irwin, Large decay of X-ray flux in 2XMM J123103.2+110648: evidence for a tidal disruption event. Mon. Not. R. Astron. Soc. **468**, 783–789 (2017b). https://doi.org/10.1093/mnras/stx489

D. Lin, J. Strader, E.R. Carrasco, D. Page, A.J. Romanowsky, J. Homan, J.A. Irwin, R.A. Remillard, O. Godet, N.A. Webb, H. Baumgardt, R. Wijnands, D. Barret, P.-A. Duc, J.P. Brodie, S.D.J. Gwyn, A luminous X-ray outburst from an intermediate-mass black hole in an off-centre star cluster. Nat. Astron. **2**, 656–661 (2018a). https://doi.org/10.1038/s41550-018-0493-1

D. Lin, J. Strader, E.R. Carrasco, O. Godet, D. Grupe, N.A. Webb, D. Barret, J.A. Irwin, Multiwavelength follow-up observations of the tidal disruption event candidate 2XMMi J184725.1-631724. Mon. Not. R. Astron. Soc. **474**, 3000–3008 (2018b). https://doi.org/10.1093/mnras/stx2940

Q.Z. Liu, J. van Paradijs, E.P.J. van den Heuvel, A catalogue of high-mass X-ray binaries. Astron. Astrophys. Suppl. Ser. **147**, 25–49 (2000). https://doi.org/10.1051/aas:2000288

Q.Z. Liu, J. van Paradijs, E.P.J. van den Heuvel, A catalogue of low-mass X-ray binaries. Astron. Astrophys. **368**, 1021–1054 (2001). https://doi.org/10.1051/0004-6361:20010075

F.K. Liu, S. Li, X. Chen, Interruption of tidal-disruption flares by supermassive black hole binaries. Astrophys. J. **706**(1), 133–137 (2009). https://doi.org/10.1088/0004-637X/706/1/L133

F.K. Liu, S. Li, S. Komossa, A milliparsec supermassive black hole binary candidate in the galaxy SDSS J120136.02+300305.5. Astrophys. J. **786**, 103 (2014). https://doi.org/10.1088/0004-637X/786/2/103

X.-L. Liu, L.-M. Dou, R.-F. Shen, J.-H. Chen, The UV/optical peak and X-ray brightening in TDE candidate AT2019azh: a case of stream-stream collision and delayed accretion. ArXiv e-prints, arXiv:1912.06081 (2019)

G. Lodato, Challenges in the modeling of tidal disruption events lightcurves, in *Eur. Phys. J. Web Conf.*. European Physical Journal Web of Conferences, vol. 39 (2012), p. 01001. https://doi.org/10.1051/epjconf/20123901001

G. Lodato, E.M. Rossi, Multiband light curves of tidal disruption events. Mon. Not. R. Astron. Soc. **410**(1), 359–367 (2011). http://doi.wiley.com/10.1111/j.1365-2966.2010.17448.x

A. Loeb, A. Ulmer, Optical appearance of the debris of a star disrupted by a massive black hole. Astrophys. J. **489**(2), 573–578 (1997). https://doi.org/10.1086/304814

M. MacLeod, J. Guillochon, E. Ramirez-Ruiz, The tidal disruption of giant stars and their contribution to the flaring supermassive black hole population. Astrophys. J. **757**(2), 134 (2012). https://doi.org/10.1088/0004-637X/757/2/134

M. MacLeod, J. Guillochon, E. Ramirez-Ruiz, D. Kasen, S. Rosswog, Optical thermonuclear transients from tidal compression of white dwarfs as tracers of the low end of the massive black hole mass function. Astrophys. J. **819**, 3 (2016). https://doi.org/10.3847/0004-637X/819/1/3

K. Makishima, Y. Maejima, K. Mitsuda, H.V. Bradt, R.A. Remillard, I.R. Tuohy, R. Hoshi, M. Nakagawa, Simultaneous X-ray and optical observations of GX 339-4 in an X-ray high state. Astrophys. J. **308**, 635–643 (1986). https://doi.org/10.1086/164534

W.P. Maksym, M.P. Ulmer, M. Eracleous, A tidal disruption flare in A1689 from an archival X-ray survey of galaxy clusters. Astrophys. J. **722**(2), 1035–1050 (2010). https://doi.org/10.1088/0004-637X/722/2/1035

W.P. Maksym, M.P. Ulmer, M.C. Eracleous, L. Guennou, L.C. Ho, A tidal flare candidate in Abell 1795. Mon. Not. R. Astron. Soc. **435**(3), 1904–1927 (2013). https://doi.org/10.1093/mnras/stt1379

W.P. Maksym, D. Lin, J.A. Irwin, RBS 1032: a tidal disruption event in another dwarf galaxy? Astrophys. J. Lett. **792**, 29 (2014a). https://doi.org/10.1088/2041-8205/792/2/L29

W.P. Maksym, M.P. Ulmer, K.C. Roth, J.A. Irwin, R. Dupke, L.C. Ho, W.C. Keel, C. Adami, Deep spectroscopy of the $M_V \sim -14.8$ host galaxy of a tidal disruption flare in A1795. Mon. Not. R. Astron. Soc. **444**(1), 866–873 (2014b). https://doi.org/10.1093/mnras/stu1485

V. Mangano, D.N. Burrows, B. Sbarufatti, J.K. Cannizzo, The definitive X-ray light curve of Swift J164449.3+573451. Astrophys. J. **817**, 103 (2016). https://doi.org/10.3847/0004-637X/817/2/103

N.J. McConnell, C.-P. Ma, Revisiting the scaling relations of black hole masses and host galaxy properties. Astrophys. J. **764**, 184 (2013). https://doi.org/10.1088/0004-637X/764/2/184

B.D. Metzger, N.C. Stone, A bright year for tidal disruptions. Mon. Not. R. Astron. Soc. **461**, 948–966 (2016). https://doi.org/10.1093/mnras/stw1394

R. Mewe, E.H.B.M. Gronenschild, G.H.J. van den Oord, Calculated X-radiation from optically thin plasmas. V. Astron. Astrophys. Suppl. Ser. **62**, 197–254 (1985). http://ui.adsabs.harvard.edu/abs/1985A%26AS...62..197M

M.J. Middleton, J.C.A. Miller-Jones, S. Markoff, R. Fender, M. Henze, N. Hurley-Walker, A.M.M. Scaife, T.P. Roberts, D. Walton, J. Carpenter, Bright radio emission from an ultraluminous stellar-mass microquasar in M 31. Nature **493**(7431), 187–190 (2013). https://doi.org/10.1038/nature11697

J.M. Miller, J.S. Kaastra, M.C. Miller, M.T. Reynolds, G. Brown, S.B. Cenko, J.J. Drake, S. Gezari, J. Guillochon, K. Gultekin, J. Irwin, A. Levan, D. Maitra, W.P. Maksym, R. Mushotzky, P. O'Brien, F. Paerels, J. de Plaa, E. Ramirez-Ruiz, T. Strohmayer, N. Tanvir, Flows of X-ray gas reveal the disruption of a star by a massive black hole. Nature **526**, 542–545 (2015). https://doi.org/10.1038/nature15708

S. Mineo, M. Gilfanov, R. Sunyaev, X-ray emission from star-forming galaxies – I. High-mass X-ray binaries. Mon. Not. R. Astron. Soc. **419**, 2095–2115 (2012). https://doi.org/10.1111/j.1365-2966.2011.19862.x

G. Miniutti, R.D. Saxton, P.M. Rodríguez-Pascual, A.M. Read, P. Esquej, M. Colless, P. Dobbie, M. Spolaor, A high Eddington-ratio, true Seyfert 2 galaxy candidate: implications for broad-line region models. Mon. Not. R. Astron. Soc. **433**, 1764–1777 (2013). https://doi.org/10.1093/mnras/stt850

B. Mockler, J. Guillochon, E. Ramirez-Ruiz, Weighing Black Holes using Tidal Disruption Events. ArXiv e-prints, arXiv:1801.08221 (2018)

K. Nandra, K.A. Pounds, GINGA observations of the X-ray spectra of Seyfert galaxies. Mon. Not. R. Astron. Soc. **268**, 405 (1994). https://doi.org/10.1093/mnras/268.2.405

H. Netzer, *The Physics and Evolution of Active Galactic Nuclei.* (Cambridge University Press, Cambridge, 2013). https://doi.org/10.1017/CBO9781139109291

H. Netzer, Bolometric correction factors for active galactic nuclei. Mon. Not. R. Astron. Soc. **488**(4), 5185–5191 (2019). https://doi.org/10.1093/mnras/stz2016

M. Nikołajuk, R. Walter, Tidal disruption of a super-Jupiter by a massive black hole. Astron. Astrophys. **552**, 75 (2013). http://adsabs.harvard.edu/abs/2013A%26A...552A..75N

J. Nishimura, K. Mitsuda, M. Itoh, Comptonization of soft X-ray photons in an optically thin hot plasma. Publ. Astron. Soc. Jpn. **38**, 819–830 (1986)

D.E. Osterbrock, *Astrophysics of Gaseous Nebulae and Active Galactic Nuclei* (University Science Books, Mill Valley, 1989)

D.R. Pasham, S.B. Cenko, A. Sadowski, J. Guillochon, N.C. Stone, S. van Velzen, J.K. Cannizzo, Optical/UV-to-X-ray echoes from the tidal disruption flare ASASSN-14li. Astrophys. J. Lett. **837**, 30 (2017). https://doi.org/10.3847/2041-8213/aa6003

Z.-K. Peng, Y.-S. Yang, R.-F. Shen, L.-J. Wang, J.-H. Zou, B.-B. Zhang, CDF-S XT1 and XT2: white dwarf tidal disruption events by intermediate-mass black holes? Astrophys. J. Lett. **884**(2), 34 (2019). https://doi.org/10.3847/2041-8213/ab481b

E.S. Phinney, Manifestations of a massive black hole in the galactic center, in *The Center of the Galaxy*, ed. by M. Morris IAU Symposium, vol. 136 (1989), p. 543. https://ui.adsabs.harvard.edu/abs/1989IAUS..136..543P

T. Piran, A. Sądowski, A. Tchekhovskoy, Jet and disc luminosities in tidal disruption events. Mon. Not. R. Astron. Soc. **453**, 157–165 (2015a). https://doi.org/10.1093/mnras/stv1547

T. Piran, G. Svirski, J. Krolik, R.M. Cheng, H. Shiokawa, Disk formation versus disk accretion—what powers tidal disruption events? Astrophys. J. **806**(2), 164 (2015b). https://doi.org/10.1088/0004-637X/806/2/164

G. Ponti, I. Papadakis, S. Bianchi, M. Guainazzi, G. Matt, P. Uttley, N.F. Bonilla, CAIXA: a catalogue of AGN in the XMM-Newton archive. III. Excess variance analysis. Astron. Astrophys. **542**, 83 (2012). https://doi.org/10.1051/0004-6361/201118326

K.A. Pounds, J.N. Reeves, A.R. King, K.L. Page, P.T. O'Brien, M.J.L. Turner, A high-velocity ionized outflow and XUV photosphere in the narrow emission line quasar PG1211+143. Mon. Not. R. Astron. Soc. **345**(3), 705–713 (2003). https://doi.org/10.1046/j.1365-8711.2003.07006.x

P. Predehl, H. Böhringer, H. Brunner, M. Brusa, V. Burwitz, N. Cappelluti, E. Churazov, K. Dennerl, M. Freyberg, P. Friedrich, eROSITA on SRG, in *X-ray Astronomy 2009; Present Status, Multi-Wavelength Approach and Future Perspectives*, ed. by A. Comastri, L. Angelini, M. Cappi American Institute of Physics Conference Series, vol. 1248 (2010), pp. 543–548. https://doi.org/10.1063/1.3475336

J.P. Pye, I.M. McHardy, The Ariel V sky survey of fast-transient X-ray sources. Mon. Not. R. Astron. Soc. **205**, 875–888 (1983). https://doi.org/10.1093/mnras/205.3.875

E. Quataert, D. Kasen, Swift 1644+57: the longest gamma-ray burst? Mon. Not. R. Astron. Soc. **419**, 1–5 (2012). https://doi.org/10.1111/j.1745-3933.2011.01151.x

M.J. Rees, Tidal disruption of stars by black holes of 10 to the 6th-10 to the 8th solar masses in nearby galaxies. Nature **333**, 523 (1988). http://adsabs.harvard.edu/abs/1988Natur.333..523R

M.J. Rees, "Dead quasars" in nearby galaxies? Science **247**(4944), 817–823 (1990). https://doi.org/10.1126/science.247.4944.817

A.E. Reines, M. Volonteri, Relations between central black hole mass and total galaxy stellar mass in the local universe. Astrophys. J. **813**, 82 (2015). https://doi.org/10.1088/0004-637X/813/2/82

S.R. Rosen, N.A. Webb, M.G. Watson, J. Ballet, D. Barret, V. Braito, F.J. Carrera, M.T. Ceballos, M. Coriat, R. Della Ceca, G. Denkinson, P. Esquej, S.A. Farrell, M. Freyberg, F. Grisé, P. Guillout, L. Heil, F. Koliopanos, D. Law-Green, G. Lamer, D. Lin, R. Martino, L. Michel, C. Motch, A. Nebot Gomez-Moran, C.G. Page, K. Page, M. Page, M.W. Pakull, J. Pye, A. Read, P. Rodriguez, M. Sakano, R. Saxton, A. Schwope, A.E. Scott, R. Sturm, I. Traulsen, V. Yershov, I. Zolotukhin, The XMM-Newton serendipitous survey. VII. The third XMM-Newton serendipitous source catalogue. Astron. Astrophys. **590**, 1 (2016). https://doi.org/10.1051/0004-6361/201526416

N. Roth, D. Kasen, What sets the line profiles in tidal disruption events? Astrophys. J. **855**(1), 54 (2018). https://doi.org/10.3847/1538-4357/aaaec6

D.M. Russell, J.C.A. Miller-Jones, T.J. Maccarone, Y.J. Yang, R.P. Fender, F. Lewis, Testing the jet quenching paradigm with an ultradeep observation of a steadily soft state black hole. Astrophys. J. Lett. **739**, 19 (2011). https://doi.org/10.1088/2041-8205/739/1/L19

P. Salucci, C. Ratnam, P. Monaco, L. Danese, The masses of black holes in the nuclei of spirals. Mon. Not. R. Astron. Soc. **317**(2), 488–496 (2000). https://doi.org/10.1046/j.1365-8711.2000.03622.x

W.L.W. Sargent, P.J. Young, A. Boksenberg, K. Shortridge, C.R. Lynds, F.D.A. Hartwick, Dynamical evidence for a central mass concentration in the galaxy M87. Astrophys. J. **221**, 731–744 (1978). https://doi.org/10.1086/156077

R.D. Saxton, A.M. Read, P. Esquej, M.J. Freyberg, B. Altieri, D. Bermejo, The first XMM-Newton slew survey catalogue: XMMSL1. Astron. Astrophys. **480**, 611–622 (2008). https://doi.org/10.1051/0004-6361:20079193

R.D. Saxton, A.M. Read, P. Esquej, S. Komossa, S. Dougherty, P. Rodriguez-Pascual, D. Barrado, A tidal disruption-like X-ray flare from the quiescent galaxy SDSS J120136.02+300305.5. Astron. Astrophys. **541**, 106 (2012a). https://doi.org/10.1051/0004-6361/201118367

R.D. Saxton, A.M. Read, S. Komossa, P. Esquej, A well-monitored, X-ray selected, tidal disruption event. Eur. Phys. J. Web Conf. **39**, 02002 (2012b). https://doi.org/10.1051/epjconf/20123902002

R.D. Saxton, S.E. Motta, S. Komossa, A.M. Read, Was the soft X-ray flare in NGC 3599 due to an AGN disc instability or a delayed tidal disruption event? Mon. Not. R. Astron. Soc. **454**, 2798–2803 (2015). https://doi.org/10.1093/mnras/stv2160

R.D. Saxton, A.M. Read, S. Komossa, P. Lira, K.D. Alexander, M.H. Wieringa, XMMSL1 J074008.2-853927: a tidal disruption event with thermal and non-thermal components. Astron. Astrophys. **598**, 29 (2017). https://doi.org/10.1051/0004-6361/201629015

R.D. Saxton, A.M. Read, S. Komossa, P. Lira, K.D. Alexander, I. Steele, F. Ocaña, E. Berger, P. Blanchard, XMMSL2 J144605.0+685735: a slow tidal disruption event. ArXiv e-prints, arXiv:1908.01065 (2019)

H. Schulz, C. Fritsch, Possible continuum-diluting processes to generate Liner-like spectra. Astron. Astrophys. **291**, 713–719 (1994). https://ui.adsabs.harvard.edu/abs/1994A%26A...291..713S

S. Sembay, R.G. West, Black hole remnants: soft X-ray flares from tidally disrupted stars. Mon. Not. R. Astron. Soc. **262**, 141–150 (1993). https://doi.org/10.1093/mnras/262.1.141

B.J. Shappee, J.L. Prieto, D. Grupe, C.S. Kochanek, K.Z. Stanek, G. De Rosa, S. Mathur, Y. Zu, B.M. Peterson, R.W. Pogge, S. Komossa, M. Im, J. Jencson, T.W.-S. Holoien, U. Basu, J.F. Beacom, D.M. Szczygieł, J. Brimacombe, S. Adams, A. Campillay, C. Choi, C. Contreras, M. Dietrich, M. Dubberley, M. Elphick, S. Foale, M. Giustini, C. Gonzalez, E. Hawkins, D.A. Howell, E.Y. Hsiao, M. Koss, K.M. Leighly, N. Morrell, D. Mudd, D. Mullins, J.M. Nugent, J. Parrent, M.M. Phillips, G. Pojmanski, W. Rosing, R. Ross, D. Sand, D.M. Terndrup, S. Valenti, Z. Walker, Y. Yoon, The man behind the curtain:

X-rays drive the UV through NIR variability in the 2013 active galactic nucleus outburst in NGC 2617. Astrophys. J. **788**, 48 (2014). https://doi.org/10.1088/0004-637X/788/1/48
R.-F. Shen, Fast, ultra-luminous X-ray bursts from tidal stripping of White Dwarfs by Intermediate-mass Black Holes. ArXiv e-prints, arXiv:1809.09359 (2018)
H. Shiokawa, J.H. Krolik, R.M. Cheng, T. Piran, S.C. Noble, General relativistic hydrodynamic simulation of accretion flow from a stellar tidal disruption. Astrophys. J. **804**, 85 (2015). https://doi.org/10.1088/0004-637X/804/2/85
X.W. Shu, S.S. Wang, L.M. Dou, N. Jiang, J.X. Wang, T.G. Wang, A long decay of X-ray flux and spectral evolution in the supersoft active galactic nucleus GSN 069. Astrophys. J. Lett. **857**, 16 (2018). https://doi.org/10.3847/2041-8213/aaba17
I.V. Strateva, S. Komossa, The X-ray point-source population of NGC 1365: the puzzle of two highly-variable ultraluminous X-ray sources. Astrophys. J. **692**, 443 (2009). https://doi.org/10.1088/0004-637X/692/1/443
N.L. Strotjohann, R.D. Saxton, R.L.C. Starling, P. Esquej, A.M. Read, P.A. Evans, G. Miniutti, Highly variable AGN from the XMM-Newton slew survey. Astron. Astrophys. **592**, 74 (2016). https://doi.org/10.1051/0004-6361/201628241
L.E. Strubbe, E. Quataert, Optical flares from the tidal disruption of stars by massive black holes. Mon. Not. R. Astron. Soc. **400**(4), 2070–2084 (2009). https://doi.org/10.1111/j.1365-2966.2009.15599.x
L. Strüder, U. Briel, K. Dennerl, R. Hartmann, E. Kendziorra, N. Meidinger, E. Pfeffermann, C. Reppin, B. Aschenbach, W. Bornemann, H. Br äuninger, W. Burkert, M. Elender, M. Freyberg, F. Haberl, G. Hartner, F. Heuschmann, H. Hippmann, E. Kastelic, S. Kemmer, G. Kettenring, W. Kink, N. Krause, S. Müller, A. Oppitz, W. Pietsch, M. Popp, P. Predehl, A. Read, K.H. Stephan, D. Stötter, J. Trümper, P. Holl, J. Kemmer, H. Soltau, R. Stötter, U. Weber, U. Weichert, C. von Zanthier, D. Carathanassis, G. Lutz, R.H. Richter, P. Solc, H. Böttcher, M. Kuster, R. Staubert, A. Abbey, A. Holland, M. Turner, M. Balasini, G.F. Bignami, N. La Palombara, G. Villa, W. Buttler, F. Gianini, R. Lainé, D. Lumb, P. Dhez, The European photon imaging camera on XMM-Newton: the pn-CCD camera. Astron. Astrophys. **365**, 18–26 (2001). https://doi.org/10.1051/0004-6361:20000066
H. Sun, B. Zhang, Z. Li, Extragalactic high-energy transients: event rate densities and luminosity functions. Astrophys. J. **812**(1), 33 (2015). https://doi.org/10.1088/0004-637X/812/1/33
Y. Terashima, N. Kamizasa, H. Awaki, A. Kubota, Y. Ueda, A candidate active galactic nucleus with a pure soft thermal X-ray spectrum. Astrophys. J. **752**, 154 (2012). https://doi.org/10.1088/0004-637X/752/2/154
L. Titarchuk, Generalized comptonization models and application to the recent high-energy observations. Astrophys. J. **434**, 570 (1994). https://doi.org/10.1086/174760
J. Truemper, The ROSAT mission. Adv. Space Res. **2**(4), 241–249 (1982). https://doi.org/10.1016/0273-1177(82)90070-9
T.J. Turner, K.A. Pounds, The EXOSAT spectral survey of AGN. Mon. Not. R. Astron. Soc. **240**, 833–880 (1989). https://doi.org/10.1093/mnras/240.4.833
A. Ulmer, Flares from the tidal disruption of stars by massive black holes. Astrophys. J. **514**, 180–187 (1999). https://doi.org/10.1086/306909
S. van Velzen, E. Körding, H. Falcke, Radio jets from stellar tidal disruptions. Mon. Not. R. Astron. Soc. **417**(1), 51–55 (2011a). https://doi.org/10.1111/j.1745-3933.2011.01118.x
S. van Velzen, G.R. Farrar, S. Gezari, N. Morrell, D. Zaritsky, L. Östman, M. Smith, J. Gelfand, A.J. Drake, Optical discovery of probable stellar tidal disruption flares. Astrophys. J. **741**(2), 73 (2011b). https://doi.org/10.1088/0004-637X/741/2/73
S. van Velzen, N.C. Stone, B.D. Metzger, S. Gezari, T.M. Brown, A.S. Fruchter, Late-time UV observations of tidal disruption flares reveal unobscured, compact accretion disks. Astrophys. J. **878**(2), 82 (2019a). https://doi.org/10.3847/1538-4357/ab1844
S. van Velzen, S. Gezari, S.B. Cenko, E. Kara, J.C.A. Miller-Jones, T. Hung, J. Bright, N. Roth, N. Blagorodnova, D. Huppenkothen, L. Yan, E. Ofek, J. Sollerman, S. Frederick, C. Ward, M.J. Graham, R. Fender, M.M. Kasliwal, C. Canella, R. Stein, M. Giomi, V. Brinnel, J. van Santen, J. Nordin, E.C. Bellm, R. Dekany, C. Fremling, V.Z. Golkhou, T. Kupfer, S.R. Kulkarni, R.R. Laher, A. Mahabal, F.J. Masci, A.A. Miller, J.D. Neill, R. Riddle, M. Rigault, B. Rusholme, M.T. Soumagnac, Y. Tachibana, The first tidal disruption flare in ZTF: from photometric selection to multi-wavelength characterization. Astrophys. J. **872**(2), 198 (2019b). https://doi.org/10.3847/1538-4357/aafe0c
S. Vaughan, R. Edelson, R.S. Warwick, Chandra observations of five X-ray transient galactic nuclei. Mon. Not. R. Astron. Soc. **349**, 1–5 (2004). https://doi.org/10.1111/j.1365-2966.2004.07615.x
W. Voges, B. Aschenbach, T. Boller, H. Bräuninger, U. Briel, W. Burkert, K. Dennerl, J. Englhauser, R. Gruber, F. Haberl, The ROSAT all-sky survey bright source catalogue. Astron. Astrophys. **349**, 389–405 (1999). https://ui.adsabs.harvard.edu/abs/1999A%26A...349..389V

J. Wang, D. Merritt, Revised rates of stellar disruption in galactic nuclei. Astrophys. J. **600**(1), 149–161 (2004). https://doi.org/10.1086/379767

T.-G. Wang, H.-Y. Zhou, L.-F. Wang, H.-L. Lu, D. Xu, Transient superstrong coronal lines and broad bumps in the galaxy SDSS J074820.67+471214.3. Astrophys. J. **740**(2), 85 (2011). https://doi.org/10.1088/0004-637X/740/2/85

T.-G. Wang, H.-Y. Zhou, S. Komossa, H.-Y. Wang, W. Yuan, C. Yang, Extreme coronal line emitters: tidal disruption of stars by massive black holes in galactic nuclei? Astrophys. J. **749**(2), 115 (2012). https://doi.org/10.1088/0004-637X/749/2/115

M.C. Weisskopf, H.D. Tananbaum, L.P. Van Speybroeck, S.L. O'Dell, Chandra X-ray Observatory (CXO): overview, in *X-Ray Optics, Instruments, and Missions III*, ed. by J.E. Truemper, B. Aschenbach Society of Photo-Optical Instrumentation Engineers (SPIE) Conference Series, vol. 4012 (2000), pp. 2–16. https://doi.org/10.1117/12.391545

T. Wevers, S. van Velzen, P.G. Jonker, N.C. Stone, T. Hung, F. Onori, S. Gezari, N. Blagorodnova, Black hole masses of tidal disruption event host galaxies. Mon. Not. R. Astron. Soc. **471**, 1694–1708 (2017). https://doi.org/10.1093/mnras/stx1703

T. Wevers, D.R. Pasham, S. van Velzen, G. Leloudas, S. Schulze, J.C.A. Miller-Jones, P.G. Jonker, M. Gromadzki, E. Kankare, S.T. Hodgkin, Ł. Wyrzykowski, Z. Kostrzewa-Rutkowska, S. Moran, M. Berton, K. Maguire, F. Onori, S. Mattila, M. Nicholl, Evidence for rapid disc formation and reprocessing in the X-ray bright tidal disruption event candidate AT 2018fyk. Mon. Not. R. Astron. Soc. **488**(4), 4816–4830 (2019a). https://doi.org/10.1093/mnras/stz1976

T. Wevers, N.C. Stone, S. van Velzen, P.G. Jonker, T. Hung, K. Auchettl, S. Gezari, F. Onori, D. Mata Sánchez, Z. Kostrzewa-Rutkowska, J. Casares, Black hole masses of tidal disruption event host galaxies II. Mon. Not. R. Astron. Soc. **487**(3), 4136–4152 (2019b). https://doi.org/10.1093/mnras/stz1602

Y.Q. Xue, X.C. Zheng, Y. Li, W.N. Brandt, B. Zhang, B. Luo, B.-B. Zhang, F.E. Bauer, H. Sun, B.D. Lehmer, X.-F. Wu, G. Yang, X. Kong, J.Y. Li, M.Y. Sun, J.-X. Wang, F. Vito, A magnetar-powered X-ray transient as the aftermath of a binary neutron-star merger. Nature **568**(7751), 198–201 (2019). https://doi.org/10.1038/s41586-019-1079-5

A. Yalinewich, J. Guillochon, R. Sari, A. Loeb, Shock Breakouts from Tidal Disruption Events. ArXiv e-prints, arxiv:1808.10447 (2018)

W. Yuan, C. Zhang, H. Feng, S.N. Zhang, Z.X. Ling, D. Zhao, J. Deng, Y. Qiu, J.P. Osborne, P. O'Brien, Einstein Probe—a small mission to monitor and explore the dynamic X-ray Universe. ArXiv e-prints, arXiv:1506.07735 (2015)

W. Yuan, S. Komossa, C. Zhang, H. Feng, Z.-X. Ling, D.H. Zhao, S.-N. Zhang, J.P. Osborne, P. O'Brien, R. Willingale, Detecting tidal disruption events of massive black holes in normal galaxies with the Einstein Probe, in *Star Clusters and Black Holes in Galaxies Across Cosmic Time*, ed. by Y. Meiron, S. Li, F.-K. Liu, R. Spurzem IAU Symposium, vol. 312 (2016), pp. 68–70. https://doi.org/10.1017/S1743921315007516

I. Zalamea, K. Menou, A.M. Beloborodov, White dwarfs stripped by massive black holes: sources of coincident gravitational and electromagnetic radiation. Mon. Not. R. Astron. Soc. **409**, 25–29 (2010). https://doi.org/10.1111/j.1745-3933.2010.00930.x

B.A. Zauderer, E. Berger, R. Margutti, G.G. Pooley, R. Sari, A.M. Soderberg, A. Brunthaler, M.F. Bietenholz, Radio monitoring of the tidal disruption event Swift J164449.3+573451. II. The relativistic jet shuts off and a transition to forward shock X-ray/radio emission. Astrophys. J. **767**(2), 152 (2013). https://doi.org/10.1088/0004-637X/767/2/152

Space Sci Rev (2020) 216:124
https://doi.org/10.1007/s11214-020-00753-z

Optical-Ultraviolet Tidal Disruption Events

Sjoert van Velzen[1,2] · Thomas W.-S. Holoien[3] · Francesca Onori[4] · Tiara Hung[5] · Iair Arcavi[6,7]

Received: 25 April 2020 / Accepted: 13 October 2020 / Published online: 27 October 2020

Abstract The existence of optical-ultraviolet Tidal Disruption Events (TDEs) could be considered surprising because their electromagnetic output was originally predicted to be dominated by X-ray emission from an accretion disk. Yet over the last decade, the growth of optical transient surveys has led to the identification of a new class of optical transients occurring exclusively in galaxy centers, many of which are considered to be TDEs. Here we review the observed properties of these events, identified based on a shared set of both photometric and spectroscopic properties. We present a homogeneous analysis of 33 sources that we classify as robust TDEs, and which we divide into classes. The criteria used here to classify TDEs will possibly get updated as new samples are collected and potential additional diversity of TDEs is revealed. We also summarize current measurements of the optical-ultraviolet TDE rate, as well as the mass function and luminosity function. Many open questions exist regarding the current sample of events. We anticipate that the search for answers will unlock new insights in a variety of fields, from accretion physics to galaxy evolution.

The Tidal Disruption of Stars by Massive Black Holes
Edited by Peter G. Jonker, Sterl Phinney, Elena Maria Rossi, Sjoert van Velzen, Iair Arcavi and Maurizio Falanga

✉ I. Arcavi
arcavi@tauex.tau.ac.il

1 Center for Cosmology and Particle Physics, New York University, New York, NY 10003, USA

2 Department of Astronomy, University of Maryland, College Park, MD 20742, USA

3 The Observatories of the Carnegie Institution for Science, 813 Santa Barbara St., Pasadena, CA 91101, USA

4 Istituto di Astrofisica e Planetologia Spaziali (INAF), via del Fosso del Cavaliere 100, Roma, 00133, Italy

5 Department of Astronomy and Astrophysics, University of California, Santa Cruz, California, CA 95064, USA

6 The School of Physics and Astronomy, Tel Aviv University, Tel Aviv 69978, Israel

7 CIFAR Azrieli Global Scholars program, CIFAR, Toronto, Canada

Keywords Tidal disruption events: optical · Tidal disruption events: ultraviolet, Bowen fluorescence

1 Introduction

The first predictions for TDE observables considered direct emission from an accretion disk. Therefore, it was expected that TDEs be seen mainly in the X-rays. Indeed, the first TDEs were detected during the *ROSAT* all sky survey as luminous, soft X-ray flares from the nuclei of otherwise quiescent galaxies (Bade et al. 1996; Komossa and Greiner 1999; Grupe et al. 1999; Greiner et al. 2000). Subsequently, similar X-ray events were discovered through dedicated searches or serendipitous discoveries with *Chandra* and *XMM-Newton* (Esquej et al. 2007, 2008; Lin et al. 2011; Saxton et al. 2012). *Swift* has been extremely useful in providing rapid follow-up observations for many of these events, and for discovering a new class of higher energy gamma-ray events. A detailed description of TDE characteristics in the X-rays and gamma-rays can be found in the X-ray Chapter and the Gamma-ray Chapter in this book, respectively.

In the last decade, several transients discovered in optical and ultraviolet wavelengths have been attributed to TDEs. The source of TDE emission at these wavelengths is currently debated to be either reprocessing of the X-ray emission from the accretion disk by optically thick material surrounding the disk (e.g. Guillochon and Ramirez-Ruiz 2013; Roth et al. 2016), or as emission from outer shocks between the debris streams as they collide (e.g. Piran et al. 2015), or perhaps some combination of both (e.g. Jiang et al. 2016; Lu and Bonnerot 2020). A detailed discussion of these emission mechanisms is presented in the Emission Mechanisms Chapter in this book. Here we focus on the observational properties of the main classes of optical and ultraviolet transients suggested in the literature to be TDEs. Additional one-of-a-kind events, whose nature is less clear, are discussed in the One-off Events and Imposters Chapter.

The main class of optical-ultraviolet TDEs contains approximately 30 events (Table 1) and is identified mainly by broad ($\sim 10{,}000\ \mathrm{km\,s^{-1}}$) H and/or He lines in their spectra (with some events showing additional features), blue-continuum ($\sim$ few $\times\ 10^4$ K) emission lasting for several months, and a $\sim$month-long rise to a peak optical absolute magnitude of approximately -20. A diversity within this class does exist, and a full unbiased view of the population is not yet available. This class of events show a surprising and still unexplained strong preference for rare post-starburst (or E+A) host galaxies (see Host Galaxies Chapter), while almost none have been found in active galactic nuclei (AGN).[1] There are obvious observational difficulties in finding TDE flares embedded in AGN light, and also in discerning TDEs from rare AGN flares (e.g. Trakhtenbrot et al. 2019a,b). Chan et al. (2019) recently suggested that TDEs in AGN will look different than those in non-active galaxies due to the interaction of the stellar debris with the pre-existing accretion disk. Distinguishing TDEs from AGN is discussed in more detail in the One-off Events and Imposters Chapter.

Here we present a current picture of the main class of established events. However, it's important to keep in mind that with new events being discovered regularly, the map of optical-ultraviolet TDEs is still being drawn. While we refer to all of the events discussed here as TDEs, some might still be considered "TDE candidates" until a better picture of the population of real events and their imposters emerges.

[1] A notable candidate exception is PS16dtm (Blanchard et al. 2017) which is discussed in the One-off Events and Imposters Chapter.

In the next section, we begin with a brief review of the history of optical-ultraviolet TDE discoveries in order to provide a picture of the different biases affecting the current sample of events. We then focus on the photometric (Sect. 3) and spectroscopic (Sect. 4) properties of this sample and discuss how a robust sample of optical-ultraviolet TDEs can be constructed (Sect. 5). We then discuss rate measurements and the luminosity function of optical-ultraviolet TDEs (Sect. 6), and end with some concluding remarks (Sect. 7).

2 A Brief History of TDE Detections in Optical and Ultraviolet Surveys

The first optical-ultraviolet TDE candidate to be discovered was GALEX-D3-13 (Gezari et al. 2006); two more events (D1-9 and D23-H1) were reported in Gezari et al. (2008, 2009). These three events were all found using multi-epoch ultraviolet images from the Galaxy Evolution Explorer (GALEX; Martin et al. 2005) which were selected to have a rich set of archival data (e.g., the Groth field). Using this archival data, recurring AGN flares could be removed.

Simultaneous optical imaging is available for two of the GALEX flares, allowing for an accurate estimate of the blackbody temperature. All three TDEs found in GALEX data have relatively high blackbody temperatures ($T \approx 5 \times 10^4$ K). Since GALEX imaging is relatively deep (limiting magnitude 23 in the near-ultraviolet band), the GALEX search for TDEs has the greatest depth of all imaging searches to date (see Table 3 in van Velzen 2018). However this search was not done in real-time, hence spectroscopy of the flares was never obtained.

The GALEX transient search (Gezari et al. 2013) remains the only ulraviolet-based TDE search to date. Below we provide an overview of the optical surveys that have led to the discovery of TDEs that can be classified based on their spectral properties (as listed in Table 1).

2.1 SDSS

The Sloan Digital Sky Survey (SDSS; York et al. 2000) provided two types of TDE candidates. First, the SDSS catalog of galaxy spectra (Strauss et al. 2002) enabled the discovery of coronal-line TDEs (e.g. Komossa et al. 2008; Wang et al. 2011, 2012; Yang et al. 2013). For these sources, the TDE is not observed directly, but rather transient narrow high-ionization emission lines[2] are seen, and are interpreted as an "echo" of a soft X-ray flare (required to produce the coronal lines) originating in a TDE. These events, as well as echoes showing broad emission lines (e.g. Komossa et al. 2008), are discussed in more detail the Echo Chapter of this book. Here we note that due to the lack of multi-epoch imaging observations that cover the flare itself, our knowledge of the photometric properties of these events (e.g., timescales, location in the host galaxy, and blackbody temperature) is limited.

Second, SDSS provided the first TDEs selected directly using optical imaging observations (van Velzen et al. 2011). These were discovered in Stripe 82 (Frieman et al. 2008), a region of ≈ 300 deg^2 that was observed with a cadence of $\sim$ few days for three months form 2005 to 2007 (lower cadence observation date back to 2000). A systematic search for nuclear flares from all galaxies in Stripe 82 yielded two events (dubbed SDSS-TDE1 and SDSS-TDE2) from otherwise quiescent galaxies. The key property distinguishing these

[2] Such as a strong Fe X λ6376 to O III λ5007 ratio, and strong Fe X λ6376, Fe XI λ7894, Fe XIV λ5304, Ar XIV λ4414, and S XII λ7612 lines.

flares from normal supernovae and AGN is a high blackbody temperature (as measured using SDSS *ugri* photometry) that remained constant for at least three months (this property later became a basis for the photometric selection of TDEs). Both events also show transient ultraviolet emission, detected in GALEX more than one year after peak. For one of the two flares, SDSS-TDE2, an optical spectrum was obtained near peak. The spectrum displays broad emission features around Hα, and Hβ and He II λ4686.

2.2 Pan-STARRS

Two TDEs (from non-AGN galaxies) were discovered by the Pan-STARRS medium-deep survey (Chambers 2007): PS1-10jh (Gezari et al. 2012) and PS1-11af (Chornock et al. 2014). Both were identified via spectroscopic follow-up of photometric transients in the medium-deep fields (Chambers et al. 2016), a set of fields that were targeted with a three-day cadence to a typical single-epoch depth of $m = 23$ (though the effective flux limit of the TDE search is likely to be shallower due to the requirement for spectroscopic classification).

Both PS1-10jh and PS1-11af show transient ultraviolet emission (detected by GALEX) and long-term high blackbody temperatures, but display different spectral properties. PS1-10jh displays a broad He II λ4686 emission feature in its optical spectrum, while PS1-11af shows two transient absorption features in the rest-frame ultraviolet, possibly attributed to Mg II. PS1-10jh is the first TDE with multiple pre-peak detections and multiple spectroscopic observations, making it the archetype for the main optical-ultraviolet class of TDEs.

While the medium-deep survey has concluded, the Pan-STARRS telescope continued to search for transients (Huber et al. 2015) and later discovered the TDEs PS17dhz (AT 2017eqx; Nicholl et al. 2019) and PS18kh (AT 2018zr; Holoien et al. 2019a). Another TDE candidate from the Pan-STARRS survey was discovered in a narrow-line Seyfert 1 galaxy (PS16dtm; Blanchard et al. 2017). This event, and others like it, for which an association with a TDE or enhanced AGN activity is not decisive, are discussed in the One-off Events and Imposters Chapter.

2.3 PTF

Three TDEs were discovered in Palomar Transient Factory (PTF: Law et al. 2009) data by searching for spectroscopically-observed nuclear flares with an R-band absolute magnitude in the range $-21 < M_R < -19$ (Arcavi et al. 2014). The three events discovered (PTF09ge, PTF09djl, and PTF09axc) all display relatively high blackbody temperatures, and have similar spectroscopic signatures as PS1-10jh and SDSS-TDE2, namely broad He and/or H emission lines. The discovery of the events established the spectroscopic sequence of TDEs (see Sect. 4), and also identified the peculiar nature of the host galaxies of such events (see Host Galaxies Chapter). For the PTF TDEs, no ultraviolet observations were obtained near peak, but two sources were detected in *HST* far-ultraviolet imaging observations, five years after peak (van Velzen et al. 2019a).

2.4 iPTF

The iPTF search for TDEs (Hung et al. 2018) used the properties of previous TDEs to identify new candidates in real time, hence more follow-up data (in particular with *HST* and *Swift*) of promising TDEs could be obtained in real time. The typical single-epoch magnitude limit of iPTF data is ≈ 20.5. No selection on the absolute magnitude was made, but instead a blue color for the flare ($g - R < -0.1$) was required to reduce the background of supernovae. Follow-up spectroscopy and *Swift*-UVOT observations were used to confirm the TDE nature. Three TDEs were found: iPTF15af (Blagorodnova et al. 2019), iPTF16fnl (Blagorodnova et al. 2017), and iPTF16axa (Hung et al. 2017). iPTF16fnl is an outlier in terms of its low luminosity and rapid decline rate. It raises the question of how many more such TDEs might have been missed due to an observational bias against finding rapidly evolving events.

2.5 ASAS-SN

To date, the All Sky Automated Survey for supernovae (ASAS-SN; Shappee et al. 2014) discovered eight TDEs: ASASSN-14ae (Holoien et al. 2014), ASASSN-14li (Holoien et al. 2016b), ASASSN-15oi (Holoien et al. 2016a), ASASSN-18pg (AT 2018dyb; Leloudas et al. 2019; Holoien et al. 2020), ASASSN-18ul (AT 2018fyk), ASASSN-18zj (AT 2018hyz; Short et al. 2020; Gomez et al. 2020; Hung et al. 2020), ASASSN-19bt (AT 2019ahk; Holoien et al. 2019b), and ASASSN-19dj (AT 2019azh). The ASAS-SN discoveries come from an all-sky search using single-band observations (either V- or g-band) with a typical limiting magnitude of 17. The TDE identification is established via spectroscopic follow-up. This shallow and wide nature of ASAS-SN means that the survey is sensitive to the lowest-redshift TDEs ($z \sim 0.03$), making them ideal for intensive followup observations. Ultraviolet detections from *Swift* have been obtained for all ASAS-SN TDEs.

ASASSN-14li is the first optical TDE with a well-sampled X-ray light curve (Holoien et al. 2016b; Miller et al. 2015), and (at the time of writing) the only published optical TDE detected in the radio (see the Radio Chapter; van Velzen et al. 2016; Alexander et al. 2016).

The nature of an additional ASAS-SN event, ASASSN-15lh, is controversial. It has been interpreted as either an extreme superluminous supernova (Dong et al. 2016; Godoy-Rivera et al. 2017) or an extreme TDE (Leloudas et al. 2016; Margutti et al. 2017; Krühler et al. 2018).

2.6 ATLAS

The Asteroid Terrestrial-impact Last Alert System (ATLAS; Tonry et al. 2018) is a survey primarily designed for discovering potentially hazardous asteroids. However, it regularly discovers transients from its two-day cadence search down to a limiting magnitude of 19–20. ATLAS has discovered three TDEs so far (van Velzen et al. 2020): ATLAS18way (AT 2018hco), ATLAS18yzs (AT 2018iih) and ATLAS19qqu (AT 2019mha).

2.7 OGLE

The Optical Gravitational Lensing Experiment (OGLE; Udalski et al. 2015) also runs a transient search (Wyrzykowski et al. 2014). The survey is focused on an area of the sky around the Magellanic Clouds, but this includes many background galaxies. Each field is observed roughly every four days with a limiting magnitude of ~ 20 in the I-band. OGLE has discovered one TDE (OGLE16aaa, Wyrzykowski et al. 2017; Kajava et al. 2020).

2.8 ZTF

The Zwicky Transient Facility (ZTF; Bellm et al. 2019) started in early 2018 and is still running. The single-epoch depth and filters are similar to iPTF, but the field of view is eight times larger. The source PS18kh was detected during the ZTF commissioning phase (van Velzen et al. 2019b), and sixteen more TDEs have been detected in ZTF data since (van Velzen et al. 2020), signaling a large increase of the TDE discovery rate.

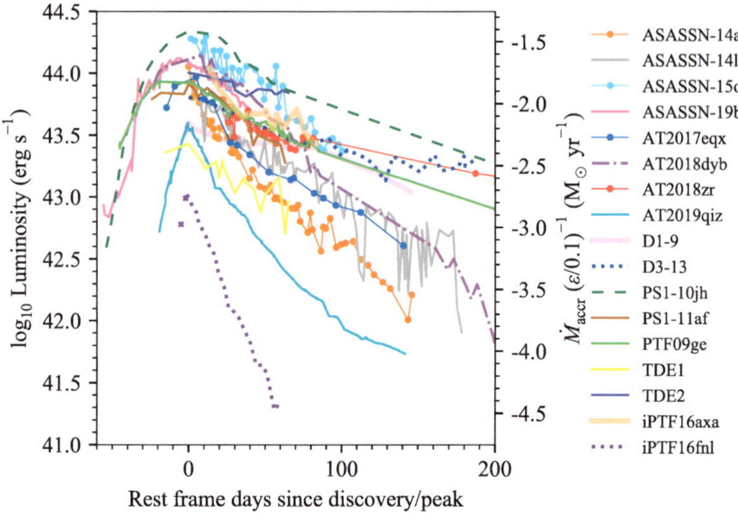

Fig. 1 A compilation of TDE light curves. The integrated optical-ultraviolet luminosity is shown as inferred from spectral energy distribution fitting. The x-axis shows the time elapsed since peak for PTF09ge, PS1-10jh, iPTF16fnl, AT 2017eqx, AT 2018dyb, AT 2018zr, and ASASSN-19bt and the time elapsed since discovery for ASASSN-14ae, ASASSN-14li, ASASSN-15oi, and iPTF16axa. The two crosses in purple are derived from pre-peak g-band data of iPTF16fnl assuming a blackbody temperature of 2×10^4 K. The mass accretion rate on the right hand side is normalized to an efficiency ε of 0.1. Figure adapted from Hung et al. (2017), data from van Velzen et al. (2011), Gezari et al. (2012), Chornock et al. (2014), Arcavi et al. (2014), Holoien et al. (2014, 2016a,b), Blagorodnova et al. (2017), Hung et al. (2017), van Velzen et al. (2019b), Nicholl et al. (2019), Leloudas et al. (2019), Holoien et al. (2019b), Nicholl et al. (2020)

3 Light Curves

Optical-ultraviolet TDEs typically rise to a peak blackbody luminosity of $L_{\rm bb} \sim 10^{43.5\text{–}44.5}$ erg s^{-1} on timescales of weeks-months with a decay broadly consistent with $t^{-5/3}$. This is the expected decline rate of the post-disruption mass return flow (Rees 1988, Phinney 1989), though the density profile of the disrupted star may influence this rate (Lodato et al. 2009; Gezari et al. 2012), as will relativistic effects (Kesden 2012a). It is also important to keep in mind that fitting a power law to TDE light curves introduces a large uncertainty on the inferred power-law index, due to the freedom in setting the time of disruption.

Peak luminosities and time scales vary between events (Fig. 1). What does not seem to vary between events is the lack of cooling of the post-peak blackbody temperature (Fig. 2), though some of this homogeneity may be artificially enhanced by the use of the non-evolving temperature as a selection criterion to discriminate TDEs from supernovae.

Perhaps the first notable outlier in its photometric properties (being still "normal" spectroscopically) is iPTF16fnl, displaying a fainter and faster light-curve evolution than the rest of the class (Fig. 1; Blagorodnova et al. 2017; Onori et al. 2019). This raised the question as to whether many more such events exist but are missed due to observational biases. Indeed, recently, more rapidly-decaying events have been found in the ZTF survey (see Table 2 and the analysis of AT 2019qiz by Nicholl et al. 2020).

Another photometric outlier is ASASSN-15lh (Leloudas et al. 2016), which displayed a double peak in the bluer bands of its light curve. This event is an outlier also spectroscopically, and its identification as a TDE is contended (Dong et al. 2016; Godoy-Rivera et al.

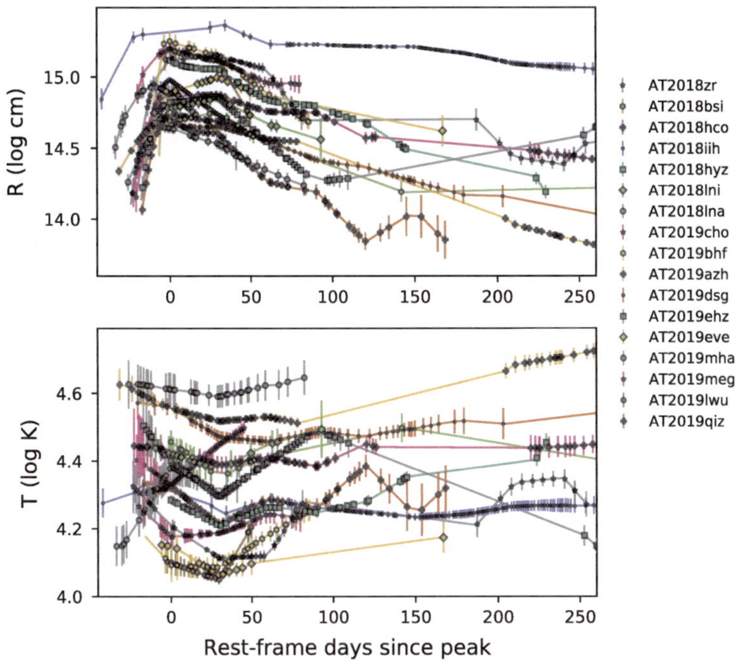

Fig. 2 A compilation of blackbody radii and temperatures obtained from optical-ultraviolet photometric observations of seventeen TDEs from van Velzen et al. (2020). All sources show a decrease of the blackbody radius after maximum light. The majority of the sources show a near-constant blackbody temperature. While some TDEs show modest temperature variability near peak or a significant temperature increase at late-time, none of the known TDEs show a monotonic decrease of the blackbody temperature. Since most of the sources in the ZTF sample were identified as TDEs within a few weeks of maximum light and the temperature is inferred from ultraviolet follow-up observations, this lack of cooling cannot be solely explained as a selection effect

2017). For this reason we do not include it in Table 1. ASASSN-15lh is discussed in more detail in the One-off Events and Imposters Chapter.

Recently, van Velzen et al. (2020) doubled the sample of known TDE light curves, and found a correlation between photometric properties and the spectroscopic types (this spectroscopic classification is introduced in the next section). Namely, that TDEs showing Bowen features have smaller blackbody radii compared to the rest of the population, while events showing only He in their spectra have the longest rise times. For the overall population, the rise timescale and fade timescale show no clear correlation, while a strong correlation would be expected if rise and fade timescales were both determined by the fallback rate. van Velzen et al. (2020) conclude that this might be an indication that the rise time of TDE light curves is not related to the fallback time but rather to a diffusion time through some extended material (which may be the source of the spectral differences as well).

Several authors have reported a (potential) correlation between the decay rate of the light curve and black-hole mass (Blagorodnova et al. 2017; Wevers et al. 2017) or total host galaxy mass (van Velzen et al. 2020). In Fig. 3 we reproduce this result using our homogeneous lightcurve analysis (see Sect. 5). Here we estimate the black-hole mass from the M-σ relation (Gültekin et al. 2009), using the velocity dispersion measurement of Wevers et al. (2017, 2019c) or SDSS (Thomas et al. 2013). The fallback timescale is estimated by fitting a power-law decay to the observed optical-ultraviolet blackbody emission, with the power-law

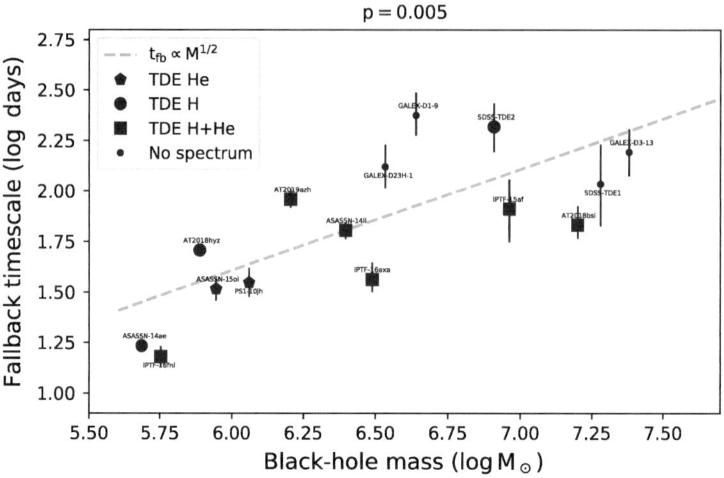

Fig. 3 Black-hole mass versus fallback time. The black-hole mass is estimated from the M-σ relation and the fallback time follows from fitting a $L \propto t^{-5/3}$ power-law decay to the optical-ultraviolet blackbody lightcurve. The dashed line indicates the predicted fallback rate for a star of one solar mass (Stone et al. 2013)

index fixed to $-5/3$. We find a statistically significant correlation between the black-hole mass and decay rate ($p = 0.005$ for a Kendall's Tau test). This correlation could indicate that the early-time decay rate of the optical-ultraviolet emission is indeed determined by the fallback rate.

X-ray detections are rare in the case of optically-discovered TDEs, and there are only a few cases showing both optical-ultraviolet and X-ray emission: ASASSN-14li (Miller et al. 2015; Holoien et al. 2016b; van Velzen et al. 2016, Fig. 4), AT 2018fyk (Wevers et al. 2019a) AT 2018zr, (van Velzen et al. 2019b), and most recently, AT 2019azh, AT 2019dsg, and AT 2019ehz (van Velzen et al. 2020).

The existence of an observational dichotomy between optical-ultraviolet and X-ray TDEs has been a topic of study. Loeb and Ulmer (1997), Strubbe and Quataert (2009), Guillochon et al. (2014), Roth et al. (2016) and Auchettl et al. (2017) invoke material surrounding the accretion disk as responsible for reprocessing of the X-ray photons from the disk into optical and ultraviolet photons. Dai et al. (2018) expand on this picture by proposing that the reprocessing material is concentrated in the equatorial direction, concluding that the observed dichotomy is a viewing angle effect: the optical and ultraviolet emission are seen from the equatorial direction, while a polar viewing angle will not cross the reprocessing material and provide a line of sight directly into the X-ray emitting accretion disk. These models are further discussed in the Emission Mechanisms Chapter of this book.

Many optical-ultraviolet TDEs with observations more than 100 days post-peak show a flattening of the light curve to almost constant luminosity. A clear example is ASASSN-14li (Brown et al. 2017). Using both *HST* and *Swift*/UVOT observations, van Velzen et al. (2019a) find a flattening of the ultraviolet light curve for 10 out of 12 TDEs with late-time observations. They show that this plateau phase can be explained by emission from an accretion disk that forms after the disruption (Cannizzo et al. 1990). Mummery and Balbus (2020) show that both the late-time ultraviolet plateau emission and the early-time X-ray emission of ASASSN-14li can be explained using an evolving accretion disk.

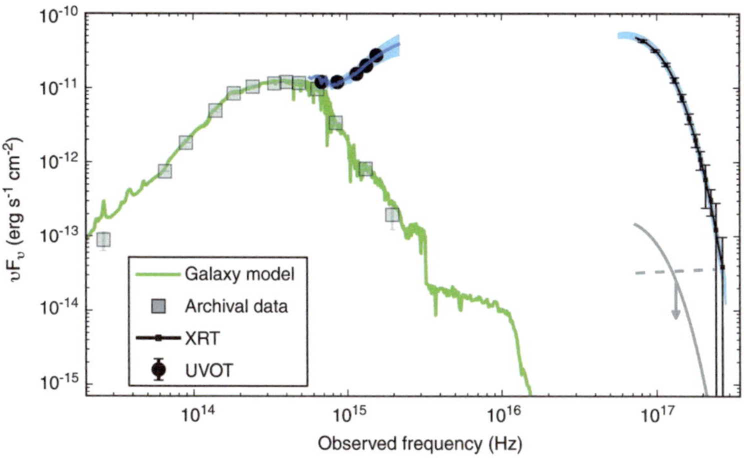

Fig. 4 *Swift* X-ray and broadband optical-ultraviolet observations of ASASSN-14li. A double-blackbody fit with temperatures of $T = 7.7 \times 10^5$ K and $T = 3.5 \times 10^4$ K is shown in blue. The pre-flare X-ray limit for the blackbody spectrum is show in grey. The SED of the host galaxy (grey squares) together with a best fit synthetic galaxy spectrum (green) are shown for comparison. Figure from van Velzen et al. (2016)

4 Spectra

Spectra are crucial for distinguishing TDEs from other nuclear activity due to AGN variability, or other transients such as supernovae. They can also provide insights into the sources of TDE emission and dynamics. The increasing sample of TDEs monitored by dense spectroscopic follow-up campaigns has revealed a set of spectroscopic classes differing in their line species, profiles and evolution.

4.1 Optical

Spectra of the first optical-ultraviolet TDEs identified were dominated by a strong and consistent blue continuum with broad (FWHM $\sim 10^4$ km s^{-1}) emission lines superimposed on it. The most common and prominent spectral feature is a broad and intense He II λ4686 line (e.g. Gezari et al. 2012, 2015), which is not seen at these widths, intensities and durations in spectra of other known transients. In some cases, broad Hα and Hβ lines are also identified (e.g. Arcavi et al. 2014). This led to an initial map of TDE spectral diversity as a continuum of H-rich to He-rich events (Fig. 5). van Velzen et al. (2020) divided the current TDE population into three spectroscopic types and showed that TDEs in these classes have significantly different photometric properties. Following their approach, in Table 1 we label events showing only He II lines as "TDE He", those showing only hydrogen lines as "TDE H" and those showing both He II and H lines as "TDE H+He".

The lack of any hydrogen signatures in the spectra of PS1-10jh was initially interpreted as evidence for the disruption of a rare helium star (Gezari et al. 2012). However, with the discovery of additional events with hydrogen-free spectra, this hypothesis became harder to justify. Guillochon et al. (2014), using hydrodynamical simulations, was able to explain the PS1-10jh spectra as produced by the disruption of a main sequence star. In this scenario, an optically thick reprocessing envelope determines the location of the line-forming region,

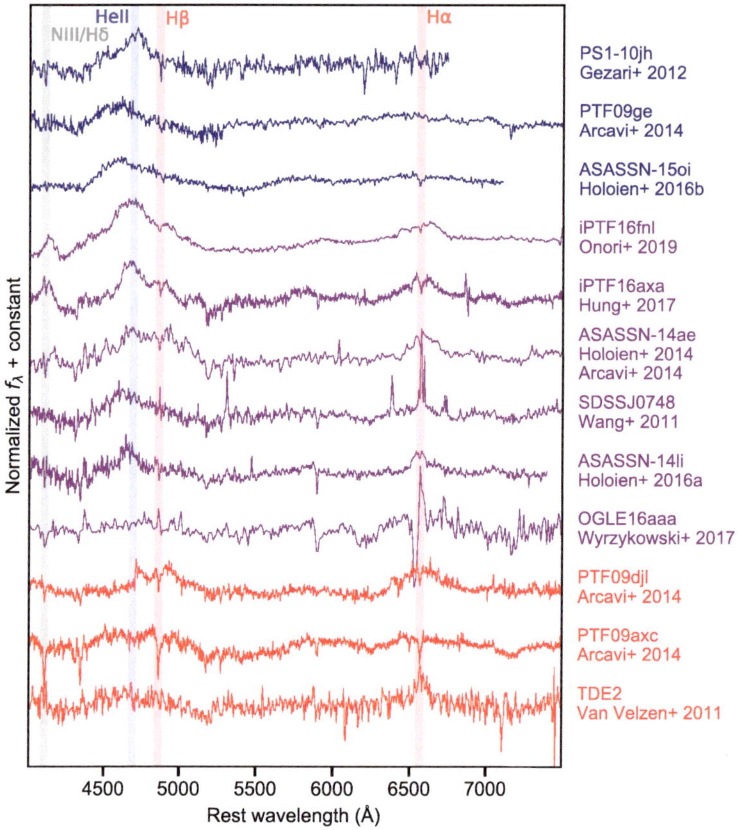

Fig. 5 Continuum-subtracted optical spectra of optically-discovered TDEs. Events in blue are dominated by a broad He II λ4686 feature with no signatures of Hα or Hβ (we denote these "TDE He"). Events in purple are characterized by the presence of both broad He II λ4686 and Hα, with variation in the intensity of these lines from one event to another, and possible broad Hβ seen in some events (we denote these"TDE H+He"). Events in red are H-dominated (we denote these "TDE H"). Additional N III features (perhaps blended with Hδ) are seen for some events in the "H+He" class. Figure adapted from Arcavi et al. (2014), data from Wang et al. (2011), van Velzen et al. (2011), Gezari et al. (2012), Arcavi et al. (2014), Holoien et al. (2014, 2016a), Hung et al. (2017), Onori et al. (2019)

and the lack of hydrogen lines is attributed to the conditions for forming the lines in the envelope, rather than the presence or absence of hydrogen. Roth et al. (2016) showed that varying the radius of the reprocessing layer (given some composition and ionizing flux) can suppress the hydrogen lines while keeping the He II lines strong, thus reproducing the observed spectral sequence displayed in Fig. 5.

In some events, the relative hydrogen to helium line intensity is seen to change with time. For instance, the early optical spectra of ASASSN-14ae (Holoien et al. 2014) are characterized by an intense and broad Hα emission line and a weak signature of a broad component in the He II λ4686 line, which developed only later. An even more extreme He II to H line-ratio evolution is seen in AT 2017eqx (Fig. 6) and interpreted, following Roth et al. (2016), as evidence for contraction of the reprocessing layer (Nicholl et al. 2019).

In Fig. 6 we show the time evolution of the He II/Hα luminosity ratio for a number of optical-ultraviolet TDEs with well-sampled spectroscopic sequences. The time evolution of

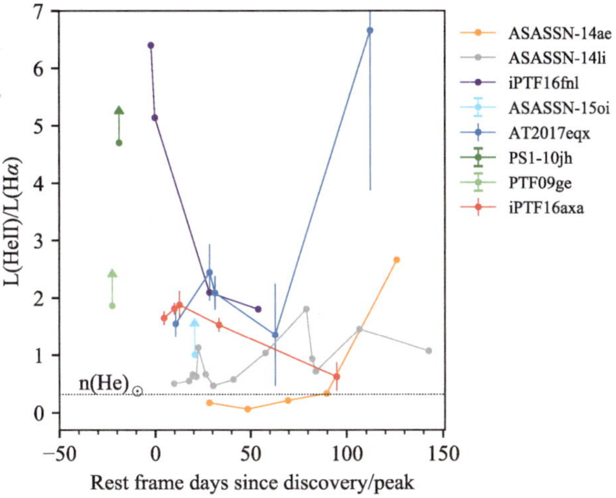

Fig. 6 He II λ4686 to Hα line luminosity ratio and its evolution with time for a sample of optical-ultraviolet TDEs with well-sampled spectroscopic sequences. The ratio changes with time in a different way for each event, yet most events show a ratio higher than the expected value for a nebular environment with solar abundance (dotted line). Figure from Hung et al. (2017), data from Gezari et al. (2012), Arcavi et al. (2014), Chornock et al. (2014), Holoien et al. (2014, 2016a,b), Blagorodnova et al. (2017)

this line ratio varies from event to event. Yet, in most, the He II/Hα luminosity ratio is above the expected value for a nebular environment with solar abundance (Hung et al. 2017, dotted line in Fig. 6). This helium enhancement has been explained as the result of the suppression of the Balmer lines (which are optically thick) by high gas densities ($> 10^{10}$ cm^{-3}) in the emitting region. This scenario was first proposed by Bogdanović et al. (2004) and further investigated using CLOUDY calculations by Gaskell and Rojas Lobos (2014) and Strubbe and Murray (2015), as well as with radiative transfer calculations by Roth et al. (2016).

The broad features observed in TDEs are, in general, pure emission line profiles, with no signs of absorption or P-cygni features. We present the line-profile evolution of Hα and He II for several TDEs in Figs. 7 and 8 respectively. Some events display evolving asymmetric profiles. For example, ASASSN-14ae (Holoien et al. 2014) shows a broad and asymmetric Hα line, with a prominent red wing and a blue-shifted centroid at early times. Later, the Hα line becomes symmetric and centered around the host rest-frame wavelength. Similar evolution is seen in the He II line of ASASSN-15oi (Holoien et al. 2016a).

Optically-thick outflows have been proposed to explain such line profiles and their evolution, especially in the case of Hα (Roth and Kasen 2018). However, in some cases, the blue-shifted component in the He II λ4686 line has been attributed to a blend with possible N III/C III lines (Gezari et al. 2015; Brown et al. 2018). Leloudas et al. (2019) found a rich set of line species in the spectra of the TDE AT 2018dyb (Fig. 9). In addition to the Balmer series and He II, they identify also He I, O III λ3760, and, indeed, N III λλ4100, 4640 (the latter blended with He II). This event demonstrates the richness of optical-ultraviolet TDE spectral diversity beyond the hydrogen to helium sequence seen initially (see also Holoien et al. 2020). Leloudas et al. (2019) identify possible N III lines also in the TDEs ASASSN-14li and iPTF16axa. Together with iPTF15af (Blagorodnova et al. 2019) and iPTF16fnl (Onori et al. 2019), they form a N-rich class of TDEs.

In addition to N III, Blagorodnova et al. (2019) identify several Bowen fluorescence lines in the optical and ultraviolet spectra of iPTF15af (Fig. 10). Bowen fluorescence (Bowen 1928; Weymann and Williams 1969) is a process by which extreme-ultraviolet and X-ray photons ionize He II, which later recombines, emitting Ly-α photons. These then excite certain O III and N III transitions with similar wavelengths, launching a cascade of transitions observed in the optical and ultraviolet regimes. In particular, strong O III λ3133, N III λ4640, and He II λ4686 have been observed in some planetary nebulae and X-ray bi-

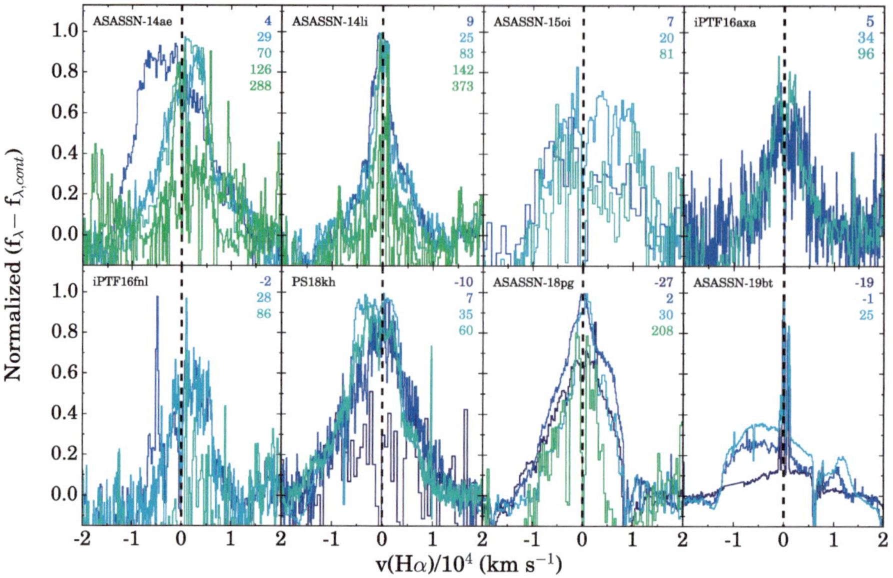

Fig. 7 Continuum-subtracted line profiles of Hα (rest-frame zero velocity marked by black dashed line) for eight optical-ultraviolet TDEs. Colors correspond to rest-frame days from peak or discovery as indicated. ASASSN-14ae can be seen to have an asymmetric profile at early times which later becomes symmetric. Data from Holoien et al. (2014, 2016a,b), Brown et al. (2016, 2017), Hung et al. (2017), Blagorodnova et al. (2017), Holoien et al. (2019a,b, 2020)

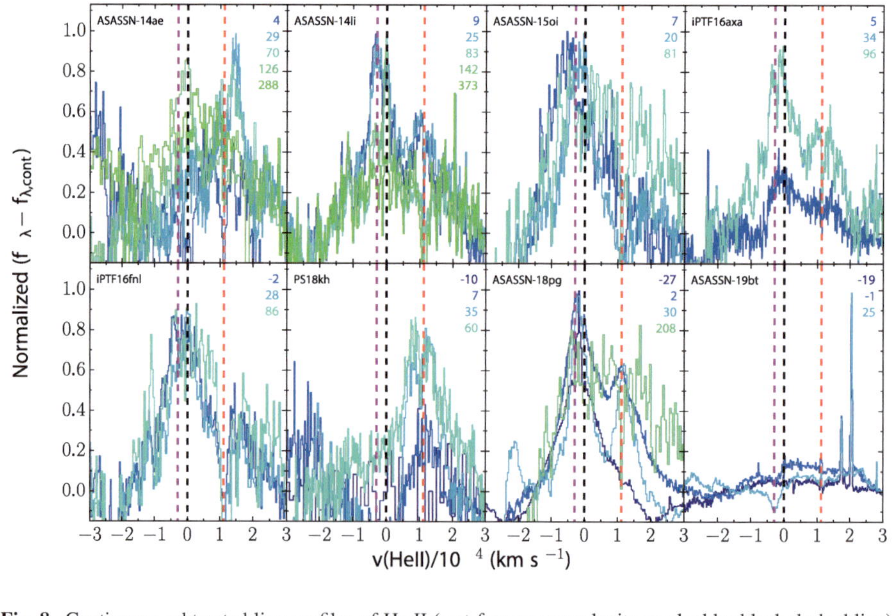

Fig. 8 Continuum-subtracted line profiles of He II (rest-frame zero velocity marked by black dashed line), N III (magenta dashed line), and Hβ (red dashed line) for eight optical-ultraviolet TDEs. Colors correspond to rest-frame days from peak or discovery as indicated. ASASSN-15oi can be seen to have an asymmetric He II line profile at early times which later becomes symmetric. Data sources are the same as in Fig. 7

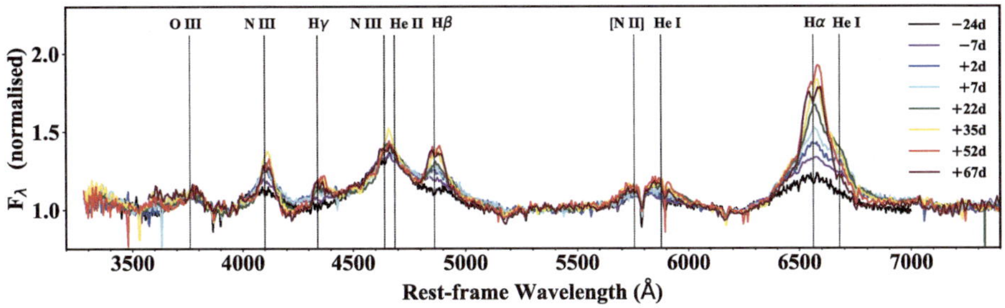

Fig. 9 Spectra of the Bowen (N-rich) TDE AT 2018dyb after normalising to the continuum. The different colors indicate different phases in rest-frame days relative to peak luminosity. A larger richness in line species compared to the initial hydrogen to helium mapping of TDE spectra (Fig. 5) can be seen. Figure from Leloudas et al. (2019)

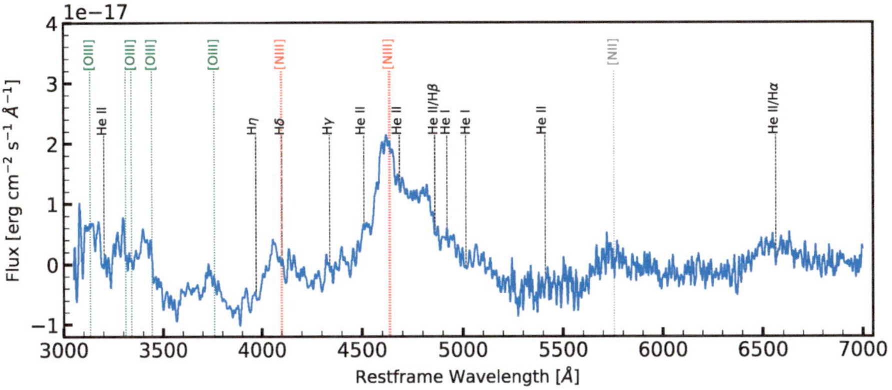

Fig. 10 Spectrum of the Bowen TDE iPTF15af. Bowen fluorescence lines can be seen ([N III] lines are marked in red and [O III] lines in green – the centroids of these forbidden lines appear shifted by $-3000\ \mathrm{km\,s^{-1}}$). This event identifies yet an additional possible layer of optical-ultraviolet TDE spectral diversity. Figure from Blagorodnova et al. (2019)

naries (Schachter et al. 1989; Kastner and Bhatia 1996). Such emission was predicted to occur around active supermassive black holes decades ago (Netzer et al. 1985), but was only recently observed for the first time in an extreme outburst of an AGN (Trakhtenbrot et al. 2019b). Its detection also in TDEs (Blagorodnova et al. 2019) may indicate the presence of similar conditions involving extreme-ultraviolet photons hitting high-density and high-optical-depth material, in both TDEs and some AGN outbursts. In Table 1 we denote both the N-rich TDEs and those with Bowen lines with a "Bowen" label (it is not yet clear whether these two spectral classes are distinct). We adopt the criterion of Leloudas et al. (2019) that use the presence of either N III $\lambda4640$ or $\lambda4100$, or O III $\lambda3760$ to identify an event as "Bowen". As first pointed out by Leloudas et al. (2019), evidence for these Bowen lines is (so far) detected only in spectra of TDEs that show both He and Balmer emission lines (i.e. TDE H+He's). In fact, almost all TDEs typed as "H+He" show one or more emission lines that might be associated with the Bowen mechanism (van Velzen et al. 2020).

Wevers et al. (2019b) identified the Fe II 37,38 multiplet in narrow emission lines in the late-time spectra of AT 2018fyk (we add a "Fe" label to this event in Table 1) and suggest similar lines might be present in SDSS J0748, PTF09ge, and ASASSN-15oi. These

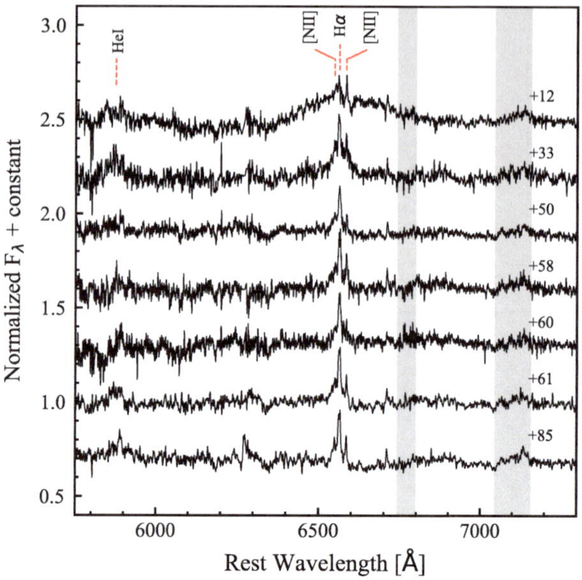

Fig. 11 A sequence of host-subtracted spectra of iPTF16fnl. The evolution of the Hα line profile is characterized by the narrowing of the broad component together with the development of a narrow core at late times. This may be an indication of decreasing optical depth, rather than of bulk velocities of the line-emitting material (Roth and Kasen 2018). The main emission lines are indicated with vertical dashed red lines, while the gray areas indicate the wavelength ranges affected by telluric lines. Days relative to peak luminosity are shown on the right. Figure adapted from Onori et al. (2019)

low-ionization lines are also found in narrow-line Seyfert 1 AGN, in some coronal-line emitters, and in flares or TDEs in active galaxies such as PS16dtm (Blanchard et al. 2017), AT 2018bcb (Neustadt et al. 2020), and AT 2018dyk (Frederick et al. 2019).

The combination of the TDE spectral features (species, line profiles and intensity) discussed so far are distinct from those of all known transients, including all types of supernovae.

In many TDEs, the widths of the lines evolve with time, starting at $\sim 10^4\ \mathrm{km\,s^{-1}}$ at early times and becoming narrower on time scales of months. The narrowing of the lines with time, as the luminosity of the event declines, is the opposite behaviour observed in the case of reverberation mapping in AGN (e.g. as noted by Holoien et al. 2016b). While the width of broad spectral lines is usually associated with the kinematics of the emitting gas, in some cases, it can be strongly affected by electron scattering instead. This can lead to an overestimation of the emitting region bulk velocity, and a mistaken interpretation of its evolution.

Roth and Kasen (2018) investigate spectral line formation due to electron scattering in an outflowing optically-thick gas. Using radiative transfer calculations they show that, indeed, non-coherent scattering of hot electrons can play a primary role in the broadening of the lines in a high-density emitting region. In this case, the line profiles are characterized by the presence of a narrow line core superimposed on a broad component at the base of the line. The broad line 'wings' are produced by photons scattered off of high-velocity electrons in the gas. The contribution of the wing to the line profiles becomes more prominent with increasing optical depth. In some cases, the broad wings may dominate the line profile completely, with the narrow core becoming visible only later as the optical depth decreases. This is seen very clearly in the case of the Hα line of iPTF16fnl (Fig. 11; Onori et al. 2019). In this scenario, the observed line profile evolution, in which a broad component becomes narrower and a central, narrow core emerges, reflects a decline in the emitting region's density, rather than a drop in its bulk velocity.

Among the current sample of optical-ultraviolet TDEs, four events exhibit broad emission lines with boxy or double-peaked profiles, which have been associated with a line-emitting disk: PTF09ge (Arcavi et al. 2014), ASASSN-14li (Liu et al. 2017), AT 2018zr (Holoien et al. 2019a), and AT 2018hyz (Short et al. 2020; Hung et al. 2020). In particular, for AT 2018zr and AT 2018hyz, the double-peaked line profile was seen to develop gradually with time, and has been interpreted as the revealing of an accretion disk through the thinning reprocessing material. Hung et al. (2020) deduce a disk-formation time scale of about a month for AT 2018hyz, implying rapid circularization of the stellar debris after disruption. Hung et al. (2020) further model the Hα emission line in AT 2018hyz as a combination of a double-peaked disk profile and a Gaussian component, which is likely produced in the outflowing gas. The Gaussian component becomes stronger over time and dominates the Hα line shape at late times. This competition with a non-disk component, and the requirement of a high inclination angle (close to edge-on) to see a disk profile may explain why double-peaked line profiles are rare in TDEs.

4.2 Ultraviolet

To date, ultraviolet spectroscopy has been obtained for five TDEs – ASASSN-14li (Cenko et al. 2016), iPTF16fnl (Brown et al. 2018), PS16dtm (Blanchard et al. 2017), iPTF15af (Yang et al. 2017; Blagorodnova et al. 2019), and AT 2018zr (Hung et al. 2019), all of which have near-ultraviolet coverage (Fig. 12). Far-ultraviolet spectra have been obtained for all but PS16dtm. While a single blackbody spectrum extrapolated from optical and near-ultraviolet photometric observations captures the overall shape of the ultraviolet spectra, the continuum often appears to be more luminous than what the blackbody predicts at the far-ultraviolet range.

Ultraviolet spectra have revealed the presence of collisionally excited broad emission lines, such as Lyα, C IV, N V and Si IV (Cenko et al. 2016; Brown et al. 2018; Blagorodnova et al. 2019), similar to those seen in QSOs. On the other hand, the nitrogen-to-carbon ratios in these TDEs are anomalously high compared to typical QSOs, thus connections of TDEs with nitrogen-rich QSOs have been suggested (Cenko et al. 2016; Kochanek 2016).

All of these TDEs, except PS16dtm, also lack the Mg II $\lambda\lambda$2796, 2803 emission that is commonly seen in QSOs. Mg II has an ionization potential of 15 eV. The Mg II $\lambda\lambda$2796, 2803 emission arises from low ionization regions where the gas is at least partially neutral. If a hard continuum is present, the majority of the Mg atoms may be photoionized to Mg III or higher states and thus lead to the weakening or absence of Mg II features. The same hard spectrum may also be responsible for the production of Bowen fluorescence emission. However, as noted by Leloudas et al. (2019), this hard ionizing source cannot be accounted for entirely by the near ultraviolet and optical blackbody with a temperature on the order of 10^4 K, as it provides too few photons at high energies.

Unlike in the optical spectra, *absorption* features have been identified in ultraviolet spectra. These are seen as broad ($\sim 10^3$–10^4 km s^{-1}) features in iPTF16fnl, iPTF15af, PS16dtm, and AT 2018zr and as narrow (few $\times$ 100 km s^{-1}) ones in ASASSN-14li. The broad absorption features are seen in the high-ionization lines C IV, N V, and Si IV, similar to what is seen in Broad Absorption Line Quasars (BALQSOs), and tend to be blueshifted by $\sim 10^3$–10^4 km s^{-1}. The absorption troughs can either be attached (e.g. iPTF16fnl and iPTF15af) or detached (e.g. AT 2018zr) from their emission peaks, which might be due to a viewing angle difference. The broad blueshifted absorption lines are interpreted as evidence for fast outflows of material accelerated by radiation. Although the driving mechanism for these outflows in the broad absorption-line TDEs is still unclear, it has been suggested that super-Eddington accretion (e.g. Strubbe and Quataert 2009, 2011), which might be common in TDEs around supermassive black holes with mass $< 10^7\ M_\odot$, could be capable of driving outflows to such high velocities ($\sim 10^4$ km s^{-1}).

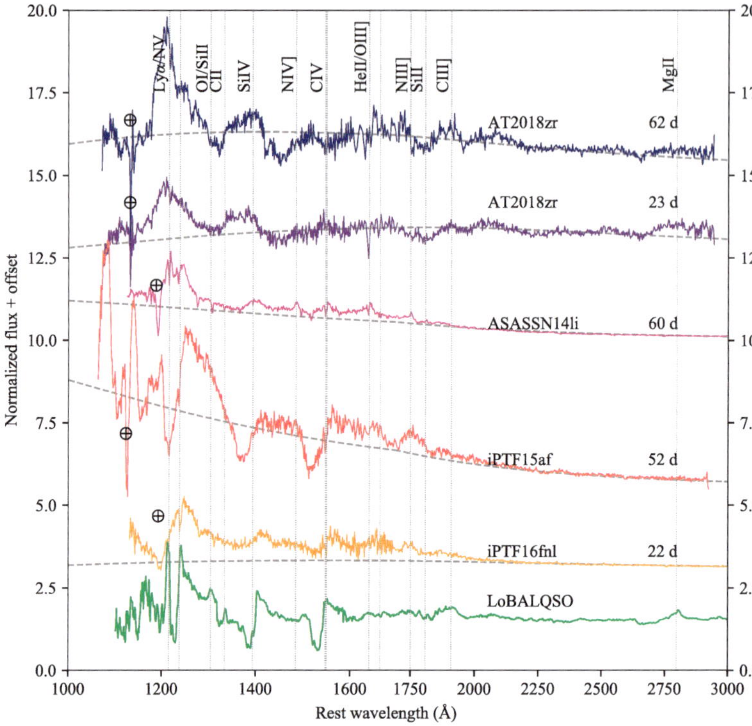

Fig. 12 Ultraviolet spectra of optical-ultraviolet TDEs (Hung et al. 2019). Grey dashed lines mark the black-body spectra with temperatures measured from ultraviolet and optical photometry. Rest-frame days from peak luminosity are marked. TDE data from Cenko et al. (2016), Brown et al. (2018), Blagorodnova et al. (2019), Hung et al. (2019). A composite spectrum of low-ionization broad absorption line QSO (LoBALQSO; from Brotherton et al. 2001) is shown for comparison

A sequence of ultraviolet spectra was obtained for two TDEs, iPTF16fnl and AT 2018zr, revealing evolution in the absorption features. In iPTF16fnl, the ultraviolet emission features are found to be redshifted relative to the systematic velocity from the earlier spectra. This apparent redshift has been interpreted as a result of fast outflows (Brown et al. 2018), where blueshifted broad absorption troughs eclipsed the bluer portion of the emission line profiles. In AT 2018zr, Hung et al. (2019) found that the high-ionization broad absorption lines only appeared ~ 2 months after maximum light (Fig. 12). Expanding the sample of TDEs with multi-epoch ultraviolet spectroscopy will test whether this timescale is linked to the timescale of super-Eddington outflows, and whether outflows are ubiquitous among TDEs.

In AT 2018zr, absorption features are also seen in the Balmer series and the metastable helium lines He I$^*\lambda 3889$ and He I$^*\lambda 5876$ at optical wavelengths, with a FWHM of $\sim 3000\ \mathrm{km\,s^{-1}}$ and displaced by the same velocity ($0.05c$) as the ultraviolet absorption features (Hung et al. 2019). Such optical absorption features are rarely detected even in BALQSOs. Photoionization models indicate both high density ($\log(n_H/\mathrm{cm}^{-3}) \gtrsim 7$) and high column density ($\log(N_H/\mathrm{cm}^{-2}) \gtrsim 23$) requirements must be met in order to produce the observed hydrogen and helium absorption lines (Hamann et al. 2019). Recent work on radiative transfer modelling of TDE outflows suggests that the column densities in TDE winds are as high as $> 10^{25}\ \mathrm{cm}^{-2}$, depending on the inclination angle of the disk (Parkinson et al. 2020). This result is in agreement with the lower limit derived from the observed ultraviolet absorption lines.

5 TDE Selection: Building a Sample of Events

We now distill the discussion of the previous sections on the photometric and spectroscopic properties into a set of selection criteria to define a sample of "secure" optical-ultraviolet TDEs (Table 1). Due to the lack of a consistent theory that explains all optical-ultraviolet TDE emission properties, our selection criteria are primarily motivated by observations. We realize that there is a somewhat circular logic here, identifying new events as optical-ultraviolet TDEs based on those previously claimed to be such. There are likely additional TDEs that do not match these criteria, and therefore, in order to truly map the population of events, we must keep an open mind regarding newly discovered types of flares (yet, we must also be cautious of calling every new type of transient a TDE). Once a large enough sample is available we can revisit the criteria. That being said, after almost one decade of collecting optical-ultraviolet data of transients found in galaxy centers, some clear patterns have emerged. Below we summarize four key properties that we use to compile the events in Table 1, which we consider to be a robust sample of optical-ultraviolet TDEs. These criteria have been applied to events claimed to be TDEs in the literature.

- **Nuclear.** That is, a small offset (d) between the host centroid and the flare. This criterion is motivated by the assumptions that SMBHs in quiescent galaxies reside at the gravitational center of their host galaxy. For the purpose of populating Table 1 we use a cut of $d < 0.3''$, which is sufficient to establish a nuclear detection for the surveys summarised in Sect. 2.
- **Spectroscopic.** As discussed in Sect. 4, nuclear flares that have blue colors, all have similar spectroscopic properties: broad helium and/or hydrogen on an otherwise predominantly featureless spectrum. In Table 1 we include also PS1-11af even though it showed no obvious features in its optical spectrum.
- **Photometric.** As discussed in Sect. 3, events that satisfy the above criteria share two key photometric properties: a high blackbody temperature and no signs of cooling after peak luminosity. For the purpose of populating Table 1 we require $T > 1.2 \times 10^4$ K near peak. We note that for almost all TDEs in Table 1, the temperature measurements include ultraviolet photometry (from *Swift* or GALEX) and in most cases multiple ultraviolet observations allow a measurement of temperature evolution (see Table 2). The three TDEs from PTF are an exception, their blackbody temperatures are measured from the optical spectrum obtained near the peak of the flare (see Arcavi et al. 2014).
- **Quiescent galaxy/not AGN.** This requirement is the most controversial because TDEs can also happen in AGN. Perhaps the TDE rate is even enhanced in AGN (Stone et al. 2018), and they might manifest themselves with distinct observational characteristics (Chan et al. 2019; Trakhtenbrot et al. 2019a). Unfortunately, the origin of large amplitude AGN variability is poorly understood and such flares can masquerade as TDEs (see the One-off Events and Imposters Chapter). For the purpose of populating Table 1 we define an event to be in an AGN if one of the following three properties is observed: (i) there is statistically-significant variability before the main flare, (ii) the signature of a dusty torus is detected using WISE colors (Stern et al. 2012), or (iii) the host is classified as a Seyfert galaxy using the BPT diagram (Baldwin et al. 1981). These criteria select a relatively pure AGN sample. We do not reject events based on observational signs that indicate recent (but not on-going) activity of the supermassive black hole, such as a LINER spectrum (Heckman 1980) or relic radio lobes.

The above selection criteria yield 33 sources presented in Table 1.

A few key photometric properties of our optical-ultraviolet TDE sample are presented in Table 2. For all sources, these features have been measured using the method presented in

Table 1 Discovery information of optically-selected TDEs with both spectroscopic and photometric observations

Discovery name	IAU name	R.A./Decl.	Peak[a] date	Redshift	Peak[b] M_g	Spectroscopic type (subtype)[c]	Host mass[d] ($\log M_\odot$)	Discovery Ref.[e]
SDSS-TDE2		23:23:48.61 −01:08:10.2	2007-06-15*	0.2515	−21.3	H	$10.57^{0.17}_{0.10}$	1
PTF-09ge		14:57:03.18 +49:36:40.9	2009-06-10	0.064	−19.8	He (Fe?)	$10.11^{0.13}_{0.12}$	2
PTF-09axc		14:53:13.07 +22:14:32.2	2009-07-04	0.1146	−19.5	H	$10.12^{0.11}_{0.17}$	2
PTF-09djl		16:33:55.97 +30:14:16.6	2009-08-05	0.184	−20.3	H	$9.95^{0.15}_{0.12}$	2
PS1-10jh		16:09:28.27 +53:40:23.9	2010-07-16	0.1696	−19.6	He	$9.61^{0.10}_{0.13}$	3
PS1-11af		09:57:26.81 +03:14:00.9	2011-01-17	0.4046	−20.2	Unknown	$10.21^{0.21}_{0.19}$	4
ASASSN-14ae		11:08:40.11 +34:05:52.2	2014-01-27*	0.0436	−19.6	H → H+He	$10.02^{0.09}_{0.17}$	5
ASASSN-14li		12:48:15.23 +17:46:26.4	2014-11-28*	0.02058	−18.1	H+He (N)	$9.69^{0.05}_{0.10}$	6
iPTF-15af		08:48:28.13 +22:03:33.4	2015-02-08	0.07897	−18.0	H+He (N+O)	$10.33^{0.11}_{0.18}$	7
ASASSN-15oi		20:39:09.14 −30:45:20.6	2015-08-13*	0.0484	−20.2	He (Fe?+O?)	$10.02^{0.04}_{0.04}$	8
OGLE16aaa		01:07:20.76 −64:16:20.4	2016-01-27	0.1655	−20.7	H+He	$10.47^{0.09}_{0.11}$	9
iPTF-16axa		17:03:34.34 +30:35:36.6	2016-05-15*	0.108	−19.0	H+He (N+O?)	$10.18^{0.10}_{0.14}$	10
iPTF-16fnl		00:29:57.05 +32:53:37.2	2016-08-26	0.0163	−17.5	H+He (N)	$9.35^{0.12}_{0.15}$	11
PS17dhz	AT 2017eqx	22:26:48.37 +17:08:52.3	2017-06-06	0.1089	−18.8	H+He → He (N+Fe?)	$9.44^{0.11}_{0.13}$	12
PS18kh	AT 2018zr	07:56:54.54 +34:15:43.6	2018-03-03	0.071	−19.9	H	$10.03^{0.09}_{0.18}$	13
ZTF18aahqkbt	AT 2018bsi	08:15:26.63 +45:35:32.0	2018-04-09*	0.051	−19.0	H+He (N)	$10.61^{0.05}_{0.06}$	14
ASASSN-18ul	AT 2018fyk	22:50:16.09 −44:51:53.5	2018-07-18*	0.059	−19.6	H+He (O+Fe)	$10.57^{0.12}_{0.21}$	15
ASASSN-18pg	AT 2018dyb	16:10:58.87 −60:55:24.3	2018-08-16	0.018	−19.8	H+He (N+O)	$10.00^{0.11}_{0.15}$	16
ATLAS18way	AT 2018hco	01:07:33.63 +23:28:34.2	2018-10-10	0.088	−19.4	H	$9.90^{0.09}_{0.18}$	14
ASASSN-18zj	AT 2018hyz	10:06:50.88 +01:41:33.9	2018-11-05*	0.04573	−20.3	H	$9.75^{0.12}_{0.26}$	14

Table 1 (*Continued*)

Discovery name	IAU name	R.A./Decl.	Peak[a] date	Redshift	Peak[b] M_g	Spectroscopic type (subtype)[c]	Host mass[d] ($\log M_{\odot}$)	Discovery Ref.[e]
ATLAS18yzs	AT 2018iih	17:28:03.92 +30:41:31.5	2018-11-20	0.212	−21.6	He	$10.76^{0.09}_{0.15}$	14
ZTF18actaqdw	AT 2018lni	04:09:37.65 +73:53:41.6	2018-12-08	0.138	−19.4	H+He (N+O)	$9.94^{0.10}_{0.15}$	14
ZTF19aabbnzo	AT 2018lna	07:03:18.64 +23:01:44.6	2019-01-25	0.091	−19.5	H+He (N?)	$9.47^{0.12}_{0.09}$	14
ZTF19aakswrb	AT 2019bhf	15:09:15.97 +16:14:22.5	2019-02-25	0.1206	−20.1	H	$10.21^{0.15}_{0.12}$	14
ASASSN-19bt	AT 2019ahk	07:00:11.39 −66:02:24.7	2019-03-06	0.0262	−20.7	H	$9.74^{0.09}_{0.10}$	17
ZTF19aakiwze	AT 2019cho	12:55:09.20 +49:31:09.9	2019-02-21	0.193	−20.7	H+He (N)	$10.10^{0.17}_{0.16}$	14
ASASSN-19dj	AT 2019azh	08:13:16.95 +22:38:53.8	2019-03-16	0.0222	−19.9	H+He	$9.86^{0.15}_{0.14}$	18
ZTF19aapreis	AT 2019dsg	20:57:02.97 +14:12:16.2	2019-04-27	0.0512	−19.2	H+He (N+O)	$10.36^{0.17}_{0.12}$	14
Gaia19bpt	AT 2019ehz	14:09:41.90 +55:29:27.7	2019-05-08	0.074	−19.6	H	$9.67^{0.09}_{0.15}$	14
ZTF19abidbya	AT 2019lwu	23:11:12.30 −01:00:10.8	2019-07-26	0.117	−19.8	H	$9.87^{0.12}_{0.15}$	14
ZTF19abhhjcc	AT 2019meg	18:45:16.19 +44:26:18.9	2019-08-01	0.152	−20.2	H	$9.69^{0.05}_{0.05}$	14
ATLAS19qqu	AT 2019mha	16:16:27.78 +56:25:56.3	2019-08-09	0.148	−19.9	H	$10.05^{0.11}_{0.15}$	14
ZTF19abzrhgq	AT 2019qiz	04:46:37.88 −10:13:34.8	2019-10-05	0.0151	−18.6	H+He (N)	$10.03^{0.10}_{0.14}$	14

[a]The peak date is obtained using a light curve model to interpolate between gaps in the data (see van Velzen et al. 2020). The sources that are only detected post-peak are indicated with an asterisk – for these we provide the date of the first detection

[b]The maximum observed optical luminosity, k-corrected to the g-band in the rest-frame of the TDE using the best-fit blackbody temperature to make the spectral correction

[c]Arrows denote a change that occurred in the spectral type with time

[d]The host stellar mass as estimated from archival optical photometry, obtained before the TDE occurred (see van Velzen et al. 2020)

[e]The discovery paper for this source (first journal article to present a classification and observed properties): 1: van Velzen et al. (2011), 2: Arcavi et al. (2014), 3: Gezari et al. (2012), 4: Chornock et al. (2014), 5: Holoien et al. (2014), 6: Miller et al. (2015), 7: Blagorodnova et al. (2019), 8: Holoien et al. (2016a), 9: Wyrzykowski et al. (2017), 10: Hung et al. (2017), 11: Blagorodnova et al. (2017), 12: (Nicholl et al. 2019), 13: Holoien et al. (2019a), 14: van Velzen et al. (2020), 15: Wevers et al. (2019b), 16: Leloudas et al. (2019), 17: Holoien et al. (2019b), 18: Liu et al. (2019)

Table 2 Light curve properties of the TDEs listed in Table 1

Survey name	IAU name	L_{bb} log erg/s	$\langle T \rangle_{100d}$ log K	dT/dt 10^2 K/day	$t_{peak/max}$ MJD	σ log day	τ log day	p	t_0 log day	$t_0\|_{p=-5/3}$ log day
SDSS-TDE2		$44.54^{0.08}_{0.06}$	$4.35^{0.03}_{0.03}$	$0.39^{0.22}_{0.25}$	54267.0	–	$2.08^{0.09}_{0.08}$	$-0.7^{0.2}_{0.2}$	$1.51^{0.35}_{0.32}$	$2.32^{0.12}_{0.13}$
PTF-09ge		$44.04^{0.01}_{0.01}$	$4.31^{0.00}_{0.00}$	–	$54992.9^{1.0}_{1.0}$	$1.3^{0.02}_{0.02}$	$1.78^{0.02}_{0.02}$	$-1.7^{0.2}_{0.2}$	$1.85^{0.08}_{0.10}$	$1.74^{0.04}_{0.03}$
PTF-09axc		$43.46^{0.03}_{0.02}$	$4.08^{0.00}_{0.00}$	–	$55016.4^{9.3}_{6.0}$	$0.9^{0.23}_{0.37}$	$2.45^{0.39}_{0.43}$	$-1.5^{0.3}_{0.3}$	$2.21^{0.14}_{0.17}$	$2.22^{0.08}_{0.07}$
PTF-09djl		$44.42^{0.04}_{0.04}$	$4.41^{0.00}_{0.00}$	–	$55048.3^{5.6}_{4.6}$	$1.1^{0.07}_{0.14}$	$2.12^{0.52}_{0.33}$	$-1.7^{0.2}_{0.3}$	$1.67^{0.11}_{0.14}$	$1.63^{0.06}_{0.05}$
PS1-10jh		$44.47^{0.07}_{0.07}$	$4.49^{0.03}_{0.03}$	$0.43^{0.43}_{0.29}$	$55393.8^{1.4}_{1.4}$	$1.3^{0.02}_{0.02}$	$1.56^{0.03}_{0.03}$	$-1.5^{0.2}_{0.6}$	$1.44^{0.41}_{0.22}$	$1.55^{0.07}_{0.07}$
PS1-11af		$44.16^{0.03}_{0.03}$	$4.31^{0.01}_{0.01}$	$-0.80^{0.39}_{0.28}$	$55578.9^{1.2}_{1.2}$	$0.9^{0.04}_{0.04}$	$1.69^{0.02}_{0.02}$	$-2.1^{0.5}_{0.5}$	$1.73^{0.16}_{0.17}$	$1.54^{0.11}_{0.09}$
ASASSN-14ae		$43.87^{0.02}_{0.01}$	$4.27^{0.01}_{0.01}$	$0.24^{0.15}_{0.13}$	56684.9	–	$1.46^{0.01}_{0.01}$	$-2.4^{0.2}_{0.1}$	$1.55^{0.05}_{0.07}$	$1.23^{0.03}_{0.03}$
ASASSN-14li		$43.66^{0.02}_{0.02}$	$4.51^{0.01}_{0.01}$	$-0.21^{0.07}_{0.07}$	56989.4	–	$1.77^{0.01}_{0.01}$	$-1.4^{0.1}_{0.1}$	$1.69^{0.09}_{0.09}$	$1.80^{0.04}_{0.04}$
iPTF-15af		$44.10^{0.10}_{0.08}$	$4.70^{0.04}_{0.03}$	$1.42^{0.43}_{0.80}$	$57061.6^{1.6}_{1.8}$	$1.5^{0.02}_{0.03}$	$1.91^{0.03}_{0.02}$	$-1.6^{0.6}_{0.6}$	$1.91^{0.22}_{0.22}$	$1.91^{0.15}_{0.16}$
ASASSN-15oi		$44.47^{0.04}_{0.04}$	$4.52^{0.02}_{0.02}$	$1.92^{0.06}_{0.14}$	57248.2	–	$1.41^{0.01}_{0.01}$	$-2.4^{0.2}_{0.2}$	$1.69^{0.08}_{0.08}$	$1.52^{0.05}_{0.06}$
OGLE16aaa		$44.22^{0.01}_{0.01}$	$4.23^{0.00}_{0.00}$	$-0.11^{0.07}_{0.06}$	$57414.3^{1.5}_{1.7}$	$1.3^{0.03}_{0.03}$	$2.01^{0.06}_{0.05}$	$-2.2^{0.3}_{0.4}$	$2.19^{0.09}_{0.09}$	$2.03^{0.03}_{0.03}$
iPTF-16axa		$43.82^{0.03}_{0.02}$	$4.37^{0.01}_{0.01}$	$-0.52^{0.29}_{0.24}$	57523.4	–	$1.73^{0.01}_{0.01}$	$-1.6^{0.3}_{0.2}$	$1.54^{0.13}_{0.16}$	$1.56^{0.08}_{0.06}$
iPTF-16fnl		$43.18^{0.03}_{0.02}$	$4.40^{0.01}_{0.01}$	$-0.14^{0.16}_{0.13}$	$57626.5^{3.6}_{11.6}$	$1.3^{0.12}_{0.18}$	$1.36^{0.01}_{0.01}$	$-2.1^{0.2}_{0.2}$	$1.35^{0.08}_{0.09}$	$1.18^{0.05}_{0.05}$
PS17dhz	AT 2017eqx	$43.82^{0.03}_{0.05}$	$4.31^{0.01}_{0.01}$	$-0.39^{0.25}_{0.23}$	$57910.4^{3.9}_{7.1}$	$1.2^{0.19}_{0.20}$	$1.59^{0.02}_{0.02}$	$-2.0^{0.3}_{0.3}$	$1.47^{0.10}_{0.13}$	$1.31^{0.09}_{0.06}$
PS18kh	AT 2018zr	$43.78^{0.02}_{0.02}$	$4.14^{0.01}_{0.01}$	$0.46^{0.05}_{0.05}$	$58180.7^{1.1}_{1.0}$	$1.0^{0.04}_{0.03}$	$1.88^{0.03}_{0.03}$	$-0.8^{0.0}_{0.1}$	$1.23^{0.15}_{0.12}$	$2.14^{0.03}_{0.03}$
ZTF18aahqkbt	AT 2018bsi	$43.87^{0.08}_{0.08}$	$4.37^{0.03}_{0.03}$	$0.21^{0.63}_{0.30}$	58217.2	–	$1.94^{0.12}_{0.10}$	$-1.9^{0.4}_{0.6}$	$1.92^{0.25}_{0.21}$	$1.83^{0.09}_{0.07}$
ASASSN-18ul	AT 2018fyk	$44.48^{0.04}_{0.03}$	$4.56^{0.02}_{0.02}$	$-0.02^{0.12}_{0.12}$	58317.5	–	$1.99^{0.02}_{0.02}$	$-1.9^{0.1}_{0.1}$	$2.14^{0.06}_{0.06}$	$1.76^{0.03}_{0.02}$
ASASSN-18pg	AT 2018dyb	$44.16^{0.01}_{0.01}$	$4.37^{0.01}_{0.01}$	$-0.25^{0.03}_{0.03}$	$58346.7^{1.4}_{1.3}$	$1.5^{0.01}_{0.03}$	$1.59^{0.01}_{0.01}$	$-1.9^{0.1}_{0.1}$	$1.49^{0.05}_{0.05}$	$1.29^{0.03}_{0.03}$
ATLAS18way	AT 2018hco	$44.25^{0.04}_{0.04}$	$4.39^{0.01}_{0.01}$	$0.05^{0.10}_{0.09}$	$58401.8^{1.7}_{1.9}$	$0.9^{0.04}_{0.04}$	$2.03^{0.04}_{0.04}$	$-1.2^{0.2}_{0.2}$	$1.73^{0.16}_{0.15}$	$2.00^{0.06}_{0.07}$
ASASSN-18zj	AT 2018hyz	$44.11^{0.01}_{0.01}$	$4.25^{0.01}_{0.01}$	$0.18^{0.05}_{0.05}$	58428.0	–	$1.72^{0.01}_{0.01}$	$-1.1^{0.1}_{0.1}$	$1.29^{0.07}_{0.06}$	$1.71^{0.01}_{0.01}$

Table 2 *(Continued)*

Survey name	IAU name	L_{bb} log erg/s	$\langle T\rangle_{100d}$ log K	dT/dt 10^2 K/day	$t_{peak/max}$ MJD	σ log day	τ log day	p	t_0 log day	$t_0\|_{p=-5/3}$ log day
ATLAS18yzs	AT 2018iih	$44.62^{0.04}_{0.03}$	$4.23^{0.01}_{0.01}$	$0.19^{0.05}_{0.05}$	$58442.2^{1.6}_{0.8}$	$1.2^{0.01}_{0.01}$	$2.05^{0.04}_{0.04}$	$-0.9^{0.1}_{0.1}$	$1.61^{0.11}_{0.07}$	$2.06^{0.06}_{0.05}$
ZTF18actaqdw	AT 2018lni	$44.21^{0.29}_{0.17}$	$4.44^{0.09}_{0.07}$	$0.48^{0.46}_{0.47}$	$58460.3^{3.6}_{5.8}$	$0.8^{0.18}_{0.50}$	$2.16^{0.16}_{0.12}$	$-1.4^{0.8}_{1.6}$	$2.46^{0.37}_{0.45}$	$2.23^{0.22}_{0.19}$
ZTF19aabbnzo	AT 2018lna	$44.56^{0.06}_{0.06}$	$4.60^{0.03}_{0.02}$	$0.93^{0.72}_{0.92}$	$58508.3^{2.2}_{2.0}$	$1.2^{0.06}_{0.07}$	$1.65^{0.03}_{0.03}$	$-2.1^{0.7}_{0.7}$	$1.79^{0.24}_{0.21}$	$1.66^{0.16}_{0.15}$
ZTF19aakswrb	AT 2019bhf	$43.91^{0.04}_{0.05}$	$4.20^{0.02}_{0.02}$	$0.74^{0.15}_{0.14}$	$58539.3^{4.4}_{5.7}$	$0.6^{0.23}_{0.35}$	$1.70^{0.04}_{0.03}$	$-1.3^{0.4}_{0.4}$	$1.70^{0.23}_{0.20}$	$1.96^{0.07}_{0.09}$
ASASSN-19bt	AT 2019ahk	$44.08^{0.01}_{0.01}$	$4.21^{0.01}_{0.01}$	$0.10^{0.03}_{0.03}$	$58548.3^{0.8}_{0.9}$	$1.3^{0.01}_{0.01}$	$1.67^{0.01}_{0.01}$	$-1.6^{0.1}_{0.1}$	$1.66^{0.05}_{0.05}$	$1.72^{0.01}_{0.01}$
ZTF19aakiwze	AT 2019cho	$43.98^{0.01}_{0.01}$	$4.19^{0.01}_{0.01}$	$0.84^{0.21}_{0.23}$	$58535.3^{3.0}_{1.6}$	$0.8^{0.10}_{0.11}$	$1.94^{0.03}_{0.03}$	$-1.0^{0.4}_{0.5}$	$1.83^{0.25}_{0.13}$	$2.19^{0.13}_{0.16}$
ASASSN-19dj	AT 2019azh	$44.50^{0.02}_{0.02}$	$4.53^{0.01}_{0.01}$	$0.98^{0.21}_{0.19}$	$58558.6^{1.5}_{1.6}$	$1.3^{0.05}_{0.04}$	$1.83^{0.01}_{0.01}$	$-1.8^{0.2}_{0.1}$	$1.87^{0.08}_{0.08}$	$1.96^{0.05}_{0.04}$
ZTF19aapreis	AT 2019dsg	$44.26^{0.04}_{0.05}$	$4.49^{0.01}_{0.01}$	$-0.07^{0.14}_{0.12}$	$58600.2^{8.2}_{10.0}$	$1.2^{0.06}_{0.12}$	$1.80^{0.02}_{0.02}$	$-1.5^{0.1}_{0.1}$	$1.51^{0.11}_{0.10}$	$1.59^{0.06}_{0.06}$
Gaia19bpt	AT 2019ehz	$44.03^{0.01}_{0.02}$	$4.33^{0.01}_{0.01}$	$-0.40^{0.03}_{0.03}$	$58611.4^{0.6}_{0.6}$	$0.9^{0.03}_{0.02}$	$1.67^{0.01}_{0.01}$	$-1.8^{0.1}_{0.1}$	$1.73^{0.06}_{0.09}$	$1.61^{0.04}_{0.04}$
ZTF19abidbya	AT 2019lwu	$43.60^{0.03}_{0.04}$	$4.09^{0.01}_{0.01}$	$0.64^{0.14}_{0.19}$	$58690.2^{1.0}_{0.7}$	$0.4^{0.17}_{0.21}$	$1.68^{0.06}_{0.05}$	$-1.5^{0.4}_{0.5}$	$1.56^{0.20}_{0.18}$	$1.67^{0.09}_{0.08}$
ZTF19abhhjcc	AT 2019meg	$44.36^{0.03}_{0.04}$	$4.44^{0.01}_{0.01}$	$1.94^{0.05}_{0.09}$	$58696.4^{0.6}_{0.6}$	$0.9^{0.02}_{0.03}$	$1.72^{0.03}_{0.02}$	$-0.2^{0.1}_{0.5}$	$2.37^{0.44}_{0.66}$	$2.56^{0.15}_{0.18}$
ATLAS19qqu	AT 2019mha	$44.05^{0.06}_{0.05}$	$4.32^{0.03}_{0.03}$	$0.36^{0.88}_{1.07}$	$58704.0^{0.7}_{0.8}$	$1.1^{0.02}_{0.02}$	$1.23^{0.03}_{0.02}$	$-4.0^{0.7}_{0.5}$	$1.63^{0.11}_{0.12}$	$1.21^{0.10}_{0.09}$
ZTF19abzrhgq	AT 2019qiz	$43.44^{0.01}_{0.01}$	$4.28^{0.01}_{0.01}$	$-0.17^{0.10}_{0.09}$	$58761.4^{0.4}_{0.4}$	$0.8^{0.01}_{0.01}$	$1.48^{0.01}_{0.01}$	$-2.0^{0.1}_{0.1}$	$1.40^{0.06}_{0.06}$	$1.26^{0.02}_{0.03}$

Notes – Light curve features are derived using the method presented in van Velzen et al. (2020). The column L_{bb} lists the peak blackbody luminosity and $\langle T\rangle_{100d}$ is the mean blackbody temperature measured during the first 100 days post maximum light. A measurement of a linear temperature change during the first year of observations is listed as dT/dt (only estimated for sources with sufficient light curve coverage). $t_{peak/max}$ lists the best-fit time of peak, or, when no pre-peak detections are available, the time of the first detection. The parameters σ and τ list the result of fitting a Gaussian rise and exponential decay, respectively, to the light curve, while p and t_0 are the free parameters of a power-law decay ($L_{bb} \propto (t/t_0)^p$). The last column lists the decay timescale obtained when the power-law index is fixed at $p = -5/3$

van Velzen et al. (2019b, 2020). Central to this approach is the use of a model that includes both the blackbody light curve shape and temperature evolution. As such, the free parameters in this model can be measured simultaneously using all available photometric observations. The blackbody temperature can be kept constant in the model in order to obtain the mean temperature ($\langle T \rangle$). Alternatively, a linear temperature change with time (dT/dt) can also be extracted. For the majority of the TDEs in our sample this approach is sufficient because the measured values of dT/dt are consistent with zero (Table 1). However to allow for more flexibility (e.g., a relatively rapid temperature increase as observed for AT 2018zr), the light curve properties reported in Table 2 are obtained using a linear interpolation of the temperature on a grid with a spacing of 30 days (see van Velzen et al. 2020 for details and Fig. 2 for examples of the resulting temperature curves). In Table 1 we report both the exponential decay rate measured for the first 100 days of post-peak observations ($L_{\rm bb} \propto e^{-t/\tau}$) and the timescale measured from a power-law decay as applied to the first year of post-peak observations ($L_{\rm bb} \propto (t/t_0)^p$). For the power-law decay, we report both the results for a fixed power-law index (at the canonical value $p = -5/3$), as well as the results obtained when this index is left as a free parameter.

6 Rate and Luminosity Function

Measurements of the TDE rate have the potential to reveal how stars are fed to the loss cone (see the Rates Chapter in this book), which in turn can teach us about stellar dynamics in galaxy centers. This measurement, however, is not trivial, because it requires both a clear set of criteria for identifying TDEs and an understanding of the selection efficiency and biases of each search.

The first comprehensive estimate of disruption rates was based on the TDE search in SDSS imaging data (van Velzen et al. 2011) and is presented by van Velzen and Farrar (2014). Since the SDSS search was based solely on photometry, the selection function was relatively straightforward. The SDSS-selected TDEs were not detected pre-peak. However, by using a light curve model based on the well-sampled TDE PS1-10jh (Gezari et al. 2012) this limitation can be addressed. The remaining systematic uncertainty due to the uncertainty of the TDE light curve was found to be a factor ≈ 2.

The approach of van Velzen and Farrar (2014) is to compute the detection efficiency of the two TDEs separately, and report the mean rate $\dot{N}$ based on the mean efficiency $\langle \epsilon \rangle$:

$$\dot{N} = \frac{2}{\langle \epsilon \rangle \, N_{\rm gal} \tau} \tag{1}$$

Here $N_{\rm gal}$ is the number of galaxies that were observed in the survey duration τ. This approach yields a rate of $(1.5-2)^{+2.7}_{-1.3} \times 10^{-5}$ galaxy^{-1} year^{-1} (the sub/super scripts denote the statistical uncertainty while the range in parentheses approximates the systematic uncertainty due to the assumption of a light curve model).

Based on two TDEs detected by the ASAS-SN survey (ASASSN-14ae and ASASSN-14li), Holoien et al. (2016b) reported a rate of (2.2–17.0) $\times 10^{-5}$ galaxy^{-1} year^{-1} (here the range in parentheses denotes the 90% confidence level statistical uncertainty). The authors used a similar approach to Eq. (1) and computed the efficiency under the assumption that the TDE absolute V-band peak magnitudes are uniformly distributed between -19 and -20. This assumption might be why Holoien et al. (2016b) obtained a slightly higher rate compared to van Velzen and Farrar (2014), who included SDSS-TDE2, an event that is brighter at peak ($M_g = -20.8$; Table 1) than the range assumed by Holoien et al. (2016b), although we note that the two estimates are consistent within their Poisson uncertainty.

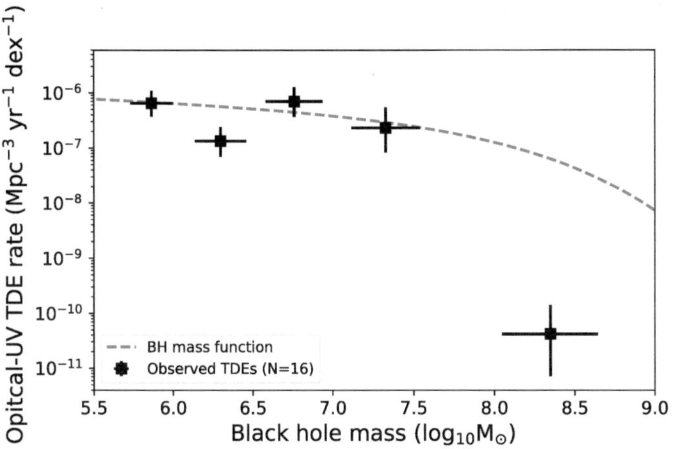

Fig. 13 The TDE black-hole mass function (updated version of Fig. 3 from van Velzen 2018). The dashed lines show the black-hole mass function (Shankar et al. 2004) multiplied by a constant factor of 6×10^{-5} yr^{-1}. The event in the highest mass bin is ASASSN-15lh (the origin of this transient, TDE or supernova, is controversial). Below the mass scale where stars can be swallowed by the black hole prior to disruption (Hills 1975; Kesden 2012b), the optical-ultraviolet TDE mass function appears to trace the black–hole mass function

Samples based on luminous events produce a lower inferred rate according to Eq. (1), since more luminous events have a higher efficiency ϵ (they can be detected to larger distance). Even when combining more luminous and less luminous events, the more luminous ones dominate the mean efficiency $\langle\epsilon\rangle$, thus Eq. (1), in general, always yields a rate that is weighted more strongly toward more luminous events. To solve this issue we have to consider the TDE luminosity function.

The fact that the small flux-limited samples of TDEs span a wide range of luminosities (for e.g. the two TDEs found in the SDSS search had a luminosity difference of almost an order of magnitude), informs us that the luminosity function can not be flat: faint events must be more common than luminous ones. This effect is quantified in van Velzen (2018), who combine events from different surveys under the assumption that each survey samples the same TDE population. This approach yields a measure of the *shape* of the luminosity function: $dN/d\log L_g \propto L_g^{-1.3\pm0.3}$ (with L_g the rest-frame g-band peak luminosity). Since the faint events dominate, van Velzen (2018) find that extrapolating the observed absolute rate from SDSS to the faintest observed TDEs increases the inferred mean rate to about 1×10^{-4} galaxy^{-1} year^{-1}. Hung et al. (2018) find a similar result of $1.7^{+2.7}_{-1.3} \times 10^{-4}$ galaxy^{-1} year^{-1} by applying the van Velzen (2018) luminosity function to compute the rate from the three TDEs discovered by the iPTF survey.

Since the g-band approaches the Rayleigh-Jeans tail of the SED, the steep L_g luminosity function does not translate directly into a steep bolometric luminosity function. For example, van Velzen et al. (2020) find that the TDE H+He and TDE H have a similar peak bolometric luminosity and occur at a similar rate in flux-limited samples, but since TDE H+He have a higher mean temperature, their optical luminosity is lower and their intrinsic rate is higher. Therefore, the temperature or radius of the optical-ultraviolet photosphere plays an important role for determining the shape of the optical luminosity function.

The work by van Velzen (2018) also enables the first measurement of the TDE black-hole mass function. Figure 13 show the TDE rate as a function of black-hole mass, as estimated based on the M-σ relation with velocity dispersion measurement from Wevers et al. (2017,

2019b). The lack of TDEs from high mass black holes is consistent with the suppression of the TDE rate by the "direct capture" of stars due to the black-hole event horizon (Hills 1975).

7 Concluding Remarks

While the set of known optical-ultraviolet TDEs is rapidly growing, many questions remain open regarding their nature, emission properties, and true diversity. Most prominent is the question of their emission source – is the emission coming entirely from outer shocks, entirely from reprocessed accretion emission, or some combination of both? Answering this question could help explain the observed emission properties and perhaps their expected diversity, thus helping to identify more events correctly.

Observationally, the origin in the diversity of TDE spectral features, such as the species seen, the line profiles measured and their evolution, and the differences between the optical and ultraviolet spectra, will likely continue to be topics of study. In addition, the scale of the photometric diversity is tied to the question of how many rapidly declining events are being missed, affecting our ability to accurately estimate TDE rates and learn about stellar dynamics in galaxy centers.

As more events are found, we will be able to continue to explore the classes of transients occurring in galaxy centers – increasing our sensitivity to new types of TDEs, mapping the diversity of the known types, and learning to distinguish between them and galaxy-central events that are not related to TDEs. Hopefully, this will help inform theoretical models on which disruption, accretion and emission mechanisms are realized in nature. In turn, such models could help find more events, and focus observational resources on the most critical phases and wavelengths.

One of the main motivations for studying TDEs has been to reveal the population of otherwise quiescent SMBHs. Some of the questions above are still hindering the realization of the full potential of TDEs on that front. However, in the mean time, optical-ultraviolet TDEs are teaching us a lot about accretion physics, emission mechanisms, galaxy dynamics, and more. Given the increasing detection rate of (optical) transients, combined with added attention and followup resources invested in transients found in galaxy centers, we will learn more about all of these topics and likely make new and unexpected discoveries along the way.

Acknowledgements We thank the International Space Science Institute for hosting us on several occasions to discuss the content and structure of this chapter, and its coordination with other chapters in this book. We are grateful to the two anonymous referees for their insightful comments and helpful suggestions. We thank Thomas Wevers for useful comments. S. van Velzen is supported by the James Arthur Postdoctoral Fellowship. F. Onori acknowledges the support of the H2020 Hemera program, grant agreement No 730970, and the support from European Research Council Consolidator Grant 647208. I. Arcavi is a CIFAR Azrieli Global Scholar in the Gravity and the Extreme Universe Program and acknowledges support from that program, from the European Research Council (ERC) under the European Union's Horizon 2020 research and innovation program (grant agreement number 852097), from the Israel Science Foundation (grant numbers 2108/18 and 2752/19), from the United States – Israel Binational Science Foundation (BSF), and from the Israeli Council for Higher Education Alon Fellowship.

References

K.D. Alexander, E. Berger, J. Guillochon, B.A. Zauderer, P.K.G. Williams, Discovery of an outflow from radio observations of the tidal disruption event ASASSN-14li. Astrophys. J. Lett. **819**, 25 (2016). https://doi.org/10.3847/2041-8205/819/2/L25

I. Arcavi, A. Gal-Yam, M. Sullivan, Y.-C. Pan, S.B. Cenko, A. Horesh, E.O. Ofek, A. De Cia, L. Yan, C.-W. Yang, D.A. Howell, D. Tal, S.R. Kulkarni, S.P. Tendulkar, S. Tang, D. Xu, A. Sternberg, J.G. Cohen, J.S. Bloom, P.E. Nugent, M.M. Kasliwal, D.A. Perley, R.M. Quimby, A.A. Miller, C.A. Theissen, R.R. Laher, A continuum of H- to He-rich tidal disruption candidates with a preference for E+A galaxies. Astrophys. J. **793**, 38 (2014). https://doi.org/10.1088/0004-637X/793/1/38

K. Auchettl, J. Guillochon, E. Ramirez-Ruiz, New physical insights about tidal disruption events from a comprehensive observational inventory at X-ray wavelengths. Astrophys. J. **838**, 149 (2017). https://doi.org/10.3847/1538-4357/aa633b

N. Bade, S. Komossa, M. Dahlem, Detection of an extremely soft X-ray outburst in the HII-like nucleus of NGC 5905. Astron. Astrophys. **309**, 35–38 (1996)

J.A. Baldwin, M.M. Phillips, R. Terlevich, Classification parameters for the emission-line spectra of extragalactic objects. Publ. Astron. Soc. Pac. **93**, 5–19 (1981). https://doi.org/10.1086/130766

E.C. Bellm, S.R. Kulkarni, M.J. Graham, R. Dekany, R.M. Smith, R. Riddle, F.J. Masci, G. Helou, T.A. Prince, S.M. Adams, C. Barbarino, T. Barlow, J. Bauer, R. Beck, J. Belicki, R. Biswas, N. Blagorodnova, D. Bodewits, B. Bolin, V. Brinnel, T. Brooke, B. Bue, M. Bulla, R. Burruss, S.B. Cenko, C.-K. Chang, A. Connolly, M. Coughlin, J. Cromer, V. Cunningham, K. De, A. Delacroix, V. Desai, D.A. Duev, G. Eadie, T.L. Farnham, M. Feeney, U. Feindt, D. Flynn, A. Franckowiak, S. Frederick, C. Fremling, A. Gal-Yam, S. Gezari, M. Giomi, D.A. Goldstein, V.Z. Golkhou, A. Goobar, S. Groom, E. Hacopians, D. Hale, J. Henning, A.Y.Q. Ho, D. Hover, J. Howell, T. Hung, D. Huppenkothen, D. Imel, W.-H. Ip, Ž. Ivezić, E. Jackson, L. Jones, M. Juric, M.M. Kasliwal, S. Kaspi, S. Kaye, M.S.P. Kelley, M. Kowalski, E. Kramer, T. Kupfer, W. Landry, R.R. Laher, C.-D. Lee, H.W. Lin, Z.-Y. Lin, R. Lunnan, M. Giomi, A. Mahabal, P. Mao, A.A. Miller, S. Monkewitz, P. Murphy, C.-C. Ngeow, J. Nordin, P. Nugent, E. Ofek, M.T. Patterson, B. Penprase, M. Porter, L. Rauch, U. Rebbapragada, D. Reiley, M. Rigault, H. Rodriguez, J. van Roestel, B. Rusholme, J. van Santen, S. Schulze, D.L. Shupe, L.P. Singer, M.T. Soumagnac, R. Stein, J. Surace, J. Sollerman, P. Szkody, F. Taddia, S. Terek, A. Van Sistine, S. van Velzen, W.T. Vestrand, R. Walters, C. Ward, Q.-Z. Ye, P.-C. Yu, L. Yan, J. Zolkower, The Zwicky transient facility: system overview, performance, and first results. Publ. Astron. Soc. Pac. **131**(1), 018002 (2019). https://doi.org/10.1088/1538-3873/aaecbe

N. Blagorodnova, S. Gezari, T. Hung, S.R. Kulkarni, S.B. Cenko, D.R. Pasham, L. Yan, I. Arcavi, S. Ben-Ami, B.D. Bue, T. Cantwell, Y. Cao, A.J. Castro-Tirado, R. Fender, C. Fremling, A. Gal-Yam, A.Y.Q. Ho, A. Horesh, G. Hosseinzadeh, M.M. Kasliwal, A.K.H. Kong, R.R. Laher, G. Leloudas, R. Lunnan, F.J. Masci, K. Mooley, J.D. Neill, P. Nugent, M. Powell, A.F. Valeev, P.M. Vreeswijk, R. Walters, P. Wozniak, iPTF16fnl: a faint and fast tidal disruption event in an E+A galaxy. Astrophys. J. **844**, 46 (2017). https://doi.org/10.3847/1538-4357/aa7579

N. Blagorodnova, S.B. Cenko, S.R. Kulkarni, I. Arcavi, J.S. Bloom, G. Duggan, A.V. Filippenko, C. Fremling, A. Horesh, G. Hosseinzadeh, E. Karamehmetoglu, A. Levan, F.J. Masci, P.E. Nugent, D.R. Pasham, S. Veilleux, R. Walters, L. Yan, W. Zheng, The broad absorption line tidal disruption event iPTF15af: optical and ultraviolet evolution. Astrophys. J. **873**(1), 92 (2019). https://doi.org/10.3847/1538-4357/ab04b0

P.K. Blanchard, M. Nicholl, E. Berger, J. Guillochon, R. Margutti, R. Chornock, K.D. Alexander, J. Leja, M.R. Drout, PS16dtm: a tidal disruption event in a narrow-line Seyfert 1 galaxy. Astrophys. J. **843**, 106 (2017). https://doi.org/10.3847/1538-4357/aa77f7

T. Bogdanović, M. Eracleous, S. Mahadevan, S. Sigurdsson, P. Laguna, Tidal disruption of a star by a black hole: observational signature. Astrophys. J. **610**, 707–721 (2004). https://doi.org/10.1086/421758

I.S. Bowen, The origin of the nebular lines and the structure of the planetary nebulae. Astrophys. J. **67**, 1–15 (1928). https://doi.org/10.1086/143091

M.S. Brotherton, H.D. Tran, R.H. Becker, M.D. Gregg, S.A. Laurent-Muehleisen, R.L. White, Composite spectra from the FIRST Bright Quasar Survey. Astrophys. J. **546**, 775–781 (2001). https://doi.org/10.1086/318309

J.S. Brown, B.J. Shappee, T.W.-S. Holoien, K.Z. Stanek, C.S. Kochanek, J.L. Prieto, Hello darkness my old friend: the fading of the nearby TDE ASASSN-14ae. Mon. Not. R. Astron. Soc. **462**, 3993–4000 (2016). https://doi.org/10.1093/mnras/stw1928

J.S. Brown, T.W.-S. Holoien, K. Auchettl, K.Z. Stanek, C.S. Kochanek, B.J. Shappee, J.L. Prieto, D. Grupe, The long term evolution of ASASSN-14li. Mon. Not. R. Astron. Soc. **466**(4), 4904–4916 (2017). https://doi.org/10.1093/mnras/stx033

J.S. Brown, C.S. Kochanek, T.W.-S. Holoien, K.Z. Stanek, K. Auchettl, B.J. Shappee, J.L. Prieto, N. Morrell, E. Falco, J. Strader, L. Chomiuk, R. Post, S. Villanueva Jr., S. Mathur, S. Dong, P. Chen, S. Bose, The ultraviolet spectroscopic evolution of the low-luminosity tidal disruption event iPTF16fnl. Mon. Not. R. Astron. Soc. **473**, 1130–1144 (2018). https://doi.org/10.1093/mnras/stx2372

J.K. Cannizzo, H.M. Lee, J. Goodman, The disk accretion of a tidally disrupted star onto a massive black hole. Astrophys. J. **351**, 38–46 (1990). https://doi.org/10.1086/168442

S.B. Cenko, A. Cucchiara, N. Roth, S. Veilleux, J.X. Prochaska, L. Yan, J. Guillochon, W.P. Maksym, I. Arcavi, N.R. Butler, A.V. Filippenko, A.S. Fruchter, S. Gezari, D. Kasen, A.J. Levan, J.M. Miller, D.R. Pasham, E. Ramirez-Ruiz, L.E. Strubbe, N.R. Tanvir, F. Tombesi, An ultraviolet spectrum of the tidal disruption flare ASASSN-14li. Astrophys. J. Lett. **818**, 32 (2016). https://doi.org/10.3847/2041-8205/818/2/L32

K.C. Chambers, The PS1 system and science mission, in *Bulletin of the AAS*, vol. 38 (2007), p. 995

K.C. Chambers, E.A. Magnier, N. Metcalfe, H.A. Flewelling, M.E. Huber, C.Z. Waters, L. Denneau, P.W. Draper, D. Farrow, D.P. Finkbeiner, C. Holmberg, J. Koppenhoefer, P.A. Price, R.P. Saglia, E.F. Schlafly, S.J. Smartt, W. Sweeney, R.J. Wainscoat, W.S. Burgett, T. Grav, J.N. Heasley, K.W. Hodapp, R. Jedicke, N. Kaiser, R.-P. Kudritzki, G.A. Luppino, R.H. Lupton, D.G. Monet, J.S. Morgan, P.M. Onaka, C.W. Stubbs, J.L. Tonry, E. Banados, E.F. Bell, R. Bender, E.J. Bernard, M.T. Botticella, S. Casertano, S. Chastel, W.-P. Chen, X. Chen, S. Cole, N. Deacon, C. Frenk, A. Fitzsimmons, S. Gezari, C. Goessl, T. Goggia, B. Goldman, E.K. Grebel, N.C. Hambly, G. Hasinger, A.F. Heavens, T.M. Heckman, R. Henderson, T. Henning, M. Holman, U. Hopp, W.-H. Ip, S. Isani, C.D. Keyes, A. Koekemoer, R. Kotak, K.S. Long, J.R. Lucey, M. Liu, N.F. Martin, B. McLean, E. Morganson, D.N.A. Murphy, M.A. Nieto-Santisteban, P. Norberg, J.A. Peacock, E.A. Pier, M. Postman, N. Primak, C. Rae, A. Rest, A. Riess, A. Riffeser, H.W. Rix, S. Roser, E. Schilbach, A.S.B. Schultz, D. Scolnic, A. Szalay, S. Seitz, B. Shiao, E. Small, K.W. Smith, D. Soderblom, A.N. Taylor, A.R. Thakar, J. Thiel, D. Thilker, Y. Urata, J. Valenti, F. Walter, S.P. Watters, S. Werner, R. White, W.M. Wood-Vasey, R. Wyse, The Pan-STARRS1 surveys. arXiv e-prints, 1612.05560 (2016)

C.-H. Chan, T. Piran, J.H. Krolik, D. Saban, Tidal disruption events in active galactic nuclei. Astrophys. J. **881**(2), 113 (2019). https://doi.org/10.3847/1538-4357/ab2b40

R. Chornock, E. Berger, S. Gezari, B.A. Zauderer, A. Rest, L. Chomiuk, A. Kamble, A.M. Soderberg, I. Czekala, J. Dittmann, M. Drout, R.J. Foley, W. Fong, M.E. Huber, R.P. Kirshner, A. Lawrence, R. Lunnan, G.H. Marion, G. Narayan, A.G. Riess, K.C. Roth, N.E. Sanders, D. Scolnic, S.J. Smartt, K. Smith, C.W. Stubbs, J.L. Tonry, W.S. Burgett, K.C. Chambers, H. Flewelling, K.W. Hodapp, N. Kaiser, E.A. Magnier, D.C. Martin, J.D. Neill, P.A. Price, R. Wainscoat, The ultraviolet-bright, slowly declining transient PS1-11af as a partial tidal disruption event. Astrophys. J. **780**, 44 (2014). https://doi.org/10.1088/0004-637X/780/1/44

L. Dai, J.C. McKinney, N. Roth, E. Ramirez-Ruiz, M.C. Miller, A unified model for tidal disruption events. Astrophys. J. **859**, 20 (2018). https://doi.org/10.3847/2041-8213/aab429

S. Dong, B.J. Shappee, J.L. Prieto, S.W. Jha, K.Z. Stanek, T.W.-S. Holoien, C.S. Kochanek, T.A. Thompson, N. Morrell, I.B. Thompson, U. Basu, J.F. Beacom, D. Bersier, J. Brimacombe, J.S. Brown, F. Bufano, P. Chen, E. Conseil, A.B. Danilet, E. Falco, D. Grupe, S. Kiyota, G. Masi, B. Nicholls, F. Olivares E., G. Pignata, G. Pojmanski, G.V. Simonian, D.M. Szczygiel, P.R. Woźniak, ASASSN-15lh: a highly super-luminous supernova. Science **351**, 257–260 (2016). https://doi.org/10.1126/science.aac9613

P. Esquej, R.D. Saxton, M.J. Freyberg, A.M. Read, B. Altieri, M. Sanchez-Portal, G. Hasinger, Candidate tidal disruption events from the XMM-Newton slew survey. Astron. Astrophys. **462**, 49–52 (2007). https://doi.org/10.1051/0004-6361:20066072

P. Esquej, R.D. Saxton, S. Komossa, A.M. Read, M.J. Freyberg, G. Hasinger, D.A. García-Hernández, H. Lu, J.R. Zaurín, M. Sánchez-Portal, H. Zhou, Evolution of tidal disruption candidates discovered by XMM-Newton. Astron. Astrophys. **489**, 543–554 (2008). https://doi.org/10.1051/0004-6361:200810110

S. Frederick, S. Gezari, M.J. Graham, S.B. Cenko, S. van Velzen, D. Stern, N. Blagorodnova, S.R. Kulkarni, L. Yan, K. De, U.C. Fremling, T. Hung, E. Kara, D.L. Shupe, C. Ward, E.C. Bellm, R. Dekany, D.A. Duev, U. Feindt, M. Giomi, T. Kupfer, R.R. Laher, F.J. Masci, A.A. Miller, J.D. Neill, C.-C. Ngeow, M.T. Patterson, M. Porter, B. Rusholme, J. Sollerman, R. Walters, A new class of changing-look LINERs. Astrophys. J. **883**(1), 31 (2019). https://doi.org/10.3847/1538-4357/ab3a38

J.A. Frieman, B. Bassett, A. Becker, C. Choi, D. Cinabro, F. DeJongh, D.L. Depoy, B. Dilday, M. Doi, P.M. Garnavich, C.J. Hogan, J. Holtzman, M. Im, S. Jha, R. Kessler, K. Konishi, H. Lampeitl, J. Marriner, J.L. Marshall, D. McGinnis, G. Miknaitis, R.C. Nichol, J.L. Prieto, A.G. Riess, M.W. Richmond, R. Romani, M. Sako, D.P. Schneider, M. Smith, N. Takanashi, K. Tokita, K. van der Heyden, N. Yasuda, C. Zheng, J. Adelman-McCarthy, J. Annis, R.J. Assef, J. Barentine, R. Bender, R.D. Blandford, W.N. Boroski, M. Bremer, H. Brewington, C.A. Collins, A. Crotts, J. Dembicky, J. Eastman, A. Edge, E. Edmondson, E. Elson, M.E. Eyler, A.V. Filippenko, R.J. Foley, S. Frank, A. Goobar, T. Gueth, J.E. Gunn, M. Harvanek, U. Hopp, Y. Ihara, Ž. Ivezić, S. Kahn, J. Kaplan, S. Kent, W. Ketzeback, S.J. Kleinman, W. Kollatschny, R.G. Kron, J. Krzesiński, D. Lamenti, G. Leloudas, H. Lin, D.C. Long, J. Lucey, R.H. Lupton, E. Malanushenko, V. Malanushenko, R.J. McMillan, J. Mendez, C.W. Morgan, T. Morokuma, A. Nitta, L. Ostman, K. Pan, C.M. Rockosi, A.K. Romer, P. Ruiz-Lapuente, G. Saurage, K. Schlesinger, S.A. Snedden, J. Sollerman, C. Stoughton, M. Stritzinger, M. Subba Rao, D. Tucker, P. Vaisanen, L.C. Watson, S. Watters, J.C. Wheeler, B. Yanny, D. York, The Sloan Digital Sky Survey-II Supernova Survey: technical summary. Astron. J. **135**, 338–347 (2008). https://doi.org/10.1088/0004-6256/135/1/338

C.M. Gaskell, P.A. Rojas Lobos, The production of strong, broad He II emission after the tidal disruption of a main-sequence star by a supermassive black hole. Mon. Not. R. Astron. Soc. **438**, 36–40 (2014). https://doi.org/10.1093/mnrasl/slt154

S. Gezari, D.C. Martin, B. Milliard, S. Basa, J.P. Halpern, K. Forster, P.G. Friedman, P. Morrissey, S.G. Neff, D. Schiminovich, M. Seibert, T. Small, T.K. Wyder, Ultraviolet detection of the tidal disruption of a star by a supermassive black hole. Astrophys. J. Lett. **653**, 25–28 (2006). https://doi.org/10.1086/509918

S. Gezari, S. Basa, D.C. Martin, G. Bazin, K. Forster, B. Milliard, J.P. Halpern, P.G. Friedman, P. Morrissey, S.G. Neff, D. Schiminovich, M. Seibert, T. Small, T.K. Wyder, UV/optical detections of candidate tidal disruption events by GALEX and CFHTLS. Astrophys. J. **676**, 944–969 (2008). https://doi.org/10.1086/529008

S. Gezari, T. Heckman, S.B. Cenko, M. Eracleous, K. Forster, T.S. Gonçalves, D.C. Martin, P. Morrissey, S.G. Neff, M. Seibert, D. Schiminovich, T.K. Wyder, Luminous thermal flares from quiescent supermassive black holes. Astrophys. J. **698**, 1367–1379 (2009). https://doi.org/10.1088/0004-637X/698/2/1367

S. Gezari, R. Chornock, A. Rest, M.E. Huber, K. Forster, E. Berger, P.J. Challis, J.D. Neill, D.C. Martin, T. Heckman, A. Lawrence, C. Norman, G. Narayan, R.J. Foley, G.H. Marion, D. Scolnic, L. Chomiuk, A. Soderberg, K. Smith, R.P. Kirshner, A.G. Riess, S.J. Smartt, C.W. Stubbs, J.L. Tonry, W.M. Wood-Vasey, W.S. Burgett, K.C. Chambers, T. Grav, J.N. Heasley, N. Kaiser, R.-P. Kudritzki, E.A. Magnier, J.S. Morgan, P.A. Price, An ultraviolet-optical flare from the tidal disruption of a helium-rich stellar core. Nature **485**, 217–220 (2012). https://doi.org/10.1038/nature10990

S. Gezari, D.C. Martin, K. Forster, J.D. Neill, M. Huber, T. Heckman, L. Bianchi, P. Morrissey, S.G. Neff, M. Seibert, D. Schiminovich, T.K. Wyder, W.S. Burgett, K.C. Chambers, N. Kaiser, E.A. Magnier, P.A. Price, J.L. Tonry, The GALEX Time Domain Survey. I. Selection and classification of over a thousand ultraviolet variable sources. Astrophys. J. **766**, 60 (2013). https://doi.org/10.1088/0004-637X/766/1/60

S. Gezari, R. Chornock, A. Lawrence, A. Rest, D.O. Jones, E. Berger, P.M. Challis, G. Narayan, PS1-10jh continues to follow the fallback accretion rate of a tidally disrupted star. Astrophys. J. Lett. **815**, 5 (2015). https://doi.org/10.1088/2041-8205/815/1/L5

D. Godoy-Rivera, K.Z. Stanek, C.S. Kochanek, P. Chen, S. Dong, J.L. Prieto, B.J. Shappee, S.W. Jha, R.J. Foley, Y.-C. Pan, T.W.-S. Holoien, T.A. Thompson, D. Grupe, J.F. Beacom, The unexpected, long-lasting, UV rebrightening of the superluminous supernova ASASSN-15lh. Mon. Not. R. Astron. Soc. **466**, 1428–1443 (2017). https://doi.org/10.1093/mnras/stw3237

S. Gomez, M. Nicholl, P. Short, R. Margutti, K.D. Alexander, P.K. Blanchard, E. Berger, T. Eftekhari, S. Schulze, J. Anderson, I. Arcavi, R. Chornock, P.S. Cowperthwaite, L. Galbany, L.J. Herzog, D. Hiramatsu, G. Hosseinzadeh, T. Laskar, T.E. Müller Bravo, L. Patton, G. Terreran, The tidal disruption event AT 2018hyz II: light curve modeling of a partially disrupted star. arXiv e-prints, 2003.05469 (2020)

J. Greiner, R. Schwarz, S. Zharikov, M. Orio, RX J1420.4+5334 – another tidal disruption event? Astron. Astrophys. **362**, 25–28 (2000)

D. Grupe, H.-C. Thomas, K.M. Leighly, RX J1624.9+7554: a new X-ray transient AGN. Astron. Astrophys. **350**, 31–34 (1999)

J. Guillochon, E. Ramirez-Ruiz, Hydrodynamical simulations to determine the feeding rate of black holes by the tidal disruption of stars: the importance of the impact parameter and stellar structure. Astrophys. J. **767**, 25 (2013). https://doi.org/10.1088/0004-637X/767/1/25

J. Guillochon, H. Manukian, E. Ramirez-Ruiz, PS1-10jh: the disruption of a main-sequence star of near-solar composition. Astrophys. J. **783**, 23 (2014). https://doi.org/10.1088/0004-637X/783/1/23

K. Gültekin, D.O. Richstone, K. Gebhardt, T.R. Lauer, S. Tremaine, M.C. Aller, R. Bender, A. Dressler, S.M. Faber, A.V. Filippenko, R. Green, L.C. Ho, J. Kormendy, J. Magorrian, J. Pinkney, C. Siopis, The M-σ and M-L relations in galactic bulges, and determinations of their intrinsic scatter. Astrophys. J. **698**, 198–221 (2009). https://doi.org/10.1088/0004-637X/698/1/198

F. Hamann, T.M. Tripp, D. Rupke, S. Veilleux, On the emergence of THOUSANDS of absorption lines in the quasar PG 1411+442: a clumpy high-column density outflow from the broad emission-line region? Mon. Not. R. Astron. Soc. **487**(4), 5041–5061 (2019). https://doi.org/10.1093/mnras/stz1408

T.M. Heckman, An optical and radio survey of the nuclei of bright galaxies – activity in normal galactic nuclei. Astron. Astrophys. **87**, 152–164 (1980)

J.G. Hills, Possible power source of Seyfert galaxies and QSOs. Nature **254**, 295–298 (1975). https://doi.org/10.1038/254295a0

T.W.-S. Holoien, J.L. Prieto, D. Bersier, C.S. Kochanek, K.Z. Stanek, B.J. Shappee, D. Grupe, U. Basu, J.F. Beacom, J. Brimacombe, J.S. Brown, A.B. Davis, J. Jencson, G. Pojmanski, D.M. Szczygieł, ASASSN-14ae: a tidal disruption event at 200 Mpc. Mon. Not. R. Astron. Soc. **445**, 3263–3277 (2014). https://doi.org/10.1093/mnras/stu1922

T.W.-S. Holoien, C.S. Kochanek, J.L. Prieto, D. Grupe, P. Chen, D. Godoy-Rivera, K.Z. Stanek, B.J. Shappee, S. Dong, J.S. Brown, U. Basu, J.F. Beacom, D. Bersier, J. Brimacombe, E.K. Carlson, E. Falco, E. Johnston, B.F. Madore, G. Pojmanski, M. Seibert, ASASSN-15oi: a rapidly evolving, luminous tidal disruption event at 216 Mpc. Mon. Not. R. Astron. Soc. **463**, 3813–3828 (2016a). https://doi.org/10.1093/mnras/stw2272

T.W.-S. Holoien, C.S. Kochanek, J.L. Prieto, K.Z. Stanek, S. Dong, B.J. Shappee, D. Grupe, J.S. Brown, U. Basu, J.F. Beacom, D. Bersier, J. Brimacombe, A.B. Danilet, E. Falco, Z. Guo, J. Jose, G.J. Herczeg, F. Long, G. Pojmanski, G.V. Simonian, D.M. Szczygieł, T.A. Thompson, J.R. Thorstensen, R.M. Wagner, P.R. Woźniak, Six months of multiwavelength follow-up of the tidal disruption candidate ASASSN-14li and implied TDE rates from ASAS-SN. Mon. Not. R. Astron. Soc. **455**, 2918–2935 (2016b). https://doi.org/10.1093/mnras/stv2486

T.W.-S. Holoien, M.E. Huber, B.J. Shappee, M. Eracleous, K. Auchettl, J.S. Brown, M.A. Tucker, K.C. Chambers, C.S. Kochanek, K.Z. Stanek, A. Rest, D. Bersier, R.S. Post, G. Aldering, K.A. Ponder, J.D. Simon, E. Kankare, D. Dong, G. Hallinan, N.A. Reddy, R.L. Sanders, M.W. Topping, Pan-STARRS, J. Bulger, T.B. Lowe, E.A. Magnier, A.S.B. Schultz, C.Z. Waters, M. Willman, D. Wright, D.R. Young, ASAS-SN, S. Dong, J.L. Prieto, T.A. Thompson, ATLAS, L. Denneau, H. Flewelling, A.N. Heinze, S.J. Smartt, K.W. Smith, B. Stalder, J.L. Tonry, H. Weiland, PS18kh: a new tidal disruption event with a non-axisymmetric accretion disk. Astrophys. J. **880** (2), 120 (2019a). https://doi.org/10.3847/1538-4357/ab2ae1

T.W.-S. Holoien, P.J. Vallely, K. Auchettl, K.Z. Stanek, C.S. Kochanek, K.D. French, J.L. Prieto, B.J. Shappee, J.S. Brown, M.M. Fausnaugh, S. Dong, T.A. Thompson, S. Bose, J.M.M. Neustadt, P. Cacella, J. Brimacombe, M.R. Kendurkar, R.L. Beaton, K. Boutsia, L. Chomiuk, T. Connor, N. Morrell, A.B. Newman, G.C. Rudie, L. Shishkovksy, J. Strader, Discovery and early evolution of ASASSN-19bt, the first TDE detected by TESS. Astrophys. J. **883**(2), 111 (2019b). https://doi.org/10.3847/1538-4357/ab3c66

T.W.-S. Holoien, K. Auchettl, M.A. Tucker, B.J. Shappee, S.G. Patel, J.C.A. Miller-Jones, B. Mockler, D.N. Groenewald, J.S. Brown, C.S. Kochanek, K.Z. Stanek, P. Chen, S. Dong, J.L. Prieto, T.A. Thompson, R.L. Beaton, T. Connor, P.S. Cowperthwaite, L. Dahmen, K.D. French, N. Morrell, D.A.H. Buckley, M. Gromadzki, R. Roy, D.A. Coulter, G. Dimitriadis, R.J. Foley, C.D. Kilpatrick, A.L. Piro, C. Rojas-Bravo, M.R. Siebert, S. van Velzen, The rise and fall of ASASSN-18pg: following a TDE from early to late times. arXiv e-prints, 2003.13693 (2020)

M. Huber, K.C. Chambers, H. Flewelling, M. Willman, N. Primak, A. Schultz, B. Gibson, E. Magnier, C. Waters, J. Tonry, R.J. Wainscoat, K.W. Smith, D. Wright, S.J. Smartt, R.J. Foley, S.W. Jha, A. Rest, D. Scolnic, The Pan-STARRS Survey for Transients (PSST) – first announcement and public release. Astron. Telegram **7153**, 1 (2015)

T. Hung, S. Gezari, N. Blagorodnova, N. Roth, S.B. Cenko, S.R. Kulkarni, A. Horesh, I. Arcavi, C. McCully, L. Yan, R. Lunnan, C. Fremling, Y. Cao, P.E. Nugent, P. Wozniak, Revisiting optical tidal disruption events with iPTF16axa. Astrophys. J. **842**, 29 (2017). https://doi.org/10.3847/1538-4357/aa7337

T. Hung, S. Gezari, S.B. Cenko, S. van Velzen, N. Blagorodnova, L. Yan, S.R. Kulkarni, R. Lunnan, T. Kupfer, G. Leloudas, A.K.H. Kong, P.E. Nugent, C. Fremling, R.R. Laher, F.J. Masci, Y. Cao, R. Roy, T. Petrushevska, Sifting for sapphires: systematic selection of tidal disruption events in iPTF. Astrophys. J. Suppl. Ser. **238**, 15 (2018). https://doi.org/10.3847/1538-4365/aad8b1

T. Hung, S.B. Cenko, N. Roth, S. Gezari, S. Veilleux, S. van Velzen, C.M. Gaskell, R.J. Foley, N. Blagorodnova, L. Yan, M.J. Graham, J.S. Brown, M.R. Siebert, S. Frederick, C. Ward, P. Gatkine, A. Gal-Yam, Y. Yang, S. Schulze, G. Dimitriadis, T. Kupfer, D.L. Shupe, B. Rusholme, F.J. Masci, R. Riddle, M.T. Soumagnac, J. van Roestel, R. Dekany, Discovery of highly blueshifted broad Balmer and metastable helium absorption lines in a tidal disruption event. Astrophys. J. **879**(2), 119 (2019). https://doi.org/10.3847/1538-4357/ab24de

T. Hung, R.J. Foley, E. Ramirez-Ruiz, J.L. Dai, K. Auchettl, C.D. Kilpatrick, B. Mockler, J. Brown, D.A. Coulter, G. Dimitriadis, T. Holoien, J. Law-Smith, A.L. Piro, A. Rest, C. Rojas-Bravo, M.R. Siebert, Prompt accretion disk formation in an X-ray faint tidal disruption event. arXiv e-prints, 2003.09427 (2020)

N. Jiang, L. Dou, T. Wang, C. Yang, J. Lyu, H. Zhou, The WISE detection of an infrared echo in tidal disruption event ASASSN-14li. Astrophys. J. Lett. **828**, 14 (2016). https://doi.org/10.3847/2041-8205/828/1/L14

J.J.E. Kajava, M. Giustini, R.D. Saxton, G. Miniutti, Rapid late-time X-ray brightening of the tidal disruption event OGLE16aaa. arXiv e-prints, 2006.11179 (2020)

S.O. Kastner, A.K. Bhatia, The Bowen fluorescence lines: overview and re-analysis of the observations. Mon. Not. R. Astron. Soc. **279**(4), 1137–1156 (1996). https://doi.org/10.1093/mnras/279.4.1137

M. Kesden, Black-hole spin dependence in the light curves of tidal disruption events. Phys. Rev. D **86**(6), 064026 (2012a). https://doi.org/10.1103/PhysRevD.86.064026

M. Kesden, Tidal-disruption rate of stars by spinning supermassive black holes. Phys. Rev. D **85**(2), 024037 (2012b). https://doi.org/10.1103/PhysRevD.85.024037

C.S. Kochanek, Tidal disruption event demographics. Mon. Not. R. Astron. Soc. **461**, 371–384 (2016). https://doi.org/10.1093/mnras/stw1290

S. Komossa, J. Greiner, Discovery of a giant and luminous X-ray outburst from the optically inactive galaxy pair RX J1242.6-1119. Astron. Astrophys. **349**, 45–48 (1999)

S. Komossa, H. Zhou, T. Wang, M. Ajello, J. Ge, J. Greiner, H. Lu, M. Salvato, R. Saxton, H. Shan, D. Xu, W. Yuan, Discovery of superstrong, fading, iron line emission and double-peaked Balmer lines of the galaxy SDSS J095209.56+214313.3: the light echo of a huge flare. Astrophys. J. Lett. **678**(1), 13 (2008). https://doi.org/10.1086/588281

T. Krühler, M. Fraser, G. Leloudas, S. Schulze, N.C. Stone, S. van Velzen, R. Amorin, J. Hjorth, P.G. Jonker, D.A. Kann, S. Kim, H. Kuncarayakti, A. Mehner, A. Nicuesa Guelbenzu, The supermassive black hole coincident with the luminous transient ASASSN-15lh. Astron. Astrophys. **610**, 14 (2018). https://doi.org/10.1051/0004-6361/201731773

N.M. Law, S.R. Kulkarni, R.G. Dekany, E.O. Ofek, R.M. Quimby, P.E. Nugent, J. Surace, C.C. Grillmair, J.S. Bloom, M.M. Kasliwal, L. Bildsten, T. Brown, S.B. Cenko, D. Ciardi, E. Croner, S.G. Djorgovski, J. van Eyken, A.V. Filippenko, D.B. Fox, A. Gal-Yam, D. Hale, N. Hamam, G. Helou, J. Henning, D.A. Howell, J. Jacobsen, R. Laher, S. Mattingly, D. McKenna, A. Pickles, D. Poznanski, G. Rahmer, A. Rau, W. Rosing, M. Shara, R. Smith, D. Starr, M. Sullivan, V. Velur, R. Walters, J. Zolkower, The Palomar Transient Factory: system overview, performance, and first results. Publ. Astron. Soc. Pac. **121**, 1395–1408 (2009). https://doi.org/10.1086/648598

G. Leloudas, M. Fraser, N.C. Stone, S. van Velzen, P.G. Jonker, I. Arcavi, C. Fremling, J.R. Maund, S.J. Smartt, T. Krìhler, J.C.A. Miller-Jones, P.M. Vreeswijk, A. Gal-Yam, P.A. Mazzali, A. De Cia, D.A. Howell, C. Inserra, F. Patat, A. de Ugarte Postigo, O. Yaron, C. Ashall, I. Bar, H. Campbell, T.-W. Chen, M. Childress, N. Elias-Rosa, J. Harmanen, G. Hosseinzadeh, J. Johansson, T. Kangas, E. Kankare, S. Kim, H. Kuncarayakti, J. Lyman, M.R. Magee, K. Maguire, D. Malesani, S. Mattila, C.V. McCully, M. Nicholl, S. Prentice, C. Romero-Cañizales, S. Schulze, K.W. Smith, J. Sollerman, M. Sullivan, B.E. Tucker, S. Valenti, J.C. Wheeler, D.R. Young, The superluminous transient ASASSN-15lh as a tidal disruption event from a Kerr black hole. Nat. Astron. **1**, 0002 (2016). https://doi.org/10.1038/s41550-016-0002

G. Leloudas, L. Dai, I. Arcavi, P.M. Vreeswijk, B. Mockler, R. Roy, D.B. Malesani, S. Schulze, T. Wevers, M. Fraser, E. Ramirez-Ruiz, K. Auchettl, J. Burke, G. Cannizzaro, P. Charalampopoulos, T.-W. Chen, A. Cikota, M. Della Valle, L. Galbany, M. Gromadzki, K.E. Heintz, D. Hiramatsu, P.G. Jonker, Z. Kostrzewa-Rutkowska, K. Maguire, I. Mandel, M. Nicholl, F. Onori, N. Roth, S.J. Smartt, L. Wyrzykowski, D.R. Young, The spectral evolution of AT 2018dyb and the presence of metal lines in tidal disruption events. Astrophys. J. **887**(2), 218 (2019). https://doi.org/10.3847/1538-4357/ab5792

D. Lin, E.R. Carrasco, D. Grupe, N.A. Webb, D. Barret, S.A. Farrell, Discovery of an ultrasoft X-ray transient source in the 2XMM catalog: a tidal disruption event candidate. Astrophys. J. **738**, 52 (2011). https://doi.org/10.1088/0004-637X/738/1/52

F.K. Liu, Z.Q. Zhou, R. Cao, L.C. Ho, S. Komossa, Disc origin of broad optical emission lines of the TDE candidate PTF09djl. Mon. Not. R. Astron. Soc. **472**(1), 99–103 (2017). https://doi.org/10.1093/mnrasl/slx147

X.-L. Liu, L.-M. Dou, R.-F. Shen, J.-H. Chen, The UV/optical peak and X-ray brightening in TDE candidate AT 2019azh: a case of stream-stream collision and delayed accretion. arXiv e-prints, 1912.06081 (2019)

G. Lodato, A.R. King, J.E. Pringle, Stellar disruption by a supermassive black hole: is the light curve really proportional to $t^{-5/3}$? Mon. Not. R. Astron. Soc. **392**, 332–340 (2009). https://doi.org/10.1111/j.1365-2966.2008.14049.x

A. Loeb, A. Ulmer, Optical appearance of the debris of a star disrupted by a massive black hole. Astrophys. J. **489**, 573–578 (1997). https://doi.org/10.1086/304814

W. Lu, C. Bonnerot, Self-intersection of the fallback stream in tidal disruption events. Mon. Not. R. Astron. Soc. **492**(1), 686–707 (2020). https://doi.org/10.1093/mnras/stz3405

R. Margutti, B.D. Metzger, R. Chornock, D. Milisavljevic, E. Berger, P.K. Blanchard, C. Guidorzi, G. Migliori, A. Kamble, R. Lunnan, M. Nicholl, D.L. Coppejans, S. Dall'Osso, M.R. Drout, R. Perna, B. Sbarufatti, X-rays from the location of the double-humped transient ASASSN-15lh. Astrophys. J. **836**, 25 (2017). https://doi.org/10.3847/1538-4357/836/1/25

D.C. Martin, J. Fanson, D. Schiminovich, P. Morrissey, P.G. Friedman, T.A. Barlow, T. Conrow, R. Grange, P.N. Jelinsky, B. Milliard, O.H.W. Siegmund, L. Bianchi, Y.-I. Byun, J. Donas, K. Forster, T.M. Heckman, Y.-W. Lee, B.F. Madore, R.F. Malina, S.G. Neff, R.M. Rich, T. Small, F. Surber, A.S. Szalay, B. Welsh, T.K. Wyder, The Galaxy Evolution Explorer: a space ultraviolet survey mission. Astrophys. J. Lett. **619**, 1–6 (2005). https://doi.org/10.1086/426387

J.M. Miller, J.S. Kaastra, M.C. Miller, M.T. Reynolds, G. Brown, S.B. Cenko, J.J. Drake, S. Gezari, J. Guillochon, K. Gultekin, J. Irwin, A. Levan, D. Maitra, W.P. Maksym, R. Mushotzky, P. O'Brien, F. Paerels,

J. de Plaa, E. Ramirez-Ruiz, T. Strohmayer, N. Tanvir, Flows of X-ray gas reveal the disruption of a star by a massive black hole. Nature **526**, 542–545 (2015). https://doi.org/10.1038/nature15708
A. Mummery, S.A. Balbus, The spectral evolution of disc dominated tidal disruption events. Mon. Not. R. Astron. Soc. **492**(4), 5655–5674 (2020). https://doi.org/10.1093/mnras/staa192
H. Netzer, M. Elitzur, G.J. Ferland, Bowen fluoresence and He II lines in active galaxies and gaseous nebulae. Astrophys. J. **299**, 752–762 (1985). https://doi.org/10.1086/163741
J.M.M. Neustadt, T.W.-S. Holoien, C.S. Kochanek, K. Auchettl, J.S. Brown, B.J. Shappee, R.W. Pogge, S. Dong, K.Z. Stanek, M.A. Tucker, S. Bose, P. Chen, C. Ricci, P.J. Vallely, J.L. Prieto, T.A. Thompson, D.A. Coulter, M.R. Drout, R.J. Foley, C.D. Kilpatrick, A.L. Piro, C. Rojas-Bravo, D.A.H. Buckley, M. Gromadzki, G. Dimitriadis, M.R. Siebert, A. Do, M.E. Huber, A.V. Payne, To TDE or not to TDE: the luminous transient ASASSN-18jd with TDE-like and AGN-like qualities. Mon. Not. R. Astron. Soc. **494**(2), 2538–2560 (2020). https://doi.org/10.1093/mnras/staa859
M. Nicholl, P.K. Blanchard, E. Berger, S. Gomez, R. Margutti, K.D. Alexander, J. Guillochon, J. Leja, R. Chornock, B. Snios, K. Auchettl, A.G. Bruce, P. Challis, D.J. D'Orazio, M.R. Drout, T. Eftekhari, R.J. Foley, O. Graur, C.D. Kilpatrick, A. Lawrence, A.L. Piro, C. Rojas-Bravo, N.P. Ross, P. Short, S.J. Smartt, K.W. Smith, B. Stalder, The tidal disruption event AT 2017eqx: spectroscopic evolution from hydrogen rich to poor suggests an atmosphere and outflow. Mon. Not. R. Astron. Soc. **488**(2), 1878–1893 (2019). https://doi.org/10.1093/mnras/stz1837
M. Nicholl, T. Wevers, S.R. Oates, K.D. Alexand er, G. Leloudas, F. Onori, A. Jerkstrand, S. Gomez, S. Campana, I. Arcavi, P. Charalampopoulos, M. Gromadzki, N. Ihanec, P.G. Jonker, A. Lawrence, I. Mandel, P. Short, J. Burke, D. Hiramatsu, D.A. Howell, C. Pellegrino, H. Abbot, J.P. Anderson, E. Berger, P.K. Blanchard, G. Cannizzaro, T.-W. Chen, M. Dennefeld, L. Galbany, S. Gonzalez-Gaitan, G. Hosseinzadeh, C. Inserra, I. Irani, P. Kuin, T. Muller-Bravo, J. Pineda, N.P. Ross, R. Roy, B. Tucker, L. Wyrzykowski, D.R. Young, An outflow powers the optical rise of the nearby, fast-evolving tidal disruption event AT 2019qiz. arXiv e-prints, 2006.02454 (2020)
F. Onori, G. Cannizzaro, P.G. Jonker, M. Fraser, Z. Kostrzewa-Rutkowska, A. Martin-Carrillo, S. Benetti, N. Elias-Rosa, M. Gromadzki, J. Harmanen, S. Mattila, M.D. Strizinger, G. Terreran, T. Wevers, Optical follow-up of the tidal disruption event iPTF16fnl: new insights from X-shooter observations. Mon. Not. R. Astron. Soc. **489**(1), 1463–1480 (2019). https://doi.org/10.1093/mnras/stz2053
E.J. Parkinson, C. Knigge, K.S. Long, J.H. Matthews, N. Higginbottom, S.A. Sim, H.A. Hewitt, Accretion disc winds in tidal disruption events: ultraviolet spectral lines as orientation indicators. Mon. Not. R. Astron. Soc. **494**(4), 4914–4929 (2020). https://doi.org/10.1093/mnras/staa1060
E.S. Phinney, Manifestations of a massive black hole in the galactic center, in *The Center of the Galaxy*, ed. by M. Morris. IAU Symposium, vol. 136 (Kluwer, Dordrecht, 1989), p. 543
T. Piran, G. Svirski, J. Krolik, R.M. Cheng, H. Shiokawa, Disk formation versus disk accretion: what powers tidal disruption events? Astrophys. J. **806**, 164 (2015). https://doi.org/10.1088/0004-637X/806/2/164
M.J. Rees, Tidal disruption of stars by black holes of 10 to the 6th-10 to the 8th solar masses in nearby galaxies. Nature **333**, 523–528 (1988). https://doi.org/10.1038/333523a0
N. Roth, D. Kasen, What sets the line profiles in tidal disruption events? Astrophys. J. **855**, 54 (2018). https://doi.org/10.3847/1538-4357/aaaec6
N. Roth, D. Kasen, J. Guillochon, E. Ramirez-Ruiz, The X-ray through optical fluxes and line strengths of tidal disruption events. Astrophys. J. **827**(1), 3 (2016). https://doi.org/10.3847/0004-637X/827/1/3
R.D. Saxton, A.M. Read, P. Esquej, S. Komossa, S. Dougherty, P. Rodriguez-Pascual, D. Barrado, A tidal disruption-like X-ray flare from the quiescent galaxy SDSS J120136.02+300305.5. Astron. Astrophys. **541**, 106 (2012). https://doi.org/10.1051/0004-6361/201118367
J. Schachter, A.V. Filippenko, S.M. Kahn, Bowen fluorescence in Scorpius X-1. Astrophys. J. **340**, 1049 (1989). https://doi.org/10.1086/167457
F. Shankar, P. Salucci, G.L. Granato, G. De Zotti, L. Danese, Supermassive black hole demography: the match between the local and accreted mass functions. Mon. Not. R. Astron. Soc. **354**, 1020–1030 (2004). https://doi.org/10.1111/j.1365-2966.2004.08261.x
B.J. Shappee, J.L. Prieto, D. Grupe, C.S. Kochanek, K.Z. Stanek, G. De Rosa, S. Mathur, Y. Zu, B.M. Peterson, R.W. Pogge, S. Komossa, M. Im, J. Jencson, T.W.-S. Holoien, U. Basu, J.F. Beacom, D.M. Szczygieł, J. Brimacombe, S. Adams, A. Campillay, C. Choi, C. Contreras, M. Dietrich, M. Dubberley, M. Elphick, S. Foale, M. Giustini, C. Gonzalez, E. Hawkins, D.A. Howell, E.Y. Hsiao, M. Koss, K.M. Leighly, N. Morrell, D. Mudd, D. Mullins, J.M. Nugent, J. Parrent, M.M. Phillips, G. Pojmanski, W. Rosing, R. Ross, D. Sand, D.M. Terndrup, S. Valenti, Z. Walker, Y. Yoon, The man behind the curtain: X-rays drive the UV through NIR variability in the 2013 active galactic nucleus outburst in NGC 2617. Astrophys. J. **788**, 48 (2014). https://doi.org/10.1088/0004-637X/788/1/48
P. Short, M. Nicholl, A. Lawrence, S. Gomez, I. Arcavi, T. Wevers, G. Leloudas, S. Schulze, J.P. Anderson, E. Berger, P.K. Blanchard, J. Burke, P. Charalampopoulos, R. Chornock, L. Galbany, M. Gromadzki, L.J. Herzog, D. Hiramatsu, K. Horne, G. Hosseinzadeh, D.A. Howell, N. Ihanec, C. Inserra, E. Kankare, K.

Maguire, C. McCully, T.E. Müller Bravo, F. Onori, J. Sollerman, D.R. Young, The tidal disruption event AT 2018hyz I: double-peaked emission lines and a flat Balmer decrement. arXiv e-prints, 2003.05470 (2020)

D. Stern, R.J. Assef, D.J. Benford, A. Blain, R. Cutri, A. Dey, P. Eisenhardt, R.L. Griffith, T.H. Jarrett, S. Lake, F. Masci, S. Petty, S.A. Stanford, C.-W. Tsai, E.L. Wright, L. Yan, F. Harrison, K. Madsen, Mid-infrared selection of active galactic nuclei with the Wide-field Infrared Survey Explorer. I. Characterizing WISE-selected active galactic nuclei in COSMOS. Astrophys. J. **753**, 30 (2012). https://doi.org/10.1088/0004-637X/753/1/30

N. Stone, R. Sari, A. Loeb, Consequences of strong compression in tidal disruption events. Mon. Not. R. Astron. Soc. **435**, 1809–1824 (2013). https://doi.org/10.1093/mnras/stt1270

N.C. Stone, A. Generozov, E. Vasiliev, B.D. Metzger, The delay time distribution of tidal disruption flares. Mon. Not. R. Astron. Soc. **480**, 5060–5077 (2018). https://doi.org/10.1093/mnras/sty2045

M.A. Strauss, D.H. Weinberg, R.H. Lupton, V.K. Narayanan, J. Annis, M. Bernardi, M. Blanton, S. Burles, A.J. Connolly, J. Dalcanton, M. Doi, D. Eisenstein, J.A. Frieman, M. Fukugita, J.E. Gunn, Ž. Ivezić, S. Kent, R.S.J. Kim, G.R. Knapp, R.G. Kron, J.A. Munn, H.J. Newberg, R.C. Nichol, S. Okamura, T.R. Quinn, M.W. Richmond, D.J. Schlegel, K. Shimasaku, M. SubbaRao, A.S. Szalay, D. Vanden Berk, M.S. Vogeley, B. Yanny, N. Yasuda, D.G. York, I. Zehavi, Spectroscopic target selection in the Sloan Digital Sky Survey: the main galaxy sample. Astron. J. **124**, 1810–1824 (2002). https://doi.org/10.1086/342343

L.E. Strubbe, N. Murray, Insights into tidal disruption of stars from PS1-10jh. Mon. Not. R. Astron. Soc. **454**, 2321–2343 (2015). https://doi.org/10.1093/mnras/stv2081

L.E. Strubbe, E. Quataert, Optical flares from the tidal disruption of stars by massive black holes. Mon. Not. R. Astron. Soc. **400**, 2070–2084 (2009). https://doi.org/10.1111/j.1365-2966.2009.15599.x

L.E. Strubbe, E. Quataert, Spectroscopic signatures of the tidal disruption of stars by massive black holes. Mon. Not. R. Astron. Soc. **415**, 168–180 (2011). https://doi.org/10.1111/j.1365-2966.2011.18686.x

D. Thomas, O. Steele, C. Maraston, J. Johansson, A. Beifiori, J. Pforr, G. Strömbäck, C.A. Tremonti, D. Wake, D. Bizyaev, A. Bolton, H. Brewington, J.R. Brownstein, J. Comparat, J.-P. Kneib, E. Malanushenko, V. Malanushenko, D. Oravetz, K. Pan, J.K. Parejko, D.P. Schneider, A. Shelden, A. Simmons, S. Snedden, M. Tanaka, B.A. Weaver, R. Yan, Stellar velocity dispersions and emission line properties of SDSS-III/BOSS galaxies. Mon. Not. R. Astron. Soc. **431**(2), 1383–1397 (2013). https://doi.org/10.1093/mnras/stt261

J.L. Tonry, L. Denneau, A.N. Heinze, B. Stalder, K.W. Smith, S.J. Smartt, C.W. Stubbs, H.J. Weiland, A. Rest, ATLAS: a high-cadence all-sky survey system. Publ. Astron. Soc. Pac. **130**(988), 064505 (2018). https://doi.org/10.1088/1538-3873/aabadf

B. Trakhtenbrot, I. Arcavi, C.L. MacLeod, C. Ricci, E. Kara, M.L. Graham, D. Stern, F.A. Harrison, J. Burke, D. Hiramatsu, G. Hosseinzadeh, D.A. Howell, S.J. Smartt, A. Rest, J.L. Prieto, B.J. Shappee, T.W.-S. Holoien, D. Bersier, A.V. Filippenko, T.G. Brink, W. Zheng, R. Li, R.A. Remillard, M. Loewenstein, 1ES 1927+654: an AGN caught changing look on a timescale of months. Astrophys. J. **883**(1), 94 (2019a). https://doi.org/10.3847/1538-4357/ab39e4

B. Trakhtenbrot, I. Arcavi, C. Ricci, S. Tacchella, D. Stern, H. Netzer, P.G. Jonker, A. Horesh, J.E. Mejía-Restrepo, G. Hosseinzadeh, V. Hallefors, D.A. Howell, C. McCully, M. Baloković, M. Heida, N. Kamraj, G.B. Lansbury, Ł. Wyrzykowski, M. Gromadzki, A. Hamanowicz, S.B. Cenko, D.J. Sand, E.Y. Hsiao, M.M. Phillips, T.R. Diamond, E. Kara, K.C. Gendreau, Z. Arzoumanian, R. Remillard, A new class of flares from accreting supermassive black holes. Nat. Astron. **3**, 242–250 (2019b). https://doi.org/10.1038/s41550-018-0661-3

A. Udalski, M.K. Szymański, G. Szymański, OGLE-IV: fourth phase of the Optical Gravitational Lensing Experiment. Acta Astron. **65**(1), 1–38 (2015)

S. van Velzen, On the mass and luminosity functions of tidal disruption flares: rate suppression due to black hole event horizons. Astrophys. J. **852**, 72 (2018). https://doi.org/10.3847/1538-4357/aa998e

S. van Velzen, G.R. Farrar, Measurement of the rate of stellar tidal disruption flares. Astrophys. J. **792**, 53 (2014). https://doi.org/10.1088/0004-637X/792/1/53

S. van Velzen, G.R. Farrar, S. Gezari, N. Morrell, D. Zaritsky, L. Östman, M. Smith, J. Gelfand, A.J. Drake, Optical discovery of probable stellar tidal disruption flares. Astrophys. J. **741**, 73 (2011). https://doi.org/10.1088/0004-637X/741/2/73

S. van Velzen, G.E. Anderson, N.C. Stone, M. Fraser, T. Wevers, B.D. Metzger, P.G. Jonker, A.J. van der Horst, T.D. Staley, A.J. Mendez, J.C.A. Miller-Jones, S.T. Hodgkin, H.C. Campbell, R.P. Fender, A radio jet from the optical and X-ray bright stellar tidal disruption flare ASASSN-14li. Science **351**, 62–65 (2016). https://doi.org/10.1126/science.aad1182

S. van Velzen, N.C. Stone, B.D. Metzger, S. Gezari, T.M. Brown, A.S. Fruchter, Late-time UV observations of tidal disruption flares reveal unobscured, compact accretion disks. Astrophys. J. **878**(2), 82 (2019a). https://doi.org/10.3847/1538-4357/ab1844

S. van Velzen, S. Gezari, S.B. Cenko, E. Kara, J.C.A. Miller-Jones, T. Hung, J. Bright, N. Roth, N. Blagorodnova, D. Huppenkothen, L. Yan, E. Ofek, J. Sollerman, S. Frederick, C. Ward, M.J. Graham, R. Fender, M.M. Kasliwal, C. Canella, R. Stein, M. Giomi, V. Brinnel, J. van Santen, J. Nordin, E.C. Bellm, R. Dekany, C. Fremling, V.Z. Golkhou, T. Kupfer, S.R. Kulkarni, R.R. Laher, A. Mahabal, F.J. Masci, A.A. Miller, J.D. Neill, R. Riddle, M. Rigault, B. Rusholme, M.T. Soumagnac, Y. Tachibana, The first tidal disruption flare in ZTF: from photometric selection to multi-wavelength characterization. Astrophys. J. **872**, 198 (2019b). https://doi.org/10.3847/1538-4357/aafe0c

S. van Velzen, S. Gezari, E. Hammerstein, N. Roth, S. Frederick, C. Ward, T. Hung, S.B. Cenko, R. Stein, D.A. Perley, K. Taggart, J. Sollerman, I. Andreoni, E.C. Bellm, V. Brinnel, K. De, R. Dekany, M. Feeney, R.J. Foley, C. Fremling, M. Giomi, V.Z. Golkhou, A.Y.Q. Ho, M.M. Kasliwal, C.D. Kilpatrick, S.R. Kulkarni, T. Kupfer, R.R. Laher, A. Mahabal, F.J. Masci, J. Nordin, R. Riddle, B. Rusholme, Y. Sharma, J. van Santen, D.L. Shupe, M.T. Soumagnac, Seventeen tidal disruption events from the first half of ZTF survey observations: entering a new era of population studies. arXiv e-prints, 2001.01409 (2020)

T.-G. Wang, H.-Y. Zhou, L.-F. Wang, H.-L. Lu, D. Xu, Transient superstrong coronal lines and broad bumps in the galaxy SDSS J074820.67+471214.3. Astrophys. J. **740**(2), 85 (2011). https://doi.org/10.1088/0004-637X/740/2/85

T.-G. Wang, H.-Y. Zhou, S. Komossa, H.-Y. Wang, W. Yuan, C. Yang, Extreme coronal line emitters: tidal disruption of stars by massive black holes in galactic nuclei? Astrophys. J. **749**(2), 115 (2012). https://doi.org/10.1088/0004-637X/749/2/115

T. Wevers, S. van Velzen, P.G. Jonker, N.C. Stone, T. Hung, F. Onori, S. Gezari, N. Blagorodnova, Black hole masses of tidal disruption event host galaxies. Mon. Not. R. Astron. Soc. **471**, 1694–1708 (2017). https://doi.org/10.1093/mnras/stx1703

T. Wevers, D.R. Pasham, S. van Velzen, G. Leloudas, S. Schulze, J.C.A. Miller-Jones, P.G. Jonker, M. Gromadzki, E. Kankare, S.T. Hodgkin, Ł. Wyrzykowski, Z. Kostrzewa-Rutkowska, S. Moran, M. Berton, K. Maguire, F. Onori, S. Mattila, M. Nicholl, Evidence for rapid disc formation and reprocessing in the X-ray bright tidal disruption event candidate AT 2018fyk. Mon. Not. R. Astron. Soc. **488**(4), 4816–4830 (2019a). https://doi.org/10.1093/mnras/stz1976

T. Wevers, D.R. Pasham, S. van Velzen, G. Leloudas, S. Schulze, J.C.A. Miller-Jones, P.G. Jonker, M. Gromadzki, E. Kankare, S.T. Hodgkin, L. Wyrzykowski, Z. Kostrzewa-Rutkowska, S. Moran, M. Berton, K. Maguire, F. Onori, S. Matilla, M. Nicholl, Evidence for rapid disk formation and reprocessing in the X-ray bright tidal disruption event AT 2018fyk. arXiv e-prints, 1903.12203 (2019b)

T. Wevers, N.C. Stone, S. van Velzen, P.G. Jonker, T. Hung, K. Auchettl, S. Gezari, F. Onori, D. Mata Sánchez, Z. Kostrzewa-Rutkowska, J. Casares, Black hole masses of tidal disruption event host galaxies II. Mon. Not. R. Astron. Soc. **487**(3), 4136–4152 (2019c). https://doi.org/10.1093/mnras/stz1602

R.J. Weymann, R.E. Williams, The Bowen fluorescence mechanism in planetary nebulae and the nuclei of Seyfert galaxies. Astrophys. J. **157**, 1201 (1969). https://doi.org/10.1086/150147

Ł. Wyrzykowski, Z. Kostrzewa-Rutkowska, S. Kozłowski, A. Udalski, R. Poleski, J. Skowron, N. Blagorodnova, M. Kubiak, M.K. Szymański, G. Pietrzyński, I. Soszyński, K. Ulaczyk, P. Pietrukowicz, P. Mróz, OGLE-IV real-time transient search. Acta Astron. **64**(3), 197–232 (2014)

Ł. Wyrzykowski, M. Zieliński, Z. Kostrzewa-Rutkowska, A. Hamanowicz, P.G. Jonker, I. Arcavi, J. Guillochon, P.J. Brown, S. Kozłowski, A. Udalski, M.K. Szymański, I. Soszyński, R. Poleski, P. Pietrukowicz, J. Skowron, P. Mróz, K. Ulaczyk, M. Pawlak, K.A. Rybicki, J. Greiner, T. Krühler, J. Bolmer, S.J. Smartt, K. Maguire, K. Smith, OGLE16aaa – a signature of a hungry supermassive black hole. Mon. Not. R. Astron. Soc. **465**, 114–118 (2017). https://doi.org/10.1093/mnrasl/slw213

C.-W. Yang, T.-G. Wang, G. Ferland, W. Yuan, H.-Y. Zhou, P. Jiang, Long-term spectral evolution of tidal disruption candidates selected by strong coronal lines. Astrophys. J. **774**(1), 46 (2013). https://doi.org/10.1088/0004-637X/774/1/46

C. Yang, T. Wang, G.J. Ferland, L. Dou, H. Zhou, N. Jiang, Z. Sheng, The carbon and nitrogen abundance ratio in the broad line region of tidal disruption events. Astrophys. J. **846**(2), 150 (2017). https://doi.org/10.3847/1538-4357/aa8598

D.G. York, J. Adelman, J.E. Anderson Jr., S.F. Anderson, J. Annis, N.A. Bahcall, J.A. Bakken, R. Barkhouser, S. Bastian, E. Berman, W.N. Boroski, S. Bracker, C. Briegel, J.W. Briggs, J. Brinkmann, R. Brunner, S. Burles, L. Carey, M.A. Carr, F.J. Castander, B. Chen, P.L. Colestock, A.J. Connolly, J.H. Crocker, I. Csabai, P.C. Czarapata, J.E. Davis, M. Doi, T. Dombeck, D. Eisenstein, N. Ellman, B.R. Elms, M.L. Evans, X. Fan, G.R. Federwitz, L. Fiscelli, S. Friedman, J.A. Frieman, M. Fukugita, B. Gillespie, J.E. Gunn, V.K. Gurbani, E. de Haas, M. Haldeman, F.H. Harris, J. Hayes, T.M. Heckman, G.S. Hennessy, R.B. Hindsley, S. Holm, D.J. Holmgren, C. Huang, C. Hull, D. Husby, S. Ichikawa, T. Ichikawa, Ž. Ivezić, S. Kent, R.S.J. Kim, E. Kinney, M. Klaene, A.N. Kleinman, S. Kleinman, G.R. Knapp, J. Korienek, R.G. Kron, P.Z. Kunszt, D.Q. Lamb, B. Lee, R.F. Leger, S. Limmongkol, C. Lindenmeyer, D.C. Long, C. Loomis, J. Loveday, R. Lucinio, R.H. Lupton, B. MacKinnon, E.J. Mannery,

P.M. Mantsch, B. Margon, P. McGehee, T.A. McKay, A. Meiksin, A. Merelli, D.G. Monet, J.A. Munn, V.K. Narayanan, T. Nash, E. Neilsen, R. Neswold, H.J. Newberg, R.C. Nichol, T. Nicinski, M. Nonino, N. Okada, S. Okamura, J.P. Ostriker, R. Owen, A.G. Pauls, J. Peoples, R.L. Peterson, D. Petravick, J.R. Pier, A. Pope, R. Pordes, A. Prosapio, R. Rechenmacher, T.R. Quinn, G.T. Richards, M.W. Richmond, C.H. Rivetta, C.M. Rockosi, K. Ruthmansdorfer, D. Sandford, D.J. Schlegel, D.P. Schneider, M. Sekiguchi, G. Sergey, K. Shimasaku, W.A. Siegmund, S. Smee, J.A. Smith, S. Snedden, R. Stone, C. Stoughton, M.A. Strauss, C. Stubbs, M. SubbaRao, A.S. Szalay, I. Szapudi, G.P. Szokoly, A.R. Thakar, C. Tremonti, D.L. Tucker, A. Uomoto, D. Vanden Berk, M.S. Vogeley, P. Waddell, S. Wang, M. Watanabe, D.H. Weinberg, B. Yanny, N. Yasuda, The Sloan Digital Sky Survey: technical summary. Astron. J. **120**, 1579–1587 (2000). https://doi.org/10.1086/301513

Space Sci Rev (2020) 216:81
https://doi.org/10.1007/s11214-020-00702-w

Radio Properties of Tidal Disruption Events

Kate D. Alexander[1] · Sjoert van Velzen[2] · Assaf Horesh[3] · B. Ashley Zauderer[4,5]

Received: 7 June 2019 / Accepted: 5 June 2020 / Published online: 29 June 2020

Abstract Radio observations of tidal disruption events (TDEs) probe material ejected by the disruption of stars by supermassive black holes (SMBHs), uniquely tracing the formation and evolution of jets and outflows, revealing details of the disruption hydrodynamics, and illuminating the environments around previously-dormant SMBHs. To date, observations reveal a surprisingly diverse population. A small fraction of TDEs (at most a few percent) have been observed to produce radio-luminous mildly relativistic jets. The remainder of the population are radio quiet, producing less luminous jets, non-relativistic outflows or, possibly, no radio emission at all. Here, we review the radio observations that have been made of TDEs to date and discuss possible explanations for their properties, focusing on detected sources and, in particular, on the two best-studied events: Sw J1644+57 and ASASSN-14li. We also discuss what we have learned about the host galaxies of TDEs from radio observations and review constraints on the rates of bright and faint radio outflows in TDEs. Upcoming X-ray, optical, near-IR, and radio surveys will greatly expand the sample of TDEs, and technologi-

The Tidal Disruption of Stars by Massive Black Holes
Edited by Peter G. Jonker, Sterl Phinney, Elena Maria Rossi, Sjoert van Velzen, Iair Arcavi and Maurizio Falanga

Kate D. Alexander is NASA Einstein Fellow.

✉ K.D. Alexander
kate.alexander@northwestern.edu

B.A. Zauderer
bezauder@nsf.gov

1 Center for Interdisciplinary Exploration and Research in Astrophysics (CIERA) and Department of Physics and Astronomy, Northwestern University, Evanston, IL 60208, USA

2 Center for Cosmology and Particle Physics, New York University, New York, NY 10003, USA

3 The Racah Institute of Physics, The Hebrew University of Jerusalem, Jerusalem 91904, Israel

4 Division of Astronomical Sciences, National Science Foundation, 2415 Eisenhower Ave., Alexandria, VA 22314, USA

5 Dark Cosmology Centre, Niels Bohr Institute, University of Copenhagen, Blegdamsvej 17, 2100 Copenhagen, Denmark

cal advances open the exciting possibility of discovering a sample of TDEs in the radio band unbiased by host galaxy extinction.

Keywords accretion, accretion disks · black hole physics · galaxies: nuclei · radiation mechanisms: non-thermal · radio continuum: galaxies · relativistic processes

1 Introduction

Radio observations have been used over the past several decades to detect and characterize outflows in extragalactic transient phenomena. These outflows can appear in different forms having a wide range of energies and velocities, ranging from narrow ultra-relativistic jets (as in gamma-ray bursts; GRBs) to sub-relativistic spherical ejecta (as in core-collapse supernovae; SNe). Some famous examples in which radio observations provided key diagnostics of the type of outflows and helped reveal the nature of the transient phenomena in question include: the associated relativistic outflow in the SN 1998bw, the first SN found to be associated with a GRB (Galama et al. 1999; Kulkarni et al. 1998); SN 2009bb, a SN with a relativistic outflow that lacks γ-ray emission (Soderberg et al. 2010); the discovery of both relativistic and sub-relativistic outflows in TDEs (Bloom et al. 2011; Levan et al. 2011; Zauderer et al. 2011; Alexander et al. 2016; van Velzen et al. 2016) and the characterization of the complex structure of the relativistic outflow in the first discovered neutron star merger GW170817 (Alexander et al. 2017b, 2018; Hallinan et al. 2017; Kim et al. 2017; Corsi et al. 2018; Dobie et al. 2018; Margutti et al. 2018; Mooley et al. 2018a,b,c; Resmi et al. 2018; Troja et al. 2018; Ghirlanda et al. 2019; Hajela et al. 2019).

It is now a common practice to perform radio observations of almost any transient phenomena discovered, especially nearby ones, as the radio emission is quite weak if the outflow is sub-relativistic. The radio observations provide key diagnostics such as calorimetry, outflow velocity, magnetic field strength, and the density of the immediate environments surrounding the transient (e.g. Chevalier 1982, 1998, Frail et al. 2000, Barniol Duran et al. 2013). In some cases these crucial pieces of information are not accessible via other wavelengths. Additionally, Very Long Baseline Interferometry (VLBI) radio observations can achieve higher resolution than available at any other wavelength, directly resolving the structure of jets and outflows for nearby events (e.g. Taylor et al. 2004, 2005, Pihlström et al. 2007, Romero-Cañizales et al. 2016, Mattila et al. 2018, Mooley et al. 2018c).

Radio observations of TDEs have revealed extremely diverse radio properties (Fig. 1). This diversity and its implications are the subject of this review. We begin with an overview of the general synchrotron mechanism for producing radio emission (Sect. 2; see also Roth et al. (2020, this issue)). To illustrate the range of radio properties observed to date, we then present a detailed discussion of the two best-studied radio TDEs, Sw J1644+57 and ASASSN-14li, and briefly discuss the other TDEs with detected radio emission (Sect. 3). We then discuss what we have learned about TDE host galaxies and TDEs' immediate environments from radio observations (Sect. 4). We finish with a discussion of occurrence rates of jets and outflows in TDEs and prospects for the future (Sect. 5).

2 Radio Emission Basics

When a fast (either relativistic or sub-relativistic) outflow interacts with the circumstellar (interstellar) medium (CSM/ISM), it drives a shockwave into that medium. As the shockwave plows through the CSM/ISM, it accelerates free electrons to relativistic velocities and enhances magnetic fields. As a result, synchrotron emission ensues. This is the basic mechanism responsible for the radio emission in all types of transients mentioned above (e.g., Chevalier 1982, 1998; Weiler et al. 2002; Granot and Sari 2002; Chevalier and Fransson 2006; Giannios and Metzger 2011; Barniol Duran et al. 2013), including TDEs. In most cases, synchrotron from the external shock between the expanding outflow and the surrounding medium dominates the observed emission, but sometimes synchrotron produced by internal shocks in the jet is also significant (e.g. in the compact cores of AGN jets, and possibly in some TDEs; see Sect. 3.2.3 and Pasham and van Velzen 2018). Roth et al. (2020, this issue) provide additional theoretical discussion of synchrotron emission in the context of TDEs; here, we review the aspects most relevant for radio observations.

The properties of the observed synchrotron emission depend on the physical properties of the system such as the CSM/ISM density, the velocity of the outflow driving the shockwave, the electron energy distribution, the total energy carried by the outflow, and other microphysical parameters such as the fractions of shockwave energy deposited in the electrons and the magnetic field, respectively. The resulting shape of the synchrotron radio spectrum is in general a multiple broken power-law, which can be fully described by its break frequencies and an overall normalization factor (e.g. Granot and Sari 2002).[1] In particular, radio observations reveal the peak of the synchrotron emission spectrum, which for TDEs is typically where the emission transitions from optically thick ($F_\nu \propto \nu^{5/2}$ at low frequencies) to optically thin ($F_\nu \propto \nu^{-(p-1)/2}$ at higher frequencies). Relativistic bulk motion of the emitting region also affects the observed luminosity of the source, due to Doppler boosting effects. In particular, a source with a simple power-law spectrum $F_\nu \propto \nu^{-\alpha}$ that emits isotropically with flux density F_0 in its rest frame, moving relativistically with Lorentz factor $\Gamma \equiv 1/\sqrt{1-\beta^2}$ at an angle θ to the observer's line of sight, will be observed to have a flux density F enhanced by an amount $\delta^{2+\alpha} < F/F_0 < \delta^{3+\alpha}$ (Condon and Ransom 2016). Here, $\delta \equiv [\Gamma(1-\beta\cos\theta)]^{-1}$ is the Doppler factor of the source. For nearly on-axis jets, the Doppler boosting can be quite large, while the flux from off-axis jets will initially be relativistically beamed away from the observer, and the jet may not become observable until it decelerates and becomes non-relativistic. This explains why on-axis relativistic jets have higher observed peak luminosities than off-axis jets or sub-relativistic sources, and thus why on-axis jetted synchrotron sources (e.g. GRBs, radio-loud TDEs) – although rarer than non-relativistic sources (e.g. SNe, some radio-quiet TDEs) – can be detected to much higher redshifts.

If the full synchrotron spectrum is observed and all of the break frequencies are measured (which typically requires coordinated radio, optical, and X-ray observations with dense temporal sampling), then all of the physical parameters listed above can be uniquely determined. If, as is often the case in TDEs, the optical and X-ray emission are dominated by other processes (see Roth et al. (2020, this issue)), then radio and millimeter observations may provide the only constraints on the synchrotron spectrum and some break frequencies will not be observable, so some parameter degeneracies may persist. In such cases, we can make the simplifying assumption that the energy in the synchrotron-emitting outflow is in equipartition between the electrons and the magnetic field (e.g. Scott and Readhead 1977, Chevalier 1998; Barniol Duran et al. 2013). By reducing the number of free parameters, this allows us

[1] We make several simplifying assumptions, most importantly that the magnetic field is constant over the emitting region, and that all electrons are accelerated into a single power-law distribution of energies, $N_e(E) \propto E^{-p}$ for $E > E_0$ (E_0 is typically taken to be $\gtrsim$ the electron rest mass energy). This is typically the case for single impulsive events like GRBs and TDE jets, but is not always true. For example, the compact jet cores of AGN typically have flat spectra because we are observing a superposition of many different synchrotron components.

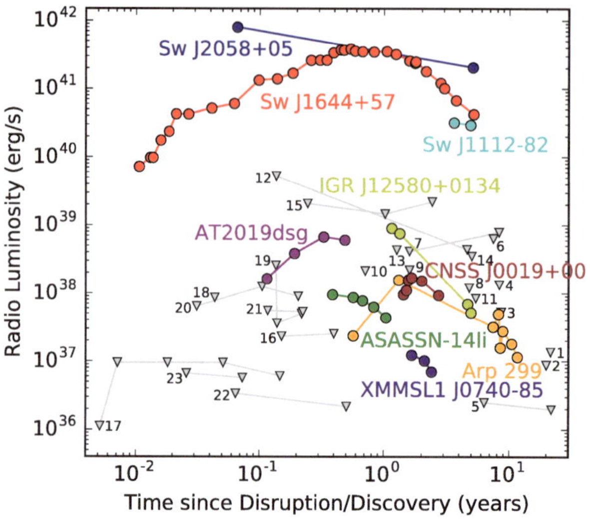

Fig. 1 Literature TDE radio observations. To date, nine TDEs have published radio detections: Sw J1644+57, Sw J2058+05, Sw J1112-82, IGR J12580+0134, ASASSN-14li, XMMSL1 J0740-85, Arp 299-B AT1, CNSS J0019+00, and AT2019dsg (colored circles; see Table 1 and references therein). Although most of the detected TDEs were observed at multiple frequencies, for simplicity we show only a single frequency for each event (8.4 GHz for Arp 299-B AT1 and AT2019dsg, 5 GHz for all others). An additional 23 events have published upper limits (gray triangles; a key to the labels is given in the first column of Table 2). When a non-detected TDE was observed at multiple frequencies on the same date, we show only the most constraining limit. All upper limits are 3σ

to estimate the physical size of the emitting region, the kinetic energy of the outflow, and other physical properties (the outflow velocity, the ambient density, the average magnetic field strength, etc.) even if only part of the synchrotron spectrum is observed (preferably including the peak). The energy thus obtained is a lower bound on the total energy, which can be much larger if the source is not exactly in equipartition, while the size of the emitting region is more robust. Multi-frequency radio observations are preferred for this technique, to constrain the peak frequency and flux density of the radio emission and their temporal evolution. Calculating the size evolution of the emitting region allows us to infer when the outflow was launched (assuming that the radio emission traces the leading edge of the outflow, as expected for external shock models). This is an important constraint for modeling TDEs, for which the time of disruption may not be known precisely (e.g. Zauderer et al. 2011, Alexander et al. 2016). For extremely nearby events, the size evolution of the outflow may also be measured directly using VLBI observations (e.g. Mattila et al. 2018).

3 Radio-Detected TDEs: Probes of Accretion and Outflow Physics

To date, several dozen TDEs have been observed in the radio, revealing a large diversity in their radio properties (Fig. 1). In particular, a few percent of TDEs are radio loud, exhibiting luminous radio emission detectable for years post-disruption, while the rest are radio quiet, with detections or upper limits constraining their radio emission to be orders of magnitude fainter than the radio-loud events. For the purpose of this review, we define a "radio-loud TDE" to have a peak radio luminosity $\nu L_\nu > 10^{40}$ erg s^{-1} and a "radio-quiet TDE" to have

Table 1 TDEs with published radio detections

Name	Redshift	Discovery method	Peak radio luminosity ($erg\,s^{-1}$)	Proposed origin(s) of radio emission (see text)	Reference(s) for radio data
Sw J1644+57	0.354	γ-rays	4×10^{41}	on-axis initially relativistic jet	Zauderer et al. (2011, 2013); Berger et al. (2012); Wiersema et al. (2012); Yang et al. (2016); Eftekhari et al. (2018)
Sw J2058+05	1.1853	γ-rays	8×10^{41}	on-axis initially relativistic jet	Cenko et al. (2012); Pasham et al. (2015); Brown et al. (2017)
Sw J1112−82	0.8901	γ-rays	3×10^{40}	on-axis initially relativistic jet	Brown et al. (2017)
IGR 12580+0134*	0.00411	X-rays	9×10^{38}	off-axis initially relativistic jet	Irwin et al. (2015); Yuan et al. (2016); Perlman et al. (2017)
AT2019dsg	0.051	optical	7×10^{38}	outflow (collimation and max velocity uncertain)	Stein et al. (2020)
Arp 299-B AT1*	0.010411	near-IR	2×10^{38}	off-axis jet	Mattila et al. (2018)
CNSS J0019+00*	0.018	radio	2×10^{38}	non-relativistic outflow	Anderson et al. (2019)
ASASSN-14li*	0.0206	optical	9×10^{37}	non-relativistic outflow, internal shocks in on-axis jet, or unbound debris	Alexander et al. (2016); van Velzen et al. (2016); Romero-Cañizales et al. (2016); Bright et al. (2018)
XMMSL1 J0740-85	0.0173	X-rays	1×10^{37}	outflow (collimation and max velocity uncertain)	Saxton et al. (2017); Alexander et al. (2017a)

*Host galaxy exhibited weak to significant AGN activity prior to the onset of the TDE

$\nu L_\nu < 10^{40}$ erg s^{-1}. We list the nine TDEs with published radio detections in Table 1, in order of decreasing peak radio luminosity. While we generally assume that both radio-loud and radio-quiet TDEs are powered by the disruption of a star, the parameters of the disruption could be different. Radio-loud TDEs also exhibit bright high-energy emission and are discussed in more detail by Zauderer et al. (2020, this issue). Their unique properties are generally attributed to the fact that they launch very energetic relativistic jets viewed on-axis, while other TDEs either launch jets viewed off-axis or do not launch energetic jets. (Off-axis jets will have a much fainter peak luminosity than on-axis jets because the radio emission is suppressed at early times by Doppler beaming, as discussed above.) Differences in parameters such as circumnuclear density, magnetic field strength or configuration, black hole spin, and disruption geometry may also contribute to the observed wide range of TDE radio luminosities (e.g. Giannios and Metzger 2011, van Velzen et al. 2011b, 2013, Krolik et al. 2016, Generozov et al. 2017, Yalinewich et al. 2019). The impact of host galaxy environment on TDE properties is discussed further in Sect. 4 and in French et al. (2020, this issue).

In contrast, radio-quiet TDEs are generally (but not always) discovered in the X-ray, optical, or UV (see Saxton et al. (2020, this issue) and van Velzen et al. (2020a, this issue)). Optical and X-ray-discovered TDEs are also called thermal TDEs because their optical, UV, and X-ray emission generally can be well-modeled as a blackbody peaking in the UV or soft X-ray. Only a handful of thermal TDEs have radio detections, but this may be partially an observational bias. Many TDEs received radio follow up that would have been too sparse or too shallow to reveal low-luminosity outflows similar to those seen in radio-detected sources (Fig. 1, gray triangles), and about half of known TDEs have not been observed in the radio at all. While the current data cannot rule out the hypothesis that radio emission with a peak luminosity $\sim 10^{38}$ erg s^{-1} may be common in radio-quiet TDEs, deep upper limits for several recent events (e.g. Blagorodnova et al. 2017, van Velzen et al. 2019, Saxton et al. 2019) hint that there may be additional diversity within this class. We have compiled a list of all published radio non-detections of TDEs available in the literature, presented in Table 2. Below, we discuss the physical insights derived from radio observations of the two best-studied radio-detected TDEs to date: the radio-loud Sw J1644+57 and the radio-quiet ASASSN-14li. We also briefly summarize the properties of other TDEs detected in the radio.

3.1 Sw J1644+57: A Relativistic Jet

Sw J1644+57 was discovered by the Neil Gehrels Swift Observatory on 28 March 2011 as an unusual γ-ray and X-ray transient coincident with the nucleus of a galaxy at redshift $z = 0.354$ (Bloom et al. 2011; Burrows et al. 2011; Levan et al. 2011; Zauderer et al. 2011). Associated radio emission was discovered within 24 hours, initiating an extensive radio observing campaign spanning $1 - 345$ GHz (Zauderer et al. 2011; Bloom et al. 2011; Levan et al. 2011; Berger et al. 2012; Wiersema et al. 2012; Zauderer et al. 2013; Yang et al. 2016; Eftekhari et al. 2018).[2] The full radio dataset includes observations extending to ≈ 2000 days post-discovery and the source continues to be detected in multi-frequency observations. Here, we focus on the radio properties of Sw J1644+57; for more discussion of this source and other TDEs with high-energy emission, see Zauderer et al. (2020, this issue).

The radio observations of Sw J1644+57 capture the peak of the synchrotron emission. The peak frequency, ν_p, decreases with time, while the peak flux density $F_{\nu,p}$ remains approximately constant (within a factor ~ 3) for ~ 400 days and then decreases (Eftekhari

[2] Sw J1644+57 was also observed at 149 MHz by LOFAR, but the resulting non-detection is not constraining (Cendes et al. 2014).

Table 2 Published radio upper limits of TDEs

Fig. 1 label	Name	Redshift	MJD (d)	Δt^1 (yr)	Frequency (GHz)	3σ upper limit (μJy)	Reference
1	RXJ1624+7554	0.06	56082.342	21.6	3.0	51	Bower et al. (2013)
2	RXJ1242-1119	0.046	56085.027	19.9	3.0	54	Bower et al. (2013)
3	SDSSJ1323+48	0.08	56082.301	8.6	3.0	102	Bower et al. (2013)
4	SDSSJ1311-01	0.156	56085.040	8.3	3.0	57	Bower et al. (2013)
5	NGC 5905	0.012	50390.786	6.3	8.46	90	Komossa (2002)
5	NGC 5905	0.012	56082.328	21.9	3.0	200	Bower et al. (2013)
6	GALEX-D1-9	0.326	55955.988	7.46	5.0	27	van Velzen et al. (2013)
7	GALEX-D3-13[2]	0.3698	53478.312	1.26	1.4	45	Bower (2011)
7	GALEX-D3-13	0.3698	55955.498	8.05	5.0	24	van Velzen et al. (2013)
8	D23H-1	0.186	55955.938	4.8	4.3	24	van Velzen et al. (2013)
9	PTF10iya	0.224	55955.479	1.6	5.0	24	van Velzen et al. (2013)
10	PS1-10jh	0.170	55649.348	0.71	5.0	45	van Velzen et al. (2013)
11	SDSS-TDE1	0.136	55955.942	5.4	5.0	30	van Velzen et al. (2013)
12	SDSS-TDE2	0.252	54407.071	0.137	8.4	255	van Velzen et al. (2011a)
12	SDSS-TDE2	0.252	55955.935	4.6	5.0	36	van Velzen et al. (2013)
13	SDSSJ1201+30	0.146	55819.767	1.267	8.3	108	Saxton et al. (2012)
13	SDSSJ1201+30	0.146	55819.779	1.267	4.8	135	Saxton et al. (2012)
13	SDSSJ1201+30	0.146	55819.788	1.267	1.4	201	Saxton et al. (2012)
14	PTF09axc	0.1146	56836.221	4.953	6.1	150	Arcavi et al. (2014)
14	PTF09axc	0.1146	56836.233	4.953	3.5	330	Arcavi et al. (2014)

Table 2 (*Continued*)

Fig. 1 label	Name	Redshift	MJD (d)	Δt^1 (yr)	Frequency (GHz)	3σ upper limit (μJy)	Reference
15	PS1-11af	0.4046	55649.053	0.186	4.9	51	Chornock et al. (2014)
15	PS1-11af	0.4046	55933.409	0.965	5.875	30	Chornock et al. (2014)
15	PS1-11af	0.4046	56443.992	2.363	5.875	45	Chornock et al. (2014)
16	PS16dtm	0.0804	57653.227	0.00610	21.8	57	Blanchard et al. (2017)
16	PS16dtm	0.0804	57653.250	0.00616	6.0	23	Blanchard et al. (2017)
16	PS16dtm	0.0804	57743.136	0.252	21.8	51	Blanchard et al. (2017)
16	PS16dtm	0.0804	57743.159	0.252	6.0	25	Blanchard et al. (2017)
17	iPTF16fnl	0.016328	57631.281	−0.00224	22	52[3]	Blagorodnova et al. (2017)
17	iPTF16fnl	0.016328	57631.293	−0.00221	6.1	34[3]	Blagorodnova et al. (2017)
17	iPTF16fnl	0.016328	57632	−0.000274	15	117	Blagorodnova et al. (2017)
17	iPTF16fnl	0.016328	57636	0.0107	15	117	Blagorodnova et al. (2017)
17	iPTF16fnl	0.016328	57648	0.0435	15	117	Blagorodnova et al. (2017)
17	iPTF16fnl	0.016328	57683	0.139	15	75	Blagorodnova et al. (2017)
18	iPTF15af	0.07897	57053.215	−0.0659	22	108	Blagorodnova et al. (2019)
18	iPTF15af	0.07897	57053.225	−0.0659	6.1	84	Blagorodnova et al. (2019)
19	AT2018zr	0.071	58205.883	0.0344	16	120	van Velzen et al. (2019)
19	AT2018zr	0.071	58207.170	0.0379	10	27	van Velzen et al. (2019)
19	AT2018zr	0.071	58236.154	0.117	10	37.5	van Velzen et al. (2019)
20	AT 2018fyk	0.059	58380.677	0.0313	18.95	38	Wevers et al. (2019)
20	AT 2018fyk	0.059	58407.490	0.1048	18.95	74	Wevers et al. (2019)
20	AT 2018fyk	0.059	58444.289	0.2055	18.95	53	Wevers et al. (2019)
21	AT 2017eqx	0.1089	57948.465	0.0736	21.7	80	Nicholl et al. (2019)
21	AT 2017eqx	0.1089	57948.477	0.0736	6.0	27	Nicholl et al. (2019)

Table 2 (*Continued*)

Fig. 1 label	Name	Redshift	MJD (d)	Δt^1 (yr)	Frequency (GHz)	3σ upper limit (μJy)	Reference
21	AT 2017eqx	0.1089	57987.240	0.180	21.7	76	Nicholl et al. (2019)
21	AT 2017eqx	0.1089	57987.252	0.180	6.0	26	Nicholl et al. (2019)
22	XMMSL2_J1446+68	0.029	57646.155	0.0643	21.7	85	Saxton et al. (2019)
22	XMMSL2_J1446+68	0.029	57646.167	0.0643	6.0	28	Saxton et al. (2019)
22	XMMSL2_J1446+68	0.029	57806.511	0.503	21.7	84	Saxton et al. (2019)
22	XMMSL2_J1446+68	0.029	57806.524	0.503	6.0	18	Saxton et al. (2019)
23	ASASSN-18pg	0.017392	58319.454	−0.06611	18.95	50	Holoien et al. (2020)
23	ASASSN-18pg	0.017392	58336.617	−0.01912	18.95	43	Holoien et al. (2020)

[1]Time relative to peak bolometric luminosity, given in the observer frame. For events with sparse multi-wavelength coverage, we list the time relative to maximum light in a single optical or X-ray observing band

[2]Based on serendipitous observations of the TDE field co-added by Bower (2011) (note: Bower 2011 reports the observation dates as 2003 Dec 2, 2005 Apr 18, and 2005 Jun 20. However, the VLA data archive records no observations within 1 degree of the TDE coordinates on Jun 20, so it is likely that one or more of the observations of the field collected under the same observing program between 2005 May 14 - 2005 Jun 7 were used instead. We use the MJD of the 2005 Apr 18 observation to compute Δt for this observation)

[3]Based on a re-reduction of the data by KDA

et al. 2018). This behavior is consistent with synchrotron emission from an initially mildly relativistic ($\Gamma \approx 3$), collimated jet that expands and slowly decelerates (Zauderer et al. 2011; Berger et al. 2012; Zauderer et al. 2013; Eftekhari et al. 2018). The jet launch date inferred from the radio observations is consistent with the onset of the high-energy emission (Zauderer et al. 2011). For equipartition models and other models where the energy carried by electrons and the energy in magnetic fields remain constant fractions of the total energy, the total energy must increase by an order of magnitude from $30 - 250$ days. This could be explained by structured ejecta, in which slower-moving material catches up to the head of the jet as it decelerates and begins contributing to the emission at later times (e.g. Berger et al. 2012, Mimica et al. 2015). An alternate possibility is that the energy distribution varies in time; e.g. most energy is initially carried by the magnetic field, and at later times shifts to being carried by the electrons (Barniol Duran and Piran 2013). In this case, it is possible for the total energy in the jet to remain constant. However, more recent work sees no evidence for such time variations (Eftekhari et al. 2018).

The electrons producing the radio emission in Sw J1644+57 are in the slow cooling regime, but at higher frequencies a cooling break in the synchrotron spectrum is expected (e.g. Granot and Sari 2002). The location of this break provides an additional constraint on the synchrotron model, breaking parameter degeneracies and removing the need to assume equipartition. The optical data alone cannot precisely constrain the position of the cooling break, as they are affected by a large but uncertain amount of host extinction (Levan et al. 2016). Until 500 days, the X-rays are dominated by emission from the base of the jet (Zauderer et al. 2011; Berger et al. 2012; Zauderer et al. 2013), rather than by the same synchrotron component producing the radio emission. At 500 days, the X-ray luminosity drops precipitously as the jet shuts off, and the residual X-rays are consistent with an extension of the radio emission spectrum (Zauderer et al. 2013). The post-500 days X-ray data, together with the optical observations, allow constraints on both the cooling break frequency and the amount of host extinction (Eftekhari et al. 2018). This reveals a magnetic field lower than the equipartition value, and correspondingly increases the total energy in the jet by a factor of 3 to $E \sim 4 \times 10^{51}$ erg (Eftekhari et al. 2018).

3.2 ASASSN-14li: A Thermal TDE with Radio Emission

ASASSN-14li was discovered by the All Sky Automated Survey for SuperNovae (ASAS-SN) on 22 November 2014, coincident with the center of the $z = 0.0206$ galaxy PGC043234 (luminosity distance $d_L \approx 90$ Mpc). It was subsequently detected across the electromagnetic spectrum (Miller et al. 2015, Holoien et al. 2016, van Velzen et al. 2016, Alexander et al. 2016, Jiang et al. 2016; Romero-Cañizales et al. 2016; Bright et al. 2018). Unlike many TDEs, it exhibited both optical and X-ray variability, along with declining radio emission and an infrared echo. ASASSN-14li has the best-studied radio emission of all thermal TDEs and has become a poster child for this class. Below, we summarize several different models that have been proposed to explain its radio emission, which is $\sim 10^3$ times less luminous than Sw J1644+57's.

3.2.1 Non-Relativistic Wind

Alexander et al. (2016) modeled ASASSN-14li's radio emission as synchrotron radiation arising from the external shock between outflowing material from the TDE and the surrounding circumnuclear medium. In this model, the peak of the radio spectral energy distribution (SED) corresponds to the synchrotron self absorption frequency, which depends on the density and the properties of the outflow. Assuming that the energy contained in the outflow is always in equipartition between the electrons and the magnetic field (and that emission from a single region dominates the radio SED at all frequencies) allows the energy and radius of the emitting region to be computed independently in each multi-frequency radio epoch (e.g. Barniol Duran et al. 2013). Thus, no dynamical assumptions are required and the derived physical properties of the emitting region don't depend on any specific launching mechanism for the outflow.

In the case of ASASSN-14li, multi-frequency radio observations with the Karl G. Jansky Very Large Array (VLA) were found to be consistent with an outflow with a constant equipartition energy of $\sim 10^{48}$ erg expanding at a constant velocity $0.04 - 0.12c$ (Alexander et al. 2016). (There is a mild degeneracy between the energy/velocity and the assumed opening angle of the outflow, with a spherical outflow moving slower than a conical one.) This method also revealed a steep circumnuclear density profile proportional to $r^{-2.5}$ (Sect. 4). These properties are consistent with expectations for a wind produced during a phase of super-Eddington accretion (Strubbe and Quataert 2009). Alexander et al. (2016) showed that the launch time of ASASSN-14li's outflow roughly coincided with the time when the accretion rate first exceeded Eddington (independently estimated from the optical, UV, and X-ray light curves), providing additional support for this picture. In this scenario, only TDEs in which the peak accretion rate exceeds the Eddington limit would produce radio emission. Alternately, recent simulations suggest that a similar outflow could also be launched by the self-intersection of the bound debris stream, expected due to relativistic apsidal precession (Lu and Bonnerot 2020). Such collision-induced outflows could produce radio emission with a wide range of peak luminosities and timescales.

3.2.2 Unbound Tidal Debris Stream

Approximately half of the stellar debris from a TDE will ultimately accrete onto the SMBH, while the rest is unbound (Rees 1988). A second possibility is that ASASSN-14li's radio emission is produced by the interaction between the unbound debris and the circumnuclear medium. In this case, synchrotron radiation is emitted in the bow shock that forms along the leading edge of the debris stream, and the emitting region is expected to expand at a velocity $\sim 0.2c$ (Krolik et al. 2016). Krolik et al. (2016) performed a similar equipartition analysis to Alexander et al. (2016) and reached similar conclusions about the velocity and energetics of the radio-emitting material in ASASSN-14li, roughly consistent with this picture. Only a tiny fraction of the unbound debris is required to produce the observed radio luminosity, consistent with a model in which the bulk of the unbound debris has not yet decelerated on the timescale of the observations.

An important prediction of this model is that all TDEs should generate radio emission from this process, with no dependence on the peak accretion rate. However, the brightness of the emission will depend on both the ambient density and the cross-section of the debris stream (which may be quite small for some observer viewing angles). Some theoretical studies suggest that in most disruptions the debris stream should remain self-gravitating until large radii (e.g. Guillochon et al. 2016, Steinberg et al. 2019); in this case, the stream may be too narrow to produce detectable radio emission. However, for disrupted stars on deeply plunging orbits, the unbound debris fan may be more dispersed (Strubbe and Quataert 2009; Yalinewich et al. 2019) and more luminous emission similar to that seen in ASASSN-14li may be possible. As only a small amount of mass is required to produce the observed radio emission in ASASSN-14li (Krolik et al. 2016; Yalinewich et al. 2019), only a small fraction of the ejecta need not be self-gravitating. Recent work by Piran et al. (in prep) suggests that in most TDEs, radio emission from the bow shock generated by the fastest unbound debris should be marginally detectable, with a peak flux density of a few μJy for a TDE at the distance of ASASSN-14li (~ 90 Mpc).

3.2.3 Jet

A third possibility to explain the radio flare of ASASSN-14li is emission from a (sub-)relativistic jet. Such jets can have two modes of emission. External emission is powered by a forward shock that the jet drives into the surrounding medium (in AGN jets these shocks power the so-called hotspots). The external mode of emission is akin to the wind and unbound-debris models discussed above. However jets can also produce synchrotron emission without interaction with a surrounding medium. This internal emission mechanism is powered by shocks inside the jet (in AGN jets, internal emission is responsible for the compact, flat-spectrum radio cores).

The first model that was proposed for radio emission of ASASSN-14li was external (i.e., shock-driven) emission from a relativistic jet (van Velzen et al. 2016). This model could explain the 16 GHz light curve, but the multi-frequency radio observations (Alexander et al. 2016) that appeared later challenged this interpretation. Romero-Cañizales et al. (2016) also suggested the possibility of a relativistic jet in ASASSN-14li, based on VLBI observations that showed two separate components of radio emission separated by ~ 2 pc (implying superluminal motion from a nearly on-axis relativistic jet with $\Gamma \gtrsim 7$). However, jets with similarly high Lorentz factors typically have much higher brightness temperatures, so they conclude that the second emission component is more likely unrelated to the TDE (Sect. 4.2).

An improvement to the jet model for ASASSN-14li is presented in Pasham and van Velzen (2018), who updated the original internal emission mechanism for TDE jets (van Velzen et al. 2011b) to include adiabatic cooling. In this model the jet consists of a superposition of synchrotron-emitting regions in a conical geometry, each with their own equipartition magnetic field. Applying this model to the observed synchrotron SEDs yield a measurement of the scaling of the equipartition magnetic field with the radius in the jet, which was found to be $B \propto R^{-1.02\pm0.05}$. This is remarkably close to the predicted power-law index of -1 for a freely expanding jet that is powered by an internal emission mechanism (Blandford and Königl 1979; Falcke and Biermann 1995).

The strongest evidence against an external emission mechanism for the radio properties of ASASSN-14li is the detection of a cross-correlation between the X-ray and radio light curves (Pasham and van Velzen 2018); the radio lags the X-rays by about 12 days (see van Velzen et al. (2020b, this issue) for details of the cross-correlation). Establishing this correlation requires a coupling between the X-ray emitting disk and the source of the radio photons. This coupling is not expected in any of the proposed external emission mechanisms (disk-wind, streams shocks, or jet hotspots) because these are all impulsive: the radio emission is governed by the evolution of the shock, which decouples from the accretion disk. However in an internal jet model, this coupling is naturally expected (and is observed for jets from active black holes; Falcke and Biermann 1995, Marscher et al. 2002). The observed 12 day lag is consistent with the time that is required for the 16 GHz radio emission to become optically thin to synchrotron self absorption.

3.3 Other Detected Sources

At the time of writing, seven other TDEs with radio detections have been reported in the literature (Fig. 1). Two of these sources, Sw J2058+05 and Sw J1112-82, were discovered in archival searches of Swift data and have similar γ-ray, X-ray, and radio properties to Sw J1644+57 (Cenko et al. 2012; Pasham et al. 2015; Brown et al. 2015, 2017). Like Sw J1644+57, they likely launched energetic relativistic jets viewed on-axis. For a more complete discussion of these sources, see Zauderer et al. (2020, this issue).

As jets in the most energetic TDEs are thought to be narrowly collimated, we expect that most jetted TDEs will be viewed off-axis. In this case, the radio emission is suppressed at early times due to relativistic beaming effects, but will appear similar to on-axis jetted events after the jet decelerates (Granot et al. 2002; van Velzen et al. 2013). However, by this point, the radio emission evolves very slowly – so it may be difficult to identify the transient nature of off-axis TDEs (Eftekhari et al. 2018). To date, two additional radio TDEs, IGR J12580+0134 and Arp 299-B AT1, have been proposed to exhibit off-axis jets (Lei et al. 2016; Mattila et al. 2018). IGR J12580+0134's radio emission was discovered serendipitously in a survey of nearby galaxies (Irwin et al. 2015), and was connected to a hard X-ray flare interpreted as a TDE that occurred in the same galaxy one year prior (Nikołajuk and Walter 2013). Its peak radio luminosity is intermediate between Sw J1644+57's and ASASSN-14li's (Fig. 1). This event occurred in a known AGN, so it is possible that the radio emission is associated with another type of AGN flaring event – nevertheless, late-time radio follow up observations show that the radio flux density has decreased by a factor $\gtrsim 10$ over five years (Perlman et al. 2017), which is extreme for typical AGN flares (Hovatta et al. 2008; Nieppola et al. 2009).

In contrast, Arp 299-B AT1 was discovered as a bright flare in the near-IR, and was luminous and long-lasting in both the IR and radio (Mattila et al. 2018). It exhibits no variability at shorter wavelengths due to extremely high host galaxy extinction (Mattila et al. 2018). Its host, Arp 299, consists of two merging galaxies, potentially resulting in a temporarily enhanced TDE rate; the nucleus in which the flare occurred is also a Compton-thick AGN with an almost edge-on torus. It has been observed extensively in the radio with VLBI, directly imaging a resolved jet in a TDE for the first time (Mattila et al. 2018). These observations initially show only a single emerging jet with no corresponding counterjet, allowing a constraint on the viewing angle ($25 - 35^\circ$). This implies that the TDE accretion disk is not aligned with the pre-existing AGN torus. It is possible that many TDEs occur in galaxy nuclei with similarly high levels of extinction, and are currently being missed by the optical and X-ray surveys that have yielded most TDE discoveries to date.

The remaining three radio-quiet TDEs also likely launched outflows, but these outflows may be quasi-spherical rather than collimated jets. XMMSL1 J0740-85 was discovered during the XMM Newton Slew Survey (Saxton et al. 2008) on 2014 April 1 as a bright X-ray transient (Saxton et al. 2017). Its X-ray spectrum contained both thermal and non-thermal components, placing it in between the jetted Swift events and TDEs not detected in γ-rays. XMMSL1 J0740-85 exhibited little to no variability in the optical and modest variability in the UV (Saxton et al. 2017). At late times ($\gtrsim 600$ days after discovery), weak, fading radio emission was observed (Alexander et al. 2017a). The origin of this radio emission is ambiguous: it is consistent with either a decelerated jet or an always non-relativistic outflow. In either model, the energy contained in the outflowing material is much less than the energy of Sw J1644+57's jet.

Further deep radio observations of TDEs are needed to determine whether weak radio emission like that seen in XMMSL1 J0740-85 and ASASSN-14li is common (Sect. 5), and to determine the physical conditions that lead to jet formation in a small subset of TDEs. Such efforts are already proving fruitful. For example, the most recently reported optical TDE with a radio detection, AT2019dsg (Stein et al. 2020), has a similar luminosity to

ASASSN-14li, but unlike ASASSN-14li it shows a linear increase of the equipartition energy with time, possibly suggesting that the outflow or jet in this source receives a constant injection of power at its origin.

Future radio facilities will have the capabilities to carry out blind searches for radio transients, and may discover many events that would have been missed by optical and X-ray surveys. Radio emission is not affected by extinction, so a uniform sample of TDEs discovered in the radio will provide an independent estimate of the TDE rate (Sect. 5.3). The final radio-detected TDE, CNSS J0019+00, was discovered in a blind search for radio transients in the SDSS Stripe 82 region, illustrating the promise of this technique (Mooley et al. 2016; Anderson et al. 2019). This TDE was in a Seyfert 2 galaxy and was localized to within 1 pc of the galaxy nucleus by VLBI observations. CNSS J0019+00's luminosity and duration make it more similar to other radio-quiet TDEs than to supernovae or AGN flares (although a reactivated AGN cannot be completely ruled out; Anderson et al. 2019). Similar to ASASSN-14li and XMMSL1 J0740-85, CNSS J0019+00's radio emission can be well-modeled as a sub-relativistic spherical outflow with an energy $\sim 10^{49}$ erg, expanding at a velocity $\sim 0.05c$ (Anderson et al. 2019).

4 Host Properties and Constraints on Nuclear Environments from Radio Observations

Radio observations of TDEs also reveal useful information about the properties of their host galaxies. The immediate vicinity of the SMBH can be probed via the interaction between outflowing debris from the TDE and the circumnuclear medium, while VLBI observations may reveal structures in the circumnuclear environment at larger scales. Processes such as star formation and AGN activity may also produce radio emission in the vicinity of the TDE; these additional sources of emission must be properly accounted for when interpreting radio observations carried out during the flare. Due to the long cooling time, relic lobes of AGN activity prior to the TDE can generate (near) nuclear radio emission that lingers for $\sim$Gyr. Additionally, AGN can exhibit radio variability on timescales ranging from hours to decades, including flares with typical timescales of a few years (Hovatta et al. 2008; Nieppola et al. 2009). The amplitude of such flares is typically much smaller than the level of variability seen during radio-detected TDEs, but more observations are needed to properly locate radio TDEs within the broader framework of SMBH accretion.

Surveying large areas of the radio sky at high resolution is time-consuming and data-intensive; until recently, only shallow extragalactic source catalogs existed for the vast majority of the radio sky. Thus, meaningful constraints on the pre-flare radio emission from TDE hosts has so far only been possible for the nearest events. This situation has started to change with the advent of the VLA Sky Survey (VLASS), which will image the entire sky north of declination -40° with a combined limiting rms of 69 μJy, and with the deployment of ASKAP and MeerKAT in the southern hemisphere, building towards the extremely sensitive Square Kilometer Array (SKA). In this section, we summarize the progress that has been made so far using radio observations to study the nuclear environments in which TDEs occur.

4.1 Density Constraints

When radio emission is produced in an external shock between a transient outflow and the surrounding medium, then we can also use radio observations of TDEs to infer the ambient

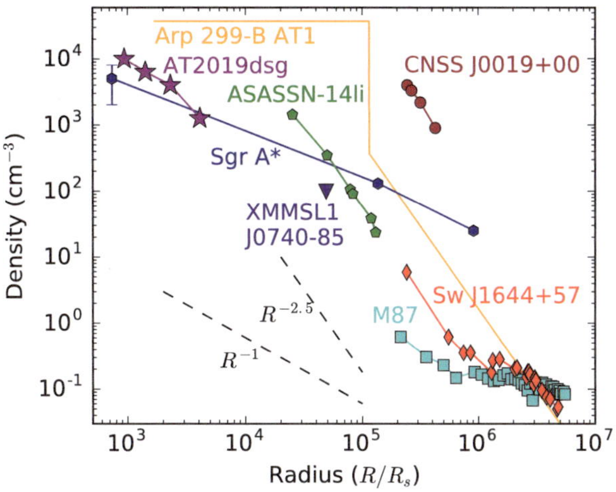

Fig. 2 The circumnuclear density profiles of TDE host galaxies inferred from their radio observations, with the distance from each SMBH given in units of its Schwarzschild radius. The normalization of the profiles for ASASSN-14li (Alexander et al. 2016), XMMSL1 J0740-85 (Alexander et al. 2017a), and AT2019dsg (Stein et al. 2020) depends on the geometry of the outflows, which is not well constrained by the data; here, we assume spherical outflows. The density profile for Sw J1644+57 is taken from the analysis by Eftekhari et al. (2018), the profile for Arp 299-B AT1 is taken from Mattila et al. (2018), and the profile for CNSS J0019+00 is taken from Anderson et al. (2019). Shown for comparison are measurements of the density profile around two other well-studied SMBHs: the nearby and more massive M87 (Russell et al. 2015), and Sagittarius A* (Baganoff et al. 2003; Gillessen et al. 2019). Sgr A* has a much shallower density profile ($\rho \propto r^{-1}$) than ASASSN-14li's SMBH ($\rho \propto r^{-2.5}$) at comparable distance scales. We also show sample r^{-1} and $r^{-2.5}$ profiles to guide the eye (black dashed lines)

density surrounding the supermassive black hole at parsec or even sub-parsec scales (e.g. Alexander et al. 2016, Krolik et al. 2016, Eftekhari et al. 2018; Mattila et al. 2018, Anderson et al. 2019, Stein et al. 2020). As the outflow expands, independent measurements of the density at different epochs can be used to construct a density profile of the circumnuclear gas. (The dust distribution at comparable distances can be indirectly probed via the detection of infrared dust echoes; see van Velzen et al. (2020b, this issue) and references therein.) Such constraints are extremely valuable, as these scales are not directly resolvable at any wavelength with current facilities at the distance of most TDE hosts.

The shape of the ambient circumnuclear density profile directly probes the history of accretion onto the SMBH in the TDE host galaxy and the circumnuclear environment. Spherically symmetric accretion results in a Bondi profile, $\rho \propto r^{-3/2}$ (Bondi 1952), while adding the effects of mass loss from massive stars produces a shallower profile (e.g. Quataert 2004, Generozov et al. 2017). A recent compilation of density measurements for Sagittarius A* reveals a density profile $\rho \propto r^{-1}$ over four orders of magnitude in distance,[3] consistent with this picture (Gillessen et al. 2019 and references therein). We compare these results with the density measurements inferred for TDE host galaxies in Fig. 2. AT2019dsg's density profile is similar to Sgr A*'s (Stein et al. 2020), while Sw J1644+57's host galaxy exhibits a Bondi-like profile at $\gtrsim 0.7$ pc (Eftekhari et al. 2018). In contrast, ASASSN-14li's and CNSS J0019+00's host galaxies exhibit a steeper profile, $\rho \propto r^{-2.5}$ (Alexander et al. 2016,

[3]We note that the available observations only sparsely sample Sgr A*'s density profile, so in principle deviations from r^{-1} over a narrower range of radii could be present.

Krolik et al. 2016; Anderson et al. 2019).[4] Arp 299-B AT1's radio emission can be best fit with a dense torus extending to 6.75×10^{17} cm, with a similarly steep $r^{-2.5}$ density profile at larger radii (Mattila et al. 2018). Such a steep profile is not expected for spherically symmetric accretion, but appears in some models of super-Eddington accretion flows (e.g. ZEBRAs; Coughlin and Begelman 2014). This may indicate that the circumnuclear environments of ASASSN-14li's, Arp 299-B AT1's, and CNSS J0019+00's host galaxies have been shaped by previous episodes of accretion onto their SMBHs. This is consistent with multi-wavelength evidence of at least a low level of AGN activity in these galaxies that predates the TDE outburst (Alexander et al. 2016; Romero-Cañizales et al. 2016; Mattila et al. 2018; Anderson et al. 2019).

4.2 Radio Evidence for AGN Activity and Complex Nuclear Environments

Interestingly, of the six published radio-quiet TDEs with detected low-luminosity radio emission, only two (XMMSL1 J0740-85 and AT2019dsg) occurred in host galaxies that showed no signs of recent or ongoing AGN activity (Table 1). This may indicate that AGN produce environmental conditions particularly favorable for the production of detectable radio emission in TDEs, but it also raises the question of possible contamination. While this is a valid concern (as noted in Sect. 3.3, AGN flares cannot be completely ruled out as an explanation for two or three of the radio-quiet TDEs discussed here), the past two decades of increasingly detailed observations have also allowed us to build up a set of multi-wavelength characteristics that differentiate AGN variability from true TDEs. The multi-wavelength properties that distinguish TDEs from "imposters" such as nuclear SNe or AGN flares are discussed in detail in Zabludoff et al. (2020, this issue). For some events whose optical and X-ray properties are ambiguous, the radio properties of the transient and its host galaxy may help to clarify its nature, as AGN variability is typically much less extreme at radio wavelengths compared to other parts of the electromagnetic spectrum (e.g. Hovatta et al. 2008, Nieppola et al. 2009).

Pre-flare radio observations of a TDE's host galaxy provide important context for interpreting radio observations collected during the TDE (e.g., Alexander et al. 2016; van Velzen et al. 2016; Romero-Cañizales et al. 2016; Mattila et al. 2018; Anderson et al. 2019). For example, ASASSN-14li's host galaxy was detected in two archival radio surveys $15-20$ years prior to the discovery of the TDE. Optical spectra of the host galaxy reveal a level of ongoing star formation an order of magnitude too low to explain the archival radio emission, so the best explanation for the archival detections is that ASASSN-14li's host galaxy hosts a weak AGN (Alexander et al. 2016; van Velzen et al. 2016). Long-term monitoring of ASASSN-14li's radio emission revealed a plateau starting ~ 1 year after discovery, indicating that the transient component of the emission has faded, and we are now viewing only the quiescent component (Bright et al. 2018).

The spectral index of the baseline radio emission in the host galaxy of ASASSN-14li is steep ($\alpha \approx -1$), suggesting that the emission is due to a "relic" population of electrons that was accelerated up to a few Gyr prior to the TDE. Integral field spectroscopy supports this evidence for recent, but not on-going, AGN activity in the host galaxy of ASASSN-14li

[4] We note that this result is only valid for models in which the radio emission arises from an external (forward) shock (Sects. 3.2.1 and 3.2.2). In the model presented by Pasham and van Velzen (2018) (Sect. 3.2.3), the radio emission arises internal to the jet and thus does not provide any information about the density profile of the external medium. However, because the data prefer a freely-expanding jet, Pasham and van Velzen (2018) infer that the external density must be low.

(Prieto et al. 2016). VLASS observations of new TDEs will help determine whether similar low levels of AGN activity are common in TDE host galaxies.

A more detailed picture of the nuclear environment may also be obtained via extremely high resolution VLBI observations. VLBI observations at early times can help pinpoint the nuclear position of the transient, confirming or ruling out a TDE origin. This is especially useful for more distant events, like Sw J1644+57 (Zauderer et al. 2011). Observations at later times can also constrain the velocity and orientation of the jet relative to the observer's line of sight (as Yang et al. 2016 do for Sw J1644+57), and can even directly resolve the jet for nearby events (as Mattila et al. 2018 do for Arp 299-B AT1).

VLBI observations of nearby events additionally reveal that many TDEs occur in complex nuclear environments. In two events, ASASSN-14li and IGR 12580+0134, VLBI observations taken after the discovery of each transient revealed multiple emission components in the nuclear vicinity, some of which were at least marginally resolved (Romero-Cañizales et al. 2016; Perlman et al. 2017). In both cases, these secondary components would require superluminal motion if associated with the TDE; more likely they are related to previous episodes of accretion activity. In the case of ASASSN-14li, it is also possible that the two observed radio components may correspond to the two components of a binary black hole (Romero-Cañizales et al. 2016). Additionally, Mattila et al. (2018) observed multiple compact radio sources in the nuclear vicinity (within a few parsecs) of Arp 299-B AT1 in pre-TDE VLBI imaging. This is not surprising, given that its host galaxy both contains an AGN and has an exceptionally high core-collapse supernova rate.

5 Upper Limits and Occurrence Rates

5.1 Radio Follow-up Observations

In 2011, the year when Sw J1644+57 was discovered, very few TDEs had received radio follow-up observations (exceptions are NGC 5905, Komossa 2002, and SDSS-TDE2, van Velzen et al. 2011a). This lack of data allowed for the exciting possibility of TDE unification via orientation, akin to the successful unification model of AGN (Bloom et al. 2011). The hope was that every TDE produces a relativistic jet, with Sw J1644+57 being the first discovered on-axis example of this new class of transient jets from SMBHs (e.g. Giannios and Metzger 2011). The first hint that this unified model of thermal and relativistic TDEs is not correct was presented by Bower (2011), who found no detections in archival radio observations available for a small number of TDEs (see Table 2). The first targeted follow-up observations of thermal TDEs were obtained using the VLA (Bower et al. 2013; van Velzen et al. 2013) and also yielded no detections.

To obtain an estimate of the off-axis radio flux of an event similar to Sw J1644+57, van Velzen et al. (2013) used the observed radio light of this source and adopted a constant late-time bulk Lorentz factor of 2; this Lorentz factor was inferred from the late-time radio light curve of Sw J1644+57 (Berger et al. 2012). Even with this conservative assumption, the radio upper limits obtained in 2013 could be used to rule out a unified model in which all thermal TDEs produce jets similar to Sw J1644+57 (van Velzen et al. 2013). A few years later, a more physical prediction of the off-axis radio luminosity of jet similar to Sw J166+57 was presented by Mimica et al. (2015). These authors tuned a hydrodynamical simulation of a fallback-powered jet interacting with a circumnuclear medium to obtain an excellent match to the observed light curve of Sw J1644+57. This simulation includes the deceleration of the jet, finding that the late-time radio emission becomes isotropic after about 3 years. At this point the luminosity at 10 GHz is predicted to be $\sim 10^{40}$ erg s^{-1}, which is larger than almost all upper limits that have been obtained. We can thus conclude that jets similar to Sw J1644+57 are rare; at the time of writing, 27 thermal TDEs (see Saxton et al. (2020, this issue) and van Velzen et al. (2020a, this issue)) have received radio follow-up, yet none show evidence for jets as powerful as Sw J1644+57.

A direct comparison of radio upper limits for known TDEs to off-axis light curves based on Sw J1644+57 has one caveat. Beside the jet power, the radio light curve depends on the circumnuclear medium. The effect of this second parameter was studied by Generozov et al. (2017), who considered a wide range of nuclear densities (as motivated by a physical model for gas supply from stellar winds; Generozov et al. 2015) to obtain an upper limit to the jet power for TDEs with radio follow-up observations. For nearly all thermal TDEs studied by Generozov et al. (2017), the upper limit to the total jet energy is an order of magnitude lower than the total of Sw J1644+57. This conclusion is consistent with the work of Mattila et al. (2018), who find that the radio transient Arp 299-B AT1 is an off-axis jet with a similar energy as Sw J1644+57. While Arp 299-B AT1 was detected in a nearby galaxy, even at the higher redshift of most thermal TDEs ($z \sim 0.1$), the isotropic radio emission of this source would exceed most of the published upper limits (Table 2). If events like Arp 299-B AT1 are common, many more should have been detected in our current follow-up observations of thermal TDEs found in blind optical/X-ray surveys, thus confirming that powerful jets following a TDE are rare.

In 2014, ASASSN-14li (Sect. 3.2) was discovered and we learned that thermal TDEs can be accompanied by a low energy radio flare ($E \sim 10^{47-48}$ erg, compared to $E \sim 4 \times 10^{51}$ erg for Sw J1644+57). At the time of its discovery, the radio luminosity of ASASSN-14li was lower than any of the upper limits obtained within one year of peak (mainly thanks to the low redshift of ASASSN-14li). Furthermore, since the radio emission was observed to fade quickly, the late-time follow-up observations of previous TDEs were also not sensitive enough to catch a radio flare similar to ASASSN-14li. Most models for the origin of the radio emission from ASASSN-14li suggested this emission could be (or should be) common to all thermal TDEs (see Sect. 3.2), which renewed the interest in rapid and sensitive radio follow-up of new TDEs.

At the time of writing, ASASSN-14li is one of two optically selected TDEs (see Table 1 of van Velzen et al. (2020a, this issue)) with published radio detections. Unlike many optical TDEs, ASASSN-14li also has bright X-ray emission, as do at least three of the other five detected radio-quiet TDEs. (Arp 299-B AT1's X-ray emission, if present, would have been masked by the large column density towards the TDE, as the AGN in Arp 299-B is Compton thick (Ptak et al. 2015), while CNSS J0019+00 was not observed in the X-ray at early times.) The first upper limits that rule out a radio flare similar to ASASSN-14li were obtained for the optical TDE iPTF-16fnl (Blagorodnova et al. 2017). However the low optical and X-ray luminosity of iPTF-16fnl ($L_X \sim 10^{37}$ erg s^{-1}), plus the rapid decay rate of its optical light curve make this source an outlier in the class of thermal TDEs—although we note that these faint TDEs dominate the volumetric rate (van Velzen 2018). The radio non-detection for iPTF-16fnl could be explained by its low luminosity. More recently, van Velzen et al. (2019) presented rapid and sensitive radio follow-up observations for optical TDE PS18kh/AT2018zr (Holoien et al. 2019) that also rule out a radio flare similar to ASASSN-14li. The optical properties of AT2018zr are similar to ASASSN-14li, and like ASASSN-14li AT2018zr also showed a soft, thermal X-ray spectrum (van Velzen et al. 2019). However the X-ray luminosity of AT2018zr is two orders of magnitude lower than ASASSN-14li's. The recent X-ray TDE XMMSL2 J1446+68 also shows no radio emission down to deep limits, suggesting that bright X-ray emission alone is not sufficient to produce detectable radio emission (Saxton et al. 2019).

In summary, at this point we can conclude that $\sim$ year-long radio flares with a peak luminosity $\sim 10^{38}$ erg s^{-1} are not common to *all* thermal TDEs. Of the six sources that have been studied with sufficient sensitivity, only ASASSN-14li and XMMSL1 0740-85 showed this feature. Somewhat brighter (peak luminosity $\sim$few 10^{38} erg s^{-1}), longer-lasting radio flares similar to Arp 299-B AT1 or IGR J12580+0134 are also not ubiquitous. If this trend is confirmed in a larger sample size, models that predict all TDEs should exhibit some level of radio emission will be disfavored (e.g. Sect. 3.2.2). While as yet there is no unifying physical characteristic that distinguishes the radio-detected thermal TDEs from those without faint radio emission, ongoing observing campaigns with the world's most sensitive radio and mm telescopes are expected to greatly expand the number of TDEs with radio detections and/or deep limits, allowing us to discriminate among the models presented in Sect. 3.2.

5.2 Rate of Sw J1644+57-Like Relativistic Jets

While the fraction of TDEs producing faint ASASSN-14li-like radio emission is still poorly constrained, existing observations do allow us to constrain the rate of radio-loud TDEs. Based on the first 10 years of *Swift*/BAT monitoring, the observed rate of events similar to Sw J1644+57 is $\sim 10^{-2}$ Gpc^{-3}yr^{-1} (Metzger et al. 2015).[5] Applying a correction for the Doppler boosting of the X-rays of $f_{\rm beam} \sim 100$ (van Velzen et al. 2018) we find $\dot{N}_{\rm J1644} \sim 1$ Gpc^{-3}yr^{-1}.

To estimate the expected fraction of observed TDEs with jets similar to Sw J166+57 we consider the ratio of the beaming-corrected rate to the rate of thermal TDEs,

$$f_{\rm J1644} \equiv \dot{N}_{\rm J1644}/\dot{N}_{\rm thermal}. \tag{1}$$

Since the volumetric rate of thermal TDEs depends on the flare's luminosity (van Velzen 2018), and most surveys are biased to the brighter events, we adopt $\dot{N}_{\rm thermal} = 10^{-4.3}$ galaxy^{-1} yr^{-1} $= 5 \times 10^{-5}$ galaxy^{-1} yr^{-1}. This rate is consistent with the observational TDE rates reported for several recent optical surveys (SDSS: $3^{+5}_{-3} \times 10^{-5}$ galaxy^{-1} yr^{-1}, van Velzen and Farrar 2014; ASAS-SN: $4.1^{+12.9}_{-1.9} \times 10^{-5}$ galaxy^{-1} yr^{-1}, Holoien et al. 2016; iPTF: $1.7^{+2.9}_{-1.3} \times 10^{-4}$ galaxy^{-1} yr^{-1}, Hung et al. 2018; see also the detailed discussion by van Velzen et al. (2020a, this issue)). The density of galaxies that can produce detectable TDEs is $\sim 10^{-2.4}$ Mpc^{-3} (for $h = 0.7$; see van Velzen and Farrar 2014), and we thus obtain $\dot{N}_{\rm thermal} \sim 10^{2.3}$ Gpc^{-3} yr^{-1}.

These two estimates of the rate imply $f_{\rm J1644} \sim 10^{-2.3}(f_{\rm beam}/100)$. One caveat to this estimate relates to the redshift evolution of the disruption rate; the jetted TDEs are found at higher redshift ($z \sim 1$) compared to the thermal TDEs ($z \sim 0.1$) and the redshift evolution of the TDE rate is unknown (depending on the loss-cone feeding mechanism the rate could both increase or decrease with redshift, see Stone et al. (2020, this issue)). At the time of writing, at least 29 TDEs have received radio follow-up observations that rule out a jet similar to Sw J1644+57 (Table 2, plus the six TDEs with faint radio detections). This lack of bright sources is consistent with our estimate of $f_{\rm J1644}$. Unless we underestimated the beaming factor, we should expect powerful jets in at most 1% of thermal TDEs. Constraints on the beaming factor can be obtained by measuring the isotropic rate of events similar to Sw J1644+57 using blind radio surveys.

[5] assuming all TDEs that launch radio-loud jets also produce bright γ-ray emission when viewed on-axis.

5.3 Blind Searches

Radio emission from TDEs has so far been detected primarily via follow-up observations (as TDEs have been discovered at other wavelengths, e.g., X-ray, UV, and optical). The sample of radio TDEs thus obtained is inevitably observationally biased, as the subset of TDEs chosen for radio follow up to date has not been uniformly selected. Direct discovery of TDEs in the radio can be achieved by blind searches in radio data. Several such searches have been conducted either by using archival data or by carrying out specifically designed time-domain radio surveys (e.g., Frail et al. 1994, Carilli et al. 2003, Bower et al. 2007, Croft et al. 2011, Ofek et al. 2011, Frail et al. 2012, Hodge et al. 2013, Mooley et al. 2016).[6] A blind search in the radio is more challenging compared, for example, to optical surveys, as the field of view of radio interferometers is much narrower and the typical sensitivity of radio facilities enables detection of radio emission from thermal TDEs only from nearby events. With time, the capabilities of numerous facilities have been upgraded, thus enabling wider and more sensitive blind searches.

Compared with theoretical expectations, most of the blind searches have not been wide and/or deep enough to discover TDEs. However, the upgrade of the VLA over the last decade has opened a window for performing effective blind surveys in which TDEs can be expected to be discovered. One such survey is the Caltech-NRAO Stripe 82 Survey (CNSS; Mooley et al. 2016). This survey was designed to cover the full $270\,\mathrm{deg}^2$ of the SDSS Stripe 82 area over multiple timescales in the $2-4$ GHz band with high spatial resolution (~ 3") and achieved a per-epoch sensitivity of ~ 80 μJy (Mooley et al. 2016). In five epochs spread over two years, one TDE was discovered, CNSS J0019+00 (Anderson et al. 2019). This implies that the rate of TDEs with radio emission brighter than 0.5 mJy is $1.8^{+5.4}_{-1.6} \times 10^{-3}$ deg^{-2} (Anderson et al. 2019). This corresponds to a volumetric rate of ~ 10 Gpc^{-3} yr^{-1} for radio TDEs with the same peak luminosity as CNSS J0019+00.

At face value, comparing this rate to the results presented above implies that TDEs with low-luminosity radio outflows similar to CNSS J0019+00 or ASASSN-14li are ~ 10 times more common than Sw J1644+57-like radio-loud TDEs, and comprise $\sim 10\%$ of thermal TDEs (Anderson et al. 2019). However, a direct comparison is not trivial due to remaining uncertainties in the TDE occurrence rate and its redshift evolution, the fraction of TDEs missed by optical/X-ray surveys, and the radio luminosity function of thermal TDEs. Additionally, identifying off-axis jetted TDEs as transient sources may be challenging because their radio emission evolves very slowly; during the 1.5 yr of the CNSS pilot survey, the 3 GHz flux density of a Sw J1644+57-like off-axis jet may not have changed sufficiently to be flagged as a variable source (Eftekhari et al. 2018). Radio surveys with more widely-separated epochs may be better suited to detect slow synchrotron transients like off-axis jetted TDEs (Metzger et al. 2015).

On the other hand, if most TDEs are thermal events that produce mostly weak radio emission, as implied by the rate from the CNSS, then a wider survey with the same sensitivity as the CNSS may be more effective to blindly detect TDEs. In fact, such a survey, the VLA Sky Survey (VLASS), has begun at the time of writing this book. The VLASS is an all sky survey that will be carried out using the VLA over the next several years. VLASS is using a drift scan technique that allows fast mapping of the sky (Mooley et al. 2019). Each piece of survey sky is expected to be observed three times during the project with each epoch separated by $\sim 18-24$ months. According to current estimates, tens of TDEs will be discovered

[6] A more complete list of radio transient surveys and additional details can be found at http://www.tauceti.caltech.edu/kunal/radio-transient-surveys/index.html.

in VLASS (Metzger et al. 2015, van Velzen et al. 2018; Anderson et al. 2019). However, this estimate is highly dependent on the true TDE rates and the TDE luminosity function, as already discussed above.

In the southern hemisphere, the best current facilities for wide-field radio transient searches are the Australian Square-Kilometre-Array Pathfinder (ASKAP; Johnston et al. 2008) and MeerKAT (Jonas 2016; Camilo et al. 2018). ASKAP utilizes a new phased array feed receiver design that effectively gives it an instantaneous field of view much larger than the VLA's (~ 30 deg^2 versus 0.25 deg^2), making it well-suited for targeted follow up of poorly-localized transients (including gravitational wave events, as recently demonstrated by Dobie et al. 2019) or for wide-field blind transient detection surveys. Several such surveys are already planned or underway (Murphy et al. 2013; Heywood et al. 2016; Hobbs et al. 2016; Bhandari et al. 2018). The largest of these is VAST (the ASKAP survey for Variables and Slow Transients), which includes a wide-field component (VAST-Wide) that will survey 10,000 deg^2 day^{-1} over a two-year period with an RMS sensitivity of 0.5 mJy. VAST-Wide is predicted to discover up to a dozen TDEs per year (van Velzen et al. 2011b; Murphy et al. 2013). MeerKAT, located in South Africa, has a smaller but still impressive field of view ($\sim 1°$ at 1.4 GHz; Driessen et al. 2020) and has achieved sub-microjansky noise in its deepest observations to date (Mauch et al. 2020). Transient science has been designated as one of the key early science priorities for MeerKAT under the ThunderKAT Large Survey Project (Fender et al. 2017), which recently published its first blind detection of a radio transient (Driessen et al. 2020).

Both ASKAP and MeerKAT are currently limited to observing frequencies ≤ 1.4 GHz, and therefore using them to blindly discover TDEs involves many of the same challenges as discovering TDEs in the CNSS and VLASS (including the slow evolution of radio TDEs at low frequencies, and the challenge of distinguishing TDEs from background variable AGN). However, both ASKAP and MeerKAT are precursors to the Square Kilometer Array (SKA; Dewdney et al. 2009), which is expected to include high-frequency sensitivity up to at least 15 GHz.[7] Radio observations with future facilities such as the SKA and the next generation VLA (ngVLA)[8] are expected to produce larger samples of TDEs (hundreds per year) that are also more distant at higher redshifts (Metzger et al. 2015; Donnarumma et al. 2015; van Velzen et al. 2018).

6 Conclusions

Radio observations of TDEs provide unique information about these rare events that cannot be gleaned from observations at other wavelengths. In particular, they probe fast-moving material ejected from the system, in contrast to the generally thermal emission seen in the optical, UV, and X-rays. Radio observations have shown that a small subset of TDEs produce relativistic jets, while a larger fraction may produce less energetic outflows. The physical conditions required for jet production and the diversity of sub-parsec scale nuclear gas profiles around SHBMs in TDE hosts will be revealed by increasing the sample size of radio-detected TDEs. Given the broad range in observed TDE radio luminosities, we suggest that

[7] https://astronomers.skatelescope.org.

[8] While the ngVLA has a much smaller field of view than the SKA under the current design, its exquisite sensitivity at high frequencies will make it a key tool for the follow up and discovery of TDEs. Additionally, its milliarcsecond resolution will be very interesting for VLBI observations of nearby events (van Velzen et al. 2018).

radio follow up of new events should optimally probe radio luminosities $\nu F_\nu \leq 10^{37}\ \mathrm{erg\,s^{-1}}$, as many existing upper limits are too shallow to detect outflows similar to those observed in e.g. ASASSN-14li. Furthermore, with the advent of ALMA, it is now possible to probe even less energetic outflows, which are expected to reach higher peak luminosities in the millimeter than in the centimeter bands probed by the VLA (Yuan et al. 2016).

The increased TDE discovery rate promised by new sensitive, wide-field optical surveys like ZTF is already being realized (van Velzen et al. 2019). Simultaneously, new and upcoming radio surveys may soon provide the first population of TDEs discovered in the radio band (Murphy et al. 2013, Metzger et al. 2015, Fender et al. 2017, Anderson et al. 2019). Within the next five years, we will have placed strong constraints on the rate of weak radio outflows in TDEs, and may discover further extremes in an already diverse radio TDE population. The sample of radio TDEs will continue to grow in the era of LSST plus SKA (Donnarumma et al. 2015) — although we note that the slow evolution of the radio light curve in low-frequency observations (1.4 GHz) will present a major challenge for TDE identification (Donnarumma and Rossi 2015). A time-domain survey at higher frequencies ($\gtrsim$ 5 GHz) with the ngVLA (van Velzen et al. 2018) or SKA band 5 would be the ultimate discovery machine for radio-emitting TDEs, finding as many as $\sim 10^2$ jetted TDEs per year.

Acknowledgements We acknowledge useful discussions with the attendees of the ISSI TDE workshop in October 2018, in particular Tsvi Piran. KDA acknowledges support provided by NASA through the NASA Hubble Fellowship grant HST-HF2-51403.001 awarded by the Space Telescope Science Institute, which is operated by the Association of Universities for Research in Astronomy, Inc., for NASA, under contract NAS5-26555. BAZ acknowledges support while serving at the National Science Foundation (NSF) and from the Dark Cosmology Centre (DARK) at the University of Copenhagen. Any opinion, findings, and conclusions expressed in this material are those of the authors and do not necessarily reflect the views of the supporting agencies.

References

K.D. Alexander, E. Berger, J. Guillochon, B.A. Zauderer, P.K.G. Williams, Discovery of an outflow from radio observations of the tidal disruption event ASASSN-14li. Astrophys. J. Lett. **819**, 25 (2016). https://doi.org/10.3847/2041-8205/819/2/L25

K.D. Alexander, M.H. Wieringa, E. Berger, R.D. Saxton, S. Komossa, Radio observations of the tidal disruption event XMMSL1 J0740-85. Astrophys. J. **837**, 153 (2017a). https://doi.org/10.3847/1538-4357/aa6192

K.D. Alexander, E. Berger, W. Fong, P.K.G. Williams, C. Guidorzi, R. Margutti, B.D. Metzger, J. Annis, P.K. Blanchard, D. Brout, D.A. Brown, H.-Y. Chen, R. Chornock, P.S. Cowperthwaite, M. Drout, T. Eftekhari, J. Frieman, D.E. Holz, M. Nicholl, A. Rest, M. Sako, M. Soares-Santos, V.A. Villar, The electromagnetic counterpart of the binary neutron star merger LIGO/Virgo GW170817. VI. Radio constraints on a relativistic jet and predictions for late-time emission from the Kilonova ejecta. Astrophys. J. Lett. **848**, 21 (2017b). https://doi.org/10.3847/2041-8213/aa905d

K.D. Alexander, R. Margutti, P.K. Blanchard, W. Fong, E. Berger, A. Hajela, T. Eftekhari, R. Chornock, P.S. Cowperthwaite, D. Giannios, C. Guidorzi, A. Kathirgamaraju, A. MacFadyen, B.D. Metzger, M. Nicholl, L. Sironi, V.A. Villar, P.K.G. Williams, X. Xie, J. Zrake, A decline in the X-ray through radio emission from GW170817 continues to support an off-axis structured jet. Astrophys. J. Lett. **863**, 18 (2018). https://doi.org/10.3847/2041-8213/aad637

M.M. Anderson, K.P. Mooley, G. Hallinan, D. Dong, E.S. Phinney, A. Horesh, S. Bourke, S.B. Cenko, D. Frail, S.R. Kulkarni, S. Myers, Caltech-NRAO Stripe 82 Survey (CNSS) Paper III: the First Radio-Discovered Tidal Disruption Event, CNSS J0019+00. arXiv e-prints (2019). arxiv:1910.11912

I. Arcavi, A. Gal-Yam, M. Sullivan, Y.-C. Pan, S.B. Cenko, A. Horesh, E.O. Ofek, A. De Cia, L. Yan, C.-W. Yang, D.A. Howell, D. Tal, S.R. Kulkarni, S.P. Tendulkar, S. Tang, D. Xu, A. Sternberg, J.G. Cohen,

J.S. Bloom, P.E. Nugent, M.M. Kasliwal, D.A. Perley, R.M. Quimby, A.A. Miller, C.A. Theissen, R.R. Laher, A continuum of H- to He-rich tidal disruption candidates with a preference for E+A galaxies. Astrophys. J. **793**, 38 (2014). https://doi.org/10.1088/0004-637X/793/1/38

F.K. Baganoff, Y. Maeda, M. Morris, M.W. Bautz, W.N. Brandt, W. Cui, J.P. Doty, E.D. Feigelson, G.P. Garmire, S.H. Pravdo, G.R. Ricker, L.K. Townsley, Chandra X-ray spectroscopic imaging of Sagittarius A* and the central parsec of the galaxy. Astrophys. J. **591**, 891–915 (2003). https://doi.org/10.1086/375145

R. Barniol Duran, T. Piran, On the origin of the radio emission of Sw 1644+57. Astrophys. J. **770**, 146 (2013). https://doi.org/10.1088/0004-637X/770/2/146

R. Barniol Duran, E. Nakar, T. Piran, Radius constraints and minimal equipartition energy of relativistically moving synchrotron sources. Astrophys. J. **772**, 78 (2013). https://doi.org/10.1088/0004-637X/772/1/78

E. Berger, A. Zauderer, G.G. Pooley, A.M. Soderberg, R. Sari, A. Brunthaler, M.F. Bietenholz, Radio monitoring of the tidal disruption event Swift J164449.3+573451. I. Jet energetics and the pristine parsec-scale environment of a supermassive black hole. Astrophys. J. **748**, 36 (2012). https://doi.org/10.1088/0004-637X/748/1/36

S. Bhandari, K.W. Bannister, T. Murphy, M. Bell, W. Raja, J. Marvil, P.J. Hancock, M. Whiting, C.M. Flynn, J.D. Collier, D.L. Kaplan, J.R. Allison, C. Anderson, I. Heywood, A. Hotan, R. Hunstead, K. Lee-Waddell, J.P. Madrid, D. McConnell, A. Popping, J. Rhee, E. Sadler, M.A. Voronkov, A pilot survey for transients and variables with the Australian Square Kilometre Array Pathfinder. Mon. Not. R. Astron. Soc. **478**(2), 1784–1794 (2018). https://doi.org/10.1093/mnras/sty1157

N. Blagorodnova, S. Gezari, T. Hung, S.R. Kulkarni, S.B. Cenko, D.R. Pasham, L. Yan, I. Arcavi, S. Ben-Ami, B.D. Bue, T. Cantwell, Y. Cao, A.J. Castro-Tirado, R. Fender, C. Fremling, A. Gal-Yam, A.Y.Q. Ho, A. Horesh, G. Hosseinzadeh, M.M. Kasliwal, A.K.H. Kong, R.R. Laher, G. Leloudas, R. Lunnan, F.J. Masci, K. Mooley, J.D. Neill, P. Nugent, M. Powell, A.F. Valeev, P.M. Vreeswijk, R. Walters, P. Wozniak, iPTF16fnl: a faint and fast tidal disruption event in an E+A galaxy. Astrophys. J. **844**, 46 (2017). https://doi.org/10.3847/1538-4357/aa7579

N. Blagorodnova, S.B. Cenko, S.R. Kulkarni, I. Arcavi, J.S. Bloom, G. Duggan, A.V. Filippenko, C. Fremling, A. Horesh, G. Hosseinzadeh, E. Karamehmetoglu, A. Levan, F.J. Masci, P.E. Nugent, D.R. Pasham, S. Veilleux, R. Walters, L. Yan, W. Zheng, The broad absorption line tidal disruption event iPTF15af: optical and ultraviolet evolution. Astrophys. J. **873**, 92 (2019). https://doi.org/10.3847/1538-4357/ab04b0

P.K. Blanchard, M. Nicholl, E. Berger, J. Guillochon, R. Margutti, R. Chornock, K.D. Alexander, J. Leja, M.R. Drout, PS16dtm: a tidal disruption event in a narrow-line Seyfert 1 galaxy. Astrophys. J. **843**, 106 (2017). https://doi.org/10.3847/1538-4357/aa77f7

R.D. Blandford, A. Königl, Relativistic jets as compact radio sources. Astrophys. J. **232**, 34–48 (1979)

J.S. Bloom, D. Giannios, B.D. Metzger, S.B. Cenko, D.A. Perley, N.R. Butler, N.R. Tanvir, A.J. Levan, P.T. O'Brien, L.E. Strubbe, F. De Colle, E. Ramirez-Ruiz, W.H. Lee, S. Nayakshin, E. Quataert, A.R. King, A. Cucchiara, J. Guillochon, G.C. Bower, A.S. Fruchter, A.N. Morgan, A.J. van der Horst, A possible relativistic jetted outburst from a massive black hole fed by a tidally disrupted star. Science **333**, 203 (2011)

H. Bondi, On spherically symmetrical accretion. Mon. Not. R. Astron. Soc. **112**, 195 (1952). https://doi.org/10.1093/mnras/112.2.195

G.C. Bower, Constraining the rate of relativistic jets from tidal disruptions using radio surveys. Astrophys. J. Lett. **732**, 12 (2011)

G.C. Bower, D. Saul, J.S. Bloom, A. Bolatto, A.V. Filippenko, R.J. Foley, D. Perley, Submillijansky transients in archival radio observations. Astrophys. J. **666**, 346–360 (2007). https://doi.org/10.1086/519831

G.C. Bower, B.D. Metzger, S.B. Cenko, J.M. Silverman, J.S. Bloom, Late-time radio emission from X-ray-selected tidal disruption events. Astrophys. J. **763**, 84 (2013). https://doi.org/10.1088/0004-637X/763/2/84

J.S. Bright, R.P. Fender, S.E. Motta, K. Mooley, Y.C. Perrott, S. van Velzen, S. Carey, J. Hickish, N. Razavi-Ghods, D. Titterington, P. Scott, K. Grainge, A. Scaife, T. Cantwell, C. Rumsey, Long-term radio and X-ray evolution of the tidal disruption event ASASSN-14li. Mon. Not. R. Astron. Soc. **475**, 4011–4019 (2018). https://doi.org/10.1093/mnras/sty077

G.C. Brown, A.J. Levan, E.R. Stanway, N.R. Tanvir, S.B. Cenko, E. Berger, R. Chornock, A. Cucchiaria, Swift J1112.2-8238: a candidate relativistic tidal disruption flare. Mon. Not. R. Astron. Soc. **452**, 4297–4306 (2015). https://doi.org/10.1093/mnras/stv1520

G.C. Brown, A.J. Levan, E.R. Stanway, T. Krühler, N.R. Tanvir, L.J.M. Davies, A. Fruchter, S.B. Cenko, B.D. Metzger, Late-time observations of the relativistic tidal disruption flare candidate Swift J1112.2-8238. Mon. Not. R. Astron. Soc. **472**, 4469–4479 (2017). https://doi.org/10.1093/mnras/stx2193

D.N. Burrows, J.A. Kennea, G. Ghisellini, V. Mangano, B. Zhang, K.L. Page, M. Eracleous, P. Romano, T. Sakamoto, A.D. Falcone, J.P. Osborne, S. Campana, A.P. Beardmore, A.A. Breeveld, M.M. Chester,

R. Corbet, S. Covino, J.R. Cummings, P. D'Avanzo, V. D'Elia, P. Esposito, P.A. Evans, D. Fugazza, J.M. Gelbord, K. Hiroi, S.T. Holland, K.Y. Huang, M. Im, G. Israel, Y. Jeon, Y.-B. Jeon, H.D. Jun, N. Kawai, J.H. Kim, H.A. Krimm, F.E. Marshall, P. Mészáros, H. Negoro, N. Omodei, W.-K. Park, J.S. Perkins, M. Sugizaki, H.-I. Sung, G. Tagliaferri, E. Troja, Y. Ueda, Y. Urata, R. Usui, L.A. Antonelli, S.D. Barthelmy, G. Cusumano, P. Giommi, A. Melandri, M. Perri, J.L. Racusin, B. Sbarufatti, M.H. Siegel, N. Gehrels, Relativistic jet activity from the tidal disruption of a star by a massive black hole. Nature **476**, 421–424 (2011). https://doi.org/10.1038/nature10374

F. Camilo, P. Scholz, M. Serylak, S. Buchner, M. Merryfield, V.M. Kaspi, R.F. Archibald, M. Bailes, A. Jameson, W. van Straten, J. Sarkissian, J.E. Reynolds, S. Johnston, G. Hobbs, T.D. Abbott, R.M. Adam, G.B. Adams, T. Alberts, R. Andreas, K.M.B. Asad, D.E. Baker, T. Baloyi, E.F. Bauermeister, T. Baxana, T.G.H. Bennett, G. Bernardi, D. Booisen, R.S. Booth, D.H. Botha, L. Boyana, L.R.S. Brederode, J.P. Burger, T. Cheetham, J. Conradie, J.P. Conradie, D.B. Davidson, G. De Bruin, B. de Swardt, C. de Villiers, D.I.L. de Villiers, M.S. de Villiers, W. de Villiers, C. De Waal, M.A. Dikgale, G. du Toit, L.J. du Toit, S.W.P. Esterhuyse, B. Fanaroff, S. Fataar, A.R. Foley, G. Foster, D. Fourie, R. Gamatham, T. Gatsi, R. Geschke, S. Goedhart, T.L. Grobler, S.C. Gumede, M.J. Hlakola, A. Hokwana, D.H. Hoorn, D. Horn, J. Horrell, B. Hugo, A. Isaacson, O. Jacobs, J.P. Jansen van Rensburg, J.L. Jonas, B. Jordaan, A. Joubert, F. Joubert, G.I.G. Józsa, R. Julie, C.C. Julius, F. Kapp, A. Karastergiou, F. Karels, M. Kariseb, R. Karuppusamy, V. Kasper, E.C. Knox-Davies, D. Koch, P.P.A. Kotzé, A. Krebs, N. Kriek, H. Kriel, T. Kusel, S. Lamoor, R. Lehmensiek, D. Liebenberg, I. Liebenberg, R.T. Lord, B. Lunsky, N. Mabombo, T. Macdonald, P. Macfarlane, K. Madisa, L. Mafhungo, L.G. Magnus, C. Magozore, O. Mahgoub, J.P.L. Main, S. Makhathini, J.A. Malan, P. Malgas, J.R. Manley, M. Manzini, L. Marais, N. Marais, S.J. Marais, M. Maree, A. Martens, S.D. Matshawule, N. Matthysen, T. Mauch, L.D. McNally, B. Merry, R.P. Millenaar, C. Mjikelo, N. Mkhabela, N. Mnyand u, I.T. Moeng, O.J. Mokone, T.E. Monama, K. Montshiwa, V. Moss, M. Mphego, W. New, B. Ngcebetsha, K. Ngoasheng, H. Niehaus, P. Ntuli, A. Nzama, F. Obies, M. Obrocka, M.T. Ockards, C. Olyn, N. Oozeer, A.J. Otto, Y. Padayachee, S. Passmoor, A.A. Patel, S. Paula, A. Peens-Hough, B. Pholoholo, P. Prozesky, S. Rakoma, A.J.T. Ramaila, I. Rammala, Z.R. Ramudzuli, M. Rasivhaga, S. Ratcliffe, H.C. Reader, R. Renil, L. Richter, A. Robyntjies, D. Rosekrans, A. Rust, S. Salie, N. Sambu, C.T.G. Schollar, L. Schwardt, S. Seranyane, G. Sethosa, C. Sharpe, R. Siebrits, S.K. Sirothia, M.J. Slabber, O. Smirnov, S. Smith, L. Sofeya, N. Songqumase, R. Spann, B. Stappers, D. Steyn, T.J. Steyn, R. Strong, A. Struthers, C. Stuart, P. Sunnylall, P.S. Swart, B. Taljaard, C. Tasse, G. Taylor, I.P. Theron, V. Thondikulam, K. Thorat, A. Tiplady, O. Toruvanda, J. van Aardt, T. van Balla, L. van den Heever, A. van der Byl, C. van der Merwe, P. van der Merwe, P.C. van Niekerk, R. van Rooyen, J.P. van Staden, V. van Tonder, R. van Wyk, I. Wait, A.L. Walker, B. Wallace, M. Welz, L.P. Williams, B. Xaia, N. Young, S. Zitha, Revival of the Magnetar PSR J1622-4950: observations with MeerKAT, Parkes, XMM-Newton, Swift, Chandra, and NuSTAR. Astrophys. J. **856**(2), 180 (2018). https://doi.org/10.3847/1538-4357/aab35a

C.L. Carilli, R.J. Ivison, D.A. Frail, Variability of submillijansky radio sources. Astrophys. J. **590**, 192–196 (2003). https://doi.org/10.1086/375005

Y. Cendes, R.A.M.J. Wijers, J.D. Swinbank, A. Rowlinson, A.J. van der Horst, D. Carbone, J.W. Broderick, T.D. Staley, A.J. Stewart, G. Molenaar, F. Huizinga, A. Alexov, M.E. Bell, T. Coenen, S. Corbel, J. Eislöffel, R. Fender, J. Grießmeier, J. Hessel, P. Jonker, M. Kramer, M. Kuniyoshi, C.J. Law, S. Markoff, M. Pietka, B. Stappers, M. Wise, P. Zarka, LOFAR Observations of Swift J1644+57 and Implications for Short-Duration Transients. arXiv:e-prints (2014). arXiv:1412.3986

S.B. Cenko, H.A. Krimm, A. Horesh, A. Rau, D.A. Frail, J.A. Kennea, A.J. Levan, S.T. Holland, N.R. Butler, R.M. Quimby, J.S. Bloom, A.V. Filippenko, A. Gal-Yam, J. Greiner, S.R. Kulkarni, E.O. Ofek, F. Olivares E., P. Schady, J.M. Silverman, N.R. Tanvir, D. Xu, Swift J2058.4+0516: discovery of a possible second relativistic tidal disruption flare? Astrophys. J. **753**, 77 (2012). https://doi.org/10.1088/0004-637X/753/1/77

R.A. Chevalier, The radio and X-ray emission from type II supernovae. Astrophys. J. **259**, 302–310 (1982). https://doi.org/10.1086/160167

R.A. Chevalier, Synchrotron self-absorption in radio supernovae. Astrophys. J. **499**, 810–819 (1998). https://doi.org/10.1086/305676

R.A. Chevalier, C. Fransson, Circumstellar emission from type Ib and Ic supernovae. Astrophys. J. **651**, 381–391 (2006). https://doi.org/10.1086/507606

R. Chornock, E. Berger, S. Gezari, B.A. Zauderer, A. Rest, L. Chomiuk, A. Kamble, A.M. Soderberg, I. Czekala, J. Dittmann, M. Drout, R.J. Foley, W. Fong, M.E. Huber, R.P. Kirshner, A. Lawrence, R. Lunnan, G.H. Marion, G. Narayan, A.G. Riess, K.C. Roth, N.E. Sanders, D. Scolnic, S.J. Smartt, K. Smith, C.W. Stubbs, J.L. Tonry, W.S. Burgett, K.C. Chambers, H. Flewelling, K.W. Hodapp, N. Kaiser, E.A. Magnier, D.C. Martin, J.D. Neill, P.A. Price, R. Wainscoat, The ultraviolet-bright, slowly declining transient PS1-11af as a partial tidal disruption event. Astrophys. J. **780**, 44 (2014). https://doi.org/10.1088/0004-637X/780/1/44

J.J. Condon, S.M. Ransom, *Essential Radio Astronomy* (2016)
A. Corsi, G.W. Hallinan, D. Lazzati, K.P. Mooley, E.J. Murphy, D.A. Frail, D. Carbone, D.L. Kaplan, T. Murphy, S.R. Kulkarni, K. Hotokezaka, An upper limit on the linear polarization fraction of the GW170817 radio continuum. Astrophys. J. Lett. **861**, 10 (2018). https://doi.org/10.3847/2041-8213/aacdfd
E.R. Coughlin, M.C. Begelman, Hyperaccretion during tidal disruption events: weakly bound debris envelopes and jets. Astrophys. J. **781**, 82 (2014). https://doi.org/10.1088/0004-637X/781/2/82
S. Croft, G.C. Bower, G. Keating, C. Law, D. Whysong, P.K.G. Williams, M. Wright, The Allen telescope array twenty-centimeter survey — a 700-square-degree, multi-epoch radio data set. II. Individual epoch transient statistics. Astrophys. J. **731**, 34 (2011). https://doi.org/10.1088/0004-637X/731/1/34
P.E. Dewdney, P.J. Hall, R.T. Schilizzi, T.J.L.W. Lazio, The square kilometre array. IEEE Proc. **97**(8), 1482–1496 (2009). https://doi.org/10.1109/JPROC.2009.2021005
D. Dobie, D.L. Kaplan, T. Murphy, E. Lenc, K.P. Mooley, C. Lynch, A. Corsi, D. Frail, M. Kasliwal, G. Hallinan, A turnover in the radio light curve of GW170817. Astrophys. J. Lett. **858**, 15 (2018). https://doi.org/10.3847/2041-8213/aac105
D. Dobie, A. Stewart, T. Murphy, E. Lenc, Z. Wang, D.L. Kaplan, I. Andreoni, J. Banfield, I. Brown, A. Corsi, K. De, D.A. Goldstein, G. Hallinan, A. Hotan, K. Hotokezaka, A.D. Jaodand, V. Karambelkar, M.M. Kasliwal, D. McConnell, K. Mooley, V.A. Moss, J.A. Newman, D.A. Perley, A. Prakash, J. Pritchard, E.M. Sadler, Y. Sharma, C. Ward, M. Whiting, R. Zhou, An ASKAP search for a radio counterpart to the first high-significance neutron star-black hole merger LIGO/Virgo S190814bv. Astrophys. J. Lett. **887**(1), 13 (2019). https://doi.org/10.3847/2041-8213/ab59db
I. Donnarumma, E.M. Rossi, Radio-X-ray synergy to discover and study jetted tidal disruption events. Astrophys. J. **803**, 36 (2015). https://doi.org/10.1088/0004-637X/803/1/36
I. Donnarumma, E.M. Rossi, R. Fender, S. Komossa, Z. Paragi, S. Van Velzen, I. Prandoni, SKA as a powerful hunter of jetted tidal disruption events, in *Advancing Astrophysics with the Square Kilometre Array (AASKA14)* (2015), p. 54
L.N. Driessen, I. McDonald, D.A.H. Buckley, M. Caleb, E.J. Kotze, S.B. Potter, K.M. Rajwade, A. Rowlinson, B.W. Stappers, E. Tremou, P.A. Woudt, R.P. Fender, R. Armstrong, P. Groot, I. Heywood, A. Horesh, A.J. van der Horst, E. Koerding, V.A. McBride, J.C.A. Miller-Jones, K.P. Mooley, R.A.M.J. Wijers, MKT J170456.2-482100: the first transient discovered by MeerKAT. Mon. Not. R. Astron. Soc. **491**(1), 560–575 (2020). https://doi.org/10.1093/mnras/stz3027
T. Eftekhari, E. Berger, B.A. Zauderer, R. Margutti, K.D. Alexander, Radio monitoring of the tidal disruption event Swift J164449.3+573451. III. Late-time jet energetics and a deviation from equipartition. Astrophys. J. **854**, 86 (2018). https://doi.org/10.3847/1538-4357/aaa8e0
H. Falcke, P.L. Biermann, The jet-disk symbiosis. I. Radio to X-ray emission models for quasars. Astron. Astrophys. **293**, 665–682 (1995)
R. Fender, P.A. Woudt, R. Armstrong, P. Groot, V. McBride, J. Miller-Jones, K. Mooley, B. Stappers, R. Wijers, M. Bietenholz, S. Blyth, M. Bottcher, D. Buckley, P. Charles, L. Chomiuk, D. Coppejans, S. Corbel, M. Coriat, F. Daigne, W.J.G. de Blok, H. Falcke, J. Girard, I. Heywood, A. Horesh, J. Horrell, P. Jonker, T. Joseph, A. Kamble, C. Knigge, E. Koerding, M. Kotze, C. Kouveliotou, C. Lynch, T. Maccarone, P. Meintjes, S. Migliari, T. Murphy, T. Nagayama, G. Nelemans, G. Nicholson, T. O'Brien, A. Oodendaal, N. Oozeer, J. Osborne, M. Perez-Torres, S. Ratcliffe, V. Ribeiro, E. Rol, A. Rushton, A. Scaife, M. Schurch, G. Sivakoff, T. Staley, D. Steeghs, I. Stewart, J. Swinbank, K. van der Heyden, A. van der Horst, B. van Soelen, S. Vergani, B. Warner, K. Wiersema, ThunderKAT: the MeerKAT Large Survey Project for Image-Plane Radio Transients. arXiv e-prints (2017). arXiv:1711.04132
D.A. Frail, S.R. Kulkarni, K.C. Hurley, G.J. Fishman, C. Kouveliotou, C.A. Meegan, M. Sommer, M. Boer, M. Niel, T. Cline, A search for the radio counterpart to the 1994 March 1 gamma-ray burst. Astrophys. J. **437**, 43 (1994). https://doi.org/10.1086/187678
D.A. Frail, E. Waxman, S.R. Kulkarni, A 450 day light curve of the radio afterglow of GRB 970508: fireball calorimetry. Astrophys. J. **537**(1), 191–204 (2000). https://doi.org/10.1086/309024
D.A. Frail, S.R. Kulkarni, E.O. Ofek, G.C. Bower, E. Nakar, A revised view of the transient radio sky. Astrophys. J. **747**, 70 (2012). https://doi.org/10.1088/0004-637X/747/1/70
K.D. French, T. Wevers, J. Law-Smith, et al., The host galaxies of tidal disruption events. Space Sci. Rev. **216**, 32 (2020, this issue). https://doi.org/10.1007/s11214-020-00657-y
T.J. Galama, P.M. Vreeswijk, J. van Paradijs, C. Kouveliotou, T. Augusteijn, F. Patat, J. Heise, J. in 't Zand, P.J. Groot, R.A.M.J. Wijers, E. Pian, E. Palazzi, F. Frontera, N. Masetti, On the possible association of SN 1998bw and GRB 980425. Astron. Astrophys. Suppl. Ser. **138**, 465–466 (1999). https://doi.org/10.1051/aas:1999311
A. Generozov, N.C. Stone, B.D. Metzger, Circumnuclear media of quiescent supermassive black holes. Mon. Not. R. Astron. Soc. **453**, 775–796 (2015). https://doi.org/10.1093/mnras/stv1607

A. Generozov, P. Mimica, B.D. Metzger, N.C. Stone, D. Giannios, M.A. Aloy, The influence of circumnuclear environment on the radio emission from TDE jets. Mon. Not. R. Astron. Soc. **464**, 2481–2498 (2017). https://doi.org/10.1093/mnras/stw2439

G. Ghirlanda, O.S. Salafia, Z. Paragi, M. Giroletti, J. Yang, B. Marcote, J. Blanchard, I. Agudo, T. An, M.G. Bernardini, R. Beswick, M. Branchesi, S. Campana, C. Casadio, E. Chassande-Mottin, M. Colpi, S. Covino, P. D'Avanzo, V. D'Elia, S. Frey, M. Gawronski, G. Ghisellini, L.I. Gurvits, P.G. Jonker, H.J. van Langevelde, A. Melandri, J. Moldon, L. Nava, A. Perego, M.A. Perez-Torres, C. Reynolds, R. Salvaterra, G. Tagliaferri, T. Venturi, S.D. Vergani, M. Zhang, Compact radio emission indicates a structured jet was produced by a binary neutron star merger. Science **363**, 968–971 (2019). https://doi.org/10.1126/science.aau8815

D. Giannios, B.D. Metzger, Radio transients from stellar tidal disruption by massive black holes. Mon. Not. R. Astron. Soc. **416**(3), 2102–2107 (2011). https://doi.org/10.1111/j.1365-2966.2011.19188.x

S. Gillessen, P.M. Plewa, F. Widmann, S. von Fellenberg, M. Schartmann, M. Habibi, A. Jimenez Rosales, M. Bauböck, J. Dexter, F. Gao, I. Waisberg, F. Eisenhauer, O. Pfuhl, T. Ott, A. Burkert, P.T. de Zeeuw, R. Genzel, Detection of a drag force in G2's orbit: measuring the density of the accretion flow onto Sgr A* at 1000 Schwarzschild radii. Astrophys. J. **871**, 126 (2019). https://doi.org/10.3847/1538-4357/aaf4f8

J. Granot, R. Sari, The shape of spectral breaks in gamma-ray burst afterglows. Astrophys. J. **568**, 820–829 (2002). https://doi.org/10.1086/338966

J. Granot, A. Panaitescu, P. Kumar, S.E. Woosley, Off-axis afterglow emission from jetted gamma-ray bursts. Astrophys. J. Lett. **570**, 61–64 (2002). https://doi.org/10.1086/340991

J. Guillochon, M. McCourt, X. Chen, M.D. Johnson, E. Berger, Unbound debris streams and remnants resulting from the tidal disruptions of stars by supermassive black holes. Astrophys. J. **822**, 48 (2016). https://doi.org/10.3847/0004-637X/822/1/48

A. Hajela, R. Margutti, K.D. Alexander, A. Kathirgamaraju, A. Baldeschi, C. Guidorzi, D. Giannios, W. Fong, Y. Wu, A. MacFadyen, A. Paggi, E. Berger, P.K. Blanchard, R. Chornock, D.L. Coppejans, P.S. Cowperthwaite, T. Eftekhari, S. Gomez, G. Hosseinzadeh, T. Laskar, B.D. Metzger, M. Nicholl, K. Paterson, D. Radice, L. Sironi, G. Terreran, V.A. Villar, P.K.G. Williams, X. Xie, J. Zrake, Two years of nonthermal emission from the binary neutron star merger GW170817: rapid fading of the jet afterglow and first constraints on the Kilonova fastest ejecta. Astrophys. J. Lett. **886**(1), 17 (2019). https://doi.org/10.3847/2041-8213/ab5226

G. Hallinan, A. Corsi, K.P. Mooley, K. Hotokezaka, E. Nakar, M.M. Kasliwal, D.L. Kaplan, D.A. Frail, S.T. Myers, T. Murphy, K. De, D. Dobie, J.R. Allison, K.W. Bannister, V. Bhalerao, P. Chandra, T.E. Clarke, S. Giacintucci, A.Y.Q. Ho, A. Horesh, N.E. Kassim, S.R. Kulkarni, E. Lenc, F.J. Lockman, C. Lynch, D. Nichols, S. Nissanke, N. Palliyaguru, W.M. Peters, T. Piran, J. Rana, E.M. Sadler, L.P. Singer, A radio counterpart to a neutron star merger. Science **358**, 1579–1583 (2017). https://doi.org/10.1126/science.aap9855

I. Heywood, K.W. Bannister, J. Marvil, J.R. Allison, L. Ball, M.E. Bell, D.C.-J. Bock, M. Brothers, J.D. Bunton, A.P. Chippendale, F. Cooray, T.J. Cornwell, D. De Boer, P. Edwards, R. Gough, N. Gupta, L. Harvey-Smith, S. Hay, A.W. Hotan, B. Indermuehle, C. Jacka, C.A. Jackson, S. Johnston, A.E. Kimball, B.S. Koribalski, E. Lenc, A. Macleod, N. McClure-Griffiths, D. McConnell, P. Mirtschin, T. Murphy, S. Neuhold, R.P. Norris, S. Pearce, A. Popping, R.Y. Qiao, J.E. Reynolds, E.M. Sadler, R.J. Sault, A.E.T. Schinckel, P. Serra, T.W. Shimwell, J. Stevens, J. Tuthill, A. Tzioumis, M.A. Voronkov, T. Westmeier, M.T. Whiting, Wide-field broad-band radio imaging with phased array feeds: a pilot multi-epoch continuum survey with ASKAP-BETA. Mon. Not. R. Astron. Soc. **457**(4), 4160–4178 (2016). https://doi.org/10.1093/mnras/stw186

G. Hobbs, I. Heywood, M.E. Bell, M. Kerr, A. Rowlinson, S. Johnston, R.M. Shannon, M.A. Voronkov, C. Ward, J. Banyer, P.J. Hancock, T. Murphy, J.R. Allison, S.W. Amy, L. Ball, K. Bannister, D.C.-J. Bock, D. Brodrick, M. Brothers, A.J. Brown, J.D. Bunton, J. Chapman, A.P. Chippendale, Y. Chung, D. DeBoer, P. Diamond, P.G. Edwards, R. Ekers, R.H. Ferris, R. Forsyth, R. Gough, A. Grancea, N. Gupta, L. Harvey-Smith, S. Hay, D.B. Hayman, A.W. Hotan, S. Hoyle, B. Humphreys, B. Indermuehle, C.E. Jacka, C.A. Jackson, S. Jackson, K. Jeganathan, J. Joseph, R. Kendall, D. Kiraly, B. Koribalski, M. Leach, E. Lenc, A. MacLeod, S. Mader, M. Marquarding, J. Marvil, N. McClure-Griffiths, D. McConnell, P. Mirtschin, S. Neuhold, A. Ng, R.P. Norris, J. O'Sullivan, S. Pearce, C.J. Phillips, A. Popping, R.Y. Qiao, J.E. Reynolds, P. Roberts, R.J. Sault, A.E.T. Schinckel, P. Serra, R. Shaw, T.W. Shimwell, M. Storey, A.W. Sweetnam, A. Tzioumis, T. Westmeier, M. Whiting, C.D. Wilson, A pilot ASKAP survey of radio transient events in the region around the intermittent pulsar PSR J1107-5907. Mon. Not. R. Astron. Soc. **456**(4), 3948–3960 (2016). https://doi.org/10.1093/mnras/stv2893

J.A. Hodge, R.H. Becker, R.L. White, G.T. Richards, Millijansky radio variability in SDSS Stripe 82. Astrophys. J. **769**, 125 (2013). https://doi.org/10.1088/0004-637X/769/2/125

T.W.-S. Holoien, C.S. Kochanek, J.L. Prieto, K.Z. Stanek, S. Dong, B.J. Shappee, D. Grupe, J.S. Brown, U. Basu, J.F. Beacom, D. Bersier, J. Brimacombe, A.B. Danilet, E. Falco, Z. Guo, J. Jose, G.J. Herczeg, F. Long, G. Pojmanski, G.V. Simonian, D.M. Szczygieł, T.A. Thompson, J.R. Thorstensen, R.M. Wagner, P.R. Woźniak, Six months of multiwavelength follow-up of the tidal disruption candidate ASASSN-14li and implied TDE rates from ASAS-SN. Mon. Not. R. Astron. Soc. **455**, 2918–2935 (2016). https://doi.org/10.1093/mnras/stv2486

T.W.-S. Holoien, M.E. Huber, B.J. Shappee, M. Eracleous, K. Auchettl, J.S. Brown, M.A. Tucker, K.C. Chambers, C.S. Kochanek, K.Z. Stanek, A. Rest, R.S. Post, J.D. Simon, E. Kankare, J. Bulger, T.B. Lowe, E.A. Magnier, A.S.B. Schultz, C.Z. Waters, M. Willman, D. Wright, D.R. Young, S. Dong, J.L. Prieto, T.A. Thompson, L. Denneau, H. Flewelling, A.N. Heinze, S.J. Smartt, K.W. Smith, B. Stalder, J.L. Tonry, H. Weiland, PS18kh: a new tidal disruption event with a non-axisymmetric accretion disk. Astrophys. J. **880**, 120 (2019). https://doi.org/10.3847/1538-4357/ab2ae1

T.W.-S. Holoien, K. Auchettl, M.A. Tucker, B.J. Shappee, S.G. Patel, J.C.A. Miller-Jones, B. Mockler, D.N. Groenewald, J.S. Brown, C.S. Kochanek, K.Z. Stanek, P. Chen, S. Dong, J.L. Prieto, T.A. Thompson, R.L. Beaton, T. Connor, P.S. Cowperthwaite, L. Dahmen, K.D. French, N. Morrell, D.A.H. Buckley, M. Gromadzki, R. Roy, D.A. Coulter, G. Dimitriadis, R.J. Foley, C.D. Kilpatrick, A.L. Piro, C. Rojas-Bravo, M.R. Siebert, S. van Velzen, The Rise and Fall of ASASSN-18pg: Following a TDE from Early to Late Times. arXiv e-prints (2020). arXiv:2003.13693

T. Hovatta, E. Nieppola, M. Tornikoski, E. Valtaoja, M.F. Aller, H.D. Aller, Long-term radio variability of AGN: flare characteristics. Astron. Astrophys. **485**, 51–61 (2008). https://doi.org/10.1051/0004-6361:200809806

T. Hung, S. Gezari, S.B. Cenko, S. van Velzen, N. Blagorodnova, L. Yan, S.R. Kulkarni, R. Lunnan, T. Kupfer, G. Leloudas, A.K.H. Kong, P.E. Nugent, C. Fremling, R.R. Laher, F.J. Masci, Y. Cao, R. Roy, T. Petrushevska, Sifting for sapphires: systematic selection of tidal disruption events in iPTF. Astrophys. J. Suppl. Ser. **238**(2), 15 (2018). https://doi.org/10.3847/1538-4365/aad8b1

J.A. Irwin, R.N. Henriksen, M. Krause, Q.D. Wang, T. Wiegert, E.J. Murphy, G. Heald, E. Perlman, CHANG-ES V: nuclear outflow in a virgo cluster spiral after a tidal disruption event. Astrophys. J. **809**, 172 (2015). https://doi.org/10.1088/0004-637X/809/2/172

N. Jiang, L. Dou, T. Wang, C. Yang, J. Lyu, H. Zhou, The WISE detection of an infrared echo in tidal disruption event ASASSN-14li. Astrophys. J. Lett. **828**, 14 (2016). https://doi.org/10.3847/2041-8205/828/1/L14

S. Johnston, R. Taylor, M. Bailes, N. Bartel, C. Baugh, M. Bietenholz, C. Blake, R. Braun, J. Brown, S. Chatterjee, J. Darling, A. Deller, R. Dodson, P. Edwards, R. Ekers, S. Ellingsen, I. Feain, B. Gaensler, M. Haverkorn, G. Hobbs, A. Hopkins, C. Jackson, C. James, G. Joncas, V. Kaspi, V. Kilborn, B. Koribalski, R. Kothes, T. Landecker, E. Lenc, J. Lovell, J.-P. Macquart, R. Manchester, D. Matthews, N. McClure-Griffiths, R. Norris, U.-L. Pen, C. Phillips, C. Power, R. Protheroe, E. Sadler, B. Schmidt, I. Stairs, L. Staveley-Smith, J. Stil, S. Tingay, A. Tzioumis, M. Walker, J. Wall, M. Wolleben, Science with ASKAP. The Australian square-kilometre-array pathfinder. Exp. Astron. **22**(3), 151–273 (2008). https://doi.org/10.1007/s10686-008-9124-7

J. Jonas (MeerKAT Team), The MeerKAT radio telescope, in *MeerKAT Science: On the Pathway to the SKA* (2016), p. 1

S. Kim, S. Schulze, L. Resmi, J. González-López, A.B. Higgins, C.H. Ishwara-Chandra, F.E. Bauer, I. de Gregorio-Monsalvo, M. De Pasquale, A. de Ugarte Postigo, D.A. Kann, S. Martín, S.R. Oates, R.L.C. Starling, N.R. Tanvir, J. Buchner, S. Campana, Z. Cano, S. Covino, A.S. Fruchter, J.P.U. Fynbo, D.H. Hartmann, J. Hjorth, P. Jakobsson, A.J. Levan, D. Malesani, M.J. Michałowski, B. Milvang-Jensen, K. Misra, P.T. O'Brien, R. Sánchez-Ramírez, C.C. Thöne, D.J. Watson, K. Wiersema, ALMA and GMRT constraints on the off-axis gamma-ray burst 170817A from the binary neutron star merger GW170817. Astrophys. J. Lett. **850**, 21 (2017). https://doi.org/10.3847/2041-8213/aa970b

S. Komossa, X-ray evidence for SMBHs in non-active galaxies: detection of X-ray flare events, interpreted as tidal disruptions of stars by SMBHs, in *Lighthouses of the Universe*, ed. by M. Gilfanov, R. Sunyeav, E. Churazov (2002), pp. 436–442

J. Krolik, T. Piran, G. Svirski, R.M. Cheng, ASASSN-14li: a model tidal disruption event. Astrophys. J. **827**, 127 (2016). https://doi.org/10.3847/0004-637X/827/2/127

S.R. Kulkarni, D.A. Frail, M.H. Wieringa, R.D. Ekers, E.M. Sadler, R.M. Wark, J.L. Higdon, E.S. Phinney, J.S. Bloom, Radio emission from the unusual supernova 1998bw and its association with the γ-ray burst of 25 April 1998. Nature **395**, 663–669 (1998). https://doi.org/10.1038/27139

W.-H. Lei, Q. Yuan, B. Zhang, D. Wang, IGR J12580+0134: the first tidal disruption event with an off-beam relativistic jet. Astrophys. J. **816**, 20 (2016). https://doi.org/10.3847/0004-637X/816/1/20

A.J. Levan, N.R. Tanvir, S.B. Cenko, D.A. Perley, K. Wiersema, J.S. Bloom, A.S. Fruchter, A. de Ugarte Postigo, P.T. O'Brien, N. Butler, A.J. van der Horst, G. Leloudas, A.N. Morgan, K. Misra, G.C. Bower, J. Farihi, R.L. Tunnicliffe, M. Modjaz, J.M. Silverman, J. Hjorth, C. Thöne, A. Cucchiara, J.M.C. Cerón, A.J. Castro-Tirado, J.A. Arnold, M. Bremer, J.P. Brodie, T. Carroll, M.C. Cooper, P.A. Curran, R.M. Cutri, J. Ehle, D. Forbes, J. Fynbo, J. Gorosabel, J. Graham, D.I. Hoffman, S. Guziy, P. Jakobsson, A.

Kamble, T. Kerr, M.M. Kasliwal, C. Kouveliotou, D. Kocevski, N.M. Law, P.E. Nugent, E.O. Ofek, D. Poznanski, R.M. Quimby, E. Rol, A.J. Romanowsky, R. Sánchez-Ramírez, S. Schulze, N. Singh, L. van Spaandonk, R.L.C. Starling, R.G. Strom, J.C. Tello, O. Vaduvescu, P.J. Wheatley, R.A.M.J. Wijers, J.M. Winters, D. Xu, An extremely luminous panchromatic outburst from the nucleus of a distant galaxy. Science **333**, 199 (2011). https://doi.org/10.1126/science.1207143

A.J. Levan, N.R. Tanvir, G.C. Brown, B.D. Metzger, K.L. Page, S.B. Cenko, P.T. O'Brien, J.D. Lyman, K. Wiersema, E.R. Stanway, A.S. Fruchter, D.A. Perley, J.S. Bloom, Late time multi-wavelength observations of Swift J1644+5734: a luminous optical/IR bump and quiescent X-ray emission. Astrophys. J. **819**, 51 (2016). https://doi.org/10.3847/0004-637X/819/1/51

W. Lu, C. Bonnerot, Self-intersection of the fallback stream in tidal disruption events. Mon. Not. R. Astron. Soc. **492**(1), 686–707 (2020). https://doi.org/10.1093/mnras/stz3405

R. Margutti, K.D. Alexander, X. Xie, L. Sironi, B.D. Metzger, A. Kathirgamaraju, W. Fong, P.K. Blanchard, E. Berger, A. MacFadyen, D. Giannios, C. Guidorzi, A. Hajela, R. Chornock, P.S. Cowperthwaite, T. Eftekhari, M. Nicholl, V.A. Villar, P.K.G. Williams, J. Zrake, The binary neutron star event LIGO/Virgo GW170817 160 days after merger: synchrotron emission across the electromagnetic spectrum. Astrophys. J. Lett. **856**, 18 (2018). https://doi.org/10.3847/2041-8213/aab2ad

A.P. Marscher, S.G. Jorstad, J.-L. Gómez, M.F. Aller, H. Teräsranta, M.L. Lister, A.M. Stirling, Observational evidence for the accretion-disk origin for a radio jet in an active galaxy. Nature **417**, 625–627 (2002)

S. Mattila, M. Pérez-Torres, A. Efstathiou, P. Mimica, M. Fraser, E. Kankare, A. Alberdi, M.Á. Aloy, T. Heikkilä, P.G. Jonker, P. Lundqvist, I. Martí-Vidal, W.P.S. Meikle, C. Romero-Cañizales, S.J. Smartt, S. Tsygankov, E. Varenius, A. Alonso-Herrero, M. Bondi, C. Fransson, R. Herrero-Illana, T. Kangas, R. Kotak, N. Ramírez-Olivencia, P. Väisänen, R.J. Beswick, D.L. Clements, R. Greimel, J. Harmanen, J. Kotilainen, K. Nandra, T. Reynolds, S. Ryder, N.A. Walton, K. Wiik, G. Östlin, A dust-enshrouded tidal disruption event with a resolved radio jet in a galaxy merger. Science **361**, 482–485 (2018). https://doi.org/10.1126/science.aao4669

T. Mauch, W.D. Cotton, J.J. Condon, A.M. Matthews, T.D. Abbott, R.M. Adam, M.A. Aldera, K.M.B. Asad, E.F. Bauermeister, T.G.H. Bennett, H. Bester, D.H. Botha, L.R.S. Brederode, Z.B. Brits, S.J. Buchner, J.P. Burger, F. Camilo, J.M. Chalmers, T. Cheetham, D. de Villiers, M.S. de Villiers, M.A. Dikgale-Mahlakoana, L.J. du Toit, S.W.P. Esterhuyse, G. Fadana, B.L. Fanaroff, S. Fataar, S. February, B.S. Frank, R.R.G. Gamatham, M. Geyer, S. Goedhart, S. Gounden, S.C. Gumede, I. Heywood, M.J. Hlakola, J.M.G. Horrell, B. Hugo, A.R. Isaacson, G.I.G. Józsa, J.L. Jonas, R.P.M. Julie, F.B. Kapp, V.A. Kasper, J.S. Kenyon, P.P.A. Kotzé, N. Kriek, H. Kriel, T.W. Kusel, R. Lehmensiek, A. Loots, R.T. Lord, B.M. Lunsky, K. Madisa, L.G. Magnus, J.P.L. Main, J.A. Malan, J.R. Manley, S.J. Marais, A. Martens, B. Merry, R. Millenaar, N. Mnyandu, I.P.T. Moeng, O.J. Mokone, T.E. Monama, M.C. Mphego, W.S. New, B. Ngcebetsha, K.J. Ngoasheng, M.T.O. Ockards, N. Oozeer, A.J. Otto, A.A. Patel, A. Peens-Hough, S.J. Perkins, A.J.T. Ramaila, Z.R. Ramudzuli, R. Renil, L.L. Richter, A. Robyntjies, S. Salie, C.T.G. Schollar, L.C. Schwardt, M. Serylak, R. Siebrits, S.K. Sirothia, O.M. Smirnov, L. Sofeya, G. Stone, B. Taljaard, C. Tasse, I.P. Theron, A.J. Tiplady, O. Toruvanda, S.N. Twum, T.J. van Balla, A. van der Byl, C. van der Merwe, V. Van Tonder, B.H. Wallace, M.G. Welz, L.P. Williams, B. Xaia, The 1.28 GHz MeerKAT DEEP2 image. Astrophys. J. **888**(2), 61 (2020). https://doi.org/10.3847/1538-4357/ab5d2d

B.D. Metzger, P.K.G. Williams, E. Berger, Extragalactic synchrotron transients in the era of wide-field radio surveys. I. Detection rates and light curve characteristics. Astrophys. J. **806**, 224 (2015). https://doi.org/10.1088/0004-637X/806/2/224

J.M. Miller, J.S. Kaastra, M.C. Miller, M.T. Reynolds, G. Brown, S.B. Cenko, J.J. Drake, S. Gezari, J. Guillochon, K. Gultekin, J. Irwin, A. Levan, D. Maitra, W.P. Maksym, R. Mushotzky, P. O'Brien, F. Paerels, J. de Plaa, E. Ramirez-Ruiz, T. Strohmayer, N. Tanvir, Flows of X-ray gas reveal the disruption of a star by a massive black hole. Nature **526**, 542–545 (2015). https://doi.org/10.1038/nature15708

P. Mimica, D. Giannios, B.D. Metzger, M.A. Aloy, The radio afterglow of Swift J1644+57 reveals a powerful jet with fast core and slow sheath. Mon. Not. R. Astron. Soc. **450**, 2824–2841 (2015). https://doi.org/10.1093/mnras/stv825

K.P. Mooley, G. Hallinan, S. Bourke, A. Horesh, S.T. Myers, D.A. Frail, S.R. Kulkarni, D.B. Levitan, M.M. Kasliwal, S.B. Cenko, Y. Cao, E. Bellm, R.R. Laher, The Caltech-NRAO Stripe 82 Survey (CNSS). I. The pilot radio transient survey in 50 deg^2. Astrophys. J. **818**, 105 (2016). https://doi.org/10.3847/0004-637X/818/2/105

K.P. Mooley, E. Nakar, K. Hotokezaka, G. Hallinan, A. Corsi, D.A. Frail, A. Horesh, T. Murphy, E. Lenc, D.L. Kaplan, K. de, D. Dobie, P. Chandra, A. Deller, O. Gottlieb, M.M. Kasliwal, S.R. Kulkarni, S.T. Myers, S. Nissanke, T. Piran, C. Lynch, V. Bhalerao, S. Bourke, K.W. Bannister, L.P. Singer, A mildly relativistic wide-angle outflow in the neutron-star merger event GW170817. Nature **554**, 207–210 (2018a). https://doi.org/10.1038/nature25452

K.P. Mooley, D.A. Frail, D. Dobie, E. Lenc, A. Corsi, K. De, A.J. Nayana, S. Makhathini, I. Heywood, T. Murphy, D.L. Kaplan, P. Chandra, O. Smirnov, E. Nakar, G. Hallinan, F. Camilo, R. Fender, S. Goedhart, P. Groot, M.M. Kasliwal, S.R. Kulkarni, P.A. Woudt, A strong jet signature in the late-time light curve of GW170817. Astrophys. J. Lett. **868**, 11 (2018b). https://doi.org/10.3847/2041-8213/aaeda7

K.P. Mooley, A.T. Deller, O. Gottlieb, E. Nakar, G. Hallinan, S. Bourke, D.A. Frail, A. Horesh, A. Corsi, K. Hotokezaka, Superluminal motion of a relativistic jet in the neutron-star merger GW170817. Nature **561**, 355–359 (2018c). https://doi.org/10.1038/s41586-018-0486-3

K.P. Mooley, S.T. Myers, D.A. Frail, G. Hallinan, B. Butler, A. Kimball, K. Golap, The Caltech-NRAO Stripe 82 Survey (CNSS). II. On-the-fly mosaicking methodology. Astrophys. J. **870**(1), 25 (2019). https://doi.org/10.3847/1538-4357/aaef7c

T. Murphy, S. Chatterjee, D.L. Kaplan, J. Banyer, M.E. Bell, H.E. Bignall, G.C. Bower, R.A. Cameron, D.M. Coward, J.M. Cordes, S. Croft, J.R. Curran, S.G. Djorgovski, S.A. Farrell, D.A. Frail, B.M. Gaensler, D.K. Galloway, B. Gendre, A.J. Green, P.J. Hancock, S. Johnston, A. Kamble, C.J. Law, T.J.W. Lazio, K.K. Lo, J.-P. Macquart, N. Rea, U. Rebbapragada, C. Reynolds, S.D. Ryder, B. Schmidt, R. Soria, I.H. Stairs, S.J. Tingay, U. Torkelsson, K. Wagstaff, M. Walker, R.B. Wayth, P.K.G. Williams, VAST: an ASKAP survey for variables and slow transients. Proc. Astron. Soc. Aust. **30**, 006 (2013). https://doi.org/10.1017/pasa.2012.006

M. Nicholl, P.K. Blanchard, E. Berger, S. Gomez, R. Margutti, K.D. Alexander, J. Guillochon, J. Leja, R. Chornock, B. Snios, K. Auchettl, A.G. Bruce, P. Challis, D.J. D'Orazio, M.R. Drout, T. Eftekhari, R.J. Foley, O. Graur, C.D. Kilpatrick, A. Lawrence, A.L. Piro, C. Rojas-Bravo, N.P. Ross, P. Short, S.J. Smartt, K.W. Smith, B. Stalder, The tidal disruption event AT2017eqx: spectroscopic evolution from hydrogen rich to poor suggests an atmosphere and outflow. Mon. Not. R. Astron. Soc. **488**(2), 1878–1893 (2019). https://doi.org/10.1093/mnras/stz1837

E. Nieppola, T. Hovatta, M. Tornikoski, E. Valtaoja, M.F. Aller, H.D. Aller, Long-term variability of radio-bright BL lacertae objects. Astron. J. **137**, 5022–5036 (2009). https://doi.org/10.1088/0004-6256/137/6/5022

M. Nikołajuk, R. Walter, Tidal disruption of a super-Jupiter by a massive black hole. Astron. Astrophys. **552**, 75 (2013). https://doi.org/10.1051/0004-6361/201220664

E.O. Ofek, D.A. Frail, B. Breslauer, S.R. Kulkarni, P. Chandra, A. Gal-Yam, M.M. Kasliwal, N. Gehrels, A very large array search for 5 GHz radio transients and variables at low galactic latitudes. Astrophys. J. **740**, 65 (2011). https://doi.org/10.1088/0004-637X/740/2/65

D.R. Pasham, S. van Velzen, Discovery of a time lag between the soft X-ray and radio emission of the tidal disruption flare ASASSN-14li: evidence for linear disk-jet coupling. Astrophys. J. **856**, 1 (2018). https://doi.org/10.3847/1538-4357/aab361

D.R. Pasham, S.B. Cenko, A.J. Levan, G.C. Bower, A. Horesh, G.C. Brown, S. Dolan, K. Wiersema, A.V. Filippenko, A.S. Fruchter, J. Greiner, P.T. O'Brien, K.L. Page, A. Rau, N.R. Tanvir, A multiwavelength study of the relativistic tidal disruption candidate Swift J2058.4+0516 at late times. Astrophys. J. **805**, 68 (2015). https://doi.org/10.1088/0004-637X/805/1/68

E.S. Perlman, E.T. Meyer, Q.D. Wang, Q. Yuan, R. Henriksen, J. Irwin, M. Krause, T. Wiegert, E.J. Murphy, G. Heald, R.-J. Dettmar, Compact resolved ejecta in the nearest tidal disruption event. Astrophys. J. **842**, 126 (2017). https://doi.org/10.3847/1538-4357/aa71b1

Y.M. Pihlström, G.B. Taylor, J. Granot, S. Doeleman, Stirring the embers: high-sensitivity VLBI observations of GRB 030329. Astrophys. J. **664**(1), 411–415 (2007). https://doi.org/10.1086/518955

J.L. Prieto, T. Krühler, J.P. Anderson, L. Galbany, C.S. Kochanek, E. Aquino, J.S. Brown, S. Dong, F. Förster, T.W.-S. Holoien, H. Kuncarayakti, J.C. Maureira, F.F. Rosales-Ortega, S.F. Sánchez, B.J. Shappee, K.Z. Stanek, MUSE reveals a recent merger in the post-starburst host galaxy of the TDE ASASSN-14li. Astrophys. J. **830**, 32 (2016). https://doi.org/10.3847/2041-8205/830/2/L32

A. Ptak, A. Hornschemeier, A. Zezas, B. Lehmer, M. Yukita, D. Wik, V. Antoniou, M.K. Argo, L. Ballo, K. Bechtol, S. Boggs, R. Della Ceca, F.E. Christensen, W.W. Craig, C.J. Hailey, F.A. Harrison, R. Krivonos, T.J. Maccarone, D. Stern, M. Tatum, T. Venters, W.W. Zhang, A focused, hard X-ray look at Arp 299 with NuSTAR. Astrophys. J. **800**(2), 104 (2015). https://doi.org/10.1088/0004-637X/800/2/104

E. Quataert, A dynamical model for hot gas in the galactic center. Astrophys. J. **613**, 322–325 (2004). https://doi.org/10.1086/422973

M.J. Rees, Tidal disruption of stars by black holes of 10 to the 6th-10 to the 8th solar masses in nearby galaxies. Nature **333**, 523 (1988). http://adsabs.harvard.edu/abs/1988Natur.333..523R

L. Resmi, S. Schulze, C.H. Ishwara-Chandra, K. Misra, J. Buchner, M. De Pasquale, R. Sánchez-Ramírez, S. Klose, S. Kim, N.R. Tanvir, P.T. O'Brien, Low-frequency view of GW170817/GRB 170817A with the Giant Metrewave Radio Telescope. Astrophys. J. **867**, 57 (2018). https://doi.org/10.3847/1538-4357/aae1a6

C. Romero-Cañizales, J.L. Prieto, X. Chen, C.S. Kochanek, S. Dong, T.W.-S. Holoien, K.Z. Stanek, F. Liu, The TDE ASASSN-14li and its host resolved at parsec scales with the EVN. Astrophys. J. **832**, 10 (2016). https://doi.org/10.3847/2041-8205/832/1/L10

Roth et al., Space Sci. Rev. (2020, this issue)

H.R. Russell, A.C. Fabian, B.R. McNamara, A.E. Broderick, Inside the Bondi radius of M87. Mon. Not. R. Astron. Soc. **451**, 588–600 (2015). https://doi.org/10.1093/mnras/stv954

R.D. Saxton, A.M. Read, P. Esquej, M.J. Freyberg, B. Altieri, D. Bermejo, The first XMM-Newton slew survey catalogue: XMMSL1. Astron. Astrophys. **480**, 611–622 (2008). https://doi.org/10.1051/0004-6361:20079193

R.D. Saxton, A.M. Read, P. Esquej, S. Komossa, S. Dougherty, P. Rodriguez-Pascual, D. Barrado, A tidal disruption-like X-ray flare from the quiescent galaxy SDSS J120136.02+300305.5. Astron. Astrophys. **541**, 106 (2012)

R.D. Saxton, A.M. Read, S. Komossa, P. Lira, K.D. Alexander, M.H. Wieringa, XMMSL1 J074008.2-853927: a tidal disruption event with thermal and non-thermal components. Astron. Astrophys. **598**, 29 (2017). https://doi.org/10.1051/0004-6361/201629015

R.D. Saxton, A.M. Read, S. Komossa, P. Lira, K.D. Alexander, I. Steele, F. Ocaña, E. Berger, P. Blanchard, XMMSL2 J144605.0+685735: a slow tidal disruption event. Astron. Astrophys. **630**, 98 (2019). https://doi.org/10.1051/0004-6361/201935650

Saxton et al., Space Sci. Rev. (2020, this issue)

M.A. Scott, A.C.S. Readhead, The low-frequency structure of powerful radio sources and limits to departures from equipartition. Mon. Not. R. Astron. Soc. **180**, 539–550 (1977). https://doi.org/10.1093/mnras/180.4.539

A.M. Soderberg, S. Chakraborti, G. Pignata, R.A. Chevalier, P. Chandra, A. Ray, M.H. Wieringa, A. Copete, V. Chaplin, V. Connaughton, S.D. Barthelmy, M.F. Bietenholz, N. Chugai, M.D. Stritzinger, M. Hamuy, C. Fransson, O. Fox, E.M. Levesque, J.E. Grindlay, P. Challis, R.J. Foley, R.P. Kirshner, P.A. Milne, M.A.P. Torres, A relativistic type Ibc supernova without a detected γ-ray burst. Nature **463**, 513–515 (2010). https://doi.org/10.1038/nature08714

R. Stein, S. van Velzen, M. Kowalski, A. Franckowiak, S. Gezari, J.C.A. Miller-Jones, S. Frederick, I. Sfaradi, M.F. Bietenholz, A. Horesh, R. Fender, S. Garrappa, T. Ahumada, I. Andreoni, J. Belicki, E.C. Bellm, M. Böttcher, V. Brinnel, R. Burruss, S.B. Cenko, M.W. Coughlin, V. Cunningham, A. Drake, G.R. Farrar, M. Feeney, R.J. Foley, A. Gal-Yam, V.Z. Golkhou, A. Goobar, M.J. Graham, E. Hammerstein, G. Helou, T. Hung, M.M. Kasliwal, C.D. Kilpatrick, A.K.H. Kong, T. Kupfer, R.R. Laher, A.A. Mahabal, F.J. Masci, J. Necker, J. Nordin, D.A. Perley, M. Rigault, S. Reusch, H. Rodriguez, C. Rojas-Bravo, B. Rusholme, D.L. Shupe, L.P. Singer, J. Sollerman, M.T. Soumagnac, D. Stern, K. Taggart, J. van Santen, C. Ward, P. Woudt, Y. Yao, A high-energy neutrino coincident with a tidal disruption event. arXiv e-prints (2020). arXiv:2005.05340

E. Steinberg, E.R. Coughlin, N.C. Stone, B.D. Metzger, Thawing the frozen-in approximation: implications for self-gravity in deeply plunging tidal disruption events. Mon. Not. R. Astron. Soc. **485**(1), L146–L150 (2019). https://doi.org/10.1093/mnrasl/slz048

N.C. Stone, E. Vasiliev, M. Kesden, et al., Rates of stellar tidal disruption. Space Sci. Rev. **216**, 35 (2020). https://doi.org/10.1007/s11214-020-00651-4

L.E. Strubbe, E. Quataert, Optical flares from the tidal disruption of stars by massive black holes. Mon. Not. R. Astron. Soc. **400**, 2070–2084 (2009). https://doi.org/10.1111/j.1365-2966.2009.15599.x

G.B. Taylor, D.A. Frail, E. Berger, S.R. Kulkarni, The angular size and proper motion of the afterglow of GRB 030329. Astrophys. J. Lett. **609**(1), 1–4 (2004). https://doi.org/10.1086/422554

G.B. Taylor, E. Momjian, Y. Pihlström, T. Ghosh, C. Salter, Late-time observations of the afterglow and environment of GRB 030329. Astrophys. J. **622**(2), 986–990 (2005). https://doi.org/10.1086/428346

E. Troja, L. Piro, G. Ryan, H. van Eerten, R. Ricci, M.H. Wieringa, S. Lotti, T. Sakamoto, S.B. Cenko, The outflow structure of GW170817 from late-time broad-band observations. Mon. Not. R. Astron. Soc. **478**, 18–23 (2018). https://doi.org/10.1093/mnrasl/sly061

S. van Velzen, On the mass and luminosity functions of tidal disruption flares: rate suppression due to black hole event horizons. Astrophys. J. **852**, 72 (2018). https://doi.org/10.3847/1538-4357/aa998e

S. van Velzen, G.R. Farrar, Measurement of the rate of stellar tidal disruption flares. Astrophys. J. **792**, 53 (2014). https://doi.org/10.1088/0004-637X/792/1/53

S. van Velzen, G.R. Farrar, S. Gezari, N. Morrell, D. Zaritsky, L. Östman, M. Smith, J. Gelfand, A.J. Drake, Optical discovery of probable stellar tidal disruption flares. Astrophys. J. **741**, 73 (2011a)

S. van Velzen, E. Körding, H. Falcke, Radio jets from stellar tidal disruptions. Mon. Not. R. Astron. Soc. **417**, 51–55 (2011b)

S. van Velzen, D.A. Frail, E. Körding, H. Falcke, Constraints on off-axis jets from stellar tidal disruption flares. Astron. Astrophys. **552**, 5 (2013). https://doi.org/10.1051/0004-6361/201220426

S. van Velzen, G.E. Anderson, N.C. Stone, M. Fraser, T. Wevers, B.D. Metzger, P.G. Jonker, A.J. van der Horst, T.D. Staley, A.J. Mendez, J.C.A. Miller-Jones, S.T. Hodgkin, H.C. Campbell, R.P. Fender, A radio jet from the optical and x-ray bright stellar tidal disruption flare ASASSN-14li. Science **351**, 62–65 (2016). https://doi.org/10.1126/science.aad1182

S. van Velzen, G.C. Bower, B.D. Metzger, in *Science with a Next Generation Very Large Array: Tidal Disruption Events*, ed. by E. Murphy. Astronomical Society of the Pacific Conference Series, vol. 517 (2018), p. 737

S. van Velzen, S. Gezari, S.B. Cenko, E. Kara, J.C.A. Miller-Jones, T. Hung, J. Bright, N. Roth, N. Blagorodnova, D. Huppenkothen, L. Yan, E. Ofek, J. Sollerman, S. Frederick, C. Ward, M.J. Graham, R. Fender, M.M. Kasliwal, C. Canella, R. Stein, M. Giomi, V. Brinnel, J. van Santen, J. Nordin, E.C. Bellm, R. Dekany, C. Fremling, V.Z. Golkhou, T. Kupfer, S.R. Kulkarni, R.R. Laher, A. Mahabal, F.J. Masci, A.A. Miller, J.D. Neill, R. Riddle, M. Rigault, B. Rusholme, M.T. Soumagnac, Y. Tachibana, The first tidal disruption flare in ZTF: from photometric selection to multi-wavelength characterization. Astrophys. J. **872**, 198 (2019). https://doi.org/10.3847/1538-4357/aafe0c

van Velzen et al., Space Sci. Rev. (2020a, this issue)

van Velzen et al., Space Sci. Rev. (2020b, this issue)

K.W. Weiler, N. Panagia, M.J. Montes, R.A. Sramek, Radio emission from supernovae and gamma-ray bursters. Annu. Rev. Astron. Astrophys. **40**, 387–438 (2002). https://doi.org/10.1146/annurev.astro.40.060401.093744

T. Wevers, D.R. Pasham, S. van Velzen, G. Leloudas, S. Schulze, J.C.A. Miller-Jones, P.G. Jonker, M. Gromadzki, E. Kankare, S.T. Hodgkin, L. Wyrzykowski, Z. Kostrzewa-Rutkowska, S. Moran, M. Berton, K. Maguire, F. Onori, S. Matilla, M. Nicholl, Evidence for rapid disk formation and reprocessing in the X-ray bright tidal disruption event AT 2018fyk. Mon. Not. R. Astron. Soc. **488**(4), 4816–4830 (2019). https://doi.org/10.1093/mnras/stz1976

K. Wiersema, A.J. van der Horst, A.J. Levan, N.R. Tanvir, R. Karjalainen, A. Kamble, C. Kouveliotou, B.D. Metzger, D.M. Russell, I. Skillen, R.L.C. Starling, R.A.M.J. Wijers, Polarimetry of the transient relativistic jet of GRB 110328/Swift J164449.3+573451. Mon. Not. R. Astron. Soc. **421**(3), 1942–1948 (2012). https://doi.org/10.1111/j.1365-2966.2011.20379.x

A. Yalinewich, E. Steinberg, T. Piran, J.H. Krolik, Radio emission from the unbound debris of tidal disruption events. Mon. Not. R. Astron. Soc. **487**(3), 4083–4092 (2019). https://doi.org/10.1093/mnras/stz1567

J. Yang, Z. Paragi, A.J. van der Horst, L.I. Gurvits, R.M. Campbell, D. Giannios, T. An, S. Komossa, No apparent superluminal motion in the first-known jetted tidal disruption event Swift J1644+5734. Mon. Not. R. Astron. Soc. **462**, 66–70 (2016). https://doi.org/10.1093/mnrasl/slw107

Q. Yuan, Q.D. Wang, W.-H. Lei, H. Gao, B. Zhang, Catching jetted tidal disruption events early in millimetre. Mon. Not. R. Astron. Soc. **461**, 3375–3384 (2016). https://doi.org/10.1093/mnras/stw1543

Zabludoff et al., Space Sci. Rev. (2020, this issue)

B.A. Zauderer, E. Berger, A.M. Soderberg, A. Loeb, R. Narayan, D.A. Frail, G.R. Petitpas, A. Brunthaler, R. Chornock, J.M. Carpenter, G.G. Pooley, K. Mooley, S.R. Kulkarni, R. Margutti, D.B. Fox, E. Nakar, N.A. Patel, N.H. Volgenau, T.L. Culverhouse, M.F. Bietenholz, M.P. Rupen, W. Max-Moerbeck, A.C.S. Readhead, J. Richards, M. Shepherd, S. Storm, C.L.H. Hull, Birth of a relativistic outflow in the unusual γ-ray transient Swift J164449.3+573451. Nature **476**, 425–428 (2011). https://doi.org/10.1038/nature10366

B.A. Zauderer, E. Berger, R. Margutti, G.G. Pooley, R. Sari, A.M. Soderberg, A. Brunthaler, M.F. Bietenholz, Radio monitoring of the tidal disruption event Swift J164449.3+573451. II. The relativistic jet shuts off and a transition to forward shock X-ray/radio emission. Astrophys. J. **767**, 152 (2013). https://doi.org/10.1088/0004-637X/767/2/152

Zauderer et al., Space Sci. Rev. (2020, this issue)

Space Science Reviews (2021) 217:63
https://doi.org/10.1007/s11214-021-00835-6

Reverberation in Tidal Disruption Events: Dust Echoes, Coronal Emission Lines, Multi-wavelength Cross-correlations, and QPOs

Sjoert van Velzen[1,2] · Dheeraj R. Pasham[3] · Stefanie Komossa[4] · Lin Yan[5] · Erin A. Kara[6]

Received: 13 June 2020 / Accepted: 2 June 2021 / Published online: 1 July 2021

Abstract
Stellar tidal disruption events (TDEs) are typically discovered by transient emission due to accretion or shocks of the stellar debris. Yet this luminous flare can be reprocessed by gas or dust that inhabits a galactic nucleus, resulting in multiple reverberation signals. Nuclear dust heated by the TDE will lead to an echo at infrared wavelengths (1-10 μm) and transient coronal lines in optical spectra of TDEs trace reverberation by gas that orbits the black hole. Both of these signal have been detected, here we review this rapidly developing field. We also review the results that have been extracted from TDEs with high-quality X-ray light curves: quasi periodic oscillations (QPOs), reverberation lags of fluorescence lines, and cross-correlations with emission at other wavelengths. The observational techniques that are covered in this review probe the emission from TDEs over a wide range of scales: from ~ 1 light year to the innermost parts of the newly formed accretion disk. They provide insights into important properties of TDEs such as their bolometric output and the geometry of the accretion flow. While reverberation signals are not detected for every TDE, we anticipate they will become more commonplace when the next generation of X-ray and infrared instruments become operational.

The Tidal Disruption of Stars by Massive Black Holes
Edited by Peter G. Jonker, Sterl Phinney, Elena Maria Rossi, Sjoert van Velzen, Iair Arcavi and Maurizio Falanga

✉ S. van Velzen
sjoert@strw.leidenuniv.nl

1 Center for Cosmology and Particle Physics, New York University, New York, NY 10003, USA

2 Leiden Observatory, Leiden University, Postbus 9513, 2300 RA Leiden, The Netherlands

3 IT Kavli Institute for Astrophysics and Space Research, Cambridge, MA 02139, USA

4 Max-Planck-Institut für Radioastronomie, Auf dem Hügel 69, 53121 Bonn, Germany

5 Caltech Optical Observatories, Cahill Center for Astronomy and Astrophysics, California Institute of Technology, Pasadena, CA 91125, USA

6 Kavli Institute for Astrophysics and Space Research, Massachusetts Institute of Technology, 77 Massachusetts Avenue, Cambridge, MA 02139, USA

1 Introduction

At the time of writing, the field of tidal disruption event (TDE) research finds itself in an interesting position: most of the new events are found in optical surveys, see the Optical Chapter (van Velzen et al. 2020) of this journal, yet the origin of optical emission from TDEs is unknown. As reviewed in the Emission Mechanisms Chapter (Roth et al. 2020) of this journal, the emission could be powered by an inner accretion disk or by shocks of the intersecting debris streams. A related open question concerns the observed fluence, which is a few orders of magnitude lower than the expected energy released when a star gets disrupted and accreted by a massive black hole (Piran et al. 2015).

The most recent models for the optical emission from TDEs all have been tailored to explain the current set of optical/X-ray properties of TDEs (e.g., peak luminosity, light curve shape, emission line widths). As such, finding new TDEs alone is not sufficient to discriminate between these models. To make progress towards understanding the emission mechanism of TDEs, probing a broader set of observables is key.

This brings us to the topic of this chapter: we will review a number of observational methods that probe beyond a single-wavelength light curve or spectrum. Each of these methods has provided valuable information about the TDE emission mechanism. However, given the requirements of high signal-to-noise or long monitoring campaigns, these observables have been measured for only a handful of events.

The techniques discussed here span a very wide range of physical scales. We start with the largest scale of ~ 1 light year, as probed by echoes of the TDE emission from dust and gas, measured using infrared imaging (Sect. 2) and coronal lines (Sect. 3), respectively. Next, we consider correlations between the emission at difference wavelengths that have been reported for the TDE ASASSN-14li (Holoien et al. 2016): a cross-correlation between the X-ray emission and radio emission (Sect. 4.2), which implies a coupling on scales of $\sim 10^{16}$ cm, and a correlation between the X-ray light curve and the UV light curves (Sect. 4.3), probing scales of $\sim 10^{14}$ cm. Finally, we consider two results from X-ray timing analysis that probes the innermost parts of the newly formed TDE accretion disks: the detection of quasi period oscillations (QPOs) for some TDEs (Sect. 5) and relativistic reverberation from the jetted TDE Swift J1644+57 (Sect. 6). We close with a brief discussion (Sect. 7).

2 Infrared Imaging: Dust Echoes

The reprocessing of TDE light by dust can yield an infrared (IR) echo that probes the environment of TDEs on a scale of ~ 0.1 pc. This subject has seen a flurry of activity in recent years. The possibility of dust reprocessing was explored for the first time by Komossa et al. (2009), who presented a *Spitzer* observation (Fig. 1) of the coronal-line selected TDE SDSS J0952+2143 (events selected by coronal lines are discussed in Sect. 3). The important theoretical paper on dust echoes by Lu et al. (2016) made a prediction that infrared emission from a few on-going TDEs could be detectable with imaging observations. Indeed a few months later, three papers that presented observations of dust echoes for different TDEs all appeared in the same week (van Velzen et al. 2016; Jiang et al. 2016; Dou et al. 2016)—although, as evidenced by van Velzen et al. (2015), some of these echoes were detected and studied prior to the forecast of Lu et al. (2016).

In the following three subsections we will review these advances in more detail. First, we give a brief overview of the theoretical background of dust echoes. Next, we present the observations of dust echoes using follow-up observations of TDEs. And third, we look at the possibility to discover new TDEs using IR observations of echoes.

2.1 Theory of Dust Reverberation

Reverberation by dust has long been used as a tool to study the dust around active galactic nuclei (AGN) (e.g., Barvainis 1987) and supernovae and supernovae remnants (e.g., Dwek 1983). The concept is relatively simple. Dust will efficiently absorb optical/UV/X-ray photons, this process heats the dust to a maximum temperature, which yields isotropic emission that peaks at a few micron. The time delay between the fluctuations in emission from the accretion disk and the IR emission yields the inner radius of the dust region (for an example of the application of this technique in AGN, see Koshida et al. 2014). In a TDE, the black hole was not actively accreting, which will modify our expectation of the dust reprocessing signal. The lack of accretion disk emission in the years prior to the TDE means the dust can survive closer to the black hole. But at the same time, the lack of gas supply to the galaxy center implies we could expect a very low dust covering factor and thus a weak reverberation signal compared to active galaxies.

To estimate relevant timescale of a dust echo we have to equate the absorbed flux at radius R from the black hole to the emission by the dust particles. The latter is relatively well-constrained, dust absorption is efficient across the wavelength range where TDEs emit most of their energy. However the dust emission is more uncertain because it depends on the model for the dust particles, which can include grain composition, size, and the shape of the emission spectrum as free parameters.

Fortunately, the uncertainty on the grain properties is reduced by the extreme conditions of a TDE, which reduce the type of dust that we expect to be visible. The grains that can survive closest to the black hole dominate the IR echoes because they radiate the received energy in the shortest amount of time. Furthermore, if the dust emission efficiency scales with the square of the dust size (Draine and Lee 1984), we expect the largest grains to dominate the reprocessing emission. Hence most authors have normalized their predictions for TDE dust emission to a single grain size, using $a \approx 0.1$ μm, the typical maximum dust size for the Milky Way (Weingartner and Draine 2001).

After restricting to "graphite-like" dust particles, because these have the highest sublimation temperature and should thus dominate the reprocessing signal, van Velzen et al. (2016) arrive at the following expression for the sublimation radius

$$R_{\rm sub} = 0.15 \left(\frac{L_{45}}{a_{0.1}^2 T_{1850}^{5.8}} \right)^{1/2} \ {\rm pc} \quad . \tag{1}$$

Here L_{45} is the bolometric luminosity of the TDE in units of 10^{45} erg s^{-1}, $a_{0.1}$ is the grain size in units of 0.1 μm, and T_{1850} is the temperature of the dust, normalized to the expected sublimation temperature of 1850 K. Equation (1) is derived with the common assumption that dust absorption is 100% efficient across the relevant wavelength range ($Q_{\rm UV} = 1$). If we use a dust model that approximates a two-component medium consisting of both graphite and silicate grains (Waxman and Draine 2000; Lu et al. 2016) the sublimation radius is

$$R_{\rm sub} = 0.12 \left(\frac{L_{45}(1 + 0.1 a_{0.1} T_{2300})}{a_{0.1} T_{2300}^5} \right)^{1/2} \ {\rm pc} \quad , \tag{2}$$

with T_{2300} the dust temperature normalized to 2300 K.

From both Eq. (1) and Eq. (2) we find a typical size scale of 0.1 pc for the emitting region, which implies a timescale of a few months for the reprocessing signal. Because this timescale can be measured from the IR light curve (Fig. 2), solving $R_{\rm sub}$ for L_{45} yields an

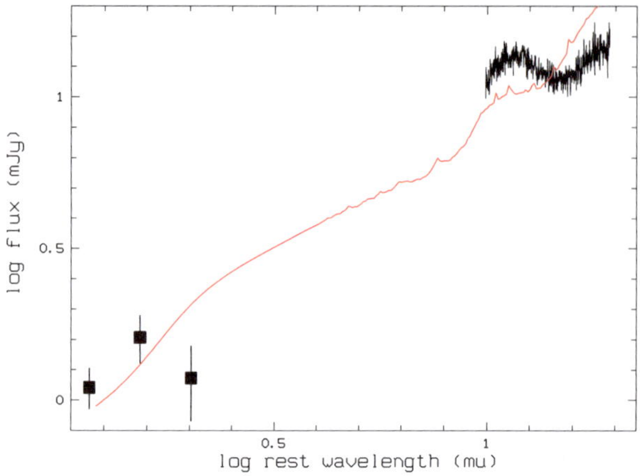

Fig. 1 IR SED of SDSS J0952+2143 observed with Spitzer (black solid line) and 2MASS (pre-flare; filled squares), reproduced from Komossa et al. (2009). The plot emphasizes the large Spitzer IR luminosity at outburst in comparison to pre-outburst NIR data, and the characteristic Spitzer spectral shape, measured for the first time in a TDE. Overplotted is the mean Spitzer IR SED of accreting extragalactic sources (solid red line; Netzer et al. 2007). Note a conversion error in Komossa et al. (2009) when plotting that comparison SED (their Figs. 7 and 8), which led to an incorrect presentation of the shape of that SED, corrected by Komossa (2010)

estimate of the bolometric luminosity of the TDE. This approach only works if the observed IR emission is dominated by a single shell that emits at the sublimation temperature. If the dust echo contains emission from larger scales (e.g., because the galaxy is devoid of dust at $R < R_{\rm sub}$), a measurement of the dust temperature is also needed to estimate $L_{\rm bol}$ (e.g., Kool et al. 2020).

While the bolometric luminosity is estimated from the *shape* of the reprocessing light curve, the covering factor (Ω_d) follows from the *amplitude* of the echo. We define $\Omega_d \equiv E_{\rm bol}/E_{\rm dust}$, with $E_{\rm dust}$ the total energy radiated by the dust and $E_{\rm bol}$ the total radiated energy of the TDE (integrated over the wavelength where dust absorption is efficient).

2.2 Infrared Echoes Detected in Follow-up Observations

The first TDE with IR follow-up observations was SDSS J0952+2143 (Komossa et al. 2008, 2009), which was selected based on variable coronal line emission and variable optical, NIR and UV continuum emission (this class of events is discussed in full detail in the next section). This source was observed once with the *Spitzer* IRS spectrograph at 10–20 μm (Fig. 1), yielding three main results (Komossa et al. 2009). First, the detection of a large MIR luminosity of $L_{10-20\,\mu\rm m} = 3.5 \times 10^{43}$ erg s^{-1} in the *Spitzer* band which implies a large X-ray luminosity (not observed directly) and efficient reprocessing. Second, a blackbody fit to the IR spectrum gave an estimate of the radius of the IR emission site $r \approx 0.5$ pc. Third, the characteristic emission features from PAHs and silicate dust were detected in the spectrum.

The single *Spitzer* observation by Komossa et al. (2009) was carried out about 4 years after the start of the optical flare (Palaversa et al. 2016). The transient nature of the IR at emission was confirmed later via WISE observations at 12 μm (the WISE W3 band), obtained 1.9 year after the *Spitzer* spectrum, showing that the flux at this wavelength has decreased by a factor of 1.8 (Dou et al. 2016).

Table 1 Events with transient infrared emission

Name	Host galaxy type	Peak IR luminosity* (erg s^{-1})	Reference
PTF-09ge	Quiescent	$10^{41.8}$	1
PTF-09axc	Quiescent (E+A)	$10^{42.6}$	1
ASASSN-14li	Quiescent (E+A)	$10^{41.6}$	2
AT2019dsg	Quiescent	$10^{43.0}$	3[†]
NGC 5092	Quiescent	$10^{42.6}$	4
IRAS F01004-2237	ULIRG	$10^{44.3}$	5
PS16dtm	AGN (Type 1)	$10^{43.0}$	6
PS1-10adi	AGN (Type 1)	$10^{44.0}$	7,8
SDSS J1657+2345	AGN (Type 2)	$10^{43.9}$	9
SDSS J0227-0420	AGN (Type 1)	$10^{43.6}$	10
Arp 299-B AT1	LIRG	$10^{43.8}$	11
AT2017gbl	LIRG	$10^{43.2}$	12
MCG-02-04-026	AGN (Type 1)	$10^{43.2}$	13
SDSS J0952+2143	Quiescent (HII-type)	$10^{43.9}$	14, 15

Notes — This table includes events published up to 2020, for more recent IR-detection from optical TDEs see Jiang et al. (2021), which appeared while this article was under review

*Peak observed luminosity (νL_ν) based on host-subtracted WISE data (except for SDSS J0952+2143 the peak flux of which is based on *Spitzer* photometry)

[†]The peak IR luminosity of AT2019dsg is based the host-subtracted neoWISE magnitude of $W1 = 13.1$ (Vega)

References: 1 (van Velzen et al. 2016), 2 (Jiang et al. 2016), 3 (Stein et al. 2020), 4 (Li et al. 2020), 5 (Dou et al. 2017), 6 (Jiang et al. 2017), 7 (Kankare et al. 2017), 8 (Jiang et al. 2019), 9 (Yang et al. 2019), 10 (Liu et al. 2020), 11 (Mattila et al. 2018), 12 (Kool et al. 2020), 13 (Sun et al. 2020) 14 (Komossa et al. 2009), 15 (Dou et al. 2016)

Infrared photometric data of optically discovered TDEs are often obtained from the Wide-field Infrared Space Explorer (WISE; Wright et al. 2010) which has been systematically scanning the whole sky in 3.4 and 4.6 μm since 2010. The primary (cryogenic, 4 band) and post-cryogenic *WISE* surveys ran from Jan 2010 to Feb 2011. Due to lack of funding, the *WISE* space telescope was placed in hibernation between 2011 to 2013. On December 13, 2013, NASA reactivated the *WISE* telescope and started an all sky survey called Near-Earth Object WISE Reactivation mission (NEOWSIE-R; Mainzer et al. 2014). NEOWISE-R is still operating since December 2013. It scans the most part of the sky at 3.4 and 4.6 μm every half year, with $10-20$ single exposures in each visit.

Based on the *WISE* archive photometry, several studies have examined the mid-IR (MIR) light curves (LC) of TDEs identified in optical imaging surveys with spectroscopic follow-up observations, see the Optical Chapter (van Velzen et al. 2020) of this journal. These include ASASSAN-14li (Jiang et al. 2016), PTF09ge, and PTF09axc (van Velzen et al. 2016). Several optical flares in active galaxies that have been interpreted as TDEs are also found to show mid-IR flares, see Table 1. Dou et al. (2016) presented WISE detections of four SDSS galaxies with variable coronal emission lines (see Sect. 3 of this chapter). It is interesting to note that the MIR LCs of the spectroscopically-selected TDEs, if detected in the first place, all have small amplitude variations with $\Delta m \sim 0.1$ magnitude. For ASASSN-14li and PTF-09ge, the peak MIR LC corresponds to infrared luminosity $L_{\rm IR} \sim 10^{41}$ erg s^{-1}

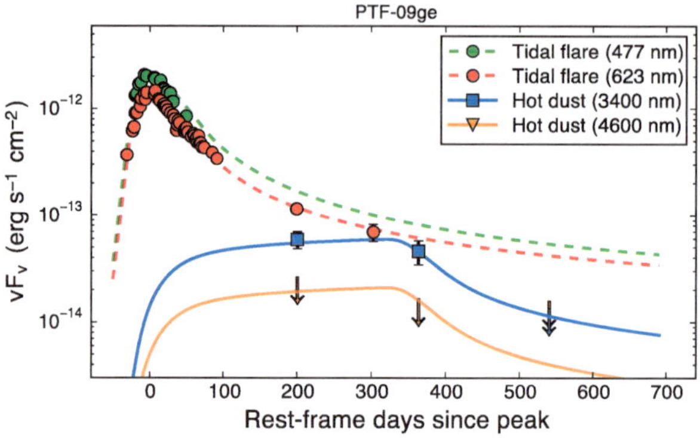

Fig. 2 The optical (Arcavi et al. 2014) and infrared (van Velzen et al. 2016) light curve of the TDE PTF-09ge. The solid lines show the dust reverberation light curve, which is derived from the optical light curve using only the radius of the dust shell and the covering factor as free parameters. The former sets the duration of the event, while the latter affects only the overall normalization. The best-fit parameters are $R = 0.15$ pc and $f_{\rm dust} = 0.01$. Figure reproduced from van Velzen et al. (2016)

and $L_{\rm IR} \sim 10^{42}$ erg s^{-1}, respectively. As shown in Table 1, this luminosity is between one and two an orders of magnitude lower than the peak IR luminosity of the TDE candidates in active galaxies.

The peak phases of the MIR LCs display time delays relative to the optical LCs, with $\Delta t \sim 100 - 200$ days. This time delay translates to the radius of dust medium of about 0.1 pc, somewhat larger than the dust sublimation radius of 0.04 pc as derived from L_{UV}. In general, these numbers are consistent within the context of MIR LCs as the result of the dust echo of the UV/optical flares. Three additional parameters – the covering factor, dust mass and geometry, and the flare bolometric luminosity (via Eq. (1)) – can be estimated from the MIR LCs, although with considerable uncertainty because dust temperatures are not well constraint with only one color ([3.4 − 4.6] μm). For PTF-09ge (Fig. 2), van Velzen et al. (2016) marginalized over the uncertainty on the dust temperature and included the measurement uncertainty on $R_{\rm sub}$ to find $\log_{10}(L_{\rm bol}) = 44.9^{+0.1}_{-0.6}$ and $\log_{10} \Omega_d = -2.0 \pm 0.2$.

Better spectroscopic coverage is one area of improvement for future near-IR follow-up observations. And with well-sampled IR light curves, we can obtain a better handle of the geometry of the dust, allowing us to differentiate between a torus model (Dou et al. 2016; Sun et al. 2020) and a polar or spherical dust geometry (van Velzen et al. 2016; Mattila et al. 2018; Kool et al. 2020).

The lower average IR luminosity of TDEs in quiescent galaxies compared to events selected by their variable coronal emission lines (ECLEs, see Sect. 3) suggests that the covering factor and the dust mass that participates in the reprocessing is larger for the ECLEs. Intriguingly, the first *WISE* detections of the MIR flare in SDSS J0952+214 is 3–5 years after the changes of coronal emission lines. Together with the high IR luminosity, this long time scale of the variation could be an indication that these do not belong to the same class as the TDEs selected based on optical photometric and follow-up spectroscopy. We discuss the ambiguity between changing-look AGNs and TDEs, and the possible ways to separate them in Sect. 2.3 below.

Infrared echo offers a unique tool to probe the dusty circumnuclear medium at close distances to the central supermassive black holes. One interesting example is PS16dtm, which could be a TDE in a Narrow Line Seyfert 1 galaxy (Blanchard et al. 2017)—however see Moriya et al. (2017) for a different interpretation. The MIR LC of PS16dtm appears to rise about 11 days before the first epoch of optical detection (Jiang et al. 2017). This can be explained by a pre-existing dusty medium with a size of 11 light days (2.8×10^{16} cm), with a sufficient high covering factor, blocking the initial optical/UV light. This size is similar to the dust sublimation radius inferred from the quiescent AGN with $L_{\rm bol} \approx 5 \times 10^{42}$ erg s^{-1}. After the TDE flare, the high flux of UV/X-ray photons destroys the dust grains and enlarges the inner radius of the dust free cavity to $\sim$ 70 light days (2×10^{17} cm). Additional Fe elements previously locked up by the dust grains are now released into the ionized gas medium, potentially explaining the abundant FeII emission lines observed in the follow-up spectra by Blanchard et al. (2017).

Infrared echoes appear to be ubiquitous to the class of highly energetic transients from the centers of active galaxies (Kankare et al. 2017). An example is PS1-10adi (Kankare et al. 2017), which has a very high MIR luminosity of $\sim 10^{44}$ erg s^{-1} (Jiang et al. 2019).

Infrared flares interpreted as dust reprocessing signals of energetic transients have also been found in star-forming and dusty luminous infrared galaxies (LIRGs; the infrared emission in these galaxies is thought to be mainly powered by star formation Genzel and Cesarsky 2000). The best-studied case is Arp 299-B AT1, which was observed with high spatial resolution radio observations combined with near- and mid-IR observations covering a period of over 10 years (Mattila et al. 2018). Based on this excellent data, a TDE is the preferred explanation proposed by Mattila et al. (2018).

Due to the small sample size of LIRGs that have been monitored for IR flares, this interpretation would imply an elevated rate of TDEs for this subset of galaxies. Likewise, a TDE interpretation of the transients found by Tadhunter et al. (2017) in a small sample of ultra-luminous infrared galaxies (ULIRGs) would imply a factor $\sim$ 100 enhancement to the TDE rate for this galaxy class (Stone et al. 2018).

2.3 Infrared Flares from Quiescent and Active Galaxies: A Tool to Find Dust-Obscured TDEs?

TDE studies using optical all-sky transient surveys face several important questions. First, tidal disruption of stars by central black-holes can in principle also occur in galaxies with active galactic nuclei (AGN) and dusty infrared luminous star-forming galaxies. Second, how do we distinguish Changing-Look (CL) AGNs from TDEs? While the underlying physics has similarities, these types of transients may probe different scales (time, size and mass) of accretion phenomena.

Some TDE candidates have been found in Seyferts and dusty LIRG/ULIRGs, as listed in see Table 1. These studies confront a common question—could these flares be due to AGN changing their states, i.e., CL AGNs?

The observational ambiguities between CL AGNs and TDEs are abundant—see the Imposters Chapter (Zabludoff et al. 2021) of this journal. Separation often requires multi-wavelength data over a long time scale of several years, as has been obtained for the systems discussed in Sect. 3. Particularly, the properties of a "turn-on" AGNs, transitioning from a quiescent to an active accreting state, can be easily confused with that of a TDE, as shown by Yan et al. (2019). Using an extensive dataset of *Swift* UV, optical, mid-IR LCs and spectroscopy over 1200 days, this study made comprehensive comparisons between CL AGN, TDE and unusual SN IIn and showed that the flare in SDSS1115+0544 is a rapid "turn-on" of a type 1 AGN, changing from its earlier quiescent state. Yan et al. (2019) highlighted that having only LCs or variable spectra, including high ionization coronal lines, is not sufficient to make conclusive classifications (see also the detailed discussion by Komossa et al. 2008). A potential solution could be collecting larger samples of the various classes of TDEs, including events in AGNs and LIRGs/ULIRGs.

Potentially, using infrared variability should offer an effective means to identify TDE candidates in dusty galaxies. Such systematic searches using *WISE* mid-IR time series photometry are being carried out and below we discuss an example of this approach.

Using only the *WISE* data between December 2009 and February 2011, Wang et al. (2018) searched for potential TDE candidates among SDSS galaxies with spectra which present no indications of AGN. Only based on their mid-IR LCs, they found a total of 14 TDE candidates from quiescent SDSS galaxies at $z < 0.22$ with estimated blackhole masses of $10^{7-8} M_{\odot}$. The estimated rate is $\sim 10^{-4}$ galaxy^{-1} yr^{-1} with large uncertainties. As concluded by these authors, these 14 candidates could contain a good fraction of CL AGNs which show short time scale episodic variability.

The second ongoing search is done by the same group but with the entire *WISE* data between December 2009 and 2018 (see Yan et al. 2019 for a brief description of the full survey). The systematic search for infrared flares was done among all SDSS spectroscopic confirmed galaxies at $z < 0.35$. A total of 137 galaxies were selected with flare amplitudes ≥ 0.5 magnitude in at least one WISE band (Jiang et al. 2020). This sample does not have complete spectroscopic classification. It contains a mixture of various types, including a small fraction of supernovae and radio loud AGNs. Jiang et al. (2020) speculates that a large fraction of this sample are possible dust echoes of TDEs and CL AGNs. One example with extensive spectroscopy is SDSS1115+0544, a rapid "turn-on" type-1 AGN (Yan et al. 2019).

Radio follow-up observations of 16 *WISE*-selected flares by Dai et al. (2020) have yielded a high detected fraction of 75% (for a typical 5σ upper limit 60 μJy at 5.5 GHz). Based on pre-flare SDSS spectroscopy, about half of targets in this sample can be classified as AGN, while most of the non-AGN show evidence for active star formation. A radio detection rate of 75% is much higher than the fraction of AGN or TDEs that are considered to be "radio-loud", see the Radio Chapter (Alexander et al. 2020) of this journal. Based on this high detection rate and similarities with other radio-detected IR flares (Mattila et al. 2018; Kool et al. 2020), one could conclude that IR-selection yields a subclass of TDEs that are often accompanied by powerful jets, with total energies of $\sim 10^{52}$ erg. Such powerful jetted TDEs have been observed before (see Bloom et al. 2011; Zauderer et al. 2011, and the Gamma-ray Chapter of this book), although the estimated volumetric rate of these events, ~ 1 Gpc^{-1} yr^{-1} or $\sim 10^{-7}$ galaxy^{-1} yr^{-1} (van Velzen et al. 2018), is three orders of magnitude lower than typical rate of WISE-selected flares from galaxies with SDSS spectra ($\sim 10^{-4}$ galaxy^{-1} yr^{-1}; Wang et al. 2018). Alternatively, one could conclude that a selection of IR flares in WISE data yields a sample of active black holes in a special phase of their evolution, during which a large change in the accretion rate triggers or rekindles jet activity, leading to a higher detection rate of radio emission.

3 Optical Spectroscopy: Emission-Line Echoes

3.1 Introduction

If TDEs occur in a gas-rich environment, their luminous electromagnetic radiation will photoionize any surrounding gas and will be reprocessed into emission lines. These are then very useful probes of the physical conditions and kinematics of the gaseous material in the SMBH vicinity, including the disrupted star itself.

Photoionization by the luminous TDE radiation, which often peaks in the EUV or soft X-rays, will then not only produce Hydrogen Balmer lines, but in particular also high-ionization lines like HeIIλ4686 and coronal lines. The ions which produce these lines have

Table 2 Extreme coronal line emitters

Full (short) name	Redshift	First reference
SDSS J095209.56+214313.3 (J0952+2143)	0.0789	Komossa et al. 2008
SDSS J074820.67+471214.3 (J0748+4712)	0.0615	Wang et al. 2011
SDSS J134244.42+053056.1 (J1342+0530)	0.0366	Wang et al. 2012
SDSS J135001.49+291609.7 (J1350+2916)	0.0777	Wang et al. 2012

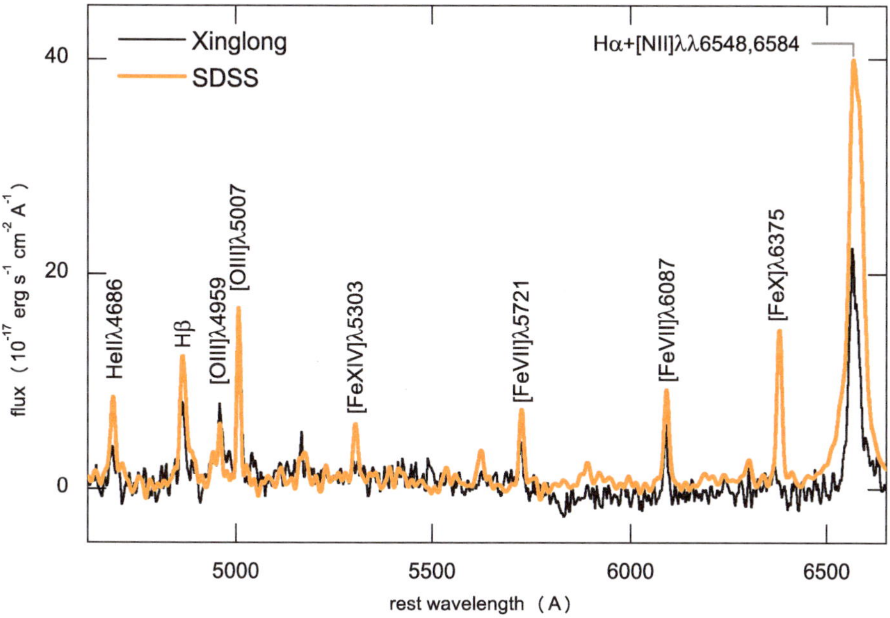

Fig. 3 Optical SDSS spectrum of SDSS J0952+2143 during the high-state (orange) from 2005 with several strong iron coronal lines and other transitions marked. A zoom on the Hα profile (shown above in degraded spectral resolution, to match the same resolution as the Xinglong spectrum) is presented in Fig. 4 in high resolution. The first optical follow-up spectrum from 2007 (taken with the Xinglong telescope, plotted in black) revealed the strong fading of the highest-ionization lines (Komossa et al. 2008)

their ionization potentials (IPs) in the EUV to soft X-ray domain. Several such events have been identified in the last decade based on Sloan Digital Sky Survey (SDSS) spectroscopy and multi-wavelength follow-up observations (Komossa et al. 2008, 2009; Wang et al. 2011, 2012; Yang et al. 2013; Palaversa et al. 2016; Dou et al. 2016) and these are discussed below.

The best-studied of these events, SDSS J0952+2143, (Table 2) was first noticed because of its unusual Balmer and HeII emission profiles and especially the exceptional strength of its high-ionization iron lines relative to [OIII]λ5007 (Komossa et al. 2008) which were, however, not permanent, but faded away dramatically on a timescale of years (Komossa et al. 2009; Yang et al. 2013), see Fig. 3. A systematic search in SDSS for similar events (Wang et al. 2012) let to the identification of six more, of which three (Yang et al. 2013) showed strongly fading coronal lines (Table 2). Key selection criterion was the presence of bright coronal line emission of [FeX]-[FeXIV] in SDSS spectra, which had dropped significantly in intensity in dedicated follow-up optical spectroscopy.

Among these extreme coronal line emitters (ECLEs hereafter), SDSS J0952+2143 at redshift $z = 0.079$ is the one with the most luminous coronal lines, and the densest multi-wavelength coverage both near the high-state and in subsequent years (Komossa et al. 2008, 2009) to which Palaversa et al. (2016) added a long-term optical light curve. Follow-up spectroscopy three years after the SDSS high-state spectrum revealed that the Balmer and coronal lines had strongly declined (Fig. 3). The fact that the outburst extended into the high-energy regime was indirectly inferred from the presence of the luminous, fading coronal lines, while at lower energies continuum variability was directly detected in the NIR, optical and UV (Komossa et al. 2008), followed by a continuous fading of the long-term optical light curve (Palaversa et al. 2016).

The host galaxy of SDSS J0952+2143 was classified as inactive, as no signs of permanent AGN activity were identified; based on line-ratios, NIR colours, and faint X-rays in low-state (Komossa et al. 2008, 2009; our Sect. 3.5). Therefore, the observations are consistent with a TDE which happened in a gas-rich environment.

In addition to the ECLs, fading broad Balmer and He line emission dominate the optical spectrum, with asymmetric profiles, further covered in the Optical Chapter (van Velzen et al. 2020) of this book.

3.2 Iron Coronal Line Diagnostics

The presence of the luminous and transient high-ionization lines in optical spectra of the galaxies immediately implied that the sources underwent luminous electromagnetic flares which extended into the soft X-ray regime – not directly observed at all wavebands, but indirectly inferred from the presence of the fading emission lines, as the ions which produce these lines have their ionization potentials (IPs) in the EUV to soft X-ray regime. The iron lines, with initial luminosities up to $L_{\rm [FeVII]-[FeXIV]} > 10^{40}$ erg s^{-1} in SDSS J0952+2143, faded away on the timescale of years, and the degree of ionization changed from higher ([FeX] to [FeXIV]) to lower ([FeVII]), and up to the complete lack of coronal lines 5-8 years after high-state.

Further, the iron lines are known to be sensitive diagnostics of the physical conditions in the line emitting gas. Individual line ratios enable us to distinguish whether the main ionization mechanism is collisional ionization or photoionization, while other line ratios are density and temperature sensitive (e.g., Nussbaumer et al. 1982; Keenan and Norrington 1987; Dere et al. 2009). Based on these diagnostics, gas temperatures of the order of $(1.5 - 3) \times 10^4$ K were inferred, consistent with a photoionization origin of the emission lines, and gas densities around $n = 10^7$ cm^{-3} (Komossa et al. 2009; Wang et al. 2012).

3.3 Oxygen Line Diagnostics

In SDSS J0952+2143, the intensity ratio of [OIII]4363/[OIII]5007 is exceptionally high (Komossa et al. 2008). [OIII]4363/[OIII]5007 $\simeq$ 0.2–0.3 is significantly above the value observed in AGN and above photoionization predictions for a large parameter range (Komossa and Schulz 1997; Groves et al. 2004). The observed value implies a high-density regime, where the ratio [OIII]4363/[OIII]5007 no longer only depends on temperature, but becomes a density diagnostic (Osterbrock 1989; Dopita and Sutherland 2003). The observed value implies $n = 10^7$ cm^{-3} (for $T = (2 - 5) \times 10^4$ K), and the [OIII]4363 emission therefore arises in a similar region as the ECL emission (Komossa et al. 2009).

Two other ECLEs, SDSS J 0748+4712 and SDSS J1342+0530, show an unusual ratio [OIII]5007/[OIII]4959 ≈ 2 in late-time spectra, strongly deviating from the canonical the-

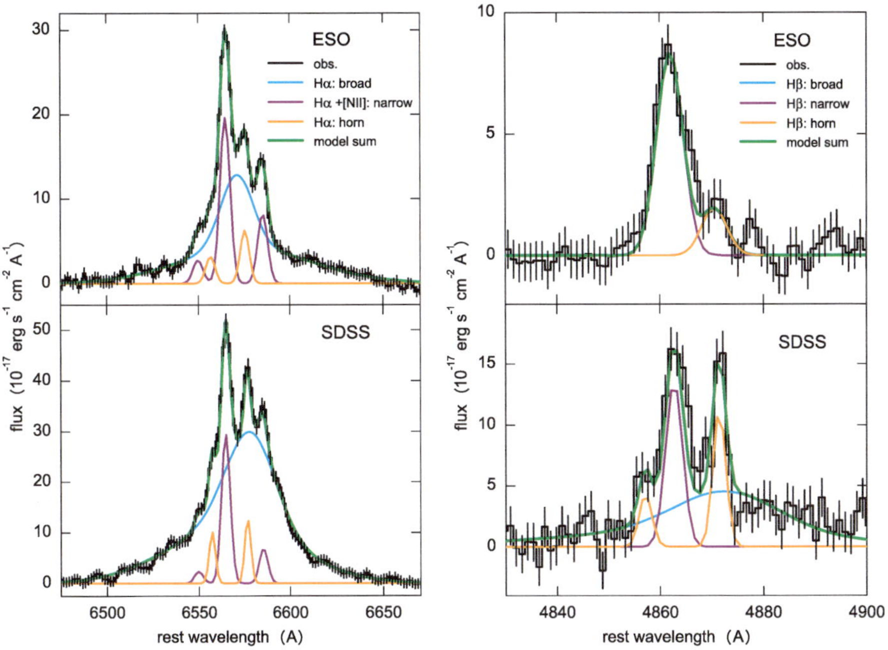

Fig. 4 Hydrogen emission-line profiles of SDSS J0952+2143 (Hα left, Hβ right) taken at SDSS high-state in 2005, and with the ESO NTT in 2008. In addition to an asymmetric, broad base (blue), two narrow horns (orange) are clearly present. Their width is unresolved. Adapted from Komossa et al. (2009)

oretical value of 3 observed in all AGN, and interpreted as sign of high density and high optical depth. Unless the metallicity is exceptionally high ($Z > 10$), the data imply that the [OIII] emission region is optically thick to electron scattering (Yang et al. 2013).

3.4 Balmer-Line Diagnostics: Double-Peaked Horns in Hα and Hβ

A narrow double-peaked structure in the Balmer lines has been detected in the optical spectra of SDSS J0952+2143 (Komossa et al. 2008), not seen in any other of the ECLEs (Wang et al. 2012), or any other source we are aware of. At high-state, the Hα and Hβ lines of SDSS J0952+2143 show a multi-component profile, which can be decomposed into at least three distinct components (Fig. 4): (1) a broad component of FWHM(Hα)=1930 km/s, asymmetric and redshifted by 570 km/s, (2) a narrow component at the same redshift as other narrow forbidden emission lines, and (3) two very narrow horns (FWHM $<$ 200 km/s, unresolved), one redshifted and the other blueshifted. While line intensities of the Balmer line components and especially the two narrow horns had faded significantly 3 years after the SDSS high-state, their velocity shifts of v=540 km/s (redshifted horn) and $v = -340$ km/s (blueshifted horn) remained remarkably constant (Komossa et al. 2009).

The exact nature of the double-horn emission remains unknown at present. The line shape, i.e., the double-peaked profile, could arise in a ring or disk geometry or a two-sided outflow. Surprisingly, the two horns do not have a counterpart in any other of the emission lines. While at first glance, this fact could imply high gas density, several of the strong

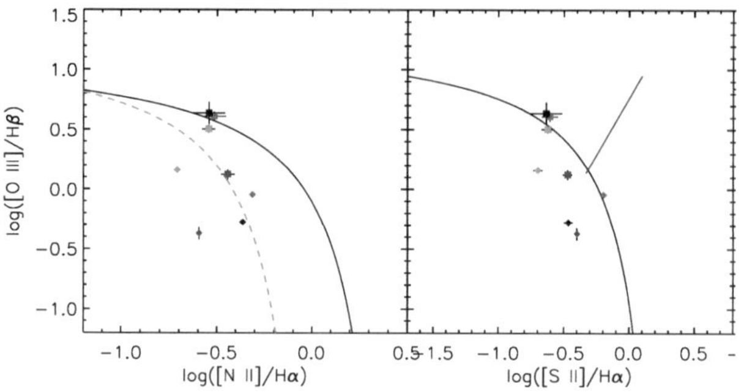

Fig. 5 Location of ECLEs in diagnostic diagrams. They are located in the starburst regime (small symbols), and shift towards higher excitation in the later optical spectra (large squares) which exclude more of the outer starburst emission. The black line marks the dividing line between starformation (lower left) and AGN (upper right), while the blue line represents the deviding line between AGN (above the line) and LINERS (below the line). The region between the green dotted line and black line represents the extreme starburst regime (see Yang et al. (2013))

coronal lines have very high critical density, too, and yet they do not show multiple horns. Komossa et al. (2009) therefore discussed the possibility that these lines form in collisionless shocks. Such shocks produce strong H lines with a two-component profile composed of a narrow component contributed by cold H atoms, and a broad component which arises from H atoms that have undergone charge transfer reactions with hot protons while other optical forbidden lines are very faint (e.g., Raymond et al. 1995; Heng and McCray 2007). Alternatively the broad underlying Balmer line with its two additional narrow horns may represent emission from the inner TDE accretion disk (Komossa et al. 2008).

3.5 A Note on the Interpretation of These Events

In summary, four events were observed which showed strong emission-line flaring in response to an optical/UV/X-ray outburst. These ECLEs reside in subL_* disk galaxies (Wang et al. 2012). A SMBH mass from stellar absorption lines of $7\times10^6\ M_\odot$ was estimated in one case (Komossa et al. 2008). Three outburst scenarios have been discussed in the literature.

The first outburst scenario involved very unusual AGN variability. However, no positive evidence for the presence of AGN has been identified: emission-line ratios are in the starburst (HII) regime of diagnostic diagrams (Fig. 5), NIR colours imply quiescent galaxies, and post-flare X-ray emission, only measured in SDSS J0952+2143, is below that of classical AGN ($L_x = 10^{41}$ erg s^{-1}) (Komossa et al. 2008, 2009; Wang et al. 2012).

Alternatively, a SN origin has been discussed (Komossa et al. 2009; Wang et al. 2011). However, the very luminous iron coronal lines, especially of SDSS J0952+2143, do imply luminous high-energy emission never observed in SNe and the coronal lines are a factor 100 more luminous than the record holder among SNe, making this event very different from any known type of SN. Further, the longevity of the UV emission, detected 1.8 years after the start of the flare (Palaversa et al. 2016), strongly disfavors an SN origin.

Finally, the events can be interpreted as TDEs in a gas-rich environment, which is preferred explanation by most authors (Komossa et al. 2008, 2009; Wang et al. 2012; Palaversa et al. 2016). In this scenario, the radiation from the temporary accretion disk photoionizes any circumnuclear gaseous material as well as the disrupted star itself.

4 Cross Correlations

4.1 Introduction and Methodology

Multi-wavelength photometric reverberation mapping using X-ray, optical, UV, and radio time series of actively accreting SMBHs, i.e., AGN, and stellar-mass black holes have yielded significant insights into the geometries of structures surrounding these systems (e.g., McHardy et al. 2018; Edelson et al. 2015; Marscher et al. 2018; Gandhi et al. 2008; O'Brien et al. 2002; Chatterjee et al. 2009). The underlying method in these studies is to first check for similarity in variability between two light curves derived from two distinct wavebands, and then measure time lags, if any, between them. The hypothesis is that if two light curves from two distinct wavebands vary in the same manner, then they are likely driven by a common physical mechanism.

Under the assumption that the emission from the reference waveband comes from very near to the black hole, any correlations and subsequently time lags give a measure of how far the reverberating regions are with respect to the black hole. The direction (leading vs lagging the emission from near the black hole) and the magnitudes of the time leads/lags vs wavelength essentially give a 1-dimensional map of the geometry of the system under consideration (e.g., Fig. 6 of Cackett et al. 2018). Thus, reverberation mapping analysis provides an invaluable tool to probe spatial scales which cannot be resolved with our current telescopes. For accreting stellar-mass black hole binaries the relevant timescales are between a fraction of a second to a few tens of seconds (e.g., Gandhi et al. 2008; O'Brien et al. 2002) while for AGN they are anywhere between a couple of minutes to a couple of months (e.g., Edelson et al. 2015).

One of the main statistical tools used in reverberation mapping analyses is the so-called cross-correlation function (or CCF). Mathematically, this defines the amount of similarity between two time series as a function of time offset (time lag/lead) between them. CCFs are usually normalized such that a value of unity means a perfect match between the two input time series while a value of negative unity implies they are exactly anti-correlated. A value near zero implies no correlation. A CCF is straightforward to compute when the two time series of interest are uniformly sampled with the same time binning. However, within the context of astronomical time series, often the data are non-uniformly sampled owing to the various observing constraints associated with telescopes. For these unevenly sampled cases, the three most widely used forms of CCF are the interpolated cross-correlation function (ICCF; Peterson et al. 2004), discrete cross-correlation function (DCF; Edelson and Krolik 1988), and the z-transformed discrete cross-correlation function (zDCF; Alexander 1997).

The ICCF is perhaps the most straightforward way to compute a CCF between two unevenly sampled light curves. In this method, for each specified value of time lag, one of the two light curves is shifted by that time lag and its flux values are interpolated (often linearly) at the times of the second light curve. Often the ICCF is computed twice, once by fixing the first light curve and shifting and interpolating the second light curve and vice versa (White and Peterson 1994). The final ICCF is an average of the two CCFs and is thus not significantly biased towards the sampling of either of the two light curves. A DCF, on the other hand, only uses the actual data points without any interpolation (see, Edelson and Krolik 1988, for more details). A zDCF is based on the DCF with several modifications. It has been argued that zDCF can tease out weaker cross-correlation signals from data than the traditional DCF (see, Alexander 1997). All three methods have been used successfully within the context of AGN reverberation mapping but only rarely for TDE studies.

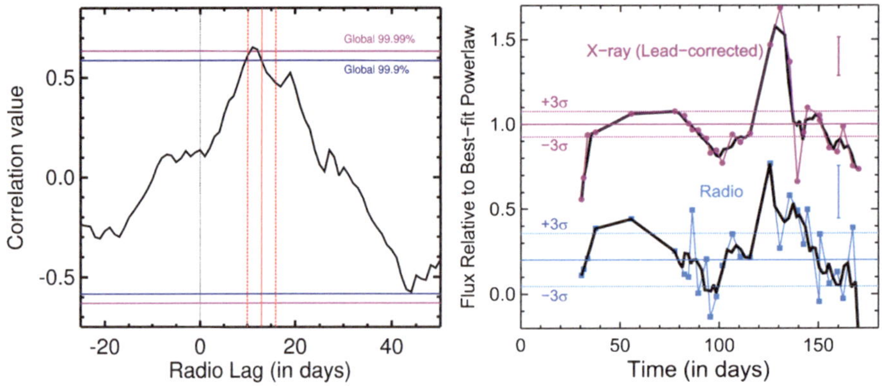

Fig. 6 *Left*: Cross-correlation function (CCF) between the soft X-ray (0.3-1 keV) and 16 GHz radio data from ASASSN-14li. The horizontal blue and magenta lines are the global 99.9% and the 99.99% white noise statistical confidence contours for a blind search between -25 and 50 days (see Sect. 3.2 of Pasham and van Velzen 2018, for more details). The solid and the dashed vertical lines are the median and the 1σ deviations (13^{+3}_{-3} days) derived by considering both the sampling and the measurement uncertainties in both the X-ray and radio light curves. *Right*: The relative X-ray (magenta) and radio (blue) light curves of ASASSN-14li obtained by dividing their corresponding best-fit power-law models. Both the light curves show the same variability pattern. The X-ray and radio fractional variability amplitudes on top of the power-law decay are $10\%\pm1\%$ and $16\%\pm1\%$, respectively. The mean flux levels are shown by the solid horizontal lines while the $\pm3\sigma$ variability contours derived using the methodology described in Romero-Cañizales et al. (2011) are indicated by dashed horizontal lines. The radio data have been vertically offset by -0.8. The solid black curves are running averages over a 10 day window (except when the gap between observations in longer than 10 days). Typical 1σ measurement uncertainties are shown as vertical bars. The X-ray data points shown here have been interpolated to match with the radio epochs (see Sect. 3.1 of Pasham and van Velzen 2018, for more analysis details) (adapted from Pasham and van Velzen 2018)

4.2 Radio and X-Ray Cross Correlations

Because of its panchromatic brightness, ASASSN-14li was monitored in the radio and the X-ray (0.3-10 keV) wavebands roughly once every 3-10 days with a temporal baseline of a few hundred days (Pasham and van Velzen 2018). ASASSN-14li's X-ray spectrum was soft with very weak emission above $\sim$1 keV. These data sets provided the first opportunity to search for reverberation signals on timescale of tens of days from a TDE. Pasham and van Velzen (2018) extracted an ICCF between the X-ray and the 16 GHz radio emission from ASASSN-14li and found that the emission at 16 GHz lagged behind the soft/thermal X-rays by roughly 13 days (see Fig. 6). Interestingly, both the fluctuations on top of the overall long-term decay and the long-term decay curves lagged by the same timescale. This suggested that the entire radio flux at 16 GHz was responding to variations in the soft X-ray (0.3-1.0 keV) brightness changes.

Based on this coupling between the radio and the soft X-rays and time-resolved radio spectral modeling, Pasham and van Velzen (2018) suggested that the radio emission from ASASSN-14li originates from a sub-relativistic jet that is regulated by the inner accretion disk which emits X-rays. Moreover, they showed using the correlation between the two wavebands and independently via modeling the time-resolved radio spectral energy distributions, that the accretion rate and the jet power are linearly coupled, i.e., the newly formed accretion disk is efficiently regulating the radio jet (for more details see Sect. 4.4 of Pasham and van Velzen 2018). Such a strong coupling between the soft X-rays, i.e., inner accretion flow, and the radio emission has never been seen before in an AGN or a supernova and thus may be a means to distinguish TDEs from TDE imposters. However in order to establish this as a common TDE phenomenon, similar studies of several events in the future are necessary. Nevertheless, this work demonstrated the use of TDE radio and X-ray observations to probe the earliest phases of jet formation which is currently only poorly constrained by observations.

4.3 Optical/UV and X-Ray Cross Correlations

In addition to radio–X-ray correlation studies, the inclusion of the optical and the UV wavebands in TDE photometric reverberation analysis can reveal insights into the physical origin of optical/UV light. Within the context of AGN, time lags between the X-ray and the optical/UV variations can arise from the following various physical scenarios (e.g., McHardy et al. 2016). Firstly, X-rays from near the black hole can scatter off material in the outer accretion disk or any other surrounding medium, lose energy, and get reprocessed into lower-energy optical/UV photons (Edelson et al. 2015). In this case, fluctuations in the X-rays would lead the optical changes by a few days to a few tens of days depending on the distance (and thus the light travel time) to the outer disk/reprocessing region (Morgan et al. 2010). Alternatively, if the optical and the UV emission is dominated by the thermal emission from the accretion disk, two kinds of correlations/lags are possible. (1) If the X-rays are produced in a Corona by Compton up-scattering of the thermal near-UV "seed" photons (Reynolds and Nowak 2003), then the changes in the X-ray flux would lag behind the near-UV fluctuations by a fraction of a day to a few days depending on the size of the corona and the inner accretion disk (which are determined by the black hole mass, e.g., Arévalo et al. 2005, and accretion rate). (2) Also in this scenario, because the optical/UV originates directly from the disk, accretion rate perturbations can manifest as variations in optical and UV emission. As these perturbations propagate inwards in the disk on the local viscous timescale (Arévalo et al. 2008), they first trigger changes in the optical band followed by the UV and finally X-rays. Thus, we may expect the optical/UV emission to lead the X-rays by a few tens to millions of days depending on the black hole mass and the accretion rate (Shakura and Sunyaev 1973), see Figs. 8 and 9.

Due to its early identification as a compelling TDE, ASASSN-14li was monitored at a cadence of once every 3-7 days in the optical/UV bands for about 250 days. Motivated by optical/UV–X-ray photometric reverberation mapping of AGN (Edelson et al. 2015; McHardy et al. 2018), Pasham et al. (2017) performed cross-correlation analysis between the soft X-ray and the several of ASASSN-14li's available optical/UV light curves. They found evidence for the presence of similar flux variations in both the X-ray and optical/UV bands. In addition, the variability pattern in X-rays lagged behind those in the optical and the UV bands by roughly 32 days (see Fig. 7).

Based on the direction of the time lag (X-rays lagging the optical/UV), it appears that X-ray reprocessing is not the dominant source of optical/UV emission in ASASSN-14li. Furthermore, the seed photon scenario described above is also unlikely because the expected UV lead times for a $10^{6.5\pm0.6}$ $M_\odot$ black hole–implied from host galaxy properties–are only a few thousand to a few tens of thousand seconds, which is an order of magnitude shorter than the observed lead time. Also, in ASASSN-14li, there is no evidence for a Corona (non-thermal emission) in the high quality X-ray spectra obtained during the first $\sim$300 d of monitoring.

The observed lags are also unlikely to be due to viscous time delays in a "standard" circular Shakura-Sunyaev accretion disk (Shakura and Sunyaev 1973). ASASSN-14li's peak bolometric luminosity of $\sim 10^{44}$ erg s^{-1} (Holoien et al. 2016) implies a peak Eddington ra-

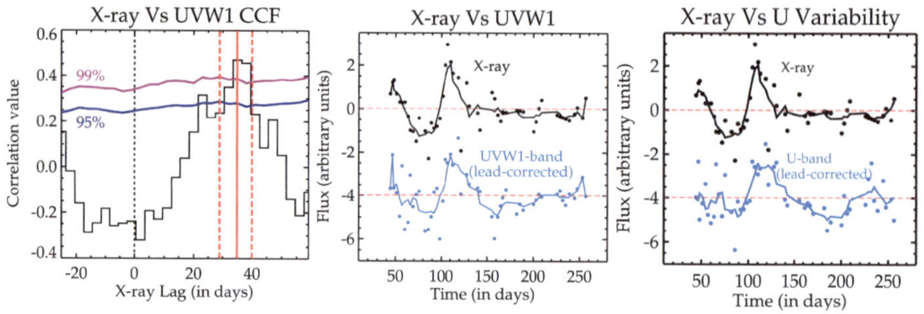

Fig. 7 *Left*: Cross-correlation function (CCF) between the soft X-ray (0.3-1 keV) and the UV (UVW1 filter of *Neil Gehrels Swift* observatory) light curves from ASASSN-14li. The blue and magenta lines are the statistical significance contours (see Pasham et al. 2017, for more details). The vertical red lines have the same meaning as in the left panel of Fig. 6. *Middle and Right*: UV and the optical (U band) variability features are compared to the X-ray variations. The UVW1/U-band light curves are offset by 35 days and the X-ray light curve is interpolated at the lead-corrected UVW1/U epochs. In both panels the observed X-ray and UVW1/U light curves were de-trended by subtracting a smooth function, leaving only the variability features. The solid curves are a running average of five neighboring points, to guide the eye (adapted from Pasham et al. 2017)

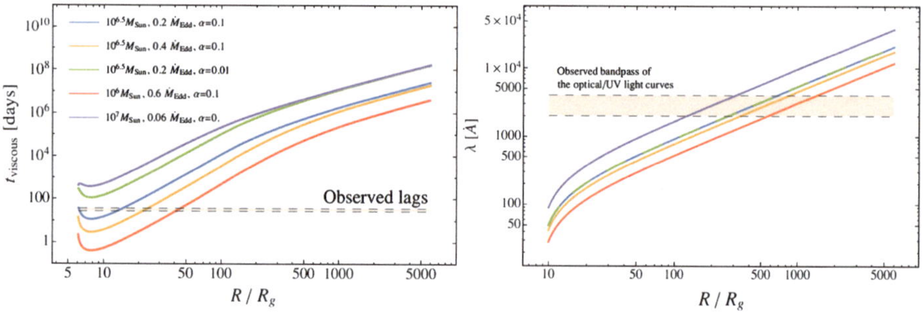

Fig. 8 The dependence of the local viscous timescale (left) and the wavelength of peak emission (right) on the radial distance from the black hole in a Shakura-Sunyaev accretion disk for a set of black hole masses, Eddington ratios and α-viscosity parameter values. The observed timescale of ~30 days is inconsistent with a circular disk and argues for an accretion flow in ASASSN-14li that is distinct from AGN accretion disks (adapted from Pasham et al. 2017)

tio of $0.2^{+0.7}_{-0.2}$ $\dot{M}_{\rm Edd}$ (where $\dot{M}_{\rm Edd}$ is the Eddington rate) for a black hole in the mass range of $10^{6.5\pm0.6}$ $M_\odot$ (e.g., Holoien et al. 2016; Pasham et al. 2019). At such sub-Eddington rates, the newly formed accretion disk can be described by a geometrically thin, optically thick disk model of Shakura and Sunyaev (1973). For the observed lag timescale of ~30 days to correspond to viscous delay in a disk the optical/UV emission should originate from within a radius of 50 R_g (see the left panel of Fig. 8). However, the bulk of the optical/UV emission in the observed *Swift* bandpass (1500-6000 Å) originates from beyond a few hundred gravitational radii (see the right panel of Fig. 8) and at these larger radii the viscous timescale and thus the time lags are orders of magnitude longer. This discrepancy between the emission radius, emission bandpass and corresponding viscous timescales rules out the standard, thin, circular disk solution.

This discrepancy can be resolved in a few of ways. First, the disrupting black hole could be a few orders of magnitude smaller than 10^6 $M_\odot$, i.e., a so-called intermediate-mass black hole. This would make the accretion flow super-Eddington and produce slim disks

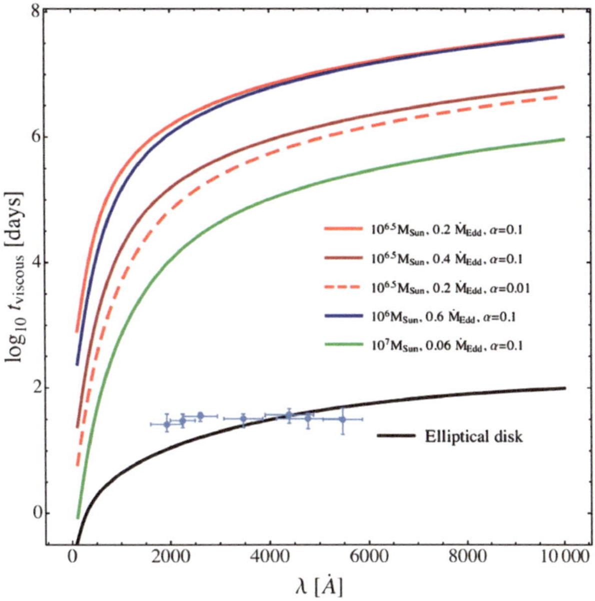

Fig. 9 The colored curves (green, blue and red) show the dependence of viscous timescale on wavelength of emission for a standard, circular thin disk. The observed time lags, shown as blue data points, are orders of magnitude faster than expected from a circular thin disk. The viscous time lags in a simplistic elliptical accretion disk described in Guillochon et al. (2014) (black curve) are shorter and are broadly consistent with the observed optical/UV–X-ray time lags in ASASSN-14li (adapted from Pasham et al. 2017)

(Abramowicz et al. 1988) which can yield faster viscous timescales, i.e., comparable to the optical/UV–X-rays lags observed in ASASSN-14li. Alternately, an elliptical accretion disk would also shorten the viscous timescales (see Fig. 9).

In any case, it is clear that at least in the quintessential TDE ASASSN-14li the optical/UV emission vary with respect to the central X-rays in a unique manner. It remains to be seen if this behavior is typical in TDEs. If so, these distinguishing "echo" signals could be a means to separate genuine TDEs from imposters, viz., AGN flares and unusual supernovae. We note that cross-correlation studies like the ones carried out for ASASSN-14li require high-cadence monitoring over a long temporal baseline (once every few days for several months) in multiple wavebands, and therefore may not be feasible in many cases.

5 X-Ray Quasi-periodic Oscillations from TDEs

Quasi-periodic oscillations (QPOs) in the X-ray flux with centroid frequencies between a few mHz to several hundreds of Hz have now been observed from several stellar-mass black hole binaries (McClintock and Remillard 2006). These are systems where the central compact object has a mass between 5 and 20 $M_\odot$ and draws material from a non-degenerate companion star. X-ray QPOs in stellar-mass black hole binaries often appear when the black hole goes into an outburst during which the overall X-ray flux can increase by orders of magnitude for a brief duration of a few weeks to months (e.g., Homan et al. 2005; Remillard et al. 2002). While the exact origin of these X-ray modulations is unclear, they are generally thought to be associated with the dynamics of hot gas in the innermost regions of the accretion flow (see McClintock and Remillard 2006 for a review). In some systems, the spins of the central black holes have been estimated under the assumption that the QPOs are associated with test particle orbits predicted from general relativity (Franchini et al. 2017; Motta et al. 2014).

Table 3 Summary of the currently known TDE quasi-periodic oscillations (QPOs)

Flare name	Centroid frequency†	Fractional rms*	Notes
SwJ1644+57	4.8±0.3 mHz	3-4%	QPO was present only within the first two weeks after the flare was discovered
2XMM J123103.2+110648	$\sim 7.3\times10^{-5}$ Hz	20-60%	The optical spectrum is unusual compared to typical TDEs. Could be an AGN flare
ASASSN-14li	7.65±0.4 mHz	3-40%	Stable QPO lasting for at least 500 d after the flare was discovered

†The centroid frequency of the X-ray QPO

*The fraction root-mean-squared amplitude of the QPO in percentage of mean flux

A wealth of X-ray timing studies now suggest that supermassive black holes (SMBHs) are scaled-up versions of stellar-mass black holes in some aspects (e.g., McHardy et al. 2006; Markowitz et al. 2003). For example, it has been shown that the X-ray power density spectra of both SMBHs in AGN and stellar-mass black holes are qualitatively similar and that certain timescales within the power spectra scale inversely with the black hole mass after correcting for the accretion rate (McHardy et al. 2006; Markowitz et al. 2003; but also see Done and Gierliński 2005). Under this black hole unification paradigm, even SMBHs should exhibit QPOs but with frequencies that are 10^{5-7} times lower depending on the exact SMBH mass. However, in spite of several searches, robust evidence for X-ray QPOs from SMBHs in AGN has been scarce (see, however, Gierliński et al. 2008; Alston et al. 2014). This could be because of a dearth of highly sensitive observations or that QPOs are intrinsically absent in steadily accreting SMBHs (see Vaughan and Uttley 2005 for details).

On the other hand, TDEs are different from steadily accreting SMBHs and to some degree are similar to stellar-mass black hole binaries when they go into outburst. This is because in both stellar-mass black hole outbursts and TDEs starving black holes are suddenly fed at a high rate. Thus, the natural expectation is that some of the same phenomena, viz., X-ray QPOs, that occur in stellar-mass black hole outbursts should also appear in TDEs. There have been three cases where X-ray QPOs have been reported from TDE (candidates). These include the relativistic TDE SwJ1644+57 (Reis et al. 2012), the super soft X-ray flare 2XMM J123103.2+110648 (Lin et al. 2015), and the thermal TDE ASASSN-14li (Pasham et al. 2019). The observed properties of these three QPOs are summarized in Table 3.

The case of ASASSN-14li is particularly interesting. Because of its widespread interest ASASSN-14li was monitored for several years at multiple wavelengths. In particular, the flare was detectable with *Swift*'s X-Ray Telescope (XRT) for multiple years after its discovery and several deeper exposures were obtained with *Chandra* and *XMM-Newton* during this time. An average power density spectrum of the *Chandra* and the *XMM-Newton* data segments spread over the first 500 d reveals a prominent feature at a frequency of 7.65±0.4 mHz (Fig. 10, left panel). The feature was surprisingly stable for over 300,000 cycles and had a fractional root-mean-squared amplitude of 40% around day 450. The stability for such large number of cycles while the flux changed by an order of magnitude suggests an origin that is tied to the fundamental properties, viz., the mass and the spin of the disrupting black hole (Pasham et al. 2019). Using a black hole mass implied from host galaxy properties, i.e.,

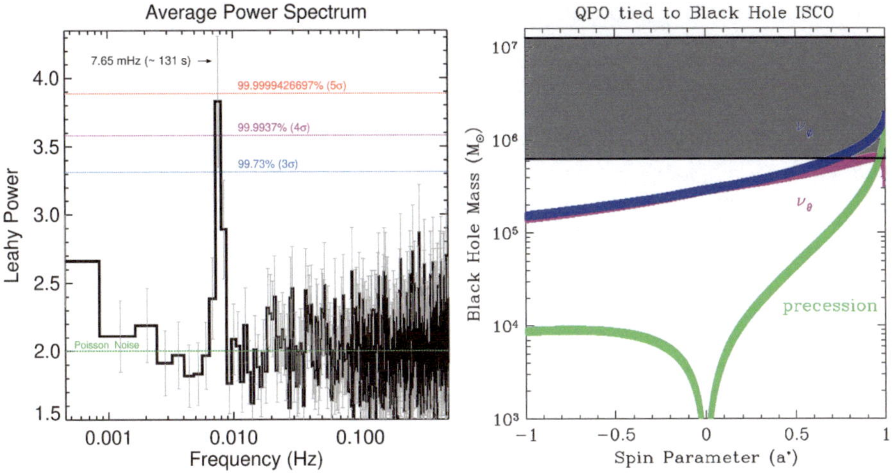

Fig. 10 *Left*: Average X-ray (0.3-1 keV) power density spectrum of ASASSN-14li using data segments spread over 500 d (adapted from Pasham et al. 2019). The global statistical confidence levels and the Poisson noise level are shown as dashed horizontal lines. The strongest features in the spectrum is at 7.65±0.4 mHz. *Right*: Black hole mass vs dimensionless spin parameter contours estimates by assuming the QPO is associated with test particle orbits at the innermost stable circular orbit (ISCO). Contours corresponding to three particle frequencies: Keplerian frequency (ν_ϕ, blue), vertical epicyclic frequency (ν_θ, magenta) and Lense-Thirring precession (ν_ϕ - ν_θ, green) at the ISCO are shown. At the ISCO the radial epicyclic frequency (ν_r) is zero and the periastron precession frequency (ν_ϕ - ν_r) is thus equal to the Keplerian frequency. The widths of these contours reflect the QPO's width of 0.7 mHz (upper limit). The shaded grey rectangular area represents ASASSN-14li's black hole mass range ($10^{5.8-7.1}$ $M_\odot$) estimated from its host galaxy scaling relations. Within this mass range, the only formal solutions are the ones that require the spin parameter to be greater than 0.7. Choosing a radius larger than the ISCO only pushes this lower limit to a higher value

based on the scaling laws between the black hole mass and the host's stellar velocity dispersion and the luminosity, and associating the QPO period to the various particle frequencies predicted from general relativity, Pasham et al. (2019) constrained the dimensionless spin parameter of the disrupting black hole to be at least 0.7 (Fig. 10, right panel). This study presented a new means to constrain spins of several SMBHs in the future when they tidally disrupt stars. On the other hand, the contours in the right panel of Fig. 10 can also be used to set an absolute upper limit on black hole mass. For ASASSN-14li this upper limit corresponds to 2×10^6 $M_\odot$. For future events with QPOs this could be a straightforward technique to identify the long-sought intermediate-mass black holes (masses in 10^{2-5} $M_\odot$) whose QPO timescales would be less than a few tens of seconds.

Historically, not many X-ray bright TDEs have been followed-up with highly sensitive X-ray observations as a majority of them were discovered in archival searches (e.g., Komossa and Bade 1999; Esquej et al. 2007 and see Komossa 2015 and references therein). But, several of the events that did have detailed X-ray observations showed QPOs. While the number of such X-ray TDEs with QPOs is currently small, it remains plausible that QPOs may have been missed in several earlier TDEs. If ASASSN-14li is indeed a "poster child" for TDEs and the observed X-ray QPOs are typical in these systems, then this opens a new opportunity to use TDEs to build a census of black hole mass. An X-ray observatory with an effective area of the order of a square meter will be transformational for this purpose. For instance, simulations using spectral responses of the Wide Field Image (WFI) and the X-ray Integral Field Unit (X-IFU) on board the anticipated meter-class X-ray observatory Athena suggest that ASASSN-14li like QPOs could be detected out to a redshift of 2 (Fig. 11).

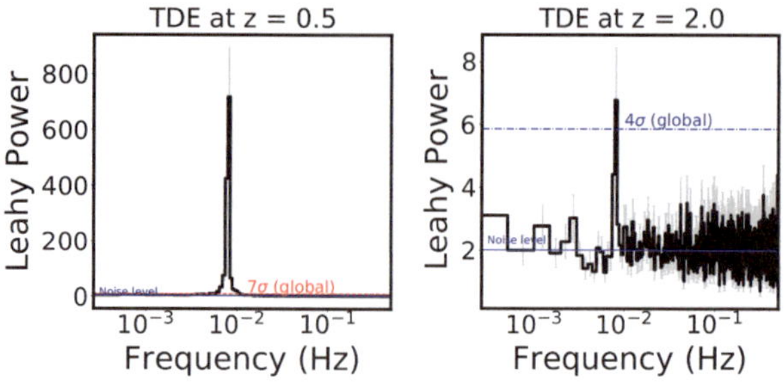

Fig. 11 Simulated power density spectra of an ASASSN-14li-like TDE at various redshifts using Athena's Wide Field Imager (WFI). For each of these simulations we used an exposure time of 32 ks and a QPO fractional rms amplitude of 40% (similar to ASASSN-14li) whose value remains constant with energy within 0.3-2 keV bandpass. Statistical confidence contours and the Poisson noise level of 2 are shown as dashed horizontal lines. The time dilation effect due to the redshift and the potential well of the source on the centroid frequency of the QPO was neglected as it will not alter the results significantly. Assuming ASASSN-14li-like behavior, TDE X-ray QPOs could be detected out to a redshift of roughly 2 with a meter-class X-ray telescope

The prospect of detecting QPOs from TDEs at cosmological distances is very exciting because it would allow us to use these signals to constrain spins of a large number of SMBHs and also construct SMBH spin distributions at various redshifts. On the other hand, the QPOs themselves are excellent probes of accretion at these redshifts. Spin distributions are especially important to constrain the models for growth of SMBHs. For example, it has been predicted that if SMBHs grew primarily via mergers with other black holes, their spin distribution should be skewed towards low spin (e.g., Berti and Volonteri 2008). This is because mergers tend to deposit angular momentum along random orientations which tends to generally spin down the remnant hole. On the other hand, growth by steady accretion tends to add angular momentum along the spin axis of the hole and thus tends to spin up the SMBH. Thus TDE QPOs and the resulting spin constraints could provide a novel opportunity to understand the dominant growth channel of SMBHs.

6 X-Ray Spectroscopy: Relativistic Reverberation

X-ray light echoes probe the smallest scales around the black hole. X-ray reverberation is commonly seen in AGN systems, most of which are sub-Eddington systems consisting of an isotropically-emitting X-ray corona that irradiates the thin accretion disc (e.g. De Marco et al. 2013; Uttley et al. 2014). This produces fluorescence lines with a time delay corresponding to a light travel distance of a few gravitational radii (e.g. Zoghbi et al. 2012; Kara et al. 2016a). The reverberation lags found in AGN can probe the extent of the corona and the inner radius of the thin accretion disc (e.g. Wilkins et al. 2016; Taylor and Reynolds 2018).

Thus far, only one TDE is bright and variable enough (on timescales of hundreds to thousands of seconds) to enable the detection of X-ray time lags. This source is Swift J1644+57, the most well-observed hard X-ray TDE. In this source, Kara et al. (2016b) discovered an emission line at 8 keV, which was seen to lag behind the continuum by 100 seconds. If associated with the Fe Kα line (the line with the highest fluorescent yield in the X-ray spectrum), then the line must be highly blueshifted.

The extremely high isotropic luminosity and the detection of strong radio emission from Swift J1644+57 suggest that this is a TDE with a jet along our line of sight, and a face-on accretion flow. The highly blueshifted iron K line and corresponding reverberation lag can be well described by reprocessing off of a Compton thick, outflowing disc wind with a velocity of $> 0.1c$ and $\sim 30\,\text{deg}$ opening angle (Kara et al. 2016b; Lu et al. 2017; Thomsen et al. 2019). The iron K lag is marginally better described ($> 98\%$ confidence) by an asymmetrically broadened line with some gravitational redshifted over a symmetric Gaussian. This suggested that the reprocessing occurred close to the black hole, and therefore at least some of the X-ray emitting continuum is isotropically emitting (Kara et al. 2016b). This was also supported by the short 100 s lag, though, Lu et al. (2017) point out that if the iron line band also contains emission from the primary continuum source (an effect known as 'dilution'), then the iron K reprocessing region could be further from the black hole, and thus do not rule out that the X-rays are produced in the slow sheath of a relativistic jet.

To conclude, we have shown that X-ray reverberation lags probe the accretion flow structure around black holes, and can be used to probe the velocity and geometrical structure of ultrafast outflows in TDEs.

7 Concluding Remarks and Outlook

Since the strength of a reverberation signal depends on the availability of material to reflect or reprocess the transient emission, we anticipate these signal are not universal in TDEs. Indeed, dust echoes are not detected for every TDE (e.g. van Velzen et al. 2016; Jiang et al. 2021) and the extreme coronal lines seen in SDSS J0952+2143 (Fig. 3) and several other SDSS sources have never been observed in follow-up optical spectra of TDEs detected in other imaging surveys. Furthermore, since not every TDE is X-ray bright (e.g. Auchettl et al. 2017), the X-ray cross-correlation and timing signatures discussed this review also remain relatively rare, with detections in just one (either ASASSN-14li or Swift J1644) or a handful of sources: see Table 3.

While they are not common to the full TDE population, the detection of reverberation signals is important since they provide a novel way to study the TDE emission mechanism and these signals can be used to select TDEs that could be missed by traditional photometric searches for TDEs. The events selected based on the detection of galaxies with extreme coronal lines are well-suited to recover a subset of TDEs that occur in gas rich circumnuclear environments, while a search for IR echoes can reveal TDEs in galaxies with a large amount of dust extinction to the center (e.g., Wang et al. 2018; Mattila et al. 2018). We should note that IR follow-up observations of TDE candidates in AGN have yielded dust echoes that are an order of magnitude more luminous compared to IR echoes in TDEs from quiescent galaxies (Table 1), implying that a blind search for IR echoes could be dominated by flares in AGN (see also the discussion at the end of 2.3).

Both the IR echoes and the detection of coronal lines can be used to probe the energy produced at EUV wavelengths (which cannot be observed directly due to absorption by neutral hydrogen). In most cases, the observed optical/UV or X-ray luminosity is lower than the energy inferred from the reverberation signals. This provides support for models that explain the optical/UV emission of TDEs with an accretion disk (Cannizzo et al. 1990; Dai et al. 2018; van Velzen et al. 2019; Mummery and Balbus 2020), since for black holes in the mass range $M_{\rm BH} \sim 10^{6-7}\,M_\odot$ these disk are expected to emit most of their energy in the EUV regime, see also the Emission Mechanisms Chapter (Roth et al. 2020) of this journal.

While the IR echoes are agnostic to the source that is heating the dust, the cross-correlation signals detected for ASASSN-14li (Sect. 4) appear to be unique to TDEs and can thus be used to help discriminate TDEs from impostors such as AGN flares, see the Imposters Chapter (Zabludoff et al. 2021) of this journal for further discussion. In particular the detection of a QPO in the X-ray light curve of ASASSN-14li is truly spectacular, providing new ways to constrain the spin of the black hole that disrupted the star. The relativistic reverberation signal (Sect. 6) detected for the jetted TDE Swift J1644+57 is another unique signature which probes the geometry of the inter accretion flow and the base of the jet.

Clearly the detection of more multi-wavelength cross-correlations and QPOs is very much anticipated. The current detection rate of such events is ~ 1 per decade. While recently the detection rate of optical TDEs has increased significantly, to dozens of new sources per year (e.g., van Velzen et al. 2021), few of these have well-sampled X-ray light curves similar to ASASSN-14li (one notable exception is AT2018fyk, see Wevers et al. 2019, 2021). This lack of useful X-ray light curves is mainly explained by the low X-ray flux of optically-selected TDEs, but is also due to a scarcity of X-ray follow-up resources. An important improvement would be a new X-ray mission that is dedicated to high-cadence observations, such as Einstein Probe (Yuan et al. 2015). As discussed in Sect. 5, X-ray observations with a meter-class X-ray observatory such as Athena would be able to detect QPOs from TDEs to $z \sim 2$, thus constraining the spin of black holes in the cosmic epoch when they are expected to be rapidly growing.

At the time of writing, the WISE satellite continues to monitor the near-IR sky (3-4 μm) with a cadence of 6 months, thus providing an excellent dataset that can be used to obtain detections (or upper limits) of dust echoes for every transient detected in the last decade. The (potential) detection of dust reprocessing using ground-based i-band observations (Stein et al. 2020) and K-band observation (Mattila et al. 2018; Kool et al. 2020) is also an encouraging result, because this allows for a much higher cadence than the scanning pattern of WISE. In the near future, observations with JWST could yield a detailed spectroscopic view of dust heated by TDEs.

References

M.A. Abramowicz, B. Czerny, J.P. Lasota, E. Szuszkiewicz, Slim accretion disks. Astrophys. J. **332**, 646 (1988). https://doi.org/10.1086/166683

T. Alexander, Is AGN variability correlated with other AGN properties? in *ZDCF Analysis of Small Samples of Sparse Light Curves*, vol. 218, ed. by D. Maoz, A. Sternberg, E.M. Leibowitz (1997), p. 163. https://doi.org/10.1007/978-94-015-8941-3_14

K.D. Alexander, S. van Velzen, A. Horesh, B.A. Zauderer, Radio properties of tidal disruption events. Space Sci. Rev. **216**(5), 81 (2020). https://doi.org/10.1007/s11214-020-00702-w

W.N. Alston, J. Markevičiūtė, E. Kara, A.C. Fabian, M. Middleton, Detection of a QPO in five XMM-Newton observations of RE J1034+396. Mon. Not. R. Astron. Soc. **445**, 16–20 (2014). https://doi.org/10.1093/mnrasl/slu127

I. Arcavi, A. Gal-Yam, M. Sullivan, Y.-C. Pan, S.B. Cenko, A. Horesh, E.O. Ofek, A. De Cia, L. Yan, C.-W. Yang, D.A. Howell, D. Tal, S.R. Kulkarni, S.P. Tendulkar, S. Tang, D. Xu, A. Sternberg, J.G. Cohen, J.S. Bloom, P.E. Nugent, M.M. Kasliwal, D.A. Perley, R.M. Quimby, A.A. Miller, C.A. Theissen, R.R. Laher, A continuum of H- to He-rich tidal disruption candidates with a preference for E+A galaxies. Astrophys. J. **793**, 38 (2014). https://doi.org/10.1088/0004-637X/793/1/38

P. Arévalo, I. Papadakis, B. Kuhlbrodt, W. Brinkmann, X-ray to UV variability correlation in MCG-6-30-15. Astron. Astrophys. **430**, 435–442 (2005). https://doi.org/10.1051/0004-6361:20041801

P. Arévalo, P. Uttley, S. Kaspi, E. Breedt, P. Lira, I.M. McHardy, Correlated X-ray/optical variability in the quasar MR2251-178. Mon. Not. R. Astron. Soc. **389**, 1479–1488 (2008). https://doi.org/10.1111/j.1365-2966.2008.13719.x

K. Auchettl, J. Guillochon, E. Ramirez-Ruiz, New physical insights about tidal disruption events from a comprehensive observational inventory at X-ray wavelengths. Astrophys. J. **838**, 149 (2017). https://doi.org/10.3847/1538-4357/aa633b

R. Barvainis, Hot dust and the near-infrared bump in the continuum spectra of quasars and active galactic nuclei. Astrophys. J. **320**, 537–544 (1987). https://doi.org/10.1086/165571

E. Berti, M. Volonteri, Cosmological black hole spin evolution by mergers and accretion. Astrophys. J. **684**, 822–828 (2008). https://doi.org/10.1086/590379

P.K. Blanchard, M. Nicholl, E. Berger, J. Guillochon, R. Margutti, R. Chornock, K.D. Alexander, J. Leja, M.R. Drout, PS16dtm: A tidal disruption event in a narrow-line Seyfert 1 galaxy. Astrophys. J. **843**, 106 (2017). https://doi.org/10.3847/1538-4357/aa77f7

J.S. Bloom, D. Giannios, B.D. Metzger, S.B. Cenko, D.A. Perley, N.R. Butler, N.R. Tanvir, A.J. Levan, P.T. O'Brien, L.E. Strubbe, F. De Colle, E. Ramirez-Ruiz, W.H. Lee, S. Nayakshin, E. Quataert, A.R. King, A. Cucchiara, J. Guillochon, G.C. Bower, A.S. Fruchter, A.N. Morgan, A.J. van der Horst, A possible relativistic jetted outburst from a massive black hole fed by a tidally disrupted star. Science **333**, 203 (2011). https://doi.org/10.1126/science.1207150

E.M. Cackett, C.-Y. Chiang, I. McHardy, R. Edelson, M.R. Goad, K. Horne, K.T. Korista, Accretion disk reverberation with Hubble space telescope observations of NGC 4593: Evidence for diffuse continuum lags. Astrophys. J. **857**, 53 (2018). https://doi.org/10.3847/1538-4357/aab4f7

J.K. Cannizzo, H.M. Lee, J. Goodman, The disk accretion of a tidally disrupted star onto a massive black hole. Astrophys. J. **351**, 38–46 (1990). https://doi.org/10.1086/168442

R. Chatterjee, A.P. Marscher, S.G. Jorstad, A.R. Olmstead, I.M. McHardy, M.F. Aller, H.D. Aller, A. Lähteenmäki, M. Tornikoski, T. Hovatta, Disk-jet connection in the radio galaxy 3C 120. Astrophys. J. **704**(2), 1689–1703 (2009). https://doi.org/10.1088/0004-637X/704/2/1689

L. Dai, J.C. McKinney, N. Roth, E. Ramirez-Ruiz, M.C. Miller, A unified model for tidal disruption events. Astrophys. J. **859**, 20 (2018). https://doi.org/10.3847/2041-8213/aab429

B.B. Dai, X.W. Shu, N. Jiang, L.M. Dou, D.Z. Liu, C.W. Yang, F.B. Zhang, T.G. Wang, Compact radio emission from nearby galaxies with mid-infrared nuclear outbursts. Astrophys. J. Lett. **896**(2), 27 (2020). https://doi.org/10.3847/2041-8213/ab97ac

B. De Marco, G. Ponti, G. Miniutti, T. Belloni, M. Cappi, M. Dadina, T. Muñoz-Darias, Time lags in the ultraluminous X-ray source NGC 5408 X-1: Implications for the black hole mass. Mon. Not. R. Astron. Soc. **436**, 3782–3791 (2013). https://doi.org/10.1093/mnras/stt1853

K.P. Dere, E. Landi, P.R. Young, G. Del Zanna, M. Landini, H.E. Mason, CHIANTI - An atomic database for emission lines. IX. Ionization rates, recombination rates, ionization equilibria for the elements hydrogen through zinc and updated atomic data. Astron. Astrophys. **498**(3), 915–929 (2009). https://doi.org/10.1051/0004-6361/200911712

C. Done, M. Gierliński, Scaling variability from stellar to supermassive black holes. Mon. Not. R. Astron. Soc. **364**, 208–216 (2005). https://doi.org/10.1111/j.1365-2966.2005.09555.x

M.A. Dopita, R.S. Sutherland, *Astrophysics of the Diffuse Universe* (2003)

L. Dou, T-g. Wang, N. Jiang, C. Yang, J. Lyu, H. Zhou, Long fading mid-infrared emission in transient coronal line emitters: Dust echo of a tidal disruption flare. Astrophys. J. **832**, 188 (2016). https://doi.org/10.3847/0004-637X/832/2/188

L. Dou, T. Wang, L. Yan, N. Jiang, C. Yang, R.M. Cutri, A. Mainzer, B. Peng, Discovery of a mid-infrared echo from the TDE candidate in the nucleus of ULIRG F01004-2237. Astrophys. J. Lett. **841**(1), 8 (2017). https://doi.org/10.3847/2041-8213/aa7130

B.T. Draine, H.M. Lee, Optical properties of interstellar graphite and silicate grains. Astrophys. J. **285**, 89–108 (1984). https://doi.org/10.1086/162480

E. Dwek, The infrared echo of a type II supernova with a circumstellar dust shell: Applications to SN 1979c and SN 1980k. Astrophys. J. **274**, 175–183 (1983). https://doi.org/10.1086/161435

R.A. Edelson, J.H. Krolik, The discrete correlation function: A new method for analyzing unevenly sampled variability data. Astrophys. J. **333**, 646 (1988). https://doi.org/10.1086/166773

R. Edelson, J.M. Gelbord, K. Horne, I.M. McHardy, B.M. Peterson, P. Arévalo, A.A. Breeveld, G. De Rosa, P.A. Evans, M.R. Goad, G.A. Kriss, W.N. Brandt, N. Gehrels, D. Grupe, J.A. Kennea, C.S. Kochanek, J.A. Nousek, I. Papadakis, M. Siegel, D. Starkey, P. Uttley, S. Vaughan, S. Young, A.J. Barth, M.C. Bentz, B.J. Brewer, D.M. Crenshaw, E. Dalla Bontà, A. De Lorenzo-Cáceres, K.D. Denney, M. Dietrich, J. Ely, M.M. Fausnaugh, C.J. Grier, P.B. Hall, J. Kaastra, B.C. Kelly, K.T. Korista, P. Lira, S. Mathur, H. Netzer, A. Pancoast, L. Pei, R.W. Pogge, J.S. Schimoia, T. Treu, M. Vestergaard, C. Villforth, H. Yan, Y. Zu, Space telescope and optical reverberation mapping project. II. Swift and HST reverberation mapping of the accretion disk of NGC 5548. Astrophys. J. **806**, 129 (2015). https://doi.org/10.1088/0004-637X/806/1/129

P. Esquej, R.D. Saxton, M.J. Freyberg, A.M. Read, B. Altieri, M. Sanchez-Portal, G. Hasinger, Candidate tidal disruption events from the XMM-Newton slew survey. Astron. Astrophys. **462**, 49–52 (2007). https://doi.org/10.1051/0004-6361:20066072

A. Franchini, S.E. Motta, G. Lodato, Constraining black hole spins with low-frequency quasi-periodic oscillations in soft states. Mon. Not. R. Astron. Soc. **467**, 145–154 (2017). https://doi.org/10.1093/mnras/stw3363

P. Gandhi, K. Makishima, M. Durant, A.C. Fabian, V.S. Dhillon, T.R. Marsh, J.M. Miller, T. Shahbaz, H.C. Spruit, Rapid optical and X-ray timing observations of GX 339-4: Flux correlations at the onset of a low/hard state. Mon. Not. R. Astron. Soc. **390**, 29–33 (2008). https://doi.org/10.1111/j.1745-3933.2008.00529.x

R. Genzel, C.J. Cesarsky, Extragalactic results from the infrared space observatory. Annu. Rev. Astron. Astrophys. **38**, 761–814 (2000). https://doi.org/10.1146/annurev.astro.38.1.761

M. Gierliński, M. Middleton, M. Ward, C. Done, A periodicity of $\sim$1 hour in X-ray emission from the active galaxy RE J1034+396. Nature **455**, 369–371 (2008). https://doi.org/10.1038/nature07277

B.A. Groves, M.A. Dopita, R.S. Sutherland Dusty, Radiation pressure-dominated photoionization. I. Model description, structure, and grids. Astrophys. J. Suppl. Ser. **153**(1), 9–73 (2004). https://doi.org/10.1086/421113

J. Guillochon, H. Manukian, E. Ramirez-Ruiz, PS1-10jh: The disruption of a main-sequence star of near-solar composition. Astrophys. J. **783**(1), 23 (2014). https://doi.org/10.1088/0004-637X/783/1/23

K. Heng, R. McCray, Balmer-dominated shocks revisited. Astrophys. J. **654**(2), 923–937 (2007). https://doi.org/10.1086/509601

T.W.-S. Holoien, C.S. Kochanek, J.L. Prieto, K.Z. Stanek, S. Dong, B.J. Shappee, D. Grupe, J.S. Brown, U. Basu, J.F. Beacom, D. Bersier, J. Brimacombe, A.B. Danilet, E. Falco, Z. Guo, J. Jose, G.J. Herczeg, F. Long, G. Pojmanski, G.V. Simonian, D.M. Szczygieł, T.A. Thompson, J.R. Thorstensen, R.M. Wagner, P.R. Woźniak, Six months of multiwavelength follow-up of the tidal disruption candidate ASASSN-14li and implied TDE rates from ASAS-SN. Mon. Not. R. Astron. Soc. **455**, 2918–2935 (2016). https://doi.org/10.1093/mnras/stv2486

J. Homan, J.M. Miller, R. Wijnands, M. van der Klis, T. Belloni, D. Steeghs, W.H.G. Lewin, High- and low-frequency quasi-periodic oscillations in the X-ray light curves of the black hole transient H1743-322. Astrophys. J. **623**, 383–391 (2005). https://doi.org/10.1086/424994

N. Jiang, L. Dou, T. Wang, C. Yang, J. Lyu, H. Zhou, The WISE detection of an infrared echo in tidal disruption event ASASSN-14li. Astrophys. J. Lett. **828**, 14 (2016). https://doi.org/10.3847/2041-8205/828/1/L14

N. Jiang, T. Wang, L. Yan, T. Xiao, C. Yang, L. Dou, H. Wang, R. Cutri, A. Mainzer, Mid-infrared flare of TDE candidate PS16dtm: Dust echo and implications for the spectral evolution. Astrophys. J. **850**(1), 63 (2017). https://doi.org/10.3847/1538-4357/aa93f5

N. Jiang, T. Wang, G. Mou, H. Liu, L. Dou, Z. Sheng, Y. Wang, Infrared echo and late-stage rebrightening of nuclear transient Ps1-10adi: Exploring the torus with tidal disruption events in active galactic nuclei. Astrophys. J. **871**(1), 15 (2019). https://doi.org/10.3847/1538-4357/aaf6b2

N. Jiang, T. Wang, L. Dou, X. Shu, X. Hu, H. Liu, Y. Wang, L. Yan, Z. Sheng, C. Yang, L. Sun, H. Zhou, Mid-InfraRed Outburst in Nearby Galaxies (MIRONG) I: Sample selection and characterization. arXiv e-prints, 2012–06806 (2020). https://doi.org/10.3847/1538-4365/abd1dc

N. Jiang, T. Wang, X. Hu, L. Sun, L. Dou, L. Xiao, Infrared echoes of optical tidal disruption events: $\sim$1% dust covering factor or less at sub-parsec scale. arXiv e-prints, 2102–08044 (2021). https://doi.org/10.3847/1538-4357/abe772

E. Kankare, R. Kotak, S. Mattila, P. Lundqvist, M.J. Ward, M. Fraser, A. Lawrence, S.J. Smartt, W.P.S. Meikle, A. Bruce, J. Harmanen, S.J. Hutton, C. Inserra, T. Kangas, A. Pastorello, T. Reynolds, C. Romero-Cañizales, K.W. Smith, S. Valenti, K.C. Chambers, K.W. Hodapp, M.E. Huber, N. Kaiser, R.-P. Kudritzki, E.A. Magnier, J.L. Tonry, R.J. Wainscoat, C. Waters, A population of highly energetic transient events in the centres of active galaxies. Nat. Astron. **1**, 865–871 (2017). https://doi.org/10.1038/s41550-017-0290-2

E. Kara, W.N. Alston, A.C. Fabian, E.M. Cackett, P. Uttley, C.S. Reynolds, A. Zoghbi, A global look at X-ray time lags in Seyfert galaxies. Mon. Not. R. Astron. Soc. **462**(1), 511–531 (2016a). https://doi.org/10.1093/mnras/stw1695

E. Kara, J.M. Miller, C. Reynolds, L. Dai, Relativistic reverberation in the accretion flow of a tidal disruption event. Nature **535**(7612), 388–390 (2016b). https://doi.org/10.1038/nature18007

F.P. Keenan, P.H. Norrington, Relative emission line strengths for Fe VII in astrophysical plasmas. Astron. Astrophys. **181**, 370–372 (1987). http://adsabs.harvard.edu/abs/1987A&A...181..370K

S. Komossa, Tidal disruption of stars by supermassive black holes: Status of observations. J. High Energy Astrophys. **7**, 148–157 (2015). https://doi.org/10.1016/j.jheap.2015.04.006

S. Komossa, N. Bade, The giant X-ray outbursts in NGC 5905 and IC 3599: Follow-up observations and outburst scenarios. Astron. Astrophys. **343**, 775–787 (1999). http://adsabs.harvard.edu/abs/1999A%26A...343..775K

S. Komossa, H. Schulz, Interpretation of the emission line spectra of Seyfert 2 galaxies by multi-component photoionization models. Astron. Astrophys. **323**, 31–46 (1997). http://adsabs.harvard.edu/abs/1997A&A...323...31K

S. Komossa, H. Zhou, T. Wang, M. Ajello, J. Ge, J. Greiner, H. Lu, M. Salvato, R. Saxton, H. Shan, D. Xu, W. Yuan, Discovery of superstrong, fading, iron line emission and double-peaked Balmer lines of the galaxy SDSS J095209.56+214313.3: The light echo of a huge flare. Astrophys. J. Lett. **678**(1), 13 (2008). https://doi.org/10.1086/588281

S. Komossa, H. Zhou, A. Rau, M. Dopita, A. Gal-Yam, J. Greiner, J. Zuther, M. Salvato, D. Xu, H. Lu, R. Saxton, M. Ajello, NTT, Spitzer, and Chandra Spectroscopy of SDSSJ095209.56+214313.3: The most luminous coronal-line supernova ever observed, or a stellar tidal disruption event? Astrophys. J. **701**, 105–121 (2009). https://doi.org/10.1088/0004-637X/701/1/105

S. Komossa, The puzzle of SDSSJ0952+2143, in *Proceedings of Frontier Objects in Astrophysics and Particle Physics*, ed. by F. Giovanelli et al. (2010), p. 10

E.C. Kool, T.M. Reynolds, S. Mattila, E. Kankare, M.A. Pérez-Torres, A. Efstathiou, S. Ryder, C. Romero-Cañizales, W. Lu, T. Heikkilä, G.E. Anderson, M. Berton, J. Bright, G. Cannizzaro, D. Eappachen, M. Fraser, M. Gromadzki, P.G. Jonker, H. Kuncarayakti, P. Lundqvist, K. Maeda, R.M. McDermid, A.M. Medling, S. Moran, A. Reguitti, M. Shahband eh, S. Tsygankov, V. U, T. Wevers, AT 2017gbl: A dust obscured TDE candidate in a luminous infrared galaxy. Mon. Not. R. Astron. Soc. **498**(2), 2167–2195 (2020). https://doi.org/10.1093/mnras/staa2351

S. Koshida, T. Minezaki, Y. Yoshii, Y. Kobayashi, Y. Sakata, S. Sugawara, K. Enya, M. Suganuma, H. Tomita, T. Aoki, B.A. Peterson, Reverberation measurements of the inner radius of the dust torus in 17 Seyfert galaxies. Astrophys. J. **788**, 159 (2014). https://doi.org/10.1088/0004-637X/788/2/159

D. Li, R.D. Saxton, W. Yuan, L. Sun, H.-Y. Liu, N. Jiang, H. Cheng, H. Zhou, S. Komossa, C. Jin, Multi-wavelength study of an X-ray tidal disruption event candidate in NGC 5092. Astrophys. J. **891**(2), 121 (2020). https://doi.org/10.3847/1538-4357/ab744a

D. Lin, P.W. Maksym, J.A. Irwin, S. Komossa, N.A. Webb, O. Godet, D. Barret, D. Grupe, S.D.J. Gwyn, An ultrasoft X-ray flare from 3XMM J152130.7+074916: A tidal disruption event candidate. Astrophys. J. **811**, 43 (2015). https://doi.org/10.1088/0004-637X/811/1/43

Z. Liu, D. Li, H.-Y. Liu, Y. Lu, W. Yuan, L. Dou, R.-F. Shen, A tidal disruption event candidate discovered in the active galactic nucleus SDSS J022700.77-042020.6. Astrophys. J. **894**(2), 93 (2020). https://doi.org/10.3847/1538-4357/ab880f

W. Lu, P. Kumar, N.J. Evans, Infrared emission from tidal disruption events - Probing the pc-scale dust content around galactic nuclei. Mon. Not. R. Astron. Soc. **458**, 575–581 (2016). https://doi.org/10.1093/mnras/stw307

W. Lu, J. Krolik, P. Crumley, P. Kumar, Radiative interaction between the relativistic jet and optically thick envelope in tidal disruption events. Mon. Not. R. Astron. Soc. **471**(1), 1141–1152 (2017). https://doi.org/10.1093/mnras/stx1668

A. Mainzer, J. Bauer, R.M. Cutri, T. Grav, J. Masiero, R. Beck, P. Clarkson, T. Conrow, J. Dailey, P. Eisenhardt, B. Fabinsky, S. Fajardo-Acosta, J. Fowler, C. Gelino, C. Grillmair, I. Heinrichsen, M. Kendall, J.D. Kirkpatrick, F. Liu, F. Masci, H. McCallon, C.R. Nugent, M. Papin, E. Rice, D. Royer, T. Ryan, P. Sevilla, S. Sonnett, R. Stevenson, D.B. Thompson, S. Wheelock, D. Wiemer, M. Wittman, E. Wright, L. Yan, Initial performance of the NEOWISE reactivation mission. Astrophys. J. **792**, 30 (2014). https://doi.org/10.1088/0004-637X/792/1/30

A. Markowitz, R. Edelson, S. Vaughan, P. Uttley, I.M. George, R.E. Griffiths, S. Kaspi, A. Lawrence, I. McHardy, K. Nandra, K. Pounds, J. Reeves, N. Schurch, R. Warwick, X-ray fluctuation power spectral densities of Seyfert 1 galaxies. Astrophys. J. **593**, 96–114 (2003). https://doi.org/10.1086/375330

A.P. Marscher, S.G. Jorstad, K.E. Williamson, A. Lähteenmäki, M. Tornikoski, J.M. Hunter, K.A. Leidig, M. Zain Mobeen, R.J.C. Vera, W. Chamani, X-ray, UV, and radio timing observations of the radio galaxy 3C 120. Astrophys. J. **867**, 128 (2018). https://doi.org/10.3847/1538-4357/aae4de

S. Mattila, M. Pérez-Torres, A. Efstathiou, P. Mimica, M. Fraser, E. Kankare, A. Alberdi, M.Á. Aloy, T. Heikkilä, P.G. Jonker, P. Lundqvist, I. Martí-Vidal, W.P.S. Meikle, C. Romero-Cañizales, S.J. Smartt, S. Tsygankov, E. Varenius, A. Alonso-Herrero, M. Bondi, C. Fransson, R. Herrero-Illana, T. Kangas, R. Kotak, N. Ramírez-Olivencia, P. Väisänen, R.J. Beswick, D.L. Clements, R. Greimel, J. Harmanen, J. Kotilainen, K. Nandra, T. Reynolds, S. Ryder, N.A. Walton, K. Wiik, G. Östlin, A dust-enshrouded tidal disruption event with a resolved radio jet in a galaxy merger. Science **361**, 482–485 (2018). https://doi.org/10.1126/science.aao4669

J.E. McClintock, R.A. Remillard, in *Black Hole Binaries*, ed. by W.H.G. Lewin, M. van der Klis (2006), pp. 157–213

I.M. McHardy, E. Koerding, C. Knigge, P. Uttley, R.P. Fender, Active galactic nuclei as scaled-up Galactic black holes. Nature **444**, 730–732 (2006). https://doi.org/10.1038/nature05389

I.M. McHardy, S.D. Connolly, B.M. Peterson, A. Bieryla, H. Chand, M.S. Elvis, D. Emmanoulopoulos, E. Falco, P. Gandhi, S. Kaspi, D. Latham, P. Lira, C. McCully, H. Netzer, M. Uemura, The origin of UV-optical variability in AGN and test of disc models: XMM-Newton and ground-based observations of NGC 4395. Astron. Nachr. **337**, 500 (2016). https://doi.org/10.1002/asna.201612337

I.M. McHardy, S.D. Connolly, K. Horne, E.M. Cackett, J. Gelbord, B.M. Peterson, M. Pahari, N. Gehrels, M. Goad, P. Lira, P. Arevalo, R.D. Baldi, N. Brandt, E. Breedt, H. Chand, G. Dewangan, C. Done, M. Elvis, D. Emmanoulopoulos, M.M. Fausnaugh, S. Kaspi, C.S. Kochanek, K. Korista, I.E. Papadakis, A.R. Rao, P. Uttley, M. Vestergaard, M.J. Ward, X-ray/UV/optical variability of NGC 4593 with Swift: Reprocessing of X-rays by an extended reprocessor. Mon. Not. R. Astron. Soc. **480**, 2881–2897 (2018). https://doi.org/10.1093/mnras/sty1983

C.W. Morgan, C.S. Kochanek, N.D. Morgan, E.E. Falco, The quasar accretion disk size-black hole mass relation. Astrophys. J. **712**, 1129–1136 (2010). https://doi.org/10.1088/0004-637X/712/2/1129

T.J. Moriya, M. Tanaka, T. Morokuma, K. Ohsuga, Superluminous transients at AGN centers from interaction between black hole disk winds and broad-line region clouds. Astrophys. J. Lett. **843**(2), 19 (2017). https://doi.org/10.3847/2041-8213/aa7af3

S.E. Motta, T.M. Belloni, L. Stella, T. Muñoz-Darias, R. Fender, Precise mass and spin measurements for a stellar-mass black hole through X-ray timing: The case of GRO J1655-40. Mon. Not. R. Astron. Soc. **437**, 2554–2565 (2014). https://doi.org/10.1093/mnras/stt2068

A. Mummery, S.A. Balbus, The spectral evolution of disc dominated tidal disruption events. Mon. Not. R. Astron. Soc. **492**(4), 5655–5674 (2020). https://doi.org/10.1093/mnras/staa192

H. Netzer, D. Lutz, M. Schweitzer, A. Contursi, E. Sturm, L.J. Tacconi, S. Veilleux, D.-C. Kim, D. Rupke, A.J. Baker, K. Dasyra, J. Mazzarella, S. Lord, Spitzer Quasar and ULIRG Evolution Study (QUEST). II. The spectral energy distributions of Palomar-Green quasars. Astrophys. J. **666**(2), 806–816 (2007). https://doi.org/10.1086/520716

H. Nussbaumer, P.J. Storey, P.J. Storey, Forbidden emission lines of Fe VII. Astron. Astrophys. **113**, 21–26 (1982). http://adsabs.harvard.edu/abs/1982A&A...113...21N

K. O'Brien, K. Horne, R.I. Hynes, W. Chen, C.A. Haswell, M.D. Still, Echoes in X-ray binaries. Mon. Not. R. Astron. Soc. **334**, 426–434 (2002). https://doi.org/10.1046/j.1365-8711.2002.05530.x

D.E. Osterbrock, *Astrophysics of Gaseous Nebulae and Active Galactic Nuclei* (1989)

L. Palaversa, S. Gezari, B. Sesar, J.S. Stuart, P. Wozniak, B. Holl, Ž. Ivezić, Revealing the nature of extreme coronal-line emitter SDSS J095209.56+214313.3. Astrophys. J. **819**, 151 (2016). https://doi.org/10.3847/0004-637X/819/2/151

D.R. Pasham, S. van Velzen, Discovery of a time lag between the soft X-ray and radio emission of the tidal disruption flare ASASSN-14li: Evidence for linear disk-jet coupling. Astrophys. J. **856**(1), 1 (2018). https://doi.org/10.3847/1538-4357/aab361

D.R. Pasham, S.B. Cenko, A. Sadowski, J. Guillochon, N.C. Stone, S. van Velzen, J.K. Cannizzo, Optical/UV-to-X-ray echoes from the tidal disruption flare ASASSN-14li. Astrophys. J. Lett. **837**, 30 (2017). https://doi.org/10.3847/2041-8213/aa6003

D.R. Pasham, R.A. Remillard, P.C. Fragile, A. Franchini, N.C. Stone, G. Lodato, J. Homan, D. Chakrabarty, F.K. Baganoff, J.F. Steiner, E.R. Coughlin, N.R. Pasham, A loud quasi-periodic oscillation after a star is disrupted by a massive black hole. Science **363**(6426), 531–534 (2019). https://doi.org/10.1126/science.aar7480

B.M. Peterson, L. Ferrarese, K.M. Gilbert, S. Kaspi, M.A. Malkan, D. Maoz, D. Merritt, H. Netzer, C.A. Onken, R.W. Pogge, Central masses and broad-line region sizes of active galactic nuclei. II. A homogeneous analysis of a large reverberation-mapping database. Astrophys. J. **613**(2), 682–699 (2004). https://doi.org/10.1086/423269

T. Piran, G. Svirski, J. Krolik, R.M. Cheng, H. Shiokawa, Disk formation versus disk accretion - What powers tidal disruption events? Astrophys. J. **806**, 164 (2015). https://doi.org/10.1088/0004-637X/806/2/164

J.C. Raymond, W.P. Blair, K.S. Long, Detection of ultraviolet emission lines in SN 1006 with the Hopkins ultraviolet telescope. Astrophys. J. Lett. **454**, 31 (1995). https://doi.org/10.1086/309772

R.C. Reis, J.M. Miller, M.T. Reynolds, K. Gültekin, D. Maitra, A.L. King, T.E. Strohmayer, A 200-second quasi-periodicity after the tidal disruption of a star by a dormant black hole. Science **337**, 949 (2012). https://doi.org/10.1126/science.1223940

R.A. Remillard, G.J. Sobczak, M.P. Muno, J.E. McClintock, Characterizing the quasi-periodic oscillation behavior of the X-ray nova XTE J1550-564. Astrophys. J. **564**, 962–973 (2002). https://doi.org/10.1086/324276

C.S. Reynolds, M.A. Nowak, Fluorescent iron lines as a probe of astrophysical black hole systems. Phys. Rep. **377**, 389–466 (2003). https://doi.org/10.1016/S0370-1573(02)00584-7

C. Romero-Cañizales, S. Mattila, A. Alberdi, M.A. Pérez-Torres, E. Kankare, S.D. Ryder, The core-collapse supernova rate in Arp 299 revisited. Mon. Not. R. Astron. Soc. **415**(3), 2688–2698 (2011). https://doi.org/10.1111/j.1365-2966.2011.18886.x

N. Roth, E.M. Rossi, J. Krolik, T. Piran, B. Mockler, D. Kasen, Radiative emission mechanisms. Space Sci. Rev. **216**(7), 114 (2020). https://doi.org/10.1007/s11214-020-00735-1

N.I. Shakura, R.A. Sunyaev, Black holes in binary systems. Observational appearance. Astron. Astrophys. **24**, 337–355 (1973). http://adsabs.harvard.edu/abs/1973A%26A....24..337S

R. Stein, S. van Velzen, M. Kowalski, A. Franckowiak, S. Gezari, J.C.A. Miller-Jones, S. Frederick, I. Sfaradi, M.F. Bietenholz, A. Horesh, R. Fender, S. Garrappa, T. Ahumada, I. Andreoni, J. Belicki, E.C. Bellm, M. Böttcher, V. Brinnel, R. Burruss, S.B. Cenko, M.W. Coughlin, V. Cunningham, A. Drake, G.R. Farrar, M. Feeney, R.J. Foley, A. Gal-Yam, V.Z. Golkhou, A. Goobar, M.J. Graham, E. Hammerstein, G. Helou, T. Hung, M.M. Kasliwal, C.D. Kilpatrick, A.K.H. Kong, T. Kupfer, R.R. Laher, A.A. Mahabal, F.J. Masci, J. Necker, J. Nordin, D.A. Perley, M. Rigault, S. Reusch, H. Rodriguez, C. Rojas-Bravo, B. Rusholme, D.L. Shupe, L.P. Singer, J. Sollerman, M.T. Soumagnac, D. Stern, K. Taggart, J. van Santen, C. Ward, P. Woudt, Y. Yao, A high-energy neutrino coincident with a tidal disruption event. arXiv e-prints, 2005–05340 (2020). https://doi.org/10.1038/s41550-020-01295-8

N.C. Stone, A. Generozov, E. Vasiliev, B.D. Metzger, The delay time distribution of tidal disruption flares. Mon. Not. R. Astron. Soc. **480**, 5060–5077 (2018). https://doi.org/10.1093/mnras/sty2045

L. Sun, N. Jiang, T. Wang, H. Zhou, L. Dou, C. Yang, X. Pan, Z. Sheng, Z. Zhong, L. Yan, G. Li, A mid-infrared flare in the active galaxy MCG-02-04-026: Dust echo of a nuclear transient event. Astrophys. J. **898**(2), 129 (2020). https://doi.org/10.3847/1538-4357/ab9f2c

C. Tadhunter, R. Spence, M. Rose, J. Mullaney, P. Crowther, A tidal disruption event in the nearby ultra-luminous infrared galaxy F01004-2237. Nat. Astron. **1**, 0061 (2017). https://doi.org/10.1038/s41550-017-0061

C. Taylor, C.S. Reynolds, Exploring the effects of disk thickness on the black hole reflection spectrum. Astrophys. J. **855**(2), 120 (2018). https://doi.org/10.3847/1538-4357/aaad63

L.L. Thomsen, J. Lixin Dai, E. Ramirez-Ruiz, E. Kara, C. Reynolds, X-ray fluorescence from super-Eddington accreting black holes. Astrophys. J. Lett. **884**(1), 21 (2019). https://doi.org/10.3847/2041-8213/ab4518

P. Uttley, E.M. Cackett, A.C. Fabian, E. Kara, D.R. Wilkins, X-ray reverberation around accreting black holes. Astron. Astrophys. Rev. **22**, 72 (2014). https://doi.org/10.1007/s00159-014-0072-0

S. van Velzen, V. Gorjian, J. Krolik, A. Mendez, Dust reprocessing of stellar tidal disruption flares. Spitzer Proposal (2015). http://adsabs.harvard.edu/abs/2015sptz.prop12046V

S. van Velzen, A.J. Mendez, J.H. Krolik, V. Gorjian, Discovery of transient infrared emission from dust heated by stellar tidal disruption flares. Astrophys. J. **829**, 19 (2016). https://doi.org/10.3847/0004-637X/829/1/19

S. van Velzen, G.C. Bower, B.D. Metzger, Science with an ngVLA: Tidal disruption events, in *Science with a Next Generation Very Large Array*, ed. by E. Murphy. Astron. Soc. Pac. Conference Series, vol. 517 (2018), p. 737. http://adsabs.harvard.edu/abs/2018ASPC..517..737V

S. van Velzen, N.C. Stone, B.D. Metzger, S. Gezari, T.M. Brown, A.S. Fruchter, Late-time UV observations of tidal disruption flares reveal unobscured, compact accretion disks. Astrophys. J. **878**(2), 82 (2019). https://doi.org/10.3847/1538-4357/ab1844

S. van Velzen, T.W.-S. Holoien, F. Onori, T. Hung, I. Arcavi, Optical-ultraviolet tidal disruption events. Space Sci. Rev. **216**(8), 124 (2020). https://doi.org/10.1007/s11214-020-00753-z

S. van Velzen, S. Gezari, E. Hammerstein, N. Roth, S. Frederick, C. Ward, T. Hung, S.B. Cenko, R. Stein, D.A. Perley, K. Taggart, R.J. Foley, J. Sollerman, N. Blagorodnova, I. Andreoni, E.C. Bellm, V. Brinnel, K. De, R. Dekany, M. Feeney, C. Fremling, M. Giomi, V.Z. Golkhou, M.J. Graham, A.Y.Q. Ho, M.M. Kasliwal, C.D. Kilpatrick, S.R. Kulkarni, T. Kupfer, R.R. Laher, A. Mahabal, F.J. Masci, A.A. Miller, J. Nordin, R. Riddle, B. Rusholme, J. van Santen, Y. Sharma, D.L. Shupe, M.T. Soumagnac, Seventeen Tidal Disruption Events from the First Half of ZTF Survey Observations: Entering a New Era of Population Studies. Astrophys. J. **908**(1), 4 (2021). https://doi.org/10.3847/1538-4357/abc258

S. Vaughan, P. Uttley, Where are the X-ray quasi-periodic oscillations in active galaxies? Mon. Not. R. Astron. Soc. **362**, 235–244 (2005). https://doi.org/10.1111/j.1365-2966.2005.09296.x

T.-G. Wang, H.-Y. Zhou, L.-F. Wang, H.-L. Lu, D. Xu, Transient superstrong coronal lines and broad bumps in the galaxy SDSS J074820.67+471214.3. Astrophys. J. **740**(2), 85 (2011). https://doi.org/10.1088/0004-637X/740/2/85

T.-G. Wang, H.-Y. Zhou, S. Komossa, H.-Y. Wang, W. Yuan, C. Yang, Extreme coronal line emitters: Tidal disruption of stars by massive black holes in galactic nuclei? Astrophys. J. **749**, 115 (2012). https://doi.org/10.1088/0004-637X/749/2/115

T. Wang, L. Yan, L. Dou, N. Jiang, Z. Sheng, C. Yang, Long-term decline of the mid-infrared emission of normal galaxies: Dust echo of tidal disruption flare? Mon. Not. R. Astron. Soc. **477**, 2943–2965 (2018). https://doi.org/10.1093/mnras/sty465

E. Waxman, B.T. Draine, Dust sublimation by gamma-ray bursts and its implications. Astrophys. J. **537**, 796–802 (2000). https://doi.org/10.1086/309053

J.C. Weingartner, B.T. Draine, Dust grain-size distributions and extinction in the Milky Way, Large Magellanic Cloud, and Small Magellanic Cloud. Astrophys. J. **548**, 296–309 (2001). https://doi.org/10.1086/318651

T. Wevers, D.R. Pasham, S. van Velzen, G. Leloudas, S. Schulze, J.C.A. Miller-Jones, P.G. Jonker, M. Gromadzki, E. Kankare, S.T. Hodgkin, Ł. Wyrzykowski, Z. Kostrzewa-Rutkowska, S. Moran, M. Berton, K. Maguire, F. Onori, S. Mattila, M. Nicholl, Evidence for rapid disc formation and reprocessing in the X-ray bright tidal disruption event candidate AT 2018fyk. Mon. Not. R. Astron. Soc. **488**(4), 4816–4830 (2019). https://doi.org/10.1093/mnras/stz1976

T. Wevers, D.R. Pasham, S. van Velzen, J.C.A. Miller-Jones, P. Uttley, K. Gendreau, R. Remillard, Z. Arzoumanian, M. Loewenstein, A. Chiti, Rapid accretion state transitions following the tidal disruption event AT2018fyk. arXiv e-prints, 2101–04692 (2021). https://doi.org/10.3847/1538-4357/abf5e2

R.J. White, B.M. Peterson, Comments on cross-correlation methodology in variability studies of active galactic nuclei. Publ. Astron. Soc. Pac. **106**, 879 (1994). https://doi.org/10.1086/133456

D.R. Wilkins, E.M. Cackett, A.C. Fabian, C.S. Reynolds, Towards modelling X-ray reverberation in AGN: Piecing together the extended corona. Mon. Not. R. Astron. Soc. **458**(1), 200–225 (2016). https://doi.org/10.1093/mnras/stw276

E.L. Wright, P.R.M. Eisenhardt, A.K. Mainzer, M.E. Ressler, R.M. Cutri, T. Jarrett, J.D. Kirkpatrick, D. Padgett, R.S. McMillan, M. Skrutskie, S.A. Stanford, M. Cohen, R.G. Walker, J.C. Mather, D. Leisawitz, T.N. Gautier III, I. McLean, D. Benford, C.J. Lonsdale, A. Blain, B. Mendez, W.R. Irace, V. Duval, F. Liu, D. Royer, I. Heinrichsen, J. Howard, M. Shannon, M. Kendall, A.L. Walsh, M. Larsen, J.G. Cardon, S. Schick, M. Schwalm, M. Abid, B. Fabinsky, L. Naes, C.-W. Tsai, The Wide-field Infrared Survey Explorer (WISE): Mission description and initial on-orbit performance. Astron. J. **140**, 1868–1881 (2010). https://doi.org/10.1088/0004-6256/140/6/1868

L. Yan, T. Wang, N. Jiang, D. Stern, L. Dou, C. Fremling, M.J. Graham, A.J. Drake, C. Yang, K. Burdge, M.M. Kasliwal, Rapid turn-on of type-1 AGN in a quiescent early-type galaxy SDSS1115+0544. Astrophys. J. **874**(1), 44 (2019). https://doi.org/10.3847/1538-4357/ab074b

C.-W. Yang, T.-G. Wang, G. Ferland, W. Yuan, H.-Y. Zhou, P. Jiang, Long-term spectral evolution of tidal disruption candidates selected by strong coronal lines. Astrophys. J. **774**, 46 (2013). https://doi.org/10.1088/0004-637X/774/1/46

Q. Yang, Y. Shen, X. Liu, X.-B. Wu, L. Jiang, J. Shangguan, M.J. Graham, S. Yao, An unusual mid-infrared flare in a type 2 AGN: An obscured turning-on AGN or tidal disruption event? Astrophys. J. **885**(2), 110 (2019). https://doi.org/10.3847/1538-4357/ab481a

W. Yuan, C. Zhang, H. Feng, S.N. Zhang, Z.X. Ling, D. Zhao, J. Deng, Y. Qiu, J.P. Osborne, P. O'Brien, R. Willingale, J. Lapington, G.W. Fraser, the Einstein Probe team, Einstein Probe - A small mission to monitor and explore the dynamic X-ray Universe. arXiv e-prints, 1506–07735 (2015)

A. Zabludoff, I. Arcavi, S. La Massa, H.B. Perets, B. Trakhtenbrot, B.A. Zauderer, K. Auchettl, J.L. Dai, K.D. French, T. Hung, E. Kara, G. Lodato, W.P. Maksym, Y. Qin, E. Ramirez-Ruiz, N. Roth, J.C. Runnoe, T. Wevers, Distinguishing tidal disruption events from impostors. arXiv e-prints, 2103–12150 (2021). https://doi.org/10.1007/s11214-021-00829-4

B.A. Zauderer, E. Berger, A.M. Soderberg, A. Loeb, R. Narayan, D.A. Frail, G.R. Petitpas, A. Brunthaler, R. Chornock, J.M. Carpenter, G.G. Pooley, K. Mooley, S.R. Kulkarni, R. Margutti, D.B. Fox, E. Nakar, N.A. Patel, N.H. Volgenau, T.L. Culverhouse, M.F. Bietenholz, M.P. Rupen, W. Max-Moerbeck, A.C.S. Readhead, J. Richards, M. Shepherd, S. Storm, C.L.H. Hull, Birth of a relativistic outflow in the unusual γ-ray transient Swift J164449.3+573451. Nature **476**, 425–428 (2011). https://doi.org/10.1038/nature10366

A. Zoghbi, A.C. Fabian, C.S. Reynolds, E.M. Cackett, Relativistic iron K X-ray reverberation in NGC 4151. Mon. Not. R. Astron. Soc. **422**(1), 129–134 (2012). https://doi.org/10.1111/j.1365-2966.2012.20587.x

Space Sci Rev (2020) 216:39
https://doi.org/10.1007/s11214-020-00661-2

Tidal Disruptions of White Dwarfs: Theoretical Models and Observational Prospects

Kate Maguire[1] · **Michael Eracleous**[2] · **Peter G. Jonker**[3,4] · **Morgan MacLeod**[5] · **Stephan Rosswog**[6]

Received: 17 May 2019 / Accepted: 16 March 2020 / Published online: 25 March 2020

Abstract White dwarf stars that enter the tidal radius of black holes with masses $\lesssim 10^5\ M_\odot$ are doomed to be ripped apart by tidal forces. Black holes in this mass range between stellar black holes and supermassive black holes have not been conclusively identified so the detection of a tidal disruption of a white dwarf would provide clear evidence for the existence of intermediate-mass black holes. In this review, we present a theoretical and observational overview of the transient events that result from the tidal disruptions of white dwarfs by intermediate-mass black holes. This includes discussion of the latest simulations and pre-

The Tidal Disruption of Stars by Massive Black Holes
Edited by Peter G. Jonker, Sterl Phinney, Elena Maria Rossi, Sjoert van Velzen, Iair Arcavi and Maurizio Falanga

✉ K. Maguire
kate.maguire@tcd.ie

M. Eracleous
mxe17@psu.edu

P.G. Jonker
p.jonker@astro.ru.nl

M. MacLeod
morgan.macleod@cfa.harvard.edu

S. Rosswog
stephan.rosswog@astro.su.se

1 School of Physics, Trinity College Dublin, The University of Dublin, College Green, Dublin 2, Ireland

2 Department of Astronomy & Astrophysics and Institute for Gravitation and the Cosmos, The Pennsylvania State University, 525 Davey Lab, University Park, PA 16802, USA

3 SRON, Netherlands Institute for Space Research, Sorbonnelaan 2, 3584 CA, Utrecht, The Netherlands

4 Department of Astrophysics/IMAPP, Radboud University Nijmegen, P.O. Box 9010, 6500 GL, Nijmegen, The Netherlands

5 Harvard-Smithsonian Center for Astrophysics, 60 Garden Street, MS-51, Cambridge, MA 02138, USA

dicted properties, the results of observational searches, as well as a summary of the potential for gravitational wave emission to be detected with upcoming missions.

Keywords White dwarf · Intermediate-mass black hole

1 Introduction

White dwarfs are the final products of the evolution of low-mass stars. They result from stars between ~ 0.1–$8\ \mathrm{M}_\odot$ that are not massive enough to ignite further burning stages of their helium or carbon-oxygen cores. Thus, these cores cool and evolve into very dense stars ($\sim 10^6\ \mathrm{g\,cm^{-3}}$) whose masses are determined by their baryon content, but the pressure necessary to maintain hydrostatic equilibrium is provided by the fermionic "degeneracy pressure" of their electrons. While under normal circumstances this is the end of their nuclear evolution, at least in principle, much more nuclear energy can be gained by transforming the white dwarfs (7.08 MeV/nucleon for helium, 7.68 MeV/nucleon for carbon, and 7.98 MeV/nucleon for oxygen) into iron group elements (8.79 MeV/nucleon): ~ 1 MeV/nucleon or $\sim 2 \times 10^{51}\ \mathrm{erg/M_\odot}$ can still be released. Perhaps the most spectacular way to trigger this release, is to send the white dwarf close a black hole (BH) and explosively ignite it by tidal compression (Luminet and Pichon 1989; Rosswog et al. 2009).

White dwarfs have typical sizes comparable to the Earth's, but following an inverted mass-radius relation they are smaller the more massive they are, as depicted in Fig. 1. Using a convenient analytic expression derived by Nauenberg (1972, see also Appendix A of Even and Tohline 2009), we can cast the mass-radius relation of a Carbon-Oxygen white dwarf as

$$R_{\mathrm{wd}} = 7.80 \times 10^8 \left(\frac{M_{\mathrm{wd}}}{M_{\mathrm{Ch}}}\right)^{-1/3} \left[1 - \left(\frac{M_{\mathrm{wd}}}{M_{\mathrm{Ch}}}\right)^{4/3}\right]^{1/2} \mathrm{cm} \tag{1}$$

where $M_{\mathrm{Ch}} = 1.435\ \mathrm{M}_\odot$ is the Chandrasekhar mass for that composition. This mass-radius relation, plotted in Fig. 1, is the reason why the average density of white dwarfs rises sharply with increasing mass, as shown by the red curve in Fig. 1.

BHs are usually classified according to their masses. The existence of stellar mass BHs ($\lesssim 60\ \mathrm{M}_\odot$) and supermassive BHs ($\sim 10^6$–$10^{10}\ \mathrm{M}_\odot$; SMBHs) is well established, but members with masses between these two regimes, so-called intermediate-mass BHs (IMBHs), have not been conclusively identified. They are thought to form via a variety of physical mechanisms, e.g., runaway stellar collisions in a dense star cluster and collapse of a massive gas cloud (see, for example, Portegies Zwart and McMillan 2002; Begelman et al. 2008; Mayer et al. 2010).

To be torn apart by the tidal forces of a BH, a white dwarf must come closer than the tidal radius

$$R_{\mathrm{t}} \simeq R_{\mathrm{wd}} \left(\frac{M_\bullet}{M_{\mathrm{wd}}}\right)^{1/3}. \tag{2}$$

For an accretion disk to form around the BH, the disruption needs to occur outside the innermost stable circular orbit at $R_{\mathrm{ISCO,S}} = 6GM_\bullet/c^2 \equiv 6R_{\mathrm{g}}$ for a Schwarzschild BH or $R_{\mathrm{ISCO,eK}} = R_{\mathrm{g}}$ for a maximally rotating Kerr BH and a prograde orbit (R_{g} is conventionally

[6] Department of Astronomy & Oskar Klein Centre, Stockholm University, AlbaNova, Roslagstullbacken 21, 10691 Stockholm, Sweden

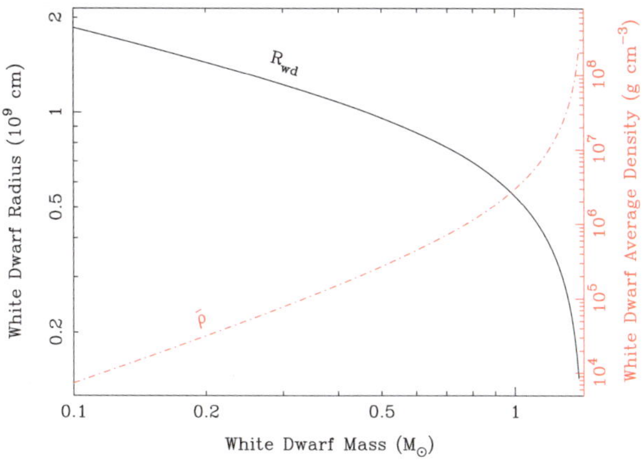

Fig. 1 Radius of a Carbon-Oxygen white dwarf as a function of its mass (in $M_\odot$). The mass-radius relation of equation (1) is adopted here, which is due to Nauenberg (1972). For reference, the red dot-dashed line (labelled $\bar{\rho}$ with corresponding scale on the right) shows the average density of the white dwarf as a function of its mass

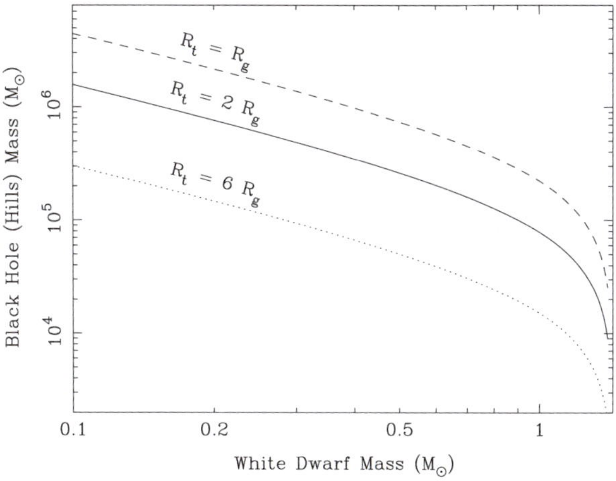

Fig. 2 The variation of the BH Hills mass with white dwarf mass. The three curves represent the boundaries below which the white dwarf is disrupted outside of the innermost stable circular orbit of a non-rotating BH (dotted line, labelled $R_t = 6R_g$), the event horizon of a non-rotating BH (solid line, labelled $R_t = 2R_g$) and the event horizon of a maximally-rotating BH, which is also its innermost stable circular orbit (dashed line, labelled $R_t = R_g$). The curves are obtained by equating the tidal disruption radius with the appropriate horizon radius and rearranging the equation to express the BH mass in terms of the white dwarf mass

referred to as the gravitational radius). If the tidal radius lies inside the event horizon ($R_{H,S} = 2R_g$ for the Schwarzschild and $R_{H,eK} = R_g$ for the extreme Kerr case), no disruption occurs and the white dwarf is "swallowed" whole. These thresholds are plotted as a function of the white white dwarf mass in Fig. 2. Since the tidal radius is proportional to $M_\bullet^{1/3}$, but the other radii are proportional to $M_\bullet$, there is a (spin-dependent) upper limit on the BH mass beyond which tidal disruption events (TDEs) are impossible. For a non-rotating black hole and a white dwarf of typical mass ($\approx 0.6\ M_\odot$; Kepler et al. 2007), this limiting BH mass is $\sim 2 \times 10^5\ M_\odot$, therefore only IMBHs can disrupt white dwarfs, see Fig. 2.

The strength of a tidal encounter is often quantified by the parameter $\beta \equiv R_t/R_p$, i.e., the inverse ratio of the pericentre distance to the tidal disruption radius. In other words, a strong encounter is one where the white dwarf penetrates deeply into the tidal radius. Figure 3

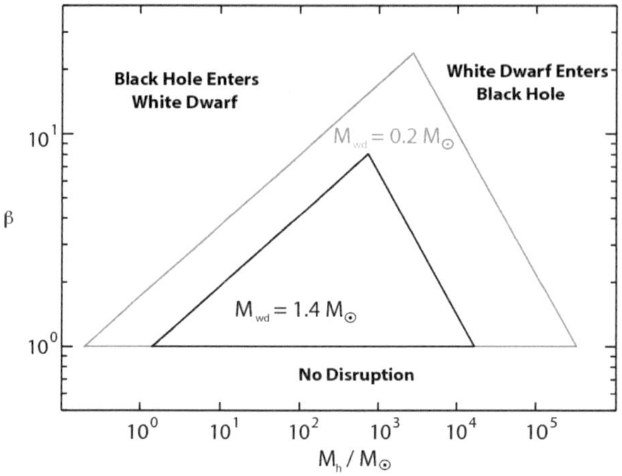

Fig. 3 The parameter space defined by the strength of the tidal encounter ($\beta \equiv R_t/R_p$) and the BH mass, M_h (assuming a non-rotating BH). This particular figure shows the interesting regions of the diagram for a 1.4 $M_\odot$ and a 0.2 $M_\odot$ white dwarf and was taken from Rosswog et al. (2009); an earlier version illustrating the same regions for a 0.6 $M_\odot$ white dwarf was shown by Luminet and Pichon (1989). In the lower region of the diagram, where $\beta < 1$, there is no disruption. If the combination of β and M_h falls within the triangle corresponding to the white dwarf's mass, the white dwarf is disrupted outside the event horizon of the BH and the post-disruption debris eventually returns to pericentre. In the upper right corner of the diagram the disruption occurs within the event horizon of the BH. In the upper left corner of the diagram, the black hole event horizon is smaller than the white dwarf, therefore the BH effectively enters the white dwarf during a strong encounter

illustrates the outcome of a tidal encounter in the β–$M_\bullet$ plane for two different white dwarf masses. For $\beta < 1$ there is no disruption since the white dwarf does not cross the tidal disruption radius. For a given white dwarf mass, the triangle encompasses the region of this parameter space where the debris eventually returns to pericentre after disruption; some of this debris can be subsequently accreted by the BH. For strong encounters outside of this triangle, the BH either enters the white dwarf or it "swallows" the white dwarf whole.

The most promising TDE locations depend on the type of disruption. TDEs of main sequence stars by SMBHs are associated with the cores of galaxies where SMBHs are generally found. However, TDEs involving white dwarfs and IMBHs may occur in a more diverse range of environments since IMBHs are predicted to occur in dwarf galaxies (e.g., Reines et al. 2013), globular clusters (Jonker et al. 2012) and in hyper-compact stellar clusters (Merritt et al. 2009), although the predicted rates for these different environments differ enormously. The discovery of a transient caused by the tidal disruption of a white dwarf by a BH would be clear evidence for the existence of an IMBH. Such a discovery would be extremely interesting for a variety of reasons:

1. IMBHs are routinely invoked as "seeds" in scenarios for the growth of SMBHs and their co-evolution with their host galaxies (see, for example, the review by Volonteri 2012). Seed masses of at least 10^3 $M_\odot$ are required in order to explain the large BH masses found in the highest-redshift quasars (e.g. Mortlock et al. 2011; Fan 2006; Bañados et al. 2018, these black holes must have grown to $\sim 10^9$ $M_\odot$ in ~ 700 Myr by Eddington-limited accretion).
2. Dwarf galaxies (with stellar masses $M_* \lesssim 10^9$ $M_\odot$) are the expected hosts of IMBHs based on an extrapolation of the relation between BH mass and host galaxy stellar

spheroid mass (e.g., McConnell and Ma 2013) and on the identification of active galactic nuclei (AGNs) in many dwarf galaxies (e.g., Reines et al. 2013; Moran et al. 2014; Baldassare et al. 2018). But the occupation fraction, which is critical in setting the white dwarf TDE rate is unknown. Circumstantial arguments by Silk (2017) suggest that an occupation fraction near unity would resolve many of the open questions in dwarf galaxy evolution.

3. Observations of flares from white dwarf disruptions would afford useful tests of models for the accretion of Hydrogen deficient matter at very high rates relative to the Eddington limit (e.g., Dai et al. 2018), the formation of jets (e.g., Krolik and Piran 2012), and the transition from super- to sub-Eddington flows. Since the BH mass is relatively low in this case, the Eddington ratio (accretion rate divided by the Eddington rate) will be much higher than in the disruption of normal stars by more massive BHs and the accretion flow will evolve on a considerably shorter time scale.
4. The event *rates* afford important tests of models for the stellar populations and dynamics in the vicinity of the BH. Since the expected rate estimates depend on a number of assumptions, as we detail later in this chapter, one could potentially eliminate entire families of models based on measurements of the event rates, which would represent substantial progress.
5. White dwarfs in bound orbits spiraling into IMBHs can produce a detectable gravitational wave signal before disruption (e.g., Sesana et al. 2008). From this signal we can potentially infer many of the fundamental properties of the system, including the masses of the white dwarf and IMBH, which is extremely useful for testing models of the post-disruption accretion flow (discussed further in Sect. 7.1, below, and in other chapters in this volume).

One can appreciate the importance of general relativistic effects during such disruptions by expressing the tidal disruption radius in units of the gravitational radius. Combining equation (2) with the definition of β and $R_{\rm g}$ one can write

$$\frac{R_{\rm p}}{R_{\rm g}} = \frac{R_{\rm t}}{\beta R_{\rm g}} \simeq \frac{80}{\beta}\left(\frac{R_{\rm WD}}{10^9\ \rm cm}\right)\left(\frac{M_{\rm wd}}{0.6\ \rm M_\odot}\right)^{-1/3}\left(\frac{M_\bullet}{10^3\ \rm M_\odot}\right)^{-2/3}. \tag{3}$$

Thus, a strong encounter of a typical white dwarf with a $10^4\ \rm M_\odot$ BH can have $R_{\rm p}/R_{\rm g} < 10$, i.e., the white dwarf gets within 5 Schwarzschild radii of the BH during its initial fly-by. Such an encounter is relativistic.

The aim of this chapter is to discuss the theoretical and observational properties of the transients that result from the tidal disruptions of white dwarfs by IMBHs. In Sect. 2, the underlying theory and simulations necessary for analysing these interactions are highlighted. In Sect. 3, the predicted nucleosynthetic yields of systems that result in the detonation of the white dwarf are discussed, while in Sect. 4 the expected optical and UV signatures of white-dwarf tidal-disruption events (TDEs) are highlighted. In Sect. 5, observational searches for the tidal disruptions of white dwarfs and candidate events are described. The intrinsic and predicted rates of white dwarf TDEs are detailed in Sect. 6. A summary of the expected gravitational wave emission from the tidal disruptions of white dwarfs by IMBHs and the cosmic ray acceleration in ensuing jets is provided in Sect. 7.

2 Theory and Simulations

Some aspects of TDEs can be described to good accuracy in a purely analytical way (Luminet and Marck 1985; Luminet and Carter 1986; Rees 1988; Stone et al. 2013), but in reality a TDE is a highly non-linear interaction between gravity (both self-gravity of the star and the "background" gravity from the BH), gas dynamics, radiation, potentially magnetic fields and – in extreme cases – thermonuclear reactions. Therefore, most astrophysically relevant questions need to be addressed via careful numerical simulations.

2.1 Geometrical Challenges

The lengths and timescales involved make TDEs formidable numerical challenges and prevent, in particular for most realistic cases, the "just-do-it" approach, where the Einstein and hydrodynamics equations are solved together in a consistent way.

The initial star far from the BH is in hydrostatic equilibrium between gas pressure gradients and self-gravity. The acceleration from self-gravity is $a_{\rm sg} \sim GM_{\rm wd}/R_{\rm wd}^2$, while the tidal acceleration at a distance d from the BH is $a_{\rm tid} \sim GM_\bullet R_{\rm wd}/d^3$, so that their ratio is

$$\frac{a_{\rm tid}}{a_{\rm sg}} \sim \left(\frac{M_\bullet}{M_{\rm wd}}\right)\left(\frac{R_{\rm wd}}{d}\right)^3 \equiv \left(\frac{R_t}{d}\right)^3, \tag{4}$$

which implies, not too surprisingly, a ratio of unity at the tidal radius given by Eq. (2). Equation (4) shows further that for the tidal acceleration to be a $< 1\%$ perturbation of the stellar equilibrium, the initial separation, d_0, for starting a numerical simulation needs to be at least

$$d_0 \sim 5 R_t, \tag{5}$$

making the initial star a tiny fraction of the simulation volume:

$$\left(\frac{R_{\rm wd}}{d_0}\right)^3 \sim 10^{-8}\left(\frac{M_{\rm wd}}{1\,{\rm M}_\odot}\right)\left(\frac{10^6\,{\rm M}_\odot}{M_\bullet}\right). \tag{6}$$

This is only dependent on the mass ratio and actually independent of the radius of the star that is disrupted. Therefore, the challenge is alleviated for an IMBH-white dwarf disruption compared to one where a main sequence star is disrupted by a SMBH. Equation (6) has repercussions for simulating TDEs. While the small volume covered by the initial star does not pose any challenge for a Lagrangian method like smooth particle hydrodynamics (SPH; Monaghan 2005; Rosswog 2009; Springel 2010; Rosswog 2015), it is a serious hurdle for Eulerian methods, where vacuum has to be treated as a low-density background gas that must be evolved during the simulation. Therefore, such simulations are often performed in the reference frame of the stellar centre of mass, with the BH being treated as a time-varying, external force (Guillochon et al. 2009; Guillochon and Ramirez-Ruiz 2013). This reduces the computational volume and avoids excessive numerical advection error due to high velocity motion of the gas with respect to the computational grid.

2.2 Hydrodynamics and Gravity Challenges and Approaches

Additional simulation challenges come from restriction of the numerical time step. For a full space-time solution approach, the relevant signal velocity that enters the Courant-Friedrichs-Lewy stability criterion (Courant et al. 1928) is the speed of light, so that the numerical time step is restricted to

$$\Delta t < 10^{-3}\,{\rm s}\left(\frac{\Delta x}{3\times 10^7\,{\rm cm}}\right), \tag{7}$$

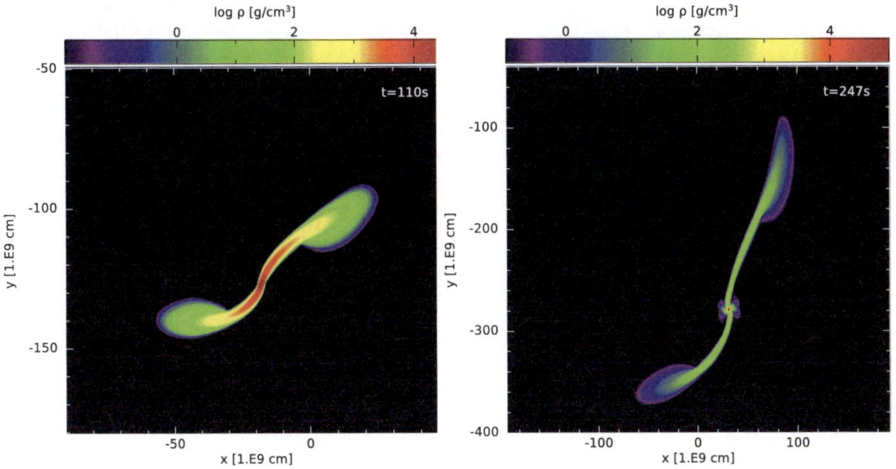

Fig. 4 Partial tidal disruption of a 0.5 $M_\odot$ white dwarf by a 1000 $M_\odot$ BH (located at the coordinate origin) illustrating a weak encounter with an impact parameter $\beta = 0.8$ (Rosswog 2019). The star is "almost completely" disrupted (left panel), but a small self-gravitating core survives within the tidal stream connecting the two tidal lobes (right panel). The simulation started at an initial separation of $5R_t$

where Δx is the resolution length and its numerical value for the scaling has been chosen so that it corresponds to resolving a typical white dwarf radius with ~ 30 resolution lengths. Self-gravity of the stellar material can play a decisive role since it, together with the BH's tidal field, determines the density structure inside the star. Since nuclear reactions are very sensitive to matter density, self-gravity is crucial for an accurate modeling of a possible explosion. For weak encounters with $\beta \lesssim 1$, also partial disruptions can occur where the star is "nearly disrupted", but its original stellar core collapses again under its own gravitational pull. Such a situation is illustrated in Fig. 4 for a 0.5 $M_\odot$ white dwarf, a 1000 $M_\odot$ BH and a penetration factor $\beta = 0.8$. Such surviving cores can have an impact on the debris properties and subsequent fallback rates (Guillochon and Ramirez-Ruiz 2013). This illustrates the serious challenge of full end-to-end simulations where a star approaches the BH from a large distance, becomes disrupted when passing the BH into a thin gas stream with a potentially self-gravitating stellar core, and returns later to the BH vicinity to finally settle into an accretion disk. Thus full space-time solution approaches are limited to the near-neighbourhood of the BH.

To date, the following strategies have been followed in numerical TDE simulations:

a) use an entirely Newtonian approach and restrict the focus to encounters that can be treated as non-relativistic with a reasonable accuracy (e.g. Guillochon and Ramirez-Ruiz 2013; Guillochon et al. 2014; Coughlin and Nixon 2015; Goicovic et al. 2019);
b) use Newtonian numerical hydrodynamics together with a pseudo-Newtonian potential for approximately capturing some relativistic effects (Rosswog et al. 2009; Tejeda and Rosswog 2013; Hayasaki et al. 2013; Bonnerot et al. 2015; Gafton et al. 2015);
c) follow a post-Newtonian approach for mildly relativistic encounters (Ayal et al. 2000, 2001; Hayasaki et al. 2016);
d) use relativistic hydrodynamics in a fixed BH space time, but neglect the self-gravity of the star (e.g. Anninos et al. 2018);
e) like d), but adding Newtonian self-gravity of the star (Kobayashi et al. 2004; Rantsiou et al. 2008);

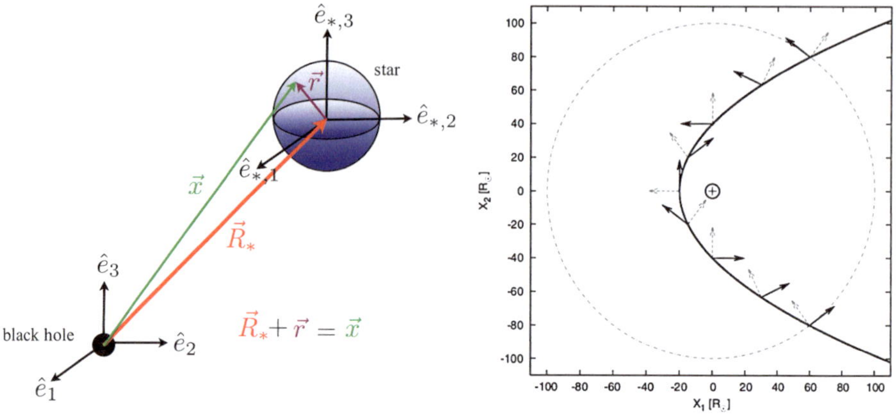

Fig. 5 Left: Reference frames used for the analysis of the tidal action of a BH on a star with quantities as discussed in Sect. 2.3. Right: Eigenvectors of the tidal tensor (shown as arrows), taken from Brassart and Luminet (2008). The circle indicates the tidal radius, the parabola the stellar orbit, the BH position is indicated by the cross. The third eigendirection (not shown) is perpendicular to the orbital plane

f) use a full numerical relativity approach by solving the Einstein equations, and restrict the attention mainly to regions near the BH (e.g. Haas et al. 2012; East 2014; Evans et al. 2015)); surviving stellar cores (as seen in Fig. 4) in partial disruptions pose a very serious challenge to such an approach;
g) use a combination of some of the above approaches, e.g. follow the gas motion far from the BH with Newtonian methods and near the BH with relativistic hydrodynamics similar to approach d) (e.g. Shiokawa et al. 2015; Sądowski et al. 2016);
h) start from a Kerr metric, but include the stellar (WD) contributions to the scalar curvature components in the metric (Anninos et al. 2019) to approximate the stellar self-gravity. This (expensive) correction is switched-off close to the black hole.
i) use the Newtonian hydrodynamics equations as the computational core, but "wrap them" into functions that contain the BH metric and its derivatives (Tejeda et al. 2017; Gafton and Rosswog 2019); this method very accurately reproduces fully relativistic approaches and includes the self-gravity of the star/its debris.

2.3 Black Hole Tides Deforming a Nearby Star

It is instructive to study how the tidal forces from a BH distort a star. As a first approximation assume that the BH can be reasonably described as a Newtonian point mass. Further assume that a coordinate system with unit vectors $\{\hat{e}_1, \hat{e}_2, \hat{e}_3\}$ is attached to the BH and another coordinate system is centred on the star with unit vectors $\{\hat{e}_{*,1}, \hat{e}_{*,2}, \hat{e}_{*,3}\}$, as sketched in Fig. 5. The center of mass of the star is displaced from the BH by $\vec{R}_*$ and we are interested in some location inside the star, displaced from the stellar center of mass by $\vec{r}$ and from the BH by $\vec{x} = \vec{R}_* + \vec{r}$. With these assumptions, the BH gravitational potential is

$$\Phi(\vec{x}) = -\frac{GM_\bullet}{x}, \tag{8}$$

where $x = |\vec{x}|$.

For a point $\vec{x} = \vec{R}_* + \vec{r}$ inside the star, Φ can be expanded around $\vec{R}_*$ into

$$\Phi(\vec{x}) = \Phi(\vec{R}_*) + (\partial_i \Phi)_{\vec{R}_*} r_i + \frac{1}{2}(\partial_{ij}\Phi)_{\vec{R}_*} r_i r_j + \frac{1}{6}(\partial_{ijk}\Phi)_{\vec{R}_*} r_i r_j r_k + O(r^4) \tag{9}$$

$$= \Phi(\vec{R}_*) + (\partial_i \Phi)_{\vec{R}_*} r_i + \Phi_{\text{tid}}(\vec{x}). \tag{10}$$

where Φ_{tid} is the tidal potential. Φ_{tid} can be further split up making use of the tidal tensor (note the sign convention)

$$\tau_{ij} = -(\partial_{ij}\Phi)_{\vec{R}_*} \tag{11}$$

and the deviation tensor

$$\Delta_{ijk} = -(\partial_{ijk}\Phi)_{\vec{R}_*}. \tag{12}$$

The deviation tensor governs the deviation of the centre of mass from a pure point mass orbit and it can be shown that

$$M_* \frac{d^2 R_{*,i}}{dt^2} = -M_* \partial_i \Phi(\vec{R}_*) + \frac{1}{2}\Delta_{ijk} \int r_j r_k \rho(\vec{r}) d^3 r + O(r^3). \tag{13}$$

Since $R_* \gg r$ the deviations from a point particle orbit are usually negligibly small. The acceleration in the reference frame of the star is then given by

$$\frac{d^2 r_i}{dt^2} = \tau_{ij}(\vec{R}_*)\, r_j \tag{14}$$

and it is, therefore, the tidal tensor τ_{ij} given by Eq. (11) that determines how the star is deformed by the BH. An eigenvector analysis of τ_{ij} delivers two negative and one positive eigenvalue, i.e. the star is compressed in two eigendirections and stretched in the third. Two eigendirections (one positive, one negative) lie in the orbital plane and change as the star approaches the BH, see Fig. 5 (taken from Brassart and Luminet (2008)). The third eigendirection, with negative eigenvalue, is perpendicular to the orbital plane. Thus, as the star approaches the BH it becomes compressed in the direction perpendicular to the orbital plane.

For relativistic BHs, an equation very similar to Eq. (14) is found, with the relativistic tidal tensor, τ_{ij}^{GR}, containing the Riemann curvature tensor. For explicit expressions in Schwarzschild space-time, see, e.g., Brassart and Luminet (2010), Cheng and Evans (2013), and Gafton et al. (2015).

2.4 Nuclear Energy Generation

Triggering dynamically important nuclear burning processes during a TDE is non-trivial since nuclear reactions are often inefficient and the compression time during the flyby is very short (Rosswog et al. 2009). An estimate of the compression time, τ_{comp}, can be made assuming the white dwarf is being squeezed with freefall velocity through a point of maximum compression ("nozzle") along its orbit:

$$\tau_{\text{comp}} \sim \frac{R_{\text{wd}}}{v_p} \sim 0.2\ \text{s} \left(\frac{M_{\text{wd}}}{0.6\ \text{M}_\odot}\right)^{-1/6} \left(\frac{R_{\text{wd}}}{10^9\ \text{cm}}\right)^{3/2} \left(\frac{M_\bullet}{10^3\ \text{M}_\odot}\right)^{-1/3}, \tag{15}$$

where the orbital velocity at periastron, v_p, is

$$v_p \sim c \left(\frac{R_g}{R_t}\right)^{1/2} \simeq 5 \times 10^9 \text{ cm s}^{-1} \left(\frac{M_{wd}}{0.6\, M_\odot}\right)^{1/6} \left(\frac{R_{wd}}{10^9 \text{cm}}\right)^{-1/2} \left(\frac{M_{BH}}{10^3\, M_\odot}\right)^{1/3}. \tag{16}$$

A simple order of magnitude estimate for the conditions needed to explode a white dwarf comes from the requirement that the nuclear energy injection time scale $\tau_{nuc} \equiv \epsilon_{nuc}/\dot{\epsilon}_{nuc}$ must be shorter than the crossing time of the star through periastron, $\tau_{passage}$, so that a substantial amount of nuclear energy can be released during the short compression time. The burning time scale also has to be substantially shorter than the dynamical time scale of the star, $\tau_{dyn} = 1/\sqrt{G\bar{\rho}_{wd}}$, where $\bar{\rho}_{wd}$ is the average white dwarf density, otherwise the star has time to react on the energy release, can expand and quench the burning. The time to pass the BH at R_t, however, is comparable to the dynamical timescale of the star,

$$\tau_{passage} \sim \frac{R_t}{v_p} \sim \frac{1}{\sqrt{GM_{wd}/R_{wd}^3}} \sim \tau_{dyn}, \tag{17}$$

so that both conditions can be summarized as $\tau_{nuc} \ll \tau_{dyn}$.

2.5 Disk Formation

The formation of an accretion disk out of the disruption debris is thought to be crucial for the electromagnetic appearance of a TDE. This process, however, is to date only incompletely understood, see Chap. "Formation of an Accretion Flow" of this book. It is, in particular, not obvious that the actual accretion rate, $\dot{M}_{acc}$, should, in all cases, follow the mass delivery/fallback rate, $\dot{M}_{fb} \propto t^{-5/3}$, that is found from simple analytical arguments and numerical simulations (Rees 1988; Phinney 1989; Evans and Kochanek 1989; Ramirez-Ruiz and Rosswog 2009; Lodato et al. 2009; Lodato and Rossi 2011; Kesden 2012; Cheng and Bogdanović 2014). For this to be true, matter must be accreted faster than it is delivered by fallback. The disk formation process depends on the relative efficiencies of circularization, viscous accretion and radiative cooling (Evans and Kochanek 1989). All of these topics currently bear a large degree of uncertainty. The self-crossing of the accretion stream can be efficient in circularizing the disruption debris into a disk-structure (of some type). It can occur for at least two reasons. First, the debris that passes the black hole has a wide range of specific energies across the stream width which translates into a "fan" spread over a range of apocentre distances (see Rees 1988; Kochanek 1994). This fan can collide and shock with the matter falling towards the black hole, as depicted in Fig. 6. Second, if strong, relativistic apsidal precession can lead to stream self-crossing with enhanced collision angles and thus efficiently circularize the debris (Bonnerot et al. 2015; Shiokawa et al. 2015; Hayasaki et al. 2016). The apsidal precession angle is approximately given by e.g. Hobson et al. (2006) and Chap. "Formation of an Accretion Flow" in this book,

$$\Delta\Phi \approx \frac{3\pi R_g}{R_p} \approx 11.5^\circ \beta \left(\frac{M_h}{10^6\, M_\odot}\right)^{2/3} \tag{18}$$

and therefore, is particularly pronounced for high-mass black holes.

Guillochon and Ramirez-Ruiz (2015) used an approximate scheme with Post-Newtonian orbital prescriptions to study the disruption of stars by spinning black holes via a Monte Carlo approach. For cases with strong relativistic effects and substantial black hole spins,

Fig. 6 The debris stream passing the black hole spreads into a "nozzle" of large apocentre distance spread. On interacting with the infalling stream a circularization shock can form. Figure from the simulations of Rosswog et al. (2009)

they found that spin-induced precession can avoid stream collisions for the times corresponding to ~ 10 orbital windings resulting in some cases in substantial "dark periods" directly after the disruptions.

They find that for cases where GR effects and precession angles are small, the streams typically self-intersect far away from the black holes. This leads to long viscous time scales that can slow down the accretion rate $\dot{M}_{\rm acc}$ compared to the fallback rate $\dot{M}_{\rm fb}$ by large factors. Their Monte Carlo studies suggest that this is mostly the case for smaller black hole masses. Cases with very long viscous time scales may only peak years after the disruption with bolometric luminosities decaying more slowly than the $t^{-5/3}$-behaviour that is often used to observationally identify TDEs. Therefore such events may have gone unnoticed.

Specifically since WDs are disrupted by IMBHs, it is currently unclear if, and in which region of the parameter space, cases with very long viscous time scales occur. This is because we have on the one hand small apsidal precession angles, see Eq. (18), but on the other hand the low black hole masses favour widely spreading "fans." This interesting question is left for future explorations.

3 Nucleosynthetic Predictions of Detonations

The tidal disruptions of white dwarfs differ from those of main sequence stars in that there can be an additional source of energy: the tidal compression during the pericentre passage can trigger thermonuclear runaway reactions (e.g. Luminet and Pichon 1989; Rosswog 2009; MacLeod et al. 2016b; Kawana et al. 2017; Tanikawa et al. 2017; Anninos et al. 2018; Tanikawa 2018a). Obviously, it is easier to ignite a He-WD than one with a CO-composition. It should be mentioned, however, that a WD below $\approx 0.46\ \rm M_\odot$ consists of a helium core engulfed in a hydrogen envelope that can extend to several core radii and therefore increase the effective tidal radius (Law-Smith et al. 2017). Alternatively tidal compression could trigger a detonation first in a He-shell on the surface of a white dwarf followed by detonation of a CO core, 'tidal double detonation' (Tanikawa 2018b).

The first three-dimensional hydrodynamic calculations involving a nuclear network were performed in 2009 (Rosswog et al. 2009) for the tidal disruption of white dwarfs by IMBHs.

Fig. 7 The mass evolution of nuclear species for simulations B3M2R09 (a 0.2 $M_\odot$ white dwarf and 10^3 $M_\odot$ BH with a β parameter of 11). Figure from Anninos et al. 2018

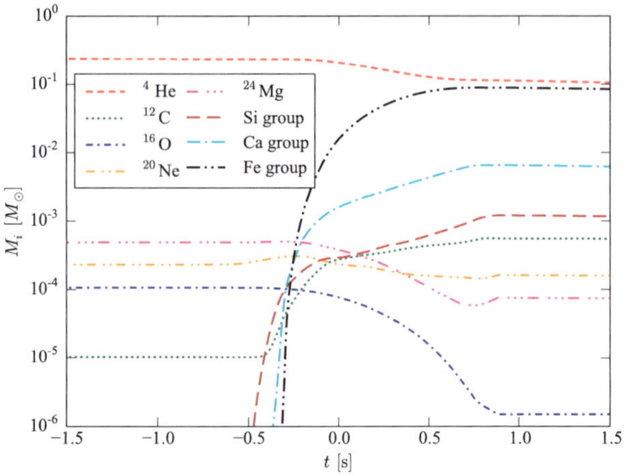

These simulations used up to 4×10^6 SPH particles and each particle solved a small nuclear reaction network (Hix et al. 1998). For deep encounters ($\beta > 3$) the released nuclear energy was found to exceed the white dwarf gravitational binding energy and to lead to an explosion (Rosswog et al. 2009; Anninos et al. 2018). A recent, systematic study (Gafton and Rosswog (2019), their Fig. 17; though focussing on disruptions of main sequence stars by SMBHs) showed that for $\beta > 3$ a large fraction of the star becomes shocked via the tidal compression described in Sect. 2.3, thus supporting the idea that the explosion is triggered by compression shocks.

Reactive flows, however, are notoriously difficult numerical problems, see the excellent review of Müller (1998). The energy release from nuclear reactions creates pressure gradients that trigger fluid motions, which again transport both fuel and ashes to and from the reaction region. In the case of detonations these regions are intrinsically three-dimensional and one needs to resolve tiny length and time scales. Tanikawa et al. (2017) recently suggested that a spatial resolution of $< 10^6$ cm (with likely $> 10^9$ particles in SPH simulations) is needed to determine if explosive burning will occur. While small nuclear reaction networks can provide an accurate estimate for the nuclear energy release, and thus guarantee a reliable dynamical evolution of the fluid, substantially larger networks are needed in a post-processing step to reliably predict nucleosynthetic yields.

The relative yields of intermediate and heavy elements depend on the density during the burning processes, iron-group elements typically require densities $\gtrsim 10^7$ g cm^{-3}. Such densities can be reached by a combination of the initial stellar mass (translating into initial density and chemical composition), the β parameter (strength of the interaction) and the BH mass (Kawana et al. 2017; Tanikawa et al. 2017; Anninos et al. 2018). In particular low-mass white dwarfs need very strong encounters to potentially produce iron group elements. Figure 7 shows the nucleosynthetic production as a function of time for a general relativistic calculation of the ignition of a 0.2 $M_\odot$ He white dwarf by approach to a BH with a mass of 10^3 $M_\odot$ and β of 11. For this model, both Fe-group and intermediate-mass elements are produced. Interestingly, the black hole spin does not have much (if any) impact on the nucleosynthesis (Anninos et al. 2019).

Some observed Ca-rich transients have been suggested to originate from the tidal disruption of white dwarf by a IMBH and it is found that Ca-rich debris can be produced for low β systems. These systems, however, also have a low burning efficiency, which results in low yields of intermediate-mass elements and so may not produce enough Ca to be the progenitors of Ca-rich transients (Kawana et al. 2017; Anninos et al. 2018).

4 Predicted Optical and UV Signatures

The tidal disruption of a white dwarf by an IMBH results in a number of separate processes that can produce observational signatures. The disruption of white dwarf material typically results in super-Eddington accretion rates which may go along with jet formation in a manner analogous to the tidal disruption of main-sequence stars. Unlike main sequence stellar disruptions, however, very strong encounters can trigger runaway nuclear burning during the first stellar passage when the star is compressed and shocked as explained in Sect. 2. This can launch a Type Ia supernova-like event, but with highly asymmetric outflow.

The early-time optical and UV spectra are expected to be featureless because of a very large electron scattering opacity in the super-Eddington outflow at this early phase (Strubbe and Quataert 2011; Roth et al. 2016). Depending on the geometry and thermal and ionization structure of the outflowing envelope, some emission and/or absorption lines may be discernible in the spectra but these are expected to fall primarily in the far-UV and soft X-ray bands. The models of Strubbe and Quataert (2011), for example, consider the transfer of continuum photons through the outflowing envelope and predict an absorption-line spectrum. The models of Roth et al. (2016) consider the random walk of continuum and emission-line photons through the envelope and predict a spectrum with a combination of emission and absorption lines that depends on the particular conditions; some of the easily discernible lines may be in the rest-frame optical part of the spectrum. Although specific calculations of the spectra from the early stages of the disruption of a white dwarf have not been carried out, we can draw an analogy with the work on main sequence stars described above. Thus we expect the rest-frame optical continuum to be largely featureless while some strong absorption lines or P-Cygni lines from elements that are abundant in a white dwarf will show up in the far-UV portion of the spectrum.

At later stages of the disruption event, i.e., after the accretion rate drops below the Eddington limit and, presumably, a conventional accretion disk forms, we may expect to see broad emission lines from the tail of returning debris. Clausen and Eracleous (2011) calculated the optical and near-UV emission-line spectrum from the photoionized debris tail and its evolution over 11 years by adopting and refining the methodology of Strubbe and Quataert (2009). They considered a 0.55 $M_\odot$ CO white dwarf (67% O, 32% C, and 1% He by mass)[1] disrupted by IMBHs of 10^2, 10^3, and 10^4 $M_\odot$ with encounter strengths of $\beta = 1$ and 3.3. They found that, the (permitted) C IV λ1550 and C III λ977 Å lines are the two brightest lines in the spectrum and that their strengths decline monotonically with time as the event evolves. During the sub-Eddington accretion rate phase, the strengths of these two lines decline *approximately* as $t^{-5/3}$, mirroring the assumed decay rate of the ionizing luminosity. The strongest lines in the rest-frame optical spectra are the (forbidden) [O III] λλ4959,5007 and [O III] λ4363 Å lines, although these are always much weaker than the near-UV C lines. This behavior is illustrated in Fig. 8. As this figure shows the O lines initially rise in strength and begin to decline after the super-Eddington phase. The rate of decline of the [O III] λ5007 line is slower than that of other lines after the first few years. As a result, this line is expected to be the strongest one about a decade after the event. Moreover,

[1] The mass fraction of H was taken to be 10^{-5} and other elements were assumed to have their Solar mass fractions.

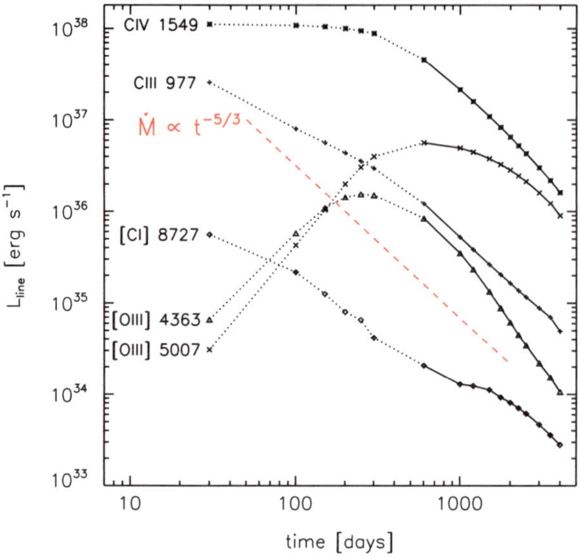

Fig. 8 Time evolution of the luminosities of the most prominent emission lines from the photoionized debris produced by the disruption of a 0.55 $M_{\odot}$ CO white dwarf by a 10^3 $M_{\odot}$ BH in a $\beta = 1$ encounter (adapted from Clausen and Eracleous 2011, see Sect. 4 of the text for the details of the models). The curves show the evolution of the luminosity of the most prominent emission lines. The dotted portion of each curve (up to 600 days) shows the evolution during and immediately after the super-Eddington accretion phase (440 days for this model), where the model predictions were deemed uncertain. The red, dashed line shows the assumed evolution of the accretion rate which sets the evolution of the strength of the ionizing continuum

since the continuum declines quickly, the observed equivalent width of the line increases in the period from a few years to a decade after the event. Clausen and Eracleous (2011) also computed the profile of the [O III] λ5007 Å line and its evolution and they found that, depending on the orientation of the observer relative to the debris tail, the profile is likely to be broad, skewed, and shifted (the shift and width can easily reach thousands of $\mathrm{km\,s^{-1}}$).

A related family of models by Clausen et al. (2012) examines the emission lines from the debris released in the disruption of the core of an evolved star, specifically a horizontal branch star. After evolving for ~ 10 Gyr, the core of a horizontal branch star that started out at 1 $M_{\odot}$, has an enhanced N abundance and suppressed C and O abundances as a result of CNO burning. Because of these modified core abundances and the depletion of hydrogen, the photoionizaiton of the debris leads to strong collisionally-excited emission lines from metals. Specifically, the [O III] λ5007 and [N II] λ6583 lines from the debris outshine the Balmer lines about a decade after the tidal disruption.

MacLeod et al. (2016b) performed radiative transfer calculations of the tidal compression of a white dwarf during a TDE to obtain predictions of their spectral and light curve properties using the hydrodynamic simulations of Rosswog et al. (2009) as input. Similarly to Type Ia supernovae, the resulting optical transients are powered by the radioactive decay of ^{56}Ni produced in the TDE. As discussed in Sect. 3, the yield of Fe-group elements will depend on the conditions of the system at the time of disruption, such as the initial white dwarf mass and composition and the β parameter. MacLeod et al. (2016b) found that the observed properties (luminosity, optical light curve decline rates, optical spectral features) of the TDEs are highly dependent on the viewing angle but in some cases have properties that are not dissimilar to unusual Type Ia supernovae but are generally less luminous (see

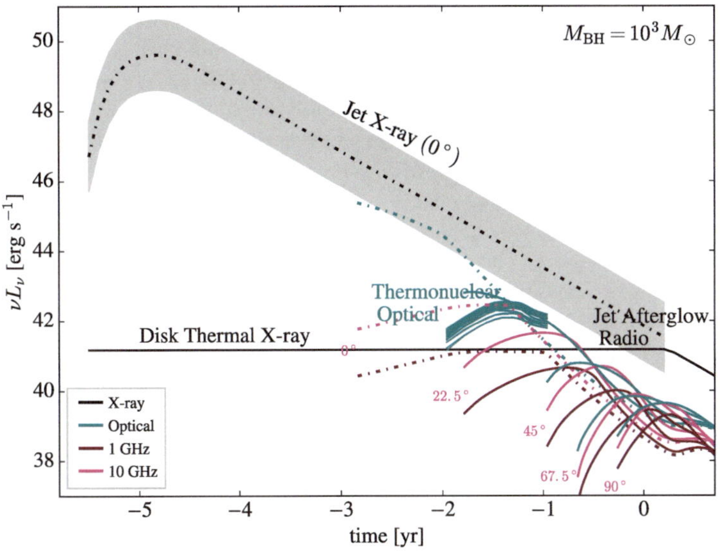

Fig. 9 The postulated multi-band signatures of a deep-passing white dwarf TDE involving a 10^3 $M_\odot$ BH and a 0.6 $M_\odot$ white dwarf. Times are from time of periapse passage of the white dwarf past the BH. This figure assumes that a jet is launched carrying kinetic energy of 6×10^{-3} times the accreted rest energy of 0.2 $M_\odot$ or 2×10^{51} erg. Line styles indicate viewing angle with respect to the jet axis: dot-dashed lines are along the jet axis, while solid lines denote an off-axis observer. Colors show different wavelengths of emission from X-ray to 1 GHz radio. At X-ray and radio wavelengths, the primary signatures are those of the jet and accretion disk. At optical wavelengths, potential signatures include the explosive transient from the thermonuclear burning of the white dwarf, depending on viewing angle, and a possible optical afterglow from the jet. For off-axis events, detection of the thermonuclear transient in the optical is a possible strategy, with accompanying X-ray emission from the accretion disk along with a radio afterglow completing the multi-wavelength picture. Adapted from MacLeod et al. (2016b, 2017)

examples in Fig. 9). This thermonuclear transient would dominate over the emission lines, such as the [O III] λλ4959,5007 and [O III] λ4363, produced by the photoionized debris tail discussed above while the transient is at its brightest but at later times these emission lines could become visible (MacLeod et al. 2016b).

5 Observational Searches and Candidates

Observational searches for the tidal disruptions of white dwarfs by IMBHs have generally taken two paths, studies looking for fast X-ray transients and optical searches for Type Ia supernova-like events. In this section we will discuss the properties of observed white dwarf TDE candidates and the link to the theoretical predictions.

5.1 Fast X-Ray Candidates

One part of the TDE transient parameter space that is still relatively unexplored is that of fast transients occurring on timescales of minutes or less. At X-ray wavelengths several events with rise times of tens of seconds have recently been discovered. Causality arguments necessitate bright transients on these timescales to be associated with compact sources, and/or jet phenomena. The relatively low BH mass involved in a tidal disruption of a white dwarf and the compactness of the white dwarf lead to short orbital time scales. Furthermore, as the white dwarf disruption must occur in a strong gravitational field, the circularisation process is probably also short (see the contribution to this Volume by Bonnerot et al.) and as the temperature of an ensuing accretion disc scales as $T \propto M_{BH}^{-0.25}$ (assuming Eddington-limited accretion) the relatively light BHs necessarily involved in white dwarf TDEs could lead to temperatures high enough for black-body radiation to peak in soft X-rays. Alternatively, given that coronal electron populations responsible for upscattering seed photons to produce hard power-law spectra are thought to be accelerated in magnetic reconnection events (Merloni and Fabian 2001), the disruption of magnetic white dwarfs also allows for hard power-law spectra. For all these reasons, white dwarf TDEs provide appealing explanations to fast X-ray transients.

In the contribution to this Volume from Saxton et al. the properties of several of the fast X-ray transients discovered are discussed. We here present a brief recap of the properties of these transients investigating if (some of) these are viable white dwarf TDE candidates such as suggested by Jonker et al. (2013) and Shen Fast (2019), for example.

The light curve properties of the observed fast X-ray events are not uniform. The rise and decay times, observed fluxes and fluences show significant diversity. For example, the main peak(s) of the flares varied in duration between ~50–700 seconds. Furthermore, for one event (XRT 000519; Jonker et al. 2013) two precursor flares were found, each 4000 s before the previous flare. The main peak of the flare was also split in two parts with significant spectral softening between the two peaks. Such precursor events were not detected in any of the other candidates but this may be explained by differences in orbital parameters of the disrupted white dwarfs (e.g. stars on eccentric orbits providing TDEs with precursor events and single-passage stars on parabolic orbits providing one-off flares; MacLeod et al. 2014). The timescale of the precursor flares is in line with the expected orbital timescale of a white dwarf in (an eccentric) orbit around an IMBH (cf. Fig. 5 in MacLeod et al. 2016a). Unfortunately, the distances to the sources responsible for these fast X-ray flares are in most cases not well constrained as no redshift information is available. However, for two flare events, there was spatial coincidence with Virgo cluster galaxies, making it probable that these flares occurred in the outskirts of these systems (Jonker et al. 2013; Irwin et al. 2016). The luminosities calculated assuming galaxy associations are consistent with the Eddington limit for IMBHs.

The X-ray spectral shape of these fast events is consistent with a power law (see the discussion and references in the contribution of Saxton et al. to this Volume). This is in contrast to the early-time soft X-ray spectra typically found in TDEs involving more massive black holes and stars other than white dwarfs (for some references and a discussion on differences between early- and late-time X-ray spectra of TDEs see for instance Jonker et al. 2019). It seems unlikely that a difference in the (Eddington ratio) mass accretion rate in white dwarf and main-sequence TDEs is the main cause of the different spectral shape.[2] A probable difference between main-sequence and white dwarf TDEs is the magnetic field of the star before disruption. The typically weak magnetic fields of main sequence stars may imply that their TDE discs start off with low levels of magnetic fields, whereas several white dwarf types host significant magnetic fields implying that accretion flows from white dwarf TDEs may start off with a higher magnetic field. Since power-law X-ray spectra are thought to

[2]Although the process of circularisation of the gas stream in white dwarf disruptions is affected by General Relativistic effects in contrast to many main sequence TDEs; see the contribution from Bonnerot et al. in this Volume on the formation of the accretion flow after disruption.

be caused by Compton up-scattering of seed disc photons by a coronal electron population, which in turn is thought to be accelerated to relativistic energies in magnetic reconnection events (Merloni and Fabian 2001), a power-law spectrum may more naturally emerge in the TDE discs with large magnetisation such as those occurring in white dwarf disruptions. Note that a main-sequence/longer duration TDEs with hard power law like spectra are also more difficult to distinguish from AGN activity and therefore observers may be biased against classifying such events as TDEs (Jonker et al. 2019).

The extreme source Swift J164449.3+573451 has been studied in detail, warranting a separate paragraph: it was discovered at the centre of a compact galaxy at $z = 0.35$ (Levan et al. 2011) due to its γ-ray emission. It displayed γ-ray emission lasting a few days (much longer than the duration of seconds to minutes for typical gamma-ray bursts, GRBs), as well as an extremely long-lived X-ray component. A number of models have been suggested to explain this event including a relativistic jet launched due to a TDE (Bloom et al. 2011) and the formation of an accretion disk after the collapse of a massive star to a BH (Quataert and Kasen 2012; Woosley and Heger 2012). Krolik and Piran (2011) suggested that jets from the tidal disruption of a white dwarf by an IMBH can explain the nature of this transient, in particular the short-term variability in and the fast rise time of its X-ray light curve. However, even though Tchekhovskoy et al. (2014) do not rule out that this event was powered by the tidal disruption of a white dwarf, this scenario is disfavoured according to their modelling.

There is some overlap in the X-ray properties of classical GRBs and the X-ray transients described above. Some X-ray transients have been interpreted as being caused by a GRB observed off-axis (e.g. Urata et al. 2015), as well as stripped envelope supernovae (Soderberg et al. 2005). Conversely, several ultra-long GRBs have been suggested to be associated with white dwarf TDEs (e.g. Levan et al. 2014), although other explanations have been proposed as well (e.g. Perets et al. 2016; Gao et al. 2010). In order to check the potential association, at least statistically, we compare the observed and predicted event rates.

The rate of Type Ibc SNe (some of which are associated with GRBs) is $0.258 \times 10^{-4}\,\mathrm{Mpc}^{-3}\,\mathrm{yr}^{-1}$ and if we include the estimate from Soderberg et al. (2004) that at most 6 per cent of Type Ibc supernovae produce a collimated jet this yields a rate of 1.55×10^{-6} Type Ibc supernovae with collimated (GRB) jets per $\mathrm{Mpc}^{-3}\,\mathrm{yr}^{-1}$. For an assumed typical X-ray luminosity of $L_X \sim 10^{48}\,\mathrm{erg\,s}^{-1}$ for the X-ray flash associated with the GRB, a distance limit of nearly $\sim$10 Gpc is implied given the observed X-ray fluxes of $\sim 10^{-10}\,\mathrm{erg\,cm}^{-2}\,\mathrm{s}^{-1}$ of the fast X-ray transients. Using this we derive that the Type Ibc SNe/GRB rate in this volume will be $\sim 1.5 \times 10^{6}\,\mathrm{yr}^{-1}$. So, comparing the observed X-ray flash rate of $1.4 \times 10^{5}\,\mathrm{yr}^{-1}$ over the whole sky as calculated by Glennie et al. (2015) with these predicted rates, we conclude that the X-ray flashes discussed in the contribution of Saxton et al. to this Volume could be due to GRBs associated with Type Ibc supernovae. We do note that the uncertainties in these rate calculations are still substantial for instance if only due to Poisson fluctuations given the low number of detected fast X-ray flashes.

In Sect. 6 we estimate the rate of X-ray flares expected from white dwarf TDEs. The end result of those deliberations on the rate R within the volume probed with a flux limit set to $\sim 10^{-10}\,\mathrm{erg\,cm}^{-2}\,\mathrm{s}^{-1}$ is $R = f_{occ} \times 10^{4}\,\mathrm{yr}^{-1}$. This value falls a factor of about $14/f_{occ}$ short of the observed rate of X-ray flashes of $1.4 \times 10^{5}\,\mathrm{yr}^{-1}$. Of course, as indicated in Sect. 6, there are significant uncertainties in these rate calculations. Nevertheless, even optimistically assuming an occupation fraction of 1 leaves a large gap between the predictions of X-ray flares from white dwarf TDEs and the observed rate of X-ray flares. Contributions from additional populations of IMBHs, such as those that have been speculated to reside in globular clusters, or hyper–compact stellar clusters, or the lighter BHs in the recently–discovered stripped nuclei (Fragione et al. 2018; Merritt et al. 2009; Voggel et al. 2019, respectively), may increase the rate somewhat. However, it is likely that only a small fraction of globular clusters host IMBHs and the feeding rate of white dwarfs into the loss cone is likely too low for the majority of these objects (see again also Sect. 6 for details). A high spin rate that increases the Hills mass for BHs with masses $< 10^6\ M_{\odot}$ could enable IMBHs to potentially disrupt more white dwarfs but overall, it is difficult to explain all the X-ray flares as solely due to white dwarf TDEs.

5.2 Optical Candidates

As discussed in Sect. 4, if the white dwarf is tidally compressed and ignites nuclear burning then the dominant powering mechanism of optical transients in the days to weeks after disruption will be the radioactive decay of Fe-group elements, provided they are produced in sufficient amounts (e.g. in very strong encounters and/or massive white dwarfs). Each Type Ia supernova typically produces $0.6\ M_{\odot}$ of Fe-group elements per explosion. MacLeod et al. (2016a,b) investigated the nucleosynthetic yields of white dwarf+IMBH TDEs and found that the amount of Fe-group elements synthesised may be significantly lower than that seen in normal Type Ia supernovae, which affects their observed light curves and spectral properties.

Ca-rich transients are a class of observed optical transients with magnitudes of ~ -16 (fainter than normal Type Ia supernovae) that have strong features of Ca in their late-time spectra, suggesting that they may have appreciable Ca in their ejecta (Perets et al. 2010; Kasliwal et al. 2012; Lunnan et al. 2017). Caution in associating strong Ca features with large quantities of Ca, however, should be exercised because Ca is a very efficient coolant and a small amount of Ca in the ejecta can produce a strong feature. The early-time spectra of Ca-rich events have optical spectra that look similar to stripped envelope Type Ib SNe that typically show strong He lines (e.g. Perets et al. 2010). Various models to explain Ca-rich transients have been suggested including TDEs involving a white dwarf disrupted by an IMBH (Sell et al. 2015), a He-shell detonation on the surface of a white dwarf ('.Ia supernova' Perets et al. 2010; Shen et al. 2010), and collisions involving a He donor and a CO or ONe white dwarf (García-Berro et al. 2017). The original white dwarf TDE models are from Rosswog (2009) but radiative transfer calculations were performed by MacLeod et al. (2016b) and some low-mass white dwarf disruptions may result in transients rich in Ca. The key distinguishing characteristic for determining if Ca-rich transients are linked to TDEs involving white dwarfs and IMBHs is the detection of high-energy emission. Supernovae originating from the thermonuclear explosion of white dwarfs have not been detected in the X-ray band but X-ray emission is a key prediction of TDE models. A search for X-ray emission was made for SN 2016hnk, a white dwarf TDE candidate, but only an upper limit was placed (Sell et al. 2018) but future searches in the X-ray band will be important for testing the potential link between Ca-rich transients and white dwarf TDEs.

5.3 Multi-Wavelength Candidates

Some candidate white dwarf and IMBH TDEs have been discovered that are bright in the optical as well as having high-energy counterparts.

AT2018cow: A rapidly rising and luminous transient was discovered in the optical by Prentice et al. (2018) and also detected in the X-ray, radio, and submillimeter (Rivera Sandoval et al. 2018; Kuin et al. 2019; Margutti et al. 2018). A TDE involving a He-star and an IMBH was suggested by Ho et al. (2019). Perley et al. (2019) suggested that

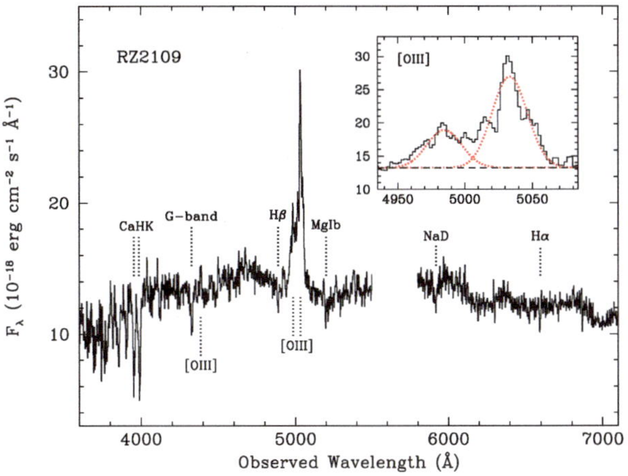

Fig. 10 The optical spectrum of the globular cluster RZ 2109 associated with the elliptical galaxy NGC 4472 in Virgo, adapted from Zepf et al. (2008). The dotted vertical lines and labels *above* the continuum identify some of the prominent absorption lines in the spectrum of an old stellar population. The dotted vertical lines and labels *below* the continuum identify three of the stronger [O III] lines expected in this range. The inset shows the profile of the [O III]4959,5007 lines with Gaussian models of FWHM ~ 2000 km s^{-1} superposed

the transient could be caused by a main-sequence star disrupted by an IMBH but this is at odds with the detection of dense circumstellar material (detected via radio measurements) that would need to be ejected from the system prior to the TDE (Margutti et al. 2018). Other suggestions for the origin of this transient include the core-collapse of a low-mass He star or a stellar collapse involving accretion onto a newly formed compact object (Prentice et al. 2018; Margutti et al. 2018).

RZ 2109: Maccarone et al. (2007) reported the discovery of an accreting BH in the globular cluster RZ 2109, associated with the elliptical galaxy NGC 4472 in the Virgo cluster, based on its X-ray luminosity and variability properties. Followup optical spectroscopy of this object by Zepf et al. (2008) revealed the remarkable spectrum illustrated in Fig. 10, which comprises a continuum from the old stellar population of the globular cluster and a very broad emission feature identified with the [O III] λλ4959,5007 doublet (see also Steele et al. 2011). The lack of any other optical emission lines prompted Clausen and Eracleous (2011) to propose that the properties of this system were the result of the tidal disruption of a CO white dwarf by an IMBH at the centre of the cluster, accretion of some of the debris, and the photoionization of remaining, returning debris by accretion-powered ionizing radiation. This proposal was bolstered by photoionization calculations that explained the emission-line spectrum (see discussion of models in Sect. 4) and a model for the profile of the [O III] λλ5007 line that reproduced the observed profile.

However, alternative explanations were also proposed, including accretion of hydrogen-deficient material from a red giant by the BH (Porter 2010), emission from the wind of an R Corona Borealis star (Maccarone and Warner 2011), and the ionization of the ejecta of a nearby nova by the accreting BH (Ripamonti and Mapelli 2012). Moreover, some of the observed properties of RZ 2109 cannot be explained by the TDE scenario, e.g., the fact that the X-ray luminosity has remained relatively steady for about 16 years (Dage et al. 2018) and the fact that the source of the emission lines is likely to be extended (Peacock et al. 2012). Thus, the nature of the X-ray and emission-line sources in RZ 2109 remains ambiguous. Another globular cluster with properties reminiscent of RZ 2109 was found in the elliptical galaxy NGC 1399 by Irwin et al. (2010, it is associated with the X-ray source CXO J033831.8−352604). The optical spectrum of this object features the [O III] λλ4959,5007 doublet as well as the [N II] λλ6548,6584 doublet but no other strong emission lines, which inspired Clausen et al. (2012) to explain it by invoking the disruption of a blue horizontal branch star. But all the alternative models proposed for RZ 2109 apply to this object as well.

6 Rates: Intrinsic and Predicted in Surveys

White dwarf TDEs, and their associated flares, are likely rare compared to TDEs of main sequence stars. There are several reasons for this. First, the compactness of white dwarfs implies that a more extreme tidal force is needed to disrupt them than is needed for a main sequence star. This implies a closer periapse approach to the BH is needed, and there is a correspondingly smaller cross section for these interactions. Further, only IMBHs with mass $M_\bullet \lesssim 10^5$ $M_\odot$, can accomodate these close periapse approaches without swallowing the white dwarf whole. Second, white dwarfs, while not rare in an old stellar population, are less common than low-mass main sequence stars. Nonetheless, the potentially luminous and distinct signatures of white dwarf disruption may separate these events from the larger number of main-sequence disruptions.

6.1 White Dwarf Tidal Disruption Sites and Dynamics

Just like stars of other stellar-evolutionary states, white dwarfs are tidally disrupted after passing at very close periapse distance to a massive BH. These events rely on the coexistence of massive BHs and surrounding dense clusters of stars. While highly uncertain at the BH masses most relevant to white dwarf tidal disruptions, in more massive BHs, these clusters are thought to contain roughly a BH mass of stars in a roughly isotropic, dynamically relaxed cluster surrounding the BH. Within this cluster, stars trace wandering orbits directed by both the gravitational attraction of the BH and interaction with all of the other stars. These wandering orbits occasionally are scattered to very high eccentricity, which causes them to plunge to close approaches to the BH (see the contribution of Stone et al. to this Volume on the processes that send stars on orbits that will lead to their tidal disruption).

Currently, predictions for the occurrence of white dwarf tidal disruptions come from two distinct methods: scaled-down models of stellar cusps present around more massive BHs, and N-body models of globular clusters containing IMBHs of hundreds to thousands of solar masses. We begin by examining the qualitative predictions of the nuclear cluster models. These models are discussed in detail by Merritt (2013, Chap. 6) and in the contribution of Stone et al. to this volume; they are applied to the white dwarf tidal disruption context by MacLeod et al. (2014, 2016b). In such clusters, a region of similar size to the BH sphere of gravitational influence $r_\mathrm{h} = GM_\bullet/\sigma^2$, where σ is the velocity dispersion of the surrounding stellar distribution, is assumed to contain approximately a BH mass of stars, arranged with a power-law profile in number density as a function of radius (Bahcall and Wolf 1976, 1977; Frank and Rees 1976). In the discussion that follows, we make the simplifying assumption that all stellar subtypes are distributed homogeneously within this power law profile. However, there are several reasons that this simplification may not occur in reality. Over a cluster relaxation time, stars of different mass segregate such that more massive objects end up more tightly bound to the BH than less tightly bound objects (See Chap. 7 of Merritt 2013).

This effect can be particularly strong if there is a significant population of stellar mass BHs, with mass much greater than the average cluster star (Alexander and Hopman 2009; Aharon and Perets 2016).

In scaling the properties of these clusters down from those of more massive BHs, MacLeod et al. (2014) applied the BH mass-velocity dispersion relation of Kormendy and Ho (2013), we note here that this has the consequence of determining the, a priori uncertain, density of the cluster that surrounds the putative BH. This point of uncertainty merits some caution, because, as we describe below, typical stellar densities in globular clusters (which could potentially harbor black holes of less than approximately $10^3 M_\odot$) have sufficiently low binding energy that the presence of a black hole modifies the entire cluster distribution, not just a central cusp.

With stellar cluster properties defined, one can compute the various "relaxation" properties that cause stellar orbits to change over time. In general, these include the stochastic scatterings of two-body gravitational encounters (two-body relaxation), and the coherent, secular torques applied to closer-in stellar orbits over many orbital cycles (resonant relaxation; see Merritt 2013). The typical magnitude of the per orbit random-walk in angular momentum space can be expressed in terms of the relevant relaxation timescale, $t_{\rm rel}$ as $\Delta J \approx J_{\rm c} (P/t_{\rm rel})^{1/2}$, where P is the orbital period and $J_{\rm c} \approx \sqrt{G M_{\rm bh} a}$ is the angular momentum of a circular orbit. This expression implies that over one relaxation time the orbital angular momentum changes by of order the circular angular momentum. Figure 11 shows the characteristic radii in such a nuclear star cluster as a function of BH mass, adapted from MacLeod et al. (2014). The line $\Delta J = J_{\rm lc}(WD)$ marks the orbital semi-major axis within the cluster where the root mean square change in orbital angular momentum over one orbit, ΔJ, equals the angular momentum of a "loss cone" orbit, $J_{\rm lc}(WD)$, that has a periapse distance similar to the white dwarf tidal radius.

At radii larger than this line, the low angular momentum phase space depleted by white dwarf tidal disruptions is repopulated each orbital period (and as a result, tidal encounters of all impact parameters occur). This is the "full loss cone" portion of the cluster phase space. At radii smaller than the $\Delta J = J_{\rm lc}(WD)$ line, the per orbit scatter is smaller than the loss cone angular momentum. This implies that the loss cone phase space is "empty" and that only grazing impact parameter encounters occur. MacLeod et al. (2014, Sect. 3) showed that most white dwarf TDEs are fed from regions where the loss cone is efficiently refilled by either two-body relaxation or mass-precession resonant relaxation. Under the assumptions of this representative cluster model, white dwarf tidal disruptions therefore occur primarily in the full loss cone limit, and therefore distribute across all periapse impact parameters.

The overall rate of white dwarf TDEs in such a model was found to be approximately $10^{-6}\ {\rm yr}^{-1}$ per BH. For comparison main-sequence star TDEs occurred with a frequency of $10^{-4}\ {\rm yr}^{-1}$ per BH in these same models. The white dwarf tidal disruption rate was therefore 1% of the overall tidal disruption occurrence rate. Although many assumptions must be made about the properties of stellar clusters surrounding BHs to estimate this rate, MacLeod et al. (2014, 2016b) have argued that the ratio of white dwarf to main sequence tidal disruptions may be robust with respect to variations in the system properties. Within a model in which stars are distributed in a homogenous, power-law profile, the unknown parameter with the strongest effect on the relative rates of main sequence versus white dwarf disruption events is the slope of stellar number density with radius in the central stellar cusp, which can affect relative rates at the factor ~ 2 level. A factor that could modify this conclusion is the differential mass segregation of stellar sub-populations. In essentially creating differently-sloped radial profiles of different stars (for example white dwarfs versus main-sequence stars) mass segregation might affect relative disruption rates. Perhaps more significantly, if

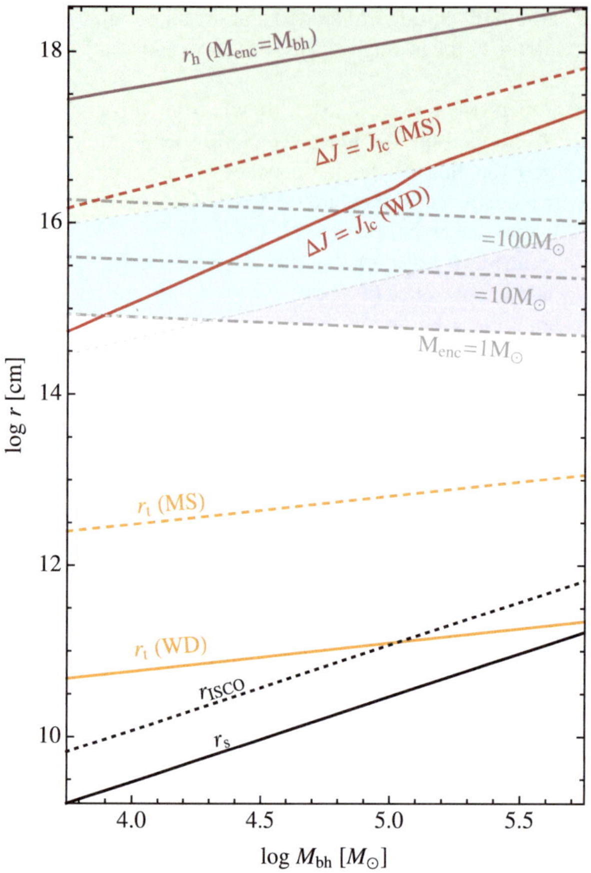

Fig. 11 Characteristic scales for interactions of white dwarfs within a central stellar cluster with the central massive BH, as a function of varying BH mass. Here the cluster is assumed to contain one BH mass of stars within the BH's sphere of gravitational influence, arranged as $n(r) \propto r^{-1.5}$. Shown, from bottom to top, are: 1) the Schwarzschild radius, r_s, and the radius of the Innermost Stable Circular Orbit $\sim 4r_s$ (black solid and dotted, respectively), 2) the tidal radius, r_t for white dwarfs ($0.5M_\odot$) and MS (sun-like) stars (yellow solid and dashed), 3) the radii that enclose 1, 10, and 100 $M_\odot$ (gray dot-dashed), 4) the characteristic orbital semi-major axis that marks the transition from the empty (smaller a) to the full (larger a) loss cone regimes (red solid and dashed lines labeled $\Delta J = J_{lc}$), and 5) the MBH radius of influence, r_h. Filled regions are, on average, populated with stars. Filling colors denote the primary orbital relaxation mechanism with general-relativistic resonant relaxation (purple), mass-precession resonant relaxation (cyan), and finally non-resonant relaxation (green) being dominant from small to large radii, respectively. Figure adapted from MacLeod et al. (2014)

a substantial population of stellar-mass BHs exists in the cluster, these could possibly or entirely occupy a portion of the tightly-bound phase space from which white dwarf disruptions are thought to arise (see, for example, the analysis of Aharon and Perets 2016, who study the related occurrence of extreme-mass ratio gravitational wave inspirals). More detailed work is needed to make quantitative statements in the context of white dwarf tidal disruptions.

The event rate of tidal disruptions by globular cluster IMBHs is thought to be considerably lower than that of galactic nuclei, on the order of 10^{-7} yr^{-1} (Ramirez-Ruiz and Rosswog 2009). In this context, N-body stellar dynamics has been the primary tool for analyzing disruption rates. We briefly review these results, with a particular focus on the ways in which they differ from the nuclear-cluster models above.

The fundamental reasons for the use of N-body dynamics simulations in the globular cluster regime is closely related to the order of magnitude lower disruption rate predicted. The IMBH and its closely bound companion objects have a binding energy comparable to the entire cluster system. Scattering encounters act to keep the cluster core inflated to higher velocity dispersion and lower density (e.g. Baumgardt et al. 2004; MacLeod et al. 2016b). As a result of this feedback process, the presumption of an homogeneous stellar surrounding, with fixed velocity dispersion σ, is typically not realized. A related aspect of cluster–IMBH dynamics is that the IMBH wanders relative to the cluster center of mass (for general discussion and a recent analysis based on N-body simulations see, respectively, Merritt 2013; de Vita et al. 2018). A priori, this is not thought to significantly impact white dwarf disruption dynamics, because disrupted white dwarfs tend to arise from orbits that are tightly bound to the IMBH and wander relative to the cluster center with the black hole (MacLeod et al. 2016b). However, more systematic investigation may be warranted, especially if black hole masses as low as $10^3 M_\odot$ are considered. Despite these numerous differences due to overall mass scale between nuclear cluster and globular cluster dynamics we again expect of the order of 1% of the disruptions that occur will be white dwarf disruptions. The overall white dwarf disruption rate may, therefore, be of the order of 1 Gyr^{-1} per IMBH hosting globular cluster. This order of magnitude event rate has been derived from N-body simulations (Baumgardt et al. 2004; MacLeod et al. 2016b) and applied to white dwarf disruption scenarios (Haas et al. 2012; Shcherbakov et al. 2013; Sell et al. 2015). Mass segregation effects may have a stronger influence on relative event rates in this context, because globular cluster relaxation times are short relative to the age of the universe. To give a particular example, Baumgardt et al. (2004) find that white dwarfs make up approximately 8% of tidal disruptions, likely because massive white dwarfs segregate into preferentially close orbits to the IMBH when few other massive remants are present.

6.2 Occurrence of Different Classes of Encounters

We have, in the preceding section, outlined a variety of classes of white dwarf tidal disruption encounters. White dwarfs have a compactness that implies that gravitational radiation and tides act with similar strengths. Their susceptibility to explosive nuclear burning under degrees of compression that correspond to the BH tidal field further diversifies the potential outcomes. The relative occurrence of these events is a problem that lies at the interface of stellar dynamics and the processes of the events themselves.

Zalamea et al. (2010) first discussed the partial tidal stripping of white dwarfs over the course of a number of orbits with corresponding gravitational radiation (in the LISA frequency range). MacLeod et al. (2014) estimated the relative occurrence rate of these events compared to single-passage disruptions to be approximately 10% of white dwarf disruptions might occur through such a channel. However, the precise proportions remain uncertain (see Fig. 7 of MacLeod et al. 2014).

Deeply-plunging encounters, which can only occur for sufficiently low BH masses, $\lesssim 10^5\ \mathrm{M}_\odot$, occur when the impact parameter is of the order of $\beta_{\mathrm{thermo}} \gtrsim 3$ (e.g. Rosswog et al. 2009; Anninos et al. 2018). By comparison, white dwarfs begin to lose mass at an impact parameter of $\beta_{\mathrm{ml}} \approx 0.5$. The fact that, for most parameters, the phase space of the white dwarf loss cone is thought to be efficiently refilled implies a distribution of impact parameters $N(>\beta) \propto \beta^{-1}$. In this limit, the fraction of disruptive events that pass deeply enough to lead to a thermonuclear explosion is $\beta_{\mathrm{ml}}/\beta_{\mathrm{thermo}} \sim 1/6$.

6.3 Volumetric Event Rates

The volumetric rate of white dwarf TDEs is very uncertain – mostly because the volumetric density of massive BHs below 10^6 M$_\odot$ is highly uncertain. Converting the specific rate per black-hole cluster system to a volumetric rate involves estimating the volume density of potential black-hole hosts and the BH occupation fraction in those hosts. Explicitly, $\dot{N}_{\rm vol} \approx \dot{N}_{\rm BH} n_{\rm BH}$, where $\dot{N}_{\rm BH}$ is the white dwarf TDE rate per BH (in a given mass range) and $n_{\rm BH} = n_{\rm host} f_{\rm BH}$ is the space density of those BHs, which is the product of the density of the hosts and the BH occupation probability for that host type. In view of the importance of the density of massive BHs, we begin the discussion of volumetric event rates by summarizing estimates of the density of potential hosts, namely dwarf galaxies and globular clusters.

An initial, theoretical estimate of $n_{\rm BH}$ can be made by a flat extrapolation of the massive BH mass function (following MacLeod et al. 2016b). Based on the results of the Illustris simulation at 10^6 M$_\odot$ (see Sijacki et al. 2015) one may infer that $n_{\rm BH} \sim 10^7$ Gpc^{-3}. Assuming a per BH rate of $\dot{N}_{\rm BH} = 10^{-6}$ yr^{-1} (see Sect. 6.1), the implied volumetric event rate is, therefore, $\dot{N}_{\rm vol} \approx 10$ yr^{-1} Gpc^{-3}. By comparison, empirical estimates based on the luminosity function or mass function of galaxies give a considerably higher density, if one assumes a high occupation fraction.

The luminosity function of galaxies can be described over the full mass range by a double Schechter function (e.g., equation (6) of Baldry et al. 2012):

$$\Phi(m)\,dm = \left[\phi_1 \left(\frac{m}{m_*}\right)^{\alpha_1} + \phi_2 \left(\frac{m}{m_*}\right)^{\alpha_2}\right] e^{-m/m_*} \frac{dm}{m_*}. \tag{19}$$

In the mass range of dwarf galaxies the exponential is approximately unity and equation (19) can be approximated as the sum of two power laws. For the purposes of our discussion, we adopt the parameters of this function reported by Wright et al. (2017), namely $\log(m_*/{\rm M}_\odot) = 10.78$, $(\alpha_1, \alpha_2) = (-0.62, -1.50)$, $(\phi_1, \phi_2) = (2.93, 0.63) \times 10^{-3}$ Mpc^{-3} (assuming a Hubble constant of $H_0 = 70$ km s^{-1} Mpc^{-1}). The density of dwarf galaxies can then be found by integrating the above function. To determine suitable limits for the integration we turn to Reines et al. (2013) who studied nuclear activity in dwarf galaxies with $\log(m/{\rm M}_\odot) = 8.5$–$9.5$ and estimated BH masses in the range $\log(M_\bullet/{\rm M}_\odot) = 4.9$–$6.5$ based on the widths of the broad emission lines. Thus we integrate between $\sim 10^7$ and $10^{9.5}$ M$_\odot$ in order to include BH masses between $\sim 10^4$ and $10^{6.5}$ M$_\odot$ and we obtain a dwarf galaxy density of $n_{\rm dwarf} \approx 9 \times 10^7$ Gpc^{-3}. An alternative approach is to integrate the luminosity function of Blanton et al. (2005, expressed in terms of absolute magnitudes in their equation 8). Carrying out the integration between absolute r magnitudes of -13 and -18 using the values of the free parameters for the "total" r-band luminosity function in Table 3 of Blanton et al. (2005), adjusted to $H_0 = 70$ km s^{-1} Mpc^{-1}, we get a dwarf galaxy density of the same order as above, $n_{\rm dwarf} \approx 3 \times 10^8$ Gpc^{-3}.

The occupation fraction of BHs in these systems is unknown, although some observational constraints have been obtained. Searches for AGNs via spectroscopy or variability yield a lower limit of $f_{\rm BH} >$ a few % (Reines et al. 2013; Moran et al. 2014; Baldassare et al. 2018) while a statistical study of the X-ray properties of nearby, low-mass galaxies yields $f_{\rm BH} > 20\%$ (Miller et al. 2015). The resulting volumetric rate, based on the latter, higher dwarf galaxy density and assuming a per BH rate of $\dot{N}_{\rm BH} = 10^{-6}$ yr^{-1} (see Sect. 6.1) can be expressed as $\dot{N}_{\rm vol} \approx 300 f_{\rm BH}$ yr^{-1} Gpc^{-3} (or $\dot{N}_{\rm vol} \gtrsim$ a few $\times$ 10 yr^{-1} Gpc^{-3}, in view of the limits on $f_{\rm BH}$).

The volume density of globular clusters can be estimated with an analogous approach. A number of recent studies (e.g., Harris et al. 2013, 2014; Zaritsky et al. 2015) show that the number of globular clusters per massive galaxy ($m > 10^{10}$ M$_\odot$) can be expressed in terms of the mass of their host galaxy as $N_{\rm glob}(m) = N_0\,(m/m_0)^\delta$. Combining this relation with the mass function of galaxies in equation (19) we get the following density of globular clusters per unit *host galaxy* mass

$$\Phi_{\rm glob}(m)\,dm = N_0 \left(\frac{m_*}{m_0}\right)^\delta \left[\phi_1 \left(\frac{m}{m_*}\right)^{\alpha_1+\delta} + \phi_2 \left(\frac{m}{m_*}\right)^{\alpha_2+\delta}\right] e^{-m/m_*}\,\frac{dm}{m_*}. \qquad (20)$$

Adopting $\log N_0 = 2.924$, $\log(m_0/{\rm M}_\odot) = 11.2$ and $\delta \approx 1$ (Harris et al. 2013), and other parameters as in equation (19) and integrating[3] for galaxy masses $m > 10^{10}$ M$_\odot$ we obtain a globular cluster volume density of $n_{\rm glob}(m > 10^{10}\ {\rm M}_\odot) = 9.5 \times 10^8\ {\rm Gpc}^{-3}$. An additional contribution may come from globular clusters associated with lower-mass galaxies. In this lower-mass range, $m = 10^8$–10^{10} M$_\odot$, the number of globular clusters per galaxy is described by a flatter power law with an index $\delta = 0.365$ and $\log N_0 = 1.274$, $\log(m_0/{\rm M}_\odot) = 9.2$ (Harris et al. 2013). Moreover, equation (20) can be simplified by neglecting the exponential, which is approximately unity for $m < 10^{10}$ M$_\odot$. Integrating $\Phi_{\rm glob}(m)$ from equation (20) in this range of galaxy masses leads to a density of $n_{\rm glob}(m = 10^8$–$10^{10}\ {\rm M}_\odot) = 4.4 \times 10^9\ {\rm Gpc}^{-3}$. But it should be borne in mind that the contribution from lower-mass galaxies is rather uncertain because the small number of globular clusters per low-mass galaxy leads to uncertain parameters for $N_{\rm glob}(m)$ in this mass range.

Searches for IMBHs have put constraints on the BH occupation fraction in globular clusters. These studies can involve measuring the dynamical properties of stars orbiting a putative IMBH in a globular cluster (e.g., Newell et al. 1976; Baumgardt et al. 2003; Gebhardt et al. 2005; Kızıltan et al. 2017; Perera et al. 2017) or measuring accretion signatures from TDEs in globular clusters, such as the IMBH candidate ESO 243-49 HLX-1 or other, similar objects (e.g., Maccarone et al. 2007; Webb et al. 2012; Lin et al. 2018). Other studies have found no signatures of accretion in large samples, suggesting that the typical masses of IMBHs in globular clusters may be lower than theoretical predictions or the occupation fraciton may be extremely small (e.g. Strader et al. 2012; Tremou et al. 2018). Using a per BH rate of $\dot{N}_{\rm BH} = 10^{-9}\ {\rm yr}^{-1}$ (see Sect. 6.1) we can express the volumetric rate as $\dot{N}_{\rm vol} \approx 4\, f_{\rm BH}\ {\rm yr}^{-1}\,{\rm Gpc}^{-3}$ or $0.2\, f_{\rm BH}\ {\rm yr}^{-1}\,{\rm Gpc}^{-3}$ depending on whether or not we include globular clusters associated with dwarf galaxies.

As the above discussion and estimates suggest, the current balance of evidence favors the cores of low-mass galaxies as the most prevalent hosts of white dwarf tidal disruptions. It is also worth emphasizing that, since BHs in globular clusters are expected to be less massive than those in dwarf galaxies the detection volume for globular cluster events should be smaller compared to dwarf galaxies because of the lower luminosities of these events.

6.4 Predicted Detectability in Surveys

In the section, we discuss the probability of the detection of white dwarf TDEs by IMBH using current X– and γ–ray missions, as well as in the future next-generation optical transient survey, the Large Synoptic Survey Telescope (LSST).

[3]The integral of the Schechter function can be expressed in terms of incomplete Γ functions, $\Gamma(p,a) = \int_a^\infty x^{p-1} e^{-x}\,dx$.

6.4.1 Detecting Beamed Emission with X– and γ–Ray Satellites

If white dwarf TDEs launch luminous jets, these will be detectable by wide-angle monitors. For example, for a jet luminosity of 10^{48} erg s^{-1}, *Swift*'s Burst Alert Telescope (BAT) will detect the event out to redshift $z \approx 1$, or a luminosity distance of 6.5 Gpc (WMAP 9 Cosmology, MacLeod et al. 2016b). The volume enclosed by redshifts $z < 1$ is approximately 150 Gpc^3. Of the population of events in this volume (see the section above) a fraction is beamed toward our observation perspective, $f_{\rm beam}$, and the BAT detects a fraction of these $f_{\rm BAT} \sim 20\%$ because of its sky coverage. The detectable rate is then

$$\dot{N}_{\rm BAT} \approx \dot{N}_{\rm vol} V f_{\rm beam} f_{\rm BAT}, \tag{21}$$

where V is the observable volume. Due to relativistic abberation, light is emitted in a narrow beam of width $1/\Gamma^2$, where Γ is the Lorentz factor. Figure 9 assumes $\Gamma \sim 7$ implying $f_{\rm beam} \sim 1/50$. However, the Lorentz factors associated with relativistic TDEs may be of the order of 2 (Zauderer et al. 2011; Cenko et al. 2012; see the contribution of Zauderer et al. to this Volume), implying a significantly larger beaming fraction of $f_{\rm beam} \sim 0.1$. Using fiducial numbers and $f_{\rm beam} \sim 1/50$ (and therefore implicitly assuming that all white dwarf TDEs lead to a jet of 10^{48} erg s^{-1}) we find $\dot{N}_{\rm BAT} \approx 6$ yr^{-1} by extrapolation of the BH mass function or $\dot{N}_{\rm BAT} \approx 180 f_{\rm BH}$ yr^{-1} from the dwarf galaxy space density.

Similarly, sensitive X–ray satellites such as *Chandra*, XMM–*Newton*, and *Swift* might serendipitously detect the X–ray flare thought to accompany a white dwarf TDE. If we again assume that the white dwarf TDE peak X-ray luminosity $L_X \sim 10^{48}$ erg s^{-1} (see Fig. 9), then the volume probed for a conservative flux limit of 10^{-10} erg cm^{-2} s^{-1} is approximately 10^3 Gpc^3. Again using the volumetric estimates of Sect. 6.3, we estimate

$$\dot{N}_{\rm X-ray} \approx \dot{N}_{\rm vol} V f_{\rm beam} f_{\rm sky}, \tag{22}$$

where $f_{\rm sky}$ is the instantaneous sky fraction observable by the X–ray satellites. This number is of the order of 6×10^{-6} for XMM–*Newton*'s pn camera with field of view of $30' \times 30'$. The number is slightly smaller for the *Swift* XRT and the *Chandra* ACIS–I detector. This fraction also assumes nearly-continuous observing. For $f_{\rm beam} = 1/50$, the implied detection rates are $\dot{N}_{\rm X-ray} \approx 1.2 \times 10^{-3}$ yr^{-1} by extrapolation of the BH mass function or $\dot{N}_{\rm X-ray} \approx 3.6 \times 10^{-2} f_{\rm BH}$ yr^{-1} from the dwarf galaxy space density. Thus, for $f_{\rm BH} \sim 1$, we could expect of the order of one detection by XMM–*Newton*, *Chandra*, or *Swift* over their roughly 20-year missions.

6.4.2 Detecting Thermonuclear Transients with LSST

Thermonuclear transients produced by the tidal disruption of a white dwarf by IMBH are well within the reach of the planned optical survey of LSST. LSST can detect a prototypical thermonuclear transient, such as that shown in Fig. 9, to $z = 0.37$ or a volume of 13.3 Gpc^3 (MacLeod et al. 2016b). We estimate the detectable rate of white-dwarf TDE thermonuclear transients as

$$\dot{N}_{\rm LSST} \approx \dot{N}_{\rm vol} V f_{\rm thermo} f_{\rm sky}, \tag{23}$$

where, $f_{\rm sky} \sim 0.5$, and thermonuclear transients are believed to occur in roughly $f_{\rm thermo} \sim 1/6$ of white dwarf TDEs, see Sect. 6.1. The resultant rates are $\dot{N}_{\rm LSST} \approx 11$ yr^{-1} based on the extrapolation of the BH mass function, or or $\dot{N}_{\rm LSST} \approx 330 f_{\rm BH}$ yr^{-1} from the dwarf galaxy space density. There is slight viewing angle dependence introduced to these baseline estimates by the asymmetric ejecta of the tidal thermonuclear transient. See MacLeod et al. (2016b) for a Monte Carlo calculation that shows the preference for certain viewing angles.

A much greater challenge than detection of tens to hundreds of events per year will be their classification. With LSST anticipated to find hundreds of thousands of supernovae per year, only the multi-wavelength signatures of these white dwarf disruptions make them truly distinctive. One might imagine that the coincidence of high-energy and optical emission (which occurs in a smaller subset of cases, but is more distinctive) might offer one of the only paths to firm classification.

7 Multi-Messenger Signatures

7.1 Gravitational Waves from White Dwarf Inspirals Before Disruption

Since the disruptions of white dwarfs are often relativistic (i.e., the white dwarf passes within a few Schwarzschild radii from the BH), we can expect substantial emission of gravitational waves. In general terms, the frequency of gravitational waves emitted during such an encounter is of the same order as the orbital/dynamical frequency at the tidal disruption radius. This time scale is given by equation (17) and is comparable to the dynamical time of the star. For the typical white dwarf parameters considered here, $\tau_{\rm dyn} \approx 3.5\ (M_{\rm wd}/0.6\ {\rm M}_{\odot})^{-1/2}\ (R_{\rm wd}/10^9\ {\rm cm})^{3/2}$ s, implying gravitational wave frequencies on the order of $f_{\rm gw} \sim 2/\tau_{\rm dyn} \sim 0.6$ Hz. This frequency is of the same order as the range accessible by the planned LASER Interferometer Space Antenna (hereafter LISA; anticipated launch date in the mid 2030s, see Danzmann 2017).

In the more likely case of a white dwarf in an unbound, parabolic orbit a burst of gravitational waves will be emitted during the initial encounter with the BH. After the initial encounter the debris will return to pericentre one dynamical time later and accretion will begin soon thereafter, leading to an electromagnetic flare. The gravitational wave strain (effectively, the observed amplitude of the wave or pulse) was computed by Rosswog et al. (2009, see their §6 for the details of the calculation and relevant formulae and references) who found it to be on the order of $h \sim 10^{-19}$, for a 0.6 $M_{\odot}$ white dwarf passing at the tidal disruption radius of a 10^3 $M_{\odot}$ BH at a distance of 10 kpc. This combination of parameters is appropriate for a Milky Way globular cluster, and suggests that the burst of gravitational waves from the disruption of a white dwarf in a nearby globular cluster may be detectable. However, more distant events, outside of the local group, are probably not detectable (see discussion in Anninos et al. 2018).

In an alternative and possibly rarer scenario the white dwarf is captured in a bound, typically eccentric, orbit around a BH and then spirals in. The per BH rate of such events is 1/10 the rate of all white dwarf disruptions (see discussion in Sect. 6.3). Such events are also referred to as "extreme mass ratio inspirals" (EMRIs, BH to white dwarf mass ratio $> 10^4$) or "intermediate mass ratio inspirals" (IMRIs, BH to white dwarf mass ratio 10^2–10^4). They are interesting because the white dwarf acts as a test particle that executes a very large number of orbits in the potential of the BH, hence the gravitational wave signal allows one to "map" the space-time around the BH and determine its fundamental properties: its mass and spin (see, for example, the review by Amaro-Seoane et al. 2007).

Since the white dwarf will complete a large number of orbital cycles around the BH before it is disrupted (10^5 revolutions, or even more) the gravitational wave signal is continuous and, in fact, the amplitude of the wave increases along with its frequency, as the white dwarf spirals in. The gravitational wave strain for an IMBH in a nearby (50–100 Mpc) dwarf galaxy is $h \sim 10^{-22}$–10^{-21}. The slow inspiral allows for a detection of the gravitational wave signal with a high signal-to-noise ratio (hereafter S/N) by accumulating a large number of orbital cycles in a long "exposure" or "integration." Sesana et al. (2008) have estimated (using the methodology described in Barack and Cutler 2004) the S/N attainable by a detector with capabilities similar to those of LISA and found that such systems may be detectable with $S/N > 30$ at a distance of 100 Mpc a few years before the disruption (with an integration time of order years). Under favourable conditions (i.e., a final orbital eccentricity of zero and an integration time of 3–5 years) detections of more distant systems, up to 450 Mpc away ($z \lesssim 0.1$), are possible, although with a lower S/N. Since the orbit is (at least initially) eccentric, the higher harmonics of the signal will be stronger than the fundamental and their detection will allow determination of the fundamental system parameters, i.e., the masses of the objects involved, the distance, and the BH spin. Knowing these parameters is extremely useful for using TDEs as tools to study BH properties (as discussed in Sect. 1).

A typical event involving the inspiral of 0.6 $M_\odot$ white dwarf into a 10^5 $M_\odot$ IMBH begins with the capture of the white dwarf in a highly eccentric, bound orbit with a pericentre distance of $\sim 15\ R_g$ ($\sim 2 \times 10^{11}$ cm), corresponding to an orbital period of 3 minutes and an initial gravitational wave frequency of $f_{\rm gw}^{\rm init} \approx 10$ mHz (see, for example, Sesana et al. 2008; MacLeod et al. 2014, 2016a). The orbit decays gradually via the emission of gravitational waves with a possible contribution from tidal heating and radiative cooling of the white dwarf (see MacLeod et al. 2016b). The tidal heating can be amplified by resonances (e.g., Rathore et al. 2005) and can lead to substantial brightening of the white dwarf and may trigger runaway fusion in a surface layer of hydrogen (Vick et al. 2017), possibly producing a soft X-ray flare. If emission of gravitational radiation is the dominant orbital decay mechanism, the decay time is approximately 3.4 years. During this period, the system is a steady source of gravitational waves whose frequency evolves to a maximum value of $f_{\rm gw}^{\rm fin} \approx 120$ mHz, as the orbital separation reaches the tidal disruption radius of $R_{\rm t} \approx 3\ R_{\rm g}$. Just before the white dwarf is disrupted, it fills its Roche lobe and goes through a phase of (stable) mass transfer onto the BH that can last days to weeks (see Zalamea et al. 2010; Dai and Blandford 2013). The accretion of this gas by the BH can lead to a substantial luminosity, on the order of 10^{43} erg s^{-1}, presumably in the X-ray band. If the orbit is still eccentric at this stage, the mass transfer, hence the X-ray emission, will also be periodic. The above scenario leads to intriguing predictions for the observational signature of this event. The first signal to be detected is the gravitational wave signal, which allows us to infer the masses of the objects involved, although the location in the sky will be fairly uncertain. This signal will also give us early warning of the TDE. The tidal heating and brightening of the white dwarf may lead to an observable, periodic signal in the optical/UV band (this is especially likely if nuclear burning is triggered on the surface of the white dwarf). Then, a few days to weeks before the ultimate disruption, we may detect steady or periodic X-ray emission as the white dwarf loses mass via Roche lobe overflow. The combination of these observational signatures holds great diagnostic power.

7.2 Ultra-High Energy Cosmic Rays from TDE Jets and Other Outflows

The jets and other outflows produced in some TDEs are also sites of particle acceleration and have been suggested as the sources of ultra-high energy cosmic rays (UHECRs, $E \gtrsim 10^{18}$ eV) and neutrinos (see, for example Farrar and Gruzinov 2009; Farrar and Piran 2014; Wang and Liu 2016; Senno et al. 2017; Lunardini and Winter 2017). Here we focus on the special conditions available in white dwarf TDEs and how they affect the process of UHECR production. The special feature of white dwarf TDEs is that they release debris that is rich in elements heavier than helium (hereafter, "metals"), which means that the resulting jets are made up of metals. Therefore, an appreciable number of metal nuclei can survive shocks internal and external to the jet and eventually arrive at the Earth and be detected by ground-based detectors as UHECRs. This particular feature of white dwarf TDE jets is essential in explaining the composition of UHECRs.

Three recent studies have computed the composition and energy spectrum of particles that are accelerated in white dwarf TDE jets and detected as UHECRs and neutrinos: Zhang et al. (2017), Alves Batista and Silk (2017), and Biehl et al. (2018). These authors considered the acceleration of the particles, the reactions that may destroy metal nuclei, the propagation of the particles from the source to the Earth and their detection by the IceCube and Pierre Auger observatories. The general conclusion is that the models can reproduce both the composition and energy spectrum of UHECRs (Zhang et al. 2017, note that O-Ne-Mg white dwarfs lead to better agreement). The implied TDE rates range between 0.1 $\mathrm{yr^{-1}\,Gpc^{-1}}$ (Biehl et al. 2018) and 10 $\mathrm{yr^{-1}\,Gpc^{-1}}$ (Alves Batista and Silk 2017) but these are inversely proportional to the baryon loading of the jets (the ratio of proton to photon luminosities) and proportional to the IMBH occupation fraction (see Sect. 6.3). In view of the uncertainties in these parameters, the rates reported in this paragraph are not incompatible with those of Sect. 6.3.

8 Conclusions

This chapter provides an overview of the tidal disruption of white dwarfs by IMBHs. While no unambiguous white dwarf TDE has yet been identified, theoretical calculations and simulations suggest that if IMBHs exist, then white dwarfs will be tidally disrupted by them if they approach close enough. The subsequent detonation of the white dwarf by tidal compression is dependent on the conditions of the system but some are expected to produce observable signatures at optical wavelengths, as well as high-energy counterparts. A number of candidate events have been suggested that match many of the predicted properties but no clear case as yet been observed. The predicted intrinsic rates of white dwarf TDEs are relatively uncertain but current and future transient surveys, such as LSST, should discover them at levels where identification is possible. This will allow the physics of these extreme systems to be studied along with the implications for populations of IMBH formation and evolution.

References

D. Aharon, H.B. Perets, The impact of mass segregation and star formation on the rates of gravitational-wave sources from extreme mass ratio inspirals. Astrophys. J. Lett. **830**(1), 1 (2016). https://doi.org/10.3847/2041-8205/830/1/L1

T. Alexander, C. Hopman, Strong mass segregation around a massive black hole. Astrophys. J. **697**(2), 1861–1869 (2009). https://doi.org/10.1088/0004-637X/697/2/1861

R. Alves Batista, J. Silk, Ultrahigh-energy cosmic rays from tidally-ignited white dwarfs. Phys. Rev. D **96**(10), 103003 (2017). https://doi.org/10.1103/PhysRevD.96.103003

P. Amaro-Seoane, J.R. Gair, M. Freitag, M.C. Miller, I. Mandel, C.J. Cutler, S. Babak, TOPICAL REVIEW: intermediate and extreme mass-ratio inspirals—astrophysics, science applications and detection using LISA. Class. Quantum Gravity **24**, 113–169 (2007). https://doi.org/10.1088/0264-9381/24/17/R01

P. Anninos, P.C. Fragile, S.S. Olivier, R. Hoffman, B. Mishra, K. Camarda, Relativistic tidal disruption and nuclear ignition of white dwarf stars by intermediate-mass black holes. Astrophys. J. **865**, 3 (2018). https://doi.org/10.3847/1538-4357/aadad9

P. Anninos, R.D. Hoffman, M. Grewal, M.J. Lavell, P.C. Fragile, Nuclear ignition of white dwarf stars by relativistic encounters with rotating intermediate mass black holes. Astrophys. J. **885**(2), 136 (2019). https://doi.org/10.3847/1538-4357/ab4ae0

S. Ayal, M. Livio, T. Piran, Tidal disruption of a solar-type star by a supermassive black hole. Astrophys. J. **545**, 772–780 (2000). https://doi.org/10.1086/317835

S. Ayal, T. Piran, R. Oechslin, M.B. Davies, S. Rosswog, Post-Newtonian smoothed particle hydrodynamics. Astrophys. J. **550**, 846–859 (2001). https://doi.org/10.1086/319769

J.N. Bahcall, R.A. Wolf, Star distribution around a massive black hole in a globular cluster. Astrophys. J. **209**, 214–232 (1976). https://doi.org/10.1086/154711

J.N. Bahcall, R.A. Wolf, The star distribution around a massive black hole in a globular cluster. II Unequal star masses. Astrophys. J. **216**, 883–907 (1977). https://doi.org/10.1086/155534

V.F. Baldassare, M. Geha, J. Greene, Identifying AGNs in low-mass galaxies via long-term optical variability. Astrophys. J. **868**, 152 (2018). https://doi.org/10.3847/1538-4357/aae6cf

I.K. Baldry, S.P. Driver, J. Loveday, E.N. Taylor, L.S. Kelvin, J. Liske, P. Norberg, A.S.G. Robotham, S. Brough, A.M. Hopkins, S.P. Bamford, J.A. Peacock, J. Bland-Hawthorn, C.J. Conselice, S.M. Croom, D.H. Jones, H.R. Parkinson, C.C. Popescu, M. Prescott, R.G. Sharp, R.J. Tuffs, Galaxy And Mass Assembly (GAMA): the galaxy stellar mass function at $z < 0.06$. Mon. Not. R. Astron. Soc. **421**, 621–634 (2012). https://doi.org/10.1111/j.1365-2966.2012.20340.x

E. Bañados, B.P. Venemans, C. Mazzucchelli, E.P. Farina, F. Walter, F. Wang, R. Decarli, D. Stern, X. Fan, F.B. Davies, J.F. Hennawi, R.A. Simcoe, M.L. Turner, H.-W. Rix, J. Yang, D.D. Kelson, G.C. Rudie, J.M. Winters, An 800-million-solar-mass black hole in a significantly neutral universe at a redshift of 7.5. Nature **553**, 473–476 (2018). https://doi.org/10.1038/nature25180

L. Barack, C. Cutler, LISA capture sources: approximate waveforms, signal-to-noise ratios, and parameter estimation accuracy. Phys. Rev. D **69**(8), 082005 (2004). https://doi.org/10.1103/PhysRevD.69.082005

H. Baumgardt, J. Makino, P. Hut, S. McMillan, S. Portegies Zwart, A dynamical model for the globular cluster G1. Astrophys. J. Lett. **589**, 25–28 (2003). https://doi.org/10.1086/375802

H. Baumgardt, J. Makino, T. Ebisuzaki, Massive black holes in star clusters. I. Equal-mass clusters. Astrophys. J. **613**, 1133–1142 (2004). https://doi.org/10.1086/423298

M.C. Begelman, E.M. Rossi, P.J. Armitage, Quasi-stars: accreting black holes inside massive envelopes. Mon. Not. R. Astron. Soc. **387**, 1649–1659 (2008). https://doi.org/10.1111/j.1365-2966.2008.13344.x

D. Biehl, D. Boncioli, C. Lunardini, W. Winter, Tidally disrupted stars as a possible origin of both cosmic rays and neutrinos at the highest energies. Sci. Rep. **8**, 10828 (2018). https://doi.org/10.1038/s41598-018-29022-4

M.R. Blanton, R.H. Lupton, D.J. Schlegel, M.A. Strauss, J. Brinkmann, M. Fukugita, J. Loveday, The properties and luminosity function of extremely low luminosity galaxies. Astrophys. J. **631**, 208–230 (2005). https://doi.org/10.1086/431416

J.S. Bloom, D. Giannios, B.D. Metzger, S.B. Cenko, D.A. Perley, N.R. Butler, N.R. Tanvir, A.J. Levan, P.T. O'Brien, L.E. Strubbe, F. De Colle, E. Ramirez-Ruiz, W.H. Lee, S. Nayakshin, E. Quataert, A.R. King, A. Cucchiara, J. Guillochon, G.C. Bower, A.S. Fruchter, A.N. Morgan, A.J. van der Horst, A possible relativistic jetted outburst from a massive black hole fed by a tidally disrupted star. Science **333**, 203 (2011). https://doi.org/10.1126/science.1207150

C. Bonnerot, E.M. Rossi, G. Lodato, D.J. Price, Disc formation from stellar tidal disruptions. ArXiv e-prints (2015). arXiv:1501.04635

M. Brassart, J.P. Luminet, Shock waves in tidally compressed stars by massive black holes. Astron. Astrophys. **481**, 259 (2008)

M. Brassart, J.-P. Luminet, Relativistic tidal compressions of a star by a massive black hole. Astron. Astrophys. **511**, 80 (2010). https://doi.org/10.1051/0004-6361/200913442

S.B. Cenko, H.A. Krimm, A. Horesh, A. Rau, D.A. Frail, J.A. Kennea, A.J. Levan, S.T. Holland, N.R. Butler, R.M. Quimby, J.S. Bloom, A.V. Filippenko, A. Gal-Yam, J. Greiner, S.R. Kulkarni, E.O. Ofek, F.E. Olivares, P. Schady, J.M. Silverman, N.R. Tanvir, D. Xu, Swift J2058.4+0516: discovery of a possible second relativistic tidal disruption flare? Astrophys. J. **753**, 77 (2012). https://doi.org/10.1088/0004-637X/753/1/77

R.M. Cheng, T. Bogdanović, Tidal disruption of a star in the Schwarzschild spacetime: relativistic effects in the return rate of debris. Phys. Rev. D **90**(6), 064020 (2014). https://doi.org/10.1103/PhysRevD.90.064020

R.M. Cheng, C.R. Evans, Relativistic effects in the tidal interaction between a white dwarf and a massive black hole in Fermi normal coordinates. Phys. Rev. D **87**(10), 104010 (2013). https://doi.org/10.1103/PhysRevD.87.104010

D. Clausen, M. Eracleous, Probing intermediate-mass black holes with optical emission lines from tidally disrupted white dwarfs. Astrophys. J. **726**(1), 34 (2011)

D. Clausen, S. Sigurdsson, M. Eracleous, J.A. Irwin, Luminous [O III] and [N II] from tidally disrupted horizontal branch stars. Mon. Not. R. Astron. Soc. **424**, 1268–1274 (2012). https://doi.org/10.1111/j.1365-2966.2012.21305.x

E.R. Coughlin, C. Nixon, Variability in tidal disruption events: gravitationally unstable streams. Astrophys. J. Lett. **808**, 11 (2015). https://doi.org/10.1088/2041-8205/808/1/L11

R. Courant, K. Friedrichs, H. Lewy, Über die partiellen Differenzengleichungen der mathematischen Physik. Math. Ann. **100**, 32–74 (1928). https://doi.org/10.1007/BF01448839

K.C. Dage, S.E. Zepf, A. Bahramian, A. Kundu, T.J. Maccarone, M.B. Peacock, X-ray variability from the ultraluminous black hole candidate X-ray binary in the globular cluster RZ 2109. Astrophys. J. **862**, 108 (2018). https://doi.org/10.3847/1538-4357/aacb2b

L. Dai, R. Blandford, Roche accretion of stars close to massive black holes. Mon. Not. R. Astron. Soc. **434**, 2948–2960 (2013). https://doi.org/10.1093/mnras/stt1209

L. Dai, J.C. McKinney, N. Roth, E. Ramirez-Ruiz, M.C. Miller, A unified model for tidal disruption events. Astrophys. J. **859**(2), 20 (2018). https://doi.org/10.3847/2041-8213/aab429

K. Danzmann, *LISA and LISA Pathfinder: Laser Interferometry in Space*. Society of Photo-Optical Instrumentation Engineers (SPIE) Conference Series, vol. 10566 (Springer, Berlin, 2017), p. 1056610. https://doi.org/10.1117/12.2308272

R. de Vita, M. Trenti, M. MacLeod, Wandering off the centre: a characterization of the random motion of intermediate-mass black holes in star clusters. Mon. Not. R. Astron. Soc. **475**(2), 1574–1586 (2018). https://doi.org/10.1093/mnras/stx3261

W.E. East, Gravitational waves from the collision of tidally disrupted stars with massive black holes. Astrophys. J. **795**, 135 (2014). https://doi.org/10.1088/0004-637X/795/2/135

C.R. Evans, C.S. Kochanek, The tidal disruption of a star by a massive black hole. Astrophys. J. Lett. **346**, 13 (1989)

C. Evans, P. Laguna, M. Eracleous, Ultra-close encounters of stars with massive black holes: tidal disruption events with prompt hyperaccretion. Astrophys. J. Lett. **805**, 19 (2015). https://doi.org/10.1088/2041-8205/805/2/L19

W. Even, J.E. Tohline, Constructing synchronously rotating double white dwarf binaries. Astrophys. J. Suppl. **184**, 248–263 (2009). https://doi.org/10.1088/0067-0049/184/2/248

X. Fan, Evolution of high-redshift quasars. New Astron. Rev. **50**, 665–671 (2006). https://doi.org/10.1016/j.newar.2006.06.077

G.R. Farrar, A. Gruzinov, Giant AGN flares and cosmic ray bursts. Astrophys. J. **693**(1), 329–332 (2009). https://doi.org/10.1088/0004-637X/693/1/329

G.R. Farrar, T. Piran, Tidal disruption jets as the source of Ultra-High Energy Cosmic Rays (2014). ArXiv e-prints

G. Fragione, N.W.C. Leigh, I. Ginsburg, B. Kocsis, Tidal disruption events and gravitational waves from intermediate-mass black holes in evolving globular clusters across space and time. Astrophys. J. **867**, 119 (2018). https://doi.org/10.3847/1538-4357/aae486

J. Frank, M.J. Rees, Effects of massive central black holes on dense stellar systems. Mon. Not. R. Astron. Soc. **176**, 633–647 (1976). https://doi.org/10.1093/mnras/176.3.633

E. Gafton, S. Rosswog, Tidal disruptions by rotating black holes: effects of spin and impact parameter. Mon. Not. R. Astron. Soc. **487**, 4790–4808 (2019)

E. Gafton, E. Tejeda, J. Guillochon, O. Korobkin, S. Rosswog, Relativistic effects on tidal disruption kicks of solitary stars. Mon. Not. R. Astron. Soc. **449**, 771–780 (2015). https://doi.org/10.1093/mnras/stv350

H. Gao, Y. Lu, S.N. Zhang, A new class of gamma-ray bursts from stellar disruptions by intermediate-mass black holes. Astrophys. J. **717**(1), 268–276 (2010). https://doi.org/10.1088/0004-637X/717/1/268

E. García-Berro, C. Badenes, G. Aznar-Siguán, P. Lorén-Aguilar, White dwarf dynamical interactions and fast optical transients. Mon. Not. R. Astron. Soc. **468**(4), 4815–4821 (2017). https://doi.org/10.1093/mnras/stx813

K. Gebhardt, R.M. Rich, L.C. Ho, An intermediate-mass black hole in the globular cluster G1: improved significance from new keck and Hubble Space Telescope observations. Astrophys. J. **634**, 1093–1102 (2005). https://doi.org/10.1086/497023

A. Glennie, P.G. Jonker, R.P. Fender, T. Nagayama, M.L. Pretorius, Two fast X-ray transients in archival Chandra data. Mon. Not. R. Astron. Soc. **450**, 3765–3770 (2015). https://doi.org/10.1093/mnras/stv801

F.G. Goicovic, V. Springel, S.T. Ohlmann, R. Pakmor, Hydrodynamical moving-mesh simulations of the tidal disruption of stars by supermassive black holes (2019). arXiv e-prints

J. Guillochon, E. Ramirez-Ruiz, Hydrodynamical simulations to determine the feeding rate of black holes by the tidal disruption of stars: the importance of the impact parameter and stellar structure. Astrophys. J. **767**, 25 (2013). https://doi.org/10.1088/0004-637X/767/1/25

J. Guillochon, E. Ramirez-Ruiz, A dark year for tidal disruption events. Astrophys. J. **809**, 166 (2015). https://doi.org/10.1088/0004-637X/809/2/166

J. Guillochon, E. Ramirez-Ruiz, S. Rosswog, D. Kasen, Three-dimensional simulations of tidally disrupted solar-type stars and the observational signatures of shock breakout. Astrophys. J. **705**, 844 (2009)

J. Guillochon, H. Manukian, E. Ramirez-Ruiz, Ps1-10jh: the disruption of a main-sequence star of near-solar composition. Astrophys. J. **783**(1), 23 (2014)

R. Haas, R.V. Shcherbakov, T. Bode, P. Laguna, Tidal disruptions of white dwarfs from ultra-close encounters with intermediate-mass spinning black holes. Astrophys. J. **749**(2), 117 (2012)

W.E. Harris, G.L.H. Harris, M. Alessi, A catalog of globular cluster systems: what determines the size of a galaxy's globular cluster population? Astrophys. J. **772**, 82 (2013). https://doi.org/10.1088/0004-637X/772/2/82

G.L.H. Harris, G.B. Poole, W.E. Harris, Globular clusters and supermassive black holes in galaxies: further analysis and a larger sample. Mon. Not. R. Astron. Soc. **438**, 2117–2130 (2014). https://doi.org/10.1093/mnras/stt2337

K. Hayasaki, N. Stone, A. Loeb, Finite, intense accretion bursts from tidal disruption of stars on bound orbits. Mon. Not. R. Astron. Soc. **434**(2), 909–924 (2013)

K. Hayasaki, N. Stone, A. Loeb, Circularization of tidally disrupted stars around spinning supermassive black holes. Mon. Not. R. Astron. Soc. **461**, 3760–3780 (2016). https://doi.org/10.1093/mnras/stw1387

W.R. Hix, A.M. Khokhlov, J.C. Wheeler, F.-K. Thielemann, The quasi-equilibrium-reduced alpha-network. Astrophys. J. **503**, 332 (1998). https://doi.org/10.1086/305968

A.Y.Q. Ho, E.S. Phinney, V. Ravi, S.R. Kulkarni, G. Petitpas, B. Emonts, V. Bhalerao, R. Blundell, S.B. Cenko, D. Dobie, R. Howie, N. Kamraj, M.M. Kasliwal, T. Murphy, D.A. Perley, T.K. Sridharan, I. Yoon, AT2018cow: a luminous millimeter transient. Astrophys. J. **871**, 73 (2019). https://doi.org/10.3847/1538-4357/aaf473

M.P. Hobson, G.P. Efstathiou, A.N. Lasenby, *General Relativity* (2006). https://doi.org/10.2277/0521829518

J.A. Irwin, T.G. Brink, J.N. Bregman, T.P. Roberts, Evidence for a stellar disruption by an intermediate-mass black hole in an extragalactic globular cluster. Astrophys. J. Lett. **712**, 1–4 (2010). https://doi.org/10.1088/2041-8205/712/1/L1

J.A. Irwin, W.P. Maksym, G.R. Sivakoff, A.J. Romanowsky, D. Lin, T. Speegle, I. Prado, D. Mildebrath, J. Strader, J. Liu, J.M. Miller, Ultraluminous X-ray bursts in two ultracompact companions to nearby elliptical galaxies. Nature **538**, 356–358 (2016). https://doi.org/10.1038/nature19822

P.G. Jonker, M. Heida, M.A.P. Torres, J.C.A. Miller-Jones, A.C. Fabian, E.M. Ratti, G. Miniutti, D.J. Walton, T.P. Roberts, The nature of the bright ULX X-2 in NGC 3921: a Chandra position and HST candidate counterpart. Astrophys. J. **758**, 28 (2012). https://doi.org/10.1088/0004-637X/758/1/28

P.G. Jonker, A. Glennie, M. Heida, T. Maccarone, S. Hodgkin, G. Nelemans, J.C.A. Miller-Jones, M.A.P. Torres, R. Fender, Discovery of a new kind of explosive X-ray transient near M86. Astrophys. J. **779**, 14 (2013). https://doi.org/10.1088/0004-637X/779/1/14

P.G. Jonker, N.C. Stone, A. Generozov, S. van Velzen, B. Metzger, Implications from Late-Time X-ray Detections of Optically Selected Tidal Disruption Events: State Changes, Unification, and Detection Rates (2019). arXiv:1906.12236

M.M. Kasliwal, S.R. Kulkarni, A. Gal-Yam, P.E. Nugent, M. Sullivan, L. Bildsten, O. Yaron, H.B. Perets, I. Arcavi, S. Ben-Ami, V.B. Bhalerao, J.S. Bloom, S.B. Cenko, A.V. Filippenko, D.A. Frail, M. Ganeshalingam, A. Horesh, D.A. Howell, N.M. Law, D.C. Leonard, W. Li, E.O. Ofek, D. Polishook, D. Poznanski, R.M. Quimby, J.M. Silverman, A. Sternberg, D. Xu, Calcium-rich gap transients in the remote outskirts of galaxies. Astrophys. J. **755**(2), 161 (2012). https://doi.org/10.1088/0004-637X/755/2/161

K. Kawana, A. Tanikawa, N. Yoshida, Tidal Disruption of a White Dwarf by a Black Hole: the Diversity of Nucleosynthesis, Explosion Energy, and the Fate of Debris Streams (2017). ArXiv e-prints

S.O. Kepler, S.J. Kleinman, A. Nitta, D. Koester, B.G. Castanheira, O. Giovannini, A.F.M. Costa, L. Althaus, White dwarf mass distribution in the SDSS. Mon. Not. R. Astron. Soc. **375**, 1315–1324 (2007). https://doi.org/10.1111/j.1365-2966.2006.11388.x

M. Kesden, Black-hole spin dependence in the light curves of tidal disruption events. Phys. Rev. D **86**(6), 064026 (2012). https://doi.org/10.1103/PhysRevD.86.064026

B. Kızıltan, H. Baumgardt, A. Loeb, An intermediate-mass black hole in the centre of the globular cluster 47 Tucanae. Nature **542**, 203–205 (2017). https://doi.org/10.1038/nature21361

S. Kobayashi, P. Laguna, E.S. Phinney, P. Mészáros, Gravitational waves and x-ray signals from stellar disruption by a massive black hole. Astrophys. J. **615**, 855 (2004)

C.S. Kochanek, The aftermath of tidal disruption: the dynamics of thin gas streams. Astrophys. J. **422**, 508 (1994). https://doi.org/10.1086/173745

J. Kormendy, L.C. Ho, Coevolution (or not) of supermassive black holes and host galaxies. Annu. Rev. Astron. Astrophys. **51**, 511–653 (2013). https://doi.org/10.1146/annurev-astro-082708-101811

J.H. Krolik, T. Piran, Swift J1644+57: a white dwarf tidally disrupted by a 10^4 $M_\odot$ black hole? Astrophys. J. **743**, 134 (2011). https://doi.org/10.1088/0004-637X/743/2/134

J.H. Krolik, T. Piran, Jets from tidal disruptions of stars by black holes. Astrophys. J. **749**(1), 92 (2012)

N.P.M. Kuin, K. Wu, S. Oates, A. Lien, S. Emery, J.A. Kennea, M. de Pasquale, Q. Han, P.J. Brown, A. Tohuvavohu, A. Breeveld, D.N. Burrows, S.B. Cenko, S. Campana, A. Levan, C. Markwardt, J.P. Osborne, M.J. Page, K.L. Page, B. Sbarufatti, M. Siegel, E. Troja, Swift spectra of AT2018cow: a white dwarf tidal disruption event? Mon. Not. R. Astron. Soc. (2019). https://doi.org/10.1093/mnras/stz053

J. Law-Smith, M. MacLeod, J. Guillochon, P. Macias, E. Ramirez-Ruiz, Low-mass white dwarfs with hydrogen envelopes as a missing link in the tidal disruption menu. Astrophys. J. **841**(2), 132 (2017). https://doi.org/10.3847/1538-4357/aa6ffb

A.J. Levan, N.R. Tanvir, S.B. Cenko, D.A. Perley, K. Wiersema, J.S. Bloom, A.S. Fruchter, A.D.U. Postigo, P.T. O'Brien, N. Butler, A.J. van der Horst, G. Leloudas, A.N. Morgan, K. Misra, G.C. Bower, J. Farihi, R.L. Tunnicliffe, M. Modjaz, J.M. Silverman, J. Hjorth, C. Thöne, A. Cucchiara, J.M.C. Cerón, A.J. Castro-Tirado, J.A. Arnold, M. Bremer, J.P. Brodie, T. Carroll, M.C. Cooper, P.A. Curran, R.M. Cutri, J. Ehle, D. Forbes, J. Fynbo, J. Gorosabel, J. Graham, D.I. Hoffman, S. Guziy, P. Jakobsson, A. Kamble, T. Kerr, M.M. Kasliwal, C. Kouveliotou, D. Kocevski, N.M. Law, P.E. Nugent, E.O. Ofek, D. Poznanski, R.M. Quimby, E. Rol, A.J. Romanowsky, R. Sánchez-Ramírez, S. Schulze, N. Singh, L. van Spaandonk, R.L.C. Starling, R.G. Strom, J.C. Tello, O. Vaduvescu, P.J. Wheatley, R.A.M.J. Wijers, J.M. Winters, D. Xu, An extremely luminous panchromatic outburst from the nucleus of a distant galaxy. Science **333**, 199 (2011). https://doi.org/10.1126/science.1207143

A.J. Levan, N.R. Tanvir, R.L.C. Starling, K. Wiersema, K.L. Page, D.A. Perley, S. Schulze, G.A. Wynn, R. Chornock, J. Hjorth, S.B. Cenko, A.S. Fruchter, P.T. O'Brien, G.C. Brown, R.L. Tunnicliffe, D. Malesani, P. Jakobsson, D. Watson, E. Berger, D. Bersier, B.E. Cobb, S. Covino, A. Cucchiara, A. de Ugarte Postigo, D.B. Fox, A. Gal-Yam, P. Goldoni, J. Gorosabel, L. Kaper, T. Krühler, R. Karjalainen, J.P. Osborne, E. Pian, R. Sánchez-Ramírez, B. Schmidt, I. Skillen, G. Tagliaferri, C. Thöne, O. Vaduvescu, R.A.M.J. Wijers, B.A. Zauderer, A new population of ultra-long duration gamma-ray bursts. Astrophys. J. **781**, 13 (2014). https://doi.org/10.1088/0004-637X/781/1/13

D. Lin, J. Strader, E.R. Carrasco, D. Page, A.J. Romanowsky, J. Homan, J.A. Irwin, R.A. Remillard, O. Godet, N.A. Webb, H. Baumgardt, R. Wijnands, D. Barret, P.-A. Duc, J.P. Brodie, S.D.J. Gwyn, A luminous X-ray outburst from an intermediate-mass black hole in an off-centre star cluster. Nat. Astron. (2018). https://doi.org/10.1038/s41550-018-0493-1

G. Lodato, E.M. Rossi, Multiband light curves of tidal disruption events. Mon. Not. R. Astron. Soc. **410**(1), 359–367 (2011). https://doi.org/10.1111/j.1365-2966.2010.17448.x

G. Lodato, A.R. King, J.E. Pringle, Stellar disruption by a supermassive black hole: is the light curve really proportional to t-5/3? Mon. Not. R. Astron. Soc. **392**, 332 (2009)

J.P. Luminet, B. Carter, Dynamics of an affine star model in a black hole tidal field. Astrophys. J. Suppl. **61**, 219 (1986)

J. Luminet, J.-A. Marck, Tidal squeezing of stars by Schwarzschild black holes. Mon. Not. R. Astron. Soc. **212**, 57 (1985)

J. Luminet, B. Pichon, Tidal pinching of white dwarfs. Astron. Astrophys. **209**, 103 (1989)

C. Lunardini, W. Winter, High energy neutrinos from the tidal disruption of stars. Phys. Rev. D **95**(12), 123001 (2017). https://doi.org/10.1103/PhysRevD.95.123001

R. Lunnan, M.M. Kasliwal, Y. Cao, L. Hangard, O. Yaron, J.T. Parrent, C. McCully, A. Gal-Yam, J.S. Mulchaey, S. Ben-Ami, A.V. Filippenko, C. Fremling, A.S. Fruchter, D.A. Howell, J. Koda, T. Kupfer, S.R. Kulkarni, R. Laher, F. Masci, P.E. Nugent, E.O. Ofek, M. Yagi, L. Yan, Two new calcium-rich gap transients in group and cluster environments. Astrophys. J. **836**(1), 60 (2017). https://doi.org/10.3847/1538-4357/836/1/60

T.J. Maccarone, B. Warner, Strong [O III] and [N II] emission lines in globular clusters from photoionized R corona borealis star winds. Mon. Not. R. Astron. Soc. **410**, 32–36 (2011). https://doi.org/10.1111/j.1745-3933.2010.00973.x

T.J. Maccarone, A. Kundu, S.E. Zepf, K.L. Rhode, A black hole in a globular cluster. Nature **445**, 183–185 (2007). https://doi.org/10.1038/nature05434

M. MacLeod, J. Goldstein, E. Ramirez-Ruiz, J. Guillochon, J. Samsing, Illuminating massive black holes with white dwarfs: orbital dynamics and high-energy transients from tidal interactions. Astrophys. J. **794**, 9 (2014). https://doi.org/10.1088/0004-637X/794/1/9

M. MacLeod, M. Trenti, E. Ramirez-Ruiz, The close stellar companions to intermediate-mass black holes. Astrophys. J. **819**, 70 (2016a). https://doi.org/10.3847/0004-637X/819/1/70

M. MacLeod, J. Guillochon, E. Ramirez-Ruiz, D. Kasen, S. Rosswog, Optical thermonuclear transients from tidal compression of white dwarfs as tracers of the low end of the massive black hole mass function. Astrophys. J. **819**, 3 (2016b). https://doi.org/10.3847/0004-637X/819/1/3

M. MacLeod, J. Guillochon, E. Ramirez-Ruiz, D. Kasen, S. Rosswog, Erratum: "Optical thermonuclear transients from tidal compression of white dwarfs as tracers of the low end of the massive black hole mass function". Astrophys. J. **843**, 154 (2017). https://doi.org/10.3847/1538-4357/aa7b74

R. Margutti, B.D. Metzger, R. Chornock, I. Vurm, N. Roth, B.W. Grefenstette, V. Savchenko, R. Cartier, J.F. Steiner, G. Terreran, G. Migliori, D. Milisavljevic, K.D. Alexander, M. Bietenholz, P.K. Blanchard, E. Bozzo, D. Brethauer, I.V. Chilingarian, D.L. Coppejans, L. Ducci, C. Ferrigno, W. Fong, D. GÖtz, C. Guidorzi, A. Hajela, K. Hurley, E. Kuulkers, P. Laurent, S. Mereghetti, M. Nicholl, D. Patnaude, P. Ubertini, J. Banovetz, N. Bartel, E. Berger, E.R. Coughlin, T. Eftekhari, D.D. Frederiks, A.V. Kozlova, T. Laskar, D.S. Svinkin, M.R. Drout, A. Macfadyen, K. Paterson, An embedded X-ray source shines through the aspherical AT2018cow: revealing the inner workings of the most luminous fast-evolving optical transients (2018). arXiv e-prints

L. Mayer, S. Kazantzidis, A. Escala, S. Callegari, Direct formation of supermassive black holes via multi-scale gas inflows in galaxy mergers. Nature **466**, 1082–1084 (2010). https://doi.org/10.1038/nature09294

N.J. McConnell, C.-P. Ma, Revisiting the scaling relations of black hole masses and host galaxy properties. Astrophys. J. **764**, 184 (2013). https://doi.org/10.1088/0004-637X/764/2/184

A. Merloni, A.C. Fabian, Accretion disc coronae as magnetic reservoirs. Mon. Not. R. Astron. Soc. **321**, 549–552 (2001). https://doi.org/10.1046/j.1365-8711.2001.04060.x

D. Merritt, *Dynamics and Evolution of Galactic Nuclei* (2013)

D. Merritt, J.D. Schnittman, S. Komossa, Hypercompact stellar systems around recoiling supermassive black holes. Astrophys. J. **699**, 1690–1710 (2009). https://doi.org/10.1088/0004-637X/699/2/1690

B.P. Miller, E. Gallo, J.E. Greene, B.C. Kelly, T. Treu, J.-H. Woo, V. Baldassare, X-ray constraints on the local supermassive black hole occupation fraction. Astrophys. J. **799**, 98 (2015). https://doi.org/10.1088/0004-637X/799/1/98

J.J. Monaghan, Smoothed particle hydrodynamics. Rep. Prog. Phys. **68**, 1703–1759 (2005). https://doi.org/10.1088/0034-4885/68/8/R01

E.C. Moran, K. Shahinyan, H.R. Sugarman, D.O. Vélez, M. Eracleous, Black holes at the centers of nearby dwarf galaxies. Astron. J. **148**, 136 (2014). https://doi.org/10.1088/0004-6256/148/6/136

D.J. Mortlock, S.J. Warren, B.P. Venemans, M. Patel, P.C. Hewett, R.G. McMahon, C. Simpson, T. Theuns, E.A. Gonzáles-Solares, A. Adamson, S. Dye, N.C. Hambly, P. Hirst, M.J. Irwin, E. Kuiper, A. Lawrence, H.J.A. Röttgering, A luminous quasar at a redshift of z = 7.085. Nature **474**, 616–619 (2011). https://doi.org/10.1038/nature10159

E. Müller, Simulation of astrophysical fluid flow, in *Computational methods for astrophysical fluid flow*, ed. by O. Steiner, A. Gautschy (Springer, Berlin, 1998)

M. Nauenberg, Analytic approximations to the mass-radius relation and energy of zero-temperature stars. Astrophys. J. **175**, 417 (1972). https://doi.org/10.1086/151568

B. Newell, G.S. Da Costa, J. Norris, Evidence for a central massive object in the X-ray cluster M15. Astrophys. J. Lett. **208**, 55 (1976). https://doi.org/10.1086/182232

M.B. Peacock, S.E. Zepf, A. Kundu, T.J. Maccarone, K.L. Rhode, J.J. Salzer, C.Z. Waters, R. Ciardullo, C. Gronwall, D. Stern, Spatially resolved spectroscopy of the globular cluster RZ 2109 and the nature of its black hole. Astrophys. J. **759**, 126 (2012). https://doi.org/10.1088/0004-637X/759/2/126

B.B.P. Perera, B.W. Stappers, A.G. Lyne, C.G. Bassa, I. Cognard, L. Guillemot, M. Kramer, G. Theureau, G. Desvignes, Evidence for an intermediate-mass black hole in the globular cluster NGC 6624. Mon. Not. R. Astron. Soc. **468**, 2114–2127 (2017). https://doi.org/10.1093/mnras/stx501

H.B. Perets, A. Gal-Yam, P.A. Mazzali, D. Arnett, D. Kagan, A.V. Filippenko, W. Li, I. Arcavi, S.B. Cenko, D.B. Fox, D.C. Leonard, D.-S. Moon, D.J. Sand, A.M. Soderberg, J.P. Anderson, P.A. James, R.J. Foley, M. Ganeshalingam, E.O. Ofek, L. Bildsten, G. Nelemans, K.J. Shen, N.N. Weinberg, B.D. Metzger, A.L. Piro, E. Quataert, M. Kiewe, D. Poznanski, A faint type of supernova from a white dwarf with a helium-rich companion. Nature **465**(7296), 322–325 (2010). https://doi.org/10.1038/nature09056

H.B. Perets, Z. Li, J.C. Lombardi Jr., S.R. Milcarek Jr., Micro-tidal disruption events by stellar compact objects and the production of ultra-long GRBs. Astrophys. J. **823**(2), 113 (2016). https://doi.org/10.3847/0004-637X/823/2/113

D.A. Perley, P.A. Mazzali, L. Yan, S.B. Cenko, S. Gezari, K. Taggart, N. Blagorodnova, C. Fremling, B. Mockler, A. Singh, N. Tominaga, M. Tanaka, A.M. Watson, T. Ahumada, G.C. Anupama, C. Ashall, R.L. Becerra, D. Bersier, V. Bhalerao, J.S. Bloom, N.R. Butler, C. Copperwheat, M.W. Coughlin, K. De, A.J. Drake, D.A. Duev, S. Frederick, J.J. González, A. Goobar, M. Heida, A.Y.Q. Ho, J. Horst, T. Hung, R. Itoh, J.E. Jencson, M.M. Kasliwal, N. Kawai, T. Khanam, S.R. Kulkarni, B. Kumar, H. Kumar, A.S. Kutyrev, W.H. Lee, K. Maeda, A. Mahabal, K.L. Murata, J.D. Neill, C.-C. Ngeow, B. Penprase, E. Pian, R. Quimby, E. Ramirez-Ruiz, M.G. Richer, C.G. Román-Zúñiga, D.K. Sahu, S. Srivastav, Q. Socia, J. Sollerman, Y. Tachibana, F. Taddia, S. Tinyanont, E. Troja, C. Ward, J. Wee, P.-C. Yu, The fast, luminous ultraviolet transient AT2018cow: extreme supernova, or disruption of a star by an intermediate-mass black hole? Mon. Not. R. Astron. Soc. **484**, 1031–1049 (2019). https://doi.org/10.1093/mnras/sty3420

E.S. Phinney, Manifestations of a massive black hole in the galactic center, in *IAU Symp. 136: The Center of the Galaxy*, ed. by M. Morris (1989), p. 543

S.F. Portegies Zwart, S.L.W. McMillan, The runaway growth of intermediate-mass black holes in dense star clusters. Astrophys. J. **576**, 899–907 (2002). https://doi.org/10.1086/341798

R.L. Porter, Broad and luminous [OIII] and [NII] in globular cluster ULXs. Mon. Not. R. Astron. Soc. **407**, 59–63 (2010). https://doi.org/10.1111/j.1745-3933.2010.00904.x

S.J. Prentice, K. Maguire, S.J. Smartt, M.R. Magee, P. Schady, S. Sim, T.-W. Chen, P. Clark, C. Colin, M. Fulton, O. McBrien, D. O'Neill, K.W. Smith, C. Ashall, K.C. Chambers, L. Denneau, H.A. Flewelling, A. Heinze, T.W.-S. Holoien, M.E. Huber, C.S. Kochanek, P.A. Mazzali, J.L. Prieto, A. Rest, B.J. Shappee, B. Stalder, K.Z. Stanek, M.D. Stritzinger, T.A. Thompson, J.L. Tonry, The cow: discovery of a luminous, hot, and rapidly evolving transient. Astrophys. J. Lett. **865**, 3 (2018). https://doi.org/10.3847/2041-8213/aadd90

E. Quataert, D. Kasen, Swift 1644+57: the longest gamma-ray burst? Mon. Not. R. Astron. Soc. **419**, 1–5 (2012). https://doi.org/10.1111/j.1745-3933.2011.01151.x

E. Ramirez-Ruiz, S. Rosswog, The star ingesting luminosity of intermediate-mass black holes in globular clusters. Astrophys. J. Lett. **697**, 77–80 (2009). https://doi.org/10.1088/0004-637X/697/2/L77

E. Rantsiou, S. Kobayashi, P. Laguna, F.A. Rasio, Mergers of black hole-neutron star binaries. I. Methods and first results. Astrophys. J. **680**, 1326–1349 (2008). https://doi.org/10.1086/587858

Y. Rathore, R.D. Blandford, A.E. Broderick, Resonant excitation of white dwarf oscillations in compact object binaries - I. The no back reaction approximation. Mon. Not. R. Astron. Soc. **357**(3), 834–846 (2005). https://doi.org/10.1111/j.1365-2966.2004.08459.x

M.J. Rees, Tidal disruption of stars by black holes of 10 to the 6th-10 to the 8th solar masses in nearby galaxies. Nature **333**, 523 (1988)

A.E. Reines, J.E. Greene, M. Geha, Dwarf galaxies with optical signatures of active massive black holes. Astrophys. J. **775**, 116 (2013). https://doi.org/10.1088/0004-637X/775/2/116

E. Ripamonti, M. Mapelli, Broad [O III] in the globular cluster RZ 2109: X-ray ionized nova ejecta. Mon. Not. R. Astron. Soc. **423**, 1144–1153 (2012). https://doi.org/10.1111/j.1365-2966.2012.20942.x

L.E. Rivera Sandoval, T.J. Maccarone, A. Corsi, P.J. Brown, D. Pooley, J.C. Wheeler, X-ray Swift observations of SN 2018cow. Mon. Not. R. Astron. Soc. **480**, 146–150 (2018). https://doi.org/10.1093/mnrasl/sly145

S. Rosswog, Astrophysical smooth particle hydrodynamics. New Astron. Rev. **53**, 78–104 (2009). https://doi.org/10.1016/j.newar.2009.08.007

S. Rosswog, SPH methods in the modelling of compact objects. Living Rev. Comput. Astrophys. **1**, 1 (2015). https://doi.org/10.1007/lrca-2015-1

S. Rosswog, The Lagrangian hydrodynamics code MAGMA2 (2019). arXiv:1911.13093

S. Rosswog, E. Ramirez-Ruiz, W.R. Hix, Tidal disruption and ignition of white dwarfs by moderately massive black holes. Astrophys. J. **695**(1), 404–419 (2009)

N. Roth, D. Kasen, J. Guillochon, E. Ramirez-Ruiz, The X-ray through optical fluxes and line strengths of tidal disruption events. Astrophys. J. **827**, 3 (2016). https://doi.org/10.3847/0004-637X/827/1/3

A. Sądowski, E. Tejeda, E. Gafton, S. Rosswog, D. Abarca, Magnetohydrodynamical simulations of a deep tidal disruption in general relativity. Mon. Not. R. Astron. Soc. **458**, 4250–4268 (2016). https://doi.org/10.1093/mnras/stw589

P.H. Sell, T.J. Maccarone, R. Kotak, C. Knigge, D.J. Sand, Calcium-rich gap transients: tidal detonations of white dwarfs? Mon. Not. R. Astron. Soc. **450**, 4198–4206 (2015). https://doi.org/10.1093/mnras/stv902

P.H. Sell, K. Arur, T.J. Maccarone, R. Kotak, C. Knigge, D.J. Sand, S. Valenti, Chandra X-ray constraints on the candidate Ca-rich gap transient SN 2016hnk. Mon. Not. R. Astron. Soc. **475**, 111–115 (2018). https://doi.org/10.1093/mnrasl/sly011

N. Senno, K. Murase, P. Mészáros, High-energy neutrino flares from X-ray bright and dark tidal disruption events. Astrophys. J. **838**(1), 3 (2017). https://doi.org/10.3847/1538-4357/aa6344

A. Sesana, A. Vecchio, M. Eracleous, S. Sigurdsson, Observing white dwarfs orbiting massive black holes in the gravitational wave and electro-magnetic window. Mon. Not. R. Astron. Soc. **391**, 718–726 (2008). https://doi.org/10.1111/j.1365-2966.2008.13904.x

R.V. Shcherbakov, A. Pe'er, C.S. Reynolds, R. Haas, T. Bode, P. Laguna, Grb060218 as a tidal disruption of a white dwarf by an intermediate-mass black hole. Astrophys. J. **769**(2), 85 (2013)

g-F. Shen Fast, Ultraluminous X-ray bursts from tidal stripping of white dwarfs by intermediate-mass black holes. Astrophys. J. Lett. **871**, 17 (2019). https://doi.org/10.3847/2041-8213/aafc64

K.J. Shen, D. Kasen, N.N. Weinberg, L. Bildsten, E. Scannapieco, Thermonuclear Ia supernovae from helium shell detonations: explosion models and observables. Astrophys. J. **715**(2), 767–774 (2010). https://doi.org/10.1088/0004-637X/715/2/767

H. Shiokawa, J.H. Krolik, R.M. Cheng, T. Piran, S.C. Noble, General relativistic hydrodynamic simulation of accretion flow from a stellar tidal disruption. Astrophys. J. **804**, 85 (2015). https://doi.org/10.1088/0004-637X/804/2/85

D. Sijacki, M. Vogelsberger, S. Genel, V. Springel, P. Torrey, G.F. Snyder, D. Nelson, L. Hernquist, The illustris simulation: the evolving population of black holes across cosmic time. Mon. Not. R. Astron. Soc. **452**, 575–596 (2015). https://doi.org/10.1093/mnras/stv1340

J. Silk, Feedback by massive black holes in gas-rich dwarf galaxies. Astrophys. J. Lett. **839**, 13 (2017). https://doi.org/10.3847/2041-8213/aa67da

A.M. Soderberg, D.A. Frail, M.H. Wieringa, Constraints on off-axis gamma-ray burst jets in type ibc supernovae from late-time radio observations. Astrophys. J. Lett. **607**, 13–16 (2004). https://doi.org/10.1086/421722

A.M. Soderberg, S.R. Kulkarni, D.B. Fox, E. Berger, P.A. Price, S.B. Cenko, D.A. Howell, A. Gal-Yam, D.C. Leonard, D.A. Frail, D. Moon, R.A. Chevalier, M. Hamuy, K.C. Hurley, D. Kelson, K. Koviak, W. Krzeminski, P. Kumar, A. MacFadyen, P.J. McCarthy, H.S. Park, B.A. Peterson, M.M. Phillips, M. Rauch, M. Roth, B.P. Schmidt, S. Shectman, An HST search for supernovae accompanying X-ray flashes. Astrophys. J. **627**, 877–887 (2005). https://doi.org/10.1086/430405

V. Springel, Smoothed particle hydrodynamics in astrophysics. Annu. Rev. Astron. Astrophys. **48**, 391–430 (2010). https://doi.org/10.1146/annurev-astro-081309-130914

M.M. Steele, S.E. Zepf, A. Kundu, T.J. Maccarone, K.L. Rhode, J.J. Salzer, Velocity structure and variability of [O III] emission in black hole host globular cluster RZ2109. Astrophys. J. **739**, 95 (2011). https://doi.org/10.1088/0004-637X/739/2/95

N. Stone, R. Sari, A. Loeb, Consequences of strong compression in tidal disruption events. Mon. Not. R. Astron. Soc. **435**(3), 1809–1824 (2013)

J. Strader, L. Chomiuk, T.J. Maccarone, J.C.A. Miller-Jones, A.C. Seth, C.O. Heinke, G.R. Sivakoff, No evidence for intermediate-mass black holes in globular clusters: strong constraints from the JVLA. Astrophys. J. Lett. **750**, 27 (2012). https://doi.org/10.1088/2041-8205/750/2/L27

L.E. Strubbe, E. Quataert, Optical flares from the tidal disruption of stars by massive black holes. Mon. Not. R. Astron. Soc. **400**, 2070 (2009)

L.E. Strubbe, E. Quataert, Spectroscopic signatures of the tidal disruption of stars by massive black holes. Mon. Not. R. Astron. Soc. **415**(1), 168–180 (2011)

A. Tanikawa, High-resolution hydrodynamic simulation of tidal detonation of a helium white dwarf by an intermediate mass black hole. Astrophys. J. **858**(1), 26 (2018a). https://doi.org/10.3847/1538-4357/aaba79

A. Tanikawa, Tidal double detonation: a new mechanism for the thermonuclear explosion of a white dwarf induced by a tidal disruption event. Mon. Not. R. Astron. Soc. **475**(1), 67–71 (2018b). https://doi.org/10.1093/mnrasl/sly006

A. Tanikawa, Y. Sato, K. Nomoto, K. Maeda, N. Nakasato, I. Hachisu, Does explosive nuclear burning occur in tidal disruption events of white dwarfs by intermediate-mass black holes? Astrophys. J. **839**, 81 (2017). https://doi.org/10.3847/1538-4357/aa697d

A. Tchekhovskoy, B.D. Metzger, D. Giannios, L.Z. Kelley, Swift J1644+57 gone MAD: the case for dynamically important magnetic flux threading the black hole in a jetted tidal disruption event. Mon. Not. R. Astron. Soc. **437**, 2744–2760 (2014). https://doi.org/10.1093/mnras/stt2085

E. Tejeda, S. Rosswog, An accurate Newtonian description of particle motion around a Schwarzschild black hole. Mon. Not. R. Astron. Soc. **433**, 1930–1940 (2013). https://doi.org/10.1093/mnras/stt853

E. Tejeda, E. Gafton, S. Rosswog, Tidal disruptions by rotating black holes: relativistic hydrodynamics with Newtonian codes (2017). ArXiv e-prints

E. Tremou, J. Strader, L. Chomiuk, L. Shishkovsky, T.J. Maccarone, J.C.A. Miller-Jones, V. Tudor, C.O. Heinke, G.R. Sivakoff, A.C. Seth, E. Noyola, The MAVERIC survey: still no evidence for accreting intermediate-mass black holes in globular clusters. Astrophys. J. **862**, 16 (2018). https://doi.org/10.3847/1538-4357/aac9b9

Y. Urata, K. Huang, R. Yamazaki, T. Sakamoto, Extremely soft x-ray flash as the indicator of off-axis orphan GRB afterglow. Astrophys. J. **806**(2), 222 (2015). https://doi.org/10.1088/0004-637x/806/2/222

M. Vick, D. Lai, J. Fuller, Tidal dissipation and evolution of white dwarfs around massive black holes: an eccentric path to tidal disruption. Mon. Not. R. Astron. Soc. **468**, 2296–2310 (2017). https://doi.org/10.1093/mnras/stx539

K.T. Voggel, A.C. Seth, H. Baumgardt, S. Mieske, J. Pfeffer, A. Rasskazov, The impact of stripped nuclei on the supermassive black hole number density in the local universe. Astrophys. J. **871**, 159 (2019). https://doi.org/10.3847/1538-4357/aaf735

M. Volonteri, The formation and evolution of massive black holes. Science **337**, 544 (2012). https://doi.org/10.1126/science.1220843

X.-Y. Wang, R.-Y. Liu, Tidal disruption jets of supermassive black holes as hidden sources of cosmic rays: explaining the IceCube TeV-PeV neutrinos. Phys. Rev. D **93**(8), 083005 (2016). https://doi.org/10.1103/PhysRevD.93.083005

N. Webb, D. Cseh, E. Lenc, O. Godet, D. Barret, S. Corbel, S. Farrell, R. Fender, N. Gehrels, I. Heywood, Radio detections during two state transitions of the intermediate-mass black hole HLX-1. Science **337**, 554 (2012). https://doi.org/10.1126/science.1222779

S.E. Woosley, A. Heger, Long gamma-ray transients from collapsars. Astrophys. J. **752**, 32 (2012). https://doi.org/10.1088/0004-637X/752/1/32

A.H. Wright, A.S.G. Robotham, S.P. Driver, M. Alpaslan, S.K. Andrews, I.K. Baldry, J. Bland-Hawthorn, S. Brough, M.J.I. Brown, M. Colless et al., Galaxy and mass assembly (gama): the galaxy stellar mass function to $z = 0.1$ from the R-band selected equatorial regions. Mon. Not. R. Astron. Soc. **470**(1), 283–302 (2017). https://doi.org/10.1093/mnras/stx1149

I. Zalamea, K. Menou, A.M. Beloborodov, White dwarfs stripped by massive black holes: sources of coincident gravitational and electromagnetic radiation. Mon. Not. R. Astron. Soc. **409**, 25–29 (2010). https://doi.org/10.1111/j.1745-3933.2010.00930.x

D. Zaritsky, M. Aravena, E. Athanassoula, A. Bosma, S. Comerón, B.G. Elmegreen, S. Erroz-Ferrer, D.A. Gadotti, J.L. Hinz, L.C. Ho et al., Globular cluster populations: first results from s4g early-type galaxies. Astrophys. J. **799**(2), 159 (2015). https://doi.org/10.1088/0004-637x/799/2/159

B.A. Zauderer, E. Berger, A.M. Soderberg, A. Loeb, R. Narayan, D.A. Frail, G.R. Petitpas, A. Brunthaler, R. Chornock, J.M. Carpenter, G.G. Pooley, K. Mooley, S.R. Kulkarni, R. Margutti, D.B. Fox, E. Nakar, N.A. Patel, N.H. Volgenau, T.L. Culverhouse, M.F. Bietenholz, M.P. Rupen, W. Max-Moerbeck, A.C.S. Readhead, J. Richards, M. Shepherd, S. Storm, C.L.H. Hull, Birth of a relativistic outflow in the unusual γ-ray transient Swift j164449.3+573451. Nature **476**(7), 425–428 (2011)

S.E. Zepf, D. Stern, T.J. Maccarone, A. Kundu, M. Kamionkowski, K.L. Rhode, J.J. Salzer, R. Ciardullo, C. Gronwall, Very broad [O III] $\lambda\lambda$4959, 5007 emission from the NGC 4472 globular cluster RZ 2109 and implications for the mass of its black hole X-ray source. Astrophys. J. Lett. **683**, 139 (2008). https://doi.org/10.1086/591937

B.T. Zhang, K. Murase, F. Oikonomou, Z. Li, High-energy cosmic ray nuclei from tidal disruption events: origin, survival, and implications. Phys. Rev. D **96**(6), 063007 (2017). https://doi.org/10.1103/PhysRevD.96.063007

Space Science Reviews (2021) 217:54
https://doi.org/10.1007/s11214-021-00829-4

Distinguishing Tidal Disruption Events from Impostors

Ann Zabludoff[1] · Iair Arcavi[2] · Stephanie La Massa[3] · Hagai B. Perets[4] · Benny Trakhtenbrot[2] · B. Ashley Zauderer[5] · Katie Auchettl[6] · Jane L. Dai[7] · K. Decker French[8] · Tiara Hung[9] · Erin Kara[10] · Giuseppe Lodato[11] · W. Peter Maksym[12] · Yujing Qin[1] · Enrico Ramirez-Ruiz[9] · Nathaniel Roth[13] · Jessie C. Runnoe[14] · Thomas Wevers[15]

Received: 1 March 2020 / Accepted: 10 April 2021 / Published online: 5 May 2021

Abstract
Recent claimed detections of tidal disruption events (TDEs) in multi-wavelength data have opened potential new windows into the evolution and properties of otherwise dormant supermassive black holes (SMBHs) in the centres of galaxies. At present, there are several dozen TDE candidates, which share some properties and differ in others. The range in properties is broad enough to overlap other transient types, such as active galactic nuclei (AGN) and supernovae (SNe), which can make TDE classification ambiguous. A further complication is that "TDE signatures" have not been uniformly observed to similar sensitivities or even targeted across all candidates. This chapter both reviews those events that are unusual relative to other TDEs, including the possibility of TDEs in pre-existing AGN, and summarises those characteristics thought to best distinguish TDEs from continuously accreting AGN, strongly flaring AGN, SNe, and Gamma-Ray Bursts (GRBs), as well as other potential impostors like stellar collisions, "micro-TDEs," and circumbinary accretion flows. We conclude that multiple observables should be used to classify any one event as a TDE. We also consider the TDE candidate population as a whole, which, for certain host galaxy or SMBH characteristics, is distinguishable statistically from non-TDEs, suggesting that at least some TDE candidates do in fact arise from SMBH-disrupted stars.

Keywords Tidal Disruption Events · Active galactic nuclei · Supernovae · Gamma-ray bursts

1 Introduction

A TDE is a star disrupted by a SMBH. The TDEs discussed in previous chapters span a range of observed characteristics, including candidates detected first or only in X-rays, optical, or UV light, and with or without broad H/He, coronal, or Bowen emission lines. All are energetic transient events consistent with arising from galactic nuclei and are not known to have re-occurred.

The Tidal Disruption of Stars by Massive Black Holes
Edited by Peter G. Jonker, Sterl Phinney, Elena Maria Rossi, Sjoert van Velzen, Iair Arcavi and Maurizio Falanga

Extended author information available on the last page of the article

Yet there are non-TDE transients—large AGN flares, SNe near or projected on the nucleus—with potentially similar features. Even some "normal," continuously accreting AGN may vary over timescales longer than some TDE candidates have so far been monitored. The range of TDE properties, which can overlap those of other transient types, suggests that there is no one observable that distinguishes TDEs unambiguously from impostors. Nor has the developing field of modelling TDE formation and emission found such a "smoking-gun."

For the time being, we must rely on the standards of jurisprudence and require a preponderance of evidence, a suite of distinguishing features. We also can employ statistical arguments to test the authenticity of at least some TDE detections. What are those features and arguments?

In this chapter, we first discuss those current TDE candidates whose properties are rare or new among claimed TDEs and examine the reasons to favour the TDE explanation. Then, for the remaining TDE candidates, we review those observables that are generally interpreted as TDE signatures, comparing them to what is known about continuously accreting AGN, AGN with strong flares arising from disk instabilities, SNe, and GRBs, as well as other potential impostors like stellar collisions, "micro-TDEs," and circumbinary accretion flows. Lastly, we explore using TDE demographics, specifically the projected offset relative to the galactic nucleus, the SMBH mass, and the stellar mass and star formation history of the host galaxy, to distinguish them statistically from non-TDE sources.

2 Unusual TDE Candidates

On-going time-domain surveys, and comprehensive follow-up campaigns, are continuously revealing new flares and transients in galactic nuclei. Some objects initially thought to be TDEs have been reclassified due to the subsequent detection of similar, non-TDE transients, e.g., F01004-2237 (Tadhunter et al. 2017), after the discoveries of OGLE17aaj (Gromadzki et al. 2019) and AT 2017bgt (Trakhtenbrot et al. 2019b, see Sect. 3.2.2), illustrating the challenge of disentangling TDE emission from other peculiar nuclear transients. Here we discuss unusual classes of objects that are still considered TDE candidates, but that are represented by only one or a handful of members. The coming era of RubinObs/LSST and *eROSITA*, when thousands of new TDE candidates will be identified, should reveal the true nature of these classes.

2.1 PS1-11af

PS1-11af is interpreted as a partial TDE (i.e., a disruption of the envelope of the star, leaving the core intact) by Chornock et al. (2014). Absorption features became apparent in the UV spectrum 24 days post flare that bear a similarity to P-Cygni troughs observed in SNe. Yet the apparent velocities of these features ($\sim 13{,}000\ \mathrm{km\,s^{-1}}$) are too high for material in homologous expansion near a SN photosphere. Furthermore, fits to the SED with a blackbody model show that the radius of the emitting ejecta does not increase and the temperature does not decrease, as would be expected for SN evolution. PS1-11af has a blue colour that evolves weakly over time, a common characteristic of optically-detected TDEs (e.g., van Velzen et al. 2011; Gezari et al. 2012). The amount of accreted mass needed to power the observed luminosity is low ($\sim 0.002\ M_\odot$), which Chornock et al. (2014) ascribe to the partial stellar tidal disruption. In this scenario, optical photons would be reprocessed from the accretion disk to higher (X-ray) energies (e.g., Strubbe and Quataert 2009; Guillochon and Ramirez-Ruiz 2013), which requires contemporaneous X-ray coverage to confirm this hypothesis. Lacking the observations in this energy range, Chornock et al. (2014) can only propose that a partial stellar tidal disruption explains the available data, but the picture remains necessarily incomplete, underscoring the importance of simultaneous multi-wavelength coverage in unveiling the physics driving transient phenomena.

2.2 ASASSN-15lh

ASASSN-15lh was first interpreted as the most luminous SN ever detected (Dong et al. 2016). Its optical spectrum is devoid of hydrogen and helium features and contains broad absorption features between 3000–4100 Å. One such feature is attributed to OII λ4100, which is also seen in hydrogen-poor superluminous SNe (e.g., Quimby et al. 2011). However, an additional OII λ4400 feature observed in SNe is not present. This discrepancy, together with the location of ASASSN-15lh in the center of a non-starforming massive galaxy, prompted Leloudas et al. (2016) to consider ASASSN-15lh as a TDE (see also Krühler et al. 2018; van Velzen 2018; Mummery and Balbus 2020). The inferred SMBH mass from simple galactic scalings exceeds $10^8\ M_\odot$, implying that a solar-mass, solar-radius star would be swallowed whole rather than disrupted, but Leloudas et al. (2016) point out that a spinning SMBH would tidally disrupt the star and produce a transient (see also Margutti et al. 2017). As a result, ASASSN-15lh, if indeed a TDE, could then be used to infer the SMBH spin, a property that is challenging to constrain with most other methods.

Another peculiar property of ASASSN-15lh is its double-peaked UV light curve (its optical light curve is single-peaked). This bimodality, also observed in the TDE candidate AT2018fyk (Wevers et al. 2019a, see below), is unusual and difficult to explain under both the SN and TDE interpretations. Leloudas et al. (2016) propose that the first peak is powered by circularisation of the debris, while the second is from accretion; the timescales are roughly consistent with those expected from a spinning SMBH. Alternatively, Coughlin and Armitage (2018) and Vigneron et al. (2018) argue that both the double peaked light curve and the apparent high SMBH mass can be explained if the TDE is due to the secondary in a SMBH binary system. Still, the nature of this TDE candidate is disputable; Godoy-Rivera et al. (2017) argue that the evolution of ASASSN-15lh's photospheric radius, its radiated energy, and the implied event rate are all more consistent with those of H-poor superluminous SNe than TDEs.

2.3 PS16dtm

The transient PS16dtm, discovered in a Narrow Line Seyfert 1 (NLSy1) galaxy, was interpreted as a TDE rather than as a SN or arising from intrinsic AGN variability (Blanchard et al. 2017). The light curve exhibits no colour evolution during the ~ 100 day plateau, similar to other optical TDEs (see van Velzen et al. 2020b, Optical Chapter). The optical spectrum has traits similar to NLSy1s, with Balmer and multi-component FeII emission lines, further arguing against a SN interpretation. Blanchard et al. (2017) rule out AGN variability, given the two orders-of-magnitude increase in optical/UV flux within ~ 50 days and the decrease in X-ray flux after the optical/UV flare; they ascribe this behaviour to obscuration of the pre-existing AGN X-ray corona by the stellar debris disk formed by the disrupted star. The rise in the light curve is followed by a plateau at roughly the Eddingtion luminosity inferred for the SMBH and then a decline.

While this source is a strong TDE candidate, its spectrum, which is AGN-like, is very different than those of many optical TDEs (see van Velzen et al. 2020b, Optical Chapter). Indeed, Moriya et al. (2017) point out that PS16dtm's flare can be explained by AGN activity: increases in the accretion disk luminosity can spur radiatively driven winds that cause shock waves to propagate within the broad line region (BLR). Interactions between the shocks and BLR clouds can convert kinetic energy of the ejecta into radiation, producing transient luminosities and timescales that match those observed in PS16dtm.

2.4 AT2018fyk

AT2018fyk is a TDE candidate with a photometric UV/optical evolution remarkably similar to ASASSN–15lh, i.e., with a secondary maximum in its light curve. Margutti et al. (2017) ascribe the second peak in ASASSN–15lh's light curve to temporal evolution in the opacity of the ejecta, which allows UV radiation to escape and produce the secondary maximum. They point out that the observations are consistent with a spinning black hole disrupting a main-sequence star as a trigger for the ASASSN-15lh flare. Despite the similarities in light curve evolution, the timescales for AT2018fyk are significantly shorter than for ASASSN–15lh, and high amplitude, erratic X-ray variability is observed for AT2018fyk from the early phases.

AT2018fyk's optical-to-X-ray luminosity ratio (L_{opt}/L_X) evolves like that of ASASSN–15oi, suggesting that similar physical processes are at play. Wevers et al. (2019a) argue that both the L_{opt}/L_X evolution and the secondary maximum in the UV/optical light curve can be explained as a tidal disruption with a relativistic pericenter, as Leloudas et al. (2016) suggested led to the double-humped light curve in ASASSN–15lh. A relativistic pericenter favours disk formation on short ($\sim$ months) timescales, compared with the typical timescale of $\sim$years (van Velzen et al. 2019). Gezari et al. (2017) note that the formation of an accretion disk on similarly short timescales may explain the peculiar L_{opt}/L_X observed in ASASSN–15oi.

Another peculiarity of AT2018fyk is the apparent decoupling of the X-ray from the UV/optical emission, about 80 days after peak. This is reminiscent of the late time X-ray detection in ASASSN–15lh, while the UV/optical lightcurve steadily declines. Margutti et al. (2017) postulate that, in ASASSN–15lh, the X-ray emission may not in fact be related to the transient and may arise instead from the host galaxy nucleus, which would favour an interpretation that the flare was caused by a stellar explosion rather than a TDE. Should the X-ray emission be due to a TDE from a massive spinning black hole, the X-ray emission would fade over time. A similar observational test can be brought to bear on AT2018fyk by monitoring its X-ray emission over the time span of years.

In the optical spectra of AT2018fyk, Wevers et al. (2019a) detect low ionisation potential Fe II emission lines like those identified in ASASSN–15oi at late times. These lines are thought to form in dense, optically thick gas in an accretion disk-like structure, favouring the rapid disk formation scenario. These lines are observed frequently in high accretion rate NLSy1s, suggesting that the physical conditions in some TDEs and AGN are similar. Arguments against the AGN interpretation for this event include the absence of galactic (narrow or broad) emission lines, a pre-flare X-ray non-detection, and IR colours consistent with a quiescent galaxy.

2.5 Summary

There is not one selection mechanism that can be used to distinguish among potential explanations for transient events. Classification instead relies on the preponderance of evidence and may still not be definitive (e.g., ASASSN-15lh). Some transient phenomena have characteristics of both SNe and TDEs (e.g., PS1-11af), AGN and TDEs (e.g., AT2018fyk), or are hosted in known active galaxies (e.g., PS16dtm), requiring care in distinguishing among flares in a pre-existing accretion disk within the high variability tail of the AGN population, the tidal disruption of a star in the vicinity of an already active black hole, or a supernova in the centre of a galaxy.

The next several sections discuss how we might differentiate TDEs from the signatures of AGN (and strongly flaring AGN), SNe, GRBs, and other potential impostors, based on photometric and spectroscopic clues as well as statistical arguments.

3 Distinguishing TDEs from AGN

Dozens of TDEs are now claimed to have been detected. The uncertainty in this number reflects the inhomogeneity of TDE definitions, incompleteness in the TDE observables, and the lack of a unifying theoretical framework. TDEs should differ from AGN in the details of their accretion, i.e., the disruption of a single star leading to the quick, inside-out formation of a small, initially inclined disk (or flow) that then disappears on a shorter timescale than typical of the more continuous nature of AGN accretion.

Yet there is much we do not know about AGN variability, particularly about the extremes of continuous variability and about instabilities in the accretion disk that may produce transient flaring. The discovery of new variable AGN classes such as hyper-variable and "changing-look" AGN, which show dramatic weakening and/or strengthening in their broad Balmer emission lines, complicates efforts to identify TDEs unambiguously. Even long-term AGN variability may be a problem; the relevant timescales for TDEs with evolved stellar progenitors (which are not the main focus here) could be far longer than for main sequence (MS) stars. Much theoretical work remains to predict TDE observational signatures and to ascertain which, if any, are unique to TDEs.

In the following discussions, we consider what may distinguish TDEs from continuously accreting AGN, including those that are highly variable, and, more problematically, from the flaring caused by AGN disk instabilities. An even bigger challenge is presented by a new class of events—combining TDE- *and* AGN-like observables—that may arise when a TDE occurs in a pre-existing AGN (e.g., Merloni et al. 2015; Chan et al. 2019; Ricci et al. 2020). In §3.2.2, we briefly discuss the few such objects detected to date, reflecting our limited knowledge at the time of this writing.

The guidelines presented below tend to err on the conservative side: we are more interested here in purity than completeness in TDE classification. As a result, TDEs that occur in galaxies with even mild signs of nuclear activity, e.g., Seyfert-like emission line ratios, otherwise strong [O III] λ5007 emission, or persistent archival X-ray emission, would be excluded by our criteria. Yet counting hybrid systems will be critical in building complete and unbiased TDE samples.

3.1 TDEs Versus Continuous AGN

The obvious difference between TDEs and steady-state AGN is that TDEs are fundamentally transient phenomena, transitioning from quiescence to near-Eddington luminosities in a few weeks, and then back to quiescence within a few years to even decades (Rees 1988; Stone et al. 2013; van Velzen et al. 2019). This opens the possibility of monitoring several state transitions in the accretion flow, which goes from near- (or super-) Eddington to sub-Eddington and eventually becomes radiatively inefficient at low accretion rates (Jonker et al. 2020). Long term monitoring of TDEs should be pursued to reveal the details of such state transitions.

Debris disks from the tidal disruption of main sequence stars are very compact, as the star is disrupted near the SMBH with low angular momentum. If the stellar debris circularises efficiently, the size of the compact disk formed is twice the tidal disruption radius, or ~ 10–$100 R_g$ where $R_g = GM_{\rm BH}/c^2$. In contrast, AGN disks are expected to be much more extended, as gas is supplied from farther distances (e.g., Alexander and Hickox 2012, and references therein).

Another difference between TDE and stable AGN is that TDE disks may be fed at super-Eddington rates, while (low-redshift) AGN are usually considered to be accreting at sub-Eddington levels. As a result, TDEs and these AGN would have different disk structures, i.e., the super-Eddington TDE disks would be geometrically and optically thick and produce optically thick winds (Strubbe and Quataert 2009; Lodato and Rossi 2011; Dai et al. 2018). As the accretion rate drops with fallback rate, the disk and wind densities will also decrease, lowering the electron scattering opacity. This behaviour can explain the narrowing of the TDE hydrogen and He II λ4686 emission lines with decreasing luminosity described below, as the line width in a scattering dominated medium scales with the opacity (Roth and Kasen 2018). The higher He II/Hα ratio observed in TDEs (also discussed below) can arise from the higher accretion levels and inner disk temperatures compared to AGN, although detailed modelling on using this ratio to directly probe the disk structure is still lacking.

The absence of hard X-ray emission in the TDEs observed so far, compared with the X-ray power-law spectrum with $\Gamma \sim 1.9$ common to AGN (as discussed below), suggests fundamental differences in the disk corona. Possible explanations include: 1) the typical duration of AGN accretion is much longer than a TDE lasts, so the corona forms only for AGN; 2) the magnetic field strength and configuration is different in TDEs than in AGN, leading to less efficient coronal production.

3.1.1 Summary of Observable Distinctions

Some continuously accreting, but variable, AGN may be identified as new, blue, and/or X-ray detected nuclear transients and thus misclassified as TDE candidates. Consequently, it is essential that we consider the breadth of known AGN properties—light curves, colors, spectral shapes and lines, and variability, across optical, UV, and X-ray wavelengths—in defining criteria that may distinguish TDEs. In what follows, we discuss possible criteria and demonstrate how they may be used to assess the data for the two best-studied TDE candidates, ASASSN-14li and ASASSN-15oi. Given that strongly flaring AGN may have properties distinct from the continuous AGN population, posing a different and perhaps greater challenge to TDE classification, we discuss other appropriate strategies in §3.2.

Features that may favour a TDE over other AGN activity include:

1. steeper (month-long) and brighter (change of several magnitudes) rise in optical/UV flux;
2. relatively narrow luminosity peak, with characteristic timescale of months;
3. smooth, power-law decline in light curve, sometimes following a $t^{-5/3}$ trend;
4. ~ 0.2 mag bluer in g-r around peak emission;
5. hot, constant $T \sim 2$–4×10^4 K blackbody in optical/UV emission;
6. absent to weak [O III] λ5007 emission, and narrow emission line ratios suggestive of star formation rather than AGN photoionisation;

7. very broad ($> 15{,}000\ \mathrm{km\,s^{-1}}$) He II $\lambda 4686$ and Balmer optical emission lines that narrow as they weaken;
8. luminous He II $\lambda 4686$ line emission, with He II/Hα flux ratio $\gtrsim 1$;
9. weak, or even absent, Mg II $\lambda 2800$ line emission;
10. softer X-ray spectrum, in terms of photon index ($\Gamma \geq 3$) and/or prominence of low-temperature emission component ($kT_{\mathrm{bb}} = 0.04$–0.12 keV; see the Saxton et al. 2020, X-ray Chapter);
11. less rapid (> hours) X-ray variability;
12. no recurrence of transient behaviour.[1]

No TDE candidate observed to date has been shown to possess all these features. Indeed, depending on the conditions, some TDEs may not generate certain features. Those TDEs with the most features above, e.g., ASASSN-14li and ASASSN-15oi, are considered the strongest TDE candidates. We discuss these two events in light of the list given above, before addressing each of the listed features in more detail.

ASASSN-14li Here is how the optical, UV, and X-ray observations of ASASSN-14li map to the criteria listed above. 1) It displayed a ΔUVW2 (*Swift*) of -4.1 and a Δg of only -0.4, where these variations in magnitudes were measured with respect to the host galaxy pre-flare archival measurements (Holoien et al. 2016b); 2) The peak of this event was not observed; 3) The UV/optical decline over the first six months of monitoring was initially fit with an exponential (Holoien et al. 2016b), but this emission was later fit with a $t^{-5/3}$ decline over a longer time interval of approximately 250 days. After this, the UV/optical light curve levelled off to a more shallow decline (Brown et al. 2017); 4) The $g - r$ colour was roughly 0.4 mag during early monitoring, but was highly affected by host contamination. The event was still quite blue, with UVM2 brighter than U (*Swift*) by a difference exceeding 0.5 mag, for at least the first 100 days of monitoring (Holoien et al. 2016a); 5) The optical/UV continuum can be fit with a blackbody with $T \approx 3.5 \times 10^4$ K, and this temperature remained almost unchanged for the first 175 days of monitoring (Hung et al. 2017).

Furthermore, 6) optical spectra indicate [O III] $\lambda 5007/\mathrm{H}\beta \ll 1$ (Holoien et al. 2016b, and see also Fig. 2 in this chapter); 7) The optical emission lines initially showed broad wings with widths of $\sim 10{,}000\ \mathrm{km\,s^{-1}}$, although these widths narrowed significantly, with only a narrow component of width $\sim 1500\ \mathrm{km\,s^{-1}}$ after 100 days (Holoien et al. 2016b). The optical emission lines in the later spectra from this event are among the narrowest seen in TDEs; 8) While He II/Hα varies, this ratio was > 1 for several epochs and generally exceeded 0.5 (Hung et al. 2017); 9) No Mg II $\lambda 2800$ emission was seen in UV spectroscopy (Cenko et al. 2016).

Lastly, 10) the X-ray spectrum was soft and could be fit with a $kT = 51$ eV blackbody (Miller et al. 2015); 11) X-ray variability *was* detected in this event: a stable quasi-periodic oscillation of roughly 131 seconds was identified (Pasham et al. 2019); 12) No recurrence has been observed.

In summary, 14li meets most of the criteria for distinguishing a TDE from other AGN activity. The most prominent exceptions are for criteria 4 and 11: the earliest (closest to peak) $g - r$ measurement was only 0.2 (but this might be due to host contamination), and the event exhibited rapid X-ray variability in the form of a QPO. Additionally, since the peak was not observed, it is not possible to assess whether the characteristic rise and fall timescales are on the order of months (criterion 2). Finally, optical emission lines were somewhat narrow

[1] One possible exception is if a TDE occurs in a binary SMBH. In this case, the TDE may be perceived as recurring transient behaviour when the X-ray light dims due to the interaction with the second SMBH.

compared to other putative TDEs (criterion 7), although these lines did narrow over time in a manner that seems characteristic of TDEs.

ASASSN-15oi Next, we consider ASASSN-15oi. 1) It displayed a ΔUVW2 (*Swift*) of -6.8 and a ΔV of only -1.2 (Holoien et al. 2016a), where once again these variations in magnitudes were measured with respect to the host galaxy pre-flare archival measurements; 2) The peak of this event was not observed; 3) A UV/optical decline similar to $t^{-5/3}$ could be fit to the first 100 days of observation, before the flux at these wavelengths dropped precipitously. While the flux initially declined steadily in all bands, the inferred bolometric flux (from a thermal fit to the optical/UV data) remained steady for approximately the first 50 days before entering a decline (Holoien et al. 2018). As with ASASSN-14li, the initial UV/optical decline could alternatively be fit with an exponential (Holoien et al. 2016a); 4) The g-r measurement was not published, but UVM2 was brighter than U (*Swift*) by at least 1.0 mag during the first 50 days of monitoring (10–60 days post-discovery) and by at least 0.5 mag for 40 days after that (Holoien et al. 2016a). 5) During roughly the first 15 days of monitoring, the optical/UV continuum could be fit with a blackbody of roughly $T \sim 2 \times 10^4$ K. The inferred temperature *increased* to about 4×10^4 K over the next 15 days and stayed at that temperature for at least the next 70 days (Hung et al. 2017).

In addition, 6) the [O III] lines do not appear prominently in the spectra, although there is an unidentified broad feature near 5000 Å, which can be seen at 21 days post-discovery in the host-subtracted spectrum. Likewise, Hβ does not appear prominently, although it may be blended with He II in the earliest spectrum (Holoien et al. 2016a, and see also Fig. 2 in this chapter). 7) The He II λ4686 emission line had a width (FWHM) of roughly 20,000 km s^{-1} in a spectrum taken seven days after discovery, which narrowed to approximately 10,000 km s^{-1} at 21 days (Holoien et al. 2016a). 8) While He II appears prominently in the spectrum, no clear detection of any hydrogen lines can be made (Holoien et al. 2016a); 9) No Mg II emission was seen in UV spectra (Alexander Dittmann et al., in prep.).

Finally, 10) the X-ray spectra were soft and could be fit with blackbodies with $kT \sim$ 40–50 eV; 11) No rapid X-ray variability has been reported; 12) No recurrence has been observed.

In summary, 15oi meets nearly all the aforementioned criteria to distinguish a TDE from other AGN activity, except for those specific to the peak of the light curve (criteria 2 and 4), which was not captured for this event.

3.1.2 UV to Optical Light Curve

The light curves of TDE candidates are characterised by a dramatic increase in optical and UV luminosity, with an observed variability of over three magnitudes (e.g., van Velzen et al. 2011), a narrow luminosity peak spanning a timescale of months (e.g., Gezari et al. 2009; Guillochon and Ramirez-Ruiz 2013), and a smooth power-law decline, which sometimes follows $t^{-5/3}$, the predicted mass fall-back rate (Rees 1988; Phinney 1989b).

In comparison, the light curves of variable AGN are dominated by more stochastic variability that lacks such dramatic increases in brightness. On timescales of months, the variability of the continuous AGN population rarely exceeds 0.1 mag (Fig. 1; see, e.g., Vanden Berk et al. 2004; MacLeod et al. 2010; van Velzen et al. 2011; MacLeod et al. 2012; Caplar et al. 2017, and references therein). Although this *typical* AGN optical variability amplitude increases towards longer timescales, it does not exceed Δmag $\sim$ 1, even over decades. Indeed, only the most extreme tail of the AGN variability distribution, with few sources in wide-field surveys, reaches Δmag $\sim$ 2 (top panels of Fig. 1; see, e.g., MacLeod et al. 2012;

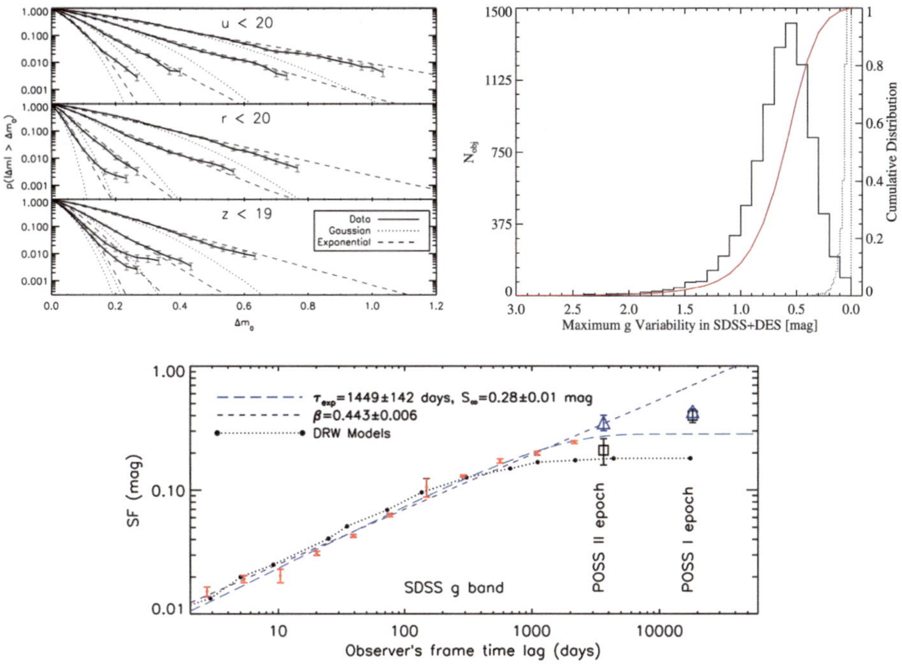

Fig. 1 Optical variability of normal, persistent (non-flaring) AGN. *Top-Left:* Distribution of UV/optical flux variations of quasars in the SDSS+POSS study of MacLeod et al. (2012). In each panel, different solid lines trace quasar variability on timescales spanning 1–30 days, 50–150 days, 200–400 days, and 1400–3000 days—the former being relevant to the rise-time of most TDEs. *Top-Right:* Distribution of optical flux variations in DES+SDSS broad-line AGN, over a period of $\sim$ 15 years (adopted from Rumbaugh et al. 2018). The cumulative distribution function (red line) indicates that only $\sim$ 10% of AGN show $|\Delta mag| > 1$ over this long period and are claimed to be highly variable on all timescales. *Bottom:* AGN variability on multiple timescales described through a structure function (SF), a measure of the *rms* variability of an AGN sample over any given time separation. This example (again from MacLeod et al. 2012) shows SDSS measurements over several years (red points) and combines them with POSS data for the longer-timescale measurements (large symbols). The different lines are phenomenological fits to the data. On timescales of months, AGN typically vary by < 0.1 mag, while TDEs show up as $\gtrsim 1$ mag transients (i.e., corresponding to SF $\gtrsim 1$ mag over < 100 days). Thus, normal, persistent (non-flaring) AGN essentially never show the month-long sharp optical flux increase seen in TDEs

Graham et al. 2017; Rumbaugh et al. 2018). Likewise, AGN typically do not show smooth and steady variability structure, such as the power-law decline seen in TDEs.

Some observed changing-look AGN stay at their peak optical luminosity for years (e.g., Runnoe et al. 2016). While there are TDE candidates detected in X-rays over a similar timescale (Lin et al. 2017; Jonker et al. 2020), the optical flare fades much more quickly. Furthermore, although the decays in some changing-look AGN light curves approximate a $t^{-5/3}$ decline (e.g., Merloni et al. 2015; Trakhtenbrot et al. 2019a) or permit a $t^{-5/3}$ solution (e.g., Runnoe et al. 2016), perhaps suggesting triggering by TDEs, many of these extremely variable AGN wane differently (Ruan et al. 2016) and/or lack the smooth decline expected from the fallback of debris from a TDE (Gezari et al. 2017).

The evolution of optical colours can also be used as a selection criterion for TDEs in ground-based optical surveys. Optically-discovered TDEs are characterised by a long-lasting blue continuum that resembles a blackbody of a few $\times 10^4$ K. Unlike SNe, which typically undergo significant colour evolution over a few weeks, TDEs and AGN can keep a constant colour for a longer period of time (∼years). The g-r colour of TDEs at peak emission (< -0.2 mag) is typically bluer than for AGN (> -0.1 mag; Fig. 12 in van Velzen et al. 2011). The observed bluer-when-brighter trend for AGN is consistent with the simple scenario of a geometrically-thin, optically-thick (i.e., Shakura-Sunyaev like) disk with variable accretion rates (e.g., Pereyra et al. 2006; Hung et al. 2016). If the observed AGN flare is an intrinsic property of the accretion disk, the classic thin disk model would predict a g-r of ~ -0.1 mag. Although intrinsic extinction in AGN is hard to estimate, it will only make the g-r colour in AGN flares redder than the predicted value, separating them further from TDEs in the optical colour space.

3.1.3 Optical Spectrum

A few key features in the optical spectra of TDE candidates can be used to differentiate them from persistent AGN. Vanden Berk et al. (2001) and the references therein provide more information about the typical UV/optical spectral energy distribution (SED) and emission lines in AGN. For complementary composite spectra in the NIR and UV, see Glikman et al. (2006) and Shull et al. (2012), respectively.

Hot, Constant Blackbody Continuum The continuum colour variability also differs between AGN and TDEs. Quasars are known to be bluer when brighter (e.g., MacLeod et al. 2010; Ruan et al. 2014). Although there is substantial scatter in this relationship, it provides additional leverage in identifying TDEs where no colour evolution is observed due to the constant blackbody temperature that produces the optical/UV continuum emission.

Weak [O III] Line The AGN narrow line region (NLR) is primarily ionised by the accreting black hole, and can span scales of order ~ 0.1–1 kpc, with some dependence on the AGN continuum luminosity (e.g., Bennert et al. 2002, 2006; Mor et al. 2009; Hainline et al. 2013, 2014). The [O III] $\lambda5007$ emission line is one of the most prominent in the NLR and has been used both to map out the size of the NLR (e.g., Schmitt et al. 2003b,a) and as a proxy of the intrinsic (bolometric) AGN luminosity (Kauffmann et al. 2003; Heckman et al. 2005; LaMassa et al. 2010; Pennell et al. 2017). Due to the much larger size scale of the [O III] $\lambda5007$ emitting region (and thus of the NLR) compared with the BLR, this line responds slower to the change in the ionising continuum than the broad emission lines (i.e., $\gg 100$ years; see, e.g., Peterson et al. 2013).

TDEs tend to have weak to no [O III] $\lambda5007$ emission, with [O III]/Hβ and [N II]/Hα emission line ratios consistent with photoionisation from star formation or LINER-like activity on the BPT (Baldwin et al. 1981; Kewley et al. 2006) diagram. Even TDE hosts with Seyfert-like line ratios, e.g., ASASSN-14ae (French et al. 2017), ASASSN-14li (French et al. 2017), and iPTF16fnl (Onori et al. 2019), generally have weak line strengths that would classify them as LINER-like on a WHAN (Cid Fernandes et al. 2010) diagram. We explore the range of AGN signatures in TDE host galaxies in the French et al. 2020, Host Galaxies Chapter.

On the other hand, galaxies with significant nuclear photometric and spectroscopic variability and strong [O III] emission are more likely to be hosting highly variable AGN, especially if the emission line ratios are within the Seyfert region of the BPT diagram. The emission line ratios of many changing-look AGN are generally consistent with those of AGN (e.g., Runnoe et al. 2016; Ruan et al. 2016). Thus, our guideline 6 in §3.1.1 would exclude most strong AGN and most changing-look AGN. We note that, while some luminous quasars have relatively weak [O III] emission (e.g., Netzer et al. 2004), their UV/optical continuum luminosities tend to be far higher than those of TDEs and could be used as a discriminant.

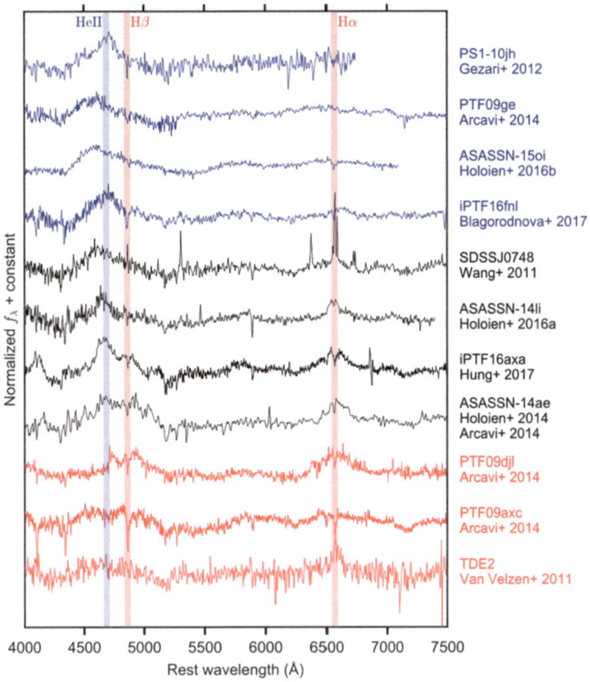

Fig. 2 Continuum-subtracted optical spectra of UV/optical-bright TDEs, most of which show He II line emission of comparable luminosity and FWHM to Hα

What do we miss with this conservative cut, where we have prioritised TDE sample purity over completeness by selecting against strong [O III] emission? We would neglect, for example, the TDE candidate PS16dtm (Blanchard et al. 2017), whose host is a NLSy1. Also excluded would be transients in the 2017-bgt class (Trakhtenbrot et al. 2019b; see below), whose nature is unclear. The presence of some activity in a galaxy does not eliminate *a priori* the possibility of a TDE, although caution should be used in such cases.

Broad, Narrowing Balmer and He II Lines The emission line velocity widths and their time-dependent changes provide another way of distinguishing between AGN and TDEs. Near peak, the typical Hα and He II λ4686 FWHM of TDEs are both of order 10^4 km s^{-1} and often exceed 15,000 km s^{-1} (Fig. 2). Only a fraction of a percent of the Hα and Hβ lines in SDSS AGN (which may have problematic spectral decompositions) are as broad as in TDEs, i.e., $\gtrsim 15{,}000$ (Fig. 3). This distinction may arise from the structure and dynamics of the BLR around SMBHs with certain masses (Laor 2003). Thus, line width is a reasonably good discriminator, but, depending on the overlap in the distributions for TDEs and AGN, may not be iron-clad.

Whenever AGN spectra do have noticeable He II λ4686 emission, the line profiles and widths are generally comparable to those of Hβ (and thus also Hα). While accurate measurements for individual AGN are often challenging, given the weakness of the He II feature and the fact that it is blended with several [Fe II] emission features, the resemblance between He II and Hβ can be seen in stacked spectra (e.g., Fig. 3 of Boroson 2002).

The temporal evolution of the velocity line widths in response to changes in the photoionising continuum provides an even better way of distinguishing between TDEs and AGN. In changing-look quasars, which are likely to contaminate TDE searches, the broad emission lines broaden as they weaken (LaMassa et al. 2015; Runnoe et al. 2016). This is the basis for

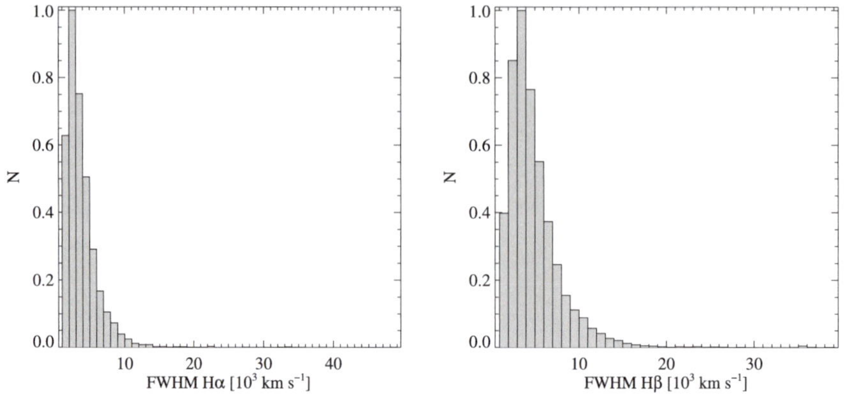

Fig. 3 The FWHM distributions for the Hα and Hβ broad emission lines in SDSS DR7 quasars, taken from the spectral decompositions of Shen et al. (2011). Only a small fraction of sources have Balmer lines that are broader than $\sim 15{,}000\ \mathrm{km\,s^{-1}}$. Given that such extremely broad Balmer lines are often seen in TDEs, Balmer line width can serve as a discriminant between TDE candidates and persistent AGN

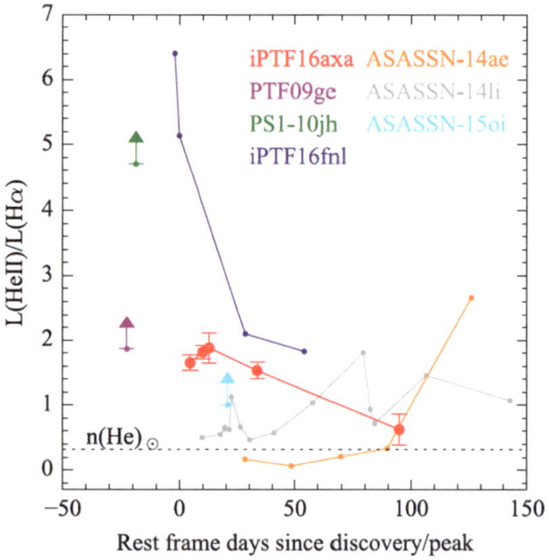

Fig. 4 The variation in the He II/Hα flux ratio over time for a collection of TDEs. Observing this flux ratio $\gtrsim 1$ in at least one spectral epoch distinguishes optical/UV TDE candidates from most AGN. However, the ratio evolves with time, and so a single spectral epoch may miss the He II line if it appears at a different time. Figure reproduced from Hung et al. (2017)

reverberation mapping in AGN (e.g., Peterson 1993) and the opposite of what is observed for TDEs (Holoien et al. 2016b).

High He II/Hα Ratio The spectra of many UV/optical-bright TDEs have prominent He II λ4686 line emission. The line luminosity is typically of order $10^{41}\ \mathrm{erg\,s^{-1}}$ at its brightest, much stronger than Hβ and comparable to (or even stronger than) Hα (Fig. 2). Therefore, a He II/Hα flux ratio $\gtrsim 1$ in at least one spectral epoch is a hallmark of UV/optical-bright TDE candidates, as is He II/H$\beta \sim 1$.

There are potential exceptions. One possible case is TDE2 (van Velzen et al. 2011, the bottom spectrum in Fig. 2), but its spectrum is low signal-to-noise. Another is PS1-11af (Chornock et al. 2014), although that event showed no emission lines at all in its spectrum.

Because the He II/Hα flux ratio is observed to evolve with time, we note that a single spectral epoch may miss the He II line if it appears at a different time (Fig. 4).

In comparison, while AGN spectra do exhibit broad He II λ4686 emission, it is typically weak compared to the Balmer lines. The He II/Hα flux ratio in the SDSS quasar composite is ~ 0.005 (Vanden Berk et al. 2001). From the theoretical side, photoionisation modelling of the BLR in AGN also gives He II/Hα < 1, although the goal of such work is usually to reproduce normal AGN spectra and not extreme outliers.

Like other high-ionisation species and transitions, He II comes from closer to the central engine than the Balmer lines (e.g., Grier et al. 2013) and is extremely responsive to continuum changes in the AGN (as in Korista and Goad 2004), whereas Hα is the least responsive of the Balmer lines. Thus, with a large (UV) flare in the AGN continuum (even if the SED shape does not change), it may be possible to boost the He II/Hα ratio temporarily. That said, while Peterson and Ferland (1986) describe a (moderate) flare in the AGN reverberation-mapping poster child, NGC 5548, that substantially boosts the He II emission, it is always weaker than Hβ. As a caveat to the above, we must mention that the measurement of the flux of the He II λ4686 line can be affected by blending with the N III λ4640 line, which can be excited by Bowen fluorescence (e.g. Leloudas et al. 2019; Onori et al. 2019; Nicholl et al. 2020). Medium resolution spectroscopy may be helpful in deblending these two components.

3.1.4 UV Spectrum

The UV spectra collected of TDEs so far (Fig. 5) have revealed striking differences with respect to AGN, although the sample is currently small. As of this writing, there are only two TDE in quiescent galaxies with published UV spectra: iPTF15af (Blagorodnova et al. 2019) and AT2018zr (Hung et al. 2019), also known as PS18kh (Holoien et al. 2019). There is at least one more event with an unpublished UV spectrum, ASASSN-15oi (Alexander Dittmann et al., in prep.). Additionally, there is a near UV spectrum of PS16dtm (Blanchard et al. 2017), but, because that event took place in a NLSy1, we exclude it for the present purpose of distinguishing TDE from AGN. ASASSN-14li (Cenko et al. 2016) and iPTF16fnl (Brown et al. 2018) also have published UV spectra, but are not in quiescent host galaxies.

These systems generally lack certain low-ionisation emission lines that are common in most AGN, in particular Mg II $\lambda\lambda$2796, 2804. The exception is AT2018zr, which displayed a broad emission feature consistent with Mg II in five spectra taken between roughly 20 to 60 days after r-band peak. However, the equivalent width of Mg II dropped steadily in time, and, by the final spectrum, the line had all but disappeared, while an absorption feature appeared at a blueshifted velocity consistent with the velocity (15,500 km s^{-1}) inferred from the Balmer lines in the optical spectrum (Hung et al. 2019). This rapid spectral variability distinguishes the event from most AGN in its own right.

Equally interesting is the general absence of C III] λ 1909 emission in the TDEs, a line seen in most AGN UV spectra. Adding to the differences is the general strength of N III] λ 1750 emission, a line which shows up prominently in only approximately 1% of SDSS AGN. Here again AT2018zr is an exception, but only in the sense that its N III does not show up clearly in emission, although it may be contributing to absorption at that wavelength. While the rare, so-called "N-rich QSOs" do have this emission line, they generally also have a strong C III] line accompanying it, as well as Mg II, which is not the case for TDEs (Jiang et al. 2008; Cenko et al. 2016).

There are broad absorption features in some UV TDE spectra, although these too distinguish themselves from broad absorption line quasars (BALQSO). TDEs with clear or

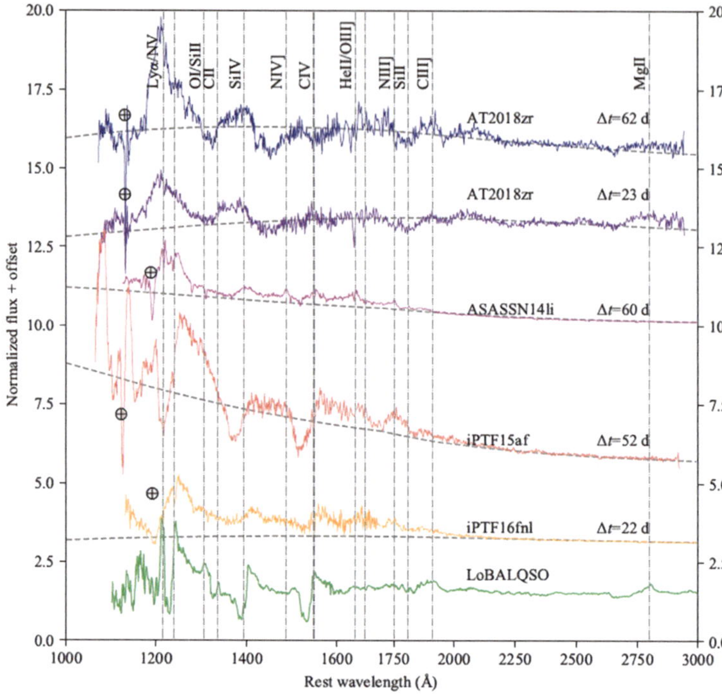

Fig. 5 Ultraviolet spectra from four TDEs, including two epochs for AT2018zr, compared to a low-ionisation broad absorption line QSO (LoBALQSO) composite spectrum. At least three of the TDE spectra lack Mg II λλ2796, 2804 emission. This figure appears in Hung et al. (2019) and is reproduced here with the permission of the American Astronomical Society

tentative absorption include iPTF16fnl, iPTF15af, and AT2018zr (PS18kh). In these cases, the absorption seems to correspond to the C IV λλ1548, 1551, Si IV λλ1394, 1403, and N V λλ1239, 1243 lines.

The FWHM of these lines is roughly in the range 5000–10,000 $\mathrm{km\,s^{-1}}$, quite similar to BALQSOs. However, in iPTF16fnl and iPTF15af, the centroid of the absorption lines is near enough to the line centre so that the absorption represents velocities from nearly 0 to 10,000 $\mathrm{km\,s^{-1}}$, blending smoothly into the emission on the red side. In contrast, the centroid of the absorption lines in BALQSOs is often blueshifted by at least 10,000 $\mathrm{km\,s^{-1}}$, and so the absorption is completely detached from the emission, with a broader wing on the higher velocity side (Blagorodnova et al. 2019). AT 2018zr (PS 18kh) once again displays its own unique behaviour, with possible broad absorption centroid velocities of 15,000 $\mathrm{km\,s^{-1}}$; these absorption lines were also highly variable, becoming increasingly prominent with time in the five spectra taken over approximately 40 days.

Of the TDEs with published UV spectra in quiescent hosts, both exhibit broad absorption lines at some point in time. This fraction is higher than that of BALQSOs. Blueshifted BALs in QSOs are thought to arise from fast-moving outflows. If the QSO/BALQSO dichotomy is largely due to viewing angle effects, then outflows in TDEs may subtend a larger solid angle than in AGN (Hung et al. 2019).

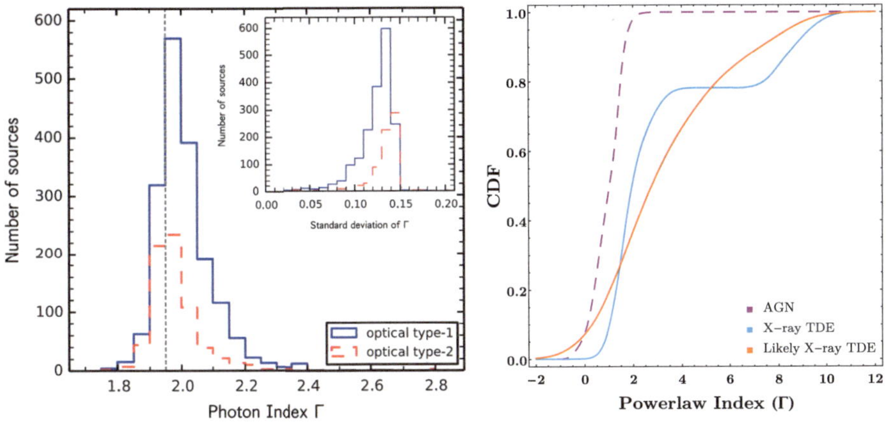

Fig. 6 Comparison of power-law photon index Γ for the X-ray spectra of AGN and TDEs. *Left:* Distribution of photon indices of ~ 2500 AGN observed with *XMM-Newton*. The AGN have non-thermal spectra with power law photon indices ranging from 1.7–2.4. Figure adapted from Liu et al. (2016). *Right:* Cumulative Γ distributions for events classified as X-ray TDEs (cyan curves) or likely X-ray TDEs (orange) by Auchettl et al. (2017) compared with AGN detected in the *Chandra* Deep Field South (magenta; Tozzi et al. 2006; Xue et al. 2011). Both TDE distributions from Auchettl et al. (2017) include jetted TDE candidates, which tend to have harder X-ray spectra than thermal TDEs and photon indices more like AGN. Observing a nuclear transient with a steeper (softer) X-ray SED ($\Gamma \gtrsim 3$) may strongly favour a TDE interpretation

3.1.5 X-Ray Spectrum

AGN activity could potentially mimic the X-ray emission arising from TDEs. In this section, we compare observations of X-ray emission from AGN and TDE candidates, focusing on those that might help us distinguish between the two types of sources. The Saxton et al. 2020, X-ray Chapter has more information on the X-ray properties of TDEs.

Thanks to its high-sensitivity and good sky coverage, the *ROSAT X-ray Observatory* (Truemper 1982) discovered the first TDE candidates. These nuclear transients had a peak X-ray luminosity of $L_x \sim 10^{44}$ erg s^{-1}, were associated with galaxies that showed no evidence of (prior) AGN emission, produced light curves that decayed following a $t^{-5/3}$ power-law, and had X-ray spectra that were best described with a $\sim 10^{5\text{–}6}$ K blackbody or with a very steep power-law index ($\Gamma = 3$–7, Fig. 6; Grupe et al. 1995; Brandt et al. 1995; Bade et al. 1996; Komossa and Bade 1999; Grupe et al. 1999; Greiner et al. 2000). Since *ROSAT*, the capabilities of the *Neil Gehrels Swift* Gamma-ray Burst Mission, the *Chandra X-ray Observatory*, and *XMM-Newton Space Observatory*, including increased effective area, spectral coverage, spatial resolution, and/or spectral resolution, have dramatically changed our ability to characterise the detailed spectral evolution of TDE candidates, leading to discoveries including rapid variability and, in Swift J1644+57, possible jet formation (Bloom et al. 2011; Burrows et al. 2011).

AGN and TDEs are quite different in the nature and appearance of their broad-band X-ray spectra. In AGN, around 10% of the bolometric luminosity is released as non-thermal X-ray continuum up to ~ 100 keV, as magnetic processes in and/or above the disk form a relativistic electron cloud that is partly cooled due to the inverse Comptonization of the thermal disk photons. In the 2–10 keV band, this continuum is typically modelled as a simple power-law, whose index is determined by the temperature distribution of the electrons and by the number of electrons that each UV seed photon encounters. In AGN, the average

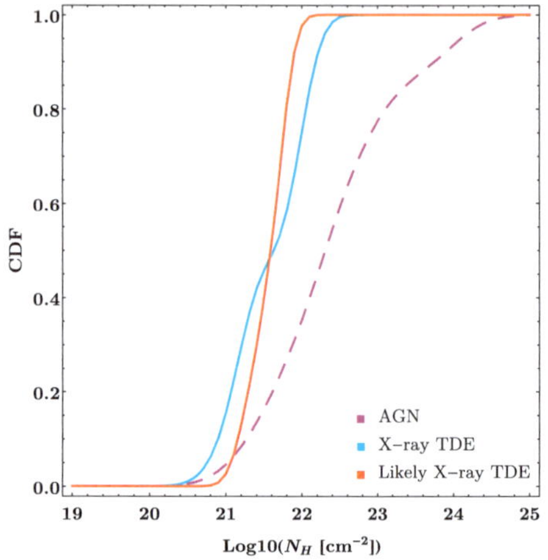

Fig. 7 Cumulative distribution function of line-of-sight hydrogen column density ($N_{\rm H}$) for TDEs classified as a X-ray TDEs (cyan curves) or likely X-ray TDEs (orange) by Auchettl et al. (2017). These TDEs are significantly less absorbed compared to AGN found at similar redshifts in the *Chandra* Deep Field South (magenta; Tozzi et al. 2006; Xue et al. 2011). It is not yet clear whether this difference is real or due to observational bias

2–10 keV spectral index is 1.9, with a spread from ~ 1.7 to 2.4 at the most extreme (Fig. 6; Liu et al. 2016).

In contrast, most thermal TDEs show X-ray emission that is dominated by a multicolor blackbody component with a temperature of 50–100 eV, e.g., ASASSN-14li (Miller et al. 2015; Holoien et al. 2016b; Brown et al. 2017). In some cases, there is an additional weak hard X-ray tail, e.g., ASASSN-14li (Kara et al. 2018) and XMMSL1 J074008.2-853927 (Saxton et al. 2017). For ASASSN-14li, Kara et al. (2018) suggest that the hard excess beyond the blackbody arises from additional inverse Compton scattering of disk photons by relativistic electrons in an X-ray corona. This additional Comptonization component appears to contribute more to the spectrum as the source evolves over time, either due to the accretion rate dropping or due to the delayed formation of an X-ray corona. Most recently, Wen et al. (2020) show that a "slim disk" accretion model adapted to SMBHs can successfully describe the multi-epoch X-ray spectra of ASASSN-14li.

Another observed distinction is that TDEs are significantly less absorbed compared to AGN found at similar redshifts (Fig. 7). We note, however, that this difference may arise from observational bias. A column density of 10^{22} cm^{-2} would completely absorb the emission of soft TDEs with a 50 eV thermal spectrum and hinder their detection.

The thermal-dominated X-ray spectra of TDEs such as ASASSN-14li differ from those of a wide range of AGN, from Type 1 (unabsorbed) to Type 1.9 (absorbed), all of which have non-thermal emission (Fig. 8). Two NLSy1s, 1H0707-495 and Ark 564, have been proposed as Eddington-limited or even super-Eddington AGN. Even these extreme accretion AGN show non-thermal X-ray emission and are distinct from ASASSN-14li.

While thermal TDEs have softer X-ray spectra than AGN (Lin et al. 2011; Auchettl et al. 2017), the seemingly separate class of jetted TDEs, such as Swift J1644+57, is characterised by strong X-ray emission up to ~ 100 keV and an AGN-like photon index (Bloom et al. 2011; Burrows et al. 2011, see also Sect. 5). Complicating the picture further are observations showing that some thermal TDEs spectrally harden as they evolve (e.g., ASASSN-14li; Kara et al. 2018), while others show little variation. For the time being, we conclude only that a nuclear transient with a steeper (softer) X-ray SED ($\Gamma \gtrsim 3$) is more likely to be a TDE than an AGN.

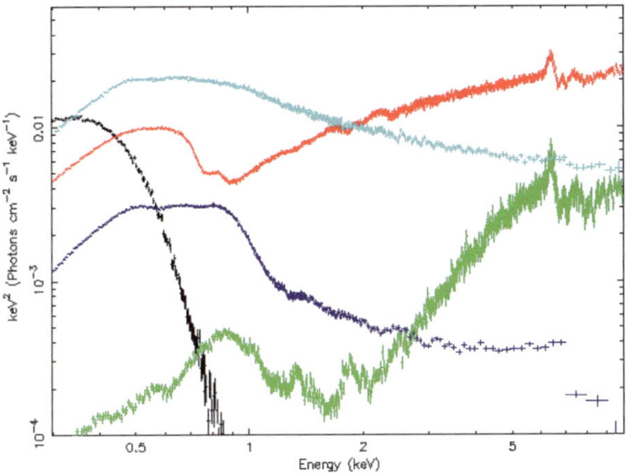

Fig. 8 Comparison between the X-ray spectra of the well-known thermal TDE ASASSN-14li (black) and AGN of different types, including MCG-6-30-15 (red), NGC 1365 (green), 1H0707-495 (blue), and Ark 454 (cyan). Even Eddington-limited NLSy1 AGN (in blue and cyan) do not show as steep an X-ray spectrum as thermal TDEs. Figure from E. Kara

3.1.6 X-Ray Variability

The variable X-ray emission observed in persistent AGN spans timescales from seconds to months and years, with an amplitude often much larger than at optical wavelengths (for the same source; e.g., Lanzuisi et al. 2014, and references therein). In contrast, the X-ray (and optical/UV) light curves of TDEs show a steady, months-long structure. Hence, measuring a transient's X-ray light curve with a short cadence and over years, and comparing it to the UV/optical light curve, may prove an important tool in identifying TDEs by excluding those AGN that vary on much shorter or longer timescales.

It is not yet clear how strongly the observed decay rate of the X-ray light curve discriminates between TDE and variable AGN. The self-regulating nature of the SMBH accretion in AGN can produce an X-ray light curve that declines as $t^{-(1.5-2.0)}$ (e.g., Hopkins and Hernquist 2009). While a power-law of $t^{-5/3}$ has been associated with some TDE optical/UV light curves and the first X-ray light curves, the observed temporal evolution of soft X-ray emission from current X-ray and likely X-ray TDE candidates (Auchettl et al. 2017) follows a wide variety of power-laws, consistent with fall-back, accretion, and disk emission (e.g., Evans and Kochanek 1989; Phinney 1989a; Cannizzo et al. 1990; Cannizzo and Gehrels 2009; Cannizzo et al. 2011; Lodato and Rossi 2011; Guillochon and Ramirez-Ruiz 2013), with the majority shallower than $t^{-5/3}$. Guillochon and Ramirez-Ruiz (2013) suggest that such shallow declines arise when TDEs are viscously delayed (i.e., the time it takes for material to accrete is slow).

Many AGN of extreme X-ray variability have been discovered thanks to the photon-counting nature of X-ray facilities, as well as the long integration times used for some sources and for certain extragalactic fields. These AGN remain poorly understood, due to the frequent lack of simultaneous data at other wavelengths. The X-ray spectra of "X-ray changing-look" AGN (Matt et al. 2003; Bianchi et al. 2012; Ricci et al. 2016) have been well modeled by occulting clouds transiting into and out of the line-of-sight towards the AGN, alternately masking and uncovering the central engine. However, the X-ray spectra of some optical changing-look AGN, with their weakening or strengthening of broad Balmer lines, cannot be ascribed to variable extinction. When observed in the dim state, the X-ray spectra do not show features of obscuration that extinguishes and reprocesses the X-ray emission, but rather are consistent with a model where the intrinsic X-ray emission diminished significantly (e.g., LaMassa et al. 2015, 2017; Husemann et al. 2016). The processes responsible for driving the X-ray variability are unknown, but there may be a link between the Eddington ratio and spectral state of the AGN due to mechanisms that operate at the scales of the inner accretion disk, similar to the spectral state transitions observed in stellar mass black hole binaries (e.g., Noda and Done 2018; Ruan et al. 2019).

TDEs tend to show relatively little variation in column density with time (Auchettl et al. 2017); after an initial drop, N_H remains roughly constant over at least several hundred days in both ASASSN-14li and -15oi (Wen et al. 2020). Thus, observing an evolving column density is a possible way of distinguishing some AGN from TDEs. We note, however, that the X-ray TDEs discovered to date tend to have lower N_H than AGN (Fig. 7), which may limit the measured N_H variation. For cases where declining absorption can be excluded as the cause of X-ray brightening, the distinction between TDE and X-ray variable AGN is less apparent. For instance, in the Seyfert 1.9 galaxy IC 3599, TDEs were alternately invoked (Campana et al. 2015) and ruled out (Grupe et al. 2015) to explain the observed, recurring X-ray flares.

3.2 Differentiating TDEs from Flaring AGN

3.2.1 Accretion due to Disk Instabilities

In contrast with steady-state AGN disk accretion, accretion events arising from AGN disk instabilities might be relatively quick and deplete only the innermost AGN disk, a scale comparable to that expected for TDEs. Indeed, fast and coherent instabilities in the innermost parts of the disk seem to be required to explain significant variability events in accreting SMBHs, as this is where most of the (continuum) radiation is produced (see, e.g., Lyubarskii 1997; King et al. 2004, and the discussion in Cannizzaro et al. 2020). Such events might even produce higher accretion rates and luminosities than during the steady-state and/or quiescent stages. In other words, AGN flares might share many similarities with TDEs.

Thermal, viscous, gravitational, and radiation-pressure instabilities have all been suggested to occur in AGN disks (e.g., Janiuk et al. 2002; Jiang et al. 2013 and references therein). Below is a brief review of the main aspects of such instabilities and their implications. We generally follow the discussion in Saxton et al. (2015).

Accretion disks may attain a limit-cycle behaviour, which can be generally divided into three phases: in the 1) *quiescent phase*, material initially accumulates at a slow rate and fills the inner region of the disk. The disk structure and properties then slowly evolve until the disk becomes unstable, leading to the 2) *rise and outburst phase*, where the instability typically leads to a runaway heating, increasing the local viscosity, scale height of the disk, and the local accretion rate (Cannizzo 1996). This process eventually changes the accretion rate and produces a rapid depletion of the unstable region, whose material then accretes onto the SMBH. Such a fast accretion episode leads to a flaring of the SMBH. As the inner disk depletes, when the matter is accreted into the black hole faster than it is replenished, the accretion flare goes through the 3) *decay phase*, leading to a new quiescent phase and the next accumulation cycle.

The disk-instability model thus predicts *repeated* flares. The possibility of observing multiple flares depends on the typical duty-cycle timescale. For an unstable disk model to explain the known TDE candidates, the timescale for the limit-cycle must be sufficiently long, given the non-detection of repeated TDEs to date.[2]

[2]The (in)ability to detect such repeated flares naturally also depends on their amplitudes, which are not necessarily as high as the flare that led to the identification of the TDE candidate.

The overall depletion of the inner unstable disk is likely to occur on the timescale $\tau_{\rm dep}$ for material to viscously accrete from the truncation radius, the outermost part of the unstable region down to innermost stable circular orbit (ISCO). The initial rise should be fast, as the accretion begins from the innermost region, and then slower, up to the timescales for the material in the outermost truncation radius region to accrete down to the SMBH. Given the mass enclosed in this region and the accretion rate, one can estimate a typical timescale for the overall rise and fall:

$$\tau_{\rm dep} = M_{\rm inner}/\dot{M}. \tag{1}$$

The enclosed mass of the inner disk is given by

$$M_{\rm inner} = \int_{R_0}^{R_{\rm trunc}} \rho(r)\, 2\pi r\, H(r)\, dr, \tag{2}$$

where R_0 is the radius of the ISCO, $R_{\rm trunc}$ is the truncation radius, $\rho(r)$ is the disk density, and $H(r)$ is its height. For a Shakura-Sunyaev thin disk (Shakura and Sunyaev 1973) and typical values, expressed in gravitational radii, the enclosed mass is

$$M_{\rm inner} = 6 \times 10^{-4} \alpha^{-8/10} M_6^{11/5} \dot{M}_{\rm Edd}^{-3/10} \left[\left(\frac{R_{\rm trunc}}{R_{\rm g}} \right)^{5/4} - \left(\frac{R_0}{R_{\rm g}} \right)^{5/4} \right] M_{\odot}, \tag{3}$$

where α is the viscosity parameter, and the accretion rate is given in units of the Eddington-limited accretion rate, $\dot{M}_{\rm Edd} \simeq 1.4 \times 10^{24} M_6\ {\rm g\,s^{-1}}$, for a SMBH mass M_6 in units of $10^6\ M_{\odot}$ (Saxton et al. 2015). The depletion time $\tau_{\rm dep}$ (with typical truncation radius of a few tens of gravitational radii) is then

$$\tau_{\rm dep} \sim 0.33\, {\alpha_{uns}}^{-8/10}\, M_6^{6/5}\, \dot{M}_{\rm Edd}^{-3/10} \left[\left(\frac{R_{\rm trunc}}{R_{\rm g}} \right)^{5/4} - \left(\frac{R_0}{R_{\rm g}} \right)^{5/4} \right] \text{months}. \tag{4}$$

In other words, $\tau_{\rm dep}$ is equivalent to the viscous timescale of a thin disk at the truncation radius, and α_{uns} is the viscosity in the unstable region, rather than the typical viscosity operating during the regular accretion phases.

There are many uncertainties both in understanding the disk instability process and in the estimates of the typical timescales and accretion rates, making this approach simplified at best. Much depends on the choice of R_{trunc} and on whether these instabilities occur there. Nevertheless, the reasoning here suggests that the timescales for the rise and fall of such flares for low mass SMBHs (up to $\sim 10^7\ M_{\odot}$) are on the order of weeks to months, while the timescales for more massive SMBHs are probably too long for the flares to be classified as fast transients. In other words, disk-instability flarings potentially masquerade as TDEs of MS stars only for lower mass SMBHs. In addition, for SMBH masses larger than $\sim 10^8\ M_{\odot}$, the ionisation region is also Toomre unstable ($Q < 1$, self-gravity).

Interestingly, the timescales of both TDEs and disk-instability flares depend on the SMBH mass, but through different scalings. This difference could potentially be used to distinguish between the models. Nevertheless, better comprehension of both the TDE process (including TDE debris disk circularisation) and disk instabilities that lead to flares is needed before reliable predictions and comparisons can be made.

As noted by Saxton et al. (2015), the overall timescale for the duty-cycle, and hence for repeating flares, is determined by the filling or viscous time at the truncation radius.

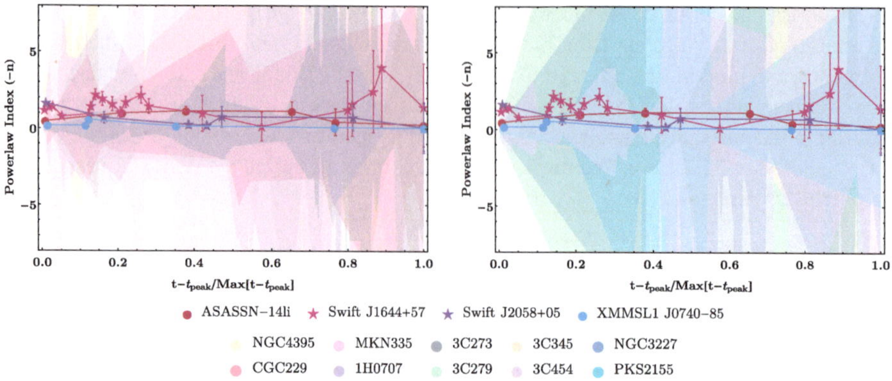

Fig. 9 Comparing the smoothness of X-ray light curve decay for TDEs and AGN flares. Auchettl et al. (2018) determine the best fit power-law index ($-n$) for TDE and AGN samples as the time of peak goes to infinity, assuming $L \propto (t - t_{peak})^{-n}$. Here we plot n as a function of $t - t_{peak}$, where t_{peak} is the time in which the peak luminosity is detected. The sources decay over different timescales, so $t - t_{peak}$ is normalised by its maximum value for each source. The derived power-law indices for AGN vary significantly over short time scales, ranging between $n \sim -10$ and $+15$. Different colour shaded bands represent the full range of these indices and their uncertainties. In contrast, the decay of a TDE flare is best fit with a power-law index between $n = 0$ and ~ -2; for most of the decay, the power-law index changes little, especially for the two non-jetted TDEs ASASSN-14li and XMMSL1 J0740-85. Figure adapted from Auchettl et al. (2018)

The viscosity in the stable region beyond the truncation radius differs from that in the inner region during the depletion, and so refilling takes far longer than the depletion and flaring timescale, i.e., decades or more for $M_{\rm BH} \sim 10^6\ M_\odot$. Repeated flares on such a timescale have been observed in the Sy 1.9 galaxy IC 3599, 20 years apart (Grupe et al. 2015).

Due to the uncertainties in the duty cycles of AGN, it is not always feasible to rule out recurring flare emission due to an AGN when seeking to confirm a TDE. Another difficulty is that, during an AGN flare or TDE, the change in source luminosity, relative to the pre-event upper-limits, is similar (Auchettl et al. 2018). However, Auchettl et al. (2018) find that TDE X-ray light curves decay much more coherently, even monotonically, while the rate of AGN decay varies widely with time (Fig. 9). Less than 4% of the coherent decay behaviour seen in their TDE sample could arise from sources like those in their AGN sample, suggesting that observing smooth decay can help to distinguish TDEs from AGN flares.

Galaxies hosting *known* AGN are typically excluded from TDE candidate selection, given the confusion arising from the expected AGN variability and flaring. However, galaxies hosting undetected, sub-luminous ("starved") AGN may present as quiescent. Given the low accretion rates in starved AGN (Saxton et al. 2018), their disk properties and variability could differ from those of persistent AGN, and their occasional flares might masquerade as TDEs. Gas-poor hosts are less likely to enable gas inflows to the nucleus, so flarings in starved AGN may favour "gas-intermediate" hosts, whose histories include higher rates of star formation in the past.

Saxton et al. (2018) suggest that the apparent preference of TDE candidates for "post-starburst" galaxies (see the French et al. 2020, Host Galaxies Chapter) naturally arises from the reasons above and that many of these events are in fact AGN flares. One prediction of the disk instability scenario is therefore that deeper observations of TDE candidate hosts will reveal weak AGN, either now or in the recent past, at higher rates than in other, comparably massive galaxies. However, at least for known low luminosity AGN (if defined by LINER

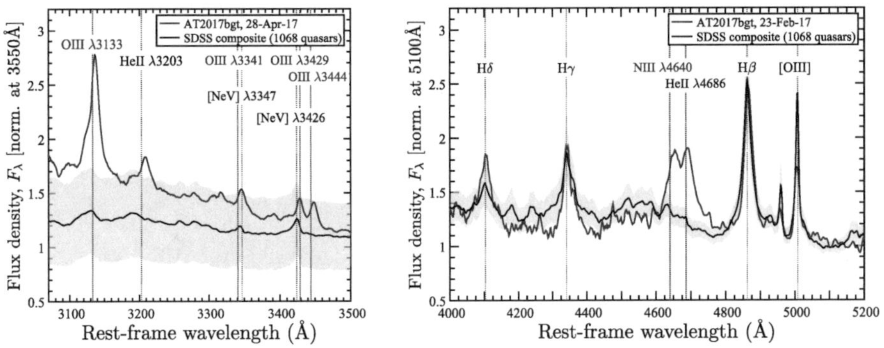

Fig. 10 Optical spectra of AT 2017bgt and unobscured AGN (adapted from Trakhtenbrot et al. 2019b). Two spectra of AT 2017bgt (blue), taken at different epochs within about two months of discovery, are compared to a composite of more than a thousand SDSS broad-line AGN (quasars) with similar hydrogen emission line widths. The broad Balmer lines and narrow forbidden [O III] λλ4959, 5007 lines of AT 2017bgt (and other events in this new class) are similar to those in the AGN. On the other hand, the prominent double-peaked emission feature near 4680 Å (*right panel*), the prominent O III λ3133 and He II λ3203 lines, and the weaker O III λλ3341, 3429, 3444 lines (*left*), are not seen in the AGN. These features arise from Bowen fluorescence, indicating an atypically strong source of high-energy (X-ray/EUV) radiation that produces intense He II emission, which in turn drives the O III and N III emission lines through multiple scatterings and excitations in an optically thick medium

emission), it is unlikely that UV/optical-bright TDE candidates arise from tail end of normal Type 2 AGN variability (Sect. 7.3).

3.2.2 Flaring of Known AGN

Trakhtenbrot et al. (2019b) identify a new class of flares from accreting SMBHs, which may be of particular importance to TDE classification. The light curves, which are exemplified by AT 2017bgt, the brightest and best-studied flare, exhibit a significant increase in UV/optical emission, followed by a long, slow decline, on timescales of a year (or more). The optical spectra show both narrow and broad emission lines, most of which resemble those of AGN, particularly NLSy1. Most importantly, AT 2017bgt-like events have a prominent double-peaked emission feature near 4680 Å, which is composed of the He II λ4686 and N III λ4640 emission lines, and several other strong O III transitions, all with widths similar to the broad Balmer lines. These lines, driven by Bowen fluorescence (BF), are not seen in normal AGN (Fig. 10), despite specific predictions (Netzer et al. 1985). Their existence in these UV-bright transients indicates that the BF process in dense gas near accreting SMBHs requires an exceptionally strong incident UV continuum.

While this new class of UV-bright flares from SMBHs was identified from only three events, including the transient in the ultra-luminous infrared galaxy F01004-2237 (Tadhunter et al. 2017) and the OGLE17aaj event (Gromadzki et al. 2019), on-going transient surveys should detect additional events of this kind. Such events are "TDE impostors," at least initially, as their early optical spectra would show a strong, broad feature near 4680 Å, which could be interpreted as He II λ4686, a common emission line in optical TDE candidates. Indeed, one of these events was originally classified as a TDE (Tadhunter et al. 2017). However, this feature, which is composed of two separate emission lines, is both narrower and weaker than the He II λ4686 emission observed in TDE candidates to date, i.e., with smaller FWHM *and* flux ratios relative to Balmer lines (see Fig. 11).

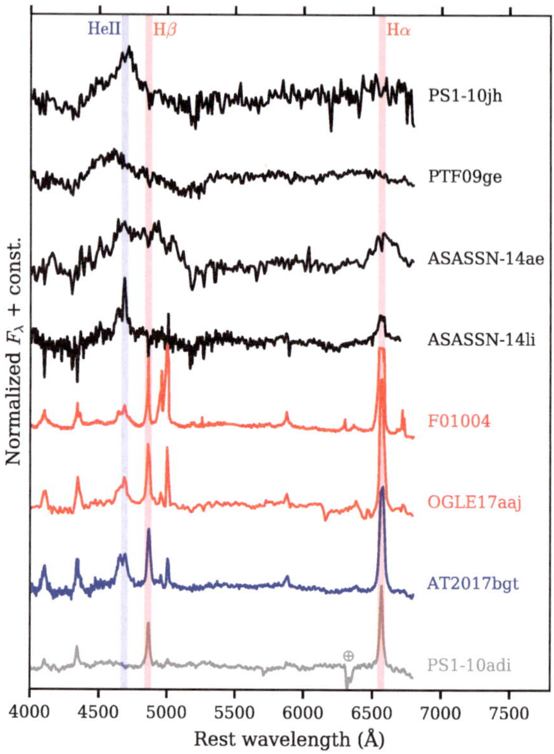

Fig. 11 Broad emission features near He II λ4686 in AT 2017bgt, and similar objects, compared to other nuclear transients (adapted from Trakhtenbrot et al. 2019b). The spectra of AT 2017bgt (blue) and the events in F01004-2237 (Tadhunter et al. 2017) and OGLE17aaj (Gromadzki et al. 2019, both in red) represent a new class of nuclear transients. Also shown are the spectra of four TDEs (from Gezari et al. 2012; Arcavi et al. 2014; Holoien et al. 2014, 2016b, in black) and of the luminous, slowly-evolving transient PS1-10adi (Kankare et al. 2017, in grey). All spectra are continuum-subtracted. The feature near 4860 Å in the AT 2017bgt-like events, which originates from He II λ4686 and the Bowen fluorescence N III λ4640 transitions, is significantly narrower than what is typically seen in most TDEs

Other recently discovered transients, some of which are strong TDE candidates, exhibit BF features in their optical spectra. TDE examples include iPTF15af (Blagorodnova et al. 2019), iPTF16fnl (Onori et al. 2019), ASASSN-18pg/AT2018dyb (Leloudas et al. 2019), and ASASSN-14li, which has some evidence for such features (Holoien et al. 2016b). Newer work shows that, in a flux-limited sample, TDEs with Bowen lines (or "TDE-Bs") are as common as TDEs with only broad hydrogen lines; in a volume-limited sample, TDE-B's are the *most* prevalent of the three TDE classes considered by van Velzen et al. (2020a).

Thus, the BF mechanism appears to frequent a range of nuclear, UV/optical-bright transients, all of which are related to enhanced accretion onto a SMBH. In some cases, the optical light curve, the peak luminosity, and the historical lack of SMBH activity, strongly favour the TDE interpretation. In other cases, where there is robust evidence for a pre-existing AGN (e.g., AT 2017bgt), we must be more prudent. While the nature of AT 2017bgt-like events remains ambiguous, the slow evolution of their light curves also argues against a "simple" TDE origin. One intriguing possibility is that these events arise from a tidal disruption stream colliding and interacting with a pre-existing AGN accretion disk. As noted by Chan et al. (2019), testing this scenario requires simulations that fully explore the geometry and orientation of the tidal stream relative to the disk, the density and velocity structure of the stream, and the properties of the unperturbed disk, as well as that predict the radiative output of the disk-stream interaction.

The flare in the well-known AGN 1ES 1927+654 (Trakhtenbrot et al. 2019a) exhibited an optical light-curve reminiscent of a TDE in rise time, peak luminosity, and decline rate. However, there were no telltale TDE features in the optical/UV spectroscopy, which instead revealed a changing-look AGN event occurring on a timescale of months. A follow-up study

(Ricci et al. 2020) speculated that the disappearance and reappearance of the X-ray emitting corona (also on timescales of months) was indeed linked to the tidal disruption of a star onto the pre-existing AGN accretion disk. This interpretation was based on qualitative agreement with models (Chan et al. 2019), which were limited in their predictive power (see more recent progress in Chan et al. 2020).

Another complex nuclear transient that occurred in a known AGN is CSS100217: 102913+404220, where the SDSS spectrum prior to the flare shows signatures of a NLSy1 galaxy (Drake et al. 2011). Drake et al. (2011) rule out a TDE based on several factors: the light curve evolution does not follow a $t^{-5/3}$ decay, the peak brightness ($M_{V,CSS} = -23$) is much higher than usually observed in TDEs ($M_V \sim -20$; see van Velzen et al. 2020b, Optical Chapter), and the fitted temperature is too low ($T = 1.5 \times 10^4$ K) compared with theoretical expectation ($T \simeq 10^5$ K). The light curve and the evolution of the narrow Balmer lines in the optical spectrum are consistent with a Type IIn supernova.

As in the case of AT 2017bgt-like events, the nature of this transient is hard to interpret. While CSS100217 could be a nuclear Type IIn SN or AGN flare, its fitted temperature is consistent with values seen in optical/UV TDEs (see van Velzen et al. 2020b, Optical Chapter), where the emission may be formed in outer shocks or reprocessing material rather than in a directly visible accretion disk. Drake et al. (2011) point out that other NLSy1 galaxies do not show the same level of optical variability and that the increase in the narrow Hα line strength after the flare died away occurred on a timescales too short to originate in the narrow line region. A focus on following up nuclear transients in AGN hosts, particularly in NLSy1 galaxies, and on quantifying the rates and range of TDEs and nuclear SNe in a control sample of quiescent galaxies, will reveal the connection between AGN and TDEs, whether it be physical or just due to mis-classification of AGN flares and/or nuclear SNe as TDEs.

3.3 Summary

Unambiguous TDE classification remains challenging, as any single observed property may be consistent with that of a persistent or flaring AGN. However, a constellation of unusual features like those cited in the previous sections and their consistency with rough expectations from theory argue that at least some TDE candidates, even unusual ones like ASASSN-15lh (Leloudas et al. 2016) and PS1-11af (Chornock et al. 2014), may in fact be tidally disrupted stars.

The observed overlap of TDE and AGN properties, particularly in the case of highly-variable AGN and AGN disk instability induced flares, complicates TDE classification and highlights the importance of archival data in constraining the level of SMBH activity prior to and long after a transient's detection. Known AGN should not be excluded from transient searches. It would be helpful to make use of existing datasets and related AGN selection criteria including: *WISE* for MIR-based AGN selection (e.g., following the criteria in Stern et al. 2012) and variability; *ROSAT* (and even *Swift*/BAT) for X-ray luminosity and X-ray SED shape; FIRST/NVSS for radio-based discrimination between galaxy star formation and AGN activity (e.g., following the SFR prescriptions of Hopkins et al. 2003); SDSS, 2dF/6dF, and earlier spectroscopic surveys, as well as relatively new efforts like GAMA and OzDES. All can be used to determine whether the transient in question is an unusual AGN or perhaps a TDE.

AGN and TDE may also be linked physically. New discoveries of dramatic, UV/optical-bright flares from accreting SMBHs, which share some TDE characteristics, raise the possibility that tidal disruption phenomena may occur in existing AGN and lead to extreme AGN variability. From a theoretical point of view, it is not yet clear how often to expect a tidal disruption event in a pre-existing AGN. Nor do we understand why some TDEs and flaring AGN both have Bowen fluorescence lines. Observationally, we do not know at present whether TDEs in pre-existing AGN are more common than those associated with dormant SMBHs or how to cleanly distinguish TDEs from the flares of known AGN.

4 Distinguishing TDEs from SNe

At first sight, optical/UV TDEs (e.g., PS1-10jh; see van Velzen et al. 2020b, Optical Chapter) can have observed properties similar to those of core collapse (CC) SNe: light curves with peak luminosities between those of "normal" and superluminous (SL) SNe (e.g., Arcavi et al. 2014), light curve timescales like those of SLSNe (e.g., Gal-Yam 2012), and broad emission line features. Indeed, ASASSN-15lh remains a debated TDE versus SN case (see Sect. 2). However, for most TDEs, a closer look at their properties reveals stark differences with SNe.

4.1 Summary of Observational Distinctions

The events in the PS1-10jh-like class of optical/UV TDEs (see van Velzen et al. 2020b, Optical Chapter) have the following properties that are not typically seen in SNe:

1. Blue, constant colour for hundreds of days;
2. Very late-time (years) UV detections;
3. Broad emission line profiles with no blueshifted absorption;
4. Strong, broad He II;
5. No late-time narrowing of emission lines in nebular phase;
6. Accompanying X-ray flare (in some cases).

4.2 Temperature Evolution

The colour of optical TDEs, implying an effective temperature of a few $\times 10^4$ K, remains constant for weeks or even months (see van Velzen et al. 2020b, Optical Chapter and references therein). While such temperatures are observed in SNe at early times, they cool within days to much lower temperatures (see Fig. 12).

In addition, some TDEs show continued UV emission years after discovery (van Velzen et al. 2019); such long-lasting UV emission is not seen in SNe.

4.3 Spectral Line Profiles

Most SNe spectra display lines with P-Cygni profiles, which originate in expanding ejecta. The line profiles of optical/UV TDEs are generally very different, showing no absorption and sometimes asymmetric emission profiles (see van Velzen et al. 2020b, Optical Chapter and Fig. 13). Even Type IIL SNe, which show weaker P-Cygni absorption (e.g., Gutiérrez et al. 2014) compared to Type IIP SNe, are still not as emission-dominated as TDEs.

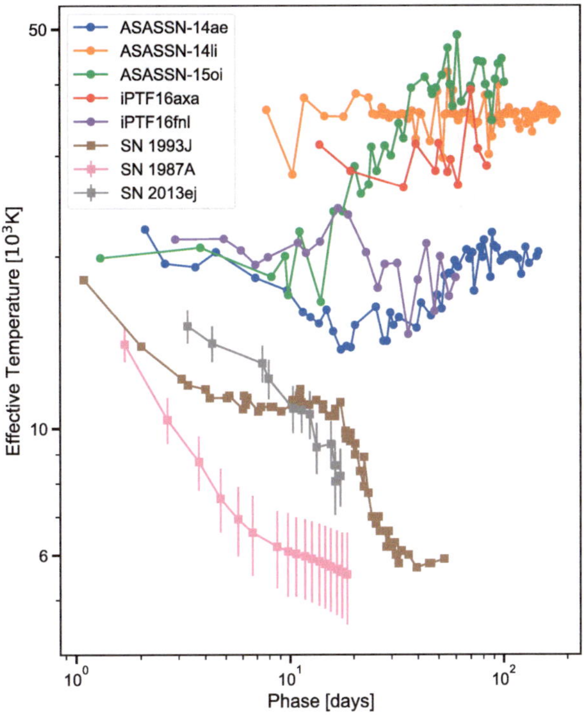

Fig. 12 Effective temperature evolution for optical/UV TDEs (circles, in days from discovery; Hung et al. 2017; Holoien et al. 2014, 2016a,b) and for hydrogen-rich core collapse SNe (squares, in days from explosion; Menzies et al. 1987; Richmond et al. 1994; Valenti et al. 2014). TDEs remain hot, while SNe cool within a few weeks. H-stripped SNe, which are not shown, cool even faster

4.4 Spectral Line Species

The spectral line species in optical TDEs differ from those of any known SN. Specifically, broad He II is not seen in any SN type at a strength comparable to H, in contrast to the broad He II and Balmer lines of some optical TDEs (Fig. 13). Ca features that are ubiquitous in SNe are not seen in TDEs at all, while only a few TDEs have shown Fe features, another element that is commonly observed in SN spectra.

4.5 Spectral Line Width Evolution

The spectral lines of optical TDEs remain relatively broad even at late times (months to years after peak); in contrast, SN spectra become "nebular" on those time scales, revealing only narrow emission lines from inner slow moving material as the ejecta expand and become transparent.

4.6 X-Ray Emission

Some optical TDEs are accompanied by X-ray emission, as expected from newly formed accretion disks. Most SNe, on the other hand, do not emit in X-rays, except for cases of obvious interaction between the SN ejecta and dense circumstellar material (CSM). Furthermore, in those cases, narrow emission lines from the unshocked CSM dominate the spectra, in stark contrast to the broad lines seen in the main class of optical TDEs.

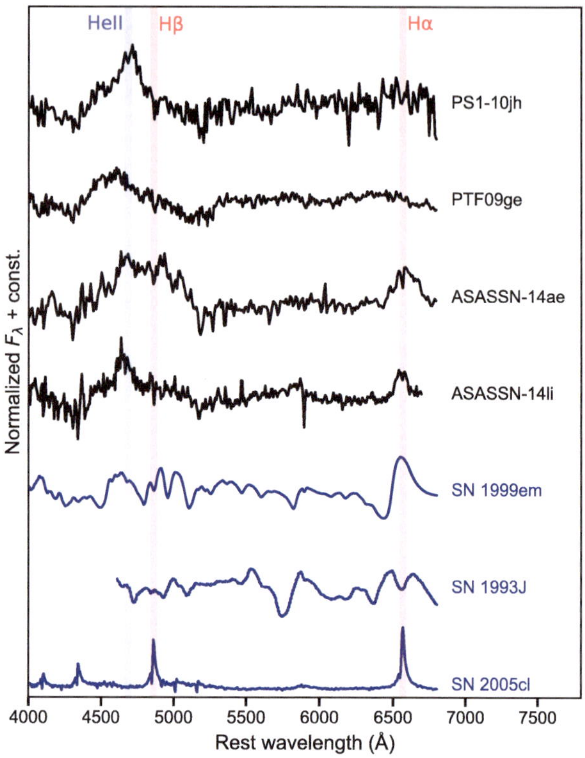

Fig. 13 Continuum subtracted spectra of optical/UV TDEs (black; Gezari et al. 2012; Arcavi et al. 2014; Holoien et al. 2014, 2016b) and of different subtypes of hydrogen-rich core collapse SNe (blue; from bottom: IIn, IIb and IIP; Kiewe et al. 2012; Barbon et al. 1995; Leonard et al. 2002). Both the spectral species and line profiles are different between TDEs and SNe

4.7 Summary

Several observational properties, which are readily measured for transients, help distinguish TDEs from SNe. This, in addition to the ample knowledge of SN populations and their typical emission properties, is the reason that there is relatively little confusion between these types of events. The notable exception is the case of ASASSN-15lh, which is a clear outlier to both known TDE and SN populations. Analysing the location of events with respect to the host galaxy centre may provide clarity; if future 15lh-like events are all found in the host centres, the TDE interpretation will be strengthened, else these sources are likely SNe.

5 Distinguishing Jetted TDEs from GRBs

The discovery of γ-ray bursts (GRBs) of cosmic origin were reported by Klebesadel et al. (1973). GRBs were a bi-modal distribution and could be divided into two categories based on the duration of their prompt emission: short and long (e.g., Kouveliotou et al. 1993). Short GRBs (SGRBs) have prompt emission (T_{90}) lasting approximately 2 sec or less, while long GRBs (LGRBs) have prompt emission from about 2 up to several thousand seconds. A more recently discovered class of "ultra-long" GRBs (GRB 111209A, GRB 121027A, and GRB 101225A) have prompt emissions at high energies from a few hundred to a few thousand seconds (e.g., Levan et al. 2014).

Jetted TDEs became a new category of TDEs after the unusual discovery of GRB 110328A by *Swift*'s γ-ray telescope (hereafter, Sw J1644+57; Burrows et al. 2011;

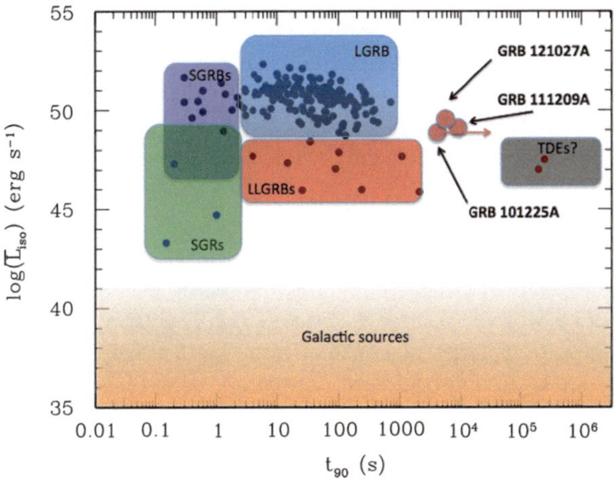

Fig. 14 Duration of burst versus approximate average luminosity over that duration for transients in the γ-ray sky. The figure compares the properties of soft-gamma repeaters (SGRs) in our own Galaxy, long- and short-duration GRBs (LGRBs and SGRBs), low-luminosity GRBs (LLGRBs), three GRB outliers (GRB 101225A, GRB 111209A, and GRB 121027A), and two very long transients thought to be jetted TDEs. Compared to GRBs, jetted TDE candidates exhibit high amplitude flaring at early times and prompt emission in X-rays lasting for weeks, i.e., t_{90} of 10^5–10^6 s. Figure reproduced from Levan et al. (2014)

Zauderer et al. 2011; Bloom et al. 2011; Levan et al. 2016). The same year, *Swift* discovered another event, Sw J2058+05 (Cenko et al. 2012), with very similar properties. In this section, we briefly describe the discovery of jetted TDEs and the primary ways in which they can be differentiated from GRBs via 1) prompt emission energy levels, 2) timescales and light curve; and 3) host galaxy association and the location of the transient within.

5.1 Prompt Emission, Timescales, and Light Curve

Jetted TDEs differ from GRBs in several ways, including high amplitude flaring at early times and prompt emission in X-rays lasting for weeks, i.e., t_{90} of 10^5–10^6 s (Fig. 14). The X-ray luminosity is also much greater (see, for example, Fig. 4 of Cenko et al. 2012). Additionally, for jetted TDEs, the X-ray light curve (after initial flaring) falls off as roughly $t^{-5/3}$. Sw J1644+57 and Sw J2058+05 have similar properties, including a X-ray light curve that slowly declined as $t^{-5/3}$ and then abruptly cut off, suggesting that the relativistic outflow shut off; see Fig. 4 of Zauderer et al. (2013) and Fig. 1 of Pasham et al. (2015).

5.2 Relationship to Host Galaxy

Another approach to distinguish jetted TDEs from GRBs is to evaluate the environment, and, if detectable, the type of host galaxy and where the transient lies within it. Long GRBs were determined to come from the deaths of massive stars based on their association with star forming galaxies, with the star-forming regions in those host galaxies, and, more specifically, with SNe themselves (see Hjorth and Bloom 2012, and references therein). Studies of the host environment (see Fong et al. 2015, and references therein) also contributed to the model of short GRBs as originating from compact object mergers, for which the discovery of GW170817 (Abbott et al. 2017) has now provided the strongest evidence. On the other hand, the jetted TDE candidates observed to-date have radio emission consistent with arising from the centres of inactive galaxies. For example, VLBA observations localised the radio afterglow of Sw 1644+57 to the host galaxy centre (Zauderer et al. 2011); multi-wavelength observations were important to 1) localise the afterglow, as the error circle for the *Swift*/XRT was more than an arcsecond; 2) obtain redshift information (e.g., from optical spectroscopy); and 3) classify the host galaxy (especially given that there were no previous observations).

5.3 Summary

For current and future wide-field surveys (e.g., VLA Sky Survey; VLASS), Metzger et al. (2015) discuss how various types of transients may be distinguishable based on emission timescales and energies. A combination of prompt emission timescales, light curves, event energetics, host galaxy properties, and transient location relative to the host centre will help to distinguish future TDEs from GRBs. In a case where the transient is detected after the high-energy emission has faded, the work may be more difficult. However, host galaxy properties will still be useful for population-wide, statistical studies.

6 Other Potential TDE Impostors

6.1 Stellar Collisions in Galactic Nuclei

Galactic nuclei hosting SMBHs are some of the most dense and collisional environments in galaxies, with stellar densities of 10^6–10^7 pc^{-3} (see Neumayer et al. 2020, for a review). Stellar collisions and mergers could thus be abundant there (see Dale et al. 2009, and references therein), making explosive transient events far more likely. Because the densest nuclear star clusters are associated with galaxies hosting relatively low mass ($< 10^8\ M_\odot$) SMBHs, stellar collisions may generate transient events in nuclei of similar masses to those expected to produce TDEs, leading to misidentifications. Moreover, mergers of binary stars can be induced by secular evolution in triple systems (Perets and Fabrycky 2009); in galactic nuclei, every binary forms a triple with the SMBH and is therefore potentially sensitive to such secular evolution. In other words, binary mergers could be triggered near SMBHs (Antonini et al. 2010; Antonini and Perets 2012; Prodan et al. 2015; Stephan et al. 2019), potentially mimicking TDEs.

Nevertheless, there are likely differences between TDEs and nuclear stellar collisions and mergers that could distinguish their populations individually and/or statistically. For example, mergers/collisions may produce reddened transients, perhaps similar to the merger in the V1309 Scorpii system (Tylenda et al. 2011). More generally, stellar collisions/mergers should be abundant in other dense stellar environments, such as globular clusters, so their expected spatial distribution and properties would need to be reconciled with the nuclear (by definition) environments of TDE candidates.

Another possibility is that stellar collisions and mergers near a SMBH could remove material from stars that is then accreted by the SMBH. For example, the rate of grazing collisions can be enhanced through the capture of stars inspiraling to the SMBH due to gravitational wave emission (Metzger and Stone 2017). Such grazing collisions can strip material from the stars, which would later be accreted by the SMBH or accumulate to form a disk around the SMBH and eventually produce a flare.

Other types of induced mergers and collisions near SMBHs could produce gas clouds that later accrete onto the SMBH and generate flares (e.g., SMBH-induced binary mergers,

Antonini and Perets 2012; Prodan et al. 2015). Such events might impersonate TDEs or disk instability flaring. It is not clear, however, why such induced stellar mergers should be preferentially observed in the hosts most favoured by TDEs, i.e., quiescent, Balmer-strong (QBS) galaxies and the subset of post-starburst (PSB, or "E+A") galaxies (Arcavi et al. 2014; French et al. 2016; Graur et al. 2018; see the French et al. 2020, Host Galaxies Chapter).

6.2 Micro-TDEs

The tidal disruption of stars by stellar-mass black holes potentially gives rise to a different class of transients. Such "micro-TDEs," especially when occurring in or near galactic nuclei, could be incorrectly interpreted as TDEs associated with the central SMBH. Perets et al. (2016) suggested that micro-TDEs arising from the disruption of MS stars or planets may appear as ultra-long GRBs ($\sim 10^3$ to a few $\times 10^4$ s). The timescales could be longer for evolved star progenitors, whose disruption might produce longer-term transient X-ray sources. Such micro-TDEs could masquerade as jetted TDEs detected in X-rays and γ-rays.

While micro-TDEs should occur across different environments, including dense non-nuclear stellar clusters and even in the field (Perets et al. 2016), they could also be frequent in galactic nuclei, given the highly collisional environment there (as discussed in the previous section).

Lin et al. (2018) reported a transient event in a star cluster offset by 12.5 kpc from the centre of its host galaxy ($z = 0.055$). Observational signatures of this event are similar to those of TDE candidates: the light curve decays by roughly $t^{-5/3}$ over 10 years, the X-ray spectrum is soft and well described by a thermal accretion disk that cools with decreasing luminosity, and the peak luminosity of the flare is 10^{43} erg s^{-1}. This event may be a micro-TDE of a post-MS star. The high disk temperature and long decay time suggest another possibility, that the TDE arose from a star shredded by an *intermediate*-mass black hole in the star cluster (see Lin et al. 2018, Wen et al. 2021), which would represent a new probe of that elusive black hole population.

6.3 Circumbinary Accretion Flows

Many studies (see De Rosa et al. 2019, for a review) explore the possibility of binary massive black holes accreting from a circumbinary disk (e.g., Ivanov et al. 1999; MacFadyen and Milosavljević 2008; Cuadra et al. 2009; Shi et al. 2012; D'Orazio et al. 2013; Tanaka 2013; Muñoz et al. 2020; Ragusa et al. 2020, and references therein). Tanaka (2013) suggests that a SMBH binary (SMBHB) can clear a central cavity in its accretion disk, which then appears dimmer and softer than an equivalent single AGN disk. Gas streams of 0.1 solar masses, which are intermittent on timescales depending on the binary period, then infall towards the SMBHB from the cavity's lip. The streams shock, giving rise to a TDE-like flare. If the recurrence period is long compared to the era of transient monitoring, a single detection might be misidentified as a TDE.

The recurrence period, debris velocities, and energy yields in a cavity flare depend on the SMBHB properties and need not correspond to the predicted tidal radii of stellar disruptions. Thus, the timescales for such flares might not be directly related to SMBH mass (cf. van Velzen et al. 2019). Subtler details might distinguish binary cavity flares: the steepness of their fading or their occurrence around a SMBH that is too massive for a TDE. For example, Leloudas et al. (2016) and Krühler et al. (2018) infer $M_{\rm BH} > 10^8\ M_\odot$ for the weak-lined TDE candidate ASASSN-15lh. If SMBH binaries are a common product of the galaxy-galaxy mergers that produce post-starburst galaxies, then the cavity flare mechanism, like

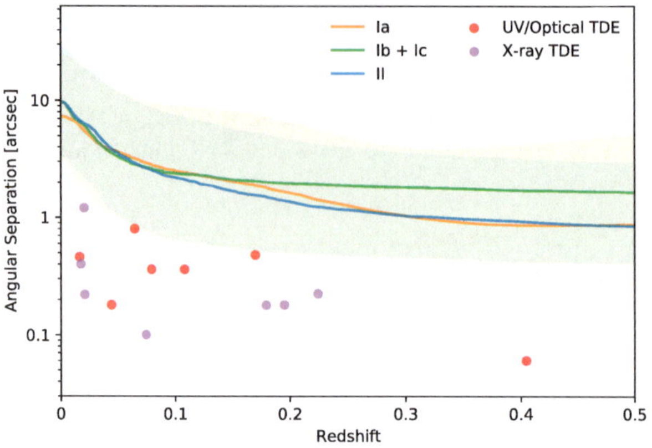

Fig. 15 Angular separation from host galaxy centre versus redshift for UV/optical and X-ray TDE samples (red and purple points, respectively) and for SNe (coloured mean lines and $\pm 1\sigma$ bands). The data are drawn from the transient-host pair catalogue of Qin et al. (in prep.). For TDEs, the angular separation uncertainties are generally unrecorded in the literature and probably dominated by the uncertainty in the position of the host's centre, especially at lower redshifts. At all redshifts, the SNe tend to lie further from the centre than TDEs, with angular offsets of over 1 dex (1.8σ) larger. This difference may arise in part from a selection effect, as certain types of SNe are likely to occur near the dense centre, but are harder to detect there, and transients may be classified as TDEs after consideration of their proximity to the nucleus

stellar tidal disruption, may favour such host galaxies; the expected rate is currently difficult to assess.

7 Distinguishing TDE Candidates Statistically

The demographics of TDE-candidates, e.g., their frequency, spatial offset from the host galaxy nucleus, and host properties, may provide evidence for their existence by *statistically* distinguishing them from impostors. The expected rates of TDEs are reviewed in the Stone et al. 2020, Rates Chapter, and TDE host galaxy properties in the French et al. 2020, Host Galaxies Chapter. Here we discuss statistical arguments suggesting that at least a fraction of TDE candidates are real, while some individual TDE detections remain ambiguous. Specifically, we compare the TDE nuclear offsets and the SMBH masses, stellar masses, and star formation histories of TDE hosts to those of AGN and SNe.

7.1 Nuclear Offsets

SNe in general have higher observed rates than TDEs (see Stone et al. 2020, Rates Chapter), and most observed SNe are significantly offset from the galactic centre (Fig. 15). However, these offsets are in part a selection effect; intrinsically, some types of SNe are more likely than others to occur near the centre. While TDEs are nuclear sources by definition (except for presumably rare cases around a recoiling BH), SNe generally follow the mass distribution of their progenitor stellar population in the host galaxy. Because TDE candidates significantly differ from typical SNe, it is hard to explain all TDE candidates as SNe, as those SNe masquerading as TDEs would need both to occur in or very near the nucleus and be unusual. Although the environments of galactic nuclei are unique, it is challenging to envision them giving rise to SNe that are rare elsewhere.

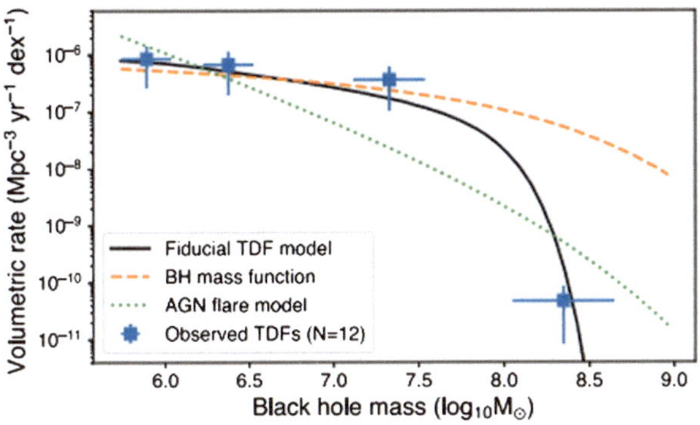

Fig. 16 Differences in expected black hole mass function for TDEs vs. AGN from van Velzen (2018). The green dotted line represents the expected volumetric rate of AGN flares, assuming that the flare reoccurrence time scales linearly with $M_{\rm BH}$. This model fails to reproduce the observed volumetric rate of observed TDE candidates. Instead, a simple model where the SMBH mass function is multiplied by a constant TDE rate (orange dashed line), and that takes into account the suppression due to direct captures, provides a better description of the observations (black solid line)

Improving the statistics of centrally located SNe requires uniform, higher spatial resolution surveys. Arguably the highest spatial resolution transient survey currently running is *Gaia*, with a typical astrometric accuracy of a few tenths of an arcsecond. Blagorodnova et al. (2016) show that a transient fainter than $G = 16$ mag (roughly equivalent to $V = 16$) can be resolved by *Gaia* if it is more than 0.2 arcsec away from the bulge of the galaxy. Within this 0.2 arcsec volume, they find that SNe and TDEs occur in roughly equal numbers. Further improvements in *Gaia* astrometry may push this ratio in favour of TDEs (see also Kostrzewa-Rutkowska et al. 2018 for a discussion on nuclear transients in *Gaia*).

Connections between extreme SNe and TDE claims (e.g., Komossa et al. 2009; Drake et al. 2011) are questioned by Gezari et al. (2009), see Sect. 4 above. Nevertheless, unusual conditions in the galactic core, including a dense interstellar medium in the circumnuclear region of a starved/dormant AGN, may boost peculiar SNe rates there. Gas confinement may alter SN shock physics and radiative efficiency, potentially producing an abnormally shallow light curve (Saxton et al. 2018). A larger and more representative census of nuclear SNe, as well as improved models, is needed to resolve this issue.

7.2 Host Galaxy SMBH and Stellar Masses

TDEs prefer hosts with SMBH masses $M_{\rm BH} < 10^8\ M_\odot$ (Wevers et al. 2017, 2019b; Mockler et al. 2019). This is not surprising, given that the tidal radius should be outside the Schwarzschild radius for TDEs to be observable. TDEs and AGN can be distinguished statistically using differences in their black hole mass functions (van Velzen 2018). The expected volumetric rate of AGN flares, assuming that the flare reoccurrence time scales linearly with $M_{\rm BH}$, does not match the observed volumetric rate of observed TDE candidates (Fig. 16). Instead, a simple model where the SMBH mass function is multiplied by a constant TDE rate, and that takes into account the suppression due to direct captures, is more successful. van Velzen (2018) argue that the strong suppression of the volumetric rate for $M_{\rm BH} > 10^8\ M_\odot$ can only be explained by the TDE scenario.

At least some AGN that arise from lower mass SMBHs may be distinguished from the TDE population by other signatures. For example, AGN driven by SMBHs with

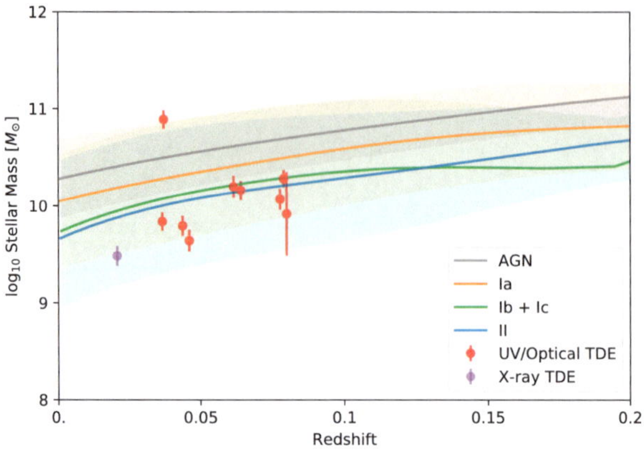

Fig. 17 Host galaxy stellar masses versus redshift for UV/optical (red points) and X-ray (purple point) TDEs, SNe (orange, green, and blue mean lines and $\pm 1\sigma$ bands), and AGN (grey mean line and $\pm 1\sigma$ band). Data are compiled from the transient-host catalogue of Qin et al. (in prep.). The high outlier event, AT2018dyk (ZTF18aajupnt), is listed as an optical/UV TDE in the French et al. 2020, Host Galaxies Chapter. For its stellar velocity dispersion of $121 \pm 3.8\ \mathrm{km\,s^{-1}}$ (SDSS DR8 spectral catalogue; Aihara et al. 2011), its SMBH mass would be roughly $1.6 \times 10^7\ M_\odot$ (from the scaling relation in McConnell et al. 2011), only marginally higher than typical of other TDE host galaxies (Wevers et al. 2019b). Frederick et al. (2019) alternately classify this source as a rapidly brightening LINER, whose later UV and optical spectra reveal a NLSy1 with strong coronal lines. The stellar masses of these TDE host galaxies are, on average, 0.74 dex ($\sim 1.8\sigma$) below those of AGN hosts at the same redshift

$10^6 \lesssim (M_{\rm BH}/M_\odot) \lesssim$ few $\times\ 10^7$ often produce relatively narrow Balmer lines (FWHM $\sim$ 1200–2000 $\mathrm{km\,s^{-1}}$) and thus are classified as NLSy1s. Such optical spectra differ from what is seen in most (optical) TDE candidates, whose lines are much broader ($\gtrsim 15{,}000\ \mathrm{km\,s^{-1}}$). Because the He II λ4686 line width is similar to those of the Balmer lines in AGN, it would be considerably more narrow than in TDEs.

By extension, the host galaxies of TDEs and AGN might be expected to have different stellar mass distributions. Data for TDE, AGN, and SNe hosts suggest that TDE host stellar masses tend to be smaller than those of AGN (Fig. 17). This offset is apparent if we consider the TDEs that satisfy many of the criteria laid out in this chapter and whose hosts have known stellar masses, i.e., the TDEs with broad optical H and He emission lines and objects with the most robust classification of "X-ray" TDEs by Auchettl et al. (2017), as in the French et al. 2020, Host Galaxies Chapter.

Furthermore, when the host properties of X-ray TDE candidates are grouped by the strength of the TDE claim (French et al. 2020, Host Galaxies Chapter), the least certain TDEs (those classified as "possible" TDEs by Auchettl et al. 2017) have on average higher SMBH masses and brighter host galaxy absolute magnitudes than "likely" or "X-ray" TDEs. While robust conclusions are limited by the small sample sizes, this result suggests that the "possible" TDEs are a different population, e.g., misidentified AGN flares, which tend to inhabit more massive and brighter host galaxies. For such flares, the lack of high quality data leads not only to the "possible" TDE label, but also makes them appear similar to TDEs.

7.3 Other Host Galaxy Properties

In addition to SMBH mass (and, by correlation, total stellar mass), other host properties can be used to distinguish TDEs statistically. TDEs prefer hosts with unusual, A-star dominated spectra, i.e., quiescent, Balmer-strong (QBS) galaxies and the subpopulation of post-starburst (PSB or "E+A") galaxies (Arcavi et al. 2014; French et al. 2016; Graur et al. 2018; Law-Smith et al. 2017; see also the French et al. 2020, Host Galaxies Chapter). The quiescent galaxies common among TDE hosts generally lack strong Hα emission, although several show broad emission lines indicative of Type 1 AGN. A higher fraction of TDE hosts exhibit weak, narrow emission lines, with line ratios indicating ionisation from sources other than star formation. Could the observed TDE population arise from the tail end of Type 2 AGN variability? We address this question here statistically.

The recent star formation histories of TDE host galaxies are different on average than Type 2 Seyfert or LINER AGN. For galaxies in the SDSS main spectroscopic sample that 1) have stellar masses typical of TDE host galaxies ($\log(M_\star/M_\odot) \simeq 9.5$–$10.5$) and 2) have all four emission lines required for AGN classification on the BPT diagram (Baldwin et al. 1981), 92% are star forming, 5% are Seyferts, and 3% are LINERs, according to the Kauffmann et al. (2003) criteria. While between 36–75% of TDE candidates with broad H/He lines are found in QBS galaxies, only 9% of Seyferts and 17% of LINERs are. If TDEs were the tail end of the Type 2 AGN LINER or Seyfert distributions, we would expect that 9–17% of the host galaxies were QBS galaxies, instead of 36–75%. Thus, it is unlikely that most of the observed broad H/He line TDEs are caused by the tail end of normal Type 2 AGN variability.

Furthermore, many galaxies with LINER-like spectra maybe be ionised by sources other than low-luminosity AGN, weakening the possibility that the observed TDE population is caused by extreme Type 2 AGN variability. Many LINER-like galaxies identified using the BPT diagram could be ionised by merger-induced shocks (Rich et al. 2015) or post-AGB stars (Yan and Blanton 2012). Both of these possible ionisation sources are expected to occur during the QBS or PSB stage, so the proportion of QBS and PSB galaxies hosting a true low-luminosity AGN will be lower than that inferred from the BPT diagram alone.

The host galaxies of the main optical and X-ray TDE classes are also distinct from those of core-collapse (CC) SNe. CC SNe originate from massive stars and thus are found almost exclusively in star forming regions. Star formation mostly ended more than ~ 100 Myr ago in QBS galaxies, the favoured hosts of TDEs, so massive stars, and thus CC SNe, in QBS galaxies are unlikely.

8 Conclusions and the Future

While TDE classification remains ambiguous—any single observed property may be found in other types of transients—a constellation of unusual features like those cited above, consistency with rough expectations from TDE theory, and the statistics of the transient and host galaxy populations argue that at least some candidates do in fact arise from stars tidally disrupted by SMBHs. The tallies of potentially distinguishing features included in this chapter are not meant to be rules, but initial guidelines, for these early days of TDE detection. In the RubinObs/LSST, SDSS-V, and *eROSITA* era, new, more extreme types of AGN and other transients will be discovered that are likely to cast doubt on some current TDE claims and to further complicate future TDE classifications. Yet, over the same period, thousands of new TDE candidates, as well as improvements to theoretical models of stellar disruption and disk accretion by SMBHs, will help us define any unique combinations of TDE signatures.

Some of the key outstanding questions and future needs are:

- Are TDEs are significantly less absorbed compared to AGN or is this difference a selection bias?

- Why do the X-ray spectra of some thermal-dominated TDEs harden as they evolve, while others do not?
- What is the physical driver of the hyper-variable and "changing-look" AGN now detected in time-domain surveys? This question is relevant here because flares from specific changing-look AGN (e.g., SDSS J015957.64+003310.5; LaMassa et al. 2015) have been attributed to the tidal disruption of a star around a previously active SMBH (Merloni et al. 2015). On the other hand, changing-look AGN may just be extreme examples of regular, continuous AGN variability. Thus, understanding the demographics and drivers of both changing-look and hyper-variable AGN may help to determine characteristics that distinguish them from TDEs. A crucial first step would be a large, fair census of such objects.
- What other kinds of flares in persistent AGN are out there and how are they related to TDEs?
- What signatures can distinguish between AGN flares due to variability in a pre-existing accretion disk and those due to TDEs?
- Are TDEs within pre-existing AGN more common than those around dormant SMBHs, suggesting a link between AGN and TDEs or similarities in the conditions that produce them?
- What does the detection of Bowen fluorescence lines, indicative of extreme UV radiation fields, in many TDE candidates and in flaring AGN tell us about the physical links between AGN and TDEs?
- To strengthen the TDE interpretation, it is critical to find additional examples of one-off events like ASASSN-15lh. If similar sources all lie in the host galaxy centres, then the case for TDEs is stronger. Otherwise, one example detected off-nucleus would suggest a new kind of superluminous SN.
- Are there other potential TDE impostors, such as stellar collisions, "micro-TDEs," or circumbinary flares, that occur more frequently than TDEs in quiescent, Balmer-strong galaxies? Or in the "post-starburst" subclass of these galaxies?
- Do any known SNe types or other potential TDE impostors occur more frequently than TDEs in galactic nuclei?
- What is the relationship between TDE claims and the high proportion of LINERs also found in post-starburst galaxies?

Acknowledgements The authors thank ISSI for their support and hospitality and the review organisers for their leadership in coordinating these reviews. This work was performed in part at the Aspen Center for Physics, which is supported by National Science Foundation grant PHY-1607611, during the January 2019 Aspen conference on *Using Tidal Disruption Events to Study Super-Massive Black Holes*. We are grateful to Sjoert van Velzen and Ryan Foley for leading a discussion there about possible differences between TDEs and supernovae. We also thank Nicholas Stone, Sixiang Wen, and Dennis Zaritsky for helpful information. AIZ acknowledges support from grant HST-GO-14717.001-A from the Space Telescope Science Institute (STScI), which is operated by the Association of Universities for Research in Astronomy (AURA), Incorporated, under NASA contract NAS5-26555. BT acknowledges support from the Israel Science Foundation (grant 1849/19). KAA is supported by the Danish National Research Foundation (DNRF132). JLD is supported by the GRF grant from the Hong Kong government under HKU 27305119. KDF is supported by Hubble Fellowship grant HST-HF2-51391.001-A from STScI, operated by AURA, Inc., under NASA contract NAS5-26555. TW is funded in part by European Research Council grant 320360 and by European Commission grant 730980. Parts of this research were supported by the Australian Research Council Centre of Excellence for All Sky Astrophysics in 3 Dimensions (ASTRO 3D), through project number CE170100013.

References

B.P. Abbott et al. (LIGO Scientific Collaboration, Virgo Collaboration), GW170817: observation of gravitational waves from a binary neutron star inspiral. Phys. Rev. Lett. **119**(16), 161101 (2017). https://doi.org/10.1103/PhysRevLett.119.161101

H. Aihara, et al., The eighth data release of the Sloan Digital Sky Survey: first data from SDSS-III. Astrophys. J. Suppl. Ser. **193**(2), 29 (2011). https://doi.org/10.1088/0067-0049/193/2/29

D.M. Alexander, R.C. Hickox, What drives the growth of black holes? Nature **56**(4), 93–121 (2012). https://doi.org/10.1016/j.newar.2011.11.003

F. Antonini, H.B. Perets, Secular evolution of compact binaries near massive black holes: gravitational wave sources and other exotica. Astrophys. J. **757**, 27 (2012). https://doi.org/10.1088/0004-637X/757/1/27

F. Antonini, J. Faber, A. Gualandris, D. Merritt, Tidal breakup of binary stars at the Galactic center and its consequences. Astrophys. J. **713**(1), 90–104 (2010). https://doi.org/10.1088/0004-637X/713/1/90

I. Arcavi, A. Gal-Yam, M. Sullivan, Y.-C. Pan, S.B. Cenko, A. Horesh, E.O. Ofek, A. De Cia, L. Yan, C.-W. Yang, D.A. Howell, D. Tal, S.R. Kulkarni, S.P. Tendulkar, S. Tang, D. Xu, A. Sternberg, J.G. Cohen, J.S. Bloom, P.E. Nugent, M.M. Kasliwal, D.A. Perley, R.M. Quimby, A.A. Miller, C.A. Theissen, R.R. Laher, A continuum of H- to He-rich tidal disruption candidates with a preference for E+A galaxies. Astrophys. J. **793**, 38 (2014). https://doi.org/10.1088/0004-637X/793/1/38

K. Auchettl, J. Guillochon, E. Ramirez-Ruiz, New physical insights about tidal disruption events from a comprehensive observational inventory at X-ray wavelengths. Astrophys. J. **838**, 149 (2017). https://doi.org/10.3847/1538-4357/aa633b

K. Auchettl, E. Ramirez-Ruiz, J. Guillochon, A comparison of the X-ray emission from tidal disruption events with those of active galactic nuclei. Astrophys. J. **852**, 37 (2018). https://doi.org/10.3847/1538-4357/aa9b7c

N. Bade, S. Komossa, M. Dahlem, Detection of an extremely soft X-ray outburst in the HII-like nucleus of NGC 5905. Astron. Astrophys. **309**, 35–38 (1996)

J.A. Baldwin, M.M. Phillips, R. Terlevich, Classification parameters for the emission-line spectra of extragalactic objects. Publ. Astron. Soc. Pac. **93**, 5–19 (1981). https://doi.org/10.1086/130766

R. Barbon, S. Benetti, E. Cappellaro, F. Patat, M. Turatto, T. Iijima, SN 1993J in M 81: one year of observations at Asiago. Astron. Astrophys. Suppl. Ser. **110**, 513 (1995)

N. Bennert, H. Falcke, H. Schulz, A.S. Wilson, B.J. Wills, Size and structure of the narrow-line region of quasars. Astrophys. J. **574**(2), 105–109 (2002). https://doi.org/10.1086/342420

N. Bennert, B. Jungwiert, S. Komossa, M. Haas, R. Chini, Size and properties of the narrow-line region in Seyfert-2 galaxies from spatially-resolved optical spectroscopy. Astron. Astrophys. **456**, 953–966 (2006). https://doi.org/10.1051/0004-6361:20065319

S. Bianchi, R. Maiolino, G. Risaliti, AGN obscuration and the unified model. Adv. Astron. **2012**, 782030 (2012). https://doi.org/10.1155/2012/782030

N. Blagorodnova, S. Van Velzen, D.L. Harrison, S. Koposov, S. Mattila, H. Campbell, N.A. Walton, Ł. Wyrzykowski, Gaia transient detection efficiency: hunting for nuclear transients. Mon. Not. R. Astron. Soc. **455**, 603–617 (2016). https://doi.org/10.1093/mnras/stv2308

N. Blagorodnova, S.B. Cenko, S.R. Kulkarni, I. Arcavi, J.S. Bloom, G. Duggan, A.V. Filippenko, C. Fremling, A. Horesh, G. Hosseinzadeh, E. Karamehmetoglu, A. Levan, F.J. Masci, P.E. Nugent, D.R. Pasham, S. Veilleux, R. Walters, L. Yan, W. Zheng, The broad absorption line tidal disruption event iPTF15af: optical and ultraviolet evolution. Astrophys. J. **873**(1), 92 (2019). https://doi.org/10.3847/1538-4357/ab04b0

P.K. Blanchard, M. Nicholl, E. Berger, J. Guillochon, R. Margutti, R. Chornock, K.D. Alexander, J. Leja, M.R. Drout, PS16dtm: a tidal disruption event in a narrow-line Seyfert 1 galaxy. Astrophys. J. **843**, 106 (2017). https://doi.org/10.3847/1538-4357/aa77f7

J.S. Bloom, D. Giannios, B.D. Metzger, S.B. Cenko, D.A. Perley, N.R. Butler, N.R. Tanvir, A.J. Levan, P.T. O'Brien, L.E. Strubbe, F. De Colle, E. Ramirez-Ruiz, W.H. Lee, S. Nayakshin, E. Quataert, A.R. King, A. Cucchiara, J. Guillochon, G.C. Bower, A.S. Fruchter, A.N. Morgan, A.J. van der Horst, A possible relativistic jetted outburst from a massive black hole fed by a tidally disrupted star. Science **333**(6039), 203 (2011). https://doi.org/10.1126/science.1207150

T.A. Boroson, Black hole mass and Eddington ratio as drivers for the observable properties of radio-loud and radio-quiet QSOs. Astrophys. J. **565**, 78–85 (2002). https://doi.org/10.1086/324486

W.N. Brandt, K.A. Pounds, H. Fink, The unusual X-ray and optical properties of the ultrasoft active galactic nucleus Zwicky 159.034 (RE J1237+264). Mon. Not. R. Astron. Soc. **273**, 47–52 (1995). https://doi.org/10.1093/mnras/273.1.47L

J.S. Brown, T.W.-S. Holoien, K. Auchettl, K.Z. Stanek, C.S. Kochanek, B.J. Shappee, J.L. Prieto, D. Grupe, The long term evolution of ASASSN-14li. Mon. Not. R. Astron. Soc. **466**, 4904–4916 (2017). https://doi.org/10.1093/mnras/stx033

J.S. Brown, C.S. Kochanek, T.W.-S. Holoien, K.Z. Stanek, K. Auchettl, B.J. Shappee, J.L. Prieto, N. Morrell, E. Falco, J. Strader, L. Chomiuk, R. Post, S. Villanueva Jr., S. Mathur, S. Dong, P. Chen, S. Bose, The ultraviolet spectroscopic evolution of the low-luminosity tidal disruption event iPTF16fnl. Mon. Not. R. Astron. Soc. **473**, 1130–1144 (2018). https://doi.org/10.1093/mnras/stx2372

D.N. Burrows, J.A. Kennea, G. Ghisellini, V. Mangano, B. Zhang, K.L. Page, M. Eracleous, P. Romano, T. Sakamoto, A.D. Falcone, J.P. Osborne, S. Campana, A.P. Beardmore, A.A. Breeveld, M.M. Chester, R. Corbet, S. Covino, J.R. Cummings, P. D'Avanzo, V. D'Elia, P. Esposito, P.A. Evans, D. Fugazza, J.M. Gelbord, K. Hiroi, S.T. Holland, K.Y. Huang, M. Im, G. Israel, Y. Jeon, Y.-B. Jeon, H.D. Jun, N. Kawai, J.H. Kim, H.A. Krimm, F.E. Marshall, P. Mészáros, H. Negoro, N. Omodei, W.-K. Park, J.S. Perkins, M. Sugizaki, H.-I. Sung, G. Tagliaferri, E. Troja, Y. Ueda, Y. Urata, R. Usui, L.A. Antonelli, S.D. Barthelmy, G. Cusumano, P. Giommi, A. Melandri, M. Perri, J.L. Racusin, B. Sbarufatti, M.H. Siegel, N. Gehrels, Relativistic jet activity from the tidal disruption of a star by a massive black hole. Nature **476**(7361), 421–424 (2011). https://doi.org/10.1038/nature10374

S. Campana, D. Mainetti, M. Colpi, G. Lodato, P. D'Avanzo, P.A. Evans, A. Moretti, Multiple tidal disruption flares in the active galaxy IC 3599. Astron. Astrophys. **581**, 17 (2015). https://doi.org/10.1051/0004-6361/201525965

G. Cannizzaro, M. Fraser, P.G. Jonker, J.E. Pringle, S. Mattila, P.C. Hewett, T. Wevers, E. Kankare, Z. Kostrzewa-Rutkowska, Ł. Wyrzykowski, F. Onori, J. Harmanen, K.E.S. Ford, B. McKernan, C.J. Nixon, Extreme variability in an active galactic nucleus: Gaia16aax. Mon. Not. R. Astron. Soc. **493**(1), 477–495 (2020). https://doi.org/10.1093/mnras/staa186

J.K. Cannizzo, The nature of the giant outbursts in the bursting pulsar GRO J1744-28. Astrophys. J. Lett. **466**, 31 (1996). https://doi.org/10.1086/310167

J.K. Cannizzo, N. Gehrels, A new paradigm for gamma-ray bursts: long-term accretion rate modulation by an external accretion disk. Astrophys. J. **700**, 1047–1058 (2009). https://doi.org/10.1088/0004-637X/700/2/1047

J.K. Cannizzo, H.M. Lee, J. Goodman, The disk accretion of a tidally disrupted star onto a massive black hole. Astrophys. J. **351**, 38–46 (1990). https://doi.org/10.1086/168442

J.K. Cannizzo, E. Troja, G. Lodato, GRB 110328A/Swift J164449.3+573451: the tidal obliteration of a deeply plunging star? Astrophys. J. **742**(1), 32 (2011). https://doi.org/10.1088/0004-637X/742/1/32

N. Caplar, S.J. Lilly, B. Trakhtenbrot, Optical variability of AGNs in the PTF/iPTF survey. Astrophys. J. **834**, 111 (2017). https://doi.org/10.3847/1538-4357/834/2/111

S.B. Cenko, H.A. Krimm, A. Horesh, A. Rau, D.A. Frail, J.A. Kennea, A.J. Levan, S.T. Holland, N.R. Butler, R.M. Quimby, J.S. Bloom, A.V. Filippenko, A. Gal-Yam, J. Greiner, S.R. Kulkarni, E.O. Ofek, F. Olivares E., P. Schady, J.M. Silverman, N.R. Tanvir, D. Xu, Swift J2058.4+0516: discovery of a possible second relativistic tidal disruption flare? Astrophys. J. **753**, 77 (2012). https://doi.org/10.1088/0004-637X/753/1/77

S.B. Cenko, A. Cucchiara, N. Roth, S. Veilleux, J.X. Prochaska, L. Yan, J. Guillochon, W.P. Maksym, I. Arcavi, N.R. Butler, A.V. Filippenko, A.S. Fruchter, S. Gezari, D. Kasen, A.J. Levan, J.M. Miller, D.R. Pasham, E. Ramirez-Ruiz, L.E. Strubbe, N.R. Tanvir, F. Tombesi, An ultraviolet spectrum of the tidal disruption flare ASASSN-14li. Astrophys. J. Lett. **818**, 32 (2016). https://doi.org/10.3847/2041-8205/818/2/L32

C.-H. Chan, T. Piran, J.H. Krolik, D. Saban, Tidal disruption events in active galactic nuclei. Astrophys. J. **881**(2), 113 (2019). https://doi.org/10.3847/1538-4357/ab2b40

C.-H. Chan, T. Piran, J.H. Krolik, Light curves of tidal disruption events in active galactic nuclei. Astrophys. J. **903**(1), 17 (2020). https://doi.org/10.3847/1538-4357/abb776

R. Chornock, E. Berger, S. Gezari, B.A. Zauderer, A. Rest, L. Chomiuk, A. Kamble, A.M. Soderberg, I. Czekala, J. Dittmann, M. Drout, R.J. Foley, W. Fong, M.E. Huber, R.P. Kirshner, A. Lawrence, R. Lunnan, G.H. Marion, G. Narayan, A.G. Riess, K.C. Roth, N.E. Sanders, D. Scolnic, S.J. Smartt, K. Smith, C.W. Stubbs, J.L. Tonry, W.S. Burgett, K.C. Chambers, H. Flewelling, K.W. Hodapp, N. Kaiser, E.A. Magnier, D.C. Martin, J.D. Neill, P.A. Price, R. Wainscoat, The ultraviolet-bright, slowly declining transient PS1-11af as a partial tidal disruption event. Astrophys. J. **780**(1), 44 (2014). https://doi.org/10.1088/0004-637X/780/1/44

R. Cid Fernandes, G. Stasińska, N. Vale Asari, A. Mateus, M.S. Schlickmann, W. Schoenell, Emission-line taxonomy and the nature of AGN-looking galaxies in the SDSS, in *Co-Evolution of Central Black Holes and Galaxies*, ed. by B.M. Peterson, R.S. Somerville, T. Storchi-Bergmann. IAU Symposium, vol. 267 (2010), pp. 65–72. https://doi.org/10.1017/S1743921310005582

E.R. Coughlin, P.J. Armitage, Tidal disruption by extreme mass ratio binaries and application to ASASSN-15lh. Mon. Not. R. Astron. Soc. **474**(3), 3857–3865 (2018). https://doi.org/10.1093/mnras/stx3039

J. Cuadra, P.J. Armitage, R.D. Alexander, M.C. Begelman, Massive black hole binary mergers within subparsec scale gas discs. Mon. Not. R. Astron. Soc. **393**(4), 1423–1432 (2009). https://doi.org/10.1111/j.1365-2966.2008.14147.x

L. Dai, J.C. McKinney, N. Roth, E. Ramirez-Ruiz, M.C. Miller, A unified model for tidal disruption events. Astrophys. J. **859**, 20 (2018). https://doi.org/10.3847/2041-8213/aab429

J.E. Dale, M.B. Davies, R.P. Church, M. Freitag, Red giant stellar collisions in the Galactic Centre. Mon. Not. R. Astron. Soc. **393**(3), 1016–1033 (2009). https://doi.org/10.1111/j.1365-2966.2008.14254.x

A. De Rosa, C. Vignali, T. Bogdanović, P.R. Capelo, M. Charisi, M. Dotti, B. Husemann, E. Lusso, L. Mayer, Z. Paragi, J. Runnoe, A. Sesana, L. Steinborn, S. Bianchi, M. Colpi, L. del Valle, S. Frey, K.É. Gabányi, M. Giustini, M. Guainazzi, Z. Haiman, N. Herrera Ruiz, R. Herrero-Illana, K. Iwasawa, S. Komossa, D. Lena, N. Loiseau, M. Perez-Torres, E. Piconcelli, M. Volonteri, The quest for dual and binary supermassive black holes: a multi-messenger view. Nature **86**, 101525 (2019). https://doi.org/10.1016/j.newar.2020.101525

S. Dong, B.J. Shappee, J.L. Prieto, S.W. Jha, K.Z. Stanek, T.W.-S. Holoien, C.S. Kochanek, T.A. Thompson, N. Morrell, I.B. Thompson, U. Basu, J.F. Beacom, D. Bersier, J. Brimacombe, J.S. Brown, F. Bufano, P. Chen, E. Conseil, A.B. Danilet, E. Falco, D. Grupe, S. Kiyota, G. Masi, B. Nicholls, F. Olivares E., G. Pignata, G. Pojmanski, G.V. Simonian, D.M. Szczygiel, P.R. Woźniak, ASASSN-15lh: a highly super-luminous supernova. Science **351**, 257–260 (2016). https://doi.org/10.1126/science.aac9613

D.J. D'Orazio, Z. Haiman, A. MacFadyen, Accretion into the central cavity of a circumbinary disc. Mon. Not. R. Astron. Soc. **436**(4), 2997–3020 (2013). https://doi.org/10.1093/mnras/stt1787

A.J. Drake, S.G. Djorgovski, A. Mahabal, J. Anderson, R. Roy, V. Mohan, S. Ravindranath, D. Frail, S. Gezari, J.D. Neill, L.C. Ho, J.L. Prieto, D. Thompson, J. Thorstensen, M. Wagner, R. Kowalski, J. Chiang, J.E. Grove, F.K. Schinzel, D.L. Wood, L. Carrasco, E. Recillas, L. Kewley, K.N. Archana, A. Basu, Y. Wadadekar, B. Kumar, A.D. Myers, E.S. Phinney, R. Williams, M.J. Graham, M. Catelan, E. Beshore, S. Larson, E. Christensen, The discovery and nature of the optical transient CSS100217:102913+404220. Astrophys. J. **735**, 106 (2011). https://doi.org/10.1088/0004-637X/735/2/106

C.R. Evans, C.S. Kochanek, The tidal disruption of a star by a massive black hole. Astrophys. J. Lett. **346**, 13–16 (1989). https://doi.org/10.1086/185567

W. Fong, E. Berger, R. Margutti, B.A. Zauderer, A decade of short-duration gamma-ray burst broadband afterglows: energetics, circumburst densities, and jet opening angles. Astrophys. J. **815**, 102 (2015). https://doi.org/10.1088/0004-637X/815/2/102

S. Frederick, S. Gezari, M.J. Graham, S.B. Cenko, S. van Velzen, D. Stern, N. Blagorodnova, S.R. Kulkarni, L. Yan, K. De, U.C. Fremling, T. Hung, E. Kara, D.L. Shupe, C. Ward, E.C. Bellm, R. Dekany, D.A. Duev, U. Feindt, M. Giomi, T. Kupfer, R.R. Laher, F.J. Masci, A.A. Miller, J.D. Neill, C.-C. Ngeow, M.T. Patterson, M. Porter, B. Rusholme, J. Sollerman, R. Walters, A new class of changing-look LINERs. Astrophys. J. **883**(1), 31 (2019). https://doi.org/10.3847/1538-4357/ab3a38

K.D. French, I. Arcavi, A. Zabludoff, Tidal disruption events prefer unusual host galaxies. Astrophys. J. Lett. **818**, 21 (2016). https://doi.org/10.3847/2041-8205/818/1/L21

K.D. French, I. Arcavi, A. Zabludoff, The post-starburst evolution of tidal disruption event host galaxies. Astrophys. J. **835**(2), 176 (2017). https://doi.org/10.3847/1538-4357/835/2/176

K.D.French, T. Wevers, J. Law-Smith, O. Graur, A.I. Zabludoff, The host galaxies of tidal disruption events. Space Sci. Rev. **216**(3), 32 (2020). https://doi.org/10.1007/s11214-020-00657-y

A. Gal-Yam, Luminous supernovae. Science **337**, 927 (2012). https://doi.org/10.1126/science.1203601

S. Gezari, T. Heckman, S.B. Cenko, M. Eracleous, K. Forster, T.S. Gonçalves, D.C. Martin, P. Morrissey, S.G. Neff, M. Seibert, D. Schiminovich, T.K. Wyder, Luminous thermal flares from quiescent supermassive black holes. Astrophys. J. **698**(2), 1367–1379 (2009). https://doi.org/10.1088/0004-637X/698/2/1367

S. Gezari, R. Chornock, A. Rest, M.E. Huber, K. Forster, E. Berger, P..J. Challis, J.D. Neill, D.C. Martin, T. Heckman, A. Lawrence, C. Norman, G. Narayan, R.J. Foley, G..H. Marion, D. Scolnic, L. Chomiuk, A. Soderberg, K. Smith, R.P. Kirshner, A.G. Riess, S.J. Smartt, C..W. Stubbs, J.L. Tonry, W..M. Wood-Vasey, W.S. Burgett, K.C. Chambers, T. Grav, J.N. Heasley, N. Kaiser, R.-P. Kudritzki, E.A. Magnier, J.S. Morgan, P.A. Price, An ultraviolet–optical flare from the tidal disruption of a helium-rich stellar core. Nature **485**(7397), 217–220 (2012). http://www.nature.com/doifinder/10.1038/nature10990

S. Gezari, S.B. Cenko, I. Arcavi, X-ray brightening and UV fading of tidal disruption event ASASSN-15oi. Astrophys. J. Lett. **851**, 47 (2017). https://doi.org/10.3847/2041-8213/aaa0c2

E. Glikman, D.J. Helfand, R.L. White, A near-infrared spectral template for quasars. Astrophys. J. **640**, 579–591 (2006). https://doi.org/10.1086/500098

D. Godoy-Rivera, K.Z. Stanek, C.S. Kochanek, P. Chen, S. Dong, J.L. Prieto, B.J. Shappee, S.W. Jha, R.J. Foley, Y.-C. Pan, T.W.-S. Holoien, T.A. Thompson, D. Grupe, J.F. Beacom, The unexpected, long-lasting, UV rebrightening of the superluminous supernova ASASSN-15lh. Mon. Not. R. Astron. Soc. **466**(2), 1428–1443 (2017). https://doi.org/10.1093/mnras/stw3237

M.J. Graham, S.G. Djorgovski, A.J. Drake, D. Stern, A.A. Mahabal, E. Glikman, S. Larson, E. Christensen, Understanding extreme quasar optical variability with CRTS—I. Major AGN flares. Mon. Not. R. Astron. Soc. **470**, 4112–4132 (2017). https://doi.org/10.1093/mnras/stx1456

O. Graur, K.D. French, H.J. Zahid, J. Guillochon, K.S. Mandel, K. Auchettl, A.I. Zabludoff, A dependence of the tidal disruption event rate on global stellar surface mass density and stellar velocity dispersion. Astrophys. J. **853**, 39 (2018). https://doi.org/10.3847/1538-4357/aaa3fd

J. Greiner, R. Schwarz, S. Zharikov, M. Orio, RX J1420.4+5334—another tidal disruption event? Astron. Astrophys. **362**, 25–28 (2000)

C.J. Grier, B.M. Peterson, K. Horne, M.C. Bentz, R.W. Pogge, K.D. Denney, G. De Rosa, P. Martini, C.S. Kochanek, Y. Zu, B. Shappee, R. Siverd, T.G. Beatty, S.G. Sergeev, S. Kaspi, C. Araya Salvo, J.C. Bird, D.J. Bord, G.A. Borman, X. Che, C. Chen, S.A. Cohen, M. Dietrich, V.T. Doroshenko, Y.S. Efimov, N. Free, I. Ginsburg, C.B. Henderson, A.L. King, K. Mogren, M. Molina, A.M. Mosquera, S.V. Nazarov, D.N. Okhmat, O. Pejcha, S. Rafter, J.C. Shields, J. Skowron, D.M. Szczygiel, M. Valluri, J.L. van Saders, The structure of the broad-line region in active galactic nuclei. I. Reconstructed velocity-delay maps. Astrophys. J. **764**(1), 47 (2013). https://doi.org/10.1088/0004-637X/764/1/47

M. Gromadzki, A. Hamanowicz, L. Wyrzykowski, K.V. Sokolovsky, M. Fraser, S. Kozłowski, J. Guillochon, I. Arcavi, B. Trakhtenbrot, P.G. Jonker, S. Mattila, A. Udalski, M.K. Szymański, I. Soszyński, R. Poleski, P. Pietrukowicz, J. Skowron, P. Mróz, K. Ulaczyk, M. Pawlak, K.A. Rybicki, J. Sollerman, F. Taddia, Z. Kostrzewa-Rutkowska, F. Onori, D.R. Young, K. Maguire, S.J. Smartt, C. Inserra, A. Gal-Yam, A. Rau, T.-W. Chen, C.R. Angus, D.A.H. Buckley, Discovery and follow-up of the unusual nuclear transient OGLE17aaj. Astron. Astrophys. **622**, 2 (2019). https://doi.org/10.1051/0004-6361/201833682

D. Grupe, K. Beuermann, K. Mannheim, N. Bade, H.-C. Thomas, D. de Martino, A. Schwope, X-ray outburst of the peculiar Seyfert galaxy IC 3599. Astron. Astrophys. **299**, 5 (1995)

D. Grupe, H.-C. Thomas, K.M. Leighly, RX J1624.9+7554: a new X-ray transient AGN. Astron. Astrophys. **350**, 31–34 (1999)

D. Grupe, S. Komossa, R. Saxton, IC 3599 did it again: a second outburst of the X-ray transient Seyfert 1.9 galaxy. Astrophys. J. Lett. **803**, 28 (2015). https://doi.org/10.1088/2041-8205/803/2/L28

J. Guillochon, E. Ramirez-Ruiz, Hydrodynamical simulations to determine the feeding rate of black holes by the tidal disruption of stars: the importance of the impact parameter and stellar structure. Astrophys. J. **767**, 25 (2013). https://doi.org/10.1088/0004-637X/767/1/25

C.P. Gutiérrez, et al., H_α spectral diversity of type II supernovae: correlations with photometric properties. Astrophys. J. Lett. **786**, 15 (2014). https://doi.org/10.1088/2041-8205/786/2/L15

K.N. Hainline, R. Hickox, J.E. Greene, A.D. Myers, N.L. Zakamska, SALT long-slit spectroscopy of luminous obscured quasars: an upper limit on the size of the narrow-line region? Astrophys. J. **774**, 145 (2013). https://doi.org/10.1088/0004-637X/774/2/145

K.N. Hainline, R.C. Hickox, J.E. Greene, A.D. Myers, N.L. Zakamska, G. Liu, X. Liu, Gemini long-slit observations of luminous obscured quasars: further evidence for an upper limit on the size of the narrow-line region. Astrophys. J. **787**, 65 (2014). https://doi.org/10.1088/0004-637X/787/1/65

T.M. Heckman, A. Ptak, A. Hornschemeier, G. Kauffmann, The relationship of hard X-ray and optical line emission in low-redshift active galactic nuclei. Astrophys. J. **634**, 161–168 (2005). https://doi.org/10.1086/491665

J. Hjorth, J.S. Bloom, *The Gamma-Ray Burst—Supernova Connection*, ed. by C. Kouveliotou, R.A.M.J. Wijers, S. Woosley (2012), pp. 169–190

T.W.-S. Holoien, J.L. Prieto, D. Bersier, C.S. Kochanek, K.Z. Stanek, B.J. Shappee, D. Grupe, U. Basu, J.F. Beacom, J. Brimacombe, J.S. Brown, A.B. Davis, J. Jencson, G. Pojmanski, D.M. Szczygieł, ASASSN-14ae: a tidal disruption event at 200 Mpc. Mon. Not. R. Astron. Soc. **445**, 3263–3277 (2014). https://doi.org/10.1093/mnras/stu1922. http://arxiv.org/abs/1405.1417v1

T.W.-S. Holoien, C.S. Kochanek, J.L. Prieto, D. Grupe, P. Chen, D. Godoy-Rivera, K.Z. Stanek, B.J. Shappee, S. Dong, J.S. Brown, U. Basu, J.F. Beacom, D. Bersier, J. Brimacombe, E.K. Carlson, E. Falco, E. Johnston, B.F. Madore, G. Pojmanski, M. Seibert, ASASSN-15oi: a rapidly evolving, luminous tidal disruption event at 216 Mpc. Mon. Not. R. Astron. Soc. **463**(4), 3813–3828 (2016a). https://doi.org/10.1093/mnras/stw2272

T.W.-S. Holoien, C.S. Kochanek, J.L. Prieto, K.Z. Stanek, S. Dong, B.J. Shappee, D. Grupe, J.S. Brown, U. Basu, J.F. Beacom, D. Bersier, J. Brimacombe, A.B. Danilet, E. Falco, Z. Guo, J. Jose, G.J. Herczeg, F. Long, G. Pojmanski, G.V. Simonian, D.M. Szczygieł, T.A. Thompson, J.R. Thorstensen, R.M. Wagner, P.R. Woźniak, Six months of multiwavelength follow-up of the tidal disruption candidate ASASSN-14li and implied TDE rates from ASAS-SN. Mon. Not. R. Astron. Soc. **455**, 2918–2935 (2016b). https://doi.org/10.1093/mnras/stv2486

T.W.-S. Holoien, J.S. Brown, K. Auchettl, C.S. Kochanek, J.L. Prieto, B.J. Shappee, J. Van Saders, The unusual late-time evolution of the tidal disruption event ASASSN-15oi. Mon. Not. R. Astron. Soc. **480**, 5689–5703 (2018). https://doi.org/10.1093/mnras/sty2273

T.W.-S. Holoien, M.E. Huber, B.J. Shappee, M. Eracleous, K. Auchettl, J.S. Brown, M.A. Tucker, K.C. Chambers, C.S. Kochanek, K.Z. Stanek, A. Rest, D. Bersier, R.S. Post, G. Aldering, K.A. Ponder, J.D. Simon, E. Kankare, D. Dong, G. Hallinan, N.A. Reddy, R.L. Sanders, M.W. Topping, Pan-STARRS, J. Bulger, T.B. Lowe, E.A. Magnier, A.S.B. Schultz, C.Z. Waters, M. Willman, D. Wright, D.R. Young, ASAS-SN, S. Dong, J.L. Prieto, T.A. Thompson, ATLAS, L. Denneau, H. Flewelling, A.N. Heinze, S.J. Smartt, K.W. Smith, B. Stalder, J.L. Tonry, H. Weiland, PS18kh: a new tidal disruption event with a non-axisymmetric accretion disk. Astrophys. J. **880**(2), 120 (2019). https://doi.org/10.3847/1538-4357/ab2ae1

P.F. Hopkins, L. Hernquist, Quasars are not light bulbs: testing models of quasar lifetimes with the observed Eddington ratio distribution. Astrophys. J. **698**(2), 1550–1569 (2009). https://doi.org/10.1088/0004-637X/698/2/1550

A.M. Hopkins, C.J. Miller, R.C. Nichol, A.J. Connolly, M. Bernardi, P.L. Gómez, T. Goto, C.A. Tremonti, J. Brinkmann, Ž. Ivezić, D.Q. Lamb, Star formation rate indicators in the Sloan Digital Sky Survey. Astrophys. J. **599**(2), 971–991 (2003). https://doi.org/10.1086/379608

T. Hung, S. Gezari, D.O. Jones, R.P. Kirshner, R. Chornock, E. Berger, A. Rest, M. Huber, G. Narayan, D. Scolnic, C. Waters, R. Wainscoat, D.C. Martin, K. Forster, J.D. Neill, The GALEX Time Domain Survey. II. Wavelength-dependent variability of active galactic nuclei in the Pan-STARRS1 Medium Deep Survey. Astrophys. J. **833**, 226 (2016). https://doi.org/10.3847/1538-4357/833/2/226

T. Hung, S. Gezari, N. Blagorodnova, N. Roth, S.B. Cenko, S.R. Kulkarni, A. Horesh, I. Arcavi, C. McCully, L. Yan, R. Lunnan, C. Fremling, Y. Cao, P.E. Nugent, P. Wozniak, Revisiting optical tidal disruption events with iPTF16axa. Astrophys. J. **842**, 29 (2017). https://doi.org/10.3847/1538-4357/aa7337

T. Hung, S.B. Cenko, N. Roth, S. Gezari, S. Veilleux, S. van Velzen, C.M. Gaskell, R.J. Foley, N. Blagorodnova, L. Yan, M.J. Graham, J.S. Brown, M.R. Siebert, S. Frederick, C. Ward, P. Gatkine, A. Gal-Yam, Y. Yang, S. Schulze, G. Dimitriadis, T. Kupfer, D.L. Shupe, B. Rusholme, F.J. Masci, R. Riddle, M.T. Soumagnac, J. van Roestel, R. Dekany, Discovery of highly blueshifted broad Balmer and metastable helium absorption lines in a tidal disruption event. Astrophys. J. **879**(2), 119 (2019). https://doi.org/10.3847/1538-4357/ab24de

B. Husemann, T. Urrutia, G.R. Tremblay, M. Krumpe, J. Dexter, G. Busch, F. Combes, S.M. Croom, T.A. Davis, A. Eckart, R.E. McElroy, M. Perez-Torres, M. Powell, J. Scharwächter, The Close AGN Reference Survey (CARS). What is causing Mrk 1018's return to the shadows after 30 years? Astron. Astrophys. **593**, 9 (2016). https://doi.org/10.1051/0004-6361/201629245

P.B. Ivanov, J.C.B. Papaloizou, A.G. Polnarev, The evolution of a supermassive binary caused by an accretion disc. Mon. Not. R. Astron. Soc. **307**(1), 79–90 (1999). https://doi.org/10.1046/j.1365-8711.1999.02623.x

A. Janiuk, B. Czerny, A. Siemiginowska, Radiation pressure instability driven variability in the accreting black holes. Astrophys. J. **576**, 908–922 (2002). https://doi.org/10.1086/341804

L. Jiang, X. Fan, M. Vestergaard, A sample of quasars with strong nitrogen emission lines from the Sloan Digital Sky Survey. Astrophys. J. **679**, 962–966 (2008). https://doi.org/10.1086/587868

Y.-F. Jiang, J.M. Stone, S.W. Davis, On the thermal stability of radiation-dominated accretion disks. Astrophys. J. **778**, 65 (2013). https://doi.org/10.1088/0004-637X/778/1/65

P.G. Jonker, N.C. Stone, A. Generozov, S. van Velzen, B. Metzger, Implications from late-time X-ray detections of optically selected tidal disruption events: state changes, unification, and detection rates. Astrophys. J. **889**(2), 166 (2020). https://doi.org/10.3847/1538-4357/ab659c

E. Kankare, R. Kotak, S. Mattila, P. Lundqvist, M.J. Ward, M. Fraser, A. Lawrence, S.J. Smartt, W.P.S. Meikle, A. Bruce, J. Harmanen, S.J. Hutton, C. Inserra, T. Kangas, A. Pastorello, T. Reynolds, C. Romero-Cañizales, K.W. Smith, S. Valenti, K.C. Chambers, K.W. Hodapp, M.E. Huber, N. Kaiser, R.-P. Kudritzki, E.A. Magnier, J.L. Tonry, R. Wainscoat, C. Waters, A population of highly energetic transient events in the centres of active galaxies. Nat. Astron. **1**(12), 865–871 (2017). https://doi.org/10.1038/s41550-017-0290-2

E. Kara, L. Dai, C.S. Reynolds, T. Kallman, Ultrafast outflow in tidal disruption event ASASSN-14li. Mon. Not. R. Astron. Soc. **474**, 3593–3598 (2018). https://doi.org/10.1093/mnras/stx3004

G. Kauffmann, T.M. Heckman, C. Tremonti, J. Brinchmann, S. Charlot, S.D.M. White, S.E. Ridgway, J. Brinkmann, M. Fukugita, P.B. Hall, Ž. Ivezić, G.T. Richards, D.P. Schneider, The host galaxies of active galactic nuclei. Mon. Not. R. Astron. Soc. **346**, 1055–1077 (2003). https://doi.org/10.1111/j.1365-2966.2003.07154.x

L.J. Kewley, B. Groves, G. Kauffmann, T. Heckman, The host galaxies and classification of active galactic nuclei. Mon. Not. R. Astron. Soc. **372**, 961–976 (2006). https://doi.org/10.1111/j.1365-2966.2006.10859.x

M. Kiewe, A. Gal-Yam, I. Arcavi, D.C. Leonard, J. Emilio Enriquez, S.B. Cenko, D.B. Fox, D.-S. Moon, D.J. Sand, A.M. Soderberg (The CCCP), Caltech Core-Collapse Project (CCCP) observations of type IIn supernovae: typical properties and implications for their progenitor stars. Astrophys. J. **744**, 10 (2012). https://doi.org/10.1088/0004-637X/744/1/10

A.R. King, J.E. Pringle, R.G. West, M. Livio, Variability in black hole accretion discs. Mon. Not. R. Astron. Soc. **348**(1), 111–122 (2004). https://doi.org/10.1111/j.1365-2966.2004.07322.x

R.W. Klebesadel, I.B. Strong, R.A. Olson, Observations of gamma-ray bursts of cosmic origin. Astrophys. J. Lett. **182**, 85 (1973). https://doi.org/10.1086/181225

S. Komossa, N. Bade, The giant X-ray outbursts in NGC 5905 and IC 3599:() hfill Follow-up observations and outburst scenarios. Astron. Astrophys. **343**, 775–787 (1999)

S. Komossa, H. Zhou, A. Rau, M. Dopita, A. Gal-Yam, J. Greiner, J. Zuther, M. Salvato, D. Xu, H. Lu, R. Saxton, M. Ajello, NTT, Spitzer, and Chandra spectroscopy of SDSSJ095209.56+214313.3: the most luminous coronal-line supernova ever observed, or a stellar tidal disruption event? Astrophys. J. **701**(1), 105–121 (2009). https://doi.org/10.1088/0004-637X/701/1/105

K.T. Korista, M.R. Goad, What the optical recombination lines can tell us about the broad-line regions of active galactic nuclei. Astrophys. J. **606**(2), 749–762 (2004). https://doi.org/10.1086/383193

Z. Kostrzewa-Rutkowska, P.G. Jonker, S.T. Hodgkin, Ł. Wyrzykowski, M. Fraser, D.L. Harrison, G. Rixon, A. Yoldas, F. van Leeuwen, A. Delgado, M. van Leeuwen, S.E. Koposov, Gaia transients in galactic nuclei. Mon. Not. R. Astron. Soc. **481**(1), 307–323 (2018). https://doi.org/10.1093/mnras/sty2221

C. Kouveliotou, C.A. Meegan, G.J. Fishman, N.P. Bhat, M.S. Briggs, T.M. Koshut, W.S. Paciesas, G.N. Pendleton, Identification of two classes of gamma-ray bursts. Astrophys. J. Lett. **413**, 101–104 (1993). https://doi.org/10.1086/186969

T. Krühler, M. Fraser, G. Leloudas, S. Schulze, N.C. Stone, S. van Velzen, R. Amorin, J. Hjorth, P.G. Jonker, D.A. Kann, S. Kim, H. Kuncarayakti, A. Mehner, A. Nicuesa Guelbenzu, The supermassive black hole coincident with the luminous transient ASASSN-15lh. Astron. Astrophys. **610**, 14 (2018). https://doi.org/10.1051/0004-6361/201731773

S.M. LaMassa, T.M. Heckman, A. Ptak, L. Martins, V. Wild, P. Sonnentrucker, Indicators of intrinsic active galactic nucleus luminosity: a multi-wavelength approach. Astrophys. J. **720**, 786–810 (2010). https://doi.org/10.1088/0004-637X/720/1/786

S.M. LaMassa, S. Cales, E.C. Moran, A.D. Myers, G.T. Richards, M. Eracleous, T.M. Heckman, L. Gallo, C.M. Urry, The discovery of the first changing look quasar: new insights into the physics and phenomenology of active galactic nucleus. Astrophys. J. **800**, 144 (2015). https://doi.org/10.1088/0004-637X/800/2/144

S.M. LaMassa, T. Yaqoob, R. Kilgard, Insight into changing-look AGN Mrk 1018 from the Fe Kα line: the reprocessing gas has yet to fully respond to the fading of the AGN. Astrophys. J. **840**, 11 (2017). https://doi.org/10.3847/1538-4357/aa68df

G. Lanzuisi, G. Ponti, M. Salvato, G. Hasinger, N. Cappelluti, A. Bongiorno, M. Brusa, E. Lusso, P.K. Nandra, A. Merloni, J. Silverman, J. Trump, C. Vignali, A. Comastri, R. Gilli, M. Schramm, C. Steinhardt, D. Sanders, J. Kartaltepe, D. Rosario, B. Trakhtenbrot, Active galactic nucleus X-ray variability in the XMM-COSMOS Survey. Astrophys. J. **781**(2), 105 (2014). https://doi.org/10.1088/0004-637X/781/2/105

A. Laor, On the nature of low-luminosity narrow-line active galactic nuclei. Astrophys. J. **590**, 86–94 (2003). https://doi.org/10.1086/375008

J. Law-Smith, E. Ramirez-Ruiz, S.L. Ellison, R.J. Foley, Tidal disruption event host galaxies in the context of the local galaxy population. Astrophys. J. **850**(1), 22 (2017). https://doi.org/10.3847/1538-4357/aa94c7

G. Leloudas, M. Fraser, N.C. Stone, S. van Velzen, P.G. Jonker, I. Arcavi, C. Fremling, J.R. Maund, S.J. Smartt, T. Krìhler, J.C.A. Miller-Jones, P.M. Vreeswijk, A. Gal-Yam, P.A. Mazzali, A. De Cia, D.A. Howell, C. Inserra, F. Patat, A. de Ugarte Postigo, O. Yaron, C. Ashall, I. Bar, H. Campbell, T.-W. Chen, M. Childress, N. Elias-Rosa, J. Harmanen, G. Hosseinzadeh, J. Johansson, T. Kangas, E. Kankare, S. Kim, H. Kuncarayakti, J. Lyman, M.R. Magee, K. Maguire, D. Malesani, S. Mattila, C.V. McCully, M. Nicholl, S. Prentice, C. Romero-Cañizales, S. Schulze, K.W. Smith, J. Sollerman, M. Sullivan, B.E. Tucker, S. Valenti, J.C. Wheeler, D.R. Young, The superluminous transient ASASSN-15lh as a tidal disruption event from a Kerr black hole. Nat. Astron. **1**, 0002 (2016). https://doi.org/10.1038/s41550-016-0002

G. Leloudas, L. Dai, I. Arcavi, P.M. Vreeswijk, B. Mockler, R. Roy, D.B. Malesani, S. Schulze, T. Wevers, M. Fraser, E. Ramirez-Ruiz, K. Auchettl, J. Burke, G. Cannizzaro, P. Charalampopoulos, T.-W. Chen, A. Cikota, M. Della Valle, L. Galbany, M. Gromadzki, K.E. Heintz, D. Hiramatsu, P.G. Jonker, Z. Kostrzewa-Rutkowska, K. Maguire, I. Mandel, M. Nicholl, F. Onori, N. Roth, S.J. Smartt, L. Wyrzykowski, D.R. Young, The spectral evolution of AT 2018dyb and the presence of metal lines in tidal disruption events. Astrophys. J. **887**(2), 218 (2019). https://doi.org/10.3847/1538-4357/ab5792

D.C. Leonard, A.V. Filippenko, E.L. Gates, W. Li, R.G. Eastman, A.J. Barth, S.J. Bus, R. Chornock, A.L. Coil, S. Frink, C.A. Grady, A.W. Harris, M.A. Malkan, T. Matheson, A. Quirrenbach, R.R. Treffers, The distance to SN 1999em in NGC 1637 from the expanding photosphere method. Publ. Astron. Soc. Pac. **114**, 35–64 (2002). https://doi.org/10.1086/324785

A.J. Levan, N.R. Tanvir, R.L.C. Starling, K. Wiersema, K.L. Page, D.A. Perley, S. Schulze, G.A. Wynn, R. Chornock, J. Hjorth, S.B. Cenko, A.S. Fruchter, P.T. O'Brien, G.C. Brown, R.L. Tunnicliffe, D. Malesani, P. Jakobsson, D. Watson, E. Berger, D. Bersier, B.E. Cobb, S. Covino, A. Cucchiara, A. de Ugarte Postigo, D.B. Fox, A. Gal-Yam, P. Goldoni, J. Gorosabel, L. Kaper, T. Krühler, R. Karjalainen, J.P. Osborne, E. Pian, R. Sánchez-Ramírez, B. Schmidt, I. Skillen, G. Tagliaferri, C. Thöne, O. Vaduvescu,

R.A.M.J. Wijers, B.A. Zauderer, A new population of ultra-long duration gamma-ray bursts. Astrophys. J. **781**, 13 (2014). https://doi.org/10.1088/0004-637X/781/1/13

A.J. Levan, N.R. Tanvir, G.C. Brown, B.D. Metzger, K.L. Page, S.B. Cenko, P.T. O'Brien, J.D. Lyman, K. Wiersema, E.R. Stanway, A.S. Fruchter, D.A. Perley, J.S. Bloom, Late time multi-wavelength observations of swift J1644+5734: a luminous optical/IR bump and quiescent X-ray emission. Astrophys. J. **819**(1), 51 (2016). https://doi.org/10.3847/0004-637X/819/1/51

D. Lin, E.R. Carrasco, D. Grupe, N.A. Webb, D. Barret, S.A. Farrell, Discovery of an ultrasoft X-ray transient source in the 2XMM catalog: a tidal disruption event candidate. Astrophys. J. **738**(1), 52 (2011). https://doi.org/10.1088/0004-637X/738/1/52

D. Lin, J. Guillochon, S. Komossa, E. Ramirez-Ruiz, J.A. Irwin, W.P. Maksym, D. Grupe, O. Godet, N.A. Webb, D. Barret, B.A. Zauderer, P.-A. Duc, E.R. Carrasco, S.D.J. Gwyn, A likely decade-long sustained tidal disruption event. Nat. Astron. **1**, 0033 (2017). https://doi.org/10.1038/s41550-016-0033

D. Lin, J. Strader, E.R. Carrasco, D. Page, A.J. Romanowsky, J. Homan, J.A. Irwin, R.A. Remillard, O. Godet, N.A. Webb, H. Baumgardt, R. Wijnands, D. Barret, P.-A. Duc, J.P. Brodie, S.D.J. Gwyn, A luminous X-ray outburst from an intermediate-mass black hole in an off-centre star cluster. Nat. Astron. **2**, 656–661 (2018). https://doi.org/10.1038/s41550-018-0493-1

Z. Liu, A. Merloni, A. Georgakakis, M.-L. Menzel, J. Buchner, K. Nandra, M. Salvato, Y. Shen, M. Brusa, A. Streblyanska, X-ray spectral properties of the AGN sample in the northern XMM-XXL field. Mon. Not. R. Astron. Soc. **459**, 1602–1625 (2016). https://doi.org/10.1093/mnras/stw753

G. Lodato, E.M. Rossi, Multiband light curves of tidal disruption events. Mon. Not. R. Astron. Soc. **410**(1), 359–367 (2011). http://doi.wiley.com/10.1111/j.1365-2966.2010.17448.x

Y.E. Lyubarskii, Flicker noise in accretion discs. Mon. Not. R. Astron. Soc. **292**(3), 679–685 (1997). https://doi.org/10.1093/mnras/292.3.679

A.I. MacFadyen, M. Milosavljević, An eccentric circumbinary accretion disk and the detection of binary massive black holes. Astrophys. J. **672**(1), 83–93 (2008). https://doi.org/10.1086/523869

C.L. MacLeod, Ž. Ivezić, C.S. Kochanek, S. Kozłowski, B. Kelly, E. Bullock, A. Kimball, B. Sesar, D. Westman, K. Brooks, R. Gibson, A.C. Becker, W.H. de Vries, Modeling the time variability of SDSS stripe 82 quasars as a damped random walk. Astrophys. J. **721**, 1014–1033 (2010). https://doi.org/10.1088/0004-637X/721/2/1014

C.L. MacLeod, Ž. Ivezić, B. Sesar, W. de Vries, C.S. Kochanek, B.C. Kelly, A.C. Becker, R.H. Lupton, P.B. Hall, G.T. Richards, S.F. Anderson, D.P. Schneider, A description of quasar variability measured using repeated SDSS and POSS imaging. Astrophys. J. **753**(2), 106 (2012). https://doi.org/10.1088/0004-637X/753/2/106

R. Margutti, B.D. Metzger, R. Chornock, D. Milisavljevic, E. Berger, P.K. Blanchard, C. Guidorzi, G. Migliori, A. Kamble, R. Lunnan, M. Nicholl, D.L. Coppejans, S. Dall'Osso, M.R. Drout, R. Perna, B. Sbarufatti, X-rays from the location of the double-humped transient ASASSN-15lh. Astrophys. J. **836**, 25 (2017). https://doi.org/10.3847/1538-4357/836/1/25

G. Matt, M. Guainazzi, R. Maiolino, Changing look: from Compton-thick to Compton-thin, or the rebirth of fossil active galactic nuclei. Mon. Not. R. Astron. Soc. **342**, 422–426 (2003). https://doi.org/10.1046/j.1365-8711.2003.06539.x

N.J. McConnell, C.-P. Ma, K. Gebhardt, S.A. Wright, J.D. Murphy, T.R. Lauer, J.R. Graham, D.O. Richstone, Two ten-billion-solar-mass black holes at the centres of giant elliptical galaxies. Nature **480**(7376), 215–218 (2011). https://doi.org/10.1038/nature10636

J.W. Menzies, R.M. Catchpole, G. van Vuuren, H. Winkler, C.D. Laney, P.A. Whitelock, A.W.J. Cousins, B.S. Carter, F. Marang, T.H.H. Lloyd Evans, G. Roberts, D. Kilkenny, J. Spencer Jones, K. Sekiguchi, A.P. Fairall, R.D. Wolstencroft, Spectroscopic and photometric observations of SN 1987a—the first 50 days. Mon. Not. R. Astron. Soc. **227**, 39–49 (1987). https://doi.org/10.1093/mnras/227.1.39P

A. Merloni, T. Dwelly, M. Salvato, A. Georgakakis, J. Greiner, M. Krumpe, K. Nandra, G. Ponti, A. Rau, A tidal disruption flare in a massive galaxy? Implications for the fuelling mechanisms of nuclear black holes. Mon. Not. R. Astron. Soc. **452**, 69–87 (2015). https://doi.org/10.1093/mnras/stv1095

B.D. Metzger, N.C. Stone, Periodic accretion-powered flares from colliding EMRIs as TDE imposters. Astrophys. J. **844**, 75 (2017). https://doi.org/10.3847/1538-4357/aa7a16

B.D. Metzger, P.K.G. Williams, E. Berger, Extragalactic synchrotron transients in the era of wide-field radio surveys. I. Detection rates and light curve characteristics. Astrophys. J. **806**, 224 (2015). https://doi.org/10.1088/0004-637X/806/2/224

J.M. Miller, J.S. Kaastra, M.C. Miller, M.T. Reynolds, G. Brown, S.B. Cenko, J.J. Drake, S. Gezari, J. Guillochon, K. Gultekin, J. Irwin, A. Levan, D. Maitra, W.P. Maksym, R. Mushotzky, P. O'Brien, F. Paerels, J. de Plaa, E. Ramirez-Ruiz, T. Strohmayer, N. Tanvir, Flows of X-ray gas reveal the disruption of a star by a massive black hole. Nature **526**, 542–545 (2015). https://doi.org/10.1038/nature15708

B. Mockler, J. Guillochon, E. Ramirez-Ruiz, Weighing black holes using tidal disruption events. Astrophys. J. **872**(2), 151 (2019). https://doi.org/10.3847/1538-4357/ab010f

R. Mor, H. Netzer, M. Elitzur, Dusty structure around type-I active galactic nuclei: clumpy torus narrow-line region and near-nucleus hot dust. Astrophys. J. **705**(1), 298–313 (2009). https://doi.org/10.1088/0004-637X/705/1/298

T.J. Moriya, M. Tanaka, T. Morokuma, K. Ohsuga, Superluminous transients at AGN centers from interaction between black hole disk winds and broad-line region clouds. Astrophys. J. Lett. **843**(2), 19 (2017). https://doi.org/10.3847/2041-8213/aa7af3

A. Mummery, S.A. Balbus, ASASSN-15lh: a TDE about a maximally rotating 10^9 $M_{\odot}$ black hole. Mon. Not. R. Astron. Soc. **497**(1), 13–18 (2020). https://doi.org/10.1093/mnrasl/slaa105

D.J. Muñoz, D. Lai, K. Kratter, R. Miranda, Circumbinary accretion from finite and infinite disks. Astrophys. J. **889**(2), 114 (2020). https://doi.org/10.3847/1538-4357/ab5d33

H. Netzer, M. Elitzur, G.J. Ferland, Bowen fluoresence and He II lines in active galaxies and gaseous nebulae. Astrophys. J. **299**, 752 (1985). https://doi.org/10.1086/163741

H. Netzer, O. Shemmer, R. Maiolino, E. Oliva, S. Croom, E. Corbett, L. di Fabrizio, Near-infrared spectroscopy of high-redshift active galactic nuclei. II. Disappearing narrow-line regions and the role of accretion. Astrophys. J. **614**, 558–567 (2004). https://doi.org/10.1086/423608

N. Neumayer, A. Seth, T. Böker, Nuclear star clusters. Astron. Astrophys. Rev. **28**(1), 4 (2020). https://doi.org/10.1007/s00159-020-00125-0

M. Nicholl, T. Wevers, S.R. Oates, K.D. Alexander, G. Leloudas, F. Onori, A. Jerkstrand, S. Gomez, S. Campana, I. Arcavi, P. Charalampopoulos, M. Gromadzki, N. Ihanec, P.G. Jonker, A. Lawrence, I. Mandel, P. Short, J. Burke, D. Hiramatsu, D.A. Howell, C. Pellegrino, H. Abbot, J.P. Anderson, E. Berger, P.K. Blanchard, G. Cannizzaro, T.-W. Chen, M. Dennefeld, L. Galbany, S. Gonzalez-Gaitan, G. Hosseinzadeh, C. Inserra, I. Irani, P. Kuin, T. Muller-Bravo, J. Pineda, N.P. Ross, R. Roy, B. Tucker, L. Wyrzykowski, D.R. Young, An outflow powers the optical rise of the nearby, fast-evolving tidal disruption event AT2019qiz. arXiv e-prints (2020). 2006.02454

H. Noda, C. Done, Explaining changing-look AGN with state transition triggered by rapid mass accretion rate drop. Mon. Not. R. Astron. Soc. **480**(3), 3898–3906 (2018). https://doi.org/10.1093/mnras/sty2032

F. Onori, G. Cannizzaro, P.G. Jonker, M. Fraser, Z. Kostrzewa-Rutkowska, A. Martin-Carrillo, S. Benetti, N. Elias-Rosa, M. Gromadzki, J. Harmanen, S. Mattila, M.D. Strizinger, G. Terreran, T. Wevers, Optical follow-up of the tidal disruption event iPTF16fnl: new insights from X-shooter observations. Mon. Not. R. Astron. Soc. **489**(1), 1463–1480 (2019). https://doi.org/10.1093/mnras/stz2053

D.R. Pasham, S.B. Cenko, A.J. Levan, G.C. Bower, A. Horesh, G.C. Brown, S. Dolan, K. Wiersema, A.V. Filippenko, A.S. Fruchter, J. Greiner, P.T. O'Brien, K.L. Page, A. Rau, N.R. Tanvir, A multiwavelength study of the relativistic tidal disruption candidate swift J2058.4+0516 at late times. Astrophys. J. **805**, 68 (2015). https://doi.org/10.1088/0004-637X/805/1/68

D.R. Pasham, R.A. Remillard, P.C. Fragile, A. Franchini, N.C. Stone, G. Lodato, J. Homan, D. Chakrabarty, F.K. Baganoff, J.F. Steiner, E.R. Coughlin, N.R. Pasham, A loud quasi-periodic oscillation after a star is disrupted by a massive black hole. Science **363**(6426), 531–534 (2019). https://doi.org/10.1126/science.aar7480

A. Pennell, J.C. Runnoe, M.S. Brotherton, Updating quasar bolometric luminosity corrections—III. [O III] bolometric corrections. Mon. Not. R. Astron. Soc. **468**, 1433–1441 (2017). https://doi.org/10.1093/mnras/stx556

H.B. Perets, D.C. Fabrycky, On the triple origin of blue stragglers. Astrophys. J. **697**, 1048–1056 (2009). https://doi.org/10.1088/0004-637X/697/2/1048

H.B. Perets, Z. Li, J.C. Lombardi Jr., S.R. Milcarek Jr., Micro-tidal disruption events by stellar compact objects and the production of ultra-long GRBs. Astrophys. J. **823**, 113 (2016). https://doi.org/10.3847/0004-637X/823/2/113

N.A. Pereyra, D.E. Vanden Berk, D.A. Turnshek, D.J. Hillier, B.C. Wilhite, R.G. Kron, D.P. Schneider, J. Brinkmann, Characteristic QSO accretion disk temperatures from spectroscopic continuum variability. Astrophys. J. **642**, 87–95 (2006). https://doi.org/10.1086/500919

B.M. Peterson, Reverberation mapping of active galactic nuclei. Publ. Astron. Soc. Pac. **105**, 247–268 (1993). https://doi.org/10.1086/133140

B.M. Peterson, G.J. Ferland, An accretion event in the Seyfert galaxy NGC 5548. Nature **324**, 345–347 (1986). https://doi.org/10.1038/324345a0

B.M. Peterson, K.D. Denney, G. De Rosa, C.J. Grier, R.W. Pogge, M.C. Bentz, C.S. Kochanek, M. Vestergaard, E. Kilerci-Eser, E. Dalla Bontà, S. Ciroi, The size of the narrow-line-emitting region in the Seyfert 1 galaxy NGC 5548 from emission-line variability. Astrophys. J. **779**(2), 109 (2013). https://doi.org/10.1088/0004-637X/779/2/109

E.S. Phinney, Cosmic merger mania. Nature **340**, 595–596 (1989a). https://doi.org/10.1038/340595a0

E.S. Phinney, Manifestations of a massive black hole in the Galactic center, in *The Center of the Galaxy*, ed. by M. Morris. IAU Symposium, vol. 136 (1989b), p. 543

S. Prodan, F. Antonini, H.B. Perets, Secular evolution of binaries near massive black holes: formation of compact binaries, merger/collision products and G2-like objects. Astrophys. J. **799**, 118 (2015). https://doi.org/10.1088/0004-637X/799/2/118

R.M. Quimby, S.R. Kulkarni, M.M. Kasliwal, A. Gal-Yam, I. Arcavi, M. Sullivan, P. Nugent, R. Thomas, D.A. Howell, E. Nakar, L. Bildsten, C. Theissen, N.M. Law, R. Dekany, G. Rahmer, D. Hale, R. Smith, E.O. Ofek, J. Zolkower, V. Velur, R. Walters, J. Henning, K. Bui, D. McKenna, D. Poznanski, S.B. Cenko, D. Levitan, Hydrogen-poor superluminous stellar explosions. Nature **474**, 487–489 (2011). https://doi.org/10.1038/nature10095

E. Ragusa, R. Alexander, J. Calcino, K. Hirsh, D.J. Price, The evolution of large cavities and disc eccentricity in circumbinary discs. Mon. Not. R. Astron. Soc. **499**(3), 3362–3380 (2020). https://doi.org/10.1093/mnras/staa2954

M.J. Rees, Tidal disruption of stars by black holes of 10^6–10^8 solar masses in nearby galaxies. Nature **333**(6173), 523–528 (1988). https://doi.org/10.1038/333523a0

C. Ricci, F.E. Bauer, P. Arevalo, S. Boggs, W.N. Brandt, F.E. Christensen, W.W. Craig, P. Gandhi, C.J. Hailey, F.A. Harrison, M. Koss, C.B. Markwardt, D. Stern, E. Treister, W.W. Zhang, IC 751: a new changing look AGN discovered by NuSTAR. Astrophys. J. **820**(1), 5 (2016). https://doi.org/10.3847/0004-637X/820/1/5

C. Ricci, E. Kara, M. Loewenstein, B. Trakhtenbrot, I. Arcavi, R. Remillard, A.C. Fabian, K.C. Gendreau, Z. Arzoumanian, R. Li, L.C. Ho, C.L. MacLeod, E. Cackett, D. Altamirano, P. Gand hi, P. Kosec, D. Pasham, J. Steiner, C.-H. Chan, The destruction and recreation of the X-ray corona in a changing-look active galactic nucleus. Astrophys. J. Lett. **898**(1), 1 (2020). https://doi.org/10.3847/2041-8213/ab91a1

J.A. Rich, L.J. Kewley, M.A. Dopita, Galaxy mergers drive shocks: an integral field study of GOALS galaxies. Astrophys. J. Suppl. Ser. **221**(2), 28 (2015). https://doi.org/10.1088/0067-0049/221/2/28

M.W. Richmond, R.R. Treffers, A.V. Filippenko, Y. Paik, B. Leibundgut, E. Schulman, C.V. Cox, UBVRI photometry of SN 1993J in M81: the first 120 days. Astron. J. **107**, 1022–1040 (1994). https://doi.org/10.1086/116915

N. Roth, D. Kasen, What sets the line profiles in tidal disruption events? Astrophys. J. **855**, 54 (2018). https://doi.org/10.3847/1538-4357/aaaec6

J.J. Ruan, S.F. Anderson, J. Dexter, E. Agol, Evidence for large temperature fluctuations in quasar accretion disks from spectral variability. Astrophys. J. **783**, 105 (2014). https://doi.org/10.1088/0004-637X/783/2/105

J.J. Ruan, S.F. Anderson, S.L. Cales, M. Eracleous, P.J. Green, E. Morganson, J.C. Runnoe, Y. Shen, T.D. Wilkinson, M.R. Blanton, T. Dwelly, A. Georgakakis, J.E. Greene, S.M. LaMassa, A. Merloni, D.P. Schneider, Toward an understanding of changing-look quasars: an archival spectroscopic search in SDSS. Astrophys. J. **826**(2), 188 (2016). https://doi.org/10.3847/0004-637X/826/2/188

J.J. Ruan, S.F. Anderson, M. Eracleous, P.J. Green, D. Haggard, C.L. MacLeod, J.C. Runnoe, M.A. Sobolewska, The analogous structure of accretion flows in supermassive and stellar mass black holes: new insights from faded changing-look quasars. Astrophys. J. **883**(1), 76 (2019). https://doi.org/10.3847/1538-4357/ab3c1a

N. Rumbaugh, Y. Shen, E. Morganson, X. Liu, M. Banerji, R.G. McMahon, F.B. Abdalla, A. Benoit-Lévy, E. Bertin, D. Brooks, E. Buckley-Geer, D. Capozzi, A. Carnero Rosell, M. Carrasco Kind, J. Carretero, C.E. Cunha, C.B. D'Andrea, L.N. da Costa, D.L. DePoy, S. Desai, P. Doel, J. Frieman, J. García-Bellido, D. Gruen, R.A. Gruendl, J. Gschwend, G. Gutierrez, K. Honscheid, D.J. James, K. Kuehn, S. Kuhlmann, N. Kuropatkin, M. Lima, M.A.G. Maia, J.L. Marshall, P. Martini, F. Menanteau, A.A. Plazas, K. Reil, A. Roodman, E. Sanchez, V. Scarpine, R. Schindler, M. Schubnell, E. Sheldon, M. Smith, M. Soares-Santos, F. Sobreira, E. Suchyta, M.E.C. Swanson, A.R. Walker, W. Wester (DES Collaboration), Extreme variability quasars from the Sloan Digital Sky Survey and the Dark Energy Survey. Astrophys. J. **854**(2), 160 (2018). https://doi.org/10.3847/1538-4357/aaa9b6

J.C. Runnoe, S. Cales, J.J. Ruan, M. Eracleous, S.F. Anderson, Y. Shen, P.J. Green, E. Morganson, S. LaMassa, J.E. Greene, T. Dwelly, D.P. Schneider, A. Merloni, A. Georgakakis, A. Roman-Lopes, Now you see it, now you don't: the disappearing central engine of the quasar J1011+5442. Mon. Not. R. Astron. Soc. **455**, 1691–1701 (2016). https://doi.org/10.1093/mnras/stv2385

R.D. Saxton, S.E. Motta, S. Komossa, A.M. Read, Was the soft X-ray flare in NGC 3599 due to an AGN disc instability or a delayed tidal disruption event? Mon. Not. R. Astron. Soc. **454**, 2798–2803 (2015). https://doi.org/10.1093/mnras/stv2160

R.D. Saxton, A.M. Read, S. Komossa, P. Lira, K.D. Alexander, M.H. Wieringa, XMMSL1 J074008.2-853927: a tidal disruption event with thermal and non-thermal components. Astron. Astrophys. **598**, 29 (2017). https://doi.org/10.1051/0004-6361/201629015

C.J. Saxton, H.B. Perets, A. Baskin, Spectral features of tidal disruption candidates and alternative origins for such transient flares. Mon. Not. R. Astron. Soc. **474**, 3307–3323 (2018). https://doi.org/10.1093/mnras/stx2928

R. Saxton, S. Komossa, K. Auchettl, P.G. Jonker, X-ray properties of TDEs. Space Sci. Rev. **216**(5), 85 (2020). https://doi.org/10.1007/s11214-020-00708-4

H.R. Schmitt, J.L. Donley, R.R.J. Antonucci, J.B. Hutchings, A.L. Kinney, J.E. Pringle, A Hubble Space Telescope survey of extended [O III] λ5007 Å emission in a far-infrared-selected sample of Seyfert galaxies: results. Astrophys. J. **597**, 768–779 (2003a). https://doi.org/10.1086/381224

H.R. Schmitt, J.L. Donley, R.R.J. Antonucci, J.B. Hutchings, A.L. Kinney, A Hubble Space Telescope survey of extended [O III] λ5007 emission in a far-infrared selected sample of Seyfert galaxies: observations. Astrophys. J. Suppl. Ser. **148**, 327–352 (2003b). https://doi.org/10.1086/377440

N.I. Shakura, R.A. Sunyaev, Black holes in binary systems. Observational appearance. Astron. Astrophys. **24**, 337–355 (1973)

Y. Shen, G.T. Richards, M.A. Strauss, P.B. Hall, D.P. Schneider, S. Snedden, D. Bizyaev, H. Brewington, V. Malanushenko, E. Malanushenko, D. Oravetz, K. Pan, A. Simmons, A catalog of quasar properties from Sloan Digital Sky Survey data release 7. Astrophys. J. Suppl. Ser. **194**, 45 (2011). https://doi.org/10.1088/0067-0049/194/2/45

J.-M. Shi, J.H. Krolik, S.H. Lubow, J.F. Hawley, Three-dimensional magnetohydrodynamic simulations of circumbinary accretion disks: disk structures and angular momentum transport. Astrophys. J. **749**(2), 118 (2012). https://doi.org/10.1088/0004-637X/749/2/118

J.M. Shull, M. Stevans, C.W. Danforth, HST-COS observations of AGNs. I. Ultraviolet composite spectra of the ionizing continuum and emission lines. Astrophys. J. **752**, 162 (2012). https://doi.org/10.1088/0004-637X/752/2/162

A.P. Stephan, S. Naoz, A.M. Ghez, M.R. Morris, A. Ciurlo, T. Do, K. Breivik, S. Coughlin, C.L. Rodriguez, The fate of binaries in the Galactic center: the mundane and the exotic. Astrophys. J. **878**(1), 58 (2019). https://doi.org/10.3847/1538-4357/ab1e4d

D. Stern, R.J. Assef, D.J. Benford, A. Blain, R. Cutri, A. Dey, P. Eisenhardt, R.L. Griffith, T.H. Jarrett, S. Lake, F. Masci, S. Petty, S.A. Stanford, C.-W. Tsai, E.L. Wright, L. Yan, F. Harrison, K. Madsen, Mid-infrared selection of active galactic nuclei with the Wide-field Infrared Survey Explorer. I. Characterizing WISE-selected active galactic nuclei in COSMOS. Astrophys. J. **753**(1), 30 (2012). https://doi.org/10.1088/0004-637X/753/1/30

N. Stone, R. Sari, A. Loeb, Consequences of strong compression in tidal disruption events. Mon. Not. R. Astron. Soc. **435**(3), 1809–1824 (2013). http://adsabs.harvard.edu/abs/2013MNRAS.435.1809S

N.C. Stone, E. Vasiliev, M. Kesden, E.M. Rossi, H.B. Perets, P. Amaro-Seoane, Rates of stellar tidal disruption. Space Sci. Rev. **216**(3), 35 (2020). https://doi.org/10.1007/s11214-020-00651-4

L.E. Strubbe, E. Quataert, Optical flares from the tidal disruption of stars by massive black holes. Mon. Not. R. Astron. Soc. **400**(4), 2070–2084 (2009). https://doi.org/10.1111/j.1365-2966.2009.15599.x

C. Tadhunter, R. Spence, M. Rose, J. Mullaney, P. Crowther, A tidal disruption event in the nearby ultra-luminous infrared galaxy F01004-2237. Nat. Astron. **1**, 0061 (2017). https://doi.org/10.1038/s41550-017-0061

T.L. Tanaka, Recurring flares from supermassive black hole binaries: implications for tidal disruption candidates and OJ 287. Mon. Not. R. Astron. Soc. **434**(3), 2275–2288 (2013). https://doi.org/10.1093/mnras/stt1164

P. Tozzi, R. Gilli, V. Mainieri, C. Norman, G. Risaliti, P. Rosati, J. Bergeron, S. Borgani, R. Giacconi, G. Hasinger, M. Nonino, A. Streblyanska, G. Szokoly, J.X. Wang, W. Zheng, X-ray spectral properties of active galactic nuclei in the Chandra Deep Field South. Astron. Astrophys. **451**, 457–474 (2006). https://doi.org/10.1051/0004-6361:20042592

B. Trakhtenbrot, I. Arcavi, C.L. MacLeod, C. Ricci, E. Kara, M.L. Graham, D. Stern, F.A. Harrison, J. Burke, D. Hiramatsu, G. Hosseinzadeh, D.A. Howell, S.J. Smartt, A. Rest, J.L. Prieto, B.J. Shappee, T.W.-S. Holoien, D. Bersier, A.V. Filippenko, T.G. Brink, W. Zheng, R. Li, R.A. Remillard, M. Loewenstein, 1ES 1927+654: an AGN caught changing look on a timescale of months. Astrophys. J. **883**(1), 94 (2019a). https://doi.org/10.3847/1538-4357/ab39e4

B. Trakhtenbrot, I. Arcavi, C. Ricci, S. Tacchella, D. Stern, H. Netzer, P.G. Jonker, A. Horesh, J.E. Mejía-Restrepo, G. Hosseinzadeh, V. Hallefors, D.A. Howell, C. McCully, M. Baloković, M. Heida, N. Kamraj, G.B. Lansbury, Ł. Wyrzykowski, M. Gromadzki, A. Hamanowicz, S.B. Cenko, D.J. Sand, E.Y. Hsiao, M.M. Phillips, T.R. Diamond, E. Kara, K.C. Gendreau, Z. Arzoumanian, R. Remillard, A new class of flares from accreting supermassive black holes. Nat. Astron. **3**, 242–250 (2019b). https://doi.org/10.1038/s41550-018-0661-3

J. Truemper, The ROSAT mission. Adv. Space Res. **2**, 241–249 (1982). https://doi.org/10.1016/0273-1177(82)90070-9

R. Tylenda, M. Hajduk, T. Kamiński, A. Udalski, I. Soszyński, M.K. Szymański, M. Kubiak, G. Pietrzyński, R. Poleski, Ł. Wyrzykowski, K. Ulaczyk, V1309 Scorpii: merger of a contact binary. Astron. Astrophys. **528**, 114 (2011). https://doi.org/10.1051/0004-6361/201016221

S. Valenti, D. Sand, A. Pastorello, M.L. Graham, D.A. Howell, J.T. Parrent, L. Tomasella, P. Ochner, M. Fraser, S. Benetti, F. Yuan, S.J. Smartt, J.R. Maund, I. Arcavi, A. Gal-Yam, C. Inserra, D. Young, The first month of evolution of the slow-rising Type IIP SN 2013ej in M74. Mon. Not. R. Astron. Soc. **438**(1), 101–105 (2014). https://doi.org/10.1093/mnrasl/slt171

S. van Velzen, On the mass and luminosity functions of tidal disruption flares: rate suppression due to black hole event horizons. Astrophys. J. **852**, 72 (2018). https://doi.org/10.3847/1538-4357/aa998e

S. van Velzen, G.R. Farrar, S. Gezari, N. Morrell, D. Zaritsky, L. Östman, M. Smith, J. Gelfand, A.J. Drake, Optical discovery of probable stellar tidal disruption flares. Astrophys. J. **741**(2), 73 (2011). https://doi.org/10.1088/0004-637X/741/2/73

S. van Velzen, N.C. Stone, B.D. Metzger, S. Gezari, T.M. Brown, A.S. Fruchter, Late-time UV observations of tidal disruption flares reveal unobscured, compact accretion disks. Astrophys. J. **878**(2), 82 (2019). https://doi.org/10.3847/1538-4357/ab1844

S. van Velzen, S. Gezari, E. Hammerstein, N. Roth, S. Frederick, C. Ward, T. Hung, S.B. Cenko, R. Stein, D.A. Perley, K. Taggart, J. Sollerman, I. Andreoni, E.C. Bellm, V. Brinnel, K. De, R. Dekany, M. Feeney, R.J. Foley, C. Fremling, M. Giomi, V.Z. Golkhou, A.Y.Q. Ho, M.M. Kasliwal, C.D. Kilpatrick, S.R. Kulkarni, T. Kupfer, R.R. Laher, A. Mahabal, F.J. Masci, J. Nordin, R. Riddle, B. Rusholme, Y. Sharma, J. van Santen, D.L. Shupe, M.T. Soumagnac, Seventeen tidal disruption events from the first half of ZTF survey observations: entering a new era of population studies. arXiv e-prints (2020a). 2001.01409

S. van Velzen, W.-S.T. Holoien, F. Onori, T. Hung, I. Arcavi, Optical-ultraviolet tidal disruption events. Space Sci. Rev. **216**(8), 124 (2020b). https://doi.org/10.1007/s11214-020-00753-z

D.E. Vanden Berk, G.T. Richards, A. Bauer, M.A. Strauss, D.P. Schneider, T.M. Heckman, D.G. York, P.B. Hall, X. Fan, G.R. Knapp, S.F. Anderson, J. Annis, N.A. Bahcall, M. Bernardi, J.W. Briggs, J. Brinkmann, R. Brunner, S. Burles, L. Carey, F.J. Castander, A.J. Connolly, J.H. Crocker, I. Csabai, M. Doi, D. Finkbeiner, S. Friedman, J.A. Frieman, M. Fukugita, J.E. Gunn, G.S. Hennessy, Ž. Ivezić, S. Kent, P.Z. Kunszt, D.Q. Lamb, R.F. Leger, D.C. Long, J. Loveday, R.H. Lupton, A. Meiksin, A. Merelli, J.A. Munn, H.J. Newberg, M. Newcomb, R.C. Nichol, R. Owen, J.R. Pier, A. Pope, C.M. Rockosi, D.J. Schlegel, W.A. Siegmund, S. Smee, Y. Snir, C. Stoughton, C. Stubbs, M. SubbaRao, A.S. Szalay, G.P. Szokoly, C. Tremonti, A. Uomoto, P. Waddell, B. Yanny, W. Zheng, Composite quasar spectra from the Sloan Digital Sky Survey. Astron. J. **122**(2), 549–564 (2001). https://doi.org/10.1086/321167

D.E. Vanden Berk, B.C. Wilhite, R.G. Kron, S.F. Anderson, R.J. Brunner, P.B. Hall, Ž. Ivezić, G.T. Richards, D.P. Schneider, D.G. York, J.V. Brinkmann, D.Q. Lamb, R.C. Nichol, D.J. Schlegel, The ensemble photometric variability of $\sim 25{,}000$ quasars in the Sloan Digital Sky Survey. Astrophys. J. **601**(2), 692–714 (2004). https://doi.org/10.1086/380563

Q. Vigneron, G. Lodato, A. Guidarelli, Tidal disruption of stars in a supermassive black hole binary system: the influence of orbital properties on fallback and accretion rates. Mon. Not. R. Astron. Soc. **476**(4), 5312–5322 (2018). https://doi.org/10.1093/mnras/sty585

S. Wen, P.G. Jonker, N.C. Stone, A.I. Zabludoff, D. Psaltis, Continuum-fitting the X-ray spectra of tidal disruption events. Astrophys. J. **897**(1), 80 (2020). https://doi.org/10.3847/1538-4357/ab9817

S. Wen, P.G. Jonker, N.C. Stone, A.I. Zabludoff, Mass spin and ultralight boson constraints from the intermediate mass black hole in the tidal disruption event 3XMM J215022.4-055108 (2021). arXiv e-prints, 2104.01498

T. Wevers, S. van Velzen, P.G. Jonker, N.C. Stone, T. Hung, F. Onori, S. Gezari, N. Blagorodnova, Black hole masses of tidal disruption event host galaxies. Mon. Not. R. Astron. Soc. **471**(2), 1694–1708 (2017). https://doi.org/10.1093/mnras/stx1703

T. Wevers, D.R. Pasham, S. van Velzen, G. Leloudas, S. Schulze, J.C.A. Miller-Jones, P.G. Jonker, M. Gromadzki, E. Kankare, S.T. Hodgkin, Ł. Wyrzykowski, Z. Kostrzewa-Rutkowska, S. Moran, M. Berton, K. Maguire, F. Onori, S. Mattila, M. Nicholl, Evidence for rapid disc formation and reprocessing in the X-ray bright tidal disruption event candidate AT 2018fyk. Mon. Not. R. Astron. Soc. **488**(4), 4816–4830 (2019a). https://doi.org/10.1093/mnras/stz1976

T. Wevers, N.C. Stone, S. van Velzen, P.G. Jonker, T. Hung, K. Auchettl, S. Gezari, F. Onori, D. Mata Sánchez, Z. Kostrzewa-Rutkowska, J. Casares, Black hole masses of tidal disruption event host galaxies II. Mon. Not. R. Astron. Soc. **487**(3), 4136–4152 (2019b). https://doi.org/10.1093/mnras/stz1602

Y.Q. Xue, B. Luo, W.N. Brandt, F.E. Bauer, B.D. Lehmer, P.S. Broos, D.P. Schneider, D.M. Alexander, M. Brusa, A. Comastri, A.C. Fabian, R. Gilli, G. Hasinger, A.E. Hornschemeier, A. Koekemoer, T. Liu, V. Mainieri, M. Paolillo, D.A. Rafferty, P. Rosati, O. Shemmer, J.D. Silverman, I. Smail, P. Tozzi, C. Vignali, The Chandra Deep Field-South Survey: 4 Ms source catalogs. Astrophys. J. Suppl. Ser. **195**, 10 (2011). https://doi.org/10.1088/0067-0049/195/1/10

R. Yan, M.R. Blanton, The nature of LINER-like emission in red galaxies. Astrophys. J. **747**(1), 61 (2012). https://doi.org/10.1088/0004-637X/747/1/61

B.A. Zauderer, E. Berger, A.M. Soderberg, A. Loeb, R. Narayan, D.A. Frail, G.R. Petitpas, A. Brunthaler, R. Chornock, J.M. Carpenter, G.G. Pooley, K. Mooley, S.R. Kulkarni, R. Margutti, D.B. Fox, E. Nakar,

N.A. Patel, N.H. Volgenau, T.L. Culverhouse, M.F. Bietenholz, M.P. Rupen, W. Max-Moerbeck, A.C.S. Readhead, J. Richards, M. Shepherd, S. Storm, C.L.H. Hull, Birth of a relativistic outflow in the unusual γ-ray transient Swift J164449.3+573451. Nature **476**(7361), 425–428 (2011). https://doi.org/10.1038/nature10366

B.A. Zauderer, E. Berger, R. Margutti, G.G. Pooley, R. Sari, A.M. Soderberg, A. Brunthaler, M.F. Bietenholz, Radio monitoring of the tidal disruption event swift J164449.3+573451. II. The relativistic jet shuts off and a transition to forward shock X-ray/radio emission. Astrophys. J. **767**(2), 152 (2013). https://doi.org/10.1088/0004-637X/767/2/152

Authors and Affiliations

Ann Zabludoff[1] · Iair Arcavi[2] · Stephanie La Massa[3] · Hagai B. Perets[4] · Benny Trakhtenbrot[2] · B. Ashley Zauderer[5] · Katie Auchettl[6] · Jane L. Dai[7] · K. Decker French[8] · Tiara Hung[9] · Erin Kara[10] · Giuseppe Lodato[11] · W. Peter Maksym[12] · Yujing Qin[1] · Enrico Ramirez-Ruiz[9] · Nathaniel Roth[13] · Jessie C. Runnoe[14] · Thomas Wevers[15]

✉ A. Zabludoff
aiz@email.arizona.edu

1 University of Arizona, Tucson, USA

2 Tel Aviv University, Tel Aviv, Israel

3 STScI, Baltimore, USA

4 Technion, Haifa, Israel

5 National Science Foundation, Arlington, USA

6 University of Melbourne, Melbourne, Australia

7 University of Hong Kong, Hong Kong, China

8 University of Illinois, Urbana, USA

9 University of California, Santa Cruz, USA

10 Massachusetts Institute of Technology, Cambridge, USA

11 University of Milano Milan, Italy

12 Harvard-Smithsonian Center for Astrophysics, Cambridge, USA

13 Lawrence Livermore National Laboratory Livermore, USA

14 Vanderbilt University, Nashville, USA

15 European Southern Observatory, Santiago, Chile

Space Sci Rev (2020) 216:32
https://doi.org/10.1007/s11214-020-00657-y

The Host Galaxies of Tidal Disruption Events

K. Decker French[1] · Thomas Wevers[2] · Jamie Law-Smith[3] · Or Graur[4] · Ann I. Zabludoff[5]

Received: 18 June 2019 / Accepted: 7 March 2020 / Published online: 18 March 2020

Abstract Recent studies of Tidal Disruption Events (TDEs) have revealed unexpected correlations between the TDE rate and the large-scale properties of the host galaxies. In this review, we present the host galaxy properties of all TDE candidates known to date and quantify their distributions. We consider throughout the differences between observationally-identified types of TDEs and differences from spectroscopic control samples of galaxies. We focus here on the black hole and stellar masses of TDE host galaxies, their star formation histories and stellar populations, the concentration and morphology of the optical light, the presence of AGN activity, and the extra-galactic environment of the TDE hosts. We summarize the state of several possible explanations for the links between the TDE rate and host galaxy type. We present estimates of the TDE rate for different host galaxy types and quantify the degree to which rate enhancement in some types results in rate suppression

The Tidal Disruption of Stars by Massive Black Holes
Edited by Peter G. Jonker, Sterl Phinney, Elena Maria Rossi, Sjoert van Velzen, Iair Arcavi and Maurizio Falanga

✉ K.D. French
kfrench@carnegiescience.edu

T. Wevers
tw@ast.cam.ac.uk

J. Law-Smith
lawsmith@ucsc.edu

O. Graur
or.graur@cfa.harvard.edu

A.I. Zabludoff
aiz@email.arizona.edu

1 Observatories of the Carnegie Institute for Science, 813 Santa Barbara St., Pasadena CA 91101, USA

2 Institute of Astronomy, University of Cambridge, Madingley Road, Cambridge CB3 0HA, UK

3 Department of Astronomy and Astrophysics, University of California, Santa Cruz, CA 95064, USA

4 Harvard-Smithsonian Center for Astrophysics, 60 Garden Street, Cambridge, MA 02138, USA

5 Steward Observatory, University of Arizona, 933 N Cherry Ave, Tucson, AZ 85721, USA

in others. We discuss the possibilities for using TDE host galaxies to assist in identifying TDEs in upcoming large transient surveys and possibilities for TDE observations to be used to study their host galaxies.

Keywords Tidal disruption events · Galaxies · Supermassive black holes

1 Introduction

Tidal Disruption Events (TDEs) are observed when a star passes close enough to a supermassive black hole (SMBH) to be disrupted and torn apart by tidal forces. The rate of TDEs and the properties of the stars and SMBHs involved depend on the nuclear conditions of the host galaxies.

The mass of the SMBH will affect whether TDEs are observed and which stars can be tidally disrupted outside the event horizon (e.g., Hills 1975; Rees 1988; MacLeod et al. 2012; Law-Smith et al. 2017a). Because the tidal radius scales as $M_{BH}^{1/3}$, while the gravitational radius scales linearly with M_{BH}, stars will be swallowed whole if the SMBH is larger than the so-called Hills mass (Hills 1975), which is $\approx 10^8$ $M_\odot$ for a non spinning black hole and a Solar type star. The Hills mass also depends on the SMBH spin, such that a faster spinning SMBH can disrupt less massive stars at a given M_{BH} (Kesden 2012). The mass of the black hole is therefore an important parameter of a TDE. It can be estimated from observations because it is closely correlated with the mass and velocity dispersion of the galaxy's stellar bulge (Magorrian et al. 1998; Ferrarese and Merritt 2000; Gültekin et al. 2009; Kormendy and Ho 2013; McConnell and Ma 2013). The mass of the stars available to be disrupted depends on the recent star formation history of the galaxy and the initial mass function. The mass of the disrupted star is much harder to infer from observations, although it has been argued that it leaves imprints on the UV/optical lightcurve (Lodato et al. 2009; Guillochon and Ramirez-Ruiz 2015; Mockler et al. 2019).

Stars will be perturbed in their orbits to pass within the tidal radius depending on the distribution function of the stars in that galaxy (Magorrian and Tremaine 1999). The parameter space of stars that can be tidally disrupted, called the loss cone, is thought to be re-filled mainly through two-body interactions, although other mechanisms may also play a non negligible role (see Stone et al. 2020, ISSI review). The stellar density profile in the vicinity of the SMBH and any deviations from an isotropic velocity/velocity dispersion field will hence affect the TDE rate (Magorrian and Tremaine 1999; Merritt and Poon 2004; Stone et al. 2018). As most TDEs are thought to be sourced from within the gravitational radius of influence of the black hole (Stone and Metzger 2016; this is typically 0.1-10 pc for galaxies with $M_{BH} \sim 10^{6-8} M_\odot$), the galaxy properties at these scales are likely to be most important in setting the TDE rate. However, the conditions in this region will be affected by the evolution and merger history of the galaxy as a whole, and may therefore be correlated with larger scale galaxy properties.

Large transient surveys such as the Palomar Transient Factory (PTF; Law et al. 2009, Rau et al. 2009), Pan-STARRS (Chambers et al. 2016), the All Sky Automated Survey for SuperNovae (ASASSN; Shappee et al. 2014), and the Zwicky Transient Facility (ZTF; Bellm et al. 2019) in the optical, as well as the Roentgen Satellite (ROSAT) and the X-ray Multi-Mirror telescope (XMM; Jansen et al. 2001) in X-rays, and *Swift* in gamma rays have enabled the detections of tens of TDEs, providing a sample large enough to study population properties. In addition to the TDE properties themselves, these new samples of TDEs also allow us to study trends in their host galaxy properties. Arcavi et al. (2014) studied the host galaxies of seven UV/optical bright TDEs with broad H/He emission lines, and found many of the hosts showed E+A, or post-starburst, spectra. Such post-starburst spectra are characterized by a lack of strong emission lines, indicating low current star formation rates, but with strong Balmer absorption, indicating a recent burst of star formation (within the last ~Gyr) that has now ended. Quiescent Balmer-strong galaxies, and the subset of post-starburst/E+A galaxies with less ambiguous star formation histories, are rare in the local universe, and yet are over-represented among TDE host galaxies (French et al. 2016, 2017; Law-Smith et al. 2017b; Graur et al. 2018).

The observed correlations between the pc-scale regions of stars which can be tidally disrupted and the kpc-scale star-formation histories and stellar concentrations are a puzzle, for which many possible solutions have been proposed. Here, we review the known host galaxy properties of TDEs observed to date and with published or archival host galaxy spectra in §2. We discuss possible drivers for the host galaxy preference in §3 and implications for the TDE rates in §4. We discuss possibilities for using the host galaxy information in future surveys to find more TDEs in §5 and study galaxy properties in §6 and conclude in §7.

1.1 TDEs Included in This Review

In this review, we have selected a list of TDEs to discuss from the sample compiled by Auchettl et al. (2017b) of X-ray and optical/UV - bright TDEs, as this sample has been used for recent host galaxy studies (Law-Smith et al. 2017b; Wevers et al. 2017; Graur et al. 2018). Given the focus of this chapter on the host galaxy properties, we only include TDEs for which a spectrum of the host galaxy has been published or is available from archival surveys. We have added three more recent TDEs for which host galaxy spectroscopy is available from before the TDE from SDSS and BOSS: AT2018dyk, AT2018bsi, and ASASSN18zj (aka AT2018hyz), as well as a new TDE with published host galaxy information (PS18kh, Holoien et al. 2018).

We note the important caveat that the classification of transient events as TDEs is complicated by the heterogeneous datasets obtained for each event. For the purposes of this review we aim to balance including a large enough sample to reflect the range of published literature in this field with giving preference to the most well-justified claims of observed TDEs. We thus preserve the classifications of Auchettl et al. (2017b) for the X-ray and optical detected TDEs that rank the likelihood an event is a TDE based on the completeness of the data, and divide the data into a number of subsets. It is important to note that the host galaxy statistics may change depending on which subset of TDE candidates are used. We comment on the differences one obtains depending on the sample used throughout, though for some subclasses we are limited by small number statistics.

We list in Table 1 the TDEs considered in this review.

We divide the TDEs into two classes—X-ray bright and optical/UV bright—with several sub-categories. The X-ray bright TDEs are subdivided further into X-ray TDEs (Holoien et al. 2016b; Levan et al. 2011; Saxton et al. 2017; Holoien et al. 2016a), likely X-ray TDEs (Saxton et al. 2012; Esquej et al. 2007; Cenko et al. 2012; Maksym et al. 2010; Lin et al. 2015, 2017), and possible X-ray TDEs (Komossa and Greiner 1999; Gezari et al. 2008; Grupe et al. 1999; Ho et al. 1995; Greiner et al. 2000; Maksym et al. 2014) as done by Auchettl et al. (2017b). TDEs with no known or observed X-ray emission are classed as optical/UV TDEs (Brown et al. 2017; Blanchard et al. 2017; Tadhunter et al. 2017; Gezari et al. 2009; Chornock et al. 2014; Komossa et al. 2009; Wang et al. 2012; van Velzen et al. 2011; Yang et al. 2013; Holoien et al. 2016b; Arcavi et al. 2014; Blagorodnova et al. 2019; Arcavi et al. 2018; Gezari et al. 2018; Dong et al. 2018). TDEs requiring re-classification

Table 1 TDEs included in this review

Name	R.A.	Dec	z	Type	BL[a]	CL[b]
ASASSN14li[c]	12:48:15.23	+17:46:26.44	0.02058	X-ray TDE	1	0
Swift J1644	16:44:49.30	+57:34:51.00	0.3534	X-ray TDE	0	0
XMM J0740	07:40:08.09	−85:39:31.30	0.0173	X-ray TDE	0	0
ASASSN15oi	20:39:09.18	−30:45:20.10	0.0484	X-ray TDE	1	0
SDSS J1201[c]	12:01:36.03	+30:03:05.52	0.146	Likely X-ray TDE	0	0
2MASX J0249	02:49:17.32	−04:12:52.20	0.0186	Likely X-ray TDE	0	0
PTF10iya	14:38:40.98	+37:39:33.45	0.22405	Likely X-ray TDE	0	0
SDSS J1311	13:11:22.15	−01:23:45.61	0.195	Likely X-ray TDE	0	0
SDSS J1323[c]	13:23:41.97	+48:27:01.26	0.0875	Likely X-ray TDE	0	0
3XMM J1521	15:21:30.73	+07:49:16.52	0.17901	Likely X-ray TDE	0	0
3XMM J1500	15:00:52.07	+01:54:53.82	0.145	Likely X-ray TDE	0	0
PS18kh	07:56:54.53	+34:15:43.63	0.074	Likely X-ray TDE	1	0
RX J1242-A	12:42:38.54	−11:19:20.85	0.05	Possible X-ray TDE	0	0
RX J1242-B	12:42:38.16	−11:19:13.62	0.05	Possible X-ray TDE	0	0
RX J1420-A	14:20:24.39	+53:34:11.14	0.148	Possible X-ray TDE	0	0
RX J1420-B	14:20:24.52	+53:34:15.72	0.147	Possible X-ray TDE	0	0
SDSS J0159	01:59:57.64	+00:33:10.49	0.31167	Possible X-ray TDE	0	0
RBS 1032[c]	11:47:26.80	+49:42:59.00	0.02604	Possible X-ray TDE	0	0
RX J1624	16:24:56.66	+75:54:56.09	0.0636	Possible X-ray TDE	0	0
NGC 5905	15:15:23.32	+55:31:01.59	0.01131	Possible X-ray TDE	0	0
GALEX D3-13	14:19:29.81	+52:52:06.37	0.3698	Possible X-ray TDE	0	0
iPTF16fnl	00:29:57.01	+32:53:37.24	0.0163	Optical/UV TDE	1	0
PS16dtm[c]	01:58:04.75	−00:52:21.87	0.0804	Optical/UV TDE	0	0
F01004	01:02:50.01	−22:21:57.22	0.1178	Optical/UV TDE	0	0
GALEX D23H-1	23:31:59.54	+00:17:14.58	0.1855	Optical/UV TDE	0	0
PS1-11af	09:57:26.82	+03:14:00.94	0.4046	Optical/UV TDE	0	0
SDSS J0952[c]	09:52:09.56	+21:43:13.24	0.0789	Optical/UV TDE	0	1
SDSS J1342[c]	13:42:44.42	+05:30:56.14	0.0366	Optical/UV TDE	0	1
SDSS J1350[c]	13:50:01.51	+29:16:09.71	0.0777	Optical/UV TDE	0	1
SDSS TDE1	23:42:01.41	+01:06:29.30	0.1359	Optical/UV TDE	0	0
SDSS TDE2	23:23:48.62	−01:08:10.34	0.2515	Optical/UV TDE	1	0
SDSS J0748[c]	07:48:20.67	+47:12:14.23	0.0615	Optical/UV TDE	1	1
ASASSN14ae[c]	11:08:40.12	+34:05:52.23	0.0436	Optical/UV TDE	1	0
PTF09axc	14:53:13.08	+22:14:32.27	0.1146	Optical/UV TDE	1	0
PTF09djl	16:33:55.97	+30:14:16.65	0.184	Optical/UV TDE	1	0
PTF09ge[c]	14:57:03.18	+49:36:40.97	0.064	Optical/UV TDE	1	0
PS1-10jh	16:09:28.28	+53:40:23.99	0.1696	Optical/UV TDE	1	0
iPTF15af[c]	08:48:28.12	+22:03:33.58	0.079	Optical/UV TDE	1	0
iPTF16axa	17:03:34.36	+30:35:36.8	0.108	Optical/UV TDE	0	0
GALEX D1-9	02:25:17.00	−04:32:59.00	0.326	Optical/UV TDE	0	0
AT2018dyk[c]	15:33:08.02	+44:32:08.20	0.037	Optical/UV TDE	1	0
AT2018bsi	08:15:26.62	+45:35:31.95	0.051	Optical/UV TDE	1	0
ASASSN18zj[c]	10:06:50.74	+01:41:34.37	0.046	Optical/UV TDE	1	0

[a]Broad H/He lines observed during TDE. [b]Coronal lines observed. [c]Host galaxies with SDSS spectroscopy. Right ascension and declination are for the host galaxies

based on new X-ray data are re-classified (ASASSN-15oi, PS18kh; K. Auchettl, private communication).

TDEs that exhibited coronal lines (Komossa et al. 2009; Wang et al. 2012) or broad H/He lines (e.g., Arcavi et al. 2014) are also indicated. Three X-ray bright TDEs (ASASSN-14li, ASASSN-15oi, and PS18kh) additionally had significant optical observations, including broad H/He lines in their spectra, and are categorized as noted in the text. D3-13 is classed as a possible X-ray TDE, and also had significant optical/UV flux, but did not show broad H/He lines. We note that these classes are based on observational distinctions, which may or may not reflect physically different phenomena. Some optical/UV TDEs may have produced significant X-ray flux which was missed because of a lack of simultaneous X-ray observations. Indeed, the optical to X-ray luminosity ratios show significant variation in the events so far detected in both X-ray and optical light, and the observations of ASASSN-15oi by Gezari et al. (2017) demonstrate that X-ray emission can even be delayed well past the peak of the optical light curve, and would have been likely missed for many optical TDEs.[1] Similarly, the coronal line detections may be a light echo from a previous TDE (Komossa and Merritt 2008), and the relation between these events and the others is still unclear. These classes represent those for which samples have been aggregated in the literature, and with adequate host galaxy localization and observations to study for the purposes of this chapter. We direct the reader to the other chapters in this review, especially those by Saxton et al., Arcavi et al., Zauderer et al., Alexander et al., and Zabludoff et al., (2020, ISSI review) for further discussion of TDE classification and the question of multiple TDE classes.

In particular, we note that the class of events including F01004 (Tadhunter et al. 2017) may be a type of nuclear phenomenon other than a TDE. Trakhtenbrot et al. (2019) argue against the TDE interpretation of this event as it has significantly narrower He lines than other broad H/He line events and the presence of Bowen fluorescence lines. However, Bowen fluorescence lines have now been found in other TDEs with broader H/He lines (Leloudas et al. 2019), indicating the space for observed TDE features may be broader than expected. The optical features of TDEs are discussed further in Arcavi et al. (2020, ISSI review), and a comparison of observational properties of observed TDE candidates with the spectrum of possible "imposters" is discussed further in Zabludoff et al. (2020, ISSI review).

We separate out 13 TDE host galaxies which are part of the SDSS main spectroscopic sample (indicated in Table 1) in some of the following analysis, as these host galaxies can be matched to the general galaxy population in a uniform way.

2 Known Host Galaxy Properties of All Events

We consider here the host galaxy properties and trends of the TDE samples discussed above. We compare the stellar mass and black hole masses to expectations given the volume-corrected mass functions and expectations from an upper cutoff in the black hole mass from event horizon suppression (§2.1). We also consider the stellar populations and inferred recent star formation histories of the TDE hosts, and discuss the observed enhancement in post-starburst and quiescent Balmer-strong galaxies (§2.2). We discuss the morphologies and concentrations of the stellar light and observed trends toward higher central concentration on kpc scales in the TDE hosts (§2.3). The presence of on-going gas accretion and AGN activity in the TDE host galaxies, as well as possible biases against identifying TDEs in such

[1] See e.g. Jonker et al. (2019) for more late-time X-ray detections of UV/optical TDEs; this article was posted to the arXiv during review of this article.

host galaxies are also discussed (§2.4). We summarize the extragalactic environments of the TDE host galaxies, given the efforts to identify TDEs in galaxy clusters (§2.5).

The redshift range of the host galaxies affects the extent to which they can be studied. Most of the TDEs discovered to date are at low redshift, such that many of the TDE host galaxies have data from the SDSS. The redshift of all of the TDEs considered in this review (see Table 1) ranges from 0.01 to 0.4. The median redshift is $z = 0.08$, and the 50 percentile range is 0.05–0.15. This redshift range is necessarily biased by the surveys which have discovered TDEs so far. Future surveys, such as LSST, may find a larger sample of higher redshift TDEs, depending on how the intrinsic TDE rate changes with redshift. A study by Kochanek (2016) predicts that the TDE rate will drop steeply with redshift between $z = 0$ and $z = 1$, based on the expected evolution of the host galaxy stellar populations, black hole masses, and merger rates. However, the rising fraction of post-starburst hosts with redshift (Yan et al. 2009; Snyder et al. 2011; Wild et al. 2016) may act to counter this effect. The blue continuum of TDE emission may result in a negative k-correction (Cenko et al. 2016), which would result in a greater number of observed TDEs at higher redshift.

2.1 Host Galaxy Stellar Mass and Black Hole Mass

The black hole mass is one of the fundamental parameters for TDE studies, as it sets both the energetics (e.g. peak luminosity, accretion efficiency) and the dynamics (e.g. orbital timescales, relativistic effects) of the disruption. While theoretical predictions (Wang and Merritt 2004) suggest that TDEs should preferentially occur in the lowest mass galaxies still hosting SMBHs (10^4–10^6 $M_\odot$), the observed distribution (using a heterogeneous set of measurements) was observed to peak around 10^7 $M_\odot$ (e.g. Stone and Metzger 2016, Kochanek 2016). More recently, Wevers et al. (2017) presented systematic measurements of black hole masses using the M-σ relation for a sample of 12 optical TDEs, and found the peak in the TDE black hole mass distribution to be significantly lower, near 10^6 $M_\odot$, consistent with theoretical predictions (taking into account the uncertainty in the calibration of the M-σ relation at the low mass end). In contrast to previous studies (e.g., Stone and Metzger 2016) which use scaling relations from photometric observations to infer black hole masses, Wevers et al. (2017) use spectroscopic observations of the bulge velocity dispersions. Black hole mass measurements from TDE light curves (Mockler et al. 2019) are consistent with measurements from galactic properties given the uncertainties in each set of measurements, but the number of TDE light curves with well-measured rises and thus more accurate black hole mass measurements is still limited.

van Velzen (2018) uses the BH masses from Wevers et al. (2017) to infer the BH mass and luminosity functions of TDEs. Correcting for selection effects such as survey depth, cadence and area, they find that the TDE rate is constant with black hole (or galaxy stellar) mass over two orders of magnitude from $M_\odot = 10^{5.5} - 10^{7.5}$. Given the uncertainties, the observed black hole mass function of TDE hosts could be consistent with either the expected black hole mass function over this mass range, or with the slightly steeper trend expected given the scaling of the TDE rate with black hole mass. The dearth of BH masses $\geq 10^8$ $M_\odot$ is consistent with the presence of BH event horizons, and the disappearance of the tidal radius for a main sequence $1 M_\odot$ star inside the event horizon.

While black hole masses are difficult and time-consuming to measure, stellar masses can be more easily measured using galaxy luminosities and stellar population estimates. The stellar masses of the host galaxies are roughly correlated with the black hole masses, via the black hole mass – bulge mass relation (e.g., McConnell and Ma 2013) and the correlation between galaxy stellar mass and bulge stellar mass (e.g., Mendel et al. 2014). For the host

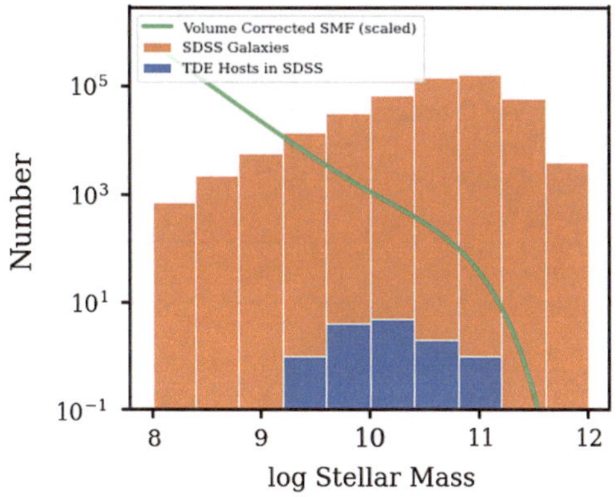

Fig. 1 Histogram of stellar masses for the SDSS main spectroscopic sample, with quality cuts as described in Law-Smith et al. (2017b) (orange), TDE host galaxies from the SDSS main spectroscopic sample (blue), and the volume corrected stellar mass function from Baldry et al. (2012). The TDE host galaxies are less massive than the typical SDSS galaxies, but with a typical stellar mass near M* (the turnover in the stellar mass function, measured by Baldry et al. (2012) to be $10^{10.66}$ $M_\odot$). This distribution is consistent with the TDE host galaxies being drawn from a volume limited sample of galaxies with stellar mass greater than 10^9 $M_\odot$

galaxies in the SDSS main spectroscopic sample, we plot a histogram of their stellar masses compared to the rest of the SDSS galaxies and the volume-corrected stellar mass function (SMF) in Fig. 1. The TDE host galaxies are less massive than the typical SDSS galaxies, but with a typical stellar mass near M*. This distribution is consistent with the TDE host galaxies being drawn from a volume limited sample of galaxies with stellar mass greater than 10^9 $M_\odot$.

Are the distributions of host galaxy stellar mass or black hole mass different for different classes of TDEs? There are several predictions in the literature, and this question depends on the details of how stars are disrupted and accreted, and the origin of the observed emission. The inverse dependence of the accretion disk temperature on black hole mass suggests X-ray TDEs should have lower black hole masses than optical TDEs (Dai et al. 2015), but if rapid circularization is required to produce X-rays, higher mass black holes may be expected to produce more X-ray emission (Guillochon and Ramirez-Ruiz 2015). Alternatively, if the difference between the classes is related to a viewing angle effect (Dai et al. 2018), no difference in the host galaxy properties would be expected.

Wevers et al. (2019a) have measured the host galaxy absolute magnitudes of a large sample of optical and X-ray TDEs using SDSS and PS1 photometry. They used the kcorrect software (Blanton and Roweis 2017) and the Petrosian or Kron magnitudes for SDSS and PS1 to estimate the host absolute magnitude as well as the galaxy stellar mass for a sample of 35 TDEs and TDE candidates. These values are presented in Table 2 for all sources in the current sample. Using different subdivisions (and a smaller sample) of host galaxies than the ones used here, Wevers et al. (2019a) found that the host galaxy absolute magnitudes, stellar masses, and black hole masses for different TDE classes are consistent with being drawn from the same parent population.

The larger sample considered here allows us to repeat the analysis in Wevers et al. (2019a) with more statistical power (Fig. 2; Table 2). We group the X-ray and likely X-ray hosts, the UV/optically discovered hosts and the possible X-ray hosts and perform pairwise Kolmogorov-Smirnov (KS) and Anderson-Darling (AD) tests for these 3 samples. For the

Table 2 Table of properties of known TDE host galaxies discussed in §2.1. $M_\star$ (stellar mass) and M_g (g-band absolute magnitude) are calculated as in Wevers et al. (2019a). σ is the measured velocity dispersion, which is used to derive the black hole mass using the M–σ relation of Ferrarese and Ford (2005). Measurement uncertainties are given between parentheses. The uncertainties in M_{BH} are the linear addition of the measurement uncertainties and the scatter in the M–σ relation. Because the scatter in the M–σ relation is a systematic uncertainty and the uncertainty in the velocity dispersion measurements are statistical, we add them linearly

Name	$\log(M_\star)$ ($M_\odot$)	M_g (mag)	σ (km s^{-1})	$\log(M_{BH})$ ($M_\odot$)
X-ray TDEs				
ASASSN14li†	10.6	−18.8	81(2)	$6.23^{+0.39}_{-0.40}$
Swift J1644	–	–	–	–
XMM J0740	–	–	–	–
ASASSN15oi†	9.9	−19.3	61(7)	$5.71^{+0.60}_{-0.57}$
Likely X-ray TDEs				
SDSS J1201	10.4	−20.6	122(4)	$7.18^{+0.41}_{-0.41}$
2MASX J0249	9.1	−17.5	43(4)	$4.93^{+0.55}_{-0.53}$
PTF10iya	9.3	−20.0	–	–
SDSS J1311	8.7	−18.6	–	–
SDSS J1323	9.8	−18.9	75(4)	$6.15^{+0.46}_{-0.45}$
3XMM J1521	9.9	−19.2	58(2)	$5.61^{+0.41}_{-0.41}$
3XMM J1500	9.3	−19.1	59(3)	$5.64^{+0.45}_{-0.45}$
PS18kh†	9.6	−19.0	–	–
Possible X-ray TDEs				
RX J1242-A	10.3	−21.0	–	–
RX J1242-B	–	–	–	–
RX J1420-A	10.3	−20.3	131(13)	$7.33^{+0.56}_{-0.54}$
RX J1420-B	–	–	–	–
SDSS J0159	10.7	−21.8	124(10)	$7.21^{+0.52}_{-0.50}$
RBS 1032	9.0	−17.7	49(7)	$5.25^{+0.67}_{-0.62}$
RX J1624	10.4	−20.8	155(9)	$7.68^{+0.45}_{-0.45}$
NGC 5905	10.0	−20.2	97(5)	$6.69^{+0.45}_{-0.44}$
GALEX D3-13	10.7	−20.8	133(6)	$7.36^{+0.43}_{-0.44}$
Optical/UV TDEs				
iPTF16fnl†	9.8	−19.8	55(2)	$5.50^{+0.42}_{-0.42}$
PS16dtm	9.6	−19.3	45(13)	$5.07^{+0.88}_{-1.06}$
F01004	9.8	−21.0	132(29)	$7.34^{+0.76}_{-0.86}$
GALEX D23H-1	10.3	−20.1	84(4)	$6.39^{+0.44}_{-0.44}$
PS1-11af	10.1	−20.1	–	–
SDSS J0952*	10.0	−20.2	–	–
SDSS J1342*	9.5	−19.0	72(6)	$6.06^{+0.51}_{-0.52}$
SDSS J1350*	–	–	–	–
SDSS TDE1	10.1	−19.2	126(7)	$7.25^{+0.45}_{-0.46}$
SDSS TDE2†	10.6	−20.6	–	–

Table 2 (*Continued*)

Name	log($M_\star$) ($M_\odot$)	M_g (mag)	σ (km s^{-1})	log(M_{BH}) ($M_\odot$)
SDSS J0748†,*	9.9	−20.0	126(7)	$7.25^{+0.45}_{-0.46}$
ASASSN14ae†	10.8	−19.2	53(2)	$5.42^{+0.46}_{-0.46}$
PTF09axc†	10.0	−20.2	60(4)	$5.68^{+0.48}_{-0.49}$
PTF09djl†	10.1	−20.0	64(7)	$5.82^{+0.56}_{-0.58}$
PTF09ge†	10.1	−19.5	82(2)	$6.31^{+0.39}_{-0.39}$
PS1-10jh†	9.5	−18.1	65(3)	$5.85^{+0.44}_{-0.44}$
iPTF15af†	10.2	−18.0	106(2)	$6.88^{+0.38}_{-0.38}$
iPTF16axa	10.1	−19.4	82(3)	$6.34^{+0.42}_{-0.42}$
AT2018dyk†	10.6	−21.4	112(4)	$7.00^{+0.41}_{-0.42}$
AT2018bsi†	10.3	−20.8	–	–
ASASSN18zj†	9.5	−19.1	60(5)	$5.68^{+0.51}_{-0.52}$

†Broad line TDEs (see Table 1)

*Coronal line TDEs (see Table 1)

X-ray and optical samples, we find that for both host galaxy stellar mass and absolute magnitude the hypothesis that they are drawn from the same parent distribution can be rejected. The KS and AD significance values are summarized in Table 3. The p-values for the X-ray and possible X-ray stellar mass comparison are higher, and we cannot reject the null hypothesis that they are drawn from the same parent distribution. For the latter, this could be due to the small size of the sample. The properties of the possible X-ray sources suggest significant contamination by AGN, which favour higher mass (both stellar mass and M_{BH}) and more luminous host galaxies. The difference with the results in Wevers et al. (2019a) can be explained by i) the larger sample considered here and ii) the different sample subdivision. In particular, the soft X-ray sample in Wevers et al. (2019a) consists of 6 likely and 6 possible X-ray TDE hosts, and the sources ASASSN–14li and ASASSN–15oi are considered as optical events.

Wevers et al. (2019a) also presented velocity dispersion measurements of an additional 19 TDE candidates, yielding a sample of 29 homogeneously measured black hole masses.[2] Fig. 2 shows a kernel density estimate (KDE) of the black hole mass distribution, divided by type. The KDE was calculated by representing each black hole mass estimate with a Gaussian function with a full width at half maximum (FWHM) equal to the measurement uncertainty (including both the velocity dispersion uncertainty and the scatter in the M–σ relation), and then summing over the respective samples.

Using statistical tests to compare the distributions between X-ray and optical samples, we find no significant differences between the black hole mass distributions. This supports the idea that the apparent dichotomy between optical and X-ray selected TDEs could be related to (for example) viewing angle or geometry (Watarai et al. 2005; Coughlin and Begelman

[2]We note that the sample presented by Wevers et al. (2019a) contains a smaller but not completely overlapping sample to that presented in this review, due to differences in the TDE selection and differences in available data.

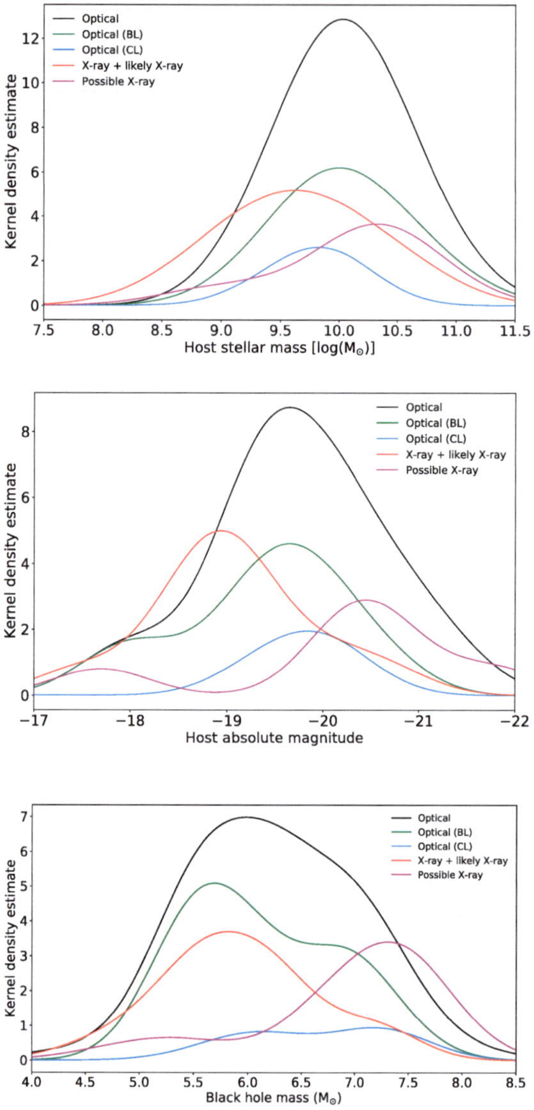

Fig. 2 Top: Kernel density estimate of TDE host galaxy stellar masses, sub-divided by TDE class. KDEs are calculated as in Wevers et al. (2019a), using Gaussians with width = 0.5 dex (the typical uncertainty on the host galaxy stellar mass measurements). Bulge to total light (B/T) corrections are applied (not individual measurements but a general prescription). The X-ray and optical/UV TDEs are consistent with being drawn from different parent samples, according to their Anderson-Darling statistic. Middle: Kernel density estimate of TDE host galaxy absolute magnitude, sub-divided by TDE class. KDEs are calculated as in Wevers et al. (2019a), using Gaussians with width = 0.5 dex (the typical uncertainty). The X-ray TDEs and optical TDEs are again consistent with being drawn from different parent samples. The possible X-ray TDE candidates are also distinct from the X-ray + likely X-ray samples as inferred from their Kolmogorov-Smirnov and Anderson-Darling tests. Bottom: Kernel density estimate of the black hole masses of X-ray and UV/optical selected TDEs, sub-divided by TDE class. The kernel size includes the velocity dispersion measurement as well as the intrinsic M–σ scatter. No statistically significant differences are observed between the TDE classes. Interestingly, while the X-ray TDEs prefer host galaxies with lower stellar mass and absolute magnitude than the optical/UV TDEs, this trend is not seen in black hole mass. More observations of black hole masses for various classes of TDEs, especially X-ray TDEs, are required to test whether this effect is physical

Table 3 Summary of statistical comparison between samples for different host properties, including host galaxy stellar mass ($M_\star$, absolute g-band magnitude (M_g) and the black hole mass (M_{BH}). We give the relevant sample sizes for each parameter. We test the hypothesis that the respective samples are drawn from the same parent distribution. The p-values of an Anderson-Darling test are given, as well as the p-values for a Kolmogorov-Smirnov test (in parentheses); values below 0.05 suggest that we can reject the hypothesis at $> 95\%$ significance. We note that these conclusions differ from those by Wevers et al. (2019a) due to the larger sample of TDEs considered here, as well as a different class division

Sample size	$M_\star$	M_g	M_{BH}
Optical	20	20	16
X-ray + likely X-ray	10	10	7
Possible X-ray	6	6	6
p-values			
Optical - X-ray	0.03 (0.09)	0.02 (0.02)	0.38 (0.42)
X-ray - pos. X-ray	0.15 (0.05)	0.03 (0.02)	0.06 (0.10)

2014; Dai et al. 2015; Metzger and Stone 2016; Roth et al. 2016; Dai et al. 2018), and that these events intrinsically belong to the same class. This is also supported by observations of UV/optical TDEs with deep X-ray upper limits: the detection of Bowen fluorescence lines in optical spectra implies that an ionizing (X-ray) radiation field exists, although no X-rays are actually observed (Leloudas et al. 2019). UV emission is not sufficient to excite the Bowen fluorescence lines for the one TDE (AT2018dyb) for which measurements are available. Emission line measurements presented by Leloudas et al. (2019) show that the Wien tail of the UV blackbody responsible for the UV/optical radiation is insufficient (by ~6 orders of magnitude) to explain the observed line fluxes. This suggests that the X-ray source is completely obscured along the line of sight.

While the black hole mass distribution is very similar, the host galaxy stellar mass and absolute magnitude are significantly lower for the X-ray sample, as we can reject the null hypothesis of a common parent sample at high significance. However, we caution that this effect could be due to the lesser number of TDE host galaxies with black hole mass measurements (Table 3). A larger sample of robust X-ray TDEs is required to draw robust statistical conclusions, and test whether optical/UV TDEs might have smaller black hole masses for their stellar masses (or whether X-ray TDEs have larger black hole masses for the stellar masses).

If the observed trends in stellar mass and absolute magnitude are driven by differences in the black hole mass distribution between the X-ray and optical samples, this suggests that smaller black holes have higher temperature accretion disks with higher X-ray luminosities, or a combination of this effect and a viewing angle effect are acting. Another possibility is that selection biases from the very different identification methods of TDEs in the optical vs. X-ray could lead to this effect. The light curve duration for TDEs may vary with black hole mass in different ways for X-ray compared to optical emission. Wen et al. (in prep) find lower mass black holes to have longer duration super-eddington plateaus in their predicted X-ray light curves. Lin et al. (2018) have found one such example of a long duration X-ray light curve from an event around a small SMBH. If it were the case that smaller black holes have longer-duration light curves in the X-ray compared to the optical, coarser cadence surveys in the X-ray would be biased against detecting TDEs in more massive black holes. Differences between TDEs found in optical vs. X-ray surveys will need to be studied further in the era of eROSITA and LSST.

Wevers et al. (2019a) also consider a class of hard X-ray selected TDE candidates (which are not included in the sample discussed here), finding these host galaxies to have significantly different black hole mass distributions, as well as absolute magnitude and stellar mass distributions. However, these conclusions are based on a sample of 5 hard X-ray TDE candidate host galaxies, and a larger sample is needed to confirm these findings and understand the cause of these potential differences.

2.2 Star Formation Rates, Star Formation Histories, and Stellar Populations

Next, we consider the current star formation rates, the past star formation history, and the stellar populations of the TDE host galaxies. While the stellar populations in the nucleus may be far removed from those of the bulk of the host galaxy, the galaxy-wide stellar populations trace the formation and evolution of the host galaxy and are closely tied to the morphologies, kinematics, interstellar medium properties, and merger histories of the host galaxies.

Star formation rates (SFRs) for host galaxies are calculated using various tracers of short-lived massive O and B stars. While many SFR tracers from the UV to the radio work well for galaxies with significant star formation, these tracers can be heavily biased in galaxies with rapidly changing SFRs (as many of the TDE host galaxies are thought to have), in dusty galaxies, or in galaxies influenced by AGN activity.

We consider here SFRs of TDE host galaxies derived using Hα luminosities from the SDSS, and correct for extinction and aperture using the corrections from the MPA-JHU SDSS catalogues (Brinchmann et al. 2004; Tremonti et al. 2004). The SFRs from this catalogue use the D4000 break to estimate the SFRs for galaxies with non-star-forming emission line ratios, like those of many of the TDE hosts (see §2.4). While the D4000-sSFR correlation will lead to accurate SFRs on average for large samples, individual galaxies will have high uncertainties. Our use of Hα here thus is more accurate for galaxies without strong AGN, but will be in general biased towards higher values for galaxies with additional Hα emission from non-star-forming sources. We convert the Hα luminosities to SFRs using $\eta = 5.4 \times 10^{-42}\ \mathrm{M_\odot\ yr^{-1}/(erg\,s^{-1})}$ (Kennicutt and Evans 2012). We compare the TDE hosts to galaxies from the SDSS main spectroscopic survey in SFR–stellar mass space in Fig. 3. While several of the TDE hosts are at the lower SFR edge of the "main sequence" of star forming galaxies, most are quiescent, with low SFRs.

We also consider the optical colours of the TDE hosts in the context of the colour magnitude relation. The $u - r$ colours and M_r absolute magnitudes are plotted in Fig. 4. The TDE hosts occupy a range of red, blue, and green-valley host galaxies. Because the colours are affected by both the current SFR and recent star formation history (SFH), we explore the physical interpretation of these quantities below.

The quiescent SFRs paired with green colours of many of the TDE hosts is suggestive of an intermediate stellar population and a recent decline in the SFH of the host. Indeed, a large number of TDEs have been observed in E+A or post-starburst galaxies. The large number of post-starburst host galaxies was first observed by Arcavi et al. (2014) in a sample of UV/optical bright TDEs with broad H/He lines. The presence of even one post-starburst galaxy amongst the hosts would be unusual given the rarity of post-starburst galaxies.

The over-representation of post-starburst galaxies in TDE hosts was then quantified by French et al. (2016), who used tracers sensitive to recent star formation on different scales to assess the recent star formation history of the host galaxies. One such comparison is to use the Hα emission as a tracer for the current star formation on ~ 10 Myr timescales and the Balmer absorption as a tracer for star formation on timescales of ~ 1 Gyr. This method of using Hα emission vs. Lick Hδ_A has been used by many (French et al. 2016; Law-Smith

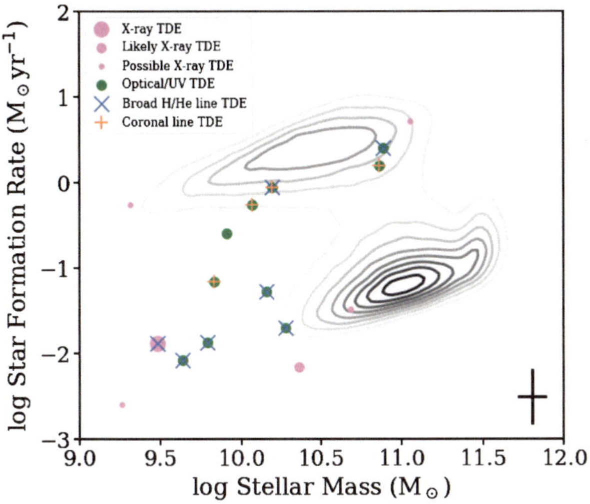

Fig. 3 Star formation rates vs. stellar masses for the SDSS main spectroscopic sample (grey contours) and the different classes of TDE hosts considered here. Stellar masses are from the MPA-JHU catalogue for all samples, as well as SFRs for the SDSS sample. We determine the TDE host SFRs based on Hα fluxes as described in §2.2. A characteristic error-bar is shown in the bottom right. While several of the TDE hosts are at the lower SFR edge of the "main sequence" of star forming galaxies, most are quiescent, with low SFRs. We note that the SDSS sample has not been volume corrected; for a more detailed analysis of the stellar mass distributions of the TDE hosts compared to a volume-limited sample, see §2.1. Thus, for the typical black hole masses of TDEs, the host galaxies lie in a relatively spare region on this diagram due to the magnitude-limited nature of the SDSS comparison sample used here

Fig. 4 Optical colour magnitude relation of TDE host galaxies and the SDSS main spectroscopic sample. All TDE hosts with SDSS photometry are plotted here. We note that the SDSS sample has not been volume corrected; for a more detailed analysis of the absolute magnitude distributions of the TDE hosts compared to a volume-limited sample, see §2.1. The TDE hosts occupy a range of red, blue, and green-valley host galaxies. Because the colours are affected by both the current SFR and recent star formation history (SFH), we explore the physical interpretation of these quantities below

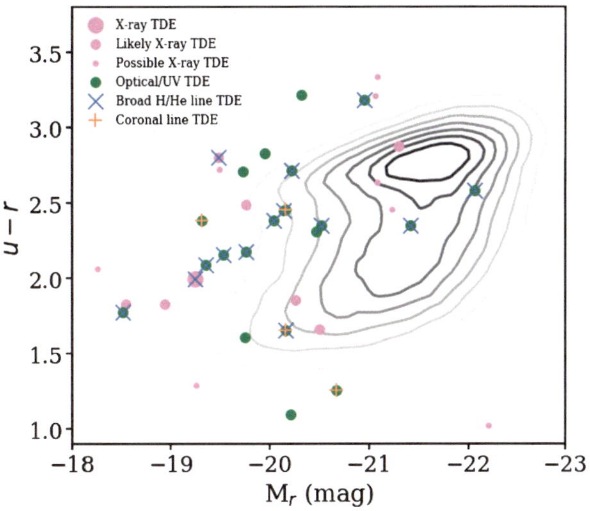

et al. 2017b; Graur et al. 2018) to study the recent star formation histories of TDE host galaxies.

There are many ways to quantify the current SFR in galaxies, as described above. Many methods that are sensitive to post-starburst galaxies use Hα emission or O[II] emission if

Table 4 Table of properties of known TDE host galaxies discussed in §2.2. Negative values indicate emission and positive values indicate absorption for the Hα and Hδ measurements. SFRs are only calculated when SDSS MPA-JHU catalogue data are available. To be classified as quiescent, quiescent Balmer-strong (QBS), or post-starburst (PSB), galaxies are required to have Hα EW > -3 Å. Galaxies with stronger Hα emission are labeled as star-forming (SF). Galaxies with strong Balmer absorption (Lick $H\delta_A > 1.31$ Å) are classified as QBS, and a subset of these are classified as PSB if they meet the threshold of $H\delta_A - \sigma(H\delta_A) > 4$ Å. 5/41 (12%) host galaxies are post-starburst galaxies and 13/41 (32%) are either QBS or PSB. Of the 4 X-ray TDEs, 3 (75%) are QBS and 1 (25%) is PSB. Of the 15 broad H/He line TDEs, 9 (60%) are QBS and 5 (33%) are PSB. We consider the effects of these choices of dividing criteria in §2.2

Name	Hα [Å]	σ(Hα) [Å]	Lick $H\delta_A$ [Å]	σ(Lick $H\delta_A$) [Å]	SFR $M_\odot$ yr^{-1}	Type
X-ray TDEs						
ASASSN14li†	−0.6	0.5	5.7	0.6	0.01	PSB
Swift J1644	−2.5	0.8	4.7	1.1	–	QBS
XMM J0740	−0.3	0.6	0.4	0.4	–	Quiescent
ASASSN15oi†	0.1	0.3	1.9	0.7	–	QBS
Likely X-ray TDEs						
SDSS J1201	0.7	0.3	−1.1	2.4	–	Quiescent
2MASX J0249	−5.7	0.6	0.4	0.5	–	SF
PTF10iya	−20.5	0.6	2.9	0.9	–	SF
SDSS J1311	−2.1	1.5	3.6	1.1	–	QBS
SDSS J1323	−0.2	0.5	−1.2	1.5	0.01	Quiescent
3XMM J1521	0.8	1.1	−1.5	2.4	–	Quiescent
3XMM J1500	−45.1	0.8	1.4	1.7	–	SF
Possible X-ray TDEs						
RX J1242-A	1.1	0.8	0.9	1.2	–	Quiescent
RX J1242-B	−0.9	0.9	−0.4	2.7	–	Quiescent
RX J1420-A	−0.2	0.9	−2.8	2.0	0.03	Quiescent
RX J1420-B	−71.6	1.7	5.8	3.4	0.54	SF
SDSS J0159	−19.9	0.8	1.7	1.0	5.25	SF
RBS 1032	−0.5	0.4	4.1	0.4	0.0	QBS
RX J1624	0.6	1.3	−1.1	2.1	–	Quiescent
NGC 5905	−28.4	0.1	––	–	–	SF
Optical/UV TDEs						
iPTF16fnl†	0.8	0.6	5.8	0.3	–	PSB
PS16dtm	−31.8	0.4	−0.2	1.1	0.26	SF
F01004	−47.0	0.2	−0.2	0.8	–	SF
GALEX D23H-1	−13.3	0.9	4.3	1.5	–	SF
PS1-11af	0.7	1.2	1.5	1.4	–	QBS
SDSS J0952*	−27.8	0.4	−2.0	1.1	1.6	SF
SDSS J1342*	−15.1	0.5	−1.0	1.3	0.07	SF
SDSS J1350*	−20.3	0.4	0.9	1.3	0.55	SF
SDSS TDE1	1.2	1.0	−1.3	1.3	–	Quiescent
SDSS TDE2†	−4.5	0.5	3.7	0.6	–	SF
SDSS J0748†,*	−11.4	1.0	1.2	0.8	0.87	SF
ASASSN14ae†	−0.7	0.4	3.4	0.8	0.01	QBS
PTF09axc†	−1.1	0.7	4.9	0.4	–	PSB
PTF09djl†	−0.3	0.7	4.7	0.5	–	PSB

Table 4 (*Continued*)

Name	Hα [Å]	σ(Hα) [Å]	Lick $H\delta_A$ [Å]	σ(Lick $H\delta_A$) [Å]	SFR $M_\odot$ yr^{-1}	Type
PTF09ge†	−1.7	0.8	0.3	0.7	0.05	Quiescent
PS1-10jh†	−0.5	0.7	1.7	0.8	–	QBS
iPTF15af†	−1.7	0.3	1.3	1.9	0.02	QBS
iPTF16axa	−1.1	1.7	0.2	1.5	–	Quiescent
PS18kh	−0.11	0.47	0.16	0.56	–	Quiescent
AT2018dyk†	−2.17	0.11	−0.27	0.48	2.54	Quiescent
AT2018bsi†	−5.25	0.69	−0.73	0.47	–	SF
ASASSN18zj†	−0.35	0.14	5.43	0.41	0.01	PSB

†Broad line TDEs (see Table 1). *Coronal line TDEs (see Table 1)

the red end of the rest-frame spectra are not available. Other methods allow for residual star formation using a PCA analysis (Wild et al. 2010) or a BPT diagram analysis (Alatalo et al. 2016). French et al. (2016) require Hα EW < 3 Å in emission in the rest frame to be considered quiescent. This corresponds to a specific SFR $\lesssim 1 \times 10^{-11}$ yr^{-1}, well below the main sequence of star-forming galaxies (e.g., Elbaz et al. 2011). The Hα emission is also corrected for stellar Balmer absorption, which is significant for post-starburst quiescent Balmer-strong galaxies.

The moderate lifetime of A stars means that the presence of a large A star population is indicative of a burst of star formation within the last Gyr. A star spectra show strong Balmer absorption, which can be best traced using the Hγ, Hδ, or Hϵ lines. French et al. (2016) use the Lick $H\delta_A$ index and its uncertainty $\sigma(H\delta_A)$, which is optimized for the stellar absorption from A stars (Worthey and Ottaviani 1997), has lower emission filling than Hβ, and smooth nearby continuum regions. The more bursty the SFH, i.e. the greater fraction of stellar mass produced over a shorter time, the higher Hδ absorption will be. A stricter cut of $H\delta_A - \sigma(H\delta_A) > 4$ Å will select galaxies with recent starbursts creating $> 3\%$ of their current stellar mass over 25–200 Myr (referred to as post-starburst galaxies throughout), and a weaker cut of $H\delta_A > 1.31$ Å (referred to as quiescent Balmer-strong galaxies throughout) will select galaxies with recent epochs of star formation which created $> 0.1\%$ of their current stellar mass over 25–1000 Myr (French et al. 2017, 2018).

We plot these SFH tracers in Fig. 5 for various subsamples of TDEs. The TDE host galaxies span a range of SFHs from star-forming galaxies, to quiescent galaxies which have been quiescent for at least the past Gyr, and galaxies which had significant star-formation within the last Gyr but are currently quiescent. This last category consists of "post-starbust" or "quiescent Balmer-strong" galaxies as defined above.

Several TDE host galaxies in the classes of the coronal line TDEs and optical/UV TDEs without coronal or broad lines have low Hδ absorption for their Hα emission compared to the rest of the galaxies in the SDSS spectroscopic sample. This may not be due to a physical difference between the host galaxies, and might instead be due to filling of the Hδ line by residual TDE emission, as discussed by Graur et al. (2018). Another possibility is contamination from the nearby Bowen fluorescence line NIII λ4100 (Blagorodnova et al. 2018; Leloudas et al. 2019; Trakhtenbrot et al. 2019).

The over-representation of a galaxy type among the TDE host galaxies can be determined using its rate in the TDE host galaxies compared to its rate in a general galaxy sample. We describe here the analyses done by various groups, and summarize in Table 5. French et al.

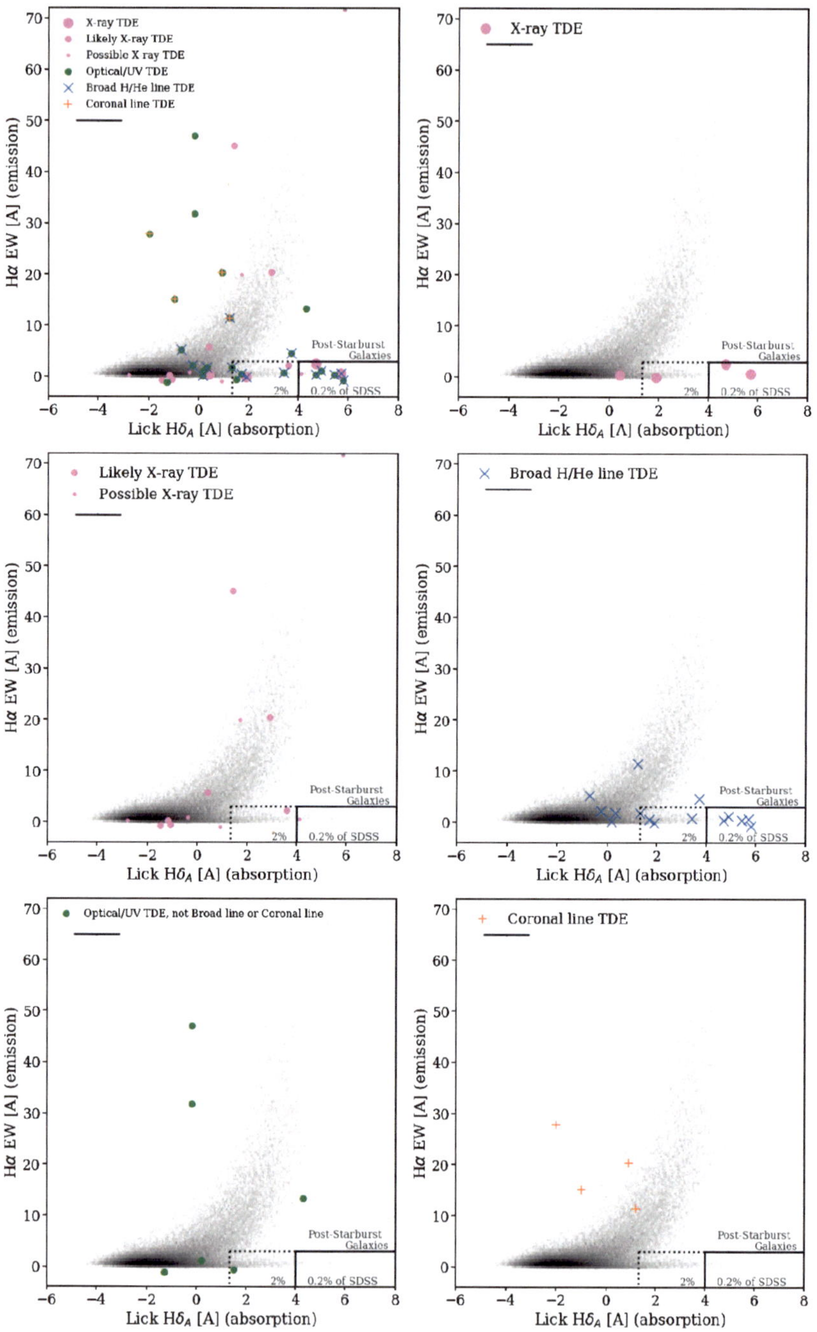

Fig. 5 Spectral indices tracing the recent star formation histories of TDE hosts and SDSS galaxies (updated from French et al. 2016; Graur et al. 2018). We plot Hα EW (sensitive to current star formation) vs. Lick $H\delta_A$ absorption (sensitive to star formation over the past Gyr) for each galaxy. Galaxies with low current star formation, yet significant star formation over the past Gyr form the post-starburst/quiescent Balmer-strong "spur" in the lower right. TDEs, especially the X-ray and broad H/He line classes, are over-represented among the post-starburst and quiescent Balmer-strong galaxies

Table 5 Summary of TDE rate overenhancement found in various samples of galaxies and TDE classifications

Overenhancement	Galaxy Sample[a]	TDE Sample[b]	Source
$33^{+7}_{-11}\times$	QBS	H/He broad line	[1]
$190^{+115}_{-100}\times$	PSB	H/He broad line	[1]
$34^{+24}_{-14}\times$	QBS	H/He broad line	[2]
$110^{+80}_{-50}\times$	PSB	H/He broad line	[2]
$18^{+13}_{-9}\times$	QBS	X-ray, likely, possible	[2]
$18^{+22}_{-18}\times$	PSB	X-ray, likely, possible	[2]
$23^{+21}_{-13}\times$	QBS	X-ray, likely	[2]
$29^{+41}_{-29}\times$	PSB	X-ray, likely	[2]
$17^{+12}_{-8}\times$	QBS	Optical	[2]
$50^{+38}_{-29}\times$	PSB	Optical	[2]
$18^{+8}_{-7}\times$	QBS	X-ray, likely, possible, optical	[2]
$35^{+21}_{-17}\times$	PSB	X-ray, likely, possible, optical	[2]
20-80×	QBS/PSB	X-ray, likely, possible, optical	[3]
40-120×	QBS/PSB	X-ray, H/He broad line	[3]

[a]We note that the definitions of Quiescent Balmer-Strong (QBS) and Post-Starburst (PSB) vary slightly between French et al. (2016) and Graur et al. (2018), and in this review we present the overenhancement as a function of the Balmer strength (see Fig. 7). [b]Similarly, the TDEs used in each classification vary. For the TDEs included in the two calculations for this review, see Table 1. [1] French et al. (2016) [2] Graur et al. (2018) [3] This review

(2016) find that 38% of a sample of eight UV/optical H/He broad line TDE hosts meet a post-starburst selection criterion with a rate of only 0.2% in the general galaxy population. Similarly, 75% of the same TDE host galaxies meet a quiescent Balmer-strong selection criterion with a rate of 2.3% in the general galaxy population. These rates imply overdensities of $33^{+7}_{-11}\times$ in quiescent Balmer-strong galaxies and $190^{+115}_{-100}\times$ in post-starburst galaxies.

Graur et al. (2018) considered the over-enhancement rates for several additional categories of observed TDEs using a similar but slightly different parent galaxy sample and post-starburst/ quiescent Balmer-strong definitions. For an updated sample of UV/optical bright H/He broad line TDEs, the over-enhancement rates are $34^{+22}_{-14}\times$ in quiescent Balmer-strong galaxies and $110^{+80}_{-50}\times$ in post-starburst galaxies, consistent with the rate enhancements found by French et al. (2016).

The over-enhancement rates in post-starburst galaxies for the X-ray bright TDEs are weaker than for the UV/optical broad line TDEs, though this comparison is limited by small number statistics. For the set of X-ray TDEs, "likely" X-ray TDEs, and "possible" X-ray TDEs identified in Auchettl et al. (2017a), Graur et al. (2018) find the over-enhancement rates to be $18^{+13}_{-9}\times$ in quiescent Balmer-strong galaxies and $18^{+22}_{-18}\times$ in post-starburst galaxies. These rates are higher once the "possible" X-ray TDEs are excluded, many of which have ambiguous light curves and may be AGN flares. Considering only the X-ray and "likely" X-ray TDEs, the over-enhancement rates are $23^{+21}_{-13}\times$ in quiescent Balmer-strong galaxies and $29^{+41}_{-29}\times$ in post-starburst galaxies. These rate enhancements for

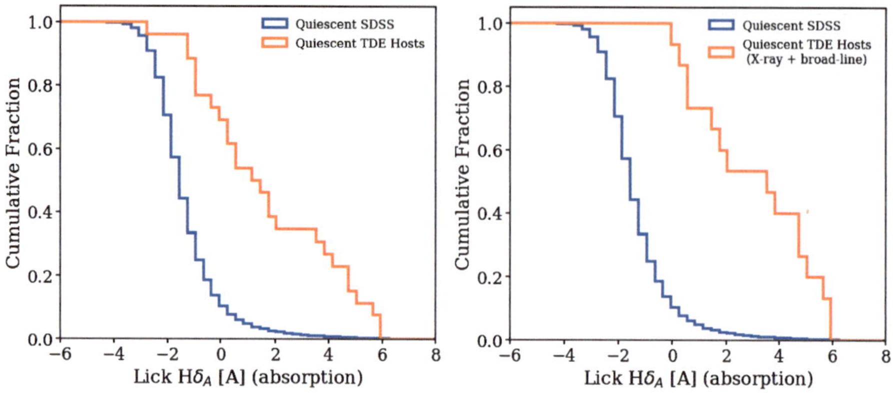

Fig. 6 Cumulative distribution of quiescent SDSS galaxies and quiescent TDE hosts with stronger Balmer absorption than the limit of the x-axis. For both the full set of TDE hosts considered here (left) as well as the subsets of the broad H/He line TDEs and the X-ray TDEs with the strongest post-starburst enhancement (right), there is a significant difference in the distributions between the quiescent SDSS galaxies and TDE hosts

the post-starburst sample are driven by the one X-ray (including the "likely" and "possible" samples) TDE that meets the strictest post-starburst criterion, ASASSN-14li.

We present classifications for the TDE hosts discussed in this review in Table 4, using the criteria described above. 5/41 (12%) host galaxies are post-starburst galaxies and 13/41 (32%) are either quiescent Balmer-strong or post-starburst. Of the 4 X-ray TDEs, 3 (75%) are quiescent Balmer-strong and 1 (25%) is post-starburst. Of the 15 broad H/He line TDEs, 9 (60%) are quiescent Balmer-strong and 5 (33%) are post-starburst.

To account for the dependence of the TDE rate enhancement on the definition of "post-starburst" or "quiescent Balmer-strong" we plot in Fig. 6 the cumulative distribution of quiescent SDSS galaxies and quiescent TDE hosts with stronger Balmer absorption than the value on the x-axis. For both the full set of TDE hosts considered here as well as the subsets of the broad H/He line TDEs and the X-ray TDEs with the strongest post-starburst enhancement, there is a significant difference in the distributions between the quiescent SDSS galaxies and TDE hosts. We also demonstrate the effect of the criteria for post-starburst or quiescent Balmer-strong on the TDE enhancement rate over normal quiescent galaxies in Fig. 7. For the full set of TDE hosts, the enhancement rate ranges between 20-80× that of normal quiescent galaxies. For the broad H/He line TDEs and the X-ray TDEs, the enhancement rate ranges from 40-120× that of normal quiescent and star-forming galaxies. We note again that these TDE classifications are tentative and subject to a number of observational biases. More observations of the host galaxies for a large sample of well-characterized TDEs are needed to overcome these uncertainties and the small-number statistics limiting our precision here.

Law-Smith et al. (2017b) consider the post-starburst and quiescent Balmer-strong galaxy over-enhancement rates in a somewhat different subset of the Auchettl et al. (2017a) X-ray and optical TDEs, but control the galaxy parent sample on different properties to test whether selection effects drive the observed rate enhancements. Controlling on bulge mass (and thus likely black hole mass), redshift, surface brightness, Sérsic index, or bulge to total light ratio can affect the relative post-starburst or quiescent Balmer-strong galaxy rate by $\leq 2\times$, within the error on the rates due to small number statistics.

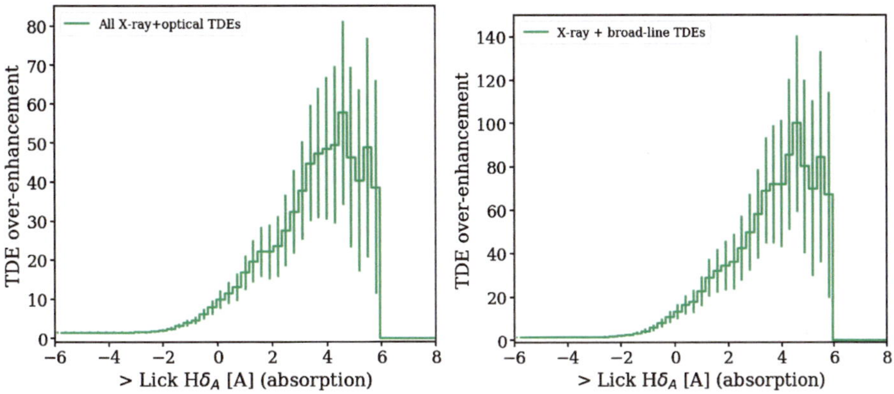

Fig. 7 Effect of the criteria for post-starburst or quiescent Balmer-strong on the TDE enhancement rate over normal quiescent and star-forming galaxies. For the full set of TDE hosts (left), the enhancement rate ranges for 20-80× that of normal quiescent and star-forming galaxies. For the broad H/He line TDEs and the X-ray TDEs (right), the enhancement rate ranges for 40-120× that of normal quiescent and star-forming galaxies. We note again that these TDE classifications are tentative and subject to a number of observational biases. More observations of the host galaxies for a large sample of well-characterized TDEs are needed to overcome these uncertainties and the small-number statistics limiting our precision here

Controlling on bulge colour has the largest possible effect on the observed post-starburst or quiescent Balmer-strong rates, decreasing the observed rate enhancements by factors of ~ 4 for the post-starburst galaxies and ~ 2.5 for the quiescent Balmer-strong galaxies, depending on what other factors are controlled for. This may be due to selection effects against finding TDEs in dustier and thus redder galaxies, or by the unique stellar populations in post-starburst galaxies, which cause them to lie in the optical green valley (Wong et al. 2012). If we control on properties which are strongly correlated with post-starburst and quiescent Balmer-strong galaxies, such as green optical colors or high central concentrations, we expect that the residual enhancement of the TDE rate in such hosts would be diminished by construction. The TDE enhancement rates in such hosts are thus consistent (given the small number statistics) between the studies of French et al. (2016), Law-Smith et al. (2017b) and Graur et al. (2018).

While the Balmer absorption is a proxy for the recent SFH, detailed stellar population fitting can be used to determine the nature of the recent SFH using the information from the full galaxy SED. French et al. (2017) fit stellar population models to the UV/optical host galaxy photometry and optical Lick indices to determine the time elapsed since the recent starburst, the fraction of mass produced in the starburst, and the duration of the recent starburst (French and Zabludoff 2018), for a sample of host galaxies of broad H/He line TDEs. While the lower Balmer absorption "quiescent Balmer-strong" galaxies could have had weaker Hδ absorption due to longer-duration bursts, older bursts, or weaker bursts, stellar population modelling by French et al. (2017) determined that this effect is driven by the fact that quiescent Balmer-strong TDE hosts had weaker starbursts than most post-starburst galaxies. While most post-starburst galaxies have formed $\sim 3-50\%$ of their stellar mass in their recent starbursts, the quiescent Balmer-strong (and non-post-starburst) hosts had burst mass fractions of $0.5-2\%$. The post-starburst and quiescent Balmer-strong TDE hosts have ages ranging from 60 Myr to 1 Gyr since the recent starbursts ended. This large range in age suggests the enhanced TDE rate is not limited to a specific time in their host's post-starburst evolution. When compared to the total sample of post-starburst and quiescent Balmer-strong galaxies, French et al. (2017) observed a statistically insignificant dearth of TDEs in older (> 600 Myr) post-starburst galaxies. The evolution of the TDE rate enhancement after a starburst implied by this early small sample is consistent with several models for explaining the enhanced TDE rate during this phase (Stone et al. 2018, discussed further in §4).

TDEs may also be over-represented in starbursting host galaxies. However, the extreme dust extinction present in the nucleus of starburst galaxies as well as the co-existence of AGN activity make detecting such TDEs difficult. Either a lucky dust-free sightline or transient detections in the NIR are required to find TDEs in starburst galaxies. Both such scenarios have been observed. Tadhunter et al. (2017) observed a light-curve over 10 years and a serendipitous appearance of broad He lines similar to those observed in other TDEs in a starburst galaxy.[3] Mattila et al. (2018) observed a jet launched from the nucleus of the starburst galaxy Arp 299 believed to be caused by a TDE, with a transient discovered in NIR AO imaging. Arp 299 has a stellar population consistent with evolving to a typical post-starburst galaxy, with an starburst age of 70-260 Myr since the starburst began and 9-29% of the total stellar mass formed in the on-going starburst (Pereira-Santaella et al. 2015).

Tadhunter et al. (2017) estimate the TDE rate to be enhanced in such galaxies by 1000-10,000×, to one per century or even one per decade per galaxy. From the observations thus far, it is unclear whether the TDE rate is enhanced during both the starburst and post-starburst phases, with selection effects biasing against observing TDEs in starburst galaxies, or if the TDE rate enhancement peak lags in time after the starburst. Upcoming infrared and radio surveys for TDEs, as well as concerted efforts to disambiguate TDEs from AGN will be necessary to resolve this question.

2.3 Concentration and Morphology of Stellar Light; Stellar Kinematics

The TDE rate is expected to depend on various physical properties of the SMBH and the stellar population in its vicinity, including the mass of the SMBH, the density of stars within its loss cone, and their velocity dispersion. Unfortunately, it is exceedingly hard to observationally probe the parsec-scale region of influence, except for the most nearby SMBHs. However, some global host-galaxy properties, on kpc scales, are known to be correlated with local properties in galactic nuclei. The most fundamental of these is the M–σ relation, which relates the mass of the SMBH, M, to the host galaxy stellar velocity dispersion, σ (e.g., Kormendy and Richstone 1995, Magorrian et al. 1998, Gebhardt et al. 2000, Tremaine et al. 2002, McConnell and Ma 2013). Moreover, these central stellar velocity dispersions have been shown to be correlated over galactic scales (Cappellari et al. 2006).

Graur et al. (2018) compared a sample of 11 TDE host galaxies with surface stellar mass densities and velocity dispersions computed from galaxy properties measured by the Sloan Digital Sky Survey (SDSS; York et al. 2000, Kauffmann et al. 2003a, Brinchmann et al. 2004) to a volume-limited sample of SDSS galaxies with galaxy properties measured by the same pipeline. Their TDE host galaxies had surface stellar mass densities in the range $\Sigma_{M_\star} = 10^{9-10}\ M_\odot\ \mathrm{kpc}^2$. The star-forming TDE hosts were significantly denser than the star-forming control sample. This effect was not significant for quiescent galaxies, which already tend to have high surface stellar mass densities. Graur et al. (2018) also measured surface stellar mass densities for a similar sample of 9 TDE host galaxies with velocity dispersions measured by Wevers et al. (2017), and found that they too had values in the range

[3]This event is included in our catalogue (F01004), although its classification as a TDE is controversial. Trakhtenbrot et al. (2019) argue it may not be a true TDE, although the space for observed TDE features may be broader than expected (Leloudas et al. 2019).

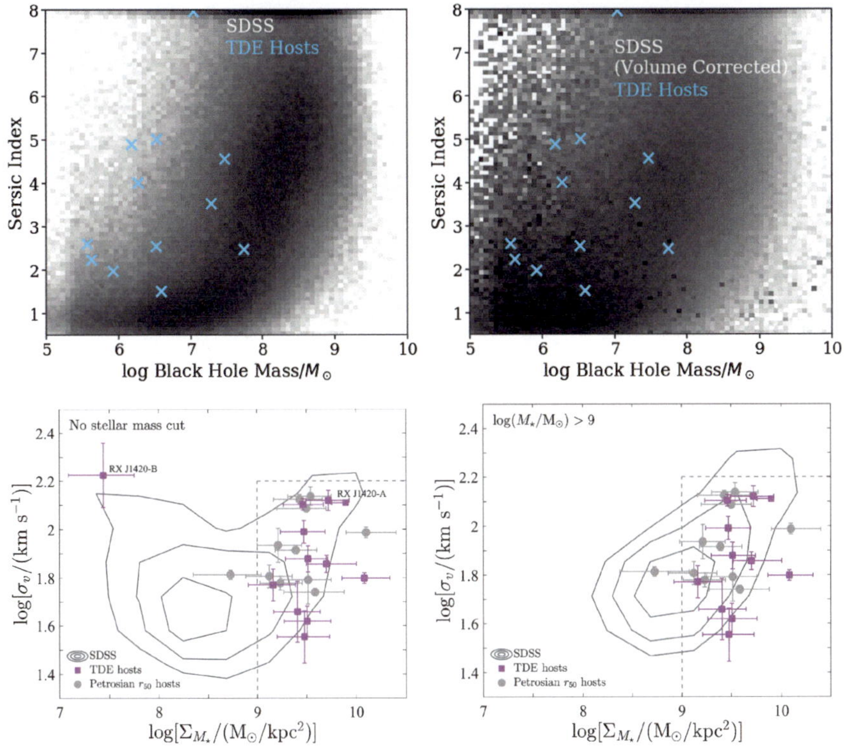

Fig. 8 Top Left: Black hole masses inferred from the $M - \sigma$ relation vs. Sérsic indices from Simard et al. (2011) for the SDSS main galaxy sample and the SDSS TDE hosts, adapted from (Law-Smith et al. 2017b). The TDE hosts have high Sérsic indices for their black hole masses. Top Right: Adapted from Law-Smith et al. (2017b) with volume corrections from Mendel et al. (2014) as done in the bottom left plot. Despite the large number of low black hole mass galaxies inferred from the volume correction, the TDE hosts still lie at high Sérsic indices for their black hole masses, compared with the rest of the galaxy sample. Bottom Left: Updated from Graur et al. (2018), TDE host galaxies (symbols) significantly prefer galaxies with high surface stellar mass densities, relative to a volume-limited SDSS sample (contours). New measurements for NGC 5905 and SDSS J1201 have been added to these plots. Contours represent the 25th, 50th, and 84th percentiles of the volume-weighted background galaxy distribution. Host galaxies marked by purple squares have galaxy properties measured by the same pipeline as the SDSS control sample. Host galaxies marked by grey circles differ from the SDSS control sample because their stellar surface densities are measured using Petrosian radii rather than Sérsic half-light radii, and some have velocity dispersions measured by Wevers et al. (2017) instead of the SDSS pipeline. The TDE host galaxies are significantly denser than the control sample (although this trend depends on whether the galaxies are star-forming or not – see text). Bottom Right: (Reproduced by permission of the AAS.) From Graur et al. (2018), same as the previous panel, but with a stellar mass cut to include only galaxies with similar stellar masses as the TDE host galaxies. Contours again represent the 25th, 50th, and 84th percentiles of the volume-weighted background galaxy distribution. As before, the TDE host galaxies have a significantly higher distribution of stellar surface mass densities, even after removing the sample of low mass, low surface density galaxies magnified by the volume correction. While the axes in these plots are not strictly analogous to those in the top plots, both analyses demonstrate that the TDE hosts have higher stellar concentrations than expected given their velocity dispersion or inferred black hole mass. Data from this figure can be found in Table 6

10^{9-10} $M_\odot$ kpc^2 with one exception: PS1-10jh, which had a surface stellar mass density of $\log(\Sigma_{M_\star}/M_\odot\ \text{kpc}^2) = 8.7^{+0.3}_{-0.4}$. Both of these samples, in purple and gray markers, respectively, are shown in Fig. 8. Because the volume-corrected quiescent galaxies have a different

Table 6 Table of properties of known TDE host galaxies discussed in §2.3

Name	Sérsic n[a]	M_{BH}[b]	A[c]	$\log(\Sigma_{M_\star})$[d]	σ_v[e] [km/s]
X-ray TDEs					
ASASSN-14li†	4.91	6.17	0.015	$10.1^{+0.2}_{-0.2}$	63 ± 3
Likely X-ray TDEs					
SDSSJ1201	5.61	–	0.008	$9.5^{+0.2}_{-0.3}$	122 ± 4
SDSSJ1323	5.03	6.52	0.002	$9.5^{+0.2}_{-0.2}$	75 ± 10
3XMM J1521	–	–	–	$9.5^{+0.2}_{-0.3}$	62 ± 14
Possible X-ray TDEs					
RX J1420-A	–	–	–	$9.7^{+0.2}_{-0.2}$	131 ± 13
RX J1420-B	–	–	–	$7.4^{+0.3}_{-0.3}$	168 ± 52
SDSSJ0159	–	–	–	< 9.9	128 ± 17
RBS1032	2.24	5.62	0.018	$9.5^{+0.2}_{-0.2}$	36 ± 9
NGC 5905	–	–	–	$10.1^{+0.3}_{-0.2}$	97 ± 5
GALEX D3-13	–	–	–	$9.4^{+0.2}_{-0.2}$	133 ± 6
Optical/UV TDEs					
iPTF16fnl†	–	–	–	$9.6^{+0.3}_{-0.3}$	55 ± 2
PS16dtm	–	–	–	$9.4^{+0.2}_{-0.2}$	45 ± 13
GALEX D23-H1	–	–	–	$9.2^{+0.2}_{-0.2}$	86 ± 14
SDSSJ0952*	7.98	7.04	0.011	< 10.4	–
SDSSJ1342*	2.56	6.51	0.012	$9.7^{+0.3}_{-0.2}$	72 ± 6
SDSSJ1350*	4.57	7.47	0.003	$9.3^{+0.3}_{-0.3}$	–
SDSS TDE1	–	–	–	$9.5^{+0.2}_{-0.2}$	137 ± 12
SDSSJ0748†,*	1.53	6.58	0.011	$9.5^{+0.2}_{-0.2}$	126 ± 7
ASASSN-14ae†	2.61	5.56	0.013	$9.5^{+0.2}_{-0.2}$	41 ± 6
PTF09axc†	–	–	–	$9.2^{+0.2}_{-0.2}$	60 ± 4
PTF09djl†	–	–	–	$9.1^{+0.3}_{-0.3}$	64 ± 7
PTF-09ge†	4.03	6.26	0.019	$9.2^{+0.2}_{-0.3}$	59 ± 9
PS1-10jh†	–	–	–	$8.7^{+0.3}_{-0.4}$	65 ± 3
iPTF15af†	3.54	7.29	0.01	$9.5^{+0.2}_{-0.2}$	98 ± 11
iPTF16axa	–	–	–	$9.4^{+0.2}_{-0.2}$	82 ± 3
AT2018dyk†	2.48	7.74	0.034	–	–
ASASSN18zj†	1.98	5.92	0.009	–	–

[a]Sérsic indices from Mendel et al. (2014) fits. [b]Black hole masses from Law-Smith et al. (2017b). [c]Residual asymmetry measures from Simard et al. (2011). [d]Surface stellar mass density, computed as $\Sigma_{M_\star} = \log[(M_\star/r_{50}^2)/(M_\odot/\mathrm{kpc}^2)]$ by Graur et al. (2018). [e]Stellar velocity dispersions from SDSS Portsmouth pipeline or Wevers et al. (2017). †Broad line TDEs (see Table 1). *Coronal line TDEs (see Table 1)

stellar mass distribution than the star-forming galaxy sample, we also compare the $\Sigma_{M_\star}$ and velocity dispersion for a comparison sample cut in stellar mass to be $M_\star > 10^9\ M_\odot$ in order to match the TDE host galaxies. Even after removing the low mass galaxies, the TDE hosts still have higher stellar surface densities than the volume-corrected comparison sample.

While the preference of TDE hosts for galaxies with high surface stellar mass densities was statistically significant (at least for star-forming galaxies), there was no significant dependence on the galaxy central stellar velocity dispersion. Only the quiescent host galaxies showed a hint that their velocity dispersions might be lower than those of the quiescent galaxies in the control sample. It remains to be seen whether this effect proves to be significant in a larger sample.

Law-Smith et al. (2017b) also compared kpc-scale indicators of stellar mass concentration for a sample of 10 TDE hosts with data from the Simard et al. (2011) and Mendel et al. (2014) SDSS catalogues, comparing the Sérsic indices of the TDE host galaxies and SDSS comparison galaxies to the black hole masses inferred from the $M - \sigma$ relation. The TDE host galaxies have Sérsic indices in the top 10–15% of the comparison sample in bins of black hole mass, indicating the TDE host galaxies have more concentrated stellar populations. In this review, we add data for two new TDEs (ASASSN-18zj and AT2018dyk) with archival SDSS information. We also perform a volume correction for the SDSS comparison sample using the volume calculations of Mendel et al. (2014) in order to compare this analysis to that of Graur et al. (2018). The updated Sérsic index–black hole mass plots are shown in Fig. 8. The volume correction accounts for the larger number of galaxies with low black hole mass, but the same trend of TDE host galaxies having higher Sérsic indices for their black hole masses is seen. We find that 50% of the TDE host galaxies have Sérsic indices in the top 20% of the volume-corrected SDSS galaxies with $M_{BH}\ 10^5 - 10^6\ M_\odot$, top 10% of the volume-corrected SDSS galaxies with $M_{BH}\ 10^6 - 10^7\ M_\odot$, and top 30% of the volume-corrected SDSS galaxies with $M_{BH}\ 10^7 - 10^8\ M_\odot$. If we only compare to the volume-corrected SDSS galaxies with quiescent levels of star formation (SFR$< 1\ M_\odot\ yr^{-1}$), we find the same result for $M_{BH}\ 10^5 - 10^7\ M_\odot$, but no significant enhancement of TDE hosts in higher Sérsic index galaxies with black hole masses $M_{BH}\ 10^7 - 10^8\ M_\odot$.

A similar trend is also seen if the bulge to total light ratio is used as a proxy for stellar concentration instead of the Sérsic index (Law-Smith et al. 2017b). However, the Sérsic index measurements from Simard et al. (2011) have lower errors and thus allow finer binning and a more detailed comparison for our analyses.

The analyses by Law-Smith et al. (2017b) and Graur et al. (2018) have established that TDE hosts are more concentrated on galaxy-wide (kpc) scales. We discuss possible mechanisms for the stellar concentration affecting the TDE rate, and the interplay between this effect and the trend with star formation history in Sect. 3.1, as post-starburst and quiescent Balmer-strong galaxies are also known to have high central concentrations of stellar light.

We also consider here the quantitative morphologies of the TDE host galaxies in asymmetry - concentration space. Spiral galaxies and elliptical galaxies separate in this space with elliptical galaxies having higher concentrations and spiral galaxies having higher asymmetries (Abraham et al. 1996). Mergers can be further identified, with higher asymmetries than individual galaxies (Conselice et al. 2003). Post-starburst galaxies often show signs of recent mergers, but at several hundred Myr past coalescence, their asymmetries as measured using HST imaging have lessened to be between those of elliptical and spiral galaxies (Yang et al. 2008, Fig. 5 reproduced in this review).

Law-Smith et al. (2017b) have compiled a sample of morphological indicators for the TDE host galaxies as well as the SDSS main spectroscopic sample, using the catalogues of Simard et al. (2011). We compare the concentration to the residual asymmetries (Simard

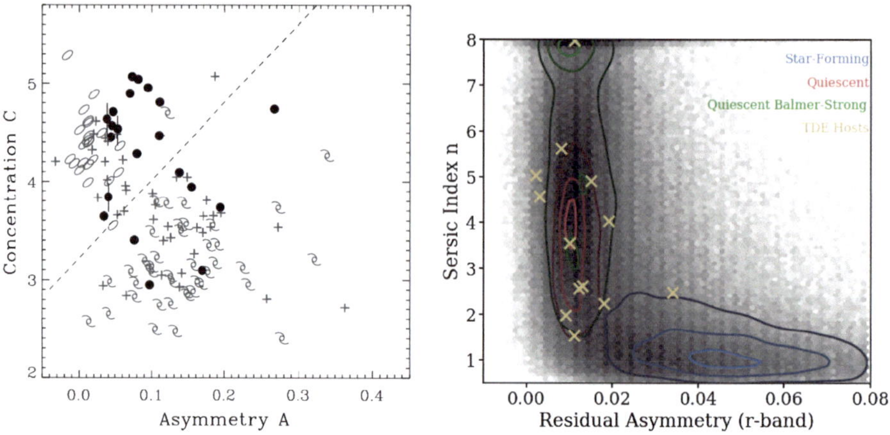

Fig. 9 Left: (Reproduced by permission of the AAS.) From Yang et al. (2008), showing the concentration and asymmetry trends for star-forming (spiral symbols), quiescent (elliptical symbols), intermediate spirals (plus signs), and post-starburst galaxies (filled circles) from HST imaging. The post-starburst sample has high concentrations of stellar light, comparable to or higher than the quiescent galaxies, and asymmetries stronger than the quiescent galaxies, but less pronounced than the star-forming galaxies. Right: Adapted from Law-Smith et al. (2017b), we plot the Sérsic index vs. the r-band residual asymmetry. In contrast to the left-hand panel, the residual asymmetries are calculated after the best-fit smooth galaxy model has already been subtracted. The residual asymmetries are calculated by Simard et al. (2002, 2011) and are denoted RA1_2 in the GIM2D output tables. Contours are plotted for the subsamples of star-forming (blue), quiescent (red), and quiescent Balmer-strong galaxies (green). TDE hosts are plotted as yellow crosses. The star-forming galaxies have high asymmetries and low Sérsic indices, whereas the quiescent galaxies have Sérsic indices of $\sim 3-5$ and low asymmetries. The quiescent Balmer-strong galaxies show high Sérsic indices, with a tail extending down to the quiescent galaxies. The TDE hosts follow the quiescent galaxy distribution, with one outlier (SDSSJ0952) at high Sérsic index ($n \sim 8$). Higher resolution imaging is required to determine whether the TDE host galaxies show the trend towards intermediate asymmetries indicative of recent mergers, as seen in the HST imaging of post-starburst galaxies. Data from this panel can be found in Table 6

et al. 2002) for the SDSS galaxies and TDE host galaxies in Fig. 9, using the Sérsic index as a proxy for concentration. We separate out star-forming, quiescent, and quiescent Balmer-strong galaxies to identify trends in this space. The star-forming galaxies have high asymmetries and low Sérsic indices, while the quiescent galaxies have Sérsic indices of $\sim 3-5$ and low asymmetries. The quiescent Balmer-strong galaxies show high Sérsic indices,[4] with a tail extending down to the quiescent galaxies, and low asymmetries. The TDE hosts are distributed like the quiescent galaxies, with one source (SDSSJ0952) having a high Sérsic index of $n \sim 8$. The shift towards higher asymmetry for the post-starburst or quiescent Balmer-strong galaxies is not observed in the residual asymmetries from the SDSS imaging as it was in the total light asymmetries from the HST imaging; this is likely due to the greater sensitivity of the HST data to low surface brightness tidal features. Yang et al. (2008) found that many of the tidal features observed with HST imaging would not be observable with ground-based imaging. Thus, the lack of high asymmetries in the TDE host sample does not rule out a recent merger, even a recent major merger. Higher resolution imaging and a variety of new measures of galaxy asymmetry (e.g., Pawlik et al. 2015) will be required to

[4] See above for a more thorough discussion of the Sérsic indices of TDE hosts, accounting for trends in stellar mass and black hole mass.

determine whether the TDE host galaxies have the trend towards intermediate asymmetries indicative of recent mergers, as seen in the HST imaging of post-starburst galaxies.

2.4 AGN Activity

We consider here the possibility that some TDEs may occur in an environment with a pre-existing accretion disk. We have compiled a BPT (Baldwin et al. 1981) diagram in Fig. 10 showing the TDE host galaxies as well as galaxies from the SDSS main spectroscopic survey. We include TDE host galaxies with SDSS spectra as described above, as well as galaxies with emission line ratios measured by French et al. (2017) and Wevers et al. (2019a). These emission line fluxes are shown in Table 7. We classify galaxies into star-forming, composite, and AGN Seyfert II or LINER based on the classifications of Kewley et al. (2001) and Kauffmann et al. (2003b). These classifications are subject to a number of caveats, and represent the likely dominant ionisation source in the aperture probed by the spectrum. However, many galaxies in the AGN region of the BPT diagram, especially those with relatively weak emission lines, may instead have ionisation consistent with an origin from shocks or evolved stars (e.g., Rich et al. 2015; Yan and Blanton 2012). "Composite" galaxies lie in between the star-forming and AGN regions and could be a mix of star-formation and other ionisation sources. The TDE host galaxies occupy a range of star-formation dominated, AGN-dominated, and ambiguous ionisation source galaxies.

Another way of identifying AGN, especially those obscured by dust, is to look for signatures of hot dust from the WISE 3.4–4.6 μm colours. Stern et al. (2012) identify a WISE colour cut of WISE 3.4–4.6$>$ 0.8 Vega mag to indicate the presence of an AGN. We present these WISE colours in Table 7, identifying four TDE hosts which meet this criterion: the hosts of F01004, SDSS J0952, SDSS J1342, and SDSS J1350. The first host galaxy is currently experiencing a starburst (Tadhunter et al. 2017). The latter three galaxies all hosted TDEs with observed coronal line emission.

The BPT analysis described above selects narrow-line AGN (Seyfert II galaxies), or obscured AGN, as does the infrared selection. When the broad-line regions of AGN are visible, this provides another way to identify them from their optical spectra. In Table 7, we also list notes from the SDSS to indicate broad-line or QSO emission. Five TDE host galaxies have such notes, with varying overlap with those galaxies classified as AGN from the BPT or WISE colour analyses. AGN selection using any of these methods is neither pure nor complete. One may note that the host galaxy of PS16dtm, a Narrow-line Seyfert I galaxy (Blanchard et al. 2017), is not selected as an AGN using a BPT or WISE colour analysis, and requires further analysis of the optical spectrum to identify the broad components of the Balmer lines. The connection between TDEs and Narrow line Seyfert I galaxies requires further study (Wevers et al. 2019b). Such galaxies make up only $\sim 15\%$ of all Seyfert I galaxies (Williams et al. 2002) and have been observed to have optical flares similar to TDEs (Kankare et al. 2017). Some of these events may even belong to a different class of transient (Frederick et al. 2019).

Caution should be taken in interpreting these results given the selection effects against identifying TDEs in AGN host galaxies. We discuss the role of AGN in either enhancing the TDE rate or as the source of selection effects against identifying TDEs in such host galaxies in §3.3. TDEs will be more difficult to identify in AGN due to selection against AGN flares and higher levels of dust obscuration. These effects may also bias the types of AGN TDEs are found in. Further study will be needed to fully understand these effects.

Furthermore, we note that spectra taken after the TDE may be contaminated by residual TDE emission, depending on how long the emission persists for. Brown et al. (2016) found

Table 7 Table of properties of known TDE host galaxies discussed in §2.4. Fluxes are in units of 10^{-17} ergs/s/cm^2

Name	Hα	Hβ	[NII]6584	[SII]6717+6731	[OIII]5007	BPT Class[a]	W1-W2[b]	SDSS Notes[c]
X-ray TDEs								
ASASSN14li[†]	47.7	22.8	42.5	17.7	89.3	AGN/Seyfert	0.01 ± 0.04	GALAXY BROADLINE
Swift J1644	–	–	–	–	–	–	–	
XMM J0740	–	–	–	–	–	–	0.01 ± 0.03	
ASASSN15oi[†]	–	–	–	–	–	–	0.06 ± 0.06	
Likely X-ray TDEs								
SDSS J1201	–	–	–	–	–	–	0.22 ± 0.07	
2MASX J0249	40.5	11.0	10.9	8.7	27.1	composite	0.05 ± 0.04	
PTF10iya	–	–	–	–	–	–	0.73 ± 0.14	
SDSS J1311	–	–	–	–	–	–	–	
SDSS J1323	2.5	4.8	–	5.3	0.6	–	0.08 ± 0.06	QSO
3XMM J1521	–	–	–	–	–	–	0.07 ± 0.26	
3XMM J1500	–	–	–	–	–	–	0.37 ± 0.46	
PS18kh[†]	–	–	–	–	–	–	-0.13 ± 0.12	
Possible X-ray TDEs								
RX J1242-A	–	–	–	–	–	–	-0.03 ± 0.04	
RX J1242-B	–	–	–	–	–	–	–	
RX J1420-A	5.0	2.5	1.2	5.1	0.4	SF	0.12 ± 0.06	ROSAT D
RX J1420-B	87.0	64.7	13.2	203.7	45.6	SF	–	ROSAT D
SDSS J0159	30.5	6.5	19.2	12.4	13.4	AGN/Seyfert	0.53 ± 0.07	QSO STARBURST BROADLINE
RBS 1032	6.7	2.3	3.6	1.7	6.3	AGN/Seyfert	0.09 ± 0.05	GALAXY
RX J1624	–	–	–	–	–	–	0.06 ± 0.05	
NGC 5905	23.6	6.4	13.5	6.4	2.9	composite	0.15 ± 0.03	

Table 7 (*Continued*)

Name	Hα	Hβ	[NII]6584	[SII]6717+6731	[OIII]5007	BPT Class[a]	W1-W2[b]	SDSS Notes[c]
Optical/UV TDEs								
iPTF16fnl[†]	–	–	–	–	–	–	0.01 ± 0.05	
PS16dtm[d]	287.3	82.5	70.9	82.5	166.4	SF	0.44 ± 0.07	GALAXY STARFORMING
F01004	–	–	–	–	–	–	*1.63 ± 0.03*	
GALEX D23H-1	21.2	5.9	15.2	7.6	2.1	composite	0.36 ± 0.13	
PS1-11af	–	–	–	–	–	–	–	
SDSS J0952*	620.7	188.4	580.4	77.9	163.4	composite	*1.02 ± 0.04*	QSO BROADLINE
SDSS J1342*	311.0	70.6	62.3	55.4	82.8	SF	*1.08 ± 0.03*	GALAXY STARFORMING
SDSS J1350*	202.7	48.4	193.1	36.0	49.3	AGN/LINER	*0.87 ± 0.03*	GALAXY BROADLINE
SDSS TDE1	–	–	–	–	–	–	0.2 ± 0.24	
SDSS TDE2[†]	–	–	–	–	–	–	0.52 ± 0.14	
SDSS J0748[†,*]	149.6	51.6	55.9	46.8	11.4	SF	0.77 ± 0.03	GALAXY STARFORMING
ASASSN14ae[†]	13.2	6.0	7.3	1.9	37.8	AGN/Seyfert	0.15 ± 0.05	GALAXY
PTF09axc[†]	–	–	–	–	–	–	0.15 ± 0.08	
PTF09djl[†]	–	–	–	–	–	–	0.33 ± 0.56	
PTF09ge[†]	17.8	1.8	16.8	13.1	12.1	AGN/Seyfert	0.13 ± 0.04	GALAXY
PS1-10jh[†]	–	–	–	–	–	–	<0.88	
iPTF15af[†]	9.0	–	–	–	2.4	–	0.18 ± 0.08	GALAXY
iPTF16axa	–	–	–	–	–	–	0.29 ± 0.18	
AT2018dyk[†]	120.4	20.3	182.2	98.4	67.6	AGN/LINER	0.03 ± 0.03	GALAXY
AT2018bsi[†]	–	–	–	–	–	–	0.09 ± 0.05	
ASASSN18zj[†]	9.3	7.9	5.4	–	8.7	composite	0.15 ± 0.05	GALAXY

[a]Classification from [NII]-BPT diagram. [b]WISE 3.4 μm - 4.6 μm colours. We *italic* those values that meet the Stern et al. (2012) WISE AGN criteria. [c]SDSS `targettype` and `subclass` notes indicating the presence of broadline AGN or QSO like spectra. [d]While the host galaxy of PS16dtm is classed as SF on the [NII]-BPT diagram, it is a Seyfert I (Blanchard et al. 2017). [†]Broad line TDEs (see Table 1). *Coronal line TDEs (see Table 1)

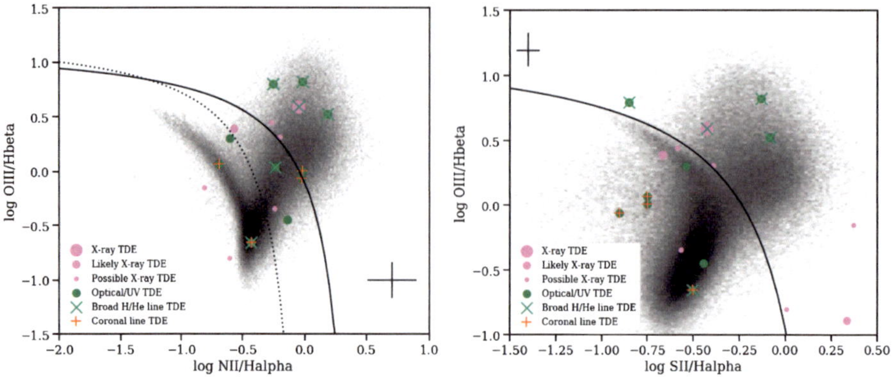

Fig. 10 BPT (Baldwin et al. 1981) diagrams of emission line ratios indicative of ionisation from AGN or star formation in the TDE host galaxies and SDSS main galaxy spectroscopic sample. Galaxies to the lower left of the dotted and solid lines have ionisation dominated by star formation. Galaxies to the upper right of the dotted lines have ionisation dominated by AGN, although those with relatively weak emission lines may instead have ionisation from shocks or evolved stars (e.g., Rich et al. 2015; Yan and Blanton 2012). The dotted line is the observed star formation - AGN separation from Kauffmann et al. (2003b) and the solid lines are the theoretical maximum starburst lines from Kewley et al. (2001). Characteristic error bars are shown in each panel. The TDE host galaxies occupy a range of star-formation dominated, AGN-dominated, and ambiguous hosts

that narrow Hα emission can persist for a year after the TDE, but after several years, Wevers et al. (2019a) find no residual narrow line emission. French et al. (2017) noted a tentative offset in Hα equivalent width and [NII]-6584/Hα emission ratio between the few events with spectroscopy before vs. after the TDE. French et al. (2017) found the host galaxies with spectroscopy from after the TDE to have higher Hα equivalent widths and [NII]-6584/Hα emission ratios than the host galaxies with spectroscopy from before the TDE, although this analysis was limited by the small number of events and the lack of events with spectra from both before and well-after the TDEs. In the sample considered in this review, we note the host galaxies with spectra taken before the TDE contain more Seyferts and the host galaxies with spectra taken after the TDE contain more star-forming and composite classifications. However, this comparison is still limited by small number statistics and selection effects between the various TDE detection methods used. A systematically collected set of follow-up spectra will be needed to better understand the presence of narrow line emission in the decade after a TDE.

Further insight into the presence or absence of AGN in TDE host galaxies requires spatially resolved emission line maps from IFU data. We present one example here; the host galaxy of AT2018dyk was observed as part of the MaNGA survey (Bundy et al. 2015). This host galaxy is in the AGN regions of the BPT diagrams in Fig. 10, and in Fig. 11 we see that there is indeed a central [OIII]-bright source, and that the outskirts of the galaxy have ionisation dominated by star formation.

The host galaxy of the TDE ASASSN-14li additionally shows evidence of past AGN activity, with large extended ionized regions visible in [OIII] 5007 and [NII] 6584 lines, seen in MUSE observations (Prieto et al. 2016, Fig. 12). Given the light travel time to these narrow-line regions, their ionisation implies strong AGN activity $10^4 - 10^5$ years in the past. Such extended ionized features have been seen around other galaxies in large imaging surveys ("voorwerps" Lintott et al. 2009; Keel et al. 2017) and around other post-starburst galaxies in narrow-band imaging (Schweizer et al. 2013; Watkins et al. 2018). These instances of

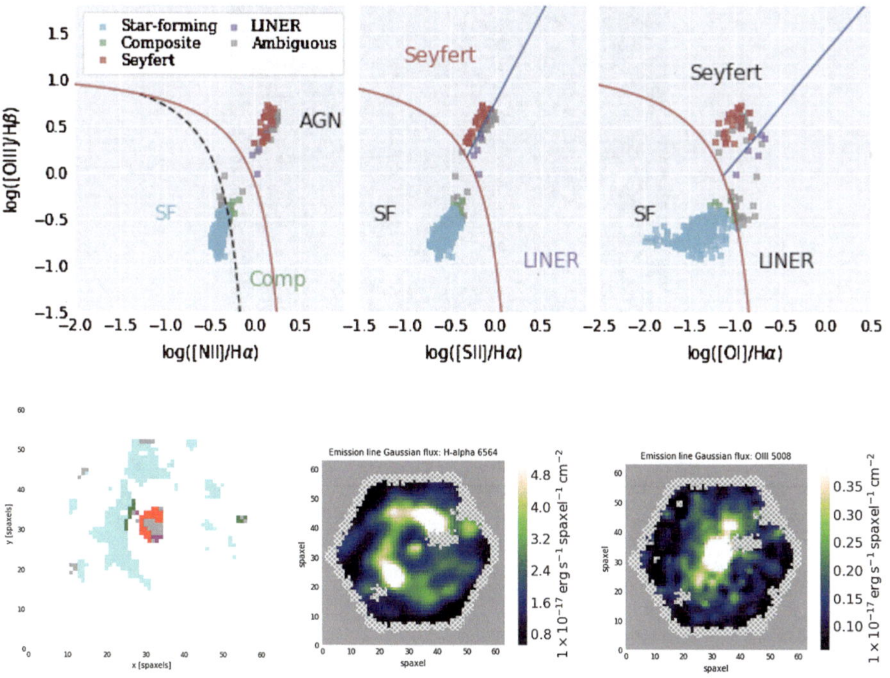

Fig. 11 Spatially resolved BPT diagram, Hα map, and [OIII]5007 map for the host galaxy of TDE AT2018dyk, made using Marvin (Cherinka et al. 2018). Points are colored blue, green, red, dark grey, or light grey based on their classification as star-forming, composite, Seyfert, LINER, or ambiguous. This host galaxy is in the AGN regions of the BPT diagrams in Fig. 10, and here we see that there is indeed a central [OIII]-bright source, and that the outskirts of the galaxy have ionisation dominated by star formation. Obtaining data like this for additional TDE hosts, especially those for which the ionisation source is ambiguous, or there is evidence for unusual AGN properties, will help further our understanding of the co-existence of TDEs and AGN

recent AGN activity in galaxies lacking strong current AGN activity further complicate our understanding of the co-existence of gas and stellar accretion by supermassive black holes in galactic nuclei, raising questions regarding the timescale for TDE rate enhancements compared to AGN duty cycles. The host galaxy of ASASSN-14li furthermore has a persistent radio source discovered in the FIRST survey which may also indicate on-going low-level AGN activity (Holoien et al. 2016a).

2.5 Environment

We consider here the extragalactic environments of the TDE host galaxies, as many mechanisms which act to change the other galaxy properties considered in this section can only act in dense cluster-scale environments. Several TDEs have been found in targeted X-ray surveys of dense galaxy clusters. J1311 (Maksym et al. 2010) was found in a search for TDEs in Abell 1689. While it is not included in our analysis due to the lack of a host galaxy spectrum, "Wings" (Maksym et al. 2013) was found in Abell 1795.

In addition to possible selection biases towards finding TDEs in clusters, post-starburst galaxies are known to lie preferentially in group environments (Zabludoff et al. 1996). While groups like the Local Group are the most common galaxy environment, the groups favored

Fig. 12 Three-colour RGB image constructed from a MUSE datacube of the host galaxy of ASASSN-14li. The red and blue components of this RGB image are from red and blue continuum regions and green is from the [OIII] 5007 line. Extended ionized features are observed via the [OIII] 5007 line extending beyond the continuum-dominated bulk of the galaxy, and are likely from past AGN activity $10^4 - 10^5$ years ago (Prieto et al. 2016). The star observed below the galaxy and the edge-on galaxy to the right are not associated with the TDE host galaxy. Recent AGN activity in galaxies lacking strong current AGN activity further complicates our understanding of the co-existence of gas and stellar accretion by supermassive black holes in galaxies, raising questions of the timescale for TDE rate enhancements compared to AGN duty cycles

by post-starburst galaxies tend to be virialized and more massive than the Local Group, with low enough velocity dispersions and high enough galaxy densities to make galaxy-galaxy mergers and tidal interactions likely (Zabludoff et al. 1996).

We cross match the catalogue in Table 1 with the Abell cluster catalogue (Abell et al. 1989), to check for host galaxies within 25 arcminutes of a cluster with a similar redshift (velocities within 3000 km/s of the host galaxy redshift). We find one additional TDE host galaxy, PTF09axc, to be associated with the cluster Abell 1986.

To test for TDE host galaxies in less rich clusters and groups, we cross match our TDE host galaxy catalogue with the group and cluster catalogue of Tempel et al. (2014). Of the 41 host galaxies, 14 are matched with the Tempel et al. (2014) catalogue. Seven are the only galaxy in their halo,[5] six are associated with groups of 1-6 additional galaxies,[6] and one is part of a cluster of 52 galaxies (SDSS J1350). Given the set of 14 TDE host galaxies in the Tempel et al. (2014) catalogue, and their full sample of galaxies, we find no evidence to suggest that the TDE host galaxies prefer different environments than the general galaxy sample. Using either a Kolmogorov-Smirnov or Anderson-Darling test, we cannot reject the hypothesis their environment richnesses are drawn from the same distribution.

3 Possible Drivers for Host Galaxy Preferences

There are several possible causes for the observed TDE rate enhancement in the host galaxy types discussed above. Many of these scenarios predict TDE enhancements in both post-starburst hosts and centrally concentrated hosts, such as a central overdensity or mechanisms related to galaxy-galaxy mergers.

[5] ASASSN14li, RBS1032, SDSS J1342, SDSS J0748, ASASSN14ae, ASASSN18zj.

[6] SDSS J1323, NGC 5905, SDSS J0952, PTF09ge, AT2018dyk, AT2018bsi.

3.1 Increased Stellar Concentration/Central Overdensities

The TDE rate depends on the number of stars which can be scattered into center-crossing orbits, and so a high central stellar density will result in a high TDE rate. For a Nuker surface-brightness profile, the inner stellar slope γ is found to correlate with the TDE rate as $\dot{N}_{\rm TDE} \propto \gamma^{0.705}$ in a sample of early type galaxies from Lauer et al. (2007) (Stone and Metzger 2016). A high central stellar concentration may be correlated with high concentrations on larger ~kpc scales. As discussed above in §2.3, TDEs are overrepresented in galaxies with high Sérsic indices for their black hole masses (Law-Smith et al. 2017b) and in galaxies with high stellar surface densities on scales of the half-light radius (Graur et al. 2018). An increased TDE rate due to a merger-induced stellar overdensity is seen in simulations by Pfister et al. (2019), although at very early stages in the merger, before the coalescence of the two black holes.

If high stellar concentrations drive the TDE rate enhancement in high Sérsic index or stellar surface density galaxies, it may also explain the rate enhancement in post-starburst galaxies. Post-starburst galaxies have high Sérsic indices (Yang et al. 2004; Quintero et al. 2004; Yang et al. 2008) as the recent starbursts are centrally concentrated, likely due to stars formed from gas infall in the recent merger, and the young/intermediate A stars dominate the light. Once the bright young stellar population in post-starburst galaxies fades, the bulge properties are consistent with evolving to normal early type galaxies, but the stellar concentrations on scales close to the black hole radius of influence have not been measured in samples of post-starburst galaxies. However, the nearby post-starburst galaxy NGC 3156 has HST imaging with high enough resolution to measure the central slope of the Nuker profile, and Stone and van Velzen (2016) find the slope to be steeper than any of the early type galaxies studied previously by Stone and Metzger (2016). Given the lack of a similar TDE rate enhancement in early type galaxies, something must change in the central galaxy concentration or dynamics in the few Gyrs after the post-starburst phase.

The evolution of a central overdensity with time was studied by Stone et al. (2018), who model the stellar density profile as $\rho \propto r^{-\gamma}$, and determine how γ changes with time. Given the post-burst ages of the post-starburst TDE hosts (French et al. 2017), the TDE rate enhancement and its tentative evolution with time could be explained if $\gamma \geq 2.5$. The predictions made by this model can be tested with larger samples of post-starburst TDE hosts in the LSST and perhaps even ZTF eras.

Based on the supposition that the TDE rate should depend on the density of stars in the SMBH loss cone, along with their velocity dispersions, Graur et al. (2018) assumed those local properties would be correlated with their global, kpc-scale counterparts, and that the TDE rate would depend on the latter as $R_{\rm TDE} \propto \Sigma_{M_\star}^{\alpha} \times \sigma_v^{\beta}$, where $\Sigma_{M_\star}$ is the surface stellar mass density on the scale of the half-light radius and σ_v is the kpc-scale velocity dispersion. By comparing their sample of TDE host galaxies with a volume-limited control sample drawn from the SDSS, Graur et al. (2018) estimated the values of the power-law indices to be $\hat{\alpha} = 0.9 \pm 0.2$ and $\hat{\beta} = -1.0 \pm 0.6$ using SDSS fiber measurements of the central few kpc, and assuming these global properties correlate with the properties on the smaller scales of the stars in the SMBH loss cone.

Wang and Merritt (2004) find that the TDE rate of an isothermal sphere ($\rho \propto r^{-2}$) depends on the SMBH mass, $M_\bullet$, and local velocity dispersion, σ, as $R_{\rm TDE} \propto M_\bullet^{-\alpha} \times \sigma^{\eta}$. The average surface stellar mass density of the stars orbiting the SMBH is $\Sigma = M_\bullet/\pi r_h^2 = \sigma^4/\pi G^2 M_\bullet$, where $r_h = GM_\bullet/\sigma^2$ is the size of the star cluster (Peebles 1972), and G is the gravitational constant. This allows us to rewrite the TDE rate as $R_{\rm TDE} \propto \Sigma^{\alpha} \times \sigma^{\eta-4\alpha}$. Using the Graur et al. (2018) estimates for α and $\beta = \eta - 4\alpha$, the values measured from the data,

$\alpha = 0.9 \pm 0.2$ and $\eta = 2.6 \pm 1.0$ are consistent with the theoretical predictions, $\alpha = 1$ and $\eta = 3.5$ (Wang and Merritt 2004). This suggests that the TDE rate is indeed driven by the dynamical relaxation of stars into the loss cone of the SMBH.

Could both the preference of TDE hosts to be in post-starburst or quiescent Balmer-strong host galaxies and the preference for host galaxies with high central concentrations be driven by the same effect? Similar galaxy over-representations can be found in both Sérsic index–black hole mass and in Hα emission–Hδ absorption, where $\geq 60\%$ of TDEs are found in $\sim 2\%$ of the parameter space (i.e., at high Hδ absorption and low Hδ emission, or at high Sérsic index and low black hole mass).[7] Of the five TDE host galaxies considered by Law-Smith et al. (2017b) with high Sérsic indices and low black hole masses, three are post-starburst or quiescent Balmer-strong, and two (PTF09ge and SDSSJ123) are not. Of the seven quiescent Balmer-strong galaxies considered by French et al. (2016), two (ASASSN-14li and ASASSN-14ae) also meet the high Sérsic index and low black hole mass criteria, and the remaining five do not have sufficient data to determine an accurate Sérsic index or bulge mass. Larger numbers of observed TDEs and more detailed analyses of low concentration post-starburst hosts or high concentration non-bursty host galaxies will be an important test of which mechanisms most affect the TDE rate.

3.2 Black Hole Binary

After a galaxy–galaxy merger, the supermassive black hole from each galaxy will inspiral and coalesce. The influence of supermassive black hole binary dynamics on TDEs is the subject of another chapter in this review (Coughlin et al. 2019, ISSI review). We summarize here the relevant points from Coughlin et al. (2019, ISSI review) for the present discussion of the host galaxies.

The TDE rate can be very high (of order 1 per year) for a short (~ 1 Myr) period during coalescence when the secondary black hole approaches the cusp of stars around the primary, at $\sim$pc scale separations. As inspiral continues, the TDE rate will then drop below that of an isolated black hole, and rise once more to a modest rate enhancement of $\sim 2 - 10\times$ that of an isolated black hole once the binary has reached mpc separations. The lightcurves of TDEs can be altered in the case of a tightly bound binary where the debris stream interacts with the companion supermassive black hole. However, if most of the TDEs around black holes in coalescing binaries happens at pc-scale separations, the debris streams will be on significantly smaller scales, and the TDE lightcurves will show no evidence of the companion supermassive black hole. Thus, we are unlikely to see observational effects from the secondary black hole at separations of the same spatial scales which will boost the TDE rate.

The main observable difference between this explanation for the TDE rate enhancement in post-starburst or centrally-concentrated galaxies and the others, is the TDE rate per galaxy. If the TDE rate is very high ($\sim 0.1 - 1$ per year per galaxy) for a Myr, we should observe a high instance of repeat TDEs per host galaxy, especially over the 10 year run of LSST. The other mechanisms for enhancing the TDE rate described in this section would act over 100 Myr - 1 Gyr, with more modest TDE rates per galaxy, and instances of repeat TDEs would be rare. No repeated TDEs have been observed to date, which means either there are no observed cases where the TDE rate is as high as 1 per several years, or systems hosting multiple TDEs within several years are obscured by dust or otherwise produce a different observational signature than the TDEs discussed in this review.

[7] Although the details of this will depend on the TDE samples and comparison samples used, see previous discussions in §2.2 and §2.3.

For now, the likelihood that supermassive black hole binary effects are driving the observed host galaxy distributions can be probed statistically. French et al. (2017) measured the star formation histories for a sample of six post-starburst TDE host galaxies to determine the time since the recent starbursts. If the starburst coincides with the coalescence of the two galaxies, this can also constrain the time since the supermassive black holes started to inspiral on kpc-scales. Most of the TDE host galaxies are less than 600 Myr since starburst. For a secondary to have in-spiraled to pc-scales in 600 Myr, that mass ratio of the two galaxies must be more equal than 12:1 given the dynamical friction timescales. This constrains the TDE rate enhancement in supermassive black hole binaries to be more strongly dependent on the mass ratio than the merger rate, since minor mergers with mass ratios less equal than 12:1 are more common than major mergers. Stone et al. (2018) also argue against the possibility of supermassive black hole binaries causing the observed rate enhancement in post-starburst galaxies by constructing an expected delay time distribution of TDEs after a starburst. Many more TDEs would be expected at times > 1 Gyr after a starburst, but the observed host galaxies have younger ages. Unless the timescale between the merger and the starburst is fine-tuned, compared to the dynamical friction timescale, the observed host galaxies are not compatible with rate enhancement from a black hole binary scenario.

3.3 Circumnuclear Gas

If a supermassive black hole is surrounded by a circumnuclear gas disk, this may act to enhance the TDE rate as stars interact with the disk. Kennedy et al. (2016) predict the TDE rate could be increased by up to $10\times$ by the presence of such a disk. This effect may coexist with the other mechanisms for affecting the TDE rate discussed here, as mergers are expected to trigger AGN activity (e.g., Treister et al. 2012) as well as bursts of centrally-concentrated star formation.

However, there are a number of selection effects which may hinder the identification of TDEs in host galaxies with pre-existing circumnuclear gas disks. As discussed by Law-Smith et al. (2017b), Wevers et al. (2019a), and others, observational searches for TDEs will select against AGN host galaxies while trying to avoid classifying AGN flares from variations in the gas accretion rate as TDEs, and many TDEs in AGN host galaxies will be heavily extincted by dust. One possibility for identifying TDEs in AGN host galaxies with heavy dust obscuration is through multi-epoch radio observations. Mattila et al. (2018) identify a TDE in Arp 299, a Seyfert II galaxy and dust-obscured LIRG (Pereira-Santaella et al. 2015), via a growing radio jet in one of the two nuclei in this merging system. Identifying the accretion of an individual star in an AGN with a high gas accretion rate may prove unfeasible, or only possible to assess statistically.

3.4 Other Dynamical or Secular Effects

There are several other effects that might increase the TDE rate in post-merger galaxies, causing the observed host galaxy preferences. A more complete description of these effects can be found in the chapter of this review by Stone et al. (2020, ISSI review); we very briefly mention two mechanisms here. First, a radial anisotropy of the orbits after a merger would increase the rate of TDEs. Stone et al. (2018) modeled the dependence of the rate enhancement on the time since the starburst, parameterizing the radial anisotropy as $\beta \equiv 1 - \frac{T_\perp}{2T_\parallel}$, where $T_\perp$ and $T_\parallel$ are the kinetic energies of the tangential and radial motion, respectively.[8]

[8] β is defined such that if all orbits are purely radial, $\beta = 1$, and if all orbits are purely tangential, $\beta = -\infty$.

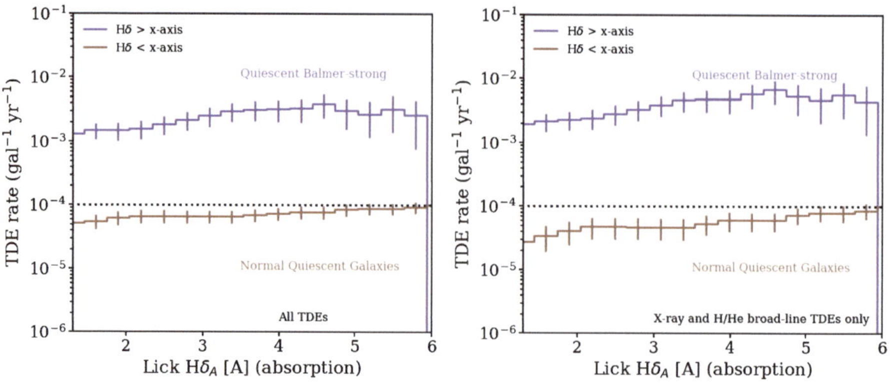

Fig. 13 TDE rate (scaled to a fiducial value of 10^{-4} per galaxy per year as measured by e.g., van Velzen 2018) in quiescent galaxies above and below each value of Hδ absorption. While the rate enhancement in quiescent Balmer-strong and post-starburst galaxies results in higher TDE rates, the resulting suppression of TDE rates in quiescent galaxies that do no meet a given Hδ absorption threshold is mild, at $< 3\times$

Stone et al. (2018) find that the observed host galaxies could be explained if $\beta > 0.6$. A triaxial nuclear potential (Merritt and Poon 2004) could have a similar effect of enhancing the TDE rate after a nuclear starburst. Second, Madigan et al. (2018) predict an enhanced TDE rate that declines with time after a nuclear starburst due to the formation of an eccentric stellar disk, where the TDE rate would be increased due to secular effects and could potentially reach $0.1 - 1$ gal^{-1} yr^{-1}. However, the timescale over which this effect might produce a high TDE rate is uncertain. The high spatial and spectral resolution of next-generation 30-m class telescopes will provide crucial tests of these mechanisms in nearby TDE hosts.

4 Implications for TDE Rates

The variation in TDE rates by host galaxy type has implications for the relative TDE rates in different types of galaxies. The TDE rate averaged across all galaxy types is observed to be $\sim 10^{-4}$ per galaxy per year (van Velzen 2018). We quantify the TDE rates in quiescent Balmer-strong and post-starburst galaxies compared to normal quiescent galaxies in Fig. 13, and how this depends on the definition of "Balmer-strong". The TDE rate in quiescent Balmer-strong and post-starburst galaxies is 1-3$\times 10^{-3}$ per galaxy per year, depending on the threshold used to define such galaxies. An enhancement in the TDE rate in certain galaxies will necessarily result in a decreased TDE rate in other galaxies given a measurement of the total TDE rate. Is this lowered rate enough to cause tension with theoretical predictions for the TDE rate in normal quiescent galaxies ($\sim$ few$\times 10^{-4}$ per galaxy per year, e.g., Stone and Metzger 2016)? Depending on the definition of "Balmer-strong" vs. "normal" quiescent galaxies, the TDE rate in normal quiescent galaxies is $0.3 - 1 \times 10^{-4}$ per galaxy per year. We stress here that despite the enhanced TDE rates observed in post-starburst and quiescent Balmer-strong galaxies, the number of TDEs in normal galaxies is still high enough to avoid a crisis in their TDE rates. The large number of galaxies which do not meet the quiescent Balmer-strong or post-starburst selection compensates for the lower TDE rate in such "normal" galaxies. This rate suppression of $\lesssim 3\times$ in normal quiescent galaxies is comparable to the uncertainties in predicting the theoretical TDE rate and determining the observed TDE rate in various surveys.

5 Using the Host Galaxy Information in Transient Surveys

5.1 Identifying TDEs Using a priori Host Galaxy Information

In addition to being used to understand what drives the TDE rates, the unique properties of TDE host galaxies may be useful for selecting candidate TDEs for spectroscopic follow-up observations in large transient surveys. The large volume of transient alerts faced by current and future surveys such as ZTF and LSST means that not all transients can be followed up for spectroscopic confirmation and further study with triggered space-based observations and high cadence photometry. Methods to flag likely TDEs will be essential for future TDE studies. One method to flag likely TDE candidates is to use a priori information about their likely host galaxies. Transients discovered in pre-identified likely TDE host galaxies, based on (for example) E+A spectroscopic signatures or concentration indices, can then be systematically classified with dedicated spectroscopic observations to efficiently select TDEs for further photometric and spectroscopic monitoring.

For transient surveys in the Northern hemisphere where transient detections will have a significant overlap with large spectroscopic surveys (especially SDSS), the detailed properties of likely host galaxies can be used to predict which candidate detections are likely TDEs, supernovae, or other transient phenomenon, and flag interesting objects for follow-up. However, a significant portion of the LSST footprint will not be covered by large spectroscopic surveys, so photometric criteria will be useful.

A technique of selecting likely TDEs for followup based on the colour (and thus star formation rate) of the host galaxy is used by iPTF/ZTF (Hung et al. 2018) in order to reduce contamination by supernovae. Further cuts on the recent star formation history or concentration of host galaxies may further refine this selection, independent of the physical reason for the enhanced TDE rate in such galaxies.

French and Zabludoff (2018) present a method for identifying likely TDE candidates using photometrically-identified quiescent Balmer-strong galaxies. A Random Forest classifier trained on spectroscopically-identified quiescent Balmer-strong galaxies can be used to detect such galaxies in LSST using LSST photometry in addition to archival photometry from *GALEX* and *WISE*. Because these galaxies have low star formation rates and thus low supernova rates, contamination from other transients will be low.

Other possible methods for identifying likely TDE hosts can use photometric information on the light concentration, given by either the Sérsic index or bulge fraction that may predict a TDE overabundance in a related set of likely host galaxies (see discussion in §3.1). For instance, choosing nuclear transients in high-Sérsic galaxies could significantly increase the success of confirming TDEs.

There are several important benefits to selecting TDEs for spectroscopic follow-up. In addition to identifying a large number of events, this method can identify TDEs with properties that diverge from the known set of TDEs. The small number of optical/UV bright TDEs studied to date already show a large variety in their peak magnitude, colour evolution, and decline timescale (Holoien et al. 2016a; Blagorodnova et al. 2017; Wevers et al. 2017). Furthermore, a priori selection of the likely host galaxies means candidate events can be followed-up early, even before a light curve is obtained. Early time light curve information will enable more detailed modeling, and may be able to be used to measure black hole masses (Mockler et al. 2019).

Any method for selecting transient events for follow-up based on the host galaxy properties will bias the host galaxy properties of the resulting sample. Thus, these selection methods are complementary to other methods which are un-targeted or use information on the transient properties alone.

6 Galaxy Studies Using TDEs

Current and upcoming transient surveys, such as LSST in the optical and e-ROSITA in the X-ray, are expected to detect hundreds to thousands of TDEs each year (van Velzen et al. 2011; Khabibullin et al. 2014). In light of the theoretical predictions and observational evidence that these TDEs should occur in the lowest mass galaxies still harbouring black holes (Wang and Merritt 2004), TDEs could become beacons for studying the black hole mass functions and spin distributions of quiescent supermassive black holes.

While measuring velocity dispersions of large samples of increasingly faint galaxies will become prohibitively expensive, alternatives are emerging to measure the black hole mass. For example, it is expected that the shape of a TDE lightcurve depends mainly on the mass of the black hole and the mass of the disrupted star. Mockler et al. (2019) exploit this by using lightcurve models to infer the black hole mass under some assumptions for the mass of the disrupted star. They show that this method yields similar uncertainties when compared to other ways of estimating the black hole mass (e.g. galaxy scaling relations), while further improvements (such as including realistic models for the stellar structure) could decrease potential systematic effects. This can provide the opportunity to measure black hole masses of galaxies that are too faint for velocity dispersion measurements, as well as large samples of galaxies, as long as the lightcurve is well sampled around peak and ideally in more then one filter.

These mass measurements will be invaluable for TDE demographics studies, but can also potentially contribute to other areas of galaxy evolution, such as the cosmic growth of SMBHs in quiescent galaxies and the BH occupation fraction and mass function of dwarf galaxies. On the high end of the mass function, observed TDE rates will be affected by the presence of an event horizon cutoff, as a consequence of the tidal radius becoming smaller than the event horizon and making disruptions invisible to outside observers. A main sequence star disrupted by a Schwarzschild black hole of mass $\geq 10^8 M_\odot$ will not produce an observable flare. However, the spin of the SMBH will affect this cutoff mass, such that TDEs would still be observable around $10^8 - 10^9 M_\odot$ SMBHs at high ($a \sim 0.9 - 0.999$) spins (Kesden 2012; Leloudas et al. 2016) A large number of TDEs with black hole masses $\geq 10^8 M_\odot$ would imply a large fraction of supermassive black holes with high spin parameters. The spin distribution for supermassive black holes has implications for their accretion and merger history, as coherent gas accretion is expected to spin up black holes while frequent mergers will spin them down (e.g., Volonteri et al. 2003; Hughes and Blandford 2003). The effect of the spin distribution on the observed TDE rate is discussed further by Stone et al. (2020, ISSI review). The possibilities for observing TDEs in dwarf galaxies with intermediate mass black holes are discussed by Maguire et al. (2020, ISSI review).

7 Summary and Discussion

We have summarized the observed host galaxy properties of X-ray bright and optical/UV bright TDEs and discussed the possible physical mechanisms that could drive the observed correlations between host galaxy properties and the TDE rate. While the black hole masses of TDEs have similar distributions between X-ray and optical/UV bright TDEs, compiling a large sample of known TDEs allows us to identify a significant shift in the stellar masses and absolute magnitude distributions between the two samples, with optical/UV TDEs having higher stellar masses than the X-ray TDEs. This may be due to an underlying trend in black hole mass we do not yet have the statistical power to resolve. Future work on determining the black hole masses of new TDEs and understanding the selection effects of TDEs identified in the X-ray or optical will be needed.

Most TDE host galaxies are quiescent, with little current star formation, and TDEs are over-represented in galaxies with post-starburst or quiescent Balmer-strong star formation histories. We present new estimates of the TDE rate enhancement in these samples depending on the definition of these classes and the types of TDEs considered. For the TDE host galaxies considered in this review, 5/41 (12%) are post-starburst galaxies and 13/41 (32%) are either quiescent Balmer-strong or post-starburst. Of the 4 X-ray TDEs, 3 (75%) are quiescent Balmer-strong and 1 (25%) is post-starburst. Of the 15 broad H/He line TDEs, 9 (60%) are quiescent Balmer-strong and 5 (33%) are post-starburst. We note again that these TDE classifications are tentative and subject to a number of observational biases. More observations of the host galaxies for a large sample of well-characterized TDEs are needed to overcome these uncertainties and the small-number statistics limiting our precision here. While controlling for some galaxy properties, notably bulge color and central concentration, can reduce the TDE enhancement rate in post-starburst or quiescent Balmer-strong galaxies, these properties are strongly correlated with galaxy SFH during the post-starburst phase. The TDE enhancement rates in such hosts are thus consistent (given the small number statistics) between the studies of French et al. (2016), Law-Smith et al. (2017b) and Graur et al. (2018). While the rate enhancement in quiescent Balmer-strong and post-starburst galaxies results in higher TDE rates, the resulting suppression of TDE rates in normal quiescent galaxies is mild, at $< 3\times$.

TDE host galaxies have higher concentrations of stellar light than expected given their stellar mass or black hole mass. We aggregate here observations from many sources of TDE host galaxy optical light concentrations, and present new versions of two analyses from the literature. We consider the effect of volume-correcting the SDSS comparison sample on the result by Law-Smith et al. (2017b) that TDE host galaxies have high Sérsic indices for their black hole masses, finding that despite the large number of low black hole mass galaxies inferred from the volume correction, the TDE hosts still lie at high Sérsic indices for their black hole masses, compared with the rest of the galaxy sample. We also add measurements for two additional TDE host galaxy velocity dispersions from Wevers et al. (2019a) to the analysis of Graur et al. (2018) that shows TDE hosts have high stellar surface densities for their velocity dispersions, and find the conclusions unchanged.

The ionisation states of TDE host galaxies span a range from star-formation dominated to AGN-dominated, and we identify many TDE host galaxies with signs of on-going gas accretion by the SMBH. However, given the significant selection biases affecting this distribution, and the uncertainties in identifying especially low-luminosity AGN or LINERs, we urge caution in its interpretation. The extra-galactic environments of TDE host galaxies show no signs of being different than the general galaxy population or the post-starburst galaxy population, but this comparison is especially limited by the small numbers of TDE hosts with well-studied extra-galactic environments.

We summarize the state of several possible explanations for the links between the TDE rate and host galaxy type, including the effect of stellar overdensities, black hole binaries, circumnuclear gas, and dynamical or secular effects. We present estimates of the TDE rate for different host galaxy types and quantify the degree to which rate enhancement in some groups results in rate suppression in others. We discuss the possibilities for using TDE host galaxies to assist in identifying TDEs in upcoming large transient surveys and possibilities for TDE observations to be used to study their host galaxies.

We note that the TDEs considered here are a relatively small number of events, with classifications subject to observational and selection biases which may change in future studies of TDEs.

We identify the following as important open questions in this field:

1. What is the primary driver of the observed correlations between galaxy-scale host properties and the TDE rate?
2. What are the distributions of stellar mass and the stellar kinematics of TDE host galaxies within the radius of gravitational influence of their supermassive black holes?
3. Do the same physical effects drive the observed enhanced TDE rates in post-starburst or quiescent Balmer-strong galaxies and galaxies with high central concentrations?
4. Are the high central concentrations at kpc scales of optical stellar light seen in TDE hosts correlated with high central concentrations of stars at the nucleus?
5. What is the unobscured TDE rate in starburst galaxies? Does the enhancement during the post-starburst phase arise from a higher absolute rate during this phase or simply a higher observed rate due to dust obscuration during the starburst phase and/or selection bias against AGN in starburst galaxies?
6. How does the presence of an existing AGN of varying accretion rates affect the rates and observability of TDEs?
7. What is the connection between TDEs and Narrow-Line Seyfert Is?
8. What is the connection between TDEs and the LINER-like emission present in most post-starburst or quiescent Balmer-strong galaxies? What can cases where a host galaxy has pre-TDE spectroscopy tell us about the evolution of narrow line emission from TDEs vs. AGN?
9. Are there differences in the host galaxies depending on the type of TDE?
10. What drives the observed difference in stellar mass between the optical and X-ray bright TDEs? Is the difference due to a difference in intrinsic black hole mass? Are the observed trends affected by selection bias in how TDEs are identified in the optical vs. X-ray?
11. How can we best use our understanding of TDE host galaxies to study supermassive black holes and their host galaxies using new observations from next-generation surveys like LSST and *eROSITA*? How can we best compare the samples of TDEs discovered using LSST vs. *eROSITA*?

Planned time-domain programs in the next decade will discover hundreds to thousands of new TDEs and will enable the detailed study of TDEs and their host galaxies with significantly greater statistical power.

Acknowledgements We thank the referees for their detailed feedback and helpful suggestions, which have improved this review.

The authors thank ISSI for their support and hospitality and the review organizers for their leadership in coordinating this set of reviews.

K.D.F. is supported by Hubble Fellowship grant HST-HF2-51391.001-A, provided by NASA through a grant from the Space Telescope Science Institute (STScI), which is operated by the Association of Universities for Research in Astronomy, Inc., under NASA contract NAS5-26555. A.I.Z. acknowledges support from NASA through STScI grant HST-GO-14717.001-A. T.W. is funded in part by European Research Council grant 320360 and by European Commission grant 730980. O.G. is supported by an NSF Astronomy and Astrophysics Fellowship under award AST-1602595.

Funding for SDSS-III has been provided by the Alfred P. Sloan Foundation, the Participating Institutions, the National Science Foundation, and the U.S. Department of Energy Office of Science. The SDSS-III website is http://www.sdss3.org/. SDSS-III is managed by the Astrophysical Research Consortium for the Participating Institutions of the SDSS-III Collaboration, including the University of Arizona, the Brazilian Participation Group, Brookhaven National Laboratory, University of Cambridge, Carnegie Mellon University, University of Florida, the French Participation Group, the German Participation Group, Harvard University, the Instituto de Astrofisica de Canarias, the Michigan State/Notre Dame/JINA Participation Group, Johns Hopkins University, Lawrence Berkeley National Laboratory, Max Planck Institute for Astrophysics, Max Planck Institute for Extraterrestrial Physics, New Mexico State University, New York University, the Ohio

State University, Pennsylvania State University, University of Portsmouth, Princeton University, the Spanish Participation Group, University of Tokyo, University of Utah, Vanderbilt University, University of Virginia, University of Washington, and Yale University.

This publication makes use of data products from the Wide-field Infrared Survey Explorer (Wright et al. 2010), which is a joint project of the University of California, Los Angeles, and the Jet Propulsion Laboratory/California Institute of Technology, funded by the National Aeronautics and Space Administration.

This work made use of the IPython package (Pérez and Granger 2007). This research made use of SciPy (Jones et al. 2001). This research made use of Astropy, a community-developed core Python package for Astronomy (Robitaille et al. 2013). This research made use of NumPy (Van Der Walt et al. 2011).

References

G.O. Abell, H.G. Corwin Jr., R.P. Olowin, A catalog of rich clusters of galaxies. Astrophys. J. Suppl. Ser. **70**, 1 (1989). https://doi.org/10.1086/191333

R.G. Abraham, S. van den Bergh, K. Glazebrook, R.S. Ellis, B.X. Santiago, P. Surma, R.E. Griffiths, The morphologies of distant galaxies. II. Classifications from the Hubble Space Telescope medium deep survey. Astrophys. J. Suppl. Ser. **107**, 1 (1996). https://doi.org/10.1086/192352

K. Alatalo, S.L. Cales, J.A. Rich, P.N. Appleton, L.J. Kewley, M. Lacy, L. Lanz, A.M. Medling, K. Nyland, Shocked POststarbust galaxy survey I: candidate poststarbust galaxies with emission line ratios consistent with shocks. Astrophys. J. Suppl. Ser. **224**, 38 (2016). https://doi.org/10.3847/0067-0049/224/2/38. arXiv:1601.05085

I. Arcavi, A. Gal-Yam, M. Sullivan, Y.-C. Pan, S.B. Cenko, A. Horesh, E.O. Ofek, A. De Cia, L. Yan, C.-W. Yang, D.A. Howell, D. Tal, S.R. Kulkarni, S.P. Tendulkar, S. Tang, D. Xu, A. Sternberg, J.G. Cohen, J.S. Bloom, P.E. Nugent, M.M. Kasliwal, D.A. Perley, R.M. Quimby, A.A. Miller, C.A. Theissen, R.R. Laher, A continuum of H- to He-rich tidal disruption candidates with a preference for E+A galaxies. Astrophys. J. **793**(1), 38 (2014). https://doi.org/10.1088/0004-637X/793/1/38

I. Arcavi, J. Burke, K.D. French, A. Zabludoff, D. Hiramatsu, C. McCully, D.A. Howell, G. Hosseinzadeh, S. Valenti, FLOYDS classification of AT 2018dyk/ZTF18aajupnt as a possible tidal disruption event. Astron. Telegram **11953**, 1 (2018)

K. Auchettl, J. Guillochon, E. Ramirez-Ruiz, New physical insights about tidal disruption events from a comprehensive observational inventory at X-ray wavelengths. Astrophys. J. **838**(2), 149 (2017a). https://doi.org/10.3847/1538-4357/aa633b. arXiv:1611.02291

K. Auchettl, E. Ramirez-Ruiz, J. Guillochon, Comparison of the X-ray emission from Tidal Disruption Events with those of Active Galactic Nuclei (2017b). arXiv:1703.06141

I.K. Baldry, S.P. Driver, J. Loveday, E.N. Taylor, L.S. Kelvin, J. Liske, P. Norberg, A.S.G. Robotham, S. Brough, A.M. Hopkins, S.P. Bamford, J.A. Peacock, J. Bland-Hawthorn, C.J. Conselice, S.M. Croom, D.H. Jones, H.R. Parkinson, C.C. Popescu, M. Prescott, R.G. Sharp, R.J. Tuffs, Galaxy And Mass Assembly (GAMA): the galaxy stellar mass function at $z < 0.06$. Mon. Not. R. Astron. Soc. **421**, 621–634 (2012). https://doi.org/10.1111/j.1365-2966.2012.20340.x

J.A. Baldwin, M.M. Phillips, R. Terlevich, Classification parameters for the emission-line spectra of extragalactic objects. Publ. Astron. Soc. Pac. **93**, 5 (1981). https://doi.org/10.1086/130766

E.C. Bellm, S.R. Kulkarni, M.J. Graham, R. Dekany, R.M. Smith, R. Riddle, F.J. Masci, G. Helou, T.A. Prince, S.M. Adams, C. Barbarino, T. Barlow, J. Bauer, R. Beck, J. Belicki, R. Biswas, N. Blagorodnova, D. Bodewits, B. Bolin, V. Brinnel, T. Brooke, B. Bue, M. Bulla, R. Burruss, S.B. Cenko, C.-K. Chang, A. Connolly, M. Coughlin, J. Cromer, V. Cunningham, K. De, A. Delacroix, V. Desai, D.A. Duev, G. Eadie, T.L. Farnham, M. Feeney, U. Feindt, D. Flynn, A. Franckowiak, S. Frederick, C. Fremling, A. Gal-Yam, S. Gezari, M. Giomi, D.A. Goldstein, V.Z. Golkhou, A. Goobar, S. Groom, E. Hacopians, D. Hale, J. Henning, A.Y.Q. Ho, D. Hover, J. Howell, T. Hung, D. Huppenkothen, D. Imel, W.-H. Ip, Ž. Ivezić, E. Jackson, L. Jones, M. Juric, M.M. Kasliwal, S. Kaspi, S. Kaye, M.S.P. Kelley, M. Kowalski, E. Kramer, T. Kupfer, W. Landry, R.R. Laher, C.-D. Lee, H.W. Lin, Z.-Y. Lin, R. Lunnan, M. Giomi, A. Mahabal, P. Mao, A.A. Miller, S. Monkewitz, P. Murphy, C.-C. Ngeow, J. Nordin, P. Nugent, E. Ofek, M.T. Patterson, B. Penprase, M. Porter, L. Rauch, U. Rebbapragada, D. Reiley, M. Rigault, H. Rodriguez, J. van Roestel, B. Rusholme, J. van Santen, S. Schulze, D.L. Shupe, L.P. Singer, M.T. Soumagnac, R. Stein, J. Surace, J. Sollerman, P. Szkody, F. Taddia, S. Terek, A. Van Sistine, S. van Velzen, W.T. Vestrand, R. Walters, C. Ward, Q.-Z. Ye, P.-C. Yu, L. Yan, J. Zolkower, The zwicky transient facility: system overview, performance, and first results. Publ. Astron. Soc. Pac. **131**(995), 018002 (2019). https://doi.org/10.1088/1538-3873/aaecbe

N. Blagorodnova, S. Gezari, T. Hung, S.R. Kulkarni, S.B. Cenko, D.R. Pasham, L. Yan, I. Arcavi, S. Ben-Ami, B.D. Bue, T. Cantwell, Y. Cao, A.J. Castro-Tirado, R. Fender, C. Fremling, A. Gal-Yam, A.Y.Q. Ho, A. Horesh, G. Hosseinzadeh, M.M. Kasliwal, A.K.H.H. Kong, R.R. Laher, G. Leloudas, R. Lunnan, F.J. Masci, K. Mooley, J.D. Neill, P. Nugent, M. Powell, A.F. Valeev, P.M. Vreeswijk, R. Walters, P. Wozniak, iPTF16fnl: a faint and fast tidal disruption event in an E+A galaxy. Astrophys. J. **844**(46), 1 (2017). https://doi.org/10.3847/1538-4357/aa7579. arXiv:1703.00965

N. Blagorodnova, S.B.C.S.R. Kulkarni, I. Arcavi, J.S. Bloom, G. Duggan, A.V. Filippenko, C. Fremling, A. Horesh, G. Hosseinzadeh, E. Karamehmetoglu, A. Levan, F.J. Masci, P.E. Nugent, D.R. Pasham, S. Veilleux, R. Walters, L. Yan, W. Zheng, The Broad Absorption Line Tidal Disruption Event iPTF15af: Optical and Ultraviolet Evolution (2018). arXiv:1809.07446

N. Blagorodnova, S.B. Cenko, S.R. Kulkarni, I. Arcavi, J.S. Bloom, G. Duggan, A.V. Filippenko, C. Fremling, A. Horesh, G. Hosseinzadeh, E. Karamehmetoglu, A. Levan, F.J. Masci, P.E. Nugent, D.R. Pasham, S. Veilleux, R. Walters, L. Yan, W. Zheng, The broad absorption line tidal disruption event iPTF15af: optical and ultraviolet evolution. Astrophys. J. **873**(1), 92 (2019). https://doi.org/10.3847/1538-4357/ab04b0

P.K. Blanchard, M. Nicholl, E. Berger, J. Guillochon, R. Margutti, R. Chornock, K.D. Alexander, J. Leja, M.R. Drout, PS16dtm: a tidal disruption event in a narrow-line Seyfert 1 galaxy. Astrophys. J. **843**(2), 106 (2017). https://doi.org/10.3847/1538-4357/aa77f7. arXiv:1703.07816

M.R. Blanton, S. Roweis, kcorrect: Calculate K-corrections between observed and desired bandpasses (2017)

J. Brinchmann, S. Charlot, S.D.M. White, C. Tremonti, G. Kauffmann, T. Heckman, J. Brinkmann, The physical properties of star-forming galaxies in the low-redshift universe. Mon. Not. R. Astron. Soc. **351**, 1151–1179 (2004). https://doi.org/10.1111/j.1365-2966.2004.07881.x

J.S. Brown, T.W.-S. Holoien, K. Auchettl, K.Z. Stanek, C.S. Kochanek, B.J. Shappee, J.L. Prieto, D. Grupe, The long term evolution of ASASSN-14li. Mon. Not. R. Astron. Soc. **446**(4), 4904–4916 (2016). arXiv:1609.04403

J.S. Brown, C.S. Kochanek, T.W.-S. Holoien, K.Z. Stanek, K. Auchettl, B.J. Shappee, J.L. Prieto, N. Morrell, E. Falco, J. Strader, L. Chomiuk, R. Post, S. Villanueva, S. Mathur, S. Dong, P. Chen, S. Bose, The Ultraviolet Spectroscopic Evolution of the Low-Luminosity Tidal Disruption Event iPTF16fnl (2017). arXiv:1704.02321

K. Bundy, M.A. Bershady, D.R. Law, R. Yan, N. Drory, N. MacDonald, D.A. Wake, B. Cherinka, J.R. Sánchez-Gallego, A.-M. Weijmans, D. Thomas, C. Tremonti, K. Masters, L. Coccato, A.M. Diamond-Stanic, A. Aragón-Salamanca, V. Avila-Reese, C. Badenes, J. Falcón-Barroso, F. Belfiore, D. Bizyaev, G.A. Blanc, J. Bland-Hawthorn, M.R. Blanton, J.R. Brownstein, N. Byler, M. Cappellari, C. Conroy, A.A. Dutton, E. Emsellem, J. Etherington, P.M. Frinchaboy, H. Fu, J.E. Gunn, P. Harding, E.J. Johnston, G. Kauffmann, K. Kinemuchi, M.A. Klaene, J.H. Knapen, A. Leauthaud, C. Li, L. Lin, R. Maiolino, V. Malanushenko, E. Malanushenko, S. Mao, C. Maraston, R.M. McDermid, M.R. Merrifield, R.C. Nichol, D. Oravetz, K. Pan, J.K. Parejko, S.F. Sanchez, D. Schlegel, A. Simmons, O. Steele, M. Steinmetz, K. Thanjavur, B.A. Thompson, J.L. Tinker, R.C.E. van den Bosch, K.B. Westfall, D. Wilkinson, S. Wright, T. Xiao, K. Zhang, Overview of the SDSS-IV MaNGA survey: mapping nearby galaxies at apache point observatory. Astrophys. J. **798**, 7 (2015). https://doi.org/10.1088/0004-637X/798/1/7

M. Cappellari, R. Bacon, M. Bureau, M.C. Damen, R.L. Davies, P.T. de Zeeuw, E. Emsellem, J. Falcón-Barroso, D. Krajnović, H. Kuntschner, R.M. McDermid, R.F. Peletier, M. Sarzi, R.C.E. van den Bosch, G. van de Ven, The SAURON project - IV. The mass-to-light ratio, the virial mass estimator and the fundamental plane of elliptical and lenticular galaxies. Mon. Not. R. Astron. Soc. **366**, 1126–1150 (2006). https://doi.org/10.1111/j.1365-2966.2005.09981.x

S.B. Cenko, J.S. Bloom, S.R. Kulkarni, L.E. Strubbe, A.A. Miller, N.R. Butler, R.M. Quimby, A. Gal-Yam, E.O. Ofek, E. Quataert, L. Bildsten, D. Poznanski, D.A. Perley, A.N. Morgan, A.V. Filippenko, D.A. Frail, I. Arcavi, S. Ben-Ami, A. Cucchiara, C.D. Fassnacht, Y. Green, I.M. Hook, D.A. Howell, D.J. Lagattuta, N.M. Law, M.M. Kasliwal, P.E. Nugent, J.M. Silverman, M. Sullivan, S.P. Tendulkar, O. Yaron, PTF10iya: a short-lived, luminous flare from the nuclear region of a star-forming galaxy. Mon. Not. R. Astron. Soc. **420**(3), 2684–2699 (2012). https://doi.org/10.1111/j.1365-2966.2011.20240.x

S.B. Cenko, A. Cucchiara, N. Roth, S. Veilleux, J.X. Prochaska, L. Yan, J. Guillochon, W.P. Maksym, I. Arcavi, N.R. Butler, A.V. Filippenko, A.S. Fruchter, S. Gezari, D. Kasen, A.J. Levan, J.M. Miller, D.R. Pasham, E. Ramirez-Ruiz, L.E. Strubbe, N.R. Tanvir, F. Tombesi, An Ultraviolet Spectrum of the Tidal Disruption Flare ASASSN-14li (2016), p. 8. arXiv:1601.03331

K.C. Chambers, E.A. Magnier, N. Metcalfe, H.A. Flewelling, M.E. Huber, C.Z. Waters, L. Denneau, P.W. Draper, D. Farrow, D.P. Finkbeiner, C. Holmberg, J. Koppenhoefer, P.A. Price, R.P. Saglia, E.F. Schlafly, S.J. Smartt, W. Sweeney, R.J. Wainscoat, W.S. Burgett, T. Grav, J.N. Heasley, K.W. Hodapp, R. Jedicke, N. Kaiser, R.-P. Kudritzki, G.A. Luppino, R.H. Lupton, D.G. Monet, J.S. Morgan, P.M. Onaka, C.W.

Stubbs, J.L. Tonry, E. Banados, E.F. Bell, R. Bender, E.J. Bernard, M.T. Botticella, S. Casertano, S. Chastel, W.-P. Chen, X. Chen, S. Cole, N. Deacon, C. Frenk, A. Fitzsimmons, S. Gezari, C. Goessl, T. Goggia, B. Goldman, E.K. Grebel, N.C. Hambly, G. Hasinger, A.F. Heavens, T.M. Heckman, R. Henderson, T. Henning, M. Holman, U. Hopp, W.-H. Ip, S. Isani, C.D. Keyes, A. Koekemoer, R. Kotak, K.S. Long, J.R. Lucey, M. Liu, N.F. Martin, B. McLean, E. Morganson, D.N.A. Murphy, M.A. Nieto-Santisteban, P. Norberg, J.A. Peacock, E.A. Pier, M. Postman, N. Primak, C. Rae, A. Rest, A. Riess, A. Riffeser, H.W. Rix, S. Roser, E. Schilbach, A.S.B. Schultz, D. Scolnic, A. Szalay, S. Seitz, B. Shiao, E. Small, K.W. Smith, D. Soderblom, A.N. Taylor, A.R. Thakar, J. Thiel, D. Thilker, Y. Urata, J. Valenti, F. Walter, S.P. Watters, S. Werner, R. White, W.M. Wood-Vasey, R. Wyse, The Pan-STARRS1 Surveys (2016). arXiv:1612.05560

B. Cherinka, B.H. Andrews, J. Sánchez-Gallego, J. Brownstein, M. Argudo-Fernández, M. Blanton, K. Bundy, A. Jones, K. Masters, D.R. Law, K. Rowlands, A.-M. Weijmans, K. Westfall, R. Yan, Marvin: a Toolkit for Streamlined Access and Visualization of the SDSS-IV MaNGA Data Set. arXiv:e-prints (2018)

R. Chornock, E. Berger, S. Gezari, B.A. Zauderer, A. Rest, L. Chomiuk, A. Kamble, A.M. Soderberg, I. Czekala, J. Dittmann, M. Drout, R.J. Foley, W. Fong, M.E. Huber, R.P. Kirshner, A. Lawrence, R. Lunnan, G.H. Marion, G. Narayan, A.G. Riess, K.C. Roth, N.E. Sanders, D. Scolnic, S.J. Smartt, K. Smith, C.W. Stubbs, J.L. Tonry, W.S. Burgett, K.C. Chambers, H. Flewelling, K.W. Hodapp, N. Kaiser, E.A. Magnier, D.C. Martin, J.D. Neill, P.A. Price, R. Wainscoat, The ultraviolet-bright, slowly declining transient PS1-11AF as a partial tidal disruption event. Astrophys. J. **780**(1), 44 (2014). https://doi.org/10.1088/0004-637X/780/1/44

C.J. Conselice, S.C. Chapman, R.A. Windhorst, Evidence for a major merger origin of high-redshift submillimeter galaxies. Astrophys. J. **596**, 5–8 (2003). https://doi.org/10.1086/379109

E.R. Coughlin, M.C. Begelman, Hyperaccretion during tidal disruption events: weakly bound debris envelopes and jets. Astrophys. J. **781**(2), 82 (2014). https://doi.org/10.1088/0004-637X/781/2/82

L. Dai, J.C. McKinney, M.C. Miller, Soft X-ray temperature tidal disruption events from stars on deep plunging orbits. Astrophys. J. **812**(2), 39 (2015). https://doi.org/10.1088/2041-8205/812/2/L39

L. Dai, J.C. McKinney, N. Roth, E. Ramirez-Ruiz, M.C. Miller, A unified model for tidal disruption events (2018). arXiv:1803.03265

S. Dong, S. Bose, P. Chen, T.G. Brink, T. de Jaeger, A.V. Filippenko, W. Zheng, Spectroscopic classification of ASASSN-18zj with the Lick 3-m Shane Telescope. Astron. Telegram **12198**, 1 (2018)

D. Elbaz, M. Dickinson, H.S. Hwang, T. Diaz-Santos, G. Magdis, B. Magnelli, D.L. Borgne, F. Galliano, M. Pannella, P. Chanial, L. Armus, V. Charmandaris, E. Daddi, H. Aussel, P. Popesso, J. Kartaltepe, B. Altieri, I. Valtchanov, D. Coia, H. Dannerbauer, K. Dasyra, R. Leiton, J. Mazzarella, V. Buat, D. Burgarella, R.-R. Chary, R. Gilli, R.J. Ivison, S. Juneau, E. LeFloc'h, D. Lutz, G.E. Morrison, J. Mullaney, E. Murphy, A. Pope, D. Scott, D. Alexander, M. Brodwin, D. Calzetti, C. Cesarsky, S. Charlot, H. Dole, P. Eisenhardt, H.C. Ferguson, N. Foerster-Schreiber, D. Frayer, M. Giavalisco, M. Huynh, A.M. Koekemoer, C. Papovich, N. Reddy, C. Surace, H. Teplitz, M.S. Yun, G. Wilson, GOODS-Herschel: an infrared main sequence for star-forming galaxies. Astron. Astrophys. **533**, 119 (2011). arXiv:1105.2537

P. Esquej, R.D. Saxton, M.J. Freyberg, A.M. Read, B. Altieri, M. Sanchez-Portal, G. Hasinger, Candidate tidal disruption events from the XMM-Newton slew survey. Astron. Astrophys. **462**(3), 49–52 (2007). https://doi.org/10.1051/0004-6361:20066072

L. Ferrarese, H. Ford, Supermassive black holes in galactic nuclei: past, present and future research. Space Sci. Rev. **116**, 523–624 (2005). https://doi.org/10.1007/s11214-005-3947-6

L. Ferrarese, D. Merritt, A fundamental relation between supermassive black holes and their host galaxies. Astrophys. J. **539**, 9–12 (2000). https://doi.org/10.1086/312838

S. Frederick, S. Gezari, M.J. Graham, S.B. Cenko, S. van Velzen, D. Stern, N. Blagorodnova, S.R. Kulkarni, L. Yan, K. De, U.C. Fremling, T. Hung, E. Kara, D.L. Shupe, C. Ward, E.C. Bellm, R. Dekany, D.A. Duev, U. Feindt, M. Giomi, T. Kupfer, R.R. Laher, F.J. Masci, A.A. Miller, C.-C. Ngeow, M.T. Patterson, M. Porter, B. Rusholme, J. Sollerman, R. Walters, A New Class of Changing-Look LINERs (2019). arXiv:1904.10973

K.D. French, A.I. Zabludoff, Identifying tidal disruption events via prior photometric selection of their preferred hosts. Astrophys. J. **868**, 99 (2018). https://doi.org/10.3847/1538-4357/aaea64

K.D. French, I. Arcavi, A. Zabludoff, Tidal disruption events prefer unusual host galaxies. Astrophys. J. **818**, 21 (2016). https://doi.org/10.3847/2041-8205/818/1/L21. arXiv:1601.04705

K.D. French, I. Arcavi, A. Zabludoff, The post-starburst evolution of tidal disruption event host galaxies. Astrophys. J. **835**, 176 (2017). https://doi.org/10.3847/1538-4357/835/2/176

K.D. French, Y. Yang, A.I. Zabludoff, C.A. Tremonti, Clocking the evolution of post-starburst galaxies: methods and first results. Astrophys. J. **862**, 2 (2018). https://doi.org/10.3847/1538-4357/aacb2d

K. Gebhardt, R. Bender, G. Bower, A. Dressler, S.M. Faber, A.V. Filippenko, R. Green, C. Grillmair, L.C. Ho, J. Kormendy, T.R. Lauer, J. Magorrian, J. Pinkney, D. Richstone, S. Tremaine, A relationship between nuclear black hole mass and galaxy velocity dispersion. Astrophys. J. Lett. **539**, 13–16 (2000). https://doi.org/10.1086/312840

S. Gezari, S. Basa, D.C. Martin, G. Bazin, K. Forster, B. Milliard, J.P. Halpern, P.G. Friedman, P. Morrissey, S.G. Neff, D. Schiminovich, M. Seibert, T. Small, T.K. Wyder, UV/optical detections of candidate tidal disruption events by GALEX and CFHTLS. Astrophys. J. **676**(2), 944–969 (2008). https://doi.org/10.1086/529008

S. Gezari, T. Heckman, S.B. Cenko, M. Eracleous, K. Forster, T.S. Gonçalves, D.C. Martin, P. Morrissey, S.G. Neff, M. Seibert, D. Schiminovich, T.K. Wyder, LUMINOUS THERMAL FLARES FROM QUIESCENT SUPERMASSIVE BLACK HOLES. Astrophys. J. **698**(2), 1367–1379 (2009). https://doi.org/10.1088/0004-637X/698/2/1367

S. Gezari, S.B. Cenko, I. Arcavi, X-ray Brightening and UV Fading of Tidal Disruption Event ASASSN-15oi (2017). arXiv:1712.03968

S. Gezari, S. van Velzen, S.B. Cenko, M. Graham, N. Blagorodnova, T. Hung, L. Yan, S. Kulkarni, ZTF discovery of a tidal disruption event at z=0.051. Astron. Telegram **12035**, 1 (2018)

O. Graur, K.D. French, H.J. Zahid, J. Guillochon, K.S. Mandel, K. Auchettl, A.I. Zabludoff, A dependence of the tidal disruption event rate on global stellar surface mass density and stellar velocity dispersion. Astrophys. J. **853**, 39 (2018). https://doi.org/10.3847/1538-4357/aaa3fd

J. Greiner, R. Schwarz, S. Zharikov, M. Orio, RX J1420.4+5334 - another tidal disruption event? Astron. Astrophys. **362**, 25–28 (2000)

D. Grupe, H.-C. Thomas, K.M. Leighly, RX J1624.9+7554: a new X-ray transient AGN. Astron. Astrophys. **350**, 31–34 (1999)

J. Guillochon, E. Ramirez-Ruiz, A dark year for tidal disruption events. Astrophys. J. **809**, 166 (2015). https://doi.org/10.1088/0004-637X/809/2/166

K. Gültekin, D.O. Richstone, K. Gebhardt, T.R. Lauer, S. Tremaine, M.C. Aller, R. Bender, A. Dressler, S.M. Faber, A.V. Filippenko, R. Green, L.C. Ho, J. Kormendy, J. Magorrian, J. Pinkney, C. Siopis, The M-σ and M-L relations in galactic bulges, and determinations of their intrinsic scatter. Astrophys. J. **698**(1), 198–221 (2009). https://doi.org/10.1088/0004-637X/698/1/198

J.G. Hills, Possible power source of Seyfert galaxies and QSOs. Nature **254**(5498), 295–298 (1975). https://doi.org/10.1038/254295a0

L.C. Ho, A.V. Filippenko, W.L. Sargent, A search for "Dwarf" Seyfert nuclei. II. An optical spectral atlas of the nuclei of nearby galaxies. Astrophys. J. Suppl. Ser. **98**, 477 (1995). https://doi.org/10.1086/192170

T.W.-S. Holoien, C.S. Kochanek, J.L. Prieto, D. Grupe, P. Chen, D. Godoy-Rivera, K.Z. Stanek, B.J. Shappee, S. Dong, J.S. Brown, U. Basu, J.F. Beacom, D. Bersier, J. Brimacombe, E.K. Carlson, E. Falco, E. Johnston, B.F. Madore, G. Pojmanski, M. Seibert, ASASSN-15oi: a Rapidly Evolving. Luminous Tidal Disruption Event at 216 Mpc., p. 17 (2016a). arXiv:1602.01088

T.W.-S. Holoien, C.S. Kochanek, J.L. Prieto, K.Z. Stanek, S. Dong, B.J. Shappee, D. Grupe, J.S. Brown, U. Basu, J.F. Beacom, D. Bersier, J. Brimacombe, A.B. Danilet, E. Falco, Z. Guo, J. Jose, G.J. Herczeg, F. Long, G. Pojmanski, G.V. Simonian, D.M. Szczygieł, T.A. Thompson, J.R. Thorstensen, R.M. Wagner, P.R. Woźniak, Six months of multiwavelength follow-up of the tidal disruption candidate ASASSN-14li and implied TDE rates from ASAS-SN. Mon. Not. R. Astron. Soc. **455**(3), 2918–2935 (2016b). https://doi.org/10.1093/mnras/stv2486

T.W.-S. Holoien, M.E. Huber, B.J. Shappee, M. Eracleous, K. Auchettl, J.S. Brown, M.A. Tucker, K.C. Chambers, C.S. Kochanek, K.Z. Stanek, A. Rest, D. Bersier, R.S. Post, G. Aldering, K.A. Ponder, J.D. Simon, E. Kankare, D. Dong, G. Hallinan, J. Bulger, T.B. Lowe, E.A. Magnier, A.S.B. Schultz, C.Z. Waters, M. Willman, D. Wright, D.R. Young, S. Dong, J.L. Prieto, T.A. Thompson, L. Denneau, H. Flewelling, A.N. Heinze, S.J. Smartt, K.W. Smith, B. Stalder, J.L. Tonry, H. Weiland, PS18kh: a New Tidal Disruption Event with a Non-Axisymmetric Accretion Disk (2018). arXiv:1808.02890

S.A. Hughes, R.D. Blandford, Black hole mass and spin coevolution by mergers. Astrophys. J. **585**, 101–104 (2003). https://doi.org/10.1086/375495

T. Hung, S. Gezari, S.B. Cenko, S. van Velzen, N. Blagorodnova, L. Yan, S.R. Kulkarni, R. Lunnan, T. Kupfer, G. Leloudas, A.K.H. Kong, P.E. Nugent, C. Fremling, R.R. Laher, F.J. Masci, Y. Cao, R. Roy, T. Petrushevska, Sifting for sapphires: systematic selection of tidal disruption events in iPTF. Astrophys. J. Suppl. Ser. **238**, 15 (2018). https://doi.org/10.3847/1538-4365/aad8b1

F. Jansen, D. Lumb, B. Altieri, J. Clavel, M. Ehle, C. Erd, C. Gabriel, M. Guainazzi, P. Gondoin, R. Much, R. Munoz, M. Santos, N. Schartel, D. Texier, G. Vacanti, XMM-Newton observatory. I. The spacecraft and operations. Astron. Astrophys. **365**, 1–6 (2001). https://doi.org/10.1051/0004-6361:20000036

E. Jones, T. Oliphant, P. Peterson, Others, SciPy: Open source scientific tools for Python (2001). http://www.scipy.org/

P.G. Jonker, N.C. Stone, A. Generozov, S. van Velzen, B. Metzger, Implications from Late-Time X-ray Detections of Optically Selected Tidal Disruption Events: State Changes, Unification, and Detection Rates (2019). arXiv:1906.12236

E. Kankare, R. Kotak, S. Mattila, P. Lundqvist, M.J. Ward, M. Fraser, A. Lawrence, S.J. Smartt, W.P.S. Meikle, A. Bruce, J. Harmanen, S.J. Hutton, C. Inserra, T. Kangas, A. Pastorello, T. Reynolds, C. Romero-Cañizales, K.W. Smith, S. Valenti, K.C. Chambers, K.W. Hodapp, M.E. Huber, N. Kaiser, R.-P. Kudritzki, E.A. Magnier, J.L. Tonry, R.J. Wainscoat, C. Waters, A population of highly energetic transient events in the centres of active galaxies. Nat. Astron. **1**, 865–871 (2017). https://doi.org/10.1038/s41550-017-0290-2

G. Kauffmann, T.M. Heckman, S.D.M. White, S. Charlot, C. Tremonti, J. Brinchmann, G. Bruzual, E.W. Peng, M. Seibert, M. Bernardi, M. Blanton, J. Brinkmann, F. Castander, I. Csábai, M. Fukugita, Z. Ivezic, J.A. Munn, R.C. Nichol, N. Padmanabhan, A.R. Thakar, D.H. Weinberg, D. York, Stellar masses and star formation histories for 10^5 galaxies from the Sloan Digital Sky Survey. Mon. Not. R. Astron. Soc. **341**, 33–53 (2003a). https://doi.org/10.1046/j.1365-8711.2003.06291.x

G. Kauffmann, T.M. Heckman, C. Tremonti, J. Brinchmann, S. Charlot, S.D.M. White, S.E. Ridgway, J. Brinkmann, M. Fukugita, P.B. Hall, Ž. Ivezić, G.T. Richards, D.P. Schneider, The host galaxies of active galactic nuclei. Mon. Not. R. Astron. Soc. **346**(4), 1055–1077 (2003b). https://doi.org/10.1111/j.1365-2966.2003.07154.x

W.C. Keel, C.J. Lintott, W.P. Maksym, V.N. Bennert, S.D. Chojnowski, A. Moiseev, A. Smirnova, K. Schawinski, L.F. Sartori, C.M. Urry, A. Pancoast, M. Schirmer, B. Scott, C. Showley, K. Flatland, Fading AGN candidates: AGN histories and outflow signatures. Astrophys. J. **835**(2), 256 (2017). https://doi.org/10.3847/1538-4357/835/2/256

G.F. Kennedy, Y. Meiron, B. Shukirgaliyev, T. Panamarev, P. Berczik, A. Just, R. Spurzem, Star-disc interaction in galactic nuclei: orbits and rates of accreted stars. Mon. Not. R. Astron. Soc. **460**, 240–255 (2016). https://doi.org/10.1093/mnras/stw908. arXiv:1604.05309

R.C. Kennicutt, N.J. Evans, Star formation in the Milky Way and nearby galaxies. Annu. Rev. Astron. Astrophys. **50**, 531–608 (2012). https://doi.org/10.1146/annurev-astro-081811-125610

M. Kesden, Tidal-disruption rate of stars by spinning supermassive black holes. Phys. Rev. D **85**(2), 024037 (2012). https://doi.org/10.1103/PhysRevD.85.024037

L.J. Kewley, M.A. Dopita, R.S. Sutherland, C.A. Heisler, J. Trevena, Theoretical modeling of starburst galaxies. Astrophys. J. **556**(1), 121–140 (2001). https://doi.org/10.1086/321545

I. Khabibullin, S. Sazonov, R. Sunyaev, SRG/eROSITA prospects for the detection of stellar tidal disruption flares. Mon. Not. R. Astron. Soc. **437**(1), 327–337 (2014). https://doi.org/10.1093/mnras/stt1889

C.S. Kochanek, Tidal Disruption Event (TDE) demographics. Mon. Not. R. Astron. Soc. **461**, 371–384 (2016). https://doi.org/10.1093/mnras/stw1290. arXiv:1601.06787

S. Komossa, J. Greiner, Discovery of a giant and luminous X-ray outburst from the optically inactive galaxy pair RX J1242.6-1119. Astron. Astrophys. **349**, 45–48 (1999)

S. Komossa, D. Merritt, Tidal disruption flares from recoiling supermassive black holes. Astrophys. J. **683**(1), 21–24 (2008). https://doi.org/10.1086/591420

S. Komossa, H. Zhou, A. Rau, M. Dopita, A. Gal-Yam, J. Greiner, J. Zuther, M. Salvato, D. Xu, H. Lu, R. Saxton, M. Ajello, NTT, Spitzer, and Chandra Spectroscopy of SDSSJ095209.56+214313.3: the most luminous coronal-line supernova ever observed, or a stellar tidal disruption event? Astrophys. J. **701**, 105–121 (2009). https://doi.org/10.1088/0004-637X/701/1/105

J. Kormendy, L.C. Ho, Coevolution (or not) of supermassive black holes and host galaxies. Annu. Rev. Astron. Astrophys. **51**(1), 511–653 (2013). https://doi.org/10.1146/annurev-astro-082708-101811

J. Kormendy, D. Richstone, Inward bound—the search for supermassive black holes in galactic nuclei. Annu. Rev. Astron. Astrophys. **33**, 581 (1995). https://doi.org/10.1146/annurev.aa.33.090195.003053

T.R. Lauer, K. Gebhardt, S.M. Faber, D. Richstone, S. Tremaine, J. Kormendy, M.C. Aller, R. Bender, A. Dressler, A.V. Filippenko, R. Green, L.C. Ho, The centers of early-type galaxies with Hubble Space Telescope. VI. Bimodal central surface brightness profiles. Astrophys. J. **664**(1), 226–256 (2007). https://doi.org/10.1086/519229

N.M. Law, S.R. Kulkarni, R.G. Dekany, E.O. Ofek, R.M. Quimby, P.E. Nugent, J. Surace, C.C. Grillmair, J.S. Bloom, M.M. Kasliwal, L. Bildsten, T. Brown, S.B. Cenko, D. Ciardi, E. Croner, S.G. Djorgovski, J. van Eyken, A.V. Filippenko, D.B. Fox, A. Gal- Yam, D. Hale, N. Hamam, G. Helou, J. Henning, D.A. Howell, J. Jacobsen, R. Laher, S. Mattingly, D. McKenna, A. Pickles, D. Poznanski, G. Rahmer, A. Rau, W. Rosing, M. Shara, R. Smith, D. Starr, M. Sullivan, V. Velur, R. Walters, J. Zolkower, The palomar transient factory: system overview, performance, and first results. Publ. Astron. Soc. Pac. **121**, 1395 (2009). https://doi.org/10.1086/648598

J. Law-Smith, M. MacLeod, J. Guillochon, P. Macias, E. Ramirez-Ruiz, Low-mass white dwarfs with hydrogen envelopes as a missing link in the tidal disruption menu. Astrophys. J. **841**(2), 132 (2017a). https://doi.org/10.3847/1538-4357/aa6ffb

J. Law-Smith, E. Ramirez-Ruiz, S.L. Ellison, R.J. Foley, Tidal disruption event host galaxies in the context of the local galaxy population. Astrophys. J. **850**(1), 22 (2017b). https://doi.org/10.3847/1538-4357/aa94c7. arXiv:1707.01559

G. Leloudas, M. Fraser, N.C. Stone, S. van Velzen, P.G. Jonker, I. Arcavi, C. Fremling, J.R. Maund, S.J. Smartt, T. Kruhler, J.C.A. Miller-Jones, P.M. Vreeswijk, A. Gal-Yam, P.A. Mazzali, A. De Cia, D.A. Howell, C. Inserra, F. Patat, A. d U, Postigo, O. Yaron, C. Ashall, I. Bar, H. Campbell, T.-W. Chen, M. Childress, N. Elias-Rosa, J. Harmanen, G. Hosseinzadeh, J. Johansson, T. Kangas, E. Kankare, S. Kim, H. Kuncarayakti, J. Lyman, M.R. Magee, K. Maguire, D. Malesani, S. Mattila, C.V. McCully, M. Nicholl, S. Prentice, C. Romero-Canizales, S. Schulze, K.W. Smith, J. Sollerman, M. Sullivan, B.E. Tucker, S. Valenti, J.C. Wheeler, D.R. Young, The Superluminous Transient ASASSN-15lh as a Tidal Disruption Event from a Kerr Black Hole (2016). arXiv:1609.02927

G. Leloudas, L. Dai, I. Arcavi, P.M. Vreeswijk, B. Mockler, R. Roy, D.B. Malesani, S. Schulze, T. Wevers, M. Fraser, E. Ramirez-Ruiz, K. Auchettl, J. Burke, G. Cannizzaro, P. Charalampopoulos, T.-W. Chen, A. Cikota, M. Della Valle, L. Galbany, M. Gromadzki, K.E. Heintz, D. Hiramatsu, P.G. Jonker, Z. Kostrzewa-Rutkowska, K. Maguire, I. Mandel, F. Onori, N. Roth, S.J. Smartt, L. Wyrzykowski, D.R. Young, The spectral evolution of AT 2018dyb and the presence of metal lines in tidal disruption events. arXiv:e-prints (2019)

A.J. Levan, N.R. Tanvir, S.B. Cenko, D.A. Perley, K. Wiersema, J.S. Bloom, A.S. Fruchter, A.D.U. Postigo, P.T. O'Brien, N. Butler, A.J. van der Horst, G. Leloudas, A.N. Morgan, K. Misra, G.C. Bower, J. Farihi, R.L. Tunnicliffe, M. Modjaz, J.M. Silverman, J. Hjorth, C. Thöne, A. Cucchiara, J.M.C. Cerón, A.J. Castro-Tirado, J.A. Arnold, M. Bremer, J.P. Brodie, T. Carroll, M.C. Cooper, P.A. Curran, R.M. Cutri, J. Ehle, D. Forbes, J. Fynbo, J. Gorosabel, J. Graham, D.I. Hoffman, S. Guziy, P. Jakobsson, A. Kamble, T. Kerr, M.M. Kasliwal, C. Kouveliotou, D. Kocevski, N.M. Law, P.E. Nugent, E.O. Ofek, D. Poznanski, R.M. Quimby, E. Rol, A.J. Romanowsky, R. Sánchez-Ramírez, S. Schulze, N. Singh, L. van Spaandonk, R.L.C. Starling, R.G. Strom, J.C. Tello, O. Vaduvescu, P.J. Wheatley, R.A.M.J. Wijers, J.M. Winters, D. Xu, An extremely luminous panchromatic outburst from the nucleus of a distant galaxy. Science **333**(6039), 199–202 (2011). https://doi.org/10.1126/science.1207143

D. Lin, P.W. Maksym, J.A. Irwin, S. Komossa, N.A. Webb, O. Godet, D. Barret, D. Grupe, S.D.J. Gwyn, An ultrasoft X-ray flare from 3XMM J152130.7+074916: a tidal disruption event candidate. Astrophys. J. **811**(1), 43 (2015). https://doi.org/10.1088/0004-637X/811/1/43

D. Lin, J. Guillochon, S. Komossa, E. Ramirez-Ruiz, J.A. Irwin, W.P. Maksym, D. Grupe, O. Godet, N.A. Webb, D. Barret, B.A. Zauderer, P.-A. Duc, E.R. Carrasco, S.D.J. Gwyn, A likely decade-long sustained tidal disruption event. Nat. Astron. **1**, 0033 (2017). https://doi.org/10.1038/s41550-016-0033

D. Lin, J. Strader, E.R. Carrasco, D. Page, A.J. Romanowsky, J. Homan, J.A. Irwin, R.A. Remillard, O. Godet, N.A. Webb, H. Baumgardt, R. Wijnands, D. Barret, P.-A. Duc, J.P. Brodie, S.D.J. Gwyn, A luminous X-ray outburst from an intermediate-mass black hole in an off-centre star cluster. Nat. Astron. **2**, 656–661 (2018). https://doi.org/10.1038/s41550-018-0493-1

C.J. Lintott, K. Schawinski, W. Keel, H. van Arkel, N. Bennert, E. Edmondson, D. Thomas, D.J.B. Smith, P.D. Herbert, M.J. Jarvis, S. Virani, D. Andreescu, S.P. Bamford, K. Land, P. Murray, R.C. Nichol, M.J. Raddick, A. Slosar, A. Szalay, J. Vandenberg, Galaxy Zoo: 'Hanny's Voorwerp', a quasar light echo? Mon. Not. R. Astron. Soc. **399**(1), 129–140 (2009). https://doi.org/10.1111/j.1365-2966.2009.15299.x

G. Lodato, A.R. King, J.E. Pringle, Stellar disruption by a supermassive black hole: is the light curve really proportional to $t^{-5/3}$? Mon. Not. R. Astron. Soc. **392**, 332–340 (2009). https://doi.org/10.1111/j.1365-2966.2008.14049.x

M. MacLeod, J. Guillochon, E. Ramirez-Ruiz, THE TIDAL DISRUPTION OF GIANT STARS AND THEIR CONTRIBUTION TO THE FLARING SUPERMASSIVE BLACK HOLE POPULATION. Astrophys. J. **757**(2), 134 (2012). https://doi.org/10.1088/0004-637X/757/2/134

A.-M. Madigan, A. Halle, M. Moody, M. McCourt, C. Nixon, H. Wernke, Dynamical properties of eccentric nuclear disks: stability, longevity, and implications for tidal disruption rates in post-merger galaxies. Astrophys. J. **853**, 141 (2018). https://doi.org/10.3847/1538-4357/aaa714

J. Magorrian, S. Tremaine, Rates of tidal disruption of stars by massive central black holes. Mon. Not. R. Astron. Soc. **309**(2), 447–460 (1999). https://doi.org/10.1046/j.1365-8711.1999.02853.x

J. Magorrian, S. Tremaine, D. Richstone, R. Bender, G. Bower, A. Dressler, S.M. Faber, K. Gebhardt, R. Green, C. Grillmair, J. Kormendy, T. Lauer, The demography of massive dark objects in galaxy centers. Astron. J. **115**, 2285–2305 (1998). https://doi.org/10.1086/300353

W.P. Maksym, M.P. Ulmer, M. Eracleous, A tidal disruption flare in A1689 from an archival X-ray survey of galaxy clusters. Astrophys. J. **722**, 1035–1050 (2010). https://doi.org/10.1088/0004-637X/722/2/1035

W.P. Maksym, M.P. Ulmer, M.C. Eracleous, L. Guennou, L.C. Ho, A tidal flare candidate in Abell 1795. Mon. Not. R. Astron. Soc. **435**, 1904–1927 (2013). https://doi.org/10.1093/mnras/stt1379

W.P. Maksym, D. Lin, J.A. Irwin, RBS 1032: a tidal disruption event in another dwarf galaxy? Astrophys. J. **792**(2), 29 (2014). https://doi.org/10.1088/2041-8205/792/2/L29

S. Mattila, M. Pérez-Torres, A. Efstathiou, P. Mimica, M. Fraser, E. Kankare, A. Alberdi, M.Á. Aloy, T. Heikkilä, P.G. Jonker, P. Lundqvist, I. Martí-Vidal, W.P.S. Meikle, C. Romero-Cañizales, S.J. Smartt, S. Tsygankov, E. Varenius, A. Alonso- Herrero, M. Bondi, C. Fransson, R. Herrero- Illana, T. Kangas, R. Kotak, N. Ramírez- Olivencia, P. Väisänen, R.J. Beswick, D.L. Clements, R. Greimel, J. Harmanen, J. Kotilainen, K. Nandra, T. Reynolds, S. Ryder, N.A. Walton, K. Wiik, G. Östlin, A dust-enshrouded tidal disruption event with a resolved radio jet in a galaxy merger. Science **361**, 482–485 (2018). https://doi.org/10.1126/science.aao4669

N.J. McConnell, C.-P. Ma, Revisiting the scaling relations of black hole masses and host galaxy properties. Astrophys. J. **764**, 184 (2013). https://doi.org/10.1088/0004-637X/764/2/184

J.T. Mendel, L. Simard, M. Palmer, S.L. Ellison, D.R. Patton, A catalog of bulge, disk, and total stellar mass estimates for the sloan digital sky survey. Astrophys. J. Suppl. Ser. **210**(1), 3 (2014). https://doi.org/10.1088/0067-0049/210/1/3

D. Merritt, M.Y. Poon, Chaotic loss cones and black hole fueling. Astrophys. J. **606**(2), 788–798 (2004). https://doi.org/10.1086/382497

B.D. Metzger, N.C. Stone, A bright year for tidal disruptions. Mon. Not. R. Astron. Soc. **461**(1), 948–966 (2016). https://doi.org/10.1093/mnras/stw1394

B. Mockler, J. Guillochon, E. Ramirez-Ruiz, Weighing black holes using tidal disruption events. Astrophys. J. **872**, 151 (2019). https://doi.org/10.3847/1538-4357/ab010f

M.M. Pawlik, V. Wild, C.J. Walcher, P.H. Johansson, C. Villforth, K. Rowlands, J. Mendez-Abreu, T. Hewlett, Shape asymmetry: a morphological indicator for automatic detection of galaxies in the post-coalescence merger stages. Mon. Not. R. Astron. Soc. **456**(3), 3032–3052 (2015). arXiv:1512.02000

P.J.E. Peebles, Star distribution near a collapsed object. Astrophys. J. **178**, 371–376 (1972). https://doi.org/10.1086/151797

M. Pereira-Santaella, A. Alonso-Herrero, L. Colina, D. Miralles-Caballero, P.G. Pérez-González, S. Arribas, E. Bellocchi, S. Cazzoli, T. Díaz-Santos, J. Piqueras López, Star-formation histories of local luminous infrared galaxies. Astron. Astrophys. **577**, 78 (2015). https://doi.org/10.1051/0004-6361/201425359

F. Pérez, B.E. Granger, IPython: a system for interactive scientific computing. J. Comput. Sci. Eng. **9**(3), 21–29 (2007). https://doi.org/10.1109/MCSE.2007.53. http://ipython.org

H. Pfister, B. Bar-Or, M. Volonteri, Y. Dubois, P.R. Capelo, Tidal disruption event rates in galaxy merger remnants (2019). arXiv:1903.09124

J.L. Prieto, T. Krühler, J.P. Anderson, L. Galbany, C.S. Kochanek, E. Aquino, J.S. Brown, S. Dong, F. Förster, T.W.-S. Holoien, H. Kuncarayakti, J.C. Maureira, F.F. Rosales-Ortega, S.F. Sánchez, B.J. Shappee, K.Z. Stanek, MUSE reveals a recent merger in the post-starburst host galaxy of the TDE ASASSN-14li. Astrophys. J. Lett. **830**(2), 32 (2016). arXiv:1609.00013

A.D. Quintero, D.W. Hogg, M.R. Blanton, D.J. Schlegel, D.J. Eisenstein, J.E. Gunn, J. Brinkmann, M. Fukugita, K. Glazebrook, T. Goto, Selection and photometric properties of K+A galaxies. Astrophys. J. **602**(1), 190–199 (2004). https://doi.org/10.1086/380601

A. Rau, S.R. Kulkarni, N.M. Law, J.S. Bloom, D. Ciardi, G.S. Djorgovski, D.B. Fox, A. Gal-Yam, C.C. Grillmair, M.M. Kasliwal, P.E. Nugent, E.O. Ofek, R.M. Quimby, W.T. Reach, M. Shara, L. Bildsten, S.B. Cenko, A.J. Drake, A.V. Filippenko, D.J. Helfand, G. Helou, D.A. Howell, D. Poznanski, M. Sullivan, Exploring the optical transient sky with the palomar transient factory. Publ. Astron. Soc. Pac. **121**, 1334 (2009). https://doi.org/10.1086/605911

M.J. Rees, Tidal disruption of stars by black holes of 106–108 solar masses in nearby galaxies. Nature **333**(6173), 523–528 (1988). https://doi.org/10.1038/333523a0

J.A. Rich, L.J. Kewley, M.A. Dopita, Galaxy mergers drive shocks: an integral field study of GOALS galaxies. Astrophys. J. Suppl. Ser. **221**, 28 (2015). https://doi.org/10.1088/0067-0049/221/2/28. arXiv:1509.08468

T.P. Robitaille, E.J. Tollerud, P. Greenfield, M. Droettboom, E. Bray, T. Aldcroft, M. Davis, A. Ginsburg, A.M. Price-Whelan, W.E. Kerzendorf, A. Conley, N. Crighton, K. Barbary, D. Muna, H. Ferguson, F. Grollier, M.M. Parikh, P.H. Nair, H.M. Unther, C. Deil, J. Woillez, S. Conseil, R. Kramer, J.E.H. Turner, L. Singer, R. Fox, B.A. Weaver, V. Zabalza, Z.I. Edwards, K. Azalee Bostroem, D.J. Burke, A.R. Casey, S.M. Crawford, N. Dencheva, J. Ely, T. Jenness, K. Labrie, P.L. Lim, F. Pierfederici, A. Pontzen, A. Ptak, B. Refsdal, M. Servillat, O. Streicher (Astropy Collaboration), Astropy: a community Python package for astronomy. Astron. Astrophys. **558**, 33 (2013). https://doi.org/10.1051/0004-6361/201322068

N. Roth, D. Kasen, J. Guillochon, E. Ramirez-Ruiz, The X-ray through optical fluxes and line strengths of tidal disruption events. Astrophys. J. **827**(1), 3 (2016). https://doi.org/10.3847/0004-637X/827/1/3

R.D. Saxton, A.M. Read, P. Esquej, S. Komossa, S. Dougherty, P. Rodriguez-Pascual, D. Barrado, A tidal disruption-like X-ray flare from the quiescent galaxy SDSS J120136.02+300305.5. Astron. Astrophys. **541**, 106 (2012). https://doi.org/10.1051/0004-6361/201118367

R.D. Saxton, A.M. Read, S. Komossa, P. Lira, K.D. Alexander, M.H. Wieringa, XMMSL1 J074008.2-853927: a tidal disruption event with thermal and non-thermal components. Astron. Astrophys. **598**, 29 (2017). https://doi.org/10.1051/0004-6361/201629015

F. Schweizer, P. Seitzer, D.D. Kelson, E.V. Villanueva, G.L. Walth, The [O III] Nebula of the merger remnant NGC 7252: a likely faint ionization echo. Astrophys. J. **773**(2), 148 (2013). https://doi.org/10.1088/0004-637X/773/2/148

B.J. Shappee, J.L. Prieto, D. Grupe, C.S. Kochanek, K.Z. Stanek, G. De Rosa, S. Mathur, Y. Zu, B.M. Peterson, R.W. Pogge, S. Komossa, M. Im, J. Jencson, T.W.-S. Holoien, U. Basu, J.F. Beacom, D.M. Szczygieł, J. Brimacombe, S. Adams, A. Campillay, C. Choi, C. Contreras, M. Dietrich, M. Dubberley, M. Elphick, S. Foale, M. Giustini, C. Gonzalez, E. Hawkins, D.A. Howell, E.Y. Hsiao, M. Koss, K.M. Leighly, N. Morrell, D. Mudd, D. Mullins, J.M. Nugent, J. Parrent, M.M. Phillips, G. Pojmanski, W. Rosing, R. Ross, D. Sand, D.M. Terndrup, S. Valenti, Z. Walker, Y. Yoon, The man behind the curtain: X-rays drive the UV through NIR variability in the 2013 active galactic nucleus outburst in NGC 2617. Astrophys. J. **788**, 48 (2014). https://doi.org/10.1088/0004-637X/788/1/48

L. Simard, C.N.A. Willmer, N.P. Vogt, V.L. Sarajedini, A.C. Phillips, B.J. Weiner, D.C. Koo, M. Im, G.D. Illingworth, S.M. Faber, The DEEP groth strip survey. II. Hubble space telescope structural parameters of galaxies in the groth strip. Astrophys. J. Suppl. Ser. **142**, 1–33 (2002). https://doi.org/10.1086/341399

L. Simard, J. Trevor Mendel, D.R. Patton, S.L. Ellison, A.W. McConnachie, VizieR Online Data Catalog: Bulge+disk decompositions of SDSS galaxies (Simard+, 2011). VizieR Online Data Catalog, 196–11 (2011)

G.F. Snyder, T.J. Cox, C.C. Hayward, L. Hernquist, P. Jonsson, K+A galaxies as the aftermath of gas-rich mergers: simulating the evolution of galaxies as seen by spectroscopic surveys. Astrophys. J. **741**(2), 77 (2011). https://doi.org/10.1088/0004-637X/741/2/77

D. Stern, R.J. Assef, D.J. Benford, A. Blain, R. Cutri, A. Dey, P. Eisenhardt, R.L. Griffith, T.H. Jarrett, S. Lake, F. Masci, S. Petty, S.A. Stanford, C.-W. Tsai, E.L. Wright, L. Yan, F. Harrison, K. Madsen, Mid-infrared selection of active galactic nuclei with the wide-field infrared survey explorer. I. Characterizing WISE-selected active galactic nuclei in COSMOS. Astrophys. J. **753**(1), 30 (2012). https://doi.org/10.1088/0004-637X/753/1/30

N.C. Stone, B.D. Metzger, Rates of stellar tidal disruption as probes of the supermassive black hole mass function. Mon. Not. R. Astron. Soc. **455**(1), 859–883 (2016). https://doi.org/10.1093/mnras/stv2281

N.C. Stone, S. van Velzen, An enhanced rate of tidal disruptions in the centrally overdense E+A galaxy NGC 3156. Astrophys. J. Lett. **825**, 14 (2016). https://doi.org/10.3847/2041-8205/825/1/L14. arXiv:1604.02056

N.C. Stone, M. Kesden, R.M. Cheng, S. van Velzen, Stellar Tidal Disruption Events in General Relativity (2018). arXiv:1801.10180

C. Tadhunter, R. Spence, M. Rose, J. Mullaney, P. Crowther, A tidal disruption event in the nearby ultra-luminous infrared galaxy F01004-2237. Nat. Astron. **1**, 0061 (2017). https://doi.org/10.1038/s41550-017-0061. arXiv:1702.02573

E. Tempel, A. Tamm, M. Gramann, T. Tuvikene, L.J. Liivamägi, I. Suhhonenko, R. Kipper, M. Einasto, E. Saar, Flux- and volume-limited groups/clusters for the SDSS galaxies: catalogues and mass estimation. Astron. Astrophys. **566**, 1 (2014). https://doi.org/10.1051/0004-6361/201423585

B. Trakhtenbrot, I. Arcavi, C. Ricci, S. Tacchella, D. Stern, H. Netzer, P.G. Jonker, A. Horesh, J.E. Mejía-Restrepo, G. Hosseinzadeh, V. Hallefors, D.A. Howell, C. McCully, M. Baloković, M. Heida, N. Kamraj, G.B. Lansbury, Ł. Wyrzykowski, M. Gromadzki, A. Hamanowicz, S.B. Cenko, D.J. Sand, E.Y. Hsiao, M.M. Phillips, T.R. Diamond, E. Kara, K.C. Gendreau, Z. Arzoumanian, R. Remillard, A new class of flares from accreting supermassive black holes. Nat. Astron. (2019). https://doi.org/10.1038/s41550-018-0661-3

E. Treister, K. Schawinski, C.M. Urry, B.D. Simmons, Major galaxy mergers only trigger the most luminous active galactic nuclei. Astrophys. J. **758**, 39 (2012). https://doi.org/10.1088/2041-8205/758/2/L39

S. Tremaine, K. Gebhardt, R. Bender, G. Bower, A. Dressler, S.M. Faber, A.V. Filippenko, R. Green, C. Grillmair, L.C. Ho, J. Kormendy, T.R. Lauer, J. Magorrian, J. Pinkney, D. Richstone, The slope of the black hole mass versus velocity dispersion correlation. Astrophys. J. **574**, 740–753 (2002). https://doi.org/10.1086/341002

C.A. Tremonti, T.M. Heckman, G. Kauffmann, J. Brinchmann, S. Charlot, S.D.M. White, M. Seibert, E.W. Peng, D.J. Schlegel, A. Uomoto, M. Fukugita, J. Brinkmann, The origin of the mass-metallicity relation: insights from 53,000 star-forming galaxies in the sloan digital sky survey. Astrophys. J. **613**(2), 898–913 (2004). https://doi.org/10.1086/423264

S. Van Der Walt, S.C. Colbert, G. Varoquaux, The numpy array: a structure for efficient numerical computation. Comput. Sci. Eng. **13**(2), 22–30 (2011)

S. van Velzen, On the mass and luminosity functions of tidal disruption flares: rate suppression due to black hole event horizons. Astrophys. J. **852**, 72 (2018). https://doi.org/10.3847/1538-4357/aa998e

S. van Velzen, G.R. Farrar, S. Gezari, N. Morrell, D. Zaritsky, L. Östman, M. Smith, J. Gelfand, A.J. Drake, Optical discovery of probable stellar tidal disruption flares. Astrophys. J. **741**(2), 73 (2011). https://doi.org/10.1088/0004-637X/741/2/73

M. Volonteri, P. Madau, F. Haardt, The formation of galaxy stellar cores by the hierarchical merging of supermassive black holes. Astrophys. J. **593**, 661–666 (2003). https://doi.org/10.1086/376722

J. Wang, D. Merritt, Revised rates of stellar disruption in galactic nuclei. Astrophys. J. **600**(1), 149–161 (2004). https://doi.org/10.1086/379767

T.-G. Wang, H.-Y. Zhou, S. Komossa, H.-Y. Wang, W. Yuan, C. Yang, Extreme coronal line emitters: tidal disruption of stars by massive black holes in galactic nuclei? Astrophys. J. **749**(2), 115 (2012). https://doi.org/10.1088/0004-637X/749/2/115

K.-Y. Watarai, K. Ohsuga, R. Takahashi, J. Fukue, Geometrical effect of supercritical accretion flows: observational implications of galactic black-hole candidates and ultraluminous X-ray sources. Publ. Astron. Soc. Jpn. **57**, 513–524 (2005). https://doi.org/10.1093/pasj/57.3.513

A.E. Watkins, J.C. Mihos, M. Bershady, P. Harding, Discovery of a vast ionized gas cloud in the M51 system. Astrophys. J. **858**(2), 16 (2018). https://doi.org/10.3847/2041-8213/aabba1

T. Wevers, S. van Velzen, P.G. Jonker, N.C. Stone, T. Hung, F. Onori, S. Gezari, N. Blagorodnova, Black hole masses of tidal disruption event host galaxies (2017). arXiv:1706.08965

T. Wevers, N.C. Stone, S. van Velzen, P.G. Jonker, T. Hung, K. Auchettl, S. Gezari, F. Onori, Black hole masses of tidal disruption event host galaxies II. arXiv:e-prints (2019a)

T. Wevers, D.R. Pasham, S. van Velzen, G. Leloudas, S. Schulze, J.C.A. Miller-Jones, P.G. Jonker, M. Gromadzki, E. Kankare, S.T. Hodgkin, L. Wyrzykowski, Z. Kostrzewa-Rutkowska, S. Moran, M. Berton, K. Maguire, F. Onori, S. Matilla, M. Nicholl, Evidence for rapid disk formation and reprocessing in the X-ray bright tidal disruption event AT 2018fyk. arXiv:1903.2203 (2019b)

V. Wild, T. Heckman, S. Charlot, Timing the starburst-AGN connection. Mon. Not. R. Astron. Soc. **405**(2), 933–947 (2010). https://doi.org/10.1111/j.1365-2966.2010.16536.x

V. Wild, O. Almaini, J. Dunlop, C. Simpson, K. Rowlands, R. Bowler, D. Maltby, R. McLure, The evolution of post-starburst galaxies from z=2 to z= 0.5. Mon. Not. R. Astron. Soc. **463**, 832–844 (2016). https://doi.org/10.1093/mnras/stw1996. arXiv:1608.00588

R.J. Williams, R.W. Pogge, S. Mathur, Narrow-line Seyfert 1 galaxies from the sloan digital sky survey early data release. Astron. J. **124**(6), 3042–3049 (2002). https://doi.org/10.1086/344765

O.I. Wong, K. Schawinski, S. Kaviraj, K.L. Masters, R.C. Nichol, C. Lintott, W.C. Keel, D. Darg, S.P. Bamford, D. Andreescu, P. Murray, M.J. Raddick, A. Szalay, D. Thomas, J. VandenBerg, Galaxy Zoo: building the low-mass end of the red sequence with local post-starburst galaxies. Mon. Not. R. Astron. Soc. **420**(2), 1684–1692 (2012). https://doi.org/10.1111/j.1365-2966.2011.20159.x

G. Worthey, D.L. Ottaviani, Hγ and Hδ absorption features in stars and stellar populations. Astrophys. J. Suppl. Ser. **111**(2), 377–386 (1997). https://doi.org/10.1086/313021

E.L. Wright, P.R.M. Eisenhardt, A.K. Mainzer, M.E. Ressler, R.M. Cutri, T. Jarrett, J.D. Kirkpatrick, D. Padgett, R.S. McMillan, M. Skrutskie, S.A. Stanford, M. Cohen, R.G. Walker, J.C. Mather, D. Leisawitz, I. Gautier, N. Thomas, I. McLean, D. Benford, C.J. Lonsdale, A. Blain, B. Mendez, W.R. Irace, V. Duval, F. Liu, D. Royer, I. Heinrichsen, J. Howard, M. Shannon, M. Kendall, A.L. Walsh, M. Larsen, J.G. Cardon, S. Schick, M. Schwalm, M. Abid, B. Fabinsky, L. Naes, C.-W. Tsai, The Wide-field Infrared Survey Explorer (WISE): mission description and initial on-orbit performance. Astron. J. **140**(6), 1868–1881 (2010). https://doi.org/10.1088/0004-6256/140/6/1868

R. Yan, M.R. Blanton, The nature of liner-like emission in red galaxies. Astrophys. J. **747**(1), 61 (2012). https://doi.org/10.1088/0004-637X/747/1/61

R. Yan, J.A. Newman, S.M. Faber, A.L. Coil, M.C. Cooper, M. Davis, B.J. Weiner, B.F. Gerke, D.C. Koo, The DEEP2 Galaxy Redshift Survey: environments of post-starburst galaxies at $z \sim 0.1$ and $\sim$0.8. Mon. Not. R. Astron. Soc. **398**(2), 735–753 (2009). https://doi.org/10.1111/j.1365-2966.2009.15192.x

Y. Yang, A.I. Zabludoff, D. Zaritsky, T.R. Lauer, J.C. Mihos, E+A galaxies and the formation of early-type galaxies at z $\sim$ 0. Astrophys. J. **607**(1), 258–273 (2004). https://doi.org/10.1086/383259

Y. Yang, A.I. Zabludoff, D. Zaritsky, J.C. Mihos, The detailed evolution of E+A galaxies into early types. Astrophys. J. **688**(2), 945–971 (2008). https://doi.org/10.1086/591656

C.-W. Yang, T.-G. Wang, G. Ferland, W. Yuan, H.-Y. Zhou, P. Jiang, Long-term spectral evolution of tidal disruption candidates selected by strong coronal lines. Astrophys. J. **774**(1), 46 (2013). https://doi.org/10.1088/0004-637X/774/1/46

D.G. York, J. Adelman, J.E. Anderson Jr., S.F. Anderson, J. Annis, N.A. Bahcall, J.A. Bakken, R. Barkhouser, S. Bastian, E. Berman, W.N. Boroski, S. Bracker, C. Briegel, J.W. Briggs, J. Brinkmann, R. Brunner, S. Burles, L. Carey, M.A. Carr, F.J. Castander, B. Chen, P.L. Colestock, A.J. Connolly, J.H. Crocker, I. Csabai, P.C. Czarapata, J.E. Davis, M. Doi, T. Dombeck, D. Eisenstein, N. Ellman, B.R. Elms, M.L. Evans, X. Fan, G.R. Federwitz, L. Fiscelli, S. Friedman, J.A. Frieman, M. Fukugita, B. Gillespie, J.E. Gunn, V.K. Gurbani, E. de Haas, M. Haldeman, F.H. Harris, J. Hayes, T.M. Heckman, G.S. Hennessy, R.B. Hindsley, S. Holm, D.J. Holmgren, C. Huang, C. Hull, D. Husby, S. Ichikawa, T. Ichikawa, Ž. Ivezić, S. Kent, R.S.J. Kim, E. Kinney, M. Klaene, A.N. Kleinman, S. Kleinman, G.R. Knapp, J. Korienek, R.G. Kron, P.Z. Kunszt, D.Q. Lamb, B. Lee, R.F. Leger, S. Limmongkol, C. Lindenmeyer, D.C.

Long, C. Loomis, J. Loveday, R. Lucinio, R.H. Lupton, B. MacKinnon, E.J. Mannery, P.M. Mantsch, B. Margon, P. McGehee, T.A. McKay, A. Meiksin, A. Merelli, D.G. Monet, J.A. Munn, V.K. Narayanan, T. Nash, E. Neilsen, R. Neswold, H.J. Newberg, R.C. Nichol, T. Nicinski, M. Nonino, N. Okada, S. Okamura, J.P. Ostriker, R. Owen, A.G. Pauls, J. Peoples, R.L. Peterson, D. Petravick, J.R. Pier, A. Pope, R. Pordes, A. Prosapio, R. Rechenmacher, T.R. Quinn, G.T. Richards, M.W. Richmond, C.H. Rivetta, C.M. Rockosi, K. Ruthmansdorfer, D. Sandford, D.J. Schlegel, D.P. Schneider, M. Sekiguchi, G. Sergey, K. Shimasaku, W.A. Siegmund, S. Smee, J.A. Smith, S. Snedden, R. Stone, C. Stoughton, M.A. Strauss, C. Stubbs, M. SubbaRao, A.S. Szalay, I. Szapudi, G.P. Szokoly, A.R. Thakar, C. Tremonti, D.L. Tucker, A. Uomoto, D. Vanden Berk, M.S. Vogeley, P. Waddell, S. Wang, M. Watanabe, D.H. Weinberg, B. Yanny, N. Yasuda, The sloan digital sky survey: technical summary. Astron. J. **120**, 1579–1587 (2000). https://doi.org/10.1086/301513

A.I. Zabludoff, D. Zaritsky, H. Lin, D. Tucker, Y. Hashimoto, S.A. Shectman, A. Oemler, R.P. Kirshner, The environment of "E+A" galaxies. Astrophys. J. **466**, 104 (1996). https://doi.org/10.1086/177495

Space Sci Rev (2020) 216:35
https://doi.org/10.1007/s11214-020-00651-4

Rates of Stellar Tidal Disruption

N.C. Stone[1] · E. Vasiliev[2] · M. Kesden[3] · E.M. Rossi[4] · H.B. Perets[5] · P. Amaro-Seoane[6]

Received: 3 September 2019 / Accepted: 24 February 2020 / Published online: 19 March 2020

Abstract Tidal disruption events occur rarely in any individual galaxy. Over the last decade, however, time-domain surveys have begun to accumulate statistical samples of these flares. What dynamical processes are responsible for feeding stars to supermassive black holes? At what rate are stars tidally disrupted in realistic galactic nuclei? What may we learn about supermassive black holes and broader astrophysical questions by estimating tidal disruption event rates from observational samples of flares? These are the questions we aim to address

The Tidal Disruption of Stars by Massive Black Holes
Edited by Peter G. Jonker, Sterl Phinney, Elena Maria Rossi, Sjoert van Velzen, Iair Arcavi and Maurizio Falanga

✉ N.C. Stone
nicholas.stone@mail.huji.ac.il

E. Vasiliev
eugvas@lpi.ru

M. Kesden
kesden@utdallas.edu

E.M. Rossi
emr@strw.leidenuniv.nl

H.B. Perets
hperets@ph.technion.ac.il

P. Amaro-Seoane
pau@ice.cat

1 Racah Institute of Physics, The Hebrew University, Jerusalem, 91904, Israel

2 Institute of Astronomy, University of Cambridge, Madingley Rd, Cambridge, CB3 0HA, UK

3 Department of Physics, University of Texas at Dallas, 800 W. Campbell Rd, Richardson, TX 75080, USA

4 Leiden Observatory, Leiden University, PO Box 9513, 2300 RA, Leiden, the Netherlands

5 Department of Physics, Technion, Haifa, Israel

6 Institut de Ciències de l'Espai, Campus UAB, Carrer de Can Magrans, Cerdanyola del Vallés, 08193, Barcelona, Spain

in this Chapter, which summarizes current theoretical knowledge about rates of stellar tidal disruption, and compares theoretical predictions to the current state of observations.

1 Introduction

The discovery and characterization of quasars in the 1960s (Schmidt 1963) was rapidly recognized as evidence for the existence of supermassive black holes (SMBHs) (Salpeter 1964). Shortly thereafter, the possibility of tidal disruption events (TDEs) was proposed by Wheeler (1971), who hypothesized that the tidal disruption of a star by a massive black hole could create a high energy transient by means of the disintegrational Penrose process. If stars were consumed frequently enough by massive black holes, the resulting accretion of their gas could, perhaps, explain the observed properties of quasars and active galactic nuclei (Hills 1975).

By the mid-1970s, the importance of *TDE rates* was evident, and rapid theoretical progress was made on understanding the dynamical processes that feed stars to SMBHs. Early work envisaged this as a diffusive process in the space of orbital energy: an isotropic distribution of stars living around a SMBH might be depleted by diffusion of stars onto more tightly bound orbits (Bahcall and Wolf 1976). In reality, however, the process of diffusion through energy space is slow and self-limiting; subsequent analytic (Frank and Rees 1976) and numerical (Lightman and Shapiro 1977) works quickly demonstrated the greater importance of velocity-space *anisotropies* in determining TDE rates. More precisely, the rate of tidal disruptions in a dense star cluster is set by diffusion through angular momentum, rather than energy, space. The stellar distribution function (DF) drains into the black hole through a "loss cone," named after the analogous phase space region in magnetic mirror fusion reactors (Rosenbluth and Post 1965).

In this Chapter, we survey the dynamical physics of dense stellar systems containing massive black holes, focusing particularly on the collisional evolution of stellar DFs in the presence of a loss cone. Our presentation will, by necessity, be rather terse and without proofs. A more detailed treatment of stellar orbits and kinetic theory can be found in several textbooks. The reader interested in going beyond this chapter may wish to consult Chap. 7 of Binney and Tremaine (2008) or Chaps. 5 and 6 of Merritt (2013).

We begin in Sect. 2, by overviewing the basic physics of stellar tidal disruption. In Sect. 3, we present the theoretical picture of the loss cone, describing both the orbital dynamics of individual stars near supermassive black holes, and the ways in which two-body relaxation and collisionless effects cause stellar populations to evolve over time. We provide both Newtonian and general relativistic treatments of loss cone dynamics, and examine more exotic types of tidal disruptions (e.g. disruption of evolved or binary stars). Next, we apply these theoretical tools to realistic astrophysical environments. In Sect. 4, we examine past efforts to estimate TDE rates by building dynamical models of nearby galactic nuclei, emphasizing both the empirically-calibrated event rate predictions and the distributions of event properties. In Sect. 5, we compare these estimates to observational inferences of the volumetric stellar disruption rate. Finally, in Sect. 6, we examine the broader importance of nuclear stellar dynamics, and describe how well-measured TDE rates may determine the uncertain bottom end of the supermassive black hole mass function, probe basic predictions of general relativity, and calibrate rate estimates for extragalactic phenomena such as mHz gravitational wave sources.

2 On the Proper Care and Feeding of Supermassive Black Holes

Modern observations demonstrate that SMBHs are ubiquitous in the nuclei of sufficiently large galaxies (e.g. Kormendy and Richstone 1995). While a minority of SMBHs live in active galactic nuclei (AGN), where they accrete steadily from long-lived and large-scale discs of interstellar gas, the majority of SMBHs are quiescent: accreting at very low rates (Heckman et al. 2004). In these quiescent galactic nuclei, it is only in the aftermath of a TDE that the SMBH may shine brightly. But what variables determine the rate of stellar tidal disruption? We can break down the controlling variables into two categories: those relevant to the *hydrodynamic* process of tidal disruption, and those related to the *orbital dynamics* of stars in galactic nuclei.

The first set of variables—those governing hydrodynamic stellar disruption—are explored in greater detail in the Disruption Chapter, whereas in this Chapter, we focus on the latter set, with the following philosophy. In quiescent galactic nuclei, hydrodynamic forces are negligible and the orbits of stars will be shaped by (i) the smooth background gravitational potential of the SMBH and the stellar population; (ii) discrete, pair-wise scatterings with other stellar-mass objects; (iii) coherent secular effects arising from correlations between the orbits of multiple stars.[1] In the continuum limit, the DF of a population of identical stars can be written in position space as $f(\vec{r}, \vec{v})$. In a sufficiently old galactic nucleus, the DF will exist in a quasi-steady state, but two-body scatterings and collisionless effects will cause it to evolve adiabatically. We call such an old nucleus "relaxed," in contrast to a nucleus with a younger stellar population that was born far from a quasi-steady state configuration; such a young nucleus would be "unrelaxed."

A full knowledge of the DF $f(\vec{r}, \vec{v})$ and the gravitational potential $\Phi(\vec{r})$ would allow us to compute osculating stellar orbits, their temporal evolution, and the rate at which stars pass near the SMBH. But in order to understand which orbits are doomed to disruption, we must first introduce at least approximate hydrodynamic criteria for the disruption process. Fortunately, these approximations are reasonably accurate.

Tidal forces increase as one approaches a black hole. The strength of the tidal field diverges as distance from the singularity $r \to 0$, so interior to some critical distance R_t (the tidal radius, Eq. (1)), any macroscopic object will be torn apart. By equating the Newtonian tidal field to the victim star's surface self-gravity, we find that a star of mass $M_\star$ and radius $R_\star$ will be torn apart by tides if its pericenter R_p is roughly[2] within a tidal radius,

$$R_t = R_\star \left(\frac{M_\bullet}{M_\star} \right)^{1/3}, \tag{1}$$

of an SMBH with mass $M_\bullet$. This Newtonian expression is reasonably accurate when $R_t \gg R_g$, where the latter is the gravitational radius $R_g \equiv GM_\bullet/c^2$. But, as we shall see in Sect. 3.5, when $R_t \sim R_g$, the tidal radius will depend significantly on SMBH spin $\chi_\bullet$, and the orbital inclination ι.

Until Sect. 3.5, however, we will treat Eq. (1) as exact. With this approximation, we can immediately note one (slightly counter-intuitive) feature of tidal disruption. Any SMBH will have an event horizon comparable in size to its gravitational radius, R_g. We may note that

[1] As we see later in Sect. 3.4, factor (iii) is generally unimportant for determining TDE rates.

[2] In reality, the exact criterion for tidal disruption of a main sequence star is that $R_p < R_t/\beta_{crit}$, where $\beta_{crit} \approx 0.95$–$1.85$ is a dimensionless constant dependent on the central concentration of the star, and can be measured precisely with numerical hydrodynamics simulations (Guillochon and Ramirez-Ruiz 2013; Mainetti et al. 2017; see also the Disruption Chapter).

$R_t \propto M_\bullet^{1/3}$, while $R_g \propto M_\bullet^1$. The weak power-law dependence of R_t on $M_\bullet$ means that, for any given stellar type, there is a *maximum black hole mass* capable of producing a TDE outside the event horizon. TDEs from SMBHs above this critical mass, often called the Hills mass (Hills 1975), will produce debris wholly swallowed by the horizon, and will fail to produce an electromagnetically luminous flare. Approximating the event horizon size as $2R_g$, we find that the Hills mass is

$$M_H = M_\star^{-1/2}\left(\frac{c^2 R_\star}{2G}\right)^{3/2} \approx 1\times 10^8 M_\odot \left(\frac{R_\star}{R_\odot}\right)^{3/2}\left(\frac{M_\star}{M_\odot}\right)^{-1/2}. \tag{2}$$

Equations (1) and (2) are fundamentally Newtonian expressions that we generalize to a relativistic context in Sect. 3.5. A relativistic treatment is necessary to account for the effects of SMBH spin, which can sizeably alter M_H (Beloborodov et al. 1992; Kesden 2012).

In both the Newtonian and relativistic regimes, it is clear that, so long as TDE rates depend on R_t (the exact dependence is non-trivial and will be quantified in Sect. 3), they will depend on the mass, radius, and (to a weaker extent) internal structure of the star. These quantities evolve over the lifetime of a star, so that the tidal radius ultimately depends on the star's zero-age main sequence (ZAMS) mass, its metallicity, its age, its spin, and even its binarity (see Sect. 3.6). In the general relativistic context (appropriate when $R_p \sim R_g$), SMBH spin $\chi_\bullet$ and spin-orbit misalignment ι may affect disruption rates as well.

If we care not only about *intrinsic* rates of tidal disruption (e.g. per-galaxy rates, volumetric rates, etc.), but rates of TDE detection in current or planned surveys, then we must consider additional properties of these events. For example, we may define the strength of the TDE with a dimensionless penetration parameter

$$\beta = R_t/R_p. \tag{3}$$

When $\beta \approx 1$, we have a relatively mild, grazing disruption; if $\beta \gg 1$, we have a more violent, deeply plunging disruption; if $\beta \lesssim 1$, a partial disruption may ensue. The observational properties of a TDE flare may depend in various ways on β (Carter and Luminet 1983; Stone et al. 2013; Dai et al. 2015), meaning that distributions of this parameter are an important prediction for event rate calculations.

As this discussion illustrates, TDE rates depend on a large number of variables related to the participating stars and SMBHs themselves. Fortunately (for the sake of minimizing astrophysical uncertainty), intrinsic TDE rates depend only weakly on stellar properties, through R_t. However, the properties of host galaxies, and their SMBHs, matter a great deal more in setting TDE rates. In the next section, we will compute first-principles rates of tidal disruption in idealized galactic nuclei, providing the theoretical framework to understand how TDE rates vary with host galaxy properties.

3 Loss-Cone Theory

In this section, we overview, from a theoretical perspective, the many ways in which stars can be fed to massive black holes. As we shall see, the dynamical evolution of stellar orbits from "safe" to "unsafe" trajectories is most easily visualized and quantified with the concept of a loss cone, which we define in Sect. 3.1. We present the physics of loss cones in spherical galactic nuclei, as well as their connections to TDE rates, in Sects. 3.2 and 3.3. The first of these sections is focused on steady-state loss-cone physics, while the second focuses on the time-dependent problem. We consider more general (aspherical) geometries in Sect. 3.4, which opens up new avenues for loss-cone repopulation. In Sect. 3.5, we generalize our treatment of the loss cone from Newtonian to general relativistic gravity. Finally, in Sect. 3.6, we briefly survey other types of SMBH loss cones, relevant for processes beyond main-sequence stellar disruption.

3.1 The Loss Cone

In Newtonian gravity, a star on a nearly parabolic orbit (eccentricity $e \approx 1$) with pericenter R_{p} will have specific orbital angular momentum $L \approx \sqrt{2GM_{\bullet}R_{\mathrm{p}}}$. This implies that stars with specific orbital angular momentum below the critical value (or, equivalently, with $\beta > 1$),

$$L_{\mathrm{t}} = \sqrt{2GM_{\bullet}R_{\mathrm{t}}}, \tag{4}$$

will be disrupted when passing within R_{t} of the SMBH. One can define analogous "loss regions" in angular momentum space for destructive processes other than tidal disruption, and we will explore these later on, in Sect. 3.6.

We are generally interested in the evolution of the stellar DF in phase space on timescales much longer than the orbital time T_{orb}, and will therefore describe populations of stars in the space of integrals of motion ("integral space"), rather than the phase space of $\vec{r}$ and $\vec{v}$. It is convenient to define a new variable $\mathcal{R} \equiv [L/L_{\mathrm{circ}}(E)]^2 \in [0, 1]$, where $L_{\mathrm{circ}}(E)$ is the specific angular momentum of a circular orbit with the given specific energy E; in a Keplerian potential, $\mathcal{R} = 1 - e^2$. The number $N(E, \mathcal{R})$ of stars per unit interval of E and $\mathcal{R}$ is related to the DF $f(E, \mathcal{R})$ as (Merritt 2013, Equation 5.166)

$$N(E, \mathcal{R}) = 4\pi^2 T_{\mathrm{orb}}(E, \mathcal{R}) L_{\mathrm{circ}}^2(E) f(E, \mathcal{R}). \tag{5}$$

Since we are interested mainly in the low-$\mathcal{R}$ region, it is convenient to ignore the dependence of orbital time on $\mathcal{R}$ (which is usually weak) and write $N(E, \mathcal{R}) \approx g(E) f(E, \mathcal{R})$, where $g(E)$ is the density of states, defined more generally as:

$$g(E) \equiv 4\pi^2 L_{\mathrm{circ}}^2(E) \int_0^1 T_{\mathrm{orb}}(E, \mathcal{R})\, \mathrm{d}\mathcal{R}. \tag{6}$$

The number of stars per unit energy is $\overline{N}(E) \equiv \int_0^1 N(E, \mathcal{R})\, \mathrm{d}\mathcal{R}$. An isotropic distribution of stars in velocity corresponds to a uniform distribution in $\mathcal{R}$: $N(E, \mathcal{R}) \approx \overline{N}(E)$.

The region of phase space with $L \le L_{\mathrm{t}}$ is colloquially called the loss cone (a term borrowed from plasma physics), which we illustrate in Fig. 1. If angular momentum were conserved for every stellar orbit, the number of stars inside the loss cone would drop to zero within one orbital period. However, L changes with time due to various processes: classical two-body and resonant relaxation, and torques induced by an aspherical stellar potential. Therefore, the time-averaged number of stars inside the loss cone is nonzero, and the tidal disruption rate per unit energy is

$$\mathcal{F}_{\mathrm{t}}(E) \equiv \frac{1}{T_{\mathrm{orb}}(E)} \int_0^{\mathcal{R}_{\mathrm{t}}(E)} N(E, \mathcal{R})\, \mathrm{d}\mathcal{R},$$
$$T_{\mathrm{orb}}(E) \equiv T_{\mathrm{orb}}(E, \mathcal{R} = 0), \qquad \mathcal{R}_{\mathrm{t}} \equiv \left(\frac{L_{\mathrm{t}}}{L_{\mathrm{circ}}(E)}\right)^2. \tag{7}$$

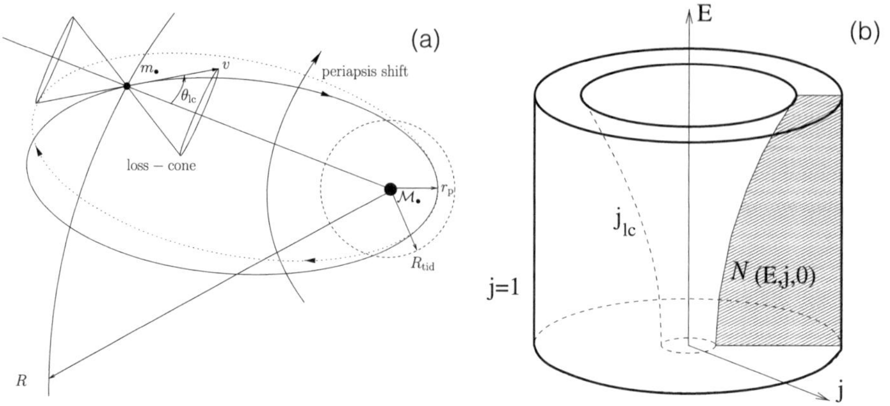

Fig. 1 Schematic representations of the loss cone in a spherical galactic nucleus. On the left (panel **a**, taken with permission from Amaro-Seoane 2018), we see the conical velocity-space loss region at an instantaneous point along a star's elliptical orbit around a supermassive black hole. The tidal radius (denoted $r_{\rm tid}$ in this image) will overlap with the star's pericenter if its instantaneous velocity vector $\vec{v}$ falls into the loss cone. On the right (panel **b**, taken with permission from Milosavljević and Merritt 2003), an orbit-averaged loss cone is depicted in the space of (negative) specific orbital energy $\mathcal{E} \equiv -E$ and angular momentum $j = \mathcal{R}^{1/2} = L/L_{\rm circ} \leq 1$. The loss cone becomes larger and larger in the space of dimensionless angular momentum as one moves to more tightly bound orbits (larger $|E|$).

Much of the remaining story revolves around estimating the equilibrium loss cone population, which is set by the efficiency of those processes that change the angular momenta of stars. As a useful benchmark, consider a situation where these processes are extremely rapid, so that the angular momentum distribution is nearly uniform (isotropic). The corresponding rate of disruptions is called the (isotropic) full loss cone flux:

$$\mathcal{F}_{\rm iso}(E) = \frac{\overline{N}(E)\mathcal{R}_{\rm t}(E)}{T_{\rm orb}(E)}. \tag{8}$$

In a more complicated situation, the loss-cone population may be either smaller or larger than this reference value, depending on both the initial conditions and the efficiency of angular momentum mixing.

3.2 Two-Body Relaxation

A mechanism that operates in all stellar systems is two-body (or collisional) relaxation, caused by the discreteness of the stellar distribution. In the classical, "Chandrasekhar," theory of two-body relaxation, the evolution of the stellar DF is driven by uncorrelated two-body encounters. We will begin, for simplicity, with a spherically symmetric nuclear star cluster surrounding a SMBH. The star cluster has local three-dimensional density $n(r)$ and velocity dispersion $\sigma(r)$ in coordinate space (here r is the distance from the SMBH), but its time evolution is most simply described by the Fokker–Planck equation in integral space, i.e. for $N(E, L, t)$. We may define a local relaxation time as

$$T_{\rm rel} \equiv \frac{\sigma^3}{G^2 n \langle m_\star^2 \rangle \ln \Lambda}, \tag{9}$$

where $\ln \Lambda \approx 10$ is the Coulomb logarithm (Chandrasekhar 1942; Spitzer and Hart 1971). This expression assumes a present-day mass function (PDMF) of stars, $\mathrm{d}N/\mathrm{d}m_\star$, defined over the interval $[m_{\min}, m_{\max}]$. The first and second moments of the PDMF are defined as

$$\begin{aligned} \langle m_\star \rangle &= \int_{m_{\min}}^{m_{\max}} m_\star \frac{\mathrm{d}N}{\mathrm{d}m_\star} \mathrm{d}m_\star \\ \langle m_\star^2 \rangle &= \int_{m_{\min}}^{m_{\max}} m_\star^2 \frac{\mathrm{d}N}{\mathrm{d}m_\star} \mathrm{d}m_\star . \end{aligned} \tag{10}$$

Notably, the rate of two-body relaxation is controlled by the second moment, $\langle m_\star^2 \rangle$, of the PDMF, indicating that for realistic stellar mass functions, it is typically the heaviest surviving species (often stellar-mass black holes) which dominate relaxation rates, especially at small radii where their relative fraction is higher due to mass segregation. For example, if one adds a population of stellar-mass black holes to a truncated Kroupa PDMF, the TDE rate can increase by up to a factor of ≈ 5 (Stone and Metzger 2016), depending on the assumed metallicity (which determines the relation between zero-age main sequence masses and compact remnant masses, e.g. Belczynski et al. 2010).

Over the characteristic relaxation timescale $T_{\rm rel} \gg T_{\rm orb}$, the typical change in energy is $\mathcal{O}(E)$ and in angular momentum is $\mathcal{O}(L_{\rm circ}(E))$. However, far smaller changes in angular momentum $\mathcal{O}(L_{\rm t})$ are needed to move stars on nearly radial orbits into or out of the loss cone: the time needed to produce characteristic changes in dimensionless angular momentum $\sim \mathcal{R}$ is $\sim \mathcal{R} T_{\rm rel}$. Because the relaxation of stars in angular momentum *near the loss-cone boundary* occurs much faster than relaxation in energy (Frank and Rees 1976), in the first approximation these types of relaxation can be considered separately. As we will discuss shortly, this two-timescale decoupling establishes a (quasi-)steady-state distribution of stars in L (or $\mathcal{R}$) at fixed E, which remains quasi-isotropic even in the presence of a loss cone (Lightman and Shapiro 1977; Cohn and Kulsrud 1978).

Over longer timescales, stellar diffusion in energy space gradually drives the stellar distribution towards a quasi-steady-state $\overline{N}(E)$ profile. If the gravitational potential is dominated by the $\Phi \propto 1/r$ potential of the SMBH, this steady-state solution is known as the Bahcall–Wolf cusp, with $f(E) \propto |E|^{1/4}$. In coordinate space, the Bahcall and Wolf (1976) distribution translates to a spherically symmetric density profile[3] $n(r) \propto r^{-7/4}$, extending out to some fraction of the SMBH influence radius $r_{\rm inf}$, which we define to be the radius enclosing a total stellar mass equal to $M_\bullet$. Notably, this is a *zero-flux solution* in energy space: in the limit as $R_{\rm t}/r_{\rm inf} \to 0$, the energy-space flux to the SMBH also goes to zero.

Accordingly, we will focus for now on the angular momentum diffusion,[4] by considering the one-dimensional, orbit-averaged Fokker–Planck equation for the evolution of the stellar distribution $N(\mathcal{R}, t)$ in the space of dimensionless angular momentum $\mathcal{R}$, at a fixed energy E (the dependence of various quantities on E is implied in the rest of this section, but is not explicitly marked):

$$\frac{\partial}{\partial t} N(\mathcal{R}, t) = \frac{\partial \mathcal{F}}{\partial \mathcal{R}}, \qquad \mathcal{F} \equiv \mathscr{D}(E, \mathcal{R}) \frac{\partial N}{\partial \mathcal{R}}, \qquad \mathscr{D}(E, \mathcal{R}) \approx \mathcal{D}(E) \mathcal{R}. \tag{11}$$

[3] If a broad spectrum of stellar masses exist, it is generally the heaviest species that relaxes to the $n(r) \propto r^{-7/4}$ profile, while lighter species will achieve shallower, $n(r) \propto r^{-3/2}$ distributions (Bahcall and Wolf 1977). However, strong mass-segregation can give rise to steeper distributions, as in Alexander and Hopman (2009), Preto and Amaro-Seoane (2010), Aharon and Perets (2016).

[4] For a less rigorous—but in some respects more intuitive—approach operating entirely in coordinate space, see the work of Syer and Ulmer (1999), which obtains qualitatively similar results to those presented here.

As before, $\mathcal{F}$ is the flux of stars through integral space in the direction of the inner boundary[5] $\mathcal{R}_t$, and the orbit-averaged diffusion coefficient $\mathcal{D}(E) \sim T_{\rm rel}^{-1}$ depends on various moments of the stellar DF (Merritt 2013, Equation 6.31)

$$\begin{aligned}
\mathcal{D}(E) &= \frac{64\pi^2 G^2 \ln\Lambda \langle m_\star^2\rangle}{3L_{\rm circ}^2(E)T_{\rm orb}(E)} \int_0^{r_{\max}(E)} \mathrm{d}r\, \frac{r^2}{\sqrt{2[E-\Phi(r)]}}(2I_0 + 3J_{1/2} - J_{3/2}), \\
I_0(E) &= \int_E^0 f(E')\,\mathrm{d}E', \qquad (12)\\
J_n(E,r) &= \int_{\Phi(r)}^E \left[\frac{E'-\Phi(r)}{E-\Phi(r)}\right]^n f(E')\,\mathrm{d}E'.
\end{aligned}$$

Here $r_{\max}(E)$, the apocenter of a radial orbit with given specific energy, is the root of $\Phi(r_{\max}) = E$. In a relatively shallow stellar density profile, or at high binding energies, $I_0(E)$ is the dominant moment of the stellar DF, and diffusion is a quasi-local process in energy space. In a relatively steep stellar density profile, or at large radii, $J_n(E,r)$ will become of greater importance, and diffusion may become strongly non-local in energy space, with very tightly bound stars playing an important role in the evolution of more loosely bound bins of E. Over long timescales, the 1D Fokker–Planck approach of Eq. (11) may break down due to changes in the energy-space distribution of stars, $\bar{N}(E)$, and this breakdown will be hastened in bins of energy whose diffusion coefficients $\mathcal{D}(E)$ are dominated by contributions from distant regions of energy-space with shorter relaxation times. In general, however, the 1D approach is a self-consistent way to determine quasi-steady state conditions near the loss cone boundary, since the time for orbits to experience a change in $\mathcal{R} \sim \mathcal{R}_t$ is $\sim T_{\rm rel}(E)\mathcal{R}_t$.

These equations are complemented with a no-flux, $\partial N/\partial\mathcal{R} = 0$ outer boundary condition (at $\mathcal{R} = 1$), and an inner boundary condition (at $\mathcal{R}_t$) of the Robin type (a linear combination of the function and its derivative):

$$N(\mathcal{R}_t, t) - \alpha\mathcal{R}_t \left.\frac{\partial N}{\partial\mathcal{R}}\right|_{\mathcal{R}=\mathcal{R}_t} = 0, \qquad \alpha \approx \left(q^2 + q^4\right)^{1/4}, \qquad q \equiv \frac{\mathcal{D}T_{\rm orb}}{R_t}. \qquad (13)$$

The dimensionless quantity $q(E)$ describes the diffusivity of relaxational loss cone refilling. If the typical change of angular momentum per orbital time is larger than the size of the loss cone ($q \gg 1$, $\alpha \approx q$), the distribution of stars near and inside the loss cone is close to uniform (its gradient is small; $\partial N/\partial\mathcal{R} \ll N/\mathcal{R}$). In this case, we speak of a "full loss cone" or a "pinhole" regime where the flux through the loss-cone boundary is

$$\mathcal{F} \approx \mathcal{F}_{\rm full} \equiv \frac{N(\mathcal{R}_t)\mathcal{R}_t}{T_{\rm orb}}. \qquad (14)$$

This flux is proportional to the size of the loss cone but independent of the relaxation rate. Note that, unlike Eq. (8) for the isotropic full loss cone flux, the above expression assumes only that the DF is *locally* well-mixed near the loss-cone boundary. In the opposite limit ($q \ll 1, \alpha \approx q^{1/2}$), we are in an "empty loss cone" or "diffusive" regime, where $N(\mathcal{R}_t) \approx 0$ and the flux is limited by the relaxation.

[5] With this definition, the flux is positive if the stars diffuse towards the loss cone, as they usually do. Note that the sign convention is the opposite in some studies.

We may consider a quasi-steady-state solution of Eq. (11) in which the shape of the DF $N(\mathcal{R}, t)$ stays the same, but its overall normalization $\overline{N} \equiv \int_0^1 N(\mathcal{R})\, \mathrm{d}\mathcal{R}$ changes with time. This solution has a nearly logarithmic profile (Cohn and Kulsrud 1978):

$$N(\mathcal{R}) \approx \overline{N} \frac{\alpha + \ln(\mathcal{R}/\mathcal{R}_\mathrm{t})}{(\alpha - 1)(1 - \mathcal{R}_\mathrm{t}) + \ln(1/\mathcal{R}_\mathrm{t})}, \tag{15}$$

and the corresponding flux is

$$\mathcal{F} = \frac{\mathcal{D}\overline{N}}{(\alpha - 1)(1 - \mathcal{R}_\mathrm{t}) + \ln(1/\mathcal{R}_\mathrm{t})} = \frac{q\mathcal{F}_\mathrm{full}}{(\alpha - 1)(1 - \mathcal{R}_\mathrm{t}) + \ln(1/\mathcal{R}_\mathrm{t})}. \tag{16}$$

As expected, in the full-loss cone limit the DF is nearly uniform ($N(\mathcal{R}) \approx \overline{N}$), and the flux approaches the isotropic full-loss-cone value (Eq. (8)), which coincides with Eq. (14) for a steady-state solution (however, this is no longer true for a time-dependent solution with strongly anisotropic initial conditions, as we will discuss in Sect. 3.3). In the opposite, empty-loss cone limit, the DF falls sharply to zero for $\mathcal{R} \lesssim \mathcal{R}_\mathrm{t}$, and the flux is

$$\mathcal{F}_\mathrm{empty} \equiv \frac{\mathcal{D}\overline{N}}{\ln(1/\mathcal{R}_\mathrm{t}) + R_\mathrm{t} - 1}. \tag{17}$$

This empty loss cone flux is proportional to the relaxation rate but only weakly dependent on the size of the loss cone. The steep drop in the DF below R_t implies that the TDE rate will be dominated by grazing encounters with $\beta \approx 1$ in the empty loss-cone regime (see Sect. 4.2).

The transition between empty and full loss cone regimes corresponds to $\mathcal{F} \approx \frac{1}{2}\mathcal{F}_\mathrm{full}$, or $q \approx \ln(1/\mathcal{R}_\mathrm{t}) \sim \mathcal{O}(10)$. This occurs at the "critical" energy E_crit, or the corresponding radius r_crit. The total TDE rate[6]

$$\dot{N} \equiv \int_{-\infty}^{0} \mathcal{F}(E)\mathrm{d}E \tag{18}$$

will contain contributions from both empty- and full-loss cone regimes. If $r_\mathrm{crit} \ll r_\mathrm{infl}$, then most of the total loss cone flux arises from $E \sim E_\mathrm{crit}$, and both the empty- and full-loss cone regimes contribute an $\mathcal{O}(1)$ fraction of the TDE rate.[7] Conversely, if $r_\mathrm{crit} \gg r_\mathrm{infl}$, then most of the integrated loss cone flux comes from $E \sim -GM_\bullet/(2r_\mathrm{infl})$, and is predominantly in the empty loss cone regime (Syer and Ulmer 1999). Astrophysical galactic nuclei typically possess $r_\mathrm{crit} \sim r_\mathrm{infl}$ (Wang and Merritt 2004), as we will discuss later in Sect. 4.1.

3.3 Anisotropic and Time-Dependent Loss Cones

So far, we have estimated the loss cone flux $\mathcal{F}(E)$ and TDE rate $\dot{N}$ in a relatively old galactic nucleus, one which has reached a quasi-steady state, quasi-isotropic distribution of

[6] In general, closed-form expressions for the variables in this section do not exist. However, for the special case of a singular isothermal sphere density profile, with $n(r) \propto r^{-2}$, Wang and Merritt (2004) provide analytic expressions for $q(E)$, $\mathcal{F}(E)$, and $\dot{N}$.

[7] An interesting caveat to this discussion concerns extremely steep stellar cusps, i.e. those with density profiles falling off faster than $n(r) \propto r^{-9/4}$. Such density profiles are not self-consistent, because they predict that, as $E \to -\infty$, $\mathcal{F}_\mathrm{empty} \to \infty$ (Syer and Ulmer 1999). Density profiles of this steepness are rarely seen in nature, with the possible exception of post-starburst galaxies, which we discuss later Sect. 5.2.1.

orbital angular momentum in each bin of orbital energy (i.e. Eq. (15)). This quasi-steady state solution is attained on a timescale $\sim T_{\rm rel}$. At earlier times, the angular momentum distribution may be quite anisotropic, and in this regime, the capture rate depends sensitively on the initial conditions. In certain types of galactic nuclei, where phenomena other than two-body relaxation are at work, it is not even clear that we can expect the angular momentum distribution to converge to Eq. (15). For example, a galactic nucleus with a long-lived SMBH *binary* will preferentially eject stars on radial orbits, continually pumping up the tangential anisotropy of the star cluster (Milosavljević and Merritt 2003; Merritt and Wang 2005). The presence of a massive gas disc (i.e. an *active* galactic nucleus) will affect stellar orbits in a more complicated way (Karas and Šubr 2007). The evolution of the nuclear stellar cusp and its build-up through star formation and/or cluster infall will also affect TDE rates (Aharon et al. 2016). In this section, we will ignore these complications, and focus on DFs $f(E, \mathcal{R}, t)$ for spherical galactic nuclei evolving only due to two-body relaxation. We consider more general loss-cone physics in aspherical potentials in Sect. 3.4.

In the limit of spherical symmetry and arbitrary initial conditions $f(E, \mathcal{R}, 0)$, it is possible to use Fourier-Bessel synthesis techniques to derive exact, time-dependent solutions to the Fokker–Planck equation in angular momentum space. These solutions can then be converted into TDE rates via Eq. (11) ($\mathcal{F} \propto \partial N/\partial \mathcal{R}$). The semi-analytic Fourier-Bessel solutions were first derived in the empty loss cone limit (Milosavljević and Merritt 2003), but can also be applied for more general inner boundary conditions (Lezhnin and Vasiliev 2015). While these solutions are useful, they are too lengthy to reproduce and explore in this review, so we will focus instead on results from the numerical solution of the time-dependent Fokker–Planck equation (Eq. (11)).

The anisotropy of a stellar distribution can be quantified with the parameter (usually called β, but here we use b to avoid confusion with Eq. (3))

$$b(E) = 1 - \frac{T_\perp(E)}{2T_\parallel(E)}, \tag{19}$$

which is a function of the total radial ($T_\parallel$) and tangential ($T_\perp$) kinetic energies of stellar orbits. This definition can be related to the DF as $f(E, \mathcal{R}) = f_E(E)\mathcal{R}^{-b(E)}$, where $f_E(E)$ is a function independent of angular momentum, although more complicated (non-separable) DFs can have the same level of anisotropy. When $b = 0$, orbital velocities and angular momenta are distributed isotropically; when $b > 0$, there is a radial orbit bias; when $b < 0$, there is a tangential orbit bias. In the case of a purely isotropic initial distribution ($b = 0$; flat in $\mathcal{R}$), the initial TDE rates $\mathcal{F}(E) \approx \mathcal{F}_{\rm iso}(E)$ (Eq. (8)), even at high binding energy where $q(E) \ll 1$. In these energy bins, stars at $\mathcal{R} < \mathcal{R}_{\rm t}$ are removed on a timescale $T_{\rm orb}$, creating a steep gradient near the loss-cone boundary. As the angular momentum distribution is progressively depleted at small but increasing $\mathcal{R}$ and approaches the Cohn–Kulsrud quasi-steady state profile (Eq. (15)), the gradient $(\partial N/\partial \mathcal{R})_{\mathcal{R}=\mathcal{R}_{\rm t}}$ will soften, and the rates will decline and approach the steady-state value (Eq. (16)). Because the Cohn–Kulsrud quasi-steady state is nearly isotropic (more specifically, a logarithmic function of $\mathcal{R}$), isotropic initial conditions produce time-dependent TDE rates that are not far from the steady-state expectation of Eq. (16).

Larger deviations from Eq. (16) will occur if the initial conditions are strongly anisotropic. For example, a galactic nucleus may inherit a strong tangential anisotropy in the aftermath of a SMBH binary merger. As two SMBHs inspiral, they eject those stars which pass within the orbit of the binary, scouring out a cavity in angular momentum space and depleting radial orbits (Milosavljević and Merritt 2003). This creates a gap in the initial distribution $N(\mathcal{R}, t=0)$ for $\mathcal{R} \le \mathcal{R}_{\rm gap} \gg \mathcal{R}_{\rm t}$. After the SMBHs have merged, the TDE rate will be suppressed until the angular momentum gap is refilled, typically over a timescale $\sim 10^{-2}$–$10^{-1} T_{\rm rel}$ (Merritt and Wang 2005; Lezhnin and Vasiliev 2015).

The opposite situation occurs when the initial distribution has an excess of stars with low angular momentum (a radially-biased velocity distribution). This may occur naturally[8] as a result of nuclear cluster buildup through infalling globular clusters (Hartmann et al. 2011, although many clusters are likely to disrupt far from the SMBH; Perets and Mastrobuono-Battisti 2014). In the case of eccentric cluster infall, stars will be left behind on preferentially radial orbits. If we assume an idealized, initial radial anisotropy of b across all bins of energy, then the TDE rate will initially be elevated and then decline with time roughly as $\mathcal{F}(E) \approx \mathcal{F}_{\rm full}(E) \times [t/T_{\rm rel}(E)]^{-b}$ (Stone et al. 2018). Despite the initially high rates of this scenario, this formula demonstrates (since $b \le 1$ definitionally) that the integrated total number of TDEs will be dominated by late times, once the distribution has become quasi-isotropic.

3.4 Asphericity and Collisionless Loss Cone Refilling

The classical loss-cone theory of Sects. 3.2 and 3.3 was developed in the 1970s for globular clusters, which are nearly spherical systems. However, galactic nuclei are, to some extent, non-spherical (see e.g. Lauer et al. 2005), and stellar orbital angular momentum is, therefore, not conserved even in the absence of two-body relaxation. Thus, even if the time-averaged angular momentum $\overline{L}$ of a star on a given orbit is large, the minimum value of angular momentum $L_{\rm min}$ reachable by this orbit may be smaller than $L_{\rm t}$, potentially bringing a much larger reservoir of stars into the loss cone.

In perfectly axisymmetric systems, one component of angular momentum (L_z) is still conserved, so $L_{\rm min} \ge L_z$; however, in triaxial or even less symmetric potentials, $L_{\rm min}$ may be zero for a significant fraction of orbits; specifically, those in "centrophilic" orbit families, such as box, pyramid or chaotic orbits (Poon and Merritt 2001). More generally, we refer to all orbits with $L_{\rm min} \le L_{\rm t}$ as the "drain region,[9]" and denote the fraction of phase space occupied by these orbits at a given energy as $\mathcal{R}_{\rm drain}(E)$. Most of these orbits have low average angular momentum, i.e. $\overline{L} \ll L_{\rm circ}(E)$, although not all orbits with $\overline{L} \ll L_{\rm circ}(E)$ are centrophilic. Nevertheless, for the sake of simplicity, we assume that the drain region is simply $\mathcal{R} < \mathcal{R}_{\rm drain}$.

There are two separate effects associated with the existence of a drain region. First, the scalar angular momentum changes significantly (by $\mathcal{O}(\epsilon L_{\rm circ})$) on a timescale $T_{\rm ang} \simeq \epsilon^{-1} T_{\rm prec}$, where ϵ is the (dimensionless) degree of non-sphericity (the relative difference between the three principal axes). Here the period of pericenter precession $T_{\rm prec} \equiv 2\pi(\Omega_r - \Omega_\phi)^{-1}$ is given by the difference between the frequencies of the radial and azimuthal motion: it is comparable to the orbital period $T_{\rm orb} \equiv 2\pi \Omega_r^{-1}$ outside the SMBH radius of influence, but is much longer than $T_{\rm orb}$ in a nearly Keplerian potential (see Equation 4.88 in Merritt 2013). In the drain region, the timescale for the scalar angular momentum to change by $\mathcal{O}(L_{\rm t})$ is typically shorter than $T_{\rm orb}$. Therefore, stars on drain orbits are in the full loss cone regime (Merritt and Poon 2004), and their capture rate is given by Eq. (14). Note that this does not imply that the capture rate is equal to the isotropic full loss cone rate of Eq. (8),

[8] But see also Arca-Sedda and Capuzzo-Dolcetta (2017) for counterexample simulations, where star cluster infall leads to a tangential bias. Ultimately, the final $b(E)$ profile is likely sensitive to the orbital properties of infalling star clusters.

[9] The drain region is often called the "loss wedge" in the axisymmetric case—e.g. Magorrian and Tremaine (1999).

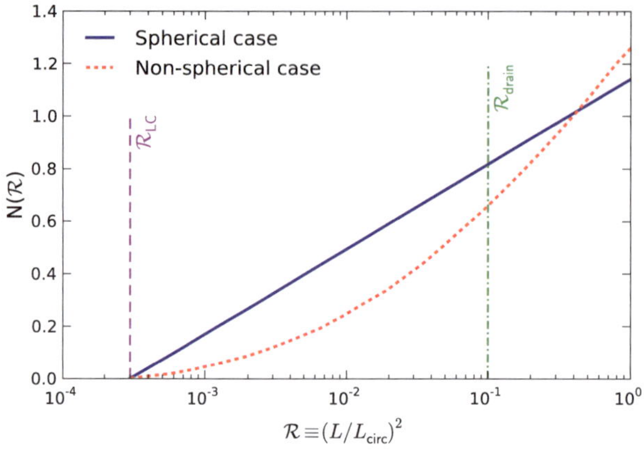

Fig. 2 Illustration of the quasi-steady-state DF in spherical (solid blue) and non-spherical (dotted red) galactic nuclei, for the case of an empty loss cone ($q = 0$ in Eq. (13)) with a fairly large loss-cone size ($\mathcal{R}_t = 3 \times 10^{-4}$). In the spherical case, the DF has a logarithmic profile (Eq. (15)), and the slope of this profile in the semi-log-scaled coordinates determines the flux. In the non-spherical case, the profile flattens out at $\mathcal{R} \lesssim \mathcal{R}_{\rm drain}$ ($= 0.1$ in this example), because the angular momenta of stars are shuffled by non-spherical torques. At the same time, the profile at $\mathcal{R} \gtrsim \mathcal{R}_{\rm drain}$ remains nearly logarithmic, as it is still determined by relaxation-driven diffusion, but it has a steeper slope (roughly a factor of two in this example), producing a correspondingly larger flux. Adapted from Fig. 5 of Vasiliev and Merritt (2013).

because the density of stars in the drain region $N(E, \mathcal{R} < \mathcal{R}_{\rm drain})$ may be different from the angular momentum-averaged density of stars $\overline{N}(E)$. We may now define the "drain timescale" needed to remove the stars from this region: $T_{\rm drain} \equiv T_{\rm orb}\mathcal{R}_{\rm drain}/\mathcal{R}_t$. It turns out that in axisymmetric systems, this time is still much shorter than the Hubble time, at least in those galactic nuclei with $M_\bullet \lesssim 10^8 M_\odot$ that are the main sources of TDEs. However, in triaxial systems, the fraction of centrophilic orbits, $\mathcal{R}_{\rm drain}$, could be $\gtrsim 0.1$ (it is proportional to the degree of non-sphericity ϵ), making the draining time longer than the Hubble time for most orbits (see Equations 6, 7 and Fig. 4 in Vasiliev 2014). During this time, the capture rate is determined by the initial conditions. In the "most neutral" case of a (nearly-)isotropic distribution in angular momentum, the TDE rate is of order the isotropic full loss cone rate of Eq. (8), which could be much higher than the relaxation-limited capture rate in purely spherical nuclei (Magorrian and Tremaine 1999).

After the initial population of stars in the drain region has been depleted, two-body relaxation again becomes the rate-limiting step. However, relaxation will still be aided by the existence of the drain region, which acts as an "extended" loss cone: one with a much larger surface area in phase space. After a star has diffused into the drain region via collisional relaxation, it will collisionlessly wander into the actual loss cone in a time $\lesssim T_{\rm drain}$ (unless it is scattered back into the higher-$\overline{L}$ region of phase space). The TDE rate from the loss wedge is technically still set by the full loss-cone rate of Eq. (14), but the phase-space density of stars in this region, $N(\mathcal{R}_t) \simeq N(\mathcal{R}_{\rm drain})$, may be much lower than $\overline{N}$, if the supply rate is diffusion-limited. In this case, the steady-state flux is given by the expression for the empty-loss cone regime of Eq. (17), but replacing $\ln(1/\mathcal{R}_t)$ with $\ln(1/\mathcal{R}_{\rm drain})$. Because of this logarithmic dependence, the actual enhancement of capture rate is at most a factor of few, as is illustrated by Fig. 2.

A different type of transient asphericity can occur in special, "degenerate" potentials where frequencies of motion are rationally commensurate with each other and orbits close. For example, in the Kepler potential of the SMBH, all three frequencies of motion are exactly equal for an individual star (so long as we neglect precession from the background stellar potential and relativistic effects). Because the stellar cusp is made of a finite number of stars, the combination of discreteness and closed orbits will create a statistical excess of time-averaged stellar mass in one direction: a temporary asphericity. An alternative view of this process is that pairwise encounters between nearby stars will be *correlated* over time, and therefore may coherently torque stellar orbits much more efficiently than do the *uncorrelated* effects from two-body relaxation. The resulting orbital evolution, known as *resonant relaxation* (Rauch and Tremaine 1996), has been proposed as a way to more efficiently refill empty loss cones (Hopman and Alexander 2006). However, general relativistic precession often prevents resonant relaxation from exciting stars to the high eccentricities needed to enter the loss cone (Merritt et al. 2011; Brem et al. 2014), and recent calculations suggest that its impact on the loss-cone flux is fairly minor in realistic galactic nuclei (e.g., Merritt 2015a, or Fig. 17 in Bar-Or and Alexander 2016).

To summarize, the TDE rate in non-spherical systems is larger than in spherical ones, if most of the flux in the latter is delivered in the diffusion-limited (empty loss cone) regime. In Sect. 4.1, we will see that in denser galactic nuclei, with $M_\bullet \lesssim 10^6 M_\odot$, this enhancement is small, while for $M_\bullet \gtrsim 10^8 M_\odot$ it could be more than an order of magnitude. However, these larger galaxies are (i) rarer and (ii) have difficulty producing luminous TDEs ($M_\bullet \sim M_{\rm H}$), and so do not dominate the cosmic TDE rate of main-sequence stars.

3.5 General Relativistic Loss Cone Theory

Einstein's theory of general relativity (GR) differs from Newtonian gravity in several respects that may carry important observational consequences for TDEs. The first of these differences concerns the nature of gravity itself in the two theories. In Newtonian gravity, the SMBH exerts an inverse-square-law force on the star, pulling more strongly on the side of the star facing the SMBH than on the stellar center of mass. Equating the difference in the acceleration experienced by the stellar center of mass and surface (i.e. the tidal acceleration) to the star's self gravity and solving for the distance from the SMBH yields the Newtonian tidal radius $R_{\rm t}$ (Eq. (1)).

General relativistic gravity is interpreted instead as a non-vanishing spacetime curvature that determines the geodesics along which test particles travel. This spacetime curvature can cause initially parallel geodesics to deviate, leading to tidal disruption if the rate of geodesic deviation exceeds the star's self-gravity. The relativistic geodesic-deviation equation is most conveniently expressed in Fermi normal coordinates $\{\tau, X^{(i)}\}$ (Manasse and Misner 1963; Marck 1983; Luminet and Marck 1983). Here τ is the proper time along the central timelike geodesic on which the star's center of mass travels, and $X^{(i)}$ are Cartesian spatial coordinates in a spatial hypersurface orthogonal to this central geodesic. In these coordinates, the geodesic-deviation equation becomes

$$\frac{d^2 X^{(i)}}{d\tau^2} = -C^{(i)}{}_{(j)} X^{(j)}, \tag{20}$$

where

$$C^{(i)}{}_{(j)} = R^{\beta}{}_{\mu\alpha\nu} \lambda_{\beta}^{(i)} \lambda^{\mu}{}_{(0)} \lambda^{\alpha}{}_{(j)} \lambda^{\nu}{}_{(0)} \tag{21}$$

is the tidal tensor, $R^{\beta}{}_{\mu\alpha\nu}$ is the Riemann curvature tensor, and $\{\lambda^{\mu}{}_{(0)}, \lambda^{\mu}{}_{(i)}\}$ is the orthonormal tetrad of 4-vectors with respect to which the Fermi normal coordinates $\{\tau, X^{(i)}\}$ are defined. The symmetries of the Riemann tensor imply that the tidal tensor is a real, symmetric 3×3

matrix with three real eigenvalues and orthogonal eigenvectors. Equation (20) implies that the star will be stretched along the eigenvector corresponding to the tidal tensor's sole negative eigenvalue, V_-; equating the tidal acceleration in this direction at pericenter to the star's self-gravity yields the relativistic generalization of the tidal radius (Kesden 2012)

$$R_{\mathrm{t,GR}} = \left[\left(\frac{|V_-|}{2GM_\bullet/r^3}\right)\left(\frac{M_\bullet}{M_\star}\right)\right]^{1/3} R_\star. \tag{22}$$

As in the Newtonian case (Eq. (1)), the exact criterion for tidal disruption is $R_{\mathrm{p}} < R_{\mathrm{t,GR}}/\beta_{\mathrm{crit}}$, where $\beta_{\mathrm{crit}} \approx 1-2$ depends on the internal structure of the star. In the non-relativistic limit ($v \ll c, r \gg R_{\mathrm{g}}$), $R_{\mathrm{t,GR}}$ reduces to R_{t}.

Although there is a superficial similarity between the Newtonian and relativistic tidal radii of Eqs. (1) and (22), there are also important differences. Because the tidal tensor $C^{(i)}_{(j)}$ and thus its negative eigenvalue V_- depend on both the Riemann tensor and the stellar geodesic, the relativistic tidal radius depends on both the spacetime metric of the SMBH and the stellar 4-position and 4-velocity at pericenter, not just the distance from the SMBH. The spacetime near SMBHs is described in GR by the Kerr metric (Kerr 1963), which depends not just on the SMBH mass $M_\bullet$, but also on its dimensionless spin, $0 \le \chi_\bullet \le 1$ (Carter 1971).

The spin dependence of the SMBH's gravity is the second important difference between Newtonian gravity and GR. The SMBH spin breaks the spherical symmetry present in Newtonian point potentials, implying that, in the commonly used Boyer-Lindquist coordinate system (Boyer and Lindquist 1967), the Riemann tensor depends on both the radial coordinate r and the polar coordinate θ. The Kerr metric is stationary and axisymmetric, implying the existence of a specific energy E and angular momentum L_z that are conserved along geodesics. In addition, the Kerr metric has a Killing tensor that provides a conserved Carter constant Q (Carter 1968; Walker and Penrose 1970). In the non-relativistic limit, the Carter constant corresponds to the square of the magnitude of the component of the orbital angular momentum in the equatorial plane of the SMBH. This allows us to define an (effective) specific orbital angular momentum $L \equiv \sqrt{L_z^2 + Q}$, and an (effective) inclination $\iota = \cos^{-1}(L_z/\sqrt{L_z^2+Q})$ that are conserved along all Kerr geodesics. These considerations indicate that the relativistic tidal radius $R_{\mathrm{t,GR}}$ can, in principle, depend on the SMBH mass $M_\bullet$ and spin $\chi_\bullet$, but also on the stellar orbital energy E, inclination ι, and argument of pericenter ω. In practice, the dependence on E and ω can be neglected to high accuracy, allowing us to define the threshold for tidal disruption $L_{\mathrm{t,GR}}(\chi_\bullet, \iota)$ as the value of L for which

$$V_-(M_\bullet, \chi_\bullet, L, \iota) = -\frac{2GM_\star}{R_\star^3} \tag{23}$$

when evaluated at pericenter.[10]

A third significant difference between Newtonian gravity and GR is that, in the latter theory, a black hole is defined as an object possessing an event horizon, a hypersurface from within which even light cannot escape (Wald 1984). The tidal force exerted by a Newtonian point mass scales $\propto r^{-3}$ and can therefore become arbitrarily large. This implies that any Newtonian point mass is capable of tidal disruption, given a small enough pericenter. However, to produce an observable TDE in GR, a SMBH's tides must overcome the star's self gravity while avoiding direct capture of the tidal debris by the event horizon. Stars on

[10]The right-hand side of Eq. (23) should be multiplied by β_{crit}^3 to account for stellar structure.

parabolic orbits (and their resulting debris) will be captured when their specific orbital angular momentum L falls below a threshold $L_{\rm cap}(\chi_\bullet, \iota)$ that depends on both the SMBH spin and orbital inclination. The rate of observable TDEs in GR will therefore be given by the rate at which two-body relaxation or other processes described above drive stars onto orbits with $L_{\rm cap}(\chi_\bullet, \iota) < L < L_{\rm t,GR}(\chi_\bullet, \iota)$. The hierarchy of distance scales, $R_{\rm g} \ll r_{\rm infl}$, between the gravitational radius and the radius of influence (from which most tidally disrupted stars are scattered into the loss cone) ensures that GR only modifies the boundaries of the loss cone, but does not otherwise affect the process responsible for refilling it.

The capture threshold $L_{\rm cap}(\chi_\bullet, \iota)$ scales linearly with $M_\bullet$ on dimensional grounds, while to lowest order in $M_\star/M_\bullet \ll 1$, the threshold for tidal disruption $L_{\rm t,GR}(\chi_\bullet, \iota) \propto M_\bullet^{2/3}$, like its Newtonian counterpart $\sqrt{2GM_\bullet R_{\rm t}}$. This implies that SMBHs more massive than the relativistic Hills mass $M_{\rm H,GR}(\chi_\bullet)$, defined such that $L_{\rm cap}(\chi_\bullet, \iota) > L_{\rm t,GR}(\chi_\bullet, \iota)$ for all inclinations ι, will be incapable of producing observable TDEs. For Schwarzschild SMBHs ($\chi_\bullet = 0$), spherical symmetry is restored and the disruption and capture thresholds are inclination-independent. The relativistic Hills mass for a Schwarzschild SMBH is

$$M_{\rm H,GR} = \left(\frac{5c^6 R_\star^3}{128\beta_{\rm crit}^3 G^3 M_\star}\right)^{1/2} = \left(\frac{5}{16\beta_{\rm crit}^3}\right)^{1/2} M_{\rm H} = 10^{7.8}\beta_{\rm crit}^{-3/2} M_\odot, \tag{24}$$

where $\beta_{\rm crit}$ is the minimum penetration factor for tidal disruption and $M_{\rm H}$ is the Newtonian Hills mass given by Eq. (2). For high-mass stars like our Sun whose equation of state can be approximated by a polytropic index $\gamma = 4/3$, Newtonian hydrodynamic simulations suggest that $\beta_{\rm crit} \simeq 1.85$ (Guillochon and Ramirez-Ruiz 2013) implying $M_{\rm H,GR} \simeq 10^{7.4} M_\odot$ (Servin and Kesden 2017). For Kerr SMBHs ($\chi_\bullet > 0$), both $L_{\rm t,GR}(\chi_\bullet, \iota)$ and $L_{\rm cap}(\chi_\bullet, \iota)$ are monotonically increasing functions of inclination ι, but the direct capture threshold has a steeper dependence. This implies that the Hills mass $M_{\rm H,GR}(\chi_\bullet)$ will be determined by the condition $L_{\rm cap}(\chi_\bullet, \iota = 0°) = L_{\rm t,GR}(\chi_\bullet, \iota = 0°)$. For maximally spinning SMBHs ($\chi_\bullet = 1$), this limit can be as large as $\sim 10^9 M_\odot$ (Sponholz 1994; Ivanov and Chernyakova 2006; Kesden 2012).

We show the reduction in the observable TDE rate due to direct capture by the event horizon in Fig. 3 (Kesden 2012). The dashed black curve shows the TDE rate as a function of SMBH mass $M_\bullet$ for galaxies with a singular isothermal sphere ($\rho \propto r^{-2}$) stellar density profile and velocity dispersions σ set by the $M_\bullet - \sigma$ relation (Ferrarese and Merritt 2000; Gebhardt et al. 2000). These rates were calculated with a Newtonian loss cone refilled by two-body relaxation as described in Sect. 3.2 (Wang and Merritt 2004). The solid colored curves (corresponding to different SMBH spins as indicated in the caption) show the suppression of this TDE rate from the fraction of stars in the tidal disruption loss cone, $L < L_{\rm t,GR}(\chi_\bullet, \iota)$, that also lie within the direct capture loss cone,[11] $L < L_{\rm cap}(\chi_\bullet, \iota)$. We see that the event horizon has a negligible effect for $M_\bullet \lesssim 10^6 M_\odot$, and that SMBHs with $M_\bullet \approx 5 \times 10^8 M_\odot$ can still produce observable TDEs provided their spins are large enough. The predicted suppression of the TDE rate by direct capture for $M_\bullet \gtrsim 10^{7.5} M_\odot$ shown in Fig. 3 is consistent with the super-exponential cutoff in the TDE rate observed in a limited sample of twelve optically-selected TDE candidates (van Velzen 2018).

Calculating observable TDE rates in a fully self-consistent manner with the asymmetric, relativistic loss cones described above remains an open problem, but we can anticipate several qualitative features of the results. For Schwarzschild SMBHs, the tidal acceleration is

[11] This calculation assumes the full loss-cone limit, in computing the relativistic correction factor, although the per-galaxy TDE rates shown in Fig. 3 are based on loss-cone calculations using contributions from both the full and empty regions.

Fig. 3 The TDE rate Γ in an idealized singular isothermal sphere ($\rho \propto r^{-2}$) galactic nucleus, shown as a function of SMBH mass M. The black dashed line shows the analytic estimate from Wang and Merritt (2004) which neglects direct capture of stars by the event horizon. Colored lines show TDE rate curves corrected for this general relativistic effect, with the red, orange, green, blue and purple curves corresponding to SMBH spin magnitudes $\chi_\bullet$ with values of 0.0, 0.5, 0.9, 0.99, and 0.999, respectively. These calculations average over an isotropic distribution of incoming stellar orbits and assume all disrupted stars are of solar mass and radius with $\beta_{\rm crit} = 2^{-1/3}$. Taken with permission from Kesden (2012).

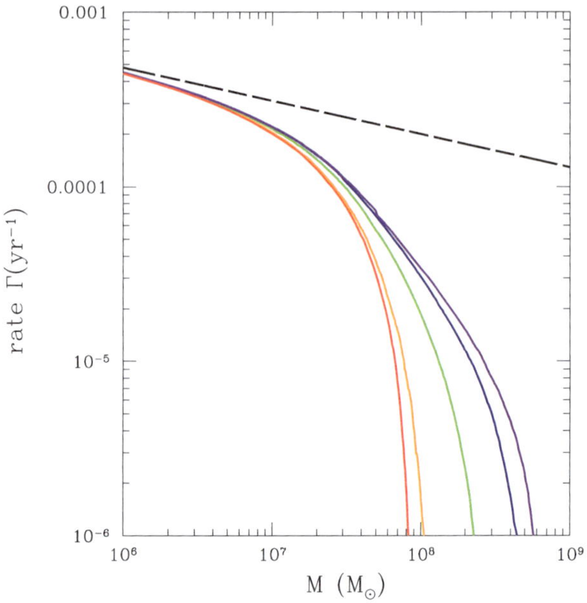

stronger in GR than on Newtonian orbits with the same angular momentum L (Servin and Kesden 2017). Equation (14) thus implies that, in the full loss-cone regime, the TDE rate will be enhanced by a factor of $(L_{\rm t,GR}/L_{\rm t})^2$ in the absence of direct capture by the event horizon. This factor is a monotonically increasing function of SMBH mass that reaches ≈ 2.7 for $M_\bullet = M_{\rm H,GR}(\chi_\bullet = 0)$, though by definition direct capture cannot be neglected for SMBHs near the Hills mass. In the *empty* loss cone regime, direct capture can be generally neglected so long as $L_{\rm t,GR}(\chi_\bullet, \iota) > L_{\rm cap,GR}(\chi_\bullet, \iota)$, but in this case the TDE rate enhancement is only by a factor $\approx \ln(L_{\rm t,GR}/L_{\rm t})$, as in Eq. (17). However, according to Eq. (13), the dimensionless factor $q(E)$ will be suppressed in GR by the same factor $(L_{\rm t,GR}/L_{\rm t})^2$, pushing more of the phase space into the empty loss-cone regime where the TDE rate is suppressed with respect to the full loss-cone regime by this factor. In either loss cone regime, these effects will be small when $M_\bullet \ll M_{\rm H,GR}$.

SMBH spin further complicates TDE rate predictions. The thresholds $L_{\rm t,GR}(\chi_\bullet, \iota)$ and $L_{\rm cap}(\chi_\bullet, \iota)$ are smaller for Kerr SMBHs than their Schwarzschild values when orbits are prograde ($\iota < 90^\circ$), and are larger when orbits are retrograde ($\iota > 90^\circ$). For isotropic distributions (flat in $\cos\iota$), even large spins have a modest $\lesssim 10\%$ effect on the $M_\bullet$-integrated TDE and capture rates in the full loss-cone regime (Young et al. 1977). However, we have already seen that spin can have *dramatic* effects on rates of observable TDEs from individual bins of SMBH mass when $M_\bullet \sim 10^8 M_\odot$ (Fig. 3). Furthermore, spin can have a significant effect on the inclination distributions of tidally disrupted and captured stars, with important observational consequences that we discuss further in Sect. 4.2, and in greater detail in the Formation of the Accretion Flow Chapter.

3.6 Other Loss Cones

So far, we have focused our attention on the loss cone relevant for standard TDEs: one centered on a massive black hole, with a boundary defined by the complete tidal disruption of a main sequence star. But the concept of a loss cone can be applied more generally to compute rates of other types of tidal disruptions. Some, such as the disruptions of binary or giant-branch stars by SMBHs, have event rates set by loss cone considerations not too dissimilar from those discussed already. More exotic types of tidal disruptions, such as "micro-TDEs" involving hyperbolic flybys of stars and stellar-mass black holes (Perets et al. 2016), or the short gamma-ray bursts produced by the quasi-circular inspiral of a neutron star into a stellar-mass black hole, are produced by quite different dynamical processes, and are therefore outside the purview of this Chapter. In this section, we briefly overview how loss cone physics is altered for different types of tidal disruptions: partial rather than full (Sect. 3.6.1), giant-branch rather than main sequence (Sect. 3.6.2), and binary rather than single (Sect. 3.6.3). Stars may also be tidally disrupted by *binary* SMBHs, but the underlying dynamics here are sufficiently complicated that they are left to the Binaries Chapter.

3.6.1 Partial Disruptions

Partial disruptions will occur when stars approach the central SMBH with $\beta < \beta_{\rm crit}$. As discussed previously, the exact value of $\beta_{\rm crit}$ is a number $\mathcal{O}(1)$ that depends on the internal structure of the victim star. For simple polytropic models, $\beta_{\rm crit}$ ranges from ≈ 0.92 (for a relatively fluffy, $n = 3/2$ polytrope, representative of lower main sequence stars) to 2.01 (for a more centrally concentrated $n = 3$ polytrope, representative of Sun-like stars). While these thresholds for full disruption are well-established for polytropic stellar models (Guillochon and Ramirez-Ruiz 2013; Mainetti et al. 2017), the corresponding $\beta_{\rm crit}$ values for more realistic stellar structures have yet to be determined. The exact threshold below which *no* mass loss occurs is also a function of stellar structure; for polytropic models, mass loss typically requires $\beta > \beta_{\rm min} \approx 0.5$ (Guillochon and Ramirez-Ruiz 2013).

Because the cross-section for partial disruption is substantially larger than that for full disruption, partial disruptions should be more common. This is clearly true in the empty loss-cone limit; when $q(E) \ll 1$, higher β values are exponentially suppressed.[12] In the full loss-cone regime, the number of stars $N(\mathcal{R})$ is roughly independent of $\mathcal{R}$ deep into the loss cone, meaning that the differential rate of disruptions $\mathrm{d}\dot{N}/\mathrm{d}\mathcal{R} \propto \mathrm{d}\dot{N}/\mathrm{d}r_{\rm p} \propto$ const. By a change of variables, this gives the differential rate $\mathrm{d}\dot{N}/\mathrm{d}\beta \propto \beta^{-2}$. The ratio of partial to full tidal disruptions will, in the $q(E) \gg 1$ limit, be

$$\frac{\dot{N}_{\rm partial}}{\dot{N}_{\rm full}} = \frac{\beta_{\rm min}^{-1} - \beta_{\rm crit}^{-1}}{\beta_{\rm crit}^{-1} - \beta_{\rm max}^{-1}}, \tag{25}$$

where $\beta_{\rm max}$ is the maximum penetration parameter that can avoid direct capture by the horizon. In Newtonian gravity, where the horizon may be approximated as an absorbing boundary at $2R_{\rm g}$,

$$\beta_{\rm max}^{\rm N} = \frac{R_\star c^2}{2GM_\bullet^{2/3} M_\star^{1/3}}. \tag{26}$$

In GR, $\beta_{\rm max} = \beta_{\rm max}^{\rm GR}(\chi_\bullet, \iota)$, and can be computed by determining both $L_{\rm t,GR}$ and $L_{\rm cap}$, although it is important to note that the definition of β becomes more ambiguous in relativistic gravity (Servin and Kesden 2017). For simplicity, we have so far discussed differential rates $\mathrm{d}\dot{N}/\mathrm{d}\beta$ in the two extreme limits of loss cone repopulation; a more sophisticated treatment of the intermediate, $q \sim 1$ case can be found in Strubbe (2011).

[12] Although we note that a power-law tail of high-β TDEs will occur, even when $q \ll 1$, due to the effects of strong scattering (Weissbein and Sari 2017).

3.6.2 Giant Stars

After a sufficient fraction of their hydrogen has been burnt, most stars will evolve off the main sequence and become giants, increasing their radial size by at least one order of magnitude. Stars with initial mass $M_\star \sim 1$–$8 M_\odot$ will spend most of their post-main sequence evolution on the red giant branch (RGB), with $R_{\rm RG} \sim 10 R_\odot$, and a smaller but dynamically important portion on the AGB branch with $R_{\rm AGB} \sim 100 R_\odot$.

Giant-branch stars are, consequently, much more vulnerable to partial tidal disruption: while their cores are no less dense than those of main sequence stars, their envelopes are distended and only weakly bound. If we reapply the results of Sect. 3.2, we therefore expect that per-star TDE rates of giants should be larger by a factor $\sim R_{\rm G}/R_\star$ in the full loss cone regime, but by a more modest factor $\sim \ln(R_{\rm G}/R_\star)$ in the empty loss cone regime. Because the total TDE rate $\dot{N}$ is an integral across these two regimes, it will typically follow a sublinear power law as shallow as $\dot{N} \propto (R_{\rm G}/R_\star)^{1/4}$ (MacLeod et al. 2012). While these arguments show that the per-star rate of giant disruption is higher than that for main sequence stars, this enhancement competes unsuccessfully with the much smaller number of giant-branch stars. As a result, the ratio between giant-branch and main sequence TDE rates is $\dot{N}_{\rm G}/\dot{N}_{\rm MS} \sim 0.1$ (Magorrian and Tremaine 1999; MacLeod et al. 2012). This ratio holds across a wide range of SMBH masses but breaks down above the main sequence Hills mass, when $M_\bullet \gtrsim M_{\rm H}$. Schwarzschild SMBHs with $10^8 M_\odot \lesssim M_\bullet \lesssim 10^9 M_\odot$ are generally unable to disrupt main sequence stars, and only produce luminous flares from giant disruptions. When $M_\bullet \gtrsim 10^9 M_\odot$, Schwarzschild SMBHs will become incapable of disrupting most giant-branch stars as well, and their only luminous flares will come from the small population of stars at the tips of the RGB and AGB (MacLeod et al. 2012).

The time-dependent radii of giants gives these stars a non-diffusive way to enter the loss cone, one that is inaccessible to main sequence stars: expanding in size until their loss cone grows to intersect their current orbit. "Growth into the loss cone" was first investigated by Syer and Ulmer (1999), who argued that this will be the dominant source of TDEs from evolved stars. However, this early work assumed that every giant disruption would be a full and highly luminous one. In reality, stars that grow onto the loss cone will, very likely, be "spoon-fed" to the SMBH in a series of tens to hundreds of very weak partial disruptions (MacLeod et al. 2013), making these events quite challenging to detect.

We note, however, that various dynamical processes can shorten relaxation times outside the influence radius (Perets et al. 2007; Hamers and Perets 2017). At these regions MS stars are typically already in the full loss cone regime and are not significantly affected by such processes. However, the empty loss cone for objects with larger tidal radii such as binary stars and giant stars extend to larger distances (as we discuss in the next section). Such processes can therefore increase the TDE rates of giants, and change the above-mentioned ratio, leading to a greater contribution of TDEs from giant stars.

3.6.3 Binary Stars

Binary stars—especially massive ones—can be as common as single stars. Their fate, when plunging along highly eccentric orbits towards a SMBH, can be richer than that of single stars. A binary star passing near a SMBH can be tidally separated with no tidal disruption of the single-star components. In this case, the individual stars may either recombine to reform a binary after pericenter passage, or undergo a three-body exchange interaction, with one star being captured around the black hole and the other being ejected with velocity in excess of the bulge escape speed (Hills 1988, 1991; Sari et al. 2010). These escapers may have already been observed as hypervelocity stars (HVSs) in the halo of our

Galaxy (e.g. Brown et al. 2018). Such outcomes most likely occur for binaries with internal semimajor axes $a_{\rm bin} \gg R_\star$, with a center-of-mass orbit just grazing the tidal separation radius $R_{\rm sep} = a(M_\bullet/m_{\rm bin})^{1/3}$ (here $m_{\rm bin}$ is the binary mass). Tidal disruption, or tidally-induced mergers, may result from either deeper encounters that cross the stellar tidal radius, $R_{\rm t} \approx (R_\star/a_{\rm bin})R_{\rm sep}$, or near-contact binaries for which $a_{\rm bin} \sim R_\star$ (Mandel and Levin 2015; Bradnick et al. 2017). For details on how the disruption and accretion processes are altered by binarity, we refer the reader to the Disruption Chapter and the Formation of the Accretion Flow Chapter. Here we will discuss properties of the stellar binaries' loss cone, the ensuing event rates, and their connections with HVSs.

As discussed in Sect. 3.2, the critical radius measures the size of the region that is in the empty loss cone regime, and it depends on the tidal radius of the system under consideration (and, strictly speaking, can be more precisely defined in energy space, a detail we neglect in this section). When the critical radius is smaller than the sphere of influence radius, $r_{\rm crit} \propto R_{\rm t}$, but in the opposite case, when $r_{\rm crit} > r_{\rm inf}$, $r_{\rm crit} \propto R_{\rm t}^{1/(4-\gamma)}$, where γ is the power-law slope of the nuclear stellar density profile (e.g. Eqs. 9–10 in Syer and Ulmer 1999). Empirically, observed nuclear density profiles typically have $0 \lesssim \gamma \lesssim 2$ (Lauer et al. 2005); in the Milky Way specifically, the central cusp appears shallower than a Bahcall–Wolf steady-state (Alexander and Hopman 2009), and recent measurements indicate $\gamma \approx 1.13$ (Schödel et al. 2018). The typical tidal separation radius of stellar binaries is larger than the single-star disruption radius by a factor $a_{\rm bin}/R_\star \sim \mathcal{O}(10)$, so we expect the critical radius for binary separation in the Milky Way to exist at several tens of parsecs (instead of the few parsecs expected for tidal disruption by Sgr A* in our Galactic Centre).

As sketched in Fig. 4, when two-body stellar relaxation is dominant and $\gamma < 9/4$, the *empty* loss cone (for binary separation) flux per unit bin of logarithmic radius increases outwards within the sphere of influence of the SMBH ($r_{\rm infl}$). On the other hand, it decreases outwards when $r > r_{\rm infl}$ (Lightman and Shapiro 1977). This implies that the loss cone flux for stellar binaries may peak around $r_{\rm infl}$, with a long tail out to ~ 100 pc. As described earlier, the peak of the loss cone flux for single-star disruption comes from $\min(r_{\rm crit}, r_{\rm infl})$, where typically $r_{\rm crit} \sim r_{\rm infl}$. Moreover, the empty loss cone flux is only logarithmically dependent on the loss cone size. These facts together might suggest that observational measurements of the rate of Galactic HVSs could be directly translated into constraints on TDE rates here.[13] This moment in time is especially propitious because of the ongoing data releases by the astrometric Galactic survey Gaia that have intensified searches for HVSs (e.g. Marchetti et al. 2017, 2018; Boubert et al. 2018; Bromley et al. 2018). Likewise, a comparison between rates of single and double TDE in other galaxies could provide an extra consistency check on the relationship between single and binary star loss cones. This comparison, however, requires more theoretical work in order to observationally distinguish these two types of transients (e.g. by the presence of a precursor in a double TDE, as in Bonnerot and Rossi 2019). In general, both comparisons need to account for the fraction of binaries being separated versus disrupted (Bradnick et al. 2017).

In principle, this is very exciting, but there are other physical ingredients—irrelevant for single-star disruption—that complicate the loss cone dynamics of binary stars. The high stellar density inside the sphere of influence means that most soft binaries will not survive "ionization" (or "evaporation") from continuous gravitational interactions with field stars around them (Perets et al. 2007; Hopman 2009; Perets 2009; Generozov et al. 2018). Hard binaries (progenitors of the fastest HVSs) can instead be driven to merger by magnetic braking. The overall result is a binary to single ratio of less than 10% at 1 pc for low mass

[13] We note that since the counterparts of HVSs can be captured around the MBH, the distribution of such stars could also reflect the processes leading to, and the rates of, TDEs (Perets and Gualandris 2010).

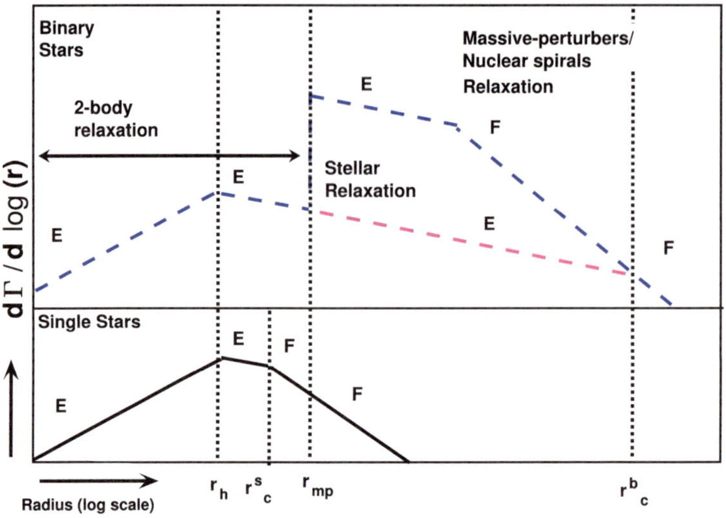

Fig. 4 A schematic representation of the local contribution to the loss-cone flux, for both binary and single stars. *Top*: the loss-cone flux for binary stars due to 2-body stellar relaxation (bottom dashed line) and due to massive perturbers (MP, top dashed line). Empty and full loss-cone regimes are denoted by "E" and "F", respectively. The radius $r_{\rm mp}$ corresponds to the region where such perturbers exist, likely just outside the innermost region of the nucleus. The critical radius for binaries and the sphere of influence radius are denoted respectively $r_{\rm c}^{\rm b}$ and $r_{\rm h}$. *Bottom*: the disruption rate of single stars. Here the presence of MPs coincides with the full loss cone regime and therefore little flux increase is expected. The critical radius for single stars is $r_{\rm c}^{\rm s} \sim r_{\rm h}$. Taken with permission from Perets et al. (2007), Fig. 3.

stars ($\le 1 M_\odot$). For more massive (but typically more rare) stars, the main sequence lifetime can be shorter than the evaporation timescale and the binary fraction would remain close to its birth value. Evaporation might therefore drastically suppress the HVS rate from $r \lesssim r_{\rm infl}$, limiting our ability to directly calibrate TDE rates from HVS or double disruption rates, with the partial exception of massive stars. On the other hand, outside the sphere of influence, dynamical relaxation can be dominated by "massive perturbers", such as giant molecular clouds, greatly shortening the relaxation timescale over that from two-body scatterings off stars. Since binary stars on loss cone orbits come from distances as far as $\sim 100 \times r_{\rm infl}$ and, unlike single stars, are in the empty loss cone regime (i.e. their flux *can* be increased), the presence of giant molecular clouds would enhance the HVS rate by a few orders of magnitude while leaving unaffected the predictions for TDEs (Perets et al. 2007; Hamers and Perets 2017). In summary, constraining the TDE rate by observing HVSs in our Galaxy or double TDEs in external galaxies is in principle possible, but not straightforward.

4 Applied Loss Cone Theory

By changing the underlying density profile $\rho(r)$ or DF $f(\epsilon)$ in a spherical, isotropic galactic nucleus, a motivated theorist can tune the TDE rate to any desired value. Asphericities and anisotropies offer further levers with which to change TDE rates. In order to produce astrophysically realistic TDE rate estimates, the underlying galaxy model must, in some way, be calibrated off observations. In this section, we will outline approximate but practical procedures for doing so. More specifically, we review how the theoretical loss cone physics of

Sect. 3 may be combined with observations to make *empirically-calibrated* TDE rate estimates in nearby galaxies (Sect. 4.1). We will then examine the implications that past rate estimates along these lines have for distributions of TDE observables (Sect. 4.2).

4.1 Simple Phase Space Modeling

In this subsection, we present a simple procedure for estimating TDE rates in individual galactic nuclei. The key assumptions of this procedure, which was first developed by Magorrian and Tremaine (1999) and Wang and Merritt (2004), are (i) spherical symmetry and (ii) nearly isotropic velocities, and this will be our starting point. Later on, we will also discuss more general prescriptions to account for geometrical asphericities or velocity anisotropies.[14] We begin with the observed 2D surface brightness profile, $I(R)$, which can be deprojected into a 3D luminosity density profile $j(r)$ using an Abel integral:

$$j(r) = \frac{-1}{\pi} \int_r^\infty \frac{\mathrm{d}I}{\mathrm{d}R} \frac{\mathrm{d}R}{\sqrt{R^2 - r^2}}. \tag{27}$$

Note that R is a projected 2D radius and r a 3D radius. We then use a mass-to-light ratio, Υ, to compute the 3D mass density $\rho(r) = \Upsilon j(r)$.

This mass density profile can be used to compute other quantities of relevance to us, such as the period of a radial orbit, $T_{\rm orb}(E)$, the stellar mass enclosed $M_\star(r)$, and the gravitational potential $\Phi(r)$:

$$M_\star(r) = \int_0^r 4\pi (r')^2 \rho(r') \mathrm{d}r' \tag{28}$$

$$\Phi(r) = -\frac{GM_\bullet}{r} - \frac{GM_\star(r)}{r} - 4\pi G \int_r^\infty \rho(r') r' \mathrm{d}r' \tag{29}$$

$$T_{\rm orb}(E) = 2 \int_0^{r_{\rm max}} \frac{\mathrm{d}r}{\sqrt{2(E - \Phi(r))}}. \tag{30}$$

Here we have also used the apocenter of a radial orbit, $r_{\rm max}(E)$. Next, under the assumption of (nearly) isotropic velocities,[15] we use Eddington's formula to compute a one-integral DF:

$$f(E) = \frac{-1}{\pi^2 \sqrt{8}} \frac{\mathrm{d}}{\mathrm{d}E} \int_E^0 \frac{\mathrm{d}n}{\mathrm{d}\Phi} \frac{\mathrm{d}\Phi}{\sqrt{\Phi - E}}. \tag{31}$$

In this integral, we convert stellar mass density to stellar number density $n(r) \equiv \rho(r)/\langle m_\star \rangle$. The difference between these two mass functions is merely the average stellar mass $\langle m_\star \rangle$, which is determined by the stellar PDMF as in Eq. (10).

With the isotropic DF $f(E)$ in hand, we may now apply the formalism of Sect. 3.2 to obtain the TDE rate, $\dot{N}$, of the galaxy in question. More specifically, we calculate the orbit-averaged diffusion coefficient $\mathcal{D}(E)$ using Eq. (12), use this to calculate the diffusivity

[14] The assumptions of spherical symmetry and quasi-isotropy are relaxed in Magorrian and Tremaine (1999), but for brevity we focus primarily on the simplest case.

[15] If the stellar density profile $n(r)$ is too shallow, the DF $f(E)$ obtained from Eq. (31) will have negative values, which is an unphysical outcome. In the limit of a Kepler potential, the shallowest self-consistently isotropic power-law density profile is $n(r) \propto r^{-1/2}$; shallower density profiles require some degree of tangential anisotropy to remain positive-definite in $f(E, \mathcal{R})$.

parameter $q(E)$ using Eq. (13), and then compute the loss cone flux $\mathcal{F}(E)$ as in Eq. (16). Integrating $\mathcal{F}(E)$ across all energies E gives the total TDE rate $\dot{N}$ (Eq. (18)).

In order to apply the above formalism to photometric observations of real galactic nuclei, we must consider a number of astrophysical uncertainties, including:

- The functional form of $I(R)$: ideally, one would operate with nonparametric data, although some smoothing may be necessary to ensure positivity of $f(E)$. However, in low-mass galaxies ($M_\bullet \sim 10^6 M_\odot$), the influence and critical radii are at best marginally resolved, so some degree of inward extrapolation in $I(R)$ is often necessary. Past works have typically employed power-law fits to the innermost isophotes (Wang and Merritt 2004; Stone and Metzger 2016), and the uncertainty produced by this extrapolation is in need of greater quantification.
- Choice of mass-to-light ratio Υ: past efforts to predict TDE rates through dynamical modeling of large samples of galaxies have made crude estimates for Υ. For example, Magorrian and Tremaine (1999) employed the scaling relationship for V-band luminosities $\Upsilon_{\rm V} = 4.9(M_\odot/L_\odot)(L_{\rm V}/10^{10}L_\odot)^{0.18}$ (Magorrian et al. 1998). Later works made virial estimates, starting with a galaxy's velocity dispersion σ, effective radius $R_{\rm eff}$, and luminosity L, then computing $\Upsilon_{\rm vir} = 2\sigma^2 R_{\rm eff}/(GL)$ (Wang and Merritt 2004; Stone and Metzger 2016). Both of these methods make the large assumption that Υ is constant throughout the galaxy; a more self-consistent method would apply simple stellar population models to multiband photometry of the galactic nucleus to estimate $\Upsilon(R)$, as was done by Stone and van Velzen (2016).
- Choice of PDMF, $\mathrm{d}N/\mathrm{d}m_\star$: as we have seen, both the first and second moments of the PDMF enter into TDE rate calculations. Since the diffusion coefficients $\langle(\Delta\mathcal{R})^2\rangle \propto \langle m_\star^2\rangle$, the heaviest surviving stellar species will generally dominate rates of two-body relaxation. For very young stellar populations this may be O stars, but for more typical stellar populations, it will be stellar-mass black holes. Depending on the mass function of stellar-mass black holes considered (Belczynski et al. 2010), their inclusion in $\mathrm{d}N/\mathrm{d}m_\star$ will enhance TDE rates by factors of ≈ 1.3–4.9 (Stone and Metzger 2016).
- Determination of SMBH mass $M_\bullet$: all the studies described so far relied on galaxy scaling relations to estimate $M_\bullet$, due to scarcity of direct SMBH mass determinations. As such, they may be biased to various extents, depending on the assumed form of these relations, for which numerous and often incompatible versions have been produced over the last two decades. See, for example, the comparison between different calibrations of the $M_\bullet - \sigma$ relation in the rate estimates of Wang and Merritt (2004).

Aside from these astrophysical uncertainties, the simplifying assumptions introduce more fundamental limitations to the validity of this formalism. So far, we have outlined a procedure for computing a one-integral, isotropic DF given only photometric information (i.e. the galaxy's surface brightness profile). If additional kinematic information is available, a two-integral DF $f(E, L_z)$ may be computed instead (Magorrian and Tremaine 1999), which will self-consistently account for the impact of flattening and orbital anisotropy on TDE rates.

The simple procedure outlined above has repeatedly been used to compute TDE rates in large galaxy samples, and at this point can be performed with the publicly available Fokker–Planck code PhaseFlow (Vasiliev 2017), as was done in, e.g. Pfister et al. (2019). This basic procedure was first used, however, by Syer and Ulmer (1999), who employed an even simpler formalism (one operating in coordinate space, not integral space) to a sample of 25 galaxies with surface brightness profiles $I(R)$ taken from Byun et al. (1996) and SMBH masses taken from Magorrian et al. (1998). The computed TDE rates were very low, with $10^{-7}\ \mathrm{yr}^{-1} \lesssim \dot{N} \lesssim 10^{-4}\ \mathrm{yr}^{-1}$ across most of the sample, although only 6 out of the 25 galaxies had SMBHs smaller than the Newtonian Hills mass. Almost simultaneously, Magorrian and Tremaine (1999) used a more sophisticated, two-integral version of the loss cone formalism to analyze a sample of 29 galaxies, taking into account multiple sources of loss cone flux: standard two-body relaxation, the draining of a loss wedge region in an axisymmetric potential, and two-body repopulation of the loss wedge. Two-integral DFs $f(E, L_z)$ were taken from Magorrian et al. (1998). This work found relatively low rates of disruption from two-body relaxation (albeit a factor of a few higher than what was found in Syer and Ulmer 1999), but once again, focused primarily on the largest galactic nuclei, with only 3 out of the 29 galaxies possessing SMBHs below the Schwarzschild Hills mass. This study was the first to point out that the observed dichotomy in nuclear density profiles $n(r)$—between steeply declining "cusp" galaxies and relatively shallow "core" galaxies (Lauer et al. 1995)—implies that the highest TDE rates will occur in lower-mass, "cuspy" galaxies. We reproduce many of the derived quantities, such as $q(E)$ and $\mathcal{F}(E)$, from Magorrian and Tremaine (1999) in Fig. 5.

A larger sample of 41 galaxies was modeled using one-integral DFs $f(E)$ by Wang and Merritt (2004), following almost exactly the procedure of this subsection. These authors estimated SMBH masses using the $M_\bullet - \sigma$ relationship of Merritt and Ferrarese (2001). This scaling relationship predicts systematically lower $M_\bullet$ values than that of Magorrian et al. (1998), and as a result this study found much higher TDE rates, with typical $\dot{N} \sim 10^{-4}\ \mathrm{yr}^{-1}$ in sub-Hills mass nuclei. This sample, which took $I(R)$ profiles from Faber et al. (1997), was more applicable to realistic TDE hosts than past studies, with 21 out of 41 galaxies possessing SMBHs below the Schwarzschild Hills mass. Wang and Merritt (2004) established much more firmly that the volumetric rate of TDEs, $\dot{n}$, should be dominated by the lowest-mass galaxy bin with a high SMBH occupation fraction. The reasons for this are (i) the greater abundance of low-mass galaxies in the Universe, (ii) the empirical preference of low-mass galaxies to have steep central stellar density cusps, which produce higher per-galaxy TDE rates $\dot{N}$, and (iii) the anticorrelation between $\dot{N}$ and SMBH mass $M_\bullet$ in a cuspy profile (see Fig. 6 for the identification of this trend in Wang and Merritt 2004).

More recently, empirically calibrated TDE rates were re-examined by Stone and Metzger (2016), who used the one-integral formalism of this section to model a sample of 144 galaxies (Lauer et al. 2007a,b), of which 42 contain SMBHs with $M_\bullet$ below the Schwarzschild Hills mass. This work was the first to estimate the fraction, f_{pin}, of TDEs from a given galaxy in the $q(E) > 1$ (pinhole) regime of disruption. Empirically, f_{pin} is $\mathcal{O}(1)$ in galaxies with $M_\bullet \lesssim 10^7 M_\odot$, but exhibits broad scatter ($10^{-2} < f_{\mathrm{pin}} < 1$) at higher SMBH masses. In any given bin of $M_\bullet$, the pinhole fraction is much higher in core galaxies than in cusp galaxies. Stone and Metzger (2016) reproduced earlier findings that per-galaxy TDE rates $\dot{N}$ are highest in smaller galaxies, and fit power-laws to their sample of dynamically modeled nuclei:

$$\dot{N} = \dot{N}_8 \left(\frac{M_\bullet}{10^8 M_\odot} \right)^B . \tag{32}$$

Here $\dot{N}_8 = 2.9 \times 10^{-5}$ yr and $B = -0.404$ when considering the entire sample of 144 modeled galaxies. For a subsample of exclusively core (cusp) galactic nuclei, $\dot{N}_8 = 1.2 \times 10^{-5}$ yr and $B = -0.247$ ($\dot{N}_8 = 6.5 \times 10^{-5}$ yr and $B = -0.223$). These results were combined with a Schechter function, the Faber–Jackson law, a recent calibration of the $M_\bullet - \sigma$ relation (McConnell and Ma 2013), and various parametrizations of the SMBH occupation fraction to estimate the volumetric TDE rate, $\dot{n}(M_\bullet)$, which is presented here in Fig. 7. These results indicate that a volume-complete sample of TDEs will be a powerful probe of the unknown bottom end of the SMBH mass function.

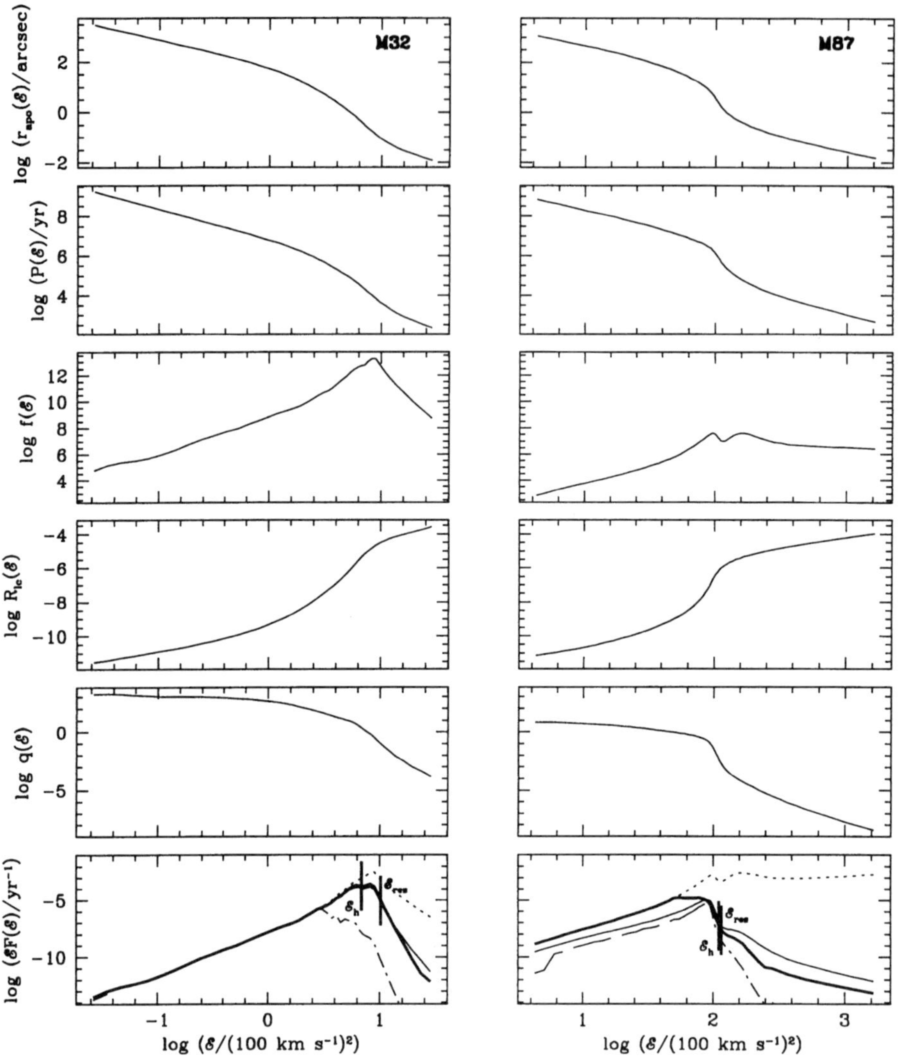

Fig. 5 Dynamically modeled photometry of two nearby galaxies, M32 (left) and M87 (right). Figures show intermediate quantities and the final product of the modeling prescription described in Sect. 4, plotted in bins of (negative) specific orbital energy $\mathcal{E} \equiv -E > 0$. The top two rows show apocentric radius $r_{\rm apo}(\mathcal{E})$ and orbital period $P(\mathcal{E})$ for radial orbits. The next two rows show the one-integral DF $f(\mathcal{E})$ (computed with an Eddington integral under the assumption of isotropic velocities) and the dimensionless size of the loss cone in angular momentum space, $\mathcal{R}_{\rm t}$ (labeled here as $\mathcal{R}_{\rm lc}$). The second-to-last row shows the diffusivity parameter $q(\mathcal{E})$ (for two-body loss cone repopulation), and the final row shows several different estimates for differential loss cone flux $\mathcal{F}(\mathcal{E})$. In the final row, the full loss cone flux is shown with a dotted line, the loss cone flux due to (spherical) two-body relaxation is a light solid line, the rate of draining of a full (axisymmetric) loss wedge is shown as a dot-dashed line, and the flux from a (axisymmetric) loss wedge being repopulated by two-body relaxation is shown as a dashed line. The thick solid line represents the total TDE rate. Taken with permission from Magorrian and Tremaine (1999).

Several astrophysical uncertainties apply to existing empirically-calibrated TDE rate estimates, as we have listed above. Aside from these caveats, it is notable that no loss cone

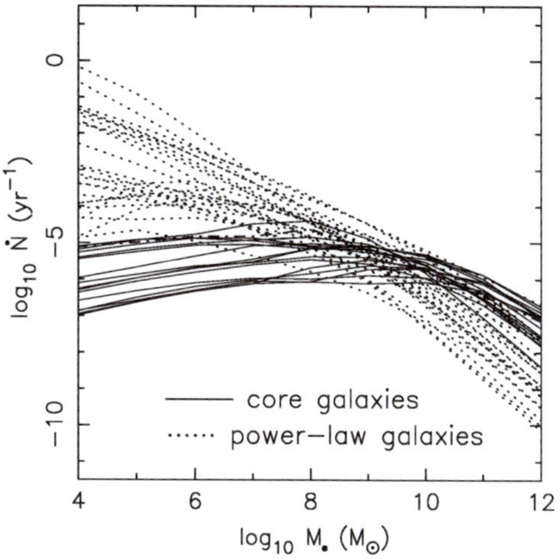

Fig. 6 The dependence of stellar consumption rates $\dot{N}$ on assumed SMBH mass $M_\bullet$ in a sample of 41 galaxies modeled with *HST* photometry (note that main-sequence stars will not produce TDEs for $M_\bullet > M_{\rm H} \sim 10^8 M_\odot$). The rates were computed assuming spherical symmetry and two-body relaxation, using the prescriptions of Sect. 4, but allowing SMBH mass to float. An interesting dichotomy emerges: for cusp ("power-law") stellar distributions, TDE rates increase with decreasing $M_\bullet$, while TDE rates are roughly independent of $M_\bullet$ in flatter, core-like nuclei. Taken with permission from Wang and Merritt (2004).

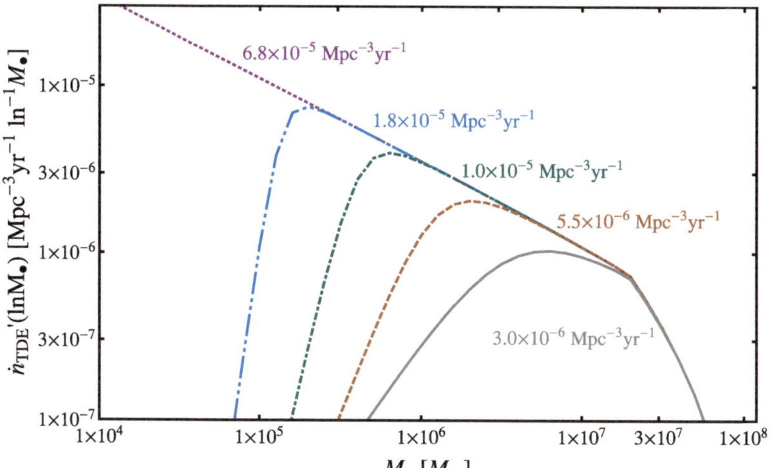

Fig. 7 The volumetric TDE rate per dex, $d\dot{n}/d\ln M_\bullet$, plotted as a function of SMBH mass $M_\bullet$. Per-galaxy TDE rates are estimated by power-law fits to the results of a large, dynamically modeled galaxy sample, and are then coupled to a Schechter function that is populated with SMBHs using the $M_\bullet - \sigma$ relation and five different models for the SMBH occupation fraction in low-mass nuclei. Specifically, the SMBH occupation fraction is assumed to resemble a step function, with SMBHs largely absent below a cutoff host stellar mass $M_{\rm cut}/M_\odot$ of 10^9 (gray solid), $10^{8.5}$ (brown dashed), 10^8 (green dot-dashed), $10^{7.5}$ (blue dot-dot-dashed), and a final scenario in which every dwarf galaxy has massive black holes down to a limiting mass of $M_\bullet = 10^4 M_\odot$ (purple dotted). Volumetric TDE rates are dominated by the smallest galaxies that typically host SMBHs. Mass-integrated rates $\dot{n}$ are labeled for each of these five models. Taken with permission from Stone and Metzger (2016).

calculations have yet been performed using the richer dynamical information available from Jeans or Schwarzschild modeling of a large sample of galaxies with (i) resolved kinematics and (ii) $M_\bullet < M_{\rm H}$. The pioneering work of Magorrian and Tremaine (1999) established the basic theoretical framework for this effort, but was applied almost exclusively to SMBHs too large to produce luminous TDEs. Applying loss cone theory to galaxies with directly measured SMBH masses and anisotropy profiles would enhance the precision of TDE rate estimates substantially, and is a logical next step forward.

4.2 Synthetic Observables

In the previous subsection, we surveyed existing efforts to empirically calibrate volumetric ($\dot{n}$) and per-galaxy ($\dot{N}$) TDE rates. However, the observable properties of TDE flares are not uniform, and can depend strongly on event parameters such as SMBH mass $M_\bullet$ or penetration parameter β. If one knows the differential distribution of TDE rates with respect to controlling parameters such as these, then it is straightforward to compute distributions of observable parameters of interest. Some of these parameters, such as the peak mass fallback rate $\dot{M}_{\rm peak}$ (or its Eddington ratio, $\dot{M}_{\rm peak}/\dot{M}_{\rm Edd}$), are very well-understood theoretically. Others, such as peak optical luminosity, are much less well-understood. Much more detailed theoretical treatments of TDE observables are offered in later Chapters of this Volume. In this section, we do not provide detailed models for TDE observables, but advise the interested reader to consult the Disruption Chapter, Formation of the Accretion Flow Chapter, Accretion Disc Chapter, and Emission Mechanisms Chapter.

Dynamical modeling of nearby galactic nuclei has already allowed us to estimate the $M_\bullet$-dependence of TDE rates, $\mathrm{d}\dot{n}/\mathrm{d}M_\bullet$ (e.g. Fig. 7). The β-dependence, $\mathrm{d}\dot{n}/\mathrm{d}\beta$, is a little more complicated, but at least in the limit of spherical loss cone repopulation, it can be estimated approximately. In the pinhole regime ($q \gg 1$), the differential distribution of penetration parameters goes as $\mathrm{d}\dot{n}/\mathrm{d}\beta \propto \beta^{-2}$, while in the diffusive regime ($q \ll 1$), $\beta \approx 1$, and higher β values are exponentially suppressed (see also the discussion surrounding Eq. (25)). Stone and Metzger (2016) find that low-mass galactic nuclei ($M_\bullet \lesssim 10^7 M_\odot$) typically produce most of their TDEs from the pinhole regime, while there is much more scatter in the "pinhole fraction" of high-mass galactic nuclei (although it is rarely less than 10%, for $M_\bullet < M_{\rm H}$). For any given bin of $M_\bullet$, the pinhole fraction is lower in cusp and higher in core nuclei.

The distribution of fallback times $t_{\rm fall}$ and the closely related peak fallback rate $\dot{M}_{\rm peak}$ may be estimated using the approximate analytic formulae (Rees 1988; Stone et al. 2013):

$$t_{\rm fall} \approx 3.5 \times 10^6\ {\rm s} \left(\frac{M_\bullet}{10^6 M_\odot}\right)^{1/2} \left(\frac{M_\star}{M_\odot}\right)^{-1} \left(\frac{R_\star}{R_\odot}\right)^{3/2} \tag{33}$$

$$\frac{\dot{M}_{\rm peak}}{\dot{M}_{\rm Edd}} \approx 130 \left(\frac{\eta}{0.1}\right) \left(\frac{M_\bullet}{10^6 M_\odot}\right)^{-3/2} \left(\frac{M_\star}{M_\odot}\right)^{2} \left(\frac{R_\star}{R_\odot}\right)^{-3/2}. \tag{34}$$

Here we have defined the Eddington-limited mass accretion rate, $\dot{M}_{\rm Edd} \equiv L_{\rm Edd}/(\eta c^2)$, in terms of a radiative efficiency $\eta < 1$ and the Eddington luminosity $L_{\rm Edd} \approx 1.5 \times 10^{46}\ {\rm erg\,s^{-1}}\ (M_\bullet/10^8 M_\odot)$. While these observables can be computed more precisely using hydrodynamical disruption simulations (Guillochon and Ramirez-Ruiz 2013), these analytic expressions were combined with TDE rate computations to predict distributions of peak fallback rates and rise times in Stone and Metzger (2016) and Kochanek (2016). We present the results of the latter paper in Fig. 8, which plots $\mathrm{d}\dot{n}/\mathrm{d}\ln M_\bullet$ in different bins of peak

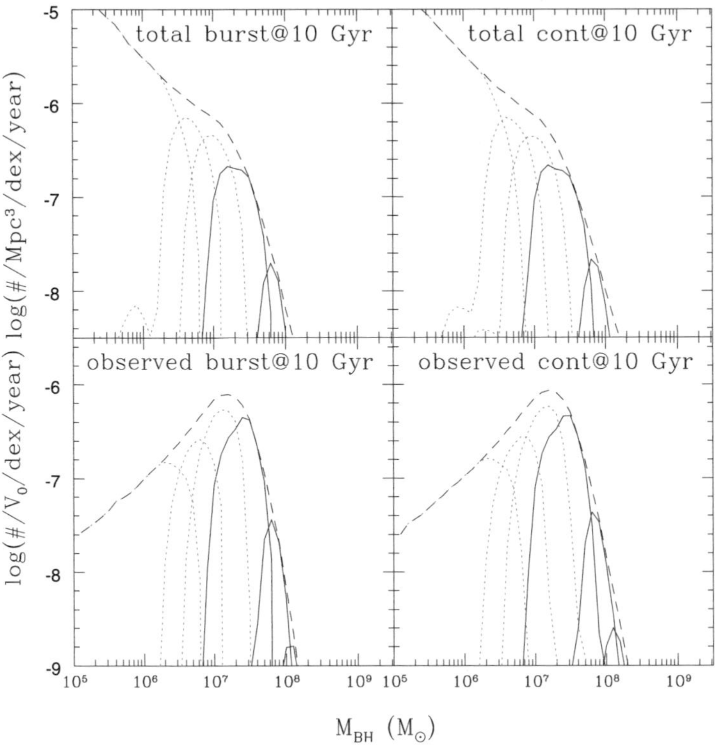

Fig. 8 The differential distribution of volumetric TDE rates (top panels) $\mathrm{d}\dot{n}/\mathrm{d}\ln M_\bullet$ and observed rates per dex $\mathrm{d}\dot{\mathcal{N}}/\mathrm{d}\ln M_\bullet$ (bottom panels). Because these TDE rates are integrated over PDMFs, they depend on assumed star formation histories; the left panels assume a single burst of star formation 10 Gyr ago, while the right panels assume a constant star formation rate over a Hubble time. Dashed black lines show total TDE rates $\mathrm{d}\dot{n}/\mathrm{d}\ln M_\bullet$ and $\mathrm{d}\dot{\mathcal{N}}/\mathrm{d}\ln M_\bullet$, while the other lines break up the rates into bins of $\dot{m}_{\rm peak} \equiv \dot{M}_{\rm peak}/\dot{M}_{\rm Edd}$. The red dotted lines correspond to super-Eddington peak fallback rate regimes of $10^{0.0} < \dot{m}_{\rm peak} < 10^{0.5}$, $10^{0.5} < \dot{m}_{\rm peak} < 10^{1.0}$, and $10^{1.0} > \dot{m}_{\rm peak}$, while the black solid lines correspond to sub-Eddington peak fallback rate regimes of $10^{-0.5} < \dot{m}_{\rm peak} < 10^{0.0}$, $10^{-1.0} < \dot{m}_{\rm peak} < 10^{-0.5}$, $10^{-1.5} < \dot{m}_{\rm peak} < 10^{-1.0}$, and $\dot{m}_{\rm peak} < 10^{-1.5}$. In translating from intrinsic rates to observed rates, the calculation assumes a flux-limited survey and that TDE peak luminosities are determined by either the peak mass fallback rate or the Eddington-limited accretion rate, whichever is smaller. Taken with permission from Kochanek (2016)

Eddington ratio $\dot{M}_{\rm peak}/\dot{M}_{\rm Edd}$. These distributions (which are integrated over different stellar PDMFs, corresponding to different star formation histories) highlight that for $M_\bullet \lesssim 10^7 M_\odot$, the large majority of TDEs have initially super-Eddington fallback rates, while the reverse is true for $M_\bullet \gtrsim 10^{7.5} M_\odot$.

Figure 8 also presents the TDE rates that will be *observed* in a flux-limited survey, $\dot{\mathcal{N}}(M_\bullet)$. These results assume a simple model relating peak luminosity to $M_\bullet$; more complicated models are explored in Stone and Metzger (2016) and Mageshwaran and Mangalam (2015). Generally, if super-Eddington fallback rates can be translated linearly into super-Eddington luminosities, then $\dot{\mathcal{N}}(M_\bullet)$ is dominated by the smallest values of $M_\bullet$ that exist with a high occupation fraction. Conversely, if TDE emission mechanisms are Eddington-limited, then $\dot{\mathcal{N}}(M_\bullet)$ is dominated by $M_\bullet \sim 10^7 M_\odot$, the characteristic SMBH mass where $\dot{M}_{\rm peak}/\dot{M}_{\rm Edd} \sim 1$. A more recent estimate of the volumetric detection rate, using a simple phenomenological model for TDE luminosities, is available in Thorp et al. (2018), which also compares standard TDE rates to those from binary SMBHs.

Other notable conclusions of efforts to produce synthetic observable distributions include: (i) regardless of star formation history, both intrinsic and observed TDE rates are dominated by the bottom end of the stellar IMF, except when $M_\bullet \sim M_{\rm H}$ (Stone and Metzger 2016; Kochanek 2016); (ii) under the assumption that galactic nuclei have similar dynamical properties at all redshifts,[16] comoving volumetric TDE rates will decline by a factor ≈ 10 from their $z = 0$ value as one moves to $z = 2$ (Kochanek 2016), largely due to the decreasing volume density of SMBHs at high z (Shankar et al. 2009); (iii) k-corrections generally aid the detectability of TDEs at cosmological distances (Strubbe and Quataert 2009). Unfortunately, the current lack of accurate, first-principles models for TDE emission mechanisms has, at the time of writing, hampered the calculation of more specific distributions of observables (e.g. emission line strengths, X-ray to optical ratios, etc.).

So far we have only discussed the observable properties of main sequence TDEs, but detailed distributions of red giant disruption properties (such as stage of post-main sequence evolution) are available in MacLeod et al. (2012). Synthetic distributions of fallback times for helium white dwarf and brown dwarf disruptions were, likewise, computed in Law-Smith et al. (2017a).

5 Comparing to Observed Rates of Tidal Disruption

For over two decades, tidal disruptions were studied from purely theoretical grounds. This situation changed in the 1990s with the discovery, in soft X-ray energies, of the first strong TDE candidates (Bade et al. 1996; Komossa and Greiner 1999, see also the X-ray Chapter). Later, wide-field ultraviolet (Gezari et al. 2006) and optical (van Velzen et al. 2011, see also the Optical Chapter) surveys discovered other classes of TDE candidates. At the time of writing, the current rate of TDE discovery stands at $\approx$ 1–3 new TDE candidates per year, and a few dozen TDE candidates are known (see e.g. the compilations of Auchettl et al. 2017; Hung et al. 2017). New and upcoming time domain surveys are poised to expand our sample by orders of magnitude (van Velzen et al. 2011; Khabibullin et al. 2014; Mageshwaran and Mangalam 2015). Clearly, it is important to assess what scientific goals can be achieved with the statistical analysis of large near-future TDE samples. A key component of any such analysis is understanding the rate of tidal disruption.

In this section, we compare the theoretical rate estimates of Sect. 4 to those inferred from the existing sample of tidal disruption flares. This type of comparison has, so far, found two interesting puzzles, which we outline in Sect. 5.1. First, there may be a broad rate discrepancy, with fewer TDEs detected than are predicted by stellar dynamics (Stone and Metzger 2016). Second, there is a more specific and definitive discrepancy related to the *overproduction* of observed TDEs in a rare galaxy subclass: so-called "E+A," or post-starburst, galaxies (Arcavi et al. 2014). We discuss these puzzles in Sect. 5.2.

[16]This assumption is unlikely to be generically true. The clearest caveat here concerns the strong preference among observed TDE flares to reside in rare E+A and, more generally, post-starburst galaxies (see discussion in Sect. 5.2.1). Because E+A and post-starburst galaxies make up very small fractions of the low-z galaxy population ($\sim 0.2\%$ and 2.3%, respectively; French et al. 2016), this implies the presence of unusual stellar dynamics enhancing TDE rates in these galaxies by at least an order of magnitude. However, the fraction of all galaxies that have a post-starburst nature increases steeply as a function of redshift. For example, going from $z \approx 0.5$ to $z \approx 2$ increases the fraction of post-starburst galaxies by a factor of ≈ 5 (Wild et al. 2016), suggesting that at high z, the decline in $\dot{n}$ due to the decreasing volume density of SMBHs may be overwhelmed by the growing abundance of this rare galaxy type.

5.1 Observationally Inferred Rates

The first TDE candidates were discovered via soft X-ray emission (Bade et al. 1996), and the first observational rate inferences were based off of these events. The pioneering work of Donley et al. (2002) analyzed three TDE candidates from the *ROSAT* All-Sky Survey and inferred a per-galaxy disruption rate $\dot{N} \approx 9 \times 10^{-6}$ yr^{-1}. Subsequent analysis of two TDE candidates from the *XMM-Newton* slew survey inferred a volumetric TDE rate $\dot{n} \approx 5.4 \times 10^{-6}$ yr^{-1} Mpc^{-3}, and a per-galaxy rate $\dot{N} \approx 2.3 \times 10^{-4}$ yr^{-1} (Esquej et al. 2008). Later, one soft X-ray TDE candidate was found by reanalysis of archival *Chandra* and *XMM-Newton* observations of galaxy clusters (Maksym et al. 2010); three more were discovered by cross-correlating *ROSAT* archives with serendipitous *XMM-Newton* pointings (Khabibullin and Sazonov 2014). The per-galaxy event rates computed from these samples were $\dot{N} \approx 1.2 \times 10^{-4}$ yr^{-1} and $\dot{N} \approx 3 \times 10^{-5}$ yr^{-1}, respectively. The sample of Khabibullin and Sazonov (2014) implies a volumetric event rate $\dot{n} \approx 4\text{–}8 \times 10^{-7}$ yr^{-1} Mpc^{-3}.

The range of inferred X-ray TDE rates spans roughly 1.5 orders of magnitude. This spread can be attributed to a number of factors. Obviously, small-number statistics play a role. Due to the poor temporal resolution of all wide-field soft X-ray surveys to date, these samples are both flux- and cadence-limited, and rate inference therefore requires assumptions about (i) the X-ray luminosity function of TDEs, and (ii) the characteristic decay time of TDEs. With these assumptions, one may estimate survey completeness as a function of source distance, and therefore infer a volumetric event rate $\dot{n}$. As a simplified but illustrative example, consider a high-cadence survey operating at low redshifts for a duration T, covering an angular area $\Delta\Omega$. In this limit, the number of detected TDEs

$$N_{\rm TDE} = \int \frac{{\rm d}\dot{n}}{{\rm d}L} \times \frac{4\pi}{3} D_{\rm max}^3(L) \times \frac{\Delta\Omega}{4\pi} \times T {\rm d}L. \tag{35}$$

Here $D_{\rm max}(L)$ is the maximum distance out to which a TDE can be detected in a flux-limited survey, and ${\rm d}\dot{n}/{\rm d}L$ is the volumetric TDE luminosity function (Hung et al. 2018). By assuming approximate functional forms for the luminosity function, one may use a measured $N_{\rm TDE}$ to estimate $\dot{n} \equiv \int ({\rm d}\dot{n}/{\rm d}L){\rm d}L$. Translating this $\dot{n}$ into a per-galaxy rate $\dot{N}$ requires further assumptions about *which* galaxies may produce TDEs (e.g. down to what limiting mass do dwarf galaxies still possess SMBHs?). Given these many uncertainties, it is perhaps unsurprising to see substantial scatter in observationally inferred X-ray event rates.

More recently, the discovery rate of TDEs has been dominated by wide-field optical/UV surveys. The first TDE candidates discovered from their thermal UV emission were found with the *GALEX* satellite (Gezari et al. 2006). The rate of TDE discovery by *GALEX* was found to be consistent (Gezari et al. 2008) with the empirical predictions of Wang and Merritt (2004). Soon afterwards, TDE candidates began to be discovered through their thermal optical emission—at first archivally (van Velzen et al. 2011), and soon afterwards in real-time surveys (Gezari et al. 2012). As the optically-selected TDE sample has grown, several groups have attempted to infer the TDE rates $\dot{N}$ and $\dot{n}$ associated with optically-bright flares. One advantage that rate inferences from optically-selected TDEs have over those from X-ray-selected TDEs is the much higher cadence of optical time-domain surveys. This means that rate estimates will be less sensitive to assumptions about the temporal evolution of TDE light curves.

Initially, rates estimated from optically-selected TDEs were low. For example, the discovery of two TDE candidates in SDSS archival data was used to compute a per-galaxy rate $\dot{N} \approx (1.5\text{–}2.0)^{+2.7}_{-1.3} \times 10^{-5}$ yr^{-1}, and a volumetric rate $\dot{n} \approx (4\text{–}8) \times 10^{-8\pm0.4}$ yr^{-1} Mpc^{-3} (van

Velzen and Farrar 2014).[17] The later discovery of two TDE candidates by the ASAS-SN survey implied a somewhat higher per-galaxy rate of $\dot{N} \approx 4.1^{+12.9}_{-1.9} \times 10^{-5}\ \mathrm{yr}^{-1}$ (Holoien et al. 2016). More recent calculations that use an *empirical* TDE luminosity function find higher event rates. For example, van Velzen (2018) analyze a sample of thirteen TDE candidates found in optical/UV surveys and find $\dot{n} = (8 \pm 4) \times 10^{-7}\ \mathrm{Mpc}^{-3}\ \mathrm{yr}^{-1}$, consistent with a per-galaxy rate $\dot{N} \approx 1 \times 10^{-4}\ \mathrm{yr}^{-1}$. Comparably high-per galaxy rates are also inferred from two TDE candidates found by iPTF ($\dot{N} \approx 1.7^{+2.9}_{-1.3} \times 10^{-4}\ \mathrm{yr}^{-1}$; Hung et al. 2018).

One particularly intriguing observational finding is an apparent overabundance of TDE candidates in quiescent galaxies with no ongoing star formation and prominent Balmer absorption features. These absorption features are generally produced by a large population of young A stars, indicating that this type of galaxy (in its extreme form, sometimes known as an "E+A" or "K+A" galaxy) recently underwent a major star formation episode, that has since ceased. Although this post-starburst preference was first noted among optically-selected TDEs (Arcavi et al. 2014), it appears to be generic across most classes of candidate TDE flares (Graur et al. 2018). While the *absolute* rate density of TDEs in post-starburst galaxies, $\dot{n}_{\rm PS}$, is subject to the same uncertainties discussed above, the relative rate enhancement factor $\mathscr{R} \equiv \dot{n}_{\rm PS}/\dot{n}$, may be computed with much less uncertainty (provided that the TDE luminosity function is nearly the same in normal and post-starburst galaxies).

Interestingly, the rate enhancement seems to depend strongly on the strength of the Balmer absorption features. The first detailed analysis of post-starburst TDE hosts found that, in the most Balmer-strong post-starbursts (classical E+As), $\mathscr{R} = 190^{+191}_{-100}$, while in less extreme Balmer-strong post-starbursts, $\mathscr{R} = 33^{+7}_{-11}$ (French et al. 2016, for a sample of 8 TDE hosts). The strength of the Balmer absorption feature represents a weighted combination of post-starburst age, fraction of the galaxy starlight produced in the starburst, and the detailed shape of star formation history during and after the starburst (French et al. 2017). A subsequent analysis (Graur et al. 2018) of a larger TDE host sample found somewhat smaller but still significant rate enhancements ($\mathscr{R} = 35^{+21}_{-17}$ and $\mathscr{R} = 18^{+8}_{-7}$ for the same categories of Balmer-line strength). The Graur et al. (2018) rate enhancements were computed from a larger but more heterogeneous sample of 33 TDE host galaxies, and vary significantly if one examines different sub-samples. TDE host galaxies display some other peculiarities as well, such as an unusually high surface brightness on $\sim$ kpc scales (Law-Smith et al. 2017b). The properties of TDE hosts are not yet fully understood, and our empirical knowledge of this subject is evolving rapidly. The interested reader is advised to consult the Host Galaxies Chapter for a more comprehensive picture.

As TDE samples grow, it will become possible to make more refined comparisons between theory and observation. For example, in Fig. 7, we can see that the predicted, volumetric TDE rate is expected to significantly depend on $M_\bullet$. Observational evidence testing this prediction is mixed, although still limited by small-number statistics. Early work by Stone and Metzger (2016) used galaxy scaling relations to find that an optical/UV selected TDE sample (consisting of 11) had a mass distribution sharply peaked near $10^{6.8} M_\odot$, with little evidence for optical/UV TDE hosts possessing SMBHs below $10^6 M_\odot$. Subsequent work by Wevers et al. (2019) improved on this by using a larger TDE sample (15 flares) and also using a more homogeneous set of SMBH mass estimates. The Wevers et al. (2019) histogram shows a more broadly peaked distribution centered on somewhat lower masses ($\approx 10^{6.2} M_\odot$), but with a qualitatively similar shape. In contrast, both of these analyses find

[17] Note that in the results of van Velzen and Farrar (2014), statistical uncertainties are denoted in superscript/subscript error ranges, while systematic uncertainties (associated with the uncertain choice of model light curve used to back out true rates from flux-limited samples) are denoted in prefactor error ranges.

a much broader distribution of host masses for TDEs selected through soft X-ray emission. In comparison to the optical/UV selected sample, the distribution of inferred SMBH masses in X-ray TDEs is less sharply peaked, with a greater fraction of these events occurring in galaxies with very high-mass ($M_\bullet \gtrsim 10^{7.5} M_\odot$) or very low-mass ($M_\bullet \lesssim 10^5 M_\odot$) SMBHs.

Neither of these analyses, however, accounted for survey selection effects, such as volume-correcting flux-limited samples. To date, the most thorough effort to back out volumetric rates $\dot{n}(M_\bullet)$ from observations comes from van Velzen (2018), who find an effectively constant TDE rate for bins of SMBH mass between $10^6 M_\odot$ and $10^{7.5} M_\odot$. At present, the two biggest obstacles to drawing inferences about the bottom end of the SMBH mass function from observed TDEs are (i) small, inhomogenous TDE samples and (ii) the uncertain luminosity function of TDEs, particularly those from smaller SMBHs where Eddington considerations may be more important.

5.2 Rate Discrepancies

As we saw in Sect. 4, there are many uncertainties associated with dynamical modeling of nearby galactic nuclei. However, a few general conclusions stand out:

1. TDE rates $\dot{N}$ are generally higher in smaller galaxies, if we consider only two-body relaxation in spherical potentials.[18] This is primarily due to the higher central densities and steeper density profiles associated with low-mass galactic nuclei.
2. The differential volumetric TDE rate $\mathrm{d}\dot{n}/\mathrm{d}\ln M_\bullet$ varies across host galaxy mass, but the integrated rate $\dot{n}$ is likely dominated by the lowest-mass range galaxies that have a high SMBH occupation fraction.
3. Averaged over the set of galaxies with a high black hole occupation fraction, per-galaxy TDE rates are $\dot{N} \gtrsim 1 \times 10^{-4}\ \mathrm{yr}^{-1}$.
4. Main sequence TDE rates should robustly cut off at the main sequence Hills mass, which, depending on the SMBH spin distribution, lies in a range $10^8 \lesssim M_\bullet/M_\odot \lesssim 10^9$.

Conclusion (3) exists in tension with a number of the lower empirical TDE rate estimates discussed in Sect. 5.1, such as those of Donley et al. (2002), Khabibullin and Sazonov (2014), and van Velzen and Farrar (2014). Another interesting, observationally motivated puzzle is the large rate enhancement seen in post-starburst galaxies (Sect. 5.1), which was not directly predicted by any pre-existing dynamical studies.

It seems, therefore, that potential tensions exist between the inferred TDE rates and those predicted from (empirically calibrated) dynamical theory. These tensions can be divided into two types: too many TDEs observed in post-starburst galaxies, and too few observed in "normal" galaxies. In the following subsections, we discuss both of these tensions. First, however, we must note one important caveat in the comparison of theory to observation: the necessity of having observed TDE samples that are both flux-complete and pure (i.e. not contaminated by a significant number of TDE impostors, such as nuclear supernovae, AGN variability, or more exotic transients). A large presence of TDE impostors in an observed sample will skew inferred event rates upwards, while a flux-incomplete TDE sample will skew inferred rates downwards. The first of these issues is discussed at length in the Imposters Chapter, while we cover the latter in some detail below.

[18]The situation becomes more complicated if galactic nuclei are significantly triaxial, in which case larger galaxies may have larger individual TDE rates $\dot{N}$. From an observational point of view, the prevalence of nuclear triaxiality remains uncertain.

5.2.1 Elevated TDE Rates in Post-Starburst Galaxies

The post-starburst preference of optical (Arcavi et al. 2014) and X-ray (Graur et al. 2018) TDEs is an interesting observational puzzle that was not predicted by theory. The tension between dynamically predicted rates and the higher rates inferred from observations is a problem that has many possible solutions. The simple (spherically symmetric, quasi-isotropic) models of Sect. 4 neglect many proposed dynamical mechanisms for increasing per-galaxy TDE rates above $\dot{N} \sim 10^{-4}\ \mathrm{yr}^{-1}$, such as binary SMBHs (Ivanov et al. 2005; Chen et al. 2011), radial anisotropies ($b > 0$; Stone et al. 2018), nuclear triaxiality (Merritt and Poon 2004), secular instabilities in stellar discs (Madigan et al. 2018), overdense central star clusters (Stone and Metzger 2016; Stone and van Velzen 2016), and rate enhancements from massive perturbers (Perets et al. 2007; Perets and Alexander 2008; Mastrobuono-Battisti et al. 2014) and/or nuclear spiral arms (Hamers and Perets 2017). In this subsection, we discuss each of these potential solutions to the discrepancy.

The first proposed explanation for the post-starburst preference invoked the correlation between starbursts and galaxy mergers (Arcavi et al. 2014). If many post-starburst galaxies are also post-merger galaxies, their nuclei may contain SMBH binaries which can increase TDE rates by many orders of magnitude (relative to galactic nuclei with solitary SMBHs) through a combination of Kozai cycles (Ivanov et al. 2005) and chaotic three-body scatterings (Chen et al. 2011; Wegg and Bode 2011). However, even though SMBH binaries may temporarily enhance TDE rates by multiple orders of magnitude, the short timescales for such enhancements (typically $\sim 10^5$ yr, e.g. Wegg and Bode 2011) may make it challenging for this mechanism to explain the global fraction of all TDEs seen in post-starburst galaxies (see also discussions in Stone and Metzger 2016; Saxton et al. 2018).

Prior to the discovery of the post-starburst preference, Perets et al. (2007), Perets and Alexander (2008) suggested that massive perturbers, such as giant molecular clouds, can greatly shorten the two-body relaxation times (since $T_{\mathrm{rel}} \propto \langle M_\star^2 \rangle^{-1}$) and enhance TDE rates as a result. Nuclear spiral arms would have a similar effect (Hamers and Perets 2017). Such enhanced rates would occur preferentially in gas-rich galaxies and, in particular, in post-merger galaxies, and would last for long timescales. However, because massive perturbers are unlikely to exist at radial scales $r \lesssim r_{\mathrm{crit}}$, these processes could possibly enhance TDE rates by a factor of two or less, and are thus unlikely explanations for the extreme, 1–2 orders of magnitude enhancements inferred for post-starburst galaxies.

Nuclear starbursts sometimes produce eccentric stellar discs in which secular effects can dramatically increase TDE rates (Madigan et al. 2018; Wernke and Madigan 2019). However, for such secular processes to operate, the nuclear cluster mass should be relatively small, so as not to give rise to mass precession that quenches coherent secular evolution. This may be problematic for this explanation of the post-starburst preference, as most low-mass SMBHs coexist with a sizeable nuclear star cluster. The more favorable environment of a disc-dominated nuclear stellar population likely exists only for more massive galaxies, which host SMBHs with $M_\bullet \gtrsim 10^8\ \mathrm{M}_\odot$ (Antonini et al. 2015) above the Hills mass M_{H} that are unable to tidally disrupt main sequence stars.

If star formation in starbursts is centrally concentrated,[19] then post-starbursts may have unusually overdense galactic nuclei. Such overdensities would result in short two-body relaxation times and high TDE rates (Stone and Metzger 2016). Nevertheless, it is not yet clear whether such extreme density nuclear clusters form in most post-starbursts (most nuclei can not be spatially resolved to assess the stellar density there, although the results of Stone and

[19] This seems to be indicated by resolved color gradients in nearby E+A galaxies, see e.g. Pracy et al. (2012).

van Velzen 2016 lend tentative support to this hypothesis in one nearby, highly overdense E+A).

A final class of dynamical explanations relies on asymmetries in stellar position and/or velocity fields. If nuclear star formation produces highly triaxial potentials (Merritt and Poon 2004) or radially anisotropic velocity distributions (Stone et al. 2018), these effects can increase TDE rates by an order of magnitude. However, both of these types of asymmetries tend to "wash out" over time due to the effects of two-body relaxation, which may pose a problem in the low-mass galaxies with short relaxation times that host most TDEs.

With so many theoretical explanations to choose from, it is important to find observational tests that can discriminate between different mechanisms. The total TDE rates from the mechanisms suggested above are sensitive to a wide variety of assumptions regarding the host galaxies and their nuclei. The statistical properties of post-starburst TDE hosts might therefore be a more useful diagnostic. For example, the observed distribution of post-starburst host masses is biased towards smaller galaxies, in agreement with the overdensity and radial orbit hypotheses, but in notable disagreement with the top-heavy distribution of host masses expected for SMBH binary-triggered TDEs (Stone et al. 2018). Likewise, the delay time distribution of TDEs in post-starbursts is highly discrepant with that of the SMBH binary explanation, generally compatible with the overdensity hypothesis, and compatible with the radial orbit hypothesis provided fairly extreme[20] ($b \gtrsim 0.5$) anisotropy parameters are chosen (Stone et al. 2018). See, for example, Fig. 9 for a comparison between observations and theoretical delay time distributions in these latter two models. Dynamical modeling of *HST* photometry of the nearby E+A galaxy NGC 3156 lends further support to the overdensity explanation, as this galaxy appears to have both an unusually steep central density profile ($\rho(r) \propto r^{-2.25}$, formally in the "ultrasteep" regime) with an unusually small SMBH influence radius (Stone and van Velzen 2016).

Finally, many of the suggested scenarios—such as SMBH binaries, eccentric nuclear discs, radial anisotropies and nuclear triaxiality—produce TDEs predominantly in the full loss-cone regime (or at the very least, with a β distribution matching that of the full loss-cone regime). TDE flares produced via these channels[21] should preferentially have a lower fraction of grazing TDEs than will TDEs arising from stellar two-body relaxation processes.

While the true explanation of the post-starburst preference remains an open question, it seems likely that (i) larger statistical samples of TDE hosts and (ii) dynamical modeling of more nearby post-starburst nuclei will allow rapid progress on this question in the near future.

5.2.2 A Possible Dearth of TDEs in "Normal" Galaxies

Very few dynamical mechanisms exist to *lower* TDE rates below the conservative floor set by two-body relaxation in a spherical and isotropic nucleus. The only mechanism that seems clearly able to do this is the presence of a strong tangential anisotropy in the stellar velocity field, which is capable of producing arbitrarily large reductions in $\dot{N}$ ($b < 0$; Merritt and Wang 2005). However, such tangential anisotropies will wash out due to two-body relaxation on a small fraction of the energy relaxation time (Lezhnin and Vasiliev 2015). The small galaxies that dominate the volumetric TDE rate (and whose relaxation times are less than $t_{\rm H}$), are therefore unlikely to have large tangential anisotropies in their nuclei unless

[20] Such large anisotropies may be vulnerable to the radial orbit instability (Polyachenko and Shukhman 1981).

[21] A notable exception to this trend is the overdensity scenario; ultrasteep density cusps will produce almost all their TDEs in the empty loss cone regime.

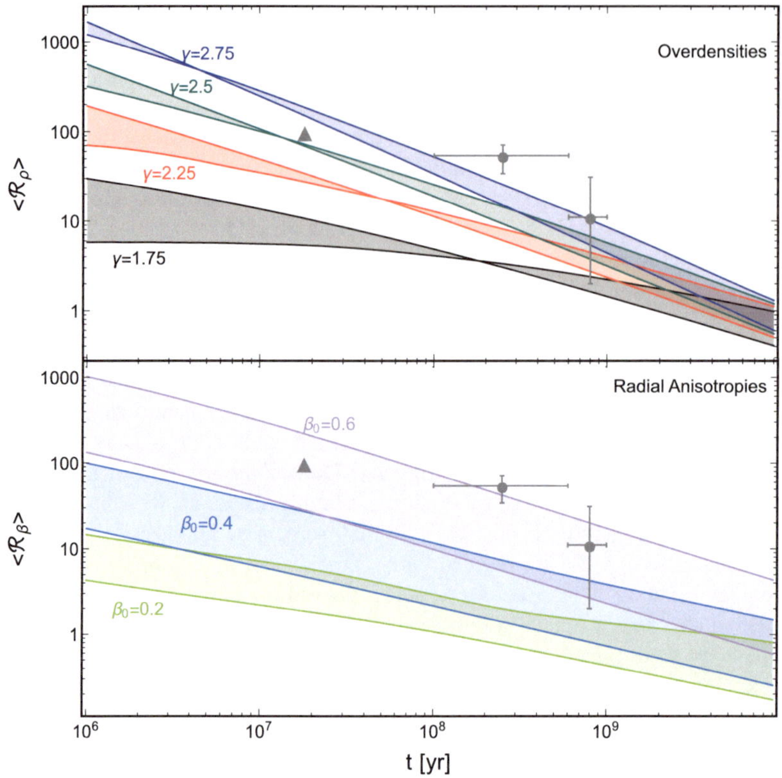

Fig. 9 Theoretical delay time distributions as functions of time t since a burst of nuclear star formation. *Top panel*: the average rate enhancement $\langle \mathscr{R}_\rho \rangle$ in *overdense* galactic nuclei, in comparison to TDE rates from a population of "normal" galaxies. The enhancement factor $\langle \mathscr{R}_\rho \rangle$ is computed by integrating across an empirical SMBH mass function (the boundaries of the shaded error regions correspond to assuming that astrophysical SMBH masses cut off either at $10^5 M_\odot$ or $10^6 M_\odot$). Theoretical models assume a range of different post-starburst density profiles $\rho(r) \propto r^{-\gamma}$, and circular data points are taken from a sample of eight TDE candidates with well-characterized host galaxies (French et al. 2016, 2017). *Bottom panel*: same as before, but now considering the average rate enhancement $\langle \mathscr{R}_\beta \rangle$ in *radially biased* galactic nuclei, with a range of different initial anisotropy parameters β_0. Taken with permission from Stone et al. (2018).

an exotic dynamical mechanism exists to pump $b < 0$ in small galactic nuclei. Long-lived SMBH binaries are expected to produce this type of tangential anisotropy (Merritt and Wang 2005), but as small galactic nuclei are the environments most favorable for solving the final parsec problem through collisional mechanisms (Begelman et al. 1980), this does not seem promising.

The lack of clear mechanisms available to decrease *dynamically-predicted* TDE rates led Stone and Metzger (2016) to identify a "rate discrepancy" between (empirically-calibrated) theory and observation.[22] Subsequent analysis of observed TDEs has identified a plausible resolution to this discrepancy: a very steep TDE luminosity function. In particular, van Velzen (2018) find that, for optically-selected TDE candidates, $\mathrm{d}\dot{N}/\mathrm{d}L \propto L^{-2.5}$, with an uncertain lower luminosity limit. This suggests that current, flux-limited TDE samples are see-

[22] Although it is worth noting that some observational rate inferences, such as Esquej et al. (2008), would not be in tension with conservative theoretical rate estimates.

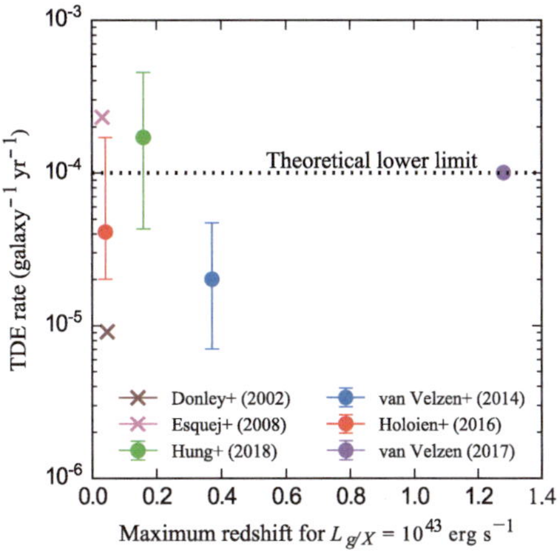

Fig. 10 Observationally inferred TDE rates from six different (but, in some cases, overlapping) TDE samples: Donley et al. (2002, brown), Esquej et al. (2008, pink), van Velzen and Farrar (2014, blue), Holoien et al. (2016, red), van Velzen (2018, purple), and Hung et al. (2018, green). Rate estimates using TDEs selected by their soft X-ray emission are indicated with "X" symbols, while rate estimates using optically-selected TDEs are shown with circles. The x-axis indicates the maximum redshift z out to which the underlying surveys would detect flares with peak g-band luminosities of $L_g = 10^{43}$ erg s^{-1} (for the X-ray samples, the equivalent condition is a 0.2–2.4 keV luminosity $L_X = 10^{43}$ erg s^{-1}). The dotted black line shows an approximate theoretical lower limit on per-galaxy TDE rates due to 2-body relaxation in isotropic galactic nuclei (Wang and Merritt 2004). Taken with permission from Hung et al. (2018).

ing only "the tip of the iceberg," in comparison to the much larger number of very dim TDEs that exist in the local Universe. While it is important to test the shape of the TDE luminosity function further, both with larger near-future TDE samples *and* with better first-principles modeling of flare emission mechanisms, this seems like a promising resolution of tensions between theoretical and observational rate estimates. The origin of the luminosity function is theoretically uncertain, and depends on the complicated physics of debris circularization (Formation of the Accretion Flow Chapter) and optical emission (Emission Mechanisms Chapter). Specific hypotheses could be tested with larger TDE samples; for example, rapid disc formation is probably disfavored for $\beta \approx 1$ TDEs around small SMBHs (Dai et al. 2015), so the smaller SMBHs that dominate volumetric event rates $\dot{n}$ might preferentially be associated with lower peak luminosities. We illustrate the current state of this rate discrepancy in Fig. 10, which shows that the most recent rate inferences from optically-selected TDE samples appear compatible with the conservative side of theoretical rate estimates.

6 Broader Implications

In the final section of this Chapter, we focus on the broader scientific importance of TDE rates. We have already explored one major astrophysical motivation for studying statistical samples of TDEs: SMBH demographic measurements. In Sect. 3.5, we saw that the statistical distribution of SMBH spins is imprinted into the mass distribution of TDE host galaxies (at least for $10^{7.5} \lesssim M_\bullet/M_\odot \lesssim 10^{8.5}$). Likewise, in Sect. 4.1, we saw that the volumetric TDE rate contains information on the uncertain bottom end of the SMBH mass function. Because these motivations have already been described in depth, we will not belabor them in this section, but instead focus on alternative applications and implications of TDE rates.

In Sect. 6.1, we discuss the potential importance of TDEs for SMBH mass, spin, and luminosity evolution. Next, in Sect. 6.2, we draw on the general relativistic analysis of Sect. 3.5 to explore how SMBH spin distributions may be encoded in volumetric disruption rates. In Sect. 6.3, we explore how TDE rates and inferences may be used to calibrate our understanding of extreme mass ratio inspirals, a related byproduct of loss cone physics in galactic nuclei. Finally, in Sect. 6.4, we offer a few speculations for what the future may hold for tidal disruption science.

6.1 Black Hole Demographics and Growth

As SMBHs feast on debris from tidally disrupted stars, they accrete both mass and angular momentum. Over a Hubble time, $t_{\rm H} \approx 1.4 \times 10^{10}$ yr, the cumulative effect of many TDEs is to increase the SMBH mass, $M_\bullet$. The net effect on SMBH angular momentum, $\chi_\bullet$, is less obvious, as individual TDEs may either spin it up or spin it down. Because TDEs usually produce transient accretion flares, the stochastic background of TDEs in otherwise quiescent (inactive) galactic nuclei produces a baseline level of accretion luminosity that may contribute in an interesting way to the bottom end of the AGN luminosity function.

The importance of all three of these effects—SMBH mass growth, SMBH spin evolution, and accretion luminosity in quiescent galaxies—depends strongly on TDE rates. These effects will be of little importance in galaxies with very low TDE rates, but can be of crucial importance in galaxies, or classes of galaxies, with higher TDE rates. Indeed, it is possible that some types of galaxies may accumulate most of their SMBH mass and/or spin angular momentum through stellar tidal disruption.

In the simplest picture, where loss cone repopulation is governed by two-body relaxation, TDE rates are decreasing functions of SMBH mass (Sect. 4), indicating that TDEs will be of the greatest importance for the growth of low-mass SMBHs (Milosavljević et al. 2006). We may compute the typical SMBH mass below which TDEs are important for growth[23] by equating $M_\bullet = \frac{1}{2}\langle M_\star\rangle \dot{N}(M_\bullet) t_{\rm H}$. We solve this equation using the empirically calibrated scaling of Eq. (32), and find $M_{\rm TDE} \approx 5 \times 10^5 M_\odot$ to be the characteristic mass below which SMBH growth should be dominated by TDEs. Larger SMBHs acquire most of their mass from other sources, primarily radiatively efficient gas accretion (Soltan 1982).

Because two-body relaxation delivers stars to the SMBH from a quasi-isotropic distribution of directions,[24] the net effect of mass growth through TDEs is to spin the SMBH down, in analogy to "chaotic accretion" of gas clouds by AGN (King and Pringle 2006), and small SMBHs that have grown primarily through tidal disruption should be spinning slowly. Interestingly, even if TDEs are a subdominant contributor to SMBH mass growth, they may still play an important role in the evolution of SMBH spin. SMBHs that grow through prolonged AGN episodes, or through short episodes with the same direction of disc angular momentum will, over a handful of mass-doubling times, spin up to very high values of $\chi_\bullet$ (Berti and Volonteri 2008). If it is only these AGN episodes that supply angular momentum to the SMBH, $\chi_\bullet$ will saturate at a value dictated by accretion physics—for example, growth through standard thin disc accretion saturates at $\chi_\bullet = 0.998$ (Thorne 1974). Because this

[23]In this calculation, we have assumed that half of the disrupted star accretes onto the SMBH. For nearly-parabolic stellar orbits, precisely half of the disrupted star is dynamically bound to the SMBH, although we caution that hydrodynamic shocks and radiation pressure in super-Eddington accretion may unbind a portion of this dynamically bound half (see the Formation of the Accretion Flow Chapter and the Accretion Disc Chapter for more discussion of these uncertainties).

[24]In principle, if TDE rates are dominated by mechanisms (such as nuclear triaxiality, or eccentric stellar discs) that preferentially supply stars from a specific orbital orientation, TDEs may act to spin up SMBHs.

dimensionless spin is so large, even a very small fraction of mass growth from (isotropically distributed) TDEs will place a lower cap on SMBH spin (Metzger and Stone 2016).

Even though TDEs are rare events, they occur in all galaxies with $M_\bullet < M_{\rm H}$, and therefore produce a minimum time-averaged accretion rate onto SMBHs in quiescent galactic nuclei. While the peak luminosity of a TDE (up to $\sim 10^{44-45}$ erg s^{-1}) is typically maintained for a period of weeks to months, there is a long, slow decay period in which the galactic nucleus may resemble a low-luminosity AGN. Recent UV observations indicate that most TDEs continue to radiate at $\sim 10^{42}$ erg s^{-1} at times ~ 10 yr post-peak (van Velzen et al. 2018). Based on simple models for TDE light curves, Milosavljević et al. (2006) estimated that main sequence TDEs could be responsible for a majority of the lowest-luminosity X-ray AGN (although it is important to note that the "AGN" produced by late stages of TDEs would probably lack standard narrow-line regions). More recent work suggests that episodic partial disruptions of giant-branch stars may be the most important type of tidal disruption for setting a floor on accretion rates in galactic nuclei (MacLeod et al. 2013). The actual luminosity floor set by the late stages of tidal disruption flares remains somewhat ambiguous, as it depends on (i) the still uncertain volumetric TDE rate, (ii) the presence (Shen and Matzner 2014) or absence (van Velzen et al. 2018) of thermal instability in late-time TDE discs, and (iii) the ability of TDE debris from red giant disruptions to efficiently accrete onto the SMBH (Bonnerot et al. 2016).

6.2 The Shadow of the Horizon

As we have seen in Sect. 3.5, the Hills mass depends sensitively on SMBH spin, and can vary by almost an order of magnitude (for prograde equatorial orbits) as one varies $\chi_\bullet$ from 0 to 1. While this has exciting astrophysical applications for measuring the distribution of SMBH spins in a mass range $10^{7.5} \lesssim M_\bullet/M_\odot \lesssim 10^{8.5}$, the hard upper limit on $M_{\rm H}$ (for lower main sequence stars) imposed by the Kerr bound ($\chi_\bullet \le 1$) raises the prospect of using TDE distributions as probes of general relativity. For example, Lu et al. (2017) demonstrate that samples of TDEs may rule out exotic alternatives to Kerr black holes, such as boson stars, which possess hard surfaces and would therefore be capable of producing limited electromagnetic emission from stellar impacts onto central massive objects with super-Hills masses. More generally, we may say that the hypothetical discovery of TDE flares from galaxies with SMBH masses $M_\bullet \gtrsim 10^9 M_\odot$ by future time-domain surveys would strongly motivate consideration of exotic SMBH alternatives.

It is interesting to note that even in a limited sample of twelve optically-selected TDE candidates, there is already statistical evidence for a super-exponential cutoff in the TDE rate near $M_\bullet \sim 10^8 M_\odot$ (van Velzen 2018) as discussed in Sect. 3.5. This tentative evidence for the Hills mass may provide support for the identification of observed TDE candidates as bona-fide stellar tidal disruption events; it likewise demonstrates how much larger near-future TDE samples may statistically test the existence of event horizons.

6.3 EMRI Rates

If a star approaching the SMBH is compact enough, the differential gravitational pull exerted on points which are diametrically separated will not lead to dangerously large tidal stresses. A compact object such as a neutron star, a white dwarf, or a stellar-mass black hole can therefore safely pass through pericenters with $R_{\rm p} \gtrsim R_{\rm mb}$[25] many times until it plunges

[25]The marginally bound radius $R_{\rm mb}$ is the minimum pericenter that avoids capture by the event horizon and is of order $R_{\rm g}$ (Bardeen et al. 1972).

through the SMBH event horizon due to energy loss via gravitational radiation. This chain of events is referred to as an "extreme mass-ratio inspiral" (EMRI, see the review of Amaro-Seoane 2018 and references therein). The number of passages is roughly proportional to the mass ratio, so that for a $10M_{\odot}$ stellar-mass black hole and a SMBH of mass $10^6 M_{\odot}$, the EMRI will complete some $\sim 10^5$ orbits before plunging.

This source of gravitational waves represents a unique probe of gravity in the strong-field regime. Generic EMRI orbits are both eccentric and inclined, and apsidal and nodal precession imply that the compact object explores the full torus between the radial and polar turning points as it inspirals towards the black hole. At every pericenter passage there is a strong burst of radiation, and with a typical total of $\sim 10^5$ such bursts, an EMRI can be used to map spacetime around SMBHs. With EMRIs one can probe the geometry of SMBHs in a regime of non-linear, strong-field and dynamical gravity. This means that one can test GR and alternative theories of gravity, but also test different questions related to stellar dynamics in galactic nuclei. The gravitational-wave radiation from these systems allows one to extract qualitative information from a regime which is inaccessible to the photon, the surface of the central supermassive dark object, and to investigate if spacetime around it differs from Kerr. In particular, in GR the Kerr solution is the unique end state of gravitational collapse. The Kerr metric depends on the mass and spin of the object only, and all higher multiple moments are related to the gravitational radiation emitted by the compact object on its orbits towards the SMBH. This can be used to test the no-hair theorem, as first put forward by Ryan (1995). Moreover, they can be used as standard sirens to measure the luminosity of the source and to test the cosmological expansion history of the Universe (Schutz 1986). See Amaro-Seoane et al. (2007, 2015) for a general description of the implications of EMRIs on cosmology and fundamental physics.

EMRIs are formed via two-body relaxation in dense stellar systems, and the theory that allows us to derive their event rate is heavily based on the concept of a loss cone. This is so because of the many similarities of the two problems. However, as we mentioned before, the compact object needs to revolve for tens or hundreds of thousands of orbits around the SMBH before disappearing into the horizon, not just one orbit, as is the case for TDEs. When an orbit's pericenter approaches $R_{\rm p} \sim$ few $\times\ R_{\rm g}$, even if its semimajor axis is $a \gg R_{\rm g}$, it starts to rapidly lose energy and progress towards smaller a while remaining at a nearly constant $R_{\rm p}$, until it is finally captured at $a \sim$ few $\times\ R_{\rm g}$ (see e.g. Fig. 43 of Amaro-Seoane 2018). However, for this process to be successful, the inspiralling object should not be disturbed by the two-body relaxation. This places an upper limit on the initial values of a for EMRIs at around 10^{-2} pc, meaning that EMRIs constitute a small fraction of all compact object capture events ($\lesssim 1\%$, see Fig. 17 in Bar-Or and Alexander 2016). The accumulation of a long number of orbits means that relativistic effects at pericenter must be taken into account.

Analogous but even stricter considerations limit the parameter space for "stellar EMRIs," which are main-sequence stars inspiraling on quasi-circular orbits. These systems are potentially interesting because of their long lives: stable Roche lobe overflow can last for millions of years, and power a weak accretion disk (Dai and Blandford 2013), though the arrival of a second stellar EMRI during the lifetime of the first may lead to a TDE impostor due to mass loss in a hypersonic stellar collision (Metzger and Stone 2017). However, such systems require fairly fine-tuned initial conditions in order to circularize through gravitational wave emission (as opposed to tidal circularization, which would destroy the star), and the most promising formation scenario may use the bound star left over after the Hills mechanism breaks up a pre-existing binary (Amaro-Seoane et al. 2012).

Unless the SMBH is Schwarzschild, the spin plays a fundamental role in the fate of a compact object EMRI (Amaro-Seoane et al. 2013), controlling the required number of pericenter passages before crossing the event horizon. Resonant relaxation can also allow EMRIs to form closer to the SMBH, where the stellar density decreases, and two-body relaxation is not efficient enough (Hopman and Alexander 2006). While the dynamics of resonant relaxation are much more complicated than that of two-body relaxation, recent studies indicate that its ultimate contribution to steady-state EMRI rates is modest (Merritt 2015b; Bar-Or and Alexander 2016; Alexander 2017).

Even though a consensus about the cusp at the Galactic Centre has recently emerged in which theory, numerical calculations and observations agree on the existence of the sub-pc stellar cusp (Gallego-Cano et al. 2018; Schödel et al. 2018; Baumgardt et al. 2018), the dynamics of distant galactic nuclei on sub-pc scales are less clear, making it challenging to predict generic EMRI rates. The first major astrophysical uncertainty concerns the massive black hole occupation fraction. If many intermediate-mass black holes exist in the nuclei of dwarf galaxies, their contribution to the volumetric EMRI rate could dominate over larger SMBHs (Babak et al. 2017). A second major uncertainty comes from the unknown distribution of stellar profiles; if cored galactic nuclei are more common than is expected, volumetric EMRI rates will be low. Finally, as noted before, the magnitude of SMBH spin is crucial in the derivation of EMRI rates (Amaro-Seoane et al. 2013), as rapidly spinning Kerr SMBHs will convert many "direct plunges" into observable EMRIs. Upcoming observations of TDEs in nuclei which harbor SMBHs in the mass range of 10^5–$10^7 M_\odot$ may help to at least partially clear up these uncertainties; using large samples of TDEs to constrain the low-mass black hole occupation fraction will be particularly valuable.

6.4 Future Prospects

Over the next five years, our sample of TDEs will expand by multiple orders of magnitude. At the time of writing, a few dozen strong TDE candidates have been discovered, primarily through thermal emission in either the soft X-ray (e.g. the compilations of Komossa 2015 and Auchettl et al. 2017, and the X-ray Chapter) or optical/UV (e.g. the compilation of Hung et al. 2017 and the Optical Chapter) bands.[26] For most of the last decade, the time-averaged TDE detection rate has been a handful of strong candidate flares per year. This situation is already starting to change in 2019, as the wide-field optical survey ZTF (Bellm et al. 2019) has begun harvesting TDE candidates, with an expected detection rate of 32^{+41}_{-25} TDEs per year (Hung et al. 2018).

While the ongoing ZTF survey is expected to increase today's TDE sample by one order of magnitude, near-future optical and X-ray time domain surveys may increase our TDE inventory even further, into the thousands and possibly tens of thousands. Two upcoming space-based X-ray surveys, *eROSITA* (Merloni et al. 2012, launched during the writing of this Chapter) and *Einstein Probe* (Yuan et al. 2015, launch date note yet determined), hold particular promise. The *Einstein Probe* is expected to detect tens to hundreds of X-ray bright TDEs per year (Yuan et al. 2015), while *eROSITA* predictions give a per-year detection rate ≈ 1000 (Khabibullin et al. 2014). On the ground, the wide-field LSST optical survey (Abell et al. 2009) has long been recognized for its ability to detect thousands of TDEs per year (Strubbe and Quataert 2009; van Velzen et al. 2011); the true detection rate depends on survey strategy, but one recent estimate predicted ≈ 3–6×10^3 TDEs per year.

It is important to consider the many science goals that can be accomplished with such large, near-future samples of TDEs. Advance planning is important both (i) to develop relevant theoretical predictions that can be tested and falsified by statistical samples of TDEs,

[26]The online "Open TDE Catalog" https://tde.space/ is a useful resource for the observationally-curious reader.

and (ii) to aid future observers in prioritizing scarce spectroscopic or multiwavelength followup resources, which will certainly be inadequate to fully characterize *every* TDE candidate found by the surveys described above. In the list below, we offer a brief (and, unavoidably, subjective) selection of important questions in physics and astrophysics that may be partially addressed by large samples of TDEs. In assembling this list, we have purposefully avoided questions focused entirely on the process and consequences of stellar tidal disruption. While this arena features many interesting open questions, they are discussed in much greater detail in the other Chapters of this Volume, and here our aim is to examine the utility of TDEs *as a tool* for answering broader questions.

- What is the distribution of SMBH masses in the Universe?
- How did SMBH seeds form at high redshift?
- What is the distribution of SMBH spins in the Universe?
- Are the "SMBHs" observed in galactic nuclei true Kerr metric black holes, or more exotic, horizonless compact objects?
- What is the rate of other processes involving loss cone physics in galactic nuclei (e.g. hypervelocity star and EMRI production)?
- What fraction of galactic nuclei host unresolved binary SMBHs?
- What exotic dynamical processes are common to post-starburst galactic nuclei?
- Are there other subclasses of galaxies with anomalous nuclear stellar dynamics?

In this Chapter, we have discussed how each of these questions may be probed—at least to some extent—with a statistical sample of TDEs. But in almost all cases, directly answering these questions with TDEs will require a firmer grasp on rates of tidal disruption, and how they depend on astrophysical variables like SMBH mass, host galaxy mass, redshift, and so on. Over the next few years, we expect progress on TDE rates to arise from three parallel channels. Better theoretical models for non-standard loss cone repopulation mechanisms (e.g. nuclear triaxiality, eccentric stellar discs) will help solidify our understanding of underlying dynamical theory. Better dynamical modeling of nearby galactic nuclei will improve our understanding of TDE rates in astrophysically realistic galaxies. And better samples of observed TDE flares—samples that are both larger and more homogeneously selected—will provide firmer observational estimates of the real volumetric TDE rate in the Universe.

Acknowledgements N.C.S. received financial support from the NASA Astrophysics Theory Research Program (Grant NNX17AK43G; PI B. Metzger). M. K. acknowledges support from NSF Grant No. PHY-1607031 and NASA Award No. 80NSSC18K0639. E.M.R. acknowledges support from NWO TOP grant Module 2, project number 614.001.401. P.A.S. acknowledges support from the Ramón y Cajal Programme of the Ministry of Economy, Industry and Competitiveness of Spain, the COST Action GWverse CA16104, and the National Key R&D Program of China (2016YFA0400702) and the National Science Foundation of China (11721303).

References

P.A. Abell, J. Allison, S.F. Anderson, J.R. Andrew, J.R.P. Angel, L. Armus, D. Arnett, S.J. Asztalos, T.S. Axelrod et al. (LSST Science Collaboration), LSST Science Book, Version 2.0. arXiv e-prints (2009)

D. Aharon, H.B. Perets, The impact of mass segregation and star formation on the rates of gravitational-wave sources from extreme mass ratio inspirals. Astrophys. J. Lett. **830**, 1 (2016). https://doi.org/10.3847/2041-8205/830/1/L1

D. Aharon, A. Mastrobuono Battisti, H.B. Perets, The history of tidal disruption events in galactic nuclei. Astrophys. J. **823**, 137 (2016). https://doi.org/10.3847/0004-637X/823/2/137

T. Alexander, EMRIs and the relativistic loss-cone: the curious case of the fortunate coincidence. J. Phys. Conf. Ser. **840**, 012019 (2017). https://doi.org/10.1088/1742-6596/840/1/012019

T. Alexander, C. Hopman, Strong mass segregation around a massive black hole. Astrophys. J. **697**, 1861–1869 (2009). https://doi.org/10.1088/0004-637X/697/2/1861

P. Amaro-Seoane, Relativistic dynamics and extreme mass ratio inspirals. Living Rev. Relativ. **21**, 4 (2018). https://doi.org/10.1007/s41114-018-0013-8

P. Amaro-Seoane, J.R. Gair, M. Freitag, M.C. Miller, I. Mandel, C.J. Cutler, S. Babak, Intermediate and extreme mass-ratio inspirals—astrophysics, science applications and detection using LISA. Class. Quantum Gravity **24**, 113–169 (2007). https://doi.org/10.1088/0264-9381/24/17/R01

P. Amaro-Seoane, M.C. Miller, G.F. Kennedy, Tidal disruptions of separated binaries in galactic nuclei. Mon. Not. R. Astron. Soc. **425**(4), 2401–2406 (2012). https://doi.org/10.1111/j.1365-2966.2012.21162.x

P. Amaro-Seoane, C.F. Sopuerta, M.D. Freitag, The role of the supermassive black hole spin in the estimation of the EMRI event rate. Mon. Not. R. Astron. Soc. **429**, 3155–3165 (2013). https://doi.org/10.1093/mnras/sts572

P. Amaro-Seoane, J.R. Gair, A. Pound, S.A. Hughes, C.F. Sopuerta, Research update on extreme-mass-ratio inspirals. J. Phys. Conf. Ser. **610**, 012002 (2015). https://doi.org/10.1088/1742-6596/610/1/012002

F. Antonini, E. Barausse, J. Silk, The coevolution of nuclear star clusters, massive black holes, and their host galaxies. Astrophys. J. **812**(1), 72 (2015). https://doi.org/10.1088/0004-637X/812/1/72

M. Arca-Sedda, R. Capuzzo-Dolcetta, The MEGaN project—I. Missing formation of massive nuclear clusters and tidal disruption events by star clusters-massive black hole interactions. Mon. Not. R. Astron. Soc. **471**, 478–490 (2017). https://doi.org/10.1093/mnras/stx1586

I. Arcavi, A. Gal-Yam, M. Sullivan, Y.-C. Pan, S.B. Cenko, A. Horesh, E.O. Ofek, A. De Cia, L. Yan, C.-W. Yang, D.A. Howell, D. Tal, S.R. Kulkarni, S.P. Tendulkar, S. Tang, D. Xu, A. Sternberg, J.G. Cohen, J.S. Bloom, P.E. Nugent, M.M. Kasliwal, D.A. Perley, R.M. Quimby, A.A. Miller, C.A. Theissen, R.R. Laher, A continuum of H- to He-rich tidal disruption candidates with a preference for E+A galaxies. Astrophys. J. **793**, 38 (2014). https://doi.org/10.1088/0004-637X/793/1/38

K. Auchettl, J. Guillochon, E. Ramirez-Ruiz, New physical insights about tidal disruption events from a comprehensive observational inventory at X-ray wavelengths. Astrophys. J. **838**, 149 (2017). https://doi.org/10.3847/1538-4357/aa633b

S. Babak, J. Gair, A. Sesana, E. Barausse, C.F. Sopuerta, C.P.L. Berry, E. Berti, P. Amaro-Seoane, A. Petiteau, A. Klein, Science with the space-based interferometer LISA. V. Extreme mass-ratio inspirals. Phys. Rev. D **95**(10), 103012 (2017). https://doi.org/10.1103/PhysRevD.95.103012

N. Bade, S. Komossa, M. Dahlem, Detection of an extremely soft X-ray outburst in the HII-like nucleus of NGC 5905. Astron. Astrophys. **309**, 35–38 (1996)

J.N. Bahcall, R.A. Wolf, Star distribution around a massive black hole in a globular cluster. Astrophys. J. **209**, 214–232 (1976). https://doi.org/10.1086/154711

J.N. Bahcall, R.A. Wolf, The star distribution around a massive black hole in a globular cluster. II. Unequal star masses. Astrophys. J. **216**, 883–907 (1977). https://doi.org/10.1086/155534

B. Bar-Or, T. Alexander, Steady-state relativistic stellar dynamics around a massive black hole. Astrophys. J. **820**, 129 (2016). https://doi.org/10.3847/0004-637X/820/2/129

J.M. Bardeen, W.H. Press, S.A. Teukolsky, Rotating black holes: locally nonrotating frames, energy extraction, and scalar synchrotron radiation. Astrophys. J. **178**, 347–370 (1972). https://doi.org/10.1086/151796

H. Baumgardt, P. Amaro-Seoane, R. Schödel, The distribution of stars around the Milky Way's central black hole. III. Comparison with simulations. Astron. Astrophys. **609**, 28 (2018). https://doi.org/10.1051/0004-6361/201730462

M.C. Begelman, R.D. Blandford, M.J. Rees, Massive black hole binaries in active galactic nuclei. Nature **287**, 307–309 (1980). https://doi.org/10.1038/287307a0

K. Belczynski, T. Bulik, C.L. Fryer, A. Ruiter, F. Valsecchi, J.S. Vink, J.R. Hurley, On the maximum mass of stellar black holes. Astrophys. J. **714**, 1217–1226 (2010). https://doi.org/10.1088/0004-637X/714/2/1217

E.C. Bellm, S.R. Kulkarni, M.J. Graham, R. Dekany, R.M. Smith, R. Riddle, F.J. Masci, G. Helou, T.A. Prince, S.M. Adams, C. Barbarino, T. Barlow, J. Bauer, R. Beck, J. Belicki, R. Biswas, N. Blagorodnova, D. Bodewits, B. Bolin, V. Brinnel, T. Brooke, B. Bue, M. Bulla, R. Burruss, S.B. Cenko, C.-K. Chang, A. Connolly, M. Coughlin, J. Cromer, V. Cunningham, K. De, A. Delacroix, V. Desai, D.A. Duev, G. Eadie, T.L. Farnham, M. Feeney, U. Feindt, D. Flynn, A. Franckowiak, S. Frederick, C. Fremling, A. Gal-Yam, S. Gezari, M. Giomi, D.A. Goldstein, V.Z. Golkhou, A. Goobar, S. Groom, E. Hacopians, D. Hale, J. Henning, A.Y.Q. Ho, D. Hover, J. Howell, T. Hung, D. Huppenkothen, D. Imel, W.-H. Ip, Ž. Ivezić, E. Jackson, L. Jones, M. Juric, M.M. Kasliwal, S. Kaspi,

S. Kaye, M.S.P. Kelley, M. Kowalski, E. Kramer, T. Kupfer, W. Landry, R.R. Laher, C.-D. Lee, H.W. Lin, Z.-Y. Lin, R. Lunnan, M. Giomi, A. Mahabal, P. Mao, A.A. Miller, S. Monkewitz, P. Murphy, C.-C. Ngeow, J. Nordin, P. Nugent, E. Ofek, M.T. Patterson, B. Penprase, M. Porter, L. Rauch, U. Rebbapragada, D. Reiley, M. Rigault, H. Rodriguez, J. van Roestel, B. Rusholme, J. van Santen, S. Schulze, D.L. Shupe, L.P. Singer, M.T. Soumagnac, R. Stein, J. Surace, J. Sollerman, P. Szkody, F. Taddia, S. Terek, A. Van Sistine, S. van Velzen, W.T. Vestrand, R. Walters, C. Ward, Q.-Z. Ye, P.-C. Yu, L. Yan, J. Zolkower, The Zwicky Transient Facility: system overview, performance, and first results. Publ. Astron. Soc. Pac. **131**(1), 018002 (2019). https://doi.org/10.1088/1538-3873/aaecbe

A.M. Beloborodov, A.F. Illarionov, P.B. Ivanov, A.G. Polnarev, Angular momentum of a supermassive black hole in a dense star cluster. Mon. Not. R. Astron. Soc. **259**, 209–217 (1992). https://doi.org/10.1093/mnras/259.2.209

E. Berti, M. Volonteri, Cosmological black hole spin evolution by mergers and accretion. Astrophys. J. **684**(2), 822–828 (2008). https://doi.org/10.1086/590379

J. Binney, S. Tremaine, *Galactic Dynamics*, 2nd edn. (Princeton University Press, Princeton, 2008)

C. Bonnerot, E.M. Rossi, Streams collision as possible precursor of double tidal disruption events. Mon. Not. R. Astron. Soc. **484**, 1301–1316 (2019). https://doi.org/10.1093/mnras/stz062

C. Bonnerot, E.M. Rossi, G. Lodato, Bad prospects for the detection of giant stars' tidal disruption: effect of the ambient medium on bound debris. Mon. Not. R. Astron. Soc. **458**(3), 3324–3330 (2016). https://doi.org/10.1093/mnras/stw486

D. Boubert, J. Guillochon, K. Hawkins, I. Ginsburg, N.W. Evans, J. Strader, Revisiting hypervelocity stars after Gaia DR2. Mon. Not. R. Astron. Soc. **479**, 2789–2795 (2018). https://doi.org/10.1093/mnras/sty1601

R.H. Boyer, R.W. Lindquist, Maximal analytic extension of the Kerr metric. J. Math. Phys. **8**, 265–281 (1967). https://doi.org/10.1063/1.1705193

B. Bradnick, I. Mandel, Y. Levin, Stellar binaries in galactic nuclei: tidally stimulated mergers followed by tidal disruptions. Mon. Not. R. Astron. Soc. **469**, 2042–2048 (2017). https://doi.org/10.1093/mnras/stx1007

P. Brem, P. Amaro-Seoane, C.F. Sopuerta, Blocking low-eccentricity EMRIs: a statistical direct-summation N-body study of the Schwarzschild barrier. Mon. Not. R. Astron. Soc. **437**, 1259–1267 (2014). https://doi.org/10.1093/mnras/stt1948

B.C. Bromley, S.J. Kenyon, W.R. Brown, M.J. Geller, Nearby high-speed stars in Gaia DR2. Astrophys. J. **868**, 25 (2018). https://doi.org/10.3847/1538-4357/aae83e

W.R. Brown, M.G. Lattanzi, S.J. Kenyon, M.J. Geller, Gaia and the galactic center origin of hypervelocity stars. Astrophys. J. **866**, 39 (2018). https://doi.org/10.3847/1538-4357/aadb8e

Y.-I. Byun, C.J. Grillmair, S.M. Faber, E.A. Ajhar, A. Dressler, J. Kormendy, T.R. Lauer, D. Richstone, S. Tremaine, The centers of early-type galaxies with HST. II. Empirical models and structural parameters. Astron. J. **111**, 1889 (1996). https://doi.org/10.1086/117927

B. Carter, Global structure of the Kerr family of gravitational fields. Phys. Rev. **174**, 1559–1571 (1968). https://doi.org/10.1103/PhysRev.174.1559

B. Carter, Axisymmetric black hole has only two degrees of freedom. Phys. Rev. Lett. **26**, 331–333 (1971). https://doi.org/10.1103/PhysRevLett.26.331

B. Carter, J.-P. Luminet, Tidal compression of a star by a large black hole. I. Mechanical evolution and nuclear energy release by proton capture. Astron. Astrophys. **121**, 97–113 (1983)

S. Chandrasekhar, *Principles of Stellar Dynamics* (1942)

X. Chen, A. Sesana, P. Madau, F.K. Liu, Tidal stellar disruptions by massive black hole pairs. II. Decaying binaries. Astrophys. J. **729**, 13 (2011). https://doi.org/10.1088/0004-637X/729/1/13

H. Cohn, R.M. Kulsrud, The stellar distribution around a black hole—numerical integration of the Fokker-Planck equation. Astrophys. J. **226**, 1087–1108 (1978). https://doi.org/10.1086/156685

L. Dai, R. Blandford, Roche accretion of stars close to massive black holes. Mon. Not. R. Astron. Soc. **434**(4), 2948–2960 (2013). https://doi.org/10.1093/mnras/stt1209

L. Dai, J.C. McKinney, M.C. Miller, Soft X-ray temperature tidal disruption events from stars on deep plunging orbits. Astrophys. J. Lett. **812**, 39 (2015). https://doi.org/10.1088/2041-8205/812/2/L39

J.L. Donley, W.N. Brandt, M. Eracleous, T. Boller, Large-amplitude X-ray outbursts from galactic nuclei: a systematic survey using ROSAT archival data. Astron. J. **124**, 1308–1321 (2002). https://doi.org/10.1086/342280

P. Esquej, R.D. Saxton, S. Komossa, A.M. Read, M.J. Freyberg, G. Hasinger, D.A. García-Hernández, H. Lu, J. Rodriguez Zaurín, M. Sánchez-Portal, H. Zhou, Evolution of tidal disruption candidates discovered by XMM-Newton. Astron. Astrophys. **489**, 543–554 (2008). https://doi.org/10.1051/0004-6361:200810110

S.M. Faber, S. Tremaine, E.A. Ajhar, Y.-I. Byun, A. Dressler, K. Gebhardt, C. Grillmair, J. Kormendy, T.R. Lauer, D. Richstone, The centers of early-type galaxies with HST. IV. Central parameter relations. Astron. J. **114**, 1771 (1997). https://doi.org/10.1086/118606
L. Ferrarese, D. Merritt, A fundamental relation between supermassive black holes and their host galaxies. Astrophys. J. Lett. **539**, 9–12 (2000). https://doi.org/10.1086/312838
J. Frank, M.J. Rees, Effects of massive central black holes on dense stellar systems. Mon. Not. R. Astron. Soc. **176**, 633–647 (1976). https://doi.org/10.1093/mnras/176.3.633
K.D. French, I. Arcavi, A. Zabludoff, Tidal disruption events prefer unusual host galaxies. Astrophys. J. Lett. **818**, 21 (2016). https://doi.org/10.3847/2041-8205/818/1/L21
K.D. French, I. Arcavi, A. Zabludoff, The post-starburst evolution of tidal disruption event host galaxies. Astrophys. J. **835**, 176 (2017). https://doi.org/10.3847/1538-4357/835/2/176
E. Gallego-Cano, R. Schödel, H. Dong, F. Nogueras-Lara, A.T. Gallego-Calvente, P. Amaro-Seoane, H. Baumgardt, The distribution of stars around the Milky Way's central black hole. I. Deep star counts. Astron. Astrophys. **609**, 26 (2018). https://doi.org/10.1051/0004-6361/201730451
K. Gebhardt, R. Bender, G. Bower, A. Dressler, S.M. Faber, A.V. Filippenko, R. Green, C. Grillmair, L.C. Ho, J. Kormendy, T.R. Lauer, J. Magorrian, J. Pinkney, D. Richstone, S. Tremaine, A relationship between nuclear black hole mass and galaxy velocity dispersion. Astrophys. J. Lett. **539**, 13–16 (2000). https://doi.org/10.1086/312840
A. Generozov, N.C. Stone, B.D. Metzger, J.P. Ostriker, An overabundance of black hole X-ray binaries in the Galactic Centre from tidal captures. Mon. Not. R. Astron. Soc. **478**, 4030–4051 (2018). https://doi.org/10.1093/mnras/sty1262
S. Gezari, D.C. Martin, B. Milliard, S. Basa, J.P. Halpern, K. Forster, P.G. Friedman, P. Morrissey, S.G. Neff, D. Schiminovich, M. Seibert, T. Small, T.K. Wyder, Ultraviolet detection of the tidal disruption of a star by a supermassive black hole. Astrophys. J. Lett. **653**, 25–28 (2006). https://doi.org/10.1086/509918
S. Gezari, S. Basa, D.C. Martin, G. Bazin, K. Forster, B. Milliard, J.P. Halpern, P.G. Friedman, P. Morrissey, S.G. Neff, D. Schiminovich, M. Seibert, T. Small, T.K. Wyder, UV/optical detections of candidate tidal disruption events by GALEX and CFHTLS. Astrophys. J. **676**, 944–969 (2008). https://doi.org/10.1086/529008
S. Gezari, R. Chornock, A. Rest, M.E. Huber, K. Forster, E. Berger, P.J. Challis, J.D. Neill, D.C. Martin, T. Heckman, A. Lawrence, C. Norman, G. Narayan, R.J. Foley, G.H. Marion, D. Scolnic, L. Chomiuk, A. Soderberg, K. Smith, R.P. Kirshner, A.G. Riess, S.J. Smartt, C.W. Stubbs, J.L. Tonry, W.M. Wood-Vasey, W.S. Burgett, K.C. Chambers, T. Grav, J.N. Heasley, N. Kaiser, R.-P. Kudritzki, E.A. Magnier, J.S. Morgan, P.A. Price, An ultraviolet-optical flare from the tidal disruption of a helium-rich stellar core. Nature **485**, 217–220 (2012). https://doi.org/10.1038/nature10990
O. Graur, K.D. French, H.J. Zahid, J. Guillochon, K.S. Mandel, K. Auchettl, A.I. Zabludoff, A dependence of the tidal disruption event rate on global stellar surface mass density and stellar velocity dispersion. Astrophys. J. **853**, 39 (2018). https://doi.org/10.3847/1538-4357/aaa3fd
J. Guillochon, E. Ramirez-Ruiz, Hydrodynamical simulations to determine the feeding rate of black holes by the tidal disruption of stars: the importance of the impact parameter and stellar structure. Astrophys. J. **767**, 25 (2013). https://doi.org/10.1088/0004-637X/767/1/25
A.S. Hamers, H.B. Perets, Relaxation near supermassive black holes driven by nuclear spiral arms: anisotropic hypervelocity stars, S-stars, and tidal disruption events. Astrophys. J. **846**, 123 (2017). https://doi.org/10.3847/1538-4357/aa7f29
M. Hartmann, V.P. Debattista, A. Seth, M. Cappellari, T.R. Quinn, Constraining the role of star cluster mergers in nuclear cluster formation: simulations confront integral-field data. Mon. Not. R. Astron. Soc. **418**, 2697–2714 (2011). https://doi.org/10.1111/j.1365-2966.2011.19659.x
T.M. Heckman, G. Kauffmann, J. Brinchmann, S. Charlot, C. Tremonti, S.D.M. White, Present-day growth of black holes and bulges: the Sloan digital sky survey perspective. Astrophys. J. **613**, 109–118 (2004). https://doi.org/10.1086/422872
J.G. Hills, Possible power source of Seyfert galaxies and QSOs. Nature **254**, 295–298 (1975). https://doi.org/10.1038/254295a0
J.G. Hills, Hyper-velocity and tidal stars from binaries disrupted by a massive galactic black hole. Nature **331**, 687–689 (1988). https://doi.org/10.1038/331687a0
J.G. Hills, Computer simulations of encounters between massive black holes and binaries. Astron. J. **102**, 704–715 (1991). https://doi.org/10.1086/115905
T.W.-S. Holoien, C.S. Kochanek, J.L. Prieto, K.Z. Stanek, S. Dong, B.J. Shappee, D. Grupe, J.S. Brown, U. Basu, J.F. Beacom, D. Bersier, J. Brimacombe, A.B. Danilet, E. Falco, Z. Guo, J. Jose, G.J. Herczeg, F. Long, G. Pojmanski, G.V. Simonian, D.M. Szczygieł, T.A. Thompson, J.R. Thorstensen, R.M. Wagner, P.R. Woźniak, Six months of multiwavelength follow-up of the tidal disruption candidate ASASSN-14li and implied TDE rates from ASAS-SN. Mon. Not. R. Astron. Soc. **455**, 2918–2935 (2016). https://doi.org/10.1093/mnras/stv2486

C. Hopman, Binary dynamics near a massive black hole. Astrophys. J. **700**, 1933–1951 (2009). https://doi.org/10.1088/0004-637X/700/2/1933

C. Hopman, T. Alexander, Resonant relaxation near a massive black hole: the stellar distribution and gravitational wave sources. Astrophys. J. **645**, 1152–1163 (2006). https://doi.org/10.1086/504400

T. Hung, S. Gezari, N. Blagorodnova, N. Roth, S.B. Cenko, S.R. Kulkarni, A. Horesh, I. Arcavi, C. McCully, L. Yan, R. Lunnan, C. Fremling, Y. Cao, P.E. Nugent, P. Wozniak, Revisiting optical tidal disruption events with iPTF16axa. Astrophys. J. **842**, 29 (2017). https://doi.org/10.3847/1538-4357/aa7337

T. Hung, S. Gezari, S.B. Cenko, S. van Velzen, N. Blagorodnova, L. Yan, S.R. Kulkarni, R. Lunnan, T. Kupfer, G. Leloudas, A.K.H. Kong, P.E. Nugent, C. Fremling, R.R. Laher, F.J. Masci, Y. Cao, R. Roy, T. Petrushevska, Sifting for sapphires: systematic selection of tidal disruption events in iPTF. Astrophys. J. Suppl. Ser. **238**, 15 (2018). https://doi.org/10.3847/1538-4365/aad8b1

P.B. Ivanov, M.A. Chernyakova, Relativistic cross sections of mass stripping and tidal disruption of a star by a super-massive rotating black hole. Astron. Astrophys. **448**, 843–852 (2006). https://doi.org/10.1051/0004-6361:20053409

P.B. Ivanov, A.G. Polnarev, P. Saha, The tidal disruption rate in dense galactic cusps containing a supermassive binary black hole. Mon. Not. R. Astron. Soc. **358**, 1361–1378 (2005). https://doi.org/10.1111/j.1365-2966.2005.08843.x

V. Karas, L. Šubr, Enhanced activity of massive black holes by stellar capture assisted by a self-gravitating accretion disc. Astron. Astrophys. **470**, 11–19 (2007). https://doi.org/10.1051/0004-6361:20066068

R.P. Kerr, Gravitational field of a spinning mass as an example of algebraically special metrics. Phys. Rev. Lett. **11**, 237–238 (1963). https://doi.org/10.1103/PhysRevLett.11.237

M. Kesden, Tidal-disruption rate of stars by spinning supermassive black holes. Phys. Rev. D **85**(2), 024037 (2012). https://doi.org/10.1103/PhysRevD.85.024037

I. Khabibullin, S. Sazonov, Stellar tidal disruption candidates found by cross-correlating the ROSAT Bright Source Catalogue and XMM-Newton observations. Mon. Not. R. Astron. Soc. **444**, 1041–1053 (2014). https://doi.org/10.1093/mnras/stu1491

I. Khabibullin, S. Sazonov, R. Sunyaev, SRG/eROSITA prospects for the detection of stellar tidal disruption flares. Mon. Not. R. Astron. Soc. **437**, 327–337 (2014). https://doi.org/10.1093/mnras/stt1889

A.R. King, J.E. Pringle, Growing supermassive black holes by chaotic accretion. Mon. Not. R. Astron. Soc. **373**(1), 90–92 (2006). https://doi.org/10.1111/j.1745-3933.2006.00249.x

C.S. Kochanek, Tidal disruption event demographics. Mon. Not. R. Astron. Soc. **461**, 371–384 (2016). https://doi.org/10.1093/mnras/stw1290

S. Komossa, Tidal disruption of stars by supermassive black holes: status of observations. J. High Energy Astrophys. **7**, 148–157 (2015). https://doi.org/10.1016/j.jheap.2015.04.006

S. Komossa, J. Greiner, Discovery of a giant and luminous X-ray outburst from the optically inactive galaxy pair RX J1242.6-1119. Astron. Astrophys. **349**, 45–48 (1999)

J. Kormendy, D. Richstone, Inward bound—the search for supermassive black holes in galactic nuclei. Annu. Rev. Astron. Astrophys. **33**, 581 (1995). https://doi.org/10.1146/annurev.aa.33.090195.003053

T.R. Lauer, E.A. Ajhar, Y.-I. Byun, A. Dressler, S.M. Faber, C. Grillmair, J. Kormendy, D. Richstone, S. Tremaine, The centers of early-type galaxies with HST.I. An observational survey. Astron. J. **110**, 2622 (1995). https://doi.org/10.1086/117719

T.R. Lauer, S.M. Faber, K. Gebhardt, D. Richstone, S. Tremaine, E.A. Ajhar, M.C. Aller, R. Bender, A. Dressler, A.V. Filippenko, R. Green, C.J. Grillmair, L.C. Ho, J. Kormendy, J. Magorrian, J. Pinkney, C. Siopis, The centers of early-type galaxies with Hubble Space Telescope. V. New WFPC2 photometry. Astron. J. **129**, 2138–2185 (2005). https://doi.org/10.1086/429565

T.R. Lauer, K. Gebhardt, S.M. Faber, D. Richstone, S. Tremaine, J. Kormendy, M.C. Aller, R. Bender, A. Dressler, A.V. Filippenko, R. Green, L.C. Ho, The centers of early-type galaxies with Hubble Space Telescope. VI. Bimodal central surface brightness profiles. Astrophys. J. **664**, 226–256 (2007a). https://doi.org/10.1086/519229

T.R. Lauer, S.M. Faber, D. Richstone, K. Gebhardt, S. Tremaine, M. Postman, A. Dressler, M.C. Aller, A.V. Filippenko, R. Green, L.C. Ho, J. Kormendy, J. Magorrian, J. Pinkney, The masses of nuclear black holes in luminous elliptical galaxies and implications for the space density of the most massive black holes. Astrophys. J. **662**, 808–834 (2007b). https://doi.org/10.1086/518223

J. Law-Smith, M. MacLeod, J. Guillochon, P. Macias, E. Ramirez-Ruiz, Low-mass white dwarfs with hydrogen envelopes as a missing link in the tidal disruption menu. Astrophys. J. **841**, 132 (2017a). https://doi.org/10.3847/1538-4357/aa6ffb

J. Law-Smith, E. Ramirez-Ruiz, S.L. Ellison, R.J. Foley, Tidal disruption event host galaxies in the context of the local galaxy population. Astrophys. J. **850**, 22 (2017b). https://doi.org/10.3847/1538-4357/aa94c7

K. Lezhnin, E. Vasiliev, Suppression of stellar tidal disruption rates by anisotropic initial conditions. Astrophys. J. Lett. **808**, 5 (2015). https://doi.org/10.1088/2041-8205/808/1/L5

A.P. Lightman, S.L. Shapiro, The distribution and consumption rate of stars around a massive, collapsed object. Astrophys. J. **211**, 244–262 (1977). https://doi.org/10.1086/154925

W. Lu, P. Kumar, R. Narayan, Stellar disruption events support the existence of the black hole event horizon. Mon. Not. R. Astron. Soc. **468**(1), 910–919 (2017). https://doi.org/10.1093/mnras/stx542

J.P. Luminet, J.A. Marck, Tidal effects in Kerr geometry, in *General Relativity and Gravitation*, vol. 1, ed. by B. Bertotti, F. de Felice, A. Pascolini (1983), p. 438

M. MacLeod, J. Guillochon, E. Ramirez-Ruiz, The tidal disruption of giant stars and their contribution to the flaring supermassive black hole population. Astrophys. J. **757**, 134 (2012). https://doi.org/10.1088/0004-637X/757/2/134

M. MacLeod, E. Ramirez-Ruiz, S. Grady, J. Guillochon, Spoon-feeding giant stars to supermassive black holes: episodic mass transfer from evolving stars and their contribution to the quiescent activity of galactic nuclei. Astrophys. J. **777**(2), 133 (2013). https://doi.org/10.1088/0004-637X/777/2/133

A.-M. Madigan, A. Halle, M. Moody, M. McCourt, C. Nixon, H. Wernke, Dynamical properties of eccentric nuclear disks: stability, longevity, and implications for tidal disruption rates in post-merger galaxies. Astrophys. J. **853**, 141 (2018). https://doi.org/10.3847/1538-4357/aaa714

T. Mageshwaran, A. Mangalam, Stellar and gas dynamical model for tidal disruption events in a quiescent galaxy. Astrophys. J. **814**, 141 (2015). https://doi.org/10.1088/0004-637X/814/2/141

J. Magorrian, S. Tremaine, Rates of tidal disruption of stars by massive central black holes. Mon. Not. R. Astron. Soc. **309**, 447–460 (1999). https://doi.org/10.1046/j.1365-8711.1999.02853.x

J. Magorrian, S. Tremaine, D. Richstone, R. Bender, G. Bower, A. Dressler, S.M. Faber, K. Gebhardt, R. Green, C. Grillmair, J. Kormendy, T. Lauer, The demography of massive dark objects in galaxy centers. Astron. J. **115**, 2285–2305 (1998). https://doi.org/10.1086/300353

D. Mainetti, A. Lupi, S. Campana, M. Colpi, E.R. Coughlin, J. Guillochon, E. Ramirez-Ruiz, The fine line between total and partial tidal disruption events. Astron. Astrophys. **600**, 124 (2017). https://doi.org/10.1051/0004-6361/201630092

W.P. Maksym, M.P. Ulmer, M. Eracleous, A tidal disruption flare in A1689 from an archival X-ray survey of galaxy clusters. Astrophys. J. **722**, 1035–1050 (2010). https://doi.org/10.1088/0004-637X/722/2/1035

F.K. Manasse, C.W. Misner, Fermi normal coordinates and some basic concepts in differential geometry. J. Math. Phys. **4**, 735–745 (1963)

I. Mandel, Y. Levin, Double tidal disruptions in galactic nuclei. Astrophys. J. Lett. **805**, 4 (2015). https://doi.org/10.1088/2041-8205/805/1/L4

T. Marchetti, E.M. Rossi, G. Kordopatis, A.G.A. Brown, A. Rimoldi, E. Starkenburg, K. Youakim, R. Ashley, An artificial neural network to discover hypervelocity stars: candidates in Gaia DR1/TGAS. Mon. Not. R. Astron. Soc. **470**, 1388–1403 (2017). https://doi.org/10.1093/mnras/stx1304

T. Marchetti, E.M. Rossi, A.G.A. Brown, Gaia DR2 in 6D: searching for the fastest stars in the galaxy. Mon. Not. R. Astron. Soc. (2018). https://doi.org/10.1093/mnras/sty2592

J. Marck, Solution to the equations of parallel transport in Kerr geometry; tidal tensor. Proc. R. Soc. Lond. Ser. A **385**, 431–438 (1983). https://doi.org/10.1098/rspa.1983.0021

A. Mastrobuono-Battisti, H.B. Perets, A. Loeb, Effects of intermediate mass black holes on nuclear star clusters. Astrophys. J. **796**, 40 (2014). https://doi.org/10.1088/0004-637X/796/1/40

N.J. McConnell, C.-P. Ma, Revisiting the scaling relations of black hole masses and host galaxy properties. Astrophys. J. **764**, 184 (2013). https://doi.org/10.1088/0004-637X/764/2/184

A. Merloni, P. Predehl, W. Becker, H. Böhringer, T. Boller, H. Brunner, M. Brusa, K. Dennerl, M. Freyberg, P. Friedrich, A. Georgakakis, F. Haberl, G. Hasinger, N. Meidinger, J. Mohr, K. Nandra, A. Rau, T.H. Reiprich, J. Robrade, M. Salvato, A. Santangelo, M. Sasaki, A. Schwope, J. Wilms, t. German eROSITA Consortium, eROSITA Science Book: Mapping the Structure of the Energetic Universe. arXiv e-prints (2012)

D. Merritt, *Dynamics and Evolution of Galactic Nuclei* (Princeton University Press, Princeton, 2013)

D. Merritt, Gravitational encounters and the evolution of galactic nuclei. II. Classical and resonant relaxation. Astrophys. J. **804**, 128 (2015a). https://doi.org/10.1088/0004-637X/804/2/128

D. Merritt, Gravitational encounters and the evolution of galactic nuclei. IV. Captures mediated by gravitational-wave energy loss. Astrophys. J. **814**, 57 (2015b). https://doi.org/10.1088/0004-637X/814/1/57

D. Merritt, L. Ferrarese, Black hole demographics from the $M_\bullet$-σ relation. Mon. Not. R. Astron. Soc. **320**, 30–34 (2001). https://doi.org/10.1046/j.1365-8711.2001.04165.x

D. Merritt, M.Y. Poon, Chaotic loss cones and black hole fueling. Astrophys. J. **606**, 788–798 (2004). https://doi.org/10.1086/382497

D. Merritt, J. Wang, Loss cone refilling rates in galactic nuclei. Astrophys. J. Lett. **621**, 101–104 (2005). https://doi.org/10.1086/429272

D. Merritt, T. Alexander, S. Mikkola, C.M. Will, Stellar dynamics of extreme-mass-ratio inspirals. Phys. Rev. D **84**(4), 044024 (2011). https://doi.org/10.1103/PhysRevD.84.044024

B.D. Metzger, N.C. Stone, A bright year for tidal disruptions. Mon. Not. R. Astron. Soc. **461**(1), 948–966 (2016). https://doi.org/10.1093/mnras/stw1394

B.D. Metzger, N.C. Stone, Periodic accretion-powered flares from colliding EMRIs as TDE imposters. Astrophys. J. **844**(1), 75 (2017). https://doi.org/10.3847/1538-4357/aa7a16

M. Milosavljević, D. Merritt, Long-term evolution of massive black hole binaries. Astrophys. J. **596**, 860–878 (2003). https://doi.org/10.1086/378086

M. Milosavljević, D. Merritt, L.C. Ho, Contribution of stellar tidal disruptions to the X-ray luminosity function of active galaxies. Astrophys. J. **652**(1), 120–125 (2006). https://doi.org/10.1086/508134

H.B. Perets, Dynamical and evolutionary constraints on the nature and origin of hypervelocity stars. Astrophys. J. **690**, 795–801 (2009). https://doi.org/10.1088/0004-637X/690/1/795

H.B. Perets, T. Alexander, Massive perturbers and the efficient merger of binary massive black holes. Astrophys. J. **677**, 146–159 (2008). https://doi.org/10.1086/527525

H.B. Perets, A. Gualandris, Dynamical constraints on the origin of the young B-stars in the galactic center. Astrophys. J. **719**, 220–228 (2010). https://doi.org/10.1088/0004-637X/719/1/220

H.B. Perets, A. Mastrobuono-Battisti, Age and mass segregation of multiple stellar populations in galactic nuclei and their observational signatures. Astrophys. J. Lett. **784**, 44 (2014). https://doi.org/10.1088/2041-8205/784/2/L44

H.B. Perets, C. Hopman, T. Alexander, Massive perturber-driven interactions between stars and a massive black hole. Astrophys. J. **656**, 709–720 (2007). https://doi.org/10.1086/510377

H.B. Perets, Z. Li, J.C. Lombardi Jr., S.R. Milcarek Jr., Micro-tidal disruption events by stellar compact objects and the production of ultra-long GRBs. Astrophys. J. **823**, 113 (2016). https://doi.org/10.3847/0004-637X/823/2/113

H. Pfister, B. Bar-Or, M. Volonteri, Y. Dubois, P.R. Capelo, Tidal disruption event rates in galaxy merger remnants. arXiv e-prints (2019)

V.L. Polyachenko, I.G. Shukhman, General models of collisionless spherically symmetric stellar systems—a stability analysis. Sov. Astron. **25**, 533 (1981)

M.Y. Poon, D. Merritt, Orbital structure of triaxial black hole nuclei. Astrophys. J. **549**, 192–204 (2001). https://doi.org/10.1086/319060

M.B. Pracy, M.S. Owers, W.J. Couch, H. Kuntschner, K. Bekki, F. Briggs, P. Lah, M. Zwaan, Stellar population gradients in the cores of nearby field E+A galaxies. Mon. Not. R. Astron. Soc. **420**, 2232–2244 (2012). https://doi.org/10.1111/j.1365-2966.2011.20188.x

M. Preto, P. Amaro-Seoane, On strong mass segregation around a massive black hole: implications for lower-frequency gravitational-wave astrophysics. Astrophys. J. Lett. **708**, 42–46 (2010). https://doi.org/10.1088/2041-8205/708/1/L42

K.P. Rauch, S. Tremaine, Resonant relaxation in stellar systems. New Astron. **1**, 149–170 (1996). https://doi.org/10.1016/S1384-1076(96)00012-7

M.J. Rees, Tidal disruption of stars by black holes of 10 to the 6th-10 to the 8th solar masses in nearby galaxies. Nature **333**, 523–528 (1988). https://doi.org/10.1038/333523a0

M.N. Rosenbluth, R.F. Post, High-frequency electrostatic plasma instability inherent to "loss-cone" particle distributions. Phys. Fluids **8**, 547–550 (1965). https://doi.org/10.1063/1.1761261

F.D. Ryan, Gravitational waves from the inspiral of a compact object into a massive, axisymmetric body with arbitrary multipole moments. Phys. Rev. D **52**, 5707–5718 (1995). https://doi.org/10.1103/PhysRevD.52.5707

E.E. Salpeter, Accretion of interstellar matter by massive objects. Astrophys. J. **140**, 796–800 (1964). https://doi.org/10.1086/147973

R. Sari, S. Kobayashi, E.M. Rossi, Hypervelocity stars and the restricted parabolic three-body problem. Astrophys. J. **708**, 605–614 (2010). https://doi.org/10.1088/0004-637X/708/1/605

C.J. Saxton, H.B. Perets, A. Baskin, Spectral features of tidal disruption candidates and alternative origins for such transient flares. Mon. Not. R. Astron. Soc. **474**, 3307–3323 (2018). https://doi.org/10.1093/mnras/stx2928

M. Schmidt, 3C 273: a star-like object with large red-shift. Nature **197**, 1040 (1963). https://doi.org/10.1038/1971040a0

R. Schödel, E. Gallego-Cano, H. Dong, F. Nogueras-Lara, A.T. Gallego-Calvente, P. Amaro-Seoane, H. Baumgardt, The distribution of stars around the Milky Way's central black hole. II. Diffuse light from sub-giants and dwarfs. Astron. Astrophys. **609**, 27 (2018). https://doi.org/10.1051/0004-6361/201730452

B.F. Schutz, Determining the Hubble constant from gravitational wave observations. Nature **323**(6086), 310–311 (1986). https://doi.org/10.1038/323310a0

J. Servin, M. Kesden, Unified treatment of tidal disruption by Schwarzschild black holes. Phys. Rev. D **95**(8), 083001 (2017). https://doi.org/10.1103/PhysRevD.95.083001

F. Shankar, D.H. Weinberg, J. Miralda-Escudé, Self-consistent models of the AGN and black hole populations: duty cycles, accretion rates, and the mean radiative efficiency. Astrophys. J. **690**(1), 20–41 (2009). https://doi.org/10.1088/0004-637X/690/1/20

R.-F. Shen, C.D. Matzner, Evolution of accretion disks in tidal disruption events. Astrophys. J. **784**(2), 87 (2014). https://doi.org/10.1088/0004-637X/784/2/87

A. Soltan, Masses of quasars. Mon. Not. R. Astron. Soc. **200**, 115–122 (1982). https://doi.org/10.1093/mnras/200.1.115

L. Spitzer Jr., M.H. Hart, Random gravitational encounters and the evolution of spherical systems. I. Method. Astrophys. J. **164**, 399 (1971). https://doi.org/10.1086/150855

H. Sponholz, Tidal processes and disruption of stars near a supermassive rotating black hole. Mem. Soc. Astron. Ital. **65**, 1135 (1994)

N.C. Stone, B.D. Metzger, Rates of stellar tidal disruption as probes of the supermassive black hole mass function. Mon. Not. R. Astron. Soc. **455**, 859–883 (2016). https://doi.org/10.1093/mnras/stv2281

N.C. Stone, S. van Velzen, An enhanced rate of tidal disruptions in the centrally overdense E+A galaxy NGC 3156. Astrophys. J. Lett. **825**, 14 (2016). https://doi.org/10.3847/2041-8205/825/1/L14

N. Stone, R. Sari, A. Loeb, Consequences of strong compression in tidal disruption events. Mon. Not. R. Astron. Soc. **435**, 1809–1824 (2013). https://doi.org/10.1093/mnras/stt1270

N.C. Stone, A. Generozov, E. Vasiliev, B.D. Metzger, The delay time distribution of tidal disruption flares. Mon. Not. R. Astron. Soc. **480**, 5060–5077 (2018). https://doi.org/10.1093/mnras/sty2045

L.E. Strubbe, Snacktime for Hungry Black Holes: Theoretical Studies of the Tidal Disruption of Stars. PhD thesis, University of California, Berkeley (2011)

L.E. Strubbe, E. Quataert, Optical flares from the tidal disruption of stars by massive black holes. Mon. Not. R. Astron. Soc. **400**, 2070–2084 (2009). https://doi.org/10.1111/j.1365-2966.2009.15599.x

D. Syer, A. Ulmer, Tidal disruption rates of stars in observed galaxies. Mon. Not. R. Astron. Soc. **306**, 35–42 (1999). https://doi.org/10.1046/j.1365-8711.1999.02445.x

K.S. Thorne, Disk-accretion onto a black hole. II. Evolution of the hole. Astrophys. J. **191**, 507–520 (1974). https://doi.org/10.1086/152991

S. Thorp, E. Chadwick, A. Sesana, Tidal disruption events from massive black hole binaries: predictions for ongoing and future surveys. arXiv e-prints (2018)

S. van Velzen, On the mass and luminosity functions of tidal disruption flares: rate suppression due to black hole event horizons. Astrophys. J. **852**, 72 (2018). https://doi.org/10.3847/1538-4357/aa998e

S. van Velzen, G.R. Farrar, Measurement of the rate of stellar tidal disruption flares. Astrophys. J. **792**, 53 (2014). https://doi.org/10.1088/0004-637X/792/1/53

S. van Velzen, G.R. Farrar, S. Gezari, N. Morrell, D. Zaritsky, L. Östman, M. Smith, J. Gelfand, A.J. Drake, Optical discovery of probable stellar tidal disruption flares. Astrophys. J. **741**, 73 (2011). https://doi.org/10.1088/0004-637X/741/2/73

S. van Velzen, N.C. Stone, B.D. Metzger, S. Gezari, T.M. Brown, A.S. Fruchter, Late-time UV observations of tidal disruption flares reveal unobscured, compact accretion disks. arXiv:e-prints (2018). arXiv:1809.00003

E. Vasiliev, Rates of capture of stars by supermassive black holes in non-spherical galactic nuclei. Class. Quantum Gravity **31**(24), 244002 (2014). https://doi.org/10.1088/0264-9381/31/24/244002

E. Vasiliev, A new Fokker-Planck approach for the relaxation-driven evolution of galactic nuclei. Astrophys. J. **848**, 10 (2017). https://doi.org/10.3847/1538-4357/aa8cc8

E. Vasiliev, D. Merritt, The loss-cone problem in axisymmetric nuclei. Astrophys. J. **774**, 87 (2013). https://doi.org/10.1088/0004-637X/774/1/87

R.M. Wald, *General Relativity* (1984)

M. Walker, R. Penrose, On quadratic first integrals of the geodesic equations for type {22} spacetimes. Commun. Math. Phys. **18**, 265–274 (1970). https://doi.org/10.1007/BF01649445

J. Wang, D. Merritt, Revised rates of stellar disruption in galactic nuclei. Astrophys. J. **600**, 149–161 (2004). https://doi.org/10.1086/379767

C. Wegg, J. Bode, Multiple tidal disruptions as an indicator of binary supermassive black hole systems. Astrophys. J. Lett. **738**, 8 (2011). https://doi.org/10.1088/2041-8205/738/1/L8

A. Weissbein, R. Sari, How empty is an empty loss cone? Mon. Not. R. Astron. Soc. **468**, 1760–1768 (2017). https://doi.org/10.1093/mnras/stx485

H.N. Wernke, A.-M. Madigan, The Effect of General Relativistic Precession on Tidal Disruption Events from Eccentric Nuclear Disks. arXiv e-prints (2019)

T. Wevers, N.C. Stone, S. van Velzen, P.G. Jonker, T. Hung, K. Auchettl, S. Gezari, F. Onori, D. Mata Sánchez, Z. Kostrzewa-Rutkowska, J. Casares, Black hole masses of tidal disruption event host galaxies II. Mon. Not. R. Astron. Soc. **487**(3), 4136–4152 (2019). https://doi.org/10.1093/mnras/stz1602

J. Wheeler, Mechanism for jets, in *Study Week on Nuclei of Galaxies*, ed. by D.J.K. O'Connell (1971), p. 539

V. Wild, O. Almaini, J. Dunlop, C. Simpson, K. Rowlands, R. Bowler, D. Maltby, R. McLure, The evolution of post-starburst galaxies from $z = 2$ to 0.5. Mon. Not. R. Astron. Soc. **463**(1), 832–844 (2016). https://doi.org/10.1093/mnras/stw1996

P.J. Young, G.A. Shields, J.C. Wheeler, The black tide model of QSOs. Astrophys. J. **212**, 367–382 (1977). https://doi.org/10.1086/155056

W. Yuan, C. Zhang, H. Feng, S.N. Zhang, Z.X. Ling, D. Zhao, J. Deng, Y. Qiu, J.P. Osborne, P. O'Brien, R. Willingale, J. Lapington, G.W. Fraser, the Einstein Probe team, Einstein Probe—a small mission to monitor and explore the dynamic X-ray Universe. arXiv e-prints (2015)

Space Sci Rev (2021) 217:40
https://doi.org/10.1007/s11214-021-00818-7

The Process of Stellar Tidal Disruption by Supermassive Black Holes

The First Pericenter Passage

E.M. Rossi[1] · N.C. Stone[2] · J.A.P. Law-Smith[3] · M. Macleod[4] · G. Lodato[5] · J.L. Dai[6] · I. Mandel[7,8]

Accepted: 9 March 2021 / Published online: 25 March 2021

Abstract Tidal disruption events (TDEs) are among the brightest transients in the optical, ultraviolet, and X-ray sky. These flares are set into motion when a star is torn apart by the tidal field of a massive black hole, triggering a chain of events which is – so far – incompletely understood. However, the disruption process has been studied extensively for almost half a century, and unlike the later stages of a TDE, our understanding of the disruption itself is reasonably well converged. In this Chapter, we review both analytical and numerical models for stellar tidal disruption. Starting with relatively simple, order-of-magnitude

The Tidal Disruption of Stars by Massive Black Holes
Edited by Peter G. Jonker, Sterl Phinney, Elena Maria Rossi, Sjoert van Velzen, Iair Arcavi and Maurizio Falanga

✉ E.M. Rossi
emr@strw.leidenuniv.nl

N.C. Stone
nicholas.stone@mail.huji.ac.il

J.A.P. Law-Smith
lawsmith@ucsc.edu

M. Macleod
morgan.macleod@cfa.harvard.edu

G. Lodato
giuseppe.lodato@unimi.it

J.L. Dai
lixindai@hku.hk

I. Mandel
ilya.mandel@monash.edu

1 Leiden Observatory, Leiden University, PO Box 9513, 2300 RA, Leiden, Netherlands

2 Racah Institute of Physics, The Hebrew University, Jerusalem, 91904, Israel

3 Department of Astronomy and Astrophysics, University of California, Santa Cruz, Santa Cruz, CA 95064, USA

4 Harvard-Smithsonian Center for Astrophysics, 60 Garden Street, Cambridge, MA, 02138, USA

5 Dipartimento di Fisica, Unversità degli Studi di Milano, Via Celoria 16, Milano, Italy

physics, we review models of increasing sophistication, the semi-analytic "affine formalism," hydrodynamic simulations of the disruption of polytropic stars, and the most recent hydrodynamic results concerning the disruption of realistic stellar models. Our review surveys the immediate aftermath of disruption in both typical and more unusual TDEs, exploring how the fate of the tidal debris changes if one considers non-main sequence stars, deeply penetrating tidal encounters, binary star systems, and sub-parabolic orbits. The stellar tidal disruption process provides the initial conditions needed to model the formation of accretion flows around quiescent massive black holes, and in some cases may also lead to directly observable emission, for example via shock breakout, gravitational waves or runaway nuclear fusion in deeply plunging TDEs.

Keywords Supermassive black holes · Transient phenomena · Stellar dynamics · Gas magnetohydrodynamics

1 Introduction

The process of tidal disruption of a star by a supermassive black hole (SMBH) was originally studied by Hills (1975) as a mechanism to fuel active galactic nuclei, whose emission had recently been associated to SMBH gas accretion by Lynden-Bell (1969). Later, however, it became clear that the stellar disruption rate may not be sufficient (e.g. Frank and Rees 1976, see also the Disruption Chapter) for producing the copious ($\sim 10\ M_\odot\ \mathrm{yr}^{-1}$) and steady accretion flows needed to explain bright quasars. Rather, Rees (1988) suggested that tidal disruption events (TDEs) could be used to identify the presence of *quiescent* SMBHs in nearby galaxies, with the distinctive signature of an accretion-powered flare lasting up to a few years. By the first decade of the 21st century, the ubiquity of SMBHs in galactic nuclei was established, with an overwhelming majority of low-redshift SMBHs being quiescent (e.g. Ferrarese and Merritt 2000). Thus, TDEs are currently regarded as a unique tool to deliver a census of SMBH properties, including mass, spin and occupation fraction up to redshifts of a few. This is vital information to unravel the galaxy formation process, which is tightly linked to cosmological evolution of SMBHs. Beside black hole demographics, the time-dependent emission of TDE flares can be exploited to understand the physics of accretion and jet launching through different accretion regimes and/or states, similar to the goal of X-ray binary observations.

This Chapter describes theoretical efforts and progress over the last 45 years to understand the (magneto-hydro) dynamics of the stellar tidal disruption process. The tidal destruction of a self-gravitating body by a denser companion is a venerable problem in astrophysics, dating back to the 19th century work of Roche. For most of its history, the problem was studied primarily in the circular-orbit limit. While stars can approach SMBHs on quasi-circular orbits, the resulting tidal mass transfer is generally stable, and therefore far less luminous than the TDEs which are our primary subject here. TDEs differ from standard Roche-lobe overflowing systems through the orbits of the disrupted stars, which are generally parabolic, or nearly so. Consequently, the entire star can be destroyed in a single pericenter passage,

[6] Department of Physics, The University of Hong Kong, Pokfulam Road, Hong Kong, Hong Kong

[7] School of Physics and Astronomy, Monash University, Clayton, Victoria 3800, Australia

[8] The ARC Centre of Excellence for Gravitational Wave Discovery – OzGrav, Birmingham Institute for Gravitational Wave Astronomy, School of Physics and Astronomy, University of Birmingham, Birmingham, B15 2TT, UK

faster even than unstable mass transfer in the circular-orbit Roche problem. Alternatively, if the star's pericenter only grazes the tidal sphere, it may suffer limited stripping of its outer envelope: a partial disruption.

While the disruption process itself is not expected to be highly luminous (although there exist some notable, albeit speculative, possibilities for observing the disruption, which we discuss later on), the dynamics of partial and full disruption set the stage for later events in a tidal disruption flare. The efficiency and the qualitative manner in which an accretion flow is formed, and the resulting light curve of the TDE, are all dictated by the rate at which tidal debris falls back to the SMBH after disruption. These later stages in the evolution of TDEs are, at the time of this writing, quite incompletely understood, and large open questions exist about the hydrodynamics of accretion disc formation and the emission mechanisms operating during TDEs. In comparison, the actual process of tidal disruption is itself reasonably well-understood. The focus of this Chapter is limited in scope[1] to events occurring in the immediate vicinity of the tidal sphere; the subsequent evolution of dynamically bound tidal debris is picked up in the Formation of the Accretion Flow Chapter, Accretion Disc Chapter, and Emission Mechanisms Chapter.

Here we review theoretical models of the tidal disruption process. In §2, we present a general theoretical framework for tides, in both the Newtonian and general relativistic regimes. We then overview both analytic and semi-analytic models for the disruption process and the dynamical properties of the stellar debris as it exits the tidal sphere. In §3, we survey the substantial literature of numerical hydrodynamic simulations of full tidal disruptions (see also the Simulation Methods Chapter for a more detailed discussion of the numerical techniques). §4 likewise surveys past numerical hydrodynamic simulations of partial tidal disruption. In §5, we explore how the disruption process depends on the detailed stellar type being examined. This section goes beyond the primarily polytropic disruption simulations of the prior sections to examine realistic models for both main sequence and giant-branch stars. In §6 we discuss the subset of highly penetrating TDEs as opposed to more common grazing disruptions, examining three as-yet unobserved signatures of deeply penetrating encounters: shock breakout, gravitational wave emission, and thermonuclear fusion. §7 reviews the fate of the $\approx 50\%$ of the star that is dynamically unbound from the black hole, and does not participate in later stages of the bound debris evolution. In §8, we explore "unusual" sub-types of TDEs, such as stable but extreme-mass-ratio Roche lobe overflow, disruption of stars on non-parabolic orbits, tidal disruption of binary stars, and repeated partial disruptions. Finally, we conclude in §9.

2 Analytical Modelling of the Process of Tidal Disruption

In Newtonian gravity, tides are a differential acceleration between two (initially) nearby points, objects, or fluid elements. If we focus on the tidal forces exerted by a massive black hole, with mass $M_{\rm BH}$, on an object at distance r away, then the "tidal approximation" will apply if the object's physical size $R_\star \ll r$. In this limit, we may Taylor expand the Newtonian gravitational field around the finite size of the object, which leads to an approximate tidal acceleration $a_{\rm t} \sim G M_{\rm BH} R_\star / r^3$, where G is the Newtonian gravitational constant. The order unity numerical prefactor on $a_{\rm t}$ varies depending on which region of the object we

[1]The one exception to this is the evolution of the star's debris which is dynamically unbound during the disruption process; because it does not participate in later stages of the bound debris evolution, we cover its evolution here.

are concerned with, but from this approximation alone, it is straightforward to define a tidal disruption radius: the distance interior to which objects are torn apart by tides from a black hole. If specifically we consider a self-gravitating star of mass $M_\star$ and radius $R_\star$, this tidal radius will be, approximately (Hills 1975),

$$R_t = R_\star \left(\frac{M_{BH}}{M_\star} \right)^{1/3}. \quad (1)$$

This equation is approximate in that it neglects a variety of effects: the internal structure of the star, the finite duration over which tides strongly perturb a star on a parabolic orbit, the positional variation of a_t across the star's surface, the stellar spin and general relativistic corrections. Ultimately, the true order unity prefactor on Eq. (1) can only be computed through (relativistic) hydrodynamic simulations of the disruption process. Fortunately, however, most of these effects are subsumed into the cube root, and Eq. (1) is therefore reasonably accurate. Note that this equation may be understood in other ways as well. For example, we see that R_t describes the orbital separation where the star of mass $M_\star$ has the same average density as a sphere of radius R_t centered on the black hole and containing its mass M_{BH}.

We may make the above dynamical arguments more mathematically rigorous by computing the exact tidal tensor C_{ij}, which describes differential accelerations experienced in a rest frame centered on the victim object. Our presentation of C_{ij} will be brief, but a more thorough treatment can be found in Brassart and Luminet (2008). Working once more in the tidal approximation ($R_\star \ll r$), the tidal acceleration may be computed via the second derivatives of the Newtonian gravitational potential $\Phi(\mathrm{r}) = -GM_{BH}/r$. If the star's position, in a lab frame centered on the SMBH, is r, then

$$C_{ij} = -\partial_{r_i}\partial_{r_j}\Phi(\mathrm{r}) = \frac{GM_{BH}}{r^3}\left(-\delta_{ij} + \frac{3r_i r_j}{r^2} \right). \quad (2)$$

Here δ_{ij} is the Kronecker delta. In principle, one may also consider[2] higher-order expansions of the Newtonian tidal potential, such as the deviation tensor $D_{ijk} = -\partial_{r_i}\partial_{r_j}\partial_{r_k}\Phi(\mathrm{r})$: a rank-3 tensor that encodes the asymmetries of the tidal field C_{ij}. However, these higher-order terms will be subdominant so long as $R_\star \ll r$, an inequality sometimes referred to as the "tidal approximation."

The symmetric, traceless tensor C_{ij} has been defined so that, in the *tidal* reference frame centered on the star's center of mass, the acceleration of a test particle with position x will be

$$\ddot{x}_i = x_j C_{ij}(\mathrm{r}). \quad (3)$$

Throughout the notation in this section, repeated indices denote summation. Because Newtonian orbits about a point mass are planar, we may specialize to a lab frame coordinate system where one of our reference axes is orthogonal to the stellar orbit, so that the matrix representation of the tidal tensor C_{ij} becomes

$$C(\mathrm{r}) = \begin{bmatrix} -1 + \frac{3r_1^2}{r^2} & \frac{3r_1 r_2}{r^2} & 0 \\ \frac{3r_1 r_2}{r^2} & -1 + \frac{3r_2^2}{r^2} & 0 \\ 0 & 0 & -1 \end{bmatrix}$$

[2]For explicit examples of higher-order tidal expansions in Newtonian and general relativistic gravity, the reader may consult Brassart and Luminet (2008) and Ishii et al. (2005), respectively.

$$= \begin{bmatrix} -1+3\cos^2 f & 3\cos f \sin f & 0 \\ 3\cos f \sin f & -1+3\sin^2 f & 0 \\ 0 & 0 & -1 \end{bmatrix}. \quad (4)$$

Here, r_1 and r_2 represent positions along rectilinear coordinate axes in the orbital plane; the tidal tensor is independent of the third, orthogonal direction, r_3. In the second line of Eq. (4), we have replaced these coordinates with the Keplerian true anomaly (azimuthal angle) f. The tidal tensor has three eigenvalues,

$$\begin{aligned} \lambda_1 &= \frac{2GM_{\rm BH}}{r^3} \\ \lambda_2 &= -\frac{GM_{\rm BH}}{r^3} \\ \lambda_3 &= -\frac{GM_{\rm BH}}{r^3}, \end{aligned} \quad (5)$$

which encode the tidal accelerations along the three principal axes (eigenvectors) of the problem, u_1, u_2, and u_3. The first two of these eigenvectors lie within the orbital plane: $\mathrm{u}_1 \parallel \mathrm{r}$, and $\mathrm{u}_2 \perp \mathrm{u}_1$. These two eigenvectors will, therefore, rotate as the star moves along any non-radial orbit. The vector u_3 is orthogonal to the orbital plane, and remains fixed in direction. Notably, $\lambda_1 > 0$, implying a "stretching" acceleration, while the negative values of λ_2 and λ_3 imply a "compressional" type of acceleration. Since in the plane the star is stretched in the radial direction, a rigorous but "generous" tidal radius could be defined by equating λ_1 to $GM_\star/R_\star^2$, i.e. $R_{\rm t} = R_\star(2M_{\rm BH}/M_\star)^{1/3}$. This is the largest radius at which any fluid elements of the star will be unable to resist the tidal pull of the black hole through self-gravity.

A similar type of estimate may be made to account for the fully general relativistic tidal field of a Schwarzschild or Kerr metric SMBH. By constructing a locally orthonormal coordinate tetrad that is parallel-propagated along the star's center of mass geodesic (a Fermi Normal Coordinate system), it is possible to create a local tidal tensor, Γ_{ij}, very analogous to the Newtonian C_{ij} we have just discussed (Marck 1983). Specifically, by projecting the Riemann curvature tensor[3] onto this coordinate tetrad, we obtain a local, Euclidean tidal tensor Γ_{ij} that nevertheless captures the leading-order effects of relativistic gravity. As was the case with C_{ij}, the indices on the Γ_{ij} tensor run from 1 to 3, omitting timelike components (an approach which is valid so long as internal motions in the center-of-mass reference frame are sub-relativistic, and the tidal approximation itself holds). Γ_{ij} describes the accelerations of particles in a small radius around the star's center of mass:

$$\ddot{x}_i = x_j \Gamma_{ij}(\mathrm{r}). \quad (6)$$

For equatorial motion in the Kerr spacetime (with a SMBH of dimensionless spin $0 \le \chi_{\rm BH} \le 1$), the matrix representation of Γ_{ij} is

$$\Gamma(\mathrm{r}) = \begin{bmatrix} -1+3(1+K/\tilde{r}^2)\cos^2\Psi & 3(1+K/\tilde{r}^2)\cos\Psi\sin\Psi & 0 \\ 3(1+K/\tilde{r}^2)\cos\Psi\sin\Psi & -1+3(1+K/\tilde{r}^2)\sin^2\Psi & 0 \\ 0 & 0 & -1+3K/\tilde{r}^2 \end{bmatrix}. \quad (7)$$

[3]In general relativity, as in Newtonian gravity, we may view tides as differential gravitational forces. The geometric nature of general relativity allows for a second, equivalent, interpretation, where tides reflect the local curvature of spacetime. This is most easily seen in the geodesic deviation equation.

The similarities with C_{ij} are self-evident, allowing for straightforward continuity with our Newtonian intuition. The azimuthal angle Ψ is analogous to, though distinct from, the Keplerian true anomaly f (both angles are 0 at pericenter). The primary difference between the two tensors is the presence of $K \equiv (L_z - \chi_{\rm BH}\varepsilon)^2 + Q$, a combination of the Kerr constants of motion: relativistic energy ε ($\varepsilon = 1$ for a parabolic orbit), z-component of angular momentum L_z, and Carter constant Q. Note that here and in Eq. (7), we have written K in geometrized ($G = c = 1$ units); we also denote the dimensionless relativistic radius $\tilde{r} = r/R_{\rm g}$, where the gravitational radius is

$$R_{\rm g} = GM_{\rm BH}/c^2. \tag{8}$$

In the Schwarzschild limit, K is just the total orbital angular momentum. For inclined orbits in the Kerr geometry, Γ_{ij} becomes considerably more complicated, and Lense-Thirring precession "mixes up" the tensor's eigenvalues[4] (Luminet and Marck 1985).

In the equatorial Kerr limit, however, we may once again compute a simple tidal radius by examining the positive eigenvalue of the tidal tensor (Beloborodov et al. 1992; Kesden 2012a). The eigenvalues of Γ_{ij} are

$$\begin{aligned} \Lambda_1 &= \frac{2GM_{\rm BH}}{r^3}(1 + 3K/\tilde{r}^2) \\ \Lambda_2 &= -\frac{GM_{\rm BH}}{r^3}(1 + 3K/\tilde{r}^2) \\ \Lambda_3 &= -\frac{GM_{\rm BH}}{r^3}, \end{aligned} \tag{9}$$

and therefore the effective tidal disruption radius[5] is (Kesden 2012a)

$$R_{\rm t}^{\rm GR} = R_\star \left(\frac{\Lambda_1}{GM_{\rm BH}/r^3} \frac{M_{\rm BH}}{M_\star} \right)^{1/3}. \tag{10}$$

The process of tidal disruption for misaligned orbits has received little analytic study, so for now we will focus mainly on the Newtonian and (to a more limited extent) planar general relativistic regimes.

As a star enters the tidal disruption radius, fluid and self-gravitational forces become subdominant to the tides from the SMBH. The process of tidal disruption can be understood through various levels of approximation. At the simplest level, we may postulate that at the moment of disruption (usually assumed to be the first moment when $r = R_{\rm t}$), the star impulsively "shatters" to pieces, with internal forces becoming negligible and each fluid element free-falling along a Keplerian trajectory (in Newtonian gravity) or timelike geodesic (in general relativity). This assumption is simplistic, but allows for exact solutions to the future evolution of the star, and provides important physical insights. Historically, analytic TDE theory based around this assumption were often referred to as "freezing" or "frozen-in" models, due to the assumption that the debris immediately freezes in to a fixed set of

[4]Speaking more rigorously, inclined Kerr orbits are no longer planar, meaning that the off-diagonal terms in Γ_{ij} are no longer equal to 0; as we shall see shortly, this complicates the dynamics of disruption.

[5]Note that $R_{\rm t}^{\rm GR}$ reduces in the Newtonian limit to the "generous" tidal radius derived from C_{ij}: $R_\star(2M_{\rm BH}/M_\star)^{1/3}$, rather than to Eq. (1). As mentioned before, the order unity prefactor on the tidal radius is uncertain, sensitive to hydrodynamic and self-gravitational effects, and best calibrated through hydrodynamical simulations.

ballistic or geodesic trajectories; in this text, instead, we will refer to this as the "impulsive disruption" approximation.

The semi-analytic "affine models" study the disruption process with a greater degree of realism, though their increased complexity requires numerical solutions. These models, first developed by Carter and Luminet (1983), couple the tensor virial theorem with strong assumptions on the geometry of the disrupting star. As long as these geometrical assumptions remain valid, the interplay between SMBH tides and weaker internal forces can be studied, and more sophisticated variants of the original affine model provide sometimes surprising degrees of physical accuracy. Of course, the greatest degree of physical realism will come from numerical hydrodynamic simulations of the disruption process, which are discussed later in this Chapter (and in the Simulation Methods Chapter). For the remainder of this section, we discuss the analytic and semi-analytic insights provided by impulsive and affine models for the disruption process.

2.1 Tidal Compression and the Affine Model

Although the semi-analytic affine model is in some ways more sophisticated than purely analytic impulse-approximation solutions, we present it first for two reasons. Most obviously, it was the earliest approach developed to studying tidal disruption, predating impulsive models by five years (Carter and Luminet 1983; Rees 1988). Second, it is focused on providing an accurate picture of the early details of disruption, while the impulse approximation is more concerned with accounting for the aftermath.

The early affine models considered the tidal disruption of a star in Newtonian gravity, and assumed that throughout the disruption process, the stellar geometry would follow nested, coaxial ellipsoids of deformable axis ratios, interacting in a way that satisfies the tensor virial theorem (Carter and Luminet 1983). More specifically, the affine model (in its simplest form) can be visualized as treating nested ellipsoids of gas that evolve due to combinations of self-gravity, external (tidal) gravity, and internal pressure. Fluid elements inside the affine star have positions

$$R_i = q_{ij}\hat{\mathrm{R}}_j, \tag{11}$$

where $\hat{\mathrm{R}}$ is the initial position of a fluid element in the unperturbed star (note that the hat notation indicates a unit vector), and q_{ij} is a deformation matrix describing the warping and rotation of the star's principal axes under tidal stress. For now, we follow the earlier implementations of the affine model and assume that q_{ij} is independent of R.

The power of the affine approximation comes from the fact that, at lowest order, Newtonian tides induce quadrupolar deformations in a spherical star (Press and Teukolsky 1977), making the ellipsoidal approximation very good for weak (non-disruptive) tidal encounters, and reasonable for the early stages of a TDE. Under these assumptions, Carter and Luminet (1983) derive a Lagrangian formulation for the process of tidal disruption, with equations of motion given by:

$$\dot{P}_i = -\,M_\star \frac{\partial \Phi(\mathrm{r})}{\partial r_i} + \frac{\mathcal{M}_\star}{2} q_{lj} q_{kj} \frac{\partial C_{lk}(\mathrm{r})}{\partial r_i} \tag{12}$$

$$\dot{p}_{ij} = \mathcal{M}_\star C_{ik}(\mathrm{r}) q_{kj} + \Pi \delta_{ik} q^{-1}_{jk} + \Omega_{ik} q^{-1}_{jk}. \tag{13}$$

Here r is the position vector of the stellar center of mass relative to the SMBH, $P_i = M_\star \dot{r}_i$ is its total ("external") momentum, and $p_{ij} = \mathcal{M}_\star \dot{q}_{ij}$ is an "internal" momentum tensor. In other words, the first of these equations describes the motion of the stellar center of mass in its orbit about the SMBH, while the second describes internal deformations of the star, which are encoded in q_{ij}. From the definitions of P_i and p_{ij}, we can re-express Eqs. (13) and (12) as sets of second-order ordinary differential equations (ODEs) for the evolution of q_{ij}. Overall, we have 12 coupled second-order ODEs, supplemented by the scalar, quadrupolar moment of inertia (evaluated for the original, unperturbed star)

$$\mathcal{M}_\star = \frac{1}{3}\iint \hat{R}_i \hat{R}_i \mathrm{d}M, \tag{14}$$

the gravitational self-energy tensor

$$\Omega_{ij} = -\frac{G}{2}\iint \frac{(R_i - R_i')(R_j - R_j')}{|\mathrm{R} - \mathrm{R}'|^3}\mathrm{d}M\mathrm{d}M', \tag{15}$$

and the volume integral of the local pressure P,

$$\Pi = \int \frac{P}{\rho}\mathrm{d}M. \tag{16}$$

In the definitions of Π and Ω_{ij}, it is useful to remember that q_{ij} can be used to relate (Lagrangian) mass coordinates to the original positions of stellar gas parcels. Finally, an equation of state is needed to relate local pressures P to local densities ρ. These definitions and equations of motion have been presented without proof or elaboration; the reader interested in a more rigorous mathematical treatment of the affine model should consult Carter and Luminet (1985); it is also covered more thoroughly in the Simulation Methods Chapter.

So far, we have written the simplest version of the affine model, and many generalizations exist that incorporate additional physical effects or, alternatively, loosen the underlying assumptions. By adding additional terms to the underlying Lagrangian, it is possible to model the effect of viscosity, other sources of internal dissipation, and internal rotation (Carter and Luminet 1985; Luminet and Carter 1986). By replacing Eq. (12) with the geodesic equations and C_{ij} with Γ_{ij}, the model can be made general relativistic (Luminet and Marck 1985). The addition of heating terms and a nuclear reaction network enable the study of nuclear fusion reactions triggered by tidal compression (Luminet and Pichon 1989b). More recent generalizations of the affine model have generalized the underlying geometry, specifically by allowing the ellipsoidal orientations and axis ratios (i.e. q_{ij}) to vary at a single moment in time as one moves from inner mass shells to outer ones (Ivanov and Novikov 2001). This generalized affine model was derived in a Newtonian context, but it has also been applied to the general relativistic tidal problem (Ivanov et al. 2003; Ivanov and Chernyakova 2006). Figure 1 illustrates results and geometrical assumptions in the extended affine model.

The most prominent application of the affine model has been to the study of tidal compression during the star's destruction. As was first noted in Carter and Luminet (1982), the decoupling of vertical from in-plane acceleration in Eq. (3) leads to a homologous collapse of the star in the direction orthogonal to the fixed orbital plane (we will refer to this as the "vertical" or z direction). The vertical deformation of the star is extremely pronounced because of the coherent effect of tidal acceleration: in Newtonian gravity,[6] $\ddot{z}$ is always negative (positive) for $z > 0$ ($z < 0$), so the star is uniformly compressed by tides in this direction. This evolution is markedly different from tidal acceleration within the orbital plane, where

[6] This statement also holds true in the Schwarzschild spacetime, and in the Kerr equatorial plane.

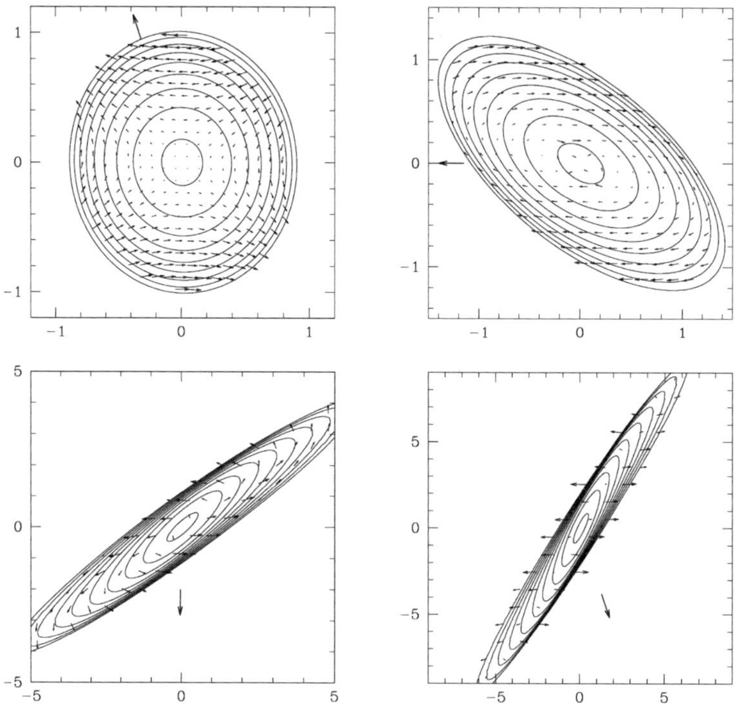

Fig. 1 The internal geometry of a star being disrupted in the framework of the extended (Ivanov and Novikov 2001), general relativistic (Ivanov et al. 2003) affine model. Panels show different moments in time in the evolution of a $\beta = 0.72$ encounter ($\beta = R_t/R_p$ is a dimensionless inverse of the orbital pericenter, R_p). The top left panel shows the star during its approach to pericenter, with $r = 1.5R_p$. The top right panel shows the star at pericenter. The bottom left and bottom right panels show the star after pericenter passage, at distances $r = 1.9R_p$ and $r = 2.73R_p$, respectively. In this figure, we can see that the extended affine model permits different axis ratios and orientations in different internal mass shells. Vectors inside the star denote internal motions, while the vector outside the star points to the SMBH. Taken from Ivanov et al. (2003), their Fig. 5

the eigenvectors of the tidal tensor C_{ij} must rotate to follow the Keplerian trajectory of the stellar center of mass. In the star's reference frame, an in-plane direction that is getting stretched at one moment in time will be squeezed at a later one, and therefore the degree of in-plane deformation during the disruption process does not exceed factors of order unity.[7]

The degree of vertical compression is thus severe, and turns out to depend strongly on the penetration factor

$$\beta = R_t/R_p, \tag{17}$$

a dimensionless inverse of the pericenter R_p. Analytic arguments (Carter and Luminet 1982) suggest that the vertical collapse velocity achieved during the star's passage through the tidal sphere is $w_c \approx \beta V_\star$ (see Stone et al. 2013 for a precise derivation of this result in the test particle limit), where we have made use of the star's natural velocity: $V_\star = \sqrt{GM_\star/R_\star}$. This velocity can be understood as the approximate average sound speed in the star (though it

[7]Note that after the stellar debris leaves the tidal sphere, it is completely deconfined along the direction of motion, and its ballistic expansion elongates the debris into a very narrow, spaghettified stream.

has other meanings as well, such as being the Keplerian speed for circular orbits around the surface of the star). If the star were made of test particles, and were collapsing uniformly everywhere, it would compress into an infinitely thin pancake somewhere in the vicinity of pericenter. However, this compression will be reversed by the buildup of internal gas pressure. If we assume the pressure increase is adiabatic, then the internal energy of the star at peak tidal compression is

$$U_c \sim \beta^2 U_\star, \tag{18}$$

its peak density is

$$\rho_c \sim \beta^{2/(\gamma-1)} \rho_\star, \tag{19}$$

and the duration of the peak compression is

$$T_c \sim \beta^{-(\gamma+1)/(\gamma-1)} T_\star. \tag{20}$$

Here we have assumed a polytropic equation of state (adiabtic index γ) and made use of other "natural" stellar variables, namely $\rho_\star = M_\star/R_\star^3$, $U_\star = GM_\star^2 R_\star^{-1}/(5\gamma - 5)$, and $T_\star = 1/\sqrt{G\rho_\star}$. Since most of the distortion of the stellar shape happens along one axis, the height at peak compression is

$$z_c \sim R_\star \rho_\star/\rho_c \sim \beta^{-2/(\gamma-1)} R_\star. \tag{21}$$

These scaling relations are simple, but match numerical integrations of the affine model (Luminet and Carter 1986) as well as 1-dimensional hydrodynamic simulations of collapsing stellar columns (Brassart and Luminet 2008). Their validity has not been explored across a wide parameter space of 3-dimensional hydrodynamic simulations.

The degree of vertical compression in a high-β TDE can be severe: if one assumes $\gamma = 5/3$, then $\rho_c \propto \beta^3$. Under a softer equation of state, such as $\gamma = 4/3$, the adiabatic compression is even more violent ($\rho_c \propto \beta^6$). This phase of stellar pancaking reverses itself rapidly, in an intense burst of hydrodynamic acceleration: for $\gamma = 5/3$ ($\gamma = 4/3$), the time of peak compression $T_c \propto \beta^{-4}$ ($T_c \propto \beta^{-7}$). The vertical compression of a $\gamma = 5/3$ star in a $\beta = 5$ TDE is illustrated in Fig. 2. Under such violent conditions, additional physics may come into play, such as shock heating or thermonuclear reactions. While these effects have been incorporated in approximate ways into the affine model (Luminet and Carter 1986; Luminet and Pichon 1989b), they are sensitive to the internal structure of the collapsing star, and are in principle more accurately treated in hydrodynamical simulations with sufficient spatial resolution. It should be noted that strong shock heating or thermonuclear detonation will cause the stellar collapse to become non-adiabatic, invalidating the assumptions behind the analytic scaling relations in Eqs. (18), (19), (20). A description of physical phenomena caused by stellar vertical collapse around the time of pericenter passage can be found in Sect. 6.

The extended affine model of Ivanov and Novikov (2001) has an additional application, which is to determine the amount of mass lost from stars in partial TDEs, with $\beta \lesssim 1$. By decoupling individual mass shells from each other, the extended affine model allows the outer shells of the star to achieve positive total energy, at which point they are treated as unbound. As we will show in §4, the predictions of the extended affine model are in fairly impressive agreement with three-dimensional hydrodynamic simulations of partial disruption.

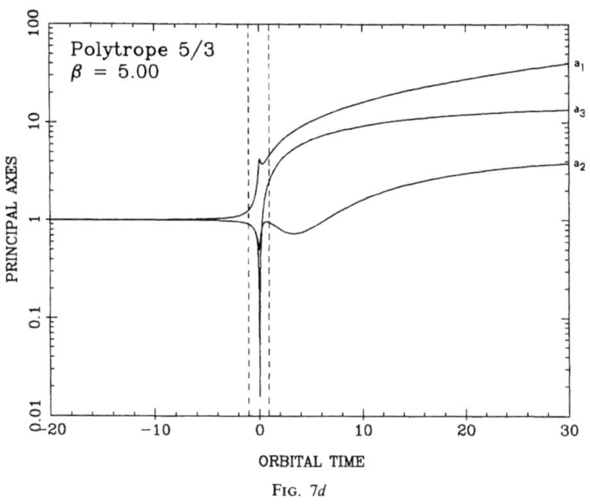

Fig. 2 The evolution of the principal axes of a star, as simulated by the simple affine model. The principal axes are shown in units of $R_\star$, and are plotted against a dimensionless time t (in units of the orbital time at $r = R_\mathrm{t}$). The disruption simulated is that of a $\gamma = 5/3$ star with $\beta = 5$, in Newtonian gravity. The duration of the encounter inside the tidal sphere is shown with vertical dashed lines. Severe compression of the second principal axis (the vertical direction) is visible at $t \approx 0$. Taken from Luminet and Carter (1986), their Fig. 7d

2.2 Impulsive Disruption Approximation

One of the earliest and most robust results in the context of tidal disruption events is without a doubt the "$t^{-5/3}$" decay of the fall-back rate after the total disruption of a star. The result is so fundamental that it is often viewed as the classical signature of TDEs, although it is non-trivial to observe directly. It was initially derived by Rees (1988), although in the original paper the result was quoted with an incorrect $t^{-5/2}$ exponent, later corrected to $t^{-5/3}$ by Phinney (1989) (interestingly, the $t^{-5/3}$ law had been independently discovered just one week after Rees 1988 in another Nature paper, by Michel 1988, to describe disc formation from supernova fall-back[8]). The basics of the argument is very simple and is based only on Kepler's third law. We will assume for now that the penetration factor $\beta = 1$ and that the disrupted star approaches on a parabolic orbit. The argument by Rees (1988) assumes that the star is almost unperturbed until it reaches pericenter, where it has an impulsive interaction with the black hole and gets torn apart. This is clearly an approximation, but, as we shall see, is not a bad one and deviations from it can be incorporated in the theory. It is straightforward, in this approximation, to compute the spread in the specific orbital energies of the debris as being due to the different depths in the potential well of the black hole across the stellar radius,

$$\Delta E = \frac{GM}{R_\mathrm{t}^2} R_\star, \tag{22}$$

which corresponds to velocities of the order of

$$v_\mathrm{ej} \sim \left(\frac{M_\mathrm{BH}}{M_\star}\right)^{1/6} V_\star, \tag{23}$$

for the highest-energy ejecta. For a typical mass ratio between the black hole and the star of 10^6, the debris can reach 10 times the stellar escape velocity $V_\star$. Interestingly, tidal forces also induce a spin in the debris, but this can only accelerate the debris up to the escape

[8] We thank Sterl Phinney for pointing out this paper during a conference.

velocity (see, for example, Sacchi and Lodato 2019 for a recent description). The argument by Rees (1988) then continues by assuming that the orbital energy E distribution is flat among the debris:

$$\frac{\mathrm{d}M}{\mathrm{d}E} = \frac{M_\star}{2\Delta E}, \tag{24}$$

and that the return time of the debris to pericenter simply follows from Keplerian dynamics:

$$T^2 = \frac{4\pi^2}{GM_{\rm BH}}\left(\frac{GM_{\rm BH}}{2|E|}\right)^3. \tag{25}$$

From the above, one can calculate the return time of the most bound debris, by setting $E = \Delta E$ in Eq. (25):

$$\begin{aligned} t_{\rm min} &= \frac{\pi}{\sqrt{2}}\left(\frac{R_\star^3}{GM_\star}\right)^{1/2}\left(\frac{M_{\rm BH}}{M_\star}\right)^{1/2} \\ &\approx 40\,\text{days}\left(\frac{M_{\rm BH}}{10^6\,M_\odot}\right)^{1/2}\left(\frac{M_\star}{M_\odot}\right)^{-1}\left(\frac{R_\star}{R_\odot}\right)^{3/2}. \end{aligned} \tag{26}$$

One can also obtain the distribution of return times (that is the fall-back rate) as:

$$\dot{M} = \frac{\mathrm{d}M}{\mathrm{d}t} = \frac{\mathrm{d}M}{\mathrm{d}E}\frac{\mathrm{d}E}{\mathrm{d}t} = \frac{M_\star}{3t_{\rm min}}\left(\frac{t}{t_{\rm min}}\right)^{-5/3}, \tag{27}$$

where in the last equation we have used Eqs. (24) and (25). For the case of a $1 M_\odot$ star disrupted by a $10^6 M_\odot$ black hole, the peak fall-back rate $M_\star/(3t_{\rm min})$ corresponds to roughly 100 times the Eddington rate.

Lodato et al. (2009) refined this calculation further by estimating the differential distribution of debris mass with respect to specific energy, $\mathrm{d}M/\mathrm{d}E$. This may be visualized as a "salami slicing" of the star at the moment of breakup: each infinitesimally thin cylindrical slice of star will have the same ϵ value. More specifically, for a spherically symmetric star with internal mass density profile $\hat{\rho} = \rho(x)/\rho_\star$, where $0 \le x = R/R_\star \le 1$,

$$\frac{\mathrm{d}M}{\mathrm{d}E} = 2\pi\frac{M_\star}{\Delta E}\int_x^1 \hat{\rho}(x')x'\mathrm{d}x'. \tag{28}$$

The cylindrical slabs of the star we integrate over are axisymmetric about a vector connecting the SMBH to the star's center of mass, and the center of each cylindrical slab is a distance R from the center of the star. Each cylinder, at the beginning of tidal free fall, freezes in to its specific orbital energy $E = x\Delta E$ (note that the approximation of constant energy across the cylindrical slab requires $R_\star \ll R_{\rm t}$). In this way, we can evaluate the distribution of debris energies that accounts for the nontrivial internal structure of the star.

It is important to note that the impulse approximation yields an accurate spread of specific energy when applied at the tidal disruption radius $r = R_{\rm t}$ as in Eq. (22) and in Eq. (28), rather than at periapsis $r = R_{\rm p}$. For high-β disruptions, $GM_{\rm BH}R_\star/R_{\rm p}^2$ may over-estimate the specific energy spread by one to two orders of magnitude (Guillochon and Ramirez-Ruiz 2013; Stone et al. 2013). The primary reason for this is the relatively short duration the star spends at radii much less than $R_{\rm t}$, which limits the amount of work internal forces (self-gravity and hydrodynamic pressure) can do to alter the energy spread that exists during the crossing of the tidal sphere. The ability of internal forces to do work on the debris is further reduced by their near-cancellation during the star's entry into the tidal sphere (when the star is not so far from hydrostatic equilibrium); later on, self-gravity will be further reduced in importance by the increasing physical size of the star.

A further elaboration of the impulse approximation was provided by Stone et al. (2013), who used the free solutions to the parabolic Hill equations (Sari et al. 2010) to write explicit orbital elements for every individual fluid element of the disrupted star (once again, under the assumption of instantaneous, impulsive freeze-in to ballistic motion once $r = R_t$). Each of the six solutions represents small perturbations of the Keplerian orbital elements around the star's parabolic (i.e. eccentricity $e = 1$) center-of-mass trajectory, and the ballistic, post-disruption orbits of the stellar debris are linear combinations of the six free solutions. One notable feature of the parabolic free solutions, already evident in Eq. (4), is the decoupling of motion within and orthogonal to the orbital plane. It is also possible to derive somewhat more complicated free solutions that allow for internal motions at the time of disruption (Stone et al. 2013), accounting for the effects of e.g. stellar rotation; for the sake of brevity we reproduce neither set here.

This geometrical picture of the disruption process is exact under the (strong) assumption of impulsive disruption and subsequent ballistic motion, and allows one to compute several quantities of interest. After entering the tidal sphere, at a true anomaly (i.e. azimuthal angle) $f_t = -\arccos(2/\beta - 1)$, the star will undergo a homologous vertical collapse. If it were made of test particles, peak vertical compression of the star would occur shortly after pericenter passage, at a true anomaly

$$f_c = \arctan(1/\sqrt{\beta - 1}). \tag{29}$$

At this point in the orbit, each component of the star will be vertically free-falling with a speed

$$w_c = -\beta \frac{z_0}{R_\star} \sqrt{\frac{GM_\star}{2R_\star}} \left(\sqrt{1 - \beta^{-1}} + 1 \right), \tag{30}$$

and the in-plane principal axes of the deformed star will have lengths

$$r_{\rm long}/R_\star \approx \frac{4}{5}\beta^{1/2} + \frac{22}{5}\beta^{-1/2} \tag{31}$$

$$r_{\rm short}/R_\star \approx 2\beta^{-1/2} - \frac{23}{2}\beta^{-3/2}. \tag{32}$$

The above expressions for $r_{\rm short}$ and $r_{\rm long}$ represent expansions in the high-β limit and will not be accurate for $\beta \approx 1$. If pressure gradients during peak compression are unable to accelerate significant motions within the orbital plane, the energy spread will not change during the compression process, and the frozen-in specific energy for each fluid element will remain

$$E = \Delta E \left(\frac{x_0}{R_\star}(1 - 2\beta^{-1}) + 2\frac{y_0}{R_\star}\sqrt{\beta^{-1} - \beta^{-2}} \right). \tag{33}$$

Here we have denoted initial positions of fluid elements inside the star as x_0, y_0, and z_0, with an origin at the stellar center of mass. By combining the free solutions with a simple approximation for the hydrodynamics of the bounce, Stone et al. (2013) argue that the frozen-in energy spread is unlikely to be altered by even severe degrees of tidal compression. This conclusion stems from the homology of the star's vertical compression, and could be altered if some source of asymmetry (e.g. a substantial stellar or SMBH spin component that is misaligned with the orbital angular momentum) breaks the homology of collapse, and increases the magnitude of in-plane pressure gradients at peak compression.

A general relativistic version of the impulse approximation was developed by Kesden (2012b), who numerically computed the spread of geodesics that stellar debris would find itself on, assuming disruption of stars on equatorial orbits in the Kerr spacetime. As with other efforts in this subsection, this work assumed that a static star shatters to pieces at $r = R_t$, although here the tidal radius in question is the general relativistic one provided by Eq. (10). Kesden (2012b) finds that the specific energy spread is not altered greatly by the relativistic disruption process, although for orbits where the gravitational radius R_g is comparable to R_p, ΔE may change at the factor of two level.

3 Numerical Simulations of the Disruption Process

The analytical picture outlined above has been confirmed numerically by various works, starting with the early simulations of Evans and Kochanek (1989), who used Smoothed Particle Hydrodynamics (SPH) to demonstrate that the fall-back rate is indeed proportional to $t^{-5/3}$, and that the energy distribution of the debris is approximately flat.[9] Later, Lodato et al. (2009) considered the effects of changing the internal structure of the star on the fall-back rate, both analytically (as described above) and in numerical hydrodynamics simulations. While Evans and Kochanek (1989) had modelled the star as a polytropic sphere with an adiabatic index $\gamma = 5/3$, Lodato et al. (2009) considered instead a range of indices, finding that the mass fallback rate can depend significantly on stellar structure.

In particular, more centrally concentrated (less compressible) stars should have a steeper energy distribution, resulting in a slower rise to the peak of the fall-back rate. This result follows from a more precise determination of the energy distribution of the debris (as is expressed analytically in Eq. (28)), rather than the simple flat distribution of Eq. (24). Note, however, that for any reasonable stellar density profile, the energy distribution of the least bound debris, which originates near the stellar center of mass and determines the late fall-back rate, should be indeed characterized by a flat $\mathrm{d}M/\mathrm{d}E$. The analytical model for $\mathrm{d}M/\mathrm{d}E$ was tested numerically with SPH in Evans and Kochanek (1989), Lodato et al. (2009), and later works, and gives good qualitative agreement (see Fig. 3 for numerical results). However, since the star is perturbed before reaching pericenter, quantitative deviations from the analytical models appear. Lodato et al. (2009) show that such deviations can be accounted for in the impulse approximation by allowing for homologous expansion of the star at pericenter, due to the reduced effective gravity. A subsequent analysis by Guillochon and Ramirez-Ruiz (2013) using the adaptive mesh grid-based code `FLASH` (simulated in the rest frame of the star's center of mass to reduce advection errors) closely confirmed this picture. Recently, Law-Smith et al. (2019) and Ryu et al. (2020a) have also studied the disruption of realistic stellar models, as opposed to simple polytropes, and discuss the differences in the resulting fall-back rates (see Sect. 5).

[9] It is worth noting that the energy distribution presented by Evans and Kochanek (1989) and some others is on a logarithmic scale, which artificially "flattens" it to the eye, but it is correct that at late times, the material falling back from a full disruption is sampled from a flat part of the $\mathrm{d}M/\mathrm{d}E$ curve.

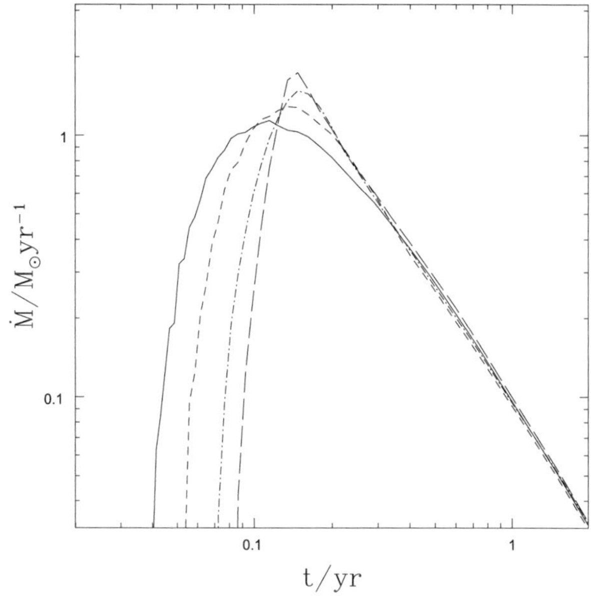

Fig. 3 Fall-back rate as a function of time from the numerical simulations of Lodato et al. (2009), for different polytropic models of the star: $\gamma = 1.4$ (solid line), $\gamma = 1.5$ (short-dashed line), $\gamma = 5/3$ (dot-dashed line), $\gamma = 1.8$ (long-dashed line). At late times, the returning mass is always drawn from the flat, central portion of the $\mathrm{d}M/\mathrm{d}E$ curve, and the fallback rate asymptotes to a $t^{-5/3}$ solution regardless of internal structure. Figure taken from Lodato et al. (2009), their Fig. 10, left panel

The above discussion considered the case $R_\mathrm{p} = R_\mathrm{t}$, or $\beta = 1$. What happens for more, or less, penetrating events? For more penetrating TDEs, the main result of numerical hydrodynamics is in agreement with the analytical arguments of §2.2: one can still use the impulse approximation to estimate ΔE, but the energy spread needs to be evaluated at the tidal radius rather than at pericenter. This was first seen numerically in the work of Guillochon and Ramirez-Ruiz (2013), and has been investigated in greater detail more recently by Steinberg et al. (2019), who simulated $\beta = 5, 6, 7$ disruptions of polytropic stars in Newtonian gravity. For this set of highly penetrating TDEs, there is little variation in the final $\mathrm{d}M/\mathrm{d}E$ for a given unperturbed stellar structure. For less concentrated $n = 3/2$ polytropes (i.e. stars with a uniform adiabatic index of $\gamma = 5/3$, representative of lower-mass main sequence stars), the assumptions of the impulse approximation work reasonably well at all β, and the final ΔE is within tens of percent of analytic predictions. For more highly concentrated $n = 3$ polytrope models (i.e. stars with a uniform adiabatic index of $\gamma = 4/3$), internal forces are seen to do substantially more work for the portion of the orbit where $r < R_\mathrm{t}$, and the final energy spread is enhanced by a factor of a few over the analytic ΔE estimate. For less penetrating encounters the disruption is only partial, as we discuss in Sect. 4 below. At this point, numerical simulations of "simple" tidal disruptions are increasingly converged in their results, but interesting open questions remain about (usually) higher-order effects in the disruption process, such as stellar spin, stellar magnetization, and general relativistic gravity. We will briefly outline the current understanding of these phenomena in the remainder of this section, along with complications that arise when the disrupting SMBH is a component of a close SMBH binary.

Until very recently, most simulations have neglected the effect of initial stellar rotation on the tidal disruption process. This is mostly due to the fact that tidal torques massively spin up the star immediately prior to disruption, so that pre-existing rotation will only be important for initial stellar spins close to break-up. The effect of stellar rotation has been studied recently by Golightly et al. (2019b) and by Sacchi and Lodato (2019), who find that prograde stellar rotation (with respect to the orbital axis) enhances the rate of mass fallback (possibly leading to a faster and more luminous flare), while the opposite occurs for stars whose spin is retrograde with the orbital plane. If the star is rapidly spinning in a retrograde sense, tidal disruption might be completely inhibited so that the outer stripped layers of the star re-accrete onto the star rather than onto the black hole, possibly giving rise to a fainter X-ray flare (Sacchi and Lodato 2019).

General relativistic effects on the disruption process have been considered analytically by Kesden (2012b) (see above), but also by a number of recent hydrodynamical simulations. Relativistic simulations of tidal disruptions by spinning black holes have been performed by Haas et al. (2012a) in the case of white dwarf disruption by intermediate black holes, by Hayasaki et al. (2013), Cheng and Bogdanović (2014), Evans et al. (2015), Hayasaki et al. (2016) for stellar disruptions, and more recently by Tejeda et al. (2017), Gafton and Rosswog (2019), Liptai et al. (2019). While in general, the frozen-in energy spread agrees with the Newtonian limit at the factor of ≈ 2 level, there are some cases where large general relativistic enhancements to the energy spread are seen for deeply plunging ($\beta \gtrsim 10$) disruptions. This has been attributed both to shock-heating during the vertical compression of the star (Tejeda et al. 2017) and also to prompt self-intersection of debris streams before they leave the region of pericenter (Evans et al. 2015).

While stellar magnetic fields are generally too weak to play a dynamically important role in the disruption process itself, they may become important later on, possibly affecting the self-gravitating equilibria of the debris streams or seeding the magneto-rotational instability in the subsequent accretion disk. During the disruption phase, any internal magnetic field in the star could in principle be amplified. This effect has been studied by Guillochon and McCourt (2017) and by Bonnerot et al. (2017). The magnetic field can be significantly amplified by at least an order of magnitude, but this is not enough to have a strong dynamical effect or modify the fall-back rate. However, the presence of a strong magnetic field can have implications for the resulting accretion flow.

Finally, the fall-back rate can be strongly affected if the stellar disruption is due to a black hole that is a member of a close binary system. The first studies of this process were by Liu et al. (2009), who used N-body simulations to predict that in this case the fall-back rate would suffer several, almost periodic interruptions. This was used by Liu et al. (2014) to argue for the presence of a hidden black hole binary system based on the light curve of an observed TDE, and motivated the first hydrodynamical simulations of TDEs in binary SMBH systems (Hayasaki and Loeb 2016). A large set of hydrodynamical simulations of this process have been performed by Coughlin et al. (2017), while a more systematic exploration of the parameter space has been provided by Vigneron et al. (2018). While in general the interruptions are very sharp, in some cases, especially if the binary orbit is perpendicular to the stellar orbit, the interruption can result in a relatively gentle decrease in the fall-back rate, which might resemble the light curve observed in ASASSN-15lh (Coughlin and Armitage 2018). More details on disruption by SMBH binaries are provided in the dedicated Binaries Chapter within this book.

4 Partial Tidal Disruptions

In star–black hole encounters with periapsis radii significantly greater than the tidal radius, non-disruptive tides can act on the star as it passes through pericenter. In this case, oscillatory motions of the star's envelope are excited by the tide and continue after the star has passed periapsis (e.g. Press and Teukolsky 1977). Deeper encounters, meaning those with higher β, lead to distortions and subsequent oscillations of progressively larger amplitude in the star.

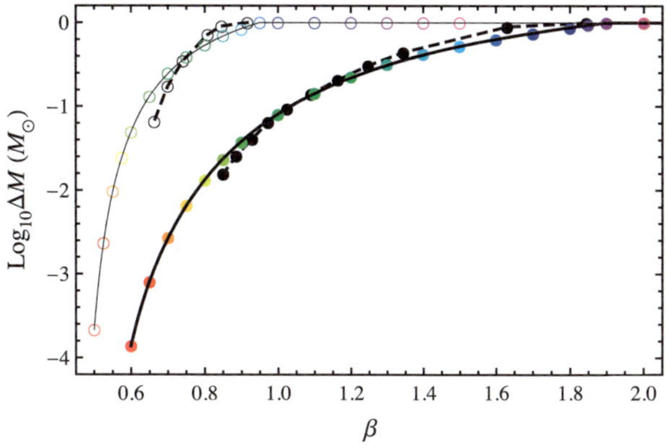

Fig. 4 The amount of mass lost from a single pericenter passage of a Solar-type star around a $10^6\ M_\odot$ SMBH with penetration parameter β. Thick curves and solid circles show $\gamma = 4/3$ models of stellar structure, while thin curves and open circles show $\gamma = 5/3$ models. The solid lines and colored data points show results from three-dimensional hydrodynamical simulations; dashed lines and black data points show results from the extended affine model of Ivanov and Novikov (2001), which is generally in good agreement. The critical β demarcating full from partial disruption is roughly $\beta = 1.875$ for $\gamma = 4/3$ ($\beta = 0.975$ for $\gamma = 5/3$). This is Fig. 4 from Guillochon and Ramirez-Ruiz (2013)

At a critical impact parameter, with β of the order of unity, a fraction of the stellar material is unbound from the star, in a partial disruption. In still-deeper encounters, the entire star is disrupted and no self-bound remnant survives.

This section focuses on the phenomenology of *partial* tidal disruptions, in which only the external layers are peeled off the star. Partial tidal disruptions occur because stars have differentiated interiors. The simple definition of the tidal radius states that the density enclosed by the tidal sphere at periapsis is equal to the mean density of the star, i.e. $M_{\rm BH}/R_{\rm t}^3 = M_\star/R_\star^3$, or $\rho_{\rm t} = \rho_\star$. If we imagine a star with a constant density interior (an $n \to 0$ polytrope), the entire interior experiences an equal ratio of tidal gravitational force to self-binding force in a given encounter. For any realistic star with a stratified interior, this statement is no longer true: the stellar density $\rho(R)$ decreases towards the surface, and consequently external layers have larger effective tidal radii, and one can define a radius-dependent tidal radius, $R_{\rm t}(R) > R_{\rm t}$ (see e.g. Ryu et al. 2020b for a more detailed example of this). Therefore, in grazing encounters with $R_{\rm p} = R_{\rm t}(R) > R_{\rm t}$, the stellar core within roughly a radius R of the center remains bound by self-gravity and proceeds on its orbit away from the black hole, stripped of its outer, more tenuous layers. Clearly, the way the density changes within the star determines the mass of the surviving core as a function of β.

More quantitative versions of these statements in the literature have relied on semi-analytic models as well as on numerical simulations. In the following, we review these efforts in chronological order. The first hydrodynamical simulations of the partial disruption process were performed by Diener et al. (1997), in Eulerian simulations that made use of a relativistic tidal tensor (the inclined generalization of Eq. (7)). While these simulations were the first to resolve the survival of a self-bound core following tidal stripping of its envelope, the computational expense limited their coverage of parameter space. Further progress originated from the semi-analytic, nested-affine model of Ivanov and Novikov (2001). By assuming shells are lost when they gain positive energy, these authors were able to estimate not just the degree of nonlinearity imparted by tides, but also fractional mass losses.

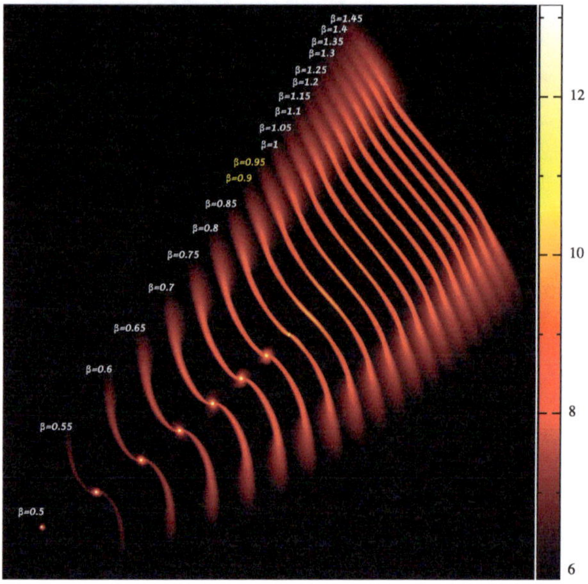

Fig. 5 Snapshots of $\gamma = 5/3$ polytropes following encounters with differing impact parameter β. These simulations are computed using the smoothed-particle hydrodynamic code GADGET2. This is Fig. 2 in Mainetti et al. (2017)

The first detailed sampling of the parameter space of partial disruptions with hydrodynamical simulations were performed in Newtonian gravity by Guillochon and Ramirez-Ruiz (2013). Figure 4 presents these results, showing also a comparison to the semi-analytic predictions of Ivanov and Novikov (2001). As a function of β, this figure shows the fraction of the star unbound in the encounter ($\Delta M/M_\star = 1$ implies a complete disruption). Figure 4 demonstrates important differences that occur for polytropes of differing internal structure. The $\gamma = 1 + 1/n = 4/3$ models transition from partial mass removal near $\beta \sim 0.6$ to full disruption near $\beta \sim 2$. The $\gamma = 5/3$ models, by contrast, are partially disrupted in a narrower range of $\beta \sim 0.5$ to $\beta \sim 0.9$. This numerical result aligns with our qualitative discussion above: the $\gamma = 4/3$ polytropic star has a wider range of internal densities and self-binding forces than the $\gamma = 5/3$ star, which is less centrally condensed.

Mainetti et al. (2017) re-examined the quantitative outcomes of partial disruption by using multiple numerical hydrodynamic methods to study the precise impact parameter that differentiates full disruption from partial disruption for $\gamma = 4/3$ and $\gamma = 5/3$ polytropes. Figure 5 shows snapshots for the $\gamma = 5/3$ stellar models. As β increases, the surviving core gets smaller and smaller while the mass and extent of the tidal tails grows. At $\beta \approx 0.9$, the star is entirely disrupted and no self-bound core remains. The results seen in this paper with both discrete-mass and discrete-volume techniques are quantitatively close to each other and those in Guillochon and Ramirez-Ruiz (2013), indicating a converged understanding of polytropic stellar disruption in Newtonian gravity.

Hydrodynamic simulations of partial tidal disruptions have revealed the morphology and dynamics of the gas around the surviving stellar core. Of the material stretched and distorted into the tidal debris streams, some remains bound to the core. The core itself is distorted by tides and may emerge from the encounter oscillating non-radially (typically dominated by an $l = m = 2$ fundamental mode). The re-accretion of bound material from the debris streams creates spiral shocks and vortices within the surviving core. These stages are clearly visualized by the magnetohydrodynamic simulations of Bonnerot et al. (2017) and Guillochon and McCourt (2017), these former of which are reproduced in Fig. 6. Both sets of

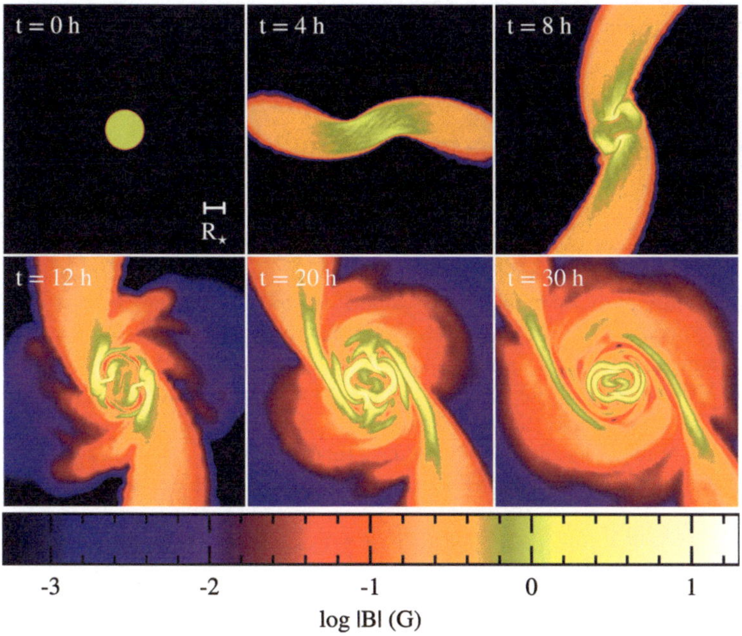

Fig. 6 The configuration of magnetic field strength (shown color-coded, with a logarithmic color scale at the bottom of the figure) in a simulated partial tidal disruption event at different snapshots in time. The top left snapshot shows the star before disruption, the top middle snapshot shows the first episode of re-accretion of the stripped gas, while later snapshots (with hours after disruption shown as white text labels) show the formation of vortices in the surviving core. Figure reproduced from Bonnerot et al. (2017), their Fig. 5

magnetohydrodynamic simulations find that the initial magnetic field strength is amplified by a factor $f_{\rm amp} \approx 10$ from vortex-driven dynamo activity, although in neither case does a self-sustaining dynamo emerge. The final degree of amplification in each study is resolution-dependent and unconverged, suggesting that these results may be lower limits on the true magnetic field strength present in a surviving core following partial disruption. Bonnerot et al. (2017) and Guillochon and McCourt (2017) both highlight the importance of repeated partial disruptions, which arise naturally for stars deep in the empty loss cone regime. For example, if a star undergoes N partial disruptions before a terminal full disruption, its initial field will be amplified by a factor $\sim f_{\rm amp}^N$, possibly producing enough magnetic flux for the final disruption to power a strong, relativistic jet.

A partial disruption produces a distinctive temporal behaviour of the fall-back rate. Guillochon and Ramirez-Ruiz (2013) find numerically that it asymptotically approaches a power-law with index $n_\infty \approx -9/4$. Interestingly, this result can also be derived and understood in the context of the impulse approximation. Coughlin and Nixon (2019) show that the "frozen-in" spread in debris energies (e.g. Eqs. (24), (28)) will be modified by the gravitational influence of a surviving core embedded within the stream. Solving the Lagrangian equation of motion for the combined stream-core-SMBH system, they find a late-time fall-back rate that is a power law with index $n_\infty \approx -1 - (\sqrt{73} - 1)/6 \approx -2.257$, and which is at leading order independent of the mass of the surviving core.

We close by noting that the processes of partial tidal disruption that we have described is quite sensitive to the interior structure of the star. The discussion above has focused on polytropic models. In the following section, we explore more closely how the basic principles of both full and partial disruption apply to realistically evolving stars with a more complex internal structure.

5 Exploring Different Stellar Types

The stellar clusters that surround galactic center black holes are believed to consist of a wide spectrum of stellar masses and evolutionary types. In our own Galactic Centre's nuclear star cluster, we observe light from giant-branch stars, substructures of young, massive stars, and diffuse light from an old population of main sequence stars (e.g. Schödel et al. 2007). In extragalactic nuclear clusters, there is also evidence for a diversity of stellar ages and types (see, for example, the work on NGC 404 of Seth et al. 2010, or the wide range of star formation histories seen in the nuclear star cluster sample of Georgiev and Böker 2014). Each of these types of stars can be scattered into orbits that lead to their disruption. In this section, we review the spectrum of possible stellar disruptions and the characteristics that relate their unique stellar evolutionary state to the outcome of a close passage by the black hole.

Differences in the disruption processes of different stellar types stem first of all from their different tidal radii. The smallest pericenter for a non-plunging parabolic orbit around a BH is the innermost bound spherical orbit (IBSO[10]). This latter depends on the BH mass, spin, and orbital inclination, is $4GM_{\rm BH}/c^2$ for a non-spinning SMBH, and can be as small as $GM_{\rm BH}/c^2$ (for prograde orbits in the equatorial plane of a maximally-spinning SMBH). Setting the tidal radius equal to this distance gives an approximate estimate of the maximum black hole mass for tidal disruption to occur. This upper limit is usually called the Hills mass (Hills 1975), and is

$$M_{\rm H} \sim \frac{R_\star^{3/2}}{M_\star^{1/2}}\left(\frac{c^2}{4G}\right)^{3/2} = 4\times 10^7 M_\odot \left(\frac{R_\star}{R_\odot}\right)^{3/2}\left(\frac{M_\star}{M_\odot}\right)^{-1/2} \propto \rho_\star^{-1/2}, \tag{34}$$

for a non-spinning black hole.[11] While the scalings in Eq. (34) are accurate, the prefactor can only be trusted to within a factor of a few, because (i) general relativistic tides differ from Newtonian tides in their strength (for example, compare C_{ij} and Γ_{ij} in §2, or see the discussion in Gafton et al. 2015), and (ii) physical radii such as the IBSO are coordinate-dependent quantities in general relativity. For these reasons, it is more accurate to perform fully general relativistic calculations using the Kerr metric tidal tensor (Eq. (7)). This can be done analytically (see Kesden 2012a, and also the discussion in the Rates Chapter) or with hydrodynamical simulations performed in a Kerr metric tidal field (e.g. Ryu et al. 2020c).

Equation (34) shows that a variety of stellar types are needed in order to probe the full range of SMBH masses. Exact relativistic calculations indicate that tidal disruptions of main sequence (MS) stars by Schwarzschild black holes[12] only happen for $M_{\rm BH} \lesssim 10^8\ M_\odot$

[10]Note that the IBSO is the tilted generalization of the familiar "marginally bound" (or unstable) circular orbit in the equatorial plane (Bardeen et al. 1972).

[11]Many calculations in the literature equate the tidal radius to the Schwarzschild horizon radius, $2GM_{\rm BH}/c^2$, rather than the IBSO radius (e.g. Hills 1975). For the reasons described above, no quasi-Newtonian calculation of the Hills mass is more trustworthy than a factor of a few, and it is better to use fully relativistic results.

[12]We note, however, that $M_{\rm H}$ is a strong function of SMBH spin $\chi_{\rm BH}$ (Kesden 2012a), and a favorably oriented orbit around a $\chi_{\rm BH} \approx 1$ SMBH can increase the MS Hills mass almost to $10^9\ M_\odot$, as in Leloudas et al. (2016).

(Kesden 2012a), while evolved stars can be disrupted by SMBHs with $M_{\rm BH} \gtrsim 10^8\ M_\odot$. Typical white dwarfs (WDs) can only be disrupted when $M_{\rm BH} \lesssim 10^5\ M_\odot$, although low-mass helium WDs with extended hydrogen envelopes can be partially disrupted so lon gas $M_{\rm BH} \lesssim 10^7\ M_\odot$. For the same stellar type, high-β events can only occur when $M_{\rm BH} \ll M_{\rm H}$. The $\beta - M_{\rm BH}$ parameter space where tidal disruptions can occur is often visualized with a "TDE triangle" diagram (see e.g. Fig. 1 in Luminet and Pichon 1989a, or Fig. 1 in Stone et al. 2019).

Figure 7 plots mass fallback rates against time seen in hydrodynamical simulations (MacLeod et al. 2012; Guillochon and Ramirez-Ruiz 2013; Law-Smith et al. 2017a) for several representative objects: a main-sequence star, a red giant at two different evolutionary stages, white dwarfs (He and CO/ONe), a brown dwarf, and a Jupiter-mass planet. The peak fallback rates and timescales span several orders of magnitude. One can understand this at the order-of-magnitude level without the need for hydrodynamical simulations, as the characteristic timescale and fallback rate for a tidal disruption scale with the BH mass, stellar mass, and stellar radius (Eqs. (26) and (27)). The fall-back rate $\dot{M}$ can range from highly super-Eddington to highly sub-Eddington, with peak timescales ranging from less than 1 day to more than 100 years. The way that the potential energy implicit in these mass return rates is converted into radiation is still highly debated (see the Formation of the Accretion Flow Chapter, Accretion Disc Chapter, and Emission Mechanisms Chapter for more details). In several—though not all—current models of accretion flow formation, the efficiency with which $\dot{M}$ is converted into radiation is a strong function of the dimensionless parameter $R_{\rm p}/R_{\rm g}$. In these models, luminous flares will arise predominantly from encounters with $R_{\rm p} \lesssim 10 R_{\rm g}$, biasing observations towards only finding TDEs from $\{M_\star, M_{\rm BH}\}$ pairs where the SMBH mass is within a factor ≈ 10 of the Hills mass (Stone and Metzger 2016); see Fig. 1 of Law-Smith et al. (2017a) for this phase space.

5.1 Main Sequence Stars

The internal structure of MS stars changes with stellar mass: more massive MS stars are more centrally concentrated. The effect of stellar structure on the disruption process was first thoroughly studied by Lodato et al. (2009) and Guillochon and Ramirez-Ruiz (2013), using polytropic stellar structure models (see Sect. 3 for discussion). While the polytropic approximation has some validity, particularly at the two extremes of the zero-age main sequence mass spectrum, it does not self-consistently account for important components of stellar physics (such as the changing structure of the star as it evolves in the MS) and has difficulty modeling stars with $0.5\ M_\odot \lesssim M_\star \lesssim 1.0\ M_\odot$. Compared to realistic stellar models, polytropic stars will have slightly different thresholds (in β) for the onset of partial mass stripping, and also for the transition between partial and full TDEs.

Aside from "bulk" questions related to total mass loss, there are more subtle (but nonetheless observationally testable) predictions that can only be made with realistic stellar models. As a star evolves along the MS, its composition profile changes, and this is reflected in the composition of the debris returning to the SMBH. Since the late-time fallback rate is dominated by material from the stellar core, and the early-time fallback rate is dominated by material from the stellar envelope, the chemical composition of fallback material (which may be reflected in emission line equivalent widths) will change over time if the progenitor star is chemically differentiated (Kochanek 2016). Using a semi-analytic fallback framework (Lodato et al. 2009, see also Eq. (28)), Gallegos-Garcia et al. (2018) calculated the time evolution of the composition of the fallback material for MS stars of varying mass and age. For most stars, they predict an enhancement in helium and nitrogen and a depletion in

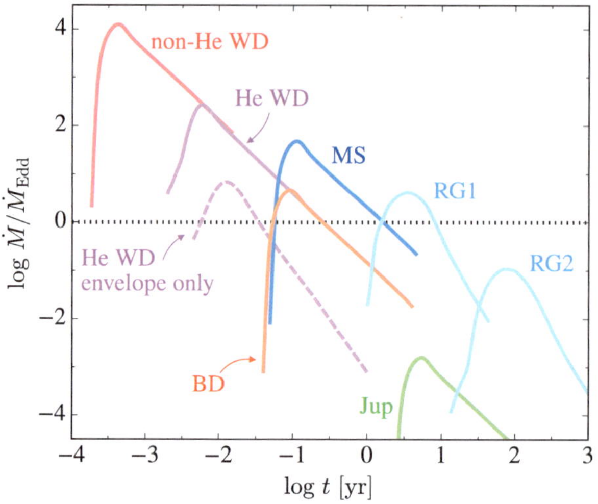

Fig. 7 Mass fallback rates $\dot{M}$ for six representative objects scaled to a single SMBH mass—$M_{\rm BH} = 10^6\ M_\odot$ is chosen for comparison, but note that a non-He WD cannot be disrupted outside the horizon of a Schwarzschild black hole for this mass. We show a 0.6 $M_\odot$ non-He WD in red, a 0.17 $M_\odot$ He WD in purple, a 0.6 $M_\odot$ MS star in blue, a 50 $M_{\rm Jup}$ brown dwarf in brown, a 1 $M_{\rm Jup}$ planet in green, and a 1.4 $M_\odot$ red giant at 10 $R_\odot$ (RG1) and at 100 $R_\odot$ (RG2) in light blue. We show a $\beta = 0.9$ encounter (full disruption) for the non-He WD, MS star, BD, and planet, and a $\beta = 1.5$ encounter for the giant stars. For the He WD, we show two curves for comparison: the solid line shows a full disruption and the dashed line shows an encounter that strips only the low-mass hydrogen envelope. Time is relative to pericenter passage. Figure from Law-Smith et al. (2017a), their Fig. 12

carbon (relative to solar) with time. The strength and timing of these abundance variations in the mass fallback depend on the mass and age of the star, and can thus help determine the properties of the victim star in an observed TDE.

Law-Smith et al. (2019) developed a simulation framework in which stars built using `MESA` are used as inputs for tidal disruption calculations in the 3D adaptive-mesh code `FLASH` (Fryxell et al. 2000) with the Helmholtz EOS. This framework uses accurate stellar density profiles and tracks the chemical abundance of the debris for 49 elements. Law-Smith et al. (2019) studied the tidal disruption of 1 $M_\odot$ and 3 $M_\odot$ stars at zero-age main sequence (ZAMS), middle-age, and terminal-age main sequence (TAMS). They find that the initial density structure of the star leads to different susceptibilities to disruption: e.g. for a ZAMS star a $\beta = 2$ encounter is a full disruption, whereas for a TAMS star this is a grazing partial disruption. In addition, significant differences in the fallback rate curves for a given stellar age and mass have been found compared to results for polytropes. This is illustrated in Fig. 8. In terms of the composition of the fallback material, the authors found that abundance anomalies in nitrogen, carbon, and helium are present before the time of peak fallback rate for some disruptions of MS stars. This proof-of-concept study was recently expanded to a large parameter-space study of main sequence tidal disruptions, in the "Stellar TDEs with Abundances and Realistic Structures" (STARS[13]) library of Law-Smith et al. (2020b).

Deviations between fallback curves for polytropic and realistic stellar models were also found in the work of Golightly et al. (2019a), which simulated the disruption of 0.3 $M_\odot$, 1 $M_\odot$ and 3 $M_\odot$ stars at three different ages (for $\beta = 3$ encounters). The authors find qualita-

[13] Tabulated results are accessible with the associated GitHub tool (Law-Smith et al. 2020a).

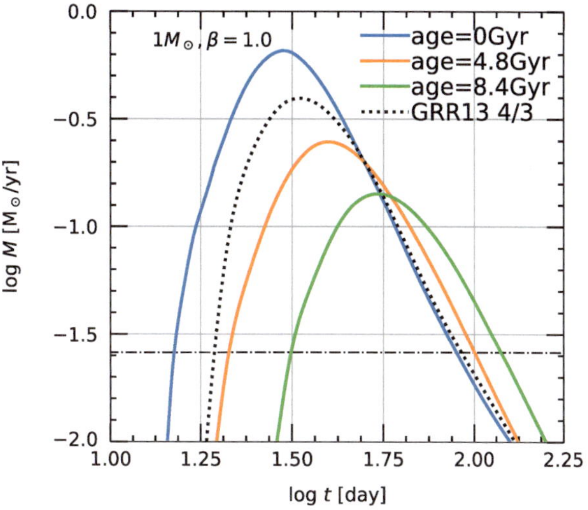

Fig. 8 Mass fall-back rate to the BH as a function of time for the disruption of a 1 $M_\odot$ star at three different ages for a $\beta = 1$ encounter with a 10^6 $M_\odot$ BH. The result for a $\gamma = 4/3$ polytrope from Guillochon and Ramirez-Ruiz (2013), scaled to the radius of a ZAMS 1 $M_\odot$ star, is in dotted black. The Eddington limit for this BH is shown by the dot-dashed line. Figure adapted from Fig. 3 in Law-Smith et al. (2019)

tive differences with polytropic TDEs, and use this comparisons to argue that determinations of SMBH mass from TDE light-curve fitting using models with polytropic stellar structures can be incorrect by as much as a factor of 5.

Goicovic et al. (2019) performed disruption simulations of a 1 $M_\odot$ ZAMS star constructed in the 1D stellar evolution code MESA (Paxton et al. 2011) for a range of β's, using the 3D moving-mesh code AREPO. Their ΔM vs. β and fallback-rate results agree relatively well with the $\gamma = 4/3$ polytrope model from Guillochon and Ramirez-Ruiz (2013), which is expected as a ZAMS 1 $M_\odot$ star is reasonably well approximated by a 4/3 polytrope. Goicovic et al. (2019) also studied the internal dynamics of the stellar remnant following a partial disruption.

Ryu et al. (2020a) very recently introduced fully relativistic simulations for a grid of stellar masses, impact parameters, and SMBH masses, at a single stellar age (halfway through the MS for each star, with models taken from MESA). They provide fitting formulae to describe the trends they find in several disruption parameters, such as the critical pericenter distance for full disruption, the mass of the remnant, and the spread in the debris energy distribution. For all partial disruptions, they find that mass loss continues for many stellar dynamical times after pericenter passage.

5.2 Giant Stars

As stars evolve off the main sequence, their radii grow by factors of tens to hundreds. Giant stars, therefore, are particularly vulnerable to tidal forces from a supermassive black hole (recall from Eq. (1) that the tidal disruption periapsis distance scales linearly with stellar radius). Also of importance in the context of TDEs is the stars' internal structure: giant stars posses a composite structure of dense core and low-density envelope.

5.2.1 Disruption of Giant Stars

The gas dynamics of the tidal disruption of giant stars was first examined extensively by MacLeod et al. (2012). While the qualitative process of tidal disruption remains the same as for main sequence stars, the composite structure of giant stars yields differing behavior for

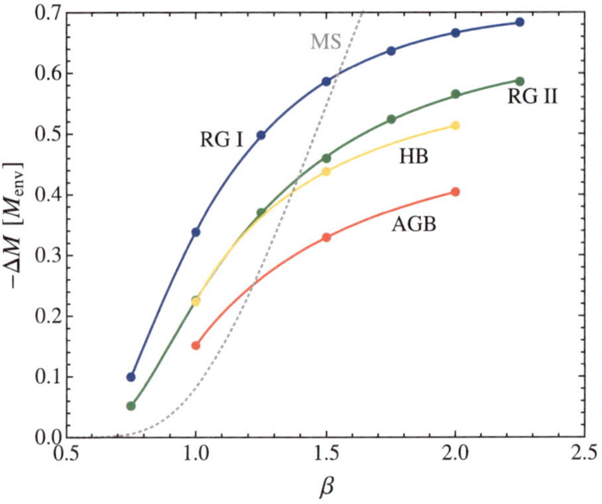

Fig. 9 Mass removed as a function of impact parameter, β, in four model giant stars. Here ΔM is normalized to the total envelope mass M_{env}. Higher impact parameter encounters remove more of the envelope material, but even relatively deeply plunging encounters with $\beta \gtrsim 2$ do not remove the entire envelope. Further, models with higher core masses retain more envelope mass, because contraction of the envelope around the core upon mass loss shelters it from complete disruption. Here MS corresponds to an $n = 3/2$ polytrope (Guillochon and Ramirez-Ruiz 2013), while RG I (ascending the red giant branch), RG II (tip of the red giant branch), HB (horizontal branch), and AGB (tip of the asymptotic giant branch) correspond to various moments in the evolution of an originally 1.4 $M_\odot$ stellar model. Adapted from MacLeod et al. (2012)'s Fig. 6, where we refer for more information

an encounter with same impact parameter. Compared to a tidal disruption of a MS star, with its less-differentiated internal structure, the tidal forces of the black hole tend to disturb only the giant star's outer envelope. Thus, the dense core generally survives the encounter, and continues on an orbit similar to that on which it first encountered the black hole.

This intuitive picture has been tested by hydrodynamical simulations that reveal how some surrounding envelope material is not lost to tides even in deeply-penetrating encounters, as is shown Fig. 9. MacLeod et al. (2012) argue that this resilience to disruption can be attributed to the adiabatic change of the inner envelope on a dynamical timescale in response to the removal of the overlying layers (Hjellming and Webbink 1987). While the outermost layers of a giant star's stellar structure tend to expand upon mass loss, when the core becomes the dominant mass component, the remaining envelope material contracts as mass is removed, self-sheltering from further mass loss in a given encounter. This change occurs because removing overlying material reduces the compression (and pressure) of stellar material at the base of the envelope, thus the pressure scale height must be smaller to achieve a pressure gradient that balances gravity in the new hydrostatic equilibrium. Another factor that inhibits mass loss is the gravitational pull of the surviving core (only very rare, deep encounters would fully disrupt a giant star) that promotes partial re-accretion of the surrounding envelope material.

In essence, this implies that all giant-star tidal disruption events are partial tidal disruptions. Depending on the orbital dynamics, the star may return for further interactions with the black hole, as was considered in detail by MacLeod et al. (2013). We discuss this possibility further in Sect. 8.3. Finally, Bogdanović et al. (2014) consider the related scenario of a giant star disruption, where a surviving degenerate compact core dynamically evolves because of tidal heating, as well as emission of gravitational waves. Its fate can be either a direct plunge into the massive $> 10^6\ M_\odot$ black hole or disruption, if the tidal heating succeeds in lifting the matter degeneracy and expanding the core.

5.2.2 *Fall-Back to the Black Hole*

One consequence of the extended radii and correspondingly large tidal disruption radii of giant stars is that the characteristic timescales for the periapsis passage and fall-back of material to the black hole are extended. Because the typical mass involved is of the same order as a lower main sequence star, this implies lower fall-back rates toward the black hole.

The characteristic timescale for the encounter itself is the stellar dynamical time, because $R_t/v_p \approx \sqrt{R_\star^3/2GM_\star} \sim T_\star/\sqrt{2}$. While a main sequence star might have a dynamical time of hours, a giant of $R_\star = 100\ R_\odot$ and $1\ M_\odot$ has a dynamical time on the order of 10^6 seconds, or a month. Thus, while the process of disruption is still rapid relative to the star, it proceeds relatively slowly from a human perspective. Likewise, from Eq. (26), we can see that the characteristic fall-back timescale $t_{\min}$ is ~ 100 yr for a solar-mass, $R_\star = 100\ R_\odot$ giant star. This also carries implications for the peak fallback rate, $\dot{M} \sim \Delta M/t_{\mathrm{fb}}$. Holding fixed other properties, $\dot{M} \propto R_\star^{-3/2}$. Therefore, TDEs of giant stars are unlikely to fuel the rapid, powerful episodes of black hole accretion that we associate with "standard" TDEs. Their characteristic properties are instead very extended duration, lower-level mass fallback toward the black hole.

The key features of red giant TDE fallback are illustrated in Fig. 7, which shows results from hydrodynamic simulations by MacLeod et al. (2012) and Law-Smith et al. (2017b). The figure compares main sequence and white dwarf tidal disruptions to two characteristic giant star phases. In comparison to their more compact counterparts, disruptions of giant stars yield fallback to the black hole at lower rates spread over much longer durations. The $10\ R_\odot$ RG I model still feeds material to a $10^6\ M_\odot$ black hole above its Eddington limit (for nearly 10 years), but the $100\ R_\odot$ RG 2 model peaks at approximately 10% of Eddington.

The dilute streams of material falling back to the black hole in the red giant TDE scenario have led Bonnerot et al. (2016a) to argue that stream interaction with the gas surrounding the quiescent supermassive black hole excites the Kelvin-Helmholtz instability. If this instability develops, the debris will fragment and dissolve into the ambient medium before they can return to the black hole, drastically decreasing the mass fall-back rate.

5.3 White Dwarfs

White dwarfs (WDs) have densities of $\approx 10^4$–10^7 g/cm^3 and can thus typically (for e.g., CO WDs) only be disrupted outside the event horizon for black holes of mass $\lesssim 10^5\ M_\odot$, but low-mass helium WDs with hydrogen envelopes can extend this limit to $\approx 10^7\ M_\odot$. The tidal disruption of WDs can thus be a unique probe of intermediate mass black holes.

The gas hydrodynamics in a WD tidal disruption is similar to the behavior of a $\gamma = 5/3$ polytrope for all but the most massive WDs. WD tidal disruptions have been studied in e.g. Rosswog et al. (2009), Cheng and Evans (2013), MacLeod et al. (2014, 2016), Law-Smith et al. (2017a). WD tidal disruptions are expected to produce flares with characteristic timescales of $\approx 10^2$–10^5 s. Cheng and Evans (2013) and Haas et al. (2012b) have closely examined the effects of passages close to the black hole horizon. Unlike in typical MS star tidal disruptions, there is the possibility for detonation due to compression in highly-penetrating encounters. The energy release in such a detonation can approach the luminosity of a Type Ia supernova. This possibility, as well as the details of the disruption dynamics and the relative rate of WD disruptions, are discussed in detail in the White Dwarf Chapter.

WDs have an inverse mass-radius relationship, which means that it is the least massive (and thus least dense) WDs that can be tidally disrupted by the highest mass BHs. Law-Smith et al. (2017a) study the disruption of helium-core hydrogen-envelope white dwarfs. These low-mass ($\lesssim 0.5\ M_{\odot}$) WDs extend the range of BH masses that can disrupt typical WDs (see above), and offer flares with peak timescales ($\sim$1–10 days) in between those of typical WDs and MS stars. Because of their unique compositional structure, these objects can also produce flares powered by hydrogen-only fall-back material for grazing encounters, or a transition from hydrogen-rich to helium-rich fall-back material for more deeply-penetrating encounters. The fallback rate from the disruption of the hydrogen envelope alone is shown in Fig. 7, and is significantly lower than that of the whole star because of this envelope's small mass fraction.

6 Phenomenology of Highly Penetrating Encounters

A TDE is typically regarded as "highly penetrating" if $\beta = R_t/R_p$ is significantly greater than one. In this case, the severe compression experienced by the star will lead to the adiabatic buildup of pressure, and the eventual reversal of the vertical collapse in a hydrodynamic rebound near pericenter. Sometimes shocks are formed during this process. In Newtonian gravity, the hydrodynamic pinch point is fixed in space at a true anomaly $f_c > 0$, as described in §2. This indicates that the collapse and rebound will occur shortly after pericenter passage.

Stellar TDEs may also be highly penetrating in a different sense, if the parameter $b = R_g/R_p \sim 1$. In this case, the star's center of mass orbit will deviate highly from that of a closed Keplerian ellipse, as relativistic effects (such as precession and, at a higher order, gravitational radiation reaction) will come into play. Unique phenomenology can emerge from the combination of high β (i.e. $\gg 1$) and high b (i.e. ~ 1). For example, in the stronger tidal field of relativistic gravity, a star with $R_p \sim R_g$ may actually undergo two or more vertical compressions and bounces, the first of which is prior to pericenter passage; Luminet and Marck (1985) provide a simple geometric proof that the number of bounces is roughly equal to the number of self-intersections of the center-of-mass geodesic inside the tidal radius, a prediction that is roughly borne out by combining a relativistic tidal field with the affine model (Luminet and Marck 1985) and one-dimensional hydrodynamic simulations (Brassart and Luminet 2010).

In this section, we explore three somewhat speculative predictions of high-β and/or high-b compression: high-energy shock breakout signals, gravitational wave emission, and runaway thermonuclear reactions. At the time of writing, none of these predictions have been clearly observed, but the detection of any would be of significant value for TDE science goals. In particular, these detections would time the disruption, providing an essential timeline for interpreting subsequent observations.

6.1 Shock Breakout and Prompt X-Rays

If an outgoing shock is launched near peak compression of the star, it may acquire a large specific energy that is at most an order unity fraction of the total kinetic energy of compression, $U_c \sim \beta^2 G M_\star^2/R_\star$. The timescale for the release of this energy is much longer than the time of peak compression ($T_c \sim T_\star \beta^{-4}$, for a $\gamma = 5/3$ polytrope) because the star does not compress simultaneously; rather, the star acquires an "hourglass" shape as it passes through the tidal pinch point, which is fixed in space, with leading portions of the star rebounding while trailing portions are still collapsing (see also §2). Assuming that each "column" of the star begins tidally free falling at the moment it crosses into the tidal sphere, the duration of shock breakout emission will be roughly the time it takes for the star to fully cross, $T_{\rm cr} \approx 2R_\star/\sqrt{2GM_{\rm BH}/R_{\rm t}} \approx \sqrt{2}T_\star(M_{\rm BH}/M_\star)^{-1/3}$. This estimate is equivalent to taking the time it takes the tidally distended stellar debris to cross pericenter: in the limit of high β, the star's long axis has a half-length $\approx \beta^{1/2}R_\star$ (Stone et al. 2013), so we again find $T_{\rm cr} \approx 2\beta^{1/2}R_\star/\sqrt{GM_{\rm BH}/R_{\rm p}} \approx \sqrt{2}T_\star(M_{\rm BH}/M_\star)^{-1/3}$.

Even if a large fraction of the total compressional energy budget $U_{\rm c}$ goes into shock-heating the star, most of this will not be promptly radiated. Because the star has an optical depth at peak compression

$$\tau_{\rm c} \sim \frac{\sigma_{\rm T}\rho_{\rm c}z_{\rm c}}{m_{\rm p}} \sim \frac{3M_\star\sigma_{\rm T}}{4\pi R_\star^2 m_{\rm p}}, \tag{35}$$

that is far greater than unity, only a small fraction of the shock heating can emerge as a prompt transient. Note that in Eq. (35), we have made use of the Thompson cross-section $\sigma_{\rm T}$, the proton mass $m_{\rm p}$, and have assumed that the star's cross-sectional area at peak compression is $\approx 4\pi R_\star^2$, as is motivated by Eqs. (31) and (32). Following Kobayashi et al. (2004), we may estimate the maximum theoretically possible bolometric luminosity by (i) assuming the shock deposits its energy uniformly through the vertical layers of the compressed star, and (ii) that during the passage of the star through the pinch point, shock-heated thermal energy diffuses out of the upper layers down to a depth

$$D \sim \sqrt{\frac{cT_{\rm cr}z_{\rm c}}{\tau_{\rm c}}}, \tag{36}$$

where we have assumed a random walk and made use of the height of the star at peak compression, which for a $\gamma = 5/3$ polytrope is $z_{\rm c} \sim R_\star\beta^{-3}$ (Eq. (21)). This is an upper limit because in a steep stellar atmosphere, most of the shock energy will be lost in deeper layers before it arrives near the dilute surface. Under these assumptions, the maximum peak luminosity will be

$$L_{\rm sh}^{\rm max} \sim \frac{U_{\rm c}D}{T_{\rm cr}z_{\rm c}} \approx 5\times 10^{42}\ {\rm erg\,s^{-1}}\ \beta^{7/2}\left(\frac{M_\star}{M_\odot}\right)^{19/12}\left(\frac{R_\star}{R_\odot}\right)^{-5/4}\left(\frac{M_{\rm BH}}{10^6\,M_\odot}\right)^{1/6}. \tag{37}$$

This is the same upper limit as in Kobayashi et al. (2004), except that it also considers the β-dependence of $U_{\rm c}$. Actual estimates for the shock breakout luminosity were performed by post-processing $\beta = 7$ hydrodynamical simulation results in Guillochon et al. (2009), taking into account the realistic density profile of a stellar atmosphere (i.e. self-consistently modeling the deposition of shock energy into layers of different density). This work found luminosities roughly one order of magnitude smaller than this upper limit. However, it is numerically challenging to resolve the breakout layer of the tidally compressed star, and underresolution may cause an overestimate of the energy in the breakout shell. To account for this, Yalinewich et al. (2019) used an analytic model incorporating a realistic stellar atmosphere profile to derive the following, far more pessimistic estimate for the shock breakout luminosity:

$$L_{\rm sh} \approx 9\times 10^{40}\ {\rm erg\ s^{-1}}\beta^{1.14}\left(\frac{M_\star}{M_\odot}\right)^{0.87}\left(\frac{R_\star}{R_\odot}\right)^{-0.34}\left(\frac{M_{\rm BH}}{10^6\,M_\odot}\right)^{0.33}, \tag{38}$$

where a $\gamma = 5/3$ polytrope has been assumed. If the shock is matter-dominated, the emitted spectrum will be quasi-thermal with a blackbody temperature $T_{\rm sh}$ in the X-rays:

$$k_{\rm B} T_{\rm sh} \approx U_{\rm c} \frac{m_{\rm p}}{M_\star} \approx 1.9\,{\rm keV}\,\beta^2 \left(\frac{M_\star}{M_\odot}\right)\left(\frac{R_\star}{R_\odot}\right)^{-1}. \tag{39}$$

However, as was pointed out by Yalinewich et al. (2019), high-β events will generally find themselves in either a radiation-dominated blackbody regime (which softens the growth of temperature to $T_{\rm sh} \propto \beta^{1/2}$) or a photon-starved regime in which the typical energy of the non-thermal emission is set by a balance between pair production and annihilation, i.e. $k_{\rm B} T \approx 50$ keV.

The short durations and low luminosities of X-ray shock breakout signals from main sequence TDEs make their detection unlikely with current instrumentation. However, the less frequent tidal disruption of red giant stars will produce optical/UV shock breakout flashes of much longer duration $T_{\rm cr}$, and LSST may detect these at a rate of $10^{-1} - 10^{1}$ yr^{-1} (Yalinewich et al. 2019).

6.2 Gravitational Wave Emission During Disruption

Gravitational waves are produced when the quadrupolar moment of the mass distribution of a source is changing with time, in an accelerated fashion. During the star's closest approach, there are contributions to the variation of the mass quadrupole moment from both the changing mass quadrupole of the star-black hole system *and* the internal mass quadrupole of the star itself.[14] In the first case, the binary components can be regarded as point masses orbiting each other, and the ultimate source of energy is their orbital energy. This signal, which is analogous to the classic "extreme mass ratio inspiral/burst" (Sigurdsson and Rees 1997), has been investigated for both main sequence stars with SMBHs (Kobayashi et al. 2004) and for white dwarfs with intermediate mass black holes (e.g. Sesana et al. 2008; Rosswog et al. 2009; Haas et al. 2012b; Anninos et al. 2018). Since the binary dynamical interaction is far from the highly relativistic merger phase, a simpler analytical description of the observed strain can be adopted:

$$h \approx \frac{1}{d}\frac{4G}{c^2}\frac{E_{\rm kin}}{c^2}, \tag{40}$$

(e.g. Thorne 1998). This signal scales with the distance d of the source to Earth and it is proportional to the non-spherical part of the source's kinetic energy $E_{\rm kin}$. Since most of the emission occurs at the closest approach $R_{\rm p} = \beta^{-1} R_{\rm t}$, and the large mass ratio leaves the black hole still at the center of mass, we can substitute $E_{\rm kin} = M_\star \left(G M_{\rm BH}/R_{\rm p}\right)$ in Eq. (40) to estimate the strength of the signal

$$\begin{aligned} h &\approx \beta \times \frac{R_{\rm s} R_{\rm s\star}}{R_{\rm t} d_{\rm L}} \\ &\approx 2\beta \times 10^{-22} \left(\frac{M_\star}{M_\odot}\right)^{4/3} \frac{R_\odot}{R_\star} \left(\frac{M_{\rm BH}}{10^6 {\rm M}_\odot}\right)^{2/3} \left(\frac{d_{\rm L}}{16{\rm Mpc}}\right)^{-1}, \end{aligned} \tag{41}$$

[14]Emission of gravitational waves during later stages of a TDE has been investigated by Toscani et al. (2019), but in this section we will only report on gravitational wave emission linked to the disruption process.

where $R_s = 2GM_{BH}/c^2$ and $R_{s\star} = 2GM_\star/c^2$. This "point particle" description of the signal was verified by numerical simulations, even in $\beta \ll 1$ encounters (Kobayashi et al. 2004). The signal duration is roughly the orbital period $T(R_p)$ at pericenter and the associated frequency is $f_{GW} \sim 1/T$,

$$f_{GW} \approx \frac{\beta^{3/2}}{2\pi}\left(\frac{GM_{BH}}{R_t^3}\right)^{1/2}$$
$$\approx 10^{-4}\,\text{Hz}\,\beta^{3/2}\left(\frac{M_\star}{M_\odot}\right)^{1/2}\left(\frac{R_\star}{R_\odot}\right)^{-3/2}. \tag{42}$$

TDEs involving WDs are necessarily associated with smaller mass black holes ($M_{BH} \leq 10^5\ M_\odot$), as otherwise disruption does not take place (see Eq. (34)). The involvement of a lower mass black hole would cause by itself a suppression of the signal (all other conditions the same), however WDs are also more compact than MS stars by a factor of ~ 100: overall the signal amplitude from a WD disruption can span from about one order of magnitude *above* to one order of magnitude *below* the value reported in Eq. (41), when considering black hole masses between $10^3\ M_\odot - 10^5\ M_\odot$, and a WD with mass $0.5\ M_\odot$ and radius $0.01\ R_\odot$. The signal duration decreases from $\sim 2.7\beta^{-3/2}$ hr for solar type stars down to $\sim 14\beta^{-3/2}$ sec or less for our fiducial WD, implying that WD signals should be expected at higher frequencies, in the range $\sim 7 \times 10^{-2} - 10$ Hz for the parameters considered here. These frequency and strain estimations suggest that disruption of WDs by intermediate mass black holes are within detection reach of future space-based interferometers like the DECI-hertz inteferometer Gravitational wave Observatory (DECIGO Sato et al. 2017) and ALIA (Baker et al. 2019), but would remain subthreshold events for the Laser Interferometer Space Antenna (LISA Amaro-Seoane et al. 2017) and TianQin (Luo et al. 2016). The much lower frequency range for TDEs involving MS stars, on the other hand, precludes detection by any planned mission.

We now turn our attention to the second source of gravitational wave emission: the time-varying quadrupole moment of the *star's* mass distribution, during the vertical compression/in-plane stretching of the star at pericenter (see Sect. 2, and Guillochon et al. 2009; Stone et al. 2013). Under the assumption of simultaneous vertical collapse of all stellar layers to a "pancake" shape, the duration of the gravitational wave burst is roughly the duration of maximal compression $T_c \approx 8.5T_\star\beta^{-4}$ (for a $\gamma = 5/3$ polytrope), with a characteristic frequency $f_{GW} \sim 1/T_c$,

$$f_{GW} \approx 4 \times 10^{-5}\ \text{Hz}\ \beta^4\left(\frac{M_\star}{M_\odot}\right)^{1/2}\left(\frac{R_\star}{R_\odot}\right)^{-3/2}. \tag{43}$$

For main sequence stars and grazing events, this frequency is too low for any planned space-based interferometers and too high for the Pulsar Timing Array.[15] However the strong dependence on β implies that nearly plunging events would emit *just* within the LIGO/VIRGO ground-based interferometer frequency band ~ 10 Hz–10 kHz: e.g. $f_{GW} \approx 15.6$ Hz for $\beta = 25$. Likewise, WDs with their smaller radius would produce events at higher frequencies spanning from the LISA band ($10^{-4} - 10^{-2}$ Hz) for $\beta \sim 1$ to the LIGO/VIRGO band for $\beta > 1$ (e.g. for[16] $\beta = 10$, $f_{GW} \approx 280$ Hz). The rapid vertical collapse ($v_z/T_c \propto$

[15]http://ipta4gw.org.

[16]Note that $\beta > 6$ for WDs is possible only for black hole masses smaller than $10^4\ M_\odot$.

$\beta^5\sqrt{GM_\star/R_\star}/T_\star$) causes an accelerated quadrupolar variation that produces a gravitational wave strain highly dependent on β, with $h \propto \beta^2$ (Stone et al. 2013). If one instead considers the less dramatic stretching of the star in the orbital plane that occurs on a slower time scale –the orbital time $T(r_\mathrm{p})$, rather than the compression timescale T_c – then the dependence of the strain on β is even steeper: $h \sim \beta^3$ (Guillochon et al. 2009). Let's now elaborate further on the former case, that produces emission at a frequency given by Eq. (43). When the strain is computed under the assumption of synchronous vertical collapse to a stellar pancake, it reads (Stone et al. 2013)

$$h \simeq 2 \times 10^{-27} \left(\frac{M_\star}{M_\odot}\right)^2 \left(\frac{R_\star}{R_\odot}\right)^{-1} \beta^2 \left(\frac{d_\mathrm{L}}{16\mathrm{Mpc}}\right)^{-1}. \tag{44}$$

This is a very weak signal for a MS star event, hardly detectable by LIGO/VIRGO even when considering highly penetrating events (e.g. $\beta = 25$, $h \sim 10^{-24}$ at $f_\mathrm{GW} \approx 15.6$ Hz). On the other hand, the same nearly plunging ($\beta \gg 1$) events involving WDs would be boosted to higher frequencies and strains (e.g. $\beta = 10$, $h \sim 5 \times 10^{-24}$ at $f_\mathrm{GW} \approx 280$ Hz), placing them above the detection threshold for future ground-based facilities, such as the Einstein Telescope.[17]

6.3 Nuclear Reactions

During high-β encounters, stars suffer large temperature and density increases from adiabatic compression. The degree of temperature and density increase can be even larger if shocks form during either the compression or early on in the rebound. In this sense, a deeply penetrating tidal disruption event is analogous to an inertial confinement fusion reactor, where the inertia of an imploding plasma can – in principle – pin the plasma in place long enough for dynamical thermonuclear fusion reactions to occur. On Earth, inertial confinement fusion is achieved by using powerful lasers to compress fuel pellets, but in galactic nuclei, the "tidal piston" of the SMBH may serve the same purpose.

This possibility, first proposed by Carter and Luminet (1982), was first investigated quantitatively (in the context of the affine model) by Luminet and Pichon (1989b). This early study found that only a small fraction of a MS star's mass could undergo fusion, even for extreme parameter choices, and that the energetics of the resulting radioisotopes would be subdominant to the impulsive disruption energy spread, ΔE. These conclusions have been qualitatively confirmed by a handful of subsequent hydrodynamical simulations that investigated thermonuclear reactions in high-β TDEs in greater detail (Guillochon et al. 2009). Thermonuclear burning does not achieve dynamical importance in MS TDEs because the primary reactions at peak compression are those on the hot CNO cycle, which is β-decay limited. The short duration of peak compression is therefore an insurmountable bottleneck to a thermonuclear runaway for stars with H-dominated interiors.

Runaway fusion appears more promising, however, in the compression of WDs, where the burning can, depending on composition, proceed via either the triple-α reaction or carbon/oxygen reactions (processes not limited by the need for β decays). Studies using both the affine model (Luminet and Pichon 1989a) and hydrodynamical simulations (Rosswog et al. 2009) have found evidence for thermonuclear detonation waves that burn an order unity fraction of the star, substantially alter the spread in debris energy, and synthesize sufficient quantities of radioisotopes to power a radioactive transient in the unbound ejecta.

[17] http://www.et-gw.eu/.

However, more recent high-resolution studies have shown that the formation of shocks that generate detonation waves during WD compression is highly sensitive to numerical resolution (Tanikawa et al. 2017), and may be artificially triggered by spurious heating produced by underresolution of the tidal compression phase. While the most recent sets of WD disruption simulations indicate a broad parameter space for ignition and thermonuclear burning (Tanikawa 2018; Anninos et al. 2018; Kawana et al. 2018), the exact parameters required to tidally detonate a WD remain contested. These issues are addressed in greater detail in the White Dwarf Chapter.

7 Unbound Debris

As mentioned earlier, part of the stellar debris acquires a positive orbital energy and leaves the vicinity of the black hole on unbound orbits. The exact unbound fraction of stellar mass depends on whether the disruption was total or partial (Sect. 4), and on the initial stellar orbit – or equivalently on the initial star's *orbital* energy $E_\star$. For a star approaching on a parabolic orbit ($E_\star = 0$) that is completely disrupted, half of the debris is expected to gain energy and be lost from the system. For the same fully disrupted star approaching instead on either an elliptical or a hyperbolic orbit, the unbound fraction would depend on the ratio $|\Delta E/E_\star|$ between the tidal energy spread ΔE (Eq. (22)) and the orbital energy. For instance, for $|\Delta E/E_\star| \ll 1$, the whole debris remains respectively bound for $E_\star < 0$ and unbound for $E_\star > 0$. These considerations have been explored both in the context of tidal separation of stellar binaries and tidal disruption of a single star (Kobayashi et al. 2012; Hayasaki et al. 2013, 2018; Park and Hayasaki 2020).

Because effectively parabolic encounters are expected to dominate overall TDE rates (see e.g. the Rates Chapter), we discuss the canonical case of a star with mass $M_\star$, on a parabolic orbit with pericenter equal to the tidal radius $R_p = R_t$, that undergoes a complete disruption. We will also first assume a constant specific energy distribution $dE/dM = \Delta E/(M_\star/2)$ between $-\Delta E$ and $+\Delta E$ (Eq. (24)). A simple estimate of the maximum terminal velocity is then

$$v_{\max} = \sqrt{2\,\Delta E} \approx 8,000\ \mathrm{km\,s^{-1}} \left(\frac{M_\star}{M_\odot}\right)^{1/3} \left(\frac{R_\star}{R_\odot}\right)^{-1/2} \left(\frac{M_{\rm BH}}{4\times 10^6\,M_\odot}\right)^{1/6}, \tag{45}$$

where we consider a SMBH with mass similar to that of Sagittarius A* (a.k.a Sgr A*), the SMBH in the Centre of our Galaxy. If not specified otherwise, these will be our fiducial parameters for this section. For such a system the tidal radius is $R_t \approx 10^{13}$ cm. Some fraction of the unbound debris then escapes at a few percent of the speed of light, comparable to the speed of a supernova blastwave. Likewise, the total available kinetic energy and momentum are $\lesssim (1/2)v_{\max}^2(M_\star/2) \approx 3\times 10^{50}$ erg and $\lesssim v_{\max}(M_\star/2) \approx 8\times 10^{41}\ \mathrm{g\,cm\,s^{-1}}$. This significant reservoir of energy and momentum stimulated the studies that we are summarizing here, as it is intriguing to explore whether they can produce observable signatures when deposited into the ambient medium. Special attention has been given to observable signatures in the Galactic Centre (Khokhlov and Melia 1996; Guillochon et al. 2016), because of the unique possibility to directly image a "remnant" of this event and because of the presence of Sagittarius A East, a radio source with unclear origin that engulfs Sgr A*'s gravitational sphere of influence.

Deceleration of the debris as it expands into the black hole surroundings and sweeps up the intervening gas is a possible way to tap its kinetic energy and convert it into radiation, in analogy with supernova remnants. The effective deceleration or "stopping" length R_{st} can be defined as the distance from the black hole at which the debris velocity is half of its initial value. This is equivalent to the distance within which an amount of gas equal to the debris's mass has been swept up along the way, i.e. $M_\star/2 = \rho_{\mathrm{ism}}(4\pi/3)R_{\mathrm{st}}^3(\Omega/4\pi)$, where ρ_{ism} is the interstellar medium mass density, which has been assumed constant. Re-arranging the mass equality to evaluate the deceleration length yields

$$\frac{R_{\mathrm{st}}}{R_{\mathrm{t}}} \approx q^{-1/3}\left(\frac{\langle\rho\rangle}{\rho_{\mathrm{ism}}}\right)^{1/3}\left(\frac{\pi}{\Omega}\right)^{1/3}, \tag{46}$$

where the mean stellar density is $\langle\rho\rangle = 3\rho_\star/4\pi$ and the mass ratio is defined as $q = M_{\mathrm{BH}}/M_*$. Assuming radial expansion into a constant solid angle equal to that under which the star is seen by the black hole $\Omega \approx \pi(R_\star/R_{\mathrm{t}})^2 \equiv \pi\theta_\star^2 \approx \pi q^{-2/3}$, we get a stopping length of ≈ 70 pc for the mean density of the Sun ($\langle\rho\rangle = 1.4\ \mathrm{g\,cm^{-3}}$) and the canonical ISM mean density (1 particle $\mathrm{cm^{-3}}$).[18]

In fact, debris with positive energy does not move radially outward but on hyperbolic orbits with a range of energies. This causes the streams to spread in the orbital plane and trace a "fan" shape. The sweeping area is therefore larger than in the radial case and the stopping length consequently smaller. We can simply estimate it by assuming that the black hole's gravity is the only force in place and ignoring relativistic effects. The true anomaly θ_∞ of the stream's orbits at $r \gg R_{\mathrm{t}}$ obeys $\cos(\theta_\infty) = -1/e$, where e is the orbital eccentricity. The maximum eccentricity of the unbound streams, corresponding to a specific energy $+\Delta E$, is $e_{\max} = 1 + 2q^{-1/3}$, therefore $\theta_\infty = \arccos(-1/(1+2q^{-1/3})) \approx \pi - \sqrt{2}q^{-1/6}$. On the other hand, the funnel of debris should continuously connect to material moving on a parabolic orbit with $e = 1$ (and $\theta_\infty = \pi$) and the overall opening angle in the orbital plane is roughly $\theta_\phi \approx \sqrt{2}q^{-1/6}$. The cross-sectional area of the "fan" is therefore an ellips and its solid angle can be estimated as $\Omega \approx \pi\theta_\phi\theta_*$. The increase in the sweeping area by $\theta_\phi/\theta_* = \sqrt{2}q^{1/6} \approx 20$ gives a shorter stopping length of a few parsecs for our fiducial parameters. So the expectation is that unbound debris streams deposit $\approx 10^{50}$ erg at and beyond the black hole's sphere of influence after $R_{\mathrm{st}}/v_{\max} \gtrsim 10^3$ yr.

This approximate picture should, however, be verified in more realistic conditions. Indeed, the above estimates neglect several physical ingredients. First of all, the shape of the unbound debris "fan" should be reconsidered to include the effect of self-gravity within the debris, which may be important in both the shallow-penetration and deep-penetration regimes (Kochanek 1994; Coughlin and Nixon 2015; Coughlin et al. 2016; Steinberg et al. 2019). The self-confinement of most of the mass in the stream may lead to a lower sweeping area and therefore to a longer deceleration scale with respect to the ballistic case. The situation may be most favourable in highly penetrating events, when an order unity fraction of the unbound debris stream remains unconstrained by self-gravity (Steinberg et al. 2019; Yalinewich et al. 2019). However, self-gravity may not play an important role in setting the deceleration scale, if suppressed at an earlier stage by the energy released in the streams from hydrogen recombination weeks to months after disruption (Kasen and Ramirez-Ruiz 2010; Guillochon et al. 2016). In addition, the shape of the energy distribution within the debris is not truly flat, but instead it depends on both the stellar internal structure and the penetration factor (see Sects. 2.2 and 3). Finally, partial disruptions would see less massive unbound ejecta, which would result in both a shorter free-expansion phase (when $r < R_{\mathrm{st}}$) and, on the other hand, less energy deposited upon deceleration.

[18] This is also the distance at which the mean density within the debris equal that of the surrounding medium in this free expanding scenario.

Other physical ingredients to reconsider pertain to the environment. The estimated stopping radius is of the same order of magnitude of the black hole sphere of influence (e.g. $\sim 1-3$ pc for Sgr A*) and therefore the stellar gravitational potential should be also be taken into account along with the debris evolution. The density profile in the sphere of influence of a SMBH is far from constant, given the deep potential well, and may also be aspherical if the SMBH harbours an accretion disc.

When analytically accounting for all of the above effects – modelling both the free expansion and the deceleration phases in a Galactic Centre type of environment – Guillochon et al. (2016) found that the typical energy and momentum deposited in the ISM are 5×10^{49} ergs and 2×10^{41} cm g s^{-1} for a MS star and an order of magnitude less for a giant star disruption. The typical stopping length is around $R_{\rm st} \sim 20$ pc for a MS star and nearly 10 times smaller for giants, while the transverse size $R_{\rm st}\theta_\phi$ is around a few parsecs. The free-expansion, mostly adiabatic phase typically lasts $\sim 10^4$ yr.

The energy and momentum deposited by the unbound debris streams can give rise to observational signatures ("unbound debris remnants"), such as X-ray and radio emission by the shocked interstellar medium, analogously to supernova remnants (Guillochon et al. 2016; Yalinewich et al. 2019; Krolik et al. 2016). The long energy deposition timescale of several thousand years implies a peak bolometric luminosity of only $\lesssim 10^{40}$ erg s^{-1}, limiting the prospects for an extragalactic detection. With radio observations of our Galactic Centre, however one may constrain the TDE rates by detecting unbound debris remnants. Guillochon et al. (2016) argue that a few such remnants should be present in our Galactic Centre within the innermost ~ 100 pc and that Sgr A East is in fact one of them. Beside the aforementioned "supernova-like" emission, other observational signatures include the optical flare from hydrogen recombination (Kasen and Ramirez-Ruiz 2010) within the unbound debris streams, and optical emission lines from the reprocessing of light coming from the accreted bound material (Strubbe and Quataert 2009). A detailed description of all these emission processes can be found in the Emission Mechanisms Chapter in this book.

8 Variations on the Classical Tune

In this section, we review works investigating variations of the classical study case explored in the preceding sections, where a single star on a nearly parabolic orbit is fully or partially tidally disrupted during a single pericenter passage. In particular, we will focus on the physical implications of elliptic and circular orbits (Sects. 8.1 and 8.2), of recursive partial disruption episodes for the same star (Sect. 8.3) and of stellar binarity (Sect. 8.4). The conditions for these different scenarios are also discussed.

8.1 Disruption of Stars on Eccentric Orbits

In the standard loss cone theory, where only scatterings between stars (in vacuum) around a SMBH are considered, a star typically enters the loss cone with marginal binding energy and therefore is disrupted on a nearly parabolic orbit, with an orbital eccentricity $1-e \sim 10^{-6}$. However, there are other channels which can supply stars to be disrupted from more tightly bound orbits. When does a finite value of $1-e$ begin to affect the dynamics of disruption? We can determine the answer to this question by comparing two specific energy scales: the frozen-in spread of debris specific energy, ΔE (see Eq. (22)), and the pre-disruption specific orbital energy of the star, $|E_\star| \approx \frac{1}{2}GM_{\rm BH}/a_0$, where a_0 is the semimajor axis of the star's final orbit. Normally, $|E_\star| \ll \Delta E$ and the orbit can be treated as effectively parabolic, but below a critical orbital eccentricity,

$$e_{\rm crit} \approx 1 - 2/\beta \times (m_\star/M_{\rm BH})^{1/3}, \tag{47}$$

$|E_\star| \gtrsim \Delta E$ (Hayasaki et al. 2013), and we are in the qualitatively different regime of an "eccentric" tidal disruption.[19] For example, $e_{\rm crit} = 0.98$ for a solar-type star disrupted by a $10^6\ M_\odot$ SMBH along an orbit with $\beta = 1$.

In an eccentric TDE, the debris fallback timescale decreases from years ($e = 1$) to weeks ($e = 0.99$), days ($e = 0.9$), or even hours or minutes ($e < 0.7$), leading to a much higher peak fallback rate and a time evolution dramatically deviating from the canonical $t^{-5/3}$ pattern (Hayasaki et al. 2013; Dai et al. 2013b). Eccentric TDEs feature steeper late-time slopes because the shift in the $\mathrm{d}M/\mathrm{d}E$ distribution means it is no longer centered on $E = 0$, and one cannot assume $\mathrm{d}M/\mathrm{d}E \to 0$ at late times (Hayasaki et al. 2018; Park and Hayasaki 2020). Therefore, it is expected that flares produced in eccentric TDEs should evolve on timescales faster than years. The accretion level should largely exceed that of standard parabolic TDEs, though the luminosity may still be regulated by the Eddington limit. Also, when the eccentricity of the initial stellar orbit is sufficiently small (less than $e_{\rm crit}$), all the debris will be bound so all of the stellar mass can, in principle, accrete onto the SMBH.

N-body simulations directly modeling stellar clusters around SMBHs generally confirm the semi-analytic expectation that most stars are disrupted from effectively parabolic orbits, if the main source of loss cone refueling is stellar two-body relaxation (Zhong et al. 2014; Hayasaki et al. 2018). However, if there exists a massive perturber deep inside the SMBH influence radius, such as an intermediate mass black hole, or if the TDE happens in a tight binary SMBH system, then one encounter between the perturber and a very tightly-bound star (one with specific orbital energy $|E_\star| \gtrsim \Delta E$) can remove enough angular momentum from the star to create an eccentric TDE. Such encounters occur naturally during the final stages of a SMBH binary inspiral (Chen et al. 2011), and the rate may be enhanced by trapping of stars in mean-motion resonances (Seto and Muto 2010, 2011). A final burst of such TDEs may occur following the merger of the SMBH with a secondary massive black hole, when anisotropic gravitational wave emission provides a recoil kick that tilts the phase space loss cone to overlap with the orbits of surviving, tightly bound stars (Stone and Loeb 2011). Eccentric TDEs may also be produced following the tidal separation of a binary stellar system by a SMBH. This process places one star on a tightly – although initially highly eccentric – bound orbit (Sari et al. 2010), while ejecting the other star on a hyperbolic (and hypervelocity) orbit. The bound star detached from its binary companion will have a much larger $|E_\star|$ than is typical in the stellar cusp, and a cluster of such stars will be more favorable for generating eccentric TDEs (Amaro-Seoane et al. 2012).

Disruption of stars on eccentric orbits have also been simulated for understanding the disk formation process in TDEs (e.g., Bonnerot et al. 2016b; Hayasaki et al. 2016). Due to the reduced dynamical spread of the debris orbital energy, the normally prohibitive computational expense of simulating the long-term aftermath of tidal disruption by SMBHs is greatly reduced (in comparison to the more astrophysically common case of a parabolic orbit TDE). In the eccentric TDE limit, the hydrodynamical dissipation or nozzle shocks associated with compression and shearing at the pericenter will be less energetically important compared

[19] For a similar discussion, impacting the amount of unbound debris, see beginning of Sect. 7. We also note that an analogous criterion to Eq. (47) can be formulated to differentiate standard from "hyperbolic" tidal disruptions, by replacing the $-$ with a $+$ (Hayasaki et al. 2018). Hyperbolic TDEs would have unusually low luminosities and energy release, if any debris at all is bound.

to standard parabolic TDEs. On the other hand, for eccentric TDEs the self-intersection of the debris stream induced by GR apsidal precession will always happen close to the SMBH, which promotes prompt disk formation, although the delays induced by the Lense-Thirring effect have not yet been thoroughly studied (Dai et al. 2013b, Hayasaki et al. 2016 – but see Liptai et al. 2019). We refer the readers to the Formation of the Accretion Flow Chapter in this book for more details.

8.2 Roche-Lobe Overflow of Stars on Nearly Circular Orbits

Under certain circumstances a star can approach a massive black hole on a close orbit and form an extreme-mass-ratio-inspiral (EMRI) system close to the SMBH (Miller et al. 2005; Amaro-Seoane et al. 2012). The orbital radius and eccentricity of the stellar orbit can both quickly decrease if gravitational radiation is sufficiently efficient, until the star eventually fills its Roche lobe on a nearly circular orbit. The star will then keep orbiting the SMBH for a long time while its envelope is steadily stripped by the tidal force from the black hole. The stripped stellar material will then accrete onto the black hole producing quasi-periodic X-ray signals.

Similar stable mass transfer processes between a star and a black hole via Roche lobe overflow have been extensively studied in the context of normal X-ray binaries (e.g., Rappaport et al. 1982; Webbink et al. 1983). The effective radius R_L of the Roche lobe of the star has been calculated in Kopal (1959), Paczyński (1971), Eggleton (1983), and is the same as the parabolic-encounter tidal radius R_t up to a factor of ≈ 2. Although much of the theoretical formalism for X-ray binaries can be adopted to study the mass-transfer between a star and a SMBH, the latter scenario is distinct in having a much more relativistic R_L (i.e. R_L is a few or a few tens of SMBH gravitational radii) due to the extreme mass ratio. Therefore, the dynamics and the evolution of the star–SMBH binary is more dramatically affected by general relativity as compared to the case of normal X-ray binaries.

Stable Roche-lobe overflow in a star-SMBH binary was first explored by King and Done (1993) and Hameury et al. (1994). Later, Dai and Blandford (2013) did a more rigorous calculation including general relativistic effects. Since the mass transfer happens on a timescale that is faster than the thermal (Kelvin–Helmholtz) timescale of the star but slower than the dynamical timescale, the interior structure of the star evolves adiabatically as its envelope is stripped (Dai et al. 2013a). The star fills its Roche lobe throughout this process, so from Kepler's law we see that its orbital period $T \propto R_t^{3/2}\, M_\mathrm{BH}^{-1/2} \propto \langle\rho\rangle^{-1/2}$, where $\langle\rho\rangle$ is the mean density of the star being tidally stripped. Therefore, the evolution of the stellar orbit is controlled by how the stellar mean density changes: low-mass stars will recede from the SMBH during Roche-lobe overflow, while high-mass stars will first continue to spiral in for some time after reaching the Roche limit, and later will spiral out.

If mass transfer is conservative (i.e. the angular momentum of the binary only changes through the torque of gravitational radiation), one can calculate the mass transfer rate to be $\sim 10^{21-23}\ \mathrm{g\,s^{-1}}$, depending on the exact black hole and stellar masses (Dai and Blandford 2013). The stripped stellar material will form an accretion disk around the SMBH, and the stream of the stripped stellar material flowing through the inner Lagrange point L1 can continuously hit this small accretion disk and produce a hot spot. The hot spot will then orbit the black hole at the orbital frequency of the star and produce quasi-periodic signals, likely in X-rays. Linial and Sari (2017) further consider the mass leakage through the outer Lagrange point L2, and analytically estimate the conditions in which it is comparable to the canonical mass transfer rate through the first Lagrange point L1, discussed so far.

Since a Roche-lobe filling star spends most of the time (at least hundreds of thousands of years) receding from the SMBH, collisions can happen between one star in the slow outward

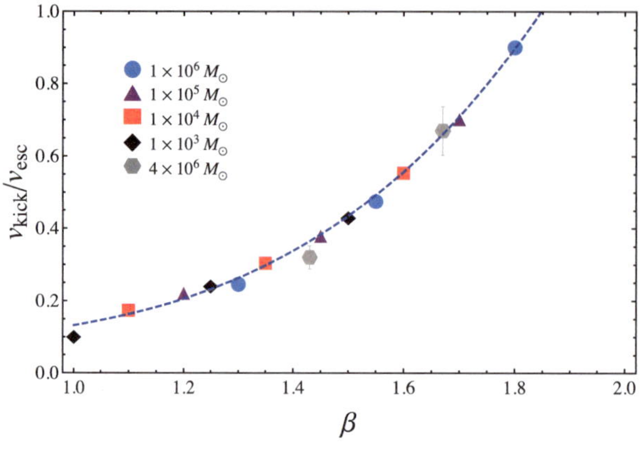

Fig. 10 The kick velocity $v_{\rm kick}$ delivered to the surviving core after partial disruption over the escape speed from the core $v_{\rm esc}$ as a function of the penetration factor β. The blue dashed line is an analytical fit to the simulation results: the kick velocity is found to be dependent on β but largely independent of the black hole to stellar mass ratio, probed here in the range $10^3 - 10^6$ (see legend). This figure is from Manukian et al. (2013) – their Fig. 3 – where a complete description of the hydrodynamical simulations that produced these results can be found

migration and another star that enters the inspiral phase at a later time. Metzger and Stone (2017) considered such colliding EMRIs and found that the timescale and the luminosity of the flare produced during the collision are similar to those of TDEs. Therefore, such EMRI collisions can serve as TDE imposters if the rate is high enough.

Gravitational waves emitted by star–SMBH Roche-lobe overflow systems would have a strain given by Eq. (40) and frequency twice the orbital one, $f_{\rm GW} = 2/T$ (Linial and Sari 2017). Contrary to classical single TDEs, these events will not be bursts of gravitational radiation but rather persistent sources, and this fact might help detection by allowing to accumulate the signal over many orbits observed with a spaced based interferometer. This leads to a signal-to-noise enhancement by a factor of $\approx \sqrt{T_{\rm obs} f_{\rm GW}}$, which can reach ~ 100 for a $T_{\rm obs} = 4$ yr mission duration (e.g. the nominal mission duration of LISA). The emission frequency for MS stars remains is rather low, $\sim 10^{-4}$ Hz. Therefore – conditional on the actual performances of LISA in the lower-frequency portion of its spectrum – the gravitational radiation produced by these systems may be detectable from the Galactic Center or from nearby galaxies, for example Andromeda.

8.3 Multiple Encounters

In certain scenarios a star may have multiple encounters with a black hole and thus experience multiple (partial) disruptions. For this to take place in practice, several conditions must be met. On the first passage, the star must be only partially disrupted, and remain in a bound orbit. This implies that any kick imparted to the core by mass-loss asymmetry (Manukian et al. 2013; Gafton et al. 2015) must be small enough that the surviving core stays bound to the black hole (see Fig. 10). The star cannot be scattered away from its disruptive orbit by dynamical processes like two-body gravitational interactions. Finally, the mass–radius relationship of the star must be such that the star becomes increasingly tidally vulnerable (less dense) following mass loss, rather than more dense.

An argument long presented in favor of multiple encounters is the possible process of capture into eccentric orbits by tidal dissipation (Fabian et al. 1975; Press and Teukolsky

1977; Lee and Ostriker 1986). The specific case of tidal capture by SMBHs has been examined by Novikov et al. (1992), Kosovichev and Novikov (1992), Diener et al. (1995) in the context of the affine model. Baumgardt et al. (2006) analyzed the related problem of tidal captures by intermediate-mass black holes in N-body dynamical models. A key point in these analyses – the spatial distribution of tidal dissipation within the star – was raised by Podsiadlowski (1996): the supplied energy can be quickly radiated away if deposited in the stellar atmosphere, while it does work and causes stellar expansion if deposited in deeper layers. In this latter scenario, the star is increasingly susceptible to tidal interaction. These details of tidal dissipation remain highly uncertain areas of active discussion, particularly in the context of possible tidal captures of Hot Jupiters by their host stars (e.g. Wu 2018).

In the limit when $R_{\rm p}$ remains too large for even partial mass loss to occur, Alexander and Morris (2003) predict the existence of "squeezars": stars on highly eccentric orbits around a SMBH, with luminosities approaching the star's Eddington limit, that are powered by tidal interactions with the black hole. These stars undergo orbital decay from tidal heating, and the inspiral terminates in a tidal disruption once the star's orbital energy exceeds its own binding energy. Squeezar formation is estimated to occur at $\sim 5\%$ of the TDE rate, but with lifetimes of $\sim 10^5$ yr, a mean number of 0.1–1 squeezars are likely orbiting our Galactic Center SMBH.

While numerous uncertainties remain, there are several situations in which conditions for repeated tidal-stripping encounters might be met. MS stars, especially low-mass stars with isentropic structures, expand upon losing mass. If these are in a sufficiently bound orbit that they are unlikely to be scattered by gravitational interactions, the star will expand following each tidal mass loss episode and will undergo runaway disruption over several orbits. Guillochon et al. (2011) discuss and simulate a very similar scenario with particular focus on the slightly different science case of giant planets scattered close to their host stars.

Because giant stars are never completely disrupted in a single passage (see e.g. §5.2), they have the opportunity to return for subsequent encounters, making them strong candidates for multiple disruptive passages. Furthermore, because the loss cone angular momenta of giant stars are larger than those of their more-compact companions (for a given semi-major axis), these objects experience less scatter in β due to two-body relaxation over the course of an orbit. This implies that they are more likely to return for a subsequent passage. The extreme limit of this occurs when the orbit of a giant star is evolving so slowly that their radial growth (from stellar evolution) controls the time evolution of $R_{\rm p}/R_{\rm t}(t)$ (Syer and Ulmer 1999; MacLeod et al. 2013; Merritt 2013). Through hydrodynamic simulations and stellar evolution modeling, MacLeod et al. (2013) find that giant stars on such orbits around a SMBH produce low-level flares that repeat on the orbital timescale. This gradual, piece-by-piece feeding of the SMBH is comparable to the feeding rate provided by stellar winds from all stars in the surrounding nuclear star cluster. We further note that the dynamics of these encounters are the high-mass ratio limit of the more-thoroughly considered scenario of Roche lobe overflow in eccentric stellar binaries (e.g. Matese and Whitmire 1983; Hamers and Dosopoulou 2019).

Finally, WDs may likewise undergo multiple passages while in an eccentric orbit (Zalamea et al. 2010; MacLeod et al. 2014). These systems can be captured or decay through a combination of gravitational radiation and tidal interaction (Vick et al. 2017). A more complete discussion of the possible phenomenology and implications for multi-messenger detection is given in the White Dwarf Chapter.

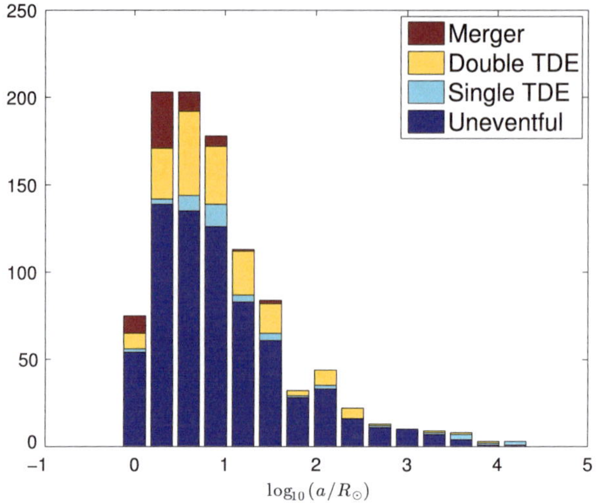

Fig. 11 The distribution – as a function of the binary semi-major axis a – of the outcomes of 1000 numerical simulations of a stellar binary encounter with an SMBH performed by Mandel and Levin (2015) in the full loss cone regime. From the top down, the outcomes are: a merger of the binary components, a double tidal disruption, a single tidal disruption, and an uneventful fly-by. Tidal disruptions of both stars become prominent for initial binary separations smaller than $\sim 100\ R_\odot$, while mergers are common only for binary separations smaller than $\sim 10\ R_\odot$. Overall among these 1000 trials, 18% produced sequential tidal disruptions of both stars, 5% resulted in single TDEs with typically the more massive star being disrupted, and binary components merged in 6% of simulations before either was disrupted. This is Fig. 2, lower panel in Mandel and Levin (2015)

8.4 Stellar Binary TDEs

When two stars in a binary of semimajor axis a and total mass M_b approach a massive black hole to within the tidal separation radius[20]

$$R_\mathrm{ts} \approx a \left(\frac{M_\mathrm{BH}}{M_\mathrm{b}}\right)^{1/3} , \tag{48}$$

several outcomes are possible: (i) both stars may be sequentially disrupted by the black hole; (ii) the stars may merge, with the merger product possibly being tidally disrupted by the black hole; (iii) one star may be gravitationally captured by the black hole (possibly to be subsequently tidally disrupted, in a TDE that may have observably non-parabolic features) with the other ejected as a hypervelocity star (Mandel and Levin 2015; Bradnick et al. 2017; Bonnerot and Rossi 2019). All three outcomes are potentially informative, with unique signatures and opportunities to extract information about nuclear cluster dynamics, tidal disruption events, and stellar evolution.

The loss cone for tidal separation is characterised by a specific angular momentum of

$$l_\mathrm{ts} \approx \sqrt{2GM_\mathrm{BH}a} \left(\frac{M_\mathrm{BH}}{M_\mathrm{b}}\right)^{1/6} . \tag{49}$$

[20] Note that this is identical to Eq. (1), but with $R_\star$ and M_* replaced by a and M_b respectively, so that binary tidal separation occurs when the binary is at least a few stellar tidal disruption radii away from the SMBH.

The stellar binary will change its orbital angular momentum over time due to relaxational interactions far from the black hole. If the binary's typical per-orbit change in angular momentum Δl is large, i.e. $\Delta l \gtrsim l_{\rm ts}$, the binary is in the full loss cone regime, meaning that the entire space of possible angular momenta is sampled. In this case, it is possible that the binary plunges toward the black hole on a trajectory with a small impact parameter, coming within the companion stars' individual tidal disruption radii. This leads to a sequential tidal disruption of two stars, outcome (i), which could produce an unusual double-peaked light-curve or at least enhance interactions between tidal streams (Mandel and Levin 2015). In fact, Bonnerot and Rossi (2019) analytically estimate that tidal streams from the two stellar disruptions following the tidal separation of a binary will collide in nearly half of such encounters. The burst of radiation associated with shock heating from these stream collisions may yield a detectable precursor to the TDE light curve (Bonnerot and Rossi 2019).

Full tidal separation loss cone encounters occasionally lead to outcome (ii), stellar mergers (Mandel and Levin 2015). Figure 11 shows quantitatively that smaller initial binary separations are more conducive to mergers and double TDEs. Mergers are even more likely as outcomes of empty loss cone encounters, in which the angular momentum is only gradually perturbed during each orbit, $\Delta l \ll l_{\rm ts}$. In this case, the binary passes by the massive black hole on a very eccentric orbit, and is tidally perturbed over many periapsis passages with $R_{\rm p} \sim \text{few} \times R_{\rm ts}$. Such encounters barely affect the semi-major axis (energy) of the stellar binary, but the exchange of angular momentum between the stellar binary and its orbit around the black hole can significantly drive up the binary's eccentricity (Bradnick et al. 2017). Stellar tides can then decrease the internal binary separation while the binary is far from the black hole. A sequence of eccentricity excitations and tidal damping further increases the likelihood of merger as a possible outcome, particularly for initially close binaries (Bradnick et al. 2017). The likely outcome of an interaction is sensitive to the initial properties of the binary (masses and separation) and the efficiency of binary tides in circularising the orbit. Meanwhile, a key environmental factor is the efficiency of relaxation of the orbit of the binary around the massive black hole.

Some of the merger products may subsequently approach the massive black hole closely enough to be tidally disrupted. A recent merger remnant could have a much larger magnetic field than would be typical for a single star; the subsequent disruption of such a magnetised object may make it easier to produce jets associated with some TDEs such as Swift J164449.3+573451 (Mandel and Levin 2015; Bradnick et al. 2017).

Meanwhile, the tidal separation of a binary, outcome (iii), generally leads to the rapid ejection of one of the companions. By analogy with Eq. (23), the velocity of the ejected star with mass $M_\star \sim M_{\rm b}/2$ is of order

$$v_{\rm ej} \sim \left(\frac{M_{\rm BH}}{M_{\rm b}}\right)^{1/6} v_{\rm bin} \,, \tag{50}$$

where $v_{\rm bin} \sim \sqrt{GM_{\rm b}/a}$ is the binary's orbital velocity, of order $100\ {\rm km\,s^{-1}}$ for a compact binary (two solar-type stars at ten solar radii). For a $10^6\ M_\odot$ black hole, the ejection velocity is then of order $v_{\rm ej,\,hvs} \sim 1000\ {\rm km\,s^{-1}}$: a hypervelocity star. Rossi et al. (2014, 2017) found that the observed population of hypervelocity stars is missing a high-velocity tail expected from models. Bradnick et al. (2017) conjectured that mergers of the most compact, high orbital velocity binaries could be responsible for the observed dearth of particularly rapid ejections.

9 Conclusions

The process of stellar tidal disruption has been studied since the pioneering work of Hills (1975). In the almost half-century of theoretical work since then, our understanding of the central features of the disruption process has largely converged. The position-dependent tidal field felt by an extended object can be written in closed form, both in Newtonian gravity and also in the general relativistic gravity of the Schwarzschild and Kerr metrics (Eqs. (4) and (7)). These tidal fields can then be used to build analytic, semi-analytic, or fully numerical models for the dynamics of tidal disruption. Because the full process of disruption is a nonlinear hydrodynamics problem, a numerical simulation – if/when adequate resolution can be achieved – provides the most precise answers, but a variety of simpler models provide useful physical intuition and some degree of accuracy.

We have reviewed these models in §2. The simplest ones can be constructed by assuming impulsive disruption of the star at the moment it enters the tidal radius, with the star's fluid elements "freezing in" to ballistic, or geodesic, trajectories in the aftermath of this disruption; this approximation can be used to make predictions about the mass fallback rate in a fully analytic way. Increasing accuracy comes from applying the free solutions to the parabolic Hill equations (Eq. (33)), which also allows limited study of the process of tidal compression; or from adding an integral equation to account for the internal structure of the star (Eq. (28)). The semi-analytic "affine model" provides significantly greater physical realism by using the tensor virial theorem (and simplifying geometrical assumptions) to model the hydrodynamics of the disruption process, rather than neglecting hydrodynamics entirely as in the impulse approximation. The affine model has been used extensively to study the tidal compression suffered by stars undergoing high-β TDEs, and makes surprisingly accurate predictions for mass lost in partial tidal disruptions.

We have also surveyed the extensive literature of numerical (magneto)-hydrodynamics simulations of TDEs, the results of which are presented in §3, §4, and §5. At this point, the tidal disruption of non-rotating polytropic stars in Newtonian gravity is a largely solved problem: at least two large parameter studies employing different numerical techniques (Guillochon and Ramirez-Ruiz 2013; Mainetti et al. 2017) are converged on the outcomes of both full and partial disruptions under these assumptions. However, there are many active frontiers of numerical disruption simulations that have not yet fully explored the parameter space of real TDEs. While tidal disruption in general relativistic gravity has been simulated since the work of Diener et al. (1997), broad parameter surveys have only recently appeared (Ryu et al. 2020c), and comparatively few simulations have been done in the Kerr (rather than Schwarzschild) metric. A full understanding of the energetics and dynamics of a star disrupted on an inclined trajectory with respect to the SMBH's spin is for instance missing. Likewise, it is only very recently that a number of groups (Golightly et al. 2019a; Law-Smith et al. 2019; Goicovic et al. 2019; Ryu et al. 2020a) have begun simulating the disruption of stars generated with fully realistic internal structure (usually employing `MESA` models). Pursuing these studies further would improve our understanding of how detectable spectral lines – their strengths and temporal behaviour – can be used to explore the disrupted star and gas thermodynamics during the time between disruption and spectral line observations. Likewise, recent – and few – investigations of the disruption of rotating (Golightly et al. 2019b; Sacchi and Lodato 2019) or magnetised (Guillochon and McCourt 2017; Bonnerot et al. 2017) stars offer leads for more in depth analysis. The rapid progress of the last two years seems promising for our future understanding of astrophysically realistic TDEs.

One of the original motivations for studying TDEs relates to the exotic phenomena that may occur during extreme tidal compression in high-β events, as we discuss in §6. The onset of runaway nuclear fusion (particularly in WD TDEs), the emission of gravitational waves, or the production of a bright shock breakout signal remain somewhat speculative possibilities, largely due to the computational challenge of resolving severe compression in high-β TDEs. If signals like these exist, however, they would pin down the precise moment of disruption, which might help to resolve open questions related to accretion flow formation. Looking into the future, observation of GW signals are particularly promising, but little work has been done so far in exploring the scientific gain from the synergy between gravitational-wave and electromagnetic observations.

We further discuss in §7 the dynamics and possible signatures of the dynamically unbound half of the star; the current assessment is that its observability remains elusive. Even if more exotic scenarios discussed in §8, such as repeated disruptions, binary disruptions, $e < 1$ TDEs, or quasi-circular Roche-lobe overflow are intrinsically rare or hard to observe, it seems plausible that the flood of data from upcoming time domain surveys may include some of these variations on more typical TDE flares.

Acknowledgements IM is a recipient of the Australian Research Council Future Fellowship FT190100574.

References

T. Alexander, M. Morris, Squeezars: tidally powered stars orbiting a massive black hole. Astrophys. J. **590**(1), 25–28 (2003). https://doi.org/10.1086/376671

P. Amaro-Seoane, M.C. Miller, G.F. Kennedy, Tidal disruptions of separated binaries in galactic nuclei. Mon. Not. R. Astron. Soc. **425**(4), 2401–2406 (2012). https://doi.org/10.1111/j.1365-2966.2012.21162.x

P. Amaro-Seoane, H. Audley, S. Babak, J. Baker, E. Barausse, P. Bender, E. Berti, P. Binetruy, M. Born, D. Bortoluzzi, J. Camp, C. Caprini, V. Cardoso, M. Colpi, J. Conklin, N. Cornish, C. Cutler, K. Danzmann, R. Dolesi, L. Ferraioli, V. Ferroni, E. Fitzsimons, J. Gair, L. Gesa Bote, D. Giardini, F. Gibert, C. Grimani, H. Halloin, G. Heinzel, T. Hertog, M. Hewitson, K. Holley-Bockelmann, D. Hollington, M. Hueller, H. Inchauspe, P. Jetzer, N. Karnesis, C. Killow, A. Klein, B. Klipstein, N. Korsakova, S.L. Larson, J. Livas, I. Lloro, N. Man, D. Mance, J. Martino, I. Mateos, K. McKenzie, S.T. McWilliams, C. Miller, G. Mueller, G. Nardini, G. Nelemans, M. Nofrarias, A. Petiteau, P. Pivato, E. Plagnol, E. Porter, J. Reiche, D. Robertson, N. Robertson, E. Rossi, G. Russano, B. Schutz, A. Sesana, D. Shoemaker, J. Slutsky, C.F. Sopuerta, T. Sumner, N. Tamanini, I. Thorpe, M. Troebs, M. Vallisneri, A. Vecchio, D. Vetrugno, S. Vitale, M. Volonteri, G. Wanner, H. Ward, P. Wass, W. Weber, J. Ziemer, P. Zweifel, Laser Interferometer Space Antenna. arXiv e-prints (2017)

P. Anninos, P.C. Fragile, S.S. Olivier, R. Hoffman, B. Mishra, K. Camarda, Relativistic tidal disruption and nuclear ignition of white dwarf stars by intermediate-mass black holes. Astrophys. J. **865**(1), 3 (2018). https://doi.org/10.3847/1538-4357/aadad9

J. Baker, S.F. Barke, P.L. Bender, E. Berti, R. Caldwell, J.W. Conklin, N. Cornish, E.C. Ferrara, K. Holley-Bockelmann, B. Kamai, S.L. Larson, J. Livas, S.T. McWilliams, G. Mueller, P. Natarajan, N. Rioux, S.R. Sankar, J. Schnittman, D. Shoemaker, J. Slutsky, R. Stebbins, I. Thorpe, J. Ziemer, Space based gravitational wave astronomy beyond LISA. arXiv e-prints (2019)

J.M. Bardeen, W.H. Press, S.A. Teukolsky, Rotating black holes: locally nonrotating frames, energy extraction, and scalar synchrotron radiation. Astrophys. J. **178**, 347–370 (1972). https://doi.org/10.1086/151796

H. Baumgardt, C. Hopman, S. Portegies Zwart, J. Makino, Tidal capture of stars by intermediate-mass black holes. Mon. Not. R. Astron. Soc. **372**(1), 467–478 (2006). https://doi.org/10.1111/j.1365-2966.2006.10885.x

A.M. Beloborodov, A.F. Illarionov, P.B. Ivanov, A.G. Polnarev, Angular momentum of a supermassive black hole in a dense star cluster. Mon. Not. R. Astron. Soc. **259**, 209–217 (1992). https://doi.org/10.1093/mnras/259.2.209

T. Bogdanović, R.M. Cheng, P. Amaro-Seoane, Disruption of a red giant star by a supermassive black hole and the case of PS1-10jh. Astrophys. J. **788**(2), 99 (2014). https://doi.org/10.1088/0004-637X/788/2/99

C. Bonnerot, E.M. Rossi, Streams collision as possible precursor of double tidal disruption events. Mon. Not. R. Astron. Soc. **484**, 1301–1316 (2019). https://doi.org/10.1093/mnras/stz062

C. Bonnerot, E.M. Rossi, G. Lodato, Bad prospects for the detection of giant stars' tidal disruption: effect of the ambient medium on bound debris. Mon. Not. R. Astron. Soc. **458**(3), 3324–3330 (2016a). https://doi.org/10.1093/mnras/stw486

C. Bonnerot, E.M. Rossi, G. Lodato, D.J. Price, Disc formation from tidal disruptions of stars on eccentric orbits by Schwarzschild black holes. Mon. Not. R. Astron. Soc. **455**(2), 2253–2266 (2016b). https://doi.org/10.1093/mnras/stv2411

C. Bonnerot, D.J. Price, G. Lodato, E.M. Rossi, Magnetic field evolution in tidal disruption events. Mon. Not. R. Astron. Soc. **469**(4), 4879–4888 (2017). https://doi.org/10.1093/mnras/stx1210

B. Bradnick, I. Mandel, Y. Levin, Stellar binaries in galactic nuclei: tidally stimulated mergers followed by tidal disruptions. Mon. Not. R. Astron. Soc. **469**, 2042–2048 (2017). https://doi.org/10.1093/mnras/stx1007

M. Brassart, J.-P. Luminet, Shock waves in tidally compressed stars by massive black holes. Astron. Astrophys. **481**(2), 259–277 (2008). https://doi.org/10.1051/0004-6361:20078264

M. Brassart, J.-P. Luminet, Relativistic tidal compressions of a star by a massive black hole. Astron. Astrophys. **511**, 80 (2010). https://doi.org/10.1051/0004-6361/200913442

B. Carter, J.P. Luminet, Pancake detonation of stars by black holes in Galactic nuclei. Nature **296**(5854), 211–214 (1982). https://doi.org/10.1038/296211a0

B. Carter, J.-P. Luminet, Tidal compression of a star by a large black hole. I. Mechanical evolution and nuclear energy release by proton capture. Astron. Astrophys. **121**, 97–113 (1983)

B. Carter, J.P. Luminet, Mechanics of the affine star model. Mon. Not. R. Astron. Soc. **212**, 23–55 (1985). https://doi.org/10.1093/mnras/212.1.23

X. Chen, A. Sesana, P. Madau, F.K. Liu, Tidal stellar disruptions by massive black hole pairs. II. Decaying binaries. Astrophys. J. **729**, 13 (2011). https://doi.org/10.1088/0004-637X/729/1/13

R.M. Cheng, T. Bogdanović, Tidal disruption of a star in the Schwarzschild spacetime: relativistic effects in the return rate of debris. Phys. Rev. D **90**(6), 064020 (2014). https://doi.org/10.1103/PhysRevD.90.064020

R.M. Cheng, C.R. Evans, Relativistic effects in the tidal interaction between a white dwarf and a massive black hole in Fermi normal coordinates. Phys. Rev. D **87**(10), 104010 (2013). https://doi.org/10.1103/PhysRevD.87.104010

E.R. Coughlin, P.J. Armitage, Tidal disruption by extreme mass ratio binaries and application to ASASSN-15lh. Mon. Not. R. Astron. Soc. **474**(3), 3857–3865 (2018). https://doi.org/10.1093/mnras/stx3039

E.R. Coughlin, C. Nixon, Variability in tidal disruption events: gravitationally unstable streams. Astrophys. J. **808**(1), 11 (2015). https://doi.org/10.1088/2041-8205/808/1/L11

E.R. Coughlin, C.J. Nixon, Partial stellar disruption by a supermassive black hole: is the light curve really proportional to $t^{-9/4}$? Astrophys. J. Lett. **883**(1), 17 (2019). https://doi.org/10.3847/2041-8213/ab412d

E.R. Coughlin, C. Nixon, M.C. Begelman, P.J. Armitage, D.J. Price, Post-periapsis pancakes: sustenance for self-gravity in tidal disruption events. Mon. Not. R. Astron. Soc. **455**(4), 3612–3627 (2016). https://doi.org/10.1093/mnras/stv2511

E.R. Coughlin, P.J. Armitage, C. Nixon, M.C. Begelman, Tidal disruption events from supermassive black hole binaries. Mon. Not. R. Astron. Soc. **465**(4), 3840–3864 (2017). https://doi.org/10.1093/mnras/stw2913

L. Dai, R. Blandford, Roche accretion of stars close to massive black holes. Mon. Not. R. Astron. Soc. **434**(4), 2948–2960 (2013). https://doi.org/10.1093/mnras/stt1209

L. Dai, R.D. Blandford, P.P. Eggleton, Adiabatic evolution of mass-losing stars. Mon. Not. R. Astron. Soc. **434**(4), 2940–2947 (2013a). https://doi.org/10.1093/mnras/stt1208

L. Dai, A. Escala, P. Coppi, The impact of bound stellar orbits and general relativity on the temporal behavior of tidal disruption flares. Astrophys. J. **775**(1), 9 (2013b). https://doi.org/10.1088/2041-8205/775/1/L9

P. Diener, A.G. Kosovichev, E.V. Kotok, I.D. Novikov, C.J. Pethick, Non-linear effects at tidal capture of stars by a massive black hole—II. Compressible affine models and tidal interaction after capture. Mon. Not. R. Astron. Soc. **275**(2), 498–506 (1995). https://doi.org/10.1093/mnras/275.2.498

P. Diener, V.P. Frolov, A.M. Khokhlov, I.D. Novikov, C.J. Pethick, Relativistic tidal interaction of stars with a rotating black hole. Astrophys. J. **479**(1), 164–178 (1997). https://doi.org/10.1086/303875
P.P. Eggleton, Approximations to the radii of Roche lobes. Astrophys. J. **268**, 368–369 (1983). https://doi.org/10.1086/160960
C.R. Evans, C.S. Kochanek, The tidal disruption of a star by a massive black hole. Astrophys. J. **346**, 13 (1989). https://doi.org/10.1086/185567
C. Evans, P. Laguna, M. Eracleous, Ultra-close encounters of stars with massive black holes: tidal disruption events with prompt hyperaccretion. Astrophys. J. **805**(2), 19 (2015). https://doi.org/10.1088/2041-8205/805/2/L19
A.C. Fabian, J.E. Pringle, M.J. Rees, Tidal capture formation of binary systems and X-ray sources in globular clusters. Mon. Not. R. Astron. Soc. **172**, 15 (1975). https://doi.org/10.1093/mnras/172.1.15P
L. Ferrarese, D. Merritt, A fundamental relation between supermassive black holes and their host galaxies. Astrophys. J. Lett. **539**(1), 9–12 (2000). https://doi.org/10.1086/312838
J. Frank, M.J. Rees, Effects of massive central black holes on dense stellar systems. Mon. Not. R. Astron. Soc. **176**, 633–647 (1976). https://doi.org/10.1093/mnras/176.3.633
B. Fryxell, K. Olson, P. Ricker, F.X. Timmes, M. Zingale, D.Q. Lamb, P. MacNeice, R. Rosner, J.W. Truran, H. Tufo, FLASH: an adaptive mesh hydrodynamics code for modeling astrophysical thermonuclear flashes. Astrophys. J. Suppl. Ser. **131**(1), 273–334 (2000). https://doi.org/10.1086/317361
E. Gafton, S. Rosswog, Tidal disruptions by rotating black holes: effects of spin and impact parameter. Mon. Not. R. Astron. Soc. , 1458 (2019). https://doi.org/10.1093/mnras/stz1530
E. Gafton, E. Tejeda, J. Guillochon, O. Korobkin, S. Rosswog, Relativistic effects on tidal disruption kicks of solitary stars. Mon. Not. R. Astron. Soc. **449**(1), 771–780 (2015). https://doi.org/10.1093/mnras/stv350
M. Gallegos-Garcia, J. Law-Smith, E. Ramirez-Ruiz, Tidal disruptions of main-sequence stars of varying mass and age: inferences from the composition of the fallback material. Astrophys. J. **857**(2), 109 (2018). https://doi.org/10.3847/1538-4357/aab5b8
I.Y. Georgiev, T. Böker, Nuclear star clusters in 228 spiral galaxies in the HST/WFPC2 archive: catalogue and comparison to other stellar systems. Mon. Not. R. Astron. Soc. **441**(4), 3570–3590 (2014). https://doi.org/10.1093/mnras/stu797
F.G. Goicovic, V. Springel, S.T. Ohlmann, R. Pakmor, Hydrodynamical moving-mesh simulations of the tidal disruption of stars by supermassive black holes. Mon. Not. R. Astron. Soc. **487**(1), 981–992 (2019). https://doi.org/10.1093/mnras/stz1368
E.C.A. Golightly, C.J. Nixon, E.R. Coughlin, On the diversity of fallback rates from tidal disruption events with accurate stellar structure. Astrophys. J. Lett. **882**(2), 26 (2019a). https://doi.org/10.3847/2041-8213/ab380d
E.C.A. Golightly, E.R. Coughlin, C.J. Nixon, Tidal disruption events: the role of stellar spin. Astrophys. J. **872**(2), 163 (2019b). https://doi.org/10.3847/1538-4357/aafd2f
J. Guillochon, M. McCourt, Simulations of magnetic fields in tidally disrupted stars. Astrophys. J. **834**(2), 19 (2017). https://doi.org/10.3847/2041-8213/834/2/L19
J. Guillochon, E. Ramirez-Ruiz, Hydrodynamical simulations to determine the feeding rate of black holes by the tidal disruption of stars: the importance of the impact parameter and stellar structure. Astrophys. J. **767**, 25 (2013). https://doi.org/10.1088/0004-637X/767/1/25
J. Guillochon, E. Ramirez-Ruiz, Hydrodynamical simulations to determine the feeding rate of black holes by the tidal disruption of stars: the importance of the impact parameter and stellar structure. Astrophys. J. **767**(1), 25 (2013). https://doi.org/10.1088/0004-637X/767/1/25
J. Guillochon, E. Ramirez-Ruiz, S. Rosswog, D. Kasen, Three-dimensional simulations of tidally disrupted solar-type stars and the observational signatures of shock breakout. Astrophys. J. **705**(1), 844–853 (2009). https://doi.org/10.1088/0004-637X/705/1/844
J. Guillochon, E. Ramirez-Ruiz, D. Lin, Consequences of the ejection and disruption of giant planets. Astrophys. J. **732**(2), 74 (2011). https://doi.org/10.1088/0004-637X/732/2/74
J. Guillochon, M. McCourt, X. Chen, M.D. Johnson, E. Berger, Unbound debris streams and remnants resulting from the tidal disruptions of stars by supermassive black holes. Astrophys. J. **822**, 48 (2016). https://doi.org/10.3847/0004-637X/822/1/48
R. Haas, R.V. Shcherbakov, T. Bode, P. Laguna, Tidal disruptions of white dwarfs from ultra-close encounters with intermediate-mass spinning black holes. Astrophys. J. **749**(2), 117 (2012a). https://doi.org/10.1088/0004-637X/749/2/117
R. Haas, R.V. Shcherbakov, T. Bode, P. Laguna, Tidal disruptions of white dwarfs from ultra-close encounters with intermediate-mass spinning black holes. Astrophys. J. **749**(2), 117 (2012b). https://doi.org/10.1088/0004-637X/749/2/117
A.S. Hamers, F. Dosopoulou, An analytic model for mass transfer in binaries with arbitrary eccentricity, with applications to triple-star systems. Astrophys. J. **872**(2), 119 (2019). https://doi.org/10.3847/1538-4357/ab001d

J.-M. Hameury, A.R. King, J.-P. Lasota, M. Auvergne, Black-hole stellar accretion in active galactic nuclei. Astron. Astrophys. **292**, 404–408 (1994)

K. Hayasaki, A. Loeb, Detection of gravitational wave emission by supermassive black hole binaries through tidal disruption flares. Sci. Rep. **6**, 35629 (2016). https://doi.org/10.1038/srep35629

K. Hayasaki, N. Stone, A. Loeb, Finite, intense accretion bursts from tidal disruption of stars on bound orbits. Mon. Not. R. Astron. Soc. **434**(2), 909–924 (2013). https://doi.org/10.1093/mnras/stt871

K. Hayasaki, N. Stone, A. Loeb, Circularization of tidally disrupted stars around spinning supermassive black holes. Mon. Not. R. Astron. Soc. **461**(4), 3760–3780 (2016). https://doi.org/10.1093/mnras/stw1387

K. Hayasaki, S. Zhong, S. Li, P. Berczik, R. Spurzem, Classification of tidal disruption events based on stellar orbital properties. Astrophys. J. **855**(2), 129 (2018). https://doi.org/10.3847/1538-4357/aab0a5

J.G. Hills, Possible power source of Seyfert galaxies and QSOs. Nature **254**, 295–298 (1975). https://doi.org/10.1038/254295a0

M.S. Hjellming, R.F. Webbink, Thresholds for rapid mass transfer in binary system. I. Polytropic models. Astrophys. J. **318**, 794 (1987). https://doi.org/10.1086/165412

M. Ishii, M. Shibata, Y. Mino, Black hole tidal problem in the Fermi normal coordinates. Phys. Rev. D **71**(4), 044017 (2005). https://doi.org/10.1103/PhysRevD.71.044017

P.B. Ivanov, M.A. Chernyakova, Relativistic cross sections of mass stripping and tidal disruption of a star by a super-massive rotating black hole. Astron. Astrophys. **448**(3), 843–852 (2006). https://doi.org/10.1051/0004-6361:20053409

P.B. Ivanov, I.D. Novikov, A new model of a tidally disrupted star. Astrophys. J. **549**(1), 467–482 (2001). https://doi.org/10.1086/319050

P.B. Ivanov, M.A. Chernyakova, I.D. Novikov, The new model of a tidally disrupted star: further development and relativistic calculations. Mon. Not. R. Astron. Soc. **338**(1), 147–155 (2003). https://doi.org/10.1046/j.1365-8711.2003.06028.x

D. Kasen, E. Ramirez-Ruiz, Optical transients from the unbound debris of tidal disruption. Astrophys. J. **714**(1), 155–162 (2010). https://doi.org/10.1088/0004-637X/714/1/155

K. Kawana, A. Tanikawa, N. Yoshida, Tidal disruption of a white dwarf by a black hole: the diversity of nucleosynthesis, explosion energy, and the fate of debris streams. Mon. Not. R. Astron. Soc. **477**(3), 3449–3460 (2018). https://doi.org/10.1093/mnras/sty842

M. Kesden, Tidal-disruption rate of stars by spinning supermassive black holes. Phys. Rev. D **85**(2), 024037 (2012a). https://doi.org/10.1103/PhysRevD.85.024037

M. Kesden, Black-hole spin dependence in the light curves of tidal disruption events. Phys. Rev. D **86**(6), 064026 (2012b). https://doi.org/10.1103/PhysRevD.86.064026

A. Khokhlov, F. Melia, Powerful ejection of matter from tidally disrupted stars near massive black holes and a possible application to Sagittarius a East. Astrophys. J. Lett. **457**, 61 (1996). https://doi.org/10.1086/309895

A.R. King, C. Done, Stellar accretion in active galactic nuclei. Mon. Not. R. Astron. Soc. **264**, 388–394 (1993). https://doi.org/10.1093/mnras/264.2.388

S. Kobayashi, P. Laguna, E.S. Phinney, P. Mészáros, Gravitational waves and X-ray signals from stellar disruption by a massive black hole. Astrophys. J. **615**(2), 855–865 (2004). https://doi.org/10.1086/424684

S. Kobayashi, Y. Hainick, R. Sari, E.M. Rossi, Ejection and capture dynamics in restricted three-body encounters. Astrophys. J. **748**(2), 105 (2012). https://doi.org/10.1088/0004-637X/748/2/105

C.S. Kochanek, The aftermath of tidal disruption: the dynamics of thin gas streams. Astrophys. J. **422**, 508–520 (1994). https://doi.org/10.1086/173745

C.S. Kochanek, Abundance anomalies in tidal disruption events. Mon. Not. R. Astron. Soc. **458**(1), 127–134 (2016). https://doi.org/10.1093/mnras/stw267

Z. Kopal, *Close Binary Systems* 1959

A.G. Kosovichev, I.D. Novikov, Non-linear effects at tidal capture of stars by a massive black hole. I—incompressible affine model. Mon. Not. R. Astron. Soc. **258**(4), 715–724 (1992). https://doi.org/10.1093/mnras/258.4.715

J. Krolik, T. Piran, G. Svirski, R.M. Cheng, ASASSN-14li: a model tidal disruption event. Astrophys. J. **827**(2), 127 (2016). https://doi.org/10.3847/0004-637X/827/2/127

J. Law-Smith, M. MacLeod, J. Guillochon, P. Macias, E. Ramirez-Ruiz, Low-mass white dwarfs with hydrogen envelopes as a missing link in the tidal disruption menu. Astrophys. J. **841**, 132 (2017a). https://doi.org/10.3847/1538-4357/aa6ffb

J. Law-Smith, E. Ramirez-Ruiz, S.L. Ellison, R.J. Foley, Tidal disruption event host galaxies in the context of the local galaxy population. Astrophys. J. **850**, 22 (2017b). https://doi.org/10.3847/1538-4357/aa94c7

J. Law-Smith, J. Guillochon, E. Ramirez-Ruiz, The tidal disruption of Sun-like stars by massive black holes. Astrophys. J. Lett. **882**(2), 25 (2019). https://doi.org/10.3847/2041-8213/ab379a

J.A.P. Law-Smith, D.A. Coulter, B. Mockler, *jamielaw-smith/STARS_library v1.0.5* (Zenodo, 2020a). https://doi.org/10.5281/zenodo.4062018

J.A.P. Law-Smith, D.A. Coulter, J. Guillochon, B. Mockler, E. Ramirez-Ruiz, Stellar TDEs with Abundances and Realistic Structures (STARS): Library of Fallback Rates. arXiv e-prints, 2007–10996 (2020b)
H.M. Lee, J.P. Ostriker, Cross sections for tidal capture binary formation and stellar merger. Astrophys. J. **310**, 176 (1986). https://doi.org/10.1086/164674
G. Leloudas, M. Fraser, N.C. Stone, S. van Velzen, P.G. Jonker, I. Arcavi, C. Fremling, J.R. Maund, S.J. Smartt, T. Krìhler, J.C.A. Miller-Jones, P.M. Vreeswijk, A. Gal-Yam, P.A. Mazzali, A. De Cia, D.A. Howell, C. Inserra, F. Patat, A. de Ugarte Postigo, O. Yaron, C. Ashall, I. Bar, H. Campbell, T.-W. Chen, M. Childress, N. Elias-Rosa, J. Harmanen, G. Hosseinzadeh, J. Johansson, T. Kangas, E. Kankare, S. Kim, H. Kuncarayakti, J. Lyman, M.R. Magee, K. Maguire, D. Malesani, S. Mattila, C.V. McCully, M. Nicholl, S. Prentice, C. Romero-Cañizales, S. Schulze, K.W. Smith, J. Sollerman, M. Sullivan, B.E. Tucker, S. Valenti, J.C. Wheeler, D.R. Young, The superluminous transient ASASSN-15lh as a tidal disruption event from a Kerr black hole. Nat. Astron. **1**, 0002 (2016). https://doi.org/10.1038/s41550-016-0002
I. Linial, R. Sari, Mass-loss through the L2 Lagrange point—application to main-sequence EMRI. Mon. Not. R. Astron. Soc. **469**(2), 2441–2454 (2017). https://doi.org/10.1093/mnras/stx1041
D. Liptai, D.J. Price, I. Mandel, G. Lodato, Disc formation from tidal disruption of stars on eccentric orbits by Kerr black holes using GRSPH. arXiv e-prints, 1910–10154 (2019)
F.K. Liu, S. Li, X. Chen, Interruption of tidal-disruption flares by supermassive black hole binaries. Astrophys. J. **706**(1), 133–137 (2009). https://doi.org/10.1088/0004-637X/706/1/L133
F.K. Liu, S. Li, S. Komossa, A milliparsec supermassive black hole binary candidate in the galaxy SDSS J120136.02+300305.5. Astrophys. J. **786**(2), 103 (2014). https://doi.org/10.1088/0004-637X/786/2/103
G. Lodato, A.R. King, J.E. Pringle, Stellar disruption by a supermassive black hole: is the light curve really proportional to $t^{-5/3}$? Mon. Not. R. Astron. Soc. **392**(1), 332–340 (2009). https://doi.org/10.1111/j.1365-2966.2008.14049.x
J.-P. Luminet, B. Carter, Dynamics of an affine star model in a black hole tidal field. Astrophys. J. Suppl. Ser. **61**, 219 (1986). https://doi.org/10.1086/191113
J.-P. Luminet, J.-A. Marck, Tidal squeezing of stars by Schwarzschild black holes. Mon. Not. R. Astron. Soc. **212**, 57–75 (1985). https://doi.org/10.1093/mnras/212.1.57
J.-P. Luminet, B. Pichon, Tidal pinching of white dwarfs. Astron. Astrophys. **209**(1-2), 103–110 (1989a)
J.-P. Luminet, B. Pichon, Tidally-detonated nuclear reactions in main sequence stars passing near a large black hole. Astron. Astrophys. **209**(1-2), 85–102 (1989b)
J. Luo, L.-S. Chen, H.-Z. Duan, Y.-G. Gong, S. Hu, J. Ji, Q. Liu, J. Mei, V. Milyukov, M. Sazhin, C.-G. Shao, V.T. Toth, H.-B. Tu, Y. Wang, Y. Wang, H.-C. Yeh, M.-S. Zhan, Y. Zhang, V. Zharov, Z.-B. Zhou, Tianqin: a space-borne gravitational wave detector. Class. Quantum Gravity **33**(3), 035010 (2016)
D. Lynden-Bell, Galactic nuclei as collapsed old quasars. Nature **223**(5207), 690–694 (1969). https://doi.org/10.1038/223690a0
M. MacLeod, J. Guillochon, E. Ramirez-Ruiz, The tidal disruption of giant stars and their contribution to the flaring supermassive black hole population. Astrophys. J. **757**, 134 (2012). https://doi.org/10.1088/0004-637X/757/2/134
M. MacLeod, E. Ramirez-Ruiz, S. Grady, J. Guillochon, Spoon-feeding giant stars to supermassive black holes: episodic mass transfer from evolving stars and their contribution to the quiescent activity of galactic nuclei. Astrophys. J. **777**, 133 (2013). https://doi.org/10.1088/0004-637X/777/2/133
M. MacLeod, J. Goldstein, E. Ramirez-Ruiz, J. Guillochon, J. Samsing, Illuminating massive black holes with white dwarfs: orbital dynamics and high-energy transients from tidal interactions. Astrophys. J. **794**(1), 9 (2014). https://doi.org/10.1088/0004-637X/794/1/9
M. MacLeod, J. Guillochon, E. Ramirez-Ruiz, D. Kasen, S. Rosswog, Optical thermonuclear transients from tidal compression of white dwarfs as tracers of the low end of the massive black hole mass function. Astrophys. J. **819**(1), 3 (2016). https://doi.org/10.3847/0004-637X/819/1/3
D. Mainetti, A. Lupi, S. Campana, M. Colpi, E.R. Coughlin, J. Guillochon, E. Ramirez-Ruiz, The fine line between total and partial tidal disruption events. Astron. Astrophys. **600**, 124 (2017). https://doi.org/10.1051/0004-6361/201630092
I. Mandel, Y. Levin, Double tidal disruptions in galactic nuclei. Astrophys. J. Lett. **805**, 4 (2015). https://doi.org/10.1088/2041-8205/805/1/L4
H. Manukian, J. Guillochon, E. Ramirez-Ruiz, R.M. O'Leary, Turbovelocity stars: kicks resulting from the tidal disruption of solitary stars. Astrophys. J. Lett. **771**(2), 28 (2013). https://doi.org/10.1088/2041-8205/771/2/L28
J. Marck, Solution to the equations of parallel transport in Kerr geometry; tidal tensor. Proc. R. Soc. Lond. Ser. A **385**, 431–438 (1983). https://doi.org/10.1098/rspa.1983.0021
J.J. Matese, D.P. Whitmire, Conservative mass transfer in close binary systems. I. Equations of motion for spin and orbital angular momenta. Astrophys. J. **266**, 776–786 (1983). https://doi.org/10.1086/160825

D. Merritt, *Dynamics and Evolution of Galactic Nuclei* (Princeton University Press, Princeton, NJ, USA, 2013)

B.D. Metzger, N.C. Stone, Periodic accretion-powered flares from colliding EMRIs as TDE imposters. Astrophys. J. **844**(1), 75 (2017). https://doi.org/10.3847/1538-4357/aa7a16

F.C. Michel, Neutron star disk formation from supernova fall-back and possible observational consequences. Nature **333**(6174), 644–645 (1988). https://doi.org/10.1038/333644a0

M.C. Miller, M. Freitag, D.P. Hamilton, V.M. Lauburg, Binary encounters with supermassive black holes: zero-eccentricity LISA events. Astrophys. J. Lett. **631**(2), 117–120 (2005). https://doi.org/10.1086/497335

I.D. Novikov, C.J. Pethick, A.G. Polnarev, Tidal capture of stars by a massive black hole. Mon. Not. R. Astron. Soc. **255**, 276–284 (1992). https://doi.org/10.1093/mnras/255.2.276

B. Paczyński, Evolutionary processes in close binary systems. Annu. Rev. Astron. Astrophys. **9**, 183 (1971). https://doi.org/10.1146/annurev.aa.09.090171.001151

G. Park, K. Hayasaki, Tidal disruption flares from stars on marginally bound and unbound orbits. Astrophys. J. **900**(1), 3 (2020). https://doi.org/10.3847/1538-4357/ab9ebb

B. Paxton, L. Bildsten, A. Dotter, F. Herwig, P. Lesaffre, F. Timmes, Modules for Experiments in Stellar Astrophysics (MESA). Astrophys. J. Suppl. Ser. **192**(1), 3 (2011). https://doi.org/10.1088/0067-0049/192/1/3

E.S. Phinney, Manifestations of a massive black hole in the Galactic center, in *The Center of the Galaxy*, ed. by M. Morris IAU Symposium, vol. 136, 1989, p. 543

P. Podsiadlowski, The response of tidally heated stars. Mon. Not. R. Astron. Soc. **279**, 1104 (1996). https://doi.org/10.1093/mnras/279.4.1104

W.H. Press, S.A. Teukolsky, On formation of close binaries by two-body tidal capture. Astrophys. J. **213**, 183–192 (1977). https://doi.org/10.1086/155143

S. Rappaport, P.C. Joss, R.F. Webbink, The evolution of highly compact binary stellar systems. Astrophys. J. **254**, 616–640 (1982). https://doi.org/10.1086/159772

M.J. Rees, Tidal disruption of stars by black holes of 10 to the 6th-10 to the 8th solar masses in nearby galaxies. Nature **333**, 523–528 (1988). https://doi.org/10.1038/333523a0

E.M. Rossi, S. Kobayashi, R. Sari, The velocity distribution of hypervelocity stars. Astrophys. J. **795**, 125 (2014). https://doi.org/10.1088/0004-637X/795/2/125

E.M. Rossi, T. Marchetti, M. Cacciato, M. Kuiack, R. Sari, Joint constraints on the Galactic dark matter halo and Galactic Centre from hypervelocity stars. Mon. Not. R. Astron. Soc. **467**, 1844–1856 (2017). https://doi.org/10.1093/mnras/stx098

S. Rosswog, E. Ramirez-Ruiz, W.R. Hix, Tidal disruption and ignition of white dwarfs by moderately massive black holes. Astrophys. J. **695**(1), 404–419 (2009). https://doi.org/10.1088/0004-637X/695/1/404

T. Ryu, J. Krolik, T. Piran, S.C. Noble, Tidal Disruptions of Main Sequence Stars – I. Observable Quantities and their Dependence on Stellar and Black Hole Mass. arXiv e-prints, 2001–03501 (2020a)

T. Ryu, J. Krolik, T. Piran, S.C. Noble, Tidal disruptions of main sequence stars – III. Stellar mass dependence of the character of partial disruptions. arXiv e-prints, 2001–03503 (2020b)

T. Ryu, J. Krolik, T. Piran, S.C. Noble, Tidal disruptions of main sequence stars – IV. Relativistic effects and dependence on black hole mass. arXiv e-prints, 2001–03504 (2020c)

A. Sacchi, G. Lodato, 'Failed' tidal disruption events and X-ray flares from the Galactic Centre. Mon. Not. R. Astron. Soc. **486**(2), 1833–1839 (2019). https://doi.org/10.1093/mnras/stz981

R. Sari, S. Kobayashi, E.M. Rossi, Hypervelocity stars and the restricted parabolic three-body problem. Astrophys. J. **708**(1), 605–614 (2010). https://doi.org/10.1088/0004-637X/708/1/605

S. Sato, S. Kawamura, M. Ando, T. Nakamura, K. Tsubono, A. Araya, I. Funaki, K. Ioka, N. Kanda, S. Moriwaki, M. Musha, K. Nakazawa, K. Numata, S.-i. Sakai, N. Seto, T. Takashima, T. Tanaka, K. Agatsuma, K.-s. Aoyanagi, K. Arai, H. Asada, Y. Aso, T. Chiba, T. Ebisuzaki, Y. Ejiri, M. Enoki, Y. Eriguchi, M.-K. Fujimoto, R. Fujita, M. Fukushima, T. Futamase, K. Ganzu, T. Harada, T. Hashimoto, K. Hayama, W. Hikida, Y. Himemoto, H. Hirabayashi, T. Hiramatsu, F.-L. Hong, H. Horisawa, M. Hosokawa, K. Ichiki, T. Ikegami, K.T. Inoue, K. Ishidoshiro, H. Ishihara, T. Ishikawa, H. Ishizaki, H. Ito, Y. Itoh, N. Kawashima, F. Kawazoe, N. Kishimoto, K. Kiuchi, S. Kobayashi, K. Kohri, H. Koizumi, Y. Kojima, K. Kokeyama, W. Kokuyama, K. Kotake, Y. Kozai, H. Kudoh, H. Kunimori, H. Kuninaka, K. Kuroda, K.-i. Maeda, H. Matsuhara, Y. Mino, O. Miyakawa, S. Miyoki, M.Y. Morimoto, T. Morioka, T. Morisawa, S. Mukohyama, S. Nagano, I. Naito, K. Nakamura, H. Nakano, K. Nakao, S. Nakasuka, Y. Nakayama, E. Nishida, K. Nishiyama, A. Nishizawa, Y. Niwa, T. Noumi, Y. Obuchi, M. Ohashi, N. Ohishi, M. Ohkawa, N. Okada, K. Onozato, K. Oohara, N. Sago, M. Saijo, M. Sakagami, S. Sakata, M. Sasaki, T. Sato, M. Shibata, H. Shinkai, K. Somiya, H. Sotani, N. Sugiyama, Y. Suwa, R. Suzuki, H. Tagoshi, F. Takahashi, K. Takahashi, K. Takahashi, R. Takahashi, R. Takahashi, T. Takahashi, H. Takahashi, T. Akiteru, T. Takano, K. Taniguchi, A. Taruya, H. Tashiro, Y. Torii, M. Toyoshima, S. Tsujikawa, Y.

Tsunesada, A. Ueda, K.-i. Ueda, M. Utashima, Y. Wakabayashi, H. Yamakawa, K. Yamamoto, T. Yamazaki, J. Yokoyama, C.-M. Yoo, S. Yoshida, T. Yoshino, The status of DECIGO. J. Phys. Conf. Ser. **840**, 012010 (2017)

R. Schödel, A. Eckart, T. Alexander, D. Merritt, R. Genzel, A. Sternberg, L. Meyer, F. Kul, J. Moultaka, T. Ott, C. Straubmeier, The structure of the nuclear stellar cluster of the Milky Way. Astron. Astrophys. **469**(1), 125–146 (2007). https://doi.org/10.1051/0004-6361:20065089

A. Sesana, A. Vecchio, M. Eracleous, S. Sigurdsson, Observing white dwarfs orbiting massive black holes in the gravitational wave and electro-magnetic window. Mon. Not. R. Astron. Soc. **391**(2), 718–726 (2008). https://doi.org/10.1111/j.1365-2966.2008.13904.x

A.C. Seth, M. Cappellari, N. Neumayer, N. Caldwell, N. Bastian, K. Olsen, R.D. Blum, V.P. Debattista, R. McDermid, T. Puzia, The NGC 404 nucleus: star cluster and possible intermediate-mass black hole. Astrophys. J. **714**(1), 713–731 (2010). https://doi.org/10.1088/0004-637X/714/1/713

N. Seto, T. Muto, Relativistic astrophysics with resonant multiple inspirals. Phys. Rev. D **81**(10), 103004 (2010). https://doi.org/10.1103/PhysRevD.81.103004

N. Seto, T. Muto, Resonant trapping of stars by merging massive black hole binaries. Mon. Not. R. Astron. Soc. **415**(4), 3824–3830 (2011). https://doi.org/10.1111/j.1365-2966.2011.18988.x

S. Sigurdsson, M.J. Rees, Capture of stellar mass compact objects by massive black holes in galactic cusps. Mon. Not. R. Astron. Soc. **284**(2), 318–326 (1997). https://doi.org/10.1093/mnras/284.2.318

E. Steinberg, E.R. Coughlin, N.C. Stone, B.D. Metzger, Thawing the frozen-in approximation: implications for self-gravity in deeply plunging tidal disruption events. Mon. Not. R. Astron. Soc. **485**(1), 146–150 (2019). https://doi.org/10.1093/mnrasl/slz048

N. Stone, A. Loeb, Prompt tidal disruption of stars as an electromagnetic signature of supermassive black hole coalescence. Mon. Not. R. Astron. Soc. **412**(1), 75–80 (2011). https://doi.org/10.1111/j.1365-2966.2010.17880.x

N.C. Stone, B.D. Metzger, Rates of stellar tidal disruption as probes of the supermassive black hole mass function. Mon. Not. R. Astron. Soc. **455**, 859–883 (2016). https://doi.org/10.1093/mnras/stv2281

N. Stone, R. Sari, A. Loeb, Consequences of strong compression in tidal disruption events. Mon. Not. R. Astron. Soc. **435**, 1809–1824 (2013). https://doi.org/10.1093/mnras/stt1270

N. Stone, R. Sari, A. Loeb, Consequences of strong compression in tidal disruption events. Mon. Not. R. Astron. Soc. **435**(3), 1809–1824 (2013). https://doi.org/10.1093/mnras/stt1270

N.C. Stone, M. Kesden, R.M. Cheng, S. van Velzen, Stellar tidal disruption events in general relativity. Gen. Relativ. Gravit. **51**(2), 30 (2019). https://doi.org/10.1007/s10714-019-2510-9

L.E. Strubbe, E. Quataert, Optical flares from the tidal disruption of stars by massive black holes. Mon. Not. R. Astron. Soc. **400**, 2070–2084 (2009). https://doi.org/10.1111/j.1365-2966.2009.15599.x

D. Syer, A. Ulmer, Tidal disruption rates of stars in observed galaxies. Mon. Not. R. Astron. Soc. **306**, 35–42 (1999). https://doi.org/10.1046/j.1365-8711.1999.02445.x

A. Tanikawa, High-resolution hydrodynamic simulation of tidal detonation of a helium white dwarf by an intermediate mass black hole. Astrophys. J. **858**(1), 26 (2018). https://doi.org/10.3847/1538-4357/aaba79

A. Tanikawa, Y. Sato, K. Nomoto, K. Maeda, N. Nakasato, I. Hachisu, Does explosive nuclear burning occur in tidal disruption events of white dwarfs by intermediate-mass black holes? Astrophys. J. **839**(2), 81 (2017). https://doi.org/10.3847/1538-4357/aa697d

E. Tejeda, E. Gafton, S. Rosswog, J.C. Miller, Tidal disruptions by rotating black holes: relativistic hydrodynamics with Newtonian codes. Mon. Not. R. Astron. Soc. **469**(4), 4483–4503 (2017). https://doi.org/10.1093/mnras/stx1089

K.S. Thorne, Probing black holes and relativistic stars with gravitational waves, in *Black Holes and Relativistic Stars*, ed. by R.M. Wald, 1998, p. 41

M. Toscani, G. Lodato, R. Nealon, Gravitational wave emission from unstable accretion discs in tidal disruption events. Mon. Not. R. Astron. Soc. **489**(1), 699–706 (2019). https://doi.org/10.1093/mnras/stz2201

M. Vick, D. Lai, J. Fuller, Tidal dissipation and evolution of white dwarfs around massive black holes: an eccentric path to tidal disruption. Mon. Not. R. Astron. Soc. **468**(2), 2296–2310 (2017). https://doi.org/10.1093/mnras/stx539

Q. Vigneron, G. Lodato, A. Guidarelli, Tidal disruption of stars in a supermassive black hole binary system: the influence of orbital properties on fallback and accretion rates. Mon. Not. R. Astron. Soc. **476**(4), 5312–5322 (2018). https://doi.org/10.1093/mnras/sty585

R.F. Webbink, S. Rappaport, G.J. Savonije, On the evolutionary status of bright, low-mass X-ray sources. Astrophys. J. **270**, 678–693 (1983). https://doi.org/10.1086/161159

Y. Wu, Diffusive tidal evolution for migrating hot jupiters. Astron. J. **155**(3), 118 (2018). https://doi.org/10.3847/1538-3881/aaa970

A. Yalinewich, E. Steinberg, T. Piran, J.H. Krolik, Radio Emission from the unbound Debris of Tidal Disruption Events. arXiv e-prints, 1903–02575 (2019)

I. Zalamea, K. Menou, A.M. Beloborodov, White dwarfs stripped by massive black holes: sources of coincident gravitational and electromagnetic radiation. Mon. Not. R. Astron. Soc. **409**(1), 25–29 (2010). https://doi.org/10.1111/j.1745-3933.2010.00930.x

S. Zhong, P. Berczik, R. Spurzem, Super massive black hole in galactic nuclei with tidal disruption of stars. Astrophys. J. **792**(2), 137 (2014). https://doi.org/10.1088/0004-637X/792/2/137

Space Sci Rev (2021) 217:16
https://doi.org/10.1007/s11214-020-00789-1

Formation of an Accretion Flow

C. Bonnerot[1] · N.C. Stone[2]

Received: 25 May 2020 / Accepted: 26 December 2020 / Published online: 25 January 2021

Abstract After a star has been tidally disrupted by a black hole, the debris forms an elongated stream. We start by studying the evolution of this gas before its bound part returns to the original stellar pericenter. While the axial motion is entirely ballistic, the transverse directions of the stream are usually thinner due to the confining effects of self-gravity. This basic picture may also be influenced by additional physical effects such as clump formation, hydrogen recombination, magnetic fields and the interaction with the ambient medium. We then examine the fate of this stream when it comes back to the vicinity of the black hole to form an accretion flow. Despite recent progress, the hydrodynamics of this phase remains uncertain due to computational limitations that have so far prevented us from performing a fully self-consistent simulation. Most of the initial energy dissipation appears to be provided by a self-crossing shock that results from an intersection of the stream with itself. The debris evolution during this collision depends on relativistic apsidal precession, expansion of the stream from pericenter, and nodal precession induced by the black hole spin. Although the combined influence of these effects is not fully understood, current works suggest that this interaction is typically too weak to significantly circularize the trajectories, with its main consequence being an expansion of the shocked gas. Global simulations of disc formation performed for simplified initial conditions find that the debris experiences additional collisions that cause its orbits to become more circular until eventually settling into a thick and extended structure. These works suggest that this process completes faster for more relativistic encounters due to the stronger shocks involved. It is instead significantly delayed if weaker shocks take place, allowing the gas to retain large eccentricities

The Tidal Disruption of Stars by Massive Black Holes
Edited by Peter G. Jonker, Sterl Phinney, Elena Maria Rossi, Sjoert van Velzen, Iair Arcavi and Maurizio Falanga

✉ C. Bonnerot
bonnerot@tapir.caltech.edu

N.C. Stone
nicholas.stone@mail.huji.ac.il

1 TAPIR, Mailcode 350-17, California Institute of Technology, Pasadena, CA 91125, USA

2 Racah Institute of Physics, The Hebrew University, Jerusalem, 91904, Israel

during multiple orbits. Radiation produced as the matter gets heated by circularizing shocks may leave the system through photon diffusion and participate in the emerging luminosity. This current picture of accretion flow formation results from recent theoretical works synthesizing the interplay between different aspects of physics. In comparison, early analytical works correctly identified the essential processes involved in disc formation, but had difficulty developing analytic frameworks that accurately combined non-linear hydrodynamical processes with the underlying relativistic dynamics. However, important aspects still remain to be understood at the time of writing, due to numerical challenges and the complexity of this process.

Keywords Tidal disruption events · Black holes · Hydrodynamics · Accretion discs

1 Introduction

In this chapter, we focus on the evolution of the debris around the black hole following stellar disruption, and most importantly the formation of an accretion flow from this matter. These two stages are described below in largely independent sections so that the reader is able to go through them in any order. While our understanding of isolated stream evolution is relatively robust, the later stage of accretion disc formation remains debated due to its greater complexity and associated numerical challenges. For the latter, we therefore present the major advances made so far while emphasizing the uncertainties of these current works.

Following the disruption of the star on its original parabolic trajectory, the debris evolves into an elongated stream due to the spread in orbital energy imparted to the stellar matter by the encounter with the black hole. While half of this gas is unbound and escapes the gravitational attraction of the compact object on hyperbolic orbits, the rest becomes bound and comes back to the stellar pericenter on highly eccentric trajectories. The hydrodynamics at play during this revolution of the stellar debris around the black hole is well understood thanks to the early work of Kochanek (1994), more recently revised by Coughlin et al. (2016a). During this phase, the stream gets stretched along its longitudinal direction due to the ballistic in-plane motion of the gas elements that evolve like test particles on a wide range of different elliptical orbits. In the other two transverse directions, the gas motion is usually specified by self-gravity that imposes a thin width such that the stream has a locally cylindrical geometry. The influence of various additional physical ingredients on this basic picture has been explored, including stream fragmentation into individual self-gravitating clumps, the recombination of hydrogen as the stream cools, the magnetic field inherited from the star and the interaction of the debris with the ambient gaseous medium. It is nevertheless fair to say that the basic picture of stream evolution described above is usually not strongly affected by this additional physics.

After the bound part of the stream has completed an entire orbit, it comes back near the black hole where it can start forming an accretion flow. As already mentioned, our understanding of this phase of evolution is less secure than the earlier one due to the more complicated mechanisms involved and the numerical challenges of studying them. The returning stream is strongly compressed at pericenter, causing a nozzle shock whose main effect is to make the gas expand. Initial dissipation is provided by a collision of the stream with itself. The parameters of this collision are specified by the combination of relativistic apsidal precession, expansion from the nozzle shock, and nodal precession due to black hole spin. Local simulations of this interaction (e.g. Jiang et al. 2016b; Lu and Bonnerot 2020) find that the ensuing self-crossing shock dissipates part of the debris kinetic energy and results in an expansion of the gas distribution around the intersection point. However, performing a global study of the disc formation process is very computationally expensive and current investigations therefore rely on simplified initial conditions to circumvent this numerical burden. Most often, this involves either decreasing the stellar eccentricity such that the debris becomes artificially more bound, or reducing the black hole mass to that of an intermediate mass black hole. These numerical works (e.g. Hayasaki et al. 2013; Bonnerot et al. 2016b; Shiokawa et al. 2015; Bonnerot and Lu 2020) find that the shocked gas experiences additional dissipation that causes the debris to move to more circular orbits and to eventually settle into a thick and extended accretion disc. Importantly, this final state may take a long time to be reached, and the gas likely retains moderate to large eccentricities even after this process is completed. The current understanding of this process presents significant improvements compared to early analytic works such as that by Rees (1988), and it significantly differs from the simple picture involving the fast formation of a compact and axisymmetric disc, often assumed in the literature. While the above qualitative description of disc formation is relatively robust, there is so far no clear consensus on the detailed hydrodynamics at play for astrophysically realistic initial conditions. Nevertheless, our understanding of this process has now reached a level sufficient to better identify the main sources of uncertainty and start developing strategies to overcome them.

This chapter is arranged as follows. We start in Sect. 2 by describing the evolution of the stream around the black hole before it comes back to pericenter. The basic gas dynamics is first presented followed by the impact of additional physical processes. In Sect. 3, we explain the current understanding of how the returning debris forms an accretion flow. We start by describing the initial sources of dissipation in a largely analytical fashion and then present the various numerical investigations of the disc formation process along with its main consequences. Finally, Sect. 4 contains a summary and our concluding remarks.

2 Stream Evolution Prior to Pericenter Return

2.1 Basic Stream Trajectory and Geometry

The trajectory of the debris stream mostly results from the disruption process discussed in the Disruption Chapter, but we start by recalling the most important properties below. The star is disrupted by the tidal force of the black hole if its pericenter distance R_p is lower than the tidal radius

$$R_\mathrm{t} = R_\star \left(\frac{M_\mathrm{h}}{M_\star}\right)^{1/3} = 0.47\,\mathrm{AU} \left(\frac{M_\mathrm{h}}{10^6\,\mathrm{M}_\odot}\right)^{1/3}, \tag{1}$$

with the depth this encounter being characterized by the penetration factor $\beta = R_\mathrm{t}/R_\mathrm{p}$. Here, M_h denotes the mass of the compact object while $M_\star$ and $R_\star$ are the stellar mass and radius, respectively. Note that the tidal radius can change by a factor of a few compared to equation (1) depending on the structure of the disrupted star. As in the above estimate, the numerical values in the remaining of the chapter are usually given assuming a solar-type star with $M_\star = \mathrm{M}_\odot$ and $R_\star = \mathrm{R}_\odot$ unless stated otherwise. The disruption imparts a spread in specific orbital energy (Rees 1988; Stone et al. 2013)

$$\Delta\varepsilon = \frac{G M_\mathrm{h} R_\star}{R_\mathrm{t}^2}, \tag{2}$$

to the debris that makes half of it unbound while the rest gets bound to the black hole. The spread in angular momentum is however negligible (see for example the Fig. 2 from Cheng and Bogdanović 2014), which implies in particular that all the stream elements keep a pericenter distance equal to that of the original star. As the gas keeps orbiting the black hole, it is usually a good approximation to consider its trajectory along the orbital plane as perfectly ballistic. This implies that the gas evolves into an elongated stream with the unbound part escaping to large distances while the bound gas returns to the stellar pericenter after a finite time. In between, a fraction of the debris is marginally-bound, remaining on the parabolic trajectory that the star followed until its disruption. The gas located at the most bound extremity of the stream has a semi-major axis

$$a_{\rm min} = \frac{R_\star}{2}\left(\frac{M_{\rm h}}{M_\star}\right)^{2/3} = 23\ {\rm AU}\left(\frac{M_{\rm h}}{10^6\ {\rm M}_\odot}\right)^{2/3}, \tag{3}$$

making use of equation (2). It is the first to come back to the black hole following the disruption after a time equal to its orbital period given by

$$t_{\rm min} = 2^{-1/2}\pi\left(\frac{GM_\star}{R_\star^3}\right)^{-1/2}\left(\frac{M_{\rm h}}{M_\star}\right)^{1/2} = 41\ {\rm d}\left(\frac{M_{\rm h}}{10^6\ {\rm M}_\odot}\right)^{1/2}, \tag{4}$$

according to Kepler's third law and equation (3). This stream element is highly-eccentric with an eccentricity $e_{\rm min} = 1 - R_{\rm p}/a_{\rm min}$ given by

$$1 - e_{\rm min} = \frac{2}{\beta}\left(\frac{M_{\rm h}}{M_\star}\right)^{-1/3} = 0.02\,\beta\left(\frac{M_{\rm h}}{10^6\ {\rm M}_\odot}\right)^{-1/3} \tag{5}$$

using equations (1) and (3). The mass fallback rate $\dot{M}_{\rm fb}$ at which the stellar matter comes back to the black hole is specified by how orbital energy is distributed among the stream. Assuming a flat distribution yields $\dot{M}_{\rm fb} = \dot{M}_{\rm p}(t/t_{\rm min})^{-5/3}$ with a peak value of

$$\dot{M}_{\rm p} = \frac{M_\star}{3\,t_{\rm min}} = 3\ {\rm M}_\odot\,{\rm yr}^{-1}\left(\frac{M_{\rm h}}{10^6\ {\rm M}_\odot}\right)^{-1/2}, \tag{6}$$

such that the integrated amount of returning mass is that $M_\star/2$ of the bound debris.

The geometry of the stream is well approximated by a cylinder whose thickness and elongation evolve in time. The latter is entirely determined by the ballistic motion of the gas along the orbital plane. In particular, one can specify the dependence of the length $l(R)$ of a stream element of fixed mass on its distance R from the black hole. At early times $t \lesssim t_{\rm min}$ when most of the gas still follows a near-parabolic orbit, this length scales as (Coughlin et al. 2016a)

$$l \propto R^2. \tag{7}$$

This result was derived by solving the equations of gas dynamics for the marginally-bound part of the stream but it can also be recovered in the following more intuitive way. Immediately after the disruption, the stream is still close to the tidal radius with a size similar to that of the original star, such that $l(R_{\rm t}) \approx R_\star$. At $t \approx t_{\rm min}$, the bulk of the stream reaches a distance of order $a_{\rm min}$ with most of its elongation being caused by the most bound debris that has returned to the black hole.[1] This implies that the length of the stream is similar to this semi-major axis with $l(a_{\rm min}) \approx a_{\rm min}$. These two estimates of the stream length can be combined to prove the quadratic dependence of the length with distance since $l(a_{\rm min})/l(R_{\rm t}) \approx (a_{\rm min}/R_{\rm t})^2$ using equations (1) and (3). The evolution of the stream length can also be expressed in terms of the time t since disruption as $l \propto t^{4/3}$ making use of the dependence $R \propto t^{2/3}$ valid for a near-parabolic orbit. After a time $t \gtrsim t_{\rm min}$, the trajectories of most gas elements start to significantly deviate from a parabolic orbit and the above scaling becomes invalid. Once they have escaped the gravity of the black hole, the unbound stream elements move at a constant speed such that their distance from the compact object evolves as $R \propto t$. Because the marginally-bound gas reaches a velocity close to zero, the length of this part of the stream follows the same scaling with $l \propto R \propto t$ (Coughlin et al. 2016a).

We now focus on the thickness of the gas distribution. Similarly to the above calculation, it is possible to determine the evolution of the width H of a stream element containing a fixed amount of mass as it orbits the black hole. While the densest stream elements have their transverse motion specified by hydrostatic equilibrium between self-gravity and gas pressure, the ones with lower densities have their width set by the shear induced by the tidal force from the black hole. The critical density separating these two regimes can be found by comparing the specific tidal and self-gravity forces acting on the stream that are given by $f_{\rm tid} \approx GM_{\rm h}H/R^3$ and $f_{\rm sg} \approx Gm/(Hl) \approx G\rho H$, respectively. To obtain the latter expression, the Gauss's theorem has been used as well as the relation $m \approx \rho H^2 l$ between the mass m and density ρ of a stream element that exploits its cylindrical geometry. Therefore, an element has its width set by hydrostatic equilibrium with $f_{\rm sg} > f_{\rm tid}$ if $\rho > \rho_{\rm c} \equiv M_{\rm h}/R^3$ where $\rho_{\rm c}$ denotes the critical density. The next step is to determine the scaling that the stream width obeys in the self-gravity and shear-dominated regimes corresponding to $\rho > \rho_{\rm c}$ and $\rho < \rho_{\rm c}$, respectively. When the tidal force dominates, the width of the stream evolves according to the homologous scaling $H \propto R$. This is because the gas parcels on its cylindrical surface experience a gravitational attraction directed in the radial direction that causes a compression of the stream element as it moves inwards. In the other regime, hydrostatic equilibrium instead imposes that the self-gravity force $f_{\rm sg} \approx G\rho H$ follows the gas pressure force $f_{\rm gas} = \nabla P/\rho \approx P/(\rho H)$, where P denotes the gas pressure that we assume to evolve according to the polytropic relation $P \propto \rho^\gamma$. Imposing $f_{\rm gas} \propto f_{\rm sg}$ then leads after some algebra to the relation $H \propto l^{(2-\gamma)/2(\gamma-1)}$. When $t \lesssim t_{\rm min}$, the stream length follows equation (7), which yields

$$H \propto R^{\frac{2-\gamma}{\gamma-1}}. \tag{8}$$

Assuming an adiabatic evolution with $\gamma = 5/3$, the width scales as $H \propto R^{1/2} \propto t^{1/3}$ that corresponds to a density evolution as $\rho \propto (H^2 l)^{-1} \propto R^{-3} \propto t^{-2}$. The unbound elements follow $l \propto R$ at late times $t \gtrsim t_{\rm min}$ such that $H \propto R^{(2-\gamma)/2(\gamma-1)} \propto R^{1/4} \propto t^{1/4}$. This latter scaling was found first by Kochanek (1994) who assumed the relation $l \propto R$ in the stream. It was later revised to $H \propto R^{1/2}$ at early times by Coughlin et al. (2016a) that makes use of the correct length evolution. Importantly, these scalings show that self-gravity confines the width of the stream such that it evolves slower with distance from the black hole than for the homologous scaling imposed by the tidal force.

For grazing encounters with $\beta \approx 1$, the stellar disruption results in a large fraction of the stream being dominated by self-gravity. This can for example be seen from Fig. 1 that shows

[1] Despite its fast outward motion, the unbound extremity of the stream remains close to the semi-major axis of the most bound gas at $t \approx t_{\rm min}$ since its velocity near this location can be approximated by $v_{\rm unb} \approx \sqrt{GM_{\rm h}/a_{\rm min}}$ such that the distance reached is $v_{\rm unb}t \approx a_{\rm min}$.

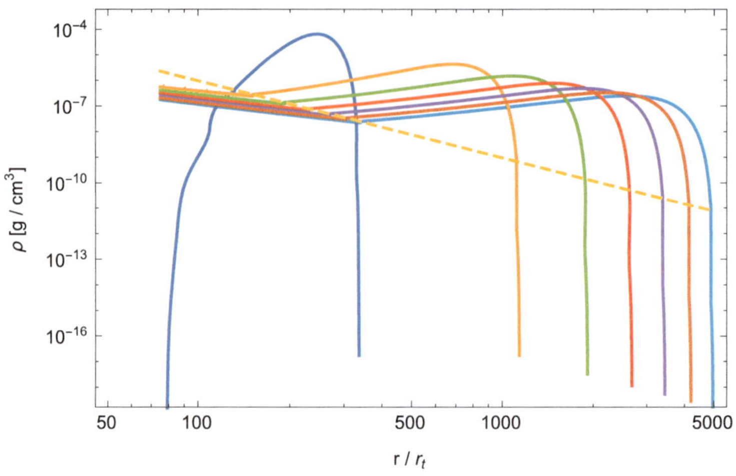

Fig. 1 Stream density as function of distance from the black hole at different times starting from $t = 32$ d (leftmost, dark blue line) and ending at $t = 620$ d (rightmost, light blue line) with an interval of 98 d between them (Coughlin et al. 2016a). The star has a solar mass and radius and is disrupted by a black hole of mass $M_{\rm h} = 10^6\ {\rm M}_\odot$ with a penetration factor of $\beta = 1$. The gas evolution is assumed to be adiabatic with $\gamma = 5/3$. The yellow dashed line shows the critical density $\rho_{\rm c} = M_{\rm h}/R^3$ that separates between the self-gravity and shear-dominated regimes

stream density profiles at different times with solid lines of different colours according to the semi-analytical model of Coughlin et al. (2016a). It considers the disruption of a solar-type star by a black hole of mass $M_{\rm h} = 10^6\ {\rm M}_\odot$ with a penetration factor $\beta = 1$, additionally assuming an adiabatic gas evolution with $\gamma = 5/3$. Most of the stream elements have a density above the dashed yellow line representing the critical value $\rho_{\rm c} = M_{\rm h}/R^3$, meaning that they are in hydrostatic equilibrium. This implies that the majority of the gas distribution is able to remain thin thanks to the confinement by self-gravity with a typical width given at $R \approx a_{\rm min}$ by $H/R_\star \approx (a_{\rm min}/R_{\rm t})^{1/2} \approx (M_{\rm h}/M_\star)^{1/6} \lesssim 10$ according to the scaling of equation (8). Shear-dominated elements are only present at the extremities of the stream due to the lower densities inherited from the original stellar density profile.[2] Numerical investigations of the stream evolution by Guillochon et al. (2014) and Coughlin et al. (2016a) confirm these expectations. It can for example be seen from Fig. 2 that shows the column density inside the stream obtained from a simulation that adopts the same parameters as the above semi-analytical study. The largest densities are reached in the central parts where the gas distribution remains thin owing to its self-gravity. Instead, the lower densities near the two ends of the stream imply that these regions are shear-dominated and therefore more extended.

Deep disruptions with $\beta \gtrsim 3$ result in heating since the star gets compressed during its disruption, as discussed in the Disruption Chapter. The ensuing expansion can cause most of the gas to have their densities decreased to $\rho < \rho_{\rm c}$. In this situation, one would expect that the stream is entirely shear-dominated with a width that evolves homologously until late times (Kochanek 1994; Coughlin et al. 2016a). This scenario is favoured by the fact that the density evolves at early times as $\rho \propto (H^2 l)^{-1} \propto R^{-4}$ in this regime such that the condition $\rho < \rho_{\rm c} = M_{\rm h}/R^3$ remains satisfied if it initially is. However, recent simulations

[2] Remarkably, the width evolution becomes homologous with $H \propto R$ near the black hole due to the increased tidal force that corresponds to an up-turn in the density profile seen at $R \lesssim 200 R_{\rm t}$ in Fig. 1.

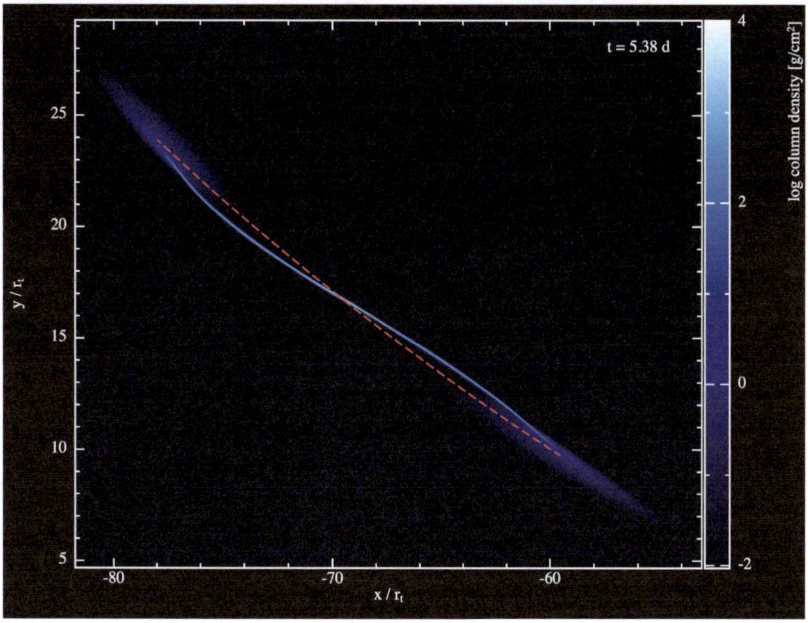

Fig. 2 Column density of the stream of debris a time of $t = 5.38$ d after the disruption of a star of solar mass and radius by a black hole of mass $M_{\rm h} = 10^6\ {\rm M}_\odot$ with a penetration factor of $\beta = 1$ (Coughlin and Nixon 2015)

by Steinberg et al. (2019) considering $\beta \geq 5$ find that a significant fraction of the stream is nevertheless able to recollapse under its self-gravity, possibly due to the presence of a weak caustic in the in-plane motion of the debris (Coughlin et al. 2016b). Although it contains more shear-dominated matter than for a grazing encounter, the resulting gas distribution remains therefore qualitatively similar to that described above for $\beta \approx 1$.

2.2 Additional Physics

2.2.1 Gravitational Fragmentation

As the stream expands, adiabatic cooling can trigger gravitational fragmentation with individual clumps forming within it. This phenomenon was first found in simulations by Coughlin and Nixon (2015) that consider a tidal disruption with $\beta = 1$ assuming an adiabatic evolution with $\gamma = 5/3$. The perturbation triggering fragmentation in this work likely has a numerical origin and the exact outcome depends the resolution used. Coughlin et al. (2016b) identified a possible physical cause for fragmentation, however, which is produced by the in-plane compression of the star as it is being disrupted by the black hole. This compression is due to the fact that, when the star is at the pericenter of its trajectory, the gas that has already passed this location decelerates while that lagging behind keeps accelerating. This results in a density increase happening a few hours after the disruption followed by oscillations as the stream settles back to hydrostatic equilibrium. The associated perturbation was proposed to destabilize the stream and trigger gravitational fragmentation but additional mechanisms likely exist such as the interaction with the surrounding medium discussed in Sect. 2.2.4.

The conditions for gravitational fragmentation have been examined more precisely by Coughlin et al. (2016a). It requires that the free-fall time $t_{\rm ff} \approx (\rho G)^{-1/2}$ is locally shorter than the dynamical time $t_{\rm dyn} \approx (G M_{\rm h}/R^3)^{-1/2}$ necessary for the stream to get stretched in the radial direction. The condition $t_{\rm ff} < t_{\rm dyn}$ then translates to $\rho \gtrsim M_{\rm h}/R^3 = \rho_{\rm c}$, which is identical to that derived in Sect. 2.1 for a stream element to be in hydrostatic equilibrium. In this regime, the density scales as $\rho \propto R^{-2/(\gamma-1)}$, implying that the condition for fragmentation remains valid at later times as long as $\gamma \geq 5/3$. This analysis suggests that clump formation is only possible for adiabatic exponents larger than this critical value and occurs faster when γ is increased. Numerical simulations by Coughlin et al. (2016b) confirm this trend by studying the stream evolution for different values of the adiabatic exponent. In reality, $\gamma > 5/3$ requires that the stream cools non-adiabatically, which can for instance be achieved through recombination when the gas becomes optically thin (see Sect. 2.2.2).

The properties of these clumps such as their preferred location along the stream are determined by the gravitational instability producing them, whose development has been studied in details by Coughlin and Nixon (2020). Due to the local density enhancements they produce, the fragments that belong to the bound part of the debris introduce variability in the fallback rate (Coughlin and Nixon 2015). The resulting emission could also be affected by the presence of these clumps, although including the impact of pressure delays their formation such that they may fall back once the emission has already faded below detection limits (Sacchi et al. 2020). The fragments forming in the unbound part of the stream instead escape the black hole, which constitutes a new kind of high-velocity objects (Coughlin et al. 2016b). Their spatial distribution has been studied more in detail by Girma and Guillochon (2018) who find that $\sim 10^7$ of them have been launched from Sgr A* within the life of our galaxy.

2.2.2 *Recombination*

Immediately after the disruption, the gas is fully ionized with a temperature of $T_\star \approx G M_\star m_{\rm p}/(k_{\rm B} R_\star) = 2 \times 10^7$ K similar to that necessary to ensure hydrostatic equilibrium inside the original star, as shown in the Disruption Chapter. Here, $k_{\rm B}$ denotes the Boltzmann constant while $m_{\rm p}$ is the proton mass. As it subsequently evolves around the black hole, this debris cools adiabatically, as its large optical depths prevent the heat from being promptly radiated. The temperature then evolves as $T \propto \rho^{2/3} \propto t^{-4/3}$ using the scaling $\rho \propto t^{-2}$ valid at early times assuming hydrostatic equilibrium (Sect. 2.1). This leads to $T \approx T_\star (t/t_{\rm str})^{-4/3}$, where $t_{\rm str} \approx (R_{\rm t}^3/G M_{\rm h})^{1/2} \approx 1$ h is the duration required for the stream to stretch by a significant amount following the disruption that is similar to the dynamical timescale near the tidal radius or, equivalently, the stellar dynamical timescale according to equation (1). Cooling continues until the temperature drops to a value of $T_{\rm rec} \approx 10^4$ K, at which hydrogen recombines. Setting $T = T_{\rm rec}$, this occurs after a time $t_{\rm rec} \approx t_{\rm str} (T_\star/T_{\rm rec})^{3/4} \approx 10$ d (Coughlin et al. 2016a).

The effect of recombination on the stream depends on its optical depth (Kochanek 1994; Coughlin et al. 2016a; Kasen and Ramirez-Ruiz 2010). By the time it occurs, the majority of the gas is still very optically thick, implying that this debris gets heated by the additional energy injection. During this process, the stream is imposed to evolve at a roughly constant temperature and becomes overpressured, causing an increase of its thickness and the end of its confinement by self-gravity. Since the stream outer layers are less dense, they are able to radiatively cool and become neutral that causes a decrease of the opacity since fewer free electrons are present. This process drives a transparency wave that propagates from the outer regions through the entire stream in a few days while producing an optical transient emission with a luminosity of $L_{\rm rec} \lesssim 10^{41}$ erg s^{-1}, as estimated by Kasen and Ramirez-Ruiz (2010). However, these values may be overestimated because this work assumes a free expansion of the stream after disruption that increases the emitting area compared to the slower thickening found in more recent investigations due to the confining effect of self-gravity.

2.2.3 Magnetic Fields

The magnetic field of the star gets transferred to the debris upon disruption and is subsequently carried by the stream in its evolution around the black hole. Its magnetic strength is then entirely specified by the stream geometry derived in Sect. 2.1 according to magnetic flux conservation. This condition imposes that the magnetic field parallel to the direction of stream elongation evolves as $B_\parallel \propto S_\parallel^{-1}$ where $S_\parallel \approx H^2$ denotes the surface the corresponding field lines go through. This results in $B_\parallel \propto R^{-1}$ according to equation (8) and assuming $\gamma = 5/3$. Similarly, the magnetic field orthogonal to the direction of stream stretching evolves as $B_\perp \propto S_\perp^{-1}$ with $S_\perp \propto Hl$ such that $B_\perp \propto R^{-5/2}$ additionally using equation (7). The fact that the parallel component decreases much slower than the perpendicular one implies that the field lines tend to align along the stream longitudinal direction.

Given this evolution, it is possible to evaluate the relative importance of magnetic and gas pressure inside the stream. Assuming that the field lines are all parallel to the direction of stream elongation, magnetic pressure follows $P_{\rm mag} \approx B_\parallel^2 \propto R^{-2}$ while the gas pressure scales as $P_{\rm gas} \propto \rho^{5/3} \propto R^{-5}$ for an adiabatic evolution. This shows that the ratio of magnetic to gas pressure increases as $P_{\rm mag}/P_{\rm gas} \propto R^3 \propto t^2$. Magnetic pressure takes over gas pressure with $P_{\rm mag}/P_{\rm gas} = 1$ after a time given by $t_{\rm mag} \approx t_{\rm str}\beta_\star^{1/2}$ where $t_{\rm str} \approx 1$ h has been defined in Sect. 2.2.2 while $\beta_\star$ denotes the initial plasma beta, defined as the ratio of gas to magnetic pressure inside the original star. This parameter can be evaluated as $\beta_\star \approx 10^{16}(B_\star/1G)^{-2}$ denoting by $B_\star$ the stellar magnetic field. Decreasing $t_{\rm mag}$ to about a month such that it becomes of order the orbital period $t_{\rm min}$ of the most bound debris defined in equation (4) therefore requires $B_\star \gtrsim 10^5 G$ that corresponds to a highly-magnetized star. In this situation, the stream gets dominated by magnetic pressure that can cause a fast increase of its width and the end of confinement by self-gravity (Guillochon and McCourt 2017; Bonnerot et al. 2017a).

2.2.4 Stream-Disc Interaction

So far, we have assumed that the stream evolves in isolation but the debris may in fact be affected by the ambient gaseous medium present around the black hole. A possible interaction relates to the Kelvin-Helmholtz instability that can develop at the interface between the debris and the surrounding matter, resulting in an efficient mixing of these two fluids. A given stream element is affected by this process if the instability has fully developed by the time the corresponding debris comes back to pericenter. Using this criterion, Bonnerot et al. (2016a) finds that the entire stream can be impacted for the disruptions of red giants by black holes of mass $M_{\rm h} \gtrsim 10^9\ {\rm M}_\odot$. In this situation, most of the debris likely gets mixed with the ambient medium that can result in a sub-luminous event. Although this type of disruption is promising to probe the high-end of the black hole mass function (MacLeod et al. 2014), the above process may significantly complicate their detection. If the galaxy contains a denser disc, tidal streams produced by main sequence stars can also be affected by the instability. This possibility has been put forward to explain the sudden drop in luminosity detected from some events (Kathirgamaraju et al. 2017) and could be involved in the capture of the disc magnetic field by the debris stream (Kelley et al. 2014).

Additional effects can take place if a TDE happens in an active galactic nucleus that contains a dense accretion flow.[3] Using hydrodynamical simulations, Chan et al. (2019) studied the interaction of a debris stream with such a disc and find two different outcomes depending on the density of the incoming stellar matter. For low densities, the stream deposits most of its kinetic energy in the disc to get promptly absorbed by it. A stream of higher density can instead pierce through the disc while significantly perturbing the inflowing matter that can result in a fraction of the stream getting unbound and a depletion of the inner part of the accretion flow.

3 Disc Formation

After a revolution around the black hole, the bound part of the stream comes back to the stellar pericenter. We are interested here in the fate of this gas and how it evolves to form an accretion flow. A simple although unrealistic scenario consists in the immediate formation of a circular disc from the infalling debris near the circularization radius $R_{\rm circ} \approx 2R_{\rm p}$ that corresponds to the lowest energy state for the gas if each fluid element conserves its specific angular momentum.[4] In practice, this full circularization of the trajectories requires that a specific energy

$$\Delta\varepsilon_{\rm circ} \approx \frac{GM_{\rm h}}{4R_{\rm p}}, \tag{9}$$

is entirely lost from the system.[5] This picture was first described in early analytical works by Rees (1988) and Ulmer (1999), which pointed out that the timescale of disc formation is uncertain but may be of order the fallback time of equation (4). They also predicted that the disc would be located near the circularization radius $R_{\rm circ}$ and have a thick vertical profile due to the impact of radiation pressure on the gas. Since these pioneering works, multiple investigations have been made that aim at improving this original picture of disc formation. An exhaustive list of these works is included in Table 1 along with the choice of parameters made and the method used. These recent studies showed that the early interactions experienced by the returning stream typically dissipate an energy much lower than $\Delta\varepsilon_{\rm circ}$, implying that disc formation could require more than a fallback time to complete. The resulting gas distribution is also found to typically reach distances much larger than the circularization radius due to angular momentum redistribution between interacting fluid elements.[6] Addi-

[3]Note that the rate of tidal disruptions in such systems may be increased due to the modification of the gravitational potential by the mass of the disc (Karas and Šubr 2007), as discussed in more details in the Rates Chapter.

[4]If all the debris have the same specific angular momentum as that of the star on its original parabolic orbit, $l_\star = \sqrt{2GM_{\rm h}R_{\rm p}}$, the lowest energy state is a circular orbit at a distance $R_{\rm circ} = l_\star^2/GM_{\rm h} = 2R_{\rm p}$ from the black hole. Although the most bound part of the stream is on an elliptical orbit, this estimate remains valid since its typical eccentricity is $e_{\rm min} = 0.98 \approx 1$ as shown in equation (5).

[5]Although the most bound part of the stream comes back to pericenter with an already negative energy given by equation (2), it is much larger than that required to reach complete circularization, since $\Delta\varepsilon/\Delta\varepsilon_{\rm circ} \approx R_{\rm p}/a_{\rm min} \ll 1$, and therefore $\Delta\varepsilon$ is irrelevant in the computation of the energy loss necessary to reach this configuration.

[6]There is also observational evidence supporting the notion that disc formation occurs differently from what early works assumed. For example, a compact accretion disc cannot reproduced the high level of optical emission detected from many events (Lodato and Price 2010; Miller 2015). Additionally, most TDEs have an integrated energy lower than $M_\star\Delta\varepsilon_{\rm circ} \approx 10^{52}$ erg, by at least an order of magnitude. This "inverse energy crisis" is discussed in Sect. 3.5.2 along with possible solutions.

tionally, we emphasize that the gas evolution is expected to be qualitatively different depending on the region of parameter space considered. Although we refer to the gas structure formed from the debris around the black hole as a "disc", it can significantly differ from the simplest disc models (e.g. Shakura and Sunyaev 1973) used in the literature.[7] Despite recent progress in the understanding of disc formation, it is important to highlight that no clear consensus has so far been reached concerning the hydrodynamics of this process, largely due to numerical limitations that have so far prevented fully realistic global simulations of disc formation. In the following, we attempt to provide a description of the current status of theoretical knowledge, remaining as agnostic as possible while also highlighting (i) the main sources of uncertainties and (ii) the areas where secure conclusions appear possible.

In Sect. 3.1, we describe the dissipation mechanisms taking place shortly after the return of the stream near the black hole. The numerical challenges inherent to a global simulation of disc formation are discussed in Sect. 3.2 along with the compromises made so far to alleviate them. Section 3.3 presents the results of these numerical works that capture the evolution of the gas as the accretion flow assembles. The radiative efficiency of the different shocks involved in this process and the resulting emission are evaluated in Sect. 3.4. Finally, Sect. 3.5 describes the consequences of disc formation on the subsequent phase of accretion onto the black hole.

3.1 Early Sources of Dissipation

We start by discussing dissipation mechanisms taking place shortly after the debris comes back near the black hole, including the nozzle shock and the self-crossing shock. The nozzle shock is due to a strong compression of the stream during pericenter passage that results in an expansion of the gas as it then moves outward. Later on, an intersection of the stream with itself causes a self-crossing shock that can induce a significant modification of the gas trajectories.

3.1.1 Nozzle Shock

An early source of dissipation occurs during the passage of the stream at pericenter. In this region, the gas trajectories are initially specified by the gravity of the black hole. A consequence is that the fluid elements furthest from the original stellar orbital plane move on inclined orbits that intersect near pericenter, resulting in a compression of the stream in the vertical direction. This effect is analogous to the compression experienced by a star if it is disrupted on a plunging trajectory with $\beta > 1$. Here, the stream is always inside its own tidal radius, with an effective penetration factor $\beta_s \gg \beta$ since its density has decreased compared to the stellar value by several orders of magnitude (see Sect. 2). The compression results in a nozzle shock, during which the kinetic energy associated with vertical gas motion near pericenter gets dissipated. Note that compression also occurs within the orbital plane due to a convergence of the trajectories along this direction. However, this effect is not guaranteed to result in an intersection of the fluid elements, which likely reduces its impact on the gas evolution compared to that of vertical compression.

As discussed in the Disruption Chapter, the vertical velocity associated to the stellar compression is given approximately by $v_z \approx \alpha\, v_p \approx \beta c_{s,\star}$ where $c_{s,\star} = \sqrt{GM_\star/R_\star}$ denotes the stellar sound speed. Here, $v_p = (GM_h/R_p)^{1/2}$ is a characteristic velocity near pericenter

[7] As will be described more precisely in Sect. 3.3, it is for example possible that this disc remains globally eccentric, or inclined with respect to the black hole spin.

while $\alpha = R_\star/\sqrt{R_t R_p}$ denotes the inclination angle between the orbital plane of the compressed gas and that of the stellar center of mass (Carter and Luminet 1982). As pointed out by Guillochon et al. (2014), this expression can be extrapolated to study the compression of the returning debris, for which the vertical velocity becomes $v_{z,s} \approx \beta_s c_{s,s}$ where $c_{s,s}$ represents the sound speed inside the stream. Denoting by $\rho_\star$ and ρ_s the stellar and stream densities, the ratio of penetration factors is given by $\beta_s/\beta \approx (\rho_\star/\rho_s)^{1/3}$ since the tidal radius scales as $R_t \propto \rho^{-1/3}$ while the pericenter distance is unchanged. Similarly, the sound speed ratio can be evaluated as $c_{s,s}/c_{s,\star} \approx (\rho_\star/\rho_s)^{-1/3}$ since $c_s \propto (P/\rho)^{1/2} \propto \rho^{1/3}$ under the legitimate assumption of an adiabatic evolution for the debris with an adiabatic exponent $\gamma = 5/3$. The density ratio therefore cancels out such that $v_{z,s} \approx v_z$.[8]

The nozzle shock dissipates a large fraction of the specific kinetic energy associated with the vertical motion of the stream,

$$\Delta\varepsilon_{no} = v_{z,s}^2 \approx \beta^2 \frac{GM_\star}{R_\star}, \tag{10}$$

where we have used the above estimates for the vertical speed of $v_{z,s} = \beta\sqrt{GM_\star/R_\star}$. It follows that $\Delta\varepsilon_{no}/\Delta\varepsilon_{circ} \approx \alpha^2 \approx 10^{-4}\beta(M_h/10^6 M_\star)^{-2/3}$, making use of equation (9). This calculation suggests that the dissipation provided by the nozzle shock is negligible compared to that necessary to completely circularize the orbits except for very low black hole masses $M_h \lesssim 10^4\ M_\odot$, where $\Delta\varepsilon_{no} \gtrsim 10^{-1}\Delta\varepsilon_{circ}$ (Guillochon et al. 2014). The internal energy injected in the shocked gas can nevertheless cause the stream to expand significantly after leaving pericenter, an effect captured by several simulations described in Sect. 3.3, and especially strong for those considering an intermediate-mass black hole (Ramirez-Ruiz and Rosswog 2009; Shiokawa et al. 2015; Guillochon et al. 2014) or a large penetration factor (Sadowski et al. 2016).

So far, the nozzle shock has not been studied in a systematic fashion for typical parameters of the problem and our understanding of it is therefore limited. In particular, it is unclear whether the resulting gas expansion significantly modifies the stream properties compared to that before pericenter passage. As discussed more in details in Sect. 3.1.2, this effect can have consequences on the later evolution of the gas. Another important uncertainty concerns the exact amount of energy dissipated by this interaction. One possibility is that a fraction of the kinetic energy associated with the orbital motion is dissipated during this process, in addition to the much lower reservoir of vertical kinetic energy we have discussed so far. Due to the strong shearing expected at pericenter between neighbouring fluid elements, additional dissipation may be caused by effective viscosity, potentially provided by magnetic fields. Alternatively, additional shocks can take place if more intersections of the trajectories occur, which could result from general relativistic effects such as apsidal and nodal precessions, the latter being due to the black hole spin. As a result, the level of heating would be higher than that predicted from equation (10), causing the gas to expand faster from pericenter.

3.1.2 Self-Crossing Shock

Another early dissipation mechanism consists of a self-crossing shock caused by the intersection between the part of the stream that has passed pericenter and the debris still

[8] This relation can also be understood from the fact that the stream has a width similar to the original stellar radius when it comes back near pericenter. Because the compression takes place near the tidal radius, it occurs on the dynamical timescale of the star, which leads to $v_{z,s} \approx \sqrt{GM_\star/R_\star}$.

approaching the black hole. The way this collision takes place depends on the following physical effects: (i) relativistic apsidal precession, which modifies the stream trajectory near pericenter to put it on a collision course with the infalling gas, (ii) the expansion resulting from the nozzle shock described in Sect. 3.1.1, which leads to an increase of the stream width after pericenter passage and (iii) nodal precession induced by the black hole spin, which changes the orbital plane of the gas during pericenter passage. The works that specifically studied this process are included in Table 1 for entries whose first letter in the last column indicating the method used is either 'L' or 'B'. We start by evaluating the influence of these three mechanisms on the self-crossing shock and then describe the gas evolution during the collision for different regimes specified by the relative importance of these effects.

(i) Relativistic Apsidal Precession When the stream passes at pericenter, its center of mass precesses by an angle that can be approximated for a non-rotating black hole as (see equation (10.2) of Hobson et al. 2006)

$$\Delta\phi \approx 3\pi R_g/R_p \approx 11.5^\circ \beta \left(\frac{M_h}{10^6\,M_\odot}\right)^{2/3}, \tag{11}$$

where $R_g = GM_h/c^2$ is the gravitational radius and the approximation $a_{min}(1-e_{min}^2) \approx 2R_p$ has been used. The precession angle decreases or increases compared to equation (11) if the black hole has a non-zero spin with angular momentum vector in the same or opposite direction to that of the star's orbit, respectively. Apsidal precession by itself can result in a collision between the part of the stream moving outward and that still approaching the black hole.[9] The self-crossing shock taking place when this mechanism dominates is illustrated in the left panel (a) of Fig. 3. Assuming that the two colliding components remain thin with widths $H_1 \approx H_2 \ll R_{int}$, the collision can be considered to happen at a single location. Its characteristics can then be derived analytically by considering two Keplerian ellipses of eccentricity e_{min} and pericenter R_p rotated according to the above precession angle, as done by several authors (Dai et al. 2015; Jiang et al. 2016b; Bonnerot et al. 2017b; Lu and Bonnerot 2020).[10] The resulting collision occurs at a true anomaly $\theta_{int} = \pi - \Delta\phi/2$ that corresponds to an intersection radius of

$$R_{int} = \frac{1}{\beta}\frac{R_t(1+e_{min})}{1-e_{min}\cos(\Delta\phi/2)}, \tag{12}$$

which decreases monotonically from the apocenter of the most bound debris to its pericenter as the relativistic apsidal angle $\Delta\phi$ increases. The two stream components collide with velocity vectors $\mathbf{v}_1^{int}$ and $\mathbf{v}_2^{int}$ of the same magnitude $v_{int} \equiv |\mathbf{v}_1^{int}| = |\mathbf{v}_2^{int}|$ but inclined by the collision angle ψ given by

$$\cos\psi = \frac{1-2e_{min}\cos(\Delta\phi/2)+e_{min}^2\cos\Delta\phi}{1-2e_{min}\cos(\Delta\phi/2)+e_{min}^2}, \tag{13}$$

[9] Another source of interactions for the stream involves the sequential tidal disruption of the two components of a binary star by a black hole (Bonnerot and Rossi 2019), during which the two debris streams can collide with each other due to the difference in their trajectory induced by the previous binary separation.

[10] Note that this treatment is approximate due to the difference in orbital energy between the two gas components involved in the collision. In reality, the part of the stream moving away from the black hole is more bound than the approaching one with a lower apocenter distance.

which can be obtained from $\cos\psi = \mathbf{v}_1^{\rm int} \cdot \mathbf{v}_2^{\rm int}/v_{\rm int}^2$ and expressing the two velocities as a function of $R_{\rm int}$. This angle is small only when apsidal precession is very weak, corresponding to low-mass black holes, i.e. $M_{\rm h} \lesssim 10^5\ \mathrm{M}_\odot$ according to equation (11). Otherwise, the velocity vectors are significantly misaligned, with directions close to being completely opposite with $\psi \approx \pi$. In the typical situation where $M_{\rm h} \approx 10^6\ \mathrm{M}_\odot$ and $\beta \approx 1$, apsidal precession is relatively weak such that the collision happens near apocenter with $R_{\rm int} \approx 2a_{\rm min}$ according to equation (12). For the two stream components to collide instead near pericenter with $R_{\rm int} \approx R_{\rm p}$, apsidal precession must be stronger, requiring that the black hole mass increases to $M_{\rm h} \gtrsim 10^7\ \mathrm{M}_\odot$ if $\beta \approx 1$ or the penetration factor to $\beta \gtrsim 5$ if $M_{\rm h} \approx 10^6\ \mathrm{M}_\odot$.

(ii) Expansion from the Nozzle Shock As described in Sect. 3.1.1, dissipation takes place during the strong compression of the stream at pericenter, leading to gas expansion as it then moves away from the black hole. As a result, this part of the stream can be thicker than that still moving inwards, so that the nozzle shock induces an asymmetry between these two components. Due to our limited understanding of the nozzle shock, its impact on the stream width past pericenter has so far not been precisely quantified. As we explained, the amount of dissipation during this interaction is uncertain, being potentially much larger than current analytical estimates. Additionally, the direction along which the gas preferentially bounces following the compression is largely unknown. It is therefore unclear whether the shocked part of the stream mainly expands within the orbital plane or perpendicular to it. This effect is particularly important since it determines the geometry of the outgoing stream component that may be involved in the self-crossing shock. To take this ambiguity into account, we allow this part of the stream to become thicker than the infalling matter, with $H_2 \geq H_1$ as a result of the nozzle shock. The width[11] H_2 can in principle take a wide range of possible values, from $H_2 = H_1$ up to a size comparable to the intersection radius, i.e. $H_2 \approx R_{\rm int}$. The left panel (a) of Fig. 3 illustrates the regime where the self-crossing shock is still mainly determined by relativistic apsidal precession, but involves stream components of slightly different widths with $H_2 \gtrsim H_1$. In the right panel (b) of Fig. 3, relativistic precession is negligible and the streams have very different widths, with $H_1 \ll H_2 \approx R_{\rm int} = 2a_{\rm min}$. The large thickness of the outgoing gas implies that the intersection takes place at a wide range of distances from the black hole. These interactions are indirectly caused by the nozzle shock that deflects a fraction of the gas in the direction of the infalling debris, even in the absence of relativistic precession. A self-crossing shock induced by this effective precession may represent an important dissipation mechanism when relativistic precession is very weak. This scenario seems particularly applicable to black holes with masses $M_{\rm h} \lesssim 10^5 M_{\rm h}$, for which the dynamical impact of the nozzle shock is additionally enhanced, as predicted by the estimates of Sect. 3.1.1. A few simulations considering intermediate-mass black holes (Ramirez-Ruiz and Rosswog 2009; Guillochon et al. 2014; Shiokawa et al. 2015) have captured this effect but its role in the disc formation process has not been further investigated in a systematic way.

(iii) Spin-Induced Nodal Precession If the black hole has a non-zero spin, the stream experiences nodal precession (dominated by the Lense-Thirring effect) near pericenter that modifies its orbital plane. As first pointed out by Cannizzo et al. (1990), this effect produces a vertical offset Δz between the infalling and outgoing debris at the intersection radius where they would otherwise collide. If $\Delta z < H_1 + H_2$, the self-crossing shock still occurs,

[11] Although for simplicity this description involves a single width H_2 for the outgoing stream component, it is not guaranteed that the gas distribution remains cylindrical following the expansion.

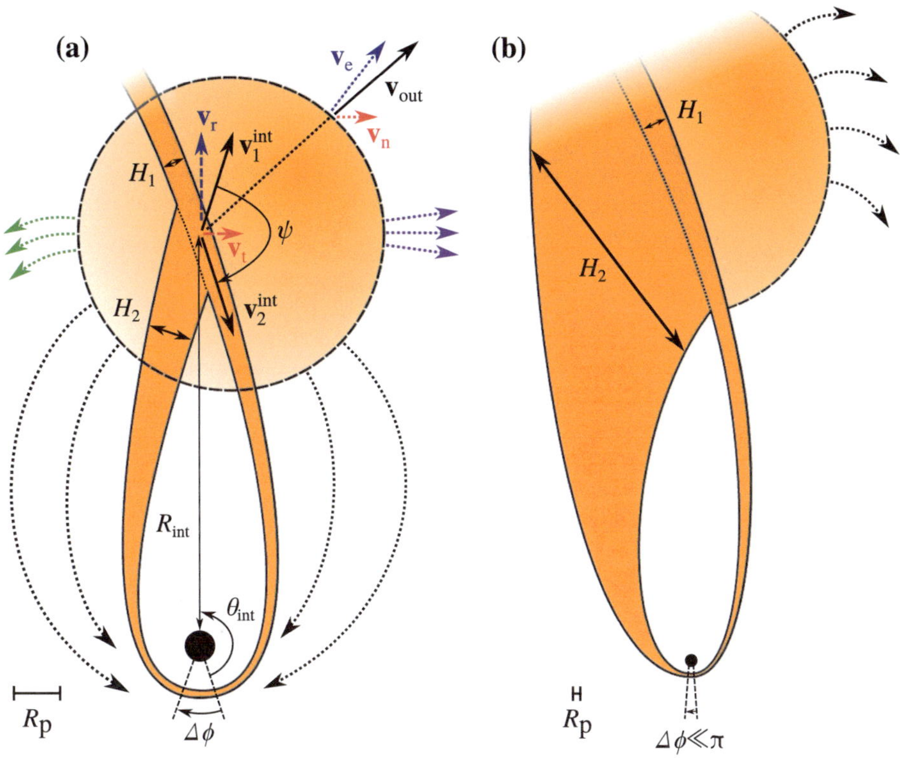

Fig. 3 Sketches representing the self-crossing shock caused by an intersection of the stream with itself, and the subsequent gas evolution in two different regimes. In the left panel (a), the collision is dominated by relativistic apsidal precession, which makes the tip of the stream precess by an angle $\Delta\phi$ during pericenter passage. The resulting intersection takes place at a distance $R_{\rm int}$ from the black hole that corresponds to a true anomaly $\theta_{\rm int} = \pi - \Delta\phi/2$ for the trajectories of the stream components that are assumed to have retained similar widths $H_1 \approx H_2 \ll R_{\rm int}$. Their velocities $\mathbf{v}_1^{\rm int}$ and $\mathbf{v}_2^{\rm int}$ are of the same magnitude but are misaligned by the collision angle ψ, such that they have identical tangential components $\mathbf{v}_{\rm t}$ but opposite radial ones $\pm\mathbf{v}_{\rm r}$. If the widths of the colliding gas are identical with $H_1 = H_2$ with no vertical offset $\Delta z = 0$ between them, all their mass passes through the self-crossing shock, resulting in the dissipation of the kinetic energy associated with the gas radial motion. This causes the formation of an outflow (orange sphere delimited by black dashes) driven by radiation pressure, whose velocity can be decomposed into a net and expansion component as $\mathbf{v}_{\rm out} = \mathbf{v}_{\rm e} + \mathbf{v}_{\rm n}$ where $|\mathbf{v}_{\rm e}| \approx |\mathbf{v}_{\rm r}|$ and $\mathbf{v}_{\rm n} \approx \mathbf{v}_{\rm t}$. Gas expanding in the direction of its net motion (dotted purple arrows) can get unbound from the black hole while that moving in the opposite way (green dotted arrows) can acquire a direction of motion opposite to that of the original star. If the stream components have a slight width difference with $H_1 \lesssim H_2$ or are weakly misaligned with $\Delta z \ll H_1 + H_2$, a small fraction of the gas can avoid the intersection while the rest experiences a collision qualitatively similar to that described above. In the right panel (b), we consider a different limit, wherein the expansion following the nozzle shock produces very different widths for the two stream components, namely $H_1 \ll H_2$. As a result, intersections occur across a broad range of radii even if relativistic precession is negligible ($\Delta\phi \ll \pi$). Due to the large density difference between the two colliding components, the infalling gas can pass through the outgoing debris with a largely unaffected trajectory. It is possible that a bow shock forms that heats the outward-moving gas causing its mild expansion (orange half-sphere delimited by black dashes) following the intersection. In both regimes, the gas that remains bound after the self-crossing shock eventually returns near the black hole (black dotted arrows) where it can experience additional interactions

but the two stream components are misaligned. The interaction is entirely avoided if $\Delta z > H_1 + H_2$: in this case, the streams are unable to impact one another, and keep following their trajectories unimpeded. Jiang et al. (2016b) carried out the following analytical estimate to determine whether this initial self-crossing shock is avoided. Nodal precession causes the gas angular momentum vector to precess around the black hole spin direction by an angle $\Delta\Omega \approx 4\pi\, 2^{-3/2} a \sin i\, (R_g/R_p)^{3/2} \approx 1° a \sin i\, \beta^{3/2} (M_h/10^6\,\mathrm{M}_\odot)$ (Merritt 2013), where a denotes the dimensionless spin parameter and i represents the angle between the black hole spin and the stellar angular momentum that is generally non-zero.[12] The resulting shift in orbital plane causes the centers of mass of the two gas components to be separated by a distance of $\Delta z \approx R_{\rm int} \Delta\Omega$ at the intersection radius given by equation (12). Additionally, the sum of the stream widths at this position is approximated by a single value $H = H_1 + H_2 \approx R_\star R_{\rm int}/R_p$ that expands homologously (see Sect. 2) from pericenter where it is assumed to be of a stellar radius. The ratio of vertical offset to stream width is then given by

$$\frac{\Delta z}{H} \approx a \sin i\, \beta^{1/2} \left(\frac{M_h}{10^6\,\mathrm{M}_\odot} \right)^{4/3}. \tag{14}$$

A prompt collision between misaligned stream components can take place if $\Delta z < H$, which is satisfied for low-mass and slowly-spinning black holes with $M_h \lesssim 10^6\,\mathrm{M}_\odot$ and $a \lesssim 1$ if $\beta \approx 1$ and $i \approx 90°$ according to this estimate. The intersection is instead completely avoided with $\Delta z > H$ if the black hole has a larger mass $M_h \gtrsim 10^7\,\mathrm{M}_\odot$ and is maximally-spinning with $a \approx 1$, keeping the other parameters fixed. If this early self-crossing shock does not occur, the stream needs to complete additional revolutions around the black hole before eventually colliding with itself. Dai et al. (2013) studied this effect first by integrating the Kerr geodesics followed by the stream. However, the impact of nodal precession was likely overestimated by assuming that the gas is still confined by self-gravity, which is unlikely to be valid near the black hole. A more extensive investigation was done by Guillochon and Ramirez-Ruiz (2015) using a more complex, semi-analytical geodesic model that assumes that the stream width evolves homologously ($H \propto R$), and also includes the additional expansion induced by the nozzle shock at each pericenter passage. In the regime where prompt self-intersection fails, streams orbiting the black hole will therefore become progressively thicker until two different windings eventually intersect with $\Delta z < H_1 + H_2$, leading to a self-crossing shock. An example of stream evolution obtained from this work is shown in the left panel of Fig. 4, while the right panel displays the time $t_{\rm delay}$ between the stellar disruption and the first successful collision as a function of black hole mass. This time delay is typically several orbital periods and increases with black hole mass to reach of order a year for $M_h = 10^7\,\mathrm{M}_\odot$. A delayed intersection is also more likely to happen between two successive windings, according to this study. In this case, the trajectories of the colliding debris may be close to that of the initial self-crossing shock shown in the left panel (a) of Fig. 3. However, the stream components may have a residual misalignment and widths larger than that of the initially returning debris due to the cumulative effect of successive nozzle shocks. Additionally, the cold streams present around the black hole due to previous missed collisions may affect the gas evolution at later times.

Depending on the importance of the effects described above, the subsequent gas evolution can differ significantly. To determine the impact of relativistic precession alone, we start by making the idealized assumption that both the nozzle shock and nodal precession have a negligible impact on the resulting self-crossing shock. The stream components can therefore be approximated as having the same width $H_1 = H_2$ and being perfectly aligned with $\Delta z =$

[12] As explained in the Rates Chapter, victim stars sourced by two-body relaxation approach the black hole with a quasi-isotropic distribution of orientations, although alternative scenarios (Wernke and Madigan 2019) are possible. As a result, the stellar angular momentum is in general not aligned with the black hole spin: $i \neq 0$.

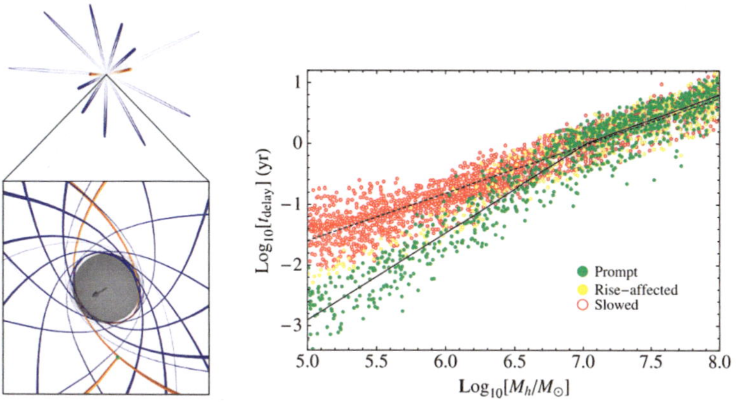

Fig. 4 Left panel: Trajectory of the stream of debris around a spinning black hole (grey sphere) obtained from the semi-analytical model of Guillochon and Ramirez-Ruiz (2015). Because of Lense-Thirring precession, the stream experiences several windings (blue curves) before eventually colliding, as shown with the orange curves that intersect at the green point. Right panel: Time $t_{\rm delay}$ elapsed between the passage of the star at pericenter and the intersection of the debris stream as a function of black hole mass. Each point corresponds to a different set of parameters assumed for the above process with the colours related to how fast the resulting disc viscously accretes with respect to the time of peak fallback rate

0. In this situation, the collision takes place in a confined region and involves all the mass of the two streams that intersect with opposite radial velocities $\pm\mathbf{v}_{\rm r}$ as indicated in the left panel (a) of Fig. 3. Because the inflow rate through the collision point is similar[13] for the two components, the specific energy dissipated by the interaction reaches its maximal value

$$\Delta\varepsilon_{\rm sc}^{\rm max} = \frac{v_{\rm r}^2}{2} \approx \frac{GM_{\rm h}}{R_{\rm int}} \sin^2(\psi/2), \tag{15}$$

since all the kinetic energy associated with radial motion gets converted into heat. In the second step, the radial velocity is expressed as $v_{\rm r} = |\mathbf{v}_{\rm r}| = v_{\rm int}\sin(\psi/2)$ and the gas velocity at the self-crossing shock is approximated by the local escape speed with $v_{\rm int} \approx (2GM_{\rm h}/R_{\rm int})^{1/2}$. Except for very low black hole masses, $\psi \approx \pi$, so that a significant fraction of the binding energy at the intersection radius gets dissipated. A dissipated energy $\Delta\varepsilon_{\rm sc} \approx \Delta\varepsilon_{\rm circ}$ requires $R_{\rm int} \approx R_{\rm p}$, which is only attained for $M_{\rm h} \gtrsim 10^7\ \mathrm{M}_\odot$ or $\beta \gtrsim 5$, as explained above. Only in this situation can we expect a fraction of the gas to reach a significant level of circularization as a result of the first collision, although even in this case, a large fraction of the gas may be unbound from the black hole. Otherwise, relativistic precession is weaker, with $\Delta\phi \ll \pi$, often implying $R_{\rm int} \approx a_{\rm min}$ and $\Delta\varepsilon_{\rm sc} \approx 10^{-2}\Delta\varepsilon_{\rm circ}$. Following this interaction away from pericenter, most of the gas remains therefore significantly eccentric.

The hydrodynamics of the self-crossing shock has been investigated through local simulations initialized with two identical thin stream components whose properties are obtained analytically (Kim et al. 1999; Jiang et al. 2016b; Lu and Bonnerot 2020). As we explain in Sect. 3.4, the collision is found to be radiatively inefficient in this limit due to the large optical depths of the confined gas such that the evolution can be accurately approximated

[13] Due to the time-dependence of the fallback rate, the inflow rate through the collision point may differ for the two stream components, especially if the intersection happens at large distances with $R_{\rm int} \gtrsim a_{\rm min}$.

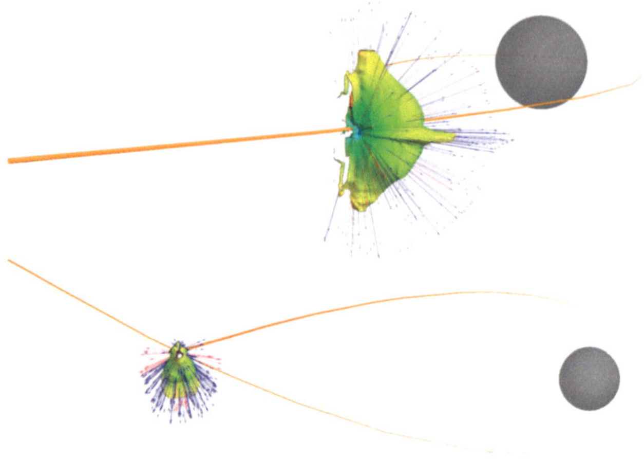

Fig. 5 Gas evolution resulting from the self-crossing shock as obtained from the local radiative transfer simulation of Jiang et al. (2016b) with the two panels adopting different points of view. The stream (orange tube) experiences apsidal precession during its passage close to the black hole (grey sphere) that results in its intersection. Due to the pressure increase, the ensuing self-crossing shock causes the formation of an outflow launched from the intersection point displayed with the blue to yellow surface. The arrows represent the gas velocity for the bound (blue) and unbound (red) matter

as adiabatic. These numerical works demonstrate that the shocked gas gets launched into an outflow driven by radiation pressure.[14] It can be seen from Fig. 5 that corresponds to the local radiation-hydrodynamics simulation of a strong intersection by Jiang et al. (2016b) where the two thin, intersecting stream components are displayed as orange tubes while the outflow is represented with the blue to yellow surface. An idealized version of this outflowing gas is also illustrated by the orange sphere delimited by black dashes in the left panel (a) of Fig. 3. In addition to this quasi-spherical expansion due the excess internal energy given by equation (15), the gas also retains a net velocity with respect to the black hole inherited from the tangential motion of the colliding streams. The velocity vector of this outflowing matter can therefore be approximately decomposed as $\mathbf{v}_{\rm out} = \mathbf{v}_{\rm e} + \mathbf{v}_{\rm n}$ where $|\mathbf{v}_{\rm e}| \approx v_{\rm r}$ and $\mathbf{v}_{\rm n} \approx \mathbf{v}_{\rm t}$, as illustrated in the left panel (a) of Fig. 3. This implies that the gas expanding near the direction of its net motion has a velocity $|\mathbf{v}_{\rm n}| + |\mathbf{v}_{\rm e}| > (\mathbf{v}_{\rm r}^2 + \mathbf{v}_{\rm t}^2)^{1/2} = v_{\rm int} \approx (2GM_{\rm h}/R_{\rm int})^{1/2}$ that is larger than the local escape speed (purple dotted arrows). The fraction of debris unbound by the collision becomes significant if the self-crossing shock is strong. For $\beta = 1$, unbinding more than 20% of the colliding gas requires a black hole mass $M_{\rm h} \gtrsim 5 \times 10^6\ {\rm M}_\odot$ according to local adiabatic simulations (Lu and Bonnerot 2020).[15] Similarly, the gas that

[14]The self-crossing shock leads to an increase of the gas temperature to $T_{\rm g} \approx m_{\rm p} v_{\rm r}^2/k_{\rm B} \approx 10^9$ K using equation (15) and a typical radial velocity $v_{\rm r} \approx 0.01\,c$ for the colliding gas. This shocked matter rapidly cools by emitting photons that results in an increase of radiation energy while that of the gas diminishes. At the end of this process, most of the energy is in the form of photons with an equilibrium temperature of $T_{\rm eq} \approx (\rho_{\rm s} v_{\rm r}^2/a)^{1/4} \approx 10^6$ K, where in the lattermost equality we have adopted a density inside the stream of $\rho_{\rm s} \approx 10^{-7}\ {\rm g\,cm^{-3}}$ as derived in Sect. 2 (see Fig. 1). The ratio of gas to radiation pressure is then $P_{\rm g}/P_{\rm r} \approx (\rho k_{\rm B} T_{\rm eq}/m_{\rm p})/(a T_{\rm eq}^4/3) \approx 10^{-3}$.

[15]For the strong self-crossing shock they consider, the radiation-hydrodynamics simulation performed by Jiang et al. (2016b) finds that about 16% of the debris gets unbound, which is lower than the value of 50% obtained from the adiabatic study of Lu and Bonnerot (2020). A possible origin for this discrepancy is that radiative losses reduce the impact of radiation pressure on the gas that therefore gets accelerated to a lower ter-

gets accelerated in the direction opposite to its net speed (green dotted arrows) experiences a change in sign of its angular momentum if $|\mathbf{v}_\mathrm{n}| < |\mathbf{v}_\mathrm{e}|$, so that it ends up rotating in the direction opposite to that of the original star around the black hole.

Most of the local studies dedicated to the self-crossing shock have been made in the idealized situation we just discussed.[16] We present now possible consequences of a more realistic configuration involving two stream components that are misaligned with different widths. For similar widths $H_1 \lesssim H_2$ and a mild vertical offset $\Delta z \ll H_1 + H_2$, only a small fraction of the colliding streams can avoid the interaction. Additionally, the two stream components may reach the intersection point with slightly different inflow rates. These effects tend to reduce the dissipated specific energy below the maximal value of equation (15), i.e. $\Delta\varepsilon_\mathrm{sc} \lesssim \Delta\varepsilon_\mathrm{sc}^\mathrm{max}$. As a result, the shocked gas will expand slower, with less unbound mass, and may evolve into an outflow that deviates from a quasi-spherical wind due to the asymmetry of the collision. Despite these possible differences, the gas evolution is likely to remain qualitatively similar to that predicted above for identical streams. On the other hand, if the widths of the two stream components are very different, with $H_1 \ll H_2$ due to a fast expansion induced by the nozzle shock, most of the outgoing gas can avoid a direct collision. Due to its much larger density, the infalling debris would be largely unaffected by the interaction and act instead as a static obstacle. As a result, a bow shock could form inside the thicker stream component that may cause a slow expansion of this gas when it continues to move outward past the intersection, as is illustrated with the orange half-sphere in panel (b) of Fig. 3. The energy dissipated during this process is likely much lower than that given by equation (15), with $\Delta\varepsilon_\mathrm{sc} \ll \Delta\varepsilon_\mathrm{sc}^\mathrm{max}$, while the radiative efficiency could be significantly higher than mentioned above, as we discuss more in detail in Sect. 3.4. If the outgoing stream component has a width $H_2 \approx R_\mathrm{int}$ similar to the intersection radius of equation (12), the self-crossing-shock takes place at a wide range of distances from the black hole due to a deflection of its trajectories induced by the nozzle shock. Despite the overall weak dissipation, a fraction of the debris could collide near pericenter due to a strong effective precession that may result in its circularization. As we mentioned before, this effect could be important for low-mass black holes with $M_\mathrm{h} \lesssim 10^5\,\mathrm{M}_\odot$. These important differences caused by the increased gas thickness suggest that the hydrodynamics in this regime qualitatively differs from the case of identical streams. However, additional work is needed to better understand the gas evolution during a realistic self-crossing shock.

As we have seen, several possible outcomes are possible for the self-crossing shock depending on how this interaction takes place. It is, however, still largely unclear for which parameters of the problem they are each realised, due to uncertainties in the mechanisms involved, most importantly the expansion at the nozzle shock. Collisions near pericenter are necessary to dissipate a specific energy $\Delta\varepsilon_\mathrm{sc} \approx \Delta\epsilon_\mathrm{circ}$ but they require either strong relativistic apsidal precession or a large expansion induced by the nozzle shock that only seems realised in specific conditions. In this case, the self-crossing shock itself could result in a significant level of circularization. More generally, interactions happen near the apocenter of the most bound debris with $\Delta\varepsilon_\mathrm{sc} \ll \Delta\epsilon_\mathrm{circ}$ such that the shocked gas retains large eccentricities. More dissipation would therefore be necessary for the gas trajectories to further circularize. In each regime, the gas remaining bound following the self-crossing shock comes

minal speed, although it may also result from differences in the numerical setup used or the initial properties of the colliding streams.

[16] Note that the self-crossing shock found in most global simulations of disc formation presented in Sect. 3.3 actually differ from the ideal situation of identical stream components, but this difference may, at least partially, originate from the simplified initial conditions used in these studies.

back towards the black hole at later times, as indicated with the dotted arrows in the two panels of Fig. 3. During its subsequent evolution, this matter is likely to experience more interactions, as will be discussed in Sect. 3.3 based on global simulations.

3.2 Computational Challenges and Treatment of General-Relativistic Effects

In Sect. 3.1, we described the early dynamics followed by the debris based on analytical arguments and local numerical studies of the self-crossing shock that apply for the realistic situation of a star disrupted on a parabolic orbit by a supermassive black hole. Studying the more complex evolution of this matter at later times requires performing a global simulation of the disc formation process. However, this task turns out to be very numerically challenging for the following reasons. The fluid elements inside the stream have semi-major axes larger than that of the most bound debris $a_{\rm min}$ while the returning matter evolves on scales similar to $R_{\rm p} \ll a_{\rm min}$ that are two to three orders of magnitude smaller. The computational cost associated with this large dynamic range is exacerbated by the high aspect ratio of the stream due to the transverse confinement by self-gravity that limits its width to $H \approx 10 R_\star \ll R_{\rm p}$ (see Sect. 2.1). This numerical challenge has long been recognized. The first hydrodynamical simulation of the disc formation process by Ayal et al. (2000) was realised for a realistic setup using a post-Newtonian smoothed-particle hydrodynamics technique. Despite the particle splitting technique used, it is now clear that the numerical resolution of this work is insufficient to accurately follow the gas evolution near pericenter, leading to important numerical artefacts such as the unbinding of most of the returning stream.[17] The numerical works investigating the global gas evolution during disc formation are included in Table 1 for entries with a first letter 'G' in the last column corresponding to the method used. The choice of parameters and the technique employed to account for general-relativistic effects are also indicated. In order to have sufficient resolution, most of these works had to adopt simplified initial conditions that are not astrophysically realistic. Two different "compromises" have been used that aim at artificially decreasing the dynamic range of the problem to make it computationally tractable. More recently, Bonnerot and Lu (2020) used a third strategy that consists in injecting gas into the computational domain to model the outflow resulting from the self-crossing shock. This method allowed them to perform for the first time a simulation of disc formation for realistic parameters of the problem. We start by describing these approaches below along with the astrophysical limitations of each and then focus on the treatment of general relativistic effects. A detailed discussion of different numerical methods can be found in the Simulation Methods Chapter while the Future Modeling Chapter describes innovative techniques that could be used in the future.

The first compromise is to simulate the disruption of a star by an intermediate-mass black hole rather than a supermassive one. Since $a_{\rm min}/R_{\rm t} \propto M_{\rm h}^{1/3}$ (see equations (1) and (3)), the dynamic range may be reduced by an order of magnitude by considering a black hole mass of $M_{\rm h} = 10^3 M_\odot$ instead of a more realistic choice of $M_{\rm h} = 10^6\ {\rm M}_\odot$. The first such simulations (Rosswog et al. 2008; Ramirez-Ruiz and Rosswog 2009; Rosswog et al. 2009) were carried out to study tidal disruptions by intermediate-mass black holes for themselves. One

[17] Analytically, this unbinding of mass at pericenter is disfavoured by the fact that the energy dissipated by the nozzle shock is much less than the binding energy of the most bound debris with a ratio $\Delta\varepsilon_{\rm no}/\Delta\varepsilon = \beta^2 (M_{\rm h}/M_\star)^{-1/3} \approx 0.01 \ll 1$ using equations (2) and (10). This effect has also not been found in any of the simulations performed at higher resolution, where the stream instead expands around its original trajectory (e.g. Guillochon et al. 2014). Its numerical origin in the low resolution simulations by Ayal et al. (2000) likely results from an inaccurate computation of pressure forces at pericenter. Because this matter moves on near-parabolic trajectories, the resulting variations in velocities may cause the artificial unbinding.

important motivation is that white dwarves can only be disrupted outside the event horizon of the black hole for $M_{\rm h} \lesssim 10^5$ $\rm M_{\odot}$. Later numerical works (Guillochon et al. 2014; Shiokawa et al. 2015) were instead performed with the intention to get insight into the case of a supermassive black hole. However, disruptions by such low-mass black holes differ from the realistic situation in three main ways that must be kept in mind when attempting to extrapolate the results: (i) the dynamical importance of energy dissipation at the nozzle shock is overestimated as shown in Sect. 3.1.1 (Guillochon et al. 2014), (ii) the pericenter distance is less relativistic since $R_{\rm t}/R_{\rm g} \propto M_{\rm h}^{-2/3}$ that can reduce the apsidal precession angle[18] and strength of the self-crossing shock (see Sect. 3.1.2) and (iii) the tidal approximation is itself less generally well-satisfied as $R_{\rm t}/R_{\star} \propto M_{\rm h}^{1/3}$. Of these three effects, (i) and (ii) may be quite important for the process of disc formation. It is less likely that (iii) has direct relevance for this phase of evolution, although it may result in a more subtle shift in the energy distribution of the debris (Brassart and Luminet 2008).

The second numerical compromise that has been made is to simulate the disruption by a supermassive black hole of a star on an initially bound rather than parabolic trajectory. The associated decrease in the semi-major axis $a_{\rm min}$ of the most-bound debris can significantly reduce the numerical cost of the simulation for stellar eccentricities of $e_{\star} \lesssim 0.95$. Stellar disruptions can occur on bound trajectories in some situations involving the previous tidal separation of a binary star (Amaro-Seoane et al. 2012), a recoiling (Stone and Loeb 2012) or binary (Coughlin et al. 2017) supermassive black hole, as described in the Binaries Chapter. However, the main motivation for carrying out such simulations is to extrapolate the results to the realistic case of an initial parabolic trajectory. This technique was first used by Hayasaki et al. (2013) and later in several other investigations (Bonnerot et al. 2016b; Hayasaki et al. 2016; Sadowski et al. 2016). Although this approach does not suffer from the issues listed above for the first method since it uses the correct black hole mass, it has its own disadvantages: (i) the stream produced by the disruption of a *bound* star is artificially truncated and, if $e_{\star}$ is too small, one winds up in the unrealistic situation where the "head" of the stream must fruitlessly chase its "tail" for several orbits before any major interaction takes place and (ii) the debris stream produced by the disruption process is artificially bound with an apocenter much less than $a_{\rm min}$, which artificially enhances the strength of the self-crossing shock.

The strategy designed by Bonnerot and Lu (2020) consists in modelling the outflow from the self-crossing shock by an injection of gas into the computational domain. This method is promising since it can be applied for both a parabolic orbit of the star and a supermassive black hole mass. Nevertheless, the idealized treatment of the self-crossing shock implies the following main caveats: (i) the properties of this interaction are fixed, which neglects any feedback resulting from the formation of the disc such as its impact on the stream trajectory near pericenter and (ii) this strategy requires that the self-crossing shock can be treated as a local interaction, which may not be possible if one of the colliding stream components has a width similar to the intersection radius (see right panel of Fig. 3).

General-relativistic effects are essential to properly describe the gas dynamics during disc formation and must therefore be taken into account to simulate this phase of evolution. In particular, this process is thought to be initiated by a self-crossing shock (see Sect. 3.1.2) that depends on both relativistic apsidal and nodal precession, the latter being induced by the black hole spin. Despite the large computational cost of this approach, some numerical

[18] Relativistic apsidal precession can be important for TDEs involving intermediate-mass black holes if $\beta \gg 1$ but this type of deeply-penetrating encounter is uncommon and poses its own computational challenges related to accurately resolving mid-plane compression in the disrupting star and the returning debris streams.

works (e.g. Shiokawa et al. 2015; Sadowski et al. 2016) adopt a fully general-relativistic treatment. To reduce this numerical expense, other investigations take these effects into account through the use of gravitational potentials that differ from the usual Keplerian one of $\Phi = -GM_{\rm h}/R$ while still using a Newtonian description of the hydrodynamics.[19] The first approach adopted by several authors (Rosswog et al. 2009; Hayasaki et al. 2013; Bonnerot et al. 2016b) is to use so called pseudo-Newtonian potentials that are designed by hand to reproduce certain features of general-relativistic gravity. Popular choices include the famous Paczynski-Wiita potential (Paczyńsky and Wiita 1980) and those designed by Wegg and Bode (2011) and Tejeda and Rosswog (2013). The latter two are especially relevant since they are tailored to reproduce apsidal precession rates in a Schwarzschild spacetime. A second option is to use post-Newtonian potentials (e.g. Blanchet 2014), which are self-consistently derived from the low-velocity limit of general relativity. This is the approach that was used in the first simulations of mass return in TDEs (Ayal et al. 2000). However, the appeal of self-consistency is often outweighed by the poor convergence of the post-Newtonian approximation as distances approach $R_{\rm g}$. Test particle dynamics in such potentials are increasingly untrustworthy at $R \lesssim 15 R_{\rm g}$, where large deviations from general-relativistic geodesics occur, although it is a good approximation at larger scales. The main advantage of this approach is that it offers a well-defined way to incorporate nodal precession due to black hole spin (Faye et al. 2006) that has been exploited by Hayasaki et al. (2016).

3.3 Global Hydrodynamics of Accretion Flow Formation

As explained in Sect. 3.2, it is not yet computationally feasible to follow the hydrodynamics of disc formation for realistic initial conditions. The global simulations listed in Table 1 therefore rely on computationally-motivated idealizations to make the problem tractable. In most of them, this is achieved by either considering an intermediate-mass black hole or an initially bound star. More recently, Bonnerot and Lu (2020) were able to consider realistic astrophysical parameters of the problem using another simplifying strategy: treating the self-crossing shock as an injection of gas into the computational domain. Most of these works adopt an adiabatic equation of state for the gas while a minority of others (e.g. Hayasaki et al. 2013) assumes that the debris evolves instead isothermally. The adiabatic choice is motivated by the radiative inefficiency of the self-crossing shock if it involves thin stream components (see Sect. 3.1.2) but it is unclear whether this assumption remains acceptable in the later stages of disc formation. As we discuss in Sect. 3.4, it is possible that the circularizing shocks become more radiatively efficient as the accretion flow starts forming, with the hydrodynamics significantly deviating from complete adiabaticity. Most of the numerical works below focus entirely on the formation of the disc and do not follow the subsequent phase of accretion onto the black hole, which is the main topic of the Accretion Disc Chapter that we also discuss separately in Sect. 3.5. The impact of magnetic fields that are expected to drive accretion is examined in only one of these investigations (Sadowski et al. 2016) while the others consider a purely hydrodynamical evolution. Accretion driven by magnetized turbulence, as well as additional dissipative processes (e.g. interactions with the stream of debris that continues impacting the accretion disc) may lead to disc evolution

[19] Note that several investigations (Guillochon et al. 2014; Bonnerot et al. 2016b) nevertheless adopt an entirely Keplerian description. This approach is legitimate when the precession angle is small but can also be useful to identify whether some features of the gas evolution occur independently of general relativistic effects.

Table 1 Chronological list of the works investigating the gas evolution leading to disc formation in TDEs and the choice of parameters adopted. The method used is specified in the last column by two sets of letters divided by a slash, with the first one corresponding to the numerical strategy used to evolve the gas and the second indicating how the black hole's gravity is treated. The first letter 'B' stands for a ballistic evolution as test particles using a simplified analytical treatment for hydrodynamics, 'L' for local hydrodynamical simulations that focus on a small domain as convenient to study the self-crossing shock and 'G' for global simulations that follow the hydrodynamics for a large fraction of the volume, as necessary to study disc formation. For the second set of letters after the slash, 'N' stands for Newtonian gravity, 'PNP' for pseudo-Newtonian potential, 'PNC' for post-Newtonian corrections, and 'GR' for methods that either fully solve the Einstein's field equations or assume a fixed background metric to evolve the gas

Reference	M_h ($M_\odot$)	a	$e_\star$	β	Method
Kim et al. (1999)	10^6	0	1	1	L/N
Ayal et al. (2000)[a]	10^6	0	1	1	G/GR
Ramirez-Ruiz et al. (2009)[b]	10^3	0	1	3	G/N
Rosswog et al. (2009)[b]	100–10^4	0	1	0.9 – 12	G/PNP
Haas et al. (2012)[b]	10^3	0–0.6	1	6–8	G/GR
Dai et al. (2013)	10^7	0–0.5	0.7–0.99	1	B/GR
Hayasaki et al. (2013)	10^6	0	0.8	5	G/PNP
Guillochon et al. (2014)	10^3	0	1	1	G/N
Guillochon et al. (2015)	10^5–10^8	0–1	1	≥ 1	B/PNC
Shiokawa et al. (2015)	500	0	1	1	G/GR
Evans et al. (2015)[b]	10^5	0	1	10–15	G/GR
Dai et al. (2015)	10^5–10^8	0	1	1–16	B/PNC
Bonnerot et al. (2016b)	10^6	0	0.8–0.95	1–5	G/PNP
Hayasaki et al. (2016)	10^6	−0.9–0.9	0.8	1–5	G/PNC
Sadowski et al. (2016)	10^5	0	0.97	10	G/GR
Jiang et al. (2016b)	10^5–10^7	0–0.8	1	0.6–1.2	L/N
Bonnerot et al. (2017b)	10^5–10^7	0	1	1–5	B/PNC
Liptai et al. (2019)	10^6	−0.99–0.99	0.95	1–5	G/GR
Lu and Bonnerot (2020)	10^5–10^8	0	1	1–2	B/GR+L/N
Bonnerot and Lu (2020)	2.5×10^6	0	1	1	G/PNP

[a]Although the early work by Ayal et al. (2000) carried out a global simulation using realistic parameters of the problem, it is now clear that most of the gas evolution they found results from numerical artefacts associated to the low numerical resolution used

[b]These global simulations explicitly aim at studying disruptions by intermediate-mass black holes while the others involving low black hole masses attempt to extrapolate their results to the case of a supermassive one

that significantly changes the disc properties subsequent to formation. In the following, we describe the ways in which the simulated debris evolves towards an accretion flow, and the properties of the newly-formed discs obtained from these simulations. All these works find that a majority of the bound gas has its average eccentricity decreased due to the interactions it experiences, resulting in a more circular gas distribution. We are particularly interested in the "circularization timescale," $t_{\rm circ}$, required for this process to complete, keeping in mind that the resulting accretion flow is in general not entirely circular with the gas retaining instead non-zero eccentricities. Because these simulations use idealized initial conditions, it is not a simple task to extrapolate from these numerical results to the astrophysically realistic situation. We attempt to highlight when this seems nevertheless possible and mention possible differences if such a generalization appears unreliable.

The first hydrodynamic simulation following accretion flow formation in its entirety is that by Hayasaki et al. (2013), who consider a bound star with $e_\star = 0.8$ and $\beta = 5$ disrupted by a supermassive black hole of mass $M_\mathrm{h} = 10^6\ \mathrm{M}_\odot$. Due to the low stellar eccentricity, all the debris is bound, unlike the situation in a parabolic encounter. As explained in Sect. 3.2, this causes the stream to chase its "tail" for a few orbits before it first intersects itself due to strong relativistic apsidal precession. Collisions near pericenter then happen at each revolution, although the first ones only involve the stream's low-density extremities. As a result, an accretion disc forms on a timescale $t_\mathrm{circ} \approx 5P_\star$ where $P_\star$ denotes the stellar period.[20] Later simulations (Bonnerot et al. 2016b; Hayasaki et al. 2016) explored the impact of the stellar trajectory and gas thermodynamics on this process. They find that an increase in eccentricity to $e_\star \approx 0.95$ allows the stream to promptly cross itself for $\beta = 5$ without the prior revolutions artificially needed for a more bound star. Due to the large precession angle, the ensuing self-crossing shock is strong, causing rapid formation of an accretion disc on a reduced timescale of $t_\mathrm{circ} \approx P_\star$. However, decreasing the penetration factor to a more common value of $\beta = 1$ results in collisions near the apocenter of the stream due to the decreased relativistic precession angle, as described in Sect. 3.1.2. As can be seen from Fig. 6, these weak interactions result in a slower disc formation with $t_\mathrm{circ} \approx 10P_\star$. As in the earlier work by Hayasaki et al. (2013), this evolution is somewhat delayed by the bound stellar trajectory considered. The gas also does not significantly expand during the circularizing shocks, causing the formation of a thin circular ring at $R_\mathrm{circ} \approx 2R_\mathrm{p}$. Importantly, this outcome directly results from the assumed isothermal equation of state, that allows the gas to effectively lose energy at shocks. As emphasized above, this assumption is unlikely to be valid in most cases due to the large optical thickness of the colliding streams. By modifying the equation of state to the more physical adiabatic limit, Hayasaki et al. (2016) and Bonnerot et al. (2016b) find that the stream expands as a result of collisions. However, this additional expansion does not unbind any gas from the black hole, which is likely due to the artificially bound stellar orbit considered. In this case, the resulting accretion disc is thick and occupies a wide range of radii centred around the stellar semi-major axis.

Sadowski et al. (2016) also studied disc formation assuming adiabaticity for an eccentric stellar trajectory, with $e_\star = 0.97$ and a large penetration factor of $\beta = 10$. As can be seen from the snapshots of Fig. 7, strong relativistic precession then results in a self-crossing shock near pericenter. This collision involves stream components of different widths with $H_2 \gtrsim H_1$ (see Fig. 3) due to a fast expansion caused by the nozzle shock during pericenter passage, as discussed in Sect. 3.1.1. Despite this asymmetry, the collision is powerful enough to cause the shocked gas to expand into an outflow whose unbound component represents about 15% of the stellar matter. While this unbinding of matter is consistent with the expectation of Sect. 3.1.2, the gas retains its original direction of rotation around the black hole, in contrast to the change of angular momentum sign predicted by local studies for a strong self-crossing shock. The bound part of this post-shock gas then returns near the black hole with a significant fraction being ballistically accreted along the funnels of the forming disc. After experiencing numerous interactions, the debris settles on a timescale $t_\mathrm{circ} \approx 5P_\star$ into an extended and thick accretion flow. As in the prior works that assume adiabaticity, this disc extends roughly to the stellar semi-major axis, with significant pressure support against gravity in its outer region. Despite its overall axisymmetric structure, the gas retains

[20]Because of the bound stellar trajectory, the difference in period between the different parts of the stream is negligible compared to the period of the star. For this reason, when discussing idealized simulations with significantly sub-parabolic stellar eccentricities, we express the circularization timescale in terms of the stellar period rather than that of the most bound debris.

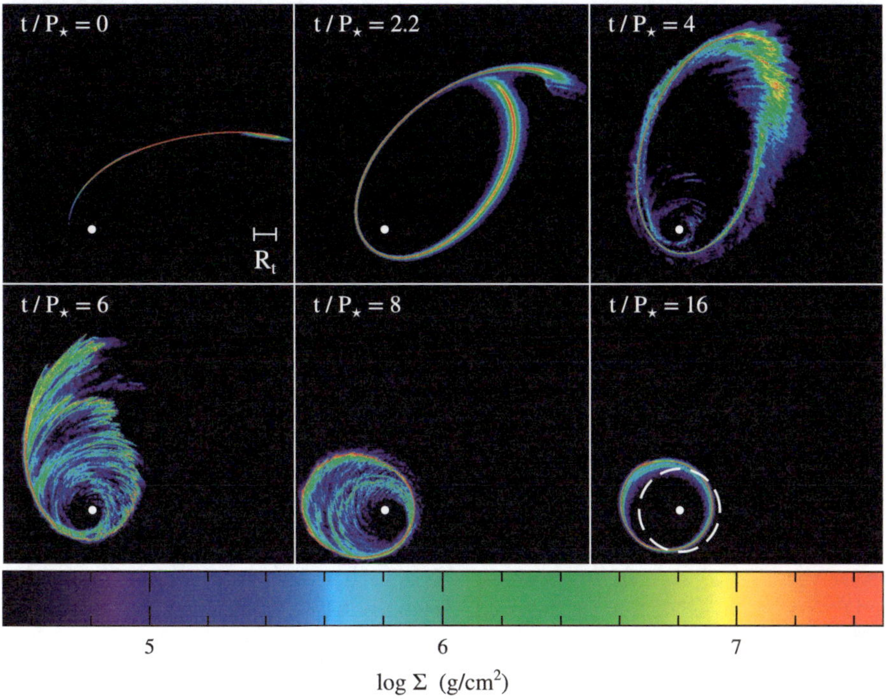

Fig. 6 Snapshots showing the evolution of column density during the disc formation process simulated by Bonnerot et al. (2016b), considering the disruption of a solar-type star on a trajectory with eccentricity $e_\star = 0.8$ and penetration factor $\beta = 1$ by a supermassive black hole of mass $M_\mathrm{h} = 10^6\ \mathrm{M}_\odot$. Due to weak relativistic precession, the stream intersects itself near apocenter, and the debris circularizes on a timescale $t_\mathrm{circ} \approx 10 P_\star$. Importantly, the gas is assumed to evolve isothermally, resulting in the formation of a thin ring located around the circularization radius $R_\mathrm{circ} \approx 2R_\mathrm{p}$ (dashed circle in the bottom right snapshot)

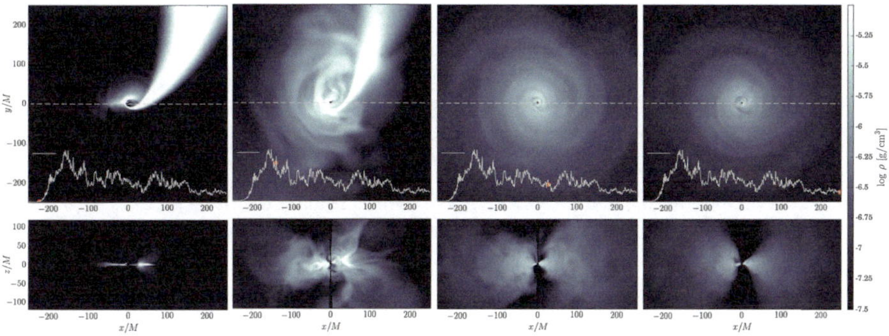

Fig. 7 Snapshots from the simulation by Sadowski et al. (2016), that studied the disc formation process for a star with eccentricity of $e_\star = 0.97$ and penetration factor $\beta = 10$, disrupted by a black hole of mass $M_\mathrm{h} = 10^5\ \mathrm{M}_\odot$. The density distribution is shown in slices parallel (upper panels) and perpendicular (lower panel) to the stellar orbital plane. Due to strong relativistic precession, a self-crossing shock takes place near pericenter, resulting in a fast outflow and the formation of an accretion flow on a timescale $t_\mathrm{circ} \approx 5 P_\star \ll t_\mathrm{min}$. Because the gas is assumed to evolve adiabatically, the resulting disc is extended and thick

moderate eccentricities with $e \approx 0.2$ on average. This work was also the first to include gas magnetic fields, but these were found not to modify the hydrodynamics, as we discuss more in detail in Sect. 3.5.3. In this study as well as the above simulations with $e \gtrsim 0.95$ and $\beta \gtrsim 5$ (e.g. Bonnerot et al. 2016b) that are able to capture a prompt self-crossing shock, the gas evolution may be comparable to that of a deep parabolic encounter, although the precise properties of the initial collision could significantly differ (see Sect. 3.1.2). Because the reduced period of the stream is washed out by the large redistribution of orbital elements induced by multiple collisions, it appears legitimate to compare the circularization timescale of a few $P_\star$ found in this regime to the period $t_{\rm min}$ of the most bound debris assuming a parabolic stellar trajectory. This comparison yields $t_{\rm circ} \lesssim t_{\rm min}$, which suggests that disc formation happens rapidly for deep encounters where a strong self-crossing shock takes place.[21] For $\beta \approx 1$, extrapolating eccentric disruption results to the parabolic case is more uncertain due to the reduced size of an entirely bound stream, which can have a larger influence on the gas evolution.

The influence of black hole spin has also been studied in simulations considering initially bound stars and different equations of state (Hayasaki et al. 2016; Liptai et al. 2019). When the gas is assumed to evolve isothermally, the energy dissipated by the nozzle shock is removed, so that there is no significant expansion of the stream during pericenter passage. In this case, nodal precession causes a vertical offset $\Delta z > H_1 + H_2$ (see Sect. 3.1.2) that prevents the stream from colliding with itself even after multiple orbits. This effect is illustrated in Fig. 8 that shows the gas evolution obtained by Hayasaki et al. (2016) for a star with $e_\star \approx 0.7$ and $\beta = 2$, disrupted by a black hole with spin $a = 0.9$, and inclined by an angle $i = 90°$ with respect to the orbital plane. A "wicker basket" configuration is created as the stream wraps around the black hole until a self-crossing shock eventually takes place causing disc formation to complete after tens of dynamical times with $t_{\rm circ} \gtrsim 10 P_\star$. These works also find that the nozzle shock makes the stream expand faster when adiabaticity is assumed, which is likely a more physical choice given the large optical depth of the confined gas. In this limit, the delay of the first collision is reduced due to the increase in stream width that can rapidly overcome the vertical offset induced by nodal precession, i.e. $\Delta z < H_1 + H_2$. As a result, no delay to the self-crossing shock is observed in the simulations by Hayasaki et al. (2016) for the parameters they consider. The more recent investigation by Liptai et al. (2019) finds a similar gas evolution, shown in Fig. 9 for a stellar trajectory with $e_\star = 0.95$ and $\beta = 5$, a black hole spin $a = 0.99$, and an inclination angle $i = 60°$. It can be seen that nodal precession prevents the stream from colliding with itself shortly after its first pericenter passage. However, the fast gas expansion caused by the nozzle shock results in a self-crossing shock during the second orbit. Later on, this interaction induces the formation of a thick accretion disc, that is only mildly delayed with $t_{\rm circ} \lesssim 5 P_\star$. This conclusion appears qualitatively different from that obtained by Guillochon and Ramirez-Ruiz (2015) in the semi-analytical work discussed in Sect. 3.1.2, which found that the first collision usually takes place after a large delay of multiple orbital timescales, even when stream expansion near pericenter is taken into account. The discrepancy could be due to an artificially small amount of stream expansion in the analytic model for the nozzle shock compared to that

[21] A more extreme situation is that of an ultra-deep encounter where the stellar pericenter is similar to the gravitational radius of the black hole. In this case, the disruption of the star is accompanied by a stretching of the debris into a elongated structure. Due to the very large values of the apsidal precession angle with $\Delta\phi \gtrsim \pi$, it is possible that this gas collides with itself during the first passage of the star near pericenter. Numerical investigations of this process (Haas et al. 2012; Evans et al. 2015; Darbha et al. 2019) find that this early self-crossing shock results in the fast formation of an accretion flow around the black hole. We note however that such relativistic pericenters are rare for black holes substantially smaller than the Hills mass.

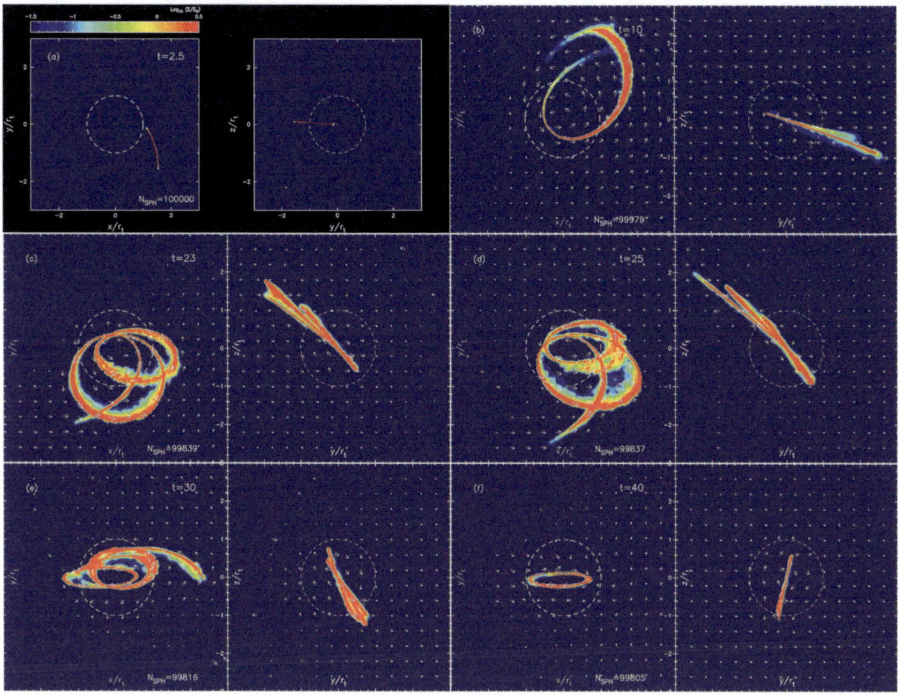

Fig. 8 Snapshots from the simulation by Hayasaki et al. (2016) who study the disc formation process for a star with initial eccentricity of $e_\star = 0.7$ disrupted by a black hole with spin $a = 0.9$ inclined with respect to the stellar orbital plane by an angle $i = 90°$. Because the gas is assumed to evolve isothermally, it does not expand significantly during pericenter passage such that nodal precession prevents a successful self-crossing shock for several orbits. Interactions eventually take place that cause the delayed formation of a narrow accretion disc on a timescale $t_{\rm circ} \gtrsim 10 P_\star$

found in simulations[22] or alternatively to the unrealistically low eccentricities of the debris resulting from the bound stellar trajectory.

The recent simulation by Bonnerot and Lu (2020) uses the outflow launched from the self-crossing shock as an initial condition by artificially injecting gas inside the computational domain from the intersection point. Using this numerical strategy, this work is able to consider realistic parameters of the problem: a parabolic stellar trajectory and a supermassive black hole of mass $M_{\rm h} = 2.5 \times 10^6\ \mathrm{M}_\odot$. The properties of the injected matter are obtained from a local numerical study of the collision (Lu and Bonnerot 2020) assuming that the two stream components have identical and thin widths with $H_1 = H_2 \ll R_{\rm int}$ (see left panel (a) of Fig. 3). For their choice of parameters, the stream crosses itself relatively close to pericenter with an intersection radius of $R_{\rm int} \approx 25 R_{\rm p}$. The ensuing self-crossing shock is strong, leading to the unbinding of about 33% of the initially bound gas via the mechanism described in Sect. 3.1.2, while most of the remaining bound matter has a sign of angular momentum opposite to that of the original star. This gas quickly expands into an envelope that completely engulfs the black hole, as can be seen from the snapshots of Fig. 10

[22] This difference could originate from the non-zero black hole spin, that may significantly increase the energy dissipated during the nozzle shock if nodal precession causes the formation of oblique collisions, as proposed in Sect. 3.1.1.

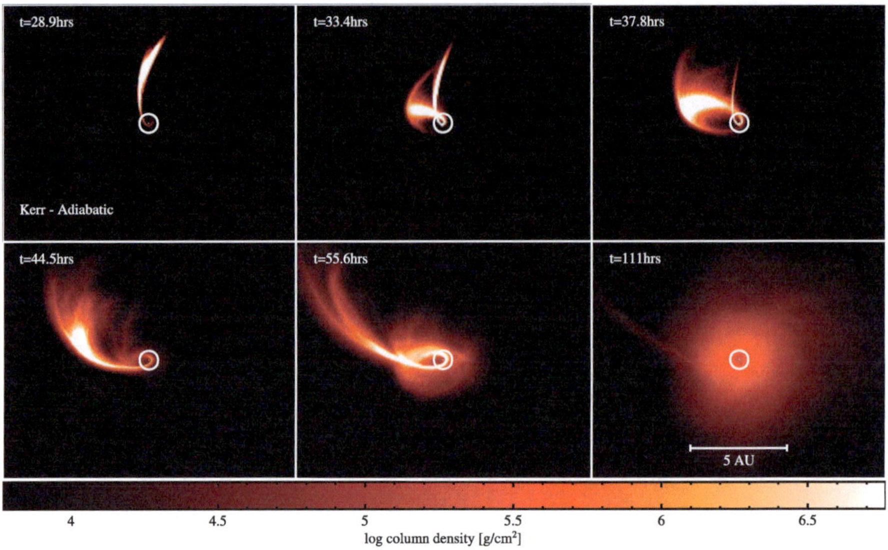

Fig. 9 Snapshots from the simulation of Liptai et al. (2019), that studies the disc formation process for a star with eccentricity $e_\star = 0.95$ and penetration factor $\beta = 5$ disrupted by a black hole with spin $a = 0.99$ inclined with respect to the stellar orbital plane by an angle $i = 60°$. Nodal precession produces a vertical offset that prevents the stream from colliding with itself promptly after its first pericenter passage. However, due to the adiabatic equation of state assumed for the gas, the nozzle shock results in a significant expansion that causes a successful self-crossing shock during the next orbit. This interaction results in the formation of a thick and extended accretion disc that is only mildly delayed, with a circularization timescale $t_{\rm circ} \lesssim 5P_\star$

obtained from this simulation. As its trajectories intersect, multiple secondary shocks take place that cause the rapid formation of an accretion disc on a timescale of $t_{\rm circ} \approx 0.3\, t_{\rm min}$. The resulting accretion disc is thick, with significant pressure support due to the adiabatic equation of state used, and retains an average eccentricity of $e \approx 0.2$. The disc is located within a radius $R_{\rm d} \approx 15 R_{\rm p}$ that contains only a small fraction of the injected gas. Due to the angular momentum sign of the injected bound debris, the accretion flow produced from this matter also rotates in the retrograde direction compared to the star. Additionally, it features two spiral shocks that drive a slow inflow along the disc mid-plane, although most of the accretion onto the black hole is due to gas falling along the funnels of the disc as it assembles. Despite the realistic astrophysical parameters it considers, this simulation remains simplified due to its idealized treatment of the self-crossing shock by an injection of gas, as explained in Sect. 3.2. In particular, it is unclear whether this collision actually results in the large outflow used as initial conditions, which closely relates to uncertainties mentioned in Sect. 3.1.2 regarding this early source of dissipation. Nevertheless, the similarity of this work with those presented above that self-consistently simulate a strong self-crossing (e.g. Sadowski et al. 2016) seems to suggest that the gas evolution is qualitatively correct in this regime.

The simulation by Shiokawa et al. (2015) considers the encounter between a white dwarf and an intermediate-mass black hole[23] with $M_{\rm h} = 500\ {\rm M}_\odot$ assuming an adiabatic evolu-

[23]Numerical studies considering intermediate-mass black holes were carried out earlier as well (Rosswog et al. 2008, 2009; Ramirez-Ruiz and Rosswog 2009; Guillochon et al. 2014). However, they do not run for long enough to capture the completion of disc formation, and we therefore do not present them in detail here.

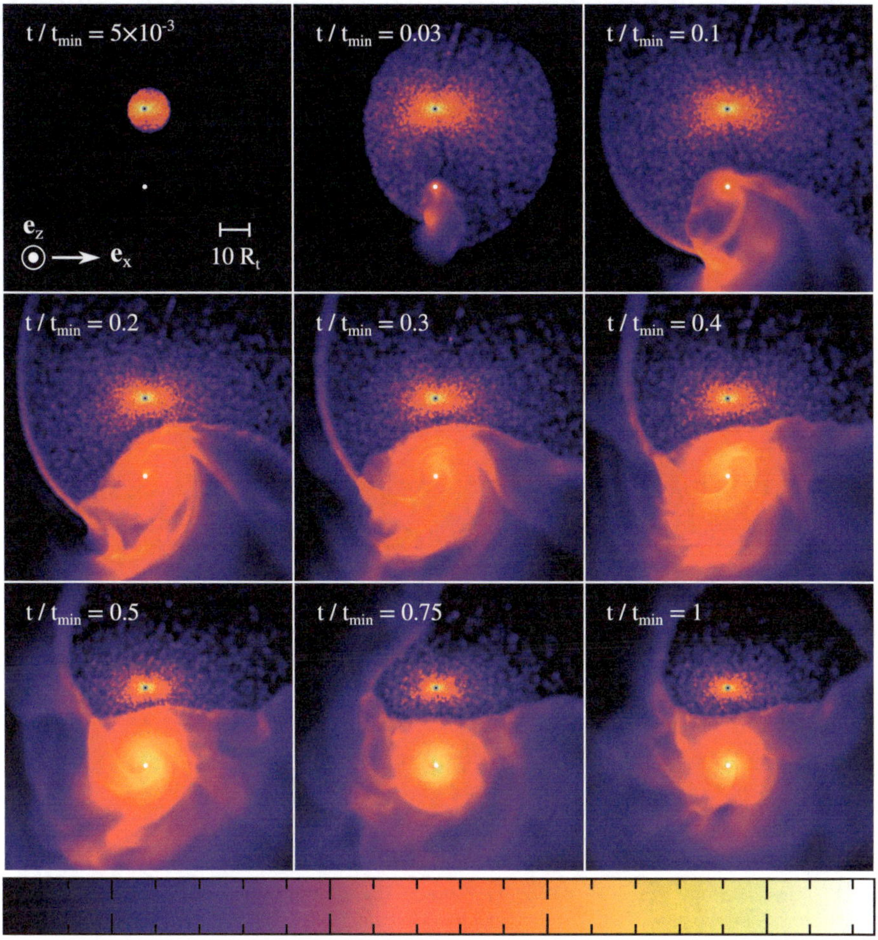

Fig. 10 Snapshots showing the gas evolution in a slice parallel to the orbital plane of the star during the disc formation process simulated by Bonnerot and Lu (2020) for a parabolic stellar trajectory and a supermassive black hole of mass $M_{\rm h} = 2.5 \times 10^6\ {\rm M}_{\odot}$. This numerical work uses as initial condition the outflow launched from a strong self-crossing shock whose properties are obtained from a local study of the collision (Lu and Bonnerot 2020). Gas is injected inside the computational domain from the intersection point (grey circle) located at $R_{\rm int} \approx 25 R_{\rm t}$. This outflowing matter rapidly engulfs the black hole and experiences multiple interactions that result in the formation of a thick and extended accretion disc on a timescale $t_{\rm circ} \approx 0.3 t_{\rm min}$

tion for the gas. Due to weak relativistic apsidal precession, this work finds that the stream intersects itself near the apocenter of its most bound part. This intersection radius is qualitatively similar to that in a large fraction of real TDEs, namely those with $\beta \approx 1$ and $M_{\rm h} \lesssim 2 \times 10^6\ {\rm M}_{\odot}$ (see Sect. 3.1.2).[24] Around this location, the two colliding components

[24]For the parameters used by Shiokawa et al. (2015), the relativistic apsidal angle is the same as for the disruption of a solar-type star by a black hole of mass $M_{\rm h} = 3 \times 10^5\ {\rm M}_{\odot}$ for $\beta = 1$.

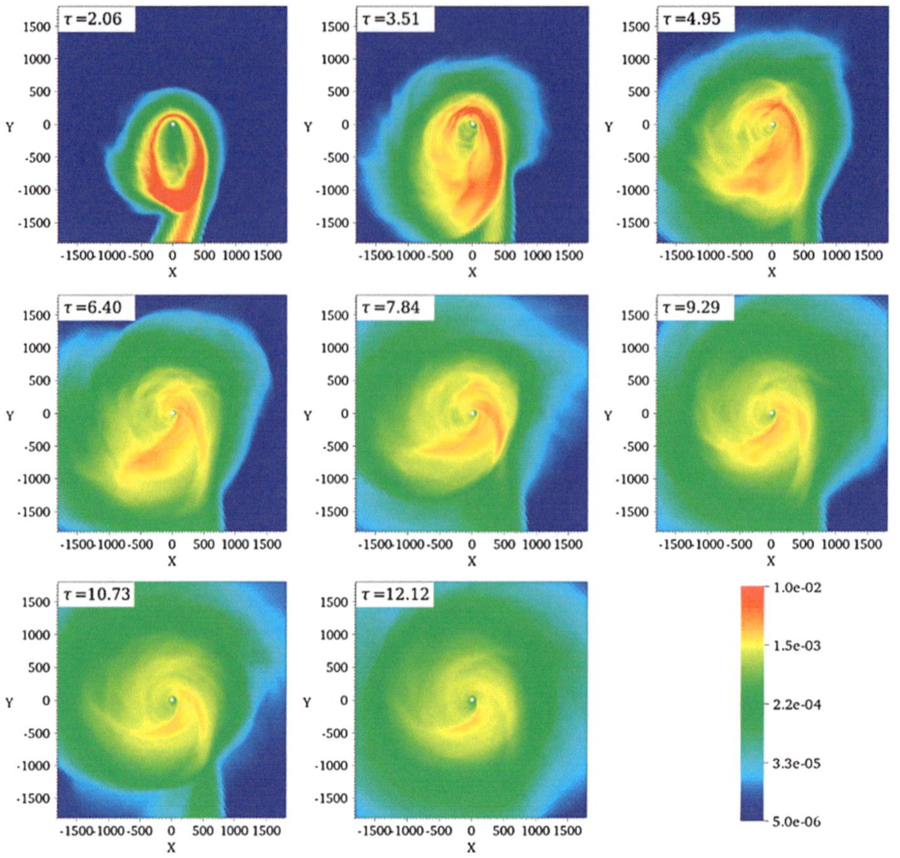

Fig. 11 Snapshots showing the column density evolution during the disc formation process in the simulation carried out by Shiokawa et al. (2015) that consider the disruption of a white dwarf on a parabolic orbit by an intermediate-mass black hole of mass $M_{\rm h} = 500\ {\rm M}_{\odot}$. In the first snapshot, the tip of the stream has already passed pericenter and intersects with the matter still moving inward that results in a self-crossing shock near apocenter. This weak collision initiates the formation of an accretion flow that remains globally eccentric for several orbital periods before eventually settling on a timescale of $t_{\rm circ} \approx 10\, t_{\rm min}$ into a more axisymmetric, thick and extended structure

are both thick, with widths $H_1 \approx H_2 \lesssim R_{\rm int} = a_{\rm min}$ (see Fig. 3) similar to the intersection radius. This is likely a consequence of the increased dynamical impact of the nozzle shock, which for low-mass black holes causes the stream to significantly expand before it first intersects itself (see Sect. 3.2). Interactions between these thick streams are therefore enhanced relative to the weaker collision that might be expected for thin streams, in a way that may compensate for the weak relativistic apsidal precession, as discussed in Sect. 3.1.2. As can be seen from the snapshots of Fig. 11, the resulting self-crossing shock initially induces mild changes in the gas trajectories associated with a slow expansion that is unable to unbind mass from the black hole. Early on, most of the dissipation still occurs at the nozzle shock, although this interaction weakens due to an expansion of the returning stream. The interactions resulting from the self-crossing shock (termed "outer shocks" in this work) are analysed in detail, and decomposed into a "forward" and "reverse" component that are initially close in space. At later times, the reverse one recedes closer to the black hole as mass accumulates near apocenter while the forward component remains around the same location. Several passages of the returning gas through this shock system progressively lead to a redistribution of the gas orbital elements accompanied by an overall decrease of its eccentricities. During this process, a significant fraction of the gas has its angular momentum diminished so that it reaches scales similar to the black hole event horizon that can dominate the early accretion, as we discuss more in detail in Sect. 3.5.2. The accretion flow remains globally eccentric for about ten dynamical times before settling into a more axisymmetric structure (i.e. $t_{\rm circ} \approx 10\, t_{\rm min}$) that is thick with internal eccentricities of $e \approx 0.3$ on average, and extends out to a radius $R_{\rm d} \approx a_{\rm min}$ similar to the semi-major axis of the most bound debris. As was also the case for the weak self-crossing shock captured by the above works (e.g. Bonnerot et al. 2016b) considering bound stars with $\beta \approx 1$, this simulation suggests that circularization is slowed when relativistic apsidal precession is reduced. However, it is unclear whether these results obtained for an intermediate-mass black hole can be extrapolated to the more common situation of a solar-type star disrupted by a supermassive black hole. An important source of uncertainty relates to the increased thickness of the debris, that may significantly modify the gas evolution compared to that involving a thin returning stream confined by self-gravity (see Sect. 2).

The above simulations aim at studying the global hydrodynamics of accretion flow formation in TDEs. Along with the works presented in Sect. 3.1, they constitute our current understanding of this complex process, that is to date still highly uncertain, although clearly different from the early prediction by Rees (1988). Despite the simplified initial conditions used, these numerical investigations allow us to gain insight into the astrophysical problem and identify the main remaining sources of uncertainty. If a powerful self-crossing shock takes place near pericenter that strongly modifies the trajectories of the colliding gas, there is convincing evidence that at least a fraction of the debris can promptly form an accretion flow with $t_{\rm circ} \lesssim t_{\rm min}$, whose properties may be qualitatively similar to those found in some simulations (e.g. Sadowski et al. 2016; Bonnerot and Lu 2020). However, the exact region of parameter space corresponding to this regime and the detailed gas evolution here are still not clearly identified. If the stream intersects itself around apocenter due to weak apsidal precession, some works above (Shiokawa et al. 2015; Bonnerot et al. 2016b) suggest a much slower formation of the accretion flow. However, it remains to be understood how sensitive these results are to parameter choices made for computational tractability. Tidal disruption by intermediate-mass black holes likely produces streams of much larger relative width due to the enhanced importance of the nozzle shock, while eccentric stellar orbits can lead to unphysical initial interactions between streams (e.g. "head chasing tail" behavior) that are probably of greater importance for very weak stream intersections. For these reasons, important uncertainties still exist in this regime, with the possibility that the gas evolves in a realistic situation very differently from what current simulations predict. For example, it is possible that the debris forms a disc that fails to circularize by retaining large eccentricities $e \approx e_{\rm min}$ for multiple orbital timescales.[25] Overall, accretion flow formation remains a very open problem, in which the hydrodynamics is not yet qualitatively understood for a large fraction of the parameter space. Progress is likely to come from a better understanding of the early sources of dissipation discussed in Sect. 3.1 combined with improved global simulations more directly applicable to the astrophysically realistic situation.

[25] Note that even though the disc could form as very eccentric, it is unclear whether this configuration can be retained due to the various dissipation processes impacting the gas evolution, such as those associated with accretion and the interaction with the stream of loosely bound debris that continues returning to interact with the disc.

Regarding the inclusion of additional physics, the influence of black hole spin on this process is also uncertain since it has so far only been studied for bound stars (Hayasaki et al. 2016; Liptai et al. 2019). The impact of nodal precession at preventing a prompt self-crossing shock remains to be evaluated for a parabolic encounter that would in particular require to estimate the expansion induced by the nozzle shock in this situation (see Sect. 3.1.1). Another important effect relates to the radiative efficiency, that may increase as the gas distribution becomes more extended due to a decrease of its optical depth, as is explained in more detail in Sect. 3.4. The gas evolution would then deviate from that assuming complete adiabaticity, which may in particular cause the newly-formed disc to get thinner and more compact as predicted from isothermal runs (Hayasaki et al. 2013; Bonnerot et al. 2016b; Hayasaki et al. 2016). Evaluating the exact level of radiative cooling requires to carry out radiation-hydrodynamics simulations of this process that has not been done so far. In order to capture early gas accretion, it may also be important to include magnetic fields, which has so far been considered in just one simulation (Sadowski et al. 2016). Despite the several uncertainties we mentioned, the results of current simulations have led to major improvements in our understanding of accretion flow formation. In the future, additional progress will come from greater computational resources and the development of innovative numerical schemes that would allow us to circumvent the computational limitations presented in Sect. 3.2. Once disc formation has been precisely understood, it will become possible to study its longer-term evolution during which most of the gas accretion takes place based on better motivated initial conditions.

3.4 Shock Radiative Efficiency and Resulting Emission

Before settling into an accretion disc, the numerical works presented in Sect. 3.3 show that the returning debris experiences a large amount of dissipation. The outcome of these shocks depends on their radiative efficiency, that is in turn determined by the optical depth of the surrounding matter. For a very optically thick gas, the photons are advected with the fluid element that created them. In this case, the interaction is radiatively inefficient and the hydrodynamics is close to completely adiabatic. In contrast, more optically thin matter allows photons to be transported away from their production site by diffusion, and to potentially leave the system entirely. The gas evolution may then significantly deviate from the adiabatic limit, and the escaping radiation can participate in the electromagnetic signal observed from TDEs, as discussed more in details in the Emission Mechanisms Chapter. We start by evaluating the radiative efficiency of the initial self-crossing shock and then focus on the later interactions that result in accretion flow formation.

As described in Sect. 3.1.2, the outcome of the self-crossing shock remains unclear since it depends on several effects whose relative importance has so far not been precisely estimated. As before, we treat first the idealized situation where the collision involves aligned stream components with the same thin width $H_1 = H_2 \ll R_{\rm int}$, and then discuss possible deviations from these strong assumptions. In this regime, the shock-heating rate is obtained from $\dot{E}_{\rm sc}^{\rm max} = \dot{M}_{\rm fb} \Delta\varepsilon_{\rm sc}^{\rm max}$ by multiplying the fallback rate at which the gas enters the collision by the specific energy dissipated. Using equation (15) then yields

$$\dot{E}_{\rm sc}^{\rm max} \approx \frac{G M_{\rm h} \dot{M}_{\rm fb}}{R_{\rm int}} \approx 7 \times 10^{43}\ {\rm erg\,s^{-1}} \left(\frac{R_{\rm int}}{a_{\rm min}}\right)^{-1} \left(\frac{M_{\rm h}}{10^6\ {\rm M_\odot}}\right)^{-1/6}, \tag{16}$$

where the numerical value assumes a collision angle $\psi \approx \pi$ and an intersection radius $R_{\rm int} \approx a_{\rm min}$ that is valid for the typical parameters $M_{\rm h} = 10^6\ {\rm M_\odot}$ and $\beta = 1$, and we have set the fallback rate equal to its peak value, with $\dot{M}_{fb} = \dot{M}_p$ using equation (6). This substantial rate of energy dissipation is comparable to the peak optical/UV luminosities observed in typical TDEs, which led Piran et al. (2015) to first propose it as the primary power source for these flares.[26] Photons are emitted by the shocked gas with the resulting radiation pressure driving the formation of an outflow that, for simplicity, we approximate as quasi-spherical (see left panel of Fig. 3) with a density profile $\rho_{out} \approx \dot{M}_{fb}/(4\pi R^2 v_{out})$ where R denotes the distance from the intersection point. The diffusion and dynamical timescales given by $t_{diff} \approx R\tau_s/c$ and $t_{dyn} = R/v_{out}$ are equal at the trapping radius R_{tr} where the scattering optical depth is $\tau_s \equiv \int_{R_{tr}}^{\infty} \kappa_s \rho_{out} \mathrm{d}R = c/v_{out}$, yielding $R_{tr} = \dot{M}_{fb}\kappa_s/(4\pi c)$, where $\kappa_s = 0.34\ \mathrm{cm^2\,g^{-1}}$ is the electron-scattering opacity. Radiation is coupled to the gas inside this radius and its energy is therefore advected outward at a rate $\dot{E} = 4\pi e_r v_{out} R^2 \propto R^{-2/3}$ where the last scaling uses the fact that the radiation energy density evolves adiabatically as $e_r \propto \rho_{out}^{4/3} \propto R^{-8/3}$. This decreasing rate implies that photons lose energy as they are transported outward due to work done on the gas through radiation pressure. After being injected by shocks near the common width H of the two colliding streams, at the luminosity given by equation (16), radiation energy is degraded in this way until photons reach the trapping radius. The emerging luminosity L_{sc}^{max} is therefore reduced by a factor

$$\frac{L_{sc}^{max}}{\dot{E}_{sc}^{max}} \approx \left(\frac{H}{R_{tr}}\right)^{2/3} \approx 0.03 \left(\frac{M_h}{10^6\ \mathrm{M_\odot}}\right)^{1/3} \left(\frac{H}{10\ \mathrm{R_\odot}}\right)^{2/3}, \tag{17}$$

where the numerical value assumes again $\dot{M}_{fb} = \dot{M}_p$ and uses a stream width $H \approx 10\ \mathrm{R_\odot}$ that corresponds to an infalling stream confined by self-gravity (see Sect. 2) (Lu and Bonnerot 2020). In this regime, the self-crossing shock is therefore radiatively inefficient, a result in agreement with radiation-hydrodynamics simulations of the collision assuming thin streams (Jiang et al. 2016b), and that justifies an assumption of adiabaticity for the gas evolution. For $M_h \approx 10^6\ \mathrm{M_\odot}$, the resulting luminosity is $L_{sc}^{max} \approx 3 \times 10^{42}\ \mathrm{erg\,s^{-1}}$, which seems too low to account for the brightest optical and UV emission detected from TDEs. If a significant fraction of the gas collides near pericenter with $R_{int} \approx R_p$, as is common for $M_h \gtrsim 10^7\ \mathrm{M_\odot}$, the heating rate of equation (16) could increase and result in a larger luminosity of $L_{sc}^{max} \gtrsim 10^{43}\ \mathrm{erg\,s^{-1}}$. As discussed in Sect. 3.1.2, the two stream components involved in the self-crossing shock may have significantly different widths, with $H_2 > H_1$. In this case, we have seen that the specific energy dissipated is likely lower than that used in equation (16), and this would tend to decrease the emerging luminosity compared to L_{sc}^{max}. However, a thicker gas distribution accompanied by a significant deviation from spherical geometry for the shocked gas could instead result in a more radiatively efficient interaction, with lower adiabatic losses than is predicted in the above case of identical streams. Our current understanding of the hydrodynamics of a realistic self-crossing shock would have to be improved to go beyond these qualitative expectations.

Following the self-crossing shock, the gas undergoes additional interactions that may circularize the orbits to form an accretion flow. We now focus on the radiation emerging during this later stage, bearing in mind that it may not always be possible to distinguish it from the initial collision, especially if a significant fraction of the gas starts accumulating near the black hole as a result of the self-crossing shock alone.[27] If these circularizing shocks can

[26] Earlier on, Lodato (2012) also estimated the energy that must be radiated for the gas to completely circularize, proposing that this could be an observable power source without further development.

[27] This situation is for example expected if early interactions take place near pericenter either as a result of strong relativistic apsidal precession or due to gas deflection induced by a fast expansion at the nozzle shock (see Sect. 3.1.2).

happen closer to the black hole, they are likely stronger than the initial collision. In the work by Bonnerot and Lu (2020) that adopts realistic astrophysical parameters, the corresponding heating rate reaches $\dot{E}_{\rm sh} \approx 10^{44}$ erg s^{-1}. Notably, Piran et al. (2015) extrapolates a similar value from the simulations by Shiokawa et al. (2015) that considers an intermediate-mass black hole. Estimating the fraction of radiation that emerges from the system requires to know the gas distribution that the corresponding photons have to go through to escape. This can be done by assuming that this matter has expanded to fill an approximately spherical region within a distance $R_{\rm d}$ from the black hole. If it contains about the mass of the original star, the optical depth of this gas may be estimated as $\tau_{\rm s} \approx \kappa_{\rm s} M_\star/(4\pi R_{\rm d}^2)$. Comparing the diffusion timescale $t_{\rm diff} \approx R_{\rm d}\tau_{\rm s}/c$ to the dynamical time $t_{\rm dyn} \approx 2\pi (G M_{\rm h}/R_{\rm d}^3)^{-1/2}$ then yields

$$\frac{t_{\rm diff}}{t_{\rm dyn}} \approx 1.5 \left(\frac{M_{\rm h}}{10^6\,{\rm M}_\odot}\right)^{-7/6} \left(\frac{R_{\rm d}}{a_{\rm min}}\right)^{-7/2}, \tag{18}$$

where the numerical value adopts a radius $R_{\rm d} = a_{\rm min}$ as motivated by the simulations described in Sect. 3.3. This simple estimate (Piran et al. 2015; Hayasaki et al. 2016) suggests that the circularizing shocks are radiatively efficient for $M_{\rm h} \gtrsim 10^6$ M$_\odot$. Based on the density distribution directly obtained from their simulation with $M_{\rm h} = 2.5 \times 10^6$ M$_\odot$, Bonnerot and Lu (2020) also find that the gas optical depths are low enough to allow the radiation generated through shocks to promptly escape. These arguments suggest that the emerging luminosity could be similar to the heating rate with $L_{\rm sh} \approx \dot{E}_{\rm sh}$, which is likely larger than that originating from the self-crossing shock in most cases. This mechanism therefore represents a plausible explanation for the observed optical and UV luminosities, although this origin may be difficult to distinguish from that associated with gas accretion (Metzger and Stone 2016). As mentioned in Sect. 3.3, this efficient loss of radiation energy from the system implies that the accretion disc may be initially thinner than expected from the numerical works that assume complete adiabaticity. Radiation-hydrodynamics simulations of this process are required to precisely evaluate the emerging electromagnetic signal and the impact of radiative cooling on the gas evolution.

3.5 Implications for the Nascent Accretion Flow

The outcome of the complex, three-dimensional hydrodynamical evolution described in Sect. 3.3 represents the initial conditions from which accretion onto the black hole will proceed. A detailed analysis of accretion processes is provided in the Accretion Disc Chapter and we only emphasize here the properties that can be extracted from the gas evolution at play during disc formation. The main reason for doing so is that most simulations of disc accretion in TDEs are, so far, initialized from an already formed disc, mostly for computational reasons that prevent us from self-consistently following the prior disk assembly process. It is therefore not guaranteed that these frequently used initial conditions possess all the properties described here, and this should be kept in mind by the reader when going through the next chapter.

3.5.1 Slow Gas Inflow: No $t^{-5/3}$ Decay of the Mass Accretion Rate?

As we have already highlighted, the current picture of disc formation has been significantly improved since the early analytical work by Rees (1988). In contrast to the simple assumption of a fast circularization on scales similar to $R_{\rm p}$, recent numerical simulations find that disc formation can be a slow process with most of the gas settling at distances $R_{\rm d} \gg R_{\rm p}$

larger than initially thought (Shiokawa et al. 2015; Sadowski et al. 2016; Bonnerot and Lu 2020). Assuming a standard α-disc, the viscous timescale $t_{\rm visc}$ at this location is given by the ratio (Frank et al. 2002)

$$\frac{t_{\rm visc}}{t_{\rm min}} \approx \frac{1}{\alpha}\left(\frac{R_{\rm d}}{a_{\rm min}}\right)^{3/2}\left(\frac{H}{R}\right)^{-2}, \tag{19}$$

from which we see that $t_{\rm visc}/t_{\rm min} \approx 10$ if $R_{\rm d} \approx a_{\rm min}$. Here, the viscous parameter and the aspect ratio have been given reasonable values of $\alpha \approx 0.1$ and $H/R \approx 1$. This implies that accretion through the disc may be slow, such that a large amount of mass is accumulated at large radii with no significant, time-averaged inflow motion even long after the disc has settled into a quasi-steady state. The early model by Rees (1988) proposed that the mass accretion rate follows the same decay law (of $t^{-5/3}$) as the fallback rate, based on the assumptions of rapid disc formation and efficient viscous processing of the freshly returning matter. According to the above analysis, these conditions may not be realized, suggesting a different evolution of the inflow rate through the disc.

3.5.2 Ballistic Accretion and Advection: Solutions to the "Inverse Energy Crisis"?

The total energy radiated in the optical, near-UV, and soft X-ray from most observed TDE candidates is almost always $\lesssim 10^{51}$ erg, a value about three orders of magnitude lower than the rest mass energy of a solar-type star. Additionally, the energy loss necessary for complete circularization is $M_\star \Delta\varepsilon_{\rm circ} \approx 10^{52}$ erg according to equation (9), which also significantly exceeds the observed values. This results in an "inverse energy crisis" where less energy is detected than one would have expected (Piran et al. 2015; Stone and Metzger 2016; Lu and Kumar 2018). We focus below on proposed solutions for this puzzle that involve the hydrodynamics of the accretion flow, but also refer to other possibilities for completeness.

The low energetics of observed events can be explained if most of the gas is ballistically accreted onto the black hole, carrying its orbital energy below the event horizon without any dissipation. As mentioned in Sect. 3.3, ballistic accretion has been found in several simulations (Shiokawa et al. 2015; Sadowski et al. 2016; Bonnerot and Lu 2020) of the disc formation process, which find that a significant fraction of the debris gets accreted with moderate eccentricities. This process is caused by angular momentum redistribution happening during the circularizing shocks and could reduce the radiative efficiency of the flow. A more extreme scenario proposed by Svirski et al. (2017) involves an highly eccentric disc with $e \approx e_{\rm min}$ that is a possible outcome of the disc formation process but has not been found in current simulations. In this situation, magnetic stresses are more efficient at transporting angular momentum near apocenter that acts to progressively reduce the pericenter of the debris. These authors propose that this mechanism can keep operating until all the gas enters the event horizon of the black hole, thereby strongly suppressing the radiation produced by either circularization or accretion. However, another effect of reducing the gas pericenter that close to the gravitational radius is to strongly enhance relativistic apsidal precession. As argued by Bonnerot et al. (2017b), the net impact could therefore be to enhance interactions between different parts of this disc, promoting dissipation. A related possibility is that significant dissipation actually occurs inside the forming disc but with a subsequent advection of the injected internal energy, as proposed by Begelman (1979) and Abramowicz et al. (1996).

Although alleviating the tension, it is not clear whether hydrodynamical models completely solve it and another perhaps more natural solution relies on the fact that the disc emits in the extreme ultraviolet (EUV) part of the spectrum that is not observable (Strubbe and Quataert 2009; Roth et al. 2016; Lu and Kumar 2018; Dai et al. 2018). This solution has gained recent support observationally through the detection of infrared dust echoes (van Velzen et al. 2016; Jiang et al. 2016a), Bowen resonance lines (Blagorodnova et al. 2019; Leloudas et al. 2019) from TDE candidates that are both known to be triggered by extreme-ultraviolet radiation, and direct fitting of disk models to observed X-ray spectra (Wen et al. 2020). Indeed, simple models for the infrared dust echoes observed in the TDE flares PTF-09ge and PTF-09axc infer EUV luminosities $\sim 10^{44-45}$ $\mathrm{erg\,s^{-1}}$ (van Velzen et al. 2016). The total EUV energy release from these flares is uncertain, but in both cases could be $\sim 10^{52}$ $\mathrm{erg\,s^{-1}}$, which would solve the inverse energy crisis if one assumes disruption of lower main sequence stars with masses $M_\star \lesssim 0.3 M_\odot$. While this could be a promising solution, we caution that the modeling of these dust echoes involves many assumptions on the interstellar dust composition and geometry.

3.5.3 Initial Source of Disc Effective Viscosity

In most accreting systems, angular momentum transport is driven by magnetic stresses resulting from the magneto-rotational instability (MRI) (Balbus and Hawley 1991) that are often parametrized in steady state as an effective viscosity. However, it is not obvious that the early phase of accretion in TDEs is produced by the same physical process. This possibility has been put forward by the magneto-hydrodynamical simulations of Sadowski et al. (2016), who found that even after the MRI has reached saturation, angular momentum transport is still dominated by purely hydrodynamical turbulence, which they attribute to either convection or the perturbation created by the matter originally ejected from the self-crossing shock and continuously joining the newly-formed disc.[28] Similarly, the simulation by Bonnerot and Lu (2020) finds that the forming disc contains spiral shocks excited by later-arriving matter that continuously strikes its outer edge. Angular momentum is transported outward at these locations, which produces a slow inflow through the mid-plane. Nealon et al. (2018) also proposed that the gas accretion at early times could be produced by angular momentum transport associated to the Papaloizou-Pringle instability (Papaloizou and Pringle 1984), whose growth timescale was evaluated to be shorter than that of the MRI owing to the low initial disc magnetic field inherited from the disrupted star. Finally, the development of MRI may differ in eccentric discs forming from debris that fails to fully circularize (Chan et al. 2018).

4 Summary and Conclusion

In this chapter, we have presented the current understanding of stream evolution prior to pericenter return (Sect. 2) and the formation of an accretion flow around the black hole from the bound part of this gas (Sect. 3).

The first phase is well-understood: different stream elements move on ballistic orbits, while their transverse motion is usually at first set by the gas self-gravity (Sect. 2.1). Several authors have also investigated the influence of various additional physical effects on the stream evolution (Sect. 2.2). Gravitational fragmentation inside this gas distribution can lead to the formation of clumps, while the confinement by self-gravity can be overcome by either magnetic pressure or through heating associated with hydrogen recombination. It

[28] We note, however, that over much of the computational domain of Sadowski et al. (2016), the magnetorotational instability is marginally or under-resolved.

is also possible that the debris experiences interactions with the surrounding medium that can enhance its mixing with this ambient gas. Nevertheless, this additional physics does not appear to drastically change the basic picture of how the stream evolves before its bound part comes back to pericenter.

Our understanding of the second phase of accretion disc formation is less robust. The initial dissipation (Sect. 3.1) is in most cases dominated by a self-crossing shock whose characteristics are specified by the combined effect of relativistic apsidal precession, expansion from pericenter due to the nozzle shock, and nodal precession produced by the black hole spin. When apsidal precession dominates, local simulations initialized with thin identical stream components find that the collision results in a quasi-spherical expansion of the gas. However, this evolution may significantly change if the nozzle shock imparts a large width difference to the streams or if nodal precession delays the first intersection and causes the collision to be misaligned. Studying the later hydrodynamics requires global simulations that are very numerically challenging owing to the large dynamic range involved, and have therefore only been carried out so far for simplified initial conditions (Sect. 3.2). These works find that the debris experiences additional interactions, causing it to progressively move to more circular orbits until eventually settling into an accretion flow (Sect. 3.3). This process appears to complete on a shorter timescale for more relativistic stellar pericenters, due to an increased strength of the initial self-crossing shock. If the gas evolves adiabatically, the final outcome is a thick and extended distribution that typically retains significant internal eccentricities. However, due to the unrealistic setup used in these numerical studies, the precise hydrodynamics at play during accretion flow formation is not yet established and it therefore cannot be excluded that its outcome is different from what current simulations predict. For example, encounters with weak apsidal precession may lead to a gas distribution that completely fails to circularize and remain instead highly eccentric. As the disc assembles, the radiation produced during circularizing shocks may leave the system to participate to the emerging luminosity from TDEs, especially in the optical/UV bands (Sect. 3.4). The outcome of the disc formation process represents the initial state for the subsequent accretion onto the black hole and the above works suggest that this later phase consists in a slow inflow of gas induced by effective viscosity with a possibly important contribution from ballistic accretion (Sect. 3.5).

Building on the progress made so far, improvements in our understanding of the complex process of accretion flow formation will come in the future from a combination of systematic studies of the different mechanisms involved and global simulations of the entire hydrodynamics applicable to an astrophysically realistic situation. Additional physics missing from most current works such as radiative diffusion, black hole spin and magnetic fields should also be incorporated in these investigations since their influence can significantly affect the gas evolution. Some of these advancements are already being undertaken and will lead to major progress in the theoretical understanding of TDEs.

Acknowledgements We gratefully acknowledge conversations with, and detailed comments from T. Piran, as well as his edits on an earlier version of this manuscript. We are also grateful to E.M. Rossi for insightful discussions during the writing of this chapter. The research of CB was funded by the Gordon and Betty Moore Foundation through Grant GBMF5076. NCS was supported by the NASA Astrophysics Theory Research program (grant NNX17AK43G; PI B. Metzger), and from the Israel Science Foundation (Individual Research Grant 2565/19).

References

M.A. Abramowicz, X.M. Chen, M. Granath, J.P. Lasota, Advection-dominated accretion flows around Kerr black holes. Astrophys. J. **10**, 762–773 (1996). https://doi.org/10.1086/178004

P. Amaro-Seoane, M.C. Miller, G.F. Kennedy, Tidal disruptions of separated binaries in galactic nuclei. Mon. Not. R. Astron. Soc. **425**(4), 2401–2406 (2012). https://doi.org/10.1111/j.1365-2966.2012.21162.x

S. Ayal, M. Livio, T. Piran, Tidal disruption of a solar-type star by a supermassive black hole. Astrophys. J. **545**, 772–780 (2000). https://doi.org/10.1086/317835

S.A. Balbus, J.F. Hawley, A powerful local shear instability in weakly magnetized disks. I - Linear analysis. Astrophys. J. **376**(9), 214 (1991). https://doi.org/10.1086/170270. ISBN 9788578110796

M.C. Begelman, Can a spherically accreting black hole radiate very near the Eddington limit. Mon. Not. R. Astron. Soc. **187**(2), 237–251 (1979). https://doi.org/10.1093/mnras/187.2.237

N. Blagorodnova, S.B. Cenko, S.R. Kulkarni, I. Arcavi, J.S. Bloom, G. Duggan, A.V. Filippenko, C. Fremling, A. Horesh, G. Hosseinzadeh, E. Karamehmetoglu, A. Levan, F.J. Masci, P.E. Nugent, D.R. Pasham, S. Veilleux, R. Walters, L. Yan, W. Zheng, The broad absorption line tidal disruption event iPTF15af: optical and ultraviolet evolution. Astrophys. J. **873**(1), 92 (2019). https://doi.org/10.3847/1538-4357/ab04b0

L. Blanchet, Gravitational radiation from post-Newtonian sources and inspiralling compact binaries. Living Rev. Relativ. **17**, 2 (2014). https://doi.org/10.12942/lrr-2014-2

C. Bonnerot, W. Lu, Simulating disc formation in tidal disruption events. Mon. Not. R. Astron. Soc. **495**(1), 1374–1391 (2020). https://doi.org/10.1093/mnras/staa1246

C. Bonnerot, E.M. Rossi, Streams collision as possible precursor of double tidal disruption events. Mon. Not. R. Astron. Soc. **484**(1), 1301–1316 (2019). https://doi.org/10.1093/mnras/stz062

C. Bonnerot, E.M. Rossi, G. Lodato, Bad prospects for the detection of giant stars' tidal disruption: effect of the ambient medium on bound debris. Mon. Not. R. Astron. Soc. **458**(3), 3324–3330 (2016a). https://doi.org/10.1093/mnras/stw486

C. Bonnerot, E.M. Rossi, G. Lodato, D.J. Price, Disc formation from tidal disruptions of stars on eccentric orbits by Schwarzschild black holes. Mon. Not. R. Astron. Soc. **455**(2), 2253–2266 (2016b). https://doi.org/10.1093/mnras/stv2411

C. Bonnerot, D.J. Price, G. Lodato, E.M. Rossi, Magnetic field evolution in tidal disruption events. Mon. Not. R. Astron. Soc. **469**(4), 4879–4888 (2017a). https://doi.org/10.1093/mnras/stx1210

C. Bonnerot, E.M. Rossi, G. Lodato, Long-term stream evolution in tidal disruption events. Mon. Not. R. Astron. Soc. **15**(August), 1–15 (2017b). https://doi.org/10.1093/mnras/stw2547

M. Brassart, J.P. Luminet, Shock waves in tidally compressed stars by massive black holes. Astron. Astrophys. **481**, 259 (2008). https://doi.org/10.1051/0004-6361:20078264

J.K. Cannizzo, H.M. Lee, J. Goodman, The disk accretion of a tidally disrupted star onto a massive black hole. Astrophys. J. **351**, 38 (1990). https://doi.org/10.1086/168442

B. Carter, J.P. Luminet, Pancake detonation of stars by black holes in galactic nuclei. Nature **296**(5854), 211–214 (1982). https://doi.org/10.1038/296211a0

C.-H. Chan, J.H. Krolik, T. Piran, Magnetorotational instability in eccentric disks. Astrophys. J. **856**(1), 12 (2018). https://doi.org/10.3847/1538-4357/aab15c

C.-H. Chan, T. Piran, J.H. Krolik, D. Saban, Tidal disruption events in active galactic nuclei. ArXiv e-prints (2019)

R.M. Cheng, T. Bogdanović, Tidal disruption of a star in the Schwarzschild spacetime: relativistic effects in the return rate of debris. Phys. Rev. D **90**(6), 064020 (2014). https://doi.org/10.1103/PhysRevD.90.064020

E.R. Coughlin, C. Nixon, Variability in tidal disruption events: gravitationally unstable streams. Astrophys. J. Lett. **808**, 11 (2015). https://doi.org/10.1088/2041-8205/808/1/L11

E.R. Coughlin, C.J. Nixon, The gravitational instability of adiabatic filaments. Astrophys. J. Suppl. **247**(2), 51 (2020). https://doi.org/10.3847/1538-4365/ab77c2

E.R. Coughlin, C. Nixon, M.C. Begelman, P.J. Armitage, On the structure of tidally-disrupted stellar debris streams. Mon. Not. R. Astron. Soc. **17**(March), 1–17 (2016a). https://doi.org/10.1093/mnras/stw770

E.R. Coughlin, C. Nixon, M.C. Begelman, P.J. Armitage, D.J. Price, Post-periapsis pancakes: sustenance for self-gravity in tidal disruption events. Mon. Not. R. Astron. Soc. **455**(4), 3612–3627 (2016b). https://doi.org/10.1093/mnras/stv2511

E.R. Coughlin, P.J. Armitage, C. Nixon, M.C. Begelman, Tidal disruption events from supermassive black hole binaries. Mon. Not. R. Astron. Soc. **465**(4), 3840–3864 (2017). https://doi.org/10.1093/mnras/stw2913

L. Dai, A. Escala, P. Coppi, The impact of bound stellar orbits and general relativity on the temporal behavior of tidal disruption flares. Astrophys. J. **775**(1), 9 (2013). https://doi.org/10.1088/2041-8205/775/1/L9

L. Dai, J.C. McKinney, M.C. Miller, Soft X-ray temperature tidal disruption events from stars on deep plunging orbits. Astrophys. J. **812**(2), 39 (2015). https://doi.org/10.1088/2041-8205/812/2/L39
L. Dai, J.C. McKinney, N. Roth, E. Ramirez-Ruiz, M.C. Miller, A unified model for tidal disruption events. Astrophys. J. **859**(2), 20 (2018). https://doi.org/10.3847/2041-8213/aab429
S. Darbha, E.R. Coughlin, D. Kasen, C. Nixon, Ultra-deep tidal disruption events: prompt self-intersections and observables. Mon. Not. R. Astron. Soc. (2019). https://doi.org/10.1093/mnras/stz1923
C. Evans, P. Laguna, M. Eracleous, Ultra-close encounters of stars with massive black holes: tidal disruption events with prompt hyperaccretion. Astrophys. J. **805**(2), 19 (2015). https://doi.org/10.1088/2041-8205/805/2/L19
G. Faye, L. Blanchet, A. Buonanno, Higher-order spin effects in the dynamics of compact binaries. I. Equations of motion. Phys. Rev. D **74**(10), 104033 (2006). https://doi.org/10.1103/PhysRevD.74.104033
J. Frank, A. King, D.s. Raine, *Accretion Power in Astrophysics*, Third edn. (Cambridge University Press, Cambridge, 2002), p. 398
E. Girma, J. Guillochon, The galactic distribution of fragments formed from tidally disrupted stars. ArXiv e-prints, 1707–03421 (2018)
J. Guillochon, M. McCourt, Simulations of magnetic fields in tidally disrupted stars. Astrophys. J. **834**(2), 19 (2017). https://doi.org/10.3847/2041-8213/834/2/L19
J. Guillochon, E. Ramirez-Ruiz, A dark year for tidal disruption events. Astrophys. J. **809**, 166 (2015). https://doi.org/10.1088/0004-637X/809/2/166
J. Guillochon, H. Manukian, E. Ramirez-Ruiz, PS1-10jh: the disruption of a main-sequence star of near-solar composition. Astrophys. J. **783**(1), 23 (2014). https://doi.org/10.1088/0004-637X/783/1/23
R. Haas, R.V. Shcherbakov, T. Bode, P. Laguna, Tidal disruptions of white dwarfs from ultra-close encounters with intermediate-mass spinning black holes. Astrophys. J. **749**(2), 117 (2012). https://doi.org/10.1088/0004-637X/749/2/117
K. Hayasaki, N. Stone, A. Loeb, Finite, intense accretion bursts from tidal disruption of stars on bound orbits. Mon. Not. R. Astron. Soc. **434**(2), 909–924 (2013). https://doi.org/10.1093/mnras/stt871
K. Hayasaki, N. Stone, A. Loeb, Circularization of tidally disrupted stars around spinning supermassive black holes. Mon. Not. R. Astron. Soc. **461**(4), 3760–3780 (2016). https://doi.org/10.1093/mnras/stw1387
M.s. Hobson, G.s. Efstathiou, A.s. Lasenby, *General Relativity: An Introduction for Physicists* (Cambridge University Press, Cambridge, 2006), p. 590
N. Jiang, L. Dou, T. Wang, C. Yang, J. Lyu, H. Zhou, The wise detection of an infrared echo in tidal disruption event ASASSN-14li. Astrophys. J. **828**(1), 14 (2016a). https://doi.org/10.3847/2041-8205/828/1/L14
Y.-F. Jiang, J. Guillochon, A. Loeb, Prompt radiation and mass outflows from the stream–stream collisions of tidal disruption events. Astrophys. J. **830**(2), 125 (2016b). https://doi.org/10.3847/0004-637X/830/2/125
V. Karas, L. Šubr, Enhanced activity of massive black holes by stellar capture assisted by a self-gravitating accretion disc. Astron. Astrophys. **19**, 11–19 (2007). https://doi.org/10.1051/0004-6361
D. Kasen, E. Ramirez-Ruiz, Optical transients from the unbound debris of tidal disruption. Astrophys. J. **714**(1), 155–162 (2010). https://doi.org/10.1088/0004-637X/714/1/155
A. Kathirgamaraju, R.B. Duran, D. Giannios, TDE fallback cut-off due to a pre-existing accretion disc. ArXiv e-prints (2017)
L.Z. Kelley, A. Tchekhovskoy, R. Narayan, Tidal disruption and magnetic flux capture: powering a jet from a quiescent black hole. Mon. Not. R. Astron. Soc. **445**(4), 3919–3938 (2014). https://doi.org/10.1093/mnras/stu2041
S.S. Kim, M. Park, H.M. Lee, The stream-stream collision after the tidal disruption of a star around a massive black hole. Astrophys. J. **519**(2), 647–657 (1999). https://doi.org/10.1086/307394
C.S. Kochanek, The aftermath of tidal disruption: the dynamics of thin gas streams. Astrophys. J. **422**, 508 (1994). https://doi.org/10.1086/173745
G. Leloudas, L. Dai, I. Arcavi, P.M. Vreeswijk, B. Mockler, R. Roy, D.B. Malesani, S. Schulze, T. Wevers, M. Fraser, E. Ramirez-Ruiz, K. Auchettl, J. Burke, G. Cannizzaro, P. Charalampopoulos, T.-W. Chen, A. Cikota, M. Della Valle, L. Galbany, M. Gromadzki, K.E. Heintz, D. Hiramatsu, P.G. Jonker, Z. Kostrzewa-Rutkowska, K. Maguire, I. Mandel, M. Nicholl, F. Onori, N. Roth, S.J. Smartt, L. Wyrzykowski, D.R. Young, The spectral evolution of AT 2018dyb and the presence of metal lines in tidal disruption events. Astrophys. J. **887**(2), 218 (2019). https://doi.org/10.3847/1538-4357/ab5792
D. Liptai, D.J. Price, I. Mandel, G. Lodato, Disc formation from tidal disruption of stars on eccentric orbits by Kerr black holes using GRSPH. Mon. Not. R. Astron. Soc., 1–12 (2020 submitted). ArXiv e-prints
G. Lodato, Challenges in the modeling of tidal disruption events lightcurves. EPJ Web Conf. **39**, 01001 (2012). https://doi.org/10.1051/epjconf/20123901001
G. Lodato, D.J. Price, On the diffusive propagation of warps in thin accretion discs. Mon. Not. R. Astron. Soc. **405**(2), 1212–1226 (2010). https://doi.org/10.1111/j.1365-2966.2010.16526.x

W. Lu, C. Bonnerot, Self-intersection of the fallback stream in tidal disruption events. Mon. Not. R. Astron. Soc. **492**(1), 686–707 (2020). https://doi.org/10.1093/mnras/stz3405

W. Lu, P. Kumar, On the missing energy puzzle of tidal disruption events. Astrophys. J. **865**(2), 128 (2018). https://doi.org/10.3847/1538-4357/aad54a

M. MacLeod, J. Goldstein, E. Ramirez-Ruiz, J. Guillochon, J. Samsing, Illuminating massive black holes with white dwarfs: orbital dynamics and high energy transients from tidal interactions. Astrophys. J. **794**(1), 9 (2014). https://doi.org/10.1088/0004-637X/794/1/9

D. Merritt, *Dynamics and Evolution of Galactic Nuclei* (Princeton University Press, Princeton, 2013). ISBN 9780691158600

B.D. Metzger, N.C. Stone, A bright year for tidal disruptions. Mon. Not. R. Astron. Soc. **461**(1), 948–966 (2016). https://doi.org/10.1093/mnras/stw1394

M.C. Miller, Disk winds as an explanation for slowly evolving temperatures in tidal disruption events. Astrophys. J. **805**(1), 83 (2015). https://doi.org/10.1088/0004-637X/805/1/83

R. Nealon, D.J. Price, C. Bonnerot, G. Lodato, On the Papaloizou-Pringle instability in tidal disruption events. Mon. Not. R. Astron. Soc. **474**(2), 1737–1745 (2018). https://doi.org/10.1093/mnras/stx2871

B. Paczyńsky, P.s.J. Wiita, Thick accretion disks and supercritical luminosities. Astron. Astrophys. **88**, 23–31 (1980)

J.C.B. Papaloizou, J.E. Pringle, The dynamical stability of differentially rotating discs with constant specific angular momentum. Mon. Not. R. Astron. Soc. **208**(4), 721–750 (1984). https://doi.org/10.1093/mnras/208.4.721. ISBN 9781479978007

T. Piran, G. Svirski, J. Krolik, R.M. Cheng, H. Shiokawa, Disk formation versus disk accretion—what powers tidal disruption events? Astrophys. J. **806**(2), 164 (2015). https://doi.org/10.1088/0004-637X/806/2/164

E. Ramirez-Ruiz, S. Rosswog, The star ingesting luminosity of intermediate-mass black holes in globular clusters. Astrophys. J. **697**(2), 77–80 (2009). https://doi.org/10.1088/0004-637X/697/2/L77

M.J. Rees, Tidal disruption of stars by black holes of 106–108 solar masses in nearby galaxies. Nature **333**(6173), 523–528 (1988). https://doi.org/10.1038/333523a0. ISBN 0028-0836

S. Rosswog, E. Ramirez-Ruiz, W.R. Hix, Atypical thermonuclear supernovae from tidally crushed white dwarfs. Astrophys. J. **679**(2), 1385–1389 (2008). https://doi.org/10.1086/528738

S. Rosswog, E. Ramirez-Ruiz, W.R. Hix, Tidal disruption and ignition of white dwarfs by moderately massive black holes. Astrophys. J. **695**(1), 404–419 (2009). https://doi.org/10.1088/0004-637X/695/1/404

N. Roth, D. Kasen, J. Guillochon, E. Ramirez-ruiz, The X-ray through optical fluxes and line strengths of tidal disruption events. Astrophys. J. **827**(1), 1–15 (2016). https://doi.org/10.3847/0004-637X/827/1/3

A. Sacchi, G. Lodato, C. Toci, V. Motta, What causes the fragmentation of debris streams in TDEs? Mon. Not. R. Astron. Soc. **495**(1), 1227–1238 (2020). https://doi.org/10.1093/mnras/staa1299

A. Sadowski, E. Tejeda, E. Gafton, S. Rosswog, D. Abarca, Magnetohydrodynamical simulations of a deep tidal disruption in general relativity. Mon. Not. R. Astron. Soc. **458**(4), 4250–4268 (2016). https://doi.org/10.1093/mnras/stw589

N.I. Shakura, R.a. Sunyaev, Black holes in binary systems. Observational appearance. Astron. Astrophys. **24**, 337–355 (1973). https://doi.org/10.1086/170270. ISBN 9810200773

H. Shiokawa, J.H. Krolik, R.M. Cheng, T. Piran, S.C. Noble, General relativistic hydrodynamic simulation of accretion flow from a stellar tidal disruption. Astrophys. J. **804**(2), 85 (2015). https://doi.org/10.1088/0004-637X/804/2/85

E. Steinberg, E.R. Coughlin, N.C. Stone, B.D. Metzger, Thawing the frozen-in approximation: implications for self-gravity in deeply plunging tidal disruption events. Mon. Not. R. Astron. Soc. Lett. **485**(1), 146–150 (2019). https://doi.org/10.1093/mnrasl/slz048

N. Stone, A. Loeb, Tidal disruption flares of stars from moderately recoiled black holes. Mon. Not. R. Astron. Soc. **422**(3), 1933–1947 (2012). https://doi.org/10.1111/j.1365-2966.2012.20577.x

N.C. Stone, B.D. Metzger, Rates of stellar tidal disruption as probes of the supermassive black hole mass function. Mon. Not. R. Astron. Soc. **455**(1), 859–883 (2016). https://doi.org/10.1093/mnras/stv2281

N. Stone, R. Sari, A. Loeb, Consequences of strong compression in tidal disruption events. Mon. Not. R. Astron. Soc. **435**(3), 1809–1824 (2013). https://doi.org/10.1093/mnras/stt1270

L.E. Strubbe, E. Quataert, Optical flares from the tidal disruption of stars by massive black holes. Mon. Not. R. Astron. Soc. **400**(4), 2070–2084 (2009). https://doi.org/10.1111/j.1365-2966.2009.15599.x

G. Svirski, T. Piran, J. Krolik, Elliptical accretion and inefficient highly eccentric accretion and the low luminosity of stellar tidal disruption events. ArXiv e-prints (2017)

E. Tejeda, S. Rosswog, An accurate Newtonian description of particle motion around a Schwarzschild black hole. Mon. Not. R. Astron. Soc. **433**(3), 1930–1940 (2013). https://doi.org/10.1093/mnras/stt853

A. Ulmer, Flares from the tidal disruption of stars by massive black holes. Astrophys. J. **514**(1), 180–187 (1999). https://doi.org/10.1086/306909. ISBN 0094-243X

S. van Velzen, A.J. Mendez, J.H. Krolik, V. Gorjian, Discovery of transient infrared emission from dust heated by stellar tidal disruption flares. Astrophys. J. **829**(1), 19 (2016). https://doi.org/10.3847/0004-637X/829/1/19

C. Wegg, J.N. Bode, Multiple tidal disruptions as an indicator of binary supermassive black hole systems. Astrophys. J. **738**(1), 8 (2011). https://doi.org/10.1088/2041-8205/738/1/L8

S. Wen, P.G. Jonker, N.C. Stone, A.I. Zabludoff, D. Psaltis, Continuum-fitting the X-ray spectra of tidal disruption events. Astrophys. J. **897**(1), 80 (2020). https://doi.org/10.3847/1538-4357/ab9817

H.N. Wernke, A.-M. Madigan, The effect of general relativistic precession on tidal disruption events from eccentric nuclear disks. ArXiv e-prints (2019)

Space Sci Rev (2021) 217:12
https://doi.org/10.1007/s11214-020-00747-x

The Physics of Accretion Discs, Winds and Jets in Tidal Disruption Events

Jane Lixin Dai[1] · **Giuseppe Lodato**[2] ·
Roseanne Cheng[3]

Received: 5 March 2020 / Accepted: 4 October 2020 / Published online: 12 January 2021

Abstract Accretion onto black holes is an efficient mechanism in converting the gas mass-energy into energetic outputs as radiation, wind and jet. Tidal disruption events, in which stars are tidally torn apart and then accreted onto supermassive black holes, offer unique opportunities of studying the accretion physics as well as the wind and jet launching physics across different accretion regimes. In this review, we systematically describe and discuss the models that have been developed to study the accretion flows and jets in tidal disruption events. A good knowledge of these physics is not only needed for understanding the emissions of the observed events, but also crucial for probing the general relativistic space-time around black holes and the demographics of supermassive black holes via tidal disruption events.

Keywords Accretion, accretion discs · Black hole physics · Galaxies: jets · Galaxies: nuclei · Gravitation · Hydrodynamics

---

The Tidal Disruption of Stars by Massive Black Holes
Edited by Peter G. Jonker, Sterl Phinney, Elena Maria Rossi, Sjoert van Velzen, Iair Arcavi and Maurizio Falanga

---

✉ J.L. Dai
lixindai@hku.hk

G. Lodato
giuseppe.lodato@unimi.it

R. Cheng
rmcheng@lanl.gov

1 Department of Physics, The University of Hong Kong, Pokfulam Road, Hong Kong

2 Dipartimento di Fisica, Università degli Studi di Milano, Via Celoria 16, Milano, Italy

3 Theoretical Division (T-3), Los Alamos National Laboratory, P.O. Box 1663, Los Alamos, NM 87545, USA

1 Introduction

A tidal disruption event (TDE) happens when a star passes by a supermassive black hole (SMBH) too close and is therefore torn apart by the tidal force. The distance at which the SMBH tidal force equals the stellar self-gravity is called the tidal radius R_t (Hills 1975), which depends on the mass $M_\star$ and radius $R_\star$ of the star as well as the mass $M_{\rm BH}$ of the SMBH: $R_t \approx R_\star\,(M_{\rm BH}/M_\star)^{1/3}$. For main-sequence stars approaching SMBHs with $M_{\rm BH} = \text{few} \times (10^6 - 10^7)\ M_\odot$, R_t is only few $\times\ (1 - 10) R_g$ with $R_g \equiv GM_{\rm BH}/c^2$ being the gravitational radius of the black hole. Therefore, general relativity (GR) naturally plays an important role in the processes happening in TDEs.

The basic theory of TDE physics has been developed since the 1970s (e.g., Hills 1975; Rees 1988; Phinney 1989). Here we give a brief description of the general picture. Following the disruption of a star approaching the black hole along a marginally bound orbit, roughly half of the stellar debris remains bound and travels along highly elliptical, ballistic orbits, while the other half escapes along hyperbolic orbits. The spread of the orbital energy determines the mass rate that bound debris returns to the pericenter as a function of time, which is called the fallback rate. Shortly after passing through the pericenter, the bound debris assembles a transient accretion disc around the SMBH through shocks and shear viscosity reducing the orbital energy of the debris. Then with the aid of magnetic fields, viscous processes transport the gas angular momentum outwards, allowing gas to accrete onto the black hole and bright flares to be produced. Under optimal conditions, winds and relativistic jets can also be launched from the transient discs formed in TDEs.

In recent years our understanding of TDE physics has been greatly pushed forward by numerical simulations. These simulations usually focus on specific phases in TDEs (e.g., the disruption of the star, the evolution of the debris stream, the formation of the disc, the accretion process and jet production, etc.) due to the large computational resources required to study the whole process. These studies have basically confirmed the disruption process described in the general picture, which is discussed in the Disruption Chapter of this book. However, up to now no self-consistent simulations have been done on the disc formation process for typical TDE parameters, which leads to uncertainties on whether the debris can quickly dissipate orbital energy to form discs as described in the general picture. We refer the readers to the Formation of the Accretion Flow Chapter for details. In this Chapter dedicated to the TDE accretion process, we will focus on the phase after stable discs have been formed and review the theoretical efforts that have been made to investigate the physics of TDE accretion discs, winds and jets.

A solid understanding of the accretion process builds foundations for understanding TDE emissions, and therefore is key for probing SMBH demographics from TDE observables. Also, TDEs give us ideal laboratories for studying black hole accretion physics in different accretion regimes. In a standard TDE (where a main-sequence star is totally disrupted by a SMBH along a parabolic orbit), the debris fallback rate that feeds the disc is expected to first quickly reach the super-Eddington regime, and then declines to the sub-Eddington regime over a few months to years (Fig. 1).

The Eddington luminosity is the theoretical maximum luminosity that a star or an accretion disc can have when there is a balance between the outward radiation force and the inward gravitational force. For the inner part of accretion discs, all of the hydrogen atoms are ionized since the temperature is higher than 10^5 K (Shakura and Sunyaev 1973; Frank et al. 2002). The Eddington luminosity for a black hole is therefore defined as:

$$L_{\rm Edd} = \frac{4\pi c G M_{\rm BH}}{\kappa_{\rm es}} \approx 1.26 \times 10^{44} \left(\frac{M_{\rm BH}}{10^6\ M_\odot}\right) \mathrm{erg\,s^{-1}}. \quad (1)$$

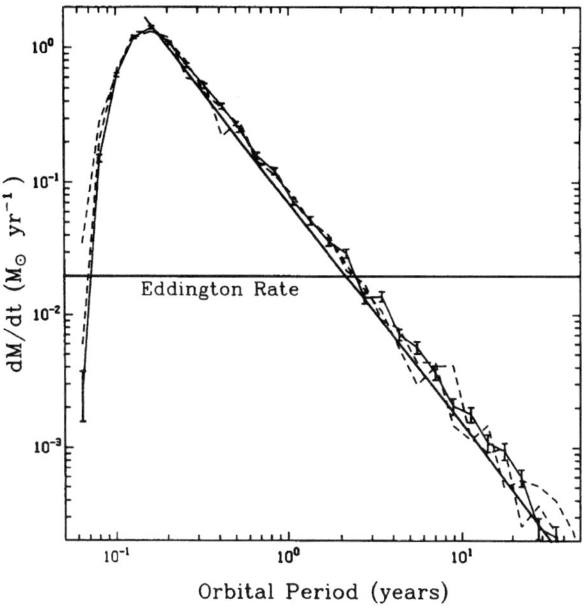

Fig. 1 The debris fallback rate as a function of time. The Eddington accretion rate shown is for a SMBH with $M_{\rm BH} = 10^6\ M_\odot$ and radiative efficiency $\eta = 0.1$. As shown the fallback rate can be one to two orders of magnitude above the Eddington accretion rate of the black hole when $M_{\rm BH} <$ few $\times\ 10^7\ M_\odot$. After the peak the fallback rate declines following a power-law function of time: $\dot{M}_{\rm fb} \propto t^{-5/3}$, sweeping through different regimes of accretion. Figure is from Evans and Kochanek (1989)

Here $\kappa_{\rm es}$ is the Thomson scattering opacity. For hydrogen $\kappa_{\rm es} = \sigma_{\rm T}/m_{\rm p} = 0.4\ {\rm cm^2 g^{-1}}$, where $m_{\rm p}$ is the proton mass and $\sigma_{\rm T}$ is the Thomson cross-section. The corresponding Eddington accretion rate is:

$$\dot{M}_{\rm Edd} = \frac{L_{\rm Edd}}{\eta c^2} \approx 0.022 \left(\frac{\eta}{0.1}\right)^{-1} \left(\frac{M}{10^6\ M_\odot}\right)\ M_\odot\ {\rm yr}^{-1}, \qquad (2)$$

where $\eta \equiv L/(\dot{M}c^2)$ is the accretion radiative efficiency. In literature η is usually given a nominal value of 10%, although its value can vary drastically as described next.

The structure of an accretion disc is primarily determined by its Eddington ratio, which is how the accretion rate compares to the Eddington accretion rate. (1) For very high, super-Eddington rates ($\dot{M}/\dot{M}_{\rm Edd} \gtrsim 1$), the disc is expected to be geometrically and optically thick. The photon diffusion time is very long and the disc is dominated by advection (the so-called slim disc solutions by Abramowicz et al. 1988) resulting in low radiative efficiency; (2) for moderate accretion rates, $0.01 \lesssim \dot{M}/\dot{M}_{\rm Edd} \lesssim 1$ (or $0.01 \lesssim L/L_{\rm Edd} \lesssim 0.3$), the disc, well described by the standard Shakura and Sunyaev (1973) solutions (or a GR prescription by Novikov and Thorne 1973), is optically thick, geometrically thin, and radiatively efficient; finally, (3) for very low accretion rates ($\dot{M}/\dot{M}_{\rm Edd} < 0.01$), the disc transitions again to an advection-dominated accretion flow (ADAF) or a radiatively inefficient accretion flow (RIAF) (Ichimaru 1977; Rees et al. 1982; Narayan and Yi 1994; Blandford and Begelman 1999, 2004) with low radiative efficiency again. Although the Eddington ratio likely plays the most important role in determining the morphology of an accretion disc, the properties of the disc are also affected other physical parameters. For example, the black hole mass, together with the accretion rate, determine the disc temperature and various opacities governing disc dynamics and radiative processes. Also, in the thin disc regime, the inner edge of the disc is set by the innermost stable circular orbit (ISCO) which has a strong dependence on the black hole spin. As a result, η varies from 6% to 42% as the black hole spin increases from 0 to 1.

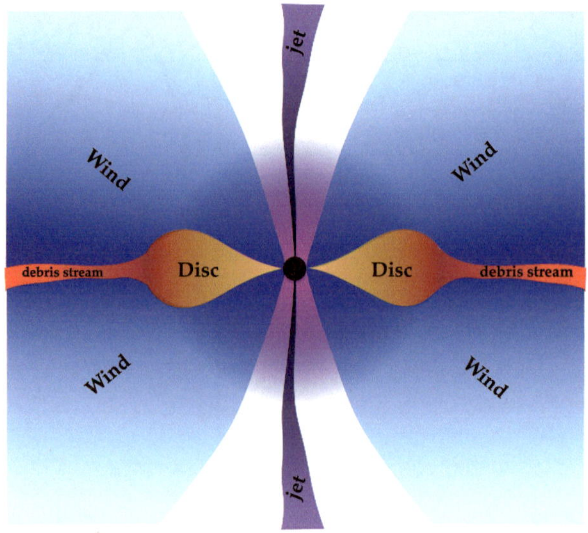

Fig. 2 A cartoon showing the geometry of a circular and aligned geometrically-thick accretion disc formed in the super-Eddington phase of a TDE, as well as the wind and jet launched from the disc. It is expected that a super-Eddington disc can generally launch optically-thick wind with large solid angle coverage. A strong relativistic jet, on the other hand, can only be produced when the black hole is spinning fast and has a large amount of ordered magnetic flux in its vicinity (Blandford and Znajek 1977). Figure is adapted from Dai et al. (2018)

Ulmer (1999) is the first to point out that different regimes of accretion should naturally happen throughout the course of a TDE, so the disc geometry should change from thick to thin as the accretion level drops from super-Eddington to sub-Eddington in a few months to years. Analytical models that encompass both the slim disc case and the standard thin disc case have been first constructed by Strubbe and Quataert (2009). We will review various disc models and simulations along this line in Sect. 2.1 (early, super-Eddington phase) and Sect. 2.2 (late time). New disc models or angular momentum transport mechanisms different from the conventional picture will be reviewed in Sect. 2.3 and Sect. 2.4.

Relativistic jets and winds have been observed in several TDEs, which cannot been explained using the standard thin disc theory. In Sect. 2.1.2 and Sect. 3, we will review works which discuss how winds and jets can be launched when TDE disc is in the super-Eddington accretion regime. A cartoon illustrating the disc-wind-jet geometry during this phase is illustrated in Fig. 2.

Most TDE literature have so far focused on modeling discs which have angular momentum aligned with the spin axis of the black hole. However, misaligned TDE discs likely can form since it is generally expected that stars are scattered into the loss cone from random orientations. We will discuss works that have been carried out to understand such discs and their jets in Sect. 4.

Lastly, a summary will be drawn in Sect. 5, where we will also raise several key open questions related to TDE accretion discs, winds and jets for future studies.

2 Models of TDE Accretion Flow

In this section, we will summarize various analytical models and numerical simulations that have been carried out to study the accretion flow formed in TDEs. While we put the focus on describing the structures of the accretion inflow/outflow and the accretion process, we will also briefly mention the properties of the emissions produced in these models. We refer the readers to the Emission Mechanisms Chapter in this book for more details of the radiative processes in TDE accretion flow.

2.1 TDE Discs with Super-Eddington Debris Fallback

During the first few weeks to years in a TDE, the debris supply rate is high and can exceed the Eddington accretion rate of the black hole if the black hole mass $M_{\rm BH} \lesssim$ few $\times 10^7\ M_\odot$. It is naturally expected that most TDEs are detected in this phase as the flares are luminous. Many theoretical studies have therefore focused on studying the properties of the TDE discs in this phase. Most of the work are semi-analytical, in which the TDE accretion flow is modeled as a steady, quasi-spherical disc (Loeb and Ulmer 1997; Ulmer 1999; Coughlin and Begelman 2014; Roth et al. 2016; Wu et al. 2018) or a disc surrounded by a quasi-spherical wind (Strubbe and Quataert 2009; Lodato and Rossi 2011; Shen and Matzner 2014; Miller 2015; Metzger and Stone 2016). Recently 3D numerical simulations of super-Eddington accretion flow have also been carried out to study TDE discs, in which the disc winds (and jets) are self-consistently produced (Dai et al. 2018; Curd and Narayan 2019).

From the observational side, the unique properties of TDE emissions also lend strong support to the notion that TDE discs should be different from the discs of local AGNs, which are believed to be sub-Eddington in most cases. Some of the most relevant observational features are: 1) Many TDE flares are strong in UV/optical wavebands but weak in X-rays, while some TDEs are strong in X-rays but not in UV/optical bands. 2) For the optically-strong TDEs, their flares are usually bluer than AGN optical emissions. There are three spectroscopic classes discovered: TDEs that produce broad H Balmer emission lines, He emission lines, or Bowen fluorescent NIII/OIII emission lines (Gezari et al. 2012; Arcavi et al. 2014; Leloudas et al. 2019; van Velzen et al. 2020). 3) For the non-jetted X-ray strong TDEs, their emissions are dominated by thermal components and do not show strong coronae which are typically seen in AGNs (Komossa 2015; Auchettl et al. 2017). 4) In some TDEs, fast winds have been reported, which will be discussed more in Sect. 2.1.2. More details on how TDE emissions are distinctive from AGN emissions can be found in the Imposters Chapter. The disc models to be discussed in this section all employ extreme debris supply or accretion level to explain these observed TDE features.

2.1.1 Quasi-Spherical Models of Steady Discs or Envelopes

Early studies (e.g., Rees 1988) assume that stellar debris can circularize fast and form a compact disc with a size $R_c = 2\ R_p$ from conservation of angular momentum, where R_p is the pericenter radius of the original parabolic orbit of the star. Due to the short viscous timescale of the disc, the debris accretion rate resembles the fallback rate. Therefore, for the first few weeks to months, the disc in the super-Eddington accretion regime is geometrically and optically thick. Ulmer (1999) calculates the effective temperature of this small, quasi-spherical disc by assuming the luminosity is limited by the Eddington luminosity $L_{\rm Edd}$ and the disc radius is comparable to the tidal radius R_t:

$$T_{\rm eff} = \left(\frac{L_{\rm Edd}}{4\pi R_t^2 \sigma}\right) \approx 5 \times 10^5\ M_6^{-1/4} {\rm K}, \tag{3}$$

where σ is the Stefan-Boltzmann constant and $M_6 = M_{\rm BH}/(10^6\ M_\odot)$.

This simple model predicts that TDEs should shine in X-ray/EUV and can therefore explain the observed X-ray strong TDEs (see the X-ray Chapter of this book). However, it cannot explain why many TDEs are only bright in optical/NUV bands but weak in X-rays (see the Optical Chapter). It has been proposed by Loeb and Ulmer (1997) that since the stellar debris is only marginally bound, radiation produced from the viscous process in the inner

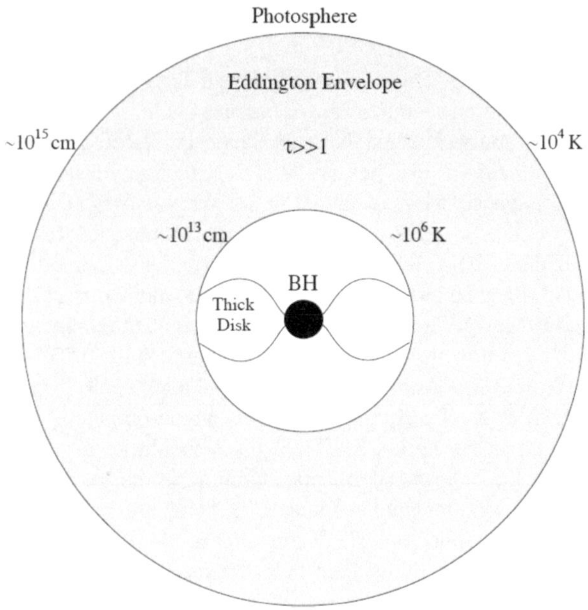

Fig. 3 Illustration of a steady, optically-thick, quasi-spherical large envelope surrounding a small accretion disc formed in TDEs. The disc X-ray/EUV emissions are reprocessed into UV/optical emissions when going through the envelope. Figure is from Loeb and Ulmer (1997)

accretion disc can easily disperse some gas into a quasi-spherical configuration at a large distance. This large, steady, optically-thick envelope surrounding the small accretion disc, as illustrated in Fig. 3, can reprocess the disc X-ray/EUV emissions into optical emissions. The envelope is supported by radiation pressure and has a density profile of $\rho(r) \propto r^{-q}$ with $q = 3$. It is assumed to be in a steady state exactly balanced by gravity and radiation pressure, which relies on a super-Eddington accretion rate with a radiative efficiency of $\gtrsim 10^{-3}$. The photosphere shines at the Eddington luminosity with a size of about 10^{15} cm and an effective temperature of around 10^4 K.

Later Coughlin and Begelman (2014) argue that the angular momentum transport between the disc and envelope in the setting of Loeb and Ulmer (1997) should further shrink the disc and flatten its density distribution, making it hard to maintain the accretion exactly at the Eddington level. Instead, they propose a model in which the whole accretion disc is inflated by radiation energy into a quasi-spherical configuration with zero Bernoulli parameter (i.e., marginally bound) everywhere (these models are called "Zero-Bernoulli accretion flows", or ZEBRAs). The disc rotates at a sub-Keplerian speed with a very small angular momentum consistent with the angular momentum of the returning stellar debris. The evolution of the accretion flow has been computed using the analytical debris fallback rate derived from the "frozen-in" approximation (Lodato et al. 2009) as well as using the numerically computed fallback rate (Wu et al. 2018). As shown in Fig. 4, the density profile of the accretion flow $\rho(r) \propto r^{-q}$ has an index q ranging between 0.5 and 1.5 during most events. The photosphere radius grows with time to $R_{\rm ph} \sim$ few $\times\ 10^{14}$ cm and the photosphere temperature drops by a few times to $T_{\rm ph} \sim$ few $\times\ 10^4$ K after a few years. In this model, most of the energy produced from accretion is actually released in the form of a jet along polar directions confined by the narrow rotational funnel.

This type of spherical and steady envelope model has also been adopted by e.g., Roth et al. (2016) to carry out detailed radiative transfer studies in TDEs.

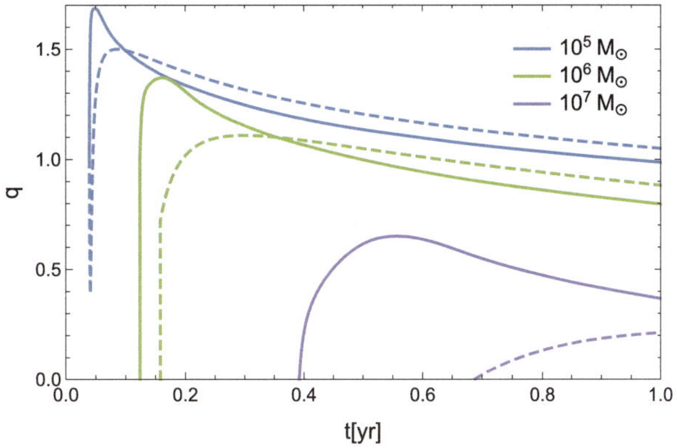

Fig. 4 Index q of the zero-Bernoulli accretion (ZEBRA) flow (Coughlin and Begelman 2014) density profile ($\rho(r) \propto r^{-q}$) as a function of time for TDEs around black holes of different masses. The solid lines show results using numerically acquired fallback rate, and the dashed lines show results using the "frozen-in" approximation at the time of disruption. Figure is from Wu et al. (2018)

2.1.2 Models of TDE Discs with Winds

A small fraction of the energy produced through accretion or other energetic processes (such as shock heating during disc formation) can in principle unbind a large fraction of returning debris and launch strong winds. Wind-launching from accretion discs can be due to: 1) large radiation pressure in super-Eddington accretion discs or other radiative transfer processes in accretion discs (e.g., Poutanen et al. 2007; Proga et al. 2000; Dotan and Shaviv 2011); 2) hydromagnetic processes (e.g. Blandford and Payne 1982); 3) in TDEs, gas heating during the processes leading to disc formation such as collisions of debris steams (Jiang et al. 2016).

Winds have been observed in several TDEs. For example, ultra-fast winds with $v \gtrsim 0.1c$ have been found through X-ray reflection signatures in Swift J1644 (Kara et al. 2016) and through X-ray absorption features or radio signals found in (soft) X-ray strong TDE ASASSN 14-li (Alexander et al. 2016; Kara et al. 2017) and 3XMM J1500 (Lin et al. 2017). Recently UV spectroscopic studies have also been fruitful in discovering winds with $v \approx 0.05c$ from optical strong TDEs (Hung et al. 2019).

In the context of TDE discs, there are several studies modeling or simulating the winds launched during the super-Eddington accretion phase (Strubbe and Quataert 2009, 2011; Lodato and Rossi 2011; Dai et al. 2018; Curd and Narayan 2019). Strubbe and Quataert (2009) consider a simple model in which the radiation energy responsible for wind-launching is generated when returning stellar debris continuously hits the already formed disc at a distance close to $R_c = 2R_p$. A constant fraction $f_{\rm out} \equiv \dot{M}_{\rm wind}/\dot{M}_{\rm fb} \approx 0.1 - 0.3$ of the fallback debris is assumed to become a spherical wind, which is launched at R_c with a velocity close to the escape velocity at R_c and adiabatically expands. Inside R_c a slim disc model similar to Abramowicz et al. (1988) is adopted to describe the disc properties. In particular, the disc temperature $T_{\rm d}$ is obtained as follows:

$$\sigma_{\rm SB} T_{\rm d}^4 = \frac{3GM_{\rm BH}\dot{M}f}{8\pi R^3}\left[\frac{1}{2} + \left(\frac{1}{4} + \left(\frac{\dot{M}}{\eta \dot{M}_{\rm Edd}}\right)^2 \left(\frac{R_{\rm s}}{R}\right)^2\right)\right]^{-1}, \tag{4}$$

where $f = 1 - \sqrt{R_{\rm in}/R}$, $R_{\rm s}$ is the Schwarzschild radius and the accretion rate in the disc is given by $\dot{M} = (1 - f_{\rm out})\dot{M}_{\rm fb}$. In this model, strong optical emissions can be produced from the wind which cools down to few $\times 10^4$ K after adiabatic expansion. The total emission is dominated by wind emission at earlier time (for $\sim 10 - 100$ days depending on the exact stellar and black hole parameters and the wind launching radius) when the wind is optically

Fig. 5 The solid line shows the wind mass rate fraction $f_{\rm out}$ as a function of the fallback rate of stellar debris in unit of the Eddington accretion rate, which is used by Lodato and Rossi (2011) to model the wind launched in TDE super-Eddington accretion discs. As the Eddington ratio increases, more fallback debris material becomes wind instead of accreting onto the black hole. The squares show the numerical results of Dotan and Shaviv (2011). Figure is from Lodato and Rossi (2011)

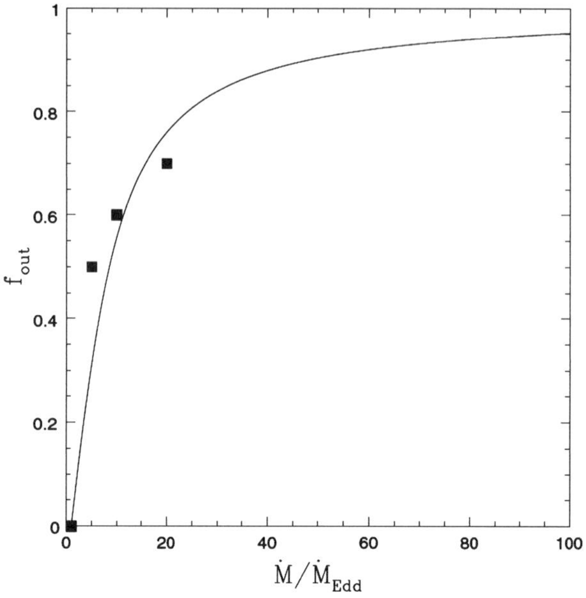

thick. At later times, the wind becomes more dilute and the photosphere recedes, revealing more disc emissions.

Lodato and Rossi (2011) consider a more realistic model for the wind produced from the radiation-pressure supported disc formed in the super-Eddington phase, based on the simulation of Dotan and Shaviv (2011). The wind mass loss fraction $f_{\rm out}$ is taken to be larger at earlier times when the debris supply rate is higher (Fig. 5). They also calculate the emission produced by the slim, compact disc, assuming that the accretion rate through the disc is a constant fraction of the fallback rate. They find that, even if the disc bolometric luminosity scales as the fallback rate, as $t^{-5/3}$, the lightcurve in specific bands scales differently with time, because the disc temperature also evolves as the accretion rate declines. In particular, at wavelengths (such as in the UV and optical) that sit on the Rayleigh-Jeans tail of the disc multi-color blackbody, the lightcurve is proportional to the temperature and therefore scales as $T_{\rm d} \propto L_{\rm bol}^{1/4} \propto t^{-5/12}$, declining much slower than the fallback rate. Conversely, for those wavelengths that are close to the peak of the blackbody at the beginning of the flare (i.e., in the soft X-rays, typically), the lightcurve initially declines as $t^{-5/3}$ (see Fig. 6). In general, the lightcurve is expected to evolve from a $t^{-5/12}$ decline, to a $t^{-5/3}$ and eventually to an exponential decline with time. For high enough photon energies (e.g., in the X-rays), the first phase could be missing, and the lightcurve would start directly from a close to $t^{-5/3}$ decline. If one also adds the wind emission in addition to the disc emission, the lightcurve gets further complicated and depends strongly on the system parameters, with the apparent lightcurve evolving coincidentially as $t^{-5/3}$ in particular regions of the parameter space.

These two models discussed previously assume that a sizable fraction of the fallback debris can make its way to the inner disc and super-Eddington accretion (onto the black hole) still happens. Instead, Miller (2015) and Metzger and Stone (2016) propose models in which most of the debris supply is launched as winds faraway from the black hole, either due to large opacities in the outer disc or due to shocks happening during the disc formation process. Under such scenarios, only a small fraction ($\lesssim 1\%$) of debris materials is accreted so the accretion rate is always sub-Eddington. The low level of accretion, together with

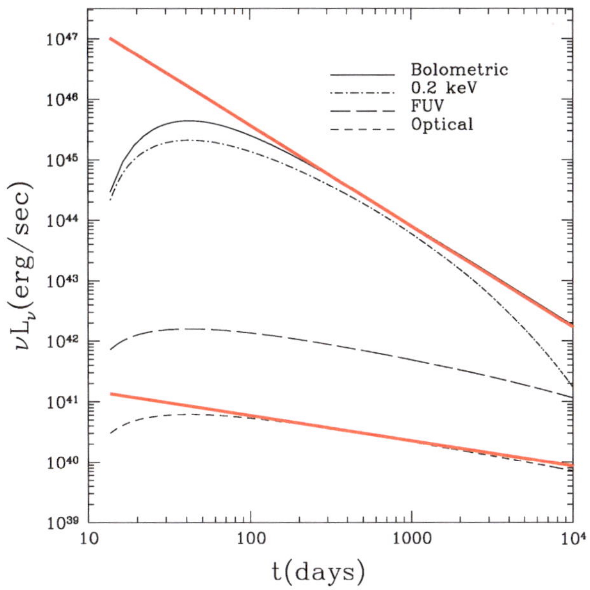

Fig. 6 Example lightcurves for the disc emission from the disruption of a solar type star by a $10^6\ M_\odot$ black hole at $\beta = 1$. The fallback rate corresponds to a $\gamma = 1.4$ polytropic model for the star (Lodato et al. 2009). The solid line shows the bolometric luminosity, the dot-dashed line shows the luminosity at 0.2 keV, the long-dashed line shows the luminosity in the FUV, at $\nu = 1.9\ 10^{15}$ Hz, while the short-dashed line shows the luminosity in the optical at $\nu = 6.3\ 10^{14}$ Hz. The two red lines mark the simple power laws expected for the bolometric luminosity ($\propto t^{-5/3}$) and for the monochromatic luminosity in the optical/UV ($\propto t^{-5/12}$). Figure is from Lodato and Rossi (2011)

the very mass-loaded wind functioning as a thick reprocessing envelope, can also reduce temperature of the TDE emission to few $\times 10^4$ K so that UV/optical emissions dominate.

Furthermore, numerical simulations of TDE discs with super-Eddington accretion rates (Dai et al. 2018; Curd and Narayan 2019) also show that strong winds can be launched in this phase, and we will discuss these numerical work in the next section.

2.1.3 Simulated Super-Eddington Accretion Flow and Inclination Dependence

Realistically, we do not expect accretion flows to be spherically symmetric. While the studies mentioned in Sect. 2.1.1 and 2.1.2 mostly adopt 1D disc or envelope models when carrying out the semi-analytical calculations, they also suggest that the structure and emission of TDE accretion flow should have a dependence on the observer's viewing-angle with respect to the disc orientation. The details, however, can only be calculated using numerical simulations due to the complexity of the problem.

Codes that can be used to calculate the dynamics of super-Eddington accretion discs and capture its basic radiative transfer processes have only been developed in the recent decade (Ohsuga et al. 2005; Ohsuga and Mineshige 2011; Jiang et al. 2014; Sądowski et al. 2013; McKinney et al. 2014). (See the Simulation Methods Chapter of this book for details.) These novel simulations reveal that super-Eddington accretion discs always launch strong, optically-thick winds due to their large radiation pressure. The winds obscure the discs from a large solid angle but has an inclination-dependent profile. Close to the pole, wind is dilute and ultra-fast speed ($v \gtrsim 0.1c$). Close to the disc, wind is dense and moves slower. Analytical models of super-Eddington accretion discs, such as the slim disc model discussed previously, however, usually do no predict super-Eddington discs launch winds.

In order to check the viewing-angle dependence of TDE super-Eddington discs, Dai et al. (2018) have carried out a 3D general-relativistic radiation magneto-hydrodynamic (GR-RMHD) simulation using a code called HARMRAD (McKinney et al. 2014, 2015). For the initial condition, they use a small, Keplerian, geometrically thick disc with mass and

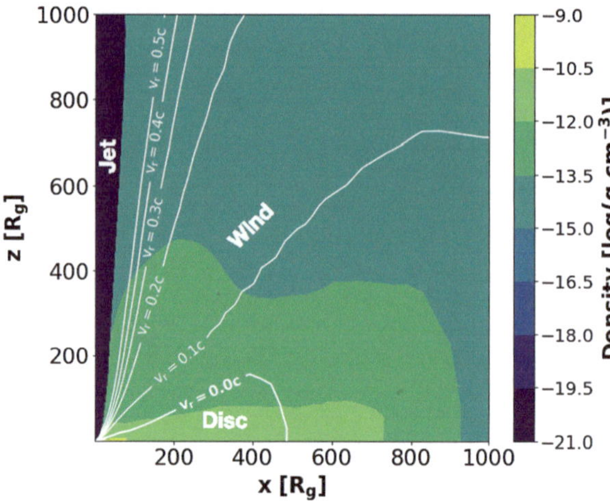

Fig. 7 Disc, wind and jet regions of the super-Eddington accretion flow simulated in Dai et al. (2018). The density of the disc and winds are shown by the background color. The white lines are contours of constant radial velocity in the wind region. Closer to the pole, the wind is more dilute and moves faster. The jet region (where magnetic pressure dominates the gas pressure) is marked using a dark blue colour. Figure is adapted from Thomsen et al. (2019)

angular momentum about one tenth of the original stellar mass and angular momentum. The simulation has been evolved to achieve an inflow equilibrium of about $100R_g$ (which is about ten times larger compared to previous super-Eddington disc simulations), in order to ensure resolving the photosphere from most of the inclination angles. The structure of the accretion disc and wind can be seen in Fig. 7. The accretion disc is geometrically and optically thick. The wind, launched by a combination of radiation and magnetic pressure, is also optically thick from most inclinations except from at angles close to the pole. A jet is also launched by the Blandford-Znajek mechanism due to the set up of the simulation with a large black hole spin and sufficient poloidal magnetic field threading the disc.

Reprocessing of the inner disc X-ray emissions happens in the wind and outer disc. At the same time, the wind also expands and cools down. Therefore, one should expect the wind emission to shine at lower energy bands due to the both adiabatic expansion and reprocessing as discussed in e.g., Strubbe and Quataert (2009), Roth et al. (2016), Metzger and Stone (2016). The simulated disc profile is then post-processed using a Monte-Carlo radiative transfer code SEDONA (Kasen et al. 2006; Roth et al. 2016) to obtain the spectra from different inclinations. It is found that UV/optical emission dominates when viewed along the disc direction, and X-ray emission dominates from directions closer to the pole. Interestingly, it is also found that most of the radiation energy will escape in the extreme ultraviolet (EUV) waveband which is heavily absorbed from extragalactic sources, and this might explain the "missing-energy" issue of the observed TDEs (Lu and Kumar 2018). Based on the results, Dai et al. (2018) propose a unified model for the observed UV/optical and X-ray TDEs based on the viewing-angle dependent structure of the accretion flow.

Curd and Narayan (2019) further simulate TDE super-Eddington discs with the GR-RMHD code KORAL (Sądowski et al. 2013), using the ZEBRA model as disc initial conditions. While the first-order disc and wind structures are similar to those of Dai et al. (2018), since the initial large-scale structure of the accretion flow used in their simulations is more isotropic, the escaped emissions from different inclinations demonstrate less variance.

Both Dai et al. (2018) and Curd and Narayan (2019) have started the simulations with already circularized geometrically-thick small discs aligned with the black hole spin. Therefore, they cannot address the concerns related to the disc formation, such as whether TDE discs can indeed circularize promptly. To our knowledge, Sądowski et al. (2016) has carried out the only MHD simulation of TDEs from the disc formation phase to the accretion phase.

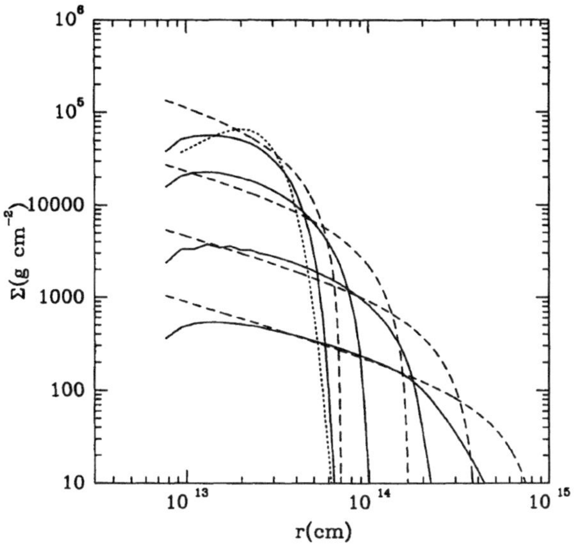

Fig. 8 Evolution of the surface density in the disc model by Cannizzo et al. (1990). The ring-structured disc has a mass of 0.5 $M_\odot$ and viscosity parameter $\alpha = 1$. The disc continues to spread out and surface density drops with the disc spreading. Curves show the surface density profile at the times of 0, 3, 30, 300,and 3000 yrs. Figure is from Cannizzo et al. (1990)

Due to the constraint on computational time, the parameters chosen (i.e., an initial eccentric stellar orbit with a penetration parameter of 10) may only represent a small fraction of TDEs. Interestingly, it is found that while the disc can circularize fast, turbulence inherited from the violent debris stream self-interaction persists in the accretion flow, sustained either by convection or by the fall-back of gas. This hydrodynamic turbulence dominates over the magnetorotational instability (MRI) in the differentially rotating flow and contributes significantly to the angular momentum transfer.

2.2 Late-Time TDE Disc Models with Disc Spreading

At late time in a TDE, as the accretion level gradually drops from super-Eddington to sub-Eddington regime, the disc thickness decreases and eventually should become a geometrically thin. The disc viscous or inflow timescale greatly increases with the reduced disc height. If the disc viscous timescale exceeds the debris fallback timescale, the draining rate of the disc through accretion will become lower than the mass supply rate of the returning debris, so material will accumulate in the disc.

Cannizzo et al. (1990) have calculated the evolution of a disc which initially has a ring structure with maximum surface density at $R \sim R_t$ and a total mass equal to a sizable fraction of the stellar mass. The disc is modeled using a standard α-disc prescription with a viscosity parameter $\alpha = 0.01 - 1$. As disc spreads both outwards and inwards due to gas pressure, its surface density decreases as shown in Fig. 8 and the disc temperate drops. While the exact accretion rate depends on the initial disc mass as well as the value of the disc viscosity and opacity, at later times the accretion rate adopts the same decline pattern with $\dot{M} \propto t^{-1.2}$, which is flatter than the canonical debris fallback rate. This model has been further revised by van Velzen et al. (2019) and Mummery and Balbus (2020) to explain the late-time light curves of TDEs observed in the UV or X-ray bands (e.g., Gezari et al. 2017; Holoien et al. 2018).

It has been a long-standing unsolved problem whether run-away thermal instability happens in radiation pressure-dominated thin accretion discs (Shakura and Sunyaev 1973). Shen

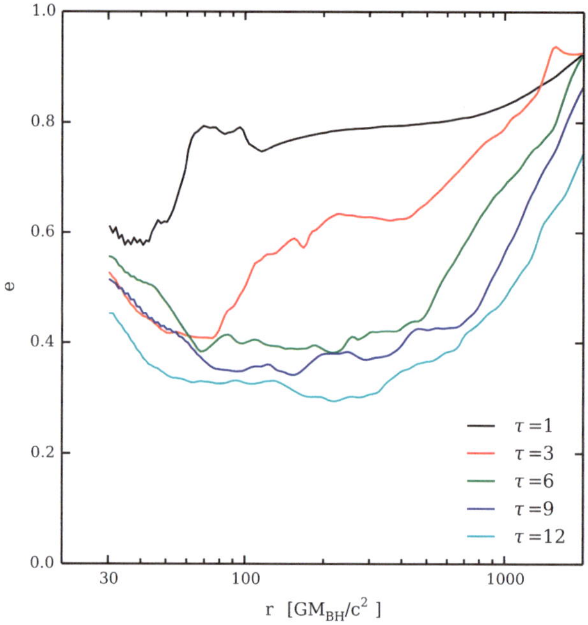

Fig. 9 Radial profile of eccentricity from the simulated accretion flow in Shiokawa et al. (2015). The time evolution of mass-weighted mean eccentricity is given in units of τ, the orbital timescale of the most tightly bound matter. Note that the inner radius in this model is truncated at $30R_g$. Figure is from Shiokawa et al. (2015)

and Matzner (2014) investigate this topic in the context of TDEs discs and predict that thermal instability, if it happens, can lead to cycles of bursts of accretion. So far observed TDE flares have not demonstrated such behaviors. This perhaps suggests that thin discs formed in TDEs can stand against thermal instability either due to the support from high iron opacity (Jiang et al. 2016) or large-scale vertical magnetic field (Liska et al. 2019a), or due to the suppression of thermal instability from large-scale toroidal magnetic field (Begelman and Pringle 2007).

2.3 Eccentric Accretion Flow Formed in TDEs

As current simulations show that nozzle shock happening at pericenter or debris stream self-crossing due to GR apisdal precession may not dissipate orbital energy fast enough, it is suggested that the just-formed disc can have an eccentric configuration. For example, for the stellar and black hole parameters in Shiokawa et al. (2015), the simulated accretion flow after several fall-back timescales is moderately eccentric. From the paper, Fig. 9 shows the time evolution of the radial profile of mass-weighted mean eccentricity, with the eccentricity settling to $e \sim 0.3 - 0.4$, extending from the inner-radius of the simulation ($\sim 30R_g$) to the minimum semi-major axis of the most bound debris ($\sim 500R_g$). Note that the tidal radius for this simulation is $\sim 100R_g$, and the general relativistic hydrodynamics simulation does not include magnetic fields. These results, large disc and slow time scale for circularization, are contrary to the canonical model, where it is assumed that the accretion flow circularizes on the order of a few times the orbital timescale of the most tightly bound matter. The distinctive structure and time scale of the accretion flow imply that optical emission in TDEs could possibly be produced by mechanisms different from accretion onto the black hole.

It is also a topic under investigation whether eccentric accretion flow around black holes can generally circularize on timescale faster than the viscous timescale. In analytic treatments to the dynamics of eccentric discs, Syer and Clarke (1992) suggest that an initially eccentric accretion flow can maintain its configuration or even become more elliptical by applying Gauss perturbation equations upon a α-viscosity disc model. Ogilvie and Barker (2014) and Barker and Ogilvie (2014) show that the eccentric disc evolution can depend on extreme, non-linear instabilities, providing a useful tool for understanding the leading mechanisms especially with follow-up numerical investigations. Based on this framework, Chan et al. (2018) show, with a toy model, how MRI growth occurs close to pericenter and declines with increasing eccentricity of the debris orbits. One remark is that the picture can become more complicated if one also considers general relativistic procession of the debris stream orbits and large-scale magnetic fields in the disc, which is beyond the capacity of analytical treatments.

In light of this, several studies have turned towards emission produced during the evolution of an elliptical accretion flow. Such mechanisms are perhaps at the location of stream collision, at apocenter $a_{\rm min}$ when the GR apsidal precession is weak (Piran et al. 2015a; Krolik et al. 2016). There are also many open questions in the evolution of eccentric accretion discs in the context of TDEs. For example, Svirski et al. (2017) address the issue of angular momentum loss by proposing a direct plunge of debris onto the black hole due to MHD stresses. We refer the readers to the Formation of the Accretion Flow Chapter in this book for more details.

Certain observational features of TDEs, especially double-peaked or flat-topped broad Hα emission lines seen in a few candidates (e.g., Holoien et al. 2019), have motivated studies adopting thin elliptical disc models to explain the line profiles (Liu et al. 2017; Cao et al. 2018). For example, Eracleous et al. (1995) give expressions for the total line flux from the disc, which depend on a relativistic Doppler factor for photons in an elliptical structure, namely the eccentricity and inclination angle. Alternatively, such line profiles may also be modeled using circular discs with optically-thick winds (Roth and Kasen 2018; Hung et al. 2019).

2.4 Angular Momentum Transport in TDE Discs

The issue of angular momentum transport in accretion discs has always been a significant source of controversy and investigations. For years, a phenomenological description based on a viscous disc model has been used, following the pioneering work of Shakura and Sunyaev (1973), until it was recognized that turbulence induced by the magneto-rotational instability (MRI) is the most likely origin for the required stress within the disc (see, e.g. Balbus and Hawley 1998 and references therein). Are discs originating from TDEs consistent with the established models for longer lived discs around Active Galactic Nuclei (AGN) and galactic black hole binaries? The question arises because the initial magnetic field of the star is expected to be relatively low, and the process of disruption itself is not expected to provide a significant enhancement in the magnetic field strength (Bonnerot et al. 2017; Guillochon and McCourt 2017).

Roughly speaking, one would imagine that, in order for the MRI to be established, the magnetic field should be strong enough that the Alfven speed is comparable to the gas sound speed. A full understanding of how MRI functions and how magnetic fields grow via MRI requires resolving the microphysical dissipation processes at small scales in the plasma. Global disc simulations have so far demonstrated that MRI can still kick in at significantly small magnetic field strengths, corresponding to a plasma β value (the ratio of thermal to magnetic pressure) of the order of $\sim 100 - 1000$ (Bugli et al. 2018; Zhu and Stone 2018). If we apply this value for the conditions of typical of TDE discs, this would correspond to $B \sim 10^{3-4}$ G, much larger than the expected 1 G field strength corresponding to the unperturbed star. Recently, Nealon et al. (2018) have demonstrated how the torus resulting from a TDE can be unstable to the Papaloizou-Pringle instability (Papaloizou and Pringle 1984), that occurs for thick torii which are supported mainly by pressure against gravity. Such instability can provide a significant means of transport of angular momentum over the first ~ 20 orbits of the torus, before the MRI eventually develops, with the effective viscosity in the range of $\alpha \sim 0.03 - 0.05$. In addition, it could also provide a time-varying quadrupole moment that can give rise to potentially detectable gravitational wave emission (Toscani et al. 2019).

3 Jets Launching in TDEs: How and When

So far three TDEs have been observed to produce powerful jets: Swift J1644+57 (Bloom et al. 2011; Burrows et al. 2011; Zauderer et al. 2011), Swift J2058+05 (Cenko et al. 2012) and Swift J1112-82 (Brown et al. 2015). Their beamed hard X-rays/soft γ-rays and the accompanied strong radio emissions have been interpreted to originate from relativistic, collimated jets formed in TDEs. This gives the most direct evidence that transient accretion discs form in TDEs. As we have discussed in Sect. 2, TDE discs transition from the super-Eddington accretion regime to the sub-Eddington thin disc regime and then eventually to the ADAF/RIAF regime, in a timescale much faster than typical AGN state-changing timescales. Therefore, TDEs can also be used to study jet physics in all three accretion regimes. In this section, we mainly focus on the jet-launching physics, and leave the jet emission physics to be discussed in details in the Emission Mechanisms Chapter, X-ray Chapter and Radio Chapter.

The best studied jet launching mechanism is the Blandford-Znajek (hereafter BZ) mechanism, in which relativistic jet can be powered by extracting the BH rotation energy via ordered magnetic field threading the ergosphere of a spinning black hole (Blandford and Znajek 1977). In the standard BZ approximation, the BH energy extraction rate scales as $a^2\Phi_B^2$ (a is the black hole spin parameter and Φ_B is the large-scale magnetic flux around the black hole). This has been demonstrated in GRMHD simulations (e.g., McKinney 2005; Tchekhovskoy et al. 2011; McKinney et al. 2012). Under certain conditions the jet power can increase even faster than a^2 for large values of a (Tchekhovskoy et al. 2012).

The accretion flow around an astrophysical BH plays a crucial role in transporting magnetic flux to its vicinity, amplifying magnetic flux via MRI or dynamo process, and confining magnetic flux from leaking (e.g., Pessah and Psaltis 2005; Beckwith et al. 2009). Since it is easier for magnetic flux to accumulate in geometrically thick discs (Lubow et al. 1994), powerful, continuous jets are expected to be launched in the super-Eddington regime or the ADAF/RIAF regime. Furthermore, the thick disc geometry can help jet collimation.

The change of accretion regime has been used to explain the X-ray light curve of the Swift J1644 event (Zauderer et al. 2013; Tchekhovskoy et al. 2014): Initially when the debris disc accretion rate was super-Eddington and the disc was geometrically thick, a relativistic, collimated jet was launched, producing strong X-ray and radio emissions. During this phase the jet power and X-ray flux roughly traced the debris fallback rate following the $t^{-5/3}$ pattern since about 10 days after the trigger. Then the X-ray emission rapidly decreased by two orders of magnitude after about 500 days, by which time the fallback rate was expected to drop to around or below the Eddington level. As a consequence, the disc morphology changed and the jet quenched. This evolution, together with the initial disc-jet alignment process which will be discussed in Sect. 4, is illustrated in Fig. 10.

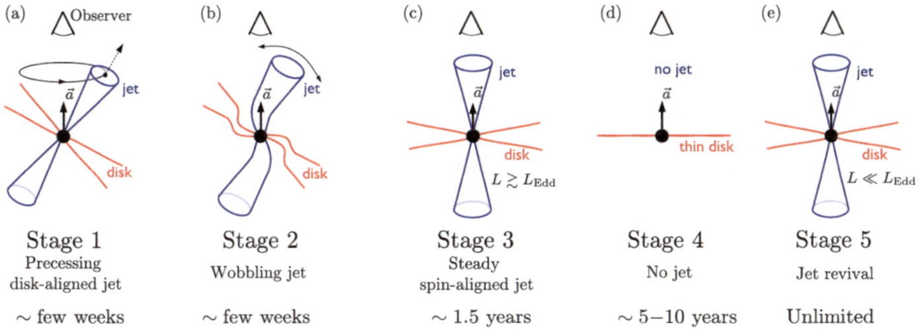

Fig. 10 Evolution of the jet and disc in Swift J1644 proposed by Tchekhovskoy et al. (2014). A misaligned jet initially forms in the super-Eddington accretion phase, which precesses and wobbles around the black hole spin axis. Within weeks the jet orientation is aligned with black hole spin axis due to the magneto-spin alignment mechanism. After about one year, the disc becomes sub-Eddington, so the jet is quenched. As the accretion continues to drop, the jet is predicted to revive when the disc eventually transitions to the ADAF/RIAF accretion regime. Figure is from Tchekhovskoy et al. (2014)

When are the most powerful jets produced? For any accretion disc there exists an upper limit of the magnetic flux that the disc can confine near the black hole. When the magnetic flux exceeds this limit, the magnetic pressure becomes stronger than the disc gravity and pushes the accretion flow away from the black hole. The type of accretion flow under the strong magnetic-field limit is called the magnetically-arrested disc (MAD) (Narayan et al. 2003). One can define a dimensionless magnetic flux parameter:

$$\Upsilon \equiv \frac{0.7\,\Phi}{\sqrt{4\pi R_g^2 \dot{M} c^2}}, \tag{5}$$

where Φ is the flux of poloidal magnetic flux threading the disc in Gaussian units and $\dot{M}$ is the accretion rate. The disc is in the MAD state when $\Upsilon \gg 1$.

GRMHD simulations show that the most powerful jets with large Lorentz factor are launched from MAD discs, which can achieve a jet efficiency $\eta_{\rm jet} \equiv L_{\rm jet}/\dot{M}_{\rm BH}c^2 \approx 10-100\%$ (Hawley and Krolik 2006; Tchekhovskoy et al. 2011, 2012; McKinney et al. 2012). On the other hand, weakly magnetized discs in the standard and normal evolution (SANE) regime (Narayan et al. 2012) can only produce low-power jets or non-relativistic outflows in the polar region (De Villiers et al. 2005; Mościbrodzka et al. 2016). Recent GR-radiation-MHD simulations of super-Eddington (TDE) discs also show consistent results (McKinney et al. 2015; Sądowski and Narayan 2016; Dai et al. 2018; Curd and Narayan 2019).

If the Swift J1644 jet is launched from a MAD disc by the BZ mechanism, then the magnetic flux threading the BH needed to explain the peak X-ray luminosity is about 10^{30} G cm^2 $\times$ $(M_{\rm BH}/10^6\ M_\odot)$ (Bloom et al. 2011; Kelley et al. 2014), which is 3–5 orders of magnitude larger than the original magnetic flux possessed by a main-sequence star. This therefore has triggered a series of investigations on how magnetic flux can be amplified during the disruption, disc forming or accretion processes in TDEs. Simulations studying the magnetic field amplification during the disruption process (Guillochon and McCourt 2017; Bonnerot et al. 2017) show that although the dynamo process in surviving cores from partial disruption events might be important, the amplification in returning debris in a total disruption event is not significant enough to supply the flux needed for the Swift J1644 jet. From the prospect of magnetic-field growth via MRI in accretion discs, Krolik and Piran

(2011) propose that Swift J1644 might have been produced by an intermediate mass black hole (IMBH) disrupting a white dwarf. In this scenario the disc orbital timescale is shorter than the case of a SMBH disrupting a main-sequence star, so the magnetic field can quickly grow via turbulence to reach saturation. (However, Kara et al. (2016) later constrains the SMBH mass in Swift J1644 to be a few $\times 10^6\ M_\odot$ from X-ray reverberation studies, which favors the disruption of a main-sequence star instead.) Kelley et al. (2014) propose another model in which extra magnetic fields can be collected by the returning debris when it sweeps through a "fossil" disc left from previous AGN activities. When the stellar orbit and the fossil disc aligns, the magnetic flux that a quiescent fossil disc can provide to the debris is enough to power the Swift J1644 jet.

Alternatively, it has been suggested that in the TDE super-Eddington accretion phase a jet can be driven by the large radiation pressure and collimated by the geometrically thick accretion flow (Coughlin and Begelman 2014; Sądowski and Narayan 2015). This jet production mechanism, however, has not been studied in details. More studies on this topic are needed to conclude whether relativistic jets can be produced purely due to radiation pressure.

4 Precession and Alignment of TDE Discs and Jets

One peculiar feature of TDEs is that in general we expect the orbit of the incoming star (and thus the angular momentum of the debris) to be randomly oriented with respect to the spin of the central supermassive black hole, with the latter determined by the merger history of the host galaxy and the accretion history of the black hole. Since the stellar pericenter is relatively close to the gravitational radius of the hole, we expect the debris to be significantly affected by relativistic effects in the form of Lense-Thirring precession. Such effects might delay the formation of a disc, by avoiding stream collisions due to a randomization of the orbital plane (Dai et al. 2013; Guillochon and Ramirez-Ruiz 2015; Hayasaki et al. 2016), or might speed-up circularization, by producing a thicker torus (Gafton and Rosswog 2019). Either way, after the disc is formed, it is expected that the initial orientation of the disc will be misaligned with the black hole spin due to conservation of angular momentum. Also, misaligned discs may also launch jets, and the jet orientation can be affected by both the disc angular momentum and the black hole angular momentum. In general, since such misalignment are specific to TDEs and might strongly affect the evolution of the system, several studies have investigated the dynamics of misaligned discs and jets in TDEs (Stone and Loeb 2012; Lei et al. 2013; Tchekhovskoy et al. 2014; Franchini et al. 2016; Xiang-Gruess et al. 2016; Ivanov et al. 2018; Zanazzi and Lai 2019).

4.1 Expected Precession Regimes in TDE Discs

The main physical effect at work in misaligned discs around a spinning black hole is Lense-Thirring precession. The precession torque at radius R is given by (see, for example, Franchini et al. 2016 and references therein):

$$\mathbf{T} = -\Sigma R^2 \Omega \left(\frac{\Omega_z^2 - \Omega^2}{\Omega^2} \right) \frac{\Omega}{2} \mathbf{e_z} \times \mathbf{l}, \tag{6}$$

where Σ is the disc surface density, $\mathbf{l}$ is the unit vector along the local disc angular momentum direction, $\mathbf{e}_z$ is the direction of the black hole spin, and Ω and Ω_z are the angular and vertical frequencies, respectively, given by:

$$\Omega = \frac{c^3}{GM} \frac{1}{r^{3/2} + a}; \tag{7}$$

$$\left(\frac{\Omega^2 - \Omega_z^2}{\Omega^2} \right) = 4ar^{-3/2} - 3a^2 r^{-2}, \tag{8}$$

where M is the black hole mass, r is the disc radius in units of the gravitational radius of the hole, and a is the black hole spin parameter.

The response of the disc to external misaligned torques has been studied extensively over the years. The pioneering work of Papaloizou and Pringle (1983) and of Papaloizou and Lin (1995) have shown that the disc responds to such perturbations either in a diffusive way (when the ratio of turbulent stresses to pressure α is larger than the disc aspect ratio H/R) or in a wave-like way (in the opposite regime, when $\alpha \lesssim H/R$). A full non-linear theory of diffusive warp propagation has been provided by Ogilvie (1999). On the other hand, a non-linear theory of bending waves has not been developed, and only a linear (Lubow and Ogilvie 2000) and quasi-linear theory (Ogilvie 2006) exist. Non-linearities might prevent the wave-like regime from happening altogether (Hawley and Krolik 2018). For thin and viscous discs, or for discs that evolve diffusively, the inner disc aligns rapidly to the black hole spin (on the warp diffusion timescale), while the outer disc retains its original direction and slowly aligns at longer times. This is known as the Bardeen-Petterson effects (Bardeen and Petterson 1975).

In the case of TDE discs, we expect initially the disc to be thick (large H/R) due to the high accretion level and with a relatively low value of internal "viscous" stresses, as the magneto-rotational instability is building up (low α), and should therefore be characterized by bending wave propagation. At later times, however, the disc becomes thinner and magnetic stresses build up, so a transition to the diffusive regime is expected to occur.

4.2 Global Disc and Jet Precession in the Thick Disc Regime

At early times in TDEs, regular global precession of the disc might occur, as shown in the GRMHD simulations of thick torii (Fragile and Anninos 2005; Fragile et al. 2007; Liska et al. 2018). For TDEs, this was initially suggested by Stone and Loeb (2012) and later confirmed with time-dependent calculations by Franchini et al. (2016). The global precessional frequency can be computed easily as a weighted average of the Lense-Thirring frequency $\Omega_{\rm LT}$ (Stone and Loeb 2012; Franchini et al. 2016):

$$\Omega_{\rm p} = \frac{\int_{R_{\rm in}}^{R_{\rm out}} \Omega_{\rm LT}(R) L(R) 2\pi R\, dR}{\int_{R_{\rm in}}^{R_{\rm out}} L(R) 2\pi R\, dR}, \tag{9}$$

where $L(R) = \Omega R^2 \Sigma$. Equation (9) is applicable when the disc is narrow (as expected for TDE) and the warp is mild (Note that a mild warp does not imply a mild misalignment). The Lense-Thirring frequency is easily derived from Eqs. (6) and (8):

$$\Omega_{\rm LT}(r) = \frac{1}{2} \frac{c^3}{GM} \frac{1}{r^{3/2} + a} (4ar^{-3/2} - 3a^2 r^{-2}). \tag{10}$$

The resulting precession periods are in the range of a few to several days, depending on black hole mass and spin, and become longer for lower spins and lower black hole masses

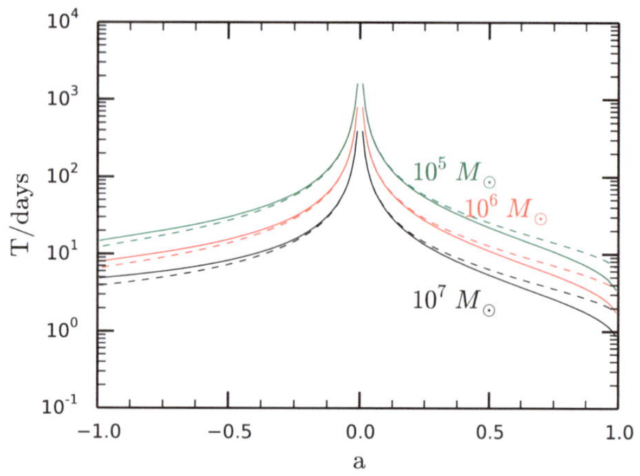

Fig. 11 Precession period expected from a TDE disc for three different choices of SMBH mass (green: $10^5\ M_\odot$, red: $10^6\ M_\odot$, black: $10^7\ M_\odot$. Solid line refer to the full relativistic calculation, while dashed line adopt a first order approximation to the Lense-Thirring frequency. These model refer to disc that extend out to $2R_t$, for $\beta = 1$. Figure is from Franchini et al. (2016)

(see Fig. 11, taken from Franchini et al. 2016). However, it should be noted that such periods strongly depend on where the disc lies, given the steep scaling of the Lense-Thirring rate with radius. If, for some reason, the disc lies much closer to the hole, the precession rate could be much higher. This will also occur if the disc is torn into smaller rings, in cases where the misalignment is very large (Nixon et al. 2012).

A few recent studies have explored the effect of the torques induced by infalling material from the debris stream onto the disc. It is found that the effect of such torques is to complicate the disc dynamics, leading to irregular precessional motion of the disc (Xiang-Gruess et al. 2016; Ivanov et al. 2018), possibly giving rise to quasi-periodic, rather than period precession.

Recent GRMHD simulations have shown that tilted discs can also launch relativistic jets (McKinney et al. 2013; Liska et al. 2018). When the disc is not highly magnetized, the jet has been shown to initially follow the disc angular momentum orientation on large scales and precess together with the disc, although the tilt angle between the black hole spin and the disc/jet decreases over longer timescales. (See Sect. 4.4 for the alternative case of highly magnetized discs.)

4.3 Viscous vs Cooling Driven Disc Alignment

Over long timescales, independently on the original misalignment, the disc is expected to settle down on the equatorial plane of the black hole due to dissipative effects (Sorathia et al. 2013b,a), although some non dissipative relaxation processes have been sometimes suggested for stellar systems (Merritt and Vasiliev 2012), the applicability of which to TDE discs has not been shown yet. Stone and Loeb (2012) argue that this would have naturally happened due to the fact that, as the accretion rate through the disc decreases (for example, following the fallback rate, as $t^{-5/3}$), the disc would become thinner and gradually move to the Bardeen-Petterson regime, at which stage rapid alignment would follow. However, even in the wave-like propagation phase, alignment is expected to happen as a consequence of viscous dissipation. Franchini et al. (2016) evaluate specifically the relevant alignment process in the presence of both cooling driven disc thinning and of viscous effects in the disc. The two possible alignment processes have a distinctly different dependence on the viscosity parameter α, whereby cooling induced alignment scales as $\alpha^{-3/5}$, while viscous induced alignment scales as α^{-1}. Franchini et al. (2016) explore a wide range of parameters and conclude that for most cases, viscously induced alignment occurs more quickly that cooling induced alignment. The only cases for which the disc aligns due to the Bardeen-Petterson effect are for large black hole mass, low spins and low values of α.

The alignment time is of the order of tens to hundreds of days for large black hole spin and grow to a few years for very low spins.

4.4 Magneto-Spin Alignment of Strongly Magnetized Discs and Jets

When large-scale electromagnetic (EM) field exists around the black hole, EM force can also affect the dynamics of jets and discs and provides another torque to affect their orientations. In the case of a highly magnetized disc, the strong coupling between the black hole spin and its magnetosphere can make not only the relativistic jet but also the inner part of a thick accretion disc align with the black hole spin axis.

One can compare the EM force torque and the torque induced by the Lense-Thirring effect to see which one dominates the disc dynamics. With simple assumptions such as that the magnetic field, when being pushed against by a disc, bends on the same scale of the disc radius r, McKinney et al. (2013) calculate that:

$$\frac{\tau_{\rm EM}}{\tau_{LT}} \approx \frac{1}{64\, a}\, \Upsilon^2\, \frac{r}{R_g}\, \alpha_{\rm eff} \left(\frac{H}{R}\right)^2. \tag{11}$$

Here a is the dimensionless black hole spin parameter, Υ is the dimensionless magnetic flux given by Eqn. 5, H/R is the disc aspect ratio, and $\alpha_{\rm eff} \equiv v_r/(v_\phi\, (H/R)^2)$ is the effective viscosity parameter. Therefore, the EM torque becomes more dominant when the disc is more magnetized, or when the disc is thicker. 3D GRMHD simulations and calculations (McKinney et al. 2013; Polko and McKinney 2017; Liska et al. 2018) have further demonstrated that for the highly magnetized thick discs (such as MAD discs) with a large range of tilt angles, the EM torque is dominant and strong enough to reorient the inner disc and inner jet to align with the BH spin axis.

Tidal disruption events likely produce thick accretion discs due to the high accretion level. These discs might also be strongly magnetized, especially when relativistic jets are produced which require large-scale magnetic fluxes to power (see Sect. 3). Therefore, EM force could play an important role in aligning the discs and jets with black hole spins in TDEs such as Swift J1644.

4.5 Warping and Tearing of Thin Accretion Discs

In this regime which is typically assumed to happen at late times in TDEs, we expect the inner part of a thin disc not to precess regularly but to simply align with the black hole spin axis due to the Bardeen-Petterson effect. This has been demonstrated in Smoothed Particle Hydrodynamics simulations by Nelson and Papaloizou (2000), Lodato and Pringle (2007), Lodato and Price (2010) as well as in GRMHD simulations by Liska et al. (2019a,b). As the disc is fed by debris on fallback timescales, the outer part of the disc may not have sufficient time to be aligned and its orientation could be still determined by the angular momentum of the falling back debris. The disc can become warped or even torn apart (Nixon et al. 2012; Liska et al. 2019b). Zanazzi and Lai (2019) compute eigenmodes for the disc warp excited by Lense-Thirring precession and apply it to disc produced in TDEs. They compute global precession frequencies finding a broad agreement with those predicted by Franchini et al. (2016), with precession periods ranging from days to weeks depending on the disc parameters.

4.6 Possible Cases of Misaligned Discs and Jets Observed in TDEs

4.6.1 QPO in ASASSN 14-li

From the observational side, ASASSN14-li interestingly shows a quasi-periodic oscillation in the soft x-ray band ($0.3 - 1$ keV) with a periodicity of only 131 seconds with $3 - 4\sigma$ significance (Pasham et al. 2019). It has been argued that this QPO is possibly associated with the global Lense-Thirring precession of a misaligned disc, though quantitative calculations pose several challenges. From Eqn. (10) such a high precession rate would require an extremely high spin for the black hole even if using the LT precession rate at the black hole ISCO radius. This case is further puzzling since the signal has been observed to be stable for roughly 450 days, which is much longer than the expected alignment process for such a high spin.

4.6.2 Behavior of the Relativistic Jet in Swift J1644

The X-ray level of Swift J1644 shows large-amplitude variability in the first few weeks (Bloom et al. 2011), which has been interpreted to be the result of jet wobbling (Tchekhovskoy et al. 2014). Then the jet and the highly magnetized disc have been aligned by the magneto-spin alignment mechanism as discussed in Sect. 4.4, as shown in Fig. 10. This model is further supported by the lack of clear evidence of jet precession during the main light curve decay phase lasting for about one year, although there is a report of a $\sim 4\sigma$ detection of a 200 second QPO in the $2 - 10$ keV X-ray band with an unknown nature (Reis et al. 2012).

5 Summary and Future Studies

The studies of accretion discs, winds and jets in TDEs discussed in this review add an important component to the studies of general accretion and jet physics around black holes. The extreme level of debris accretion and the fast transition across different accretion regimes offer us unique opportunities of studying physics that do not in happen other accretion phenomena such as AGNs or X-ray binaries.

The masses and spins of the SMBHs contain valuable information about their growth as well as the assembling and merger history of the galaxies since the early universe. TDEs are important probes to the demographics of SMBHs, since they can illuminate dormant SMBHs, which represent the majority of the SMBH population at low redshifts, with bright flares lasting for years. Also, while AGN observations tend to select luminous, massive targets, TDEs allow us to study less massive galaxies and their SMBHs, since stars approaching black holes with $M \gtrsim 10^8\ M_\odot$ will typically plunge into the event horizon before being tidally disrupted (unless the SMBH is spinning very rapidly, see Kesden 2012).

A good understanding of the accretion and jet physics in TDEs will help us independently constrain SMBH parameters from TDE observables. For example, the power of a magnetically driven jet strongly correlates with the value of the black hole spin. Also, while the debris mass fallback rate has a dependence on the SMBH mass (e.g., $\dot{M}_{\rm fb,peak} \propto M_{\rm BH}^{-1/2}$, as in Guillochon and Ramirez-Ruiz 2013) and therefore the TDE light curve can in principle be used to constrain the SMBH mass (Mockler et al. 2019), the luminosity (in monochromatic bands) may not exactly scale with the fallback rate, since the overall radiative efficiency or the SED of the emission can change with accretion rate (Lodato and Rossi 2011; Piran et al. 2015b), letting alone the uncertainties and delays related to the disc formation process.

As of now most of the studies of TDE accretion flow have focused on the ideal case of circular and aligned discs. They have shown that winds and relativistic jets can be launched at early times when the accretion is in the super-Eddington regime and the disc geometry is thick. These studies also offer explanations for the basic emission properties of TDEs, such as how X-ray emission produced close to the black hole can be reprocessed into UV/optical emissions in optically-thick winds or outer accretion flow. Studies of TDE discs at late times in the sub-Eddington phase show that such unobscured thin discs should mostly produce X-ray emissions. They have long viscous timescales so they can keep accumulating mass and significantly spread outwards on debris fallback timescales. Recently, more explorations have been made into less understood regimes such as misaligned discs/jets or eccentric accretion discs. A lot of the progress as discussed in this chapter has been made possible largely thanks to various novel numerical simulations carried out recently to investigate accretion flow and jets which are too complex to be studied analytically.

At the end of this review, we pose a few open questions to motivate future studies:

* Recently more TDEs have been followed up for several years past the peak of the flare. This offers a good opportunity for further studying thin accretion discs and especially the transition between the super-Eddington regime and the thin disc regime. For example, do TDE emissions at late times share similar properties as AGN emissions which are assumed to be produced from thin accretion discs?
* If, in a TDE, the disc is initially highly eccentric, how does such a disc evolve? Can it maintain an eccentric configuration for a long time or circularize quickly? How does MRI work in such discs? Can eccentric discs also generate winds and jets?
* The basic emission properties of TDEs and some of their spectroscopic features can be qualitatively explained using current accretion and emission models. For example, from the observed Bowen fluorescent lines in several optically strong TDEs, we know from first order calculations that there exists an optically-thick reprocessing layer that converts X-ray/EUV emissions into optical emissions (Leloudas et al. 2019). Also the asymmetric, broad emission line profiles in some TDEs (Hung et al. 2019; Holoien et al. 2019) indicate that these discs either produce winds or have an eccentric geometry. In the future more quantitative studies along these lines will be very useful for studying the properties of the accretion flow.
* How can we apply TDE accretion and jet physics to study other accretion phenomena, such as AGNs, Galactic X-ray binaries, ultra-luminous X-ray sources and gamma-ray bursts? And vice versa?
* The emission from a typical X-ray strong AGN always has a power-law hard X-ray component, which is believed to be produced by the Compotonization of the cold disc photons by a hot "corona" above the disc. However, (non-jetted) X-ray strong TDEs only produce soft thermal emissions. What makes coronae disappear in TDE discs?
* If most TDE discs are surrounded by large, optically-thick reprocessing layers (in the form of steady envelopes or winds), then it may not be straightforward to directly probe the space-time very close to black holes with TDE observables. Perhaps X-ray strong TDEs and jetted TDEs give us the best hope for constraining black hole spins.

In the future, detailed models and simulations of disc formation, accretion and emission processes will help us better answer these questions in our quest of probing the SMBH demographics, black hole accretion/jet physics and general relativity with potential samples of thousands of tidal disruption events.

Acknowledgements The authors thank the International Space Science Institute (ISSI) for their support and hospitality and the review organisers for their leadership in coordinating these reviews. The authors also thank the Yukawa Institute for Theoretical Physics (YITP) at Kyoto University for hosting the International Molecule-type Workshop "Tidal Disruption Events: General Relativistic Transients" (YITP-T-19-07), where the authors held useful discussions to complete this work. The authors are grateful to the referees for providing constructive suggestions and comments. JLD is supported by the GRF grant from the Hong Kong government under HKU 27305119. The work of RMC is funded by the Advanced Simulation Computing Physics and Engineering Program under the auspices of Los Alamos National Laboratory, operated by Triad National Security, LLC, for the National Nuclear Security Administration of U.S. Department of Energy (Contract No. 89233218CNA000001).

References

M.A. Abramowicz, B. Czerny, J.P. Lasota, E. Szuszkiewicz, Slim accretion disks. Astrophys. J. **332**, 646–658 (1988). https://doi.org/10.1086/166683

K.D. Alexander, E. Berger, J. Guillochon, B.A. Zauderer, P.K.G. Williams, Discovery of an outflow from radio observations of the tidal disruption event ASASSN-14li. Astrophys. J. Lett. **819**, 25 (2016). https://doi.org/10.3847/2041-8205/819/2/L25

I. Arcavi, A. Gal-Yam, M. Sullivan, Y.-C. Pan, S.B. Cenko, A. Horesh, E.O. Ofek, A. De Cia, L. Yan, C.-W. Yang, D.A. Howell, D. Tal, S.R. Kulkarni, S.P. Tendulkar, S. Tang, D. Xu, A. Sternberg, J.G. Cohen, J.S. Bloom, P.E. Nugent, M.M. Kasliwal, D.A. Perley, R.M. Quimby, A.A. Miller, C.A. Theissen, R.R. Laher, A continuum of h- to He-rich tidal disruption candidates with a preference for E+A galaxies. Astrophys. J. **793**, 38 (2014). https://doi.org/10.1088/0004-637X/793/1/38

K. Auchettl, J. Guillochon, E. Ramirez-Ruiz, New physical insights about tidal disruption events from a comprehensive observational inventory at X-ray wavelengths. Astrophys. J. **838**(2), 149 (2017). https://doi.org/10.3847/1538-4357/aa633b

S.A. Balbus, J.F. Hawley, Instability, turbulence, and enhanced transport in accretion disks. Rev. Mod. Phys. **70**, 1–53 (1998). https://doi.org/10.1103/RevModPhys.70.1

J.M. Bardeen, J.A. Petterson, The Lense-Thirring effect and accretion disks around Kerr black holes. Astrophys. J. Lett. **195**, 65 (1975). https://doi.org/10.1086/181711

A.J. Barker, G.I. Ogilvie, Hydrodynamic instability in eccentric astrophysical discs. Mon. Not. R. Astron. Soc. **445**(3), 2637–2654 (2014). https://doi.org/10.1093/mnras/stu1939

K. Beckwith, J.F. Hawley, J.H. Krolik, Transport of large-scale poloidal flux in black hole accretion. Astrophys. J. **707**(1), 428–445 (2009). https://doi.org/10.1088/0004-637X/707/1/428

M.C. Begelman, J.E. Pringle, Accretion discs with strong toroidal magnetic fields. Mon. Not. R. Astron. Soc. **375**(3), 1070–1076 (2007). https://doi.org/10.1111/j.1365-2966.2006.11372.x

R.D. Blandford, M.C. Begelman, On the fate of gas accreting at a low rate on to a black hole. Mon. Not. R. Astron. Soc. **303**(1), 1–5 (1999). https://doi.org/10.1046/j.1365-8711.1999.02358.x

R.D. Blandford, M.C. Begelman, Two-dimensional adiabatic flows on to a black hole—I. Fluid accretion. Mon. Not. R. Astron. Soc. **349**(1), 68–86 (2004). https://doi.org/10.1111/j.1365-2966.2004.07425.x

R.D. Blandford, D.G. Payne, Hydromagnetic flows from accretion disks and the production of radio jets. Mon. Not. R. Astron. Soc. **199**, 883–903 (1982). https://doi.org/10.1093/mnras/199.4.883

R.D. Blandford, R.L. Znajek, Electromagnetic extraction of energy from Kerr black holes. Mon. Not. R. Astron. Soc. **179**, 433–456 (1977). https://doi.org/10.1093/mnras/179.3.433

J.S. Bloom, D. Giannios, B.D. Metzger, S.B. Cenko, D.A. Perley, N.R. Butler, N.R. Tanvir, A.J. Levan, P.T. O'Brien, L.E. Strubbe, F. De Colle, E. Ramirez-Ruiz, W.H. Lee, S. Nayakshin, E. Quataert, A.R. King, A. Cucchiara, J. Guillochon, G.C. Bower, A.S. Fruchter, A.N. Morgan, A.J. van der Horst, A possible relativistic jetted outburst from a massive black hole fed by a tidally disrupted star. Science **333**, 203 (2011). https://doi.org/10.1126/science.1207150

C. Bonnerot, D.J. Price, G. Lodato, E.M. Rossi, Magnetic field evolution in tidal disruption events. Mon. Not. R. Astron. Soc. **469**(4), 4879–4888 (2017). https://doi.org/10.1093/mnras/stx1210

G.C. Brown, A.J. Levan, E.R. Stanway, N.R. Tanvir, S.B. Cenko, E. Berger, R. Chornock, A. Cucchiaria, Swift J1112.2-8238: a candidate relativistic tidal disruption flare. Mon. Not. R. Astron. Soc. **452**, 4297–4306 (2015). https://doi.org/10.1093/mnras/stv1520

M. Bugli, J. Guilet, E. Müller, L. Del Zanna, N. Bucciantini, P.J. Montero, Papaloizou-Pringle instability suppression by the magnetorotational instability in relativistic accretion discs. Mon. Not. R. Astron. Soc. **475**, 108–120 (2018). https://doi.org/10.1093/mnras/stx3158

D.N. Burrows, J.A. Kennea, G. Ghisellini, V. Mangano, B. Zhang, K.L. Page, M. Eracleous, P. Romano, T. Sakamoto, A.D. Falcone, J.P. Osborne, S. Campana, A.P. Beardmore, A.A. Breeveld, M.M. Chester, R. Corbet, S. Covino, J.R. Cummings, P. D'Avanzo, V. D'Elia, P. Esposito, P.A. Evans, D. Fugazza, J.M. Gelbord, K. Hiroi, S.T. Holland, K.Y. Huang, M. Im, G. Israel, Y. Jeon, Y.-B. Jeon, H.D. Jun, N. Kawai, J.H. Kim, H.A. Krimm, F.E. Marshall, P. Mészáros, H. Negoro, N. Omodei, W.-K. Park, J.S. Perkins, M. Sugizaki, H.-I. Sung, G. Tagliaferri, E. Troja, Y. Ueda, Y. Urata, R. Usui, L.A. Antonelli, S.D. Barthelmy, G. Cusumano, P. Giommi, A. Melandri, M. Perri, J.L. Racusin, B. Sbarufatti, M.H. Siegel, N. Gehrels, Relativistic jet activity from the tidal disruption of a star by a massive black hole. Nature **476**, 421–424 (2011). https://doi.org/10.1038/nature10374

J.K. Cannizzo, H.M. Lee, J. Goodman, The disk accretion of a tidally disrupted star onto a massive black hole. Astrophys. J. **351**, 38–46 (1990). https://doi.org/10.1086/168442

R. Cao, F.K. Liu, Z.Q. Zhou, S. Komossa, L.C. Ho, A large accretion disc of extreme eccentricity in the TDE ASASSN-14li. Mon. Not. R. Astron. Soc. **480**(3), 2929–2938 (2018). https://doi.org/10.1093/mnras/sty1997

S.B. Cenko, H.A. Krimm, A. Horesh, A. Rau, D.A. Frail, J.A. Kennea, A.J. Levan, S.T. Holland, N.R. Butler, R.M. Quimby, J.S. Bloom, A.V. Filippenko, A. Gal-Yam, J. Greiner, S.R. Kulkarni, E.O. Ofek, F. Olivares E., P. Schady, J.M. Silverman, N.R. Tanvir, D. Xu, Swift J2058.4+0516: discovery of a possible second relativistic tidal disruption flare? Astrophys. J. **753**, 77 (2012). https://doi.org/10.1088/0004-637X/753/1/77

C.-H. Chan, J.H. Krolik, T. Piran, Magnetorotational instability in eccentric disks. Astrophys. J. **856**(1), 12 (2018). https://doi.org/10.3847/1538-4357/aab15c

E.R. Coughlin, M.C. Begelman, Hyperaccretion during tidal disruption events: weakly bound debris envelopes and jets. Astrophys. J. **781**, 82 (2014). https://doi.org/10.1088/0004-637X/781/2/82

B. Curd, R. Narayan, GRRMHD simulations of tidal disruption event accretion discs around supermassive black holes: jet formation, spectra, and detectability. Mon. Not. R. Astron. Soc. **483**(1), 565–592 (2019). https://doi.org/10.1093/mnras/sty3134

L. Dai, A. Escala, P. Coppi, The impact of bound stellar orbits and general relativity on the temporal behavior of tidal disruption flares. Astrophys. J. Lett. **775**, 9 (2013). https://doi.org/10.1088/2041-8205/775/1/L9

L. Dai, J.C. McKinney, N. Roth, E. Ramirez-Ruiz, M.C. Miller, A unified model for tidal disruption events. Astrophys. J. **859**(2), 20 (2018). https://doi.org/10.3847/2041-8213/aab429

J.-P. De Villiers, J.F. Hawley, J.H. Krolik, S. Hirose, Magnetically driven accretion in the Kerr metric. III. Unbound outflows. Astrophys. J. **620**(2), 878–888 (2005). https://doi.org/10.1086/427142

C. Dotan, N.J. Shaviv, Super-Eddington slim accretion discs with winds. Mon. Not. R. Astron. Soc. **413**(3), 1623–1632 (2011). https://doi.org/10.1111/j.1365-2966.2011.18235.x

M. Eracleous, M. Livio, J.P. Halpern, T. Storchi-Bergmann, Elliptical accretion disks in active galactic nuclei. Astrophys. J. **438**, 610 (1995). https://doi.org/10.1086/175104

C.R. Evans, C.S. Kochanek, The tidal disruption of a star by a massive black hole. Astrophys. J. Lett. **346**, 13–16 (1989). https://doi.org/10.1086/185567

P.C. Fragile, P. Anninos, Hydrodynamic simulations of tilted thick-disk accretion onto a Kerr black hole. Astrophys. J. **623**, 347–361 (2005). https://doi.org/10.1086/428433

P.C. Fragile, O.M. Blaes, P. Anninos, J.D. Salmonson, Global general relativistic magnetohydrodynamic simulation of a tilted black hole accretion disk. Astrophys. J. **668**, 417–429 (2007). https://doi.org/10.1086/521092

A. Franchini, G. Lodato, S. Facchini, Lense-Thirring precession around supermassive black holes during tidal disruption events. Mon. Not. R. Astron. Soc. **455**, 1946–1956 (2016). https://doi.org/10.1093/mnras/stv2417

J. Frank, A. King, D.J. Raine, *Accretion Power in Astrophysics: Third Edition* 2002

E. Gafton, S. Rosswog, Tidal disruptions by rotating black holes: effects of spin and impact parameter. arXiv e-prints (2019)

S. Gezari, R. Chornock, A. Rest, M.E. Huber, K. Forster, E. Berger, P.J. Challis, J.D. Neill, D.C. Martin, T. Heckman, A. Lawrence, C. Norman, G. Narayan, R.J. Foley, G.H. Marion, D. Scolnic, L. Chomiuk, A. Soderberg, K. Smith, R.P. Kirshner, A.G. Riess, S.J. Smartt, C.W. Stubbs, J.L. Tonry, W.M. Wood-Vasey, W.S. Burgett, K.C. Chambers, T. Grav, J.N. Heasley, N. Kaiser, R.-P. Kudritzki, E.A. Magnier, J.S. Morgan, P.A. Price, An ultraviolet-optical flare from the tidal disruption of a helium-rich stellar core. Nature **485**, 217–220 (2012). https://doi.org/10.1038/nature10990

S. Gezari, S.B. Cenko, I. Arcavi, X-ray brightening and UV fading of tidal disruption event ASASSN-15oi. Astrophys. J. Lett. **851**(2), 47 (2017). https://doi.org/10.3847/2041-8213/aaa0c2

J. Guillochon, M. McCourt, Simulations of magnetic fields in tidally disrupted stars. Astrophys. J. Lett. **834**(2), 19 (2017). https://doi.org/10.3847/2041-8213/834/2/L19

J. Guillochon, E. Ramirez-Ruiz, Hydrodynamical simulations to determine the feeding rate of black holes by the tidal disruption of stars: the importance of the impact parameter and stellar structure. Astrophys. J. **767**, 25 (2013). https://doi.org/10.1088/0004-637X/767/1/25

J. Guillochon, E. Ramirez-Ruiz, A dark year for tidal disruption events. Astrophys. J. **809**, 166 (2015). https://doi.org/10.1088/0004-637X/809/2/166

J.F. Hawley, J.H. Krolik, Magnetically driven jets in the Kerr metric. Astrophys. J. **641**, 103–116 (2006). https://doi.org/10.1086/500385

J.F. Hawley, J.H. Krolik, Sound speed dependence of alignment in accretion disks subjected to Lense–Thirring torques. Astrophys. J. **866**, 5 (2018). https://doi.org/10.3847/1538-4357/aadf90

K. Hayasaki, N. Stone, A. Loeb, Circularization of tidally disrupted stars around spinning supermassive black holes. Mon. Not. R. Astron. Soc. **461**, 3760–3780 (2016). https://doi.org/10.1093/mnras/stw1387

J.G. Hills, Possible power source of Seyfert galaxies and QSOs. Nature **254**, 295–298 (1975). https://doi.org/10.1038/254295a0

T.W.-S. Holoien, J.S. Brown, K. Auchettl, C.S. Kochanek, J.L. Prieto, B.J. Shappee, J. Van Saders, The unusual late-time evolution of the tidal disruption event ASASSN-15oi. Mon. Not. R. Astron. Soc. **480**(4), 5689–5703 (2018). https://doi.org/10.1093/mnras/sty2273

T.W.-S. Holoien, M.E. Huber, B.J. Shappee, M. Eracleous, K. Auchettl, J.S. Brown, M.A. Tucker, K.C. Chambers, C.S. Kochanek, K.Z. Stanek, A. Rest, D. Bersier, R.S. Post, G. Aldering, K.A. Ponder, J.D. Simon, E. Kankare, D. Dong, G. Hallinan, N.A. Reddy, R.L. Sanders, M.W. Topping (Pan-STARRS), J. Bulger, T.B. Lowe, E.A. Magnier, A.S.B. Schultz, C.Z. Waters, M. Willman, D. Wright, D.R. Young (ASAS-SN), S. Dong, J.L. Prieto, T.A. Thompson (ATLAS), L. Denneau, H. Flewelling, A.N. Heinze, S.J. Smartt, K.W. Smith, B. Stalder, J.L. Tonry, H. Weiland, PS18kh: a new tidal disruption event with a non-axisymmetric accretion disk. Astrophys. J. **880**(2), 120 (2019). https://doi.org/10.3847/1538-4357/ab2ae1

T. Hung, S.B. Cenko, N. Roth, S. Gezari, S. Veilleux, S. Van Velzen, C.M. Gaskell, R.J. Foley, N. Blagorodnova, L. Yan, Discovery of Highly Blueshifted Broad Balmer and metastable helium absorption lines in a tidal disruption event. arXiv e-prints, 1903–05637 (2019)

S. Ichimaru, Bimodal behavior of accretion disks: theory and application to Cygnus X-1 transitions. Astrophys. J. **214**, 840–855 (1977). https://doi.org/10.1086/155314

P.B. Ivanov, V.V. Zhuravlev, J.C.B. Papaloizou, The dynamics of a twisted disc formed after the tidal disruption of a star by a rotating black hole. Mon. Not. R. Astron. Soc. **481**, 3470–3496 (2018). https://doi.org/10.1093/mnras/sty2493

Y.-F. Jiang, J.M. Stone, S.W. Davis, A global three-dimensional radiation magneto-hydrodynamic simulation of super-Eddington accretion disks. Astrophys. J. **796**, 106 (2014). https://doi.org/10.1088/0004-637X/796/2/106

Y.-F. Jiang, S.W. Davis, J.M. Stone, Iron opacity bump changes the stability and structure of accretion disks in active galactic nuclei. Astrophys. J. **827**(1), 10 (2016). https://doi.org/10.3847/0004-637X/827/1/10

Y.-F. Jiang, J. Guillochon, A. Loeb, Prompt radiation and mass outflows from the stream-stream collisions of tidal disruption events. Astrophys. J. **830**, 125 (2016). https://doi.org/10.3847/0004-637X/830/2/125

E. Kara, J.M. Miller, C. Reynolds, L. Dai, Relativistic reverberation in the accretion flow of a tidal disruption event. Nature **535**, 388–390 (2016). https://doi.org/10.1038/nature18007

E. Kara, L. Dai, C.S. Reynolds, T. Kallman, Ultrafast Outflow in Tidal Disruption Event ASASSN-14li. arXiv e-prints (2017)

D. Kasen, R.C. Thomas, P. Nugent, Time-dependent Monte Carlo radiative transfer calculations for three-dimensional supernova spectra, light curves, and polarization. Astrophys. J. **651**, 366–380 (2006). https://doi.org/10.1086/506190

L.Z. Kelley, A. Tchekhovskoy, R. Narayan, Tidal disruption and magnetic flux capture: powering a jet from a quiescent black hole. Mon. Not. R. Astron. Soc. **445**, 3919–3938 (2014). https://doi.org/10.1093/mnras/stu2041

M. Kesden, Tidal-disruption rate of stars by spinning supermassive black holes. Phys. Rev. D **85**(2), 024037 (2012). https://doi.org/10.1103/PhysRevD.85.024037

S. Komossa, Tidal disruption of stars by supermassive black holes: status of observations. J. High Energy Astrophys. **7**, 148–157 (2015). https://doi.org/10.1016/j.jheap.2015.04.006

J.H. Krolik, T. Piran, Swift J1644+57: a white dwarf tidally disrupted by a 10^4 $M_\odot$ black hole? Astrophys. J. **743**(2), 134 (2011). https://doi.org/10.1088/0004-637X/743/2/134

J. Krolik, T. Piran, G. Svirski, R.M. Cheng, ASASSN-14li: a model tidal disruption event. Astrophys. J. **827**(2), 127 (2016). https://doi.org/10.3847/0004-637X/827/2/127

W.-H. Lei, B. Zhang, H. Gao, Frame dragging, disk warping, jet precessing, and dipped X-ray light curve of Sw J1644+57. Astrophys. J. **762**(2), 98 (2013). https://doi.org/10.1088/0004-637X/762/2/98

G. Leloudas, L. Dai, I. Arcavi, P.M. Vreeswijk, B. Mockler, R. Roy, D.B. Malesani, S. Schulze, T. Wevers, M. Fraser, E. Ramirez-Ruiz, K. Auchettl, J. Burke, G. Cannizzaro, P. Charalampopoulos, T.-W. Chen, A.a. Cikota, M. Della Valle, L. Galbany, M. Gromadzki, K.E. Heintz, D. Hiramatsu, P.G. Jonker, Z. Kostrzewa-Rutkowska, K. Maguire, I. Mandel, M. Nicholl, F. Onori, N. Roth, S.J. Smartt, L. Wyrzykowski, D.R. Young, The spectral evolution of AT 2018dyb and the presence of metal lines in tidal disruption events. Astrophys. J. **887**(2), 218 (2019). https://doi.org/10.3847/1538-4357/ab5792

D. Lin, J. Guillochon, S. Komossa, E. Ramirez-Ruiz, J.A. Irwin, W.P. Maksym, D. Grupe, O. Godet, N.A. Webb, D. Barret, A likely decade-long sustained tidal disruption event. Nat. Astron. **1**, 0033 (2017). https://doi.org/10.1038/s41550-016-0033

M. Liska, C. Hesp, A. Tchekhovskoy, A. Ingram, M. van der Klis, S. Markoff, Formation of precessing jets by tilted black hole discs in 3D general relativistic MHD simulations. Mon. Not. R. Astron. Soc. **474**(1), 81–85 (2018). https://doi.org/10.1093/mnrasl/slx174

M. Liska, A. Tchekhovskoy, A. Ingram, M. van der Klis, Bardeen-Petterson alignment, jets, and magnetic truncation in GRMHD simulations of tilted thin accretion discs. Mon. Not. R. Astron. Soc. **487**(1), 550–561 (2019a). https://doi.org/10.1093/mnras/stz834

M. Liska, C. Hesp, A. Tchekhovskoy, A. Ingram, M. van der Klis, S.B. Markoff, M. Van Moer, Disc Tearing and Bardeen-Petterson Alignment in GRMHD simulations of highly tilted thin accretion discs. arXiv e-prints, 1904–08428 (2019b)

F.K. Liu, Z.Q. Zhou, R. Cao, L.C. Ho, S. Komossa, Disc origin of broad optical emission lines of the TDE candidate PTF09djl. Mon. Not. R. Astron. Soc. Lett. **472**(1), 99–103 (2017). https://doi.org/10.1093/mnrasl/slx147.

G. Lodato, D.J. Price, On the diffusive propagation of warps in thin accretion discs. Mon. Not. R. Astron. Soc. **405**(2), 1212–1226 (2010). https://doi.org/10.1111/j.1365-2966.2010.16526.x

G. Lodato, J.E. Pringle, Warp diffusion in accretion discs: a numerical investigation. Mon. Not. R. Astron. Soc. **381**(3), 1287–1300 (2007). https://doi.org/10.1111/j.1365-2966.2007.12332.x

G. Lodato, E.M. Rossi, Multiband light curves of tidal disruption events. Mon. Not. R. Astron. Soc. **410**(1), 359–367 (2011). https://doi.org/10.1111/j.1365-2966.2010.17448.x

G. Lodato, A.R. King, J.E. Pringle, Stellar disruption by a supermassive black hole: is the light curve really proportional to $t^{-5/3}$? Mon. Not. R. Astron. Soc. **392**, 332–340 (2009). https://doi.org/10.1111/j.1365-2966.2008.14049.x

A. Loeb, A. Ulmer, Optical appearance of the debris of a star disrupted by a massive black hole. Astrophys. J. **489**, 573–578 (1997). https://doi.org/10.1086/304814

W. Lu, P. Kumar, On the missing energy puzzle of tidal disruption events. Astrophys. J. **865**(2), 128 (2018). https://doi.org/10.3847/1538-4357/aad54a

S.H. Lubow, G.I. Ogilvie, On the tilting of protostellar disks by resonant tidal effects. Astrophys. J. **538**, 326–340 (2000). https://doi.org/10.1086/309101

S.H. Lubow, J.C.B. Papaloizou, J.E. Pringle, Magnetic field dragging in accretion discs. Mon. Not. R. Astron. Soc. **267**(2), 235–240 (1994). https://doi.org/10.1093/mnras/267.2.235

J.C. McKinney, Total and jet Blandford-Znajek power in the presence of an accretion disk. Astrophys. J. Lett. **630**, 5–8 (2005). https://doi.org/10.1086/468184

J.C. McKinney, A. Tchekhovskoy, R.D. Bland ford, General relativistic magnetohydrodynamic simulations of magnetically choked accretion flows around black holes. Mon. Not. R. Astron. Soc. **423**(4), 3083–3117 (2012). https://doi.org/10.1111/j.1365-2966.2012.21074.x

J.C. McKinney, A. Tchekhovskoy, R.D. Bland ford, Alignment of magnetized accretion disks and relativistic jets with spinning black holes. Science **339**(6115), 49 (2013). https://doi.org/10.1126/science.1230811

J.C. McKinney, A. Tchekhovskoy, A. Sadowski, R. Narayan, Three-dimensional general relativistic radiation magnetohydrodynamical simulation of super-Eddington accretion, using a new code HARMRAD with M1 closure. Mon. Not. R. Astron. Soc. **441**, 3177–3208 (2014). https://doi.org/10.1093/mnras/stu762

J.C. McKinney, L. Dai, M.J. Avara, Efficiency of super-Eddington magnetically-arrested accretion. Mon. Not. R. Astron. Soc. **454**, 6–10 (2015). https://doi.org/10.1093/mnrasl/slv115

D. Merritt, E. Vasiliev, Spin evolution of supermassive black holes and galactic nuclei. Phys. Rev. D **86**(10), 102002 (2012). https://doi.org/10.1103/PhysRevD.86.102002

B.D. Metzger, N.C. Stone, A bright year for tidal disruptions. Mon. Not. R. Astron. Soc. **461**, 948–966 (2016). https://doi.org/10.1093/mnras/stw1394

M.C. Miller, Disk winds as an explanation for slowly evolving temperatures in tidal disruption events. Astrophys. J. **805**, 83 (2015). https://doi.org/10.1088/0004-637X/805/1/83

B. Mockler, J. Guillochon, E. Ramirez-Ruiz, Weighing black holes using tidal disruption events. Astrophys. J. **872**(2), 151 (2019). https://doi.org/10.3847/1538-4357/ab010f

M. Mościbrodzka, H. Falcke, S. Noble, Scale-invariant radio jets and varying black hole spin. Astron. Astrophys. **596**, 13 (2016). https://doi.org/10.1051/0004-6361/201629157

A. Mummery, S.A. Balbus, The spectral evolution of disc dominated tidal disruption events. Mon. Not. R. Astron. Soc. **492**(4), 5655–5674 (2020). https://doi.org/10.1093/mnras/staa192
R. Narayan, I. Yi, Advection-dominated accretion: a self-similar solution. Astrophys. J. Lett. **428**, 13 (1994). https://doi.org/10.1086/187381
R. Narayan, I.V. Igumenshchev, M.A. Abramowicz, Magnetically arrested disk: an energetically efficient accretion flow. Publ. Astron. Soc. Jpn. **55**, 69–72 (2003). https://doi.org/10.1093/pasj/55.6.L69
R. Narayan, A. SÄ dowski, R.F. Penna, A.K. Kulkarni, GRMHD simulations of magnetized advection-dominated accretion on a non-spinning black hole: role of outflows. Mon. Not. R. Astron. Soc. **426**(4), 3241–3259 (2012). https://doi.org/10.1111/j.1365-2966.2012.22002.x
R. Nealon, D.J. Price, C. Bonnerot, G. Lodato, On the Papaloizou-Pringle instability in tidal disruption events. Mon. Not. R. Astron. Soc. **474**, 1737–1745 (2018). https://doi.org/10.1093/mnras/stx2871
R.P. Nelson, J.C.B. Papaloizou, Hydrodynamic simulations of the Bardeen-Petterson effect. Mon. Not. R. Astron. Soc. **315**(3), 570–586 (2000). https://doi.org/10.1046/j.1365-8711.2000.03478.x
C. Nixon, A. King, D. Price, J. Frank, Tearing up the disk: how black holes accrete. Astrophys. J. Lett. **757**(2), 24 (2012). https://doi.org/10.1088/2041-8205/757/2/L24
I.D. Novikov, K.S. Thorne, Astrophysics of black holes, in *Black Holes (Les Astres Occlus)*, ed. by C. Dewitt, B.S. Dewitt, 1973, pp. 343–450
G.I. Ogilvie, The non-linear fluid dynamics of a warped accretion disc. Mon. Not. R. Astron. Soc. **304**, 557–578 (1999). https://doi.org/10.1046/j.1365-8711.1999.02340.x
G.I. Ogilvie, Non-linear bending waves in Keplerian accretion discs. Mon. Not. R. Astron. Soc. **365**, 977–990 (2006). https://doi.org/10.1111/j.1365-2966.2005.09776.x
G.I. Ogilvie, A.J. Barker, Local and global dynamics of eccentric astrophysical discs. Mon. Not. R. Astron. Soc. **445**(3), 2621–2636 (2014). https://doi.org/10.1093/mnras/stu1795
K. Ohsuga, S. Mineshige, Global structure of three distinct accretion flows and outflows around black holes from two-dimensional radiation-magnetohydrodynamic simulations. Astrophys. J. **736**, 2 (2011). https://doi.org/10.1088/0004-637X/736/1/2
K. Ohsuga, M. Mori, T. Nakamoto, S. Mineshige, Supercritical accretion flows around black holes: two-dimensional, radiation pressure-dominated disks with photon trapping. Astrophys. J. **628**, 368–381 (2005). https://doi.org/10.1086/430728
J.C.B. Papaloizou, D.N.C. Lin, On the dynamics of warped accretion disks. Astrophys. J. **438**, 841–851 (1995). https://doi.org/10.1086/175127
J.C.B. Papaloizou, J.E. Pringle, The time-dependence of non-planar accretion discs. Mon. Not. R. Astron. Soc. **202**, 1181–1194 (1983). https://doi.org/10.1093/mnras/202.4.1181
J.C.B. Papaloizou, J.E. Pringle, The dynamical stability of differentially rotating discs with constant specific angular momentum. Mon. Not. R. Astron. Soc. **208**, 721–750 (1984). https://doi.org/10.1093/mnras/208.4.721
D.R. Pasham, R.A. Remillard, P.C. Fragile, A. Franchini, N.C. Stone, G. Lodato, J. Homan, D. Chakrabarty, F.K. Baganoff, J.F. Steiner, E.R. Coughlin, N.R. Pasham, A loud quasi-periodic oscillations after a star is disrupted by a massive black hole. Science **363**, 531–534 (2019)
M.E. Pessah, D. Psaltis, The stability of magnetized rotating plasmas with superthermal fields. Astrophys. J. **628**, 879–901 (2005). https://doi.org/10.1086/430940
E.S. Phinney, Manifestations of a massive black hole in the galactic center, in *The Center of the Galaxy*, ed. by M. Morris IAU Symposium, vol. 136, 1989, p. 543
T. Piran, G. Svirski, J. Krolik, R.M. Cheng, H. Shiokawa, Disk formation versus disk accretion – what powers tidal disruption events? Astrophys. J. **806**, 164 (2015a). https://doi.org/10.1088/0004-637X/806/2/164
T. Piran, A. Sądowski, A.e. Tchekhovskoy, Jet and disc luminosities in tidal disruption events. Mon. Not. R. Astron. Soc. **453**(1), 157–165 (2015b). https://doi.org/10.1093/mnras/stv1547
P. Polko, J.C. McKinney, Electromagnetic versus Lense-Thirring alignment of black hole accretion discs. Mon. Not. R. Astron. Soc. **464**(3), 2660–2671 (2017). https://doi.org/10.1093/mnras/stw1875
J. Poutanen, G. Lipunova, S. Fabrika, A.G. Butkevich, P. Abolmasov, Supercritically accreting stellar mass black holes as ultraluminous X-ray sources. Mon. Not. R. Astron. Soc. **377**(3), 1187–1194 (2007). https://doi.org/10.1111/j.1365-2966.2007.11668.x
D. Proga, J.M. Stone, T.R. Kallman, Dynamics of line-driven disk winds in active galactic nuclei. Astrophys. J. **543**(2), 686–696 (2000). https://doi.org/10.1086/317154
M.J. Rees, Tidal disruption of stars by black holes of 10 to the 6th-10 to the 8th solar masses in nearby galaxies. Nature **333**, 523–528 (1988). https://doi.org/10.1038/333523a0
M.J. Rees, M.C. Begelman, R.D. Blandford, E.S. Phinney, Ion-supported tori and the origin of radio jets. Nature **295**(5844), 17–21 (1982). https://doi.org/10.1038/295017a0
R.C. Reis, J.M. Miller, M.T. Reynolds, K. Gültekin, D. Maitra, A.L. King, T.E. Strohmayer, A 200-second quasi-periodicity after the tidal disruption of a star by a dormant black hole. Science **337**(6097), 949 (2012). https://doi.org/10.1126/science.1223940

N. Roth, D. Kasen, What sets the line profiles in tidal disruption events? Astrophys. J. **855**(1), 54 (2018). https://doi.org/10.3847/1538-4357/aaaec6
N. Roth, D. Kasen, J. Guillochon, E. Ramirez-Ruiz, The X-ray through optical fluxes and line strengths of tidal disruption events. Astrophys. J. **827**, 3 (2016). https://doi.org/10.3847/0004-637X/827/1/3
A. Sądowski, R. Narayan, Powerful radiative jets in supercritical accretion discs around non-spinning black holes. Mon. Not. R. Astron. Soc. **453**(3), 3213–3221 (2015). https://doi.org/10.1093/mnras/stv1802
A. Sądowski, R. Narayan, Three-dimensional simulations of supercritical black hole accretion discs—luminosities, photon trapping and variability. Mon. Not. R. Astron. Soc. **456**, 3929–3947 (2016). https://doi.org/10.1093/mnras/stv2941
A. Sądowski, R. Narayan, A. Tchekhovskoy, Y. Zhu, Semi-implicit scheme for treating radiation under M1 closure in general relativistic conservative fluid dynamics codes. Mon. Not. R. Astron. Soc. **429**(4), 3533–3550 (2013). https://doi.org/10.1093/mnras/sts632
A. Sądowski, E. Tejeda, E. Gafton, S. Rosswog, D. Abarca, Magnetohydrodynamical simulations of a deep tidal disruption in general relativity. Mon. Not. R. Astron. Soc. **458**(4), 4250–4268 (2016). https://doi.org/10.1093/mnras/stw589
N.I. Shakura, R.A. Sunyaev, Black holes in binary systems. Observational appearance. Astron. Astrophys. **24**, 337–355 (1973)
R.-F. Shen, C.D. Matzner, Evolution of accretion disks in tidal disruption events. Astrophys. J. **784**, 87 (2014). https://doi.org/10.1088/0004-637X/784/2/87
H. Shiokawa, J.H. Krolik, R.M. Cheng, T. Piran, S.C. Noble, General relativistic hydrodynamic simulation of accretion flow from a stellar tidal disruption. Astrophys. J. **804**, 85 (2015). https://doi.org/10.1088/0004-637X/804/2/85
K.A. Sorathia, J.H. Krolik, J.F. Hawley, Magnetohydrodynamic simulation of a disk subjected to Lense-Thirring precession. Astrophys. J. **777**(1), 21 (2013a). https://doi.org/10.1088/0004-637X/777/1/21
K.A. Sorathia, J.H. Krolik, J.F. Hawley, Relaxation of warped disks: the case of pure hydrodynamics. Astrophys. J. **768**(2), 133 (2013b). https://doi.org/10.1088/0004-637X/768/2/133
N. Stone, A. Loeb, Observing Lense-Thirring precession in tidal disruption flares. Phys. Rev. Lett. **108**(6), 061302 (2012). https://doi.org/10.1103/PhysRevLett.108.061302
L.E. Strubbe, E. Quataert, Optical flares from the tidal disruption of stars by massive black holes. Mon. Not. R. Astron. Soc. **400**, 2070–2084 (2009). https://doi.org/10.1111/j.1365-2966.2009.15599.x
L.E. Strubbe, E. Quataert, Spectroscopic signatures of the tidal disruption of stars by massive black holes. Mon. Not. R. Astron. Soc. **415**(1), 168–180 (2011). https://doi.org/10.1111/j.1365-2966.2011.18686.x
G. Svirski, T. Piran, J. Krolik, Elliptical accretion and low luminosity from high accretion rate stellar tidal disruption events. Mon. Not. R. Astron. Soc. **467**(2), 1426–1432 (2017). https://doi.org/10.1093/mnras/stx117
D. Syer, C.J. Clarke, The viscous evolution of elliptical accretion discs. Mon. Not. R. Astron. Soc. **255**, 92–104 (1992). https://doi.org/10.1093/mnras/255.1.92
A. Tchekhovskoy, R. Narayan, J.C. McKinney, Efficient generation of jets from magnetically arrested accretion on a rapidly spinning black hole. Mon. Not. R. Astron. Soc. **418**(1), 79–83 (2011). https://doi.org/10.1111/j.1745-3933.2011.01147.x
A. Tchekhovskoy, J.C. McKinney, R. Narayan, General relativistic modeling of magnetized jets from accreting black holes, in *Journal of Physics Conference Series*. Journal of Physics Conference Series, vol. 372, 2012, p. 012040. https://doi.org/10.1088/1742-6596/372/1/012040
A. Tchekhovskoy, B.D. Metzger, D. Giannios, L.Z. Kelley, Swift J1644+57 gone MAD: the case for dynamically important magnetic flux threading the black hole in a jetted tidal disruption event. Mon. Not. R. Astron. Soc. **437**, 2744–2760 (2014). https://doi.org/10.1093/mnras/stt2085
L.L. Thomsen, J. Lixin Dai, E. Ramirez-Ruiz, E. Kara, C. Reynolds, X-ray fluorescence from super-Eddington accreting black holes. Astrophys. J. Lett. **884**(1), 21 (2019). https://doi.org/10.3847/2041-8213/ab4518
M. Toscani, G. Lodato, R. Nealon, Gravitational wave emission from unstable accretion discs in tidal disruption events. Mon. Not. R. Astron. Soc. **489**(1), 699–706 (2019). https://doi.org/10.1093/mnras/stz2201
A. Ulmer, Flares from the tidal disruption of stars by massive black holes. Astrophys. J. **514**, 180–187 (1999). https://doi.org/10.1086/306909
S. van Velzen, N.C. Stone, B.D. Metzger, S. Gezari, T.M. Brown, A.S. Fruchter, Late-time UV observations of tidal disruption flares reveal unobscured, compact accretion disks. Astrophys. J. **878**(2), 82 (2019). https://doi.org/10.3847/1538-4357/ab1844
S. van Velzen, S. Gezari, E. Hammerstein, N. Roth, S. Frederick, C. Ward, T. Hung, S.B. Cenko, R. Stein, D.A. Perley, K. Taggart, J. Sollerman, I. Andreoni, E.C. Bellm, V. Brinnel, K. De, R. Dekany, M. Feeney, R.J. Foley, C. Fremling, M. Giomi, V.Z. Golkhou, A.Y.Q. Ho, M.M. Kasliwal, C.D. Kilpatrick, S.R. Kulkarni, T. Kupfer, R.R. Laher, A. Mahabal, F.J. Masci, J. Nordin, R. Riddle, B. Rusholme, Y. Sharma, J. van Santen, D.L. Shupe, M.T. Soumagnac, Seventeen Tidal Disruption Events from the first

half of ZTF survey observations: entering a new era of population studies. arXiv e-prints, 2001–01409 (2020)

S. Wu, E.R. Coughlin, C. Nixon, Super-Eddington accretion in tidal disruption events: the impactof realistic fallback rates on accretion rates. Mon. Not. R. Astron. Soc. **478**(3), 3016–3024 (2018). https://doi.org/10.1093/mnras/sty971

M. Xiang-Gruess, P.B. Ivanov, J.C.B. Papaloizou, On the formation of a quasi-stationary twisted disc after a tidal disruption event. Mon. Not. R. Astron. Soc. **463**, 2242–2264 (2016). https://doi.org/10.1093/mnras/stw2130

J.J. Zanazzi, D. Lai, Tidal disruption event disks around supermassive black holes: disk warp and inclination evolution. arXiv e-prints (2019)

B.A. Zauderer, E. Berger, A.M. Soderberg, A. Loeb, R. Narayan, D.A. Frail, G.R. Petitpas, A. Brunthaler, R. Chornock, J.M. Carpenter, G.G. Pooley, K. Mooley, S.R. Kulkarni, R. Margutti, D.B. Fox, E. Nakar, N.A. Patel, N.H. Volgenau, T.L. Culverhouse, M.F. Bietenholz, M.P. Rupen, W. Max-Moerbeck, A.C.S. Readhead, J. Richards, M. Shepherd, S. Storm, C.L.H. Hull, Birth of a relativistic outflow in the unusual γ-ray transient Swift J164449.3+573451. Nature **476**(7361), 425–428 (2011). https://doi.org/10.1038/nature10366

B.A. Zauderer, E. Berger, R. Margutti, G.G. Pooley, R. Sari, A.M. Soderberg, A. Brunthaler, M.F. Bietenholz, Radio monitoring of the tidal disruption event Swift J164449.3+573451. II. The relativistic jet shuts off and a transition to forward shock X-ray/radio emission. Astrophys. J. **767**(2), 152 (2013). https://doi.org/10.1088/0004-637X/767/2/152

Z. Zhu, J.M. Stone, Global evolution of an accretion disk with a net vertical field: coronal accretion, flux transport, and disk winds. Astrophys. J. **857**(1), 34 (2018). https://doi.org/10.3847/1538-4357/aaafc9

Space Sci Rev (2020) 216:63
https://doi.org/10.1007/s11214-020-00697-4

Simulations of Tidal Disruption Events

Giuseppe Lodato[1] · **Roseanne M. Cheng**[2] · **Clément Bonnerot**[3] · **Jane Lixin Dai**[4]

Received: 23 May 2019 / Accepted: 17 May 2020 / Published online: 2 June 2020

Abstract Numerical simulations have historically played a major role in understanding the hydrodynamics of the tidal disruption process. Given the complexity of the geometry of the system, the challenges posed by the problem have indeed stimulated much work on the numerical side. Smoothed Particles Hydrodynamics methods, for example, have seen their very first applications in the context of tidal disruption and still play a major role to this day. Likewise, initial attempts at simulating the evolution of the disrupted star with the so-called affine method have been historically very useful. In this Chapter, we provide an overview of the numerical techniques used in the field and of their limitations, and summarize the work that has been done to simulate numerically the tidal disruption process.

Keywords Black hole physics · Galaxies: nuclei · Hydrodynamics · Methods: numerical

The Tidal Disruption of Stars by Massive Black Holes
Edited by Peter G. Jonker, Sterl Phinney, Elena Maria Rossi, Sjoert van Velzen, Iair Arcavi and Maurizio Falanga

✉ G. Lodato
giuseppe.lodato@unimi.it

R.M. Cheng
rmcheng@lanl.gov

C. Bonnerot
bonnerot@tapir.caltech.edu

J.L. Dai
lixindai@hku.hk

1 Dpartimento di Fisica, Università degli Studi di Milano, Via Celoria 16, Milano, Italy

2 Theoretical Division (T-1, T-3), Los Alamos National Laboratory, P.O. Box 1663, Los Alamos, NM 87545, USA

3 TAPIR, California Institute of Technology, Mailcode 350-17, Pasadena, CA 91125, USA

4 Department of Physics, The University of Hong Kong, Pok-fulam Road, Hong Kong, Hong Kong

1 Introduction

The dynamics of tidal disruption events is relatively complex. It involves treating the hydrodynamics of a self-gravitating fluid (the star), subject to a general relativistic force provided by the black hole. A full treatment of the problem would thus involve a general relativistic (magneto)-hydrodynamics code including heating and radiation. In some respects, this is what is also required when describing accretion flows around black holes in other systems, such as Active Galactic Nuclei, X-ray Binaries, and Ultra Luminous X-ray sources. Here the situation is made more complex because of the rapid variability of the system and the process of disruption, which is composed of several stages that are best described separately.

In a nutshell, we can describe a TDE as being composed of three separate phases: (a) the disruption phase, (b) the evolution of the disrupted stream leading to disc formation, and (c) the accretion phase. In the disruption phase, the relevant physics is essentially the stellar self-gravity, fluid dynamics, and the tidal field of the black hole. For highly penetrating events, the tidal compression can lead to nuclear detonation (at least in the case of the disruption of a white dwarf), introducing complex thermodynamics in the problem. This phase is the simplest one, providing the only firm analytical prediction for TDE, the $t^{-5/3}$ fallback rate predicted by Rees (1988) and Phinney (1989). However, when treating it numerically, the problem arises that, in the inertial frame of the black hole, the action is spread over a very large volume, which is mostly composed of "empty space", and the tidal debris only occupies a relatively small portion of it. This is best treated either with a grid following the star for an Eulerian hydrodynamics code, or by using a Lagrangian hydrodynamics code. The second phase has been long thought to be relatively straightforward, with the debris essentially orbiting in Keplerian orbits until they fall back to pericenter. Recent investigations have shown complications as the flow can become unstable to self-gravity effects (Coughlin and Nixon 2015). Subsequently, it is thought that the debris falls back and forms an accretion disc, although the process of disc formation is much less well understood (see the dedicated Chapter by Bonnerot et al. within this book). The process of disc formation can be strongly affected by both relativistic (such as apsidal and Lense-Thirring precession) (Dai et al. 2015; Guillochon and Ramirez-Ruiz 2015; Hayasaki et al. 2016; Bonnerot et al. 2017b) and magnetic effects (Bonnerot et al. 2017a). Finally, as mentioned above, the accretion phase is in some respects similar to the case of accretion tori in other black hole systems except that the conditions are extreme in terms of fallback rate. Thus, we expect the formation of winds, outflows, and jets because this rate is strongly super-Eddington.

While no attempt has been made to simulate the whole process in a single framework, there has been significant progress over the years to simulate the individual stages independently. In this Chapter, we focus on the techniques that have been used in this context and on their applications to the various phases of TDE. The Chapter is organised as follows. In Sect. 2, we describe the various kinds of codes that have been used for TDE: from Smoothed Particles Hydrodynamics, to grid-based methods, to the affine model that was specifically developed in the '80s to treat this problem. In Sect. 3, we describe the efforts that have been done to treat the important relativistic effects at play in the process. In Sect. 4, we describe the attempts at including more complex thermal and radiation physics. In Sect. 5, we review the numerical work that has been done to simulate the various phases of a TDE: disruption, stream evolution and accretion. Finally, in Sect. 6, we conclude and provide an outlook for the future simulations.

2 Numerical Techniques Used in TDE Simulations

2.1 Smoothed Particle Hydrodynamics

Smoothed Particles Hydrodynamics (SPH, Lucy 1977, Gingold and Monaghan 1977) is a mesh-less Lagrangian algorithm that has been widely used, especially in cosmological simulations and hydrodynamic simulation of accretion discs and star formation. Its application to simulations of TDE is convenient especially in view of its Lagrangian approach, given the large dynamical range of the disruption phase of a TDE and since most of the computational domain is actually "empty", with the stellar debris occupying only a limited volume.

It is interesting to note that some of the earliest application of SPH in astrophysics were indeed in the field of TDE, indicating that it was immediately apparent that a Lagrangian method such as SPH was particularly suited for this kind of problem. Looking back at the historical first papers, Nolthenius and Katz (1982) used ~ 40 SPH particles to simulate the disruption of a $1 M_\odot$ star by a $10^4 M_\odot$ black hole. Bicknell and Gingold (1983) used 500 SPH particles to simulate a highly penetrating encounter of a $1 M_\odot$ star by a $10^5 M_\odot$ black hole, in order to test the possibility of tidal detonation of the star–a fashionable topic at the time. It is quite impressive to see how these early results were obtained with such limited numerical resolution (by comparison, modern codes use $\sim 10^6$ particles, a factor of ten thousand more than these seminal studies).

In SPH, the fluid is discretized into finite mass elements (called "particles"), whose properties are computed by suitable averages between neighbouring particles that lie within a "smoothing region" around it. Within this region, particles that lie close to the edge have a progressively lower weight. For example, the density ρ at the position of particle a is given by:

$$\rho(\boldsymbol{r}_a) = \sum_{b=1}^{N} m_b W\left(\boldsymbol{r}_a - \boldsymbol{r}_b, h\right), \tag{1}$$

where W is the so-called "smoothing kernel", which is typically a bell-shaped function with compact support and a size defined by the parameter h, called the "smoothing length". N is the number of particles within the kernel, which are called the "neighbours" of particle a. Each particle *must* have their own smoothing length, determined in such a way to keep the number of neighbours roughly constant and avoid unwanted sampling error fluctuations between particles. A typical choice is:

$$h\left(\boldsymbol{r}_a\right) = \eta \left(\frac{m_a}{\rho_a}\right)^{1/3}, \tag{2}$$

where, with $\eta = 1.2$ (for the standard choice of a cubic spline kernel, see Price 2012) one has roughly 60 neighbours per particle in three dimensions. Thus, SPH naturally has the advantage that resolution follows density, so that high density regions have relatively small smoothing length (and thus have a higher spatial resolution). SPH is naturally an "adaptive" method.

The equation of motion of the particles is derived from the Euler-Lagrange equations obtained from a variational principle formulation of fluid dynamics (Eckart 1960). The fluid Lagrangian L is:

$$L = \int \left[\rho v^2 - u(\rho, s) - \Phi\right] dV, \tag{3}$$

where ρ is the density, v is the fluid velocity, u is the internal energy, that depends only on density and on the specific entropy s, and Φ is the gravitational potential. This can be discretized immediately, giving:

$$L = \sum_b m_b \left[\frac{1}{2} v_b^2 - u_b(\rho_b, s_b) - \Phi_b \right], \tag{4}$$

from which the Euler-Lagrange equation of motion follows, after a little algebra and by using the first principle of thermodynamics for a dissipationless fluid (see, for example, Price 2012 for details):

$$\frac{\mathrm{d}\mathbf{v}_a}{\mathrm{d}t} = -\sum_b m_b \left[\frac{P_a}{\rho_a^2} + \frac{P_b}{\rho_b^2} \right] \frac{\partial W_{ab}}{\partial \mathbf{r}_a} - \frac{\partial \Phi}{\partial \mathbf{r}_a}. \tag{5}$$

The above equation is valid only for the simple case in which the smoothing length is assumed to be constant (a more complete equation can be found in Price 2012), but it shows that the SPH equations indeed recover the fluid equations appropriately, whereby in the left-hand side we find the Lagrangian derivative of the velocity and on the right-hand side we find the pressure gradient and the gravitational force. Since they follow the Euler-Lagrange equations, the set of SPH particles represent a Hamiltonian system and as such conserve exactly and simultaneously linear and angular momentum, as well as energy (see Price 2012 and Springel 2010 for recent reviews).

The derivation described above, however, is only valid for a dissipationless fluid and requires all fluid properties to be continuous and differentiable. Particular care should thus be given to the handling of shocks and discontinuities in SPH (Price 2008). In particular, to resolve a shock, SPH typically uses an artificial viscosity. It can be shown (Lodato and Price 2010), in the absence of switches, the artificial viscosity scales as $\propto c_s h$, where c_s is the gas sound speed. Thus, low density regions, which are characterized by a large smoothing length h, also suffer from, in principle, unwanted, large artificial viscosity. In most cases, however, such dissipation can be effectively limited by using suitable switches, such as the Morris and Monaghan (1997) switch, or the Cullen and Dehnen (2010) switch. This can be particularly important for simulations that try to follow the formation of a disc after a TDE because, at the beginning of the fallback phase, the gas density is bound to be low. Great care should be taken in ensuring that the results are not affected by excessive numerical dissipation.

Another interesting aspect to consider is the inclusion of relativistic effects. SPH naturally lends itself to a fully general relativistic implementation (in a fixed metric), as shown by Monaghan and Price (2001). Progress is in underway to implement this feature in available codes. The PHANTOM code (Price et al. 2018) already implements several GR effects as corrections to Newtonian dynamics and very recently a full general relativistic implementation has been implemented (Liptai and Price 2019). These include Lense-Thirring precession around a spinning black hole (following the approach of Nelson and Papaloizou 2000) and apsidal precession around a Schwarzschild black hole (Tejeda and Rosswog 2013), which has been already used in simulations of TDE (Bonnerot et al. 2016b). These aspects will be discussed more extensively in Sect. 3 below.

2.2 Grid Based Codes

An alternative to SPH is solving the fluid equations on a mesh. In such grid based codes, the physical domain is discretized into volumes. In each volume, the conservative equations of mass, momentum, and energy are evolved in time as

$$\frac{\partial \rho}{\partial t} + \vec{\nabla} \cdot (\rho \vec{v}) = 0,$$
$$\frac{\partial \rho \vec{v}}{\partial t} + \vec{\nabla} \cdot (\rho \vec{v} \vec{v}) + \vec{\nabla} P = 0,$$
$$\frac{\partial E}{\partial t} + \vec{\nabla} \cdot [(E + P)\vec{v}] = 0, \tag{6}$$

for density ρ, velocity $\vec{v}$, gas pressure P, and total energy density $E = u + \frac{1}{2}\rho|\vec{v}|^2$, where the internal energy density u is determined by the choice in equation of state (Landau and Lifshitz 1959). For an ideal gas of adiabatic index γ, this is $u = P/(\gamma - 1)$. Extending to ideal magnetohydrodynamics, for magnetic field **B**, the evolution equations are also in conservative form,

$$\frac{\partial \rho}{\partial t} + \vec{\nabla} \cdot (\rho \vec{v}) = 0,$$
$$\frac{\partial \rho \vec{v}}{\partial t} + \vec{\nabla} \cdot (\rho \vec{v} \vec{v} - \vec{B}\vec{B} + \vec{P}^*) = 0,$$
$$\frac{\partial E}{\partial t} + \vec{\nabla} \cdot [\,(E + P^*)\vec{v} - \vec{B}(\vec{B} \cdot \vec{v})] = 0,$$
$$\frac{\partial \vec{B}}{\partial t} - \vec{\nabla} \times (\vec{v} \times \vec{B}) = 0, \tag{7}$$

where $\vec{P}^*$ is a diagonal tensor with components $P* = P + B^2/2$ and total energy density $E = u + \frac{1}{2}\rho|\vec{v}|^2 + \frac{1}{2}\vec{B} \cdot \vec{B}$ (Gardiner and Stone 2008; Stone et al. 2008). Note that special numerical treatment is necessary in order to enforce the solenoidal constraint $\vec{\nabla} \cdot \vec{B} = 0$ on the computational domain (Evans and Hawley 1988). For a fixed region, decreasing the volume size increases the resolution of the simulation. This method is useful in providing a high-resolution shock capturing technique to model, with high-accuracy, the dynamical response of the star to the black hole's tidal field as well as the formation and evolution of an accretion disc from the stellar debris. It is inefficient when there is a significant amount of "empty-space" in the simulation.

Unfortunately, this is the case for a significant phase of the TDE, namely when the star initially passes by the black hole. Specifically, if one used a computational domain as large as the size of the expected stream, a majority of the region would be modeled as vacuum until a significant time after the disruption, when the thin stream of debris returned to the black hole, forming a disc. By focusing the simulation on the star, adopting a reference frame with respect to its center, modeling "empty-space" is avoided. Also, since the computational error scales linearly with the fluid velocity with respect to the grid, such a reference frame also helps in keeping the accuracy of the simulation high. This type of calculation is useful for studying the dynamics of tidal compression and calculating quantities such as the return rate of debris to the black hole. Inefficiencies aside, grid based techniques are useful for modeling the disc formation and accretion not only because of the accuracy in capturing shocks, but also because implementing additional physics such as general relativity, magnetic fields, and radiation is straightforward. Furthermore, many of these numerical tools are available to the astrophysics community due to the effort in solving related complex multi-physics problems in stellar structure and accretion discs.

To solve the basic equations of hydrodynamics, finite volume methods are often implemented because they are conservative by construction (Toro 1999). The numerical scheme is based on the Godunov update from time level t^n to t^{n+1}, where the conservative equations of hydrodynamics Eq. (6) and magnetohydrodynamics Eq. (7) are discretized as

$$\mathbf{U}_i^{n+1} = \mathbf{U}_i^n + \frac{\Delta t}{\Delta x}\left(\mathbf{F}_{i-1/2}^n - \mathbf{F}_{i+1/2}^n\right), \tag{8}$$

in one dimension over a zone of length Δx within a time step Δt. Mass, momentum, and energy $\mathbf{U} = \{\rho, \rho v^x, E\}$ are volume-averaged and defined at zone centers i and fluxes $\mathbf{F} = \{\rho v^x, \rho v^x v^x + P, (E+P)v^x\}$ are area-averaged and defined at zone faces $i \pm 1/2$. At t^n, $\mathbf{U}$ is given. The fluxes are time-averaged and depend on the solution to the Riemann problem or the jump discontinuity conditions of the conservative hydrodynamics equations at the zone faces. High spatial accuracy of these schemes depends on the order chosen in solving the Riemann problem given the data defined at the zone-center and approximated at the zone face. The choice of numerical integration determines the temporal accuracy, where more sophisticated updates at intermediate time-levels may be implemented instead of Eq. (8), which is a first-order scheme.

For local simulations, the gravitational effects of the black hole are treated with respect to the local reference frame of the star. Numerically, calculations of the disruption phase in this frame have greater accuracy than those with computational domains in the black hole frame. This is because, when the star is tidally deformed, changes in the fluid velocity are much smaller than the orbital velocity of all of the debris with respect to the black hole. These small changes are easily tracked in a local frame while they can be lost as numerical errors reduce the precision of calculations in the black hole frame. Local, Newtonian treatments model the disruption of the star with Newtonian stellar and black hole gravity and ideal gas hydrodynamics (Khokhlov et al. 1993a,b; Brassart and Luminet 2008; Guillochon and Ramirez-Ruiz 2013). Post-disruption, the debris expands significantly and, as the density decreases, the hydrodynamical behaviour becomes unimportant. In this limit, the debris can be treated as ballistic particles. If the local simulation follows the debris to this point, then an estimate of the return rate of debris to the black hole can be derived from the numerical results (Guillochon and Ramirez-Ruiz 2013). While the star is typically modeled as a polytrope given by a single adiabatic index, the use of stellar profiles derived from the open source code MESA, Modules for Experiments in Stellar Astrophysics, enables a study of the effects of tidal disruptions on stellar cores and tidal stripping of stellar envelopes (MacLeod et al. 2012, 2013; Law-Smith et al. 2017; Anninos et al. 2018; Law-Smith et al. 2019; Goicovic et al. 2019). It is interesting to note that for some partial disruptions, the remnant core receives a "kick" into an unbound orbit (Manukian et al. 2013). Furthermore, with a magnetized star, one can see the magnetic field amplification and structure in the debris field due to the tidal distortion (Guillochon and McCourt 2017).

In global simulations with respect to the black hole reference frame, the returning debris stream wraps around the black hole, forming a disc. If the debris reaches within a radius significantly close to the black hole, then the fluid velocity is a significant fraction of the speed of light. In this limit, it is useful to implement a fully general relativistic hydrodynamics (GRHD) code in order to accurately capture important general relativistic effects (e.g., Haas et al. 2012). Furthermore, it is necessary to follow the interaction of gas with electromagnetic fields in the strong gravitational field around the black hole. Plasma often possesses relatively strong magnetic fields. Modeling their interaction is important for accretion and relativistic jet production. Simulations of magnetic fields in the strong gravity regime are performed with general relativistic magnetohydrodynamic (GRMHD) codes (e.g., Gammie et al. 2003). These codes are usually simplified with the use of an ideal gas equation of state, ideal MHD, and a static metric around the black hole. For the latter, this works well for TDEs because the accretion of the debris from a star only trivially changes the mass and angular momentum of a supermassive black hole. The equations evolved are similar to Eq. (7) but with relativistic stress-energy tensors, which we will explain in detail in Sect. 3. Due to the set up, GRMHD codes are more efficient when following the evolution of the gas after it has already formed a somewhat axisymmetric structure around the black hole (e.g., Sądowski et al. 2016). Under certain circumstances the interaction between strong radiation and gas also needs to be well captured, such as when studying the super-Eddington accretion phase of TDE discs (e.g., Dai et al. 2018). Such simulations can be performed using general relativistic radiation magnetohydrodynamic (GRRMHD) codes (e.g., Sądowski et al. 2014; McKinney et al. 2014), which evolve radiation in parallel with gas while including basic scattering and absorption physics.

We briefly discuss the current state-of-the art. For local simulations, finite volume method codes such as FLASH (Fryxell et al. 2000) and VH-1 (Hawley et al. 2012) follow the star through its disruption phase in the Newtonian (Guillochon et al. 2009) and post-Newtonian (Cheng and Evans 2013) regimes. Before the disruption point, the star is a compact ball, but once it is disrupted, the gas expands, following trajectories around the black hole that further spread out the debris. To follow the debris long enough until the hydrodynamic effects are insignificant (at least for the unbound portion of the debris), all of the gas must remain inside the grid. For a fixed Cartesian grid, one needs to start with a very large and computationally inefficient domain which models the "empty-space" around the compact ball for a significant, and costly, amount of time. In order to ameliorate this problem, Guillochon et al. (2009), Guillochon and Ramirez-Ruiz (2013) use an adaptive mesh refinement (AMR) technique (Berger and Colella 1989) which provides overlaying levels of increasing grid resolutions. Much of the debris is captured with a reduction of cost by placing high resolution on a high density regions of the debris and low resolution on "empty-space." However, this technique introduces uncertainties in numerical errors through interpolation between fine and coarse levels of refinement and relies on arbitrary conditions to distinguish between low-density debris and numerical atmosphere. Such AMR techniques on a Cartesian grid have also been implemented in GRHD simulations of the disruption and disc formation such as with MAYA in (Haas et al. 2012), to reduce the problem of computational inefficiency. It is worth noting that the problem of maintaining the solenoidal constraint for magnetohydrodynamics is non-trivial with AMR. Furthermore, Anninos et al. (2018) use a combination of AMR and moving-mesh techniques to model the disruption phase with COSMOS++ (Fragile et al. 2012, 2014). Alternatively, one can re-express the spatial coordinate system used in evolving the hydrodynamic equations in terms of modified coordinates which provide logarithmic spacing in radius (grid zones increase with increasing radius) along with an increase in resolution along the equatorial plane (Gammie et al. 2003; Noble et al. 2009; Sądowski et al. 2015). This is done with HARM3D in Shiokawa et al. (2015), with HARMRAD in Dai et al. (2018) and with KORAL in Sądowski et al. (2016).

In the following sections, we discuss grid-based general relativistic treatments to the black hole and hydrodynamics (Haas et al. 2012; Cheng and Evans 2013; East and Pretorius 2013; Shiokawa et al. 2015), GRMHD (Sądowski et al. 2016), the use of nuclear networks (Anninos et al. 2018), and studies with radiation (Jiang et al. 2016) as well as with both radiation and general relativity (Dai et al. 2018; Curd and Narayan 2019). While these multiphysics simulations begin to address questions about the leading physical mechanisms, the problem of computational expense limits the use of grid-based codes.

2.3 Moving Mesh Codes

Recently, moving mesh Lagrangian codes have been developed, where the volume is discretized, as in grid-based methods, but the cells are allowed to move, thus including Galilean invariance, as in SPH methods (Springel 2010; Hopkins 2015). The volume discretization is performed via a Voronoi tessellation, where the various cells are allowed to move and adapt to the flow. While not perfectly Lagrangian in nature, the method maintains Galilean invariance, unlike other grid-based methods. At the same time, it allows one to solve the Riemann problem at the cell interfaces, providing better shock capturing. In some sense, such moving-mesh codes combine the best properties of both grid-based and SPH methods.

Such methods have been rarely used to simulate TDE. Mainetti et al. (2017) compare moving-mesh codes to SPH and grid-based codes for the problem of partial tidal disruptions, finding a relatively good agreement between the three methods. Anninos et al. (2018) combine moving mesh techniques with general relativistic hydrodynamics and thermonuclear reactions. More recently, Steinberg et al. (2019) have used moving mesh codes for the study of deeply plunging events, while Goicovic et al. (2019) have also considered the evolution of partially disrupted stars, focusing on the evolution of the stellar remnant after pericenter passage.

Another interesting application of moving mesh codes has been done by Yalinewich et al. (2019), who simulate the evolution of the unbound debris and the associated radio emission.

2.4 The Affine Model

To model the tidal disruption process, it is necessary to solve complex systems of non-linear equations for the fluid and self-gravity of the star as well as the gravity due to the black hole. Analytical models provide reasonable estimates under basic assumptions. A useful approximation is the limit where the pericenter of the star's orbit is much larger than the radius of the star ($R_p \gg R_*$). If the orbital timescale is much shorter than the stellar dynamical timescale, then the tidal forces dominate over the internal forces of self-gravity and pressure gradients, compressing the star significantly. An interesting observational consequence of severe tidal compression was suggested by Carter and Luminet (1982). For stellar encounters with pericenters that are much smaller than the tidal radius of the black hole ($\beta = R_t/R_p \gg 1$), a short phase of high compression or "pancaking" leads to orders of magnitude increases in density and temperature, enough to detonate a significant fraction of thermonuclear fuel. For a main sequence star, such a temperature increase would set off helium combustion by a triple-α reaction. More nuclear reactions are possible (Carter and Luminet 1983). While gas cools as it expands after the compression, β-decay of by-products can reheat and eject the gas. Potentially, this reheating leads to yet another stage of energy release by self-sustaining hydrogen combustion through a hot C-N-O cycle or rapid proton capture process, assuming that helium combustion is efficient. Note, however, that such results of the affine model are not reproduced in 3D hydrodynamical simulations of the disruption of main sequence stars, and nuclear detonation has only been demonstrated for the case of white dwarf disruption.

Coupling a dynamical model with nuclear reactions would reveal promising observational signatures. A useful, analytic description is given by the affine star model, where layers of constant density are followed as ellipsoids in a frame with respect to the center of mass of the star (Carter and Luminet 1982, 1983, 1985) and is described, as follows. Assume that general relativistic effects are negligible such that $R_p \gg GM_{\rm bh}/c^2$. The star is defined in terms of 12 variables: 9 parameters for a 3×3 deformation matrix q_{ij} and the position of the center of mass in the black hole's reference frame $\vec{X}_0$. Consider the quantities in the local frame of the star. Let $\hat{r}$ be the position vector of a fluid element in the initial configuration of the star, in spherical equilibrium. Let $\vec{r}$ be the position vector of a fluid element during the evolution. In the affine limit, the star is constrained by the linear relation,

$$r_i = q_{ij}\hat{r}_j \tag{9}$$

The mechanics is governed by a Lagrangian function $\mathcal{L}$, with $\| = q_{ij}$,

$$\mathcal{L} = \mathcal{L}_I(\|) + \mathcal{L}_C(\vec{X}_0) + \mathcal{L}_E(\vec{X}). \tag{10}$$

The external term is $\mathcal{L}_E = T_E - M_*\Phi_E$, which depends on the external coordinates of the stellar fluid elements with respect to the black hole frame $\vec{X}$. Specifically, for mass of the star M_*, $T_E = \frac{1}{2}M_*\dot{X}_i\dot{X}_j$ is the kinetic energy and $\Phi_E(\vec{X}_0)$ is the Newtonian gravitational field of the black hole evaluated at the center of mass of the star. The coupling term is $\mathcal{L}_C = \frac{1}{2}\mathcal{M}_*C_{ij}q_{ik}q_{jk}$, where $C_{ij}(\vec{X}_0) = \partial_i\partial_j\Phi_E$ is defined in terms of derivatives with respect to the black hole frame coordinates $\vec{X}$ and evaluated at the center of mass of the star $\vec{X}_0$. This trace-free, symmetric tensor represents the tidal force due to the black hole. The quantity $\mathcal{M}_* = \frac{1}{3}\int \hat{r}_i\hat{r}_j dM$ is the scalar quadrupole moment of the initial configuration for the star. The internal term is $\mathcal{L}_I = T_I - \Omega - U$, where the kinetic energy $T_I = \frac{1}{2}\mathcal{M}_*\dot{q}_{ij}\dot{q}_{ij}$, the self-gravitational potential $\Omega = -\frac{1}{2}\iint dMdM'[(r_i - r_i')(r_i - r_i')]^{-1/2}$, and the total internal compression energy U are all in terms of the internal coordinates, only. The Lagrangian equations of motion for the fluid elements are decomposed into evolution equations for the external momentum $P_i = M_*\dot{X}_i$,

$$\dot{P}_i = -M_*\frac{\partial\Phi_E}{\partial X_i} + \frac{1}{2}\mathcal{M}_*q_{lk}q_{jk}\frac{\partial C_{lk}}{\partial X_i}, \tag{11}$$

and for the internal momentum $p_{ij} = \mathcal{M}_*\dot{q}_{ij}$,

$$\dot{p}_{ij} = \mathcal{M}_*C_{ik}q_{kj} + \Pi q_{ji}^{-1} + \Omega_{ik}q_{jk}^{-1}, \tag{12}$$

where $\Pi = \int PdV$ is the volume integral over pressure. These evolution equations depend on the assumptions for the trajectory of the star $\vec{X}_0$, the equation of state which determines the total internal compression energy U, and the total power input due to nuclear reactions $\dot{Q}$. In this model, the timescales are too short for radiative transfer. Assuming a polytropic gas with index γ, the evolution of q_{ij} is specified by

$$\ddot{q}_{ij} = C_{ik}q_{kj} + (\gamma - 1)\Psi\|q\|^{1-\gamma}q_{ji}^{-1} - \frac{3}{2}(\gamma - 1)\Psi_*\int_0^\infty \frac{du}{\Delta}(\mathbb{S} + u\mathbb{1})_{ki}^{-1}q_{kj}, \tag{13}$$

where $S_{ij} = q_{ik}q_{jk}$, $\Delta = \|\mathbb{S} = u\mathbb{1}\|^{1/2}$, $\Psi_* = \mathcal{M}^{-1}U_*$ is the initial (equilibrium) state of the quantity that couples the thermonuclear release and evolves as

$$\dot{\Psi} = \mathcal{M}_*^{-1}\|q\|^{\gamma-1}\dot{Q}. \tag{14}$$

Without nuclear reactions, $\dot{Q} = 0$.

Consider the Carter & Luminet model for an adiabatic gas and in the absence of nuclear reactions. Given the analytic form of the evolution for the star in Eq. (13), it is straightfoward to apply approximations in regions where the star is close to pericenter. For encounters where $\beta \gg 1$, when the star reaches a pericenter much closer to the black hole than the tidal radius, the tidal forces are much greater than those due to the pressure and self-gravity of the gas. The star is approximately particles in free-fall in the black hole's gravitational field, evolved as

$$\ddot{q}_{ij} \approx C_{ik} q_{kj}. \tag{15}$$

Neglecting the minor corrections due to the star's initial rotation, the motion is roughly invariant under reflection in the orbital plane. Specifically, the motion is decoupled between the action orthogonal and within the orbital plane. For orthogonal motion, Eq. (15) is $\ddot{q} = -Cq$, where $q = q_{33}$, and the star "pancakes", compresses and flattens in the orbital plane. Once sufficiently flattened, the pressure forces dominate over the tidal forces, reversing the compression. At this point, under the adiabatic approximation, where there is no nuclear energy release, all of the internal energy of the gas is converted from the kinetic energy of the vertical, pancake motion. From this, the maximum temperature $\Theta_{\rm max}$ increases from its initial temperature Θ_* by $\Theta_{\rm max} \approx \beta^2 \Theta_*$. Assuming a polytropic equation of state with adiabatic index $\gamma = 5/3$, we have the following. The maximum increase in the initial central density ρ_* is given by $\rho_{\rm max} \approx \beta^3 \rho_*$. The timescale for maximum compression $\tau_{\rm max}$, in terms of stellar dynamical timescale τ_*, is $\tau_{\rm max} \approx \beta^{-4} \tau_*$. The time delay $t_{\rm max}$ between the instant of passage through pericienter and the instant of maximum compression is $t_{\rm max} \approx \beta^{-2} \tau_*$. Extending the affine model for nuclear reactions, Luminet and Pichon (1989b,a) explore the process of nucleosynthesis in deep, plunging disruptions and its impact on the dynamics of the stellar debris and the isotropic enrichment of the interstellar medium.

The limitations of the affine model are due to its idealized hydrodynamical description of the star and linearization of the properties of the black hole's tidal field. It is appropriate for the central bulk of the star, but not the outer layers (which are less dense than the center). Because the hydrodynamic description of compression is simple, it does not account for non-linear terms in the tide, which will produce shock waves, reducing the increase in density and temperature. Nonetheless, the affine model is qualitatively useful for nuclear reactions and, as we discuss in the next section, extensions to this model can deliver interesting input in the relativistic corrections to the tide and orbit.

3 Treatment of Relativistic Effects in Simulations

Numerical techniques used in simulating the relativistic effects of the black hole are generally divided into two categories: local simulations centered on the star and global simulations with a computational domain that includes the black hole. For local simulations, the general relativistic formalism was first presented in the context of the affine model (Luminet and Marck 1983, 1985; Luminet and Carter 1986; Marck et al. 1996). This framework was later used in grid-based simulations of the disruption phase (Diener et al. 1995; Frolov et al. 1994; Cheng and Evans 2013; Cheng and Bogdanović 2014). In particular, by assuming that the star's radius is much smaller than the separation from the black hole, the stellar trajectory is approximately a geodesic. For a non-spinning black hole of mass M described by the Schwarzschild spacetime with coordinates $X^\mu = \{t, r, \theta, \phi\}$, to first order, the equations of motion are

$$\frac{dt}{d\tau} = \frac{\epsilon}{1 - \frac{2M}{r}},$$

$$\frac{d\phi}{d\tau} = \frac{l}{r^2},$$
$$\left(\frac{dr}{d\tau}\right)^2 = \epsilon^2 - \left(1 - \frac{2M}{r}\right)\left(1 + \frac{l^2}{r^2}\right), \tag{16}$$

for specific orbital energy ϵ and angular momentum l. Within this approximation, a local inertial frame, sometimes called the Fermi normal coordinate frame, is constructed with respect to the star's center of mass (Marck 1983). The tidal potential due to the black hole is represented as

$$\Phi_i^{\text{tidal}} = \frac{1}{2}C_{ij}x^i x^j + \frac{1}{6}C_{ijk}x^i x^j x^k + \frac{1}{24}C_{ijkl}x^i x^j x^k x^l + \cdots, \tag{17}$$

where C_{ij}, C_{ijk}, and C_{ijkl} are the quadrupole, octupole, and hexadecapole tidal tensors etc. and $x^i = \{\tau, x, y, z\}$ represents the coordinates in the star's local reference frame. This is a convenient form because the black hole tide is simply applied to the star through external accelerations in a locally flat spacetime. This enables the use of a Newtonian hydrodynamics code with Newtonian self-gravity and high-order relativistic corrections to the black hole tide. The tidal tensors are obtained from the curvature or Riemann tensor $R_{\mu\nu\alpha\beta}$ in the black hole frame. For a known spacetime such as Schwarzschild or Kerr, $R_{\mu\nu\alpha\beta}$ is given. Using the Fermi normal frame vectors or tetrad $\lambda^{\mu}_{\ i}$, the Riemann tensor in the black hole frame is transformed into the local inertial frame, where the tidal tensors can be calculated and added to the tidal potential. For example, for the quadrupole tensor C_{ij} we have

$$C_{ij} \equiv R_{\tau i \tau j} = R_{\mu\nu\alpha\beta}\lambda^{\mu}_{\ \tau}\lambda^{\nu}_{\ i}\lambda^{\alpha}_{\ \tau}\lambda^{\beta}_{\ j}, \tag{18}$$

where R_{ijkl} is the Riemann tensor transformed into the local star's frame. Extensions to higher-order tides and spin effects are straightforward in this Fermi normal frame (Ishii et al. 2005; Cheng and Evans 2013).

Relativistic corrections to the tides lead to more disruptive encounters than the Newtonian approximation for a given encounter strength (Frolov et al. 1994; Diener et al. 1995). The relativistic effects are seen in the return rate through a slower rise and lower peak with respect to the Newtonian encounters (Cheng and Bogdanović 2014). Local simulations of the disruption phase are especially useful because at the end of the simulation, the orbital parameters of the debris are on-hand to either make predictions about the disc or use as initial conditions for a fully global simulation of the interacting debris streams and the formation of the disc (Shiokawa et al. 2015).

The full general relativistic form of the basic conservative hydrodynamics equations in a black hole spacetime defined by coordinate X^{μ} and black hole metric $g_{\mu\nu}$ is given by

$$\partial_t(\sqrt{-g}\rho u^t) + \partial_i(\sqrt{-g}\rho u^i) = 0,$$
$$\partial_t(\sqrt{-g}T^t_{\ \mu}) + \partial_i(\sqrt{-g}T^i_{\ \mu}) = \sqrt{-g}\Gamma^{\sigma}_{\ \rho\mu}T^{\rho}_{\ \sigma}, \tag{19}$$

in terms of four-velocity u^{μ}, stress-energy tensor $T^{\mu\nu}$, connection coefficient $\Gamma^{\sigma}_{\ \rho\mu}$, and determinant of the black hole metric g. Extending to ideal GRMHD (Gammie et al. 2003; White et al. 2016), the stress-energy tensor is written in terms of the magnetic field components B^i, where

$$T^{\mu\nu} = (\rho h + b_\lambda b^\lambda)u^\mu u^\nu + \left(P + \frac{1}{2}b_\lambda b^\lambda\right)g^{\mu\nu} - b^\mu b^\nu, \tag{20}$$

for specific enthalpy h and contravariant magnetic field components $b^0 = g_{i\mu} B^i u^\mu$ and $b^i = (B^i + b^0 u^i)/u^0$, which define the dual of the electromagnetic field tensor as

$$^*F^{\mu\nu} = b^\mu u^\nu - b^\nu u^\mu. \tag{21}$$

Then, in addition to Eq. (19), the conservative equations for evolving the hydrodynamic variables as well as the magnetic field are

$$\partial_t(\sqrt{-g} B^i) + \partial_j(\sqrt{-g}\, ^*F^{ij}) = 0. \tag{22}$$

Note that while the magnetic field B^i is evolved, the electric field E^i is inferred from B^i and the fluid velocities u^μ through Eq. (21) and the definition,

$$^*F^{ij} = \begin{pmatrix} 0 & -B^1 & -B^2 & -B^3 \\ B^1 & 0 & E^3 & -E^2 \\ B^2 & -E^3 & 0 & E^1 \\ B^3 & E^2 & -E^1 & 0 \end{pmatrix}. \tag{23}$$

To-date, there are few tidal disruption event simulations that solve the full general relativistic set of equations in the black hole frame. This is because the computational expense for the hydrodynamics scales with black hole to star mass ratio. Most grid-based black hole frame codes leverage the computational expense with a small black hole to star mass ratio. In particular, white dwarf disruptions by intermediate mass black holes are chosen because of the computationally tractable length and timescale as well as the fact that the parameters are well within the relativistic regime without the use of a large penetration factor β. To further decrease the computational expense, Shiokawa et al. (2015) consider the problem in a two-step process where post-disruption simulation data from Cheng and Evans (2013) is used as initial data for a global GRHD simulation of the accretion flow of the bound stream. This technique is also implemented in Sądowski et al. (2016), with initial data from Tejeda and Rosswog (2013), in a study with GRMHD. Both studies use a fixed Schwarzschild spacetime. The initial data represents the results of a hydrodynamic treatment of the star under the influence of its own gravity and the black hole.

While the stellar self-gravity is important for the disruption process, it is much less so very close to the black hole. A few grid-based studies which focus on deeply relativistic encounters neglect self-gravity in order to save the computational expense for different physical effects (Haas et al. 2012; East 2014; Anninos et al. 2018). Haas et al. (2012) use a fully general relativistic hydrodynamics code with an adaptive mesh refinement technique where both the star and the black hole share the computational domain. Several refinement levels follow the star as it passes by a spinning black hole. It is found that after disruption, the black hole spin disperses the debris scattering it far from the orbital plane. East (2014) models a more extreme mass ratio while making use of an error-correcting technique which reduces the expense of calculating the black hole and star's gravity. Both codes used in Haas et al. (2012) and East (2014) directly calculate the gravitational wave form. The former provides the waveform for a white dwarf disrupted by an intermediate mass black hole. The latter study focuses on generating the waveform for a main sequence star disrupted by a 10^6 solar mass black hole. Anninos et al. (2018) use a fully general relativistic hydrodynamics code in the black hole frame, but make use of a moving mesh technique as well as adaptive mesh refinement in order to focus the computational domain on the star only. This decrease in computational expense allows more resources dedicated to resolving the star and coupling a nuclear network to the hydrodynamic evolution. The results of this study, on thermonuclear reactions in tidal disruption events is discussed below.

The question of what gravitational wave form is generated from a tidal disruption event has been the focus of a wide variety of studies using the affine model (Casalvieri et al. 2006), smoothed-particle hydrodynamics (Rosswog 2009; Kobayashi et al. 2004), and full general relativistic hydrodynamics (Haas et al. 2012; East 2014; Anninos et al. 2018). Tidal disruption events generate a burst-like gravitational wave with low frequencies of $0.1 - 0.7$ mHz and strain amplitudes of $h \sim 10^{-22}$ at a source distance of 10 Mpc. More recently, Toscani et al. (2019) have also studied the gravitational wave emission generated by hydrodynamical instabilities in the accreting torus, after the disruption proper. The typical frequency in this case is similar to the case of disruption (in the mHz regime), the strain is lower $h \sim 10^{-24}$, but rather than an instantaneous burst, in this case the emission is expected to be produced over a few cycles, which may help detection.

Following the disruption of a star, the debris orbits around the black hole with apsidal and nodal precession due to general relativity (Merritt 2013), the rate of which dictates how fast the debris stream can form a disc from self-intersection. As discussed previously, simulating the motion and evolution of debris stream around the black hole with hydrodynamical codes is computationally inefficient even when such codes are Newtonian. However, under the assumption that hydrodynamical effects are unimportant for the debris orbital motion, the debris can also be approximated to follow geodesics around the black hole until debris stream self-intersection happens. In many cases, even semi-analytical calculations only considering the first-order precession rate are sufficient (Dai et al. 2015; Guillochon and Ramirez-Ruiz 2015; Bonnerot et al. 2017b). One can also more accurately calculate the geodesics using a numerical approach. For example, Dai et al. (2013) calculated the debris motion in full general relativity around a spinning black hole by solving the geodesic equations of motions from Fuerst and Wu (2004).

A computationally efficient way of treating general-relativistic effects in SPH simulations consists in using a gravitational potential that incorporates the most important ones. One possibility is to use pseudo-Newtonian potentials that reproduce the rate of relativistic apsidal precession around the black hole. For instance, the potential developed by Wegg (2012) reproduces this feature accurately for highly-eccentric orbits while that of Tejeda and Rosswog (2013) provides an exact treatment of the gas trajectories around a non-rotating black hole. These approaches have been exploited to study partial stellar disruptions (Gafton et al. 2015) and disc formation (Hayasaki et al. 2013; Bonnerot et al. 2016b). In the latter case, a correct evaluation of the apsidal angle is fundamental since it is at the origin of the self-crossing shock that initiates this process. Spin-related effects can be taken into account in a similar way by adding post-Newtonian acceleration terms to the Keplerian component as described by Blanchet (2006). This method has been used to study the impact of Lense-Thirring precession on the stream-stream collision and ensuing circularization (Hayasaki et al. 2016). In the above approaches, note that general relativity is only taken into account to specify the gravity of the black hole while the hydrodynamics still assumes a flat space-time.

A full general-relativistic treatment of the gas dynamics is also possible using SPH (see the recent implementation in the PHANTOM code, Liptai and Price 2019), but it requires a rewrite of the equations of gas dynamics to account for the space-time curvature, as detailed in Rosswog (2009). The earliest application of fully general relativistic SPH to tidal disruption events was provided by Laguna et al. (1993a), using the code developed by Laguna et al. (1993b) based on the approach of Kheyfets et al. (1990). This formulation, however, was not in conservative form (as the one described above in Sect. 2.1) and had problems in resolving shocks efficiently. More modern SPH versions have been developed (Rosswog 2009; Liptai and Price 2019) that are fully conservative (thus exploiting the major advantage of an SPH formulation) and are able to handle shocks much more efficiently. The application of such methods to TDE has been limited, though. Note that this approach assumes that the metric is entirely specified by the black hole and not affected by the mass of the gas that evolves around it. Of course, this is a very good approximation for tidal disruptions where the compact object is always much more massive that the stellar debris. This technique has been used by Tejeda et al. (2017) and more recently by Gafton and Rosswog (2019) and Liptai et al. (2019) to simulate the disruption of a star by a Schwarzschild and a rotating black hole. In these investigations, note that the gas self-gravity remains treated in a Newtonian way that, although generally being a excellent approximation, is strictly speaking not entirely self-consistent.

4 Treatment of Thermodynamics, Radiation and Nuclear Reactions

In the description of simulation methods above, the primary goal of previous work has been to study the dynamics of the stellar gas through gravitational interactions. Generally, the stellar model is assumed to be a non-rotating polytrope (except in recent papers, such as Sacchi and Lodato 2019 and Golightly et al. 2019a), where the distinguishing characteristic between main sequence stars and white dwarfs, for example, is the difference in adiabatic index. Radiation has not been included in most simulations although the importance of its role is still an open question. While the interaction of the debris streams may reduce the high rate of the funneling of gas around the black hole, the efficiency of accretion is not well understood. As new multi-wavelength observations are likely to deliver detailed light curves, it is increasingly important to have theoretical models which explain the spectral features. This calls for new attention to the description of the stars themselves, their layers and nuclear constituents, as well as a direct treatment of radiative processes helpful in understanding the nature of multi-wavelength emission in TDE.

The state-of-the art includes few studies of the thermodynamics beyond ideal gases. Differences in the choice of adiabatic index of the gas lead to changes in the density structure of the star, which in turn affects the shape of the debris return rate (Lodato et al. 2009; Guillochon and Ramirez-Ruiz 2013). If the light curve of TDE traces the return rate, then these theoretical models could help determine the types of stars that are disrupted. In detailed stellar structure studies, MacLeod et al. (2012, 2013) use initial configurations from a MESA stellar evolution code while evolving the star as a composite polytrope. This enables a simulation code to model tidal stripping of material from envelopes and partially disrupt cores.

Local radiation-hydrodynamics simulations have been carried out to evaluate the amount of radiative losses experienced by the heated gas after it has passed through the self-crossing shock at the origin of disc formation. In the early investigation of Kim et al. (1999), radiation transport is approximated by incorporating a volume cooling rate. The more recent work by Jiang et al. (2016) makes use of a more accurate treatment that consists in solving the frequency-independent radiation-hydrodynamics equations.

Furthermore, for the evolution of the debris disc, choices in the adiabatic index of the gas lead to dramatically different configurations: thin, circularized (isothermal) and puffy, elliptical (adiabatic) (Bonnerot et al. 2016b; Hayasaki et al. 2016). For GRMHD codes used to simulate the formation of accretion flow (Shiokawa et al. 2015; Sądowski et al. 2016), the gas is usually assumed to have an adiabatic index of 4/3 (relativistic) or 5/3 (Newtonian).

However, when simulating accretion discs that are optically thick where the dynamics of the gas and radiation are coupled, such as in the super-Eddington accretion regime, the sensitive interaction between gas and radiation needs to be captured. In order to address this, GR-radiation-MHD (GRRMHD) codes have been developed in recent years. The state-of-the art codes use the flux-limited diffusion approximation which only allows isotropic emission relative to the fluid frame (e.g., Ohsuga et al. 2005) or the more accurate M1 closure (Sądowski et al. 2014; McKinney et al. 2014; Dai et al. 2018; Curd and Narayan 2019) for the gas-radiation coupling. The stress-energy tensor evolved in the equations in these codes does not only contain the electromagnetic energy component and the matter component as in Eq. (20), but also a radiation stress-energy component. The dynamics of gas and radiation are evolved in parallel and coupled to each other using opacity terms addressing electron Thomson and Compton scattering as well as various absorption processes.

Motivated by Carter and Luminet (1982) and Luminet and Pichon (1989b,a), several numerical studies investigate the interesting observational consequences once the star reaches pericenter. While most probe the effects at very close encounters with the black hole, a few focus on weakly disruptive encounters in the relativistic regime (Wilson and Mathews 2004; Dearborn et al. 2005). The main focus of studies is modeling the nuclear reactions triggered by severe compression in the orbital plane leading to an energy release comparable to Type Ia supernovae, particularly with white dwarf disruptions by intermediate mass black holes. The threshold for nuclear ignition is set by the star's gravitational binding energy. Given a rapid increase in temperature and energy above this threshold within a short timescale, the star is considered ignited. Several numerical studies model the temperature increase due to the shock after severe compression (Kobayashi et al. 2004; Brassart and Luminet 2008; Rosswog et al. 2009; Guillochon et al. 2009; Haas et al. 2012; Tanikawa 2018b,a; Kawana et al. 2018; Anninos et al. 2018; Kagaya et al. 2019). Rosswog et al. (2009) model the compression and expansion during the disruption phase while tracking the energy and nucleosynthesis of the debris. To track nucleosynthesis, they implement a HELMHOLTZ equation of state with flexible specification of the chemical composition of the gas and couple an α-network with seven nuclear species to the hydrodynamic evolution. Several models meet the conditions for nuclear ignition with sizeable iron-group injection into the outflow. However, Tanikawa (2018a) finds a numerical limit to accurately resolve nuclear flows and notes that ignition is sensitive to resolution. In a detailed follow-up investigation with general relativistic hydrodynamics, initial data from MESA, a HELMHOLTZ equation of state, and a large nuclear network, Anninos et al. (2018) show ignition with diverse thermonuclear environments and in particular, show calcium-group injection in the outflow in addition to the iron-group.

5 Simulating the Various Stages of TDE

5.1 Disruption

In this section, we provide an overview of the evolution of simulations of the disruption phase in TDE, emphasizing the numerical advances and innovations obtained in the various studies. Not all of the papers investigating the disruption phase are reported here (this is the subject of the Disruption Chapter in this book), but only those that have provided significant improvements from the point of view of the numerical techniques.

After the seminal simulations by Nolthenius and Katz (1982) and Bicknell and Gingold (1983), the first simulations of the disruption process with reasonable resolution were performed by Evans and Kochanek (1989). This paper is particularly important, as it sets the standard for the numerical simulations of the disruption process. They simulate the disruption of a $1 M_\odot$ star by a $10^6 M_\odot$ black hole, with a penetration factor $\beta = 1$, on a parabolic orbit. They were the first investigators to derive the distribution of mechanical energies of the debris, apparently confirming spectacularly the prediction of a flat energy distribution by Rees (1988) (although their plots were in log scale, which hides possible smaller changes in the distribution) and they obtain the predicted fallback rate, again confirming Rees' prediction of a $t^{-5/3}$ decline. From the numerical point of view, they use $\sim 10^4$ particles and perform a convergence test, showing that for this stage of the process, the simulations are converged in SPH.

Laguna et al. (1993a) were the first to introduce relativistic effects in SPH simulations of the disruption process (see Sect. 3) and simulated strongly penetrating disruptions of a $1 M_\odot$ star by a $10^6 M_\odot$ Schwarzschild black hole. Their numerical resolution, however, was quite low, even for the standards of those days (they used 10^3 particles).

In a series of relatively poorly known, but pioneering papers, Khokhlov et al. (1993b), Khokhlov et al. (1993a) and Diener et al. (1997) perform some of the first Eulerian simulations of TDE. They simulate the disruption of polytropic stars, with different polytropic indices, on a fixed and on a moving mesh. Khokhlov et al. (1993b) simulate weak encounters that do not lead to disruption. Khokhlov et al. (1993a) simulate partial and total disruptions and estimate the critical penetration parameter to have disruption as a function of stellar structure. They find $\beta_{\rm crit} \sim 0.65 - 0.75$ for $\gamma = 5/3$, and $\beta_{\rm crit} \sim 1.4 - 1.6$ for $\gamma = 4/3$. Subsequent simulations with much higher resolution and AMR techniques by Guillochon and Ramirez-Ruiz (2013) later corrected these values to $\beta_{\rm crit} = 0.9$ and $\beta_{\rm crit} = 1.85$, for the two cases, respectively. Diener et al. (1997) introduce full relativistic effects on a fixed metric in their Eulerian code, such that the debris evolves following Kerr geodesic. Interestingly, despite their pioneering quality, such papers did not achieve significant success in the community, probably because their main outcome was essentially an estimate of the amount of mass, energy and angular momentum transferred in the encounter.

Ayal et al. (2000) try to follow the evolution of the tidal debris both during the disruption phase and during the subsequent stream evolution. They use a post-Newtonian approach to include some relativistic effects into an SPH method. However, their simulations suffer significantly from poor resolution. While the number of particles used is similar, or slightly lower than those of their contemporaries (they employ $\sim 10^3$ particles), the resolution issue comes about because they want to study the post-disruption phase, where the debris significantly spread and reach very low densities, thus requiring a very large smoothing length in SPH. They attempt to resolve this issue by employing the so-called "particle splitting" technique that however introduces significant noise in the particle distribution and has been mostly abandoned in more modern SPH studies.

Two "modern" papers investigating the disruption phase are those of Lodato et al. (2009) and of Guillochon and Ramirez-Ruiz (2013). Both papers investigate the disruption of a solar mass star by a $10^6 M_\odot$ black hole essentially using advanced versions of the codes employed in the past. Lodato et al. (2009) use a modern SPH code with a full Lagrangan formulation, employing adequate viscosity switches and a variable smoothing length (that was not always used in the past) and employ a reasonably large number of particles (10^5), thus reaching adequate numerical resolution. Guillochon and Ramirez-Ruiz (2013) use the FLASH code, with AMR. Both use Newtonian gravity. Their main results can be summarized as follows. Lodato et al. (2009) revisit the conclusions of Evans and Kochanek (1989) noting that the energy distribution of the debris is actually not exactly flat for the most bound debris, leading to deviations to the $t^{-5/3}$ law at early times in ways that depend on the internal structure of the star. Such deviations can be even predicted analytically, although the analytical calculations need some adjustments in order to reproduce the numerical results. Guillochon and Ramirez-Ruiz (2013) essentially confirm the numerical results of Lodato et al. (2009) but extend their analysis to different penetration factors (Lodato et al. 2009 only considered the $\beta = 1$ case). They also simulate partial disruptions, and give more solid estimates of the amount of mass loss and on the critical penetration for disruption, with respect to the earlier analysis of Khokhlov et al. (1993a) and Khokhlov et al. (1993b). The close agreement of the results of these two papers despite the different numerical technique used, and the convergence tests presented in them, allow to put solid constraints on the outcome of the disruption phase (which is anyway the simplest phase to simulate numerically).

More recently, Mainetti et al. (2017) compare different codes to determine the critical penetration for total disruption. They use two different implementations of SPH, a moving mesh code and also compare their results to those obtained with FLASH by Guillochon and Ramirez-Ruiz (2013). They find a broad agreement between the various codes and set $\beta_{\rm crit} = 0.92$ and 2.01 for the two cases $\gamma = 5/3$ and $\gamma = 4/3$, respectively.

From the numerical point of view, a couple of papers have recently included the stellar magnetic field in MHD simulations. This has been done both with grid based codes by Guillochon and McCourt (2017) and with SPH (Bonnerot et al. 2017a). Interestingly, unlike earlier calculations of magnetic field amplification during tidal encounters, the results of these two papers are consistent with each other, probably due to the enhanced divergence cleaning method used in the SPH code used by Bonnerot et al. (2017a).

Finally, some simulations now go beyond simple polytropic models for the star and use instead MESA based stellar structures. One example is Golightly et al. (2019b) who introduced MESA based density profiles in their SPH simulations, emphasising their role in the shape of the fallback rate. Law-Smith et al. (2019) introduce realistic stellar models in their FLASH simulations, which allows to track the compositional evolution of the debris stream.

For deeply penetrating events, the tidal compression can induce nuclear reactions in the disrupted star and many studies have studied this problem, as discussed above in Sect. 4.

5.2 Disc Formation

Following the stellar disruption, the debris evolves into an elongated stream that keeps orbiting the black hole with approximately half of it being bound and falling back towards the compact object while the rest gets unbound and escapes. When the bound gas returns to pericenter, it starts forming an accretion disc around the black hole. This process is initiated by a shock occurring when the tip of the stream intersects with the later-arriving material due to a modification of its trajectory by relativistic apsidal precession. The physics at play in the phases of stream evolution and subsequent disc formation are described in detail in the Formation of the Accretion Flow Chapter in this book and we focus here only on the various numerical methods used to study them.

Numerical studies of the stream evolution have been carried out until the first debris falls back to pericenter using both SPH (Coughlin and Nixon 2015) and grid-based codes (Guillochon et al. 2014). The former investigation found that the stream can fragment into self-gravitating clumps. While the perturbation at the origin of this collapse is likely numerical in this work, several processes exist that can sustain this fragmentation including periodic density variations inherited from the passage of the star at pericenter (Coughlin et al. 2016). Such simulations do not require a large computational cost.

Following the process of disc formation from this stream when it comes back to pericenter is, however, very numerically challenging. Most importantly, an accurate numerical treatment of the passage of the stream at pericenter requires one to properly resolve its thin transverse profile. Reaching such a resolution for each part of the stream represents in itself a large computational overhead because of its very elongated profile. Also, because of such elongation, the evolutionary time at the head and tail of the stream differ substantially, so that while the time-step criterion is usually determined at pericenter, the system needs to be evolved over the long dynamical time-scale at apocenter. In SPH, such disparity of time-scales is alleviated substantially by the employment of individual particle time-steps (Bate et al. 1995). Remarkably, the issue of numerical resolution appears to be present independently of the numerical method used. Using SPH to explore this phase of evolution requires an unmanageably large number of particles. It has been tackled early on by Ayal et al. (2000), but it is now clear that their main results are flawed by a resolution much lower than would be required to accurately follow this process. Perhaps the most important artifact is that the majority (75%) of the stellar mass gets unbound during the passage of the stream at pericenter. This effect is likely due to artificial dissipation taking place at this location due to the unresolved stream thickness. It is also unsurprising because they used a few thousand SPH particles while more recent (unpublished) investigations find that a thousand times more particles is still not enough to get rid of similar spurious numerical effects. Grid-based codes encounter the same type of numerical issues. While this numerical method could in principle study this process by using a more refined grid near pericenter, such a simulation has not yet been carried out, possibly due to the too large computational cost required. This issue may be resolved by the advent of new dedicated computational methods such as those discussed in the Future Modeling Chapter by Krolik et al in this book but this has so far not been the case.

One method used to gain insight into the disc formation process while avoiding the above computational burden was to study the self-crossing shock by means of local simulations. The passage of the stream near pericenter is followed analytically assuming that its orbit remains ballistic, which can be carried out by taking into account relativistic apsidal precession but also spin-related effects. This preliminary calculation allows to determine the properties of the two components of the stream when they reach the collision point, which is then used to initialize the local simulation of the self-crossing shock. This approach was used early on by Kim et al. (1999), who used an approximate treatment of radiation to evaluate the amount of thermal energy injected into the shocked gas that is able to diffuse out. More recently, this technique has been employed assuming gas adiabaticity (Lu and Bonnerot 2020) and making use of a more accurate algorithm for radiation transport (Jiang et al. 2016).

Global simulations of the disc formation process have been performed but all of them rely on important simplifications made in order to alleviate the computational burden discussed above that arises in the physically realistic case. Most of these simplifications aim at artificially reducing the size of the stream, either by decreasing the black hole mass to $M_{\rm bh} \approx 10^3 M_\odot$ or considering stars on elliptical rather than parabolic orbits with eccentricities in the range $0.8 \le e \le 0.97$. Such simulations have been carried out using both grid-based codes and the SPH technique and using various treatments of general relativistic effects from full GR to simpler gravitational potentials (as discussed in more details in Sect. 3). Early simulations, including those by Rosswog (2009), have investigated the early phases of disc formation, although not in a detailed manner. The first one to follow the disc formation process in its entirety has been carried out by Hayasaki et al. (2013) who consider an eccentric stellar trajectory with $e = 0.8$ and $\beta = 5$ and assumed an isothermal evolution for the gas to approximate efficient gas cooling. This work was improved by Bonnerot et al. (2016b) who focused on more eccentric stellar orbits with $e = 0.95$ and a grazing penetration factor $\beta = 1$. In addition, this study investigated the influence of inefficient gas cooling corresponding to an adiabatic evolution of the gas. These works treat relativistic apsidal precession through the use of pseudo-Newtonian potentials. The hydrodynamics of disc formation has also been investigated with a grid-based code that treats general-relativistic effects exactly by Shiokawa et al. (2015) considering a parabolic stellar trajectory around a black hole of mass $M_{\mathrm{bh}} = 500 M_{\odot}$ and assuming adiabaticity. The results obtained are similar to the earlier ones concerning the timescale and final outcome of the disc formation process. More recently, Bonnerot and Lu (2019) have used another simplifying strategy that consists of modelling the self-crossing shock by an injection of outflowing gas inside their computational domain, which allows them to follow disc formation for astrophysically realistic parameters of the problem, most importantly a black hole mass $M_{\mathrm{bh}} = 2.5 \times 10^6 M_{\odot}$ and a parabolic stellar orbit.

Two further improvements have been made more recently. The first shows that the influence of the black hole spin can delay the initial self-crossing shock through Lense-Thirring precession. This additional effect has been investigated by Hayasaki et al. (2016) who consider bound stars and a treatment of spin-related effects through a gravitational potential with post-Newtonian corrections. The second advance is the inclusion of magnetic fields inside the stellar debris, which was carried out in the numerical work by Sądowski et al. (2016). These authors use a hybrid approach that consists of simulating the stellar disruption with SPH and the ensuing disc formation process using a grid-based code after adding magnetic fields to the debris. Most importantly, they were able to study the relative influence of magnetic fields and gas dynamics at producing the turbulence at the origin of angular momentum transport through the newly-formed disc.

Recently, Chan et al. (2019) have simulated the interaction of the debris stream with a pre-existing accretion disc, when the disruption occurs in an AGN. They have used the grid-based code Athena++ and find that the stream can excite significant inflow in the disc (strongly super-Eddington), while the emergent radiation is expected to be close to the Eddington value.

5.3 Accretion

After a disc is formed out of stellar debris, viscous or magnetic processes in the disc can effectively transport the angular momentum of gas outwards (Shakura and Sunyaev 1973; Balbus and Hawley 1991), allowing it to accrete onto the black hole and produce emission. If the black hole mass is smaller than $\sim 10^8\ M_{\odot}$ and the disc formation process is prompt, the accretion rate is expected to resemble the fallback rate, which exceeds the Eddington accretion rate for $\sim 1\,\mathrm{yr}$ around peak and then decays to below the Eddington level. We refer to the Chapter on Accretion Physics in Tidal Disruption Events by Dai et al in this book for the details of physics involved in this process and only focus on the development of numerical studies on TDE accretion here.

While various hydrodynamical simulations have been employed to improve our understanding on the TDE disruption and disc formation processes as discussed in 5.1 and 5.2, there have been much fewer numerical studies on the TDE accretion process. One main reason for this is that we do not know the exact initial disc configuration to use, since currently we still cannot fully follow the long-term debris stream evolution until a disc is fully assembled without making simplifications. For example, in the simulations by Shiokawa et al. (2015) and Bonnerot et al. (2016b), the assembled discs have different eccentricities and structures, due to the different initial set-up used to reduce the dynamical range. Also, while local simulations of debris stream-stream collision such as Jiang et al. (2016) show that such collisions can induce outflows and therefore the accretion rate could be lower than the fallback rate, an accurate measure of the lost debris fraction during the disc formation process can only be obtained using global simulations which include the black hole gravity.

The other reason for the slow progress is that numerical tools needed to study super-Eddington accretion, which is assumed to happen around the peak of most TDEs, have not been developed until a few years ago (e.g., Ohsuga et al. 2005; Sądowski et al. 2014; Jiang et al. 2014; McKinney et al. 2014). Unlike the accretion discs at low accretion rates, super-Eddington accretion discs, puffed up by large radiation pressure, are geometrically and optically thick. Therefore, the disc gas dynamics is affected by radiation, and the radiation emitted by disc inner region is heavily reprocessed in the outer disc and optically thick wind. While magnetohydrodynamical (MHD) codes are needed to study optically thin accretion discs, more advanced radiation-MHD (RMHD) codes are further needed to study optically thick super-Eddington discs. There are a few differences between these codes. Some codes are pseudo-Newtonian (Ohsuga et al. 2005; Jiang et al. 2014) while others are general-relativistic (Sądowski et al. 2014; McKinney et al. 2014). Also the radiative transport scheme used in these codes are different (see Sect. 4). However, the simulations of super-Eddington discs using these codes have produced qualitatively consistent results, showing that strong outflows can be produced and the escaped luminosity can greatly exceed the Eddington limit (e.g., Jiang et al. 2014; McKinney et al. 2015), which are very different from the predictions from analytical modelings (e.g., Abramowicz et al. 1988).

Recently Dai et al. (2018) perform the simulation of a circular, super-Eddington disc mimicking discs formed in TDE using the code HARMRAD (McKinney et al. 2014), assuming that such discs can circularize promptly and strong magnetic field threads the disc. The simulation results, post processed with another state-of-the-art Monte-Carlo radiative transfer code SEDONA(Kasen et al. 2006; Roth et al. 2016), show that the anisotropic super-Eddington disc-wind structure can naturally lead to X-ray dominant or optical dominant emission from different viewing angles. Curd and Narayan (2019), using the code KORAL (Sądowski et al. 2014), further study TDE super-Eddington discs around black holes with different spin parameters or magnetic flux strength.

It is also worth noting that the current generation of GRRMHD codes are extremely computationally expensive. Therefore, it is inefficient to use such codes to follow the TDE disc evolution over the fallback time scale. Also, as the accretion rate approaches the Eddington accretion rate, the disc structure will become thin and the viscous timescale will grow longer, putting more challenges to simulating the transition from the super-Eddington regime to the thin disc regime. Here, one may instead just adopt the analytical solution of standard thin discs, although the classical question on whether or how such discs withstand thermal instability still remains to be solved by future numerical work.

6 Conclusions and Outlook

Numerical simulations have played a major role over the years in sharpening our understanding of the dynamics and hydrodynamics of tidal disruption events. Here, we have presented the various methods that have been employed for this task, from Lagrangian particle based methods, such as SPH, to grid-based codes. We have discussed how such different methods have been applied to the various stages of the process of tidal disruption, and emphasised their successes and their limitations. Table 1 shows a "map" of the current effort in simulating TDEs in terms of different codes, treatment of thermodynamics and general relativity, etc. We hope this map can be useful to guide the reader into the field of numerical TDEs.

Table 1 Map of numerical methods. I: *Method*, A: Affine, SPH, MM: Moving Mesh, G: Fixed Grid, AMR: Fixed grid w/ AMR, Log: Fixed Grid w/ Logarithmic Radius; II: *Frame*, Loc: Local, Gl: Global; III: *BH Gravity*, Newt: Newtonian, GR: General relativistic (either approximate or exact); IV: *EOS*, Id: Ideal gas, MESA, H: Helmholtz, O: Other; V: *MHD*, Y: Yes, N: No; VI: *Radiation*, Y: Yes, N: No; VII: *Nuclear reactions*, Y: Yes, N: No; VIII: *Self-gravity*, Y: Yes, N: No; IX: *Gravitational Waves*, Y: Yes, N: No; X: *Trajectory*, P: Parabolic, E: Elliptical, Var: Variety, NR: not relevant

Paper	I	II	III	IV	V	VI	VII	VIII	IX	X
Carter and Luminet (1982, 1983, 1985)	A	Loc	Newt	Id	N	N	Y	Y	N	P
Luminet and Pichon (1989b,a)	A	Loc	GR	O	N	N	Y	Y	N	P
Casalvieri et al. (2006)	A	Loc	Newt	Id	N	N	N	N	Y	Var
Nolthenius and Katz (1982)	SPH	Gl	Newt	Id	N	N	N	Y	N	P
Evans and Kochanek (1989)	SPH	Gl	Newt	Id	N	N	N	Y	N	P
Laguna et al. (1993a)	SPH	Gl	GR	Id	N	N	N	Y	N	P
Ayal et al. (2000)	SPH	Gl	GR	Id	N	N	N	Y	N	P
Lodato et al. (2009)	SPH	Gl	Newt	Id	N	N	N	Y	N	P
Bonnerot et al. (2016b)	SPH	Gl	GR	Id	N	N	N	Y	N	E
Bonnerot et al. (2017b)	SPH	Gl	Newt	Id	Y	N	N	Y	N	P
Coughlin and Nixon (2015)	SPH	Gl	Newt	Id	N	N	N	Y	N	P
Liptai and Price (2019)	SPH	Gl	GR	Id	N	N	N	Y	N	E
Anninos et al. (2018)	MM	Gl	GR	O	N	N	Y	Y	Y	P
Goicovic et al. (2019)	MM	Gl	Newt	Id/MESA	N	N	N	Y	N	P
Marck et al. (1996)	G	Loc	GR	Id	N	N	Y	Y	N	P
Khokhlov et al. (1993a,b)	G	Loc	Newt	Id	N	N	N	Y	N	P
Frolov et al. (1994)	G	Loc	GR	Id	N	N	N	Y	N	P
Diener et al. (1997)	G	Loc	GR	Id	N	N	N	Y	N	P
Brassart and Luminet (2008)	G	Loc	Newt	N	N	N	N	N	N	P
Cheng and Evans (2013)	G	Loc	GR	Id	N	N	N	Y	Y	P
Cheng and Bogdanović (2014)	G	Loc	GR	Id	N	N	N	Y	N	P
Guillochon et al. (2009)	AMR	Loc	Newt	Id	N	N	N	Y	Y	P
Guillochon and Ramirez-Ruiz (2013)	AMR	Loc	Newt	Id	N	N	N	Y	N	P
Manukian et al. (2013)	AMR	Loc	Newt	Id	N	N	N	Y	N	P
MacLeod et al. (2012)	AMR	Loc	Newt	Id/MESA	N	N	N	Y	N	P
MacLeod et al. (2013)	AMR	Loc	Newt	Id/MESA	N	N	N	Y	N	E
Guillochon and McCourt (2017)	AMR	Loc	Newt	Id	Y	N	N	Y	N	P
Law-Smith et al. (2017)	AMR	Loc	Newt	Id/MESA	N	N	N	Y	N	P
Law-Smith et al. (2019)	AMR	Loc	Newt	MESA/H	N	N	N	Y	N	P
Haas et al. (2012)	AMR	Gl	GR	Id	N	N	N	N	Y	P
East (2014)	AMR	Gl	Newt	N	N	N	N	Y	Y	P
Shiokawa et al. (2015)	Log	Gl	GR	Id	N	N	N	N	N	P
Sądowski et al. (2016)	Log	Gl	GR	Id	Y	N	N	N	N	E
Bonnerot and Lu (2019)	SPH	Gl	GR	Id	N	N	N	N	N	P
Jiang et al. (2016)	G	Loc	Newt	Id	N	Y	N	N	N	P
Lu and Bonnerot (2020)	G	Loc	Newt	Id	N	N	N	N	N	P
Roth et al. (2016)	G	Gl	Newt	O	N	Y	N	N	N	NR
Dai et al. (2018)	Log	Gl	GR	Id	Y	Y	N	N	N	NR
Curd and Narayan (2019)	Log	Gl	GR	Id	Y	Y	N	N	N	NR
Tanikawa et al. (2017), Tanikawa (2018a, 2018b)	SPH	Gl	Newt	H	N	N	Y	Y	N	P
Kawana et al. (2018)	SPH	Gl	Newt	H	N	N	Y	Y	N	P
Mainetti et al. (2017)	SPH	Gl	Newt	Id	N	N	N	Y	N	P
Steinberg et al. (2019)	MM	Loc	Newt	Id	N	N	N	Y	N	P
Yalinewich et al. (2019)	MM	Gl	Newt	Id	N	N	N	Y	N	P
Gafton and Rosswog (2019)	SPH	Gl	GR	Id	N	N	N	Y	N	P
Tejeda et al. (2017)	SPH	Gl	GR	Id	N	N	N	Y	N	P

For what concerns the disruption phase, it is fair to say that current codes behave reasonably well. Numerical convergence is reached easily (already using $\sim 10^5$ particles, for SPH simulations) and it is encouraging to see that the results obtained with different methods agree well (Lodato et al. 2009; Guillochon and Ramirez-Ruiz 2013; Mainetti et al. 2017). Care should be taken in this context only for very deeply penetrating events, for which relativistic effects and thermonuclear detonation (for white dwarfs disruption) could play an important role. As a result, most of the numerical work done in this context, rather than improving the numerics, is aimed at including progressively more complex geometries, for example exploring disruption by binary black holes (Coughlin et al. 2017; Vigneron et al. 2018) or the role of stellar spin (Golightly et al. 2019a; Sacchi and Lodato 2019), or that of realistic stellar structure (Law-Smith et al. 2019).

The subsequent evolution of the debris appears to be more problematic. For several years, it has been assumed that the debris would evolve according to simple Keplerian dynamics, with effects due to pressure and self-gravity being negligible. However, it has been recently shown that this may not be the case (Coughlin and Nixon 2015; Coughlin et al. 2016) and the stream may develop (in the case of the disruption of giant stars) hydrodynamic instability (Bonnerot et al. 2016a) (although note that the instability criterion used by Bonnerot et al. 2016a assumed an incompressible flow). The extent to which numerical effects can affect the evolution in this regime is still unclear: for example, Coughlin et al. (2016) find, rather surprisingly, that increasing the resolution of their simulations (and thus improving the ability to resolve small scale fragmentation) actually reduces the tendency of the stream to fragment due to gravitational instability.

The formation of a disc or an accretion flow from the fallback material does present difficulties, especially for parabolic stellar orbits. This is because the gas density at the beginning of the fallback is very low. Application of adaptive methods (such as SPH) would result in very low spatial resolution, leading to unphysical behaviour. In some sense, it is disappointing that no simulation to date has been able to follow the whole process of stellar disruption and disc formation for the standard case of a solar mass star on a parabolic orbit, being disrupted by a SMBH. Additionally, this phase is expected to strongly depend on relativistic effects, such as apsidal and Lense-Thirring precession. These effects still need to be properly accounted for in numerical schemes, although recent work has extended SPH to a general relativistic frame, by solving the equation of motion on a curved, static metric (Liptai and Price 2019).

Finally, the proper accretion phase is most naturally treated in the context of GR (radiation) MHD simulations. Simulations such as Dai et al. (2018) have illustrated the basic properties of TDE discs in the super-Eddington phase. However, if one wants to fully understand the dynamical evolution of the disc as well as the reprocessed emission, there are a few remaining bottlenecks. 1) The current approach of treating the disc formation and accretion stages separately introduces an intrinsic uncertainty for the proper initialization of the disc when studying the TDE accretion phase. 2) Close to the black hole, the disc structure is regulated by the accretion process. Typically the disc inflow equilibrium can only be established for a few tens of gravitational radii around the black hole in GRRMHD simulations which are computationally expensive. (In Dai et al. 2018, this equilibrium can be established for hundreds of gravitational radii because the viscous timescale is shorter for a disc threaded by stronger magnetic flux.) This problem will only become more severe when studying the disc transitioning from the super-Eddington regime to the sub-Eddington regime, when the disc structure becomes thinner and its viscous timescale grows longer. Therefore, the dynamical evolution of the disc over the whole fallback timescale can be hard to simulate. 3) The current generation of GRRMHD codes all employ frequency-averaged absorption opacity to calculate the gas-radiation coupling. In order to make this calculation more accurate and especially to obtain TDE spectra directly from disc simulations, frequency-dependent radiative transfer physics (such as line transitions) should be incorporated into the GRRMHD codes, which remains as a goal for future studies.

In this Chapter, we have concentrated on current efforts in simulating TDE, specifically from the algorithmic point of view, and certainly more work is required. We identify the most pressing area of further progress to be: (a) a proper inclusion of general relativistic effects, and in particular of the role of black hole spin; (b) a better treatment of dissipative effects, with the aim of resolving the dynamics that leads to disc formation; (c) a more detailed inclusion of radiative effects, that have been rarely included in simulations. Additionally, improved hardware, such as GPU computing, will definitely offer new resources for future work. A detailed discussion of such issues can be found in the dedicated Chapter within this book, by Krolik et al.

Acknowledgements The authors thank the Yukawa Institute for Theoretical Physics at Kyoto University. Discussions during the YITP workshop YITP-T-19-07 on International Molecule-type Workshop "Tidal Disruption Events: General Relativistic Transients" were useful to complete this work. The work of RMC was funded by a Nicholas C. Metropolis Postdoctoral Fellowship and the Advanced Simulation Computing Physics and Engineering Program under the auspices of Los Alamos National Laboratory, operated by Triad National Security, LLC, for the National Nuclear Security Administration of U.S. Department of Energy (Contract No. 89233218CNA000001). The research of CB was funded by the Gordon and Betty Moore Foundation through Grant GBMF5076. JLD is supported by the GRF grant from the Hong Kong government under HKU 27305119.

References

M.A. Abramowicz, B. Czerny, J.P. Lasota, E. Szuszkiewicz, Slim accretion disks. Astrophys. J. **332**, 646–658 (1988). https://doi.org/10.1086/166683

P. Anninos, P.C. Fragile, S.S. Olivier, R. Hoffman, B. Mishra, K. Camarda, Relativistic tidal disruption and nuclear ignition of white dwarf stars by intermediate-mass black holes. Astrophys. J. **865**, 3 (2018). https://doi.org/10.3847/1538-4357/aadad9

S. Ayal, M. Livio, T. Piran, Tidal disruption of a solar-type star by a supermassive black hole. Astrophys. J. **545**, 772–780 (2000). https://doi.org/10.1086/317835

S.A. Balbus, J.F. Hawley, A powerful local shear instability in weakly magnetized disks. I. Linear analysis. Astrophys. J. **376**, 214 (1991). https://doi.org/10.1086/170270

M.R. Bate, I.A. Bonnell, N.M. Price, Modelling accretion in protobinary systems. Mon. Not. R. Astron. Soc. **277**(2), 362–376 (1995). https://doi.org/10.1093/mnras/277.2.362

M.J. Berger, P. Colella, Local adaptive mesh refinement for shock hydrodynamics. J. Comput. Phys. **82**, 64–84 (1989). https://doi.org/10.1016/0021-9991(89)90035-1

G.V. Bicknell, R.A. Gingold, On tidal detonation of stars by massive black holes. Astrophys. J. **273**, 749–760 (1983). https://doi.org/10.1086/161410

L. Blanchet, Gravitational radiation from post-Newtonian sources and inspiralling compact binaries. Living Rev. Relativ. **9**, 4 (2006). https://doi.org/10.12942/lrr-2006-4

C. Bonnerot, W. Lu, Simulating realistic disc formation in tidal disruption events. arXiv e-prints (2019). arXiv:1906.05865

C. Bonnerot, E.M. Rossi, G. Lodato, Bad prospects for the detection of giant stars' tidal disruption: effect of the ambient medium on bound debris. Mon. Not. R. Astron. Soc. **458**(3), 3324–3330 (2016a). https://doi.org/10.1093/mnras/stw486

C. Bonnerot, E.M. Rossi, G. Lodato, D.J. Price, Disc formation from tidal disruptions of stars on eccentric orbits by Schwarzschild black holes. Mon. Not. R. Astron. Soc. **455**, 2253–2266 (2016b). https://doi.org/10.1093/mnras/stv2411

C. Bonnerot, D.J. Price, G. Lodato, E.M. Rossi, Magnetic field evolution in tidal disruption events. Mon. Not. R. Astron. Soc. **469**(4), 4879–4888 (2017a). https://doi.org/10.1093/mnras/stx1210

C. Bonnerot, E.M. Rossi, G. Lodato, Long-term stream evolution in tidal disruption events. Mon. Not. R. Astron. Soc. **464**, 2816–2830 (2017b). https://doi.org/10.1093/mnras/stw2547

M. Brassart, J.-P. Luminet, Shock waves in tidally compressed stars by massive black holes. Astron. Astrophys. **481**, 259–277 (2008). https://doi.org/10.1051/0004-6361:20078264

B. Carter, J.P. Luminet, Pancake detonation of stars by black holes in galactic nuclei. Nature **296**, 211–214 (1982). https://doi.org/10.1038/296211a0

B. Carter, J.-P. Luminet, Tidal compression of a star by a large black hole. I Mechanical evolution and nuclear energy release by proton capture. Astron. Astrophys. **121**, 97–113 (1983)

B. Carter, J.P. Luminet, Mechanics of the affine star model. Mon. Not. R. Astron. Soc. **212**, 23–55 (1985). https://doi.org/10.1093/mnras/212.1.23

C. Casalvieri, V. Ferrari, A. Stavridis, Gravitational signals due to tidal interactions between white dwarfs and black holes. Mon. Not. R. Astron. Soc. **365**, 929–938 (2006). https://doi.org/10.1111/j.1365-2966.2005.09773.x

C.-H. Chan, T. Piran, J.H. Krolik, D. Saban, Tidal disruption events in active galactic nuclei. Astrophys. J. **881**(2), 113 (2019). https://doi.org/10.3847/1538-4357/ab2b40

R.M. Cheng, T. Bogdanović, Tidal disruption of a star in the Schwarzschild spacetime: relativistic effects in the return rate of debris. Phys. Rev. D **90**(6), 064020 (2014). https://doi.org/10.1103/PhysRevD.90.064020

R.M. Cheng, C.R. Evans, Relativistic effects in the tidal interaction between a white dwarf and a massive black hole in Fermi normal coordinates. Phys. Rev. D **87**(10), 104010 (2013). https://doi.org/10.1103/PhysRevD.87.104010

E.R. Coughlin, C. Nixon, Variability in tidal disruption events: gravitationally unstable streams. Astrophys. J. Lett. **808**, 11 (2015). https://doi.org/10.1088/2041-8205/808/1/L11

E.R. Coughlin, C. Nixon, M.C. Begelman, P.J. Armitage, D.J. Price, Post-periapsis pancakes: sustenance for self-gravity in tidal disruption events. Mon. Not. R. Astron. Soc. **455**, 3612–3627 (2016). https://doi.org/10.1093/mnras/stv2511

E.R. Coughlin, P.J. Armitage, C. Nixon, M.C. Begelman, Tidal disruption events from supermassive black hole binaries. Mon. Not. R. Astron. Soc. **465**, 3840–3864 (2017). https://doi.org/10.1093/mnras/stw2913

L. Cullen, W. Dehnen, Inviscid smoothed particle hydrodynamics. Mon. Not. R. Astron. Soc. **408**, 669–683 (2010). https://doi.org/10.1111/j.1365-2966.2010.17158.x

B. Curd, R. Narayan, GRRMHD simulations of tidal disruption event accretion discs around supermassive black holes: jet formation, spectra, and detectability. Mon. Not. R. Astron. Soc. **483**(1), 565–592 (2019). https://doi.org/10.1093/mnras/sty3134

L. Dai, A. Escala, P. Coppi, The impact of bound stellar orbits and general relativity on the temporal behavior of tidal disruption flares. Astrophys. J. Lett. **775**, 9 (2013). https://doi.org/10.1088/2041-8205/775/1/L9

L. Dai, J.C. McKinney, M.C. Miller, Soft X-ray temperature tidal disruption events from stars on deep plunging orbits. Astrophys. J. Lett. **812**, 39 (2015). https://doi.org/10.1088/2041-8205/812/2/L39

L. Dai, J.C. McKinney, N. Roth, E. Ramirez-Ruiz, M.C. Miller, A unified model for tidal disruption events. Astrophys. J. **859**, 20 (2018). https://doi.org/10.3847/2041-8213/aab429

D.S.P. Dearborn, J.R. Wilson, G.J. Mathews, Relativistically compressed exploding white dwarf model for Sagittarius A East. Astrophys. J. **630**, 309–320 (2005). https://doi.org/10.1086/431200

P. Diener, A.G. Kosovichev, E.V. Kotok, I.D. Novikov, C.J. Pethick, Non-linear effects at tidal capture of stars by a massive black hole - II. Compressible affine models and tidal interaction after capture. Mon. Not. R. Astron. Soc. **275**, 498–506 (1995). https://doi.org/10.1093/mnras/275.2.498

P. Diener, V.P. Frolov, A.M. Khokhlov, I.D. Novikov, C.J. Pethick, Relativistic tidal interaction of stars with a rotating black hole. Astrophys. J. **479**, 164–178 (1997). https://doi.org/10.1086/303875

W.E. East, Gravitational waves from the collision of tidally disrupted stars with massive black holes. Astrophys. J. **795**, 135 (2014). https://doi.org/10.1088/0004-637X/795/2/135

W.E. East, F. Pretorius, Simulating extreme-mass-ratio systems in full general relativity. Phys. Rev. D **87**, 101502 (2013). https://doi.org/10.1103/PhysRevD.87.101502

C. Eckart, Variation principles of hydrodynamics. Phys. Fluids **3**, 421–427 (1960). https://doi.org/10.1063/1.1706053

C.R. Evans, J.F. Hawley, Simulation of magnetohydrodynamic flows - a constrained transport method. Astrophys. J. **332**, 659–677 (1988). https://doi.org/10.1086/166684

C.R. Evans, C.S. Kochanek, The tidal disruption of a star by a massive black hole. Astrophys. J. Lett. **346**, 13–16 (1989). https://doi.org/10.1086/185567

P.C. Fragile, A. Gillespie, T. Monahan, M. Rodriguez, P. Anninos, Numerical simulations of optically thick accretion onto a black hole. I. Spherical case. Astrophys. J. Suppl. Ser. **201**, 9 (2012). https://doi.org/10.1088/0067-0049/201/2/9

P.C. Fragile, A. Olejar, P. Anninos, Numerical simulations of optically thick accretion onto a black hole. II. Rotating flow. Astrophys. J. **796**, 22 (2014). https://doi.org/10.1088/0004-637X/796/1/22

V.P. Frolov, A.M. Khokhlov, I.D. Novikov, C.J. Pethick, Relativistic tidal interaction of a white dwarf with a massive black hole. Astrophys. J. **432**, 680–689 (1994). https://doi.org/10.1086/174607

B. Fryxell, K. Olson, P. Ricker, F.X. Timmes, M. Zingale, D.Q. Lamb, P. MacNeice, R. Rosner, J.W. Truran, H. Tufo, FLASH: an adaptive mesh hydrodynamics code for modeling astrophysical thermonuclear flashes. Astrophys. J. Suppl. Ser. **131**, 273–334 (2000). https://doi.org/10.1086/317361

S.V. Fuerst, K. Wu, Radiation transfer of emission lines in curved space-time. Astron. Astrophys. **424**, 733–746 (2004). https://doi.org/10.1051/0004-6361:20035814

E. Gafton, S. Rosswog, Tidal disruptions by rotating black holes: effects of spin and impact parameter. arXiv e-prints (2019). arXiv:1903.09147

E. Gafton, E. Tejeda, J. Guillochon, O. Korobkin, S. Rosswog, Relativistic effects on tidal disruption kicks of solitary stars. Mon. Not. R. Astron. Soc. **449**(1), 771–780 (2015). https://doi.org/10.1093/mnras/stv350

C.F. Gammie, J.C. McKinney, G. Tóth, HARM: a numerical scheme for general relativistic magnetohydrodynamics. Astrophys. J. **589**, 444–457 (2003). https://doi.org/10.1086/374594

T.A. Gardiner, J.M. Stone, An unsplit Godunov method for ideal MHD via constrained transport in three dimensions. J. Comput. Phys. **227**, 4123–4141 (2008). https://doi.org/10.1016/j.jcp.2007.12.017

R.A. Gingold, J.J. Monaghan, Smoothed particle hydrodynamics - theory and application to non-spherical stars. Mon. Not. R. Astron. Soc. **181**, 375–389 (1977). https://doi.org/10.1093/mnras/181.3.375

F.G. Goicovic, V. Springel, S.T. Ohlmann, R. Pakmor, Hydrodynamical moving-mesh simulations of the tidal disruption of stars by supermassive black holes. arXiv e-prints (2019). arXiv:1902.08202

E.C.A. Golightly, E.R. Coughlin, C.J. Nixon, Tidal disruption events: the role of stellar spin. Astrophys. J. **872**, 163 (2019a). https://doi.org/10.3847/1538-4357/aafd2f

E.C.A. Golightly, C.J. Nixon, E.R. Coughlin, On the diversity of fallback rates from tidal disruption events with accurate stellar structure. Astrophys. J. Lett. **882**(2), 26 (2019b). https://doi.org/10.3847/2041-8213/ab380d

J. Guillochon, M. McCourt, Simulations of magnetic fields in tidally disrupted stars. Astrophys. J. Lett. **834**, 19 (2017). https://doi.org/10.3847/2041-8213/834/2/L19

J. Guillochon, E. Ramirez-Ruiz, Hydrodynamical simulations to determine the feeding rate of black holes by the tidal disruption of stars: the importance of the impact parameter and stellar structure. Astrophys. J. **767**, 25 (2013). https://doi.org/10.1088/0004-637X/767/1/25

J. Guillochon, E. Ramirez-Ruiz, A dark year for tidal disruption events. Astrophys. J. **809**, 166 (2015). https://doi.org/10.1088/0004-637X/809/2/166

J. Guillochon, E. Ramirez-Ruiz, S. Rosswog, D. Kasen, Three-dimensional simulations of tidally disrupted solar-type stars and the observational signatures of shock breakout. Astrophys. J. **705**, 844–853 (2009). https://doi.org/10.1088/0004-637X/705/1/844

J. Guillochon, H. Manukian, E. Ramirez-Ruiz, PS1-10jh: the disruption of a main-sequence star of near-solar composition. Astrophys. J. **783**, 23 (2014). https://doi.org/10.1088/0004-637X/783/1/23

R. Haas, R.V. Shcherbakov, T. Bode, P. Laguna, Tidal disruptions of white dwarfs from ultra-close encounters with intermediate-mass spinning black holes. Astrophys. J. **749**, 117 (2012). https://doi.org/10.1088/0004-637X/749/2/117

J. Hawley, J. Blondin, G. Lindahl, E. Lufkin, *VH-1: Multidimensional Ideal Compressible Hydrodynamics Code* (2012)

K. Hayasaki, N. Stone, A. Loeb, Finite, intense accretion bursts from tidal disruption of stars on bound orbits. Mon. Not. R. Astron. Soc. **434**, 909–924 (2013). https://doi.org/10.1093/mnras/stt871

K. Hayasaki, N. Stone, A. Loeb, Circularization of tidally disrupted stars around spinning supermassive black holes. Mon. Not. R. Astron. Soc. **461**, 3760–3780 (2016). https://doi.org/10.1093/mnras/stw1387

P.F. Hopkins, A new class of accurate, mesh-free hydrodynamic simulation methods. Mon. Not. R. Astron. Soc. **450**, 53–110 (2015). https://doi.org/10.1093/mnras/stv195

M. Ishii, M. Shibata, Y. Mino, Black hole tidal problem in the Fermi normal coordinates. Phys. Rev. D **71**(4), 044017 (2005). https://doi.org/10.1103/PhysRevD.71.044017

Y.-F. Jiang, J.M. Stone, S.W. Davis, A global three-dimensional radiation magneto-hydrodynamic simulation of super-eddington accretion disks. Astrophys. J. **796**, 106 (2014). https://doi.org/10.1088/0004-637X/796/2/106

Y.-F. Jiang, J. Guillochon, A. Loeb, Prompt radiation and mass outflows from the stream-stream collisions of tidal disruption events. Astrophys. J. **830**, 125 (2016). https://doi.org/10.3847/0004-637X/830/2/125

K. Kagaya, S. Yoshida, A. Tanikawa, Tidal disruptions of rotating stars by a supermassive black hole. arXiv e-prints (2019). arXiv:1901.05644

D. Kasen, R.C. Thomas, P. Nugent, Time-dependent Monte Carlo radiative transfer calculations for three-dimensional supernova spectra, light curves, and polarization. Astrophys. J. **651**, 366–380 (2006). https://doi.org/10.1086/506190

K. Kawana, A. Tanikawa, N. Yoshida, Tidal disruption of a white dwarf by a black hole: the diversity of nucleosynthesis, explosion energy, and the fate of debris streams. Mon. Not. R. Astron. Soc. **477**(3), 3449–3460 (2018). https://doi.org/10.1093/mnras/sty842

A. Kheyfets, W.A. Miller, W.H. Zurek, Covariant smoothed particle hydrodynamics on a curved background. Phys. Rev. D **41**(2), 451–454 (1990). https://doi.org/10.1103/PhysRevD.41.451

A. Khokhlov, I.D. Novikov, C.J. Pethick, Strong effects during close encounters of a star with a massive black hole. Astrophys. J. **418**, 181 (1993a). https://doi.org/10.1086/173380

A. Khokhlov, I.D. Novikov, C.J. Pethick, Weak tidal encounters of a star with a massive black hole. Astrophys. J. **418**, 163 (1993b). https://doi.org/10.1086/173379

S.S. Kim, M. Park, H.M. Lee, The stream-stream collision after the tidal disruption of a star around a massive black hole. Astrophys. J. **519**(2), 647–657 (1999). https://doi.org/10.1086/307394

S. Kobayashi, P. Laguna, E.S. Phinney, P. Mészáros, Gravitational waves and X-ray signals from stellar disruption by a massive black hole. Astrophys. J. **615**, 855–865 (2004). https://doi.org/10.1086/424684

P. Laguna, W.A. Miller, W.H. Zurek, M.B. Davies, Tidal disruptions by supermassive black holes - hydrodynamic evolution of stars on a Schwarzschild background. Astrophys. J. Lett. **410**, 83–86 (1993a). https://doi.org/10.1086/186885

P. Laguna, W.A. Miller, W.H. Zurek, Smoothed particle hydrodynamics near a black hole. Astrophys. J. **404**, 678 (1993b). https://doi.org/10.1086/172321

L.D. Landau, E.M. Lifshitz, *Fluid Mechanics* (Pergamon Press, Oxford, 1959)

J. Law-Smith, M. MacLeod, J. Guillochon, P. Macias, E. Ramirez-Ruiz, Low-mass white dwarfs with hydrogen envelopes as a missing link in the tidal disruption menu. Astrophys. J. **841**, 132 (2017). https://doi.org/10.3847/1538-4357/aa6ffb

J. Law-Smith, J. Guillochon, E. Ramirez-Ruiz, The tidal disruption of sun-like stars by massive black holes. Astrophys. J. Lett. **882**(2), 25 (2019). https://doi.org/10.3847/2041-8213/ab379a

D. Liptai, D.J. Price, General relativistic smoothed particle hydrodynamics. Mon. Not. R. Astron. Soc. **485**, 819–842 (2019). https://doi.org/10.1093/mnras/stz111

D. Liptai, D.J. Price, I. Mandel, G. Lodato, Disc formation from tidal disruption of stars on eccentric orbits by Kerr black holes using GRSPH. arXiv e-prints (2019). arXiv:1910.10154

G. Lodato, D.J. Price, On the diffusive propagation of warps in thin accretion discs. Mon. Not. R. Astron. Soc. **405**, 1212–1226 (2010). https://doi.org/10.1111/j.1365-2966.2010.16526.x

G. Lodato, A.R. King, J.E. Pringle, Stellar disruption by a supermassive black hole: is the light curve really proportional to $t^{-5/3}$?. Mon. Not. R. Astron. Soc. **392**, 332–340 (2009). https://doi.org/10.1111/j.1365-2966.2008.14049.x

W. Lu, C. Bonnerot, Self-intersection of the fallback stream in tidal disruption events. Mon. Not. R. Astron. Soc. **492**(1), 686–707 (2020). https://doi.org/10.1093/mnras/stz3405

L.B. Lucy, A numerical approach to the testing of the fission hypothesis. Astron. J. **82**, 1013–1024 (1977). https://doi.org/10.1086/112164

J.-P. Luminet, B. Carter, Dynamics of an affine star model in a black hole tidal field. Astrophys. J. Suppl. Ser. **61**, 219–248 (1986). https://doi.org/10.1086/191113

J.P. Luminet, J.A. Marck, Tidal effects in Kerr geometry, in *General Relativity and Gravitation*, vol. 1, ed. by B. Bertotti, F. de Felice, A. Pascolini (1983), p. 438

J.-P. Luminet, J.-A. Marck, Tidal squeezing of stars by Schwarzschild black holes. Mon. Not. R. Astron. Soc. **212**, 57–75 (1985). https://doi.org/10.1093/mnras/212.1.57

J.-P. Luminet, B. Pichon, Tidal pinching of white dwarfs. Astron. Astrophys. **209**, 103–110 (1989a)

J.-P. Luminet, B. Pichon, Tidally-detonated nuclear reactions in main sequence stars passing near a large black hole. Astron. Astrophys. **209**(1–2), 85–102 (1989b)

M. MacLeod, J. Guillochon, E. Ramirez-Ruiz, The tidal disruption of giant stars and their contribution to the flaring supermassive black hole population. Astrophys. J. **757**(2), 134 (2012). https://doi.org/10.1088/0004-637X/757/2/134

M. MacLeod, E. Ramirez-Ruiz, S. Grady, J. Guillochon, Spoon-feeding giant stars to supermassive black holes: episodic mass transfer from evolving stars and their contribution to the quiescent activity of galactic nuclei. Astrophys. J. **777**, 133 (2013). https://doi.org/10.1088/0004-637X/777/2/133

D. Mainetti, A. Lupi, S. Campana, M. Colpi, E.R. Coughlin, J. Guillochon, E. Ramirez-Ruiz, The fine line between total and partial tidal disruption events. Astron. Astrophys. **600**, 124 (2017). https://doi.org/10.1051/0004-6361/201630092

H. Manukian, J. Guillochon, E. Ramirez-Ruiz, R.M. O'Leary, Turbovelocity stars: kicks resulting from the tidal disruption of solitary stars. Astrophys. J. Lett. **771**(2), 28 (2013). https://doi.org/10.1088/2041-8205/771/2/L28

J.-A. Marck, Solution to the equations of parallel transport in Kerr geometry; tidal tensor. Proc. R. Soc. Lond. Ser. A **385**, 431–438 (1983). https://doi.org/10.1098/rspa.1983.0021

J.A. Marck, A. Lioure, S. Bonazzola, Numerical study of the tidal interaction of a star and a massive black hole. Astron. Astrophys. **306**, 666 (1996)

J.C. McKinney, A. Tchekhovskoy, A. Sadowski, R. Narayan, Three-dimensional general relativistic radiation magnetohydrodynamical simulation of super-Eddington accretion, using a new code HARMRAD with M1 closure. Mon. Not. R. Astron. Soc. **441**, 3177–3208 (2014). https://doi.org/10.1093/mnras/stu762

J.C. McKinney, L. Dai, M.J. Avara, Efficiency of super-Eddington magnetically-arrested accretion. Mon. Not. R. Astron. Soc. **454**, 6–10 (2015). https://doi.org/10.1093/mnrasl/slv115

D. Merritt, *Dynamics and Evolution of Galactic Nuclei* (Princeton University Press, Princeton, 2013). ISBN 9780691158600

J.J. Monaghan, D.J. Price, Variational principles for relativistic smoothed particle hydrodynamics. Mon. Not. R. Astron. Soc. **328**, 381–392 (2001). https://doi.org/10.1046/j.1365-8711.2001.04742.x

J.P. Morris, J.J. Monaghan, A switch to reduce SPH viscosity. J. Comput. Phys. **136**, 41–50 (1997). https://doi.org/10.1006/jcph.1997.5690

R.P. Nelson, J.C.B. Papaloizou, Hydrodynamic simulations of the Bardeen-Petterson effect. Mon. Not. R. Astron. Soc. **315**, 570–586 (2000). https://doi.org/10.1046/j.1365-8711.2000.03478.x

S.C. Noble, J.H. Krolik, J.F. Hawley, Direct calculation of the radiative efficiency of an accretion disk around a black hole. Astrophys. J. **692**, 411–421 (2009). https://doi.org/10.1088/0004-637X/692/1/411

R.A. Nolthenius, J.I. Katz, The passage of a star by a massive black hole. Astrophys. J. **263**, 377–385 (1982). https://doi.org/10.1086/160511

K. Ohsuga, M. Mori, T. Nakamoto, S. Mineshige, Supercritical accretion flows around black holes: two-dimensional, radiation pressure-dominated disks with photon trapping. Astrophys. J. **628**, 368–381 (2005). https://doi.org/10.1086/430728

E.S. Phinney, Manifestations of a massive black hole in the galactic center, in *The Center of the Galaxy*, ed. by M. Morris. IAU Symposium, vol. 136 (1989), p. 543

D.J. Price, Modelling discontinuities and Kelvin Helmholtz instabilities in SPH. J. Comput. Phys. **227**, 10040–10057 (2008). https://doi.org/10.1016/j.jcp.2008.08.011

D.J. Price, Smoothed particle hydrodynamics and magnetohydrodynamics. J. Comput. Phys. **231**, 759–794 (2012). https://doi.org/10.1016/j.jcp.2010.12.011

D.J. Price, J. Wurster, T.S. Tricco, C. Nixon, S. Toupin, A. Pettitt, C. Chan, D. Mentiplay, G. Laibe, S. Glover, C. Dobbs, R. Nealon, D. Liptai, H. Worpel, C. Bonnerot, G. Dipierro, G. Ballabio, E. Ragusa, C. Federrath, R. Iaconi, T. Reichardt, D. Forgan, M. Hutchison, T. Constantino, B. Ayliffe, K. Hirsh, G. Lodato, Phantom: a smoothed particle hydrodynamics and magnetohydrodynamics code for astrophysics. Publ. Astron. Soc. Aust. **35**, 031 (2018). https://doi.org/10.1017/pasa.2018.25

M.J. Rees, Tidal disruption of stars by black holes of 10 to the 6th-10 to the 8th solar masses in nearby galaxies. Nature **333**, 523 (1988). https://doi.org/10.1038/333523a0

S. Rosswog, Astrophysical smooth particle hydrodynamics. New Astron. **53**, 78–104 (2009). https://doi.org/10.1016/j.newar.2009.08.007

S. Rosswog, E. Ramirez-Ruiz, W.R. Hix, Tidal disruption and ignition of white dwarfs by moderately massive black holes. Astrophys. J. **695**(1), 404–419 (2009). https://doi.org/10.1088/0004-637X/695/1/404

N. Roth, D. Kasen, J. Guillochon, E. Ramirez-Ruiz, The X-ray through optical fluxes and line strengths of tidal disruption events. Astrophys. J. **827**, 3 (2016). https://doi.org/10.3847/0004-637X/827/1/3

A. Sacchi, G. Lodato, 'Failed' tidal disruption events and X-ray flares from the galactic centre. Mon. Not. R. Astron. Soc. **486**, 1833–1839 (2019). https://doi.org/10.1093/mnras/stz981

A. Sądowski, R. Narayan, J.C. McKinney, A. Tchekhovskoy, Numerical simulations of super-critical black hole accretion flows in general relativity. Mon. Not. R. Astron. Soc. **439**(1), 503–520 (2014). https://doi.org/10.1093/mnras/stt2479

A. Sądowski, R. Narayan, A. Tchekhovskoy, D. Abarca, Y. Zhu, J.C. McKinney, Global simulations of axisymmetric radiative black hole accretion discs in general relativity with a mean-field magnetic dynamo. Mon. Not. R. Astron. Soc. **447**, 49–71 (2015). https://doi.org/10.1093/mnras/stu2387

A. Sądowski, E. Tejeda, E. Gafton, S. Rosswog, D. Abarca, Magnetohydrodynamical simulations of a deep tidal disruption in general relativity. Mon. Not. R. Astron. Soc. **458**, 4250–4268 (2016). https://doi.org/10.1093/mnras/stw589

N.I. Shakura, R.A. Sunyaev, Reprint of 1973A&A....24..337S. Black holes in binary systems. Observational appearance. Astron. Astrophys. **500**, 33–51 (1973)

H. Shiokawa, J.H. Krolik, R.M. Cheng, T. Piran, S.C. Noble, General relativistic hydrodynamic simulation of accretion flow from a stellar tidal disruption. Astrophys. J. **804**, 85 (2015). https://doi.org/10.1088/0004-637X/804/2/85

V. Springel, Smoothed particle hydrodynamics in astrophysics. Annu. Rev. Astron. Astrophys. **48**, 391–430 (2010). https://doi.org/10.1146/annurev-astro-081309-130914

E. Steinberg, E.R. Coughlin, N.C. Stone, B.D. Metzger, Thawing the frozen-in approximation: implications for self-gravity in deeply plunging tidal disruption events. Mon. Not. R. Astron. Soc. **485**(1), 146–150 (2019). https://doi.org/10.1093/mnrasl/slz048

J.M. Stone, T.A. Gardiner, P. Teuben, J.F. Hawley, J.B. Simon, Athena: a new code for astrophysical MHD. Astrophys. J. Suppl. Ser. **178**, 137–177 (2008). https://doi.org/10.1086/588755

A. Tanikawa, High-resolution hydrodynamic simulation of tidal detonation of a helium white dwarf by an intermediate mass black hole. Astrophys. J. **858**(1), 26 (2018a). https://doi.org/10.3847/1538-4357/aaba79

A. Tanikawa, Tidal double detonation: a new mechanism for the thermonuclear explosion of a white dwarf induced by a tidal disruption event. Mon. Not. R. Astron. Soc. **475**(1), 67–71 (2018b). https://doi.org/10.1093/mnrasl/sly006

A. Tanikawa, Y. Sato, K. Nomoto, K. Maeda, N. Nakasato, I. Hachisu, Does explosive nuclear burning occur in tidal disruption events of white dwarfs by intermediate-mass black holes? Astrophys. J. **839**(2), 81 (2017). https://doi.org/10.3847/1538-4357/aa697d

E. Tejeda, S. Rosswog, An accurate Newtonian description of particle motion around a Schwarzschild black hole. Mon. Not. R. Astron. Soc. **433**, 1930–1940 (2013). https://doi.org/10.1093/mnras/stt853

E. Tejeda, E. Gafton, S. Rosswog, J.C. Miller, Tidal disruptions by rotating black holes: relativistic hydrodynamics with Newtonian codes. Mon. Not. R. Astron. Soc. **469**, 4483–4503 (2017). https://doi.org/10.1093/mnras/stx1089

E.F. Toro, *Riemann Solvers and Numerical Methods for Fluid Dynamics* (Springer, Berlin, 1999)

M. Toscani, L. Lodato, R. Nealon, Gravitational wave emission from unstable accretion discs in tidal disruption events. Mon. Not. R. Astron. Soc. **489**, 699–706 (2019). https://doi.org/10.1093/mnras/stz2201

Q. Vigneron, G. Lodato, A. Guidarelli, Tidal disruption of stars in a supermassive black hole binary system: the influence of orbital properties on fallback and accretion rates. Mon. Not. R. Astron. Soc. **476**, 5312–5322 (2018). https://doi.org/10.1093/mnras/sty585

C. Wegg, Pseudo-Newtonian potentials for nearly parabolic orbits. Astrophys. J. **749**, 183 (2012). https://doi.org/10.1088/0004-637X/749/2/183

C.J. White, J.M. Stone, C.F. Gammie, An extension of the Athena++ code framework for GRMHD based on advanced Riemann solvers and staggered-mesh constrained transport. Astrophys. J., Suppl. **225**, 22 (2016). https://doi.org/10.3847/0067-0049/225/2/22

J.R. Wilson, G.J. Mathews, White dwarfs near black holes: a new paradigm for type I supernovae. Astrophys. J. **610**, 368–377 (2004). https://doi.org/10.1086/421449

A. Yalinewich, E. Steinberg, T. Piran, J.H. Krolik, Radio emission from the unbound debris of tidal disruption events. Mon. Not. R. Astron. Soc. **487**(3), 4083–4092 (2019). https://doi.org/10.1093/mnras/stz1567

Space Sci Rev (2020) 216:114
https://doi.org/10.1007/s11214-020-00735-1

Radiative Emission Mechanisms

Nathaniel Roth[1] · Elena Maria Rossi[2] · Julian Krolik[3] · Tsvi Piran[4] · Brenna Mockler[5] · Daniel Kasen[6]

Received: 16 December 2019 / Accepted: 9 September 2020 / Published online: 6 October 2020

Abstract We describe how the various outcomes of stellar tidal disruption give rise to observable radiation. We separately consider the cases where gas circularizes rapidly into an accretion disc, as well as the case when shocked debris streams provide the observable emission without having fully circularized. For the rapid circularization case, we describe how outflows, absorption by reprocessing layers, and Comptonization can cause the observed radiation to depart from that of a bare disc, possibly giving rise to the observed optical/UV emission along with soft X-rays from the disc. If, instead, most of the debris follows highly eccentric orbits for a significant time, many properties of the observed optical/UV emission can be explained by the scale of those eccentric orbits and the shocks embedded in the debris flow near orbital apocenter. In this picture, soft X-ray emission at early times results from the smaller amount of debris mass deflected into a compact accretion disc by weak shocks near the stellar pericenter. A general proposal for the near-constancy of the ultraviolet/optical color temperatures is provided, by linking it to incomplete thermalization of radiation in the atmosphere of the emitting region. We also briefly discuss the radio signals from the interaction of unbound debris and jets with the black hole environment.

Keywords Black holes · Thermal radiation · Non-thermal radiation · Accretion

The Tidal Disruption of Stars by Massive Black Holes
Edited by Peter G. Jonker, Sterl Phinney, Elena Maria Rossi, Sjoert van Velzen, Iair Arcavi and Maurizio Falanga

✉ N. Roth
nroth@astro.umd.edu

1 Joint Space Science Institute, University of Maryland, College Park, MD 20742, USA

2 Leiden Observatory, Leiden University, PO Box 9513, 2300 RA, Leiden, the Netherlands

3 Dept. of Physics and Astronomy, Johns Hopkins University, Baltimore, MD 21218, USA

4 Racah Institute of Physics, Hebrew University, Jerusalem 91904, Israel

5 Dept. of Astronomy and Astrophysics, University of California, Santa Cruz, CA 95064, USA

6 Depts. of Physics and Astronomy, University of California, Berkeley, Berkeley, CA 94720 USA

1 Introduction

In this chapter we describe how the various components of the disrupted star emit the radiation we observe. We immediately face the difficulty that, despite much progress, the underlying hydrodynamics are not fully understood and remain the subject of active research (see for example the Formation of the Accretion Flow Chapter and Accretion Disc Chapter). We also expect the disruption outcome to vary substantially among separate events. We therefore describe the emission that might result from a variety of post-disruption scenarios, with the understanding that no single picture is capable of explaining every feature of the ever-growing observational dataset.

A summary of the TDE components which may give rise to detectable radiation is provided in Table 1. As indicated, a subset of these are discussed in detail in this chapter. For the others, we have referred readers to relevant chapters in this review collection, or to specific papers.

In Sect. 2 we summarize the emission from accretion discs (and their outflows) that may form from TDEs, which are continuously fed and potentially super-Eddington at the outset. In Sect. 3 we discuss the emission from shock-heated and elliptically orbiting tidal debris. In Sect. 4 we consider general principles governing optical and ultraviolet (UV) line and continuum emission, which may apply to several of the above scenarios. Finally, Sect. 5 addresses the radio emission from the interaction of unbound stellar debris and/or accretion disc jets with the surrounding circum-nuclear environment.

2 Emission from Accretion Discs and Their Outflows, for Material That Rapidly Circularizes

In this section we will consider the radiative emission that is produced if a large fraction of the bound stellar material is capable of rapidly dissipating its orbital energy and entering a circularized accretion flow around the black hole. In Sect. 2.1.1 we first consider the classical picture for the observational electromagnetic properties, dating back to Rees (1988) and Phinney (1989), where thermal emission peaking in soft X-rays is emitted by a circularised accretion disc fed by the continuous stream of stellar debris. We will track the emission properties of such an unobscured accretion disc with an accretion rate that is initially tied to the post-disruption mass fallback rate. In Sect. 2.1.2 we examine how the accretion flow is expected to behave several years after disruption, when the accretion rate is governed by viscous processes within the disc, rather than the mass fallback rate. While late-time disc emission models have been successfully matched to UV and X-ray data (van Velzen et al. 2019; Jonker et al. 2019), the bare discs of Sect. 2.1.1 generally do not fit observational data well on their own for the first few years after detection by optical surveys. In particular, the bare disc models often over-predict the amount of soft x-rays emitted when compared to observed limits, and vastly under-predict the observed optical and UV emission at early times. Thus, in Sect. 2.2 we will describe a number of factors that may cause the observed radiation to depart from the bare disc at early times, which may shift the peak of the emitted spectrum away from the soft X-rays, towards UV and optical wavelengths. These include adiabatic reprocessing of radiation in outflows (Sect. 2.2.1); absorption, re-thermalization and re-emission of inner disc emission by material farther out (Sect. 2.2.2); and Comptonization (Sect. 2.2.3). In Sect. 2.3, we discuss a model for fitting multi-band TDE light curves based on tying the light curves to the post-disruption mass fallback rate and allowing for a dynamically adjusting photosphere.

Table 1 A listing of sources of radiative emission from tidal disruption events by super-massive black holes

Components of TDEs posited to give rise to radiative emission		
Emission source	Primary detection bands	Locations discussed
Flash from vertical compression shock during first passage	Soft X-ray, Hard X-ray, γ-ray	Kobayashi et al. (2004), Brassart and Luminet (2008), Brassart and Luminet (2010), Guillochon et al. (2009), Stone et al. (2013), Yalinewich et al. (2019b), Gafton and Rosswog (2019), and Disruption Chapter
Nuclear burning and/or detonation following vertical compression at pericenter	Optical, UV	*Main sequence stars:* Carter and Luminet (1982, 1983), Bicknell and Gingold (1983), Pichon (1985), Luminet and Pichon (1989), Brassart and Luminet (2008). *White dwarfs:* White Dwarf Chapter
Electron recombination and cooling in unbound debris	Optical, UV	Kochanek (1994), Kasen and Ramirez-Ruiz (2010)
Accretion flow and associated thermal or quasi-thermal emission	Soft X-ray, UV	Sect. 2.1 and references therein, Accretion Disc Chapter, X-ray Chapter
Outflow launched from location of first stream collision	Optical, UV	Sect. 2.2.1, Formation of the Accretion Flow Chapter
Disc outflow launched by radiation pressure	Optical, UV, Soft X-ray	Sect. 2.2.1, Sect. 2.3, Accretion Disc Chapter
Reprocessing of inner accretion emission by material farther out (e.g. thick disc, outflowing disc material, or incompletely circularized bound material)	Optical, UV	Sect. 2.2.2, Sect. 2.3, Sect. 4.1, Optical Chapter
Shock-heated, incompletely circularized stellar debris	Optical, UV	Sect. 3, Formation of the Accretion Flow Chapter, Optical Chapter
Inverse Comptonization of photons by thermal and/or non-thermal electrons in a jet	Soft X-ray, Hard X-ray, γ-ray	Sect. 2.1.1, Sect. 2.2.3, Sect. 5.2, X-ray Chapter
Forward shock driven by jets encountering circum-nuclear material	Radio	Sect. 5.2, Radio Chapter
Mildly relativistic outflow interacting with circum-nuclear material	Radio	Sect. 5.1, Radio Chapter
Unbound debris interacting with circum-nuclear material	Radio	Sect. 5.1, Radio Chapter
Dust heated by UV and X-rays from TDE	IR	Sect. 2.2, Echo Chapter

2.1 Bare Disc Emission

2.1.1 Disc Properties at Early Times (< a Few Years After Disruption)

The prototype system we adopt consists of a main sequence star with mass M_* and radius R_* disrupted on an orbit with pericenter radius r_p equal to the tidal radius $r_\mathrm{t} \equiv R_* \, (M_\mathrm{BH}/M_*)^{1/3}$ where M_BH is the mass of the black hole (i.e. we set the so-called "penetration factor" $\beta \equiv r_\mathrm{t}/r_\mathrm{p}$ to 1 where r_p is the pericenter distance from the black hole of the initial stellar orbit). For scaling relations we will adopt the short-hand $M_\mathrm{BH,6} \equiv M_\mathrm{BH}/(10^6 \, M_\odot)$. We neglect the role of black hole spin unless stated otherwise.

After stellar disruption, if self-gravity in the stellar debris can be ignored, then the debris move ballistically, i.e. under the sole influence of the black hole's gravity. The portions of the debris that remain bound follow highly eccentric ($1 - e \leq 0.02$) elliptical orbits returning to the site of the star's disruption at a time-varying rate. We let $t = 0$ refer to the time of pericenter passage of the initial stellar orbit. Assuming for the moment a flat distribution of orbital energy for the stellar debris, the ballistic dynamics implies a fallback rate of

$$\dot{M}_\mathrm{fb} = \frac{1}{3}(M_*/t_\mathrm{fb})(t/t_\mathrm{fb})^{-5/3} \qquad \text{for } t \geq t_\mathrm{fb} \,, \; 0 \text{ otherwise} \tag{1}$$

where t_fb is the orbital period of the most bound debris (thus the first to come back to the tidal radius), and evaluates to

$$t_\mathrm{fb} \approx 40 \text{ days } (M_\mathrm{BH,6})^{1/2} \left(\frac{M_*}{M_\odot}\right)^{-1} \left(\frac{R_*}{R_\odot}\right)^{3/2} . \tag{2}$$

This approximation for $\dot{M}_\mathrm{fb}$ is taken for simplicity and the implications of this simplification will also be discussed below.

In this scenario, the timescale for the stellar debris to lose the necessary energy to circularize is assumed to be short compared to t_fb (see Sect. 3 for challenges to this assumption). Within the newly formed disc, most of the mass is initially deposited at a distance $r_d \approx 2r_t$, as demanded by the conservation of the star's specific angular momentum $j^2 = (2GM_\mathrm{BH}r_t)$ on its originally parabolic orbit with pericenter r_t where G is Netwon's gravitational constant. This evaluates to

$$\begin{aligned} r_d &\approx 100 \, r_g \left(M_\mathrm{BH,6}\right)^{-2/3} \left(\frac{M_*}{M_\odot}\right)^{-1/3} \left(\frac{R_*}{R_\odot}\right) \\ &\approx 10^{13} \text{ cm } \left(M_\mathrm{BH,6}\right)^{1/3} \left(\frac{M_*}{M_\odot}\right)^{-1/3} \left(\frac{R_*}{R_\odot}\right), \end{aligned} \tag{3}$$

where $r_g \equiv GM_\mathrm{BH}/c^2$. Meanwhile, the high rate of mass accretion through the disc may lead it to be geometrically thick, with a vertical scale height to radius ratio $h/r \sim 1$ (we will elaborate on this point shortly). In this case the viscous (disc inflow) timescale t_ν at r_d is initially much shorter than the fallback timescale:

$$\begin{aligned} \frac{t_\nu(r_d)}{t_\mathrm{fb}} &\approx \alpha^{-1} \left(\frac{h}{r}\right)^{-2} \frac{P(r_d)}{2\pi \, t_\mathrm{fb}} \\ &\approx 10^{-2} \left(\frac{\alpha}{0.1}\right)^{-1} \left(\frac{h}{r}\right)^{-2} \left(M_\mathrm{BH,6}\right)^{-1/2} \left(\frac{M_*}{M_\odot}\right)^{1/2} , \end{aligned} \tag{4}$$

where $P(r_d) \approx 8$ hours $\ll t_{fb}$ is the orbital period at r_d and α is the standard Shakura & Sunyaev viscosity parameter (Shakura and Sunyaev 1973) with a value appropriate for thick discs. As a result of this timescale ordering, material is immediately accreted through the disc. The rate at which mass feeds the black hole, $\dot{M}_{acc}$, is then given by

$$\dot{M}_{acc} = f_{in} \dot{M}_{fb} \,, \tag{5}$$

where the factor $f_{in} \leq 1$ accounts for mass launched in an outflow during circularization and/or accumulating in the disc. Generally we will expect f_{in} to be a function of time; however, for the purpose of deriving an approximate time dependence for $\dot{M}_{acc}$ we will take f_{in} to be constant with time so that $\dot{M}_{acc} \propto \dot{M}_{fb}$.

Accretion through the disc will cause orbital energy to be dissipated as heat, and some fraction of this dissipated energy will escape as radiation. We will label the radially-integrated bolometric luminosity of this radiation, with contributions from both sides of the disc, as L_{bol}, and we will often make reference to the instantaneous radiative efficiency $\eta = L_{bol}/(\dot{M}_{acc} c^2)$ as a function of time.

Notably, the fallback rate implies accretion rates that may exceed a critical value defined as the rate that would lead to the Eddington luminosity L_{Edd} if the radiative efficiency were set to a constant reference value η_{ref}:

$$\dot{M}_{cr} \equiv \frac{L_{Edd}}{\eta_{ref} c^2}. \tag{6}$$

We will be using the definition $L_{Edd} \equiv 4\pi G M_{BH}\, c/\kappa_{es}$ where κ_{es} is the electron scattering opacity. It will be useful to consider the time-dependent ratio f_{Edd} of the instantaneous value of $\dot{M}_{acc}$ over $\dot{M}_{cr}$:

$$\begin{aligned} f_{Edd} \equiv \frac{\dot{M}_{acc}}{\dot{M}_{cr}} &= \frac{f_{in}\, \dot{M}_{fb}\, \eta_{ref} c^2}{L_{Edd}} \\ &\approx 100\, f_{in} \left(\frac{\eta_{ref}}{0.1}\right) \left(\frac{\kappa_{es}}{0.34 \text{ cm}^2 \text{ g}^{-1}}\right) \\ &\times \left(M_{BH,6}\right)^{-3/2} \left(\frac{M_*}{M_\odot}\right)^2 \left(\frac{R_*}{R_\odot}\right)^{-3/2} \left(\frac{t}{t_{fb}}\right)^{-5/3} . \end{aligned} \tag{7}$$

We can use equation (7) to provide an estimate of the time after disruption t_{cr} when the mass fallback rate transitions from super-critical to sub-critical, which is when f_{Edd} drops to 1:

$$\begin{aligned} t_{cr} &\approx 700 \text{ days } (f_{in})^{3/5} \left(\frac{\eta_{ref}}{0.1}\right)^{3/5} \left(\frac{\kappa_{es}}{0.34 \text{ cm}^2 \text{ g}^{-1}}\right)^{3/5} \\ &\times \left(M_{BH,6}\right)^{-2/5} \left(\frac{M_*}{M_\odot}\right)^{1/5} \left(\frac{R_*}{R_\odot}\right)^{3/5} . \end{aligned} \tag{8}$$

The time-dependence in equation (7) follows from the approximations stated at the beginning of this section, in addition to the assumption that f_{in} does not vary with time. The power-law time-dependence for $\dot{M}_{fb}$ is closest to reality after a few t_{fb} from the return of the most bound debris, and the actual power-law may deviate from $-5/3$ (see the Disruption Chapter). At earlier times the star's structure imprints a characteristic rising-to-a-peak shape, with some dependence on the impact parameter of the initial stellar orbit. Hydrodynamic simulations show that the peak $\dot{M}_{\rm fb}$ remains super-critical for our fiducial parameters but it is suppressed by a factor of a few compared to the peak value stated in equation (1) (e.g Evans and Kochanek 1989). Interestingly, since $\dot{M}_{\rm fb}/\dot{M}_{\rm cr} \propto M_{\rm bh}^{-3/2}$, only black holes with masses below $\sim 10^7\ M_\odot$ may present these super-critical mass accretion flows (neglecting the effects of black hole spin). For other considerations concerning how the transition between super- and sub-critical accretion flow depends on $M_{\rm BH}$, see Wu et al. (2018). We also note that the super-criticality of $\dot{M}_{\rm acc}$ as we have discussed it here depends crucially on $f_{\rm in}$ not being too small, and on our initial assumption of rapid circularization.

Super-critical accretion is a topic of ongoing research, and the most recent results on such accretion flows in the context of TDEs is presented in the Accretion Disc Chapter. Here we summarize some analytic results that are at least qualitatively in line with more detailed numerical calculations. At the highest accretion rates ($f_{\rm Edd} \gtrsim 100$), the energy transport through the disc is dominated by advection of trapped radiation, as the large optical depth of the disc leads to the radiative cooling timescale is much longer than the timescale that trapped radiation is advected inward (e.g. Begelman 1978; Paczyński and Wiita 1980; Narayan and Yi 1994). This trapping of radiative energy leads to a large disc scale height, justifying our earlier decision to set $h/r \sim 1$ in equation (4). For more near-critical accretion rates ($0.1 \lesssim f_{\rm Edd} \lesssim 100$) the trapping of radiation is reduced yet still important. Potential complications in this regime are thermal and viscous instabilities (Shakura and Sunyaev 1976; Lightman and Eardley 1974). Some of the consequences of these instabilities for accretion following tidal disruption were investigated by Shen and Matzner (2014). However, recent global radiation magneto-hydrodynamic simulations of sub-critical ($f_{\rm Edd} \simeq 0.1$) accretion discs around super-massive black holes have found that, for certain magnetic field configurations, these instabilities are not triggered (Jiang et al. 2019b). Another important consideration is that outflows are generally expected in advection-dominated flows, as discussed in the literature concerning advection-dominated inflow-outflow solutions (Blandford and Begelman 1999, 2004) and in numerical calculations (e.g. Sądowski et al. 2014; Jiang et al. 2019a). We will have more to say about the role of outflows in Sect. 2.2.1.

The near-critical regime is of the greatest interest to us. Putting aside the issue of stability, we may make use of a formula for the steady-state effective temperature (a measure of the radiative flux) of disc annuli as a function of radius and $\dot{M}_{\rm acc}$ provided by Strubbe and Quataert (2009) (henceforth SQ09):

$$\sigma_{\rm SB} T_{\rm eff}^4 = \frac{3GM_{\rm BH}\dot{M}_{\rm acc} f_{\rm nt}}{8\pi r^3}\left[\frac{1}{2} + \left\{\frac{1}{4} + 6 f_{\rm nt}\left(\frac{\dot{M}_{\rm acc}c^2}{L_{\rm Edd}}\right)^2\left(\frac{r}{r_g}\right)^{-2}\right\}^{1/2}\right]^{-1}, \tag{9}$$

where $\sigma_{\rm SB}$ is the Stefan-Boltzmann constant and $f_{\rm nt} \equiv 1 - \sqrt{r_{\rm ISCO}/r}$, which comes from imposing a no-torque boundary condition at the innermost stable circular orbit (ISCO) (this may be compared to more recent work by Mummery and Balbus 2020, who allow for finite ISCO stress and a fully relativistic, time-dependent treatment). Note that $\dot{M}_{\rm acc}c^2/L_{\rm Edd}$ is equal to $f_{\rm Edd}/\eta_{\rm ref}$ as we have defined it. The temperature profile in equation (9) becomes identical to the thin disc result of Shakura and Sunyaev (1973) for $f_{\rm Edd} \ll 1$. The assumptions underlying equation (9) are related to the so-called "slim disc" models of Abramowicz et al. (1988) which apply to discs with accretion rates that are close to the critical value, although in deriving the SQ09 formula, general relativistic modifications to the black hole's gravitational potential are ignored when computing the orbital velocities of the gas.

Equation (9) is a simplified version of what would be computed from a detailed hydrodynamic simulation of the accretion process, which would compute physical stresses driving the accretion and account for cooling using radiative transfer. We provide this formula to give a concrete example of how $L_{\rm bol}$ and η can be computed once $T_{\rm eff}(r)$ has been determined: we integrate the flux over disc annuli, accounting for both sides of the disc (again ignoring relativistic effects):

$$L_{\rm bol} = 2 \int \sigma_{\rm SB} T_{\rm eff}^4(r) 2\pi r dr \tag{10}$$

For $f_{\rm Edd} \ll 1$, then for this simplified disc model, as for Shakura and Sunyaev (1973), $L_{\rm bol}$ for a disc of a given size varies linearly with $\dot{M}_{\rm acc}$. In other words, η is constant in this regime for fixed disc size. On the other hand, for $f_{\rm Edd} \gg 1$, equation (9) indicates that the radiative efficiency drops with increasing $\dot{M}_{\rm acc}$, so that $L_{\rm bol}$ varies sub-linearly with $\dot{M}_{\rm acc}$. This is the result of increased trapping of radiation at larger values of $f_{\rm Edd}$. As a consequence, while the total radiative luminosity of the disc may exceed $L_{\rm Edd}$, the rate at which the luminosity grows beyond this value is limited. For rough approximations, a common practice is to allow $L_{\rm bol}$ to grow with $\dot{M}$ according to the thin-disc result until $L_{\rm Edd}$ is reached, and then to cap $L_{\rm bol}$ at $L_{\rm Edd}$ for larger values of $\dot{M}_{\rm acc}$. This approximation, in addition to neglecting the more detailed growth of $L_{\rm bol}$ near and above $L_{\rm Edd}$, also neglects how the change in the disc size with time will affect $L_{\rm bol}$, in addition to a changing $\dot{M}_{\rm acc}$.

In order to get closer to observations, the luminosity in a given band, rather than the bolometric one, needs to be derived. If the radiation can effectively thermalize in the disc atmosphere, the disc radiates as a superposition of blackbody spectra from disc annuli, with a temperature profile equal to the effective temperature profile specified in equation (9), with generally outwards-decreasing peak frequency and flux. However, if the ionization state of the disc becomes high enough, electron scattering opacity is expected to be comparable to or exceed absorption opacities that thermalize the radiation, preventing local thermodynamic equilibrium (LTE) from being established in the disc atmosphere. This may lead to several departures from the full thermalization prediction, the simplest of which is a color correction $f_{\rm col}(r)$ that measures the degree to which the local color temperature of the radiation exceeds the local effective temperature. While such color corrections are most often applied to X-ray binary accretion discs (Shakura and Sunyaev 1973; Shimura and Takahara 1995; Davis et al. 2005), recent work also indicates they may apply to some accreting super-massive black holes such as those of narrow-line Seyfert 1 galaxies, with values of $f_{\rm col}$ reaching 2 – 3 at small radii (Done et al. 2012). We will have more to say about non-LTE effects on the emission observed from TDEs, including emitting components other than the disc, in Sects. 2.2.2 and 4.1. For now, we simply incorporate these effects on the disc continuum emission by implicitly allowing for a color correction whenever a temperature is specified below.

Following Miller (2015), it is useful to determine the maximum value of $T_{\rm eff}$ that is reached in the hottest disc annulus, as a function of $\dot{M}_{\rm acc}$. Using equation (9), we find that near the ISCO,

$$\begin{aligned} T_{\rm max} &\approx 4.3 \times 10^5\ {\rm K} \left(M_{\rm BH,6}\right)^{-1/4} \left(\frac{\kappa_{\rm es}}{0.34\ {\rm cm^2\ g^{-1}}}\right)^{-1/4} \\ &\quad \text{for } f_{\rm Edd} \gg 1, \\ &\approx 2.4 \times 10^5\ {\rm K} \left(\frac{f_{\rm Edd}}{0.1}\right)^{1/4} \left(M_{\rm BH,6}\right)^{-1/4} \\ &\quad \times \left(\frac{\eta_{\rm ref}}{0.1}\right)^{-1/4} \left(\frac{\kappa_{\rm es}}{0.34\ {\rm cm^2\ g^{-1}}}\right)^{-1/4} \end{aligned}$$

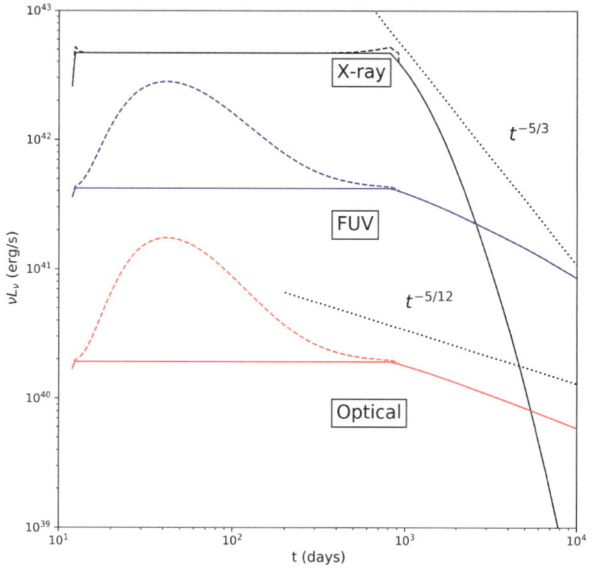

Fig. 1 Specific luminosity in soft X-rays, far ultraviolet (FUV) and optical radiation as a function of time for an accretion disc-plus-outflow as discussed in Sects. 2.1.1 and 2.2.1. Solid lines are light curves from an analytic model of a super-critical, Eddington limited accretion disc (equation (9)) with $\dot{M}_{\rm acc}$ proportional to the fallback $\dot{M}$, while the dashed lines include the emission from a wind component, as described in Lodato and Rossi (2011). We here consider a $10^6\ M_\odot$ black hole, a solar mass star, a pericenter equal the tidal radius, and we assume that 10% of the accretion rate is lost from the disc into the wind

$$\text{for}\quad f_{\rm Edd} \ll 1. \tag{11}$$

Allowing for finite fluid stress at the ISCO can lead to higher maximum disc temperatures in the case of sub-critical accretion onto a black hole of a given mass (Mummery and Balbus 2020). Note that for $f_{\rm edd} \gg 1$, the maximum disc temperature approaches a value that it is independent of $f_{\rm edd}$. Meanwhile, an approximate scaling for $T_{\rm eff}$ of the outer disc, which we call $T_{\rm out}$, based on equation (9) for $r_d/r_g \gg 1$, is

$$\begin{aligned}
T_{\rm out} &\approx 1.6\times 10^5\ {\rm K}\left(M_{\rm BH,6}\right)^{-1/4}\left(\frac{\kappa_{\rm es}}{0.34\ {\rm cm^2\ g^{-1}}}\right)^{-1/4}\left(\frac{r_d}{100\, r_g}\right)^{-1/2} \\
&\quad \text{when}\ \frac{f_{\rm Edd}}{\eta_{\rm ref}}\frac{r_g}{r_d} \gg 1, \\
&\approx 6.1\times 10^4\ {\rm K}\left(\frac{f_{\rm Edd}}{0.1}\right)^{1/4}\left(M_{\rm BH,6}\right)^{-1/4} \\
&\quad \times \left(\frac{\eta_{\rm ref}}{0.1}\right)^{-1/4}\left(\frac{\kappa_{\rm es}}{0.34\ {\rm cm^2\ g^{-1}}}\right)^{-1/4}\left(\frac{r_d}{100\, r_g}\right)^{-3/4} \\
&\quad \text{when}\ \frac{f_{\rm Edd}}{\eta_{\rm ref}}\frac{r_g}{r_d} \ll 1.
\end{aligned} \tag{12}$$

These temperatures are high enough that the optical luminosity mainly comes from the outer radii of the disc and sits in the Rayleigh-Jeans portion of the spectrum for emission at $T_{\rm out}$. In the case $(f_{\rm Edd}/\eta)(r_{\rm g}/r_{\rm d}) \gg 1$, the optical luminosity scales roughly as $\nu L_{\nu,{\rm opt}} \propto T_{\rm out}$, which depends only on how far the disc extends. Once the mass accretion rate has fallen low enough so that $(f_{\rm Edd}/\eta)(r_{\rm g}/r_{\rm d}) \ll 1$ we then have $T_{\rm out} \propto \dot{M}_{\rm acc}^{1/4}$, and so for the approximate scaling $\dot{M}_{\rm acc} \propto t^{-5/3}$ we arrive at $\nu L_{\nu,{\rm opt}} \propto t^{-5/12}$ (see Fig. 1, and Strubbe and Quataert 2009; Lodato and Rossi 2011). The duration of this temporal behaviour at optical wavelengths depends on several factors, such as whether $t_\nu/t_{\rm fb}$ at r_d approaches unity (see equation (4)), or the black hole accretion state changes in other ways at lower $\dot{M}_{\rm acc}$, or whether the TDE emission is outshone by that of the host galaxy.

On the other hand, observed soft X-ray frequencies (photon energies $\gtrsim 0.2$ keV) may initially fall close to the peak of the disc emission spectrum, in which case the X-ray light curve should more closely track the bolometric luminosity. As previously indicated, the bolometric luminosity may be capped near Eddington initially, and later decline as $\nu L_{\nu,{\rm X}} \propto t^n$ ($n \approx -5/3$). As shown in Fig. 1, for a $10^6\ M_\odot$ black hole, this power-law phase – following the Eddington limited stage – precedes an exponential decline when the spectrum softens to the point that the observed soft X-ray band overlaps only with the Wien tail of the spectrum emitted by the innermost parts of the disc (Lodato and Rossi 2011). For yet another possibility, Mummery and Balbus (2020) considered the case where the soft X-ray band is initially located on the Wien tail of the spectrum for an initially *sub-critical* disc, and the disc inflow time remains non-negligible compared to the duration of the observations. They found that the X-ray light curve is then described by a power-law multiplied by an exponential, with the power-law being tied to the viscous evolution of the disc, rather than the mass fallback rate (see Sect. 2.1.2). We noted that Mummery and Balbus (2020) initialized their disc as a ring of material at radius 15 r_g from the black hole, which if interpreted as the initial circularization radius, corresponds to a very close pericenter passage for the star ($r_p = 7.5 r_g$).

Observationally, some X-ray TDE candidates seem to show an approximate $t^{-5/3}$ decline in the soft X-ray band (X-ray Chapter), which is consistent with the disc models discussed in this chapter during the *sub-critical* mass inflow phase, for both the luminosity and temperature. The X-ray light curve of the event ASASSN-14li (Jose et al. 2014; Holoien et al. 2016) can be fit by the bare disc model of Mummery and Balbus (2020), which was also sub-critical. In contrast, the event discussed in Lin et al. (2017) does suggest an Eddington-limited plateau in the early X-ray emission. Aside from that event, however, there is little evidence for sustained, Eddington-limited phases in the X-ray data lasting as long as predicted by equation (8) with $f_{\rm in} \sim 1$. Moreover, optically selected TDEs have constraining X-ray upper limits, at least in the first few months following detection, and their optical specific luminosities are at least an order of magnitude larger than predicted from bare disc models. This can be interpreted in several ways. In addition to explanations for this fact described in this chapter (Sects. 2.2 and 3), another possibility is that the low temperatures of TDE flares result from line-driven winds from an extended accretion disc (along the lines of Laor and Davis 2014), which suppresses the fraction of the accreting gas which reaches the hotter inner disc (Miller 2015). In this case, in order to explain the stringent X-ray upper limits (sometimes $L_X < 10^{41}\ {\rm erg\,s^{-1}}$), the fraction of gas that reaches the inner disc would need to be as low as 10^{-4}.

We leave this section by noting an additional puzzle when comparing X-ray TDEs to steadily accreting super-massive black holes in active galactic nuclei (AGN). With the exception of the rare TDEs that have produced relativistic jets, the early X-ray spectra of TDEs are systematically "softer", i.e. weighted toward lower photon energies, than a broad sample of flaring AGN. Early TDE spectra also do not harden with decreasing flare luminosity, as is typically the case for AGN flares (Auchettl et al. 2018). This soft emission is indeed expected for the thermal emission from a disc as described above. However, AGN spectra typically also possess an emission component of non-thermal, higher energy photons, which are attributed to a hot "corona" which may produce high-energy emission through inverse Compton scattering (e.g. Haardt and Maraschi 1991). Some unspecified mechanism seems to inhibit the growth of such coronae during the first few years of TDE accretion disc evolution. Some authors (e.g. Wevers et al. 2019) have suggested a connection between the soft

X-ray spectra of TDEs in their first few years following disruption and the X-ray properties of the AGN class known as narrow-line Seyfert 1 galaxies, which have systematically higher accretion rates (in the sense of $f_{\rm Edd}$) and systematically softer X-ray spectra than the AGN population at large. However, there are limits to how far this analogy can be taken (see the Imposters Chapter for more details).

2.1.2 *Disc Emission at Later Times (at Least Several Years After Disruption)*

After $t_{\rm cr}$, as the accretion rate has decreased and dropped below $\dot{M}_{\rm cr}$, a state transition takes place when the thick disc we have postulated becomes geometrically thin ($h/r \ll 1$). As a consequence, the viscous timescale for the outer disc increases and eventually becomes longer than the fallback timescale at the outer edge of the disc (equation (4)), where material increasingly accumulates. If the disc remains viscously and thermally stable, the expectation is that internal viscous processes now determine the mass accretion flow, flattening the bolometric light curve to $L_{\rm bol} \propto \dot{M}_{\rm acc} \propto t^{-1.2}$ when the disc evolution is calculated in the Newtonian limit for the gas orbital motion and using a no-torque inner boundary condition (Cannizzo et al. 1990; Shen and Matzner 2014). Allowing for finite stress at the ISCO, and analyzing the thin disc equations using full general relativity, yields a shallower asymptotic decline, $L_{\rm bol} \propto t^{-n}$ with $n \approx 0.6$–0.7 (Balbus and Mummery 2018), and X-ray emission on the Wien tail then continues to follow the power-law-times-exponential decline discussed earlier (Mummery and Balbus 2020). The predicted light curve for frequency bands close to the Rayleigh-Jeans part of the spectrum is $L_\nu \propto \dot{M}_{\rm acc}^{1/4} \propto t^{-0.3}$ for the Newtonian, vanishing-inner-stress predictions, and $\approx t^{-0.16}$ for the updated treatment by Balbus and Mummery (2018). In all cases, the viscous spreading of the disc increases the emitting area of the outer disc annuli and causes the actual Rayleigh-Jeans band light curves to decline even more gradually before they approach the asymptotic power-law decays quoted here.

Indeed, slow (nearly-flat) temporal decay in all three bands of the Swift UVOT instrument is seen in a sample of TDEs followed up 5–10 years after first detection (van Velzen et al. 2019). The characteristic photospheric radii are close to the tidal disruption radii, as expected for discs that have spread out by just a factor of a few. Late-time X-ray detections of TDEs, particularly those without early X-ray detections, also support the interpretation that a bare accretion disc is responsible for the late-time emission (Jonker et al. 2019). Interestingly, the spectral energy distribution of the disc model that can explain the late-time observations suggest that most of the radiation is emitted in the hardly observable extreme-ultraviolet frequency band. This is consistent with some observed TDE dust echoes (Echo Chapter). The energy radiated over time is estimated to be a few percent of the available accretion energy, a cumulative radiative efficiency that is not much lower than many AGN, providing a possible solution to one aspect of what was dubbed "the missing energy problem" prior to these late-time observations (Piran et al. 2015a; Stone and Metzger 2016; Lu and Kumar 2018).

2.2 Reprocessing by Outflows, Radiative Absorption, and Comptonization That Affect the Early UV/Optical Emission

As discussed elsewhere in this book (e.g. the Optical Chapter), many TDEs exhibit optical and UV emission at early times (weeks to months after first detection) far in excess of what is predicted from the preceding description of bare disc emission. Similarly, their x-ray emission is, at least initially, much lower than the bare disc predictions for the case of rapid inflow and circularization. Thus, in the case that a circularized accretion disc has formed at those times, the story cannot be as simple as previously described if such accretion flows are responsible for the optically bright events.

There are three primary mechanisms that might convert the early bare disc emission to that with an SED peaking closer to the UV or optical. These are 1) Trapping of radiation in outflowing material by electron scattering, and the subsequent adiabatic losses of radiative energy as the photons do work on the outflowing material 2) absorption of disc photons by material at larger distances, which emits primarily at longer wavelengths 3) Compton scattering of photons off thermal electrons (here the electron thermal motions take the place of bulk kinetic motion that applied to point 1).

All three ought to be happening to some extent in the picture of a rapidly circularized accretion flow, and the manner in which they combine depends on the underlying hydrodynamics assumed.

2.2.1 Adiabatic Trapping of Radiation in Expanding Flows

As discussed earlier, when mass is brought into the vicinity of a black hole at a super-critical rate, an expected outcome is the formation of outflows that carry away mass, energy, and angular momentum that cannot be captured by the black hole. These outflows may be powered by the energy released during the initial collision of the bound debris with itself (e.g. Jiang et al. 2016; Sądowski et al. 2016; Lu and Bonnerot 2019), or by energy released during the subsequent circularization and accretion of the circularized gas. In some cases the accretion-related outflows may take the form of a relativistic jet, either magnetically launched (e.g. Tchekhovskoy et al. 2014; Sądowski and Narayan 2015; Dai et al. 2018; Curd and Narayan 2019) or radiatively launched (e.g. Sądowski and Narayan 2015; Coughlin and Begelman 2014). We do not focus on emission relating to the jet here, although we do discuss the radio afterglow signal associated with it in Sect. 5.2. Instead, here we focus on wild-angle, mildly relativistic or non-relativistic outflows.

The fundamental ingredients of the emission corresponding to these wide-angle outflows in the context of TDEs were laid out in SQ09, drawn in part from the treatment by Rossi and Begelman (2009) for the disruptive tidal interaction of neutron star binaries by Rossi and Begelman (2009). The TDE outflow model has been subsequently refined by Lodato and Rossi (2011), Metzger and Stone (2016), and Lu and Bonnerot (2019) (henceforth LR11, MS16, and LB19, respectively), and others. In short, the energy released from circularization and/or accretion goes into radiation that is trapped in a promptly launched, quasi-spherical envelope of outflowing material (see Fig. 2). In the process of escaping, the radiation does work adiabatically on the gas, transferring energy from the radiation to the kinetic energy of the outflow. This causes the radiation that finally escapes to have a lower overall flux and peak emission frequency than it did at the point where the orbital energy of the gas was first converted to radiation.

The various papers mentioned above differ on details such as the fraction of the falling back stellar debris that goes into the outflow, the location at which the outflow is launched, and whether the energy carried by the outflow comes only from the circularization process, or also from energy released from the accretion disc. SQ09's and LR11's outflow models are driven primarily by circularization, and they predict the resulting bolometric light curve to rise as $t^{11/9}$, followed by a $t^{-5/9}$ decline that lasts until $t_{\rm cr}$ (equation (8)).

It is useful to keep track of a "temperature" $T_{\rm rad}$ corresponding to the radiation at all depths within the outflow. Here we use the term temperature not in a strict thermodynamic sense, but simply as a measure of the energy density of the radiation field: $T_{\rm rad} \equiv (E_{\rm rad}/a_{\rm rad})^4$, where $E_{\rm rad}$ is the radiation energy per unit volume, and $a_{\rm rad} \equiv 4\sigma_{\rm SB}/c$.

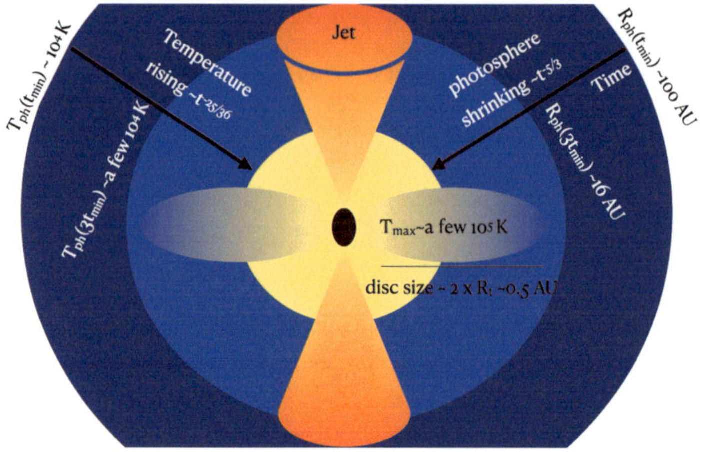

Fig. 2 Schematic of the disc-wind-jet structure that may arise following rapid circularisation of the stellar debris. The stated time dependencies of the shrinking photospheric radius and increasing photospheric temperature of the wind come from the adiabatic trapping model by Lodato and Rossi (2011)

Of particular interest is the value of $T_{\rm rad}$ at the electron scattering photosphere, which we will label as $T_{\rm ph}$, since this will be related to the mean energy of the photons that escape the system. From the analysis of SQ09 and LR11, $T_{\rm ph}$ evolves gradually but does not stay constant, first declining as $t^{-7/36}$ and then, past the peak of the light curve, rising as $t^{25/36}$.

The predictions of MS16 are slightly different for several reasons, but most relevant here is that they keep track of when the radiation will no longer be sufficiently trapped in the outflow to continue to adiabatically transfer its energy to the gas. This estimate is made in terms of the so-called trapping radius $r_{\rm tr}$, the location in the outflow where the radiative diffusion time through the outlying material first equals the expansion time of the gas. This leads to the estimate that the optical depth beyond the trapping radius, $\tau(r_{\rm tr})$, equals $c/v_{\rm out}$, where $v_{\rm out}$ is the speed of the outflow, and where τ can be estimated by assuming that electron scattering is taken to be the dominant process that may trap the radiation. For a steady-state spherical outflow ejecting mass at a rate $\dot{M}_{\rm out}$, MS16 show that this leads to $r_{\rm tr} = \dot{M}_{\rm out}\kappa_{\rm es}/(4\pi c)$, which evaluates to

$$r_{\rm tr} = 5.7 \times 10^{13}\ {\rm cm} \left(\frac{\dot{M}_{\rm out}}{1\ M_\odot\ {\rm yr}^{-1}}\right)\left(\frac{\kappa_{\rm es}}{0.34\ {\rm cm}^2\ {\rm g}^{-1}}\right). \tag{13}$$

This estimate assumes that the outflow corresponding to this steady $\dot{M}$ effectively extends to infinity - accounting for its truncation at finite radius, and the presence of material launched earlier at a different $\dot{M}_{\rm out}$, will modify the true location of $r_{\rm tr}$. Indeed, while the mass outflow rate may remain approximately steady over several time intervals of $r_{\rm tr}/v_{\rm out}$, it will change over longer time scales. As the mass fallback rate drops, eventually $\dot{M}_{\rm out}$ in this model drops as well, and the trapping radius shrinks and eventually approaches the launching radius, at which time the radiation no longer loses a significant amount of energy radiative energy to adiabatic work. MS16 estimate this time, labeled $t_{\rm tr}$, by assuming that $\dot{M}_{\rm out}$ is a constant, near-unity fraction of the total mass fallback rate $\dot{M}$. They find $t_{\rm tr}$, to be only a factor of a few times $t_{\rm fb}$. Notice that the location of $r_{\rm tr}$ is important for determining how $L_{\rm bol}$ compares to the output from the bare disc.

By keeping track of how r_{tr} evolves over time, MS16 predict that after its initial rise, L_{bol} experiences a shallow decline similar to the $t^{-5/9}$ predicted by the previous authors, while adiabatic losses are still important. Beginning around t_{tr}, however, this then transitions to a decline tied to the standard approximation for the mass fallback rate, $t^{-5/3}$, as the accretion luminosity escapes without being adiabatically adjusted. Meanwhile, T_{ph} declines gradually at first, but after reaching its minimum when L_{bol} is at peak, T_{ph} rises as $\approx t^{1/2}$, which is more gradual than the $t^{25/36}$ rise in T_{ph} found by the previous authors.

Once again, to connect to observations what is really needed is a prediction for the light curve in a particular band. Sometimes (e.g. LR11, MS16), the assumption is made that the radiation spectrum emitted from the scattering photosphere is fully thermalized and so can be represented as a blackbody emitting at T_{ph}. This is not always justified. Even though the radiation is optically thick to electron scattering out to the photosphere by definition, that does not guarantee that it is optically thick to the absorption and emission processes that thermalize the radiation spectrum, as it would need to be at every wavelength of interest.

We now list the predictions that can be made for these outflow models under the assumption that the radiation is emitted as a blackbody at T_{ph} after all. For LR11, for any band on the Rayleigh-Jeans tail corresponding to T_{ph}, then in this case L_ν will first rise as $t^{65/36}$ and then decline as $t^{-95/36}$. For MS16, for bands on the Rayleigh-Jeans tail, L_ν will initially rise and then remain nearly constant during the adiabatically trapped phase, but eventually transition to the rapid $t^{-95/36}$ shortly after t_{tr}. In both cases, T_{ph} is typically high enough that optical bands lie in this Rayleigh-Jeans regime.

In Sects. 2.2 and 4.1 we will take a closer look at the requirements for re-thermalization of the radiation and the consequences of partial thermalization. However, we will now briefly consider how the radiation field might evolve in the absence of *any* absorption, and is instead affected only through trapping in the fast-moving outflow by scattering. Likewise, we will momentarily take this scattering to be coherent in the rest frame of the fluid - the effects of thermal electron motion on the photon scattering process will be discussed in Sect. 2.2.3 on thermal Comptonization.

In these pure-scattering conditions, adiabatic losses will reduce the mean energy of the photons as they scatter through the outflow. This results in a "stretching" of the spectrum so that it extends to lower energies, making the spectral shape wider than that of a blackbody with the same peak energy (Roth and Kasen 2018; Dai et al. 2018). The amount by which the mean photon energy shifts can be estimated by determining the fraction of radiative energy that is lost up to the trapping radius, although in reality photons continue to lose energy as they scatter even beyond that radius. Another way to do the estimate, valid in the regime when the radiative diffusion time is short compared to the timescales over which the outflow is evolving, is to use an integral of the velocity divergence (equation 20 of Roth and Kasen 2018).

In summary, the radiation escaping from such promptly launched, quasi-spherical outflows might be bright enough to match the peak optical luminosities of observed TDEs (i.e. peak νL_ν in the approximate range of 10^{42} to 10^{43} erg s^{-1} in optical bands), although the optical luminosity predicted by some versions of these models is too faint by about an order of magnitude when compared to many observed events and for typical parameter choices (see Fig. 1). The challenge of producing bright enough optical emission from these outflows is exacerbated when the radiation is not efficiently re-thermalized in the outflow. In any case, after several t_{fb} the optical light-curves from these outflows are expected to drop rapidly and become subdominant compared to more direct disc emission.

2.2.2 *Reprocessing via Absorption and Re-Emission, and the Requirements for Re-Thermalization of Radiation*

In the last section we saw that material launched in outflows can alter the radiation from the bare disc by adiabatically converting radiative energy to gas kinetic energy. Throughout that discussion, coherent electron scattering was the only interaction between the radiation and the stellar material. Now we consider the role of absorption and re-emission on re-thermalizing the radiation at a lower temperature.

The idea that a "reprocessing envelope", consisting of disrupted stellar material at large radii, might intercept X-rays produced from the inner accretion and re-radiate at lower temperatures to produce bright UV / optical emission was first put forth at least as early as Loeb and Ulmer (1997). The outflows discussed in the last section are one potential origin of such a reprocessing envelope. A thick accretion torus (e.g. Coughlin and Begelman 2014) might also serve this purpose. Yet another case, not discussed here, is reprocessing by the tail of bound and unbound debris. (Bogdanović et al. 2004).

Dust that is present in the host galaxy or formed in the reprocessing envelope might also reprocess a large portion of the emitted radiation to infrared wavelengths (first noted in Kormendy and Richstone 1995). We defer that topic to the Echo Chapter and instead we focus on reprocessing envelopes generated in some manner by the stellar debris without *in situ* dust formation, and we focus on the UV/optical emission.

In all cases, we are considering a large fraction of the material from the disrupted star concentrated within $\lesssim 1000 r_g$. This is irradiated by a hot inner source of temperature $\gtrsim 10^5$ Kelvin and peak luminosity in the range of 10^{44} to 10^{45} $\mathrm{erg\,s^{-1}}$. In these conditions, we expect the continuum UV/optical spectrum that escapes to be formed in regions of appreciable scattering depth and high ionization. This influences the re-thermalization of the radiation. Let α^{es} and α_ν^{abs} respectively denote the opacities from electron scattering and from all absorption processes for frequencies between ν and $\nu + d\nu$. We define an opacity ratio ϵ_ν as

$$\epsilon_\nu = \frac{\alpha_\nu^{abs}}{\alpha^{es} + \alpha_\nu^{abs}} \tag{14}$$

An elementary result of radiative transfer theory is that the *effective* optical depth to absorption is then given by $\tau_{\nu,\mathrm{tot}}\sqrt{\epsilon_\nu}$. Said another way, photons emitted at optical depths above the thermalization depth

$$\tau_{\nu,\mathrm{therm}} \equiv 1/\sqrt{\epsilon_\nu} \tag{15}$$

have an exponentially increasing chance of being absorbed rather than eventually escaping to the observer by scattering their way out. We define the thermalization radius $r_{\nu,\mathrm{therm}}$ as the radius corresponding to the thermalization depth, or equivalently the radius where the *effective* optical depth to absorption for the given frequency interval is 1. An analytic expression for $r_{\nu,\mathrm{therm}}$ in the simplified case where the only absorption process is free-free absorption is presented in Sect. 4.1.

If and when the absorption opacities are large enough that $r_{\nu,\mathrm{therm}}$ coincides with the true photosphere (the average surface of last photon interaction for any process), and if this is true for *all* frequencies of interest, then we may apply the properties of blackbody emission. If the blackbody photosphere is spherically symmetric, then we have

$$L_{\mathrm{bol}} = 4\pi r_{\mathrm{ph}}^2 \sigma_{\mathrm{SB}} T_{\mathrm{eff}}^4 , \tag{16}$$

where $L_{\rm bol}$ here refers to the luminosity escaping from the system (not necessarily equal to the $L_{\rm bol}$ from the bare disc), and $r_{\rm ph}$ is the photospheric radius. In this situation, $T_{\rm eff}$ will also correspond to the observed color temperature. Near the peak of the light curve, and for typical color temperatures associated with optically bright TDEs (Optical Chapter), the spherical blackbody model puts the photospheric radius at

$$r_{\rm ph} = 4.2 \times 10^{14}\ {\rm cm} \left(\frac{L_{\rm bol}}{10^{44}\ {\rm erg\,s^{-1}}}\right)^{1/2} \left(\frac{T_{\rm eff}}{3 \times 10^4\ {\rm K}}\right)^{-2}. \tag{17}$$

If $r_{\nu,{\rm therm}}$ lies beneath the scattering photosphere but the effective photon path to absorption is sufficiently small compared to $r_{\nu,{\rm therm}}$ for all ν of interest, then we can treat the system using equation (16) along with a color correction, as discussed in Sect. 2.1.1. For even lower values of ϵ_ν, $r_{\nu,{\rm therm}}$ becomes significantly different for each ν, or does not exist at all for some ν. In this case the spectrum will no longer have a blackbody shape, as discussed in the last section and also in Sect. 4.1.

In TDE models with rapidly circularized accretion discs, the assumption that there is a large effective optical depth to absorption at all UV/optical wavelengths poses theoretical challenges. On the one hand, the radiative transfer calculations of *static* reprocessing envelopes performed by Roth et al. (2016) suggested that there can be sufficiently high effective absorption optical depth to provide the observed optical flux, although even then the radiation was not completely thermalized. This calculation also assumed that a large fraction of the stellar debris ($\gtrsim 0.1\ M_\odot$) can be present all at once in a roughly spherical distribution contained within a radius of $\lesssim 10^{15}$ cm. What remains to be seen is whether the same required absorption optical depth is present in more realistic, time-dependent, hydrodynamically motivated situations that assume rapid circularization. If there is an accretion disc involved, it will be crucial to track how that disc builds up, spreads out and evolves in thickness over time, or in the case of an outflow, whether the required effective optical depth can be sustained for a long enough time before the debris becomes too optically thin.

MS16 have provided estimates of the "ionization breakout" time, when the envelope becomes optically thin to the absorption of soft X-rays from the inner disc,[1] and find that for $M_{\rm BH} \lesssim 10^7\ M_\odot$, the ionizing radiation may take several months to escape. Such an X-ray breakout may have been observed in ASASSN-15oi, although another interpretation is slow circularization leading to delayed disc formation (Gezari et al. 2017). In any case, if X-ray breakout is to happen at all, it must occur in such a way that the envelope remains effectively optically thick at UV/optical wavelengths at least until the time of breakout. This is because for UV/optical emission to be coming from re-thermalized radiation, we will need the effective optical depth to be large for *both* the soft X-rays (to completely or partially absorb them), and for the UV/optical wavelengths (to produce a thermal or near-thermal spectrum at those wavelengths). Otherwise, we would also expect to see the optical flux drop off faster than usual around the time of X-ray breakout (in such a scenario the peak of the re-emitted emission would mostly likely be moving farther into the extreme ultraviolet). In ASASSN-15oi, a smooth power-law decline could connect the optical flux before and after the X-ray flare. This suggests either that the system did in fact remain

[1] The analytic estimates for soft X-ray absorption in MS16 and Roth 2016 were based on the He II photoionization opacity. However, further examination of the Roth et al. (2016) results by those authors have indicated that the dominant absorption process for the soft X-rays was the photoionization of oxygen, the only metal included in that calculation. This suggests that improved X-ray breakout estimates must account for opacities from oxygen, nitrogen, carbon, and possibly other species, rather than only helium.

effectively optically thick to optical wavelengths past the time the X-rays could escape, or that something other than the model described here is responsible for the optical emission.

In practice, UV/optical emission from TDEs is almost always fit to a spherical blackbody with temperature and radius specified by equation (16). It is usually possible to produce a satisfactory match to the emission in the observed bands this way, despite the theoretical challenges we have described for equation (16) to apply. However, it is usually the case that the optical bands do not cover the peak of the spectrum for the assumed blackbody, so it is impossible to know whether the true spectrum deviates from a blackbody at extreme ultra-violet wavelengths.

A feature of such fits for UV/optical blackbody temperatures is that they either remain constant over several months, or to the extent that the temperature changes it seems to do so slowly (Optical Chapter). Again, this statement is subject to some uncertainty given that we usually do not observe the peak of the spectral emission. This behavior of the fitted temperature can be contrasted with supernovae, which tend to exhibit temperatures that decrease with time as the ejecta expands in cools (Imposters Chapter). The fact that TDEs do not seem to cool in the same way has become an important feature used to identify promising TDE candidates in optical surveys (e.g. Hung et al. 2018) The consequence of the slowly changing temperature and the rising then falling $L_{\rm bol}$ is that $r_{\rm ph}$ is inferred to move out and then in. Connecting this to the underlying hydrodynamics is an important task for models seeking to explain optical emission this way. Some authors use this $r_{\rm ph}$ evolution as evidence for an outflow (e.g. Strubbe and Murray 2015), and we will see similar behavior in the models discussed in Sect. 2.3.

2.2.3 Comptonization

In many accreting black hole systems, inverse Comptonization of soft photons by hot electrons transfers energy from the electrons to the radiation field and produces the harder X-rays that are observed. Such processes are likely happening in TDEs with observed hard X-ray emission such as Swift J1644. For example, in Bloom et al. (2011), an external Comptonization model was suggested to explain the hard x-ray emission from this event, in which softer photons produced primarily by the accretion disc and/or circularization shock are up-scattered by relativistic electrons in the jet. It was later suggested that the hard x-ray power-law in Swift J1644 could also be produced by mildly-relativistic, more thermalized electrons in the jet (Lu et al. 2017). Bright, beamed hard X-ray emission from inverse compton scattering is also produced in simulations with a relativistic jet and radiative transfer post-processing (Dai et al. 2018; Curd and Narayan 2019, hereafter C19). In fact, even in cases in which no highly relativistic outflow was produced, and only thermal electrons were considered, the C19 radiative post-processing suggests that a tail of hard X-ray emission will be produced and potentially detected by viewers at a wide range of inclinations, which is line with Lu et al. (2017).

It is also important to consider the role that direct (as opposed to inverse) Comptonization will play in shaping the spectrum of TDEs at a range of lower energies including soft X-ray and UV, which will often involve the transfer of energy from the photons to the electrons. While the bound electrons responsible for most of the absorption and re-emission of light beyond the accretion disc may be far out of LTE, it is generally reasonable to conclude that at the relevant densities the free electrons responsible for scattering the light are in LTE. This means they possess a velocity distribution described by a Maxwell-Boltzmann distribution and an associated temperature T_e, which in general might differ from other characteristic temperatures assigned to a region of gas, such as the radiation color temperature there.

As such, we can use the Compton y heuristic for determining the importance of Comptonization. This quantity if formed by multiplying the average change in photon energy expected for each scatter by the expected number of times the photon will scatter before escaping the system:

$$y \equiv \frac{4kT_e}{m_e c^2}\tau^2 \qquad \text{when } \tau \gg 1 \tag{18}$$

Values of T_e outside a jet/funnel take on values that range from several 10^4 to 10^5 Kelvin, while the photons from the inner disc may be more energetic. For T_e of 10^5 Kelvin, y evaluates to 0.7 for τ of 100. This implies a significant shift in the photon energy due to thermal motion of electrons during repeated scattering.

In addition to computing y, it is equally important to identify the thermalization depth for a particular wavelength, since the relevant quantity is really y above that depth. In other words, photon absorption competes with Comptonization. For example, as seen in the radiative post-processing of Dai et al. (2018), the majority of the soft X-rays produced in the disc close to the black hole were affected by Comptonization and shifted to lower energies. However, when viewed at disc inclinations close to edge-on, the optical portion of the spectrum remained relatively unaltered between test cases where Comptonization was included or not. This can be understood because the optical thermalization depths tended to be closer to the scattering photosphere than for the soft X-rays, as treated in this calculation. Therefore there was a relatively small value for y for the optical photons emitted at larger radii; that is, those photons were able to escape with fewer scatterings.

2.3 Simplified TDE Models for Light Curve Fitting

We conclude this section with one approach to connect TDE light curves to the mass fallback rate (Mockler et al. 2019, hereafter M19), which has been implemented using the MOSFiT astronomical transient fitting framework (Guillochon et al. 2018). Along with everything that has been written in this chapter up to this point, this model assumes that circularization is rapid for the portion of the debris that forms an accretion disc. This model invokes the presence of a reprocessing layer (Sect. 2.2.2), emitting as a spherical blackbody, to convert radiative energy emitted from the disc at primarily soft X-ray wavelengths into UV/optical emission, at least for several $t_{\rm fb}$ while the reprocessing layer remains optically thick.

2.3.1 Motivation and Method

Among the motivations for such an approach, we consider again simple estimates of the stellar mass fallback rate (see also Sect. 2.1.1):

$$\text{peak } \dot{M}_{\rm fb} \propto M_{\rm BH}^{-1/2} M_*^2 R_*^{-3/2},$$

$$t_{\rm fb} \propto M_{\rm BH}^{1/2} M_*^{-1} R_*^{3/2}.$$

The black hole mass affects both the mass fallback rate and the timescale to the same extent. However, the mass-radius relationship of solar metallicity main-sequence stars is such that the effects of the star's mass and radius largely cancel in the equation for $t_{\rm fb}$. The timescale of mass fallback curves is therefore highly sensitive to the mass of the black hole, as shown in Fig. 3, even when simulations of the disruption process provide more accurate information than the simple scalings listed above. To the extent that the luminosity of observed TDEs is directly tied to the mass fallback rate as computed from hydrodynamics simulations (e.g.

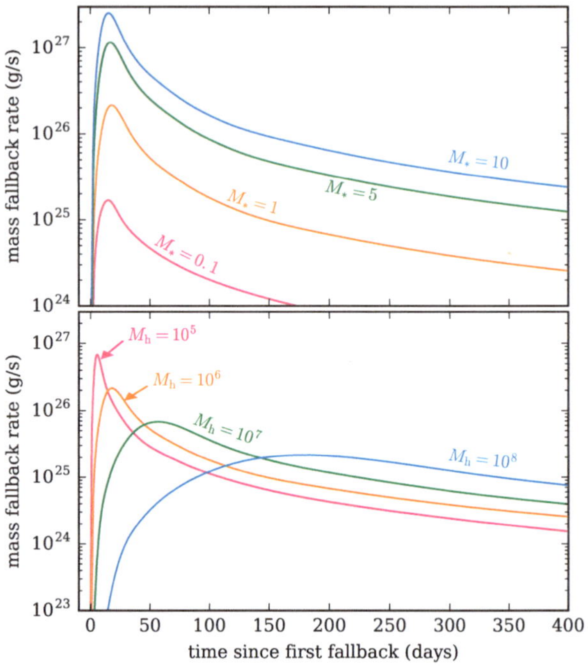

Fig. 3 Top panel: The mass of the disrupted star has a large effect on the overall normalization of the fallback curve, but a relatively minor effect on its shape. Bottom panel: The black hole mass, on the other hand, is largely responsible for setting the timescale t_{fb} of the fallback rate. If the observed light curves have a direct relation to the mass fallback rate, it may be possible to estimate M_{BH} based on the shape of the light curves

Guillochon and Ramirez-Ruiz 2013), fitting the light curves of TDEs may provide an indirect measurement of the mass of the disrupting black hole (M19). The mass of the star does have a second-order effect on the peak timescale and the power-law decline, and in some cases this is enough to fit for the stellar mass as well.

As mentioned earlier in this chapter, the observed temperatures of TDEs do not change dramatically even while the luminosities change by many orders of magnitude, implying that such a blackbody photosphere is dynamic, growing and shrinking in radius in a manner that tracks the light curve. The precise physical origin of the reprocessing layer is left unspecified in what follows. However, given the dynamic nature of the photosphere, simulations of outflows launched during the circularization and/or accretion processes (Sect. 2.2.1) are especially relevant. In particular, Jiang et al. (2016) found that a solution of the form $R_{phot} \propto L^{\xi}$ fit their simulations for a power law index of $\xi \sim 1$, although it must be kept in mind that these simulations treated the stream intersection as an isolated system that did not interact with other stellar debris that might influence the evolution of the outflow over the course of the event. The same general power-law formula is used to model the photosphere in the MOSFiT model. M19 found that the best-fit power-law indices for a sample of 14 optical TDEs with bright optical and UV emission varied between $\xi \sim 0.5$ and $\xi \sim 4$, with the majority of the fits preferring ξ-values between 1 and 2. For a blackbody photosphere to emit at a constant temperature as the luminosity increases, the power-law index ξ would have to equal 0.5. Both the simulations from Jiang et al. (2016) and the fitting results from M19 found $\xi > 0.5$, implying that the blackbody temperature decreases slightly near peak, and increases at late times as the effective photosphere radii decrease (Fig. 4). This behavior qualitatively resembles the temperature evolution for the models discussed in Sect. 2.2.1.

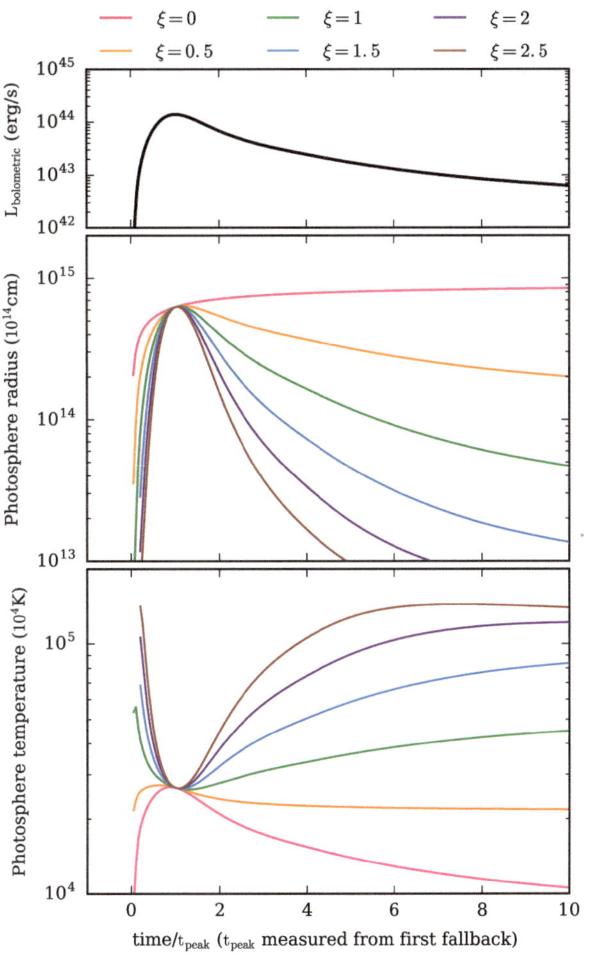

Fig. 4 Top panel: A reference bolometric light curve fit following the procedure of Mockler et al. (2019), which assumes that it closely tracks mass fallback rates found in hydrodynamic simulations. Middle panel: the time evolution of the blackbody photospheric radius that is assumed to give rise to the UV/optical emission, for various choices of the tuning parameter ξ. Bottom panel: the evolution of the corresponding blackbody temperature. For $\xi \gtrsim 0.5$, the temperature first declines and then rises, as was the case for the outflow models discussed in Sect. 2.2.1

2.3.2 Sources of Modeling Uncertainty

As discussed above, the effects of the stellar mass and radius on the TDE light curve largely cancel out when the disrupted star is young and has metallicity close to solar. This allows the peak timescale to be mostly sensitive to the mass of the black hole. However, varying the age and metallicity of the star changes the mass-radius relationship (Choi et al. 2016), allowing the mass and radius of the star to influence the peak timescale. This uncertainty can be expressed as a systematic error, as described in M19. They found that the error in the black hole mass due to this uncertainty is ± 0.2 dex.

In this model it is often difficult to disentangle the influence of the mass of the star from that of the radiative efficiency. The main effect both of these parameters have on the light curve is to increase (or decrease) the bolometric luminosity, and therefore their degeneracy has very little impact on the uncertainty in the measurement of the mass of the black hole.

Another source of uncertainty relates to the choice to set the radiative efficiency to a constant value for each event. Since the mass fallback rate spans a large range of values, from possibly super-critical at early times to sub-critical at late times, we might expect the radiative efficiency to change significantly over time if the accretion rate is tied directly to mass fallback rate.

Most difficult to quantify are the other systematic uncertainties in the values of physical parameters derived such as the black hole mass using this method. These systematic uncertainties relate to the underlying assumptions that have been made to connect the fitting parameters to the underlying physical processes. Purely phenomenological models can be declared successes if they describe the data to within an acceptable level using some goodness-of-fit metric such as χ^2. In fact, it is common to add pieces to such models until they meet this standard. The parameters may then be interpreted as actual physical measurements, but only when there is strong evidence for the underlying model.

As this book illustrates, there are many competing models for most aspects of these events. For this reason, a high priority of empirical studies will be to test and distinguish these models, in addition to exploring the values for physical parameters that are returned using an assumed model. Thus, it is important to keep in mind that even when a model may be able to reproduce the measurements at a statistical error level consistent within the error level of the data, there is more work that must be done to reduce the aforementioned systematic uncertainties.

3 Emission from Shocked Tidal Debris That Does Not Rapidly Circularize

We now discuss a post-disruption hydrodynamic scenario that differs qualitatively from most of the models discussed in Sect. 2, which generally had assumed that an accretion disc containing a large fraction of the bound stellar material could form rapidly (compared to $t_{\rm fb}$) upon fallback. In Sect. 3.1 we briefly review the underlying hydrodynamics that motivates a picture where the bound stellar material remains incompletely circularized for several months following the initial fallback. Sections 3.2 and 3.3 describe the inner and outer portions of this flow, which may give rise to prompt emission of soft X-rays and UV/optical emission, respectively. Section 3.4 then applies these principles to infer the physical parameters of the event ASASSN-14li based on its multi-band emission.

3.1 Underlying Hydrodynamics

It is common to pose the entire problem in terms of a characteristic scale, the nominal tidal radius $r_t \equiv R_*(M_{\rm BH}/M_*)^{1/3}$, defined by the distance from the black hole at which tidal gravity is comparable to the star's self-gravity. However, this number is only an order-of-magnitude estimate of the radius at which strong tidal effects occur because its derivation is on the basis of dimensional analysis rather than actual dynamical calculation.

As first indicated by Newtonian simulations using stars with polytropic internal structures disrupted at a systematically adjusted range of impact parameters (Guillochon and Ramirez-Ruiz 2013), and refined by more recent fully-relativistic simulations using stars with realistic main-sequence internal structures (Ryu et al. 2020a,b,c), the critical radius within which a star is completely disrupted is in general a number of order unity times r_t (see also Golightly et al. 2019; Goicovic et al. 2019; Law-Smith et al. 2019, for non-relativistic treatments of the disruption of realistic stellar models). Because, somewhat fortuitously, this critical distance is nearly independent of stellar mass for all masses $\lesssim 3\ M_\odot$ (Ryu et al. 2020a,b), the dependence of this order-unity correction Ψ on stellar mass and black hole mass can be factored into one piece that is a function only of M_* and another that is a function only of $M_{\rm BH}$ (Ryu et al. 2020a).

Similarly, the nominal tidal radius may be used to make an order-of-magnitude estimate of the maximum binding energy of the tidal debris, $\Delta\epsilon \equiv GM_{\rm BH}R_*/r_t^2$, but both structural and relativistic effects can case the actual maximum binding energy ΔE to differ from $\Delta\epsilon$ by an order-unity factor $\Xi \equiv \Delta E/\Delta\epsilon$. Like Ψ, Ξ can also be factored into one portion dependent on M_* and another dependent on $M_{\rm BH}$.

When matter is initially torn off the star, its orbital eccentricity lies in the range $1 \pm 2\Xi(M_*/M_{\rm BH})^{1/3}$ and its semi-major axis is at least that of the most tightly-bound orbit,

$$\begin{aligned} a_{\rm min} &= [1/(2\Xi)]r_t(M_{\rm BH}/M_*)^{1/3} = [1/(2\Xi)]R_*(M_{\rm BH}/M_*)^{2/3} \\ &\approx 3.5\times 10^{14}\,\Xi^{-1}(M_*/M_\odot)^{0.22}M_{\rm BH,6}^{2/3}\ {\rm cm} \\ &\approx 2400\,\Xi^{-1}(M_*/M_\odot)^{0.22}M_{\rm BH,6}^{-1/3}\,r_g. \end{aligned} \tag{19}$$

Here we have adopted a single power-law fit to the main-sequence radius-mass relation $R_* \simeq 0.9R_\odot(M_*/M_\odot)^{0.88}$ (Ryu et al. 2020b). The orbital period of this most tightly-bound matter sets the characteristic timescale for the evolution of the system, $t_{\rm fb}$ (Sect. 2.1.1), but note that the expression given in that section requires a correction factor $\Xi^{-3/2}$.

To join a nearly-circular accretion flow at radii $\sim r_t$ therefore requires, at the very least, a substantial loss of orbital energy. As simulations have shown, the "nozzle" shock, where the returning streams converge near the stellar pericenter, is usually a rather weak shock because the convergence speed is a fraction $\sim (M_*/M_{\rm BH})^{1/3} \sim 10^{-2}$ times the orbital speed (Guillochon et al. 2014; Shiokawa et al. 2015, the latter henceforth S15). It can eventually create larger deflections, but only for a minority of the returning mass (S15). Additional shocks can be generated by relativistic apsidal precession, but for many parameter combinations the precession angle is small enough that these shocks occur near the orbital apocenters, where there is little kinetic energy available for dissipation (S15, Dai et al. 2015). By converting some orbital energy into heat, these shocks can increase $1-e$ by factors of a few, but that still leaves the orbits strongly eccentric (for related results, see also Hayasaki et al. 2016; Bonnerot et al. 2016). In contrast, for stellar pericenters closer than $\approx 10r_g$, where the precession angle is large enough to bring the shock close to the pericenter scale (Dai et al. 2015), the orbital energy can be dissipated much more rapidly. However, when the pericenter is that small, the probability of direct capture is larger than the probability of a tidal disruption, so such events are relatively rare (Krolik et al. 2020). Another possibility is that, even for somewhat larger initial pericenters ($r_p \gtrsim 15r_g$), gas is launched from the initial stream intersection point and sent on converging trajectories toward the black hole, leading to secondary shocks and increased dissipation (Bonnerot and Lu 2020). For a detailed review, see the Formation of the Accretion Flow Chapter. For the remainder of this section we focus on the case where these secondary shocks are not present and the rate of orbital energy dissipation is too low to promptly circularize most of the bound stellar debris falling back until several intervals of $t_{\rm fb}$ have elapsed.

3.2 Inner Flow

Although the shocks resulting from apsidal precession in this case cannot deflect material onto orbits with a scale $\sim r_t$, there are other mechanisms to deposit mass in such a compact flow. In particular, as just mentioned, S15 found that the nozzle shock can do so for $\sim 1/3$ of the bound gas. However, this does not necessarily happen immediately upon the debris'

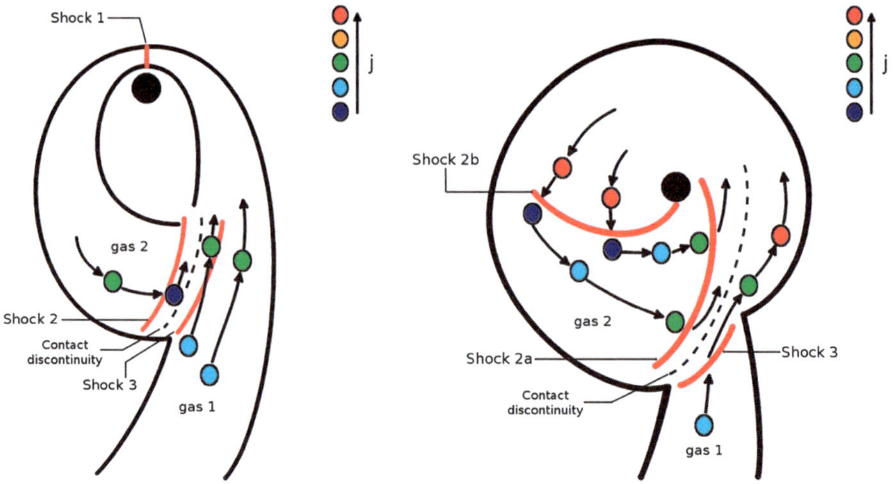

Fig. 5 A schematic for the bound stellar mass returning to pericenter and shocking with itself, as discussed in sections Sect. 3.2 and 3.3. This figure originally appeared in Shiokawa et al. (2015) and has been reproduced with permission. The "outer flow" consists of the majority of this bound stellar material that has returned and that remains on moderately elliptical orbits with semi-major axes on the order of $a_{\rm min}$ (equation (20)). The "inner flow" consists of the subset of the material depicted in the right panel that has been deflected by shocks into orbits of lower angular momentum and has dissipated a substantial fraction of its orbital energy. The color coding represents the specific angular momentum of the sample streamlines that have been drawn with arrows. For illustrative purposes, not all aspects of the diagram are drawn to scale

first return from apocenter. In the simulation of S15, it took $\approx 3t_{\rm fb}$ for the nozzle shock to build in strength before it deflected significant amounts of gas toward smaller radii. Matter drawn from the larger-scale accretion flow was then pushed inward at a roughly constant rate, $\sim 0.1\times$ the maximum mass-return rate, until $\approx 10t_{\rm fb}$, after which the inflow rate fell.

The duration of these different periods and the fraction of the mass that quickly enters a small accretion disc depends upon M_* and $M_{\rm BH}$. In particular, it should be noted that S15 used a mass ratio $M_* / M_{\rm BH}$ of 780, which is small compared to more common values of $\sim 10^6$ for the tidal disruption of a main-sequence star of a super-massive black hole, and this mass ratio influences the hydrodynamics of the nozzle shock and the stream intersection shock. Nevertheless, it is reasonable to expect that whenever the apsidal precession angle per pericenter passage is sufficiently small (the precise value and its related parameter dependences still requiring further investigation), the timescale to divert mass to smaller radii are generically a few $t_{\rm fb}$ and that only a minority of the bound gas enters an accretion disc quickly.

At later times, inflow into this inner region depends increasingly upon the level of internal stresses in gas at larger radii, presumably due to the stirring of MHD turbulence by the magnetorotational instability (Balbus and Hawley 1998).

3.3 Outer Flow

By contrast, the fate of most of the bound mass after it has returned from its first apocenter passage is a very extended, highly eccentric, and rather asymmetric flow, as depicted in Fig. 5. "Very extended" means that its radial scale is comparable to that of the most-bound debris, $a_{\rm min} \sim 10^2 r_t$. Spreading half a stellar mass over an area $\sim \pi a_{\rm min}^2$ produces a Thomson optical depth (measured from the midplane vertically outward)

$$\tau_T \sim 500\,\Xi^2 (M_*/M_\odot)^{0.56} M_{\mathrm{BH},6}^{-4/3}. \tag{20}$$

If the shocks, taken all together, have been able to magnify the streams' binding energy by a factor of a few, the internal energy density of the gas is comparable to the binding energy. In the outer portion of the orbit, where the matter spends most of its time, the vertical scale height would then be comparable to the local radius relative to the black hole. As a result, the characteristic volume density is

$$\rho_* \sim 4 \times 10^{-12}\,\Xi^3 (M_*/M_\odot)^{0.35} M_{\mathrm{BH},6}^{-2}\ \mathrm{gm\,cm^{-3}}, \tag{21}$$

while the vertical component of gravity is

$$g_z \sim 1000\,\Xi^2 (h/r)(M_*/M_\odot)^{-0.44} M_{\mathrm{BH},6}^{-1/3}\ \mathrm{cm\,s^{-2}}. \tag{22}$$

Thus, the conditions at the surface of the bound stellar mass are much like those of stellar atmospheres. The interior is sufficiently optically thick to be close to thermodynamic equilibrium, and the volume density near the photosphere is a few orders of magnitude smaller than in main sequence stars, while the photospheric surface gravity is roughly in the range of evolved massive stars (Langer and Kudritzki 2014).

Also like stars, a heat flux emerges through the surface of this flow. However, the energy supply for this heat flux is not derived from nuclear reactions, but from shocks between the debris streams and, potentially, from the dissipation of MHD turbulence. As shown by Piran et al. (2015b), the total heating rate due to shocks near the apocenter scales like $GM_{\mathrm{BH}}\dot{M}_0/a_{\mathrm{min}}$, where $\dot{M}_0$ is the peak mass-return rate. Calibrated by the simulations of S15, the peak expected heating rate is

$$\dot{E} \simeq 1 \times 10^{43}\,\Xi^{5/2} (M_*/M_\odot)^{0.44} M_{\mathrm{BH},6}^{-1/6}\ \mathrm{erg\,s^{-1}}. \tag{23}$$

In the simulation of S15, it lasted from $\approx 3t_{\mathrm{fb}}$ to $\approx 8t_{\mathrm{fb}}$ after the star's pericenter passage. Although $\dot{E}$ is $\sim 0.1L_{\mathrm{Edd}}$ for fiducial values of the scaling parameters, it is important to recognize that this is a small fraction of the luminosity expected from radiatively efficient accretion onto a black hole at the peak of the mass-return rate, only $\sim 10^{-3}\,\Xi(M_*/M_\odot)^{-0.22}(\eta/0.1)^{-1} M_{\mathrm{BH},6}^{1/3}$. Like the durations of peak shock heating, the calibration to the simulation of S15 may also be subject to some dependence on M_* and M_{BH}. If, for a different choice of system parameters, secondary shocks are formed at radii smaller than a_{min} such as those found by Bonnerot and Lu (2020), this would raise the heating rate above this estimate.

Spread over both sides of a flow occupying an area $\sim \pi a_{\mathrm{min}}^2$, this heating rate corresponds to a mean effective temperature

$$T_{\mathrm{eff}} \simeq 2.2 \times 10^4\,\Xi^{9/8} M_{\mathrm{BH},6}^{-3/8}\ \mathrm{K}. \tag{24}$$

This temperature is quite insensitive to parameters; by numerical coincidence, the main-sequence mass-radius relation leads to it being essentially *independent* of M_*, while it depends only modestly on M_{BH}. If the absorption opacity in the gas is high enough that the effective mean free path to absorption at all UV and optical frequencies is small compared to a_{min}, this effective temperature becomes a good estimator of the actual thermodynamic temperature of the surface. The case in which thermalization is not complete is treated in Sect. 4.1.

Whether the heating emerges with a smooth surface brightness depends on whether the photon diffusion time from deep inside the flow is long or short compared to the time required for fluid elements to travel far from a shock front. The latter time is generically $\sim t_{\rm fb}/2$; the former is

$$t_{\rm cool} \sim \tau_*(a_{\rm min}/c)(h/r) \simeq 6 \times 10^6\, \Xi(h/r)(M_*/M_\odot)^{0.78} M_{\rm BH,6}^{-2/3}\ {\rm s}. \tag{25}$$

Thus, the cooling time and the gas orbital time are generally similar, with the orbital time perhaps a bit shorter; the ratio

$$t_{\rm cool}/t_{\rm fb} \sim 2\Xi^{5/2}(M_*/M_\odot)^{-0.13} M_{\rm BH,6}^{-1/6}. \tag{26}$$

As a result, the surface brightness (or, equivalently, the effective temperature) may or may not be thoroughly smoothed out by slow photon diffusion, but the degree to which it is smoothed is dependent primarily on M_* through its determination of Ξ. Whatever its distribution, the profile of effective temperature due to shock heating does not resemble what might be expected from a classical accretion disc: there is no reason to expect it to be a declining power-law in radius, nor is it likely to be axisymmetric.

3.4 Multi-Wavelength Properties and Parameter Inference: ASASSN-14li, a Worked Example

As previously argued, when relativistic apsidal precession is too weak to cause a sizable amount of the debris' orbital kinetic energy to be dissipated, the matter spreads out into a large ($r \sim (M_{\rm BH}/M_*)^{1/3} r_t$), eccentric, and asymmetric disc (S15). In this flow, the initial heating is predominantly due to shocks, both because the dynamics naturally lead to shocks and because MHD turbulence requires ~ 10 orbits to grow. The optical/UV lightcurve during this initial phase is determined by the heating rate due to these shocks convolved with delays caused by photon diffusion through the matter. For times beyond $\simeq 5$–$10 t_{\rm fb}$ after disruption, the shock heating should decay $\propto t^{-5/3}$ because its ultimate energy source is the returning matter. At later times, accretion from the apocenter region down to the black hole grows in importance to optical/UV emission on a timescale $\sim 10 t_{\rm fb} \sim 1\, \Xi^{-3/2}(M_*/M_\odot)^{0.32} M_{\rm BH,6}^{1/2}$ yr.

Although the simulation of S15 describes only a single, rather special, choice of parameters, its results in regard to the time evolution of the heating rate in these shocks can be checked for qualitative consistency with a large sample of observed events. For example, long-term HST monitoring of 12 TDEs demonstrated that in nearly all cases, the total energy radiated in the UV at times more than 1 yr after event-discovery was comparable to the energy radiated in that band within the first month (van Velzen et al. 2019). It immediately follows that the long-term lightcurve in this band is much better described by t^{-1} than $t^{-5/3}$, which is closely consistent with the prediction of the previous paragraph, taken directly from a scaling of the S15 simulation.

To illustrate how to use the theory described in Sect. 3.3 for parameter inference, it is easiest to work through an example. ASASSN-14li is especially suitable for this purpose (Krolik et al. 2016) because it is the best-observed TDE so far, with extensive X-ray and radio observations augmenting optical/UV data. The availability of this multi-wavelength data provides both additional consistency checks and a broader view of what may be possible in a TDE.

We begin with the shape of the optical/UV spectrum. In ASASSN-14li, it was well-fit by a blackbody spectrum with a temperature 3.5×10^4 K for as long as 600 d after this event's discovery (Cenko et al. 2016; Holoien et al. 2016; Brown et al. 2017). For fiducial parameter values this is less than a factor of 2 from the prediction of the apocenter shock model Piran et al. (2015b) given in Sect. 3.3 that $T_{\rm eff} \approx 2 \times 10^4 \Xi^{9/8} M_{\rm BH,6}^{-3/8}$ K. Because the temperature depends on M_* only through Ξ, the temperature is primarily a constraint on $M_{\rm BH}$.

A second constraint may be obtained from the total optical/UV luminosity, likewise estimated in Sect. 3.3 on the basis of heating in apocenter shocks. Integrating over the full blackbody spectrum, even though only part was directly observed, Holoien et al. (2016) found that the luminosity from ASASSN-14li at discovery was $\approx 6 \times 10^{43}$ erg s^{-1}, again quite similar to the fiducial prediction, $\simeq 1 \times 10^{43} \Xi^{5/2} (M_*/M_\odot)^{0.44} M_{\rm BH,6}^{-1/6}$ erg s^{-1}. As this expression demonstrates, if the observed luminosity is close to the peak luminosity, it is primarily a constraint on M_*, except for the $M_{\rm BH}$-dependence hidden in Ξ.

A different measure of this consistency is the effective area required to radiate the observed optical/UV luminosity at the observed temperature. Describing this area in terms of a double-sided disc of radius $r_{\rm eff}$, Brown et al. (2017) showed that $r_{\rm eff}$ peaked at $\simeq 1600 M_{\rm BH,6}^{-1} r_g$ and slowly declined to $\simeq 500 M_{\rm BH,6}^{-1} r_g$ over a period of ~ 2 yr. These values compare well with a picture in which the matter orbits on a scale similar to the semi-major axis of the most-bound matter's orbit $\approx 2400\, \Xi^{-1} (M_*/M_\odot)^{0.22} M_{\rm BH,6}^{-1/3} r_g$, especially when allowance is made for the facts that the shocks reduce its orbital energy by order unity and not all of the area of the eccentric accretion flow may contribute to the observed light.

A third arises from the emission line profiles. In terms of full width at half maximum, the various UV lines seen in ASASSN-14li ranged from 1700 km s^{-1} to 7700 km s^{-1} (Cenko et al. 2016), while the optical lines began with full width at zero intensity $\simeq$ 10,000 km s^{-1} but these narrowed to $\simeq 5000$ km s^{-1} at late times (Holoien et al. 2016). Given this range, one cannot define constraints too tightly; instead one can merely test whether the posited source region would possess a range of velocities consistent with what is seen at the factor of 2 level. In the apocenter region, and assuming that forces from radial pressure gradients do not dominate over centrifugal support, the characteristic orbital speed is

$$v_{\rm orb}(r) \approx 6000\, \Xi^{1/2} (M_*/M_\odot)^{-0.11} M_{\rm BH,6}^{1/6} (2a_{\rm min}/r - 1)^{1/2}\ {\rm km\,s^{-1}}, \tag{27}$$

squarely in the observed range. The last term in the estimate incorporates the variation in speed around an eccentric orbit of semi-major axis $a_{\rm min}$ if the instantaneous radius of a fluid element is r. Comparison to the observed profile width requires specifying both the range in orbital speeds corresponding to the range in orbital semimajor axes and radius contributing, and allowing for inclination.

Because the dependence of $v_{\rm orb}$ on M_* and $M_{\rm BH}$ is so weak (and is largely through the factor $\Xi^{1/2}$), equation (27) gives an extremely robust prediction of the scale of line-widths to be expected. That it is so readily consistent with the magnitude of the observed line-widths, despite the almost complete lack of parameter-dependence, is also evidence on behalf of the underlying model. For the same reason, however, the line-width is not very useful for parameter inference because it is more sensitive to factors beyond the scope of the model (extent of the radiating region relative to the semimajor axis, orbital inclination) than to the parameters one would wish to constrain.

Nonetheless, detailed fitting of the optical emission line profiles to a model in which they are radiated by a highly elliptical disc (a more specific application of the apocenter shock model) found that in the case of PTF09djl they were well described by a disc with semi-major axis $1700 r_g$ and eccentricity 0.97 (Liu et al. 2017; Cao et al. 2018), which would be well in line with the fiducial estimate for $a_{\rm min}/r_g$.

The next step is to examine the light curve. In the optical/UV, the flux was roughly constant for the first ~ 25 d and then began a decline that roughly followed an exponential form for the next ~ 150 d (Holoien et al. 2016). At still later times, the decline decelerated: from ~ 250 d until ~ 550 d post-discovery, the flux fell by only a factor $\simeq 3$. An alternative description for the period comprising the exponential decline and the first ~ 50 d of the later epoch is $L_{\rm opt,UV} \propto (t + 29)^{-5/3}$ if t is the time in days since discovery (Brown et al. 2017), but it fits the data better than the exponential only for the period 175–250 d after discovery. In the UV, the late-time light curve was even shallower: from $\simeq 400$ d to at least $\simeq 1250$ d, there was no change in flux to within measurement error, and this flux was only a factor ~ 30 below the flux at 60 d (van Velzen et al. 2019).

In many TDEs, only optical/UV data are available. However, for ASASSN-14li, radio and X-ray data were also obtained. Remarkably, equipartition analysis of the many epochs of radio data showed that the length scale of the synchrotron-emitting gas grew at a very nearly constant rate per unit time: $\sim 12{,}000$ km s^{-1} (van Velzen et al. 2016; Krolik et al. 2016; Alexander et al. 2016). Moreover, this speed matches the predicted speed of the fastest edge of the unbound ejecta (see Sect. 5.1) quite well. Having this constant speed allows one to project back to the date at which the outflow was launched: 70 days before discovery. If one further identifies the outflow with the unbound ejecta, this launch date is also the date of the actual disruption.

It was also possible to detect the flare in X-rays, enabling a much improved estimate of the bolometric luminosity. Brown et al. (2017) showed that for the first 450 d after discovery, the luminosity in soft X-rays stayed very nearly equal to the optical/UV if both are estimated by integrating over their respective blackbody fits (for the X-rays, the temperature was $\simeq$ 68 eV for the first ~ 200 d, but fell to $\simeq 56$ eV for the next ~ 200 d). Thus, for that period the bolometric luminosity was roughly double either the optical/UV or the X-ray luminosity separately. Interestingly, however, the peak value of $L_{\rm bol}$ corresponded to almost exactly the Eddington luminosity of a $10^6\ M_\odot$ black hole, and it fell below that level within a few weeks after discovery. If the bolometric luminosity were primarily due to accretion onto such a black hole at a rate given by the mass-return rate, this would require the accretion to be extremely radiatively inefficient (see Sect. 2.1.1).

However, if the optical/UV luminosity is generated in the apocenter region as a result of shock heating, one would expect its power to be roughly steady for a period of several $t_{\rm fb}$ beginning $\sim t_{\rm fb}$ after the actual disruption (although the mean flux may be more or less steady over this time period, the details of shock evolution may create fluctuations at the tens of percent level). At later times, as the source of energy to the apocenter shocks, the infall of late-arriving debris, diminishes, so would $L_{\rm opt/UV}$. At still later times ($>\sim 10 t_{\rm fb}$), accretion inflow supersedes shocks as a source of heat in this region and the zone just interior to it. Once this happens, $L_{\rm opt,UV}$ would decline as the mass reservoir in the apocenter region is depleted. Because the inflow time at these distances is $\sim \alpha^{-1}(h/r)^{-2} t_{\rm fb}$, the decay in $L_{\rm opt,UV}$ would occur on timescales at least $\sim O(10) t_{\rm fb}$, consistent with the noticeable flattening of the optical/UV light curve after ~ 250 d.

Combining the optical/UV peak luminosity and $T_{\rm eff}$ constraints leads to

$$\Xi^2 (M_*/M_\odot)^{0.44} \simeq 4.6 \tag{28}$$

$$\Xi^3 M_{\rm BH,6}^{-1} \simeq 3.4 \tag{29}$$

Accounting for how Ξ implicitly depends on M_* and $M_{\rm BH}$ (equations 10 and 11 of Ryu et al. 2020a) leads to $M_* \simeq 2.8\ M_\odot$, and $M_{\rm BH,6} \simeq 1.4$. The corresponding $t_{fb} \simeq 28$ d. If so, discovery took place when the system was $\simeq 2.5 t_{\rm fb}$ past disruption; the optical and X-ray plateau lasted another $\simeq 1 t_{\rm fb}$. Because the black hole mass is derived entirely from TDE properties, it is very encouraging that the value derived from the apocenter shocks model

($1.4 \times 10^6\ M_\odot$) is close to the value ($\simeq 1.7 \times 10^6\ M_\odot$) estimated from its bulge velocity dispersion (Wevers et al. 2017).

4 Additional Considerations for the UV/Optical/Near-IR Continuum and Line Emission

In this section we discuss general principles relating to the UV, optical, and near-IR emission from TDEs, regardless of whether this emission originates from a circularized accretion disc (Sect. 2.2) or shock-heated streams in non-circularized flows (Sect. 3.3).

The state of the emitting gas in TDEs requires a separate set of assumptions than those commonly used in modeling either the broad-line region (BLR) of active galactic nuclei, or the simplest treatment of supernova spectra. Unlike the BLR, the gas is dense enough to be far from nebular conditions. By this we mean that the system is electron scattering dominated, and line photons, even from non-resonance lines, may be reabsorbed after many scatterings. For these reasons, a photoionization calculation which does not make use of a solution of the radiative transfer calculation will face difficulties in self-consistently exploring the conditions of interest. At the same time, the strong radiation field generated by the shocked and/or accreting gas prohibits the establishment of local thermodynamic equilibrium (LTE), an assumption that is often made to simplify the calculation of early-phase supernova spectra. Finally, the production of X-rays leads to Comptonization that is not often important in the context of supernovae (Sect. 2.2.3). The situation for radiative transfer in TDEs is most analogous to non-LTE treatments of AGN accretion disc emission (e.g. Hubeny et al. 2000, 2001), stellar atmospheres (e.g. Mihalas 1978; Hubeny 1988), and super-luminous supernovae driven by central engines (e.g. Dessart et al. 2012).

In Sect. 4.1 we consider how these conditions affect the properties of the UV, optical, and IR continuum from TDEs. In Sect. 4.2 we perform a similar analysis for the UV and optical emission and absorption lines, focusing on what sets the strengths of these features with respect to the continuum emission at neighboring frequencies.

4.1 Constancy of the UV/Optical Color Temperature: the Potential Role of Partial Thermalization of Radiation

As mentioned in Sects. 2.2.2 and 3.3, if the effective *absorption* optical depth through the emitting stellar debris is low, the continuum emission might not completely thermalize, even when the gas has high total scattering depth. Here we examine this situation in more detail, and we begin with the case where the dominant absorption opacity is due to free-free processes, which are expected to dominate for TDEs at near-infrared (NIR) and longer wavelengths. Later we will make qualitative statements about how a partially thermalized spectrum at UV and optical wavelengths might help to explain the constancy of fitted blackbody temperatures over several months in TDEs, although at those wavelengths other absorption opacities dominate over free-free.

The depth-dependent thermalization we will describe is related to continuum formation in Wolf-Rayet wind atmospheres (e.g. Wright and Barlow 1975), but here we are in the scattering-dominated regime. A similar analysis to what follows was performed by Illarionov and Sunyaev (1972) and more recently by Shussman et al. (2016), Lu and Bonnerot (2019), and Margutti et al. (2019). This particular derivation follows the latter reference most closely.

Neglecting gaunt factors, the free-free emissivity ($\mathrm{erg\,cm^{-2}\,s^{-1}\,Hz^{-1}\,Sr^{-1}}$) is

$$j_\nu^{ff} \propto T_e^{-1/2} Z^2 n_e n_i \exp\left(-h\nu/kT_e\right), \tag{30}$$

where T_e is the temperature of the free electrons. Near the surface of the emitting region, the electron temperature tends to level off (Hubeny et al. 2000; Roth et al. 2016), so that it can be absorbed into the constant. In the NIR, $h\nu/kT_e \ll 1$ and so the exponential factor is approximately constant. Thus, the free-free emissivity will vary with position roughly according to $j_\nu \propto n_e n_i \propto \rho^2$, with the assumption that the material is highly ionized. While the bound electrons are coupled to the radiation field and are likely to be out of the thermal equilibrium, the *free* electrons and the ions should have a thermal velocity distribution given the densities of interest, so that $j_\nu^{ff} = \alpha_\nu^{ff} B_\nu$. Expanding the Planck function to first order in $h\nu/kT_e$, we find $\alpha_\nu^{ff} \propto \rho^2 \nu^{-2}$.

While the discussion up to this point has been independent of an assumed gas geometry, at this point we will assume spherical symmetry, and future work may extend this argument to other geometries. We assume that near the surface of the emitting material the density is dropping as a power-law, r^{-n}, for some $n > 1$. Due to the ionization from the engine, electron scattering dominates the opacity. The electron scattering optical depth (integrated from the outside in) is then wavelength-independent and given at radius r by

$$\tau_{\mathrm{es}} = \left(\frac{1}{n-1}\right) \rho_0 \kappa_{\mathrm{es}}\, r_0^n\, r^{1-n}, \tag{31}$$

where r_0 is some reference radius within the region where this power-law expression for the density holds, and ρ_0 is the density at that location. This expression assumes the material in the power-law effectively extends to infinity.

We seek an expression for the thermalization radius $r_{\nu,\mathrm{therm}}$ (see Sect. 2.2.2), which we recall is the radius where the *effective* optical depth for frequency ν equals 1. Labeling the associated total scattering optical depth as $\tau_{\nu,\mathrm{therm}}$, and making use of our expression for α_ν^{ff}, we have $\tau_{\nu,\mathrm{therm}} \approx 1/\sqrt{\epsilon_\nu} \propto \rho^{-1/2}\nu^1$. Substituting this into equation (31), solving for the radius, and restoring the proportionality factors for a pure hydrogen gas, we have

$$r_{\nu,\mathrm{therm}} = \left[\frac{2\,(2\pi)^{1/4}}{n-1} \left(3k_B m_e\right)^{-3/4} e^3 \left(m_p \sigma_T c\right)^{-1/2} \rho_0^{3/2} \kappa_{\mathrm{es}} r_0^{3n/2} T_e^{-3/4} \nu^{-1}\right]^{\frac{2}{3n-2}}, \tag{32}$$

where m_e is the electron mass, m_p is the proton mass, k_B is the Boltzmann constant, σ_T is the Thomson electron scattering cross section, and e is the elementary charge in Gaussian-cgs units. When $n = 2$ this yields

$$\begin{aligned} r_{\nu,\mathrm{therm}} = 2.2 \times 10^{14}\ \mathrm{cm} \left(\frac{\rho_0}{10^{-12}\ \mathrm{cm}}\right)^{3/4} \left(\frac{\kappa_{\mathrm{es}}}{0.4\ \mathrm{cm^2\,g^{-1}}}\right)^{1/2} \left(\frac{r_0}{10^{14}\ \mathrm{cm}}\right)^{3/2} \\ \times \left(\frac{T_e}{5 \times 10^4\ \mathrm{Kelvin}}\right)^{-3/8} \left(\frac{\nu}{3 \times 10^{14}\ \mathrm{Hz}}\right)^{-1/2}. \end{aligned} \tag{33}$$

Accounting for mixed gas composition is straightforward, but changes the result only slightly.

We can obtain an estimate of the observed spectrum by integrating the emissivity

$$L_\nu \approx 4\pi \int_{r_{\nu,\mathrm{therm}}}^{\infty} j_\nu \left(4\pi r^2\right) dr. \tag{34}$$

Using our assumed density power-law, and our scaling for j_ν^{ff}, we then have

$$L_\nu \approx \frac{(4\pi)^2}{2n-3}\, j_\nu\left(r_{\nu,\mathrm{therm}}\right) r^3_{\nu,\mathrm{therm}} \propto r^{3-2n}_{\nu,\mathrm{therm}}. \tag{35}$$

Using equation (32), we finally have $L_\nu \propto \nu^{(6-4n)/(2-3n)}$, and for convenience we define $\beta \equiv (3-2n)/(3n-2)$. The full result is

$$\begin{aligned} L_\nu = {} & \frac{4\pi}{2n-3}\left(6.8416\times 10^{38}\ \mathrm{erg\,cm^3\,K^{1/2}}\right)\left(0.0177\ \mathrm{cm^5\,K^{3/2}\,s^{-2}}\right)^\beta \\ & \times \left(\frac{1}{m_p^3\sigma_T}\right)^\beta \rho_0^{2+3\beta}\,\kappa_{\mathrm{es}}^{2\beta}\, r_0^{n(2+3\beta)}\, T_e^{-(1+3\beta)/2}\, \nu^{-2\beta} \exp\left(\frac{-h\nu}{kT_e}\right). \end{aligned} \tag{36}$$

For $n = 2$, and on the Rayleigh-Jeans tail where $h\nu/kT_e \ll 1$, we have

$$\begin{aligned} \nu L_\nu = {} & 1.5\times 10^{41}\ \mathrm{erg\,s^{-1}}\left(\frac{\rho_0}{10^{14}\ \mathrm{cm}}\right)^{5/4}\left(\frac{\kappa_{\mathrm{es}}}{0.4\ \mathrm{cm^2\,g^2}}\right)^{-1/2}\left(\frac{r_0}{10^{14}\ \mathrm{cm}}\right)^{5/2} \\ & \times \left(\frac{T_e}{5\times 10^4\ \mathrm{Kelvin}}\right)^{-1/8}\left(\frac{\nu}{3\times 10^{14}\ \mathrm{Hz}}\right)^{3/2}. \end{aligned} \tag{37}$$

The $L_\nu \propto \nu^{1/2}$ and $T_e^{-1/8}$ dependencies for $n = 2$ were also found by LB19. For shallow density profiles with n approaching 2/3, this result predicts $L_\nu \propto \nu^\alpha$ with $\alpha \to -\infty$; α equals 0 at $n = 3/2$, and for $n \to \infty$, $\alpha \to 4/3$. In all these cases ($n > 2/3$), this result predicts a slope that is shallower than the Rayleigh-Jeans law $\alpha = 2$.

Equation (37) can only be directly applied to NIR wavelengths where free-free opacity dominates. Nevertheless, we now consider the possibility that the UV and optical continuum spectra of TDEs might be described by an expression similar to equation (37) in the sense that 1) F_ν is a power-law in ν or a sequence of power-laws over frequency intervals, where the power-laws exponents are not directly connected to the overall UV/optical flux, and 2) the overall flux normalization is sensitive to the gas density ρ but not very sensitive to the gas temperature or bolometric luminosity. This is not the standard interpretation for the UV and optical spectra of TDEs. Instead, they are generally fit to portions of thermal blackbodies (see Sect. 2.2.2) because these yield better fits, in the sense of a lower reduced chi-squared statistic, than a single power-law as suggested by equation (37).

However, the peaks of thermal spectra at the fitted temperatures of roughly 2 to 4 $\times 10^4$ Kelvin lie at unobservable extreme ultraviolet wavelengths. So, it is not yet entirely clear that a thermal fit is appropriate, either, because the observations are not probing the portions of the spectra that would be most constraining of their thermal nature. Purely thermal UV/optical spectra also raise other theoretical puzzles. As discussed in Sect. 2.2.2, for the highly irradiated gas in a reprocessing region of a TDE accretion disc, it can be hard to explain how a high enough optical depth to absorption can be maintained to re-thermalize the radiation. Perhaps more importantly, whether or not the stellar debris circularizes rapidly or not, it is difficult to explain why the blackbody temperatures for the thermal fits tend to remain nearly constant over several months in many events, even as the overall UV/optical flux drops by an order of magnitude or more. This would require the position of the photosphere to co-evolve with the UV/optical luminosity in such a way as to maintain the nearly constant color temperature. Such behavior is possible, but currently there is no hydrodynamic explanation for it.

On the other hand, a continuum spectrum resulting from partial thermalization, similar to the simple one derived here using only free-free opacity, might pose a solution to some of these theoretical challenges. It accounts for the low absorption opacities at optical wavelengths that might apply in TDE conditions. It also predicts a spectral slope that is independent of the overall UV/optical flux as a function of time, because in this case the shape of the continuum only depends on the overall luminosity only to the extent that a high ionization state is maintained. Meanwhile, the normalization of the spectrum (i.e. the total UV/optical flux) is most sensitive to gas density, which might be expected to drop in time, but the power-law slope should mostly persist unchanged, modulo some weak dependence on T_e and the shifting of the exponential break at $h\nu/kT_e$.

If the partial thermalization models are ever to succeed to the point where they can be used to fit spectra, they will need to be improved beyond their current state. To make quantitative time-dependent predictions, they will need to be tied to a hydrodynamic model. Additionally, they will need a more complete treatment of the relevant absorption opacities at bluer optical and UV wavelengths where bound-free opacities from metals (especially oxygen, nitrogen, carbon and iron) are expected to dominate over free-free. Bound-free opacity is more difficult to treat because it depends sensitively on the precise ionization state of the gas, which is coupled directly to the radiation. The calculations described in Roth et al. (2016) included bound-free opacities from hydrogen, helium, and oxygen, and found that the optical continuum generally took the form of a power-law with $\nu L_\nu \propto \nu^2$ ($\lambda L_\lambda \propto \lambda^{-2}$), that gradually steepened moving from optical to UV wavelengths. Additionally, as the mass in the reprocessing envelope varied, this would change the UV and optical flux but not the spectral shape. However, these calculations were not tied to a hydrodynamically motivated, time-dependent model for the reprocessing envelope, and did not include opacities from other metals which are likely to be important. Time will tell whether the partial thermalization explanation of the UV and optical continuum spectra is correct, or whether the purely thermal explanation can succeed after all.

4.2 Line Strengths with Respect to Continuum

Similar principles of emissivity and thermalization depth set the strength of TDE emission lines with respect to the continuum. Consequently, inferences of the relative abundances of elements in the stellar debris based on emission line ratios, as is done in nebular spectroscopy, must be treated with caution. As one example, Roth et al. (2016) showed how the ratio of the He II / Hα emission line ratio in optical TDE spectra can vary substantially as the density and spatial extent of the line-emitting gas are varied, while keeping the composition unchanged.

To gain more understanding of what sets these line strengths, we may perform the following toy calculation, which will culminate with Fig. 6. We consider a plane-parallel atmosphere with surface area A and we focus on a narrow range of wavelengths centered on a line. For concreteness we will choose $H\alpha$ and its nearby continuum, although the formalism presented here also applies to other lines.

We adopt a Gaussian absorption profile $\phi(\lambda)$ for the line with Doppler width corresponding to a velocity of 10^4 km/s. At the center of the line, we will denote the ratio of the line absorption opacity to the electron scattering opacity by ϵ_l, and we will assume this line-center value is constant at all atmospheric depths we are considering. Similarly, the continuum has an opacity ratio denoted by ϵ_c that is taken to be constant with respect to both depth in the atmosphere and the range of wavelengths under consideration.

We specify the source functions for these processes in terms of "temperatures" so that $S_\lambda(T) = B_\lambda(T)$ where B_λ is the Planck function. "Temperature" is in quotes because in the radiation- and scattering-dominated environment of a TDE atmosphere the gas will not be in local thermodynamic equilibrium, so here temperature is merely serving as a convenient way to quantify the source function. The line and continuum each have their own source function temperatures, T_l and T_c. These quasi-temperatures are distinct from the quasi-temperature $T_{\rm rad}$ introduced in Sect. 2.2.1. However, they come about for the same reason - the system is not in LTE. The closer the system gets to LTE, the closer all of these quasi-temperatures will come to equalling to the shared thermodynamic temperature.

In the limit $\epsilon_l, \epsilon_c \ll 1$, the combined source function for the line and continuum processes may be written as

$$S_\lambda = \frac{\epsilon_l \, \phi(\lambda) B_\lambda(T_l) + \epsilon_c B_\lambda(T_c)}{\epsilon_l + \epsilon_c} \tag{38}$$

As discussed earlier, photons of frequency λ that reach the observer will mostly be emitted from radii that satisfy $\tau_{es} < 1/\sqrt{\epsilon_\lambda}$ (here ϵ should be taken as the opacity ratio with contributions from both line and absorption processes). From equation (34), with the understanding that the total emissivity requires the sum of the line and continuum emissivities, and that the thermalization depth depends on opacity contributions from both lines and continuum, this leads to the following estimate for the escaping luminosity (c.f. Rybicki and Lightman 1986, equation 1.103)

$$L_\lambda \approx 4\pi \sqrt{\epsilon_l \, \phi(\lambda) + \epsilon_c} \, S_\lambda A. \tag{39}$$

Figure 6 displays the combined line and continuum emission as each of the parameters ϵ_l, ϵ_c, T_l, and T_c are varied. Adjusting any of these affects the relative strength of the line with respect to the continuum. Additionally, these values will differ for each line in the spectrum and the continuum at wavelengths near that line. The values of these parameters can also be determined by a simultaneous solution of the radiative transfer equation and the non-LTE rate equations for the gas state. For rough comparison to observations, we have taken A to equal $(4\pi)\left(10^{15}\ \text{cm}\right)^2$.

There are some general statements we can make. Ordinarily, the line-center opacity for Hα (as measured by ϵ_l) is greater than the continuum opacity, so that the line thermalizes closer to the surface than the continuum. Additionally, the continuum tends to rise at longer wavelengths in the optical due to the increasing free-free opacity at those wavelengths. However, to quantitatively determine the line ratios, the detailed calculation is needed. The range of values of ϵ_l, ϵ_c, T_l and T_c displayed in Fig. 6 share a broad overlap with the range of outcomes considered in Roth et al. (2016) for Hα; those calculations considered used reprocessing envelopes with total masses in the range 0.1 to 0.5 $M_\odot$, outer edges at radii between 5×10^{14} and 2×10^{15} cm, and bolometric luminosities between 5×10^{43} and 10^{45} erg s^{-1}. However, it should be noted that the Roth et al. (2016) calculations tended to favor combinations with lower values of ϵ_c near Hα (usually < 0.001) and with high enough values of T_c to keep the line from contrasting too strongly with the continuum. Once again, it will be important to improve upon the assumptions and simplifications that went into those calculations before being able to use such a model to infer physical parameters from fitted spectra.

5 Black Hole Environment: Radio Signals from Unbound Debris and Jets

While the majority of the emitted radiation from TDEs arises from matter infalling onto the black hole, some radiation arises due to the interaction of unbound stellar debris with pre-existing gas in the circum-nuclear environment of the galaxy. There are two broad categories

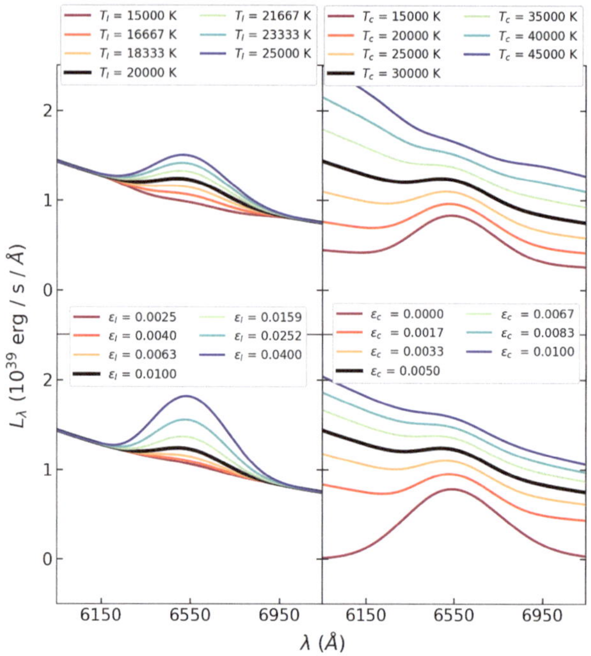

Fig. 6 Line and continuum emission from the simplified plane-parallel atmosphere model described in this section. The emitting area is taken to be $(4\pi)(10^{15}\ \text{cm})^2$. Depending on the opacities and emissivities of the line and nearby continuum, the equivalent width of the line may vary substantially. These effects must be considered when attempting to infer gas composition from emission line ratios in TDEs. The black emission line is the same in all four panels, and in each panel one parameter relating to the opacity or emissivity of the line or the continuum is varied

of this unbound material to consider: 1) the portion of the star that is unbound during its initial disruption, and 2) material that remains bound during disruption but is then launched and unbound during the subsequent circularization and/or accretion processes. This second class can be further classified into relativistic jets on the one hand, and the sub-relativistic outflows mentioned in Sect. 2.2.1 on the other. The material that is unbound during disruption will be present in every TDE where the star is fully disrupted, while the question of whether one or all of the other outflows are launched depends on the details of the subsequent hydrodynamics.

At least in two cases, the observed γ- and X-ray emission clearly indicate that a relativistic jet emerged (Swift J1644: Bloom et al. 2011; Burrows et al. 2011; Zauderer et al. 2011), (Swift J2058: Cenko et al. 2012). In these cases, aside from the prompt and beamed high-energy emission, the interaction of the jet with the surrounding environment is expected to be a source of radio emission (Giannios and Metzger 2011). Upper limits on this radio emission from samples of TDEs (van Velzen et al. 2013; Alexander et al. 2017) indicate that in many cases there is no evidence for environmental interaction with a jet. The expected properties of the radio emission produced in the forward shock of a relativistic jet when it *is* launched, such as for the two cases highlighted above, will be discussed in Sect. 5.2.

Aside from the events with jets, prompt radio emission has also been detected for a handful of other TDEs (Radio Chapter). Section 5.1 will present an argument for how interaction of the highest velocity portion of the tail of unbound debris with the circum-nuclear environment might be responsible for these radio signals. Alternatively, these radio detections might be related to interaction of wide-angle, sub-relativistic outflows discussed in Sect. 2.2.1 with the circum-nuclear environment. For more details regarding that interpretation, please refer to the Radio Chapter.

5.1 Radio Emission from Material Unbound During Disruption

The interaction of the unbound matter with the surrounding environment resembles a supernova remnant, although the geometry is very different (Guillochon et al. 2016; Krolik et al. 2016). Depending on the adiabatic index, the unbound material may or may not fragment and collapse into dense blobs (Coughlin et al. 2016), or it many remain roughly wedge-shaped in the stellar orbital plane. In either case, it drives a bow shock in the ambient gas. Calculations such as those performed by Yalinewich et al. (2019a) provide more details on the shape of the shock created by the unbound debris. Similar to a supernova remnant shock, this bow shock both accelerates electrons to relativistic velocities and amplifies the ambient magnetic field, producing a synchrotron-radiating region. This radio emission can vary significantly from one event to another as it depends sensitively on how much gas the debris plows through and the viewing angle. While self-gravity may play some role in confining the unbound debris and influencing its subsequent environmental interaction, this effect can be undercut when the ionized gas recombines, counteracting cooling (Kochanek 1994).

Originally, Guillochon et al. (2016) estimated that in typical galactic environments the unbound stream must travel a distance > 1 pc, and possibly much farther, before sweeping up mass comparable to itself and appreciably decelerating. In this case the associated radio emission would be delayed by $\gtrsim 100$ years or longer. However, the hydrodynamic simulations of realistic stellar models performed by Ryu et al. (2020a) found that for a large range of relevant values for $M_{\rm BH}$ and M_*, there is a larger spread in energy in the high-velocity edge of the unbound debris than previous estimates had indicated. This raises the possibility that a small amount of mass in the most energetic portion of the unbound debris stream can give rise to prompt radio synchrotron emission.

The characteristic speed of the unbound ejecta at $R \gg a_{\rm min}$ is

$$v_{\rm out} \simeq 6000 \text{ km s}^{-1} M_{\rm BH,6}^{1/6} (M_*/M_\odot)^{-0.11} \Xi^{1/2}, \tag{40}$$

where once again we have used the approximate main-sequence mass-radius relation of $R_* = 0.9 R_\odot (M_*/M_\odot)^{0.88}$, as in Sect. 3. This outflow carries a kinetic energy[2] of

$$E_k \simeq 2 \times 10^{50} \text{ erg } M_{\rm BH,6}^{1/3} (M_*/M_\odot)^{-1/3} \Xi \ . \tag{41}$$

As the ejecta continue to move out at nearly constant velocity they drive a bow shock in the ambient medium (at the relevant distances, deceleration from the gravity of the black hole is negligible). The front edge of this outflow is determined by the small, but non-zero, fraction of the ejecta with greater energy: for stars with $M_* \gtrsim 1\ M_\odot$, $\sim 3 \times 10^{-4}$ of the unbound ejecta have speeds at least $2\times$ the fiducial value given in equation (40) (Ryu et al. 2020b).

To make a simple estimate of the synchrotron emission from the shock one can follow the same procedure as for calculating synchrotron emission from a supernova remnant (Chevalier 1998), which is a Newtonian version of the calculation used for gamma-ray burst afterglows (Sari et al. 1998). At frequencies above the self-absorption frequency ν_A, the spectrum will take on a power-law $F_\nu \propto \nu^{-(p-1)/2}$ where p is the power-law index for the electron energy distribution, and often $p \sim 3.5$. Below this frequency, the spectrum will have slope $5/2$.

To present an estimate for ν_A and the emitted flux, we must define a number of quantities. The parameters ϵ_e and ϵ_B will represent the fraction of dissipated energy that goes into

[2]See Krolik et al. (2016) for more refined expressions.

electrons and magnetic fields, respectively, while γ_m represents the minimum Lorentz factor for the electron energy distribution. The geometric parameters f_A and f_V relate to the area and volume of the emitting region (see Barniol Duran et al. 2013, for details). The external medium into which the shock is expanding is taken to have a power-law density profile of the form $n(r) = n_0(r/r_0)^k$. The near-constant velocity of the ejecta (prior to deceleration, which will occur when the shock sweeps up enough material to be comparable to its own mass) is represented by v_0. Then at time t after the disruption, we find

$$\begin{aligned} \nu_a &= 0.4\ \text{GHz}\ f_A^{-2/7} f_V^{2/7} (\epsilon_e/0.1)^{2/7} (\epsilon_B/0.1)^{5/14} (\gamma_m/2)^{2/7} \\ &\times (n_0/1500\ \text{cm}^{-3})^{9/14} (r_0/10^{16}\ \text{cm})^{9k/14} (t/100\ \text{d})^{(4-9k)/14} \\ &\times (v_o/11{,}000\ \text{km/s})^{(14-9k)/14}, \end{aligned} \tag{42}$$

and the corresponding flux measured at distance d from the black hole, is

$$\begin{aligned} F_\nu(\nu_a) &= 3.4\ \mu\text{Jy}\ f_A^{2/7} f_V^{5/7} (\epsilon_e/0.1)^{5/7} (\epsilon_B/0.1)^{9/14} (\gamma_m/2)^{5/7} \\ &\times (n_0/1500\ \text{cm}^{-3})^{19/14} (r_0/10^{16}\ \text{cm})^{19k/14} (t/100\ \text{d})^{19(2-k)/14} \\ &\times (v_o/11{,}000\ \text{km/s})^{(56-19k)/14} (d/2.7 \times 10^{26}\ \text{cm})^2 . \end{aligned} \tag{43}$$

(Krolik et al. 2016).

These estimates made use of the faster velocities that are more likely to correspond to the fastest edge of the unbound debris, as discussed earlier. We also note that the flux may be higher for deeply penetrating (high β) events (Yalinewich et al. 2019a).

5.2 Radio Emission from a Relativistic Jet and the Radio Emission from Swift J1644

The estimates of the radio emission due to the interaction of a relativistic jet with the surrounding matter are similar to those derived for the radio afterglow from GRB jets (Sari et al. 1998). Like in the Newtonian case these estimates assume that a fraction of the blast wave energy is converted to magnetic energy and another constant fraction is converted to accelerated electrons. Once the kinetic energy of the jet and the external density are given the calculation of the synchrotron emission is straightforward. An essential generalization is the consideration of a power-law density profile, as expected around a SMBH in a galactic center. Giannios and Metzger (2011) find that for typical parameters the synchrotron frequency is:

$$\nu_m = \begin{cases} 0.1\ \text{GHz}\ \left(\frac{\epsilon_e}{0.1}\right)^2 \left(\frac{\epsilon_B}{0.01}\right)^{1/2} \left(\frac{n_0}{10\ \text{cm}^{-3}}\right)^{1/2} & \text{forward shock,} \\ 25\ \text{GHz}\ \left(\frac{\epsilon_e}{0.1}\right)^2 \left(\frac{\epsilon_B}{0.01}\right)^{1/2} \left(\frac{n_0}{10\ \text{cm}^{-3}}\right)^{1/2} \left(\frac{\Gamma_j}{10}\right)^2 & \text{reverse shock,} \end{cases} \tag{44}$$

where Γ_j is the Lorentz factor of the jet. The corresponding maximal flux is:

$$F_{\nu_m} = 2\ \text{mJy}\ (\epsilon_j/0.01)(\epsilon_B/0.01)^{1/2}(n_0/10\ \text{cm}^{-3})^{1/2}(\Gamma_j/10)^{-1}(D/\text{Gpc})^{-2}, \tag{45}$$

where ϵ_j is the ratio of the total kinetic energy of the jet to the total energy liberated by accretion.

The initial γ-rays and X-rays detected by Swift revealed the existence of a relativistic jet in Swift J1644 (Bloom et al. 2011; Burrows et al. 2011). As expected (Giannios and

Metzger 2011) a few days after the event a radio signature was detected (Zauderer et al. 2011). However, the temporal evolution of the radio signal was not the expected one. While the initial expectation was that the energy involved would be roughly constant, it became clear that as time passed, more and more energy was added to the emitting region. This was clear from either a blast wave calculation of the kind mentioned above (Berger et al. 2012; Zauderer et al. 2013; Eftekhari et al. 2018) or when an equipartition analysis was carried out (Barniol Duran and Piran 2013). This latter analysis indicated that quite early on the source had slowed down and became mildly relativistic or even Newtonian. At the same time it showed that the energy within the emitting region increased by a factor of ~ 10 to 20 over ~ 200 days. While the total amount of energy was within the total energy available in TDEs, it was not clear why the total energy increased as a function of time. One possible explanation was provided by Kumar et al. (2013), who proposed that X-ray photons from close to the black hole are inverse-Compton scattered on the hot electrons within the radio emitting region, thereby cooling the electrons. As the X-ray flux decreased, this effect gradually disappeared, and more energy was available to the electrons to emit in the radio.

More information about radio emission from TDEs possessing relativistic jets can also be found in the Radio Chapter.

Acknowledgements We thank Leiden Observatory BSc students Rens Verkade and Ashmara Wederfoort for providing Fig. 1. We thank the anonymous referees for providing feedback which greatly improved the quality of this chapter. NR acknowledges the support from the University of Maryland through the Joint Space Science Institute Prize Postdoctoral Fellowship.

References

M.A. Abramowicz, B. Czerny, J.P. Lasota, E. Szuszkiewicz, Slim accretion disks. Astrophys. J. **332**, 646–658 (1988). https://doi.org/10.1086/166683

K.D. Alexander, E. Berger, J. Guillochon, B.A. Zauderer, P.K.G. Williams, Discovery of an outflow from radio observations of the tidal disruption event ASASSN-14li. Astrophys. J. Lett. **819**, 25 (2016). https://doi.org/10.3847/2041-8205/819/2/L25

K.D. Alexander, M.H. Wieringa, E. Berger, R.D. Saxton, S. Komossa, Radio observations of the tidal disruption event XMMSL1 J0740-85. Astrophys. J. **837**, 153 (2017). https://doi.org/10.3847/1538-4357/aa6192

K. Auchettl, E. Ramirez-Ruiz, J. Guillochon, A comparison of the X-ray emission from tidal disruption events with those of active galactic nuclei. Astrophys. J. **852**(1), 37 (2018). https://doi.org/10.3847/1538-4357/aa9b7c

S.A. Balbus, J.F. Hawley, Instability, turbulence, and enhanced transport in accretion disks. Rev. Mod. Phys. **70**(1), 1–53 (1998). https://doi.org/10.1103/RevModPhys.70.1

S.A. Balbus, A. Mummery, The evolution of Kerr discs and late-time tidal disruption event light curves. Mon. Not. R. Astron. Soc. **481**(3), 3348–3356 (2018). https://doi.org/10.1093/mnras/sty2467

R. Barniol Duran, T. Piran, On the origin of the radio emission of Sw 1644+57. Astrophys. J. **770**(2), 146 (2013). https://doi.org/10.1088/0004-637X/770/2/146

R. Barniol Duran, E. Nakar, T. Piran, Radius constraints and minimal equipartition energy of relativistically moving synchrotron sources. Astrophys. J. **772**(1), 78 (2013). https://doi.org/10.1088/0004-637X/772/1/78

M.C. Begelman, Black holes in radiation-dominated gas: an analogue of the Bondi accretion problem. Mon. Not. R. Astron. Soc. **184**, 53–67 (1978). https://doi.org/10.1093/mnras/184.1.53

E. Berger, A. Zauderer, G.G. Pooley, A.M. Soderberg, R. Sari, A. Brunthaler, M.F. Bietenholz, Radio monitoring of the tidal disruption event Swift j164449.3+573451. I. Jet energetics and the pristine parsec-scale environment of a supermassive black hole. Astrophys. J. **748**(1), 36 (2012)

G.V. Bicknell, R.A. Gingold, On tidal detonation of stars by massive black holes. Astrophys. J. **273**, 749 (1983)

R.D. Blandford, M.C. Begelman, On the fate of gas accreting at a low rate on to a black hole. Mon. Not. R. Astron. Soc. **303**(1), 1–5 (1999). https://doi.org/10.1046/j.1365-8711.1999.02358.x

R.D. Blandford, M.C. Begelman, Two-dimensional adiabatic flows on to a black hole - I. Fluid accretion. Mon. Not. R. Astron. Soc. **349**(1), 68–86 (2004). https://doi.org/10.1111/j.1365-2966.2004.07425.x

J.S. Bloom, D. Giannios, B.D. Metzger, S.B. Cenko, D.A. Perley, N.R. Butler, N.R. Tanvir, A.J. Levan, P.T. O'Brien, L.E. Strubbe, F. De Colle, E. Ramirez-Ruiz, W.H. Lee, S. Nayakshin, E. Quataert, A.R. King, A. Cucchiara, J. Guillochon, G.C. Bower, A.S. Fruchter, A.N. Morgan, A.J. van der Horst, A possible relativistic jetted outburst from a massive black hole fed by a tidally disrupted star. Science **333**, 203 (2011)

T. Bogdanović, M. Eracleous, S. Mahadevan, S. Sigurdsson, P. Laguna, Tidal disruption of a star by a black hole: observational signature. Astrophys. J. **610**(2), 707–721 (2004)

C. Bonnerot, W. Lu, Simulating disc formation in tidal disruption events. Mon. Not. R. Astron. Soc. **495**(1), 1374–1391 (2020). https://doi.org/10.1093/mnras/staa1246

C. Bonnerot, E.M. Rossi, G. Lodato, D.J. Price, Disc formation from tidal disruptions of stars on eccentric orbits by Schwarzschild black holes. Mon. Not. R. Astron. Soc. **455**, 2253–2266 (2016). https://doi.org/10.1093/mnras/stv2411

M. Brassart, J.P. Luminet, Shock waves in tidally compressed stars by massive black holes. Astron. Astrophys. **481**, 259 (2008)

M. Brassart, J.-P. Luminet, Relativistic tidal compressions of a star by a massive black hole. Astron. Astrophys. **511**, 80 (2010). https://doi.org/10.1051/0004-6361/200913442

J.S. Brown, T.W.-S. Holoien, K. Auchettl, K.Z. Stanek, C.S. Kochanek, B.J. Shappee, J.L. Prieto, D. Grupe, The long term evolution of ASASSN-14li. Mon. Not. R. Astron. Soc. **466**(4), 4904–4916 (2017). https://doi.org/10.1093/mnras/stx033

D. Burrows, J. Kennea, G. Ghisellini, V. Mangano, B. Zhang, K. Page, M. Eracleous, P. Romano, T. Sakamoto, A. Falcone, J. Osborne, S. Campana, A. Beardmore, A. Breeveld, M. Chester, R. Corbet, S. Covino, J. Cummings, P. D'avanzo, V. D'elia, P. Esposito, P. Evans, D. Fugazza, J. Gelbord, K. Hiroi, S. Holland, K. Huang, M. Im, G. Israel, Y. Jeon, H. Jun, N. Kawai, J. Kim, H. Krimm, F. Marshall, P. Mészáros, H. Negoro, N. Omodei, W. Park, J. Perkins, M. Sugizaki, H. Sung, G. Tagliaferri, E. Troja, Y. Ueda, Y. Urata, R. Usui, L. Antonelli, S. Barthelmy, G. Cusumano, P. Giommi, A. Melandri, M. Perri, J. Racusin, B. Sbarufatti, M. Siegel, N. Gehrels, Relativistic jet activity from the tidal disruption of a star by a massive black hole. Nature **476**, 421–421424424 (2011)

J.K. Cannizzo, H.M. Lee, J. Goodman, The disk accretion of a tidally disrupted star onto a massive black hole. Astrophys. J. **351**, 38 (1990)

R. Cao, F.K. Liu, Z.Q. Zhou, S. Komossa, L.C. Ho, A large accretion disc of extreme eccentricity in the TDE ASASSN-14li. Mon. Not. R. Astron. Soc. **480**(3), 2929–2938 (2018). https://doi.org/10.1093/mnras/sty1997

B. Carter, J. Luminet, Pancake detonation of stars by black holes in galactic nuclei. Nature **296**, 211 (1982)

B. Carter, J. Luminet, Tidal compression of a star by a large black hole. I Mechanical evolution and nuclear energy release by proton capture. Astron. Astrophys. **121**, 97 (1983)

S.B. Cenko, H.A. Krimm, A. Horesh, A. Rau, D.A. Frail, J.A. Kennea, A.J. Levan, S.T. Holland, N.R. Butler, R.M. Quimby, J.S. Bloom, A.V. Filippenko, A. Gal-Yam, J. Greiner, S.R. Kulkarni, E.O. Ofek, F.E. Olivares, P. Schady, J.M. Silverman, N.R. Tanvir, D. Xu, Swift J2058.4+0516: discovery of a possible second relativistic tidal disruption flare? Astrophys. J. **753**, 77 (2012). https://doi.org/10.1088/0004-637X/753/1/77

S.B. Cenko, A. Cucchiara, N. Roth, S. Veilleux, J.X. Prochaska, L. Yan, J. Guillochon, W.P. Maksym, I. Arcavi, N.R. Butler, A.V. Filippenko, A.S. Fruchter, S. Gezari, D. Kasen, A.J. Levan, J.M. Miller, D.R. Pasham, E. Ramirez-Ruiz, L.E. Strubbe, N.R. Tanvir, F. Tombesi, An ultraviolet spectrum of the tidal disruption flare ASASSN-14li. Astrophys. J. Lett. **818**, 32 (2016). https://doi.org/10.3847/2041-8205/818/2/L32

R.A. Chevalier, Synchrotron self-absorption in radio supernovae. Astrophys. J. **499**(2), 810–819 (1998). https://doi.org/10.1086/305676

J. Choi, A. Dotter, C. Conroy, M. Cantiello, B. Paxton, B.D. Johnson, Mesa isochrones and stellar tracks (MIST). I. Solar-scaled models. Astrophys. J. **823**(2), 102 (2016). https://doi.org/10.3847/0004-637X/823/2/102

E.R. Coughlin, M.C. Begelman, Hyperaccretion during tidal disruption events: weakly bound debris envelopes and jets. Astrophys. J. **781**(2), 82 (2014)

E.R. Coughlin, C. Nixon, M.C. Begelman, P.J. Armitage, D.J. Price, Post-periapsis pancakes: sustenance for self-gravity in tidal disruption events. Mon. Not. R. Astron. Soc. **455**, 3612–3627 (2016). https://doi.org/10.1093/mnras/stv2511

B. Curd, R. Narayan, GRRMHD simulations of tidal disruption event accretion discs around supermassive black holes: jet formation, spectra, and detectability. Mon. Not. R. Astron. Soc. **483**(1), 565–592 (2019). https://doi.org/10.1093/mnras/sty3134

L. Dai, J.C. McKinney, M.C. Miller, Soft X-ray temperature tidal disruption events from stars on deep plunging orbits. Astrophys. J. Lett. **812**, 39 (2015). https://doi.org/10.1088/2041-8205/812/2/L39

L. Dai, J.C. McKinney, N. Roth, E. Ramirez-Ruiz, M.C. Miller, A unified model for tidal disruption events. Astrophys. J. Lett. **859**(2), 20 (2018). https://doi.org/10.3847/2041-8213/aab429

S.W. Davis, O.M. Blaes, I. Hubeny, N.J. Turner, Relativistic accretion disk models of high-state black hole X-ray binary spectra. Astrophys. J. **621**, 372–387 (2005). https://doi.org/10.1086/427278

L. Dessart, D.J. Hillier, R. Waldman, E. Livne, S. Blondin, Superluminous supernovae: ^{56}Ni power versus magnetar radiation. Mon. Not. R. Astron. Soc. **426**(1), 76–80 (2012). https://doi.org/10.1111/j.1745-3933.2012.01329.x

C. Done, S.W. Davis, C. Jin, O. Blaes, M. Ward, Intrinsic disc emission and the soft X-ray excess in active galactic nuclei. Mon. Not. R. Astron. Soc. **420**, 1848–1860 (2012). https://doi.org/10.1111/j.1365-2966.2011.19779.x

T. Eftekhari, E. Berger, B.A. Zauderer, R. Margutti, K.D. Alexander, Radio monitoring of the tidal disruption event Swift J164449.3+573451. III. Late-time jet energetics and a deviation from equipartition. Astrophys. J. **854**(2), 86 (2018). https://doi.org/10.3847/1538-4357/aaa8e0

C.R. Evans, C.S. Kochanek, The tidal disruption of a star by a massive black hole. Astrophys. J. Lett. **346**, 13 (1989)

E. Gafton, S. Rosswog, Tidal disruptions by rotating black holes: effects of spin and impact parameter. Mon. Not. R. Astron. Soc. **487**(4), 4790–4808 (2019). https://doi.org/10.1093/mnras/stz1530

S. Gezari, S.B. Cenko, I. Arcavi, X-ray brightening and UV fading of tidal disruption event ASASSN-15oi. Astrophys. J. Lett. **851**, 47 (2017). https://doi.org/10.3847/2041-8213/aaa0c2

D. Giannios, B.D. Metzger, Radio transients from stellar tidal disruption by massive black holes. Mon. Not. R. Astron. Soc. **416**(3), 2102–2107 (2011). https://doi.org/10.1111/j.1365-2966.2011.19188.x

F.G. Goicovic, V. Springel, S.T. Ohlmann, R. Pakmor, Hydrodynamical moving-mesh simulations of the tidal disruption of stars by supermassive black holes. Mon. Not. R. Astron. Soc. **487**(1), 981–992 (2019). https://doi.org/10.1093/mnras/stz1368

E.C.A. Golightly, C.J. Nixon, E.R. Coughlin, On the diversity of fallback rates from tidal disruption events with accurate stellar structure. Astrophys. J. Lett. **882**(2), 26 (2019). https://doi.org/10.3847/2041-8213/ab380d

J. Guillochon, E. Ramirez-Ruiz, Hydrodynamical simulations to determine the feeding rate of black holes by the tidal disruption of stars: the importance of the impact parameter and stellar structure. Astrophys. J. **767**, 25 (2013). https://doi.org/10.1088/0004-637X/767/1/25

J. Guillochon, E. Ramirez-Ruiz, S. Rosswog, D. Kasen, Three-dimensional simulations of tidally disrupted solar-type stars and the observational signatures of shock breakout. Astrophys. J. **705**, 844 (2009)

J. Guillochon, H. Manukian, E. Ramirez-Ruiz, Ps1-10jh: the disruption of a main-sequence star of near-solar composition. Astrophys. J. **783**(1), 23 (2014). http://adsabs.harvard.edu/abs/2014ApJ...783...23G

J. Guillochon, M. McCourt, X. Chen, M.D. Johnson, E. Berger, Unbound debris streams and remnants resulting from the tidal disruptions of stars by supermassive black holes. Astrophys. J. **822**, 48 (2016). https://doi.org/10.3847/0004-637X/822/1/48

J. Guillochon, M. Nicholl, V.A. Villar, B. Mockler, G. Narayan, K.S. Mandel, E. Berger, P.K.G. Williams, MOSFiT: modular open source fitter for transients. Astrophys. J. Suppl. Ser. **236**(1), 6 (2018). https://doi.org/10.3847/1538-4365/aab761

F. Haardt, L. Maraschi, A two-phase model for the X-ray emission from Seyfert galaxies. Astrophys. J. Lett. **380**, 51 (1991). https://doi.org/10.1086/186171

K. Hayasaki, N. Stone, A. Loeb, Circularization of tidally disrupted stars around spinning supermassive black holes. Mon. Not. R. Astron. Soc. **461**, 3760–3780 (2016). https://doi.org/10.1093/mnras/stw1387

T.W.-S. Holoien, C.S. Kochanek, J.L. Prieto, K.Z. Stanek, S. Dong, B.J. Shappee, D. Grupe, J.S. Brown, U. Basu, J.F. Beacom, D. Bersier, J. Brimacombe, A.B. Danilet, E. Falco, Z. Guo, J. Jose, G.J. Herczeg, F. Long, G. Pojmanski, G.V. Simonian, D.M. Szczygieł, T.A. Thompson, J.R. Thorstensen, R.M. Wagner, P.R. Woźniak, Six months of multiwavelength follow-up of the tidal disruption candidate ASASSN-14li and implied TDE rates from ASAS-SN. Mon. Not. R. Astron. Soc. **455**, 2918–2935 (2016). https://doi.org/10.1093/mnras/stv2486

I. Hubeny, A computer program for calculating non-LTE model stellar atmospheres. Comput. Phys. Commun. **52**, 103–132 (1988). https://doi.org/10.1016/0010-4655(88)90177-4

I. Hubeny, E. Agol, O. Blaes, J.H. Krolik, Non-LTE models and theoretical spectra of accretion disks in active galactic nuclei. III. Integrated spectra for hydrogen-helium disks. Astrophys. J. **533**, 710–728 (2000). https://doi.org/10.1086/308708

I. Hubeny, O. Blaes, J.H. Krolik, E. Agol, Non-LTE models and theoretical spectra of accretion disks in active galactic nuclei. IV. Effects of compton scattering and metal opacities. Astrophys. J. **559**, 680–702 (2001). https://doi.org/10.1086/322344

T. Hung, S. Gezari, S.B. Cenko, S.van Velzen, N. Blagorodnova, L. Yan, S.R. Kulkarni, R. Lunnan, T. Kupfer, G. Leloudas, A.K.H. Kong, P.E. Nugent, C. Fremling, R.R. Laher, F.J. Masci, Y. Cao, R. Roy, T. Petrushevska, Sifting for sapphires: systematic selection of tidal disruption events in iPTF. Astrophys. J. Suppl. Ser. **238**(2), 15 (2018). https://doi.org/10.3847/1538-4365/aad8b1

A.F. Illarionov, R.A. Sunyaev, The thermal radiation spectra of supermassive stars and X-ray sources. Astrophys. Space Sci. **19**(1), 61–74 (1972). https://doi.org/10.1007/BF00643167

Y.-F. Jiang, J. Guillochon, A. Loeb, Prompt radiation and mass outflows from the stream-stream collisions of tidal disruption events. Astrophys. J. **830**, 125 (2016). https://doi.org/10.3847/0004-637X/830/2/125

Y.-F. Jiang, J.M. Stone, S.W. Davis, Super-Eddington accretion disks around supermassive black holes. Astrophys. J. **880**(2), 67 (2019a). https://doi.org/10.3847/1538-4357/ab29ff

Y.-F. Jiang, O. Blaes, J.M. Stone, S.W. Davis, Global radiation magnetohydrodynamic simulations of sub-Eddington accretion disks around supermassive black holes. Astrophys. J. **885**(2), 144 (2019b). https://doi.org/10.3847/1538-4357/ab4a00

P.G. Jonker, N.C. Stone, A. Generozov, S.van Velzen, B. Metzger, Implications from late-time X-ray detections of optically selected tidal disruption events: state changes, unification, and detection rates. ArXiv e-prints (2019). arXiv:1906.12236

J. Jose, Z. Guo, F. Long, G. Herczeg, S. Dong, T.W.-S. Holoien, J.L. Prieto, D. Grupe, B.J. Shappee, K.Z. Stanek, C.S. Kochanek, A.B. Davis, G. Simonian, U. Basu, J.F. Beacom, D. Bersier, J. Brimacombe, D. Szczygiel, G. Pojmanski, ASAS-SN discovery of an unusual nuclear transient in PGC 043234. Astron. Telegram **6777**, 1 (2014)

D. Kasen, E. Ramirez-Ruiz, Optical transients from the unbound debris of tidal disruption. Astrophys. J. **714**, 155 (2010)

S. Kobayashi, P. Laguna, E.S. Phinney, P. Mészáros, Gravitational waves and x-ray signals from stellar disruption by a massive black hole. Astrophys. J. **615**, 855 (2004)

C.S. Kochanek, The aftermath of tidal disruption: the dynamics of thin gas streams. Astrophys. J. **422**, 508 (1994)

J. Kormendy, D. Richstone, Inward bound—the search for supermassive black holes in galactic nuclei. Annu. Rev. Astron. Astrophys. **33**, 581 (1995). https://doi.org/10.1146/annurev.aa.33.090195.003053

J. Krolik, T. Piran, G. Svirski, R.M. Cheng, ASASSN-14li: a model tidal disruption event. ArXiv e-prints (2016)

J. Krolik, T. Piran, T. Ryu, Tidal disruptions of main sequence stars – V. The varieties of disruptions. ArXiv e-prints (2020). arXiv:2001.03234

P. Kumar, R. Barniol Duran, Ž. Bošnjak, T. Piran, A model for the multiwavelength radiation from tidal disruption event Swift J1644+57. Mon. Not. R. Astron. Soc. **434**(4), 3078–3088 (2013). https://doi.org/10.1093/mnras/stt1221

N. Langer, R.P. Kudritzki, The spectroscopic Hertzsprung-Russell diagram. Astron. Astrophys. **564**, 52 (2014). https://doi.org/10.1051/0004-6361/201423374

A. Laor, S.W. Davis, Line-driven winds and the UV turnover in AGN accretion discs. Mon. Not. R. Astron. Soc. **438**(4), 3024–3038 (2014). https://doi.org/10.1093/mnras/stt2408

J. Law-Smith, J. Guillochon, E. Ramirez-Ruiz, The tidal disruption of sun-like stars by massive black holes. Astrophys. J. Lett. **882**(2), 25 (2019). https://doi.org/10.3847/2041-8213/ab379a

A.P. Lightman, D.M. Eardley, Black holes in binary systems: instability of disk accretion. Astrophys. J. Lett. **187**, 1 (1974). https://doi.org/10.1086/181377

D. Lin, J. Guillochon, S. Komossa, E. Ramirez-Ruiz, J.A. Irwin, W.P. Maksym, D. Grupe, O. Godet, N.A. Webb, D. Barret, B.A. Zauderer, P.-A. Duc, E.R. Carrasco, S.D.J. Gwyn, A likely decade-long sustained tidal disruption event. Nat. Astron. **1**, 0033 (2017). https://doi.org/10.1038/s41550-016-0033

F.K. Liu, Z.Q. Zhou, R. Cao, L.C. Ho, S. Komossa, Disc origin of broad optical emission lines of the TDE candidate PTF09djl. Mon. Not. R. Astron. Soc. **472**(1), 99–103 (2017). https://doi.org/10.1093/mnrasl/slx147

G. Lodato, E.M. Rossi, Multiband light curves of tidal disruption events. Mon. Not. R. Astron. Soc. **410**(1), 359–367 (2011). http://doi.wiley.com/10.1111/j.1365-2966.2010.17448.x

A. Loeb, A. Ulmer, Optical appearance of the debris of a star disrupted by a massive black hole. Astrophys. J. **489**, 573 (1997). https://doi.org/10.1086/304814

W. Lu, C. Bonnerot, Self-intersection of the fallback stream in tidal disruption events. ArXiv e-prints (2019)

W. Lu, P. Kumar, On the missing energy puzzle of tidal disruption events. Astrophys. J. **865**(2), 128 (2018). https://doi.org/10.3847/1538-4357/aad54a

W. Lu, J. Krolik, P. Crumley, P. Kumar, Radiative interaction between the relativistic jet and optically thick envelope in tidal disruption events. Mon. Not. R. Astron. Soc. **471**(1), 1141–1152 (2017). https://doi.org/10.1093/mnras/stx1668
J. Luminet, B. Pichon, Tidal pinching of white dwarfs. Astron. Astrophys. **209**, 103 (1989)
R. Margutti, B.D. Metzger, R. Chornock, I. Vurm, N. Roth, B.W. Grefenstette, V. Savchenko, R. Cartier, J.F. Steiner, G. Terreran, B. Margalit, G. Migliori, D. Milisavljevic, K.D. Alexand er, M. Bietenholz, P.K. Blanchard, E. Bozzo, D. Brethauer, I.V. Chilingarian, D.L. Coppejans, L. Ducci, C. Ferrigno, W. Fong, D. Götz, C. Guidorzi, A. Hajela, K. Hurley, E. Kuulkers, P. Laurent, S. Mereghetti, M. Nicholl, D. Patnaude, P. Ubertini, J. Banovetz, N. Bartel, E. Berger, E.R. Coughlin, T. Eftekhari, D.D. Frederiks, A.V. Kozlova, T. Laskar, D.S. Svinkin, M.R. Drout, A. MacFadyen, K. Paterson, An embedded X-ray source shines through the aspherical AT 2018cow: revealing the inner workings of the most luminous fast-evolving optical transients. Astrophys. J. **872**(1), 18 (2019). https://doi.org/10.3847/1538-4357/aafa01
B.D. Metzger, N.C. Stone, A bright year for tidal disruptions. Mon. Not. R. Astron. Soc. **461**, 948–966 (2016). https://doi.org/10.1093/mnras/stw1394
D. Mihalas, *Stellar Atmospheres*, 2nd edition (Freeman, New York, 1978)
M.C. Miller, Disk winds as an explanation for slowly evolving temperatures in tidal disruption events. Astrophys. J. **805**, 83 (2015). https://doi.org/10.1088/0004-637X/805/1/83
B. Mockler, J. Guillochon, E. Ramirez-Ruiz, Weighing black holes using tidal disruption events. Astrophys. J. **872**(2), 151 (2019). https://doi.org/10.3847/1538-4357/ab010f
A. Mummery, S.A. Balbus, The spectral evolution of disc dominated tidal disruption events. Mon. Not. R. Astron. Soc. **492**(4), 5655–5674 (2020). https://doi.org/10.1093/mnras/staa192
R. Narayan, I. Yi, Advection-dominated accretion: a self-similar solution. Astrophys. J. Lett. **428**, 13 (1994). https://doi.org/10.1086/187381
B. Paczyński, P.J. Wiita, Thick accretion disks and supercritical luminosities. Astron. Astrophys. **88**, 23–31 (1980)
E.S. Phinney, Manifestations of a massive black hole in the galactic center, in *The Center of the Galaxy*, ed. by M. Morris IAU Symposium, vol. 136 (1989) , p. 543
B. Pichon, Helium detonation in pancake stars. Astron. Astrophys. **145**(2), 387–390 (1985)
T. Piran, A. Sądowski, A. Tchekhovskoy, Jet and disc luminosities in tidal disruption events. Mon. Not. R. Astron. Soc. **453**, 157–165 (2015a). https://doi.org/10.1093/mnras/stv1547
T. Piran, G. Svirski, J. Krolik, R.M. Cheng, H. Shiokawa, ′ disk formation versus disk accretion–what powers tidal disruption events? Astrophys. J. **806**(2), 164 (2015b). https://doi.org/10.1088/0004-637X/806/2/164
M.J. Rees, Tidal disruption of stars by black holes of 10 to the 6th-10 to the 8th solar masses in nearby galaxies. Nature **333**, 523 (1988). http://adsabs.harvard.edu/abs/1988Natur.333..523R
E.M. Rossi, M.C. Begelman, Delayed X-ray emission from fallback in compact-object mergers. Mon. Not. R. Astron. Soc. **392**(4), 1451–1455 (2009). https://doi.org/10.1111/j.1365-2966.2008.14139.x
N. Roth, D. Kasen, What sets the line profiles in tidal disruption events? Astrophys. J. **855**, 54 (2018). https://doi.org/10.3847/1538-4357/aaaec6
N. Roth, D. Kasen, J. Guillochon, E. Ramirez-Ruiz, The X-ray through optical fluxes and line strengths of tidal disruption events. Astrophys. J. **827**, 3 (2016). https://doi.org/10.3847/0004-637X/827/1/3
G.B. Rybicki, A.P. Lightman, *Radiative Processes in Astrophysics* (Wiley/VCH, Weinham, 1986)
T. Ryu, J. Krolik, T. Piran, S.C. Noble, Tidal disruptions of main sequence stars – I. Observable quantities and their dependence on stellar and black hole mass. ArXiv e-prints (2020a). arXiv:2001.03501
T. Ryu, J. Krolik, T. Piran, S.C. Noble, Tidal disruptions of main sequence stars – II. Simulation methodology and stellar mass dependence of the character of full tidal disruptions. ArXiv e-prints, (2020b). arXiv:2001.03502
T. Ryu, J. Krolik, T. Piran, S.C. Noble, Tidal disruptions of main sequence stars – IV. Relativistic effects and dependence on black hole mass. ArXiv e-prints (2020c). arXiv:2001.03504
A. Sądowski, R. Narayan, Powerful radiative jets in supercritical accretion discs around non-spinning black holes. Mon. Not. R. Astron. Soc. **453**(3), 3213–3221 (2015). https://doi.org/10.1093/mnras/stv1802
A. Sądowski, R. Narayan, J.C. McKinney, A. Tchekhovskoy, Numerical simulations of super-critical black hole accretion flows in general relativity. Mon. Not. R. Astron. Soc. **439**(1), 503–520 (2014). https://doi.org/10.1093/mnras/stt2479
A. Sądowski, E. Tejeda, E. Gafton, S. Rosswog, D. Abarca, Magnetohydrodynamical simulations of a deep tidal disruption in general relativity. Mon. Not. R. Astron. Soc. **458**, 4250–4268 (2016). https://doi.org/10.1093/mnras/stw589
R. Sari, T. Piran, R. Narayan, Spectra and light curves of gamma-ray burst afterglows. Astrophys. J. Lett. **497**(1), 17–20 (1998). https://doi.org/10.1086/311269
N.I. Shakura, R.A. Sunyaev, Black holes in binary systems. Observational appearance. Astron. Astrophys. **24**, 337–355 (1973)

N.I. Shakura, R.A. Sunyaev, A theory of the instability of disk accretion on to black holes and the variability of binary X-ray sources, galactic nuclei and quasars. Mon. Not. R. Astron. Soc. **175**, 613–632 (1976). https://doi.org/10.1093/mnras/175.3.613

R.-F. Shen, C.D. Matzner, Evolution of accretion disks in tidal disruption events. Astrophys. J. **784**, 87 (2014). https://doi.org/10.1088/0004-637X/784/2/87

T. Shimura, F. Takahara, On the spectral hardening factor of the X-ray emission from accretion disks in black hole candidates. Astrophys. J. **445**, 780–788 (1995). https://doi.org/10.1086/175740

H. Shiokawa, J.H. Krolik, R.M. Cheng, T. Piran, S.C. Noble, General relativistic hydrodynamic simulation of accretion flow from a stellar tidal disruption. Astrophys. J. **804**, 85 (2015). https://doi.org/10.1088/0004-637X/804/2/85

T. Shussman, R. Waldman, E. Nakar, Type II supernovae Early Light Curves. ArXiv e-prints (2016). arXiv: 1610.05323

N.C. Stone, B.D. Metzger, Rates of stellar tidal disruption as probes of the supermassive black hole mass function. Mon. Not. R. Astron. Soc. **455**, 859–883 (2016). https://doi.org/10.1093/mnras/stv2281

N. Stone, R. Sari, A. Loeb, Consequences of strong compression in tidal disruption events. Mon. Not. R. Astron. Soc. **435**(3), 1809–1824 (2013). http://adsabs.harvard.edu/abs/2013MNRAS.435.1809S

L.E. Strubbe, N. Murray, Insights into tidal disruption of stars from PS1-10jh. Mon. Not. R. Astron. Soc. **454**, 2321–2343 (2015). https://doi.org/10.1093/mnras/stv2081

L.E. Strubbe, E. Quataert, Optical flares from the tidal disruption of stars by massive black holes. Mon. Not. R. Astron. Soc. **400**, 2070 (2009)

A. Tchekhovskoy, B.D. Metzger, D. Giannios, L.Z. Kelley, Swift j1644+57 gone mad: the case for dynamically important magnetic flux threading the black hole in a jetted tidal disruption event. Mon. Not. R. Astron. Soc. **437**(3), 2744–2760 (2014)

S.van Velzen, D.A. Frail, E. Körding, H. Falcke, Constraints on off-axis jets from stellar tidal disruption flares. Astron. Astrophys. **552**, 5 (2013). https://doi.org/10.1051/0004-6361/201220426

S.van Velzen, G.E. Anderson, N.C. Stone, M. Fraser, T. Wevers, B.D. Metzger, P.G. Jonker, A.J.van der Horst, T.D. Staley, A.J. Mendez, J.C.A. Miller-Jones, S.T. Hodgkin, H.C. Campbell, R.P. Fender, A radio jet from the optical and x-ray bright stellar tidal disruption flare ASASSN-14li. Science **351**(6268), 62–65 (2016). https://doi.org/10.1126/science.aad1182

S.van Velzen, N.C. Stone, B.D. Metzger, S. Gezari, T.M. Brown, A.S. Fruchter, Late-time UV observations of tidal disruption flares reveal unobscured, compact accretion disks. Astrophys. J. **878**(2), 82 (2019). https://doi.org/10.3847/1538-4357/ab1844

T. Wevers, S.van Velzen, P.G. Jonker, N.C. Stone, T. Hung, F. Onori, S. Gezari, N. Blagorodnova, Black hole masses of tidal disruption event host galaxies. ArXiv e-prints (2017)

T. Wevers, D.R. Pasham, S.van Velzen, G. Leloudas, S. Schulze, J.C.A. Miller-Jones, P.G. Jonker, M. Gromadzki, E. Kankare, S.T. Hodgkin, Ł. Wyrzykowski, Z. Kostrzewa-Rutkowska, S. Moran, M. Berton, K. Maguire, F. Onori, S. Mattila, M. Nicholl, Evidence for rapid disc formation and reprocessing in the X-ray bright tidal disruption event candidate AT 2018fyk. Mon. Not. R. Astron. Soc. **488**(4), 4816–4830 (2019). https://doi.org/10.1093/mnras/stz1976

A.E. Wright, M.J. Barlow, The radio and infrared spectrum of early type stars undergoing mass loss. Mon. Not. R. Astron. Soc. **170**, 41–51 (1975). https://doi.org/10.1093/mnras/170.1.41

S. Wu, E.R. Coughlin, C. Nixon, Super-Eddington accretion in tidal disruption events: the impact of realistic fallback rates on accretion rates. Mon. Not. R. Astron. Soc. **478**(3), 3016–3024 (2018). https://doi.org/10.1093/mnras/sty971

A. Yalinewich, E. Steinberg, T. Piran, J.H. Krolik, Radio emission from the unbound debris of tidal disruption events. Mon. Not. R. Astron. Soc. **487**(3), 4083–4092 (2019a). https://doi.org/10.1093/mnras/stz1567

A. Yalinewich, J. Guillochon, R. Sari, A. Loeb, Shock breakouts from tidal disruption events. Mon. Not. R. Astron. Soc. **482**(3), 2872–2877 (2019b). https://doi.org/10.1093/mnras/sty2809

B.A. Zauderer, E. Berger, A.M. Soderberg, A. Loeb, R. Narayan, D.A. Frail, G.R. Petitpas, A. Brunthaler, R. Chornock, J.M. Carpenter, G.G. Pooley, K. Mooley, S.R. Kulkarni, R. Margutti, D.B. Fox, E. Nakar, N.A. Patel, N.H. Volgenau, T.L. Culverhouse, M.F. Bietenholz, M.P. Rupen, W. Max-Moerbeck, A.C.S. Readhead, J. Richards, M. Shepherd, S. Storm, C.L.H. Hull, Birth of a relativistic outflow in the unusual γ-ray transient Swift j164449.3+573451. Nature **476**(7), 425–428 (2011)

B.A. Zauderer, E. Berger, R. Margutti, G.G. Pooley, R. Sari, A.M. Soderberg, A. Brunthaler, M.F. Bietenholz, Radio monitoring of the tidal disruption event Swift J164449.3+573451. II. The relativistic jet shuts off and a transition to forward shock X-ray/radio emission. Astrophys. J. **767**(2), 152 (2013). https://doi.org/10.1088/0004-637X/767/2/152

Space Sci Rev (2019) 215:45
https://doi.org/10.1007/s11214-019-0612-z

The Influence of Black Hole Binarity on Tidal Disruption Events

Eric R. Coughlin[1,2] · **Philip J. Armitage**[3,4,5] · **Giuseppe Lodato**[6] · **C.J. Nixon**[7]

Received: 9 April 2019 / Accepted: 27 September 2019 / Published online: 9 October 2019

Abstract Mergers are fundamental to the standard paradigm of galaxy evolution, and provide a natural formation mechanism for supermassive black hole binaries. The formation process of such a binary can have a direct impact on the rate at which stars are tidally disrupted by one or the other black hole, and the luminous signature of the tidal disruption itself can have distinct imprints of a binary companion. In this article we review our current understanding of the influence of black hole binarity on the properties of tidal disruption events. We discuss the rates of tidal disruption by supermassive black hole binaries, the impact of a second black hole on the fallback of debris and the formation of an accretion flow, and the prospects for detection of tidal disruption events by supermassive black hole binaries.

Keywords Binaries: general · Black hole physics · Galaxies: nuclei · Hydrodynamics · Methods: numerical

The Tidal Disruption of Stars by Massive Black Holes
Edited by Peter G. Jonker, Sterl Phinney, Elena Maria Rossi, Sjoert van Velzen, Iair Arcavi and Maurizio Falanga

✉ E.R. Coughlin
eric.r.coughlin@gmail.com

1 Department of Astrophysical Sciences, Princeton University, Princeton, NJ 08544, USA

2 Columbia Astrophysics Laboratory, Columbia University, New York, NY 10027, USA

3 Department of Physics and Astronomy, Stony Brook University, Stony Brook, NY 11790, USA

4 Center for Computational Astrophysics, Flatiron Institute, 162 Fifth Avenue, New York, NY 10010, USA

5 JILA, University of Colorado & NIST, 440 UCB, Boulder, CO 80309, USA

6 Dipartimento di Fisica, Università degli Studi di Milano, Via Celoria 16, 20133 Milano, Italy

7 Department of Physics and Astronomy, University of Leicester, Leicester, LE1 7RH, UK

1 Introduction

Supermassive black hole (SMBH) binaries are a byproduct of the hierarchical picture of galactic evolution. When two galaxies merge, the constituent black holes inspiral over the course of Myrs–Gyrs (see Eq. (1)) through dynamical friction (Chandrasekhar 1960), whereby the "wake" of stars generated by the gravitational focusing of the black holes saps their kinetic energy. This process continues to reduce the separation between the SMBHs until the SMBH binary circular velocity exceeds the stellar velocity dispersion of the stars within the merged galaxy, at which point it becomes a bound system in the sense that the two SMBHs feel each other's gravitational presence. The binary becomes "hard" and only individual stellar interactions—which eject stars from the galaxy at high velocities—continue to shrink the binary separation once it reaches a critical separation $a_{\rm h} \sim GM_2/4\sigma^2$, where M_2 is the mass of the secondary and σ is the velocity dispersion (e.g., Begelman et al. 1980; Quinlan 1996; Milosavljević and Merritt 2001; Sesana et al. 2006; Kelley et al. 2017). Depending on the specific properties of the SMBH binary and the structure of the galactic nucleus, the binary can stall at parsec-scale separations for Gyr before gravitational radiation results in their coalescence, though gas dynamics (e.g., Dunhill et al. 2014; Goicovic et al. 2016) and third bodies (e.g., Ryu et al. 2018; Bonetti et al. 2019) may alleviate this "last parsec problem."

Theoretical predictions for how SMBH binaries form and coalesce within merged galaxies need to be tested across a large range of spatial separations, where different physical processes dominate. On kpc-scales there is a growing body of informative observational data from the detection of dual Active Galactic Nuclei (AGN) (e.g., Comerford et al. 2013). In this article we argue that observing the tidal disruption of stars by one black hole in a binary system—on scales roughly a million times smaller than dual AGN—could yield valuable constraints on merger physics immediately prior to (i.e., when the SMBH binary is within the gravitational-wave inspiral regime) the final coalescence detectable in gravitational waves. However, realizing this opportunity to probe the SMBH binary population on small scales is only achievable once the effects of binarity on the evolution of tidal disruption events (TDEs) are understood from theoretical grounds.

Here we discuss theoretical progress toward understanding the impact of black hole binarity on the properties of TDEs. In Sect. 2, we outline some very basic arguments concerning the SMBH binary separations and mass ratios at which we expect certain physical mechanisms to alter the rate and appearance of TDEs. In Sect. 3, we discuss the recent work that has been done on the rates of TDEs by black holes in binaries, and Sect. 4 describes the influence of a binary companion on the fallback and accretion of debris following a TDE. Section 5 gives a discussion of the prospects of detecting TDEs from binaries, and we summarize in Sect. 6.

2 Elementary Considerations

The qualitative evolution of supermassive black hole binaries that form following galactic mergers follows the scenario described by Begelman et al. (1980). There are three distinct phases.

Immediately following the merger, the two black holes orbit in the gravitational potential of the merged galaxy, with a separation that may initially be of the order of a few kpc. Dynamical friction against stars (and gas, if present) causes the separation to shrink. If the galaxy has a core containing N stars within radius r, with one-dimensional velocity dispersion σ, the time scale for orbital decay for a black hole of mass M is[1] (Yu 2002),

$$t_{\rm fric} \sim \frac{4\times 10^6}{\log N}\left(\frac{\sigma}{200\ {\rm km\,s^{-1}}}\right)\left(\frac{r}{100\ {\rm pc}}\right)^2\left(\frac{M}{10^8\ M_\odot}\right)^{-1}\ {\rm yr}. \tag{1}$$

Dynamical friction is efficient enough to bring binaries to separations of the order of $a \sim$ 10–100 pc (depending on the galaxy), *unless* the secondary is of very low mass. Very low mass secondaries formed from mergers will instead continue to orbit with minimal energy loss in the merged galaxy.

A bound binary forms at $a \sim \sqrt{GM/\sigma^2}$, when the binary separation matches the sphere of influence for the total mass M. Further orbital decay is accompanied by an increase in the orbital velocity and reduced effectiveness of dynamical friction. By the time the binary passes the threshold separation to be hard (in the sense of Heggie 1975; Binney and Tremaine 1987; Quinlan 1996; Yu 2002),

$$a_{\rm h} = \frac{GM_2}{4\sigma^2}, \tag{2}$$

the dominant energy loss process is not dynamical friction but rather the scattering of individual stars. Scattering by a hard black hole binary generally removes the intruding stars from the vicinity of the binary, meaning that stars that possess pericenters within a few binary separations belong to a *loss cone* within the stellar energy-angular momentum phase space distribution; if the refilling of this cone by two-body relaxation does not occur quickly enough, this region of the phase space is empty (Frank and Rees 1976). The rate of decay is then determined by whatever processes bring fresh stars into dynamical contact with the binary (Alexander 2017). For spherical or axisymmetric stellar systems in massive galaxies the maximum predicted inspiral time can exceed the Hubble time (Yu 2002), leading to the prediction that some binaries might stall at roughly 0.1–1 pc scales (the "last parsec problem"; Begelman et al. 1980). There is no last parsec problem in triaxial galactic nuclei (Berczik et al. 2006; Holley-Bockelmann and Sigurdsson 2006; Gualandris et al. 2017), or in general for low mass black holes in cuspy stellar distributions. Angular momentum can also be exchanged with gas disks if there is sufficient gas present on sub-pc scales (Armitage and Natarajan 2002; Dotti et al. 2007; Cuadra et al. 2009; Lodato et al. 2009; Nixon et al. 2011; Tang et al. 2017).

Finally, energy loss via gravitational waves controls the final phase of inspiral. At fixed mass ratio and eccentricity the merger time due to gravitational waves scales as $t_{\rm gr} \propto a^4/M^3$, so there is a sharp transition between the stellar (or gas) dynamical phase of merger and that dominated by gravitational radiation.

Depending upon how far a binary has advanced through the above sequence we can identify distinct effects that binarity may have on TDEs. For "wide" binaries—where wide here typically means any separation greater than $\sim 10^{-3}$ pc—neither the dynamics of the disruption nor that of the fallback is appreciably altered by the presence of the second black hole. The presence of a binary may be inferred indirectly, either by observing TDEs that are displaced from the center of galaxies, or by measuring a different *rate* of TDEs in post-merger galaxies compared to similar galaxies that have not recently undergone merger. In galaxies with only one black hole, the majority of tidally disrupted stars originate from the

[1] We note, however, that at early times each black hole "sees" the other galactic nucleus as a single entity with a mass much greater than its mass, which enhances the inspiral rate at early times.

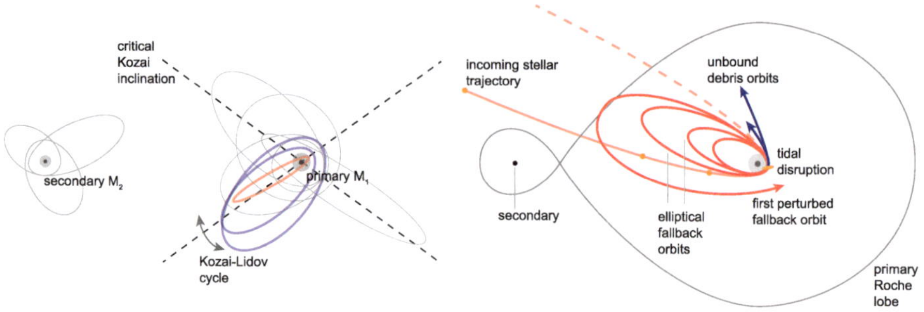

Fig. 1 Illustration of two of the ways by which supermassive black hole binaries impact the physics of tidal disruption events. At relatively large binary separations (*left*) the gravitational potential of the companion can excite Kozai-Lidov oscillations among bound stars whose orbits are inclined at $i > i_{\rm crit}$. The cycles increase the orbital eccentricity at fixed semi-major axis, potentially reducing the pericenter distance to within the tidal radius. This can increase the *rate* of TDEs. At typically smaller binary separations (*right*) the binary potential can affect the *dynamics* of the debris stream. Fallback orbits whose apocenter distance approaches the size of the disrupting black hole's Roche lobe will be significantly perturbed, modifying the fallback rate

boundary between the full and empty loss cone, which corresponds roughly to the sphere of influence (Alexander and Hopman 2003; Stone and Metzger 2016), so at a minimum we expect that changes to the rate of TDEs will occur around the time that a bound supermassive black hole binary forms. The rate can be altered either by changes in how unbound stars are scattered into the loss cones of the black holes, or by dynamics that the binary companion induces in the bound stars surrounding either hole. Kozai-Lidov oscillations, illustrated in Fig. 1, are one binary specific process that can affect the rate of bound star disruptions. Galactic mergers may also lead to the formation of new populations of stars, for example in eccentric disks, whose dynamics can have a large impact on TDE rates (Madigan et al. 2018). If the rates are increased dramatically, a single galaxy may also exhibit multiple flares within $\sim$ year timescales, which would also be a good, indirect indicator of the presence of a SMBH binary (Wegg and Bode 2011; Thorp et al. 2018).

For close binaries, the dynamics of the debris stream that provides the fallback material can be perturbed by the gravitational influence of the other black hole. Let us sketch out the simplest estimate of when such perturbations have an observable effect on TDEs (via, e.g., altering the fallback rate, initiating accretion on to *both* black holes, or changing how streams collide and circularize). We consider a binary of total mass M, made up of two equal mass black holes with $M_1 = M_2 = M/2$. The binary is on a circular orbit, with separation $a \gg r_{\rm t}$, the tidal disruption radius. After a star has been tidally disrupted by one of the holes, the debris stream follows a set of highly eccentric orbits with $e \simeq 1$. At time T after the disruption, the relation between the time and the apocenter of the debris that is only then returning to the vicinity of the black hole is,

$$T^2 = \frac{\pi^2 a_{\rm apo}^3}{GM}. \tag{3}$$

Clearly, the fallback dynamics will be drastically perturbed if $a_{\rm apo} > a/2$. (For general mass ratios, as illustrated in Fig. 1, strong perturbations occur if the debris, at apocenter, leaves the Roche lobe of the disrupting black hole.) Imposing this condition we find that fallback in a binary system differs strongly from the single black hole case, at time T, if the binary has separation,

$$a \le (GM)^{1/3} P^{2/3},$$
$$= 3 \times 10^{-4} \left(\frac{M}{10^6 \, M_\odot} \right)^{1/3} \left(\frac{T}{1 \text{ month}} \right)^{2/3} \text{pc.} \tag{4}$$

(Note that we have dropped unimportant numerical factors from this estimate.) This separation is interestingly small. If we observe TDEs whose dynamics is directly influenced by the binary potential, we will be probing an unexplored regime of supermassive black hole binaries much tighter than any securely identified to date.

At the small separations where we expect perturbations to the fallback to be observable, the binary will be decaying due to gravitational wave emission. This imposes a strong bias toward seeing binary-modified TDEs around the lowest mass black holes. Evaluating the gravitational wave merger time at the critical separation given by Eq. (4) we find,

$$t_{\text{merge}} \sim \frac{5c^5}{64G^{5/3}} M^{-5/3} T^{8/3},$$
$$\sim 2 \times 10^7 \left(\frac{M}{10^6 \, M_\odot} \right)^{-5/3} \left(\frac{T}{1 \text{ month}} \right)^{8/3} \text{yr.} \tag{5}$$

The persistence of low mass binaries close enough to appreciably affect TDE fallback dynamics is much greater than for higher mass systems. We therefore expect there to be a strong bias toward seeing "binary TDEs" around low mass supermassive black holes, *in addition to* the usual bias toward low mass black holes that arise even for normal TDEs simply because low mass systems are more numerous. The peak fallback rate $\dot{M}$ onto the black hole is also expected to scale with the black hole mass M as $\dot{M} \sim M^{-1/2}$ (Rees 1988), making TDEs by larger black holes less luminous overall and harder to detect, though we note that the more relativistic tidal radii of larger black holes enhances general relativistic precession and should facilitate disc formation.

Finally, when $a \sim r_t$ the physics of the disruption itself will be sensitive to the presence of a binary potential. The merger time scale at this separation is very short, so absent some special process that synchronizes disruption with merger it is unlikely that TDEs will be observed in this regime.

3 TDE Rates in Black Hole Binary Systems

When the galactic merger that initiates the formation of the SMBH binary is still in its early stages and the SMBHs are at kpc separations, the TDE rate—onto one or the other black hole—is likely not greatly enhanced above the typical value that results from loss cone refilling (though triaxiality and chaotic stellar orbits induced by the merger could conceivably enhance the rate; Merritt and Poon 2004). Once the binary separation becomes comparable to the sphere of influence of the primary, the secondary encounters the Bahcall-Wolf "cusp" around the primary (Bahcall and Wolf 1976), and the presence of the secondary can start to significantly modify the stellar dynamics and, correspondingly, the rate at which stars are tidally disrupted.

As mentioned in Sect. 2, one modification to the TDE rate arises from Kozai-Lidov oscillations (Lidov 1962; Kozai 1962), whereby the secondary causes circumprimary stars on highly-inclined orbits (relative to the binary orbital plane) to be driven to low-angular momentum orbits and disrupted by the primary; see the left panel of Fig. 1. This stage of the problem was analyzed in detail by Ivanov et al. (2005), who demonstrated analytically that the rate of TDEs—exclusively by the primary—can be increased to $\dot{N}_{\mathrm{TDE}} \simeq 0.1\text{–}1\ \mathrm{gal}^{-1}\ \mathrm{yr}^{-1}$ when the black hole mass ratio is fairly small (such that one can approximate the gravitational potential as being dominated by the primary for those stars undergoing Kozai oscillations and contributing to the increased rate). The timescale over which the orbits of stars are altered by the Kozai process, and thereby the time over which the rate of TDEs is expected to be enhanced by this process, is (Ivanov et al. 2005)

$$T_{\mathrm{K}} \simeq \frac{1}{q}\left(\frac{a}{D}\right)^{-3/2}\sqrt{\frac{D^3}{GM}}, \tag{6}$$

where q is the binary mass ratio, a is the distance of a star from the primary, D is the binary separation, and M is the mass of the primary (assumed to be much more massive than the secondary). Taking $D = 1$ pc, $a/D = 0.5$, $M = 10^7 M_{\odot}$, and $q = 0.1$, this timescale is $T_{\mathrm{K}} \simeq 10^5$ yr. Therefore, the enhancement to the TDE rate resulting from Kozai oscillations induced by the secondary is very short-lived.

An additional modification to the TDE rate arises from direct interactions between the secondary and individual, bound stars. Chen et al. (2009) demonstrated that these interactions generate highly complex, restricted three-body orbits as stars are scattered off of the secondary, which—-similar to the findings of Ivanov et al. (2005)—generates a short-lived ($\lesssim 10^6$ yr) burst of TDEs. However, in contrast to the mechanism proposed by Ivanov et al. (2005), which relied on Kozai-Lidov oscillations to fill the loss cone of the primary, Chen et al. (2009) found that the majority of disrupted stars entered the loss cone (of one or the other black hole) from the chaotic orbits induced by close interactions with the secondary. Wegg and Bode (2011) and Chen et al. (2011) validated these findings and additionally accounted for the inspiral of the secondary, and also found that the importance of Kozai oscillations is *reduced* by general relativistic precession and precession induced by the gravitational field of the cusp itself. Lezhnin and Vasiliev (2019) also recently confirmed this increase in the rate of TDEs, and accounted for the dynamical evolution of the distribution function of the stellar cluster, with Monte Carlo methods, and Li et al. (2017) found a similar boost to the TDE rate by employing a direct, N-body integration of the inspiral phase. Interestingly, even though this phase is very short-lived, the rate of TDEs can be so substantially increased that up to 10% of all TDEs can be from this stage of the inspiral of the SMBH binary (Chen et al. 2011).

As the secondary continues to inspiral through the cusp, the population of bound stars is depleted by three-body ejections and the binary continues to "harden"; these ejections occur primarily from the encounters between slowly-intruding (relative to the binary orbital speed) stars and the binary, and hence the separation at which the binary enters this hard phase is $a_{\mathrm{h}} \sim G\mu/\sigma^2$, where $\mu = M_1 M_2/(M_1 + M_2)$ (see also Eq. (2); Quinlan 1996). For mass ratios of order unity, this separation is comparable to the sphere of influence of the pimary, but can be significantly smaller for very unequal-mass binaries. Chen et al. (2008) studied the interactions between stars encountering a hardened binary, and showed that, because of the destruction of the stellar cusp during the hardening phase of the binary, the rate of tidal disruptions can actually fall significantly below that of an isolated black hole.

Finally, as the binary reaches semimajor axes such that the observed TDEs are expected to appear qualitatively differently from those by isolated SMBHs (semimajor axes of the order mpc and below; see Sects. 2 and 4), the speed of the binary greatly exceeds the velocity dispersion of the stars incident on the binary. In this case, the stars that encounter the binary have a binding energy of effectively zero at the large distances (relative to the binary separation) from which they originate, and the physical setup of the problem is similar to that of an isolated SMBH. However, the "cross section" of the binary is enhanced by a factor $a/r_{\rm t}$ (assuming that the stars entering the loss cone are in the pinhole regime), meaning that—for the same diffusion rate of stars into the loss cone—the number of stars incident on the binary is larger than that of an isolated black hole.

Coughlin et al. (2017) investigated the rate of tidal disruptions by a SMBH binary in this regime, focusing on binaries with a mass ratio between 0.1 and 1 (in increments of 0.1) and a separation of $a/r_{\rm t} = 100$; for a $10^6 M_\odot$ primary, this separation is $a \simeq 0.1$ mpc. They found that the vast majority of stars encountering such a small-separation binary are ejected on hypervelocity orbits, with only $\sim 2\%$ being tidally disrupted by either the primary or the secondary; the total rate of disruption is also almost completely independent of the mass ratio (though the rate of disruption by the secondary scales almost exactly linearly with the mass ratio). However, the increased cross section of the binary implies that the rate of disruption *relative to an isolated black hole* is $\dot{N}_{\rm TDE} = 2\% \times a/r_{\rm t} \simeq 2$. Thus, in this regime, the rate of disruption by the binary is augmented by a modest factor, which is due to the competition between the increased binary cross section and the prevalence of stellar ejection over disruption. Darbha et al. (2018) expanded upon the analysis of Coughlin et al. (2017) by studying the three-body interactions between incident, parabolic stars and binaries with a wide array of separations and mass ratios, and found that—for the majority of mass ratios and separations—the relative rate is augmented by a factor of ~2–5. However, for certain combinations of binary properties that favor relatively wide separations and small mass ratios, the rate of disruption reaches a relative maximum that can be of the order 10.

In summary, binary black holes can generate a short-lived, $\lesssim 1$ Myr burst of TDEs as the secondary first encounters the cusp of stars around the primary and reaches the hardening radius. As the binary continues to harden, the rate of TDEs can drop below the single-black hole value owing to the scouring out of the cusp and the potentially slow rate of repopulation of the loss cone. Just prior to coalescence, when the binary reaches $\sim$ mpc separations, the relative rate of tidal destruction approaches and modestly exceeds (by a factor of 2–10) that of an isolated black hole.

4 Fallback Dynamics in Black Hole Binaries

As we noted in Sect. 2, there is expected to be a critical separation where the disrupted debris stream becomes significantly perturbed by the presence of the companion black hole on the orbital timescale of the binary. Within this separation, one expects the return rate of the debris to the disrupting hole to exhibit some modulation that is imparted by the non-Keplerian potential of the binary. As shown in Sect. 2, this separation can be found by equating the Roche radius of the non-disrupting hole with the apocenter of the disrupted debris after the binary period P, which yields

$$\frac{a}{r_{\rm t,1}} \simeq N(q)\left(\frac{GM_*}{4\pi^2 R_*^3}\right)^{1/3} P^{2/3}, \tag{7}$$

where $N(q)$ is a numerical factor, of order unity, that depends weakly on the mass ratio (see Eq. (2) of Coughlin et al. (2017) for the full equation). For periods on the order of months—within which time we would have some chance of detecting the variability—this expression yields separations on the order of hundreds of tidal radii.

After similarly motivating an appropriate selection of binary parameters, Liu et al. (2009) (see also Liu et al. 2014) used restricted three-body integrations to quantitatively assess the impact of the secondary on the fallback rate of debris to the primary (the disrupting black hole for their setup). They found that the fallback rate is indeed modified on timescales consistent with their analytic arguments, and for their specific setup (consisting of a $10^7 M_\odot$ primary) it exhibited dramatic "dips" around five to ten years. These dips were displayed when the secondary came close enough to the disrupted debris stream to significantly alter its dynamics, thereby deflecting it from its return path to the primary.

Using similar numerical techniques, Ricarte et al. (2016) extended the work of Liu et al. (2009) by performing a more expansive parameter space study, and investigated the response of the fallback rate—and in particular the presence and longevity of the interruptions—on the binary mass ratio, the primary mass, and the binary separation. They also accounted for the fallback rate onto the secondary. While certain binary parameters exhibited unambiguous dips in the total accretion rate (primary plus secondary), in others, and in particular for more equal mass ratios, the reduction in the fallback rate of the primary was supplemented by a nearly equal increase in the fallback onto the secondary.

While these first pioneering studies validated the basic intuition regarding the impact of the secondary on the fallback, a number of aspects of the fallback remained unanswered. In particular, stream self-intersections should, non-trivially, influence and likely augment the accretion rate onto both black holes, as suggested by Ricarte et al. (2016), if the hydrodynamics of the gas is taken into account. Likewise, at small scales ($\lesssim r_t$) around the black holes the debris should circularize viscously into discs; therefore, the abrupt and complete shutoff of the fallback of debris found by Chen et al. (2009) should be less pronounced if the inner regions around the black holes are resolved and the material is allowed to self-interact. Owing to its reduced Hill sphere, disruptions by the secondary, while statistically less likely (Chen et al. 2008, 2009), should also exhibit a greater degree of irregularity in the fallback rate and debris morphology.

Additionally, Liu et al. (2009) and Ricarte et al. (2016) assumed that the frozen-in approximation was upheld by the disrupted star prior to reaching pericenter, and thereby let the stellar center of mass be described by a parabolic orbit up until the moment at which the star entered the tidal sphere around the disrupting SMBH. However, the star can interact strongly with, and enter into a long-lived, chaotic orbit about, the binary for many orbital periods *prior* to disruption. In these instances, the specific energy and angular momentum of the stellar COM could be modified significantly from their fiducial values (being 0 and $\sqrt{2GMr_p}$ respectively, where r_p is the pericenter distance of the stellar center of mass to the disrupting black hole of mass M), which would correspondingly introduce variations in the accretion rates.

Coughlin et al. (2017) performed a study of both the three-body (i.e., stellar COM specific energy and angular momentum-altering) and hydrodynamic effects of the binary on the fallback of tidally-disrupted debris to both the SMBHs. By performing millions of restricted three-body integrations by binaries with mass ratio in the range 0.1–1 (in increments of 0.1) and a separation $a/r_{t,1} = 100$, they found that the stellar specific energy could be significantly modified from its original, parabolic value. The distribution of $\beta = r_t/r_p$ for disrupted stars was found to be very well reproduced by a $1/\beta^2$ falloff, indicating that the angular momentum distribution of disrupted stars is roughly uniform. Some ($\gtrsim 10\%$) tidally disrupted stars were also found to have one or more "close encounters" with one or the other black hole, defined to be instances in which the star came within three tidal radii of either hole. In these cases, partial disruptions could occur or significant rotation could be induced to the star prior to its full disruption (for the impact of stellar rotation on the disruption process see Golightly et al. 2019).

To investigate the effects of hydrodynamics on the evolution of the tidally-disrupted debris, Coughlin et al. (2017) randomly selected 120 of the three-body encounters that resulted in the disruption of the star, and used their orbital parameters to initialize smoothed-particle hydrodynamics (SPH) simulations (with the code PHANTOM; Price et al. 2017) of the disruption of a Solar-like star by a binary with primary mass $M_1 = 10^6 M_\odot$, secondary mass $M_2 = 2 \times 10^5 M_\odot$, and separation $a = 100\, r_{t1} \simeq 0.2$ mpc. They found that the modifications to the specific energy of the center of mass could have significant effects on the return of the debris, and these effects were isolated from the post-disruption influence of the companion black hole by performing "control" simulations; in the controls, the non-disrupting black hole was removed, but the same initial conditions (in the rest frame of the disrupting hole) were kept, thereby singling out the contributions to the fallback from the change in COM properties. In extreme cases, all of which were disruptions by the secondary, the stream was completely ejected from the binary and resulted in no accretion onto either black hole owing to the large (positive) shift in the COM energy. In other instances where the stellar COM was placed on a bound orbit prior to disruption, the return of the most-bound debris could be significantly shorter than the parabolic value, with an appropriately increased peak fallback rate.

In general agreement with Liu et al. (2009) and Ricarte et al. (2016), the accretion rates measured by Coughlin et al. (2017) exhibited dips after a characteristic timescale comparable to a fraction of an orbital period (being $T_{\rm orb} \simeq 106$ d). However, the relative minima attained by these dips only amounted to fractional reductions of 1 to 10%, due to the fact that small-scale accretion disks around the individual black holes continued to supply them with a finite amount of gas. Moreover, there were some rarer instances in which the accretion rate actually *increased* relative to the control simulation, and these cases tended to occur when the returning debris stream, the disrupting black hole, and the binary center of mass were all roughly colinear (see also Coughlin and Armitage 2018).

In some instances, the morphology of the tidally-disrupted debris resembled a "cloud" that enshrouded the binary itself, and extended out to a large number of binary semimajor axes; there were also a modest number of disruptions—especially those that were ejected nearly perpendicular to the binary orbital plane—that were far more "tame" in appearance, with debris primarily accreting onto one or the other black hole in a manner roughly consistent with the isolated black hole picture. In line with the expectations of Ricarte et al. (2016), a large number of self-intersections occurred when the stream was confined to within a small angle of the binary orbital plane, which likely contributed an additional source of energy dissipation and enhanced the accretion rates of the holes. Figure 2 shows six examples of the morphology of the disrupted debris that resulted from the simulations in Coughlin et al. (2017).

Vigneron et al. (2018) performed a more controlled study of the effects of several physical and geometrical parameters involved in the disruption by a binary black hole, with the aim to use TDE fallback rates as probes of hidden sub-pc black hole binaries. The main binary parameters are the mass of the primary black hole (held constant to $10^6 M_\odot$ by Vigneron et al. 2018), the mass ratio q, the binary separation a and the eccentricity of the binary orbit (assumed to be equal to zero). The stellar orbit is characterized by its eccentricity, assumed to be equal to 1 (i.e. a parabolic orbit), its pericenter distance (assumed to be equal to the tidal radius with respect to the primary black hole) and the three angles shown in Fig. 3, indicating the inclination of the stellar orbit with respect to the binary orbit θ, the angle of the line of the nodes Ω and the apsidal angle of the pericenter ω. Vigneron et al. (2018)

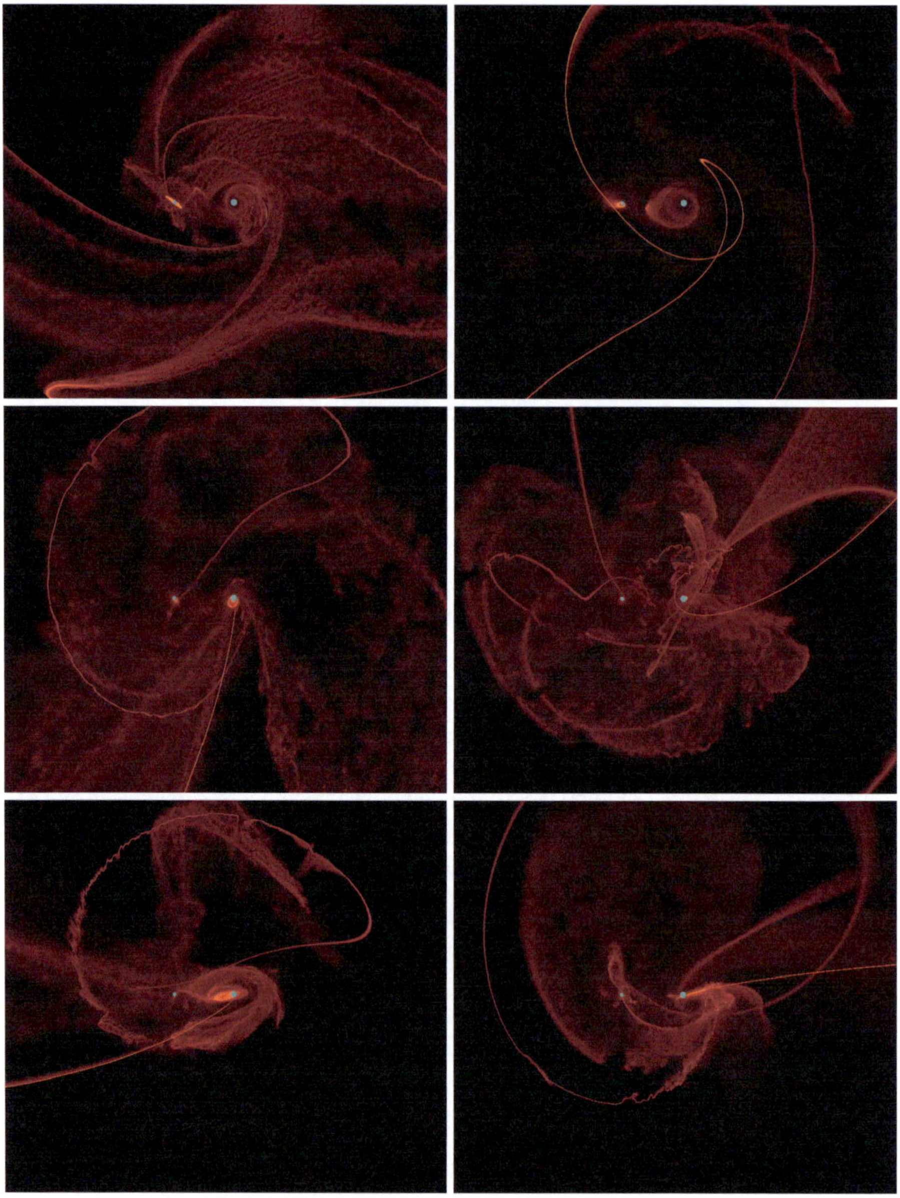

Fig. 2 The morphology of the tidally-disrupted debris resulting from six different tidal disruptions of stars by a supermassive black hole binary (adapted from Coughlin et al. 2017); bright (dark) colors indicate regions of increased (reduced) column density integrated normal to the plane of the binary, and the blue circles indicate the supermassive black holes. It can be seen that accretion onto one or the other black hole can occur, and there are a number of "gaps" where the stream self-intersects. The debris also expands to very large radii and angles and surrounds the SMBH binary

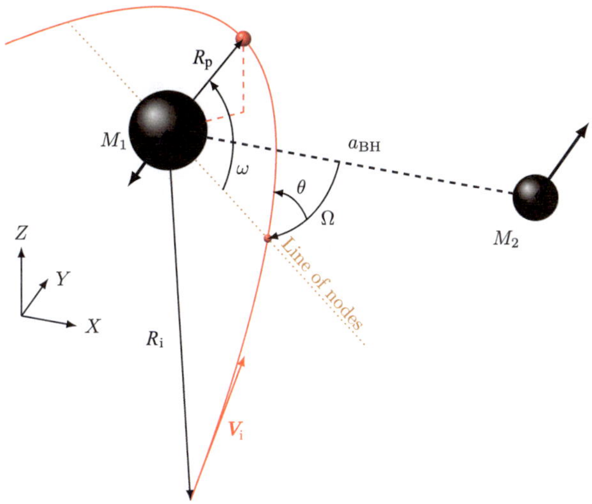

Fig. 3 Schematic of the geometry of the disruption of a star by a binary black hole. From Vigneron et al. (2018)

performed simulations by varying the mass ratio, the binary separation and the angles θ, Ω and ω. Only disruptions from the primary black hole were considered. Firstly, it is important to define two fundamental scales for the binary separation. For very close binaries, the apocenter of the most bound debris after disruption will be larger than the Roche lobe of the primary. In this case, all the debris will be perturbed by the binary and the fallback rate will not show a power-law decay at any time. Conversely, for large enough binary separation, the debris will be affected by the presence of the binary only at very late times, when the fallback is unlikely to be observable. If we assume that this happens when the fallback rate has reached 1% of its peak value, the two critical binary separations are given by:

$$a_{\min} = \frac{0.6q^{2/3} + \ln(1+q^{1/3})}{0.49q^{2/3}} R_\star \left(\frac{M_1}{M_\star}\right)^{2/3}, \tag{8}$$

$$a_{\max} = \epsilon^{-2/5} a_{\min} \tag{9}$$

where $\epsilon = 0.01$. For typical parameters, $a_{\min} \approx 0.4$–0.6 mpc and $a_{\max} \approx 2$–4 mpc. For binary separations in between the two critical values, we thus expect a TDE to initially show the standard $t^{-5/3}$ fallback curve, but to then experience dips and interruptions as found by previous studies. The numerical simulations were performed using the same PHANTOM code (Price et al. 2017) as Coughlin et al. (2017). The results confirm the expectations, in that for separations smaller than $a_{\min}$ the fallback curve shows a series of quasi-periodic dips (with a period of the order of the binary period) and no immediately recognizable power-law decay. On the contrary, for separations of the order of $a_{\max}$ the fallback proceeds undisturbed, and follows the trend of a TDE by a single black hole. Figure 4 shows the fallback rates of several simulations, where $\theta = 0$ (disruption in the binary plane). The top left panel refers to a separation that is just above $a_{\max}$, while the bottom right is just within $a_{\min}$. The red curve refers to a TDE from a single black hole, and the various curves refer to different choices of Ω.

An interesting behaviour is seen when comparing disruptions in the plane of the binary (that is, with $\theta = 0$) to disruptions perpendicular to the binary orbit ($\theta = \pi/2$). For perpendicular stellar orbits, rather than a series of periodic dips, the fallback rate shows a smooth decline below the expected, $t^{-5/3}$ curve, with a possible recovery only at late times and again

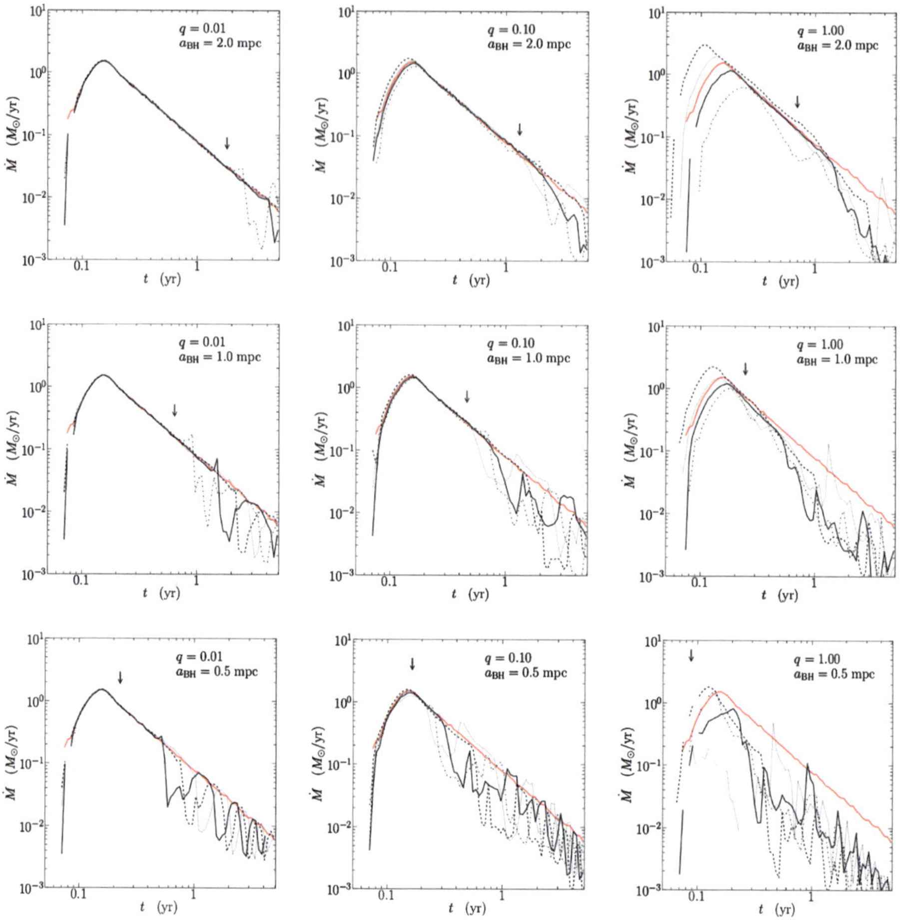

Fig. 4 Fallback rate of TDE from a binary black hole system. The red curve shows the "control" run of a TDE from a single black hole, while the various lines refer to different values of the angle Ω. In all these simulations the stellar orbit was in the same plane as the binary orbit. See Vigneron et al. (2018) for details

in a relatively smooth fashion, with respect to the sudden interruptions seen in the in-plane disruption. These are shown in Fig. 5, which shows the same fallback rates as Fig. 4, but for a polar stellar orbit. Such smooth decline and re-brightening of the fallback luminosity has been observed in the lightcurve of ASASSN-15lh (Leloudas et al. 2016), which has indeed been suggested to be a TDE from a binary black hole (Coughlin and Armitage 2018).

Finally, we note that while we focused on the case of fallback dynamics generated by tidal disruption events generated by supermassive black hole binaries, there has also been interest in the effects of tidal disruption events by *stellar mass* black holes in binary systems. Lopez et al. (2018) investigated the hydrodynamics of these events numerically, and found that such disruptions can potentially alter the mass and spin distributions of stellar mass black holes. Samsing et al. (2019) also showed that the electromagnetic signatures of TDEs by stellar mass black hole binaries in globular clusters could probe the evolution of the binary fraction, and the orbital properties of the binaries themselves, within those clusters.

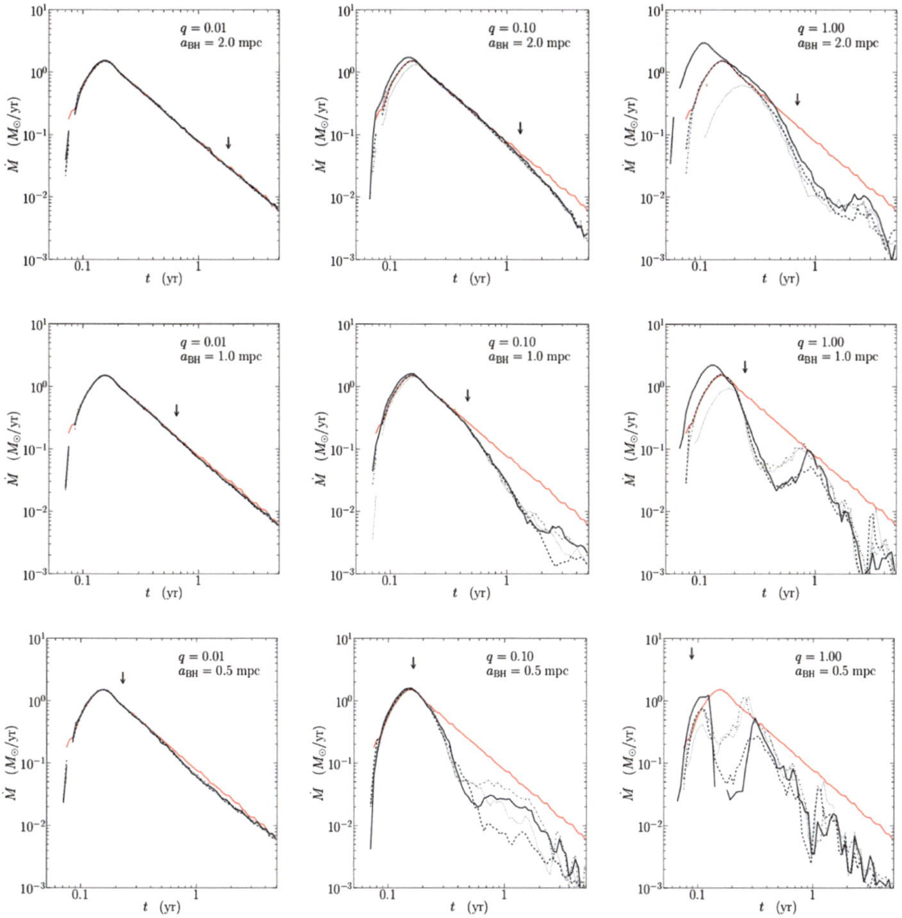

Fig. 5 Same as Fig. 4, but for a polar stellar orbit. See Vigneron et al. (2018) for details

5 Prospects for Detection

From Sect. 3, the most pronounced effects of the binary in terms of the rates of disruption occur when the secondary first encounters the cusp of stars around the primary. For typical primary masses, this initial, burst of TDEs will occur as the binary spirals in to the hardening radius, which is of the order $\sim 0.01 - 10$ pc separations depending on the black hole mass ratio and the stellar velocity dispersion (see Eq. (2)). Unfortunately, since the tidal disruption radius of either black hole is on the order of tens of gravitational radii, such separations are millions of tidal radii. Therefore, even though one could detect more than one tidal disruption event from the same galaxy during this phase (over the course of a human lifetime, especially during the upcoming age of LSST; see Table 1 in Wegg and Bode 2011), which would be good indirect evidence for the existence of a black hole binary, the influence of the binary companion on the fallback of debris—which would enable the more direct detection of the second black hole—will be completely negligible.

As noted in Sects. 2 and 4, the binary companion is expected to, and has been verified numerically to, modify the accretion process onto the disrupting black hole when the fallback time of the most-bound debris is comparable to the binary orbit, and typical black hole masses then yield a requisite binary separation on the order of $\sim$ mpc. At these separations, the binary has entered the gravitational-wave inspiral regime (i.e., the emission of gravitational waves will cause the binary to merge within the Hubble time), and the time to merger is (Peters 1964)

$$T_{\mathrm{GW}} = \frac{5}{256}\frac{c^5}{G^3}\frac{a^4}{M_1 M_2 (M_1+M_2)} \simeq 5\left(\frac{a}{0.23\ \mathrm{mpc}}\right)^4\left(1+\left(\frac{q}{0.2}\right)^{-1}\right)\left(\frac{M}{10^6 M_\odot}\right)^{-3}\ \mathrm{Myr} \tag{10}$$

(Note that 0.23 mpc is roughly 100 times the tidal radius of a $10^6 M_\odot$ SMBH). Thus, by the time the binary reaches the separations at which TDEs are expected to appear distinct from single-SMBH disruptions, the lifetime of the binary is cosmologically short. Therefore, the vast majority of TDEs that are caused by binary black holes will occur when the binary separation is too wide to permit direct detection (though, as noted above, multiple TDEs in the same galaxy would be very good, indirect evidence for the existence of a SMBH binary, and it is predicted that LSST may observe such multiple-TDE galaxies at a rate comparable to one per year; Wegg and Bode 2011; Thorp et al. 2018).

Nevertheless, there have been suggestions that some observed transients originate from tidal disruptions by black holes in small-separation binaries. For example, the lightcurve from the galaxy SDSS J1201 + 30, which was previously classified as a candidate TDE, exhibited dramatic reductions in luminosity over a timescale of years. From the results of Liu et al. (2009), Liu et al. (2014) argued that these sudden dips could indicate a binary companion at a separation of $\sim$1 mpc. However, the sparse coverage of the source does not rule out variability on smaller timescales, which could be due to inherent AGN activity (e.g., Grupe et al. 2015).

It has also been suggested that the event ASASSN-15lh, originally described by Dong et al. (2016) as one of the most luminous supernovae ever observed, was caused by the disruption of a star by a black hole in a binary system. In particular, its puzzling, "double-humped" nature in the UV can be explained by a secular change in the accretion rate onto the disrupting black hole, that change induced by the perturbing gravitational presence of a second black hole. If one requires that the time to rebrightening of $\sim$ 100 days equal the time for the disrupted debris to reach the Hill sphere of the primary, the mass of which being $\sim 10^8 M_\odot$ from the M-σ relation of the host galaxy, then one finds that a secondary black hole with mass $\sim 5\times 10^5 M_\odot$ at a separation of $\sim$ 0.5 mpc can reproduce the observed variation (and was reproduced with the simulations of Coughlin and Armitage 2018). However, there have also been other interpretations of this event, including the tidal disruption of a star by a rapidly-rotating black hole (Leloudas et al. 2016), fallback onto a newly-formed magnetar (Moriya et al. 2016; Metzger et al. 2018; Margalit et al. 2018), and the sequential, tidal disruption of two stars previously in a stellar binary (Coughlin et al. 2018).

While the likelihood of there being a tidal disruption in the final Myr of the existence of a supermassive black hole binary may be relatively low, there are other features of such tidal disruption events that may make them easier to detect. For one, there is an additional, apsidal precession of the returning debris stream that is caused by the gravitational influence of the second black hole, and this deflection—which can exceed the general relativistic one—creates more efficient dissipation of the kinetic energy of the debris. Indeed, this self-intersection was observed in the simulations of Coughlin et al. (2017), in spite of the fact that general relativity was not included. As a consequence, disc formation and the brightening of the source may be less adversely affected by Lense-Thirring precession, which torques the stream out of the orbital plane and causes the two streams to miss one another (Hayasaki et al. 2013), and the "dark years" that may plague low-β disruptions (Guillochon and Ramirez-Ruiz 2015) can be more easily avoided.

In addition, there are often multiple, distinct accretion episodes that occur from the disruption of a star by a black hole in a binary, as both black holes can actively accrete the stellar debris and the gravitational field of the second black hole induces variation in the binding energy of the disrupted material. Furthermore, the timescales associated with the brightenings of each flare are not necessarily limited by the intrinsic fallback time to the disrupting black hole, as shown in the simulations of Coughlin et al. (2017), and the luminosities can—in some instances—exceed the one predicted for an isolated SMBH. Thus, while a single flare may be too dim or fade too rapidly for the cadence of current and even upcoming surveys, the recurrent and larger flares induced by the second hole can be more readily detected.

As noted in Sect. 4, the pre-disruption orbit of a star in the binary potential can also induce large changes in the orbital properties of the center of mass of the star. As shown by Darbha et al. (2018), the greatest changes to the binding energy occur when the binary is at its smallest separations, and in the most extreme cases can result in completely bound debris streams. In these scenarios, the accretion rate is greatly increased above the standard, isolated-black hole rate, which makes such systems much more luminous. In other instances the center of mass receives a large, positive-energy kick, which can result in completely *unbound streams*. While there is no accretion luminosity from such TDEs, the radio emission from the unbound ejecta (as it shocks against the circumnuclear medium; Guillochon et al. 2016; Yalinewich et al. 2019) will be more luminous owing to its increased kinetic energy budget, which may also enhance the detectability (albeit outside the context of optical surveys).

Finally, even if the tidal disruption event from a small-separation, SMBH binary is not detected, the tidally-disrupted debris can provide an additional reservoir of gas in the vicinity of the binary at the time of merger. As shown by Coughlin and Armitage (2017), circumbinary rings can form out of the debris streams for binaries with small mass ratios, which then spread viscously into a disc over many binary orbits. This disc of gas then responds dynamically to the mass lost to gravitational waves at the time of merger, the electromagnetic signature from which being potentially detectable in nearby galaxies (Rossi et al. 2010; Corrales et al. 2010; Rosotti et al. 2012).

6 Summary

In this article we presented an overview of the current understanding of the influence of black hole binarity on the properties of tidal disruption events. As described in Sect. 2, the presence of a second black hole can affect both the rate at which TDEs occur and the evolution of the fallback. The magnitude of each effect is, however, highly-dependent on the separation of the binary.

Past work on the rates of TDEs by black holes in binary systems (Sect. 3) has indicated that, as the secondary spirals in through the cusp of the primary and approaches the hardening radius, the rate can be appreciably augmented (by 2–3 orders of magnitude above the isolated-black hole rate) owing to Kozai oscillations and direct interactions between the secondary and bound stars. As the binary continues to harden, the depletion of the population of bound stars and the potentially-slow rate of refilling of the loss cone can cause the TDE rate to plummet below the single-black hole case. Once the binary reaches very small separations, such that the binary orbital velocity greatly exceeds the stellar velocity dispersion, the rate increases slightly above (by a factor of a few to 10 at most) that of an isolated black hole with the same rate of refilling of the loss cone.

While the TDE rate is modified once the binary separation is comparable to the sphere of influence of the primary, being $\sim$ pc scales for typical galaxy types and black hole masses, the fallback of debris onto a black hole in a binary system (see Sect. 4) is only dramatically altered once the binary shrinks to $\sim$ mpc separations. As argued in Sect. 2, at these small separations the binary orbital time is comparable to the fallback time of the most bound debris, meaning that the dynamics of the returning gas stream will be affected by the secondary in the early stages of the accretion process. This basic argument has been substantiated with numerical simulations that investigated both the gravitational and hydrodynamical evolution of TDEs from binary SMBHs.

Because of the small separations required to alter the fallback of tidally-disrupted debris, SMBH binaries that yield directly observable deviations from the single-black hole predictions are well into the gravitational-wave inspiral regime and are correspondingly short-lived; the overall rate of detection of TDEs by binary SMBHs (Sect. 5) is therefore reduced by this relatively narrow window during which we can observe such binaries. However, the features of the TDEs themselves, including much greater variability and increased accretion rates (and likely luminosities), likely render them easier to detect over single-SMBH TDEs, and there have been some claims of detections. Moreover, the extremely high rate of TDEs by binaries at larger separations implies that multiple TDEs may be discovered in the same galaxy, especially in the current era of wide-field optical surveys. While observing multiple TDEs from the same galaxy would (likely) not yield direct evidence in the form of variation in the accretion luminosity, it would provide good, indirect evidence of the existence of SMBH binaries.

Acknowledgements ERC acknowledges support from NASA through the Einstein Fellowship Program, grant PF6-170170, and the Hubble Fellowship, grant #HST-HF2-51433.001-A awarded by the Space Telescope Science Institute, which is operated by the Association of Universities for Research in Astronomy, Incorporated, under NASA contract NAS5-26555. PJA acknowledges support from NASA through grant NNX16AI40G. CJN is supported by the Science and Technology Facilities Council (STFC) (grant number ST/M005917/1)

References

T. Alexander, Stellar dynamics and stellar phenomena near a massive black hole. Annu. Rev. Astron. Astrophys. **55**, 17–57 (2017). https://doi.org/10.1146/annurev-astro-091916-055306

T. Alexander, C. Hopman, Orbital in-spiral into a massive black hole in a galactic center. Astrophys. J. Lett. **590**(1), 29–32 (2003). https://doi.org/10.1086/376672

P.J. Armitage, P. Natarajan, Accretion during the merger of supermassive black holes. Astrophys. J. Lett. **567**, 9–12 (2002). https://doi.org/10.1086/339770

J.N. Bahcall, R.A. Wolf, Star distribution around a massive black hole in a globular cluster. Astrophys. J. **209**, 214–232 (1976). https://doi.org/10.1086/154711

M.C. Begelman, R.D. Blandford, M.J. Rees, Massive black hole binaries in active galactic nuclei. Nature **287**, 307–309 (1980). https://doi.org/10.1038/287307a0

P. Berczik, D. Merritt, R. Spurzem, H.-P. Bischof, Efficient merger of binary supermassive black holes in nonaxisymmetric galaxies. Astrophys. J. Lett. **642**, 21–24 (2006). https://doi.org/10.1086/504426

J. Binney, S. Tremaine, *Galactic Dynamics* (1987)

M. Bonetti, A. Sesana, F. Haardt, E. Barausse, M. Colpi, Post-Newtonian evolution of massive black hole triplets in galactic nuclei – IV. Implications for LISA. Mon. Not. R. Astron. Soc. **486**(3), 4044–4060 (2019). https://doi.org/10.1093/mnras/stz903

S. Chandrasekhar, *Principles of Stellar Dynamics* (1960)

X. Chen, F.K. Liu, J. Magorrian, Tidal disruption of stellar objects by hard supermassive black hole binaries. Astrophys. J. **676**, 54–69 (2008). https://doi.org/10.1086/527412

X. Chen, P. Madau, A. Sesana, F.K. Liu, Enhanced tidal disruption rates from massive black hole binaries. Astrophys. J. Lett. **697**, 149–152 (2009). https://doi.org/10.1088/0004-637X/697/2/L149

X. Chen, A. Sesana, P. Madau, F.K. Liu, Tidal stellar disruptions by massive black hole pairs. II. Decaying binaries. Astrophys. J. **729**, 13 (2011). https://doi.org/10.1088/0004-637X/729/1/13

J.M. Comerford, K. Schluns, J.E. Greene, R.J. Cool, Dual supermassive black hole candidates in the AGN and galaxy evolution survey. Astrophys. J. **777**, 64 (2013). https://doi.org/10.1088/0004-637X/777/1/64

L.R. Corrales, Z. Haiman, A. MacFadyen, Hydrodynamical response of a circumbinary gas disc to black hole recoil and mass loss. Mon. Not. R. Astron. Soc. **404**, 947–962 (2010). https://doi.org/10.1111/j.1365-2966.2010.16324.x

E.R. Coughlin, P.J. Armitage, Circumbinary discs from tidal disruption events. Mon. Not. R. Astron. Soc. **471**, 115–119 (2017). https://doi.org/10.1093/mnrasl/slx114

E.R. Coughlin, P.J. Armitage, Tidal disruption by extreme mass ratio binaries and application to ASASSN-15lh. Mon. Not. R. Astron. Soc. **474**, 3857–3865 (2018). https://doi.org/10.1093/mnras/stx3039

E.R. Coughlin, P.J. Armitage, C. Nixon, M.C. Begelman, Tidal disruption events from supermassive black hole binaries. Mon. Not. R. Astron. Soc. **465**, 3840–3864 (2017). https://doi.org/10.1093/mnras/stw2913

E.R. Coughlin, S. Darbha, D. Kasen, E. Quataert, Stellar binaries incident on supermassive black hole binaries: implications for double tidal disruption events, calcium-rich transients, and hypervelocity stars. Astrophys. J. Lett. **863**, 24 (2018). https://doi.org/10.3847/2041-8213/aad7bd

J. Cuadra, P.J. Armitage, R.D. Alexander, M.C. Begelman, Massive black hole binary mergers within subparsec scale gas discs. Mon. Not. R. Astron. Soc. **393**, 1423–1432 (2009). https://doi.org/10.1111/j.1365-2966.2008.14147.x

S. Darbha, E.R. Coughlin, D. Kasen, E. Quataert, Gravitational interactions of stars with supermassive black hole binaries – I. Tidal disruption events. Mon. Not. R. Astron. Soc. **477**, 4009–4034 (2018). https://doi.org/10.1093/mnras/sty822

S. Dong, B.J. Shappee, J.L. Prieto, S.W. Jha, K.Z. Stanek, T.W.-S. Holoien, C.S. Kochanek, T.A. Thompson, N. Morrell, I.B. Thompson, U. Basu, J.F. Beacom, D. Bersier, J. Brimacombe, J.S. Brown, F. Bufano, P. Chen, E. Conseil, A.B. Danilet, E. Falco, D. Grupe, S. Kiyota, G. Masi, B. Nicholls, F. Olivares E., G. Pignata, G. Pojmanski, G.V. Simonian, D.M. Szczygiel, P.R. Woźniak, ASASSN-15lh: a highly super-luminous supernova. Science **351**, 257–260 (2016). https://doi.org/10.1126/science.aac9613

M. Dotti, M. Colpi, F. Haardt, L. Mayer, Supermassive black hole binaries in gaseous and stellar circumnuclear discs: orbital dynamics and gas accretion. Mon. Not. R. Astron. Soc. **379**, 956–962 (2007). https://doi.org/10.1111/j.1365-2966.2007.12010.x

A.C. Dunhill, R.D. Alexander, C.J. Nixon, A.R. King, Misaligned accretion on to supermassive black hole binaries. Mon. Not. R. Astron. Soc. **445**, 2285–2296 (2014). https://doi.org/10.1093/mnras/stu1914

J. Frank, M.J. Rees, Effects of massive central black holes on dense stellar systems. Mon. Not. R. Astron. Soc. **176**, 633–647 (1976). https://doi.org/10.1093/mnras/176.3.633

F.G. Goicovic, J. Cuadra, A. Sesana, F. Stasyszyn, P. Amaro-Seoane, T.L. Tanaka, Infalling clouds on to supermassive black hole binaries – I. Formation of discs, accretion and gas dynamics. Mon. Not. R. Astron. Soc. **455**, 1989–2003 (2016). https://doi.org/10.1093/mnras/stv2470

E.C.A. Golightly, E.R. Coughlin, C.J. Nixon, Tidal disruption events: the role of stellar spin. Astrophys. J. **872**, 163 (2019). https://doi.org/10.3847/1538-4357/aafd2f

D. Grupe, S. Komossa, R. Saxton, IC 3599 did it again: a second outburst of the X-ray transient Seyfert 1.9 Galaxy. Astrophys. J. Lett. **803**, 28 (2015). https://doi.org/10.1088/2041-8205/803/2/L28

A. Gualandris, J.I. Read, W. Dehnen, E. Bortolas, Collisionless loss-cone refilling: there is no final parsec problem. Mon. Not. R. Astron. Soc. **464**, 2301–2310 (2017). https://doi.org/10.1093/mnras/stw2528

J. Guillochon, E. Ramirez-Ruiz, A dark year for tidal disruption events. Astrophys. J. **809**, 166 (2015). https://doi.org/10.1088/0004-637X/809/2/166

J. Guillochon, M. McCourt, X. Chen, M.D. Johnson, E. Berger, Unbound debris streams and remnants resulting from the tidal disruptions of stars by supermassive black holes. Astrophys. J. **822**, 48 (2016). https://doi.org/10.3847/0004-637X/822/1/48

K. Hayasaki, N. Stone, A. Loeb, Finite, intense accretion bursts from tidal disruption of stars on bound orbits. Mon. Not. R. Astron. Soc. **434**, 909–924 (2013). https://doi.org/10.1093/mnras/stt871

D.C. Heggie, Binary evolution in stellar dynamics. Mon. Not. R. Astron. Soc. **173**, 729–787 (1975). https://doi.org/10.1093/mnras/173.3.729
K. Holley-Bockelmann, S. Sigurdsson, A Full Loss Cone For Triaxial Galaxies. arXiv e-prints, 0601520 (2006)
P.B. Ivanov, A.G. Polnarev, P. Saha, The tidal disruption rate in dense galactic cusps containing a supermassive binary black hole. Mon. Not. R. Astron. Soc. **358**, 1361–1378 (2005). https://doi.org/10.1111/j.1365-2966.2005.08843.x
L.Z. Kelley, L. Blecha, L. Hernquist, Massive black hole binary mergers in dynamical galactic environments. Mon. Not. R. Astron. Soc. **464**, 3131–3157 (2017). https://doi.org/10.1093/mnras/stw2452
Y. Kozai, Secular perturbations of asteroids with high inclination and eccentricity. Astron. J. **67**, 591 (1962). https://doi.org/10.1086/108790
G. Leloudas, M. Fraser, N.C. Stone, S. van Velzen, P.G. Jonker, I. Arcavi, C. Fremling, J.R. Maund, S.J. Smartt, T. Krìhler, J.C.A. Miller-Jones, P.M. Vreeswijk, A. Gal-Yam, P.A. Mazzali, A. De Cia, D.A. Howell, C. Inserra, F. Patat, A. de Ugarte Postigo, O. Yaron, C. Ashall, I. Bar, H. Campbell, T.-W. Chen, M. Childress, N. Elias-Rosa, J. Harmanen, G. Hosseinzadeh, J. Johansson, T. Kangas, E. Kankare, S. Kim, H. Kuncarayakti, J. Lyman, M.R. Magee, K. Maguire, D. Malesani, S. Mattila, C.V. McCully, M. Nicholl, S. Prentice, C. Romero-Cañizales, S. Schulze, K.W. Smith, J. Sollerman, M. Sullivan, B.E. Tucker, S. Valenti, J.C. Wheeler, D.R. Young, The superluminous transient ASASSN-15lh as a tidal disruption event from a Kerr black hole. Nat. Astron. **1**, 0002 (2016). https://doi.org/10.1038/s41550-016-0002
K. Lezhnin, E. Vasiliev, Evolution of supermassive black hole binaries and tidal disruption rates in non-spherical galactic nuclei. Mon. Not. R. Astron. Soc. **484**, 2851–2865 (2019). https://doi.org/10.1093/mnras/stz172
S. Li, F.K. Liu, P. Berczik, R. Spurzem, Boosted tidal disruption by massive black hole binaries during galaxy mergers from the view of N-body simulation. Astrophys. J. **834**(2), 195 (2017). https://doi.org/10.3847/1538-4357/834/2/195
M.L. Lidov, The evolution of orbits of artificial satellites of planets under the action of gravitational perturbations of external bodies. Planet. Space Sci. **9**, 719–759 (1962). https://doi.org/10.1016/0032-0633(62)90129-0
F.K. Liu, S. Li, X. Chen, Interruption of tidal-disruption flares by supermassive black hole binaries. Astrophys. J. **706**, 133–137 (2009). https://doi.org/10.1088/0004-637X/706/1/L133
F.K. Liu, S. Li, S. Komossa, A milliparsec supermassive black hole binary candidate in the Galaxy SDSS J120136.02+300305.5. Astrophys. J. **786**, 103 (2014). https://doi.org/10.1088/0004-637X/786/2/103
G. Lodato, S. Nayakshin, A.R. King, J.E. Pringle, Black hole mergers: can gas discs solve the 'final parsec' problem? Mon. Not. R. Astron. Soc. **398**, 1392–1402 (2009). https://doi.org/10.1111/j.1365-2966.2009.15179.x
M. Lopez Jr., A. Batta, E. Ramirez-Ruiz, I. Martinez, J. Samsing, Tidal Disruptions of Stars by Binary Black Holes: Modifying the Spin Magnitudes and Directions of LIGO Sources in Dense Stellar Environments. arXiv e-prints (2018)
A.-M. Madigan, A. Halle, M. Moody, M. McCourt, C. Nixon, H. Wernke, Dynamical properties of eccentric nuclear disks: stability, longevity, and implications for tidal disruption rates in post-merger galaxies. Astrophys. J. **853**, 141 (2018). https://doi.org/10.3847/1538-4357/aaa714
B. Margalit, B.D. Metzger, E. Berger, M. Nicholl, T. Eftekhari, R. Margutti, Unveiling the engines of fast radio bursts, superluminous supernovae, and gamma-ray bursts. Mon. Not. R. Astron. Soc. **481**, 2407–2426 (2018). https://doi.org/10.1093/mnras/sty2417
D. Merritt, M.Y. Poon, Chaotic loss cones and black hole fueling. Astrophys. J. **606**, 788–798 (2004). https://doi.org/10.1086/382497
B.D. Metzger, P. Beniamini, D. Giannios, Effects of fallback accretion on protomagnetar outflows in Gamma-ray bursts and Superluminous Supernovae. Astrophys. J. **857**, 95 (2018). https://doi.org/10.3847/1538-4357/aab70c
M. Milosavljević, D. Merritt, Formation of galactic nuclei. Astrophys. J. **563**, 34–62 (2001). https://doi.org/10.1086/323830
T.J. Moriya, B.D. Metzger, S.I. Blinnikov, Supernovae powered by magnetars that transform into black holes. Astrophys. J. **833**, 64 (2016). https://doi.org/10.3847/1538-4357/833/1/64
C.J. Nixon, P.J. Cossins, A.R. King, J.E. Pringle, Retrograde accretion and merging supermassive black holes. Mon. Not. R. Astron. Soc. **412**, 1591–1598 (2011). https://doi.org/10.1111/j.1365-2966.2010.17952.x
P.C. Peters, Gravitational radiation and the motion of two point masses. Phys. Rev. **136**, 1224–1232 (1964). https://doi.org/10.1103/PhysRev.136.B1224
D.J. Price, J. Wurster, T.S. Tricco, C. Nixon, S. Toupin, A. Pettitt, C. Chan, D. Mentiplay, G. Laibe, S. Glover, C. Dobbs, R. Nealon, D. Liptai, H. Worpel, C. Bonnerot, G. Dipierro, G. Ballabio, E. Ragusa, C. Federrath, R. Iaconi, T. Reichardt, D. Forgan, M. Hutchison, T. Constantino, B. Ayliffe, K. Hirsh,

G. Lodato, Phantom: A smoothed particle hydrodynamics and magnetohydrodynamics code for astrophysics. ArXiv e-prints (2017)

G.D. Quinlan, The dynamical evolution of massive black hole binaries I. Hardening in a fixed stellar background. New Astron. **1**, 35–56 (1996). https://doi.org/10.1016/S1384-1076(96)00003-6

M.J. Rees, Tidal disruption of stars by black holes of 10^6–10^8 solar masses in nearby galaxies. Nature **333**(6173), 523–528 (1988). https://doi.org/10.1038/333523a0

A. Ricarte, P. Natarajan, L. Dai, P. Coppi, Tidal disruption events by a massive black hole binary. Mon. Not. R. Astron. Soc. **458**, 1712–1727 (2016). https://doi.org/10.1093/mnras/stw355

G.P. Rosotti, G. Lodato, D.J. Price, Response of a circumbinary accretion disc to black hole mass loss. Mon. Not. R. Astron. Soc. **425**, 1958–1966 (2012). https://doi.org/10.1111/j.1365-2966.2012.21488.x

E.M. Rossi, G. Lodato, P.J. Armitage, J.E. Pringle, A.R. King, Black hole mergers: the first light. Mon. Not. R. Astron. Soc. **401**, 2021–2035 (2010). https://doi.org/10.1111/j.1365-2966.2009.15802.x

T. Ryu, R. Perna, Z. Haiman, J.P. Ostriker, N.C. Stone, Interactions between multiple supermassive black holes in galactic nuclei: a solution to the final parsec problem. Mon. Not. R. Astron. Soc. **473**, 3410–3433 (2018). https://doi.org/10.1093/mnras/stx2524

J. Samsing, T. Venumadhav, L. Dai, I. Martinez, A. Batta, M. Lopez Jr., E. Ramirez-Ruiz, K. Kremer, Probing the Black Hole Merger History in Clusters using Stellar Tidal Disruptions. arXiv e-prints (2019)

A. Sesana, F. Haardt, P. Madau, Interaction of massive black hole binaries with their stellar environment. I. Ejection of hypervelocity stars. Astrophys. J. **651**, 392–400 (2006). https://doi.org/10.1086/507596

N.C. Stone, B.D. Metzger, Rates of stellar tidal disruption as probes of the supermassive black hole mass function. Mon. Not. R. Astron. Soc. **455**(1), 859–883 (2016). https://doi.org/10.1093/mnras/stv2281

Y. Tang, A. MacFadyen, Z. Haiman, On the orbital evolution of supermassive black hole binaries with circumbinary accretion discs. Mon. Not. R. Astron. Soc. **469**, 4258–4267 (2017). https://doi.org/10.1093/mnras/stx1130

S. Thorp, E. Chadwick, A. Sesana, Tidal disruption events from massive black hole binaries: predictions for ongoing and future surveys. arXiv e-prints (2018)

Q. Vigneron, G. Lodato, A. Guidarelli, Tidal disruption of stars in a supermassive black hole binary system: the influence of orbital properties on fallback and accretion rates. Mon. Not. R. Astron. Soc. **476**, 5312–5322 (2018). https://doi.org/10.1093/mnras/sty585

C. Wegg, J. Bode, Multiple tidal disruptions as an indicator of binary supermassive black hole systems. Astrophys. J. Lett. **738**, 8 (2011). https://doi.org/10.1088/2041-8205/738/1/L8

A. Yalinewich, E. Steinberg, T. Piran, J.H. Krolik, Radio Emission from the unbound Debris of Tidal Disruption Events. arXiv e-prints (2019)

Q. Yu, Evolution of massive binary black holes. Mon. Not. R. Astron. Soc. **331**, 935–958 (2002). https://doi.org/10.1046/j.1365-8711.2002.05242.x

Space Sci Rev (2020) 216:88
https://doi.org/10.1007/s11214-020-00680-z

Future Simulations of Tidal Disruption Events

Julian H. Krolik[1] · **Philip J. Armitage**[2] ·
Yanfei Jiang[3] · **Giuseppe Lodato**[4]

Received: 22 July 2019 / Accepted: 23 April 2020 / Published online: 2 July 2020

Abstract Tidal disruption events involve numerous physical processes (fluid dynamics, magnetohydrodynamics, radiation transport, self-gravity, general relativistic dynamics) in highly nonlinear ways, and, because TDEs are transients by definition, frequently in non-equilibrium states. For these reasons, numerical solution of the relevant equations can be an essential tool for studying these events. In this chapter, we present a summary of the key problems of the field for which simulations offer the greatest promise and identify the capabilities required to make progress on them. We then discuss what has been—and what cannot be—done with existing numerical methods. We close with an overview of what methods now under development may do to expand our ability to understand these events.

1 Pressing Unsolved Problems Requiring Simulation

As the preceding chapters have illustrated in many ways, the field of TDEs remains filled with important unsolved problems. For many, the physics is so complex that analytic meth-

The Tidal Disruption of Stars by Massive Black Holes
Edited by Peter G. Jonker, Sterl Phinney, Elena Maria Rossi, Sjoert van Velzen, Iair Arcavi and Maurizio Falanga

✉ J.H. Krolik
jhk@jhu.edu

P.J. Armitage
philip.armitage@stonybrook.edu

Y. Jiang
yjiang@flatironinstitute.org

G. Lodato
giuseppe.lodato@unimi.it

[1] Department of Physics and Astronomy, Johns Hopkins University, Baltimore, MD 21218, USA

[2] Department of Physics and Astronomy, Stony Brook University, Stony Brook, NY 11794, USA

[3] Center for Computational Astrophysics, Flatiron Institute, New York, NY 10010, USA

[4] Dipartimento di Fisica, Universita di Milano, Via Celoria 16, 20133 Milano, Italy

ods have little power: although it is possible in many cases to study idealized problems analytically, it is often at the cost of ignoring mechanisms of comparable influence to the one treated, or of imposing artificial symmetries or boundary conditions, or of being restricted to implausible corners of parameter space (most often by being limited to linear perturbation theory). By contrast, numerical methods can permit the simultaneous consideration of multiple mechanisms, are capable of dealing with highly asymmetric geometries, and make no distinction between small and large amplitude fluctuations.

In this chapter, we will begin by posing a number of TDE problems for which numerical work promises to yield major rewards in understanding. In the remainder of the chapter, we will critically discuss the simulation methods (i.e., codes) currently available in terms of how they measure up to the demands of these questions, and then apply the same standards to methods currently under development.

1.1 Magnetic Seeding of Bound Debris

Once the bound remains of a destroyed star return from the apocenter of their first orbit, they join an accretion flow around the black hole, as discussed in the contribution of Bonnerot et al. in this issue. In most accretion flows, inward motion is limited by angular momentum transport, and the dominant mechanism providing that transport is correlated magnetic stresses in MHD turbulence driven by the magnetorotational instability (MRI) (Balbus and Hawley 1998). In TDE accretion flows, although stream deflection in oblique shocks initially dominates angular momentum transport, these shocks decay over a span of 5–10 orbital periods of the most-bound debris (Shiokawa et al. 2015). Any further inward flow must be due to some other mechanism, perhaps a version of the correlated MHD turbulence familiar from other accretion problems.

However, little is known about how MHD turbulence might develop in these circumstances. Presumably, the magnetic field of the star is bequeathed to the bound debris because the matter formerly in the star should stay highly-conductive, consistently supporting flux-freezing, but it is unclear what intensity or geometry it will have once the debris settles into the accretion flow. There are more questions about the development of the MRI because our entire knowledge-base hitherto has been built upon the assumption of circular orbits, while the debris orbits of a TDE are highly elliptical, with initial eccentricities e within a few percent of unity. Only recently has it been shown that the MRI remains linearly unstable in elliptical disks (Chan et al. 2018) and grows with a rate comparable to that found in circular-orbit disks, but to date there has been no work on its nonlinear saturation.

Such knowledge is critical for understanding the longer-term evolution of the accretion flow. It is also critical for understanding the possible launch of relativistic jets. The phenomenology of several TDEs points strongly to the existence of relativistic jets (Burrows et al. 2011; Bloom et al. 2011; Cenko et al. 2012; Brown et al. 2015), but the magnetic flux required to support a jet of sufficient strength through the Blandford-Znajek mechanism (or some variant) is $>\sim 10^{29} M_6 L_{46}$ G cm^2, considerably greater than the magnetic flux of typical stars, $\sim 10^{22}$ G cm^2. Thus, field evolution after disruption is also of great interest.

These desiderata point strongly to the need for simulations able to follow the magnetic field from an origin in the star, through its expulsion into the much larger volume around the black hole occupied by the debris, and continuing over the course of whatever subsequent events may amplify (or diminish) its intensity, shape its geometry, or reconfigure its flux. These simulations will not be easy because they must span a very large dynamic range in lengthscale, from within the star to the span of the debris orbits: the radius of a $1 M_\odot$

main sequence star is $\simeq 0.5M_6^{-1}r_g$, while the orbital semi-major axes are larger by a factor of at least $(M_{\rm BH}/M_*)^{2/3} \sim 10^4 M_6^{2/3}$ (here $M_6 \equiv M_{\rm BH}/10^6 M_\odot$ and $r_g \equiv GM/c^2$ is the gravitational radius corresponding to the black hole's mass). Any such simulation must also ensure that the magnetic field is kept divergence-free to very high accuracy in order to avoid unphysical magnetic forces.

1.2 Radiation Flow and Forces

The only information we have on TDEs (until they can be detected as gravitational wave events, as suggested by Kobayashi et al. (2004)) is through photons. If we're to claim any real understanding of them, we must therefore be able to show how the dynamics we think control them lead to the electromagnetic spectra and lightcurves we observe. To accomplish this will demand a greate deal of effort because the EM radiation properties span a wide range of wavelengths, vary in time (these are, after all, flare events), and likely depend on viewing-angle (e.g., relative to the star's orbital axis or the black hole's spin axis). Tracing radiation properties is also a vital part of understanding the dynamics themselves because at many locations and times during a TDE radiation forces can be important. This is the case within the initial star if its mass is $>\sim 10M_\odot$, in shocked debris (Shiokawa et al. 2015; Jiang et al. 2016), and in the accretion flow close to the black hole because accretion rates can easily be in the Eddington range (Rees 1988; Ulmer 1999; Krolik et al. 2016). Even when the radiation fluxes are too small to cause significant forces, photons carrying off heat frequently play a significant role in determining the gas's equation of state, and therefore have indirect dynamical influence. This last role is of special importance to the flow near the black hole because photon trapping may be substantial when the mass accretion rate is comparable to or greater than Eddington. Thus, for all these reasons, incorporating radiation transport (at least at the level of gray opacity) into the dynamical simulations is another important goal.

There is, however, good reason why as of yet this has only rarely been done—simultaneous solution of the radiation transfer problem and the equations of hydrodynamics can be both complex and computationally very expensive. Although the diffusion approximation, or extensions of it like flux-limited diffusion (FLD) and the M1 closure, are excellent assumptions where the optical depth to the surface is very large, they can produce spurious effects when the optical depth becomes order unity or less (McKinney et al. 2014). For this reason, diffusion-based schemes are not reliable near or outside an external surface, and this is, of course, precisely where the radiation we see is determined. Moreover, over time, a bad photospheric boundary condition (which is what FLD and M1 provide) can influence the intensity of radiation deep inside the system where locally the diffusion approximation is entirely valid. Credible results may, in the end, rest upon genuine multi-angle radiation transfer solutions—which may also need to be time-dependent—and possibly their extension to multi-frequency calculations. Such calculations can be extremely computationally expensive both because of the large number of equations solved and because time-dependent transfer can force time-steps considerably smaller than those required for the fluid motion.

Further complexity and cost come from spectral considerations. Spectral properties are often determined by the interaction of many individual atomic processes because the total opacity of an astrophysical gas depends on a sum over its many line and edge features, and these in turn depend on the gas's ionization balance and excited populations (which, in regions of line formation, are frequently not in thermodynamic equilibrium). Still further complexity is introduced because the portion of the flow near the black hole is necessarily relativistic. Radiation-matter interactions are almost always most simply described in the rest-frame of the matter, but in a relativistic flow, there are significant frame-shifts between each pair of adjacent cells in a simulation.

1.3 The Influence of Black Hole Spin

For main sequence stars suffering tidal disruption, the tidal radius is never more than several tens of gravitational radii. General relativistic effects can therefore be important, and in more than one way. One effect whose influence has only begun to be explored is the gravitomagnetic (Lense-Thirring) torque exerted by the black hole on orbiting matter. If the black hole has any spin at all, and the stellar orbital plane does not lie exactly in the equatorial plane of the Kerr spacetime, gravitomagnetic torque drives a precession of the orbital plane at a rate $2(a/M)(r/r_g)^{-3/2}\times$ the orbital frequency (here a/M is the black hole spin parameter and r is the radius of a fluid element relative to the black hole). Because it is hard to imagine any reason for the stellar orbit's orientation to be correlated with the black hole's spin direction, this precession should occur generically in TDEs. Moreover, it should do so throughout the event: while the star passes through its orbital pericenter and begins to fall apart, when the bound debris returns to the pericenter region, and during the evolution of the accretion flow created by debris return. Much interesting behavior could result from this precession (Guillochon and Ramirez-Ruiz 2015).

Ever since the work of Bardeen and Petterson (1975), it has been generally believed that, granted sufficient time, the inner regions of an accretion flow oriented obliquely relative to its central black hole should become aligned with the black hole's spin. However, there remains much controversy about how rapidly this takes place, under which circumstances, and by what specific mechanisms (Hatchett et al. 1981; Papaloizou and Pringle 1983; Nelson and Papaloizou 2000; Lodato and Price 2010; Sorathia et al. 2013; Zhuravlev et al. 2014; Krolik and Hawley 2015; Hawley and Krolik 2018; Liska et al. 2018b). The issue of evolutionary timescale is especially important for TDEs, which are intrinsically transient events; because they may evolve on timescales that are not a great many orbital timescales, while the precession time can be a large multiple of an orbital period, non-equilibrium is a very real issue in this context. Thus, even the extensive (and disputatious) literature on this topic, which has generally focused on longer-lived systems, gives only limited guidance.

This is a problem that positively calls out for numerical simulation: it involves nonlinear, general relativistic hydrodynamics in intrinsically 3-d geometry. The need for simulation is underlined by the fact that this problem is all about angular momentum transport and in accretion disks that means MHD turbulence must be included. Although capable general relativistic MHD codes exist, it remains a challenging problem to treat because the precession timescale is long compared to the orbital timescale and because very good resolution is necessary to ensure that numerical diffusion on the gridscale does not masquerade as physical angular momentum transport.

1.4 Self-Gravity

The star's self-gravity is, of course, essential to determining its structure in isolation as well as the trade-off between this force and the external tidal gravity of the black hole when the star approaches sufficiently close. Where its effect remains poorly understood is in the debris. Kochanek (1994) pointed out that self-gravity could cause confinement of the debris in the plane perpendicular to the stream's extension; at the same time, however, he also noted that modest injections of entropy, e.g., from H recombination or internal shocks, could substantially counteract its action (Kochanek 1994). More recently, Coughlin et al. (2016)

argued that if the stream evolution is perfectly adiabatic, whether self-gravity is vital or negligible depends on the value of $d \ln p / d \ln \rho$. It is also possible that the innermost portion of the debris is subject to significant self-gravity, while a lower-density halo is not (Yalinewich et al. 2019; Steinberg et al. 2019).

The degree to which self-gravitational confinement occurs might be important to later stages of the event because compression of the stream cross-section could affect the deceleration of returning streams when they strike the accretion flow created by earlier-returning debris. However, the epoch of its possible importance ends once most of the returning mass has suffered at least one shock; at that point, the additional entropy created overwhelms any self-gravity because these shocks have speeds $\sim 10^{3\pm0.5}$ km s^{-1}, comparable to stellar interior sound speeds, but occur in gas of far lower density.

Once again, the nonlinearity and geometrical complexity of the problem make it intractable to anything but computational methods. Also once again, there are technical challenges. Several of them stem from the long, narrow geometry of the debris stream. This shape automatically implies that self-gravity is important in only a small fraction of the volume around the black hole. It also means that high spatial resolution is necessary because the stream width is often a very small fraction of its distance from the black hole. It also raises the importance of local methods. In the early stages of a disruption, when the material is still close enough to the black hole for relativistic effects to be significant, any global self-gravity calculation would have to be framed in terms of the Einstein Field Equations rather than the Poisson Equation, although it might be possible to treat the Field Equations in a perturbative fashion. Even in a local solution (e.g., within a few stellar radii of the star's center-of-mass) it is still necessary to be careful about relativistic considerations (for one method, see Ryu et al. (2020)). Later, when the matter is farther away, only the matter within a few stream thicknesses of a point contributes significantly to the local self-gravity.

Other complications arise because of the sensitivity to entropy in the balance between pressure and self-gravity. In other words, it is necessary to be careful about radiative cooling or phase changes (like atomic recombination).

2 Prerequisite Capabilities

This survey of major TDE problems for which simulations would be valuable reveals that numerous physics elements beyond classical hydrodynamics are demanded: general relativity; MHD; radiation transport and radiation forces; self-gravity (Newtonian locally, but requiring reconciliation with the relativistic background); non-adiabatic equations of state. None of the scientific problems posed demands all of these, but all require at least several of them.

Progress on these scientific problems through simulations also requires overcoming a number of technical difficulties: large dynamic range in lengthscales and timescales; incorporating a phase space (photon direction and energy) beyond the configuration space of fluid dynamics; and a physics repertory that varies with position and time. Particularly the first two difficulties can pose severe practical problems due to their magnification of the computing time needed.

3 Currently-Available Tools

We next turn to an evaluation of how existing simulation methods measure up the prerequisites just listed.

3.1 The *HARM* Family

A number of codes in current use for tidal disruption problems all descend from *HARM-2D* (Gammie et al. 2003). They share a number of characteristics: all are general relativistic, treat 3 spatial dimensions, employ an intrinsically-conservative fluid algorithm using the Lax-Friedrichs approximation to the Riemann problem, and use a constrained transport (CT) algorithm to update the magnetic field and preserve $\nabla \cdot \vec{B} = 0$. Because they are relativistic, a fully self-consistent calculation of gas self-gravity would require the solution of the Einstein Field Equations, but the associated computational load is so great that none of them attempts it.

Although sharing a great deal, they differ from one another in several ways. *HARM3D* (Noble et al. 2009, 2012) offers complete flexibility in coordinate and spacetime definition, including time-dependent spacetimes. Its fundamental coordinate system is "index-space" for the spatial arrays, permitting grid-cells whose physical dimensions can vary in any fashion prescribed by the user. *H-AMR* (Liska et al. 2018a), on the other hand, is restricted to the Kerr spacetime in Kerr-Schild spherical coordinates. However, because it runs on GPUs rather than CPUs, its processing speed is at least an order of magnitude greater than *HARM3D*. In addition, it has the capacity for adaptive mesh refinement (AMR) on blocks of cells. Both *HARM3D* and *H-AMR* offer comparatively primitive thermodynamics: heat can be lost by radiation only through optically-thin cooling functions ($\nabla_\mu T^{\mu\nu} = -\mathcal{L}u^\nu$, for stress-energy trensor $T^{\mu\nu}$, 4-velocity u^ν, and cooling function $\mathcal{L}$). *KORAL* (Sądowski et al. 2013, 2014) shares the spacetime restrictions of *H-AMR*, but introduces quasi-diffusive radiation transfer through the "M1 closure" approximation, as well as the forces exerted by radiation fluxes. The M1 closure's essential assumption is that the intensity distribution is axisymmetric around the direction of the mean flux.

3.2 *Flash*

Flash (Fryxell et al. 2010) differs from the *HARM* family in a number of respects. It is Newtonian (or special relativistic), but not general relativistic, and it offers a number of Riemann solution choices (Roe, HLL, HLLC, and HLLD in addition to Lax-Friedrichs). Most significantly, because its primary version is Newtonian, it can compute self-gravity in terms of the Poisson equation (a number of different algorithms are implemented: multigrid, multipole, FFT, and "tree"). Like *KORAL*, it can solve radiation transfer problems and thereby account for radiation forces, but does so via flux-limited diffusion. Like *H-AMR*, it offers AMR.

3.3 *Athena*

Athena (Stone et al. 2008) is a finite-volume grid based code that solves the Newtonian magneto-hydrodynamic (MHD) equations. It uses constrained transport (CT) to evolve the magnetic fields and a Godunov scheme for shock capturing. In addition to the ideal MHD equations, other terms can be included to describe different physical processes such as resistivity, viscosity, ambipolar diffusion, Hall effects and self-gravity. The radiative transfer equation can be solved based on either the Variable Eddington Tensor method (Jiang et al. 2012, Davis et al. 2012) or an algorithm that solves the time-dependent equation for specific intensities directly (Jiang et al. 2014). Coupled with the MHD equations, either radiation module can be used to determine the thermal properties of the gas self-consistently and predict the observed bolometric lightcurve. The original *Athena* code has only Cartesian and cylindrical (Skinner and Ostriker 2010) coordinate systems with a fixed resolution. It is useful for studying the local physics of TDE streams (Salbi et al. 2014, Jiang et al. 2016), but it would be difficult to use it for a global TDE simulation.

Some of these limitations are removed by *Athena++*, which solves the same set of equations as the original *Athena* code, but also offers Cartesian, cylindrical and spherical polar coordinate systems with adaptive mesh refinement. A constant time step is currently adopted for all refinement levels. The radiation modules written for the original version of *Athena* also work with *Athena++* (Jiang et al. 2017).

In addition, *Athena++* provides two new physics modules. One supports general relativistic magneto-hydrodynamics. It is based on advanced Riemann solvers and staggered-mesh constrained transport (White et al. 2016) and can be used to study the effects of general relativity in TDEs (note that the radiation modules are not consistent with relativistic dynamics). A second (appropriate only when the spacetime is nearly flat) computes self-gravity in the interior of a bounded region in cylindrical coordinates using an eigenfunction expansion for the azimuthal and vertical dependence and a tridiagonal matrix solver for the radial dependence (the radial grid can be either uniform or logarithmic). To create appropriate boundary conditions for such a finite region, it iterates between the interior solution and an exterior solution that calculates the "image charge" necessary to make the interior solution reach the appropriate boundary condition at infinity (Moon et al. 2019).

3.4 Moving Mesh Codes

"Moving-mesh" is an umbrella term for a class of methods that includes both fully unstructured Lagrangian hydrodynamics schemes such as *AREPO* (Springel 2010) and codes that implement a moving, but structured mesh, that is tailored to a specific problem geometry. All such schemes potentially reduce advection errors as compared to using the same solver on a fixed grid, while the fully general codes additionally offer an adaptive capability that can be coupled to high-order Godunov schemes. Examples include *AREPO*, *RICH* (Yalinewich et al. 2015) and *ChaNGa* (Chang et al. 2017), all of which are based on a Voronoi tessellation of the simulation domain.

A unique numerical consideration relevant to moving mesh codes is the presence of "mesh noise" that originates from the remapping operations needed as the flow evolves. In addition, current codes that implement general moving meshes do not conserve angular momentum to machine precision, and the adequacy of angular momentum conservation would therefore need to be monitored for long term simulations where it is important (e.g., accretion disks). However, in a situation such as a Keplerian disk where a high degree of symmetry is guaranteed, a moving mesh whose motion is constrained to follow a known, dominant component of the motion has advantages over a fully unstructured one (Duffell 2016) because it reduces these noise sources. In common with particle-based codes where forces are computed on a pairwise basis, angular momentum is typically well-conserved globally. Local conservation of specific angular momentum, on the other hand, is dependent on the time integration scheme, and was relatively poor for first generation moving mesh codes. Pakmor et al. (2016) discuss one approach to obtaining improved angular momentum conservation within *AREPO*.

Existing codes include different subsets of the physics needed for TDE simulations. When implemented, self-gravity is typically calculated using a tree algorithm (e.g. Barnes and Hut 1986). A constrained transport scheme for MHD is possible (Mocz et al. 2016), as is a method based instead on the vector potential (Fragile et al. 2018). Most do not include radiation physics or relativity (even in a fixed metric formulation), although these capabilities do not pose intrinsic problems.

Another existing option for TDE studies is *GIZMO* (Hopkins 2015), which offers a choice between SPH and a mesh-less scheme that implements an arbitrary Lagrangian-Eulerian finite-volume method based on Riemann solvers. *GIZMO* conserves angular momentum to machine precision, and uses divergence cleaning for MHD. However, there are indications that both options suffer from excessive numerical dissipation and incomplete divergence-cleaning when treating nonlinear MHD turbulence (Deng et al. 2019).

These codes have been employed on TDE problems only within the past year (Goicovic et al. 2019; Yalinewich et al. 2019; Steinberg et al. 2019), even though their key strengths—the ability to model a range of scales adaptively, follow supersonic flows with reduced diffusion, and capture shocks—are in principle well-matched to the problem.

3.5 SPH

Smoothed Particle Hydrodynamics (SPH, Lucy 1977, Gingold and Monaghan 1977) is a mesh-less Lagrangian algorithm that has been widely used, especially in cosmological simulations and hydrodynamic simulation of accretion disks and star formation. Its application to simulations of TDEs is especially convenient in view of its Lagrangian approach, given the large dynamic range in lengthscales between the stellar interior and the extended debris flow. It is also computationally efficient in the sense that most of the computational domain is "empty", with the stellar debris occupying only a limited volume. In fact, some of the earliest applications of the SPH method were in the context of TDEs. Nolthenius and Katz (1982) used ~ 40 SPH particles to simulate the disruption of a $1 M_\odot$ star by a $10^4 M_\odot$ black hole, and Bicknell and Gingold (1983) used 500 SPH particles to simulate a highly penetrating encounter of a $1 M_\odot$ star with a $10^5 M_\odot$ black hole. The goal of the latter simulation was to test the possibility of tidal detonation of the star, a fashionable topic at the time.

In SPH, the fluid is discretized into finite mass elements (called "particles") that evolve according to the Euler-Lagrange equations obtained from a variational principle formulation of fluid dynamics (Eckart 1960). Thus, the set of SPH particles represent a Hamiltonian system and as such exactly conserve linear and angular momentum, as well as energy (see Price 2012 and Springel 2010 for recent reviews). The fluid properties at one particle's position are computed by averaging the properties of neighbouring particles that lie within a "smoothing region" around it; various weighting functions are used, but in general they decline with distance from the particle. The typical size of this region is called the "smoothing length" h. Typically, the smoothing length is chosen such as to have a constant mass (i.e., number of particles) inside the smoothing sphere. Thus, in SPH resolution automatically follows density and in this sense is a naturally "adaptive" method; this adaptivity does not, however, automatically recognize local gradient scales. It is also important to note that the smallest "resolved" mass is the mass within a smoothing volume, not the mass of an individual particle; moreover, the mass, energy, and momentum of a particular smoothing volume are subject to Poisson fluctuations due to the finite number of particles within that volume (often chosen to be a few dozen).

Particular care should be given to the handling of shocks and discontinuities in SPH (Price 2008). In particular, to resolve a shock, SPH typically requires an artificial bulk viscosity. It can be shown (Lodato and Price 2010) that, in the absence of switches, the artificial viscosity scales $\propto c_s h$, where c_s is the gas sound speed. Thus, low density regions, which also have large smoothing lengths h, can suffer from large artificial viscosity. In most cases, however, such dissipation can be effectively limited to the shock region by using suitable switches, such as the Morris and Monaghan (1997) switch, or the Cullen and Dehnen (2010) switch. This can be particularly important for simulations that try to follow the formation of a disc after a TDE, since at the beginning of the fallback phase, the gas density is bound to be low, and great care should be taken in ensuring that the results are not affected by excessive numerical dissipation.

Another interesting aspect to consider is the inclusion of relativistic effects. SPH naturally lends itself to a fully general relativistic implementation (in a fixed metric), as shown by Monaghan and Price (2001), but it has taken a number of years to complete development of this feature. This effort began with Hayasaki et al. (2013), who used the modified pseudo-Newtonian potential of Wegg (2012) to include apsidal precession for particles with nearly-parabolic orbits, and Tejeda and Rosswog (2013), who introduced a "generalized Newtonian potential" that reproduces test-particle motion in a Schwarzschild spacetime very well for orbits that are identically parabolic, but has unspecified errors for orbits with non-zero binding energy, such as the tidal debris. This treatment of particle motion as well as a first-order post-Newtonian approximation to Lense-Thirring torque have been added to the *PHANTOM* code (Price et al. 2017) and used in TDE simulations (Bonnerot et al. 2016); a second-order pN approximation was developed by Hayasaki et al. (2016). Very recently, a fully relativistic version has been published, using a formalism based on entropy conservation rather than energy conservation (Liptai and Price 2019). However, no provision for relativistically-consistent calculation of stellar self-gravity in SPH simulations has yet been made.

An algorithm for computing 3D radiation transfer in the flux-limited diffusion approximation exists (Whitehouse et al. 2005), but has not yet been used on TDE problems.

4 Tools Under Development

We turn next to new tools on the horizon, comparing them as well to the prerequisites for progress on TDE problems.

4.1 *Patchwork*

Many physical systems, including many in astrophysics, contain local inhomogeneities where the local geometric symmetry, or physical lengthscales, or relevant physical processes are different from those elsewhere in the system. These contrasts pose great difficulties to simulations in which the entire problem is assigned to a single program. Graded meshes, whether determined in advance or adaptively, can help with contrasting lengthscales, but it is very difficult to adjust the equations to be solved or the symmetry of the grid within a single program.

"Multipatch" systems enable a number of independent programs, each governing a particular region, to solve a unified physical problem by exchanging boundary condition data. A particularly flexible example is the *Patchwork* infrastructure (Shiokawa et al. 2018). With this infrastructure, built to be intrinsically general relativistic, multiple patches, which may be stationary with respect to one another or moving, can have wholly independent grids, coordinate systems, and physical equations. For consistency, they must be regarded as moving through the same spacetime and must share the same time coordinate. When boundary condition data are exchanged, they are therefore transformed by coordinate (not Lorentz) transformations; the transformations for scalar, 4-vector, and tensor quantities follow the standard rules. In its published form, *Patchwork* was restricted to problems without magnetic fields because interpolation of magnetic field data onto a new grid as part of inter-patch boundary data exchange creates inter-patch monopoles. Since then, it has been extended to include algorithms that remove these monopoles (Avara et al., in preparation).

Patchwork presents great promise for tidal disruption simulations because TDEs are a prime example of strong inhomogeneities within a single physical system. They exhibit extreme contrasts in lengthscale: the radius of a $1 M_\odot$ main-sequence star is only $\approx 0.5 r_g$ when the black hole mass is $10^6 M_\odot$, while the debris orbits extend as far as $\sim 10^4 r_g$. They have strong contrasts in grid symmetry: the natural symmetry for orbital motion around a black hole implies the use of polar coordinates with an origin at the center of the black hole, while the natural symmetry for the star's self-gravity argues for an origin at its center-of-mass—and this moves relative to the black hole. They involve contrasts in relevant physics: self-gravity is, of course, essential to a star and can be significant for the initial evolution of the tidal debris, but elsewhere and later in the development of the event it is irrelevant. In fact, the challenges of TDEs were one of the prime motivations for development of *Patchwork.*

It is also worth noting that multipatch systems can present computational as well as physical advantages. When the coordinates are chosen with reference to local dynamical symmetries, numerical diffusion can be minimized, and it is often possible to design grids with far fewer total cells while maintaining high resolution where it is needed. Different patches can also have different time-steps, potentially offering avenues for improved load-balancing.

4.2 General Relativistic Radiation Athena

Extending the current radiative transfer module for *Athena* to be fully compatible with relativistic magneto-hydrodynamics will be a major next step; problems in flat spacetime and curved spacetime will be handled by separate modules. This extension will make it possible to study the dynamics of streams from TDEs as they travel around the black hole with self-consistent thermodynamics and radiation forces. Although the first use of such a module will undoubtedly assume gray opacity, increasing computing power would permit use of a more general version incorporating frequency-dependent radiation transport, as well as scattering angular distributions sensitive to background physical properties such as magnetic field direction.

The ability to include radiation transport is greatly enhanced by improvements in MHD computing speed. The MHD module of *Athena++* has already been converted to a GPU version and preliminary testing indicates substantial acceleration: it can update as many as 10^8 cells per second on a single Nvidia Volta GPU, while maintaining 75% parallelization efficiency on as many as 250 GPUs. Although initially written for conventional parallelization, the plan is to write both relativistic radiation modules to be compatible with future conversion to GPUs.

4.3 Moving Mesh and SPH Codes

The intrinsic potential of moving mesh codes for TDE problems has not been fully realized for a number of reasons, including the fact that first-generation moving mesh codes typically implemented a more limited set of physical processes than more mature methods. Restricted access to some codes, together with an initially small base of experienced users, may also have been a factor.

Near-future improvements should remove these limitations. In particular, the core TDE physics of coupling radiation transport to hydrodynamics is expected to be available soon in several moving mesh or mesh-less schemes. *GIZMO*, for example, now includes a number of radiation transport methods, including flux-limited diffusion, M1, and Monte Carlo ray-tracing (Hopkins and Grudić 2019). Monte Carlo radiation transport is being implemented in the SPH code PHANTOM, and there is ongoing work to include similar radiation transport capabilities within *ChaNGa*, which is in addition being extended to relativistic computations within a fixed metric.

4.4 New Fluid Algorithms

Astrophysical simulation codes most commonly implement one of two families of algorithms for solving the fluid equations: either finite-volume methods based on Godunov schemes, or SPH. Finite difference and spectral methods are also in use. Many other algorithms have been developed in the fluid dynamics community, some of which may offer advantages for astrophysical problems including TDEs. A number of groups, for example, have recently developed astrophysical simulation codes based on discontinuous Galerkin methods (e.g. Schaal et al. 2015; Kidder et al. 2017; Anninos et al. 2017). Discontinuous Galerkin schemes can provide high order accuracy—approaching that of spectral methods—without the non-local communication costs that are typically the price that high-order accuracy demands.

4.5 Programming Model Advances

In addition to improvements in physics capability and numerical algorithms, advances in computer architecture and operating systems also promise greater capability in the near future.

The most mature of these new programming models is the use of GPUs, whose rapid growth in use is due to the fact that GPU performance has increased at a faster rate than CPU performance in recent years. The GPU version of *HARM*, *H-AMR*, outperforms its CPU cousin by a factor of between 9 and 14 on the general relativistic MHD disk simulation problem considered by Liska et al. (2018a). In comparing codes and methods it must be remembered that it is the overall compute time needed to obtain a solution of a given fidelity that matters, and that an impressive speed-up on one problem does not necessarily generalize to other problems of interest (for example those where other physical effects dominate the computation). Nonetheless, it is clear that large GPU speed-ups are possible for some problems of interest to the TDE community, and it is likely that codes other than *HARM* could benefit from the use of GPU technology.

Whatever algorithm is employed, the goal is to obtain an appropriate balance of absolute code performance (on a small number of cores or GPUs) together with good scaling to large numbers of processing elements. This requires identifying and taking advantage of whatever parallelism the problem offers, while minimizing overheads in communication between processors or in breaking down the work into pieces that can be executed independently. The traditional approach to parallelism in astrophysical simulation codes has been based on *data decomposition* using MPI. In this approach, each core executes the same code on a different piece of the data, which is usually sub-divided into spatial domains. The domains can be fixed in advance (static domain decomposition), or adjusted dynamically in cases where the workload is expected to vary as the simulation progresses. Data decomposition often suffices to provide good weak scaling, i.e. ever larger numbers of processors can be efficiently deployed to solve ever larger instances of the same problem. There are, however, drawbacks. The inherent near-synchronicity of the model does not always make full use of the available communication between processors, leading to less efficient strong scaling. This is a weakness because ideally we would like larger computer systems to be able to solve "small" problems faster, as well as being able to tackle problems that were previously too large to attempt.

Task-based parallelism is a programming model that can overcome some of the problems of the data-based approach. The basic idea is that the overall problem is divided into *tasks*, which can be quite distinct (for example, computing fluxes of energy and momentum across boundaries might be one task, while another might be determining the next time step to take). Each task comes with a list of dependencies (other tasks whose output is needed before this task can be executed) and conflicts. A scheduler assigns tasks dynamically as the simulation proceeds, taking note of the dependencies and with the goal of minimizing the number of idle cores at any one time. Adopting this more asynchronous model, in which computation tasks are heterogeneous and executed in a dynamically-determined order, has many advantages (see, e.g., Kidder et al. 2017). It is more robust against failures and can potentially "hide" more of the inevitable communication behind useful computation.

With the exception of GPU computing, the benefit that these and other "next-generation" methods offer for TDE simulations is largely unquantified. That said, the overall TDE problem can be broken down into several sub-problems (initial disruption, circularization, accretion, generation of emission in different bands, jet formation, etc) that have quite distinct numerical requirements. It is highly likely that there are large gains to be found from the adoption of new algorithms and parallelism models in at least some of those sub-problems.

Acknowledgements This work was partially supported by: NSF Grant AST-1715032 and Simons Foundation Grant 559794 (JHK); by NSF Grant PHY-1748958 (YJ); and by NASA Grant NNX16AI40G (PJA). We also thank the International Space Science Institute for hospitality.

References

P. Anninos, C. Bryant, P.C. Fragile, A.M. Holgado, C. Lau, D. Nemergut, CosmosDG: an hp-adaptive discontinuous Galerkin code for hyper-resolved relativistic MHD. Astrophys. J. Suppl. Ser. **231**, 17 (2017). https://doi.org/10.3847/1538-4365/aa7ff5

S.A. Balbus, J.F. Hawley, Instability, turbulence, and enhanced transport in accretion disks. Rev. Mod. Phys. **70**, 1–53 (1998). https://doi.org/10.1103/RevModPhys.70.1

J.M. Bardeen, J.A. Petterson, The Lense-Thirring effect and accretion disks around Kerr black holes. Astrophys. J. Lett. **195**, L65 (1975). https://doi.org/10.1086/181711

J. Barnes, P. Hut, A hierarchical O(N log N) force-calculation algorithm. Nature **324**, 446–449 (1986). https://doi.org/10.1038/324446a0

G.V. Bicknell, R.A. Gingold, On tidal detonation of stars by massive black holes. Astrophys. J. **273**, 749–760 (1983). https://doi.org/10.1086/161410

J.S. Bloom, D. Giannios, B.D. Metzger, S. Bradley Cenko, D.A. Perley, N.R. Butler, N.R. Tanvir, A.J. Levan, P.T. O'Brien, L.E. Strubbe, F. De Colle, E. Ramirez-Ruiz, W.H. Lee, S. Nayakshin, E. Quataert, A.R. King, A. Cucchiara, J. Guillochon, G.C. Bower, A.S. Fruchter, A.N. Morgan, A.J. van der Horst, A possible relativistic jetted outburst from a massive black hole fed by a tidally disrupted star. Science **333**, 203 (2011)

C. Bonnerot, E.M. Rossi, G. Lodato, D.J. Price, Disc formation from tidal disruptions of stars on eccentric orbits by Schwarzschild black holes. Mon. Not. R. Astron. Soc. **455**, 2253–2266 (2016). https://doi.org/10.1093/mnras/stv2411

G.C. Brown, A.J. Levan, E.R. Stanway, N.R. Tanvir, S.B. Cenko, E. Berger, R. Chornock, A. Cucchiaria, Swift J1112.2-8238: a candidate relativistic tidal disruption flare. Mon. Not. R. Astron. Soc. **452**, 4297–4306 (2015). https://doi.org/10.1093/mnras/stv1520

D. Burrows, J. Kennea, G. Ghisellini, V. Mangano, B. Zhang, K. Page, M. Eracleous, P. Romano, T. Sakamoto, A. Falcone, J. Osborne, S. Campana, A. Beardmore, A. Breeveld, M. Chester, R. Corbet, S. Covino, J. Cummings, P. D'avanzo, V. D'elia, P. Esposito, P. Evans, D. Fugazza, J. Gelbord, K. Hiroi, S. Holland, K. Huang, M. Im, G. Israel, Y. Jeon, H. Jun, N. Kawai, J. Kim, H. Krimm, F. Marshall, P. Mészáros, H. Negoro, N. Omodei, W. Park, J. Perkins, M. Sugizaki, H. Sung, G. Tagliaferri, E. Troja, Y. Ueda, Y. Urata, R. Usui, L. Antonelli, S. Barthelmy, G. Cusumano, P. Giommi, A. Melandri, M. Perri, J. Racusin, B. Sbarufatti, M. Siegel, N. Gehrels, Relativistic jet activity from the tidal disruption of a star by a massive black hole. Nature **476**, 421 (2011)

S.B. Cenko, H.A. Krimm, A. Horesh, A. Rau, D.A. Frail, J.A. Kennea, A.J. Levan, S.T. Holland, N.R. Butler, R.M. Quimby, J.S. Bloom, A.V. Filippenko, A. Gal-Yam, J. Greiner, S.R. Kulkarni, E.O. Ofek, F. Olivares E., P. Schady, J.M. Silverman, N.R. Tanvir, D. Xu, Swift J2058.4+0516: discovery of a possible second relativistic tidal disruption flare? Astrophys. J. **753**, 77 (2012). https://doi.org/10.1088/0004-637X/753/1/77

C.-H. Chan, J.H. Krolik, T. Piran, Magnetorotational instability in eccentric disks. Astrophys. J. **856**, 12 (2018). https://doi.org/10.3847/1538-4357/aab15c

P. Chang, J. Wadsley, T.R. Quinn, A moving-mesh hydrodynamic solver for ChaNGa. Mon. Not. R. Astron. Soc. **471**, 3577–3589 (2017). https://doi.org/10.1093/mnras/stx1809

E.R. Coughlin, C. Nixon, M.C. Begelman, P.J. Armitage, On the structure of tidally disrupted stellar debris streams. Mon. Not. R. Astron. Soc. **459**, 3089–3103 (2016). https://doi.org/10.1093/mnras/stw770

L. Cullen, W. Dehnen, Inviscid smoothed particle hydrodynamics. Mon. Not. R. Astron. Soc. **408**, 669–683 (2010). https://doi.org/10.1111/j.1365-2966.2010.17158.x

S.W. Davis, J.M. Stone, Y.-F. Jiang, A radiation transfer solver for Athena using short characteristics. Astrophys. J. Suppl. Ser. **199**, 9 (2012). https://doi.org/10.1088/0067-0049/199/1/9

H. Deng, L. Mayer, H. Latter, P.F. Hopkins, X.-N. Bai, Local simulations of MRI turbulence with meshless methods. ArXiv e-prints (2019)

P.C. Duffell, DISCO: a 3D moving-mesh magnetohydrodynamics code designed for the study of astrophysical disks. Astrophys. J. Suppl. Ser. **226**, 2 (2016). https://doi.org/10.3847/0067-0049/226/1/2

C. Eckart, Variation principles of hydrodynamics. Phys. Fluids **3**, 421–427 (1960). https://doi.org/10.1063/1.1706053

P.C. Fragile, D. Nemergut, P.L. Shaw, P. Anninos, Divergence-free magnetohydrodynamics on conformally moving, adaptive meshes using a vector potential method. ArXiv e-prints (2018)

B. Fryxell, K. Olson, P. Ricker, F.X. Timmes, M. Zingale, D.Q. Lamb, P. MacNeice, R. Rosner, J.W. Truran, H. Tufo, FLASH: adaptive mesh hydrodynamics code for modeling astrophysical thermonuclear flashes. Astrophysics Source Code Library (2010)

C.F. Gammie, J.C. McKinney, G. Tóth, HARM: a numerical scheme for general relativistic magnetohydrodynamics. Astrophys. J. **589**, 444–457 (2003). https://doi.org/10.1086/374594

R.A. Gingold, J.J. Monaghan, Smoothed particle hydrodynamics - theory and application to non-spherical stars. Mon. Not. R. Astron. Soc. **181**, 375–389 (1977). https://doi.org/10.1093/mnras/181.3.375

F.G. Goicovic, V. Springel, S.T. Ohlmann, R. Pakmor, Hydrodynamical moving-mesh simulations of the tidal disruption of stars by supermassive black holes. ArXiv e-prints (2019)

J. Guillochon, E. Ramirez-Ruiz, A dark year for tidal disruption events. Astrophys. J. **809**, 166 (2015). https://doi.org/10.1088/0004-637X/809/2/166

S.P. Hatchett, M.C. Begelman, C.L. Sarazin, A new look at the dynamics of twisted accretion disks. Astrophys. J. **247**, 677–685 (1981). https://doi.org/10.1086/159079

J.F. Hawley, J.H. Krolik, Sound speed dependence of alignment in accretion disks subjected to Lense-Thirring torques. Astrophys. J. **866**, 5 (2018). https://doi.org/10.3847/1538-4357/aadf90

K. Hayasaki, N. Stone, A. Loeb, Finite, intense accretion bursts from tidal disruption of stars on bound orbits. Mon. Not. R. Astron. Soc. **434**(2), 909–924 (2013). https://doi.org/10.1093/mnras/stt871

K. Hayasaki, N. Stone, A. Loeb, Circularization of tidally disrupted stars around spinning supermassive black holes. Mon. Not. R. Astron. Soc. **461**(4), 3760–3780 (2016). https://doi.org/10.1093/mnras/stw1387

P.F. Hopkins, A new class of accurate, mesh-free hydrodynamic simulation methods. Mon. Not. R. Astron. Soc. **450**, 53–110 (2015). https://doi.org/10.1093/mnras/stv195

P.F. Hopkins, M.Y. Grudić, Numerical problems in coupling photon momentum (radiation pressure) to gas. Mon. Not. R. Astron. Soc. **483**, 4187–4196 (2019). https://doi.org/10.1093/mnras/sty3089

Y.-F. Jiang, J.M. Stone, S.W. Davis, A Godunov method for multidimensional radiation magnetohydrodynamics based on a variable Eddington tensor. Astrophys. J. Suppl. Ser. **199**, 14 (2012). https://doi.org/10.1088/0067-0049/199/1/14

Y.-F. Jiang, J.M. Stone, S.W. Davis, An algorithm for radiation magnetohydrodynamics based on solving the time-dependent transfer equation. Astrophys. J. Suppl. Ser. **213**, 7 (2014). https://doi.org/10.1088/0067-0049/213/1/7

Y.-F. Jiang, J. Guillochon, A. Loeb, Prompt radiation and mass outflows from the stream-stream collisions of tidal disruption events. Astrophys. J. **830**, 125 (2016). https://doi.org/10.3847/0004-637X/830/2/125

Y.-F. Jiang, J. Stone, S.W. Davis, Super-Eddington accretion disks around supermassive black holes. ArXiv e-prints (2017)

L.E. Kidder, S.E. Field, F. Foucart, E. Schnetter, S.A. Teukolsky, A. Bohn, N. Deppe, P. Diener, F. Hébert, J. Lippuner, J. Miller, C.D. Ott, M.A. Scheel, T. Vincent, SpECTRE: a task-based discontinuous Galerkin code for relativistic astrophysics. J. Comput. Phys. **335**, 84–114 (2017). https://doi.org/10.1016/j.jcp.2016.12.059

S. Kobayashi, P. Laguna, E.S. Phinney, P. Mészáros, Gravitational waves and X-ray signals from stellar disruption by a massive black hole. Astrophys. J. **615**, 855–865 (2004). https://doi.org/10.1086/424684

C.S. Kochanek, The aftermath of tidal disruption: the dynamics of thin gas streams. Astrophys. J. **422**, 508 (1994)

J.H. Krolik, J.F. Hawley, A steady-state alignment front in an accretion disk subjected to Lense-Thirring torques. Astrophys. J. **806**, 141 (2015). https://doi.org/10.1088/0004-637X/806/1/141

J. Krolik, T. Piran, G. Svirski, R.M. Cheng, ASASSN-14li: a model tidal disruption event. ArXiv e-prints (2016)

D. Liptai, D.J. Price, General relativistic smoothed particle hydrodynamics. Mon. Not. R. Astron. Soc. (2019). https://doi.org/10.1093/mnras/stz111

M. Liska, C. Hesp, A. Tchekhovskoy, A. Ingram, M. van der Klis, S. Markoff, Formation of precessing jets by tilted black hole discs in 3D general relativistic MHD simulations. Mon. Not. R. Astron. Soc. **474**, L81–L85 (2018a). https://doi.org/10.1093/mnrasl/slx174

M. Liska, A. Tchekhovskoy, A. Ingram, M. van der Klis, Bardeen-Petterson alignment, jets and magnetic truncation in GRMHD simulations of tilted thin accretion discs. ArXiv e-prints (2018b)

G. Lodato, D.J. Price, On the diffusive propagation of warps in thin accretion discs. Mon. Not. R. Astron. Soc. **405**, 1212–1226 (2010). https://doi.org/10.1111/j.1365-2966.2010.16526.x

L.B. Lucy, A numerical approach to the testing of the fission hypothesis. Astron. J. **82**, 1013–1024 (1977). https://doi.org/10.1086/112164

J.C. McKinney, A. Tchekhovskoy, A. Sadowski, R. Narayan, Three-dimensional general relativistic radiation magnetohydrodynamical simulation of super-Eddington accretion, using a new code HARMRAD with M1 closure. Mon. Not. R. Astron. Soc. **441**, 3177–3208 (2014). https://doi.org/10.1093/mnras/stu762

P. Mocz, R. Pakmor, V. Springel, M. Vogelsberger, F. Marinacci, L. Hernquist, A moving mesh unstaggered constrained transport scheme for magnetohydrodynamics. Mon. Not. R. Astron. Soc. **463**, 477–488 (2016). https://doi.org/10.1093/mnras/stw2004

J.J. Monaghan, D.J. Price, Variational principles for relativistic smoothed particle hydrodynamics. Mon. Not. R. Astron. Soc. **328**, 381–392 (2001). https://doi.org/10.1046/j.1365-8711.2001.04742.x

S. Moon, W.-T. Kim, E.C. Ostriker, A fast Poisson solver of second-order accuracy for isolated systems in three-dimensional Cartesian and cylindrical coordinates. ArXiv e-prints (2019)

J.P. Morris, J.J. Monaghan, A switch to reduce SPH viscosity. J. Comput. Phys. **136**, 41–50 (1997). https://doi.org/10.1006/jcph.1997.5690

R.P. Nelson, J.C.B. Papaloizou, Hydrodynamic simulations of the Bardeen-Petterson effect. Mon. Not. R. Astron. Soc. **315**, 570–586 (2000). https://doi.org/10.1046/j.1365-8711.2000.03478.x

S.C. Noble, J.H. Krolik, J.F. Hawley, Direct calculation of the radiative efficiency of an accretion disk around a black hole. Astrophys. J. **692**, 411–421 (2009). https://doi.org/10.1088/0004-637X/692/1/411

S.C. Noble, B.C. Mundim, H. Nakano, J.H. Krolik, M. Campanelli, Y. Zlochower, N. Yunes, Circumbinary magnetohydrodynamic accretion into inspiraling binary black holes. Astrophys. J. **755**, 51 (2012). https://doi.org/10.1088/0004-637X/755/1/51

R.A. Nolthenius, J.I. Katz, The passage of a star by a massive black hole. Astrophys. J. **263**, 377–385 (1982). https://doi.org/10.1086/160511

R. Pakmor, V. Springel, A. Bauer, P. Mocz, D.J. Munoz, S.T. Ohlmann, K. Schaal, C. Zhu, Improving the convergence properties of the moving-mesh code AREPO. Mon. Not. R. Astron. Soc. **455**, 1134–1143 (2016). https://doi.org/10.1093/mnras/stv2380

J.C.B. Papaloizou, J.E. Pringle, The time-dependence of non-planar accretion discs. Mon. Not. R. Astron. Soc. **202**, 1181–1194 (1983). https://doi.org/10.1093/mnras/202.4.1181

D.J. Price, Modelling discontinuities and Kelvin Helmholtz instabilities in SPH. J. Comput. Phys. **227**, 10040–10057 (2008). https://doi.org/10.1016/j.jcp.2008.08.011

D.J. Price, Smoothed particle hydrodynamics and magnetohydrodynamics. J. Comput. Phys. **231**, 759–794 (2012). https://doi.org/10.1016/j.jcp.2010.12.011

D.J. Price, J. Wurster, T.S. Tricco, C. Nixon, S. Toupin, A. Pettitt, C. Chan, D. Mentiplay, G. Laibe, S. Glover, C. Dobbs, R. Nealon, D. Liptai, H. Worpel, C. Bonnerot, G. Dipierro, G. Ballabio, E. Ragusa, C. Federrath, R. Iaconi, T. Reichardt, D. Forgan, M. Hutchison, T. Constantino, B. Ayliffe, K. Hirsh, G. Lodato,

Phantom: a smoothed particle hydrodynamics and magnetohydrodynamics code for astrophysics. ArXiv e-prints (2017)
M.J. Rees, Tidal disruption of stars by black holes of 10 to the 6th-10 to the 8th solar masses in nearby galaxies. Nature **333**, 523 (1988). http://adsabs.harvard.edu/abs/1988Natur.333..523R
T. Ryu, J. Krolik, T. Piran, S.C. Noble, Tidal disruptions of main sequence stars – II. Simulation methodology and stellar mass dependence of the character of full tidal disruptions. ArXiv e-prints arXiv:2001.03502 (2020)
A. Sądowski, R. Narayan, A. Tchekhovskoy, Y. Zhu, Semi-implicit scheme for treating radiation under M1 closure in general relativistic conservative fluid dynamics codes. Mon. Not. R. Astron. Soc. **429**, 3533–3550 (2013). https://doi.org/10.1093/mnras/sts632
A. Sądowski, R. Narayan, J.C. McKinney, A. Tchekhovskoy, Numerical simulations of super-critical black hole accretion flows in general relativity. Mon. Not. R. Astron. Soc. **439**, 503–520 (2014). https://doi.org/10.1093/mnras/stt2479
P. Salbi, C.D. Matzner, S. Ro, Y. Levin, Oblique shock breakout in supernovae and gamma-ray bursts. II. Numerical solutions for non-relativistic pattern speeds. Astrophys. J. **790**, 71 (2014). https://doi.org/10.1088/0004-637X/790/1/71
K. Schaal, A. Bauer, P. Chandrashekar, R. Pakmor, C. Klingenberg, V. Springel, Astrophysical hydrodynamics with a high-order discontinuous Galerkin scheme and adaptive mesh refinement. Mon. Not. R. Astron. Soc. **453**, 4278–4300 (2015). https://doi.org/10.1093/mnras/stv1859
H. Shiokawa, J.H. Krolik, R.M. Cheng, T. Piran, S.C. Noble, General relativistic hydrodynamic simulation of accretion flow from a stellar tidal disruption. Astrophys. J. **804**, 85 (2015). https://doi.org/10.1088/0004-637X/804/2/85
H. Shiokawa, R.M. Cheng, S.C. Noble, J.H. Krolik, PATCHWORK: a multipatch infrastructure for multiphysics/multiscale/multiframe fluid simulations. Astrophys. J. **861**, 15 (2018). https://doi.org/10.3847/1538-4357/aac2dd
M.A. Skinner, E.C. Ostriker, The Athena astrophysical magnetohydrodynamics code in cylindrical geometry. Astrophys. J. Suppl. Ser. **188**, 290–311 (2010). https://doi.org/10.1088/0067-0049/188/1/290
K.A. Sorathia, J.H. Krolik, J.F. Hawley, Magnetohydrodynamic simulation of a disk subjected to Lense-Thirring precession. Astrophys. J. **777**, 21 (2013). https://doi.org/10.1088/0004-637X/777/1/21
V. Springel, Smoothed particle hydrodynamics in astrophysics. Annu. Rev. Astron. Astrophys. **48**, 391–430 (2010). https://doi.org/10.1146/annurev-astro-081309-130914
E. Steinberg, E.R. Coughlin, N.C. Stone, B.D. Metzger, Thawing the frozen-in approximation: implications for self-gravity in deeply plunging tidal disruption events. ArXiv e-prints (2019)
J.M. Stone, T.A. Gardiner, P. Teuben, J.F. Hawley, J.B. Simon, Athena: a new code for astrophysical MHD. Astrophys. J. Suppl. Ser. **178**, 137–177 (2008). https://doi.org/10.1086/588755
E. Tejeda, S. Rosswog, An accurate Newtonian description of particle motion around a Schwarzschild black hole. Mon. Not. R. Astron. Soc. **433**, 1930–1940 (2013). https://doi.org/10.1093/mnras/stt853
A. Ulmer, Flares from the tidal disruption of stars by massive black holes. Astrophys. J. **514**, 180 (1999)
C. Wegg, Pseudo-Newtonian potentials for nearly parabolic orbits. Astrophys. J. **749**(2), 183 (2012). https://doi.org/10.1088/0004-637X/749/2/183
C.J. White, J.M. Stone, C.F. Gammie, An extension of the Athena++ code framework for GRMHD based on advanced Riemann solvers and staggered-mesh constrained transport. Astrophys. J. Suppl. Ser. **225**, 22 (2016). https://doi.org/10.3847/0067-0049/225/2/22
S.C. Whitehouse, M.R. Bate, J.J. Monaghan, A faster algorithm for smoothed particle hydrodynamics with radiative transfer in the flux-limited diffusion approximation. Mon. Not. R. Astron. Soc. **364**, 1367–1377 (2005). https://doi.org/10.1111/j.1365-2966.2005.09683.x
A. Yalinewich, E. Steinberg, R. Sari, RICH: open-source hydrodynamic simulation on a moving Voronoi mesh. Astrophys. J. Suppl. Ser. **216**, 35 (2015). https://doi.org/10.1088/0067-0049/216/2/35
A. Yalinewich, E. Steinberg, T. Piran, J.H. Krolik, Radio emission from the unbound debris of tidal disruption events. ArXiv e-prints (2019)
V.V. Zhuravlev, P.B. Ivanov, P.C. Fragile, D. Morales Teixeira, No evidence for Bardeen-Petterson alignment in GRMHD simulations and semi-analytic models of moderately thin, prograde, tilted accretion disks. Astrophys. J. **796**, 104 (2014). https://doi.org/10.1088/0004-637X/796/2/104

GPSR Compliance

The European Union's (EU) General Product Safety Regulation (GPSR) is a set of rules that requires consumer products to be safe and our obligations to ensure this.

If you have any concerns about our products, you can contact us on ProductSafety@springernature.com

In case Publisher is established outside the EU, the EU authorized representative is:

Springer Nature Customer Service Center GmbH
Europaplatz 3
69115 Heidelberg, Germany

Batch number: 10777430

Printed by Printforce, the Netherlands